COMPREHENSIVE MEDICINAL CHEMISTRY

IN 6 VOLUMES

COMPREHENSIVE MEDICINAL CHEMISTRY

The Rational Design, Mechanistic Study & Therapeutic Application of Chemical Compounds

Chairman of the Editorial Board
CORWIN HANSCH
Pomona College, Claremont, CA, USA

Joint Executive Editors
PETER G. SAMMES
Brunel University of West London, Uxbridge, UK

JOHN B. TAYLOR
Rhône-Poulenc Ltd, Dagenham, UK

Volume 1
GENERAL PRINCIPLES

Volume Editor
PETER D. KENNEWELL
Roussel Laboratories Ltd, Swindon, UK

PERGAMON PRESS
Member of Maxwell Macmillan Pergamon Publishing Corporation
OXFORD • NEW YORK • BEIJING • FRANKFURT
SÃO PAULO • SYDNEY • TOKYO • TORONTO

U.K.	Pergamon Press plc, Headington Hill Hall, Oxford OX3 0BW, England
U.S.A.	Pergamon Press, Inc., Maxwell House, Fairview Park, Elmsford, New York 10523, U.S.A.
PEOPLE'S REPUBLIC OF CHINA	Pergamon Press, Room 4037, Qianmen Hotel, Beijing, People's Republic of China
FEDERAL REPUBLIC OF GERMANY	Pergamon Press GmbH, Hammerweg 6, D-6242 Kronberg, Federal Republic of Germany
BRAZIL	Pergamon Editora Ltda, Rua Eça de Queiros, 346, CEP 04011, Paraiso, São Paulo, Brazil
AUSTRALIA	Pergamon Press Australia Pty Ltd., P.O. Box 544, Potts Point, N.S.W. 2011, Australia
JAPAN	Pergamon Press, 5th Floor, Matsuoka Central Building, 1-7-1 Nishishinjuku, Shinjuku-ku, Tokyo 160, Japan
CANADA	Pergamon Press Canada Ltd., Suite No. 241, 253 College Street, Toronto, Ontario, Canada M5T 1R5

First edition 1990

Library of Congress Cataloging in Publication Data

Comprehensive medicinal chemistry: the rational design, mechanistic study & therapeutic application of chemical compounds/ chairman of the editorial board, Corwin Hansch; joint executive editors, Peter G. Sammes, John B. Taylor. — 1st ed.
p. cm.
Includes index.
1. Pharmaceutical chemistry. I. Hansch, Corwin. II. Sammes, P. G. (Peter George) III. Taylor, J. B. (John Bodenham), 1939– .
[DNLM: 1. Chemistry, Pharmaceutical. QV 744 C737]
RS402.C65
615'.19—dc20
DNLM/DLC 89–16329

British Library Cataloguing in Publication Data

Hansch, Corwin
Comprehensive medicinal chemistry
1. Pharmaceutics
I. Title
615'.19

ISBN 0–08–037057–8 (Vol. 1)
ISBN 0–08–032530–0 (set)

Printed in Great Britain by
BPCC Hazell Books, Aylesbury, Bucks, England
Member of BPCC Ltd.

Contents

Preface

Medicinal chemistry is a subject which has seen enormous growth in the past decade. Traditionally accepted as a branch of organic chemistry, and the near exclusive province of the organic chemist, the subject has reached an enormous level of complexity today. The science now employs the most sophisticated developments in technology and instrumentation, including powerful molecular graphics systems with 'drug design' software, all aspects of high resolution spectroscopy, and the use of robots. Moreover, the medicinal chemist (very much a new breed of organic chemist) works in very close collaboration and mutual understanding with a number of other specialists, notably the molecular biologist, the genetic engineer, and the biopharmacist, as well as traditional partners in biology.

Current books on medicinal chemistry inevitably reflect traditional attitudes and approaches to the field and cover unevenly, if at all, much of modern thinking in the field. In addition, such works are largely based on a classical organic structure and therapeutic grouping of biologically active molecules. The aim of *Comprehensive Medicinal Chemistry* is to present the subject, the modern role of which is the understanding of structure–activity relationships and drug design from the mechanistic viewpoint, as a field in its own right, integrating with its central chemistry all the necessary ancillary disciplines.

To ensure that a broad coverage is obtained at an authoritative level, more than 250 authors and editors from 15 countries have been enlisted. The contributions have been organized into five major themes. Thus Volume 1 covers general principles, Volume 2 deals with enzymes and other molecular targets, Volume 3 describes membranes and receptors, Volume 4 covers quantitative drug design, and Volume 5 discusses biopharmaceutics. As well as a cumulative subject index, Volume 6 contains a unique drug compendium containing information on over 5500 compounds currently on the market. All six volumes are being published simultaneously, to provide a work that covers all major topics of interest.

Because of the mechanistic approach adopted, Volumes 1–5 do not discuss those drugs whose modes of action are unknown, although they will be included in the compendium in Volume 6. The mechanisms of action of such agents remain a future challenge for the medicinal chemist.

We should like to acknowledge the way in which the staff at the publisher, particularly Dr Colin Drayton (who initially proposed the project), Dr Helen McPherson and their editorial team, have supported the editors and authors in their endeavour to produce a work of reference that is both complete and up-to-date.

Comprehensive Medicinal Chemistry is a milestone in the literature of the subject in terms of coverage, clarity and a sustained high level of presentation. We are confident it will appeal to academic and industrial researchers in chemistry, biology, medicine and pharmacy, as well as teachers of the subject at all levels.

CORWIN HANSCH
Claremont, USA

PETER G. SAMMES
Uxbridge, UK

JOHN B. TAYLOR
Dagenham, UK

Contributors to Volume 1

Dr S. P. Adams
Biological Sciences Division, Monsanto Company, 700 Chesterfield Village Parkway, St Louis, MO 63198, USA

Dr N. Anand
Central Drug Research Institute, Chattar Manzil, PO Box 173, Lucknow-226 001, India

Dr P. Bost
Direction Strategie Recherche et Development, Rhône-Poulenc Santé, 20 Avenue Raymond Aron, F-92165 Antony Cedex, France

Dr G. Bourat
Direction Strategie Recherche et Development, Rhône-Poulenc Santé, 20 Avenue Raymond Aron, F-92165 Antony Cedex, France

Dr L. A. Brown
Mill Leat, West Gomeldon, Salisbury, Wilts SP4 6JY, UK

Professor A. Burger
510 Wiley Drive, Charlottesville, VA 22901, USA

Professor W. R. Butt
36 Avonside, Mill Lane, Stratford-upon-Avon CV37 6BJ, UK

Professor P. S. Callery
School of Pharmacy, Department of Medicinal Chemistry, University of Maryland, 20 North Pine Street, Baltimore, MD 21201, USA

Professor G. deStevens
Department of Chemistry, Drew University, Madison, NJ 07940, USA

Dr K. Drlica
The Public Health Research Institute of the City of New York, 455 First Avenue, New York, NY 10016, USA

Dr J. E. Fincham
School of Pharmacy, Samford University, Birmingham, AL 35229, USA

Dr S. C. Gilman
Cytogen Corporation, 201 College Road East, Princeton, NJ 08540, USA

Dr L. Hodes
Drug Synthesis and Chemistry Branch, Developmental Therapeutics Program, Division of Cancer Treatment, Department of Health and Human Services, National Cancer Institute, Landow Building, Room 5C-19, Bethesda, MD 20892, USA

Dr E. A. Horak
Ciba-Geigy Ltd, Postfach, CH-4002 Basle, Switzerland

Professor Huang Liang
The Institute of Materia Medica, Chinese Academy of Medical Sciences, 1 Xian Nong Tan Street, Beijing, People's Republic of China

Dr A.-C. Jouanneau
22 rue de Bièvre, F-75005 Paris, France

Dr T. P. Kenakin
Department of Molecular Pharmacology, Glaxo Research Laboratories, Glaxo Inc., Five Moore Drive, PO Box 13438, Research Triangle Park, NC 27709, USA

Dr P. D. Kennewell
Roussel Laboratories Ltd, Kingfisher Drive, Covingham, Swindon, Wilts SN3 5BZ, UK

Dr T. Laird
2 Richmead Gardens, Mayfield, East Sussex TN20 6DE, UK

Dr A. J. Lewis
Division of Experimental Therapeutics, Wyeth Laboratories Inc., Research and Development Division, PO Box 8299, Philadelphia, PA 19101, USA

Dr J. Liebenau
Department of Information Systems, The London School of Economics and Political Science, Houghton Street, London WC2A 2AE, UK

Dr R. L. Lipnick
5308 Pender Court, Alexandria, VA 22304, USA

Dr J. R. McClintic
28135 Burrough North Road, Tollhouse, CA 93667, USA

Dr P. Miller
Roussel Laboratories Ltd, Kingfisher Drive, Covingham, Swindon, Wilts SN3 5BZ, UK

Dr S. A. North
Information Services Department, Glaxo Group Research Ltd, Park Road, Ware, Herts SG12 0DP, UK

Mr G. E. Powderham
Roussel Laboratories Ltd, Broadwater Park, North Orbital Road, Denham, Uxbridge, Middlesex UB9 5HP, UK

Dr N. S. B. Rawson
Box 92, University Hospital, Saskatoon, S7N 0X0, Canada

Dr J. H Seamer
Mill Leat, West Gomeldon, Salisbury, Wilts SP4 6JY, UK

Mr J. Sharp
Mill Close, Sandford Manor, Woodley, Reading RG5 4SY, UK

Dr J. Skidmore
Worldwide Information, Smith Kline & French Research Ltd, The Frythe, Welwyn, Herts AL6 1AR, UK

Dr W. Sneader
Department of Pharmacy, Royal College, University of Strathclyde, 204 George Street, Glasgow G1 1XW, UK

Dr B. Spilker
Department of Project Coordination, Burroughs Wellcome Co., 3030 Cornwallis Road, Research Triangle Park, NC 27709, USA

Dr J. A. Sutton
Knoll Ltd, Fleming House, 71 King Street, Maidenhead, Berks SL6 1EU, UK

Mr I. J. Tarr
18–20 Hill Rise, Richmond, Surrey TW10 6US, UK

Mr D. Taylor
ABPI, 12 Whitehall, London SW1A 2DY, UK

Dr J.-C. Vieillefosse
Departement de Brevets, Centre de Recherches, Roussel-Uclaf, 111 Route de Noisy, F-93230 Romainville, France

Dr S. E. Ward
Information Services Department, Glaxo Group Research Ltd, Park Road, Ware, Herts SG12 0DP, UK

Mr N. E. J. Wells
Glaxo Pharmaceuticals Ltd, 891–995 Greenford Road, Greenford, Middlesex UB6 0HE, UK

Dr P. H. Wooley
Department of Immunology, Ayerst Laboratories, Box CN 8000, Princeton, NJ 08540, USA

Professor Zhou Jin
The Institute of Materia Medica, Chinese Academy of Medical Sciences, 1 Xian Nong Tan Street, Beijing, People's Republic of China

Contents of All Volumes

Volume 1 General Principles

Historical Perspective

Targets of Biologically Active Molecules

Bioactive Materials

Socio-economic Factors of Drug Development

Volume 2 Enzymes and Other Molecular Targets

Volume 3 Membranes and Receptors

Membranes, Membrane Receptors and Second Messenger Pathways

Neurotransmitter and Autocoid Receptors

Peptidergic Receptors

Drugs Acting on Ion Channels and Membranes

Lymphokines and Cytokines

Intracellular Receptors

Volume 4 Quantitative Drug Design

Volume 5 Biopharmaceutics

1.1
Medicinal Chemistry: A Personal View

ALFRED BURGER
University of Virginia, Charlottesville, VA, USA

A treatise on medicinal chemistry is a survey of the research and accomplishments of thousands of medicinal chemists, past and present, who have furnished the observations recorded in these volumes. These data have been obtained not only from factual experiments and measurements but are also the distillation of the dreams and the imagination of different scientists who have explored the origins and the preparation of drugs and the fate and mode of action of biologically active substances. Thus, biologists and chemists of every ilk, biostatisticians, experimental and clinical pharmacologists, microbiologists, geneticists, experts in organic synthesis and fermentation processes, and instrument-minded spectroscopists have combined forces and zeroed in on the problems of medicinal agents. My own professional history reflects the interests that create a medicinal chemist and is probably similar to the evolution of many medicinal chemists now active in our science.

Formal courses in medicinal chemistry—still called pharmaceutical chemistry in some institutions—are offered in colleges of pharmacy for undergraduates, and later on a biomedical level for graduate students. Few chemistry departments list such topics. My first course in this subject was a simple recital of alkaloid chemistry that included the botanical sources, structural proof and synthesis of such compounds. Several of my teachers were natural products chemists, and my doctoral dissertation was entitled 'Syntheses of Benzylisoquinoline Alkaloids'. It included partial syntheses of papaverine,[1] laudanine, laudanidine and (±)-codamine.[2] This background in opium alkaloids stood me in good stead when a year after graduation, in 1929, a postdoctoral position became available at the University of Virginia.

In the mid-1920s a wave of addiction to morphine and heroin swept the United States, not as widespread as the present similar dependence liability, but alarming enough to require federal intervention. A narcotic prison-hospital was erected in Lexington, Kentucky, and addicts were treated by psychiatrists and by controlled withdrawal procedures. Soon it became apparent that prevention would be preferable to therapy; therefore chemists and pharmacologists were called in to take up a study of chemical modification of the morphine structure and to elucidate the contribution of its structural moieties to the problem of separating analgesic properties from euphoric and other side effects. A similar approach had succeeded in the elaboration of procaine and other local anesthetics from cocaine, and of antispasmodic-type amino esters from atropine. It required the rudiments of drug design which, at that time, was almost nonexistent. The chemical project at Virginia was headed by Lyndon Small and Erich Mosettig;[3] the pharmacological evaluation of our compounds was studied by Nathan B. Eddy at the University of Michigan.

A medicinal chemist studying a problem as a member of a team can only contribute but so much to the attainment of the project's goal. My corner of the task was the synthesis of series of amino alcohols derived from partial structures discernible in morphine and codeine. The sterically twisted piperidine ring of these alkaloids contains the amino function which we tried to emulate, and the secondary alcohol group is located in the cyclohexene portion. Since amino ketones obtainable from the naturally occurring secondary alcohols also had some analgesic properties, we included both amino alcohols and amino ketones in our plans.

Morphine can be dissected in many ways. One-ring systems in this molecule could be an aromatic ring, cyclohexane, cyclohexene, piperidine and furan. Condensed ring systems that could be visualized are hydronaphthalene, partially hydrogenated phenanthrenes, isoquinolines, benzo- and dibenzo-furans, *etc.* We did what most people would do, *i.e.* start with a simple one-ring system and build on to it our β-amino alcohol groups.[4] Our first attempt was to treat cyclohexanone with diazomethane and to convert the resulting epoxide to amino alcohols. The real bonanza consisted of a ring enlargement of cyclohexanone to cycloheptanone and cyclooctanone. Our method made cycloheptanone more easily accessible than before.

We soon chose as partial structures the hydrophenanthrene, furan, dihydrobenzofuran and piperidine systems and attached amino alcohol groups to these rings.[5–11] We also built steric models of morphine but missed the phenylpiperidine structure, discovered later by Eisleb and Schaumann,[12,13] that is present in meperidine and related analgesics. As a consolation to us, these investigators arrived at the analgesic properties of meperidine by the back door and not by design. They had conceived meperidine as a reversed ester structure of aminoalkyl antispasmodics and discovered the analgesic action by comprehensive screening. The real biochemical events which distinguish opiate analgesic from psychotomimetic effects are only now beginning to be unravelled.[14] These precipitous discoveries probably account for the fact that graduate students, sent on a search of the literature, rarely look up any references older than three or four years.

Working day after day on amino alcohols derived from the wrong ring systems, I myself must have made 150 such compounds, when I could no longer suppress a nagging question. With analgesic activities of only a few percent of those of codeine, were we really on the right track? Should we stick to our program or branch out to something more unorthodox? A few tetrahydroisoquinolino alcohols had a slightly higher analgesic index but we barely suspected that the increased lipophilicity may have had something to do with it; we really did not know how to grope for a rationale. Such questions still plague organic chemists who vary 'lead' structures systematically and cannot discern clear trends in the biological activity of the derivatives and analogs.

I have often felt that, had I been born in another age, I still would have wanted to work and study in the same general field of science I was lucky enough to spend my life in. However, methodology would have been different. Biochemical considerations would have been paramount, with molecular modification of natural biochemically active metabolites as a main goal.

In 1932, the seminal papers by Hans Erlenmeyer of the University of Basel began to appear,[15,16] which heralded a turning point in rational drug design. Erlenmeyer transposed Langmuir's concept of isosterism,[17] Grimm's rules of hydride displacement,[18] and Hinsberg's ring equivalents[19] to structures of biological interest. Briefly, it was assumed that atoms and groups that are spatially and electronically similar could replace each other and thereby open the way to new structural analogs with biological activities similar to those of the prototype compounds. Erlenmeyer's criteria for steric and electronic similarity were based on valid but narrowly circumscribed properties such as isomorphism in the crystalline state, and were therefore restricted to a few examples of isosteres. By giving these physical criteria wider play, and especially by including biochemical or biological analogy, Harris L. Friedman[20] enlarged the scope of these comparisons and called them bioisosterism. This widely used practical concept has been reviewed and expanded repeatedly[21–25] and has become the mainstay of chemical approaches to drug design. Only more than 30 years later did the computer age make it possible to base the possibilities of singling out peaks in potency or biological suitability to smaller numbers of analogs in some cases.[26,27]

When in 1938 my first graduate students chose to work with me, we pursued isosteric replacements in the search for alternate drug structures.[28,29] These efforts brought us friendships and support in the pharmaceutical industry, where similar studies had taken a foothold. Analogs of classical aralkylamines were then in vogue, stimulated by the discoveries of norepinephrine as a neurotransmitter and of amphetamine[30,31] as a simplification of the CNS stimulatory alkaloid, ephedrine. We designed an analog, 2-phenylcyclopropylamine, as a more rigid spatial analog of amphetamine; this compound also represented a minimal molecular modification of amphetamine, lacking only two of its hydrogen atoms.[32] We did not know that its cyclopropane ring structure would predispose it to become an inhibitor of monoamine oxidase (MAO) and a clinically useful antidepressant.[33,34] These properties were discovered eight years after the compound had been tested for amphetamine-like cardiovascular and CNS stimulatory activities. When independent studies of MAO inhibitors in other structural series had awakened interest in such compounds,[35] the slight MAO-inhibitory properties of amphetamine were remembered and all available analogs of amphetamine, including our cyclopropylamines, were fed into the test program. We had originally submitted 8 g of our *trans* isomer, 6 g of which were wasted on tests for amphetamine-like pharmacology and toxicology. The anti-MAO tests which revealed the useful properties of our drug had only 2 g to work with.

Similar discoveries with leftover substances have been experienced by other investigators.[36] No wonder that an authoritative textbook of pharmacology[37] states that 'new therapeutic agents are discovered by screening, by structural modification of established drugs or endogenous substances, or by accident.' All claims about rational drug design to the contrary, this statement is still woefully correct. Structural classes of 'lead' compounds can now be found by rational design based on biochemical relationships but the final compound that is eligible for clinical testing and likely to weather the exigencies of this process is still only obtained by screening. Nevertheless, the narrowing down to several structural types in a structurally related series of compounds is aided immeasurably by the principles of bioisosterism and by comparing and computing as many trends of pertinent physical properties as possible.

In 1950, few, if any, books on medicinal chemistry were available to readers in this emerging science. There were textbooks of organic chemistry for undergraduate students that illustrated organic reactions using compounds of therapeutic interest. I tried to fill this need with a treatise that concentrated on drug design, drug metabolism where known, and other biochemical aspects of medicinal agents. A few years later the literature problem of medicinal chemistry became aggravated when strictures of available journal space closed several major journals to articles of our hybrid science. In fact, most chemists published their chemical work and the biological evaluation of their test compounds was often lost or hidden in references not usually read by chemists. The *Journal of Medicinal Chemistry* was founded by Arnold H. Beckett and myself in 1959, and edited by me until 1971. It combines chemical, biochemical and biological aspects of medicinal science and has become the principal journal in our field of study.

It is often assumed that academic research scientists can dream up anything they want and work up such ideas in the laboratory without any restrictions. The real world looks different, however. Even the smallest research projects consume materials and require apparatus and instruments of ever increasing complexity, sensitivity and cost. Both the professor who initiates the research, and the postgraduate assistants, want to eat, drive a car and engage in other costly activities. All this needs direct financial support plus administrative overhead for the university that provides water, gas, electricity, a laboratory, a library and perhaps a secretary. The time of the Medicis is over, especially since the sum total of the required funds is almost always greater than that affordable by private largesse. This means that in selected cases private industry, but otherwise almost always governmental agencies, have to be the grantors of research funds. These agencies have their own spheres of interest which have been assigned to them by the elected representatives which created them. Since the mid-1950s, the main support of biomedical research in the United States of America has come from the NIH, the NIMH and to some extent from the National Science Foundation. The decade 1955–1965 was the most prosperous for federal grants. Since then it has become increasingly more hazardous to become eligible for such research grants. Yet, the peer control system by which an application is judged and graded by committees of experts in the field has remained the most just and most respected means of selecting grant applications for financial support.

I served on five such committees over a period of 20 years and found this activity an opportunity for learning about exciting new thoughts in medicinal chemistry and other areas of biomedical research. My own research was supported by federal grants for 15 years and enabled me to engage coworkers from all over the world. The topics we worked on were affected by national needs and current research policies, but left enough latitude to explore novel approaches. During World War II and the Vietnamese War we made, like so many others, quinoline antimalarials including close precursors of mefloquine. In between, we ran across aromatic amino ethers with potent antituberculous activity.[38–41] Among other topics we worked on, the Virginia tobacco industry helped us in syntheses of *Nicotiana* alkaloids and their metabolites[42–44] and a pesticide company sponsored work on phosphonate analogs of bioactive phosphate esters.[45–53] After the selection of *trans*-2-phenylcyclopropylamine as an antidepressant (tranylcypromine), most of our work turned to psychopharmacological agents. It was both the high tide of research in this field and an opportunity to get our compounds tested meaningfully.

In all this work we shared the fate of other medicinal chemists in that we had to rely on hunches and often semiquantitative test data in planning our next step in drug design. I have often compared this to shadow-boxing, trying to hit a target without even recognizing its outlines. The existence of drug receptors had been assumed for almost a century but their probable chemical classification had remained shrouded in mystery. As analytical and spectroscopic methodology improved, it became possible to isolate and study a few of these elusive biochemicals. Among the earliest ones were the cholinergic receptors.[54] In some cases where receptor-containing tissues were abundant, such as the nicotinic acetylcholine receptor,[55] the receptor material could be fractionated and purified to some extent. In most other cases, the very presence of a receptor was identified only on the basis of the pharmacokinetic behavior of drugs and substrates. Thus, receptors remained pharmacological rather than biochemical concepts, and did not contribute much to drug design where minor steric or small substituent effects could increase or abolish drug activity without a satisfactory chemical explanation. This led increasingly to drawing general receptor areas as circles or rectangles, superimposing such hypothetical outlines on ligating groups of known drug molecules. These unrealistic geometrical representations contributed little to the stepwise design of new active structures. Molecular graphics analysis has refined the circles and rectangles,[56] but is still beset with major uncertainties. Even the inclusion of X-ray crystallographic data of enzymes inhibited by a given drug has not solved all these problems[57] but points more clearly to the direction drug design should take in future research.[58, 59] Electron microscopy of biomaterials, recognized autoradiographically as drug receptors, has revealed the participation of receptors in facilitating the flow of drug molecules across cell membranes and will thereby change our ideas of receptors as static molecules to functional participants on a molecular biological level. In selected cases, such as for MAO inhibitors, a detailed molecular interpretation of their mode of biochemical action has become possible.[60]

Such research, on which the future of drug design depends, is being, and will be, carried out mostly in university laboratories. In the meantime, new drugs for disease conditions will continue to be demanded by the medical profession, and the burden of developing them will rest on the pharmaceutical industry. The two lines of research, for receptor-based drug design and the largely empirical emergence of new bioactive and therapeutic agents, will run parallel and complement and support each other. The discovery of 'lead' compounds will remain a province of pharmacologists who do not mind screening or who devise novel methods of testing for biological activities.

Nature offers us an abundance of compounds as yet unexplored, and observations of side effects of test chemicals can often be boosted to become major activities by molecular modification. There are thousands of metabolites and natural substrates; the modification of their structures is bound to yield metabolite antagonists that may have a potential as drugs. This remains the most viable chemical approach to finding 'lead' structures; among peptides this has barely begun to unfold.

Medicinal scientists have embraced with enthusiasm studies of peptides and especially those protein problems that can be initiated by recombinant DNA techniques. The peptides and proteins clarified by these researches have already made possible new therapies of genetic and hormonal diseases that have not been approachable by conventional drugs. However, this exciting area has been emphasized in some instances over drug design. It should be remembered that recombinant DNA work and straightforward peptide chemistry yields mostly peptides, biologically and catalytically active macromolecules, or peptide-like compounds. The hundreds of drugs now in use have every kind of other structure, aliphatic, aromatic, heterocyclic, *etc.* They are often beset with side effects, are not as specific as peptide drugs and need improvement. Nature did not have human medicine in mind when it directed plants and microbes to metabolize their nutrients to odd 'natural products' that we can isolate from their tissues. Molecular modification of natural 'lead' compounds will always be necessary, although its vicissitudes and ennui will need to be ironed out.

Medicinal chemistry remains a challenging science, which provides profound satisfaction to its practitioners. It intrigues those of us who like to solve problems posed by Nature. It verges increasingly on biochemistry and on all the physical, genetic and chemical riddles in animal physiology which bear on medicine. Medicinal chemists have a chance to participate in the fundamentals of prevention, therapy and understanding of diseases and thereby to contribute to a healthier and happier life.

REFERENCES

1. E. Späth and A. Burger, *Chem. Ber.*, 1927, **60**, 704.
2. E. Späth and A. Burger, *Monatsh. Chem.*, 1926, **47**, 733.
3. L. F. Small, N. B. Eddy, E. Mosettig and C. K. Himmelsbach, 'Studies on Drug Addiction', Supplement No. 138 to the Public Health Reports, U.S. Government Printing Office, Washington, DC, 1938.
4. E. Mosettig and A. Burger, *J. Am. Chem. Soc.*, 1930, **52**, 3456.
5. E. Mosettig and A. Burger, *J. Am. Chem. Soc.*, 1931, **53**, 2295; 1935, **57**, 2189.
6. A. Burger and E. Mosettig, *J. Am. Chem. Soc.*, 1934, **56**, 1745; 1936, **58**, 1570, 1857.
7. J. Van de Kamp, A. Burger and E. Mosettig, *J. Am. Chem. Soc.*, 1938, **60**, 1321.
8. A. Burger, *J. Am. Chem. Soc.*, 1938, **60**, 1533.
9. A. Burger and G. H. Harnest, *J. Am. Chem. Soc.*, 1943, **65**, 2382.
10. A. Burger, R. W. Alfriend and A. J. Deinet, *J. Am. Chem. Soc.*, 1944, **66**, 1327.
11. A. Burger and A. J. Deinet, *J. Am. Chem. Soc.*, 1945, **67**, 566.

12. O. Eisleb and O. Schaumann, *Dtsch. Med. Wochenschr.*, 1939, **65**, 967.
13. O. Eisleb (Winthrop Chemical Co.), *US Pat.* 2 167 351 (1939) (*Chem. Abstr.*, 1939, **33**, 8923); *Chem. Ber.*, 1941, **74**, 1433.
14. A. Pfeiffer, V. Brantl, A. Hertz and H. M. Emrich, *Science (Washington, D.C.)*, 1986, **233**, 774.
15. H. Erlenmeyer and M. Leo, *Helv. Chim. Acta*, 1932, **15**, 1171.
16. H. Erlenmeyer and M. Leo, *Helv. Chim. Acta*, 1933, **16**, 897, 1381.
17. I. Langmuir, *J. Am. Chem. Soc.*, 1919, **41**, 1543.
18. H. G. Grimm, *Z. Elektrochem.*, 1925, **31**, 474; 1928, **34**, 430; *Angew. Chem.*, 1929, **42**, 367; 1934, **47**, 53, 594; *Naturwissenschaften*, 1929, **17**, 535, 557.
19. O. Hinsberg, *J. Prakt. Chem.*, 1916, **93**, 302.
20. H. L. Friedman, National Academy of Sciences—National Research Council, Publication No. 206, Washington, DC, 1951, p. 295.
21. A. Burger, '*Medicinal Chemistry*', Interscience, New York, 1951, vol. I, chap. 4; in '*Medicinal Chemistry*', 3rd edn., ed. A. Burger, Wiley-Interscience, New York, 1970, part 1, chap. 6.
22. C. Hansch, in '1972 Intrascience Symposium', Intrascience Research Foundation, Santa Monica, CA; *Intra-Science Chem. Rep.*, 1974, **8**, 17.
23. C. W. Thornber, *Chem. Soc. Rev.*, 1979, **8**, 563.
24. C. A. Lipinski, *Annu. Rep. Med. Chem.*, 1986, **21**, 283.
25. A. Burger, 'A Guide to the Chemical Basis of Drug Design', Wiley-Interscience, New York, 1983, pp. 28–30, 84–87.
26. C. Hansch and T. Fujita, *J. Am. Chem. Soc.*, 1964, **86**, 1616.
27. C. Hansch, *J. Med. Chem.*, 1976, **19**, 1.
28. A. Burger, W. B. Wartman, Jr. and R. E. Lutz, *J. Am. Chem. Soc.*, 1938, **60**, 2628.
29. A. Burger and S. Avakian, *J. Am. Chem. Soc.*, 1940, **62**, 226.
30. G. A. Alles, *J. Pharmacol. Exp. Ther.*, 1927, **32**, 121; 1933, **47**, 339.
31. G. A. Alles and M. Prinzmetal, *J. Pharmacol. Exp. Ther.*, 1933, **48**, 161.
32. A. Burger and W. L. Yost, *J. Am. Chem. Soc.*, 1948, **70**, 2198.
33. C. L. Zirkle, C. Kaiser, D. H. Tedeschi, R. E. Tedeschi and A. Burger, *J. Med. Pharm. Chem.*, 1962, **5**, 1265.
34. A. Burger and S. Nara, *J. Med. Chem.*, 1965, **8**, 859.
35. P. Zeller, A. Pletscher, K. F. Gey, H. Gutmann, B. Hegedüs and O. Straub, *Ann. N.Y. Acad. Sci.*, 1959, **80**, 555.
36. L. H. Sternbach, *J. Med. Chem.*, 1980, **22**, 1.
37. E. Fingle and D. M. Woodbury, in 'The Pharmacological Basis of Therapeutics', 5th edn., ed. L. S. Goodman and A. Gilman, 1975, Macmillan, New York, p. 41.
38. L. Long, Jr. and A. Burger, *J. Am. Chem. Soc.*, 1941, **63**, 1586.
39. A. Burger, E. L. Wilson, C. O. Brindley and F. Bernheim, *J. Am. Chem. Soc.*, 1945, **67**, 1416.
40. A. K. Saz, F. R. Johnston, A. Burger and F. Bernheim, *Am. Rev. Tuberc.*, 1943, **48**, 40.
41. A. Burger, *J. Am. Pharm. Assoc.*, 1947, **36**, 372.
42. A. Burger (Tobacco Byproducts and Chemical Corporation), *US Pat.* 2 315 314 (1943). (*Chem. Abstr.*, 1943, **37**, 5556).
43. R. N. Castle and A. Burger, *J. Am. Pharm. Assoc.*, 1954, **43**, 163.
44. M. L. Stein and A. Burger, *J. Am. Chem. Soc.*, 1957, **79**, 154.
45. N. D. Dawson and A. Burger, *J. Am. Chem. Soc.*, 1952, **74**, 5312.
46. B. E. Smith and A. Burger, *J. Am. Chem. Soc.*, 1953, **75**, 5891.
47. A. Burger, J. B. Clements, N. D. Dawson and R. B. Henderson, *J. Org. Chem.*, 1955, **20**, 1383.
48. J. R. Parikh and A. Burger, *J. Am. Chem. Soc.*, 1955, **77**, 2386.
49. B. S. Griffin and A. Burger, *J. Am. Chem. Soc.*, 1956, **78**, 2336.
50. A. Burger and J. J. Anderson, *J. Am. Chem. Soc.*, 1957, **79**, 3575.
51. M. E. Wolff and A. Burger, *J. Am. Chem. Soc.*, 1957, **79**, 1970.
52. R. D. Bennett, A. Burger and W. A. Volk, *J. Org. Chem.*, 1958, **23**, 940.
53. M. E. Wolff and A. Burger, *J. Am. Pharm. Assoc.*, 1959, **48**, 56.
54. B. M. Conti-Tronconi, C. M. Gotti, M. W. Hunkapiller and M. A. Raftery, *Science*, 1982, **218**, 1227.
55. H. G. Mautner, in 'New Methods in Drug Research', ed. A. Makriyannis, Prous, Barcelona, 1985, chap. 5, p. 83.
56. C. Hansch, T. Klein, J. McClarin, R. Langridge and N. W. Cornell, *J. Med. Chem.*, 1986, **29**, 615.
57. C. D. Selassie, Z.-X. Fang, R. Li, C. Hansch, T. Klein, R. Langridge and B. T. Kaufman, *J. Med. Chem.*, 1986, **29**, 621.
58. U. Rickenbacher, J. D. McKinney, S. J. Oatley and C. C. F. Blake, *J. Med. Chem.*, 1986, **29**, 641.
59. M. Williams and S. J. Enna, *Annu. Rep. Med. Chem.*, 1986, **21**, 211.
60. L. E. Richards and A. Burger, *Prog. Drug Res.*, 1986, **30**, 205.

1.2
Chronology of Drug Introductions

WALTER SNEADER
University of Strathclyde, Glasgow, UK

1.2.1 INTRODUCTION

1.2.1.1 Drug Prototypes[1]

There are more than 1000 organic compounds in regular use around the world as therapeutic agents. These have been developed from a few score of prototypes, of which only a small minority were suitable for clinical purposes without further development. The structure of this chapter should indicate to the reader how these prototypes served as the lead compounds from which many valuable medicines were derived.

Until the advent of synthetic organic chemistry in the 19th century, the plant kingdom remained the major source of complex chemicals which could interfere with biological processes.[2,3] Nature never generated such compounds with the intention that they would be exploited for medicinal purposes. On the contrary, some are likely to have been synthesized by plants as poisons to repel foraging predators. For this reason, it is hardly surprising that few of the active principles from plants have been able to compete with those modern synthetic drugs which have been selected from scores, hundreds or sometimes even thousands of analogues before being released for use in man.[4] The 1980 'British Pharmacopoeia' lists only four active principles from plants which have been introduced during the past half century.

Plants and other natural products containing pharmacologically active constituents have been used by man from the dawn of history, but only since the 18th century have there been critical attempts to assess whether these confer therapeutic benefit. Notable events in that century were the testing of different products by Lind[5] in an attempt to evaluate antiscorbutic activity, the assessment of the foxglove by Withering[6] with a view to establishing its true value in the clinic, and the comparison by Pringle[7] of the preservative powers of different salts. These particular studies affirmed the value of some of the materials under investigation, but most other studies achieved the opposite. A vast range of animal and herbal products with which the recently introduced pharmacopoeias had been crammed were steadily eliminated from therapeutic practice by enlightened practitioners, so that by the beginning of the 19th century there was a void waiting to be filled. The successful isolation of the active principles from many plants at first held out the prospects of filling that void with more potent drugs of consistent quality. What actually happened was quite different, for instead long-cherished remedies were exposed as near worthless once their active principles had been found wanting when tested by pharmacologists or clinicians. However, there were some developments which helped to counter the inevitable mood of therapeutic nihilism during the first half of the 19th century. The demonstration of the beneficial properties of the pure alkaloids from cinchona bark and ipecacuanha root showed that drugs from the New World could rival those from the ancient—a harbinger of things to come!

What of the traditional remedies from the Ancient World? The isolation of morphine in 1817 confirmed the narcotic and analgesic powers of opium, the dried latex exuded by the poppy *Papaver somniferum*, whilst the separation of colchicine[8,9] from the autumn crocus *Colchicum autumnale*, shortly after, confirmed its value in gout. Both had been known to the Greeks, as had many of the other herbal remedies still being prescribed. Why did so few retain their place in therapeutic practice? The answer seems to be that their reputations were, for the most part, founded either on magico-religious belief or unsubstantiated empiricism. That they had survived since ancient times is a reflection of the psychotherapeutic aura generated by physicians over the centuries

1.2.1.2 Drugs in the Ancient World[10,11]

The earliest documentary record of drug therapy is to be found on a Sumerian clay tablet from around 2100 BC, which records several recipes without indicating for what they were used. However, our major source of information prior to the Greco-Roman period comes from several Egyptian medical papyri, especially the Ebers papyrus,[12] dated around 1550 BC. The latter contains more than 800 prescriptions, many of which are accompanied by ritual incantations requesting divine intervention to alter the course of the disease.

1.2.1.2.1 Egyptian medicine

The selection of drugs to counter diseases often reflects the current appreciation of their aetiology. In the absence of any obvious cause of internal diseases, the Egyptians believed that supernatural forces were at work. This inevitably resulted in the introduction of magical and religious strategies in treating patients.[13] The actual remedies were drawn from animal, vegetable and sometimes mineral sources, many being household remedies borrowed from the kitchen or plucked from the garden. Where the cause of a complaint was recognizable, as in wounds and swelling, the Egyptians exercised good common sense in their selection of soothing balms and lotions prepared with materials such as frankincense, myrrh and acacia. There is little evidence that their use of vegetable purgatives such as colcynth, senna and castor oil was to expel demons, as has been the case in some primitive societies. Purgatives were apparently administered to ease constipation, just as vegetable diuretics were chosen to relieve oedema.

1.2.1.2.2 Greek Medicine[14]

The trade and cultural links which developed between Greece and Egypt ensured that there would be a constant exchange of medical knowledge between the two countries. The Greeks incorporated many of the Egyptian remedies into their own herbals, in the process often seeking to rationalize their use of these drugs in terms of their own humoral theories of disease rather than rely on religious or magical interpretations. An excess or deficiency of any of the four humours (blood, phlegm, black bile and yellow bile), so it was claimed, could be countered by the administration of a concoction of herbs with opposing properties. This reliance on humoral theories resulted in secularization of therapeutic practice, but it certainly did not result in the rejection of those herbs to which had formerly been attributed supernatural qualities. Nevertheless, there were those who questioned the value of such theories. Most notable amongst these critics was the first century physician Dioscorides,[15] who deplored the vain speculation of his contemporaries about the reasons as to why drugs worked. Instead, he urged them to concentrate their attention on what actually happened when drugs were administered. His five volume treatise 'De Materia Medica' was by far the largest and most authoritative pharmaceutical guide in antiquity.[16] In discussing over 600 plants, 35 animal products and 90 minerals, it added considerably to the knowledge of drugs in the Ancient World. It was widely circulated and passed down through the generations,[17] exercising a major influence on the Arabian physicians and, ultimately, on Europe during the Renaissance. A direct consequence of this is that the system of plant nomenclature still used at present was largely determined by the influence of the writings of Dioscorides on the 16th century medical botanists. Many of the products cited in his works are familiar to modern ears, *e.g.* almond oil, aloes, belladonna, calamine, cherry syrup, cinnamon, coriander, galbanum, galls, ginger, juniper, lavender, lead acetate, marjoram, mastic, mercury, olive oil, opium, pepper, pine bark, storax, sulfur, terebinth, thyme and wormwood.

The other dominant figure from the same era was Galen (129–199).[18] He was an outstanding physician, but his enthusiasm for humoral medicine ensured its perpetuation through the reproduction of his extensive writings over the next 1600 years. His major work dealing with drug therapy was 'On the Art of Healing'. The writings of Galen constituted the basis of the curricula of medical schools throughout the Middle Ages. This had a stultifying influence on the development of therapeutics until well into the 17th century since the emphasis was in correcting an imagined humoral imbalance in the patient rather than seeking an external cause of disease. Galen was probably the most influential physician of all times.

1.2.1.3 Arabic Medicine[19,20]

The Muslim conquerers of much of the former Roman Empire preserved and nourished Greco-Roman science and medicine, ensuring that the writings of Dioscorides, Galen and others would not disappear under the strictures of a Church which construed the practice of the healing art to be an attempt to thwart divine providence. Many medical texts were translated into Arabic in the middle of the 9th century at Baghdad under the supervision of Hunayn (809–873), and resulted in the spread of humoral medicine throughout the territories administered by the enlightened Eastern Caliphate, and beyond. In the process there was added the medical lore of India and China, gleaned through extensive trading contacts.

An important contribution of Arabic physicians was the introduction of many mineral products previously ignored as medicinal agents. Whilst most were applied externally, a few were taken by mouth. One of the leading exponents of the medicinal use of minerals was al-Razi (865–925), who employed salts of copper, mercury, arsenic and gold. Another was Abulcasis, who lived in Spain in the second half of the 10th century and made extensive use of minerals which were readily obtained from local mines. He described in his 'Liber Servitoris' how raw minerals could be purified for medicinal application. It is not without relevance in this present chapter to note that this constituted one of the earliest references to the use of pure chemicals in therapy, and might be held to represent the dawn of medicinal chemistry.

1.2.1.4 Post-Renaissance European Medicine[21]

The medical knowledge of Greece was preserved in the Arab world during what was, for Europe, the Dark Ages. In time, Jewish physicians travelling around the courts of Europe reintroduced the old medical lore, but it was the invention of the printing press in the 15th century which made the works of Dioscorides, Galen and their contemporaries readily available throughout Europe. The first printed formularies and pharmacopoeias clearly reflected the influence of Dioscorides. Inevitably, there was an immediate resurgence of interest in medicinal herbs. However, the use of herbal remedies was challenged by Paracelsus (1493–1541)[22] and his followers. They urged the adoption of the chemical remedies, in particular preparations of antimony, gold and mercury.[23] Several of the Paracelsan physicians attained positions of considerable influence, with the result that the mineral salts began to appear in the new pharmacopoeias that were being published. Nevertheless, the herbal remedies continued to be looked upon with favour throughout Europe.

The wide availability of the old medical texts during the 16th and 17th centuries may have brought herbal medicine to its zenith, but the dawn of scientific enlightenment soon took its toll of many worthless remedies. The advent of modern chemistry ultimately resulted in the demonstration of the merit or otherwise of ancient herbs as their long-hidden constituents were finally extracted and at last exposed to critical examination. The first steps in this direction were taken by Scheele (1742–1786),[24] the father of plant chemistry, who had isolated tartaric acid in his pharmacy by 1768 and went on to do likewise with gallic, oxalic, uric, lactic, mucic, citric and malic acids. None of these were physiologically active. It was not until early in the next century that the first active principle was to be isolated.[25]

1.2.2 ACTIVE PRINCIPLES FROM PLANTS, AND THEIR DERIVATIVES

1.2.2.1 Alkaloids[26]

1.2.2.1.1 Opium

The earliest attempts to isolate the active principles in opium[27] were inspired by a desire amongst pharmacists to find a chemical method of distinguishing authentic from adulterated material. Although Derosne obtained what appears to have been the antitussive alkaloid noscapine (**1**) in 1804, the credit for isolating the first physiologically active substance from a plant is usually attributed to Sertürner. He isolated morphine (**2**), probably in an impure form, in 1805. His detailed report of 1817 was republished in a prominent French journal where it received wide attention and stimulated workers in Paris to seek out other alkaloids.[28–30]

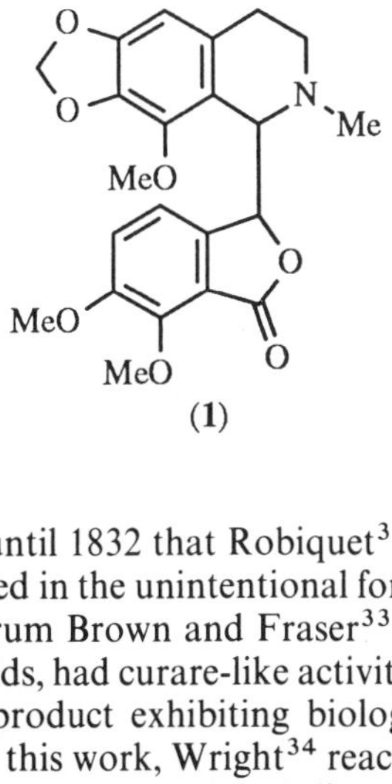

(**1**) 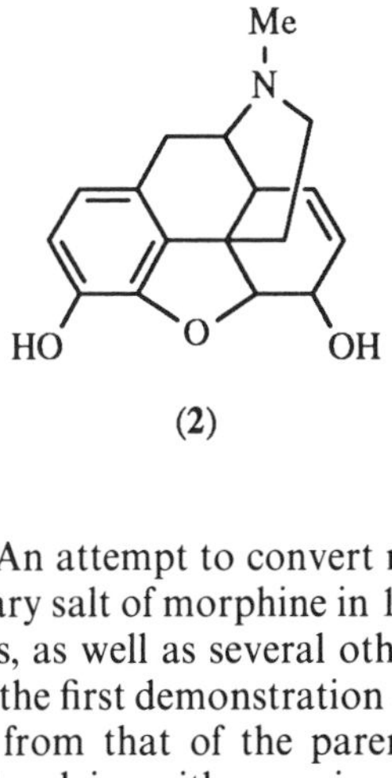 (**2**)

It was not until 1832 that Robiquet[31] isolated codeine (**3**). An attempt to convert morphine into codeine resulted in the unintentional formation of the quaternary salt of morphine in 1853.[32] Fifteen years later, Crum Brown and Fraser[33] demonstrated that this, as well as several other quaternary salts of alkaloids, had curare-like activity. This was apparently the first demonstration of a derivative of a natural product exhibiting biological activity different from that of the parent compound. Stimulated by this work, Wright[34] reacted both morphine and codeine with organic acids. Amongst the derivatives that resulted was diacetylmorphine (**4**) (diamorphine), the properties of which were first examined in 1874. Almost a quarter of a century later this was introduced commercially by Dreser[35] as a morphine substitute with allegedly diminished respiratory depressant properties. It followed close in the wake of the introduction by a rival manufacturer of the ethyl ether of morphine as a codeine-like cough suppressant. The only morphine ether still frequently used is pholcodine (**5**). The addictive liability of diamorphine discouraged the preparation of other similar compounds, but a series of hydrogenated derivatives including oxycodone (**6**) and hydrocodone (dihydrocodeineone) (**7**) were synthesized from the opium alkaloid thebaine during the years 1916–1920 by Freund and Speyer.[36] Hydromorphone (**8**) followed in 1923. During the 1930s, a programme of research led by Small and Eddy and sponsored by several national agencies in the United States failed to produce a non-addictive morphine substitute. The most promising compound resulting from that effort was methyldihydromorphinone (**9**),[37] which received the approved name of metopon. It was about three times as potent as morphine and less likely to produce drowsiness or nausea.

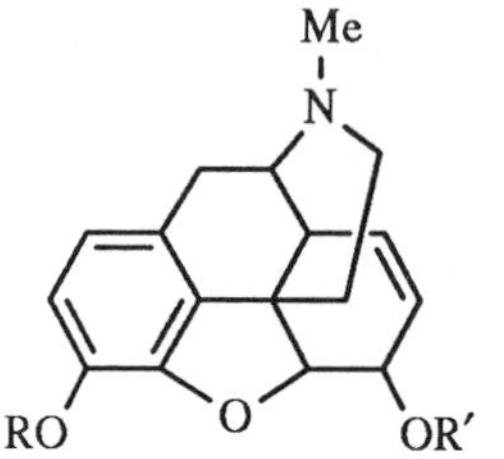

(**3**) R = Me, R′ = H
(**4**) R = COMe, R′ = COMe
(**5**) R = CH_2CH_2-N(morpholino)O, R′ = H

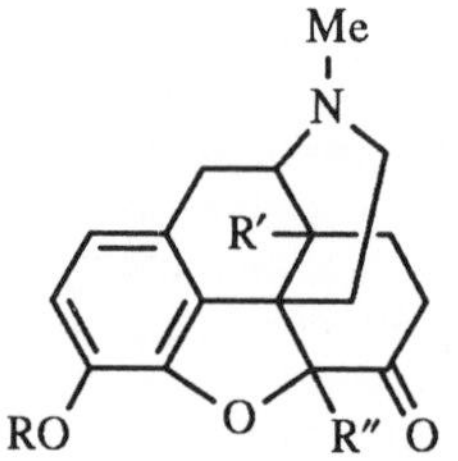

(**6**) R = Me, R′ = OH, R″ = H
(**7**) R = Me, R′ = H, R″ = H
(**8**) R = H, R′ = H, R″ = H
(**9**) R = H, R′ = H, R″ = Me

Until the correct structure of morphine was determined by Gulland and Robinson[38] in 1923, its analogues necessarily consisted of simple derivatives such as ethers, esters and oxidation or reduction products. Just before the outbreak of the Second World War, Grewe[39] synthesized *N*-methylmorphinan, the first structurally simplified analogue of morphine to exhibit analgesic activity. The hydroxy derivative levorphanol (**10**) proved to be particularly effective as an oral analgesic. The dextrorotatory isomer of this latter compound's methyl ether was devoid of analgesic activity, but retained useful antitussive activity. Many other analogues of levorphanol have been synthesized, most being modelled on the known variants of morphine and codeine.

In 1954, Lasagna and Beecher reported that the morphine antagonist nalorphine was also a non-addictive analgesic. Its hallucinatory activity precluded clinical use. This arose from a lack of selectivity for the appropriate type of opioid receptors, hence a search began for more selective agents.[40] It took 10 years before Archer *et al.*[41] introduced pentazocine (**11**), but this still retained hallucinatory activity at effective analgesic dose levels. The first non-addictive analgesic devoid of hallucinatory properties, nalbuphine (**12**), was patented by Pachter[42] in 1968. It was followed by butorphanol (**13**),[43] which likewise was only effective when administered parenterally. Also in 1968, Bentley[44] prepared buprenorphine (**14**), which was formulated for sublingual administration so as to avoid rapid hepatic metabolism. None of the recent non-addictive analgesics seem to have yet satisfied all the clinical requirements for complete control of severe pain.

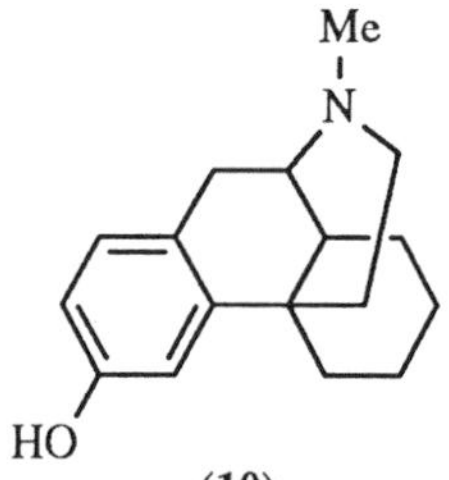

(**10**) 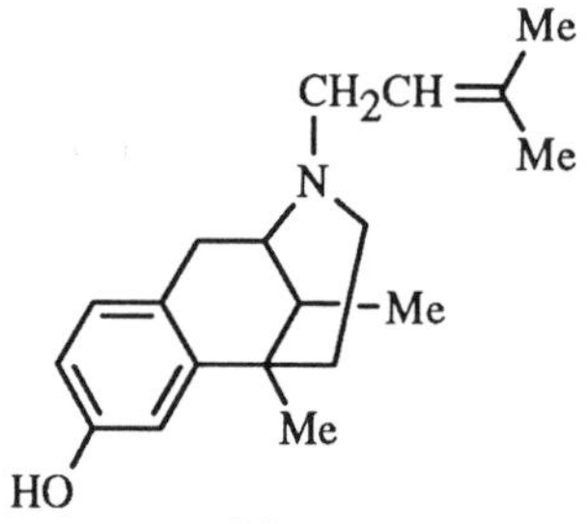 (**11**) 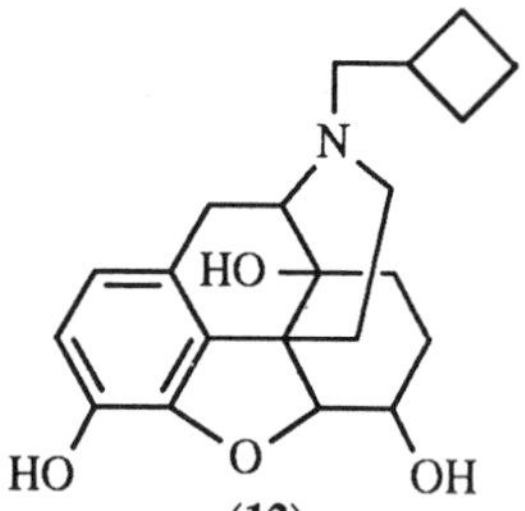 (**12**)

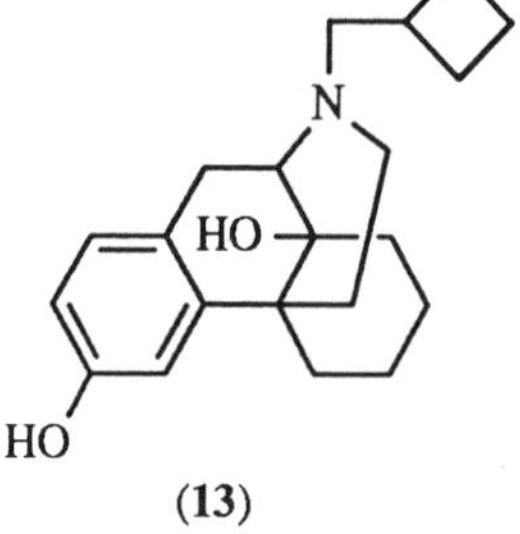

(**13**) 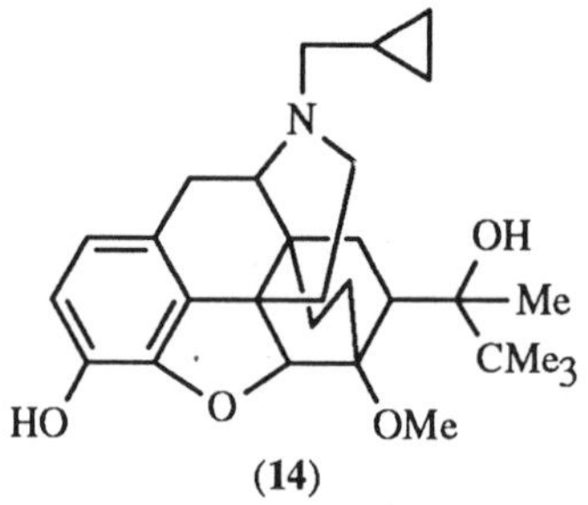 (**14**)

In 1848, George Merck recovered the benzylisoquinoline alkaloid papaverine (**15**) from the mother liquors remaining after the extraction of morphine from opium. It was largely ignored for many years since it exhibited only feeble analgesic activity. Not until after Macht,[45] in 1917, had described its spasmolytic action on smooth muscle were attempts made to employ papaverine in the clinic. From then it was widely used until the introduction of synthetic atropine analogues in the 1930s. When supplies were threatened through the introduction of more stringent legislation against the uncontrolled production of opium, attempts were immediately made to synthesize the alkaloid by a commercially viable process. This was duly achieved by Mannich[46] in 1927 and by others during the next few years. Modifications to these syntheses of papaverine soon permitted the preparation of closely related analogues, such as eupapverin (**16**) and ethaverine (**17**). Their popularity stimulated the introduction of open ring analogues such as alverine[47] (**18**) in 1935, cyverine[48] (**19**) in 1939 and mebeverine (**20**) in 1962. The antianginal drug verapamil (**21**), also introduced in 1962, is another open chain analogue of papaverine.

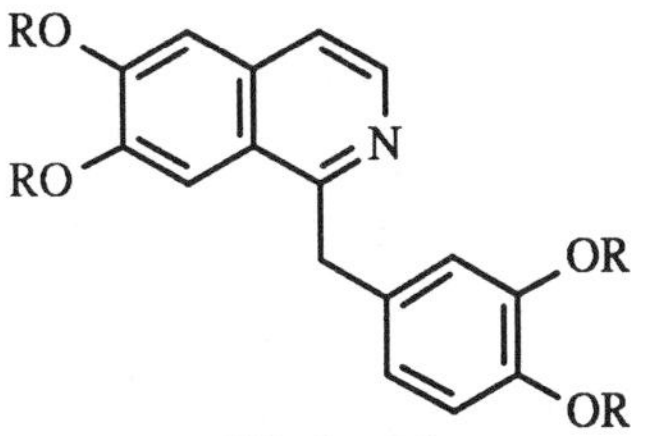

(**15**) R = Me
(**17**) R = Et

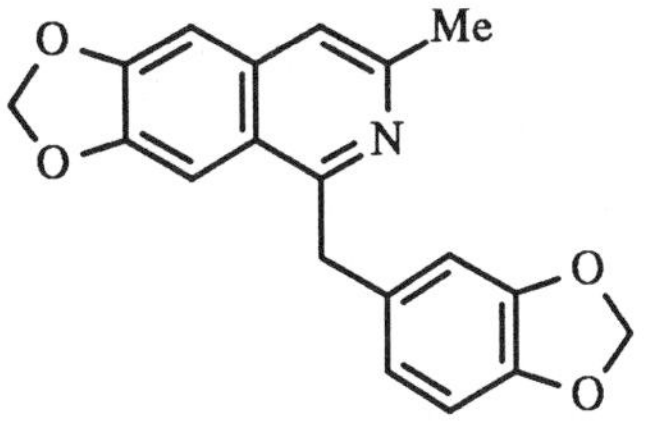

(**16**)

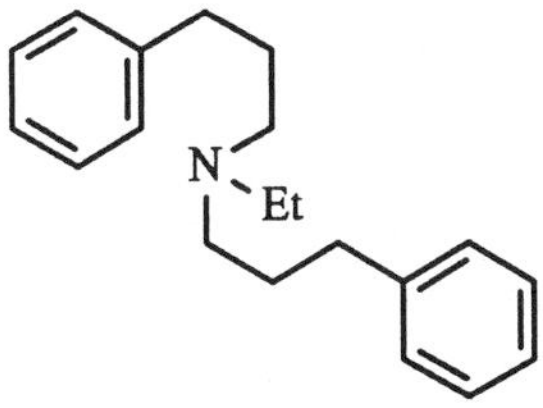

(**18**)

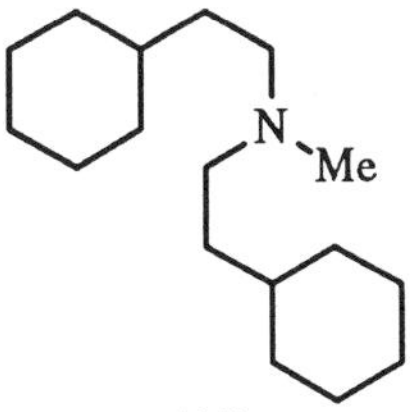

(**19**)

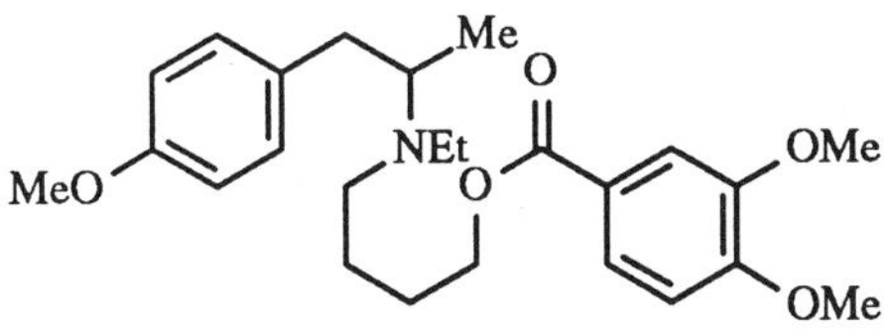

(**20**)

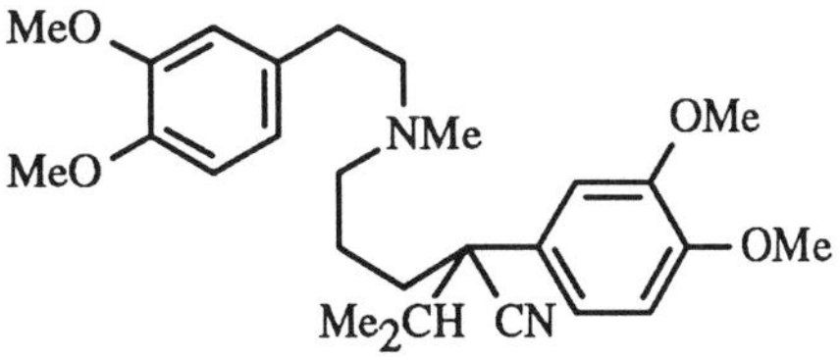

(**21**)

1.2.2.1.2 Cinchona

The isolation of emetine (**22**) from ipecacuanha root, *Cephaelis ipecacuanha*, by Pelletier and Magendie in 1817[49–51] was followed a year later by that of strychnine (**23**) from *Strychnos nux vomica* by Pelletier and Caventou. In 1819, Runge isolated caffeine (**24**) from coffee beans and quinine (**25**) from cinchona bark (naming it China base).[52] The discovery of the latter is often attributed to Pelletier and Caventou who, in fact, discovered it shortly after Runge.[53,54] Its optical isomer quinidine was isolated in 1833 by Henry and Delondre,[55] being generally considered somewhat inferior as an antimalarial. However, it was eventually applied in the treatment of cardiac arrhythmias in 1918.

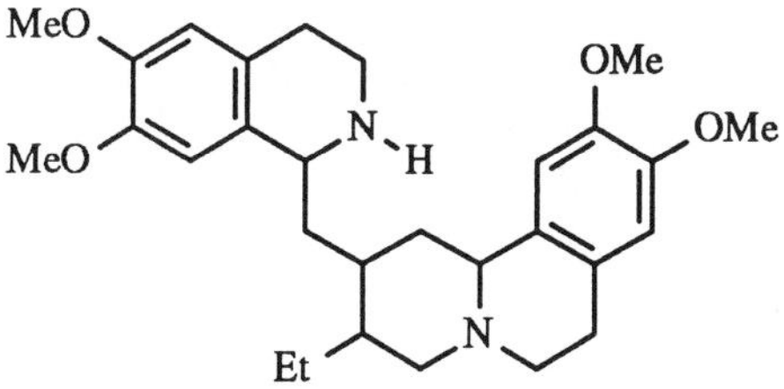

(**22**)

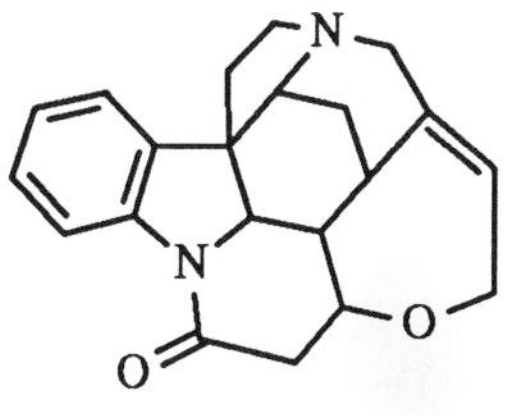

(**23**)

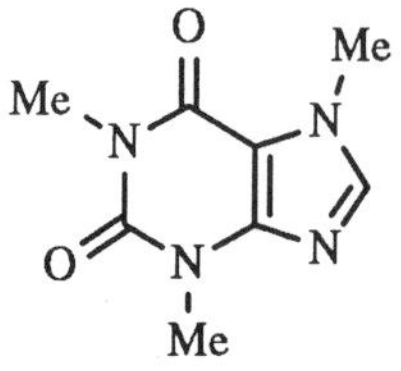

(24)

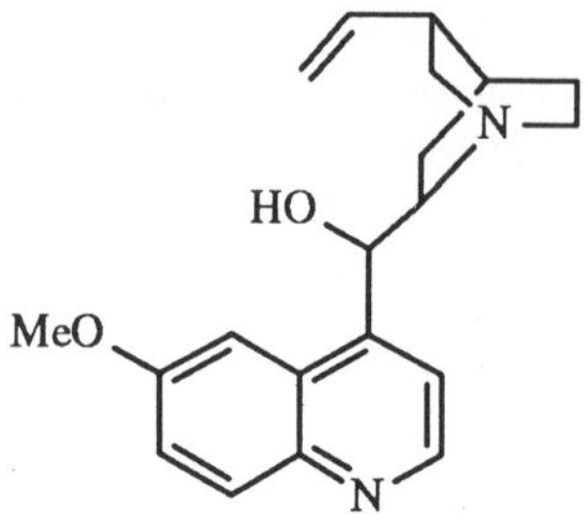

(25)

Acting on the mistaken impression that quinine was a tetrahydroquinoline alkaloid, Koenigs and Fischer prepared a series of tetrahydroquinolines for pharmacological evaluation. In 1881, one of these was marketed as an antipyretic under the proprietary name of kairin (**26**),[56] soon to be superseded by the related kairoline A (**27**) and thalline (**28**). The toxicity of these compounds rendered them inferior to phenazone (**29**), a drug which was assumed to be a tetrahydroquinoline when first synthesized by Knorr in 1884.[57] Tested by Filehne, as had been the earlier compounds, it was seen to be safer and more palatable. Its outstanding commercial success did much to stimulate interest in synthetic drugs. Amidopyrine (**30**), one of its analogues prepared by Knorr, was shown to be three times as potent and was marketed 12 years later.[59] Widely used not only as antipyretics, but also as analgesics and antirheumatics, these drugs were incriminated from the 1930s onwards as having caused agranulocytosis, a fate which also befell the still more potent and highly popular phenylbutazone (**31**). This was introduced in 1952, retaining a significant market share for nearly 30 years until regulatory authorities began to demand its withdrawal on the grounds that the hazards, which also included a slight risk of aplastic anaemia, were no longer acceptable when compared with certain more modern agents. Oxyphenbutazone (**32**), its active metabolite, and feprazone (**33**) have also been withdrawn.

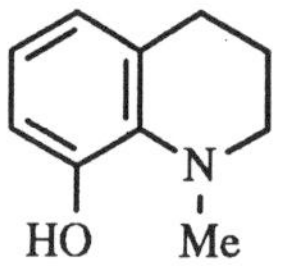

(26) 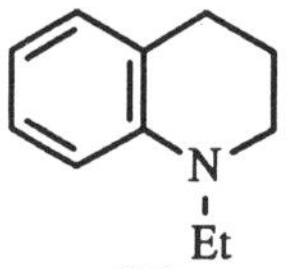 (27) 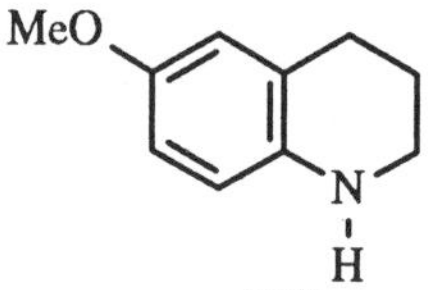 (28)

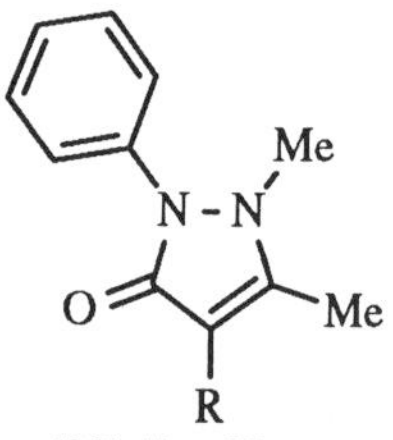

(29) R = H
(30) R = NMe_2

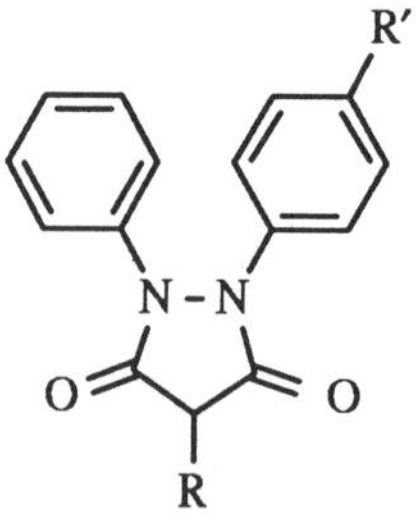

(31) R = $(CH_2)_3Me$, R′ = H
(32) R = $(CH_2)_3Me$, R′ = OH
(33) R = $CH_2CH{=}CMe_2$, R′ = H

1.2.2.1.3 Colchicum

The isolation of colchicine (**34**) from *Colchicum autumnale* in 1820 by Pelletier and Caventou, has already been mentioned (Section 1.2.1.1). It represented an important advance in the treatment of gout since the pure alkaloid was thereby freed from contamination by the cardiotoxic veratrine present in the colchicum corm.

1.2.2.1.4 Calabar bean

Jobst and Hesse isolated physostigmine (**35**) in a pure form from Calabar bean[59] (*Physostigma venenosum*) in 1864.[60] The elucidation of its structure by Stedman and Barger in 1925 enabled the former to determine which molecular features were essential for miotic activity. In turn, this facilitated the synthesis of simpler molecules that retained biological activity. Aeschlimann[61] introduced the widely used analogue neostigmine (**36**) in 1931. Its enhanced resistance to chemical hydrolysis when compared to physostigmine was advantageous. Subsequent studies revealed neostigmine was also superior to the natural product in the treatment of the muscular disease myasthenia gravis. During the 1950s, similar analogues with diminished cholinergic side-effects were developed, *viz.* pyridostigmine (**37**), distigmine (**38**) and ambenonium (**39**).

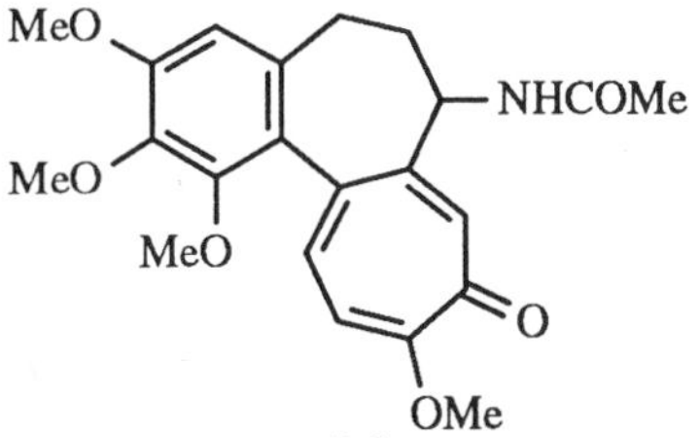

(**34**) 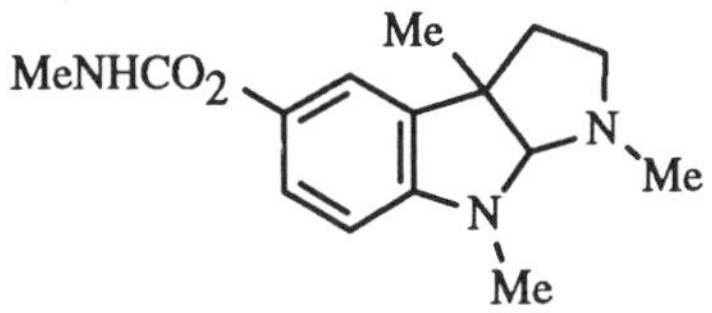 (**35**) 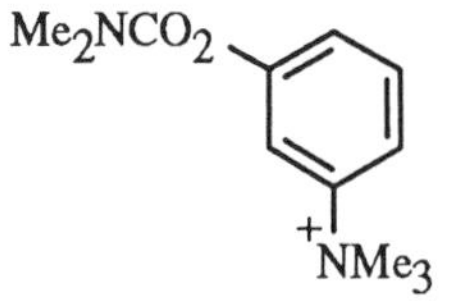 (**36**)

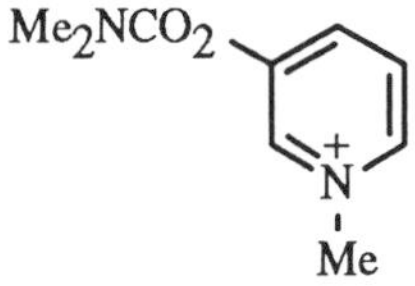

(**37**) 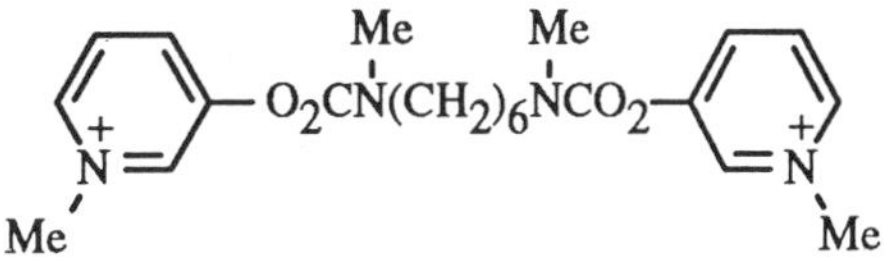 (**38**)

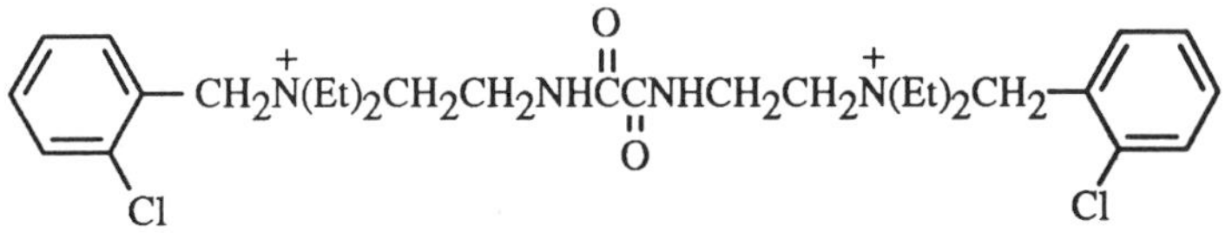

(**39**) 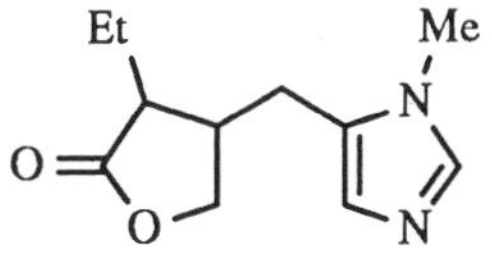 (**40**)

1.2.2.1.5 Pilocarpus

Another alkaloid capable of mimicking the action of acetylcholine on the eye is pilocarpine (**40**). It was isolated from *Pilocarpus jaborandi* in 1875 by both Hardy[62] and Gerrard.[63] Initially it was used as a diaphoretic to eliminate excess fluid from the body by increased sweating, but the development of effective diuretics has rendered this obsolete. Pilocarpine, like physostigmine, is now prescribed solely for the reduction of increased intraocular pressure in glaucoma.

1.2.2.1.6 Solanaceous alkaloids

Shortly after its isolation, physostigmine was shown to antagonize the action of atropine (**41**) on the pupil of the eye. This latter alkaloid had itself been isolated in 1831 by Mein, but it was not until almost half a century later, after Buchheim had suspected a second alkaloid might be present in belladonna extracts, that Ladenburg separated hyoscine (scopolamine) (**42**).

Following his discovery that atropine could be synthesized from its components tropine and tropic acid, Ladenburg[64] proceeded to prepare a series of physiologically active analogues by combining a variety of aromatic acids with tropine. One of these was homatropine (**43**), synthesized in 1880, which has retained its place as an alternative to atropine in ophthalmology due to its shorter duration of action. Another variant was its quaternary ammonium salt, atropine methonitrate. Originally introduced in 1902 as a mydriatic for use by ophthalmologists, it is today employed as an antispasmodic since its polar nature lessens the likelihood of side effects on the central nervous system.

The most important development relating to the synthesis of atropine analogues arose from von Braun's exploratory studies[65] on the transposition of functional groups within drug molecules. These revealed the wide scope for structural variation without loss of anticholinergic activity. This afforded pharmaceutical companies considerable freedom to introduce analogues without infringing the patents held by their rivals! The first commercially important agent was marketed in the late 1920s under the name navigan (**44**), being promoted as an alternative to hyoscine for sea-sickness.

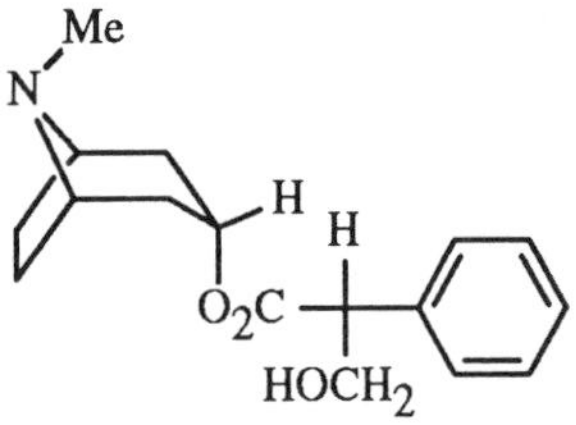

(41) 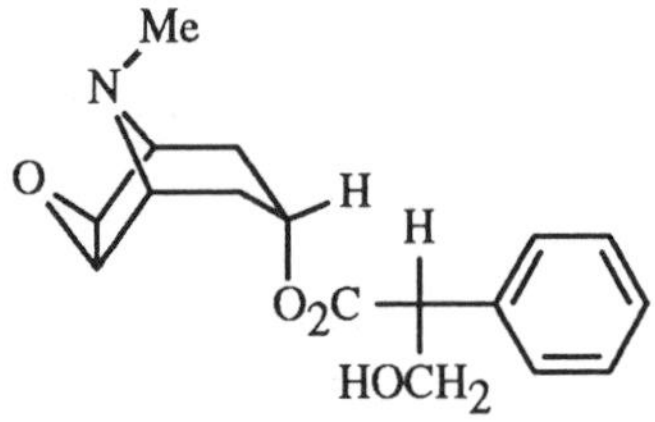 (42) 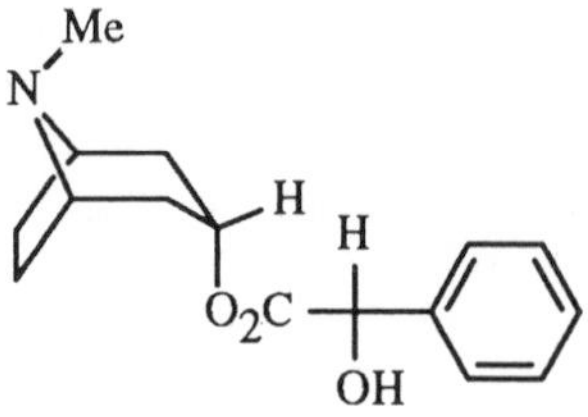 (43)

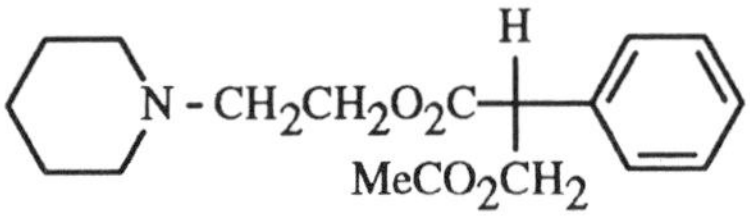

(44)

The recognition of the structural similarity between atropine and cocaine was responsible for the grafting of a typical synthetic local anaesthetic side chain on to a tropine substitute to produce amprotropine (**45**). The modest reduction in its antisecretory as compared with its antispasmodic activity stimulated researchers to seek an as yet still unattained complete separation of these two activities.[66] Variants of amprotropine include agents such as lachesine (**46**), developed during the Second World War as an antidote to nerve gas poisoning. Others are pipenzolate (**47**), piperidolate (**48**) and mepenzolate (**49**). Amongst the most important analogues of amprotropine are pethidine (meperidine) (**50**) and methadone (**51**). The former was unexpectedly found, in 1937, to exhibit analgesic activity when tested in animals,[67] and it is often used as an alternative to morphine. Soon after, methadone[68] was found to be an even stronger analgesic. Both, however, were addictive.

One of the analogues of pethidine is diphenoxylate (**52**). This has proved to be a useful antidiarrhoeal agent as it retains morphine-like activity on the gut, but after absorption is enzymatically destroyed in the liver before it can exert systemic effects. However, phenoperidine (**53**) and fentanyl (**54**) are metabolically much more stable and thus retain analgesic activity, as do the

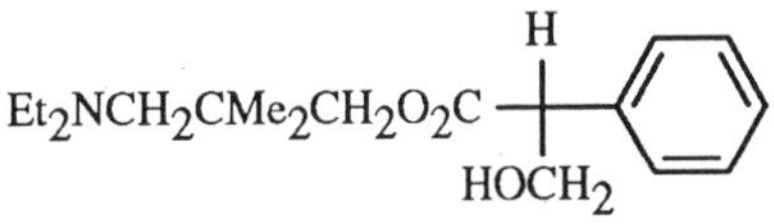

(45) 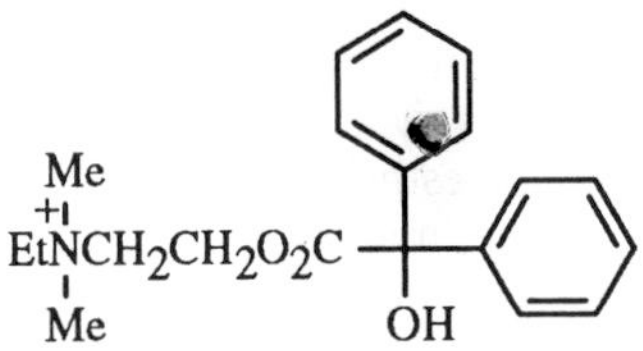 (46)

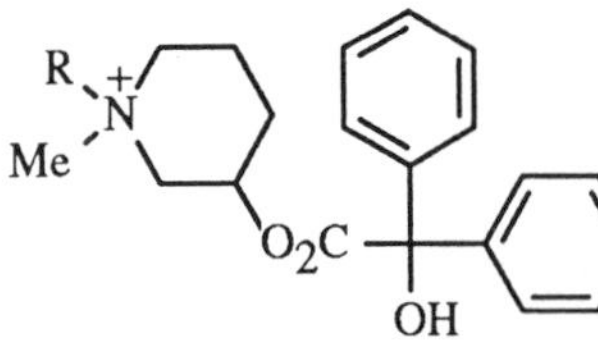

(47) R = Et
(49) R = Me

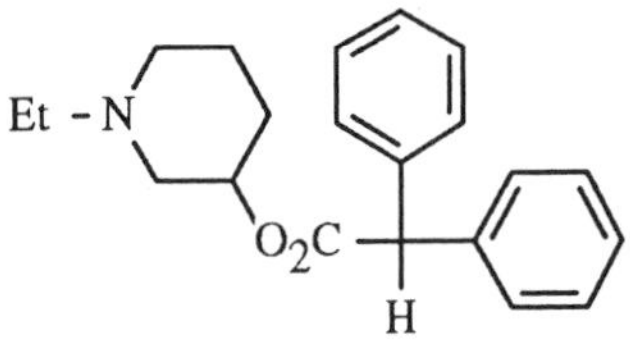

(48)

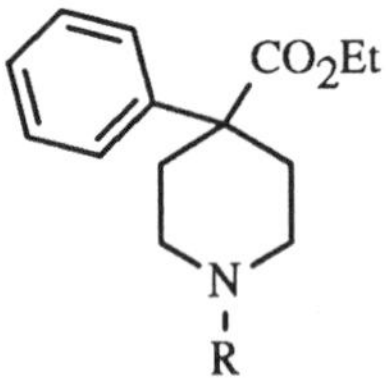

(50) R = Me
(53) R = $CH_2CH_2CH(OH)Ph$

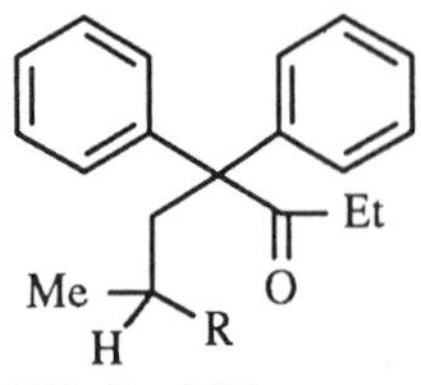

(51) R = NMe_2
(55) R = —N(piperidino)

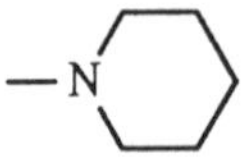

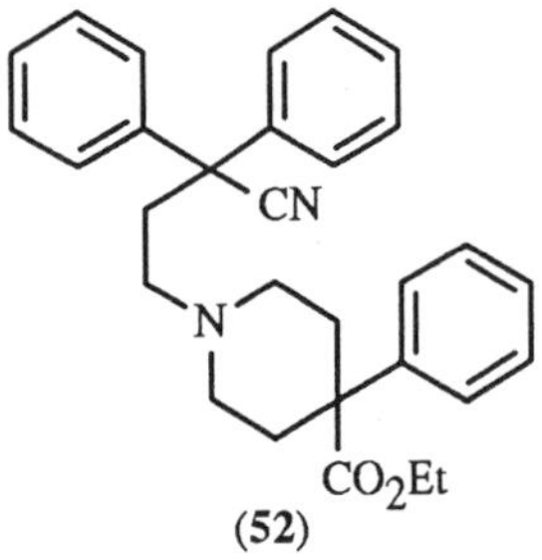

(52) 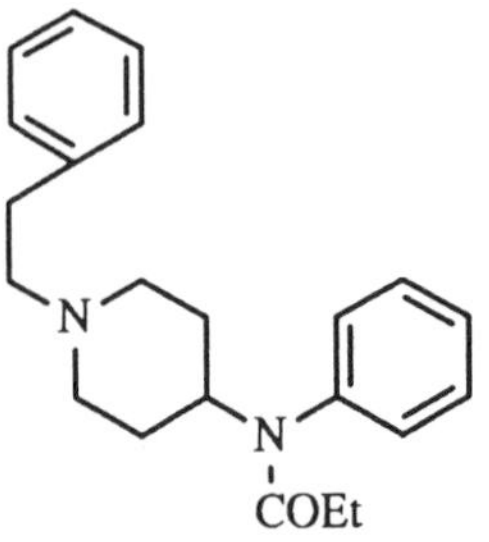 (54) 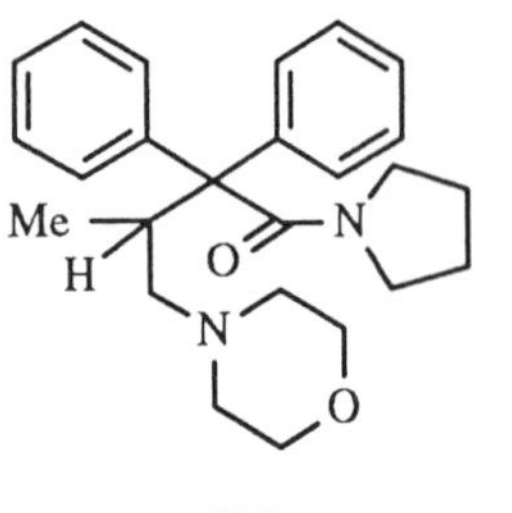 (56)

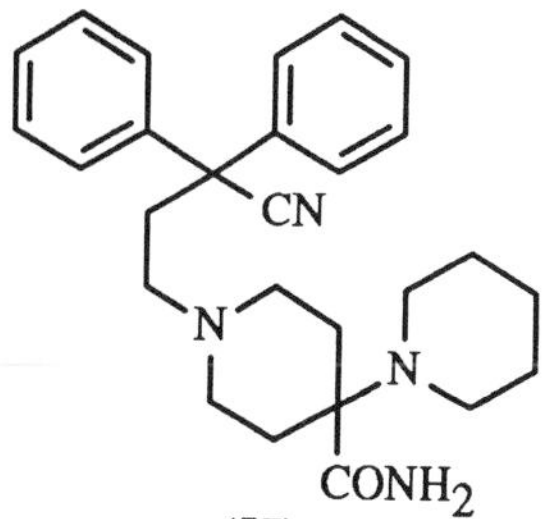

(57) 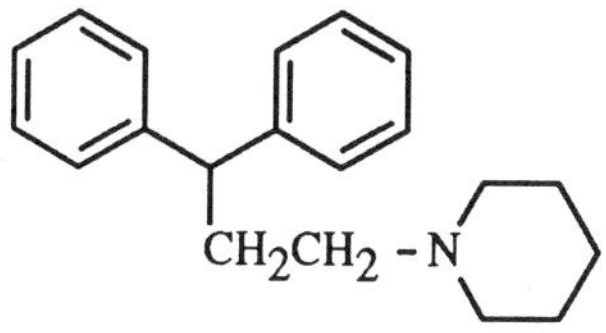 **(58)** 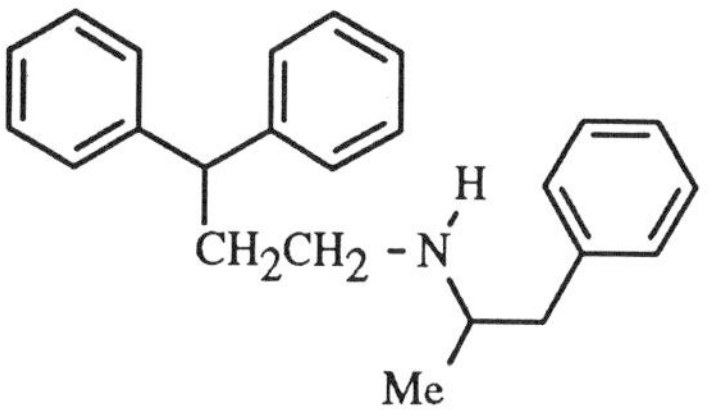 **(59)**

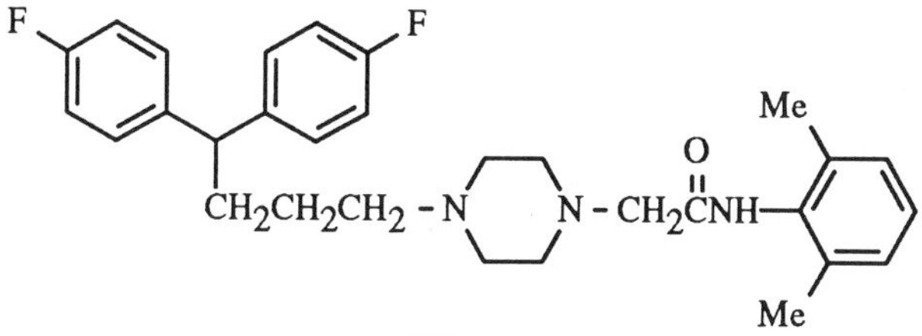

(60) 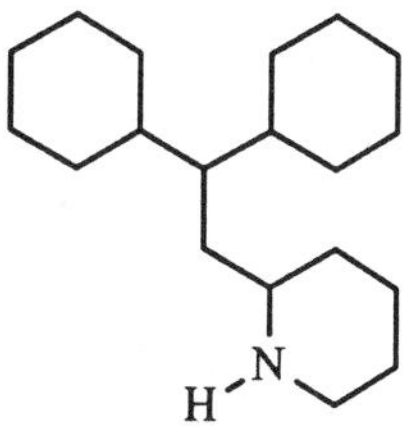 **(61)**

highly potent methadone analogues dipipanone (**55**), dextromoramide (**56**) and piritramide (**57**). There are also methadone analogues that were introduced as antispasmodics free of analgesic actions. These include fenpiprane (**58**), developed as a papaverine substitute during the Second World War. Its analogues were later designed as coronary dilators for use in the treatment of angina. The most successful of these, however, owe their beneficial actions to their ability to act as calcium antagonists. These include prenylamine (**59**), lidoflazine (**60**) and perhexiline (**61**).

1.2.2.1.7 Coca leaf

Cocaine (**62**), another medicinally important tropane alkaloid, was obtained as pure crystals from coca leaf (*Erythroxylon coca*) by Niemann[69] in 1860. A report published in 1885 to the effect that there were chemical similarities between cocaine and atropine led Filehne[70] to test atropine and its tropeine analogues as potential local anaesthetics. Benzoyltropine turned out to be the most potent of these, but its irritancy to the eyes of animals precluded it from clinical application. The apparent importance of the benzoyl residue stimulated further investigations into benzoate esters of a variety of alkaloids and cyclic amines. Filehne published the results of his studies in 1887, but it was another nine years before Merling[71] introduced α-eucaine (**63**) as the first cocaine substitute for clinical use. It was quickly superseded by its analogue β-eucaine (**64**), a less irritant compound. The eucaines were based upon the incorrect interpretation of the structure of cocaine and they were the last of the cocaine substitutes to be based on the wrongly perceived structure of the alkaloid.

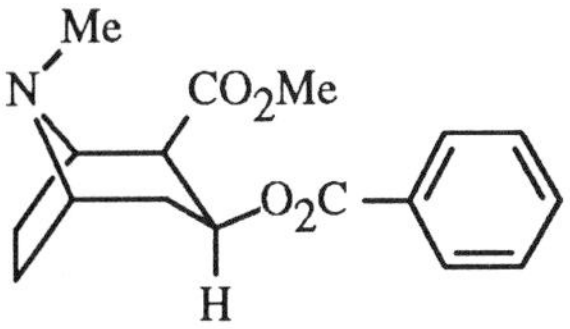

(62) 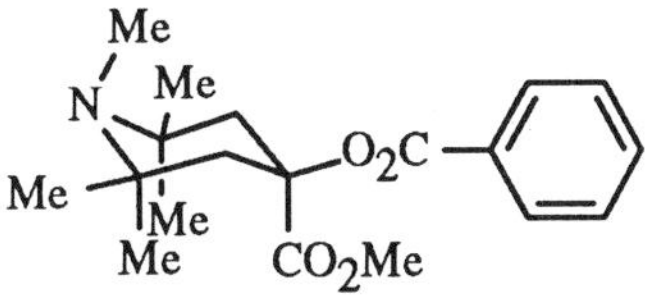 **(63)** 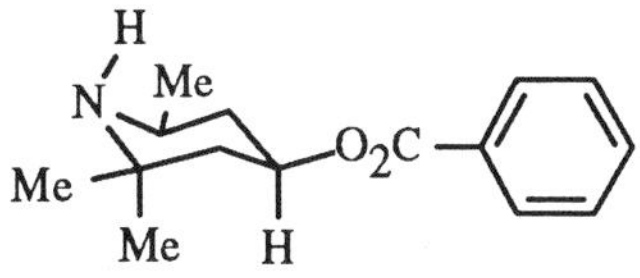 **(64)**

Whilst seeking to synthesize further analogues of cocaine, Einhorn ran into difficulties and was left with several aromatic esters and the phenols from which these had been prepared. Although the phenolic compounds bore no resemblance to the alkaloid, he arranged for them to be routinely tested along with the esters. The phenols turned out to be active although quite unsuitable for anything other than topical application since they were too insoluble. Yet it was one of these compounds, orthocaine (**65**),[72] that provided the prototype from which amylocaine (**66**) and procaine (**67**) were to be derived as the first cocaine substitutes that were superior local anaesthetics so far as clinical application was concerned. These were respectively synthesized by Fourneau in 1903,[73] and Einhorn shortly after.[74] Both were capable of being formulated as water soluble salts.

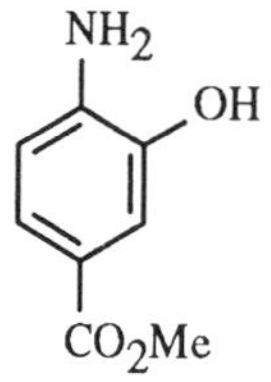

(65) 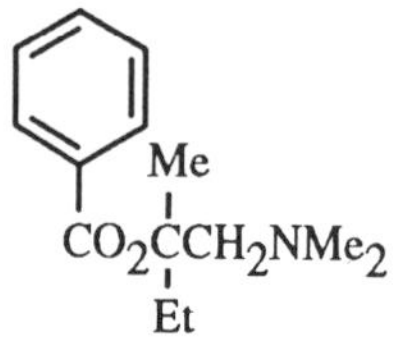 **(66)** 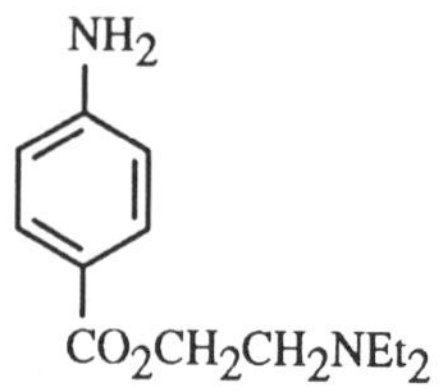 **(67)**

1.2.2.1.8 Ergot[75]

Ergot (*Claviceps purpurea*) proved to be one of the most difficult traditional products from which to extract active principles. In 1905, Barger and Carr separated a crystalline complex called ergotoxine, which was later shown to be a mixture consisting mainly of ergocornine in association with small quantities of ergocristine and ergocryptine. Pure ergotamine (**68**) was isolated by Stoll in 1918, but the most important alkaloid for obstetric purposes, ergometrine (ergonovine) (**69**), was not isolated until 1932.[76] It is used both to prevent and treat uterine haemorrhage. Five years later, Stoll and Hofmann synthesized ergometrine from lysergic acid. Their chosen route permitted the preparation of analogues such as methylergometrine (**70**) and lysergic acid diethylamide (**71**). The latter was found, in 1943, to be an exceedingly potent hallucinogen after Hofmann[77] had inadvertantly ingested traces of it. Several hydrogenated analogues have been prepared from the ergot alkaloids.

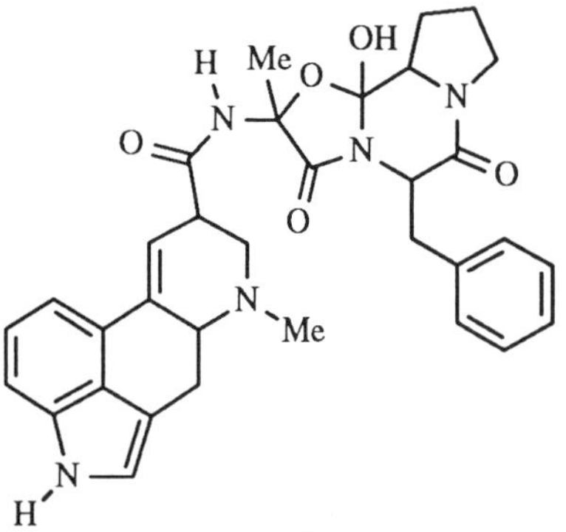

(68) 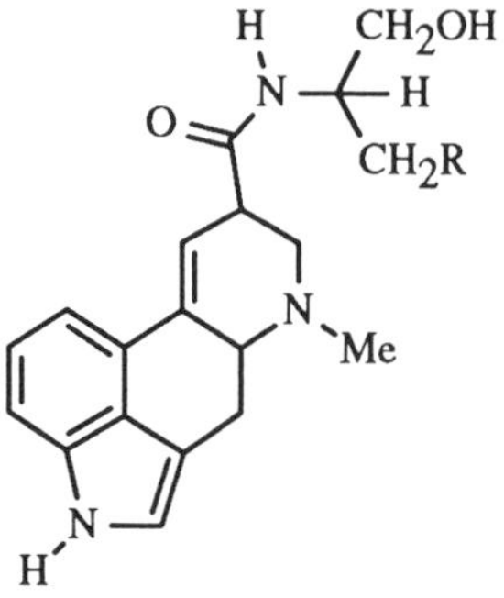 **(69)** R = H **(70)** R = Me **(71)**

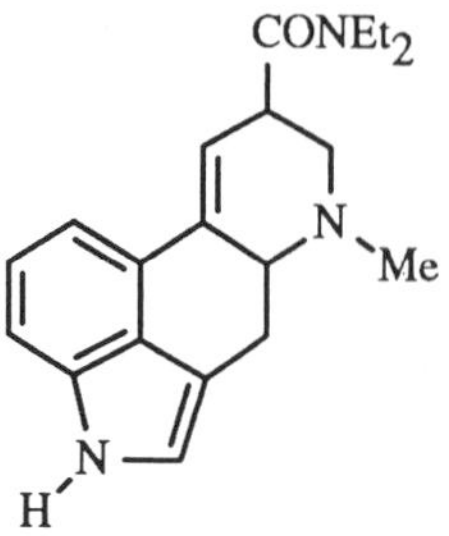

1.2.2.1.9 Curare[78-80]

The elusiveness of the active principles in curare were more to do with the uncertainty surrounding the exact source of this important substance than any particular chemical difficulties. Indeed, tubocurarine (**72**) was first obtained from unidentified material by King in 1935,[81] several years before it was shown to be present in extracts that were being tested in the clinic. Following its clinical introduction,[82-84] there were extensive studies on similar substances isolated from the barks of varieties of *Strychnos* plants. Alcuronium (**73**), introduced in 1961, was a derivative of one of the many alkaloids thus isolated.

The desire to obtain simplified analogues of tubocurarine led Bovet to investigate diquinoline ethers that were converted to their quaternary ammonium salts. The tubocurarine-like specificity of action of one of these ethers focused attention on phenolic ethers, resulting in the synthesis of gallamine (**74**) in 1947.[85] This compound was remarkable for its rapid onset and brief duration of action.

Another approach based on the diquaternary structure of tubocurarine proposed by King was that of Barlow and Ing.[86] By varying the interonium distance in a series of polymethylene dionium compounds, they established that decamethonium (**75**) was a potent neuromuscular blocking agent. However, its resistance to metabolism made decamethonium unacceptably long-acting. The problem was overcome in 1949, when the analogue which incorporated two ester functions was shown to be rapidly metabolized in most patients by esterases. This was given the approved name of

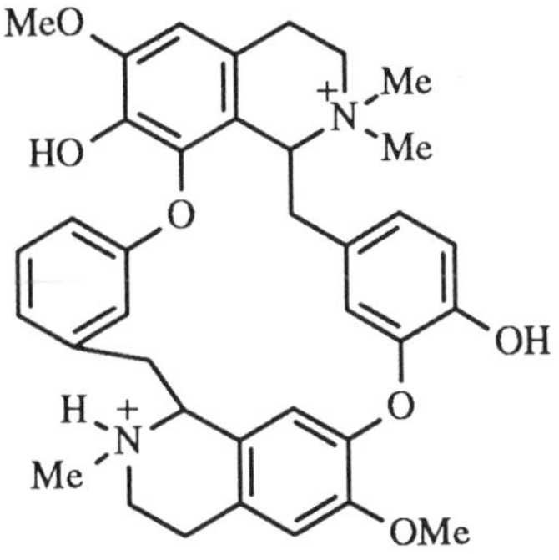

(72) 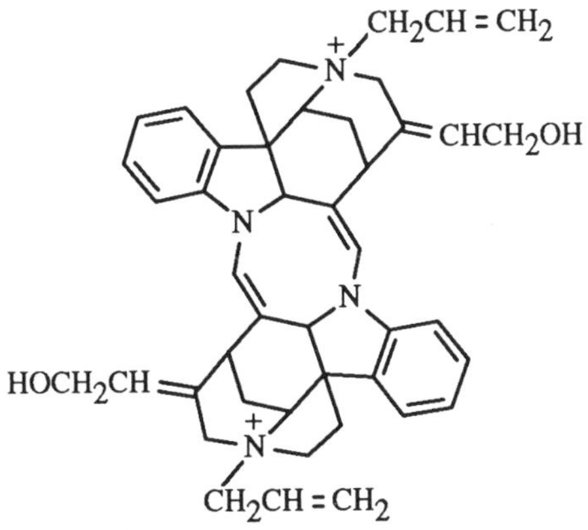 (73) 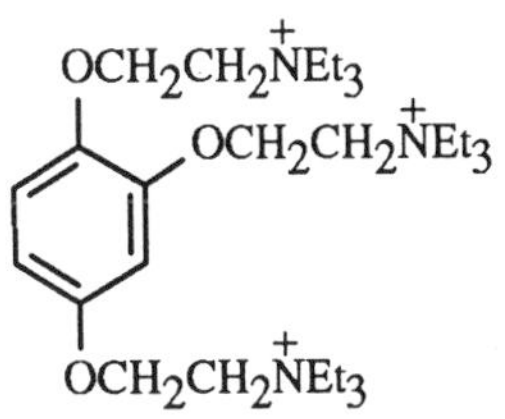 (74)

$Me_3\overset{+}{N}(CH_2)_{10}\overset{+}{N}Me_3$

(75)

$Me_3\overset{+}{N}CH_2CH_2OCOCH_2CH_2COOCH_2CH_2\overset{+}{N}Me_3$

(76)

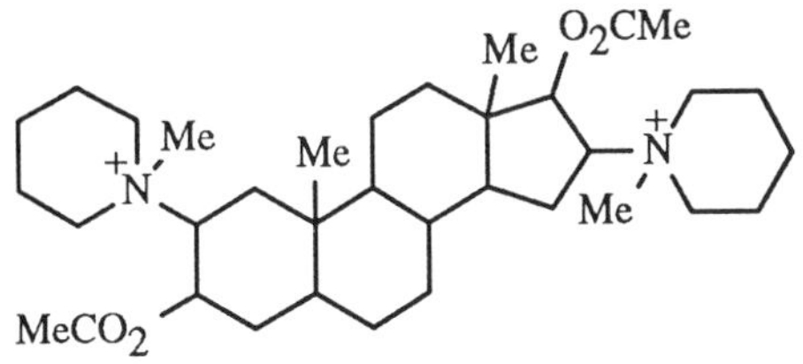

(77) 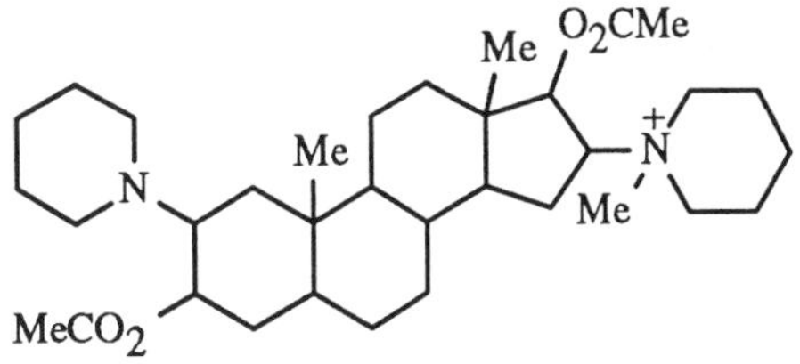 (78)

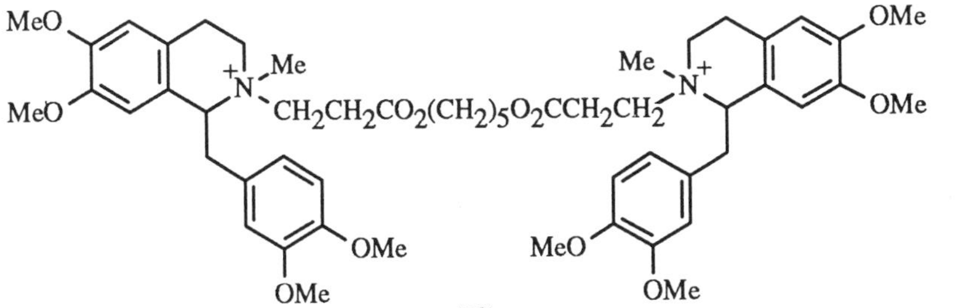

(79)

suxamethonium (**76**). Unfortunately, its action being different from that of tubocurarine, the paralysis could not be antagonized by physostigmine in those patients in whom its effects were long-lasting due to enzymic insufficiency.

In 1966, Savage and his colleagues[97] introduced an analogue of tubocurarine in which the key features seen in King's structure were incorporated into a steroid nucleus. This compound, pancuronium (**77**), was particularly successful clinically, but it was not until 1983, after it had been shown that tubocurarine was actually a monoquaternary alkaloid,[88] that the monoquaternary analogue of pancuronium was found to be superior. This was marketed as vecuronium (**78**) at the same time as atracurium (**79**), another diquaternary analogue of suxamethonium. By incorporating the quaternary ammonium groups into tetrahydroquinoline rings in such a way that the molecule would rapidly decompose at physiological pH, Stenlake *et al.*[89] overcame the dependence of suxamethonium on enzymatic hydrolysis.

1.2.2.1.10 Ephedra

Although ephedrine (**80**) was isolated by Nagai from the Chinese plant Ma Huang (*Ephedra vulgaris*) in 1897, it was not until the mid-1920s that its clinical value as an orally active sympathomimetic was recognized. Initial high cost of the alkaloid persuaded Alles[90] and his colleagues to examine structurally related compounds such as amphetamine (**81**), a compound originally synthesized in 1887. Its enhanced volatility ensured its commercial success when it was marketed in 1932 in the form of a decongestant inhaler. Amphetamine and its analogues were subsequently found to be potent central stimulants and appetite suppressants. As such they became widely abused during and after the Second World War.

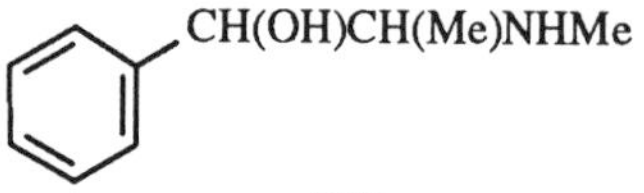

$CH(OH)CH(Me)NHMe$

(80)

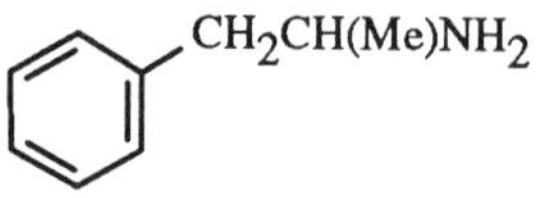

$CH_2CH(Me)NH_2$

(81)

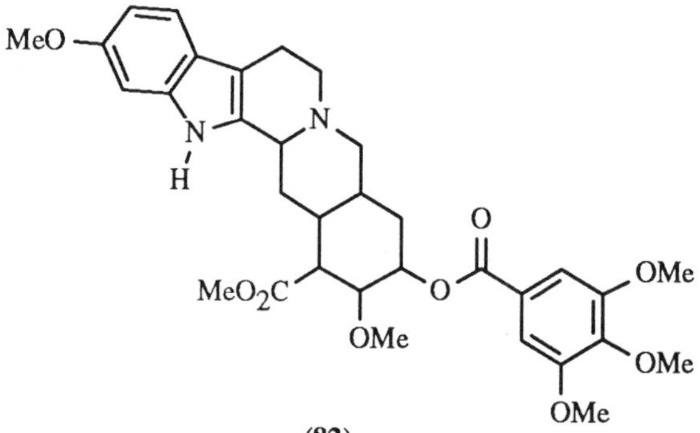

(82)

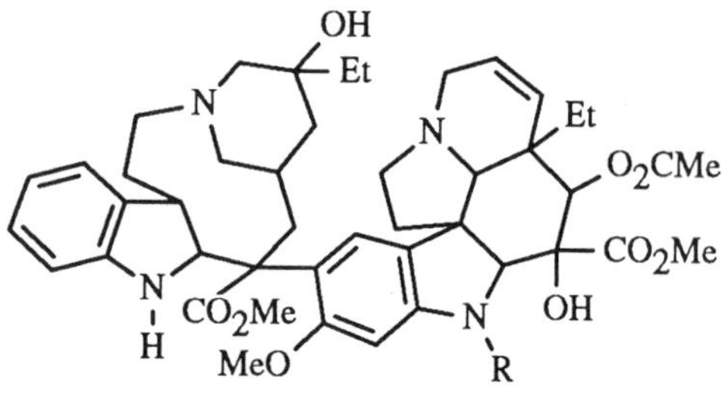

(83) R = Me
(84) R = CHO

1.2.2.1.11 Rauwolfia[91]

Enthusiastic claims for the hypotensive and sedative properties of the Indian snakeroot plant, *Rauwolfia serpentina*, led to intensive pharmacological investigations that culminated in the isolation of reserpine (**82**) in 1952.[92] Shortly after, the term 'tranquillizer' was used to describe its clinical action. The early commercial success of reserpine stimulated considerable research into natural products as sources of new medicines, but the truly successful outcome of this was the introduction of two alkaloids from *Catharanthus rosea* as cytotoxic agents with some use in cancer chemotherapy. These were vinblastine (**83**) and vincristine (**84**). The former was also discovered serendipitously during an independent investigation by Noble, which started in 1949, into the alleged hypoglycaemic activity of the periwinkle. The first announcement of the isolation of an active cytotoxic agent was made nine years later.[93]

1.2.2.2 Plant Acids

1.2.2.2.1 Tars

Coal being a fossil fuel derived from the plant kingdom, it is appropriate to consider the medicinal application of coal tar in this section. In 1859, an antiseptic powder prepared from coal tar and plaster was evaluated in several hospitals in Paris. The tendency for this to 'cake' was avoided by the use of an emulsified coal tar preparation introduced later that year by LeBeuf and Lemaire. The latter considered the success of this formulation to be due to the presence of phenol, an acid first isolated from coal tar by Runge in 1833 and subsequently purified by Laurent seven years later. Lemaire published extensively on the medicinal uses of phenol,[94,95] but it is Lister to whom the credit for recognizing its particular value in antiseptic (as opposed to aseptic) surgery must be attributed.[96]

The necrosive action of phenol on tissues stimulated the search for milder substitutes amongst its chemical derivatives. The first of these was salicylic acid, introduced in 1874. Many others have found a useful role as antiseptics, perhaps the most important being chloroxylenol (**85**). In the early 1930s, Colebrook pioneered the introduction of chloroxylenol solution as a non-irritant antistreptococcal topical antiseptic for use in obstetric practice.

Creosote was first distilled from beechwood tar in 1830. Subsequently, it acquired a reputation as a much less irritant antiseptic than coal tar. When Behal and Choay isolated guaiacol (**86**) from beechwood tar in 1887, this phenolic compound was assured of widespread popularity since it could be cheaply synthesized. Numerous derivatives were marketed in attempts to eliminate the remaining irritancy. That so many were devised is adequate testimony to the essential futility of the exercise.

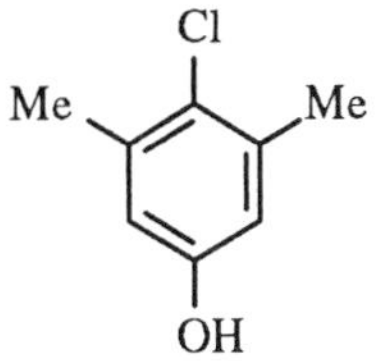

(85) 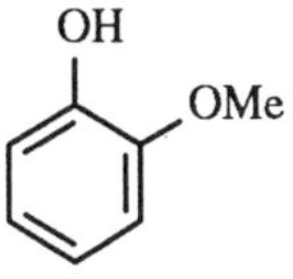 (86) 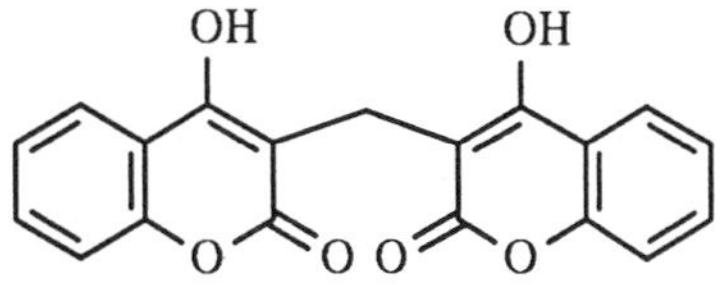 (87)

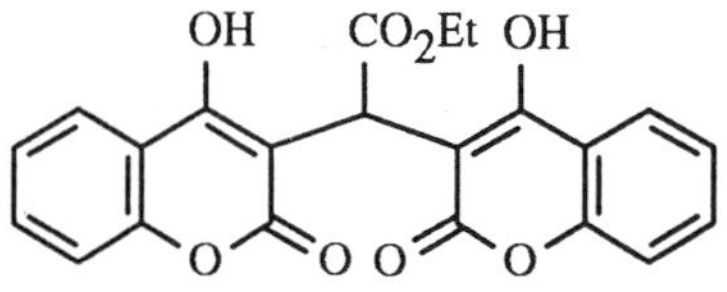

(88) 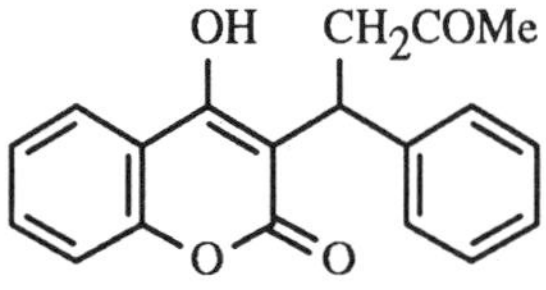 (89) 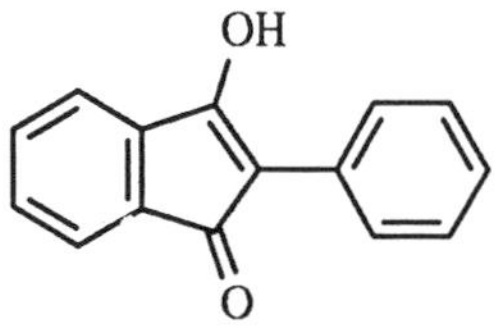 (90)

1.2.2.2.2 Dicoumarol

Apart from the phenols derived from coal and beechwood tars, the only plant acid devoid of basic properties which has had an unquestioned role in therapy is dicoumarol (**87**). In 1941, Link[97] found this to be the toxic compound responsible for inducing haemorrhages in cattle that had grazed in fields contaminated with sweet clover. It was immediately investigated for its potential as an orally administered anticoagulant in thromboembolic disease. Dicoumarol and its synthetic analogues such as ethyl biscoumacetate (**88**), warfarin (**89**) and phenindione (**90**) are now considered to have a limited role in clinical practice.

1.2.2.3 Glycosides

1.2.2.3.1 Digitalis

The failure of early attempts to isolate any active principle from digitalis led the Société de Pharmacie in Paris to offer a prize to whomsoever could achieve this. In 1841, this was awarded to Homolle and Quevenne for isolating from the leaves of *Digitalis purpurea* a somewhat impure crystalline material consisting mainly of digitoxin (**91**).[98] Purer samples of this were isolated by Nativelle in 1869 and Schmiedeberg in 1875, the latter also obtaining the less potent digitalin (**92**). The structures of these cardiac glycosides were elucidated just over half a century later by Windaus. However, in 1933, Stoll[99] established that all the glycosides that had been reported in the literature were artefacts formed by decomposition of digitoxin during extraction. Nevertheless, a genuine analogue with greater and more controllable clinical activity was isolated by Smith from *Digitalis lanata* in 1930. Given the approved name of digoxin (**93**), it is now the most widely used cardiac glycoside.

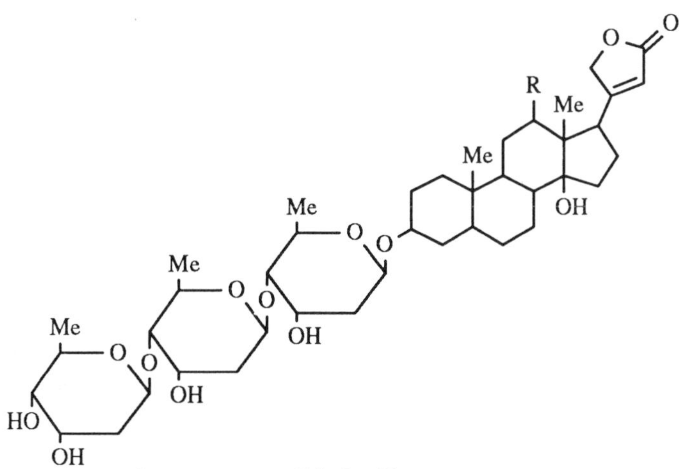

(**91**) R = H
(**93**) R = OH

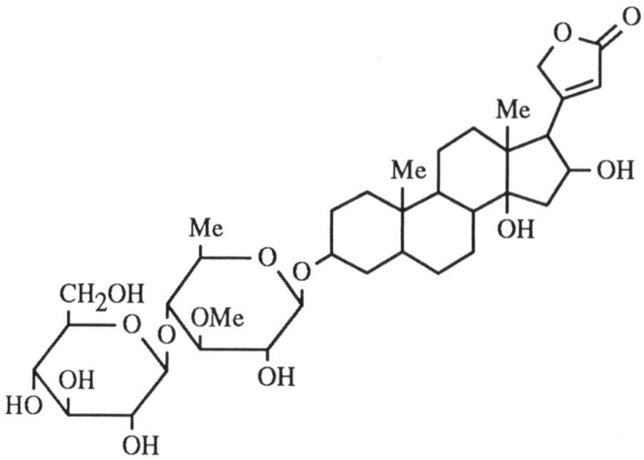

(92)

1.2.2.3.2 *Salicin*

Salicin (**94**) was isolated from willow bark, *Spirea ulmaria*, by Leroux in 1829. No truly useful therapeutic application was found for this glycoside until 1874, when Maclagan[100] introduced it in the treatment of rheumatic fever. Subsequently, he found its metabolite, salicylic acid, to be more efficacious in the treatment of a variety of rheumatic conditions. By this time, its antipyretic properties had also been recognized.

The widespread use of salicylic acid naturally exposed its irritant nature to the gastric mucosa, so it is hardly surprising that attempts were made to find more acceptable derivatives. One of the first was salol (**95**), formed by esterification of salicylic acid with phenol in a misconstrued effort to exploit both compounds as internal antiseptics for use in combatting intestinal infections. By far the most successful derivative, of course, has been aspirin (**96**). This was selected in 1898 by Hoffmann from a range of known salicylates.[101,102] Although more acceptable to patients, the gastric irritancy persisted. A variety of substitutes for salicylic acid have been introduced since the early 1960's in an effort to find a widely acceptable antirheumatic agent. These include mefenamic acid (**97**), flufenamic acid (**98**), diclofenac (**99**), benorylate (**100**) and diflunisal (**101**). The structural relationships between these and the salicylates is apparent. The belief that antiinflammatory activity was connected with the presence of the carboxylic acid group lay behind a screening programme which led to the introduction of ibuprofen (**102**) in 1964 after more than 600 phenoxyalkanoic acids had been evaluated.[103] Several analogues of ibuprofen have been developed, *e.g.* naproxen (**103**), fenoprofen (**104**), ketoprofen (**105**), flurbiprofen (**106**) and benoxaprofen (**107**). The latter was marketed in 1980 as suitable for once daily dosage due to its slow elimination from the body. Tragically, the rate of elimination was prolonged in elderly patients, resulting in scores of deaths or permanent disabilities arising from allergic reactions to the drug. It was withdrawn from the market in 1982.

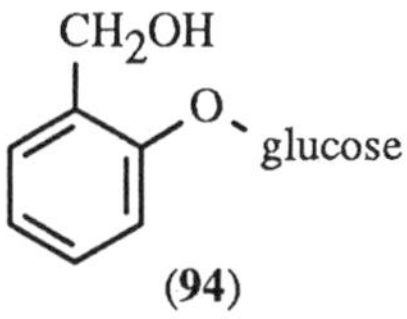

(94) 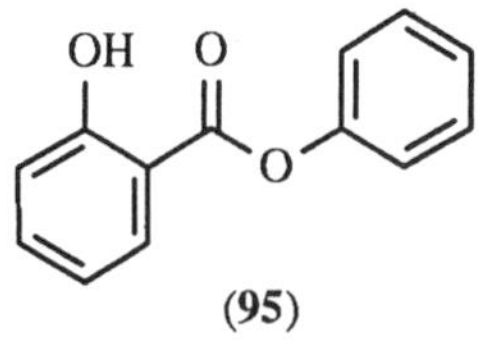 (95) 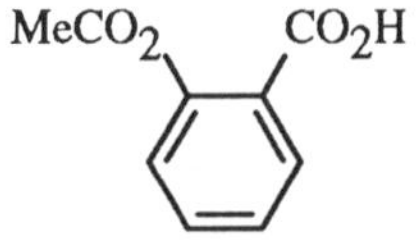 (96) 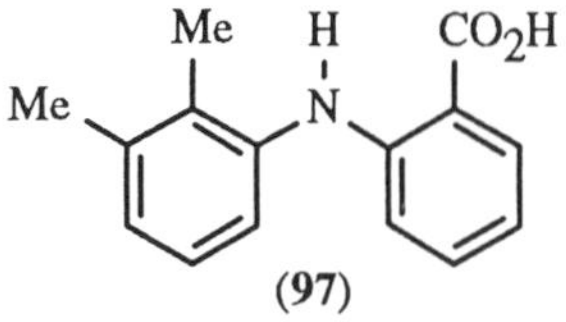 (97)

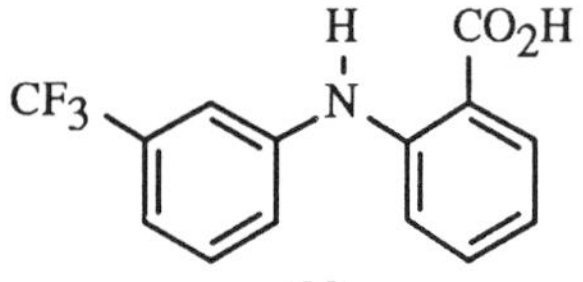

(98) 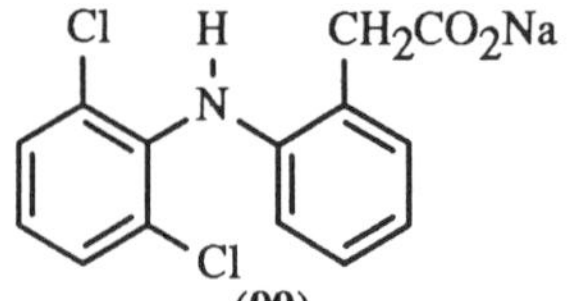 (99) 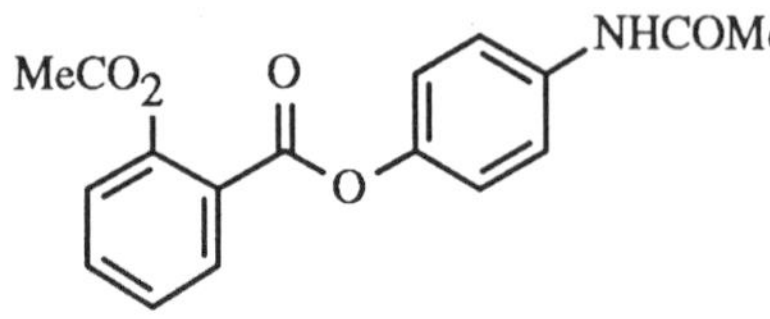 (100)

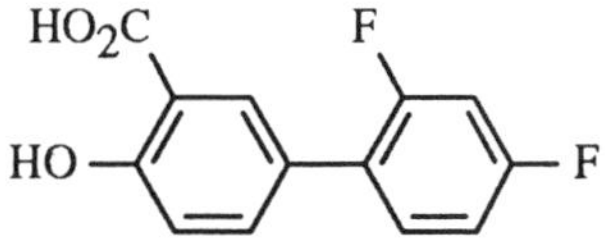

(101) 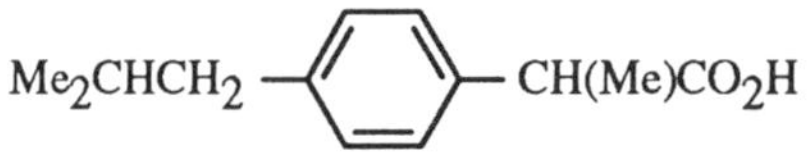 (102) 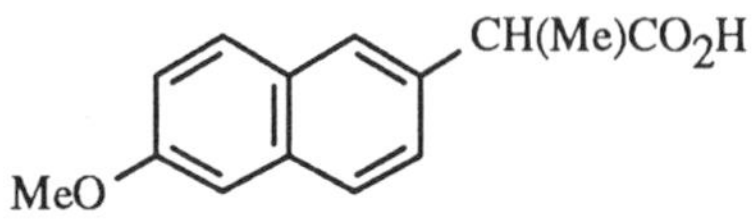 (103)

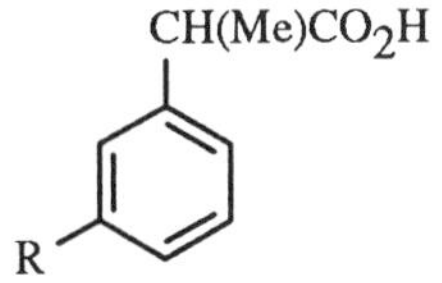

CH(Me)CO_2H

R

(**104**) R = OPh
(**105**) R = COPh

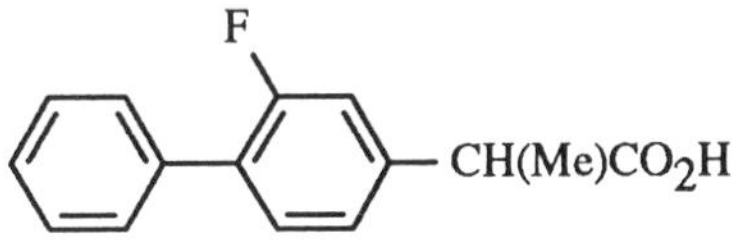

F
CH(Me)CO_2H

(**106**)

N
CH(Me)CO_2H
Cl
O

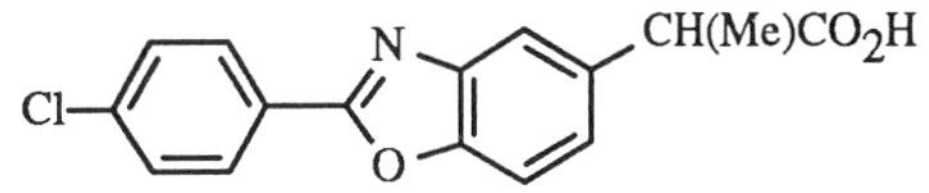

(**107**)

1.2.2.4 Miscellaneous

1.2.2.4.1 Cannabis

Mechoulam and Gaoni[104] isolated tetrahydrocannabinol (**108**) from *Cannabis sativa* in 1964, but despite many interesting enquiries, so far no generally accepted clinical application has been found for this euphoriant. Its ability to prevent nausea in leukaemic patients undergoing intensive chemotherapy with cytotoxic drugs can only be achieved at dose levels which produce a high incidence of central and cardiovascular side effects. Nabilone (**109**) is a synthetic analogue which was introduced in 1982 as an antinauseant and unlike the natural product is active when given by mouth.

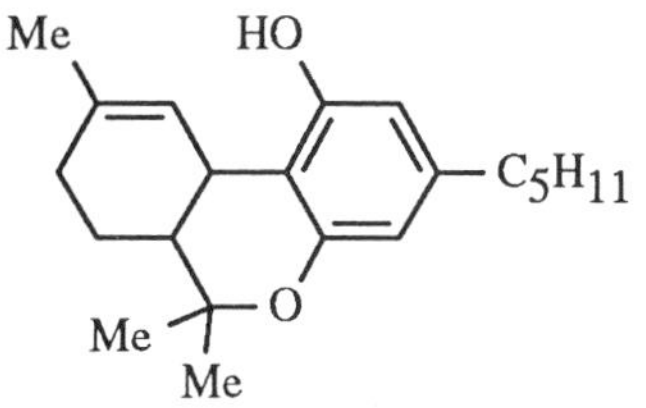

Me HO C_5H_{11} Me O Me

(**108**)

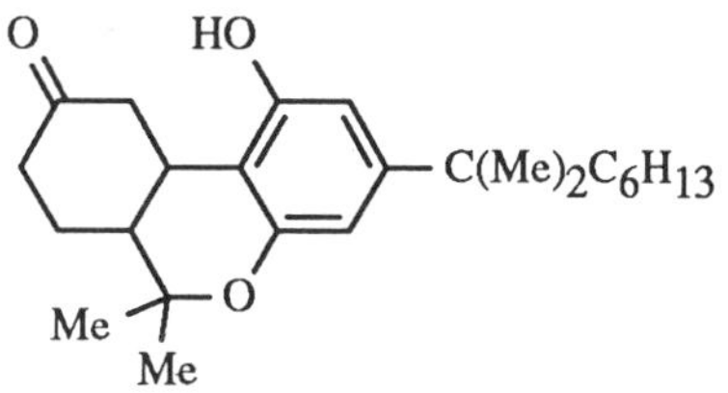

O HO $C(Me)_2C_6H_{13}$ Me O Me

(**109**)

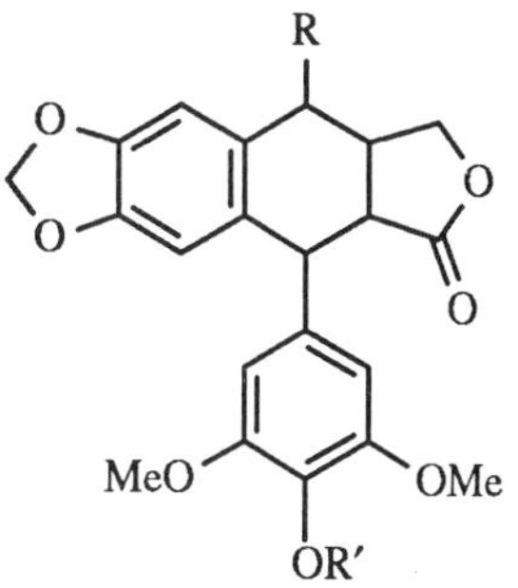

R O O O O MeO OMe OR'

(**110**) R = OH, R′ = Me

(**111**) R = O (sugar: O, O, Me, O, OH, OH), R′ = H

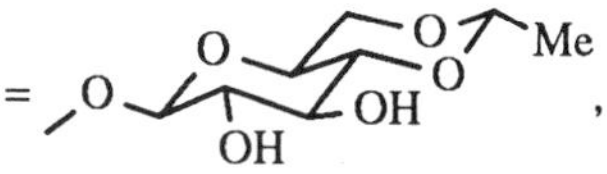

1.2.2.4.2 Podophyllum

The juice expressed from the root of the May Apple (American mandrake), *Podophyllum peltatum*, has long been employed by the Wyandotte Indians as a purgative and anthelmintic. In 1835, King pioneered the use of the highly potent podophyllum resin and ultimately saw it become one of the most widely used purgatives throughout the world. The irritant action on the bowel led to its demise, yet was exploited in the use of the powdered root as an escharotic for the treatment of warts. In recent years, the recognition of the ability of the active principle, podophyllotoxin (**110**), to act as a mitotic poison has resulted in the investigation of this substance and its derivatives in the treatment of cancer. In 1973, Stahelin[105] reported that etoposide (**111**) was active against experimental tumours. It has shown promise in the treatment of non-small-cell carcinoma of the lung.

1.2.3 MAMMALIAN HORMONES AND THEIR ANALOGUES[106,107]

Mammalian hormones constitute one of the major groups of drug prototypes that have been successfully exploited to provide useful therapeutic agents. Difficulties have been encountered in their isolation because in the body they often exist in trace amounts in order to fulfil their role as chemical messengers. A further complication has often been their lack of selectivity when administered as drugs, again a reflection of their natural role, whereby they may be released only in the vicinity of target organs. However, medicinal chemists have often been able either to hinder or enhance the fit of analogues to hormone receptors in order to achieve a degree of selectivity of action that is acceptable for clinical purposes.

When Brown-Séquard reported in 1889 that he had rejuvenated himself with injections of an aqueous extract of testicles from guinea pigs, there were many who were quick to profit from his unsubstantiated claim.[108] The resulting notoriety of organotherapy dissuaded most researchers from entering such a controversial field, but a few ignored the disdain of colleagues and proceeded with their own investigations. Thus it was that the closing years of the 19th century witnessed early experiments with physiologically active extracts of the thyroid gland, the adrenal medulla, the posterior pituitary and the ovaries.

1.2.3.1 Thyroid

In 1873, Gull suggested that myxoedema was associated with atrophy of the thyroid gland, but it was another 18 years before surgeons were able to remove the gland successfully and then confirm this. In 1891, Murray[109] injected a glycerol extract of sheep thyroid into a patient desperately ill with severe thyroid insufficiency. Her complete recovery after receiving regular maintenance doses of the extract stimulated other investigators to administer similar extracts by mouth.

Kendall[110] isolated crystals of thyroid hormone in 1914, but three further years were required before he was able to amass enough material for clinical studies. The efficiency of the isolation process was later increased, but the cost of the hormone remained prohibitive so far as its use in therapy was concerned. Although Harrington and Barger devised a synthesis of thyroxine (**112**) in 1927, which confirmed the former's proposed structure for the hormone, it was not until the mid-1950's that a commercially viable eight-stage synthesis by Hems and his colleagues permitted the routine use of the synthetic hormone.[111] This synthesis also permitted the manufacture of the more potent hormone liothyronine (**113**), formed in the gland from thyroxine. It had first been isolated in 1952 by Gross and Pitt-Rivers.

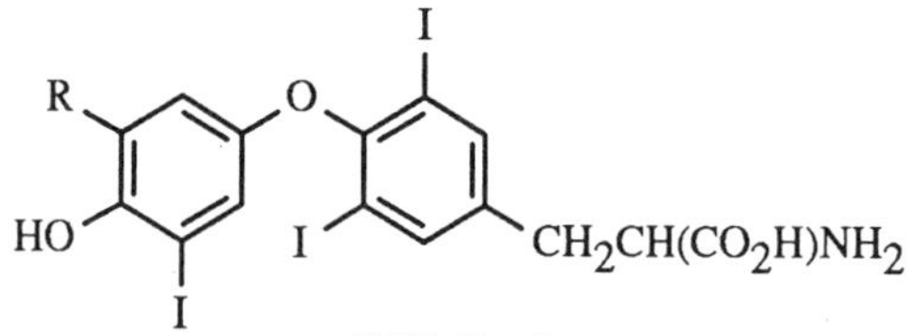

(**112**) R = I
(**113**) R = H

1.2.3.2 Adrenal Medulla

In 1893, Oliver[112] carried out experiments on volunteers with a view to establishing the effects of the oral administration of glycerol extracts of glands on the diameter of the radial artery. These experiments revealed the presence of a physiologically active substance in extracts of sheep adrenal glands. In 1901, crystalline adrenaline (epinephrine) (**114**) was isolated by Takamine, this being the first hormone to be obtained in a pure state. A commercially viable synthesis was achieved within five years, permitting the synthetic hormone to compete on the market with the natural product. It was initially used as a haemostatic and also as a vasoconstrictor to prolong the action of procaine, a synthetic local anaesthetic.

Only after the introduction of ephedrine and its synthetic analogues as orally active sympathomimetic agents was it realized that adrenaline analogues might have similar therapeutic value. One of the first of these was phenylephrine (**115**), marketed as a decongestant in the 1930s. A few years later, the superiority of inhaled isoprenaline (**116**) over adrenaline was noted,[113] although this was not disclosed until after the Second World War. It did not stimulate the α-adrenoceptors in the vascular system, thus avoiding the unwelcome pressor effects hitherto associated with the administration of adrenaline. The success of isoprenaline eventually inspired a search for orally active analogues possessing greater resistance to metabolic deactivation. Notable amongst these were orciprenaline (**117**) and soterenol (**118**), which could be administered by mouth for the prophylaxis of asthmatic attacks. Being more polar than ephedrine and its analogues, these new bronchodilators were devoid of the undesirable central stimulating activity.

In the course of seeking novel isoprenaline analogues based on the saligenin nucleus, it was unexpectedly found that salbutamol (**119**) was selective for the β-adrenoceptors of the lung.[114] That is to say, therapeutic doses had a negligible effect on the heart. The announcement of this in 1967 was rapidly followed by that of the similarly selective orciprenaline analogue known as terbutaline (**120**).

HO, HO, CH(OH)CH_2NHMe

(114)

HO, CH(OH)CH_2NHMe

(115)

HO, HO, CH(OH)CH_2NHCHMe_2

(116)

HO, OH, CH(OH)CH_2NHCRMe_2

(117) R = H

(120) R = Me

$MeSO_2NH$, HO, CH(OH)CH_2NHCHMe_2

(118)

$HOCH_2$, HO, CH(OH)CH_2NHCMe_3

(119)

HO, OH, CH(OH)CH_2NHR

(121) R = —CH(Me)CH_2—C_6H_4—OH

(124) R = —$CH_2CH_2CH_2$— (1,3-dimethylxanthin-7-yl: N, N, N—Me, N—Me, O, O)

HO, HO, CH(OH)CH(Et)NHCHMe_2

(122)

$HOCH_2$, N, HO, CH(OH)CH_2NHCMe_3

(123)

HO, HO, H, OH, H, N

(125)

Other drugs of this type are fenoterol (**121**), isoetharine (**122**), pirbuterol (**123**), reproterol (**124**) and rimiterol (**125**).

The quest for isoprenaline analogues took an unexpected twist in 1957 when it was found that dichloroisoprenaline (**126**) antagonized the action of adrenaline on the β-adrenoceptors of both the heart and lung. Since dichloroisoprenaline itself stimulated these receptors (*i.e.* it was a partial agonist), it had no clinical value. The recognition that a potent β-adrenoceptor antagonist could have far-reaching clinical consequences in patients with cardiovascular disease spurred Black to seek such an agent. He and his colleagues introduced pronethalol[115] (**127**) in 1960, but this was quickly superseded by the more efficacious propranolol (**128**) after toxicity tests had indicated pronethalol was carcinogenic.[116] A range of similar compounds reached the market from the mid-1960s onwards. Amongst these was sotalol (**129**), an analogue of which was found by Dunlop and Shanks to have some selectivity for the β-adrenoceptors of the heart in preference to those of the lung. The potential of this in asthmatic patients was recognized, leading to the development of practolol (**130**). Unfortunately, this produced an unanticipated range of severe hypersensitivity reactions in a very small proportion of patients. Similar compounds, including acebutolol (**131**), atenolol (**132**) and metoprolol (**133**) have proved to be safe.

The β-adrenoceptor antagonists were long preceded by antagonists of the α-adrenoceptors, although the distinction between these types of receptors was not recognized until 1948. It was, in fact, during the first decade of this century that Dale established that ergotoxine could not only stimulate smooth muscle, but could also block the action of adrenaline at what were later to be recognized as α-adrenoceptors.

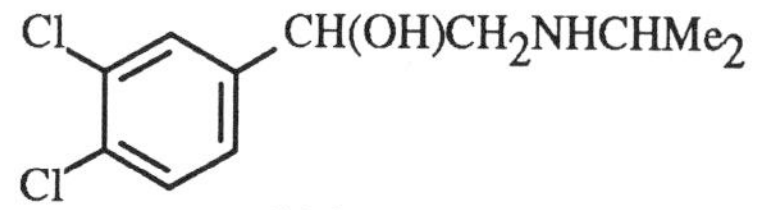

(126)

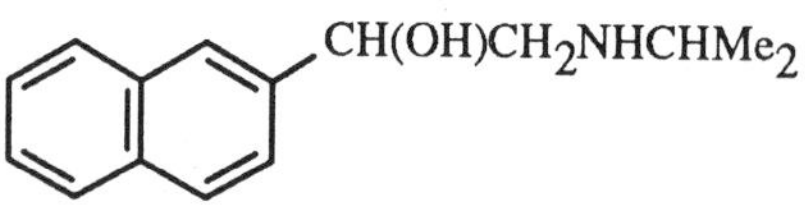

(127)

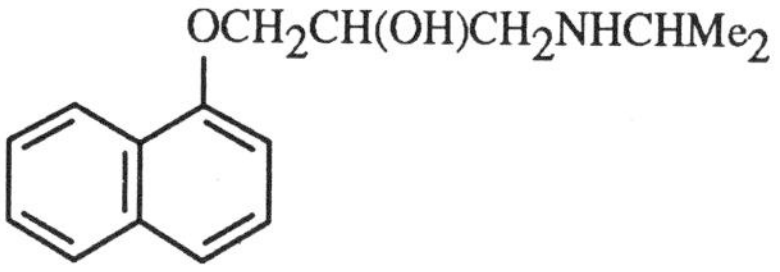

(128)

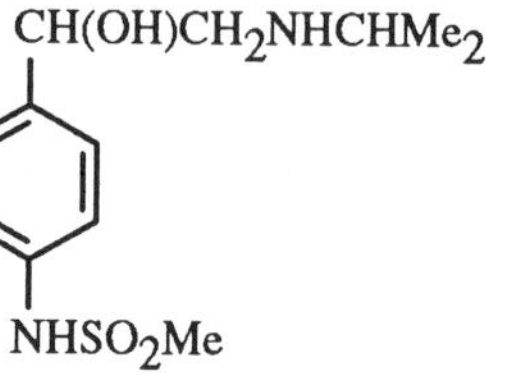

(129)

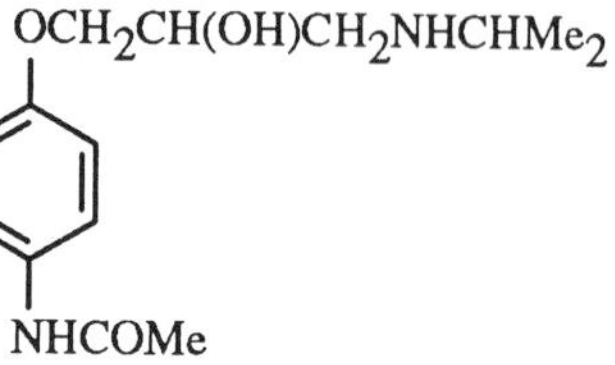

(130)

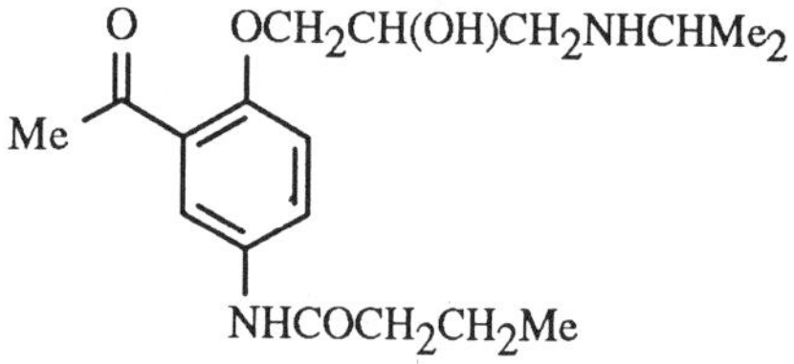

(131)

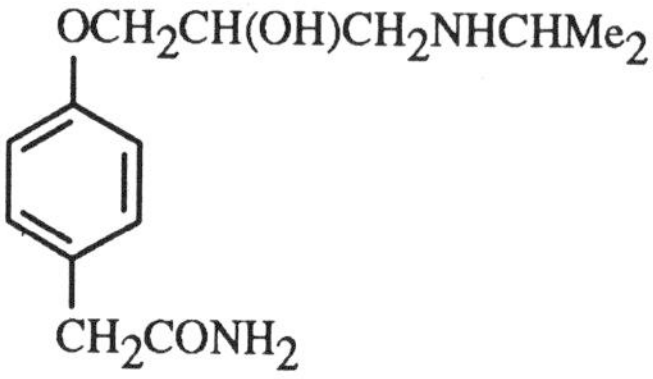

(132)

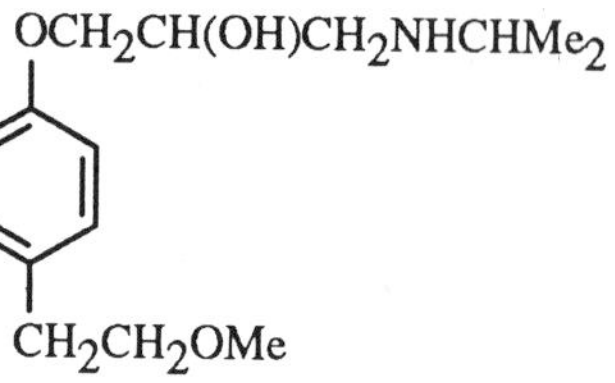

(133)

Amongst the earliest synthetic compounds that were found to exhibit adrenergic blocking activity were a series of benzodioxans prepared as analogues of a uterine stimulant known as gravitol (**134**). In 1933, Fourneau and Bovet synthesized further analogues, of which piperoxan (**135**) was the most effective adrenaline antagonist, though its duration of action was too brief for most clinical purposes. Nevertheless, piperoxan served as the prototype from which the first antihistamines were subsequently derived.

Confusing reports in the literature led Uhlmann to investigate the activity of substituted imidazolines during the early 1940s. This resulted in the introduction not only of the sympathomimetic drug naphazoline (**136**) and its analogues xylometazoline (**137**) and oxymetazoline (**138**) used as nasal decongestants, but also of the α-adrenoceptor antagonist tolazoline (**139**).[117] A more potent antagonist, phentolamine (**140**), was subsequently developed in the 1950s.

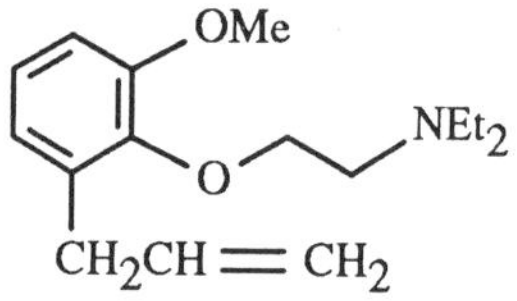

(134)

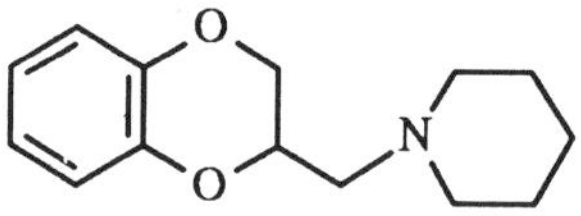

(135)

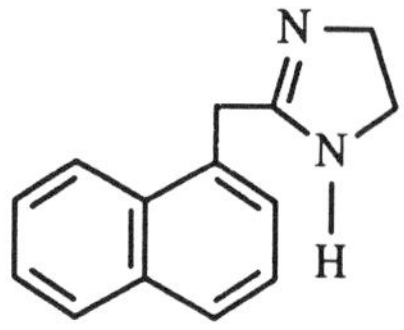

(136)

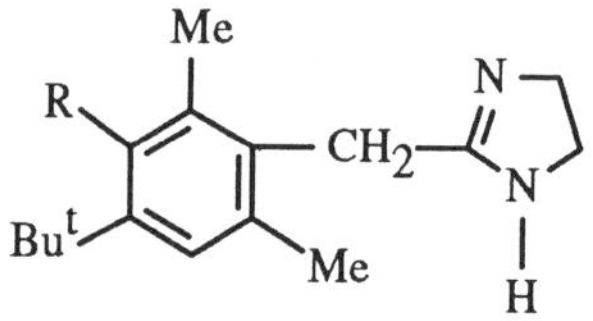

(137) R = H
(138) R = OH

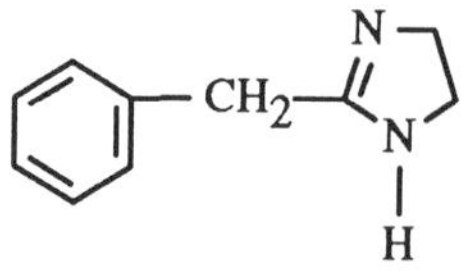

(139)

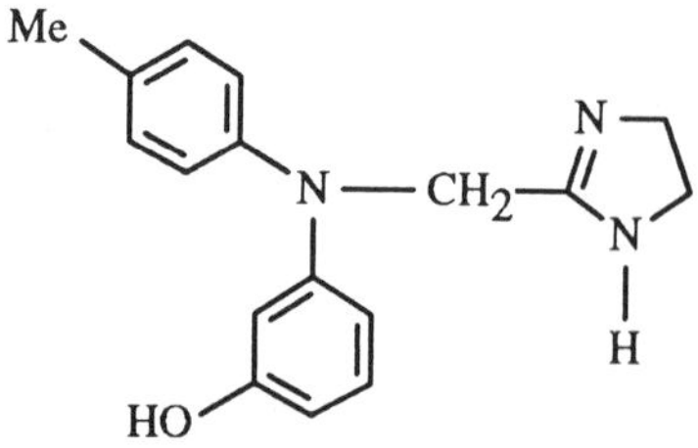

(140)

1.2.3.3 Pancreas[118,119]

In 1889, Minkowski and von Mering demonstrated that excision of its pancreas caused the death of a dog from diabetes. Six years later, Schafer proposed that pathological changes in the region of the pancreas known as the islets of Langerhans induced diabetes mellitus in humans. In 1906, Cohnheim found that during attempts to isolate it, the pancreatic hormone was being destroyed by the digestive enzymes in the gland. To avoid this, he found it necessary to deactivate these by boiling prior to alcoholic extraction. Two years later, Zuelzer kept the urinary sugar levels of a depancreatized dog under control by administering daily injections of an alcoholic extract. His trial of this extract on eight patients was initially promising, but had to be abandoned through the occurrence of fever due to contamination with pyrogens and the shortage of supplies. The subsequent history of attempts to treat diabetics with insulin is characterized by difficulties in overcoming production problems. These were ultimately overcome in 1921 by the industry and persistence of the Canadian researchers Banting and Best, who utilized the latest microchemical techniques to obtain rapid assays of blood sugar levels in dogs treated with pancreatic extracts. By the summer of the following year, commercial production of insulin had begun. This was undoubtedly the most significant event in the development of modern pharmaceuticals until the introduction of penicillin. It is noteworthy that both involved the tackling of production problems by academics who made the successful breakthroughs before freely handing over all their information to the pharmaceutical industry so that literally millions of patients could gain the benefits of these discoveries as soon as possible. It is to the lasting credit of industry and academe that the timescales involved in both instances almost defy belief when the nature of the inherent problems is fully understood.

Abel was able to obtain insulin in crystalline form in 1926. Human insulin (**141**), which has a slight difference in its amino acid sequence from bovine or porcine material, is now commercially prepared from bacterial sources through the use of genetic-engineering processes.

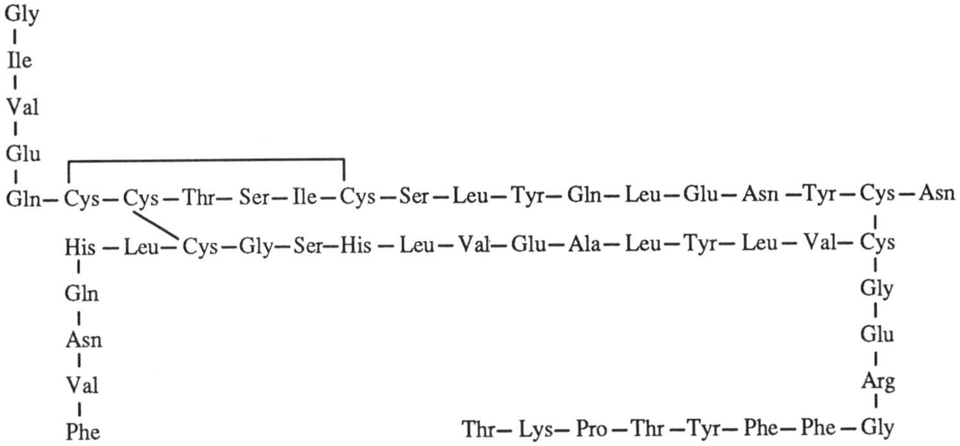

Gly–Ile–Val–Glu–Gln–Cys–Cys–Thr–Ser–Ile–Cys–Ser–Leu–Tyr–Gln–Leu–Glu–Asn–Tyr–Cys–Asn

Phe–Val–Asn–Gln–His–Leu–Cys–Gly–Ser–His–Leu–Val–Glu–Ala–Leu–Tyr–Leu–Val–Cys–Gly–Glu–Arg–Gly–Phe–Phe–Tyr–Thr–Pro–Lys–Thr

(141)

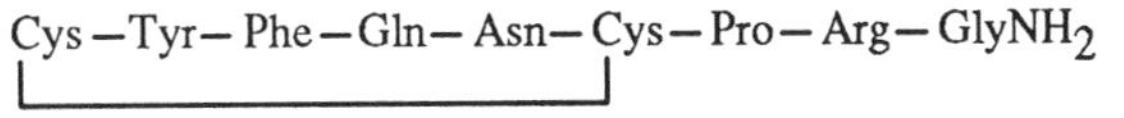

Cys–Tyr–Phe–Gln–Asn–Cys–Pro–Arg–GlyNH_2

(142)

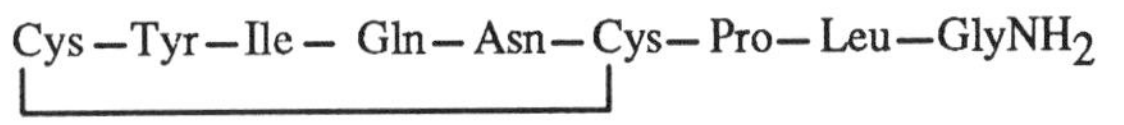

Cys–Tyr–Ile–Gln–Asn–Cys–Pro–Leu–GlyNH_2

(143)

1.2.3.4 Pituitary Hormones

1.2.3.4.1 Posterior pituitary

Oliver and Schafer prepared active extracts of the pituitary gland in 1894, and four years later Howell confirmed that the pressor activity was only present in material obtained from the posterior part of the gland. The first clinical application of posterior pituitary extract followed Dale's discovery that it had the ability to induce contraction of the uterus in a manner akin to that of ergot. From 1909 onwards, it was regularly employed in obstetric practice. Another successful clinical application of the extract was in the treatment of dangerously excessive urinary production in patients with diabetes insipidus. This followed upon van den Velden's discovery that the principal role of the posterior pituitary was to produce antidiuretic hormone.

The isolation of insulin served as a powerful stimulus for investigations on hormones. In 1928, Kamm separated two still impure hormones from posterior pituitary extract, namely the pressor agent vasopressin (**142**) and the uterine stimulant oxytocin (**143**). The latter replaced the use of the crude extract. However, it was only after applying the sophisticated counter-current extraction and preparative chromatographic techniques developed during the Second World War for the isolation of penicillin that du Vigneaud[120] was eventually able to obtain these hormones as pure octapeptides. He then established their correct structures and confirmed these by syntheses in 1953. Today, synthetic posterior pituitary hormones and their analogues are routinely used in therapy.

1.2.3.4.2 Anterior pituitary

Following Minkowski's observation, in 1887, that pituitary tumours were associated with the characteristic overgrowth of the hands and feet in patients with acromegaly, evidence accrued to support the belief that a growth hormone was produced by the pituitary. However, when Evans and Long injected an extract of the anterior lobe of the gland more than 30 years later, they were surprised to detect not only stimulation of growth, but also of a similar effect on the development of the gonads. This influence of the pituitary was confirmed later by their colleague, Smith, who studied the consequences of ablation of the gland, only to detect still more hormonal effects. This time atrophy of both the thyroid and adrenal glands was observed. The activity of these glands was restored by implants prepared from the anterior pituitary. These findings led Collip to introduce an adrenotropic extract for clinical use in 1933. Following the clinical success of cortisone in the treatment of leukaemia and arthritis, a more refined extract was used for several years until the isolation of pure adrenocorticotrophin in 1956. Clinicians now prefer to use adrenocortical steroids.

Early attempts to treat pituitary dwarfism with extracts of the pituitary gland were unsuccessful, even when Evans and Li isolated pure growth hormone from ox pituitary in 1944. Only after it was realized that there was marked species sensitivity were attempts made to utilize human growth hormone. Cadaver hormone was first used in 1957, with dramatic results. Pure human growth hormone is now obtained from genetically engineered bacteria. Another human pituitary hormone that is occasionally used is menotrophin. This is the follicle stimulating hormone, first obtained from sheep pituitary in the early 1950s by Li. It is administered to women who cannot ovulate because of pituitary deficiency.

1.2.3.5 Sex Hormones

1.2.3.5.1 Female hormones[121]

Around the turn of the century, Knauer[122] revealed the existence of an ovarian hormone by hastening the onset of sexual maturity in young animals through the transplantation of ovaries removed from mature animals. Amongst the numerous ovarian extracts subsequently administered to treat menopausal disorders during the years leading up to the First World War, only those prepared with fat solvents were effective. One such extract was marketed in 1913. However, the problem with this and other similar preparations was the lack of standardization. This was resolved in 1923, when Allen and Doisy[123] introduced an assay procedure based on earlier reports that oestrogenic material induced consistent changes in the appearance of cells lining the vaginal wall in rodents. The assay also cleared the way for the isolation of female hormone after Ascheim and Zondek[124] had used it to demonstrate that pregnancy urine was an exceptionally rich source. Doisy was the first to report the isolation of oestrone (**144**), in August 1929, only to be followed by others within months.[125] Early suspicions that oestrone might be a steroid were confirmed by Butenandt, and six years later the structure which Rosenheim and King proposed in 1932 was shown to be correct. The more potent oestradiol (**145**) was isolated from four tons of sows' ovaries by Doisy in 1935, although it had already been synthesized in the laboratory from oestrone.

Dodds and his colleagues detected weak oestrogenic activity in several phenanthrene compounds modelled upon oestrone. Unexpectedly, they then found activity was retained even in dihydroxystilbene. In 1938, they reported[126] that stilboestrol (diethylstilboestrol) (**146**), dienoestrol (**147**) and hexoestrol (**148**) were as potent as oestradiol when injected. However, unlike the natural hormone, these cheaply synthesized analogues were resistant to metabolic degradation and hence retained high potency when administered by mouth. Robson and Schonberg obtained similar results with a series of halogen-substituted triphenylmethanes based on Dodds' original observations. Their work was to lead others to develop drugs such as chlorotrianisene (**149**) and clomiphene (**150**). The latter compound, which was introduced in 1962, was found unexpectedly to induce ovulation in some women whose infertility was due to ovulatory failure. This was not the only instance of a serendipitous advance in this series of compounds, for the related tamoxifen (**151**)[127] turned out to bind so firmly to oestrogen receptors that it acted as an antioestrogen. Since its first clinical trial in the treatment of breast cancer in 1971, impressive evidence has accumulated to testify to its outstanding value in this disease.

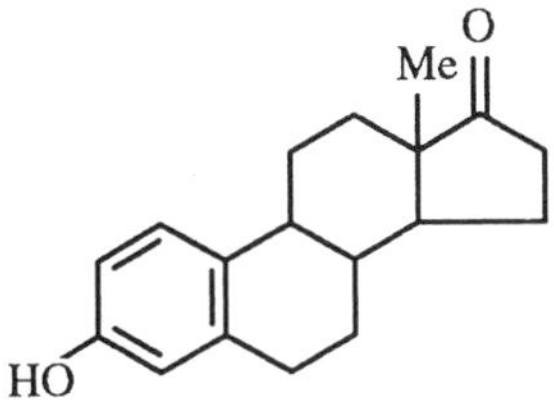

(144) 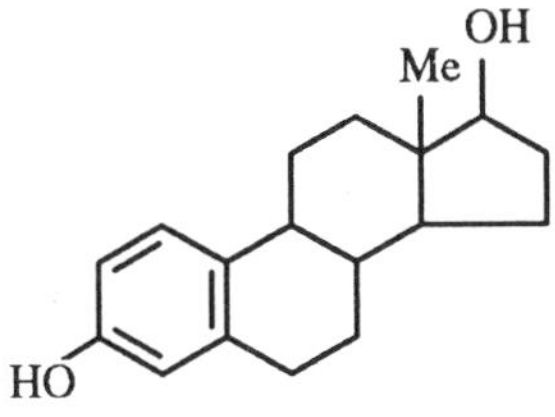 **(145)** 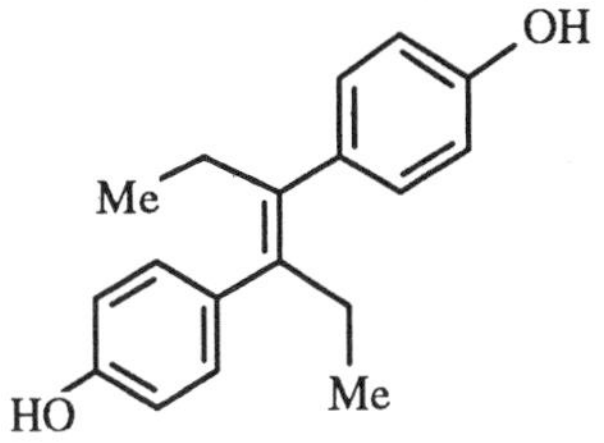 **(146)**

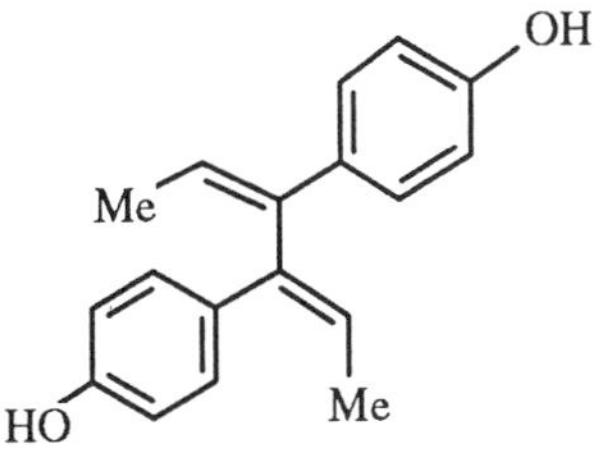

(147) 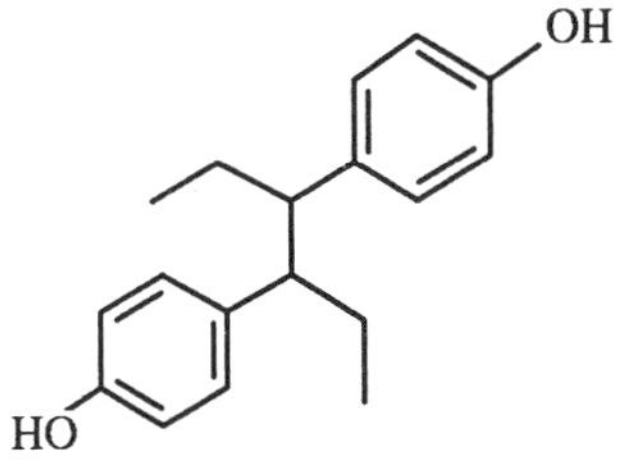 **(148)** 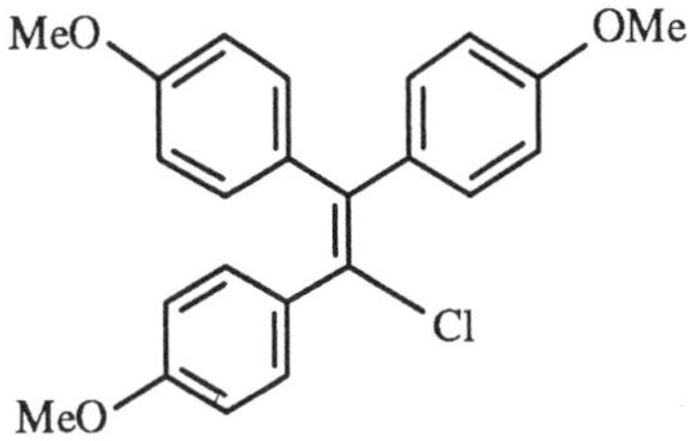 **(149)**

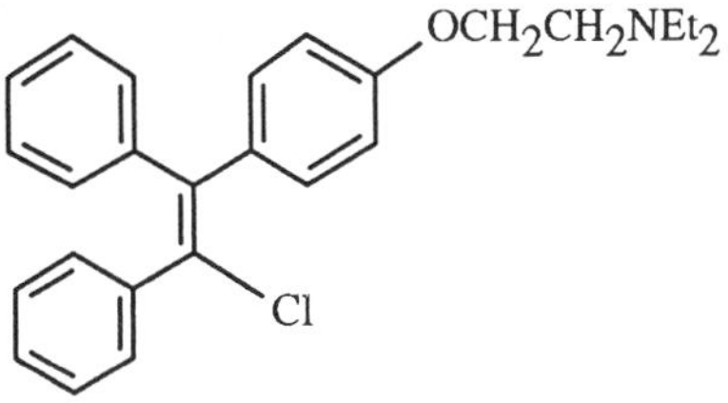

(150) 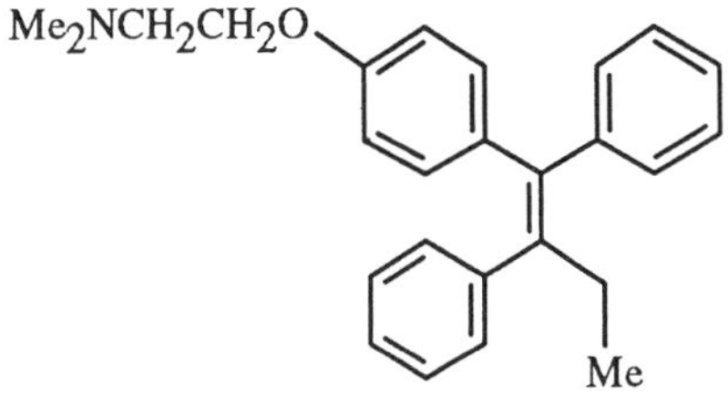 **(151)**

In 1903, Fraenkel[128] observed that a yellow substance formed in the ruptured egg sack after the release of ova during the oestrus cycle. Later, it was established that this corpus luteum somehow enabled the uterus to maintain pregnancy. The isolation of oestrone encouraged Allen and Corner[129] to use similar methods to prepare an active luteal extract. In 1934, three independent groups of researchers isolated progesterone (**152**). It was shown to prevent miscarriages in the early months of pregnancy so long as it was administered by injection. The extensive metabolic degradation of progesterone in the liver following oral administration could, however, be avoided by the use of its analogue ethisterone (**153**). When this was introduced in 1938 it marked an important advance, but the virilizing (androgenic) action on the foetus due to structural similarity to testosterone eventually gave rise to concern when it was used in pregnant women. This problem was tackled in the mid-1950s[130] when progestogens were rendered more resistant to metabolism by the introduction of small blocking groups at positions in the steroid nucleus that were sensitive to attack by oxidative enzymes. Ethisterone was thus converted to dimethisterone (**154**) to effect a twelve-fold enhancement of potency. When the 17-acetoxy ester of progesterone (prepared for use in oily depot injections) was unexpectedly found to be orally active, several pharmaceutical companies competed to be the first to patent the 6-methyl analogue that subsequently became known as medroxyprogesterone acetate (**155**). It was 25 times as potent as ethisterone whilst its analogue with a double bond at the 6-position, megestrol acetate (**156**), was even more potent. In the same year as this was introduced, 1958, chlormadinone acetate (**157**) appeared, featuring a chlorine atom instead of a methyl group to block oxidation at the 6-position.

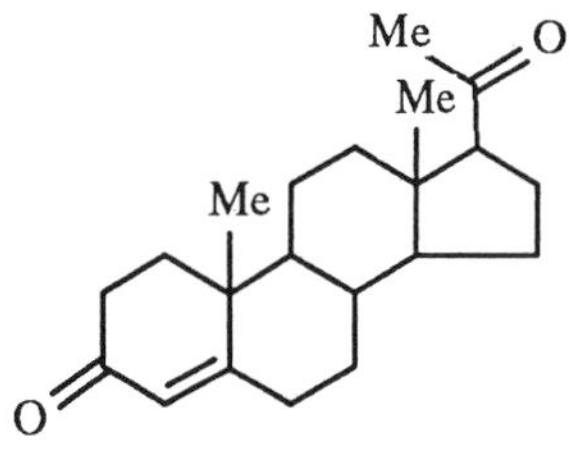

(152) 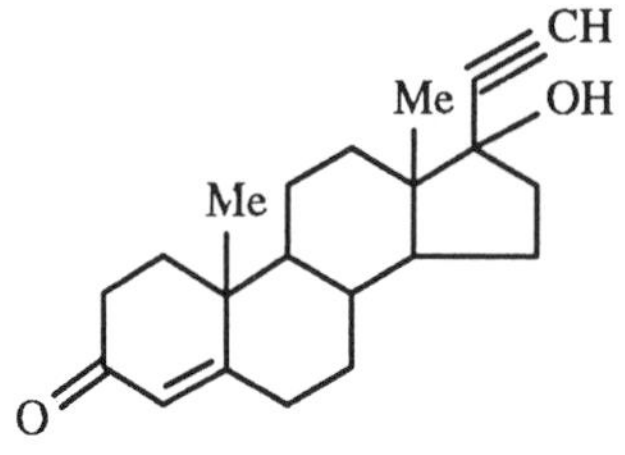 (153) 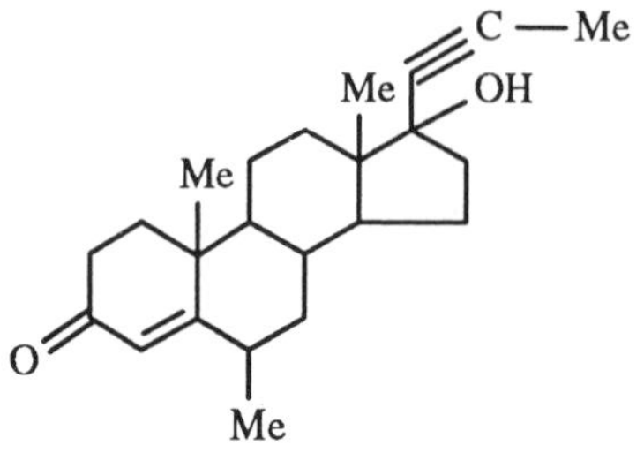 (154)

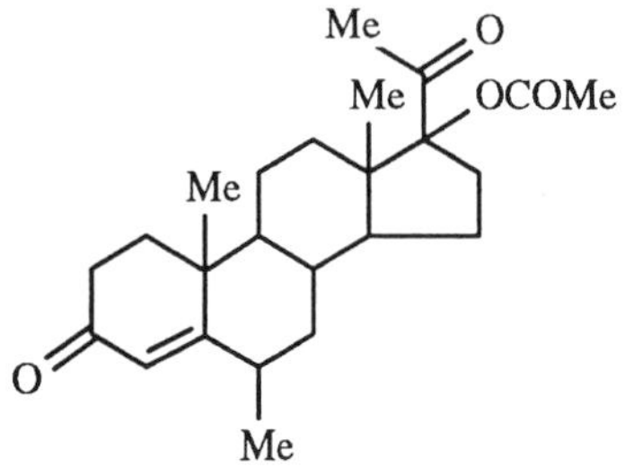

(155)

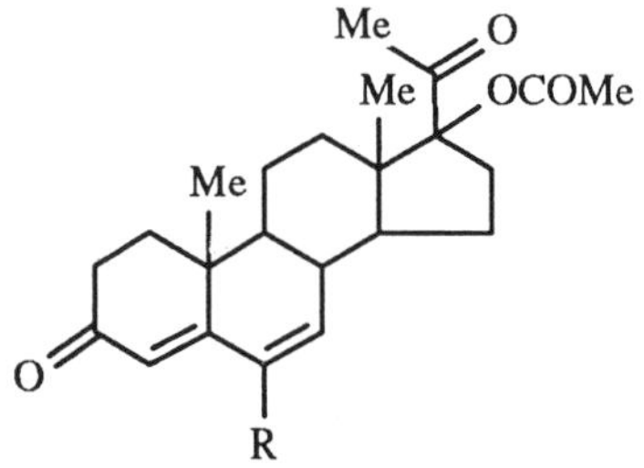

(156) R = Me
(157) R = Cl

The main stimulus for the development of potent, orally active progestational agents came from the pioneering studies of Pincus and Rock on oral contraception through the cyclical administration of such compounds.[131] They screened 200 steroids supplied by pharmaceutical companies, thereby establishing that the most potent progestogens were all 19-norsteroids. Of these, norethynodrel (**158**) was selected in 1956 for their first large-scale trial of oral contraception. It was known to be free of the androgenic activity associated with the use of ethisterone. Interestingly, norethisterone (**159**) was one of the first norsteroids prepared as an analogue of progesterone. It was synthesized in 1951 and was subsequently to rival norethynodrel as an oral contraceptive.

The universal popularity of oral contraception offered high financial rewards to the developers of safer progestogens. The search for such drugs ultimately manifested itself in the production of increasingly more potent compounds, despite the fact that enhanced potency in itself did not necessarily confer greater safety. Initially, the aim had been to increase progestational activity without affecting residual androgenic potency. Norethisterone, having proved to be one of the most successful progestogens, served as the prototype for several of the newer progestogens introduced in the 1960s. These included norgestrel (**160**), lynoestrenol (**161**) and ethynodiol diacetate (**162**). Fears over the safety of prolonged administration of progestogens dissuaded manufacturers from investing

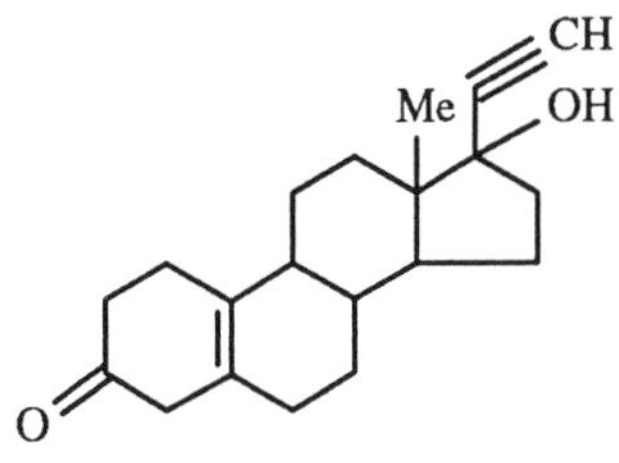

(158) 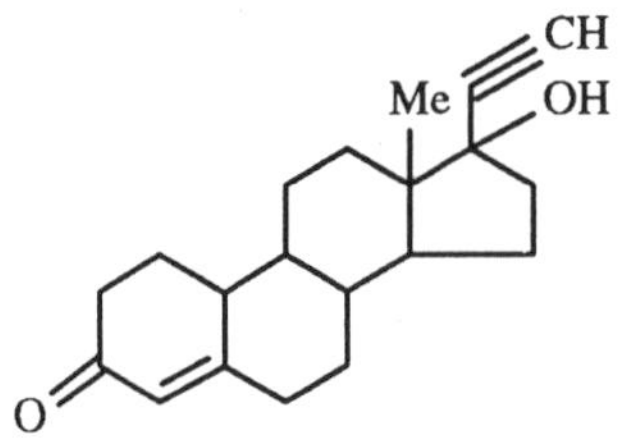 (159) 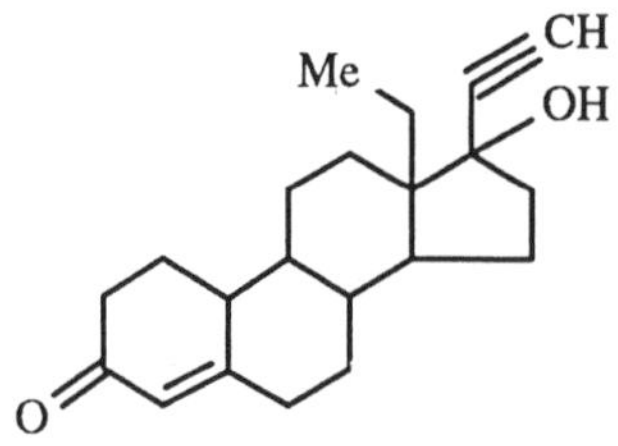 (160)

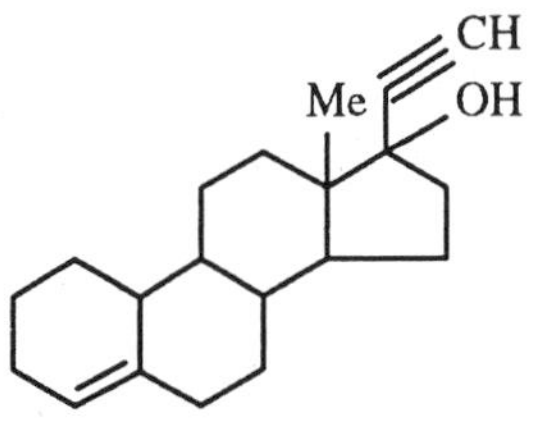

(161) 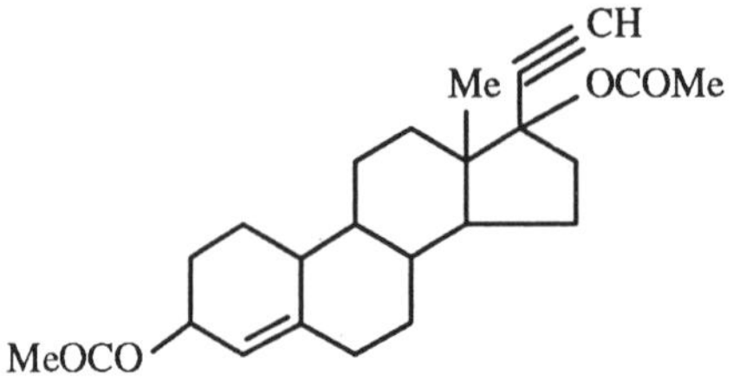 (162) 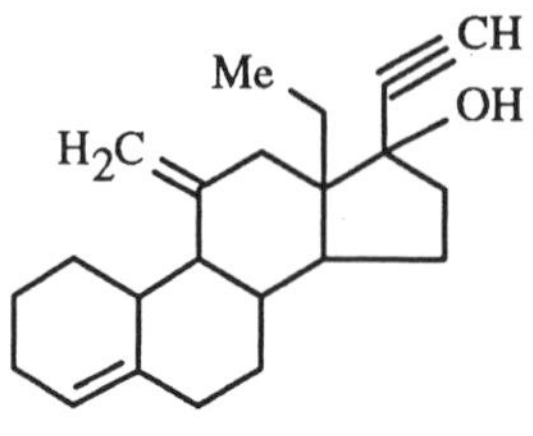 (163)

the massive funds needed to develop still more of these drugs. The only new progestogen to be introduced in the United Kingdom since the 1960s has been the norgestrel analogue known as desogestrel (**163**).

1.2.3.5.2 Male hormones

The techniques responsible for the isolation of female hormones were promptly applied to that of the male hormone. Thus the success of the Allen and Doisy assay for oestrogenic activity encouraged Koch and his colleagues[132] to develop a bioassay for androgenic activity. They were able to measure the potency of extracts of bull testicles by their ability to induce growth of the capon's comb. This resulted in the preparation of a potent extract, but the real breakthrough came after it was discovered that male hormone was present in urine. Butenandt, in 1931, isolated 15 mg of androsterone (**164**) from 15 000 L of urine. Two years later, Laqueur[133] isolated 5 mg of the much more potent testosterone (**165**) from nearly one ton of bulls' testicles. Androsterone proved to be a metabolite of testosterone, the true male hormone. The latter was synthesized from cholesterol in 1935 by Ruzicka.

Soon after its isolation, testosterone was shown to exhibit muscle-building properties. In 1948, Saunders and Drill embarked on an ambitious screening programme which examined approximately 1000 steroids in an attempt to identify one which retained anabolic activity whilst devoid of hormonal activity. Seven years passed before Colton's norethandrolone (**166**) (prepared by reducing the 17-ethinyl group of norethisterone) was found to have only one-sixteenth the androgenic activity of testosterone whilst exhibiting the full anabolic action.[134] It had the added advantage of being orally active. Its analogue which lacks the carbonyl at the 3-position, ethyloestrenol (**167**), proved to be 10 times as potent. Other anabolic steroids are analogues of methyltestosterone (**168**), a derivative of the male hormone synthesized by Ruzicka in 1935 and which owes its oral activity to the ability of the 17-methyl group to block metabolic oxidation in the liver. Structural modifications to methyltestosterone have interfered with its ability to fit those receptors which mediate the undesired hormonal activity. Compounds such as methandienone (methandrostenolone) (**169**), oxymetholone (**170**), and stanozolol (**171**) have thus been introduced as anabolic steroids.

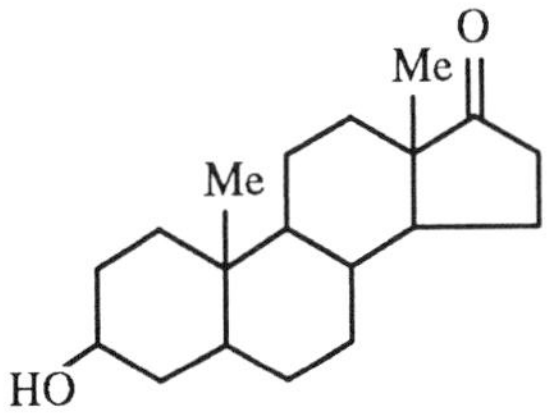

(164) 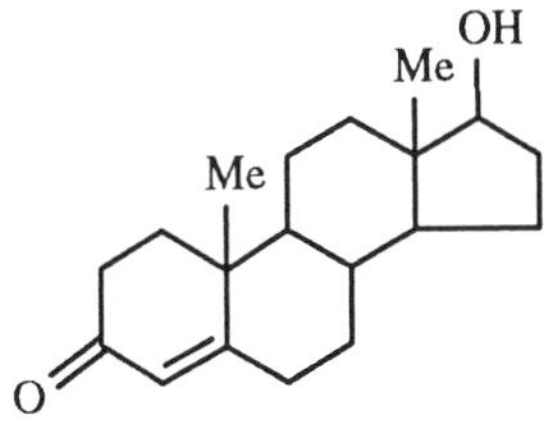 **(165)** 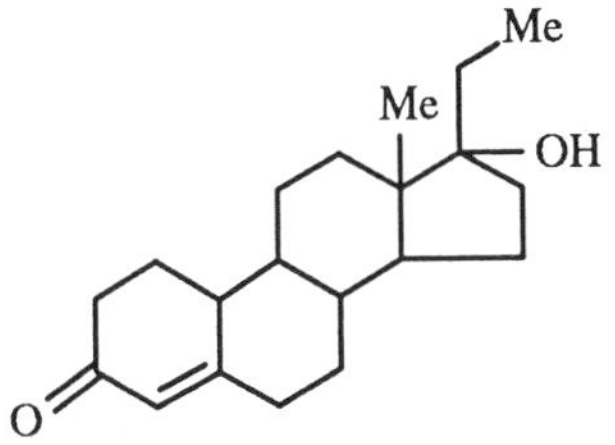 **(166)**

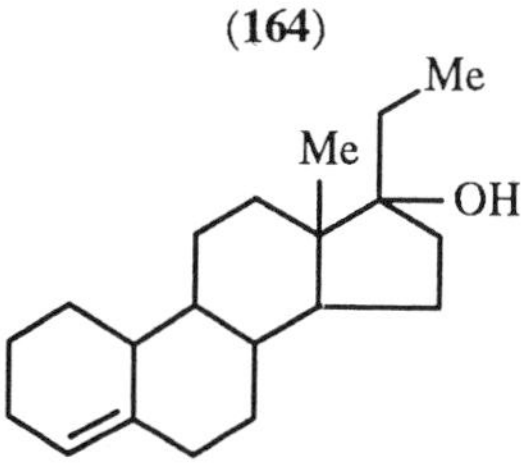

(167) 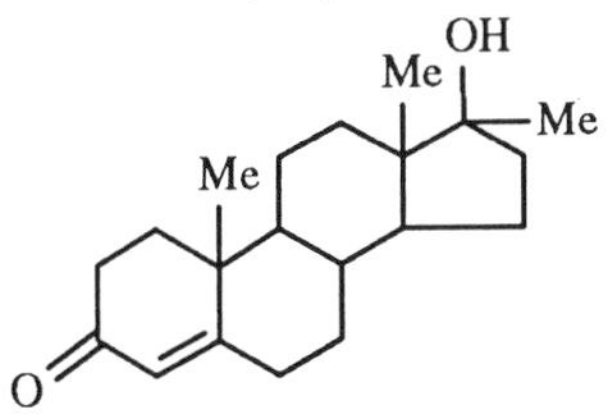 **(168)** 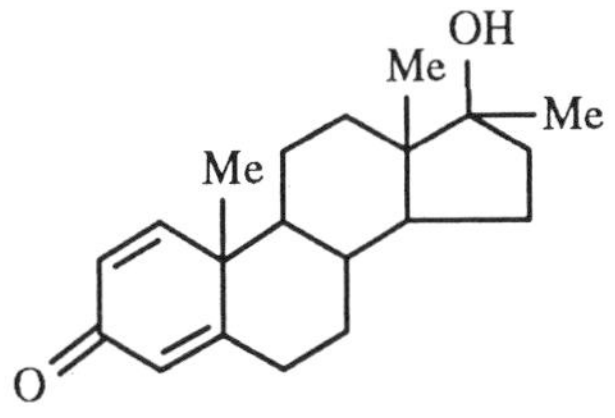 **(169)**

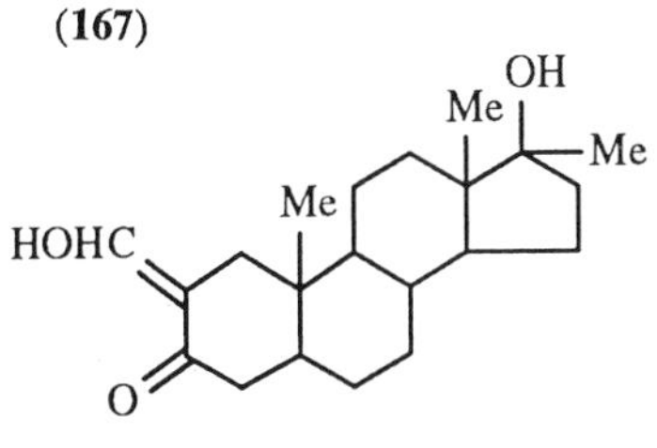

(170) 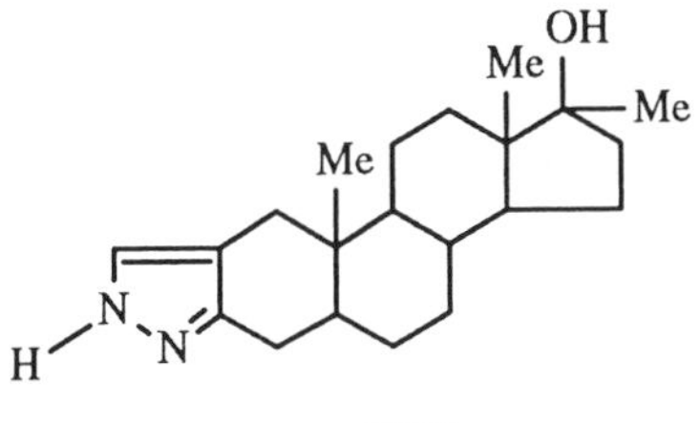 **(171)**

1.2.3.6 Adrenal Cortex

In 1927, Rogoff and Stewart demonstrated that dogs from which the outer cortex of the adrenal gland had been excised could be kept alive by injections of a canine adrenal cortical extract. By utilizing benzene instead of aqueous extraction, Swingle and Pfiffner obtained a much more potent extract, enabling commercial production to be undertaken so that patients with Addison's disease could be treated for their life-threatening hormone deficiency state from 1930 onwards. By 1935, adrenocortical extracts were free from contamination with adrenaline. This encouraged investigators to attempt the isolation of the hormone or hormones concerned. Within a year or two Pfiffner, Wintersteiner and Vars, and also Kendall working independently, had isolated six steroid hormones. By the end of the decade, even that achievement had been surpassed by Reichstein who isolated no less than 19 steroids with hormonal activity! His source had been extracts prepared from 20 000 cattle adrenal glands. He also managed to synthesize one of these hormones, deoxycorticosterone (**172**), from a readily available plant product. This was marketed in the form of an oily injection, in 1939.

Wartime interest in the potential of adrenocortical steriods to reduce high altitude stress (based on an unfounded rumour), led to extensive investigations in the United States and resulted in the synthesis of cortisone (**173**) by Sarett. In 1948, Hensch demonstrated the remarkable ability of this hormone to ameliorate severe arthritis. Initially, there was excessive enthusiasm amongst clinicians and public alike for the use of cortisone and its analogues, resulting in extensive investigations by the pharmaceutical industry.

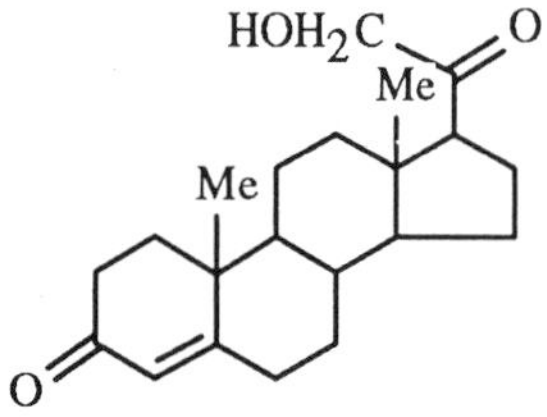

(**172**)

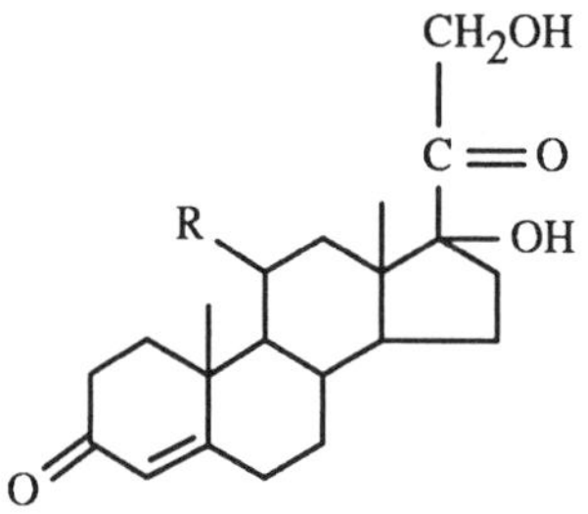

(**173**) R = O
(**174**) R = OH

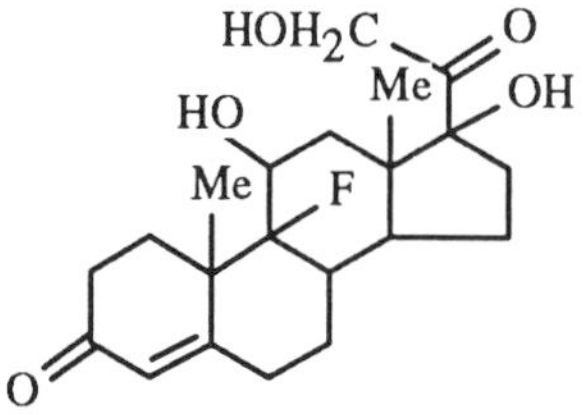

(**175**)

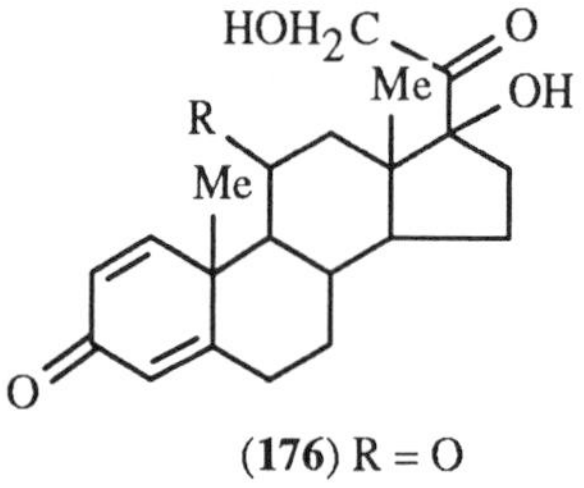

(**176**) R = O
(**177**) R = OH

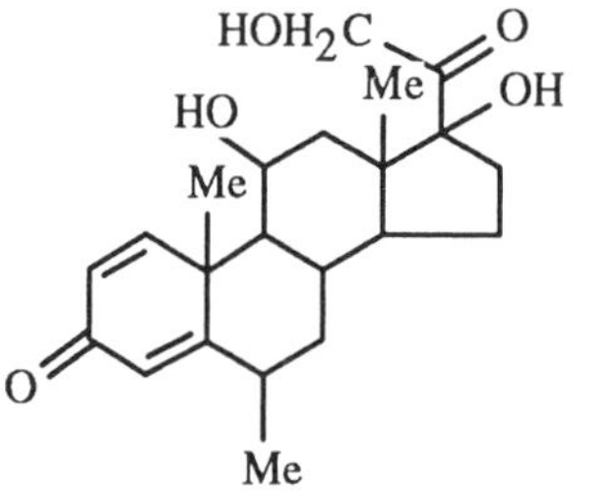

(**178**)

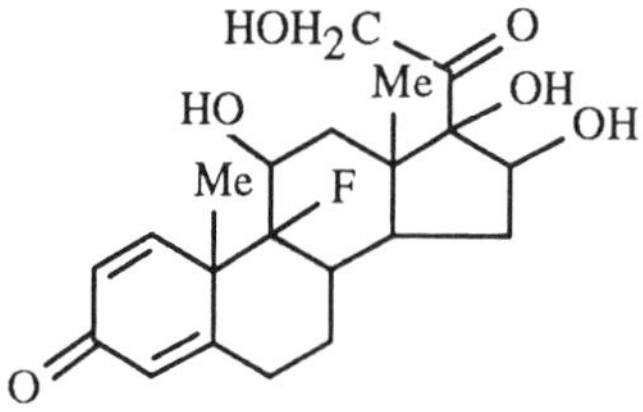

(**179**)

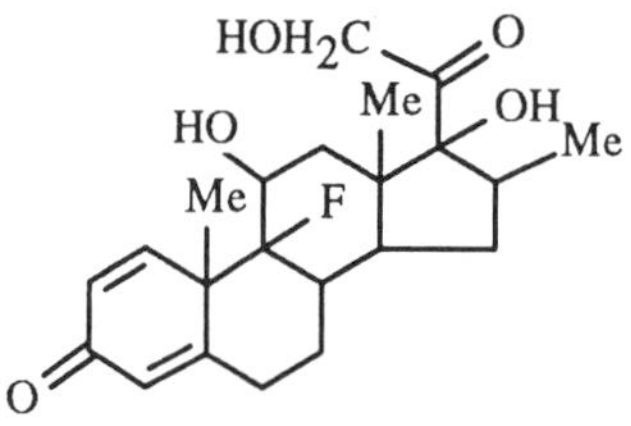

(**180**)

In the course of an attempted synthesis of the hormone hydrocortisone (**174**), a derivative of epicortisol substituted with a bromine atom at the 9-position of the steroid nucleus, was routinely screened. As this proved to be much more potent than epicortisol itself, Fried synthesized a series of halogen-substituted steroids. Tests then revealed that 9-fluoro steroids were highly potent. This resulted in the introduction of fludrocortisone (**175**) in 1954. Although this compound was about 10 times as potent as cortisone with regard to its antiinflammatory (glucocorticoid) activity, its (mineralocorticoid) ability to induce salt retention was enhanced by a factor of more than 300. This afforded an early indication that mineralocorticoid and glucocorticoid activity could be separated by appropriate molecular manipulations. This view was reinforced when it was revealed that prednisone (**176**) and prednisolone (**177**), analogues obtained by mould fermentation from cortisone, exhibited a selective fivefold enhancement of glucocorticoid activity. Derivatives of prednisolone that were substituted at the 6-position with either methyl groups or halogens, *e.g.* methylprednisolone (**178**), were slightly more potent due to enhanced resistance to metabolic deactivation.

Following the isolation of a 16-hydroxy-substituted steroid from a patient with an adrenal tumour, investigations into the biological activity of this type of compound disclosed reduced mineralocorticoid activity. By combining the introduction of a 16-hydroxy substituent with that of a double bond in the same position as in prednisolone, Bernstein was able to prepare a fludrocortisone analogue with very weak mineralocorticoid activity but high glucocorticoid activity. It was named triamcinolone (**179**). Its ability to cause nausea and dizziness limited its usefulness to topical application in the form of its lipophilic acetonide. This problem did not arise when the 16-hydroxy substituent was replaced by a methyl group as in dexamethasone (**180**). In this instance, the substituent had been introduced solely with the intention of hindering metabolism at the adjacent 17-position. The apparently unanticipated enhancement of glucocorticoid activity to make the analogue at least six times as potent as prednisolone ensured its clinical success and that of its equiactive stereoisomer betamethasone. Both compounds exhibited high antiinflammatory activity and minimal salt-retaining activity. Still more potent analogues have appeared since the introduction of these two drugs in 1958, but whether there is any real clinical need for them depends on whether or not the mineralocorticoid activity had been completely eliminated. This does not seem to have been the case.

1.2.3.7 Neurohormones

Although the physiological activity of acetylcholine (**181**) had been observed by Hunt and Traveau[135] in 1906 and later extensively studied by Dale[136] after he detected it in ergot extracts, the hormone was not isolated from mammalian tissue until 1926 when Loewi was able to avoid its endogenous destruction by utilizing physostigmine as a cholinesterase inhibitor. Loewi and Navratil[137] proved that acetylcholine was the neurohormone responsible for neurotransmission in the parasympathomimetic nervous system. Its similar role to that of adrenaline as a chemotransmitter encouraged investigations into its analogues, leading to the introduction of methacholine (**182**) in 1931. These drugs were used as parasympathomimetic stimulants to treat conditions such as post-operative intestinal spasm. In keeping with a prediction by Simonart,[138] the placing of a methyl substituent adjacent to the metabolically labile ester linkage stabilized this molecule somewhat towards attack by esterase enzyme. This permitted clinical use of the analogue, which had not been feasible with the evanescent natural hormone. Another stable analogue introduced around the same time was carbachol (**183**), soon to be followed by bethanechol (**184**).

$MeCO_2CH_2CH_2\overset{+}{N}Me_3$ (**181**)

$MeCO_2CH(Me)CH_2\overset{+}{N}Me_3$ (**182**)

$H_2NCO_2CH(R)CH_2\overset{+}{N}Me_3$ (**183**) R = H; (**184**) R = Me

Noradrenaline (**185**) was not used clinically until after the demonstration, in 1946, by von Euler[139] that it was the principal chemotransmitter in the sympathetic nervous system. Within two years, it was marketed for the treatment of clinical shock. This use has since been abandoned as the resultant rise in blood pressure is usually gained at the expense of perfusion of vital organs. Unlike adrenaline, noradrenaline has no bronchodilator action or any other important clinical application.

In 1959, Carlsson proposed that dopamine was not simply the metabolic precursor of noradrenaline, but was actually an important neurohormone. The following year, Hornykiewicz produced evidence of depletion of dopamine reserves in patients with Parkinson's disease in the region of their brains known to be implicated in its aetiology. The inability of dopamine to enter the brain from the general circulation was overcome by the administration of its metabolic precursor, DOPA. Large doses had to be administered to secure clinical improvement, but the use of the laevorotatory isomer known as levodopa (**186**) was advantageous.[140] Further dosage reduction has been attained by the concurrent administration of drugs that were found to inhibit the destructive enzyme DOPA decarboxylase, *e.g.* benzserazide (**187**) or carbidopa (**188**).

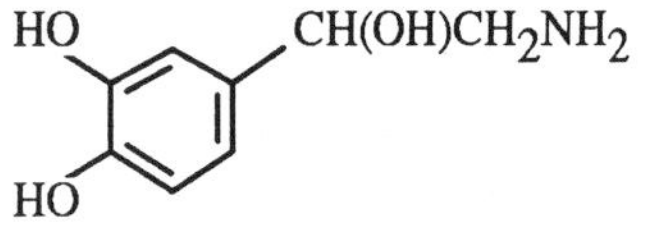

(185) 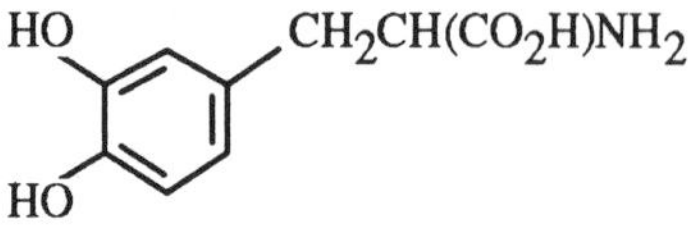 (186)

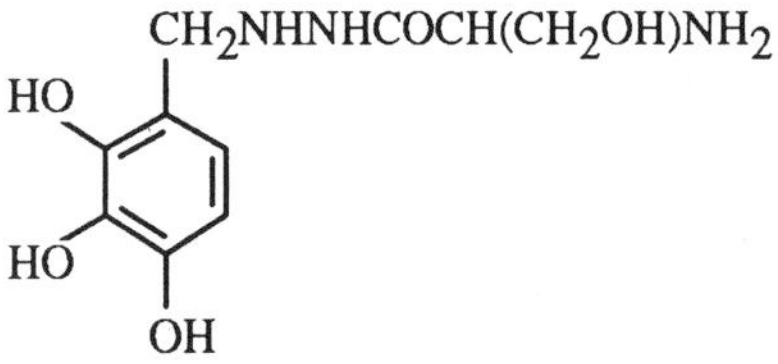

(187) 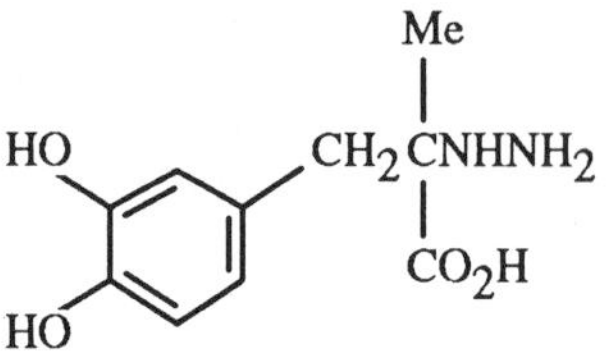 (188)

The indolic compound serotonin (5-hydroxytryptamine) (**189**) was isolated from serum in 1948 and subsequently became recognized as a neurohormone. Implicated in both inflammatory processes and as a chemotransmitter in the central nervous system, pharmaceutical companies showed much interest in its potential for therapeutic exploitation. The consequences of this have proved less far-reaching than had been anticipated. The serotonin analogue oxypertine (**190**) was introduced as a tranquillizer in 1962. It was followed the next year by indomethacin (**191**),[141] one of more than 350 indole derivatives that had been screened for ability to interfere with the inflammatory process. Its somewhat high incidence of side effects led to the development of analogues such as sulindac and tolmetin.

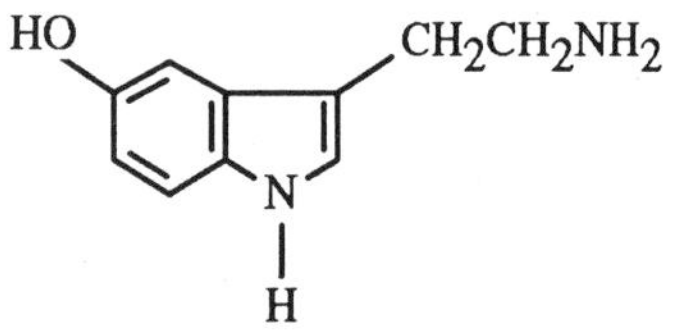

(189) 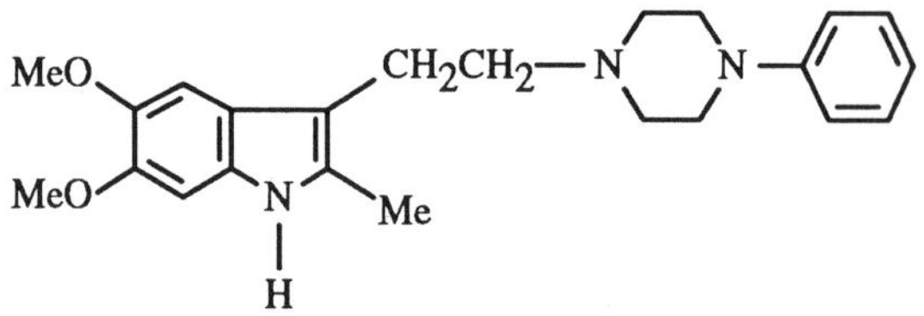 (190)

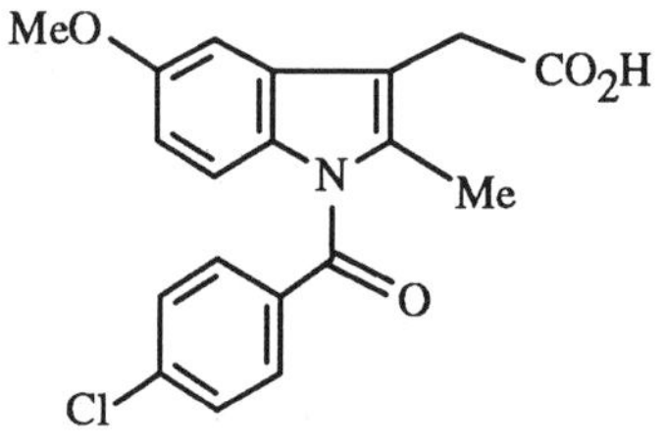

(191)

1.2.4 MISCELLANEOUS MAMMALIAN PRODUCTS

1.2.4.1 Histamine and Antihistaminic Drugs

In 1910, it was shown that histamine (**192**) was formed by putrefaction of proteins containing the amino acid histidine. Despite its stimulant effect on the uterus, the compound could not be used clinically since it had so many other actions. Nor was its presence in the body recognized until 1926, although Dale and Laidlaw[142] had previously noted that its effects on animals bore a strong resemblance to the intense allergic reaction described as anaphylactic shock. When antagonists of histamine first became available for clinical use, it was this latter role that was considered for exploitation.

Following the discovery of the occurrence of histamine in the body, efforts were made to elucidate its role. Bovet believed these could be facilitated by the preparation of antagonists that would do for histamine what atropine and ergotamine had done for acetylcholine and adrenaline respectively. The sole lead for Bovet in seeking an effective antihistamine was a series of reports indicating that some adrenaline analogues and antagonists seemed to interfere with the action of histamine during pharmacological experiments on the isolated intestine. He was able to confirm this by using his colleague Fourneau's adrenergic blocker piperoxan. Its failure to protect guinea pigs from the lethal effects of injected histamine led him and Staub to seek more effective analogues of piperoxan amongst the compounds already examined for adrenergic blocking activity in their laboratory.[143] Much progress was made during the late 1930s,[144] but the first antihistamine potent enough for clinical application was not introduced until 1942.[145] This was phenbenzamine (**193**), an analogue of the compounds prepared by Bovet and Staub. Bovet introduced mepyramine two years later (**194**), this being rapidly followed by similar compounds from other laboratories, *e.g.* tripelennamine (**195**), diphenhydramine (**196**), triprolidine (**197**) and cyclizine (**198**). All of these were developed before 1950, and they have served as models for the many more that have been synthesized since then. Their principal value is in the alleviation of allergies. However, their residual anticholinergic properties conferred decongestant activity and this permitted extravagant claims to be made for their role in dealing with the common cold.

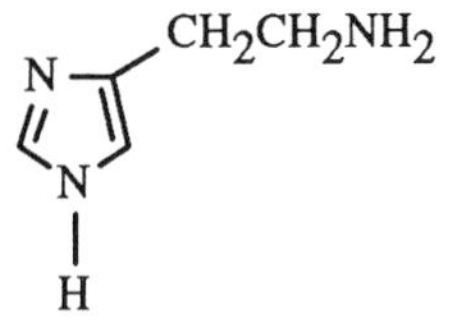

(192) 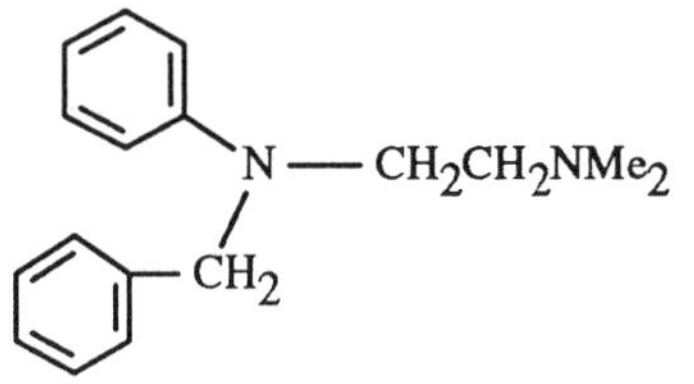 **(193)** 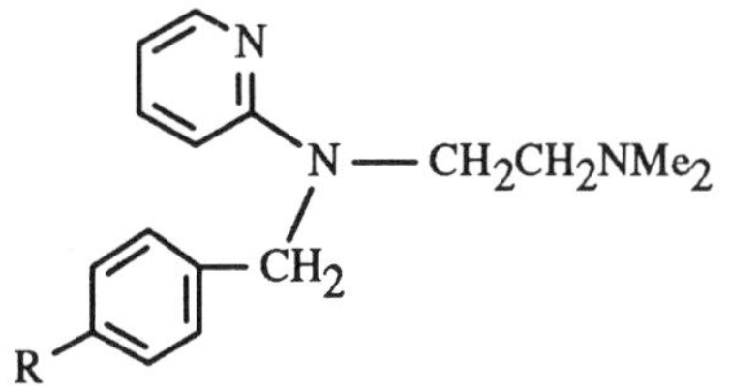 **(194)** R = OMe
(195) R = H

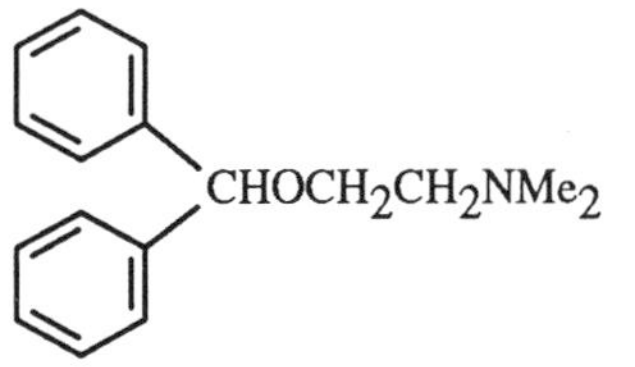

(196) 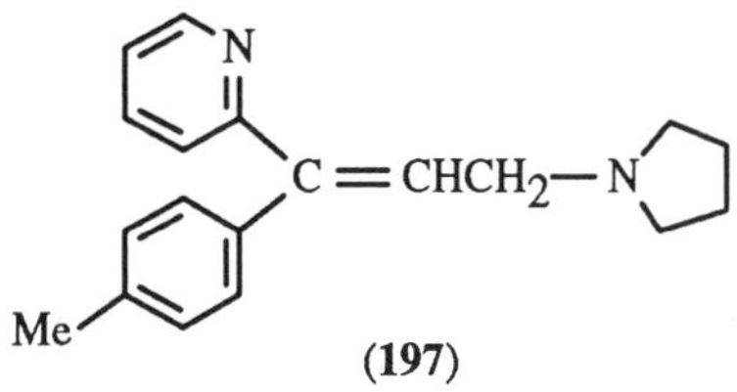 **(197)**

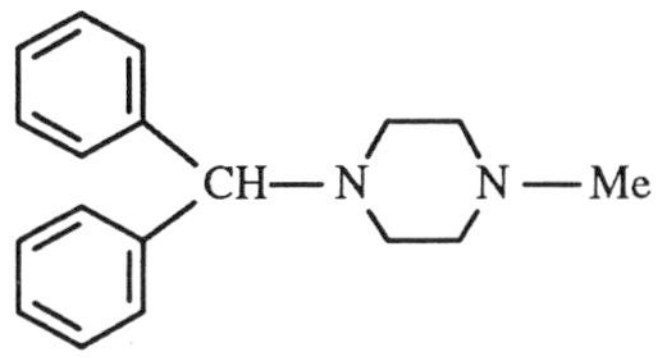

(198) 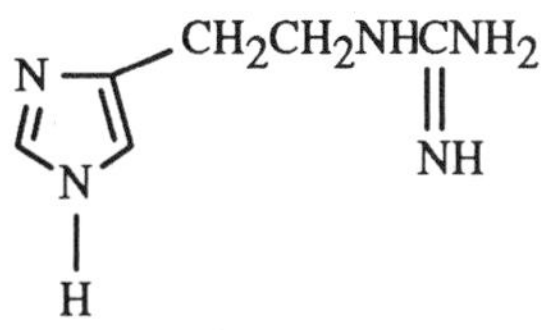 **(199)**

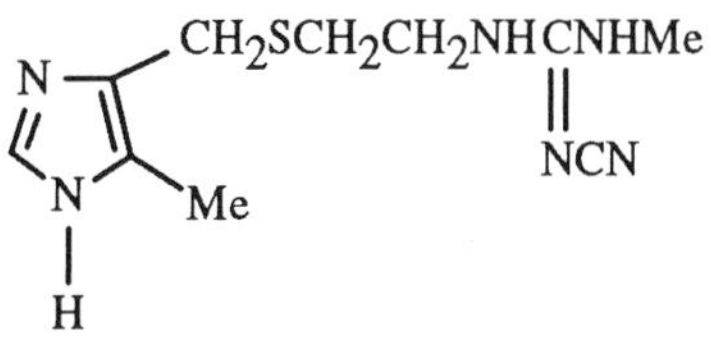

(200) 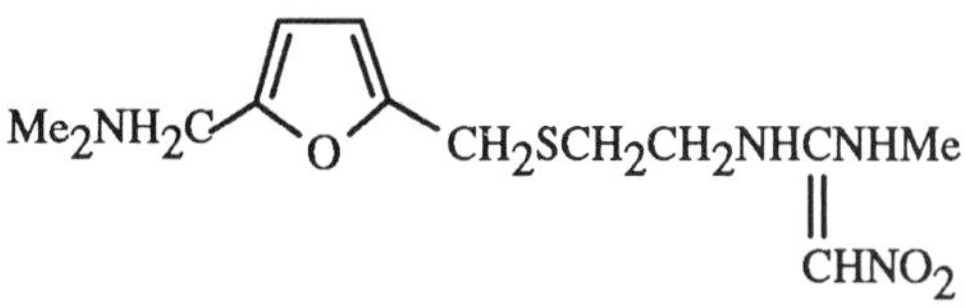 **(201)**

The early antihistamines were unable to antagonize the release of gastric acid caused by the action of histamine. In 1964, Black and Ganellin began their search for an antihistamine that would do this and thus possibly be of value in the treatment of gastric ulcers. With no clear lead to follow, the only way for them to proceed was to prepare analogues of histamine and subject them to screening. Over 200 compounds were examined during the next four years before guanylhistamine (**199**) was shown to have slight activity. Considerable effort was required to exploit this through highly sophisticated medicinal chemistry techniques, but cimetidine (**200**) satisfied all the expectations when it was introduced into clinical practice in 1976.[146] Ten years later it and the closely related ranitidine (**201**) had become the two best-selling drugs in the world.

1.2.4.2 Prostaglandins and Prostanoids

The prostaglandins were discovered in 1934 by von Euler,[147] who detected their ability to reduce blood pressure and to stimulate smooth muscle. In the late 1950s, Bergstrom was able to apply combined gas chromatography/mass spectrometry techniques so as to establish their structures, thereby clearing the way for the first synthesis by Beal in 1965. This permitted clinical investigations to be conducted, resulting in the marketing in 1972 of dinoprost (**202**) for obstetric use, and dinoprostone (**203**) for induction of abortion. Four years later, Vane isolated the prostanoid epoprostenol (prostacyclin) (**204**) from the walls of blood vessels where it produces vasodilation and prevents blood clotting. Like the closely related prostaglandins, it was rapidly metabolized within minutes of administration as a drug. However, this was successfully exploited in high risk patients to whom it has been administered during renal dialysis or heart–lung by-pass surgery as an alternative to the more persistent anticoagulant heparin.

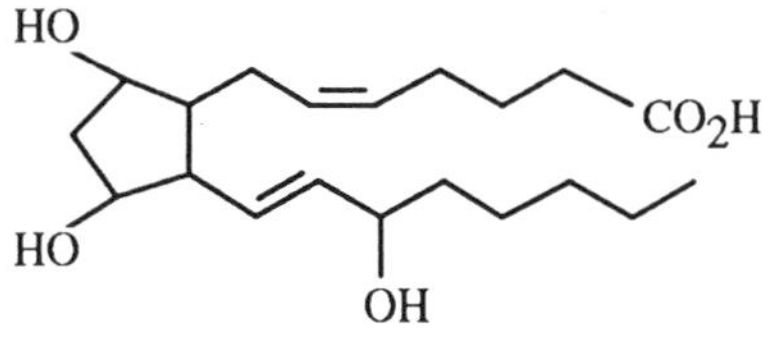

(**202**) 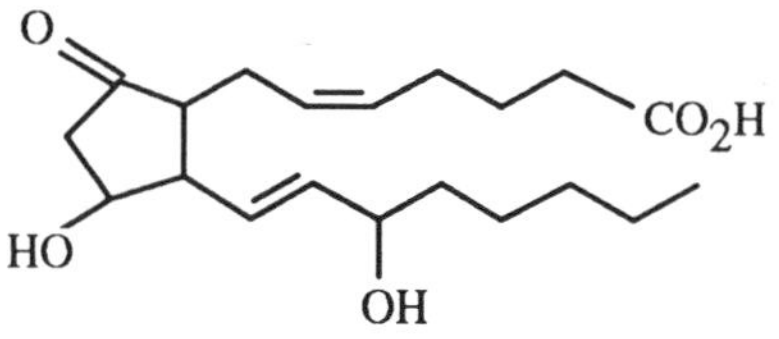 (**203**)

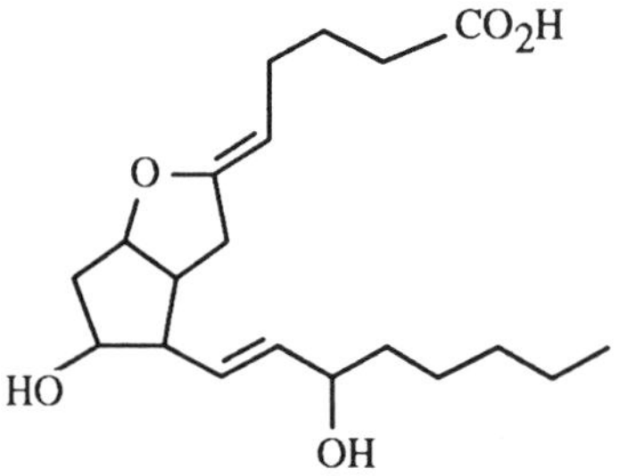

(**204**)

1.2.4.3 Vitamins[148,149,150,151]

McCollum and Davis reported, in 1913, that they had noted disruption of normal growth in mice being fed on artificial diets for experimental purposes. This was attributed to deficiency of an essential fat soluble substance found in natural diets. Osborne and Mendel then observed that rats fed on a diet deficient in the fat soluble substance developed an eye disease similar to the xerophthalmia which was known to respond to treatment with cod liver oil. The eye disorder in the rats was alleviated with cod liver oil, and during the First World War a similar condition in undernourished Dutch children was successfully treated in the same manner.

Prior to its characterization, the fat soluble vitamin was known as vitamin A. During the 1920s, steam distillation processes yielded the vitamin from cod liver oil in almost pure form. The structure of the pure vitamin, now known as retinol (**205**), was determined in 1933 by Karrer. A commercial synthesis initiated in 1947 rendered the former fish oil extraction procedures obsolete.

In 1901, Grijns came to the conclusion that, in order to prevent polyneuritis in fowls and beriberi in humans, a healthy diet had to contain an unidentified substance known to be present in the husk of unmilled rice as well as meat and vegetables. His findings had been inspired by the work of his colleague Eijkman, who subsequently established that the essential food factor, later to be described as vitamin B, had a low molecular weight and was water soluble. The isolation of this depended on the development of a reliable assay by Jansen who, with Donath, was able to isolate crystals of pure vitamin in 1926.[152] A commercially viable extraction procedure was introduced seven years later by Williams, who established the structure and synthesized the vitamin in 1936. Two other syntheses were independently developed around the same time. In 1951, the approved name of thiamine (**206**) was universally adopted.

In 1919, Mitchell[153] questioned the contention that vitamin B was a single substance. Following the isolation of pure thiamine in 1926, a distinction was drawn between it and a more heat stable vitamin, these then being called vitamin B_1 and vitamin B_2 respectively. Experiments on rats fed on artificial diets to which vitamin-containing extracts were added soon revealed the presence of several more water soluble vitamins. These were then described as the vitamin B complex. The following were subsequently isolated from this complex: riboflavine (**207**),[154] nicotinamide (**208**),[155] pyridoxine (**209**),[156] and biotin (**210**).[157]

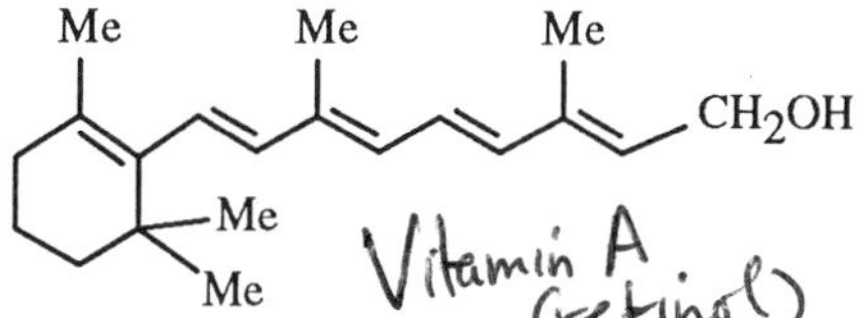

(205) Vitamin A (retinol)

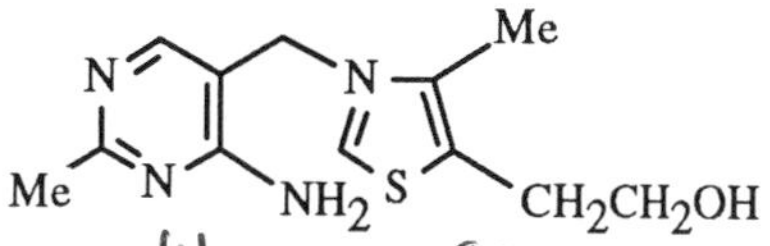

(206) Vitamin B (thiamine)

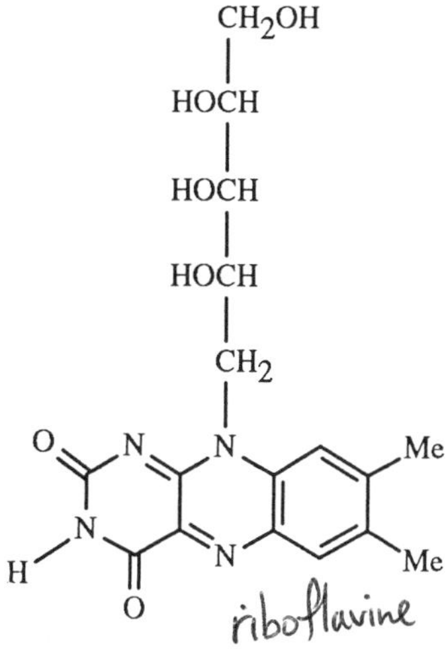

(207) riboflavine

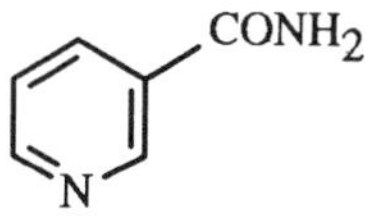

(208) nicotinamide

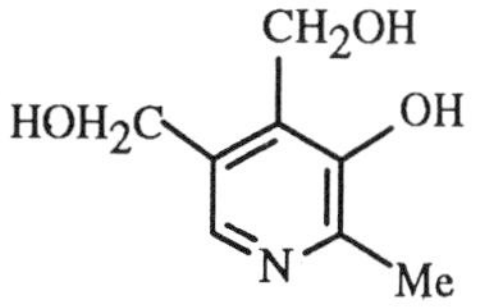

(209) pyridoxine

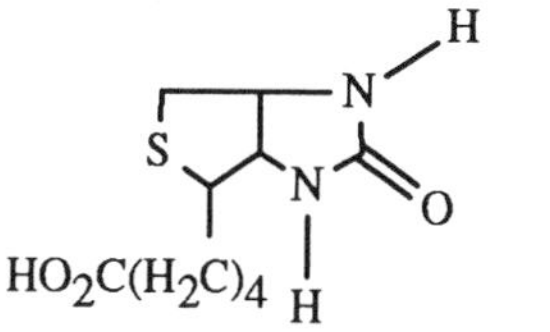

(210) biotin

Having observed that diet influenced the rate of hemoglobin formation, Whipple's interest in liver disease persuaded him to examine the effect of feeding liver to anaemic dogs.[158] This revealed that its influence on hemoglobin formation was greater than that of any other food. In 1924, Minot and Murphy initiated a clinical trial to assess the value of feeding liver to patients with pernicious anaemia. Remarkable improvements were noted within weeks or less, although as much as half a kilogram of liver had to be taken daily. Potent liver extracts were introduced later in the decade, and it became possible to treat the disease effectively with injections of what became known as vitamin B_{12}, given once every three weeks. In 1948, Smith and Folkers independently isolated pure cyanocobalamin (**211**), its structure being determined seven years later in a collaborative project headed by Smith, Todd and Hodgkin.

Wills speculated that a pernicious anaemia-like disease (tropical macrocytic anaemia) in undernourished mothers of premature babies might have been due to a dietary deficiency of the vitamin B complex. Unlike pernicious anaemia, it responded to administration of a yeast extract known to be a rich source of the vitamin B complex. Injections of liver extract had no beneficial value. These observations, supported by controlled studies on animals, were reported in an Indian journal in 1930 but were largely ignored. Later in the decade, feeding experiments on experimental animals suggested that deficiency of an unknown B complex vitamin could be responsible for the development of an anaemic condition which responded favourably to either yeast extract or liver. The vitamin was isolated from liver in 1942 and found to be identical to an essential growth factor for *Lactobacillus casei* that had just been isolated from yeast, liver and spinach.[159] It was given the name folic acid (**212**) and a commercial synthesis was reported three years later.

From the 16th century onwards numerous reports testified to the ability of citrus and certain other fruits to prevent or cure scurvy. When the first reports suggesting dietary deficiency was the cause of beriberi were published around the turn of the century, Holst initiated feeding experiments on guinea pigs as there appeared to be similarities between this disease and scurvy. Fed on diets of rice from which the husk had been removed by polishing, the guinea pigs developed scurvy. This was not alleviated by adding rice polishings to the diet, as would have happened in the case of beriberi, but those fruits and vegetables known to alleviate scurvy cured the deficiency.

This led Holst to conclude, in 1907, that there must be an antiscorbutic factor as well as the antineuritic factor already postulated by Grijns. Subsequently, he demonstrated that it was a low molecular weight, water soluble compound. In 1920, it was given the tentative name of vitamin C until it could be identified. The vitamin was finally isolated by King 11 years later,[160] and given the name ascorbic acid (**213**). Its structure was determined by Hirst in 1933. Within months it had

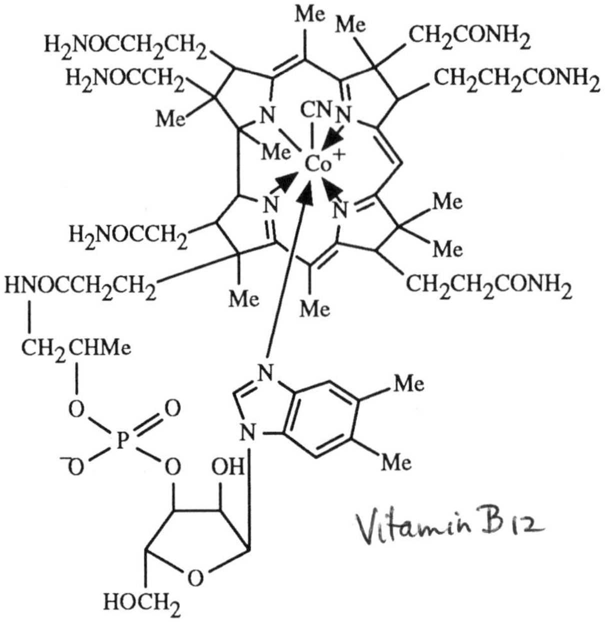

Vitamin B12

(211)

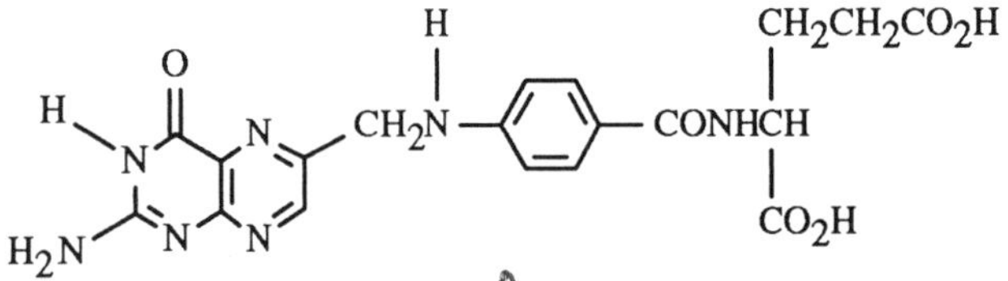

(Vitamin B complex) **(212)** folic acid

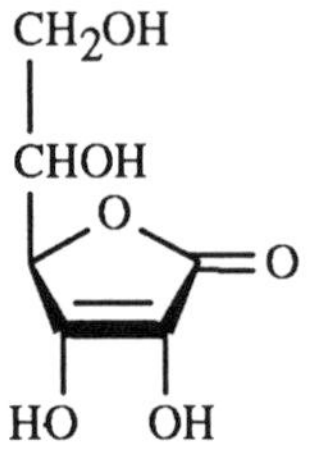

(213)

Vitamin C (ascorbic acid)

been synthesized by Haworth and Hirst, and also by Reichstein whose synthesis was exploited commercially.

In 1912, Hopkins suggested that rickets could be a dietary deficiency disease. Mellanby confirmed this by feeding dogs on restricted diets and then demonstrated that the disease could be cured by the addition of animal fats such as butter, suet or cod liver oil. Isolated reports of the use of the latter, a traditional Northern European folk remedy for a variety of ailments, had made little impact prior to this. Suffice it to say that Mellanby's findings enabled health authorities to eradicate rickets in London by the early 1930s.

McCollum found that cod liver oil heated for a prolonged period in a current of air retained antirachitic activity, thus confirming that this could not be attributable to vitamin A. The results of this experiment were published in 1922; three years later the unidentified second fat soluble vitamin was designated vitamin D. In the meantime, Hess suggested that the ability of UV exposure of the skin to cure rickets might be due to the presence of a provitamin. In 1924, Hess and Steenbock independently demonstrated that irradiation of certain foods conferred potent antirachitic activity. Three years later, in collaboration with Windaus and Pohl, Hess established that provitamin A was ergosterol (**214**), a known substance. Windaus isolated the product formed by irradiation of ergosterol in 1932, naming it vitamin D_2 (**215**) to distinguish it from the previously isolated complex of it and lumisterol once thought to have been the pure vitamin. This complex was renamed vitamin D_1, having once being called calciferol. Confusingly, the 'British Pharmacopoeia' retained this term to describe the pure vitamin D_2, a substance described as ergocalciferol in the 'United States Pharmacopoeia'. Within four years, Windaus and also Heilbron and Spring had established the chemical structure of vitamin D_2. Windaus and Bock also found that the antirachitic activity conferred by exposure of skin to UV irradiation was actually due to the conversion of 7-dehydrocholesterol to vitamin D_3, cholecalciferol.

In 1922, Evans and Bishop reported that the addition of all vitamins then known could not remedy the markedly diminished ability of rats fed on artificial diets to maintain normal pregnancies. It was later suggested that an unidentified vitamin E was required in the diet and this was

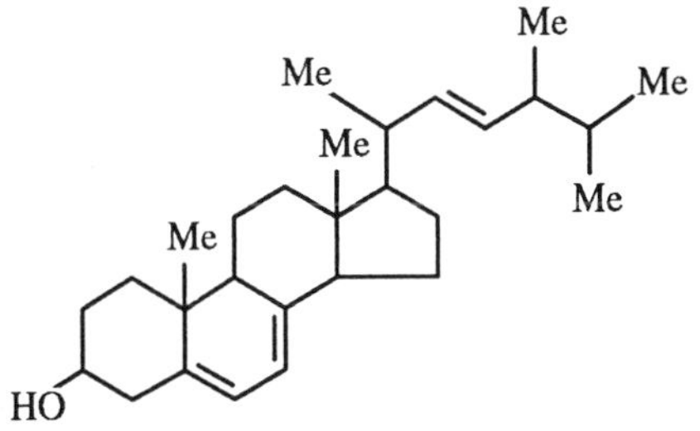

(214)

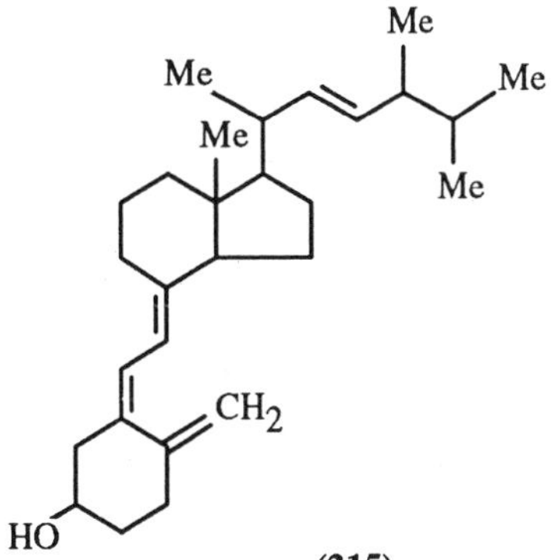

(215)

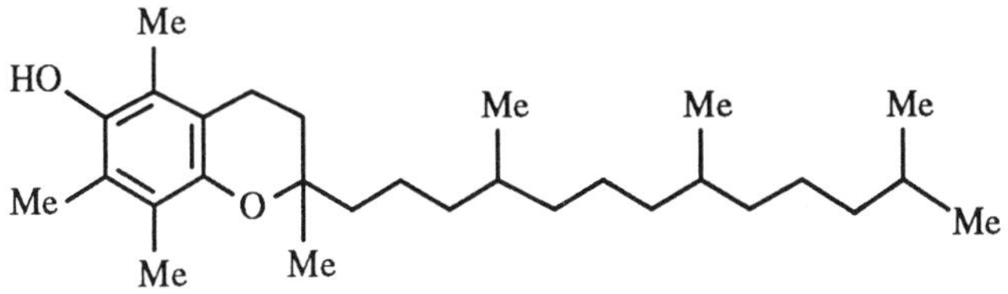

(216)

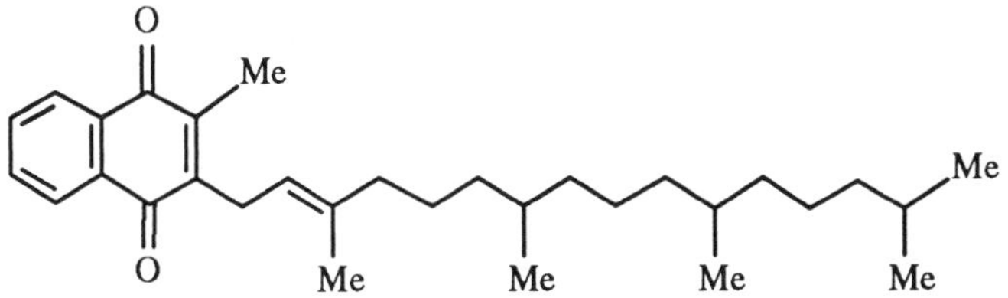

(217)

shown to be present in wheat germ oil. It was isolated by Evans in 1936 and named α-tocopherol (**216**). Its structure was established by Fernholz and then synthesized by Karrer. Despite numerous unsubstantiated claims to the contrary, its therapeutic role in humans has never been established. Nevertheless, it is widely used by the food and pharmaceutical industries as a safe, fat soluble antioxidant.

During the course of feeding chickens on fat free diets as part of an investigation into cholesterol metabolism, Dam found that despite the addition of all known fat soluble vitamins the chickens began to haemorrhage after only two or three weeks. In 1935, he reported that the hitherto unrecognized fat soluble vitamin K was essential for normal coagulation of the blood to proceed. Almquist detected the vitamin in alfalfa meal, from which source Dam and Karrer isolated the pure vitamin in 1939.[161] Later that year, Doisy determined its structure and it was promptly synthesized by Fieser. It was given the name phytomenadione (**217**).

1.2.5 ANTIBIOTICS[162]

It would be quite wrong to believe that Fleming's detection of the antibiotic action of a *Penicillium* mould was the first observation of such a phenomenon. More than 300 examples of antibiosis had been observed prior to Fleming's 1928 discovery. Just over half a century earlier, Pasteur had observed how the growth of one group of microorganisms in a nutrient medium was able to inhibit that of another group.[163]

1.2.5.1 Bacterial Antibiotics

A protein-free extract of *Bacillus brevis* was shown by Dubos,[164] in 1939, to have bactericidal activity. It was given the name tyrothricin and was soon shown to consist largely of a cyclic peptide called tyrocidine (**218**) together with an open chain peptide, gramicidin (**219**), which was 50 times as potent. Although both components were too toxic for systemic administration, tyrothricin became the first antibiotic of known constitution to be marketed when it was introduced in 1942 for the topical treatment of Gram-positive infections. It has remained a popular ingredient of non-prescription throat lozenges.

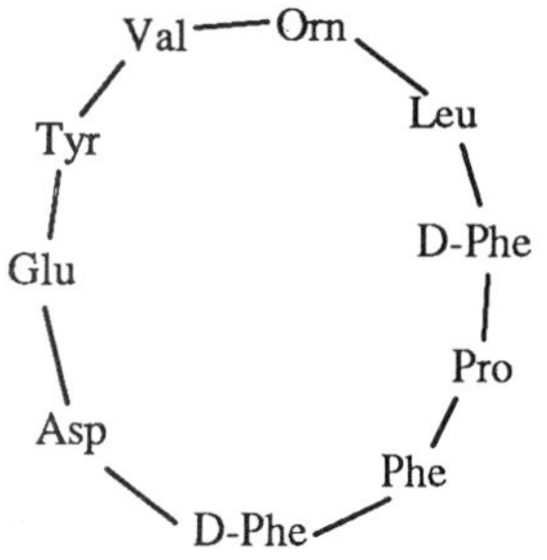

Val—Orn—Leu—D-Phe—Pro—Phe—D-Phe—Asp—Glu—Tyr (cyclic)

(218)

Val—Gly—Ala—Leu—Ala—Val—Val—Val—Trp—Leu—Trp—Leu—Trp—Leu—Trp—NHCH$_2$OH; CHO on Val

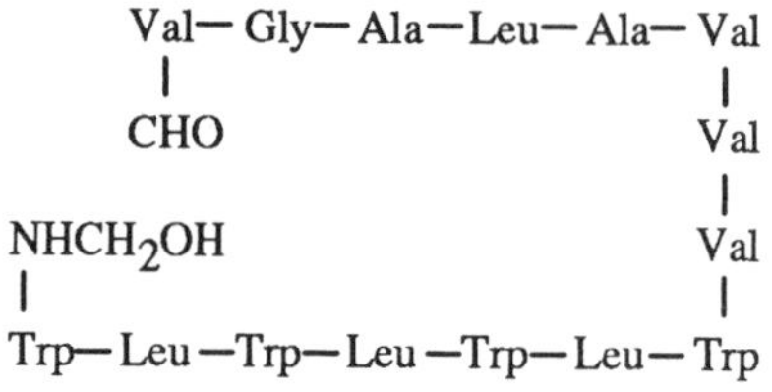

(219)

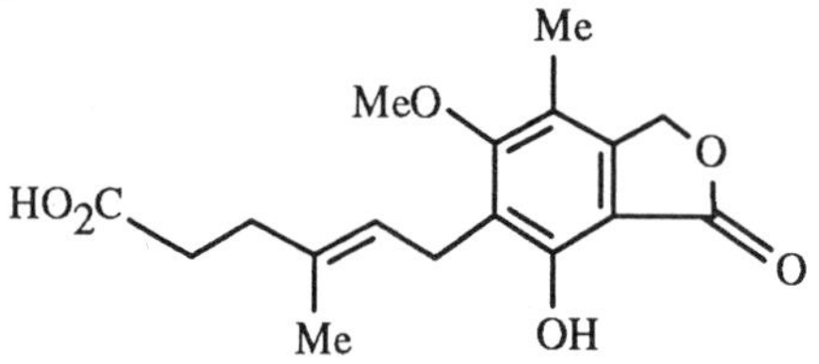

(220)

1.2.5.2 Fungal Antibiotics[165,166,167,168,169]

Antibiotic activity has not infrequently been reported to be present in various preparations derived from *Penicillium* moulds. It is most likely to have been due to the presence of mycophenolic acid (**220**), a somewhat toxic antiseptic compound first obtained in crystalline form by Gosio in 1896. When its structure was elucidated in 1952, it was seen to be chemically unrelated to that of the penicillins.

Hare[170] has published a description of the probable sequence of remarkable coincidences that permitted Fleming to detect the inhibitory action of a *Penicillium* mould on bacterial growth. Very few varieties of *Penicillium* actually secrete penicillin, but in a laboratory on the floor beneath his an associate was working with a rare strain of *Penicillium notatum*, which fortuitously produced large amounts of the antibiotic. It seems this wafted through the air on to an exposed culture plate already impregnated with staphylococci. The plate was set aside and left undisturbed during an unseasonal cool spell during which Fleming was on holiday. This provided ideal conditions for sufficient mould growth to ensure destruction of bacteria in the immediate vicinity of the *Penicillium* colony. Records confirm that before Fleming returned, the weather improved sufficiently to favour extensive growth of the staphylococci that had not been affected. This permitted him to observe a clear zone of inhibition of bacterial growth in the centre of the culture plate.

Fleming recognized the significance of the zone of bacterial inhibition surrounding the *Penicillium* mould. He prepared subcultures and gave the name penicillin to the filtrate of the broth in which these were allowed to grow for one or two weeks. Assistants were given the task of preparing sufficient quantities of the antibiotic and were instructed to attempt to isolate it. Lack of adequate facilities thwarted them, but they did devise an effective extraction procedure that supplied Fleming with enough pencillin for him to draw the conclusion that it was not a particularly useful antiseptic. He confined his efforts to a study of the action of penicillin on bacteria cultured in the presence of blood or serum, a procedure he had developed to assess the clinical potential of antiseptics. At no time did he carry out the crucial test for a true chemotherapeutic agent, namely administering it systemically to a live animal infected with a pathogenic organism. It is noteworthy that had atoxyl and the first active antibacterial sulfonamide, the other two major chemotherapeutic prototypes, been similarly neglected by being tested only against test-tube cultures then the chemotherapeutic revolution might never have occurred.

The credit for bringing penicillin to the clinic must surely go to Florey and Chain who, at the very outbreak of the Second World War, recognized its potential and single mindedly strove to produce sufficient quantities of refined material for animal and clinical studies to be conducted. Ably assisted by their colleagues at Oxford, an entire university department was effectively transformed into a factory to produce penicillin despite the absence of convincing evidence that it would be effective in the clinic.[171] In May 1940, this evidence was forthcoming when the first test on mice was completed. The following February, the life of a policeman dying from a mixed staphylococcal and streptococcal infection was almost saved before supplies of penicillin ran out. Encouraging results were obtained shortly after in other patients, enabling Florey to approach the American scientific community for it to facilitate the large-scale production of the antibiotic to meet wartime demands.[172] The collaborative response of federal, academic and industrial institutions in the United States was unprecedented, surpassing all expectations. It was, of course, firmly underpinned by outstanding progress at Oxford. When the D-Day landings took place in June 1944, sufficient penicillin was available to meet all the requirements of the Allied Forces. This could never have been achieved in wartime Britain, especially as British manufacturers had no experience of the deep culture fermentation process that marked the turning point in penicillin production. Perhaps the best measure of the progress that took place in the United States is the fact that in 1940 the early batches of mould juice being collected at Oxford contained 1–2 units ml^{-1}. Four years later, the leading American manufacturer was producing 100 000 million units per month!

Penicillin was first obtained as pure crystals by Wintersteiner in 1943. The Oxford workers then discovered that the crystals of their penicillin were different, indicating for the first time that variant forms of the antibiotic existed. Their material, 2-pentenylpenicillin (**221**), became known as penicillin F, whilst that being produced in massive quantities in the United States was described as penicillin G, although it subsequently received the approved name of benzylpenicillin (**222**). By the end of 1943, scientists at Oxford and at the Merck laboratories in the United States had concluded that benzylpenicillin was one of two possible structures. The oxazolone–thiazolidine structure (**223**) was finally rejected as an option in 1945 after X-ray crystallographic studies by Hodgkin confirmed the presence of the rare β-lactam ring.

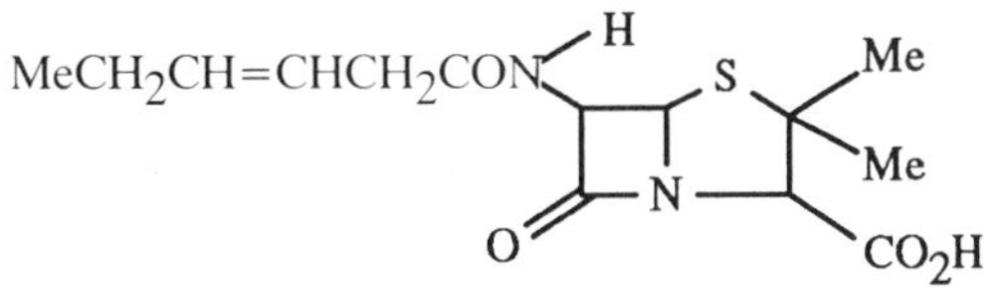

(**221**) 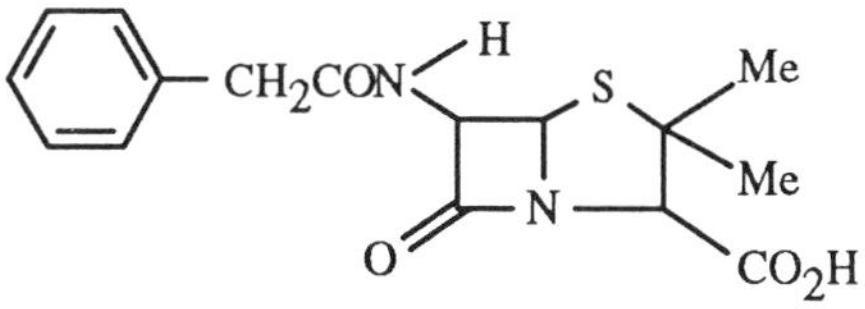 (**222**)

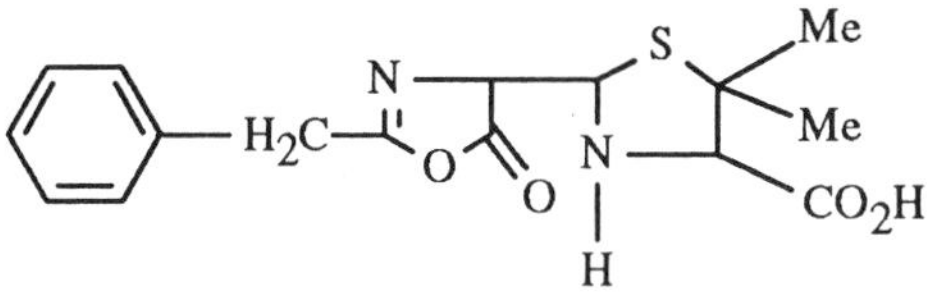

(**223**) 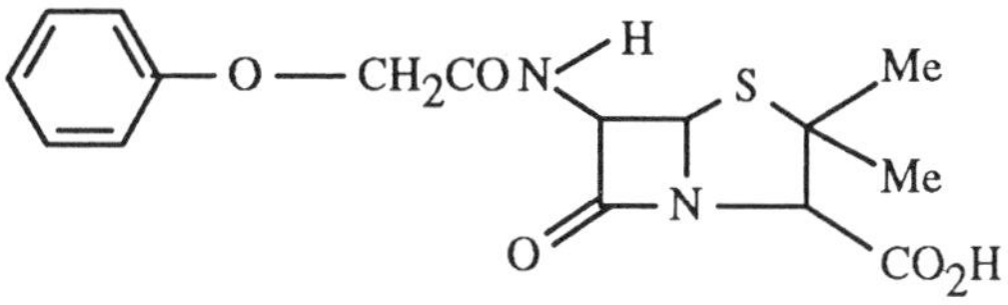 (**224**)

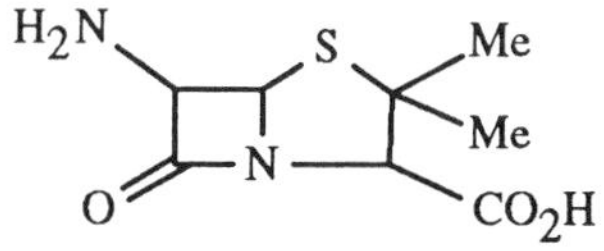

(**225**)

Sheehan synthesized phenoxymethylpenicillin (**224**), the variant also known as penicillin V, in 1957. His synthetic route afforded a superior method for obtaining novel variants to that formerly employed with only limited success, *i.e.* by the addition of precursors to the liquor in which the *Pencillium* mould was grown. Sheehan demonstrated that a key intermediate, 6-aminopenicillanic acid (**225**), could be acylated with acid chlorides. Although Sheehan managed to improve the yield from his synthesis to around 60%, it could never rival what was to become the universal industrial method for producing novel penicillins and, later, cephalosporins. This was based on the surprising discovery by Batchelor *et al.*[173] in 1958 that 6-aminopenicillanic acid was always present in the fermentation liquor from penicillin production, and was chemically stable. By appropriate manipulation of the fermentation process, large quantities of this key intermediate became available.

The relative ease with which novel semisynthetic penicillins could be made from 6-aminopenicillanic acid was a turning point in antibiotic chemotherapy, for it became easier to meet the demands of clinical progress by examining novel penicillins rather than indulging in expensive programmes based on screening thousands upon thousands of soil samples brought from the four corners of the earth in the hope of finding that exceedingly rare product of nature—a non-toxic antibiotic. Most of the new antibiotics introduced since the early 1960s have been employed for the treatment of conditions that could not be expected to respond to penicillin therapy, *viz.* non-bacterial infections and cancer.

Phenoxymethylpenicillin had been isolated in 1948, but its unique stability in acid conditions was overlooked until five years later. Only then was its potassium salt (**226**) marketed as the first reliable orally active penicillin. Its success influenced the selection of its close analogue phenethicillin (**227**) as the first semisynthetic penicillin to be put into commercial production, in 1959. It had no significant clinical advantage over phenoxymethylpenicillin since the higher blood levels from equivalent doses were offset by the price disadvantage. On the other hand, when methicillin (**228**) was marketed a few months later it offered the real advantage of being stable in the presence of the penicillinase enzyme produced by resistant staphylococci, especially the troublesome *S. aureus*. Its insensitivity to the destructive action of the enzyme was attributable to the presence of substituents on the benzene ring serving to interfere with the fit to the active site. This pointed the way forward to antibiotics that had the additional advantage of being acid stable and hence orally active, *e.g.* oxacillin (**229**), cloxacillin (**230**), dicloxacillin (**231**) and flucloxacillin (**232**).

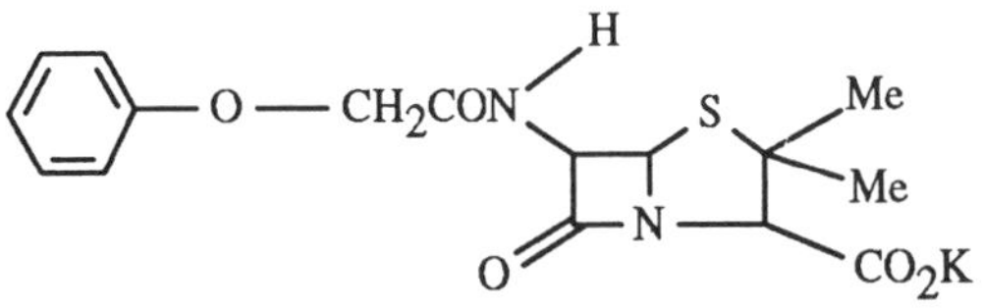

(226)

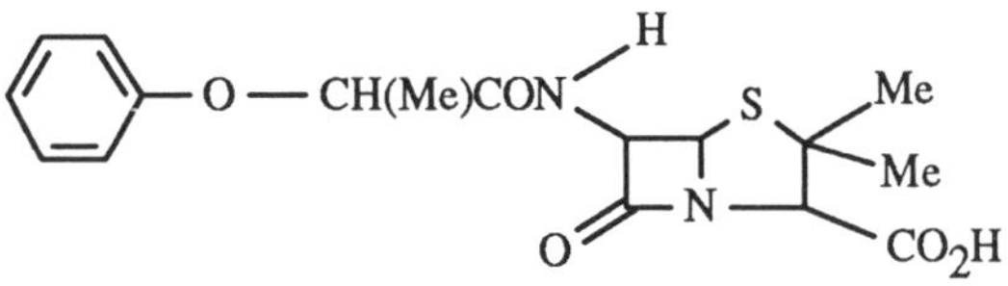

(227)

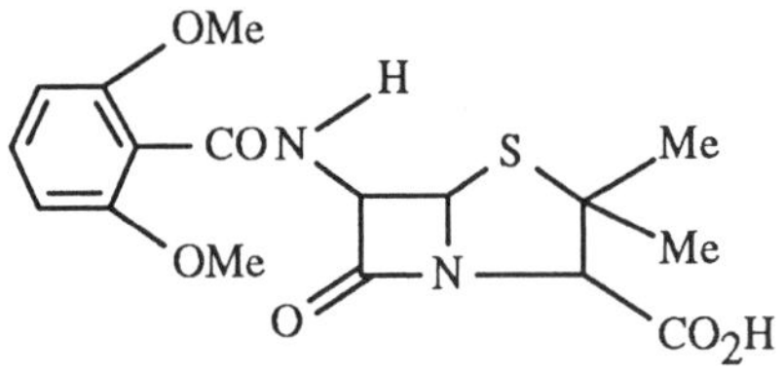

(228)

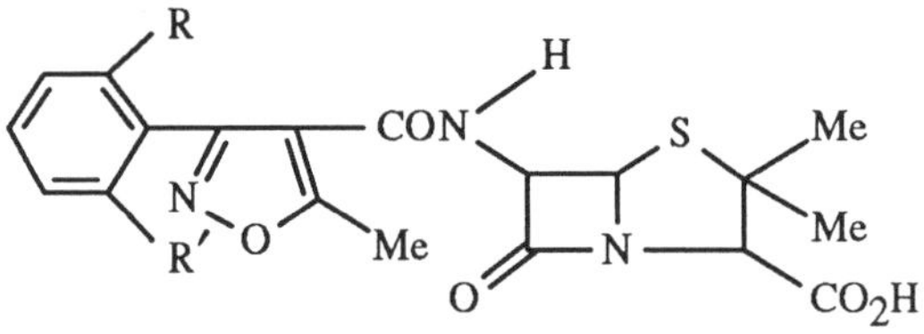

(229) R = R′ = H
(230) R = Cl; R′ = H
(231) R = R′ = Cl
(232) R = Cl; R′ = F

An analysis of the factors influencing acid stability in semisynthetic penicillins led Doyle *et al.*[174] to conclude that the incorporation of an electron-withdrawing amino group in the side chain of benzylpenicillin might produce an orally active antibiotic which retained its broad spectrum of activity against both Gram-positive and Gram-negative bacteria. The latter had proved rather insensitive to the semisynthetic penicillins. The objective of this exercise was actually surpassed when the orally active ampicillin (**233**) was found to have stronger activity against Gram-negative bacteria than had benzylpenicillin, although that against Gram-positive organisms was somewhat diminished. Ampicillin proved to be an outstanding success, becoming the most frequently prescribed drug in most countries for many years after its introduction in 1961. This, however, was to highlight its major drawback. Unlike its immediate forerunners, ampicillin was sensitive to penicillinase and hence many bacterial strains eventually acquired resistance to it by evolving to produce the enzyme.

One problem with ampicillin that was effectively dealt with was its poor absorption from the gut caused by its dipolar nature. The introduction of its phenolic analogue amoxycillin (**234**), in 1964, largely overcame the problem. Five years later saw the marketing of pivampicillin (**235**), the first of several prodrugs in which the carboxylic acid function was esterified to eliminate the dipolar nature of the molecule and thus ensure good absorption. Similar products followed, such as talampicillin (**236**) and bacampicillin (**237**). In all of these prodrugs, the ester was decomposed to release ampicillin after absorption from the gut.

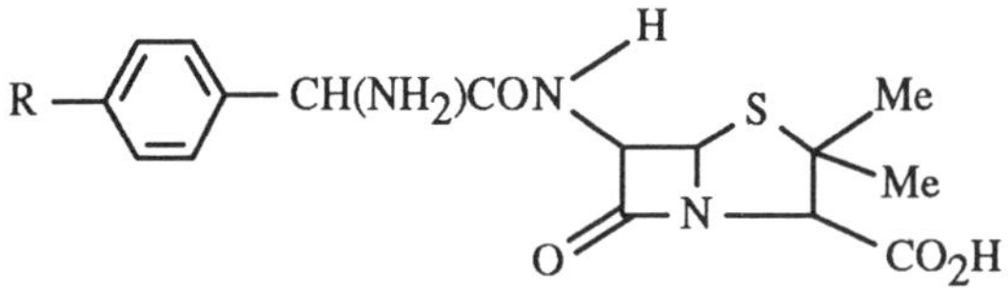

(233) R = H
(234) R = OH

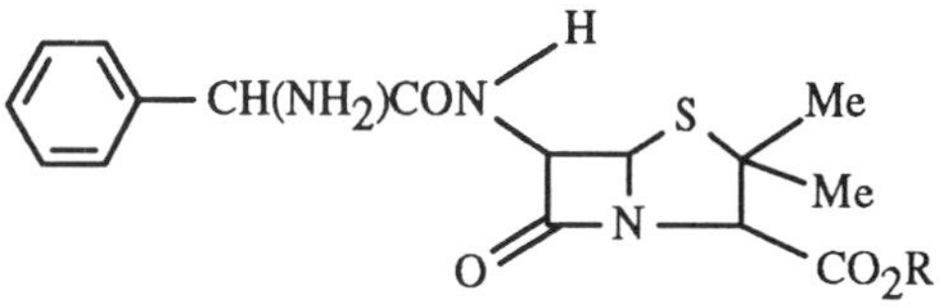

(235) R = CH_2OCOBu^t
(236) R =

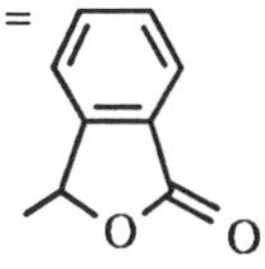

(237) R = $CH(Me)OCO_2Et$

Although ampicillin was prepared as an analogue of benzylpenicillin by incorporating a basic amino group into the side chain, carbenicillin (**238**) featured an acidic carboxyl group when it was introduced in 1964. This, and the similar ticarcillin (**239**), which was introduced at the same time, turned out to have unprecedented activity amongst the penicillins against the dangerous pathogen *Pseudomonas aeruginosa.* The polar nature of the side chain carboxyl group was masked in carfecillin (**240**), a prodrug of carbenicillin designed for oral administration. The ampicillin analogue azlocillin (**241**) was later also found to have powerful antipseudomonal activity.

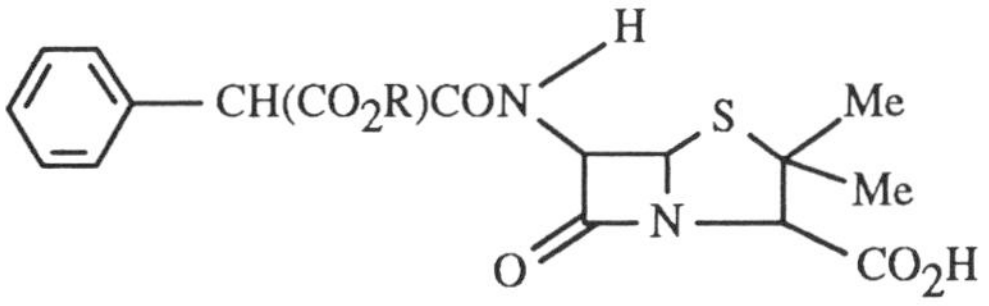

(238) R = H
(240) R =

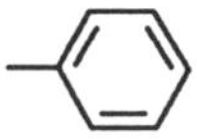

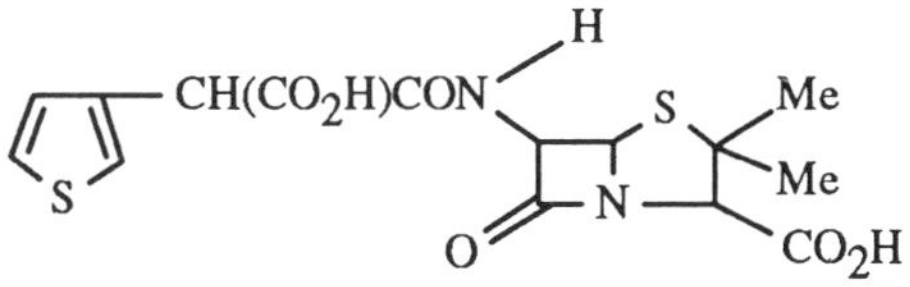

(239)

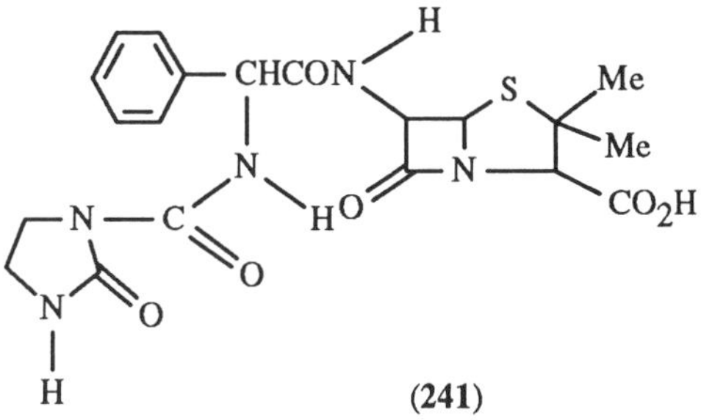

(241)

During the immediate post-war years, microbiologists conducted an extensive global search for novel antibiotic-producing organisms. One of the most far-reaching successes of that quest took many years to bear fruit. This was Brotzu's isolation from a sewer in Sardinia of *Cephalosporium acremonium*, a mould that inhibited the growth of typhoid bacilli on culture plates. From it was obtained a concentrate with a wider spectrum of antibiotic action than penicillin, and this was examined by researchers at Oxford. Several antibiotic compounds were isolated, including cephalosporin C (**242**). This was structurally related to the penicillins, but exhibited resistance to penicillinase. The introduction of methicillin undermined the significance of this, but once Morin *et al.* had devised a method for industrial production of 7-aminocephalosporanic acid from cephalosporin C, the way cleared for the development of semisynthetic cephalosporins.[175] In 1964, cephalothin (**243**) and cephaloridine (**244**) were marketed as broad spectrum injectable antibiotics, soon to be followed by the orally active analogue of the recently developed ampicillin. Known as cephaloglycin (**245**), the latter required frequent dosage since it retained the metabolically labile acetyl ester function originally present in cephalosporin C from which it had been synthesized. This disadvantage was overcome by replacing the entire ester function with either a hydrogen or chlorine atom to form cephalexin (**246**) and cefaclor (**247**), in 1967 and 1974 respectively. By analogy with the ampicillin analogue amoxycillin, cephalexin was also converted to the phenolic derivative cefadroxil (**248**). Several so-called second generation cephalosporins with a wider spectrum of activity were also

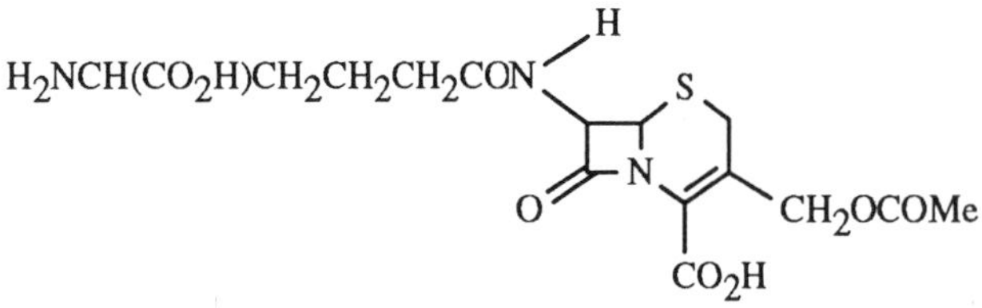

H
$H_2NCH(CO_2H)CH_2CH_2CH_2CON$
S
O N
CH_2OCOMe
CO_2H

(**242**)

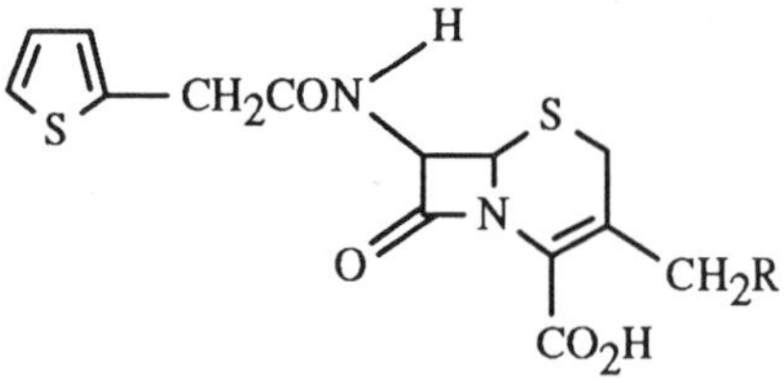

H
S CH_2CON
S
O N
CH_2R
CO_2H

(**243**) R = OCOMe

(**244**) R = —N

H
R′ $CH(NH_2)CON$
S
O N
CH_2R
CO_2H

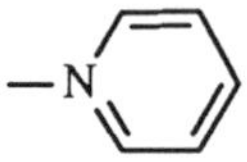

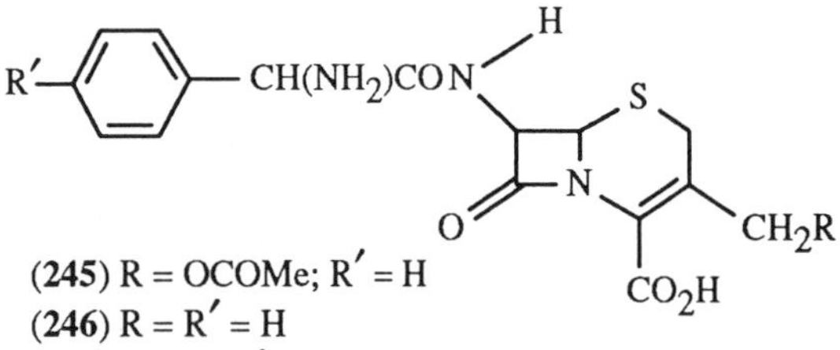

(**245**) R = OCOMe; R′ = H
(**246**) R = R′ = H
(**248**) R = H; R′ = OH

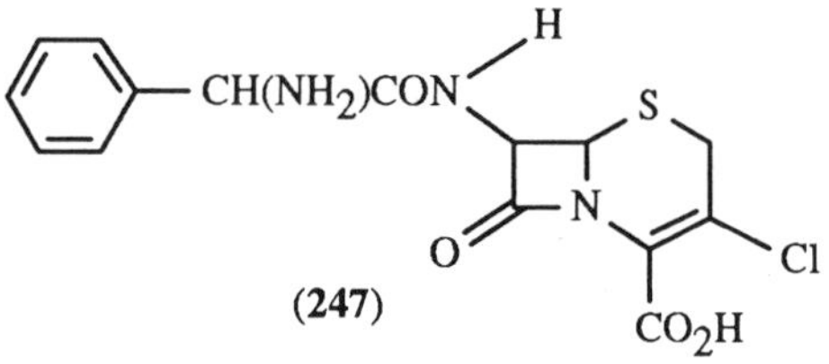

H
$CH(NH_2)CON$
S
O N
Cl
CO_2H

(**247**)

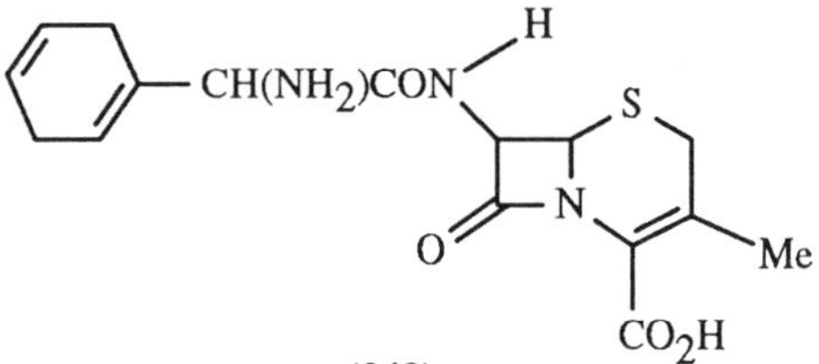

H
$CH(NH_2)CON$
S
O N
Me
CO_2H

(**249**)

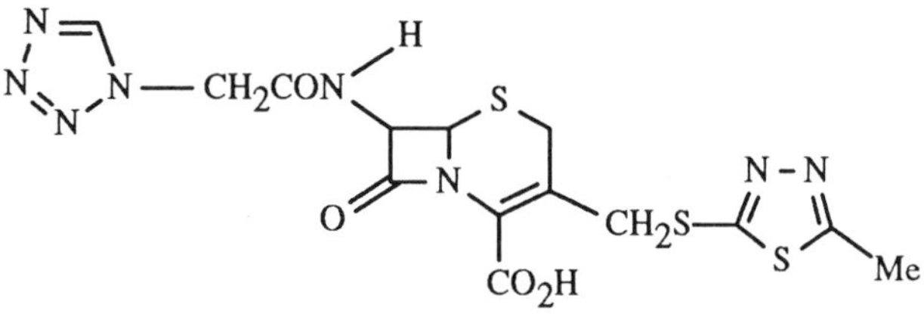

N N N N
H
N — CH_2CON
S
O N
N - N
CH_2S
S
Me
CO_2H

(**250**)

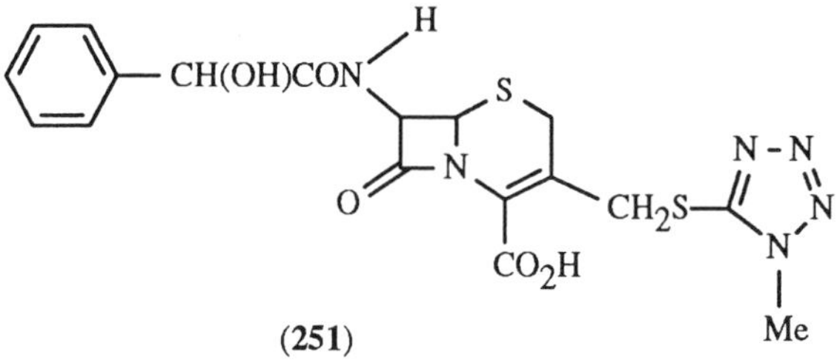

H
CH(OH)CON
S
O N
N - N
CH_2S
N N
N
Me
CO_2H

(**251**)

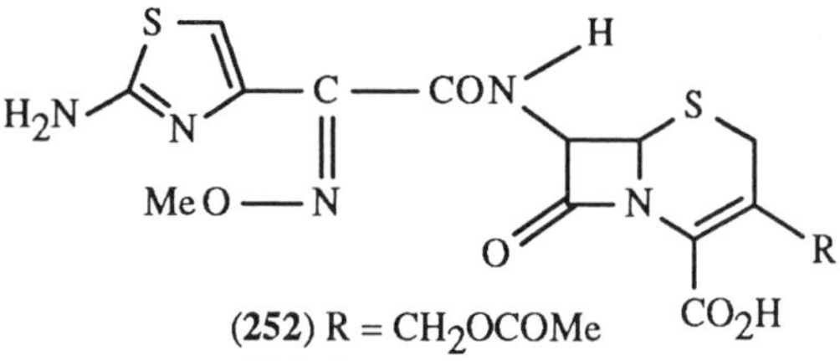

S
H
H_2N N C — CON
MeO — N
S
O N
R
CO_2H

(**252**) R = CH_2OCOMe
(**255**) R = H

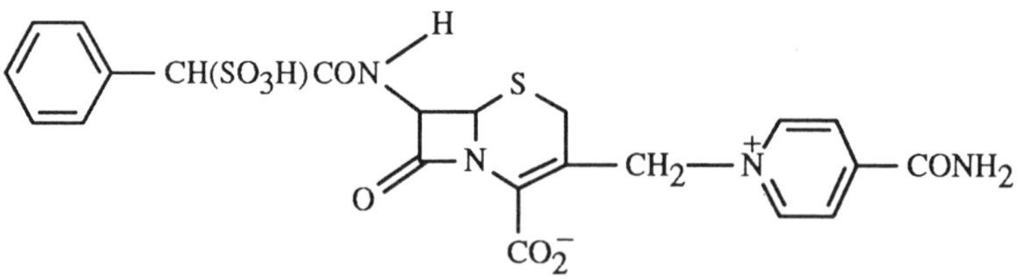

H
$CH(SO_3H)CON$
S
O N
CH_2—$\overset{+}{N}$
$CONH_2$
CO_2^-

(**253**)

S
H
H_2N N C — CON
N
O
$C(Me)_2CO_2H$
S
O N
CH_2—$\overset{+}{N}$
CO_2^-

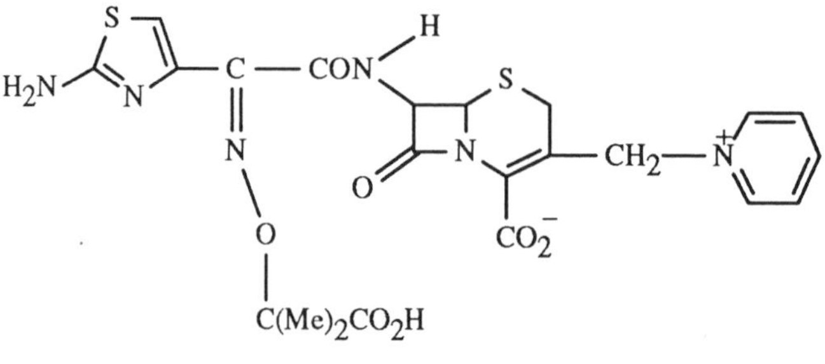

(**254**)

developed, but they had to be injected as they were unstable in acid. Activity against Gram-negative organisms and some penicillin resistant organisms was the important feature of these drugs, which included cephradine (**249**), cephazolin (**250**) and cefamandole (**251**). The succeeding generation of cephalosporins exhibited a restricted spectrum of activity, but were highly effective against some of the most dangerous pathogens such as *Pseudomonas aeruginosa*. Examples of such drugs are cefotaxime (**252**), cefsulodins (**253**), ceftazidime (**254**) and ceftizoxime (**255**). The variant spelling of the names of the cephalosporins affords a clue to the chronology of their introduction.

A metabolite first isolated from *Penicillium griseofulvum* by Raistrick in 1939 was later found in extracts of *P. janczewskii* in which Brian *et al.* detected antibiotic activity directed against soil fungi that was required to be present for normal tree growth. Grove determined the structure of the metabolite in 1952. Hopes that it could be introduced as a horticultural antifungal agent were not realized, but Gentles[176] found it effective by mouth when administered to guinea pigs infected with a severe fungal infection. In 1959, under the name griseofulvin (**256**), it was introduced into medicine as an antifungal antibiotic.

Cyclosporin A (**257**) was one of several antifungal antibiotics isolated from certain varieties of fungi imperfecti in 1969. Although it proved too toxic for clinical use as an antibiotic, Borel[177] discovered it to act as an immunosuppressant agent. Its early promise has been fully realized, cyclosporin having made a major impact on transplant surgery since the results of the first clinical trials were published in 1978.

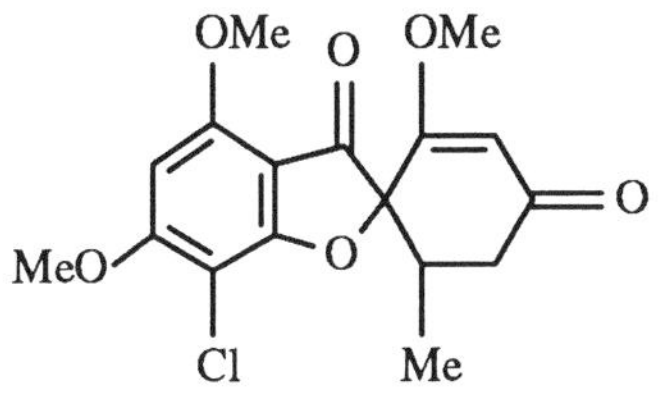

OMe O OMe
MeO O O
Cl Me

(**256**)

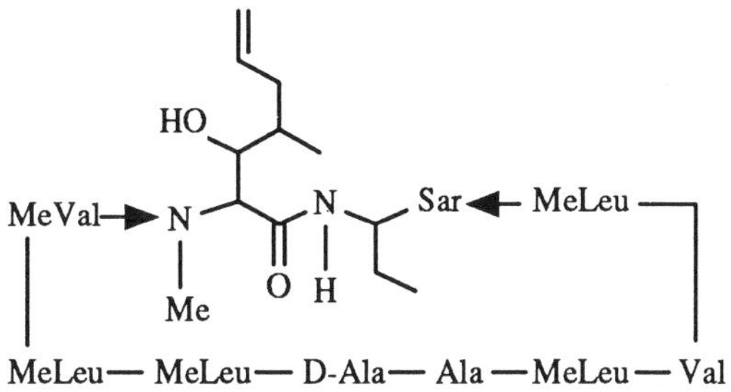

HO
MeVal → N N Sar ← MeLeu
Me O H
MeLeu— MeLeu— D-Ala— Ala —MeLeu— Val

(**257**)

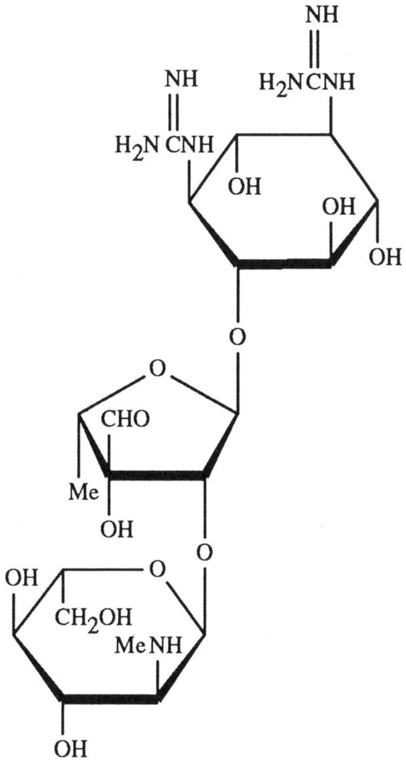

NH
NH H_2NCNH
H_2N CNH
OH OH
OH
O
O
CHO
Me
OH
O
OH O
CH_2OH
MeNH
OH

(**258**)

1.2.5.3 Actinomycetal Antibiotics

Waksman held the view that the destruction of pathogenic microorganisms deposited in soil could be due to their destruction by chemicals secreted by normal soil organisms. Following the isolation of tyrothricin from a soil bacterium by Dubos in 1939, he initiated a programme to seek further antibiotic-producing soil microorganisms.[178] Studies on actinomycetes initiated by Gratia in 1923 had already furnished extensive evidence of their capacity to secrete substances that lyzed other microorganisms. It was, therefore, hardly surprising that a preliminary survey showed Waksman that out of 244 soil cultures, more than 100 exhibited antimicrobial activity.

In 1940, Waksman isolated actinomycin A from *Actinomyces antibioticus*, but after extensive animal studies this crystalline antibiotic was judged to lack sufficient efficacy or safety for clinical application. The following year, a further two actinomycetal antibiotics were isolated, but again proved too toxic. As the early clinical results from the ever-expanding penicillin programme were indicating that there would be a need for an antibiotic with activity against Gram-negative bacteria, Waksman made this his prime objective. It appeared he had finally acheived this when streptothricin was isolated from *Actinomyces lavendulae*, but chronic renal toxicity was detected during animal studies. Restricting his objective still further by seeking specifically for an antibiotic capable of destroying the causative organism of tuberculosis, *Mycobacterium tuberculosis*, Waksman finally isolated streptomycin (**258**) early in 1944 from *Streptomyces griseus*. Successful studies in infected animals were rapidly followed by most encouraging trials in the clinic, confirming streptomycin as the first drug to be proven effective against tuberculosis. Although it is a somewhat toxic drug, having caused deafness and kidney damage, it remains a first line drug in the combination chemotherapy of tuberculosis. Like streptomycin, other aminoglycoside antibiotics are bactericidal, particularly towards Gram-negative bacteria. They also share its toxicity profile to a greater or lesser degree.

The significance of the isolation of a major antibiotic like streptomycin from an actinomycete in the soil was not overlooked by the pharmaceutical industry. Extensive research programmes were initiated by most major companies, involving the screening of thousands of soil samples culled from all over the globe. The financial rewards for success were unprecedented, but although scores of promising antibiotics were isolated, only a handful retain a major role in therapeutics. One of the first successes was the discovery by Duggar[179] in 1945 of chlortetracycline (**259**), an antibiotic secreted by *Streptomyces aureofaciens*. After its isolation, tests confirmed that it was an orally active, broad spectrum antibiotic. Industrial production by the deep fermentation process began late in 1948. The following year, oxytetracycline (**260**) was isolated from *Streptomyces rimosus*. A semi-synthetic analogue, tetracycline (**261**), was first prepared in 1952 by catalytic hydrogenation of chlortetracycline, before subsequently being isolated from the same actinomycete as chlortetracycline itself! The commercial success of these tetracycline antibiotics was based on their oral activity, despite the fact that the efficiency of their absorption from the gut was really quite poor. Erythromycin (**262**), another actinomycetal antibiotic that was suitable for oral administration, was well absorbed. Isolated from a strain of *Streptomyces erythreus* in 1952, it had a spectrum of activity almost identical to that of penicillin. Had the oral penicillins not become available around the same time as it was first marketed, it could have become one of the most frequently prescribed antibiotics. Its radically different chemical structure has ensured that it retains an important role in the therapy of infections where resistance is due to pencillinase production, as well as in treating patients who are allergic to penicillin.

Illustrative of what can happen when aggressive marketing brings a new drug to the fore too speedily is the fate of chloramphenicol (**263**), an antibiotic isolated from *Streptomyces venezuelae* in 1947. Within months of its isolation, it was used successfully in an epidemic of typhus in Bolivia. Shortly after, it was found to be a broad spectrum antibiotic, active when administered by mouth. It appeared to have a high margin of safety, reminiscent of penicillin. Its structure was rapidly elucidated and a synthesis was developed, enabling industrial production to begin in 1949.[180] Within two or three years, 8 000 000 patients had received chloramphenicol, the vast majority experiencing no adverse effects. Tragically, somewhere between one in 20 000 and one in 100 000 of those taking the drug developed a delayed onset aplastic anaemia, with four out of five dying. The drug never regained its popularity, but it remains of considerable value in the treatment of typhoid, salmonella, meningitis and typhus.

Nystatin (**264**) is an antifungal antibiotic produced by various species of *Streptomyces*, including *S. noursei* and *S. aureus*. It was isolated in 1950. Severe toxicity has limited its clinical application to topical use. The closely related polyene antibiotic amphotericin (**265**)[181] is somewhat less toxic and thus can be administered intravenously to combat life-threatening fungal infection. It was isolated in 1953 from a strain of *Streptomyces nodosus*.

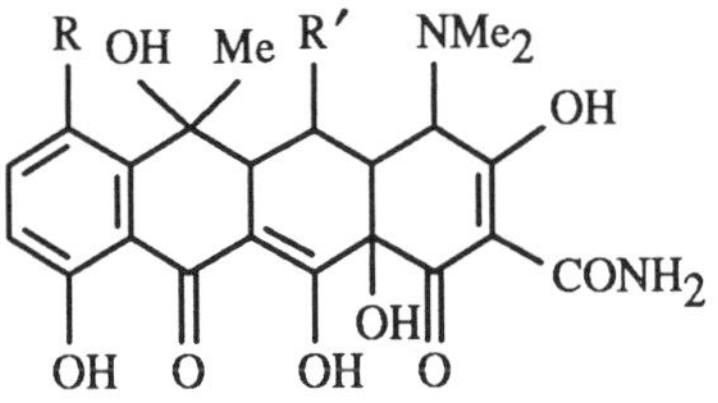

(259) R = Cl; R′ = H
(260) R = H; R′ = OH
(261) R = R′ = H

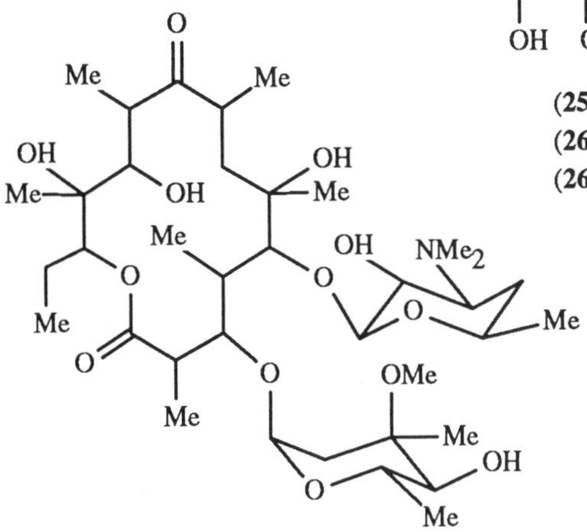

(262)

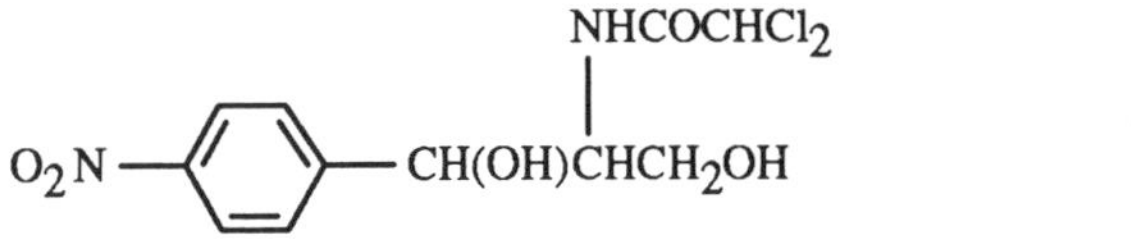

(263)

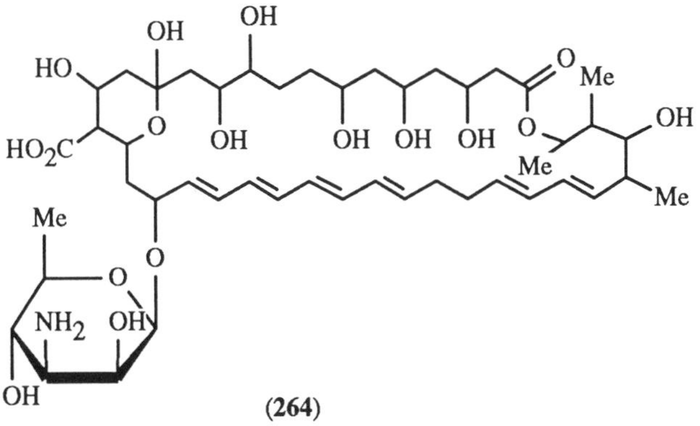

(264)

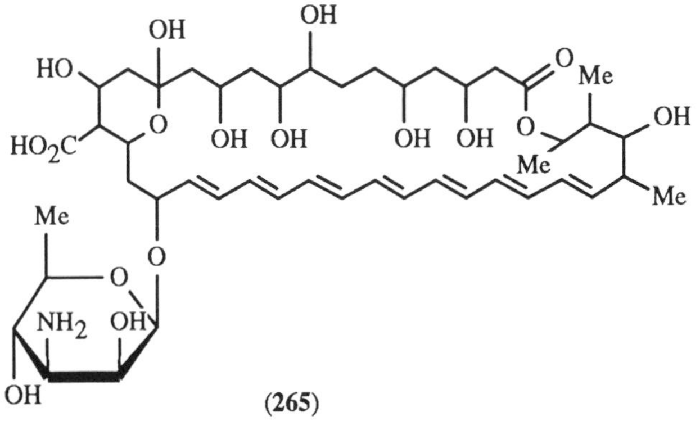

(265)

A group of antibiotics collectively described as the rifamycins were isolated by Sensi[182] from cultures of *Streptomyces mediterranei* in 1959. One of these antibiotics, rifamycin B (**266**), was obtained in a crystalline form and its structure was determined in 1963. In solution, it oxidized to form a more potent product that was named rifamycin S (**267**). The potency of this, in turn, was enhanced by reducing it to rifamycin SV (**268**), a compound with antituberculous activity. Its early promise was not fulfilled due to rapid metabolism when administered to patients. Only after an extensive screening programme, involving the synthesis of hundreds of derivatives, was a satisfactory drug found. The effort proved to be wholly justified and should serve to silence those whose ignorance of the nature of pharmaceutical research leads them to criticize the industry for placing so much reliance on this type of approach. The drug that emerged from the screening programme, in

1966, was rifampicin (**269**). Used as the mainstay of combination chemotherapy, it has revolutionized the treatment of tuberculosis to such an extent that patients no longer spend months being treated in hospitals or sanitoria.

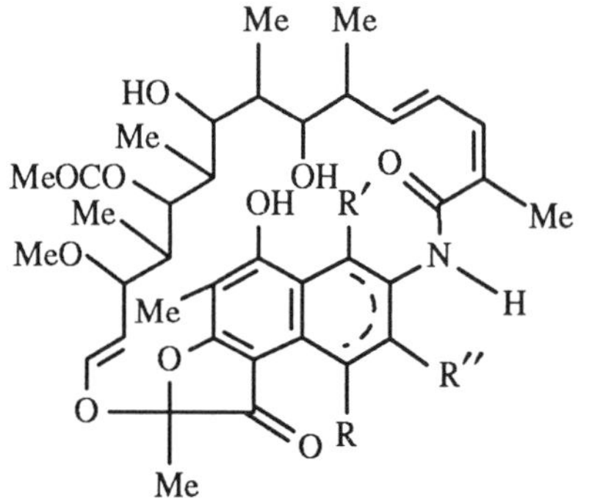

(**266**) R = OCH_2CO_2H; R′ = OH; R″ = H
(**267**) R = R′ = O; R″ = H
(**268**) R = R′ = OH; R″ = H
(**269**) R = R′ = OH; R″ = CH=N—N(piperazine)N—Me

Actinomycetes have proved to be an exclusive source of antibiotics for use in cancer chemotherapy. A report that actinomycin A had some activity against a transplanted tumour in mice encouraged Brockmann to submit actinomycin C, a complex of related antibiotics from *Streptomyces chrysomallus*, for screening. This was introduced clinically in 1952 for the treatment of Hodgkin's disease of the lymphatic system. Results were unconvincing until Waksman introduced actinomycin D (**270**), a compound isolated from *Streptomyces parvullus* corresponding to one of the components of Brockmann's antibiotic complex. This has proved to be a life-saving drug in children with Wilm's tumour of the kidney, as well as in several other rare malignancies. It is now generally known as dactinomycin.[183]

Further screening programmes have led to the introduction of mitomycin C (**271**), isolated from *Streptomyces caespitosus* in 1956, and of mithramycin (aureolic acid) (**272**), isolated from *Streptomyces argillaceus*, the antitumour activity of which was not recognized until 1962. The toxicity of these drugs has severely restricted their clinical application. Bleomycin (**273**), isolated from *Streptomyces verticullus* in 1962, has proved of much more value. Daunorubicin (daunomycin, rubidomycin; **274**) was isolated from *Streptomyces peucetius* in that same year. Although cardiotoxic, it could be used with caution in combination with other drugs to treat acute leukaemia. The closely related doxorubicin (**275**), isolated from the same source in 1967, remained cardiotoxic but proved much more successful in the clinic. Its isomer, epirubicin (**276**), is much less toxic and promises to become a major chemotherapeutic agent. A synthetic analogue of doxorubicin, mitozantrone (**277**), was designed to retain the anticancer activity whilst avoiding the cardiotoxicity.

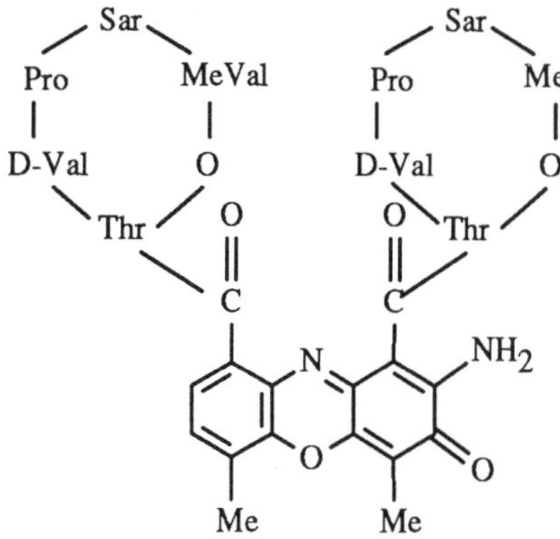

(**270**)

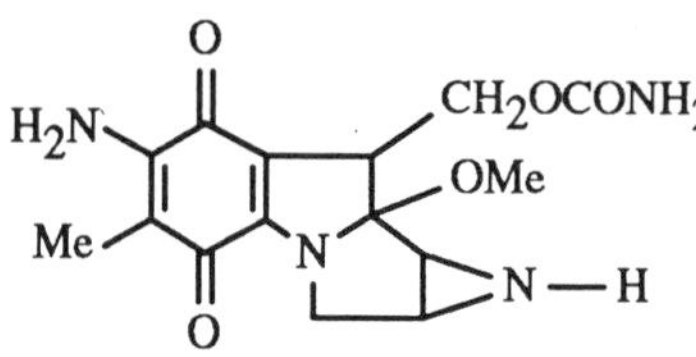

(**271**)

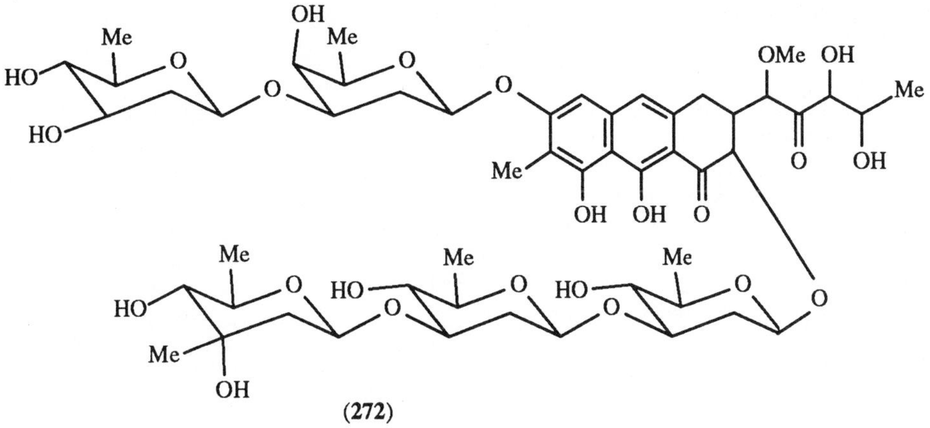

(272)

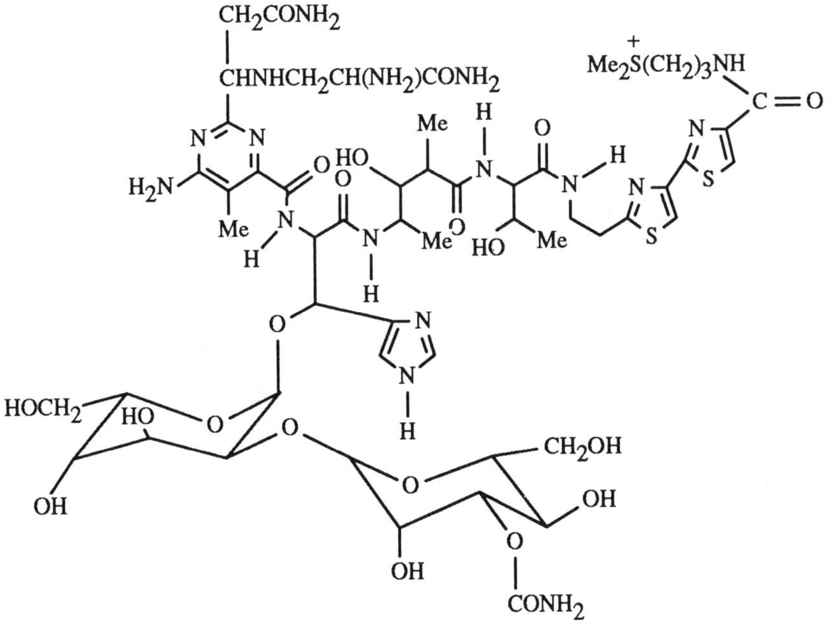

(273)

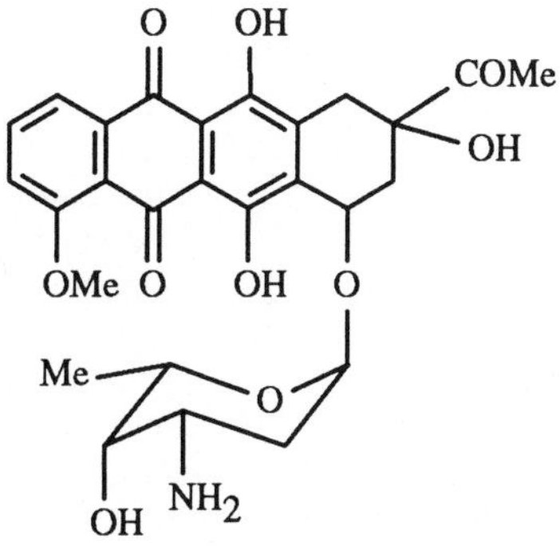

(274)

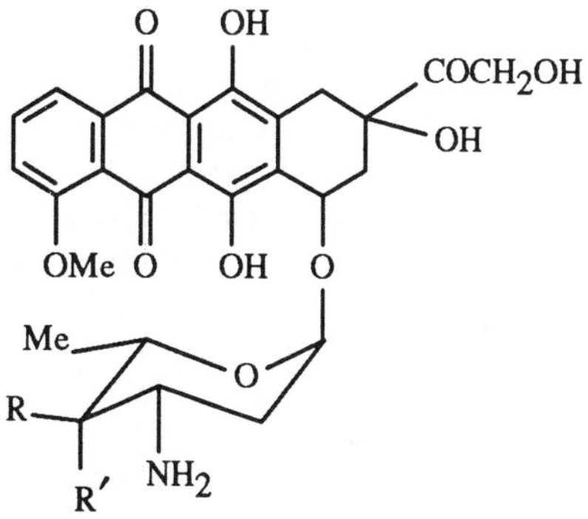

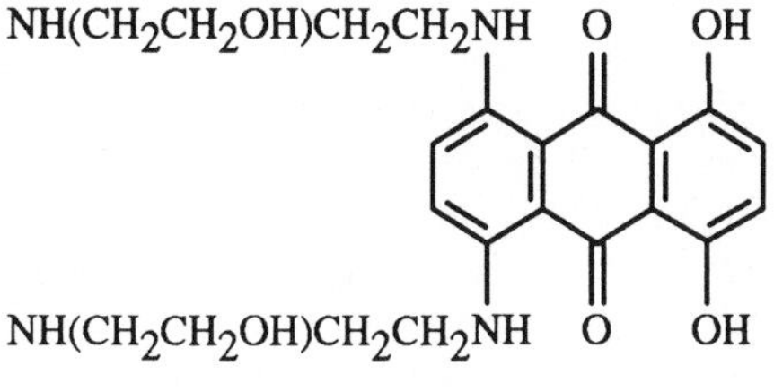

(275) R = H; R′ = OH
(276) R = OH; R′ = H

(277)

1.2.6 INORGANIC AND ORGANOMETALLIC COMPOUNDS

1.2.6.1 Mercurials

In 1881, Koch tested some 70 antiseptics for their ability to kill anthrax spores adhering to silk thread. When he found that mercury(II) chloride was the only one with sporicidal activity, it appeared that the age old belief in the medicinal value of mercury and its salts had finally found justification. During the same decade, several insoluble organomercury(II) salts were introduced, including the benzoate, phenate and salicylate. They were selected on the basis of their diminished astringency when compared with inorganic mercurials. Around the turn of the century, the range of applications for mercury preparations was widened by the introduction of derivatives containing sulfonic acid functions to confer water solubility. Despite the introduction of organomercurials such as mercurophen (**278**), mercurochrome (**279**), phenylmercuric nitrate (**280**), nitromersal (**281**) and thiomersal (**282**), the early hopes for these have never been fulfilled. They are used solely as preservatives and antiseptics.

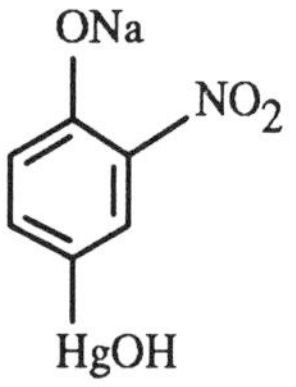

ONa, NO_2, HgOH
(278)

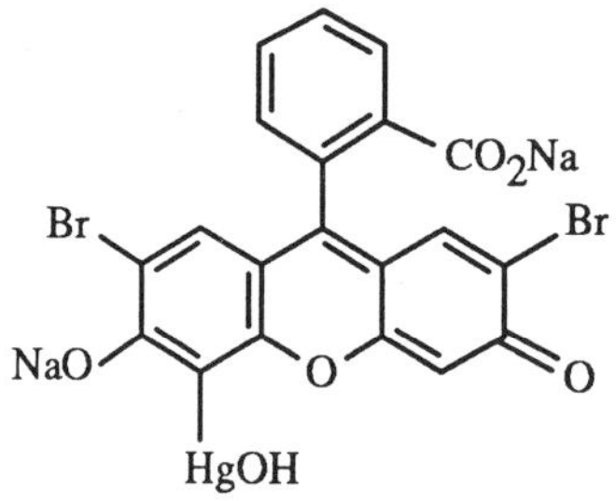

CO_2Na, Br, Br, NaO, O, O, HgOH
(279)

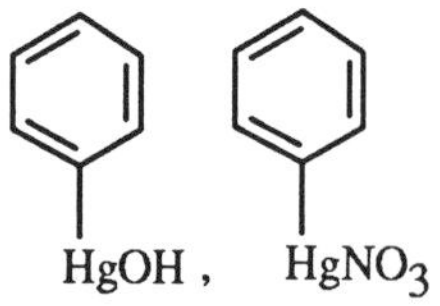

HgOH , $HgNO_3$
(280)

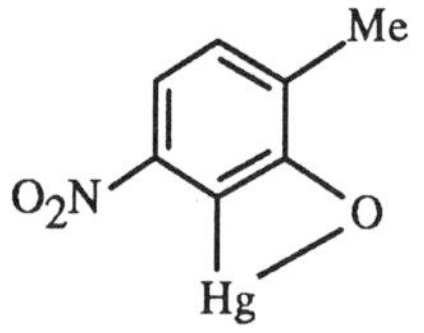

Me, O_2N, O, Hg
(281)

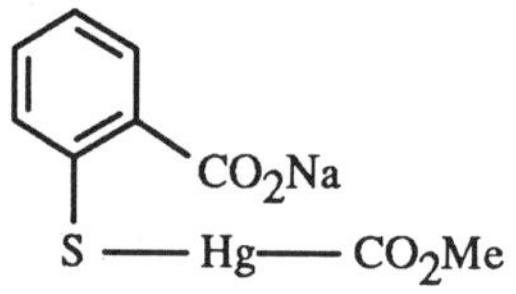

CO_2Na, S —— Hg—— CO_2Me
(282)

1.2.6.2 Arsenicals

The reputation of Fowler's solution (potassium arsenite solution) as an alternative, albeit of questionable efficacy, to quinine in the treatment of malaria led to its use in Africa as a remedy in sleeping sickness. In 1894, following upon his demonstration of the presence of trypanosomes in the blood of infected cattle, Bruce found these protozoa could be temporarily eliminated by administration of Fowler's solution. Eleven years later, Thomas[184] described how a proprietary organic arsenical known as atoxyl (**283**) could cure animals experimentally infected with trypanosomes. Unfortunately, the drug could not live up to its name when tested in the clinic. It was, however, to provide the prototype which the genius of Ehrlich[185] successfully exploited through his painstaking screening of literally hundreds of analogues. The ratio of the toxic dose to the therapeutic dose of each and every one of these analogues was carefully determined, the results being used to direct the design of subsequent analogues towards the ultimate goal of a truly safe and efficacious cure for sleeping sickness. Although the attainment of that specific objective using organic arsenicals was achieved by others when tryparsamide (**284**) and acetarsol (**285**) were introduced after his death, Ehrlich[186] had the satisfaction of introducing arsphenamine (**286**) and the more easily formulated neoarsphenamine (**287**) to treat the causative organism of syphilis, the *Treponema pallidum.* His screening programme had been expanded to include this organism because of its close similarity to those he was already employing. The patent for arsphenamine was applied for in the summer of 1909, only weeks after Ehrlich and Hata had begun to screen hundreds of the arsenicals against *T. pallidum.* When it was introduced into the clinic the following year, arsphenamine was an outstanding success; it truly merited the title of being the first man-made chemotherapeutic agent and Ehrlich equally deserves to be remembered as the father of chemotherapy.[187] His approach may have been empirical, but whenever possible he sought a theoretical basis on which to base the design of the many analogues that had to be screened.[188] His utilization of Langley's receptor theory illustrates this. Langley had promulgated his theory of 'receptive substances' in 1905 in order to explain the antagonism between drugs such as nicotine and curare or pilocarpine and atropine. The conclusion he had reached was that a receptive substance could either respond to the stimulant or be blocked by the antagonizing drug. Ehrlich's adoption of the concept ensured its wider acceptance. In it, he saw a parallel with the binding between enzymes and their substrates which could fit the active sites of the former. Thus, he began to refer to drugs acting at receptors, a concept which still underpins much of current thinking in medicinal chemistry.

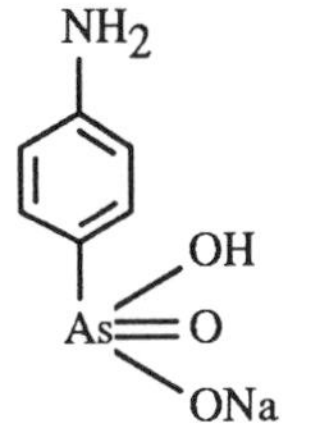

NH_2 / OH / As=O / ONa

(283)

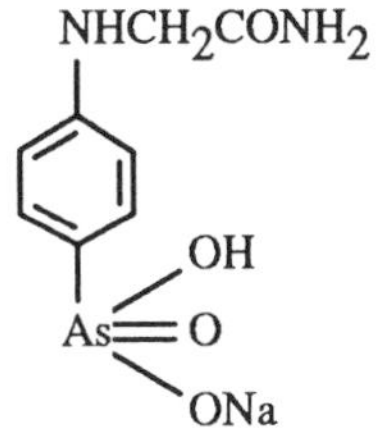

$NHCH_2CONH_2$ / OH / As=O / ONa

(284)

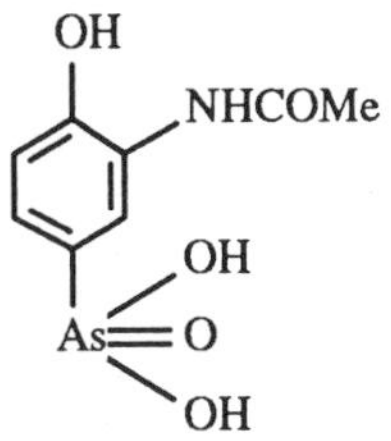

OH / NHCOMe / OH / As=O / OH

(285)

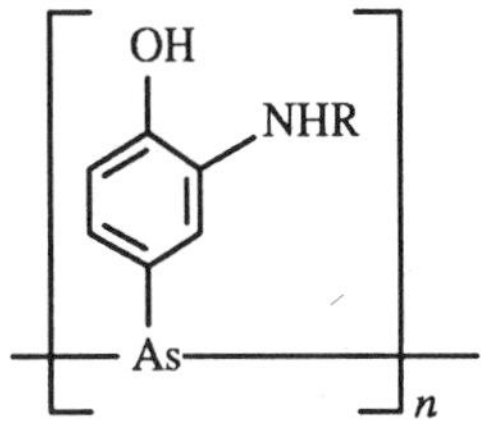

OH / NHR / As / n

(286) R = H
(287) R = H or CH_2SO_2Na

1.2.6.3 Antimonials

In 1905, Laveran demonstrated that Fowler's solution could destroy trypanosomes in the blood of experimentally infected mice. The next year, his colleagues Nicolle and Mesnil obtained similar results with intravenous injections of tartar emetic (antimony potassium tartrate). Although subsequently used in cattle, the drug was considered unacceptably toxic for the treatment of sleeping sickness in humans. Notwithstanding this, in 1912 Vianna reported his successful use of it in the treatment of leishmaniasis, another protozoal disease. The impact of his pioneering studies has been underestimated all too often. Simply put, by the administration of intravenous injections of the sometimes lethal tartar emetic, the mortality rate for this infectious disease which has afflicted many millions of people in the tropics has been reduced from around nine out of ten to the present figure of about one-tenth of that! Tartar emetic was subsequently found to be effective against schistosomiasis, a protozoal disease afflicting an estimated 200 000 000 people. If ever an example had to be cited to justify the importance of having a clear understanding of disease processes in order to find curative drugs, none better could be found than that of tartar emetic. Although known since biblical times, its successful application to save more lives than any other drug has probably ever done had to await the advent of modern microbiology. That its high toxicity remains even in analogues such as the organic antimonial stibophen (**288**),[189] developed from a screening programme in the 1920s, should not obscure the remarkable results that have been achieved.

1.2.6.4 Gold Compounds

In 1890, Koch disclosed that he had found gold–cyanide complexes to be more effective than other antiseptics when tested in high dilution against cultures of *Mycobacterium tuberculosis*. Disappointing results led to the abandonment of this approach, but in 1924 a report was published claiming that gold sodium thiosulfate had a beneficial effect against bovine tuberculosis. Results in humans were disappointing, and although subsequent laboratory studies revealed that the compound protected animals against lethal streptococcal infections, the risk of severe kidney damage was unacceptable in the clinic. However, during experimental clinical studies, patients with rheumatism reported relief of joint pain. The use of this and other gold compounds as antirheumatic agents persisted during the 1930s, but their value was not generally accepted until after the Second World War.

1.2.6.5 Platinum Compounds

In 1964, during a study of the effects of an electric current on the growth of bacterial cultures, Rosenberg *et al.*[190] observed unusual elongation of cells of *Escherichia coli*. Investigation of this phenomenon revealed it to have been caused by the release of a soluble platinum complex formed at the platinum electrode. The effect on bacterial cultures was reproduced merely by adding similar complexes with the *cis* configuration. Tests against transplanted tumours confirmed that cisplatin (**289**) had cytotoxic effects and it was then subjected to a thorough evaluation before being introduced into the clinic in the early 1970s. It has since proved to be an outstanding drug in the treatment of testicular cancer and some other malignancies. Its major drawback of causing severe renal damage appears to have been overcome with the introduction of carboplatin (**290**), a compound selected from extensive screening of analogues of cisplatin.

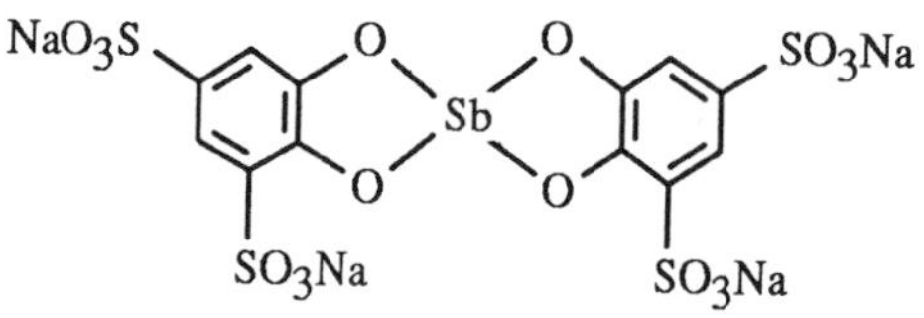

(**288**)

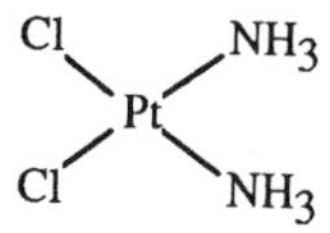

(**289**)

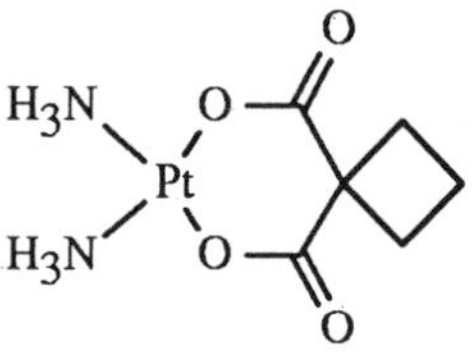

(**290**)

1.2.6.6 Halogens and their Derivatives

The ability of chlorinated water to eliminate the foul odour of decaying organic matter was noted by Berthollet in 1788, and this led him to recommend it as a disinfectant. Labarraque, in 1825, introduced a chlorinated solution for wound disinfection. The preparation which bore his name was subsequently employed in France to treat gangrene and to disinfect the hands of medical attendants, a use exploited first by Holmes and later by Semmelweis for the prevention of the spread of childbirth fever. These applications, of course, preceded the discovery by Pasteur that infectious disease was caused by the spread of microorganisms. The irritancy of chlorinated solutions when applied to wounds was overcome to some extent when Dakin[191] introduced the stable aromatic chloramines B (**291**) and T (**292**) in 1915.

The known depressant action of potassium bromide on the nervous system induced Locock to study its action in epilepsy. However, it was not until 11 years later, in 1868, that Clouston[192] was able to demonstrate convincingly that this salt could reduce the number of fits so long as it was administered chronically. This improvement could only be achieved at highly depressant dose levels. The introduction of phenobarbitone and modern drugs has resulted in the withdrawal of potassium bromide from the therapeutic armamentarium, although it is a matter of concern that in some countries it is still sold as a non-prescription sedative.

Burnt sea sponge was used as a remedy for goitre from the Middle Ages onwards. In 1820, Coindet[193] came to the conclusion that the only ingredient that was likely to have any therapeutic effect was iodine, which he promptly substituted for the traditional remedy. A high incidence of overdosing resulted in disillusionment, with the result that many patients were denied this simple remedy for a dietary deficiency of iodine. Almost exactly a century later, the work of Marine[194] resulted in the introduction of iodine supplements for inhabitants of goitrogenic areas.

Tincture of iodine was introduced as an antiseptic in 1836, but the demonstration of its ability to destroy pathogens had to await the development of appropriate test techniques in the 1870s. In the interim, its irritancy to wounds led surgeons to prefer iodoform, a simple organic compound that had been used in the treatment of goitre. The popularity of iodoform in hospitals persisted until the mid-20th century. Such was the unjustified faith in it that the pharmaceutical industry vigorously sought still less irritant analogues. These mild disinfectants owed their activity to their propensity to release small quantities of iodine into solution. Early examples included iodol (**293**), soziodol (**294**) and chiniofon (**295**).

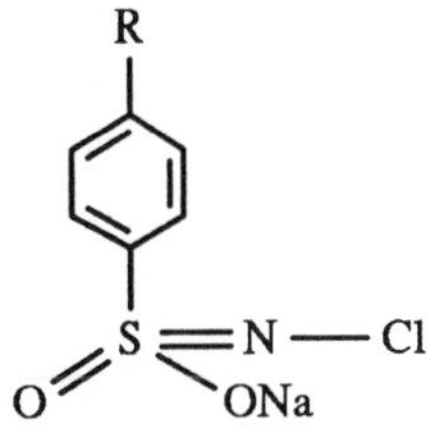

R

S=N—Cl, O, ONa

(291) R = H
(292) R = Me

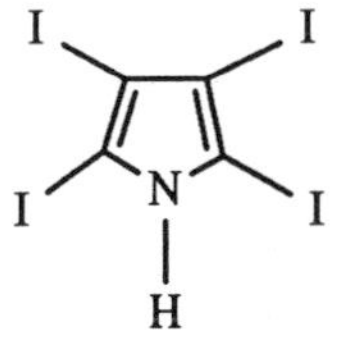

I I I I N H

(293)

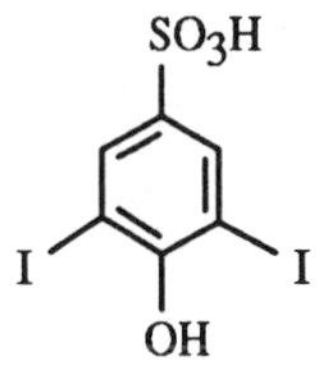

SO_3H I I OH

(294)

SO_3H I N OH

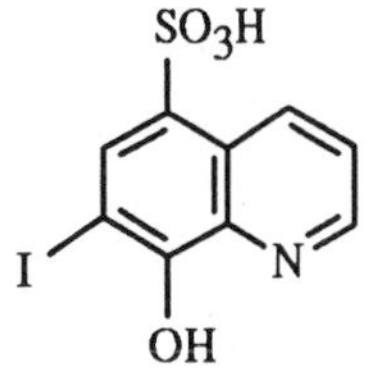

(295)

1.2.7 SERENDIPITOUSLY DISCOVERED DRUG PROTOTYPES AND THEIR DERIVATIVES

The term *serendipity* was coined by Horace Walpole in 1754 after he had read a poem about three princes of Serendip (Sri Lanka) who were 'always making discoveries, by accidents and sagacity, of things they were not in quest of'. There can be no doubt that so far as drug research is concerned, our pharmacopoeias would be much slimmer were it not for those whose sagacity exploited those accidental discoveries that less alert minds would have overlooked. Louis Pasteur clearly recognized the importance of serendipity in the development of science when he stated that 'chance favours the prepared mind'.

1.2.7.1 Perceived Effects of Chemicals

1.2.7.1.1 Central nervous system depressants[195,196]

The profound effects of certain chemicals on the nervous system could not but be noticed, although the relevance of these to clinical practice had to be recognized by those with prepared minds. Nitrous oxide acquired its name of 'laughing gas' after Humphry Davy described its euphoriant properties in 1799, yet it was not until 1844 that Wells was inspired to evaluate it as a dental anaesthetic after he had noted that a volunteer had not responded on injuring his leg during a public demonstration of the effects of the gas. Much the same happened with ether when Long[197] employed it as an alternative to nitrous oxide at a boisterous party. Noting that neither he nor his companions could account for their bruises after the effects of inhaling the solvent had worn off, he concluded that it might be of value in his clinical work. This was to result in his excising a growth from the neck of one of the party guests who volunteered to inhale ether. This took place in 1842, although it was not reported until after others had claimed to have been the first to introduce ether anaesthesia.

Surgical anaesthesia with ether was universally accepted after the celebrated demonstration in 1846 by Morton before the leading American surgeons. Nevertheless, there were those who sought alternatives. Simpson believed that the inflammability and the need to carry large volumes of ether were particularly disadvantageous in his domiciliary obstetric practice. He and his friends tested several volatile solvents as substitutes before settling on chloroform in 1847. It was widely used until well into the next century, eventually falling into disrepute as safer alternatives became available. Of particular relevance in the present context is that it inspired Liebreich to seek a derivative which would release small amounts of chloroform over a period of time and thus act as a hypnotic.

One of the early approaches towards the development of novel synthetic drugs was to devise means of obtaining a sustained release of an active component in the body. Liebreich believed that as chloral hydrate (**296**) liberated chloroform in alkaline solution, it would do likewise in the blood.[198] At that time, 1868, there was no true perception of the nature of alkalinity, but this did not prevent administration of chloral hydrate from inducing sleep even though it was too polar to enter the central nervous system. In fact, it formed an active metabolite, *viz.* trichloroethanol (**297**). Thus, Liebreich serendipitously discovered a prodrug form of this alcohol whilst intent upon finding a prodrug of chloroform! Within months of the disclosure of its hypnotic activity, chloral hydrate was in universal demand. As is frequently the case with any popular drug, its disadvantages eventually became apparent. The most notable of these was its irritancy to the stomach. Numerous attempts were made to design a less irritant analogue that would release chloral in the body. Early examples which were introduced between 1888 and 1890 included chloral ammonia (**298**), chloral formamide (**299**), chloralose (**300**) and dichloralphenazone (**301**). Trichloroethanol was far too irritant to the stomach, but its phosphate ester, triclofos sodium (**302**), was introduced in 1962.

$Cl_3CCH(OH)_2$ (**296**)

Cl_3CCH_2OH (**297**)

$Cl_3CCH(OH)NHR$ (**298**) R = H; (**299**) R = CHO

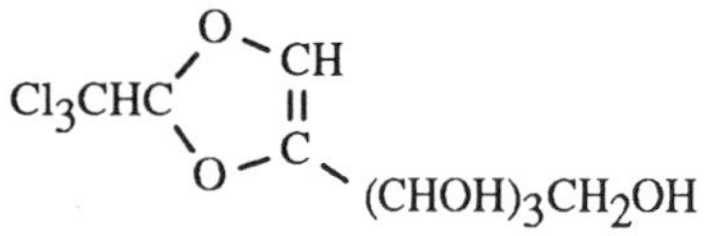

(**300**)

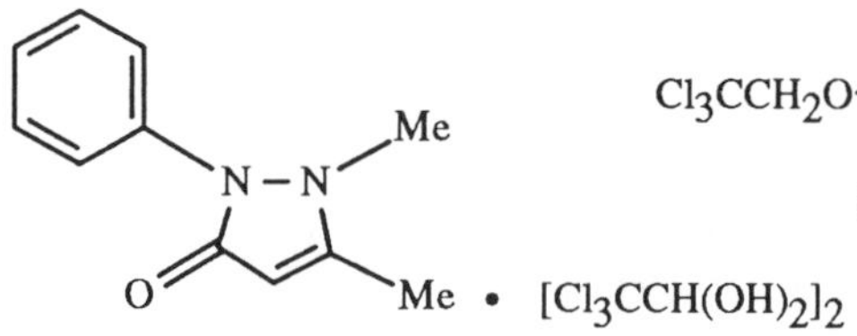

(**301**) $\cdot\ [Cl_3CCH(OH)_2]_2$

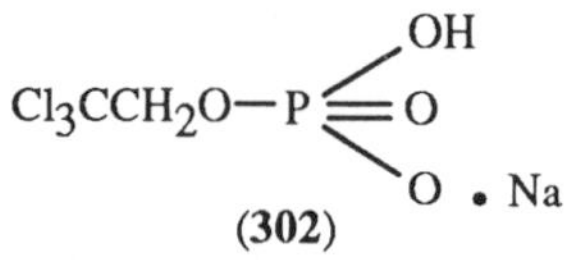

$Cl_3CCH_2O{-}P(=O)(OH)O \cdot Na$ (**302**)

The search for alternatives to chloral hydrate during the last two decades of the 19th century culminated in the introduction of the first barbiturate, barbitone (**303**), by von Mering and his colleagues in 1903. This cyclic ureide was based on an open chain precursor, diethylacetylurea (**304**), which itself had been modelled on existing hypnotics, *viz.* sulphonal and hedonal (**305**). The latter was the most successful of a series of analogues of urethane, a hypnotic tested with success on animals in 1885 by Schmiedeberg. He had believed that this would decompose in the body to liberate ethanol and thereby induce sleep, this being accompanied by release of carbon dioxide and ammonia which would act as respiratory stimulants. At that time there was increasing concern over the dangerous depression of the respiratory centre which had been observed following overdosage with chloral hydrate. In the event, the intact urethane molecule itself possessed the hypnotic activity, but results in humans were disappointing. The subsequent screening programme to find an effective analogue, which culminated in the introduction of hedonal in 1899, was possibly the first major one to be conducted within the pharmaceutical industry.

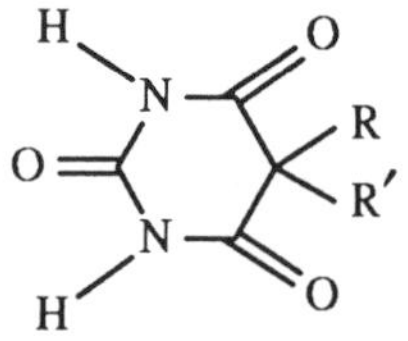

	R	R′
(**303**)	Et	Et
(**306**)	Et	Ph
(**307**)	Et	Bu
(**308**)	Et	$CH_2CH_2Pr^i$
(**309**)	$CH_2CH{=}CH_2$	CH(Me)Pr
(**310**)	Et	CH(Me)Pr

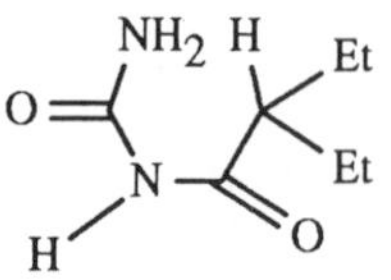

(**304**)

$NH_2CO_2CH_2CH_2Pr^i$ (**305**)

Barbitone was initially so successful that it was not challenged for some years. Phenobarbitone (**306**) was introduced in 1911 by Fischer, who had synthesized barbitone for von Mering. Being more rapidly excreted, it was less likely to leave patients with a hangover on awakening. During the 1920s, barbiturates such as butobarbitone (**307**), pentylbarbitone (**308**), quinalbarbitone (**309**) and pentobarbitone (**310**) were introduced following the discovery that an increase in the chain length of the 5-alkyl substituent significantly reduced their duration of action. This can be explained by the increased lipophilicity of these drugs ensuring their ability to enter liver cells where they were deactivated by oxidative enzymes. The even greater lipophilicity of hexobarbitone (**311**) and thiopentone (**312**) was responsible for their ultra-short duration of action which enabled them to be evaluated as intravenous anaesthetics in the early 1930s. Few other barbiturates were introduced after the 1930s. At first, this was due to the existing drugs having captured such a firm hold on the market. However, in the 1950s mounting discontent and their hazardous properties resulted in a search for alternatives, leading to the introduction of analogues that the cynic would claim usually bore some structural resemblance to the barbiturates but which could legitimately be promoted by their vendors as non-barbiturate hypnotics. Examples include glutethimide (**313**), and thalidomide (**314**). The notoriety of the latter[199] is surely so well known that little need be written here other than to state, for the record, that it caused lasting misery for thousands of otherwise healthy infants born with deformed or absent limbs as a consequence of their mothers having consumed a drug once promoted for its claimed safety in pregnancy. The enormity of the thalidomide disaster was revealed late in 1961. Its consequences for the pharmaceutical industry were immense, with governments throughout the world following the example of the United States (where thalidomide was never licensed for use) in formulating strict legislative controls over the testing and marketing of drugs.

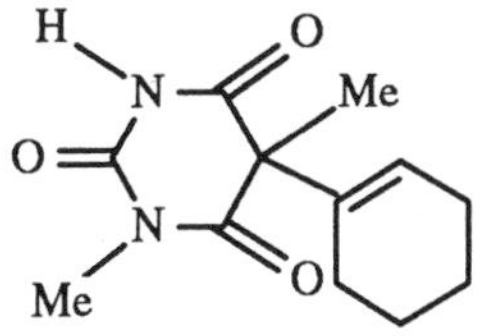

(311)

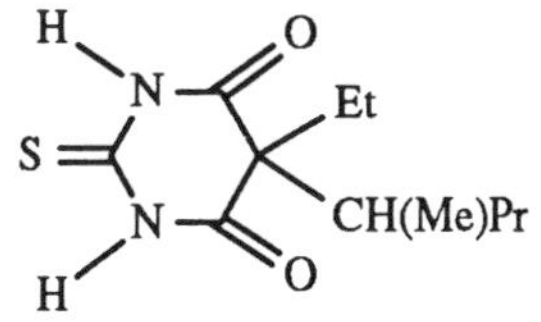

(312)

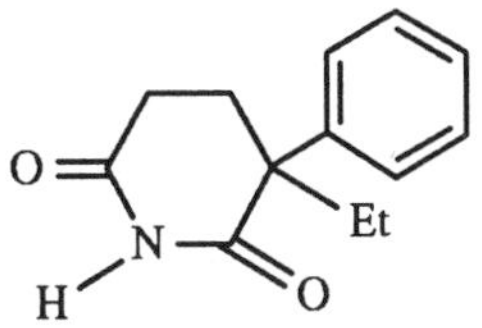

(313)

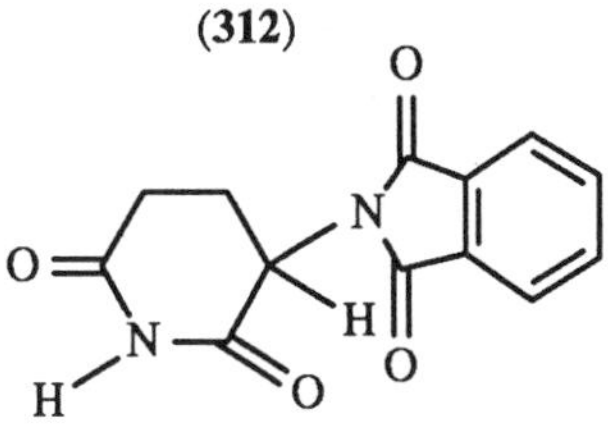

(314)

The major importance of phenobarbitone lay in its anticonvulsant activity. This was discovered serendipitously by Hauptmann[200] shortly after the drug was released for clinical evaluation. He noticed a general reduction in the frequency of daytime seizures in those of his patients who had been receiving the new drug as a hypnotic, then drew the appropriate conclusion. A quarter of a century later, Putnam[201] screened a variety of non-barbiturate analogues of phenobarbitone to obtain phenytoin (**315**), the next major anticonvulsant to be introduced. Carrington likewise used phenobarbitone as the prototype from which he developed primidone (**316**) in 1952. The succinimide anticonvulsants phensuximide (**317**), methsuximide (**318**) and ethosuximide (**319**) were developed during the mid-1950s as a consequence of an extensive screening programme designed to find safe analogues of troxidone (**320**), a somewhat toxic drug that controlled petit mal absence seizures. These did not respond to therapy with other drugs. Troxidone itself had been synthesized in 1943 as a potential mild analgesic, but whilst being evaluated in combination with an antispasmodic which turned out to have convulsant properties, its hitherto unsuspected anticonvulsant action was revealed.

The introduction of both nitrous oxide and ether anaesthesia into medicine could be described as exploitation of solvent abuse, however, this does not detract from what was gained. A third instance of an anaesthetic being found in this manner is afforded by trichloroethylene (**321**), introduced after apprentices were observed inhaling it from degreasing vats used to clean aircraft parts during the Second World War.

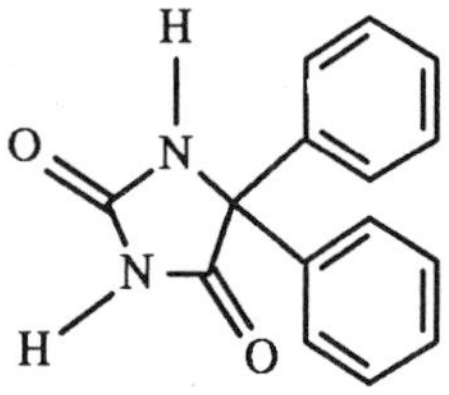

(315)

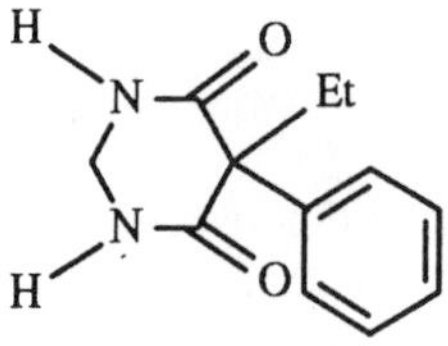

(316)

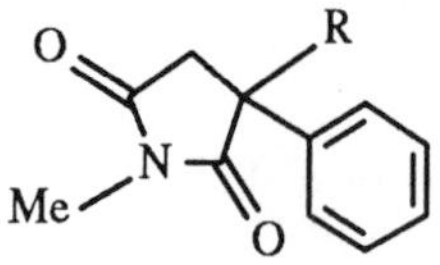

(317) R = H
(318) R = Me

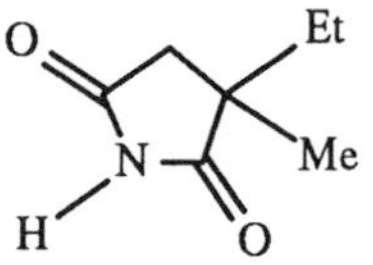

(319)

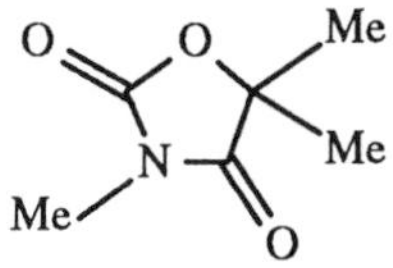

(320)

$ClCH{=}CCl_2$

(321)

1.2.7.1.2 Pentyl nitrite and organic nitrates

In 1844, Balard synthesized pentyl nitrite (**322**) and noted that its fumes caused severe headache. Investigations into its physiological actions continued over the next 20 years, revealing that it caused a drop in blood pressure through dilation of the capillaries. This seemed to have no therapeutic application until Brunton had pioneered the use of the sphygmograph to demonstrate a pronounced rise in blood pressure accompanying attacks of angina. Simply by giving pentyl nitrite to patients to inhale from a cloth soaked with the liquid, he produced rapid relief of their agonizing chest pain.[202] Brunton confirmed that other organic nitrites had similar actions, but it was Murrell[203] who recognized the superiority of nitroglycerine (glyceryl trinitrate; **323**) 12 years later, in 1879. The brevity of its action led to the introduction of pentaerythritol tetranitrate (**324**) in 1896, and other organic nitrates since then. Some clinicians have disputed the value of these longer-acting nitrates.

$C_5H_{11}ONO$

(322)

CH_2ONO_2–$CHONO_2$–CH_2ONO_2

(323)

O_2NOCH_2–$C(CH_2ONO_2)_2$–CH_2ONO_2

(324)

1.2.7.1.3 Local anaesthetics

Isogramine (**325**) was synthesized in 1935 by Erdtman in conjunction with research into the chemistry of gramine, a toxic alkaloid. On testing a trace of the isogramine, he noted that it numbed his tongue. Further investigation revealed similar local anaesthetic activity existed in the open chain synthetic precursor, and this persuaded Erdtman and Lofgren to seek less irritant analogues.[204] Over the next seven years, 57 compounds were examined before lignocaine (**326**) emerged as an excellent local anaesthetic.[205] Several analogues of lignocaine have also found clinical application, including the longer-acting bupivacaine (**327**) introduced in 1957. This was synthesized to assess

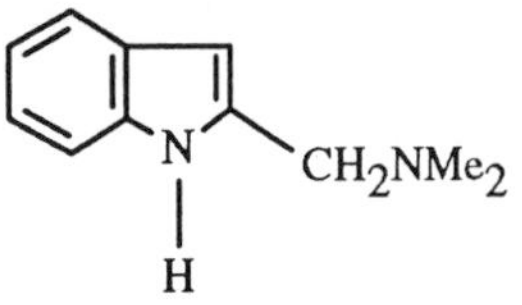

(325)

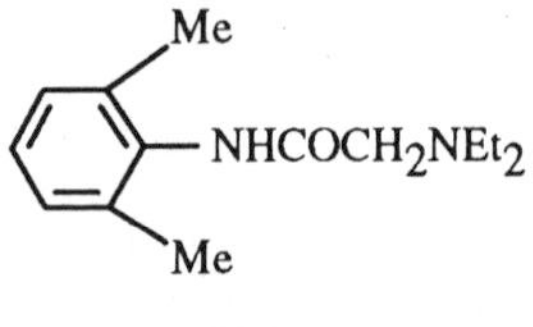

(326)

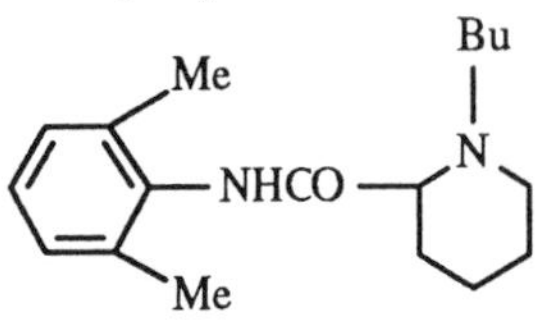

(327)

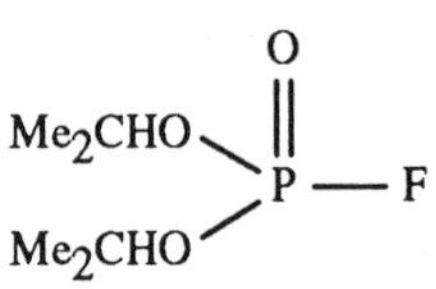

(328)

whether a cyclic analogue offered any advantage over lignocaine. It has been of considerable value to obstetricians as an epidural anaesthetic.

1.2.7.1.4 Organophosphorus anticholinesterases[206]

Lange and von Kreuger[207] synthesized the first phosphorus–fluorine compounds in 1932. Whilst handling these, they experienced pressure in the larynx, breathlessness, blurring of vision and clouding of consciousness. Subsequently, this observation was exploited by the preparation of over 2000 analogues as potential insecticides and chemical warfare agents. These were synthesized by Schrader; their anticholinesterase activity was confirmed by Gross, although not disclosed because of their military significance. However, the characteristic prolonged miotic action on the eye of the most toxic of the compounds described in the original publication of Lange and von Kreuger was shown by Adrian, in 1941, to be due to its potent anticholinesterase activity. After the war, it was given the name dyflos (**328**) and found occasional use in the hands of ophthalmologists.

1.2.7.2 Observations During Routine Laboratory Studies

The greatest serendipitous discovery of all—the discovery of penicillin—which was later to cause a revolution in antibacterial chemotherapy, was made in the laboratory. It was, however, by no means the sole discovery of that type.

1.2.7.2.1 Alkylhydrocupreines

In 1911, the accidental substitution of trypanosomes for pneumococci in a laboratory procedure revealed that they too could be solubilized by means of bile salts. Wondering whether this similarity in behaviour between the organisms extended to sensitivity towards quinine and its derivatives which he was then screening, Morgenroth[208] was able to confirm this to be the case so far as the hydrogenated derivative of quinine, methylhydrocupreine, was concerned. Further tests confirmed that its ethyl homologue was particularly effective in preventing the spread of infection in mice, guinea pigs and rabbits inoculated with pneumococci. Despite the early recognition of the hazard of damage to the optic and acoustic nerves, ethylhydrocupreine (**329**) was marketed as a remedy for pneumonia. It was the first antibacterial chemotherapeutic agent. Some patients were blinded by it, yet it remained in use until the introduction of sulphapyridine in 1938. Morgenroth introduced isopentylhydrocupreine (**330**) and isooctylhydrocupreine (**331**) during the First World War for direct application to deep wounds infected with streptococci or staphylococci.

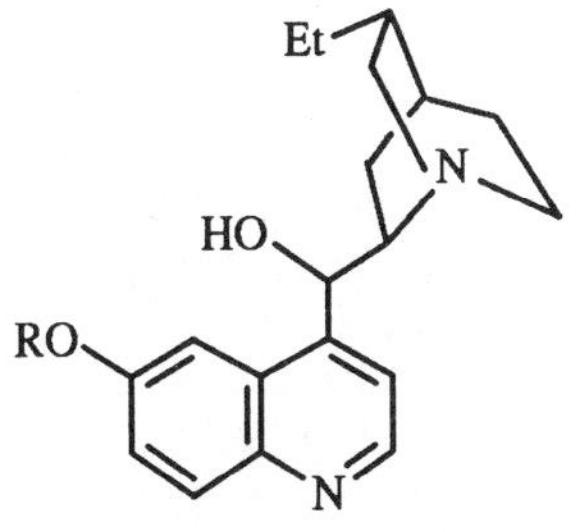

(329) R = Et
(330) R = $CH_2CH_2Pr^i$
(331) R = $(CH_2)_5Pr^i$

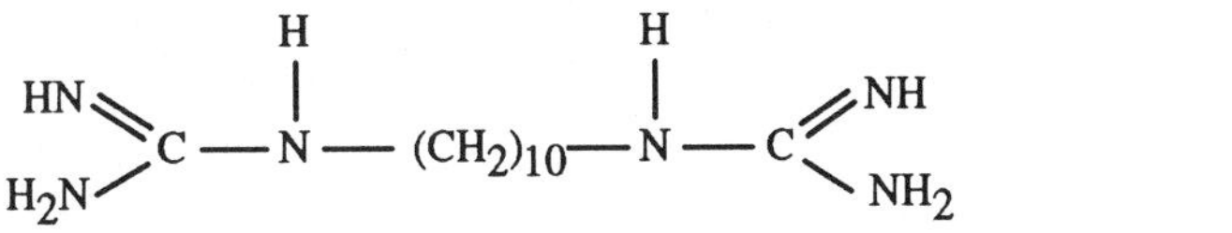

(332) $HN=C(NH_2)-NH-(CH_2)_{10}-NH-C(=NH)NH_2$

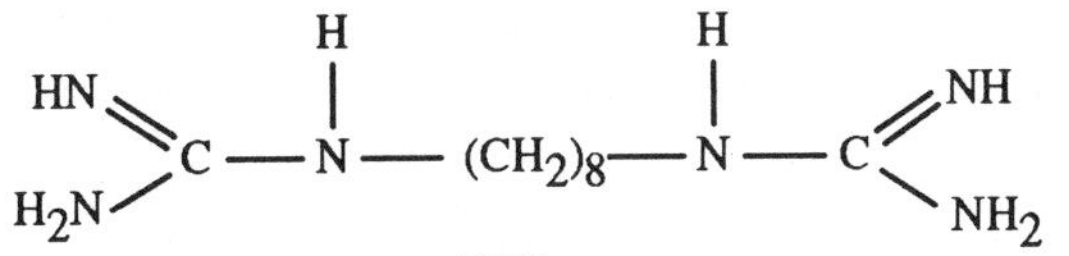

(333) $HN=C(NH_2)-NH-(CH_2)_8-NH-C(=NH)NH_2$

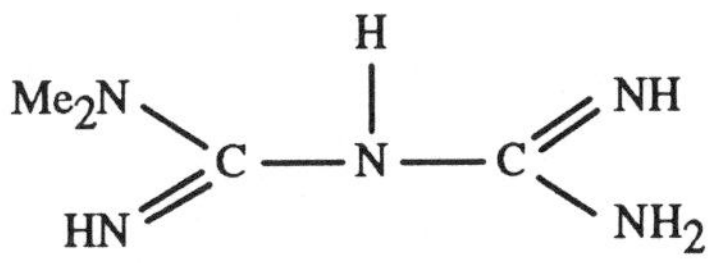

(334) $Me_2N-C(=NH)-NH-C(=NH)NH_2$

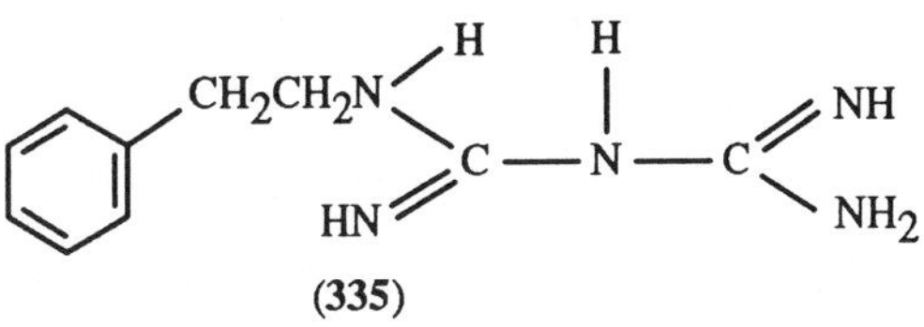

(335) $PhCH_2CH_2NH-C(=NH)-NH-C(=NH)NH_2$

1.2.7.2.2 Guanidines

In 1918, Watanabe noted a drop in blood sugar levels following extirpation of the parathyroid gland. He attributed this to the presence of large amounts of guanidine in the blood, but subsequent tests of this metabolite as an antidiabetic agent showed that it was too toxic. Substituted guanidines seemed more promising, so after the structure of the alkaloid galegine (from *Galega officinalis*) was shown by Barger and White in 1923 to be of this type, it was tested clinically. Three years later, one of the series of diguanidines synthesized by Slotta and Tschesche[209] was marketed under the name synthalin (**332**) as an oral hypoglycaemic agent for patients with mild diabetes. Its toxicity required its early replacement by synthalin B (**333**), but this had to be withdrawn in the early 1940s after numerous reports of its damaging effects on the liver. It was not until 1957 that further development took place, when Shapiro *et al.*[210] introduced the biguanides metformin (**334**) and phenformin (**335**) for the oral treatment of diabetes.

A valuable group of antitrypanosomal agents owe their introduction to work done by Jancso and Jancso[211] in 1935. They followed up reports that trypanosomes required large amounts of glucose in order to reproduce, and that the survival of animals infected with these protozoa could be prolonged merely by injecting them with insulin to maintain low blood sugar levels. The obvious alternative was to inject infected mice with synthalin and its analogues. The Jancsos' established these to be trypanocidal, although subsequent studies by Yorke and Lourie[212] revealed that synthalin was effective in doses which did not affect blood sugar levels. The action was shown to be a hitherto unrecognized direct trypanocidal one. Ewins then synthesized a range of analogues for screening, from which emerged stilbamidine (**336**) and pentamidine (**337**). Extensive trials of these two amidines confirmed their value in the chemotherapy of sleeping sickness and leishmaniasis.

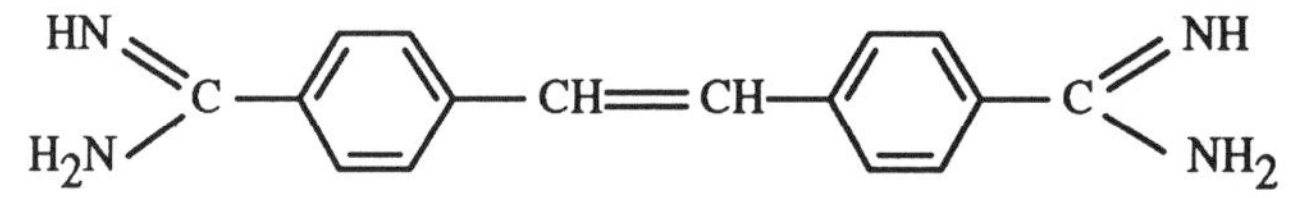

(336)

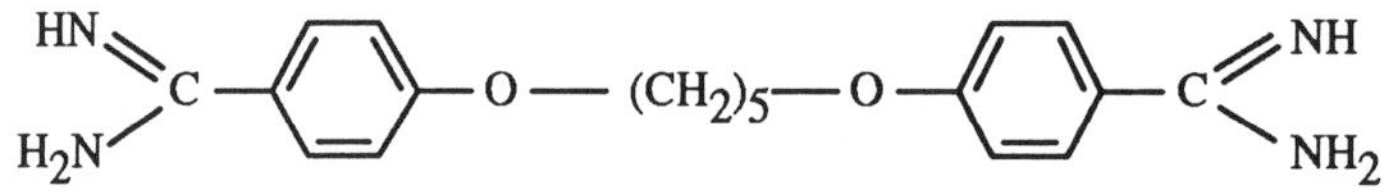

(337)

1.2.7.2.3 Lithium salts[213]

Seeking evidence for hormonal imbalance as a possible cause of manic depressive disease, Cade, in 1948, injected guinea pigs with concentrates of urine collected from patients and healthy volunteers. He noted that convulsions caused by the urea that was present were more likely to occur when concentrates from depressive patients were injected, despite the absence of significantly increased amounts of this metabolite. Cade considered the possibility that the toxicity of the urea might be augmented by the presence of either uric acid or some other substance present in the concentrates from depressives. In order to test this hypothesis, he had to use the lithium salt of uric acid in order to achieve the high concentrations that were required. When urea dissolved in a saturated solution of lithium urate was injected into guinea pigs, the expected toxicity was absent. Cade established that the protective action was due not to the uric acid, but rather to the lithium ion. Further experiments with lithium carbonate revealed its sedating effect on animals. This was followed by administration of it by mouth to a manic depressive patient, a procedure that could be accepted as lithium salts had been widely used last century for treating gout (in the belief that uric acid deposits would be dissolved). That patient's remarkable restoration to normality so long as the treatment was continued was to be the forerunner of a major advance in psychiatry. Although requiring regular monitoring for toxicity and effectiveness, lithium carbonate is now a standard form of treatment for manic depression.

1.2.7.2.4 Asparaginase

In 1953, Kidd[214] observed that the growth of transplanted tumours in mice could be inhibited by guinea pig serum, but not by serum from either horses or rabbits. As it had been known for over 30 years that unusually high levels of the enzyme asparaginase were present in guinea pig serum, Broome[215] conducted tests that confirmed the dependency of the tumour cells on asparagine. Subsequent investigations confirmed that enzyme extracted from *E. coli*, known as colaspase, can be a useful addition to the combination of drugs that are administered to initiate the chemotherapy of acute leukaemia in children. The value of asparaginase in other malignancies is questionable.

1.2.7.3 Unanticipated Consequences of Therapy

It is a fact of life that the observant clinician is just as likely to be responsible for the discovery of a novel type of drug response as any medicinal chemist. Indeed, one wonders whether too much emphasis is now placed on being alert to adverse reactions rather than exploitable side effects!

1.2.7.3.1 Salicylic acid

Salicylic acid was introduced in 1874 as an antiseptic suitable for internal use.[216] In the course of a trial of its efficacy against typhoid, nursing staff maintained fever charts which revealed a drop in temperature after oral dosing. Buss recognized that as there was no remission of the infection the drug was acting as an antipyretic. When the report of this appeared, Stricker immediately tested salicylic acid for its ability to control the temperature of his patients with rheumatic fever. To his surprise, it proved to be of quite definite value as a specific antirheumatic agent. His findings were published in 1876, shortly before the antirheumatic activity of salicin was discovered by MacLagan (see Section 1.2.2.3.2).

1.2.7.3.2 Acetanilide

In 1886, Cahn and Hepp[217] attempted unsuccessfully to eradicate worms in patients by prescribing oral doses of naphthalene, which was then under investigation as an antiseptic for internal use. Eventually, they traced the failure of the treatment to a dispensing error which had resulted in acetanilide being supplied instead of naphthalene. Before this was revealed, however, they confirmed that the medication had reduced the fever of a patient suffering from a variety of complaints besides worms. No time was wasted in publishing a report confirming the value of acetanilide as an antipyretic, and commercial production began later in the year. Shortly after, Hinsberg introduced the somewhat less toxic 4-ethoxy derivative as phenacetin (**338**). Its metabolite, now known as paracetamol (**339**; acetaminophen), was tested clinically in 1893, but its superiority over phenacetin was not recognized until 60 years later.

1.2.7.3.3 Hexamine

Following Ladenburg's observation, in 1894, that uric acid formed a soluble salt with piperazine, attempts were made to treat gout with this base. Nicolaier examined hexamine (**340**) as an alternative, but found it no better. Unexpectedly, however, it did have a beneficial effect on urinary tract infections. This was later attributed to the decomposition of hexamine in acid urine to release formaldehyde as the active antiseptic.

1.2.7.3.4 Mercurial diuretics

It was not until 1919, seven years after its introduction, that the diuretic action of the antisyphilitic drug merbaphen (**341**) was recognized. Vogl[218] had administered it by injection to a patient whose progress was meticulously monitored by nursing staff. As a consequence, the striking increase in urine production was brought to his attention. Detailed investigations then established the outstanding value of the drug in patients with congestive heart failure. No previous diuretic had comparable value, and until the development of the thiazides, mercurial diuretics were widely used in such patients. The use of merbaphen was abandoned when drugs less damaging to the kidney became available, such as mersalyl (**342**), which was introduced in 1924. Like merbaphen, it had been developed earlier for use in syphilis.

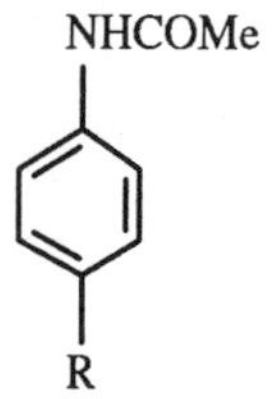

(338) R = OEt
(339) R = OH

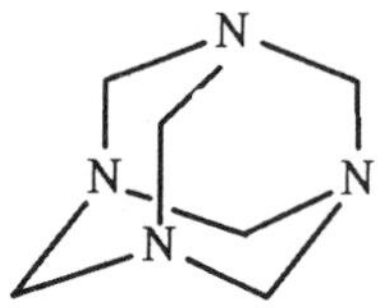

(340)

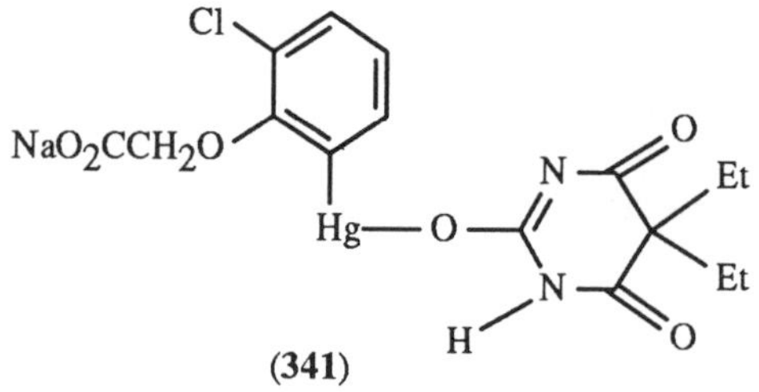

(341)

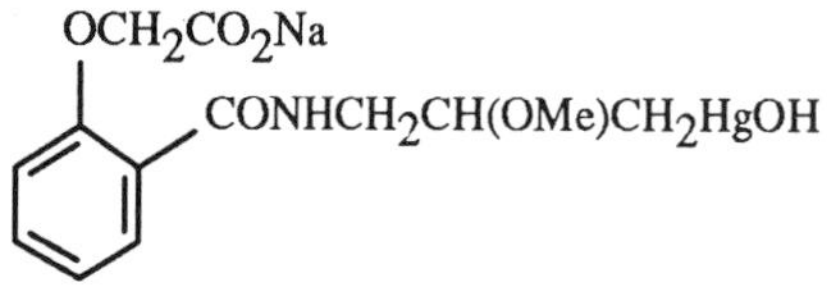

(342)

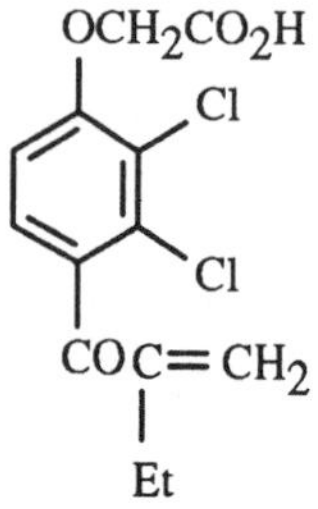

(343)

The mercurial diuretics were not absorbed from the gut. Being somewhat toxic, they were given in intermittent courses. These factors, and the fact that they were often life-saving drugs, encouraged research into the development of safer alternatives. This objective seemed unattainable until it was shown, in 1941, that the mercurials probably interfered with the movement of ions across the kidney tubules by inhibiting dehydrogenase enzymes. A prolonged search for mercury free enzyme inhibitors resulted in the introduction of ethacrynic acid (**343**) in 1962.[219] This had been modelled on the molecular characteristics of merbaphen and mersalyl.

1.2.7.3.5 Sulfonamide diuretics

Not long after the introduction of the antibacterial sulfonamides in the late 1930s, clinicians became aware that patients who had received massive doses of some of these drugs produced a large volume of slightly alkaline urine. It was suggested that this could be accounted for by the increased excretion of sodium bicarbonate arising from the inhibition of carbonic anhydrase. This inhibition of the enzyme had been shown by Mann and Keilin[220] to account for the fall in carbon dioxide binding power of the blood that accompanied the administration of some sulfonamides. This serendipitous observation provided the basis for the subsequent use, in 1949, of large oral doses of sulfanilamide as a diuretic in patients with congestive heart failure. As this produced too high an incidence of toxic reactions, Roblin and Clapp[221] screened some 20 heterocyclic sulfonamides and found that acetazolamide (**344**) was over 300 times as potent as sulfanilamide as a carbonic anhydrase inhibitor. It was introduced as a diuretic in 1952.

Acetazolamide was non-specific in its action, inhibiting the enzyme not only in the kidney but also in other parts of the body. This had the advantage of enabling it to reduce intraocular pressure in glaucoma in much the same way as it acted on the kidney, but it also resulted in complications when used as a diuretic and hence it could only be taken intermittently by patients. In an attempt to avoid this, Beyer[222] screened analogues in a search for one which would only inhibit the enzyme in the proximal portion of the kidney tubules and thus produce excretion of sodium chloride rather than the bicarbonate. He was convinced that the enhanced excretion of salt, accompanied by a diuresis, would be of particular value in treating hypertension. During the screening, dichlorphenamide (**345**) emerged as having similar properties to acetazolamide. The surprising discovery was then made that introduction of an amino group into the benzene ring of compounds like dichlorphenamide reduced their carbonic anhydrase inhibitory activity yet still permitted them to induce increased renal secretion of chloride. During the search for more potent diuretics of this type, an unintended ring closure produced the very potent chlorothiazide (**346**). Clinical tests confirmed that it produced a marked increase in the excretion of sodium chloride. When introduced in 1957, its oral activity immediately rendered the injectable mercurial diuretics obsolete in the treatment of congestive heart failure. Many analogues have been marketed, these either being more potent or producing their effects more quickly. Examples include hydrochlorothiazides (**347**), hydroflumethazide (**348**), mefruside (**349**), frusemide (furosemide) (**350**), bumetanide (**351**) and piretanide (**352**).

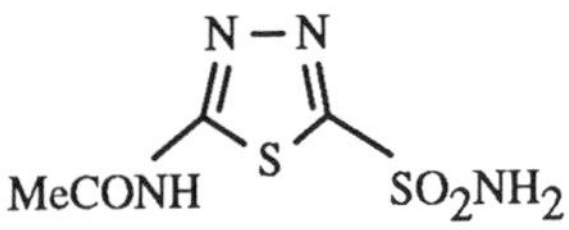

(344) 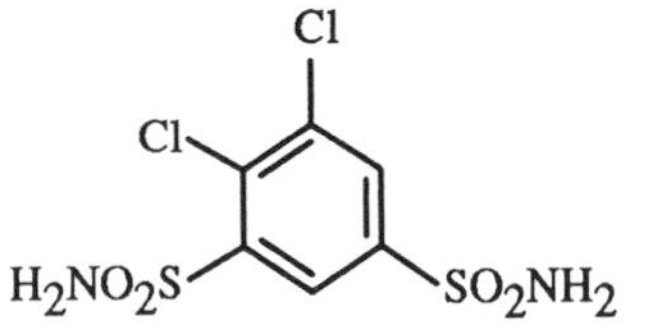 (345) 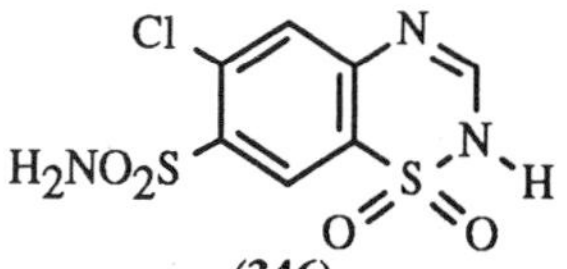 (346)

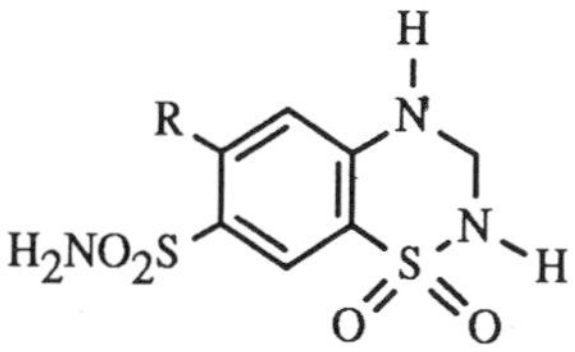

(347) R = Cl
(348) R = CF_3

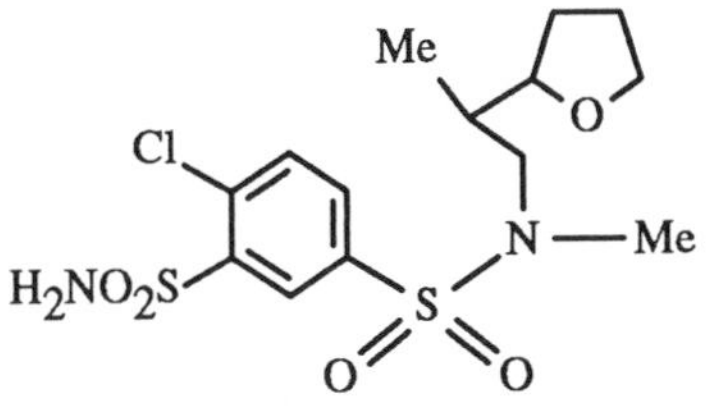

(349) 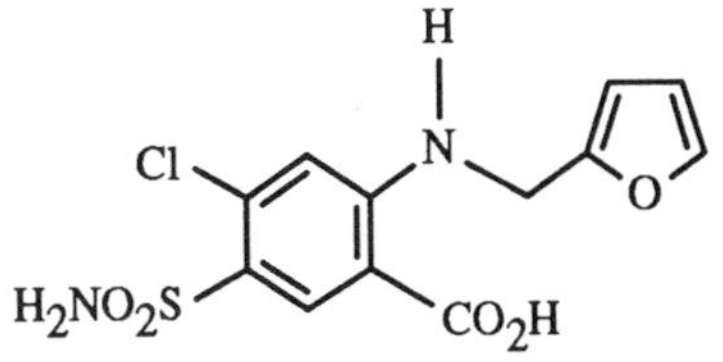 (350)

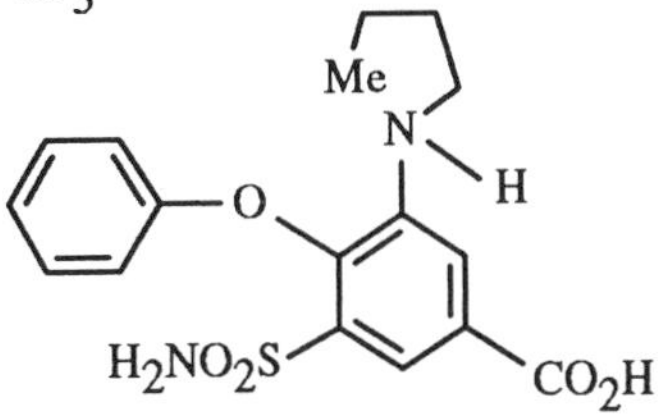

(351) 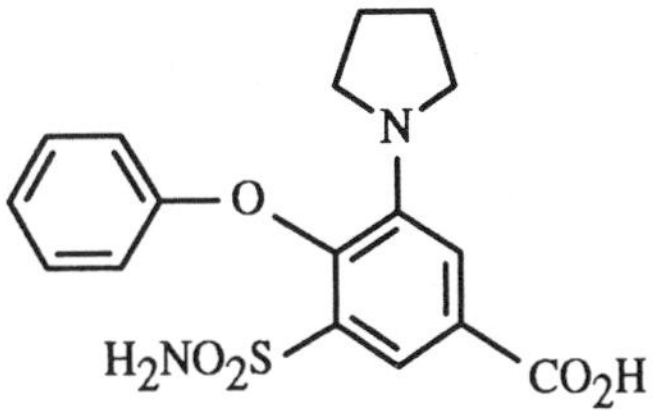 (352)

1.2.7.3.6 *Antihistaminic travel sickness remedies*

In an attempt to counteract the unwelcome depressant side effects of antihistamines, diphenhydramine was formulated as a salt with a caffeine analogue, 8-chlorotheophylline, this preparation being known as dimenhydrinate. Samples of the drug being evaluated clinically by Gay and Carling[223] were supplied to a patient troubled with urticaria. A week later, she reported disappointment over the failure of dimenhydrinate to provide relief of her skin condition. However, she added the comment that since regularly taking the drug her persistent sickness when travelling on Baltimore streetcars had disappeared! This chance observation was seized upon and led to a full-scale clinical trial involving hundreds of marines aboard the troopship 'General Ballou' crossing the Atlantic in 1947. The results removed any doubt as to the efficacy of the remedy. Subsequently, it was established that most antihistamines possessed similar properties.

1.2.7.3.7 *Antifolates*

During a preliminary clinical trial designed to establish their safety and suitability for use in the treatment of cancer, Farber[224] found the folic acid derivatives pteroyldiglutamic acid (**353**) and pteroyltriglutamic acid (**354**) were accelerating the growth of malignant cells in the bone marrow of patients with acute leukaemia. He obtained the newly synthesized antifolate pteroylaspartic acid from SubbaRow, this having been prepared speculatively for investigative purposes. During 1947, encouraging results were obtained in children with acute leukaemia, but much better responses were subsequently observed with the more potent aminopterin (**355**) and methotrexate (**356**). Injections of the latter used on its own induced temporary remissions in 30% of children, but Farber and his colleagues later combined it with drugs such as cortisone and mercaptopurine to achieve results which paved the way towards the current position where most children can be cured of the disease if given the appropriate treatment.

The precise mode of action of the antifolates became apparent in 1952. This was the inhibition of dihydrofolate reductase, the enzyme that catalyzed the transformation of folic acid so that it could be utilized for the synthesis of thymine for incorporation into DNA. Surprisingly, no other antifolate has emerged to rival methotrexate in the treatment of either acute leukaemia or cancers. Nonetheless, analogues which are selective in their action towards the dihydrofolate reductase enzymes of the malarial parasite and many bacteria have been developed following the early recognition of

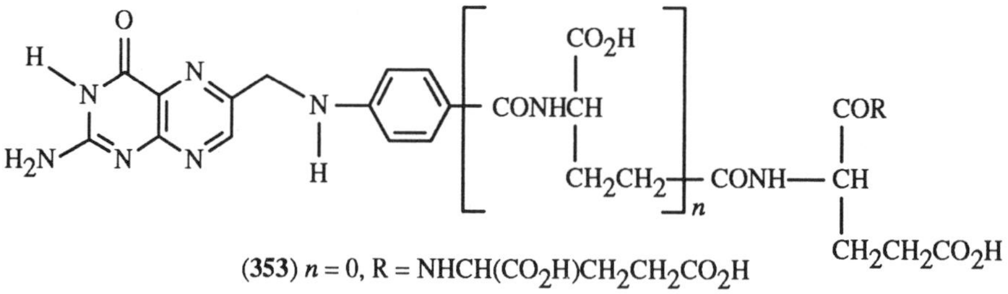

(353) $n = 0$, R = $NHCH(CO_2H)CH_2CH_2CO_2H$

(354) $n = 2$, R = OH

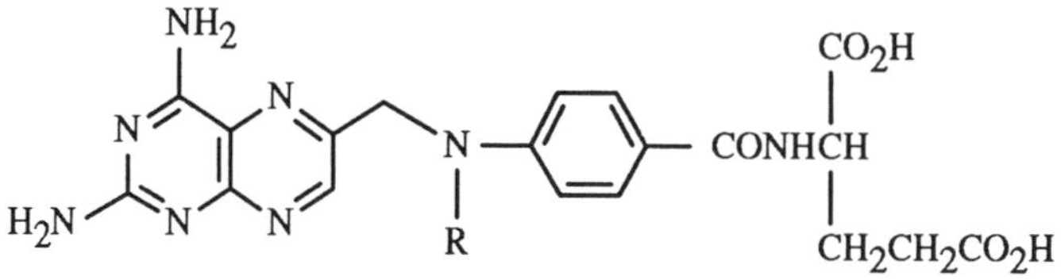

(355) R = H

(356) R = Me

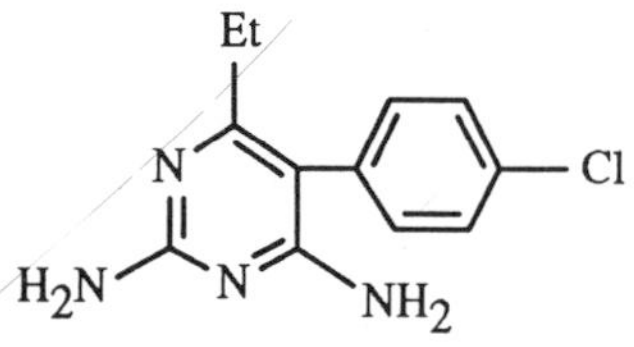

(357)

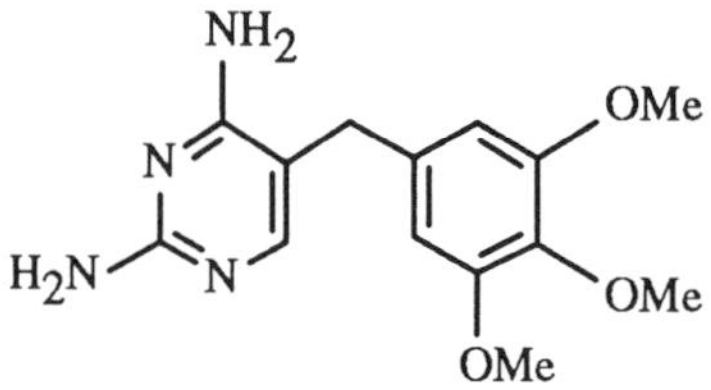

(358)

species sensitivity due to slight variations in enzymic structure. These include the antimalarial pyrimethamine (**357**) and the antibacterial trimethoprim (**358**).[225]

1.2.7.3.8 Tricyclic psychopharmacological agents[226]

Seeking a drug to prevent surgical shock during anaesthesia, Laborit established that promethazine was not only superior to other antihistamines in regular clinical use, but appeared to possess unusual central actions. In conjunction with Courvoisier, the central effects of the full series of phenothiazines originally synthesized by Charpentier as potential antihistamines were then examined. When it became apparent that promazine came closest to meeting the requirements for antishock activity, new analogues were synthesized by Charpentier. Chlorpromazine, (**359**) prepared in 1950, was found to have outstanding activity and low toxicity. In 1951, Laborit not only confirmed its safety in the clinic, but observed that patients appeared exceptionally relaxed and unconcerned prior to and after surgery. He recognized the significance of this and persuaded psychiatrists to evaluate the drug in disturbed patients. In 1952, the full tranquillizing action of the drug was demonstrated when a severely agitated patient received it by injection. Subsequent studies confirmed its early promise, ushering in a new era for psychiatry.

The outstanding clinical and commercial success of chlorpromazine encouraged pharmaceutical companies to synthesize analogues. Some of these were made by introducing alternative substituents to replace the chlorine atom, whilst others involved variation of the side chain. Room for manoeuvre being limited, alternative ring systems were examined. Thioxanthene tranquillizers such as chlorprothixene (**360**),[227] introduced in 1958, retained the overall shape and dimensions of the phenothiazines and thus possessed much the same type of activity. However, this was by no means the case with certain other tricyclic compounds. Imipramine (**361**),[228] introduced in 1956, turned out to be an antidepressant rather than a tranquillizer. When amitriptyline (**362**) was synthesized in 1960 as an analogue of the thioxanthene tranquillizers, it not only retained tranquillizing activity, but also exhibited antidepressant properties. Molecular modification of amitriptyline in turn produced drugs with similar activity, namely doxepin (**363**) and dothiepin (**364**). Analogues of all these tricyclic drugs have been marketed.

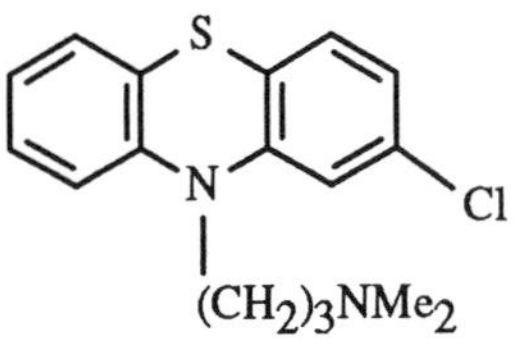

(359) 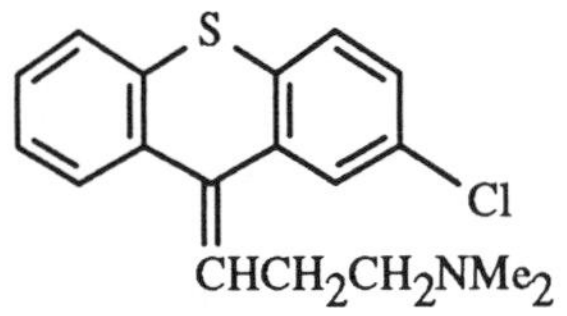 (360) 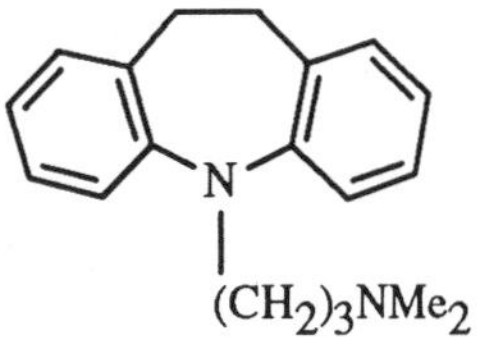 (361)

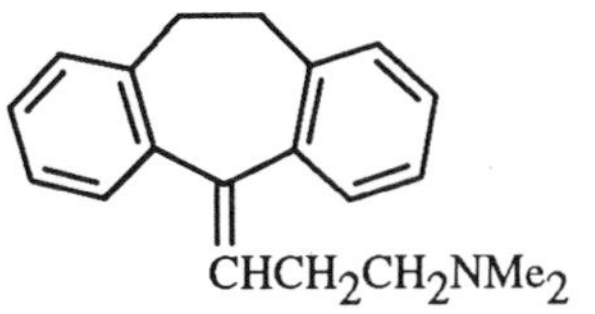

(362) 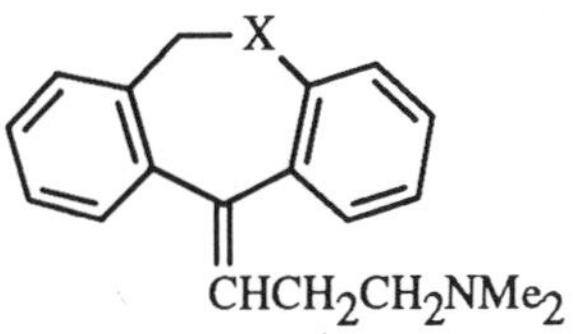 (363) X = S
(364) X = O

1.2.7.3.9 Antidiabetic sulfonylureas

In an attempt to find out more about a severe toxic reaction caused by a sulfonamide antibacterial undergoing clinical trial in 1954, Fuchs tested the drug on himself. He recognized the ensuing symptoms as those of severe hypoglycaemia. Consequently, the drug was thoroughly investigated and then introduced as an oral antidiabetic agent under the name of carbutamide (**365**).[229] It was quickly superseded by less toxic analogues, including tolbutamide (**366**), which was synthesized in 1956. Although several others have been developed, it took 10 years and the screening of thousands of compounds before the first of these, glibenclamide (**367**), was able to challenge tolbutamide.

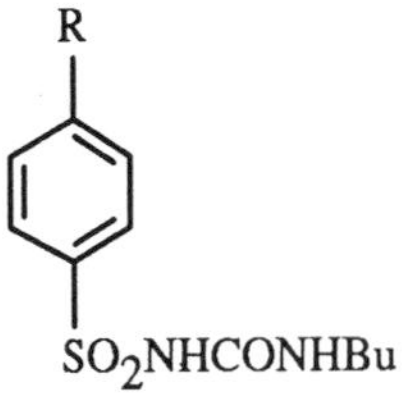

(365) R = NH_2
(366) R = Me

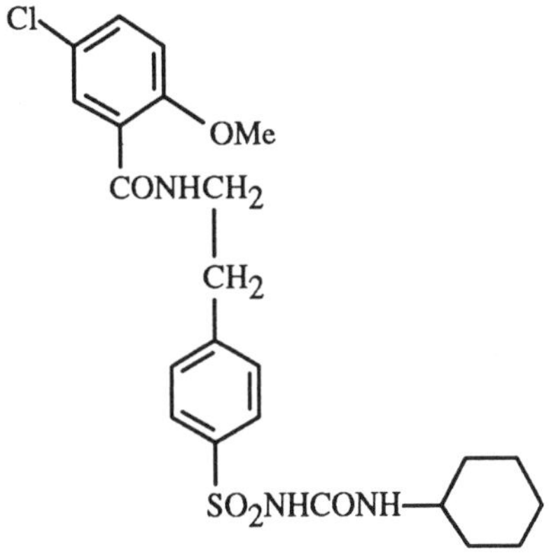

(367)

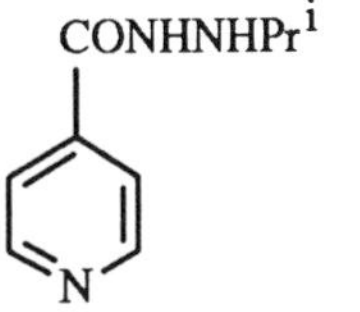

(368) 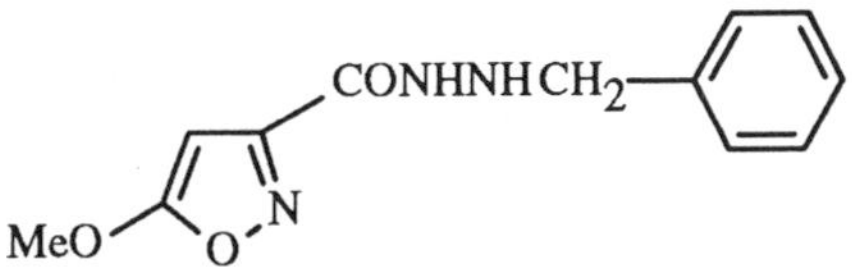 (369) 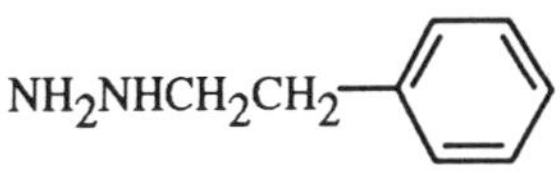 (370)

1.2.7.3.10 Monoamine oxidase inhibitors[230]

At a meeting of the American Psychiatric Association in 1957, delegates heard several reports of the mood-elevating effect of iproniazid (**368**), an antituberculous drug that had been introduced five years earlier. One of the reports testified to the specific value of iproniazid in chronically depressed psychotic patients. As a result the drug was prescribed for hundreds of thousands of patients during the next four years before it had to be withdrawn because of an unacceptable incidence of jaundice. As its ability to inhibit monoamine oxidases, known since its original introduction in 1952, was considered to be the basis of its antidepressant activity, it was superseded by less toxic monoamine oxidase inhibitors such as isocarboxazid (**369**) and phenelzine (**370**).

1.2.7.3.11 Methyldopa

Methyldopa (**371**) was synthesized in 1953[231] and shown to be an *in vitro* inhibitor of the enzyme DOPA decarboxylase, which was considered to play a major role in catecholamine biosynthesis. Supplies of the drug were sent to Udenfriend who considered its pharmacological profile to justify a clinical trial in hypertensive patients. In a report published in 1960, he confirmed that methyldopa was capable of reducing blood pressure.[232] Initially, it was assumed that this must be attributable to its metabolic effects, but subsequent studies revealed that the hypotensive action was due to a direct effect on the brain. Hitherto, there had been no reason to suspect this.

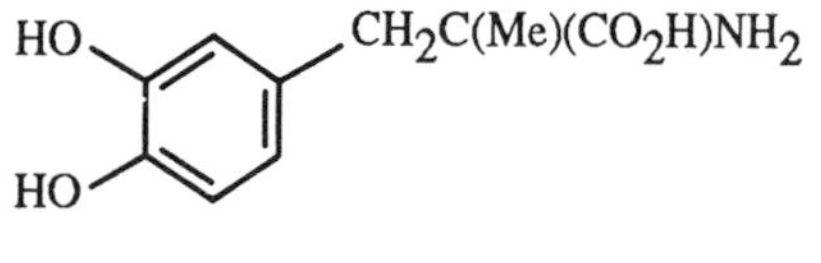

HO, HO, $CH_2C(Me)(CO_2H)NH_2$

(371)

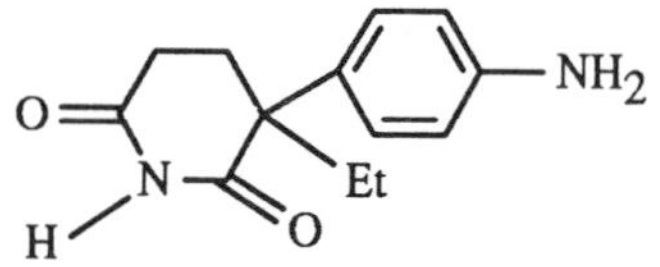

O, N, H, O, Et, NH_2

(372)

1.2.7.3.12 Aminoglutethimide

In 1963, Cash detected signs of Addison's disease (adrenal insufficiency) in a young girl who had been taking aminoglutethimide (**372**) to control her epilepsy. The drug had been on the market for less than three years. When others confirmed the existence of the problem, the drug was withdrawn. However, once it had been shown that it blocked steroid biosynthesis, tests were carried out on patients with Cushing's disease in an attempt to interfere with corticosteroid synthesis. Although results were unsatisfactory in this instance,[233] encouraging responses were later obtained when aminoglutethimide was given by mouth to women with hormone dependent breast cancer. Together with tamoxifen, aminoglutethimide represents a major advance in the treatment of this disease.

1.2.7.4 Effects Observed During Pharmacological Experiments

1.2.7.4.1 Central nervous system depressants

During the 1950s, the concern over the use of barbiturates in self-poisoning attempts created a climate conducive to the introduction of new hypnotics. There was also an interest in developing novel intravenous anaesthetics which were devoid of the hazardous cumulative properties of thiopentone. Pharmacologists were not slow to exploit the serendipitous opportunities that sometimes offered a way to develop such drugs.

Whilst screening analogues of febrifugine, an antimalarial quinazolone alkaloid found in the common hydrangea, Gujral *et al.*[234] found several to have hypnotic activity in rats. In 1955, they reported methaqualone (**373**) to be a promising non-barbiturate hypnotic. Although it lived up to its promise in some ways, the drug was subject to abuse and was withdrawn in some countries in the early 1980s.

Curiosity about the convulsant activity observed when thiamine was injected intravenously led Charonnat *et al.*[235] to examine the fragments obtained by chemical degradation of this vitamin. They found that the thiazole component antagonized the convulsant action of the pyrimidine fragment. Replacement of the primary alcohol in this thiazole had already been achieved 20 years earlier, in 1935, when the compound now known as chlormethiazole (clomethiazole) (**374**) had been screened as a potential vitamin antagonist. When tested for hypnotic activity it was found to offer a useful alternative to the barbiturates.

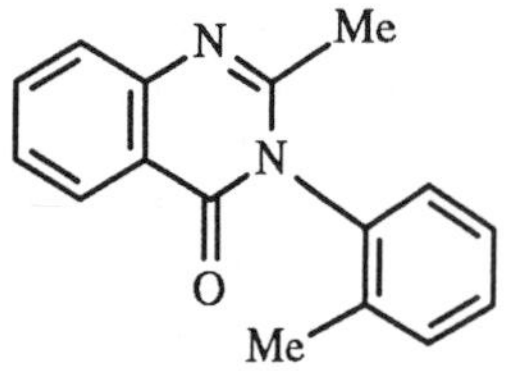

N, Me, N, O, Me

(373)

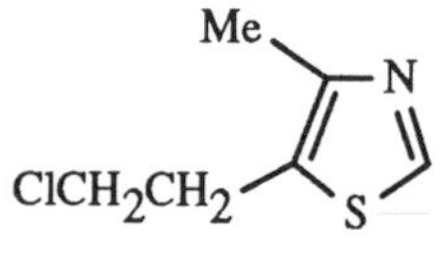

Me, N, $ClCH_2CH_2$, S

(374)

During a programme of research into ethisterone analogues, an alkynic intermediate subjected to routine screening was discovered to have hypnotic activity. It was introduced in 1956 under the approved name of methylpentynol (meparfynol) (**375**). It had first been synthesized in 1913, hence could not be patented. The analogues ethchlorvynol (**376**) and ethinamate (**377**) were also marketed.

Analogues of eugenol (**378**), the topical anaesthetic present in oil of cloves, were evaluated in the early 1950s. Several unexpectedly turned out to possess general anaesthetic properties, estil being the most interesting of these. Attempts to find a water soluble derivative for use as an intravenous anaesthetic were not entirely successful, but propanidid (**379**) was introduced in 1964 in a formulation that contained a surfactant.[236] Twenty years later, mounting concern about allergic responses to the surfactant resulted in the withdrawal of propanidid.

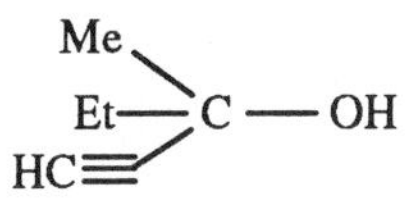

Me, Et, HC≡ — C — OH

(**375**)

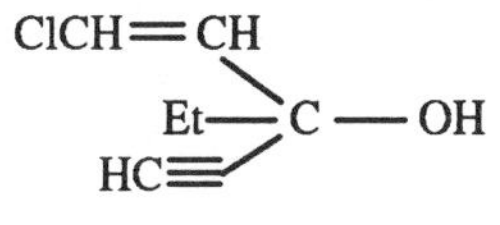

ClCH=CH, Et, HC≡ — C — OH

(**376**)

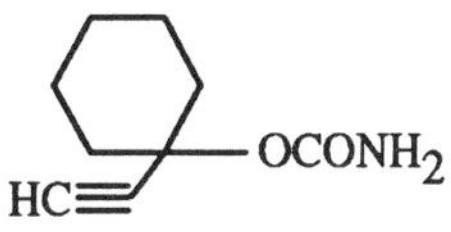

HC≡, $OCONH_2$

(**377**)

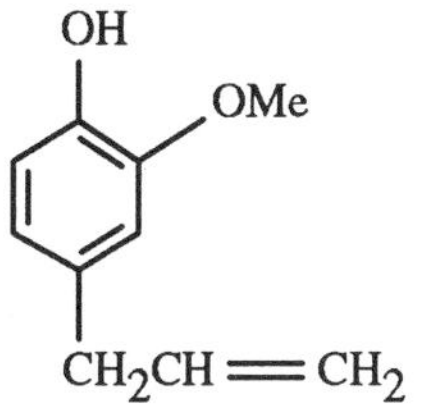

OH, OMe, $CH_2CH{=}CH_2$

(**378**)

OCH_2CONEt_2, OMe, CH_2CO_2Pr

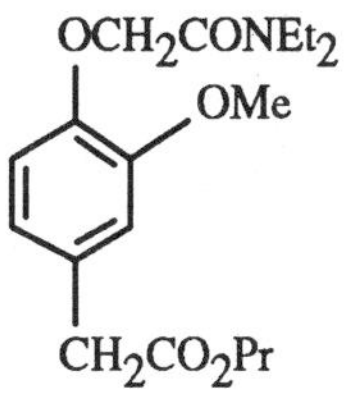

(**379**)

The outstanding impact of chlorpromazine on psychiatric practice ensured that compounds found to possess unusual sedating properties would be thoroughly investigated. When Janssen observed mice becoming calm and sedated after being injected with a pethidine (meperidine) analogue, he set out to find a derivative of this which was devoid of analgesic activity.[237] In 1958, he reported that the screening of hundreds of butyrophenone analogues of pethidine had shown haloperidol (**380**) to be the most potent tranquillizer known to medicine. During the following 10 years, a further 5000 analogues were examined, a dozen or so finding a place in clinical practice.

The benzodiazepines are mild tranquillizers which owe their discovery to what might strictly be described as luck rather than serendipity. Following the introduction of chlorpromazine, Sternbach[238] decided to synthesize a series of tricyclic compounds derived from a novel ring system he had investigated 20 years earlier. All but one of the new series were screened and found to be inactive. It was only after a routine clearing out of the laboratory that the remaining compound was sent for screening. It turned out to have outstanding properties as an antianxiety agent. Further investigation revealed that this particular compound, chlordiazepoxide (**381**) had undergone a chemical rearrangement to form a novel ring system that conferred upon it the interesting biological properties. It was patented in 1958. Subsequently, it was established that the *N*-oxide function was superfluous and even reduced potency, hence a whole range of new analogues were screened. Chlordiazepoxide itself reached the market in 1960, to be followed by many other benzodiazepine tranquillizers, *e.g.* diazepam (**382**) and bromazepam (**383**). Some members of the series were eminently suitable as hypnotics due to their high safety margin, the first of these being nitrazepam (**384**). More rapidly metabolized compounds, less likely to produce drowsiness on awakening, were later introduced, *e.g.* lormetazepam (**385**), temazepam (**386**) and triazolam (**387**).

Although benzodiazepines may be of use in controlling epileptic fits, they are too sedating for prophylactic use. However, there are two important anticonvulsants which were serendipitously discovered and are now frequently prescribed. These are carbamazepine (**388**) and sodium valproate (**389**). The former was prepared as an analogue of chlorpromazine in 1953, but it was another 10 years before its anticonvulsant properties were discovered and exploited clinically. As for the latter, valproic acid was used to dissolve a somewhat insoluble khellin derivative that was subjected to a battery of pharmacological tests. Only when the same solvent was used to dissolve an unrelated compound was it realized that the anticonvulsant action was attributable to the solvent.[239] The sodium salt was marketed in 1967 after extensive clinical studies had confirmed its suitability.

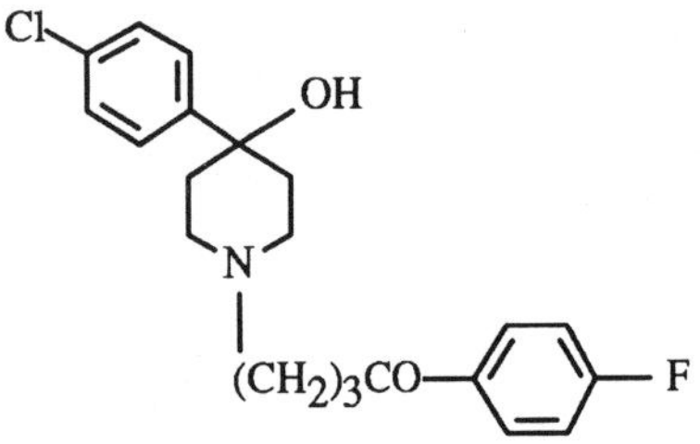

(380) 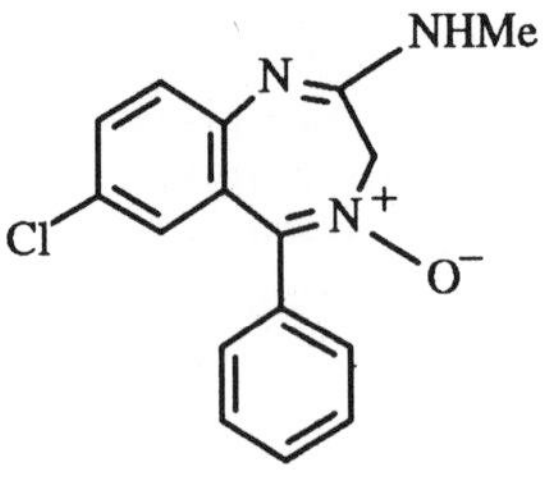 (381) 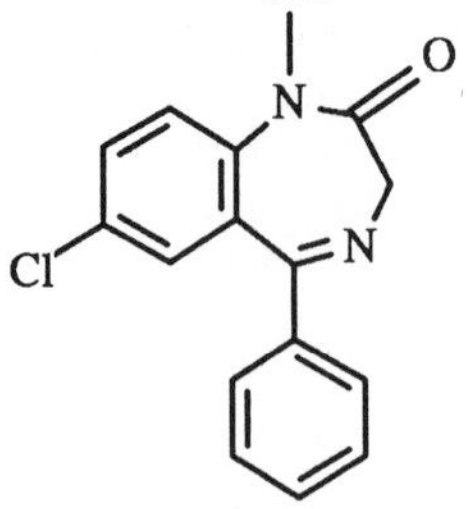 (382)

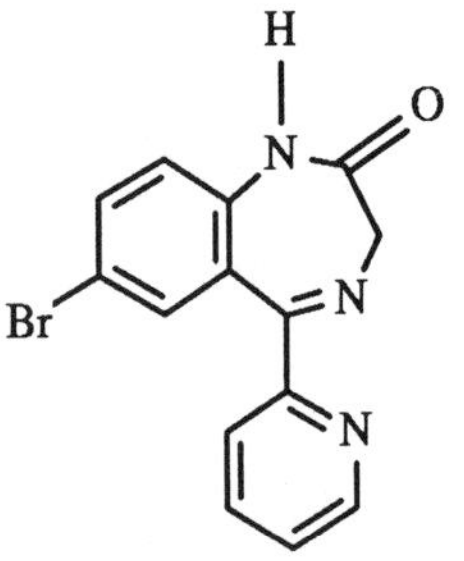

(383) 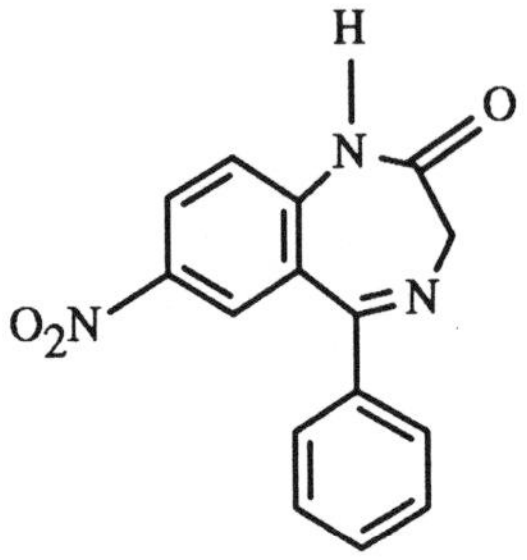 (384) 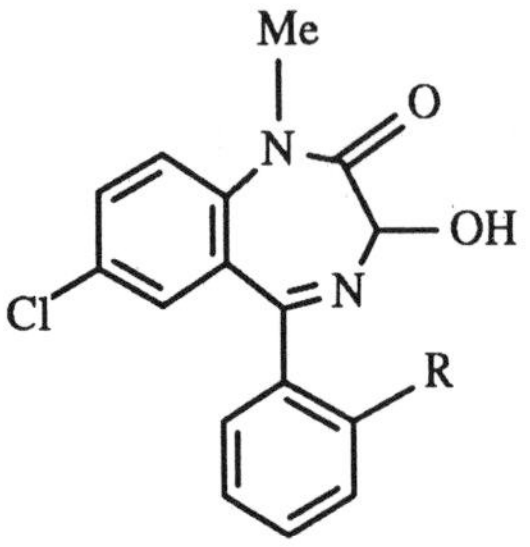 (385) R = Cl
(386) R = H

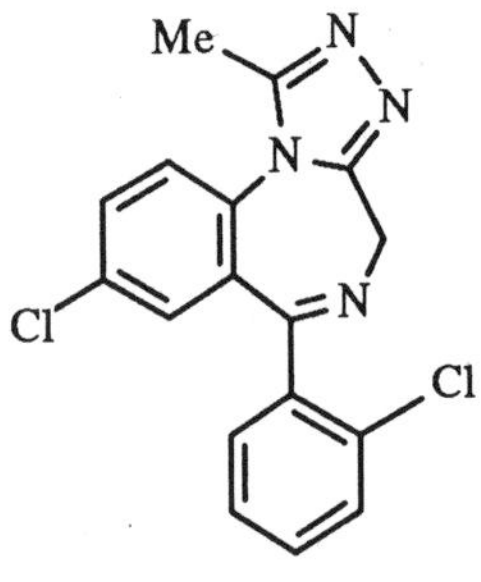

(387) 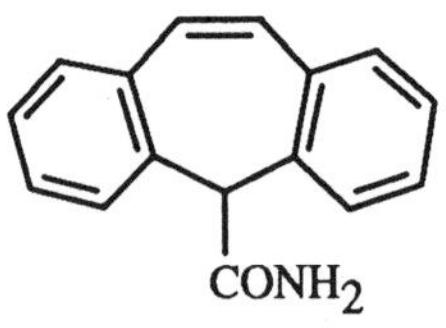 (388)  (389)

1.2.7.4.2 Cardiovascular drugs

In 1949, whilst testing a series of dionium polymethylene compounds with a view to ascertaining the influence of chain length on neuromuscular blocking activity, Paton and Zaimis[240] noted that injection of hexamethonium (**390**) caused flushing of a rabbit's ears, accompanied by a drop in blood pressure. This was attributed to a blockade of the ganglia that mediated chemotransmission of electrical impulses in the autonomic nervous system. It thus afforded a novel approach to the treatment of high blood pressure, but its poor absorption from the gut necessitated parenteral administration. Furthermore, hexamethonium blocked all autonomic ganglia, producing unwanted side effects such as dry mouth, visual disturbance and fainting. The general lack of potency and duration were to some extent overcome when pentolinium tartrate (**391**) was introduced in 1952. This was followed three years later by the first orally active analogue, mecamylamine (**392**). This was an unexpected product obtained when camphene had been reacted with hydrogen cyanide. Routine screening revealed it to be capable of blocking ganglia, despite the absence of a quaternary ammonium function. Its analogue known as pempidine (**393**), introduced in 1958, had the important clinical advantage of being non-cumulative and thus gave more consistent control of blood pressure. Like the earlier drugs, it lacked selectivity.

$Me_3\overset{+}{N}-(CH_2)_6-\overset{+}{N}Me_3$

(390) (391) 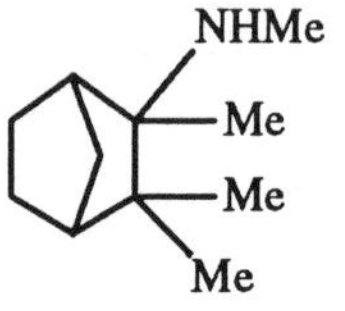 (392) 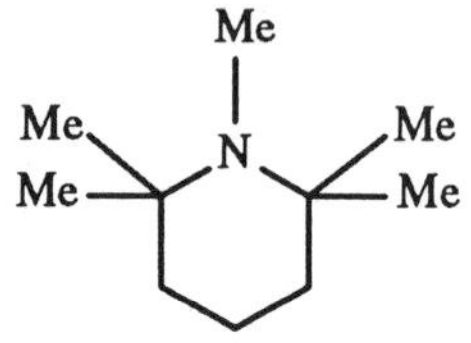 (393)

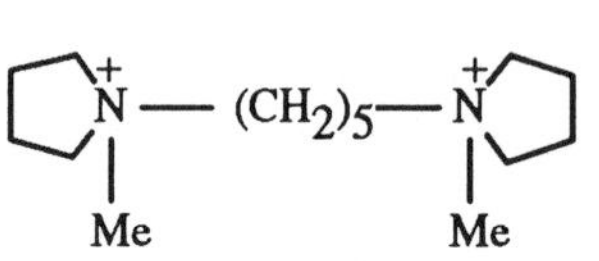

It was serendipity again which led to the introduction of selective ganglion blockers. One of a series of new cholinergic agents was found to be a monoamine oxidase inhibitor. Analogues were then prepared, including xylocholine (**394**). On investigation, this was found to have the unique property of blocking adrenergic neurones by preventing the release of noradrenaline at nerve endings. Following the reporting of this in 1956, a search was initiated for a similar compound devoid of the cholinergic activity present in xylocholine. Three years later, bretylium (**395**) was marketed.[241] It was soon challenged by guanethidine (**396**), a derivative of a potential chemotherapeutic agent that had unexpectedly been found to reduce blood pressure. Similar compounds were marketed in the mid-1960s, *viz.* bethanidine (**397**), debrisoquine (**398**), guanoclor (**399**) and guanoxan (**400**).

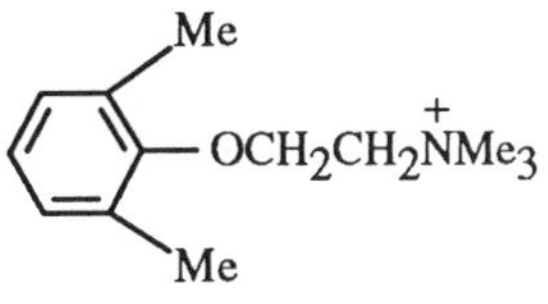

Me
$OCH_2CH_2\overset{+}{N}Me_3$
Me
(394)

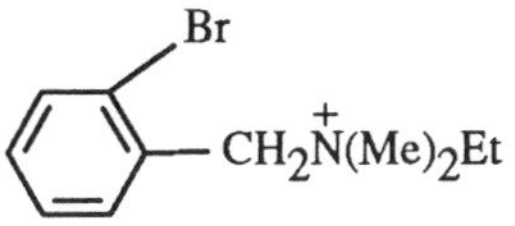

Br
$CH_2\overset{+}{N}(Me)_2Et$
(395)

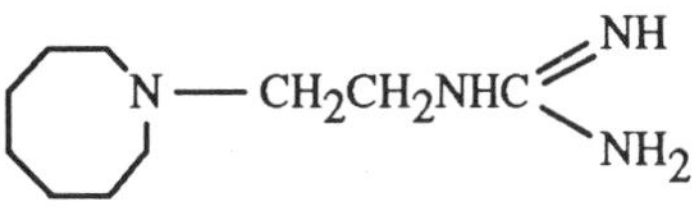

$N-CH_2CH_2NHC(=NH)NH_2$
(396)

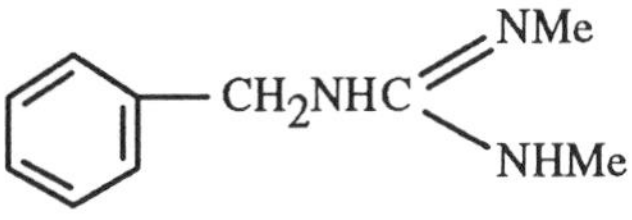

$CH_2NHC(=NMe)NHMe$
(397)

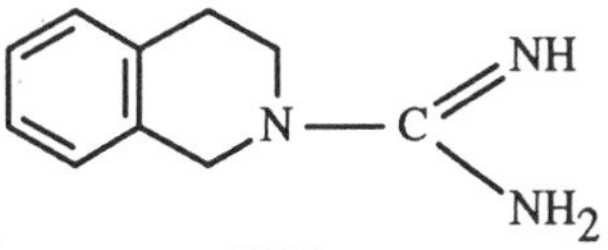

$N-C(=NH)NH_2$
(398)

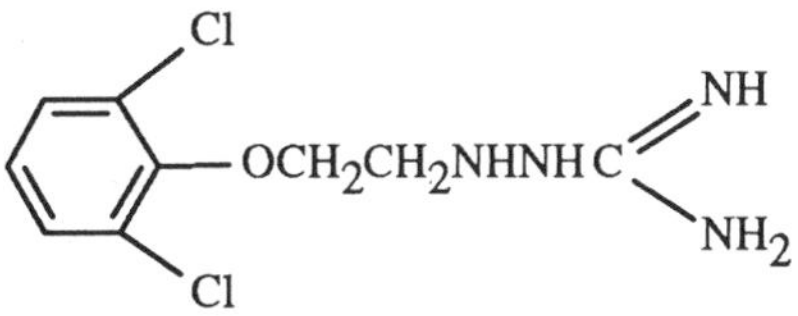

Cl
$OCH_2CH_2NHNHC(=NH)NH_2$
Cl
(399)

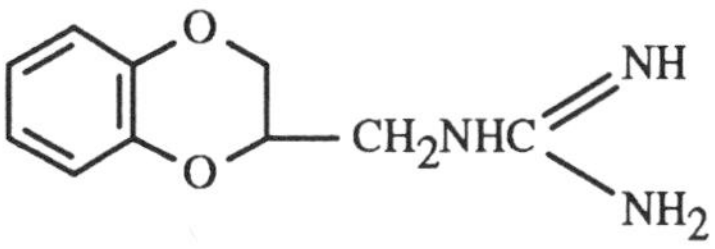

O
O
$CH_2NHC(=NH)NH_2$
(400)

1.2.7.5 Effects Observed During Toxicological Experiments

During the toxicological testing of drugs and chemicals, animals are exposed to unusually large doses. On several occasions, these doses have revealed hitherto unrecognized pharmacological properties of the compounds concerned.

1.2.7.5.1 Anaesthetics

Ethylene was investigated by Nunnely as a possible inhalational anaesthetic in 1849, but was discarded on the grounds that it had a narrow safety margin and would be difficult to manufacture. Its reinstatement provides a classic example of serendipity, beginning in 1908 with the discovery that the failure of carnations to open up when stored in greenhouses was attributable to the presence of ethylene in the gas used for illumination at night. It transpired that ethylene was toxic to a variety of plants, so it was tested on laboratory animals by Luckhardt and Thompson. Unaware of the studies conducted by Nunnely, they were surprised to observe the anaesthetic action of the gas when inhaled. Extensive animal and clinical tests established that ethylene induced aneasthesia remarkably smoothly, with an equally uneventful recovery. In 1923 it became the first new general anaesthetic to be introduced this century.

Investigations into propylene, a homologue of ethylene, indicated that it could also be a valuable anaesthetic if it were not for the formation of a toxic impurity (now considered to be a mixture of hexenes) when stored in pressurized steel cylinders. Lucas thus arranged for cyclopropane to be tested as it seemed likely to be the impurity. When kittens were exposed to a mixture of cyclopropane and oxygen in a bell jar, they quickly fell asleep and soon recovered without mishap! Yet again, serendipity had resulted in the discovery of a new anaesthetic. It was subjected to extensive studies before coming into general use in 1934. It retains its place in contemporary anaesthetic practice.

The availability of large amounts of a variety of steroid hormones by 1941 permitted Selye to study the consequences of administering large doses to animals. To his surprise, rats injected intraperitoneally were rendered unconscious by several of the compounds, only to recover uneventfully after a short period. Being aware that steroids were often inactivated in the liver, Selye then avoided this by injecting the hormones intravenously. The doses required to produce anaesthesia were now considerably reduced. The study was subsequently extended and confirmed that steroids without hormonal activity were more potent intravenous anaesthetics, the most active being pregnanedione. Its lack of water solubility was probably the main reason why Selye did not consider it should be investigated for clinical application. However, in 1955, this problem was overcome with the introduction by Laubach *et al.*[242] of hydroxydione sodium succinate (**401**). The rate of onset proved to be unacceptably slow, due to the requirement for the drug to be metabolized to form an active metabolite. Nevertheless, the wide safety margin of this type of drug encouraged others to seek alternatives. In 1969, a mixture of alphaxolone (**402**) and alphadolone acetate (**403**) was found to give excellent results in trials,[243] but several years of clinical experience pointed to an unacceptable incidence of allergic responses to the surfactant used in its formulation. It was withdrawn from the market along with the similarly formulated propanidid in 1984.

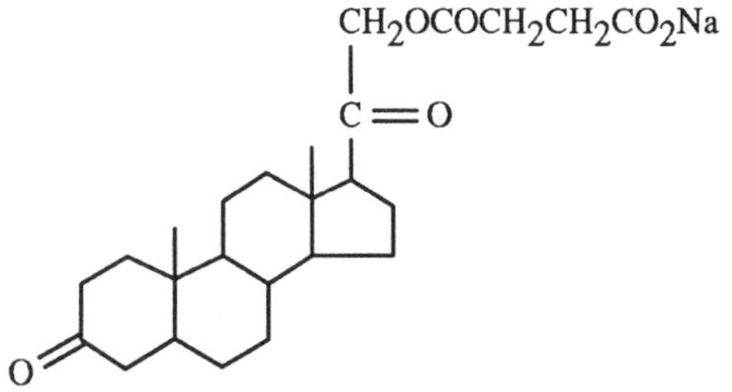

$CH_2OCOCH_2CH_2CO_2Na$ | C=O ; O

(**401**)

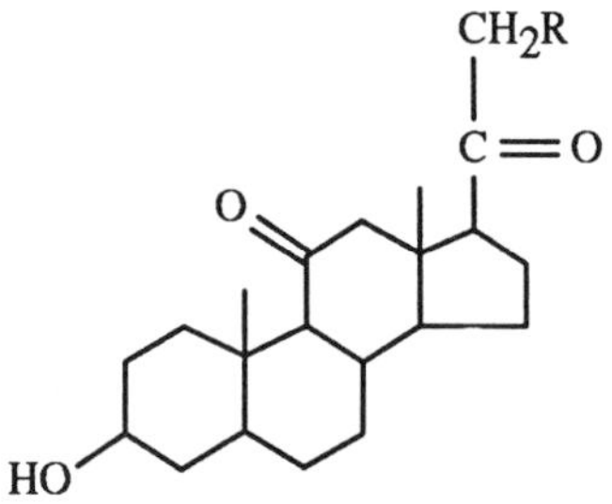

CH_2R | C=O ; O ; HO

(**402**) R = H
(**403**) R = OCOMe

1.2.7.5.2 *Antianxiety agents*[244]

In 1946, Berger reported that whilst carrying out toxicological studies on analogues of the antiseptic agent phenoxyethanol he observed a temporary flaccid paralysis of the limbs of mice. Smaller doses had a quietening effect on the demeanour of the animals. This he described as 'tranquillization'. Over 140 analogues were screened before mephenesin (**404**) emerged as a potential substitute for tubocurarine, but anaesthetists found its lack of consistency of action to be unacceptable. Nevertheless, its ability to ease symptoms of anxiety without clouding consciousness made it popular with patients in an era when the alternative was to administer small doses of hypnotics as sedatives. The principal drawback was that the drug was rapidly oxidized, necessitating frequent dosing. Finding superior activity in a related series of substituted 1,3-propanediols, Berger was able to prolong duration by blocking the sensitive alcohol functions by forming the carbamate esters. Meprobamate (**405**), synthesized in 1950 and marketed five years later, proved an outstanding success until the appearance of the benzodiazepines in the following decade. The related carisoprodol (**406**) was recommended for relief of muscular spasm due to injury.

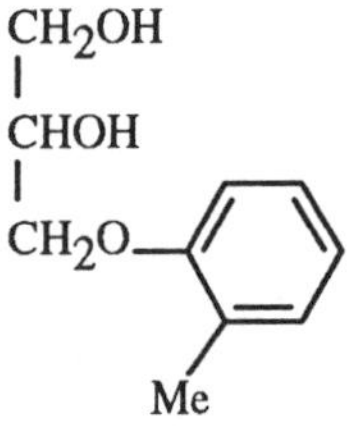

CH_2OH | CHOH | CH_2O– ; Me

(**404**)

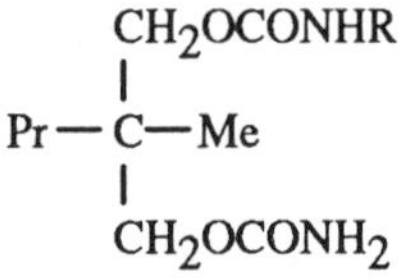

$CH_2OCONHR$ | Pr—C—Me | CH_2OCONH_2

(**405**) R = H
(**406**) R = Pr^i

1.2.7.5.3 *Antithyroid agents*

In 1941, toxicological studies on the recently developed sulfonamide antibacterial agent sulphaguanidine (**407**) revealed that it had a profound effect on the thyroid of rats. Several thiourea derivatives were found to do likewise. Two years later, Astwood reported[245] that thiouracil had emerged from the screening of over 100 compounds as the most potent inhibitor of thyroid function. It was used with success in two patients with hyperactive thyroids, but in a third it produced a severe blood dyscrasia known as agranulocytosis. By 1945, Anderson *et al.*[246] had introduced the much safer propylthiouracil (**408**). Methimazole (**409**) and carbimazole (**410**) were developed in 1949 and 1954 respectively, and were much more potent.

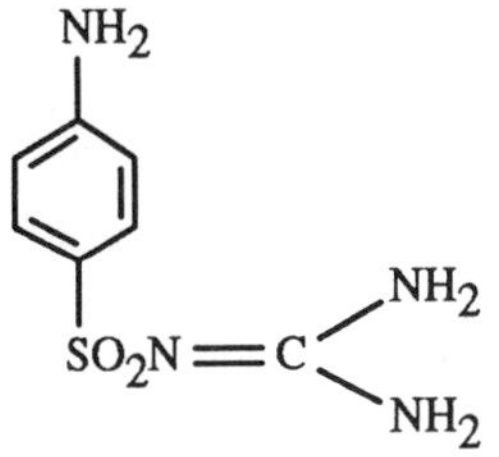

(**407**)

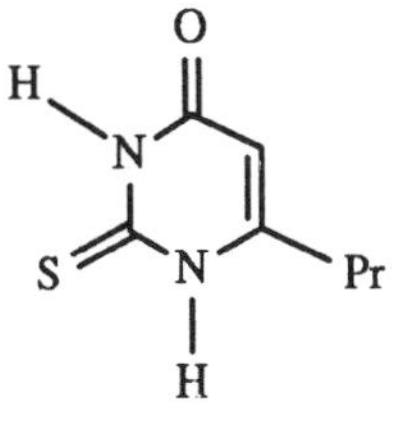

(**408**)

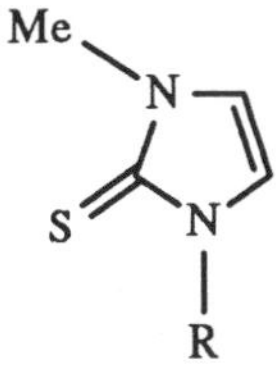

(**409**) R = H
(**410**) R = CO_2Et

1.2.7.5.4 Biological alkylating agents

In 1942, Goodman and Gilman initiated a study of the properties of the chemical warfare agents known as nitrogen mustards. They soon established that amongst the consequences arising from exposure to these noxious materials was not only blistering of the skin, oedema of the lung and blindness, but also serious systemic effects. The latter were seen to be directed at rapidly proliferating cells, especially the blood-forming elements of the bone marrow and lymphoid tissue. The significance of this was fully appreciated, and consequently Dougherty was asked to examine the effect of a nitrogen mustard on a mouse bearing a transplanted tumour. Such was the unprecedented response that an extensive investigation was initiated amidst full secrecy because of the wartime situation. Only after the war had ended was it disclosed that cancer patients had been receiving nitrogen mustard therapy since the autumn of 1942.[247] The most successful drug to emerge from this was mustine (**411**), one of the first to have been tried in animal tests. It had been of value in patients with Hodgkin's disease, lymphomas and chronic leukaemias.

In 1949, it was suggested that nitrogen mustards acted as bifunctional alkylating agents which cross-linked strands of DNA, thereby interfering with cell replication. This has led to the introduction of a variety of different types of cross-linking agents, *e.g.* tretamine (**412**), thiotepa (**413**), busulphan (**414**), ethoglucid (**415**) and treosulphan (**416**). Recognizing the need for greater specificity of action than could be afforded merely by relying on the sensitivity of rapidly-dividing cancer cells to undergo attack, Everett *et al.*[248] introduced chlorambucil (**417**). The polar nature of this zwitterionic compound restricted its ability to penetrate tissues indiscriminately, affording an agent which has consistently proved its worth in the chemotherapy of chronic lymphocytic leukaemia. A similar concept lay behind the design of melphalan (**418**) by Bergel and Stock[249] in 1954. This time, the passage of the drug through membranes was dependent upon the presence of active transport mechanisms for the amino acid phenylalanine. Melphalan has been of considerable value in treating myelomas. Other approaches to the design of selective biological alkylating agents have attempted to incorporate carrier moieties from which the drug separates at or near the intended site of action.

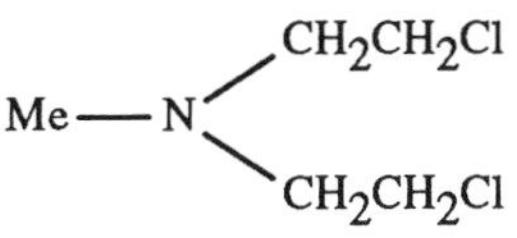

(**411**)

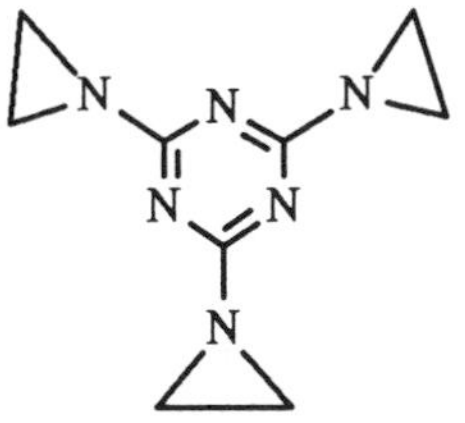

(**412**)

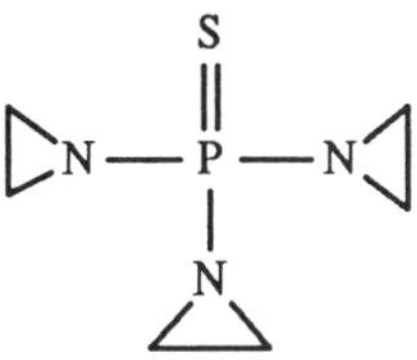

(**413**)

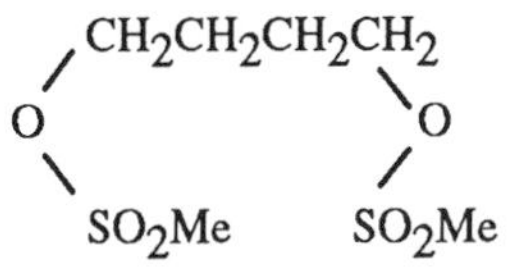

(**414**)

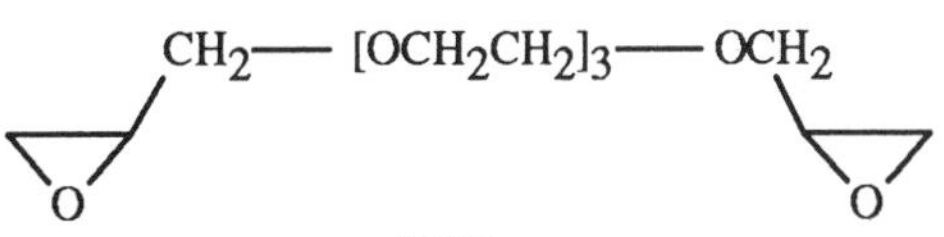

(**415**)

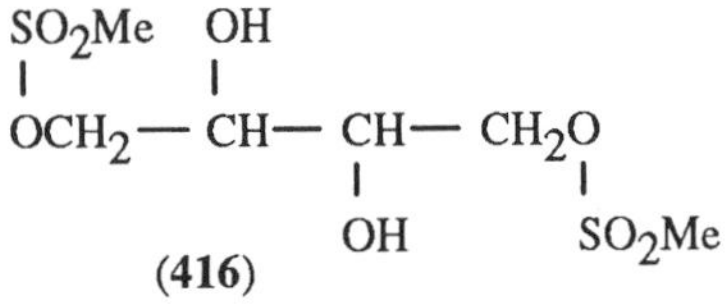

(416) 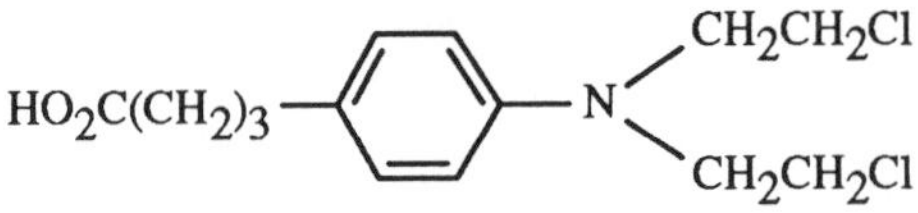 **(417)** 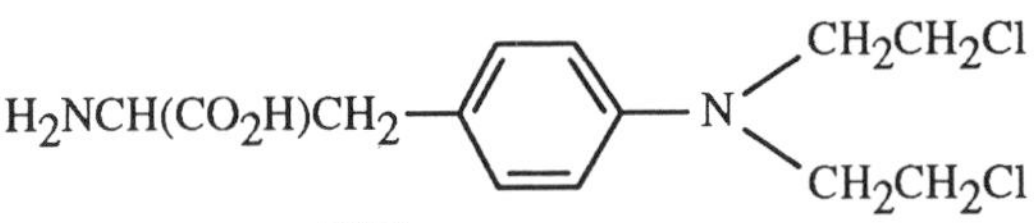 **(418)** 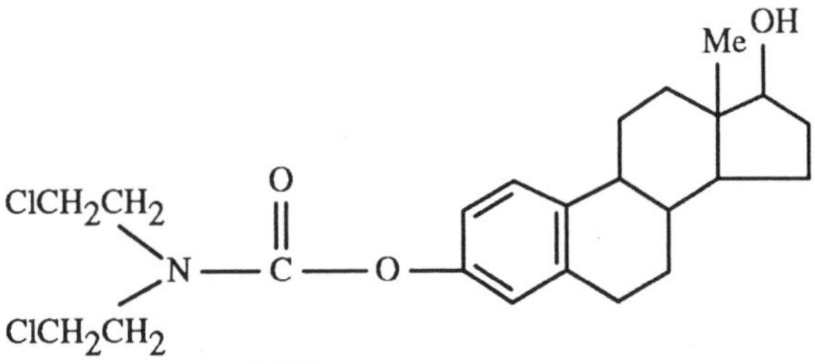 **(419)** 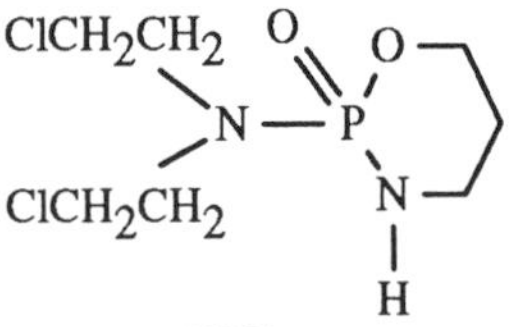 **(420)**

This has not proved particularly successful, but the oestradiol derivative estramustine (**419**) has been used in cancer of the prostate. The most successful of the nitrogen mustards, cyclophosphamide (**420**), was introduced in 1956 by Arnold *et al.*[250] It was intended to be selectively activated by enzymes secreted by prostatic tumours, but it was actually oxidized by liver enzymes to form a chemically unstable metabolite which released normustine.

1.2.8 DRUG PROTOTYPES DEVELOPED THROUGH SCREENING PROGRAMMES

Many examples of valuable therapeutic agents which were discovered by screening analogues of prototype drugs have been cited in the preceding pages. Systematic screening programmes have been employed principally for this purpose. The question that must still be answered is whether or not drug prototypes have been found by speculatively screening large numbers of compounds. The answer is a qualified yes, for screening has only occasionally proved to be an effective means of generating novel drug prototypes other than antibiotics isolated from soil samples. There are a variety of reasons why this is not always appreciated. Firstly, it has frequently been necessary to have screened vast numbers of compounds before finding analogues with superior activity to the prototype, a feature which can easily give the mistaken impression that screening was random. A second reason for overestimating the value of screening as a means of generating lead compounds is that prototypes have occasionally been discovered whilst testing drugs for completely unrelated activities. Such cases must be considered as examples of serendipity. Another reason is that a few of the early successes of modern drug research arose from the screening of dyes for potential chemotherapeutic activity. Also, the hypnotic action of the now obsolete sulphonal (**421**) was revealed in 1887 when a test dose was injected into a dog simply to see what would happen. The commercial success of this drug may have encouraged hope that many more might be discovered simply by injecting novel chemicals into animals!

1.2.8.1 Azo Dyes

In 1903, Ehrlich screened over 100 diverse dyes to determine their capacity to bring about the disappearance of trypanosomes from the blood of infected mice. This led him to select a benzopurpurin azo dye for further investigation. He named the dye nagana red (**422**) in view of its activity against *Trypanosoma brucei*, the protozoan which caused nagana in cattle.

Ehrlich had chosen to screen dyes because of his prior experience with methylene blue (**423**). In 1888, he had tested it on patients in the hope that it might act as an analgesic by interfering with nerve conduction; he knew it to have a high affinity for nerves. The results did not match his expectations, but three years later he cured two patients suffering from a mild form of malaria by the oral administration of methylene blue. This followed his discovery that it stained specimens of the plasmodia which caused the disease. It was the first time a synthetic chemical had cured a specific

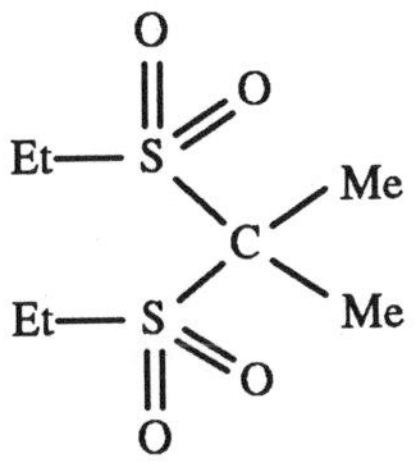

(421)

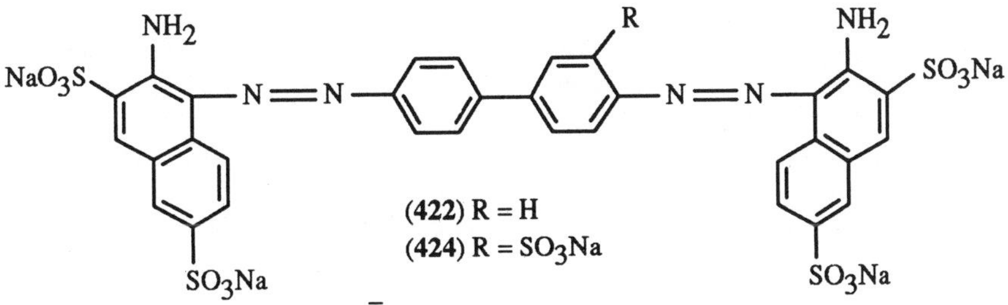

(422) R = H
(424) R = SO_3Na

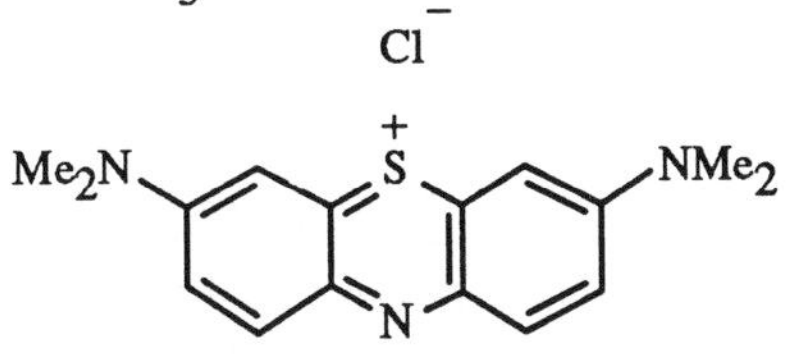

(423)

disease, so it is hardly surprising that when Ehrlich had the opportunity to seek a cure for trypanosomiasis he chose to screen dyes.

Ehrlich sought a more soluble analogue of nagana red which would enhance its ability to enter the circulation after subcutaneous injection. He found that trypan red (**424**) suited his purpose, producing such greatly improved results that it was even tested clinically. Unfortunately, in order to effect a cure for sleeping sickness the dose that was required was likely to cause blindness in many patients. However, the drug subsequently proved to be of value in veterinary practice, where it was used as an antiprotozoal agent for many years. It can be seen that the introduction of trypan red was a landmark in the history of chemotherapy since it represented the first successful attempt to design a molecule that would enhance antiprotozoal activity.

Ehrlich examined another 50 analogues of trypan red, but the analogue which eventually proved to be a major success in both preventing and treating sleeping sickness was not discovered until after his death. This was a colourless compound called suramin (**425**), introduced clinically around 1920 after more than a thousand similar compounds had been synthesized and screened in infected mice by Ehrlich's former associate Roehl.[251] It still remains an important drug in the prevention and treatment of sleeping sickness.

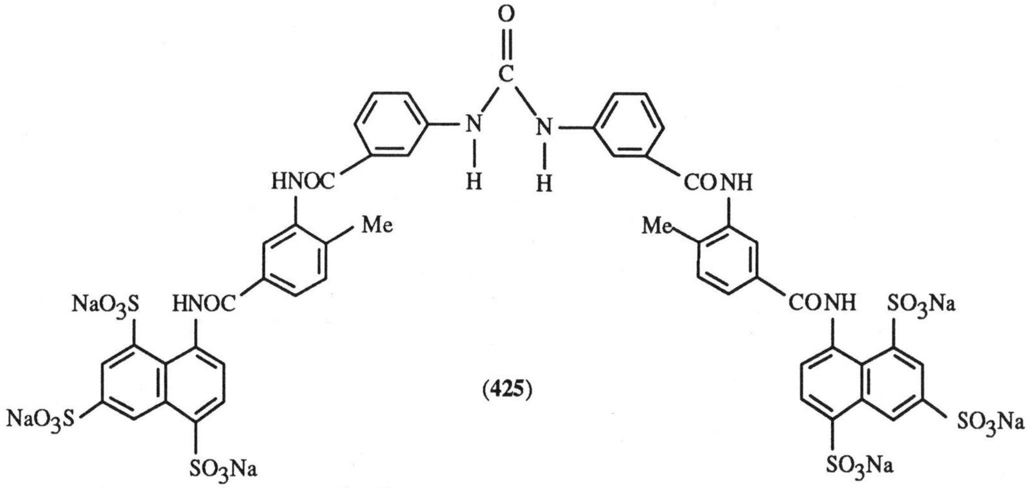

(425)

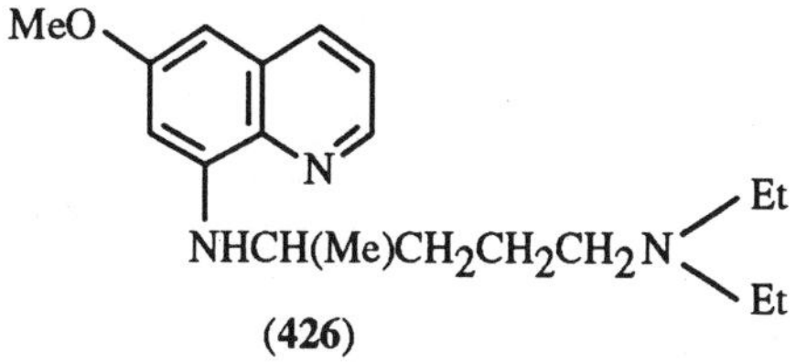

(426)

The success of suramin encouraged Schulemann and his colleagues to screen derivatives of methylene blue as potential antimalarials, using canaries infected with plasmodia.[252] Having established that activity was enhanced if a diethylaminoethyl side chain was inserted in place of one of the amine methyl substituents, they then examined alternative heterocyclic ring systems. An 8-aminoquinoline derivative gave better results, but many more compounds had to be synthesized before pamaquin (**426**) emerged as an effective substitute for quinine. As it features part of the quinine structure, it could possibly be considered as having been derived from the natural product rather than from methylene blue.

1.2.8.2 Acridine Dyes[253]

In 1913, Browning[254] screened a variety of dyes for their ability to kill bacteria. Trypaflavine, a dye found by Ehrlich to have trypanocidal activity, emerged from the screen as being effective against a wide range of bacteria. Further investigation revealed it to share with Morgenroth's ethylhydrocupreine, the unique property of being effective against a variety of pathogens even in the presence of serum. This was fully exploited by Browning and his colleagues[255] during the First World War when the drug was renamed acriflavine and saw extensive use in preventing wound sepsis. It was not until 1934 that acriflavine was found to be a mixture of (**427**) and (**428**). The more active component (**427**) was introduced as proflavine. During the Second World War, a non-staining analogue was developed, *viz.* aminacrine (**429**). These compounds were intended solely for topical application.

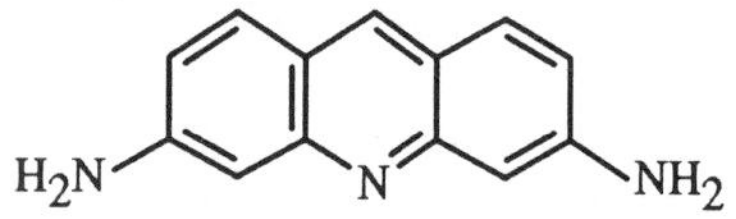

(427) 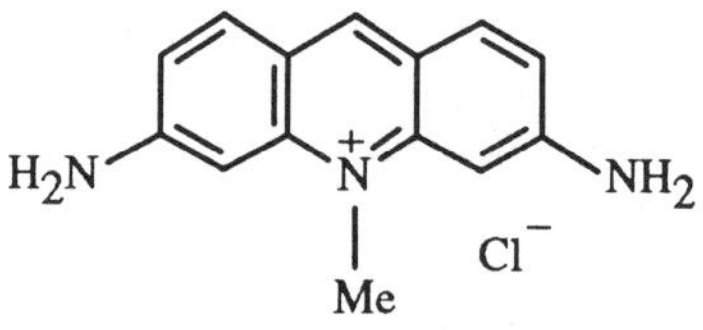 **(428)** 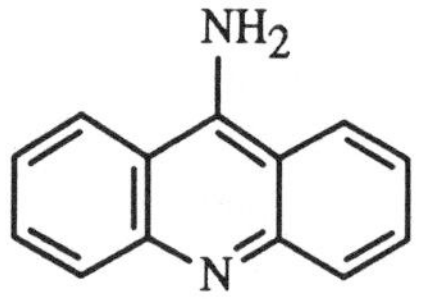 **(429)**

Morgenroth and his colleagues[256] began their search for an acridine derivative that was safe enough to be injected by incorporating part of the ethylhydrocupreine structure to form ethacridine (**430**), which was introduced in 1920. It proved to have similar clinical properties to acriflavine, being unable to cure generalized infections. Schnitzer[257] then examined hundreds of analogues of ethacridine during the next seven to eight years. These were routinely screened for trypanocidal as well as antibacterial activity, and from 1924 onwards also in canaries infected with malaria. Several nitroacridines were found to control streptococcal infections in mice, but were not potent enough for clinical application. In 1928, however, one was considered to be promising enough to be marketed under the proprietary name of 'Entozon' (**431**). Unfortunately, it turned out to be inadequate for therapeutic purposes and also caused unacceptable irritation at the site of i njection. The search for an antimalarial acridine was much more successful, resulting in the discovery by Kikuth, around 1930, of high activity in mepacrine (**432**). As many as 12 000 compounds may have been synthesized by Mietzch and Mauss before it emerged as having outstanding activity. This remarkable achievement by a small group of researchers working in the German pharmaceutical industry was followed by still further screening of analogues to provide chloroquine in 1934 (**433**). Its outstanding value was not recognized at that time. During the Second World War, over 15 000 more compounds were screened in the UK and USA during a collaborative programme of research involving many universities and industrial companies. This resulted in the introduction of amodiaquine (**434**) and primaquine (**435**), as well as the recognition of the true value of chloroquine.

Many of the compounds synthesized in the programme of research that led to the introduction of mepacrine by the German pharmaceutical industry were subsequently screened in mice infected with schistosomiasis. This resulted in the introduction of lucanthone (**436**) by Mauss, and it was used in Africa during the war. Its active metabolite was marketed as hycanthone (**437**). Analogues of these compounds include the highly successful and relatively non-toxic oxamniquine (**438**),[258] discovered in 1968.

Despite the failure of nitroacridines to cure bacterial infections in patients, the quest for antibacterial agents was continued in Germany by Domagk and his colleagues. Recognizing that they had failed in the clinic despite their apparent effectiveness in mice, he introduced more rigorous

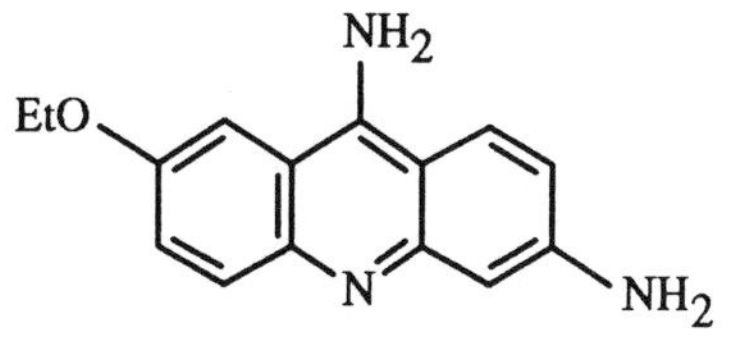

(430) 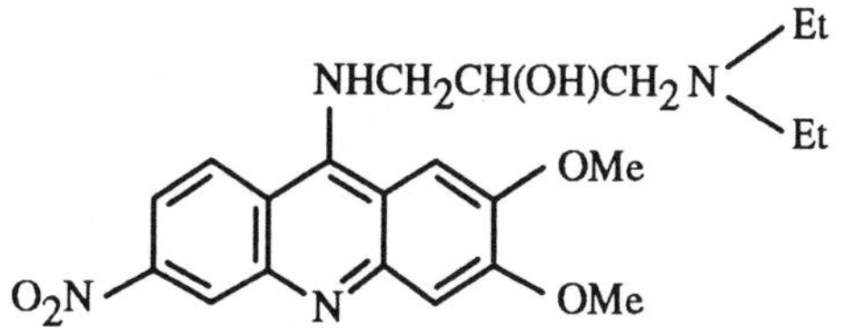 **(431)**

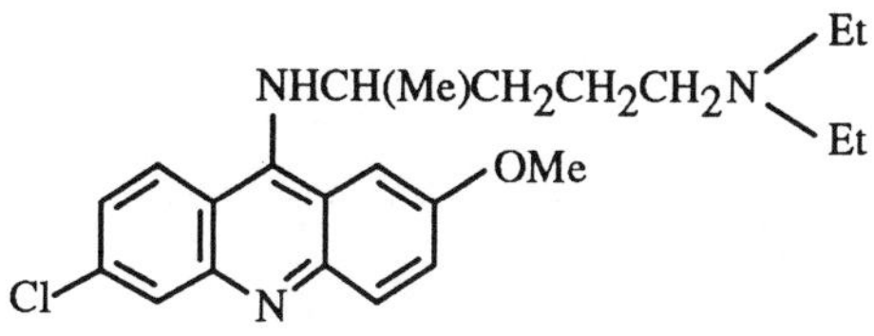

(432) 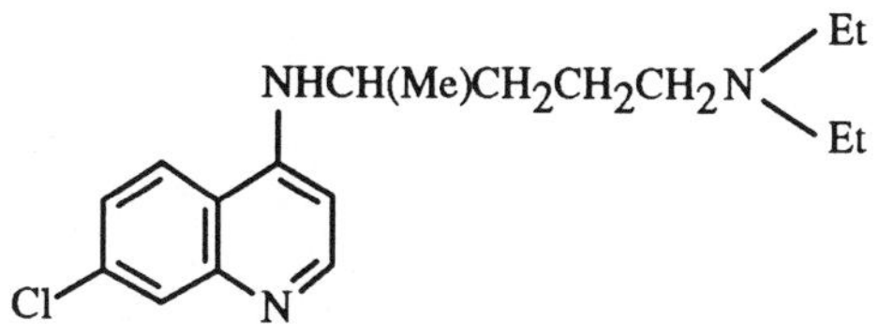 **(433)**

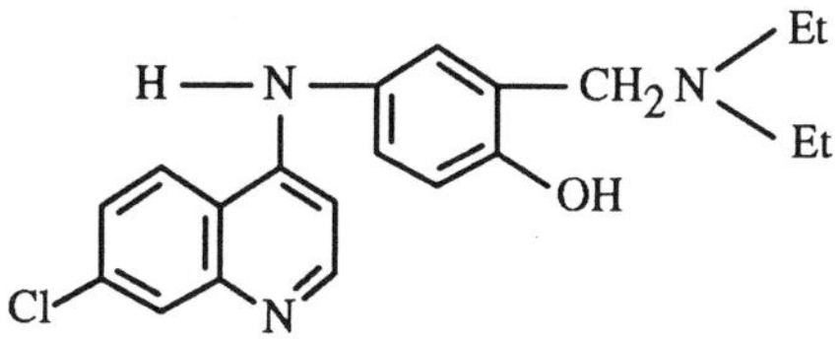

(434) 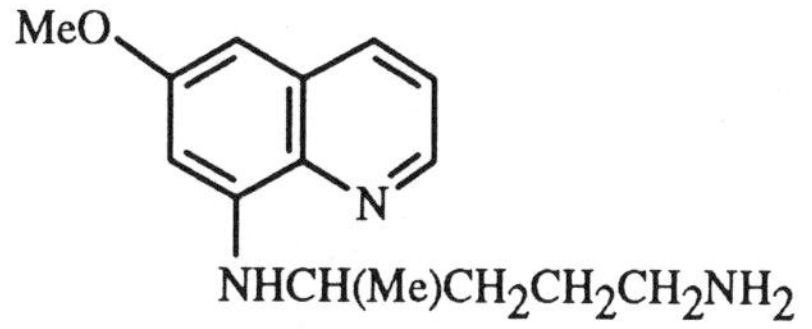 **(435)**

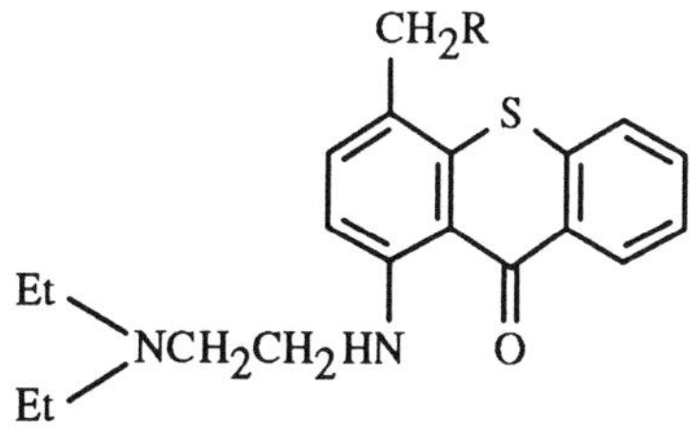

(436) R = H
(437) R = OH

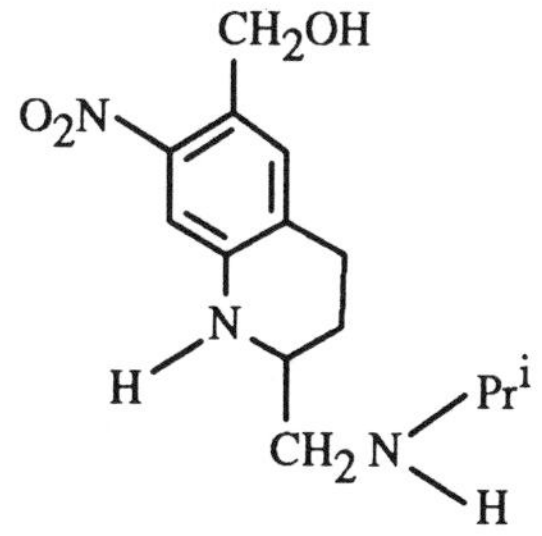

(438)

testing protocols, including the use of a particularly virulent strain of haemolytic streptococcus. As well as examining acridines, Domagk screened a series of azo dyes containing side chains which had previously been found to confer antistreptococcal activity when present on acridines. One of these (**439**) had strong activity against cultures of streptococci, but failed to cure infected mice. However, when a sulfonamide group was introduced into this compound, the first truly promising results were obtained. Mietzsch and Klarer then synthesized a vast range of analogues for Domagk to test, resulting in the discovery of outstanding activity in sulphamidochrysoidine (**440**),[259] for which a patent was sought in 1932. Clinical results were unprecedented, with patients being cured of life-threatening streptococcal and staphylococcal infections. The drug was not put on the market until 1935,[260] and later in that year workers in France disclosed that the antibacterial activity was entirely due to the release of sulphanilamide (**441**) *in vivo*. This opened the door for pharmaceutical companies to synthesize their own analogues of sulphanilamide. Amongst the earliest was sulphapyridine (**442**), synthesized in 1937 and shown the following year to have a wider spectrum of antibacterial activity than sulphamidochrysoidine and, in particular, to be highly effective against lobar pneumonia.[261] Until then, this had been a common cause of death in the elderly and also otherwise healthy people.

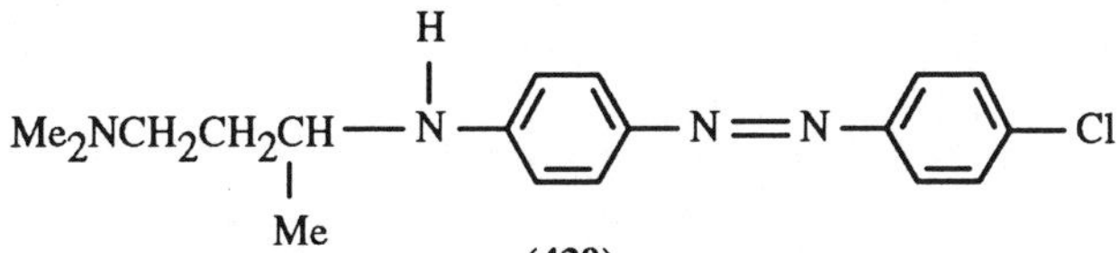

(439)

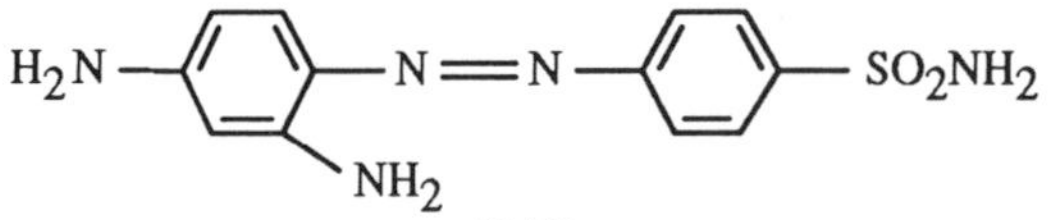

(440)

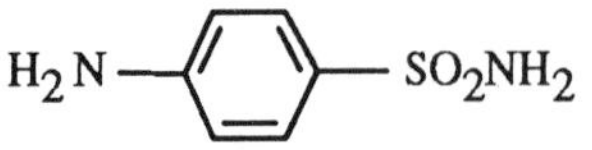

(441)

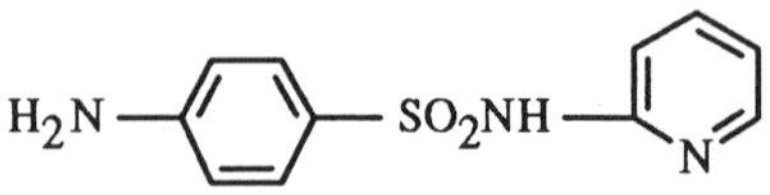

(442)

There can be no doubt that the remarkable success of sulphapyridine and other sulfonamides altered thinking in medical circles with regard to the possibility of treating bacterial infections by drug therapy. It was this change in the climate of opinion which led Florey to give serious consideration to the possible use of penicillin as a systemic antibacterial, whereas only 10 years earlier Fleming had merely assessed it as an antiseptic for topical application.

Numerous antibacterial sulfonamides have been marketed since sulphapyridine was introduced. It is a sobering thought that they, together with the related antileprotic sulfones and several antituberculous agents, retain a unique position in therapeutics as synthetic antibacterial drugs of proven clinical value in systemic infections. The sulfones were originally examined as analogues of sulphanilamide and found to be more potent. Unfortunately, they were also much more toxic. Various analogues were synthesized in an attempt to reduce this toxicity, amongst which was glucosulphone (**443**). Screening against *Mycobacterium leprae* in 1941 revealed it to have antileprotic activity. During the remainder of the decade, clinical trials confirmed its value and that of the parent compound, dapsone (**444**), as well as of solapsone (**445**).

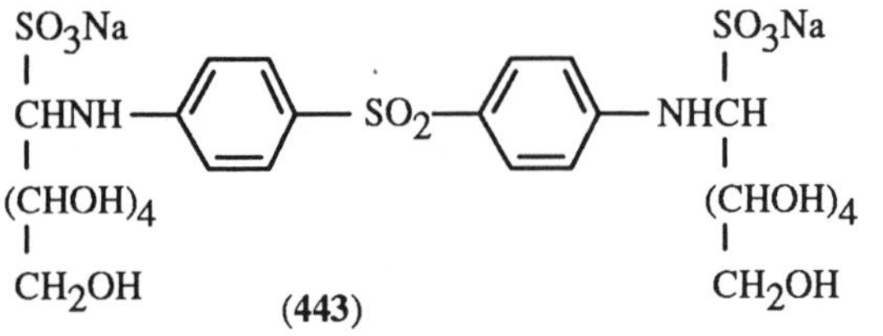

(443)

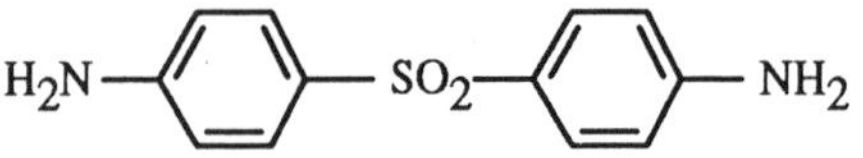

(444)

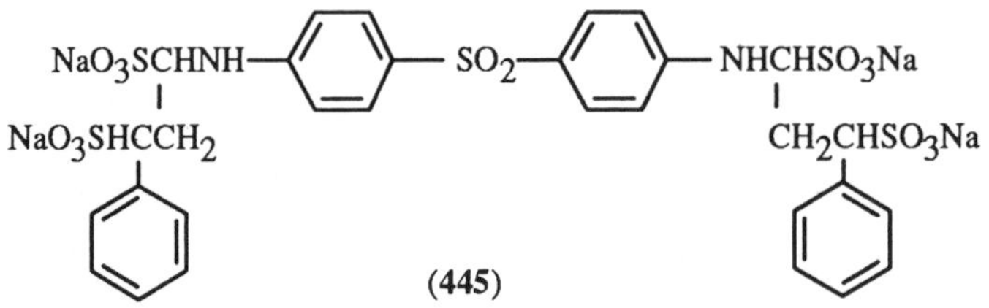

(445)

1.2.8.3 Antituberculous Drugs

By the early 1940s, Domagk was screening compounds against an ever widening range of organisms. When supplied with a series of novel chemical intermediates used in the synthesis of potential sulfonamide antibacterials, he found them to possess stronger antituberculous activity than did any sulfonamide. More analogues were then synthesized, leading to the selection of thiacetazone (**446**) for clinical trials during an epidemic of tuberculosis at the end of the war.[262] Despite early enthusiasm for the drug, it eventually became clear that it was unacceptably toxic to the liver. Bernstein and his colleagues then set up their own programme and screened some 5000 analogues of thiacetazone. In 1951, they detected strong activity once more in a synthetic intermediate, *viz.* isonicotinic acid hydrazide (**447**). It proved to be 15 times as potent as streptomycin! After clinical trials revealed it to be superior to any other antituberculous drug, it was introduced early in the following year under the name isoniazid. A few days before this, however, the same compound was reported to be an outstanding antituberculous drug by Schnitzer and Fox, and Domagk did likewise shortly after. Their independent discoveries had been based on following up a report that the B complex vitamin nicotinamide had mild tuberculostatic activity. The isoniazid analogue pyrazinamide (**448**) is also used clinically. Another extensive screening programme revealed antituberculous activity in *N,N'*-diisopropylethylenediamine. When this was shown to be somewhat toxic, many more analogues were examined. The outcome of this was the introduction of ethambutol (**449**) in 1961.

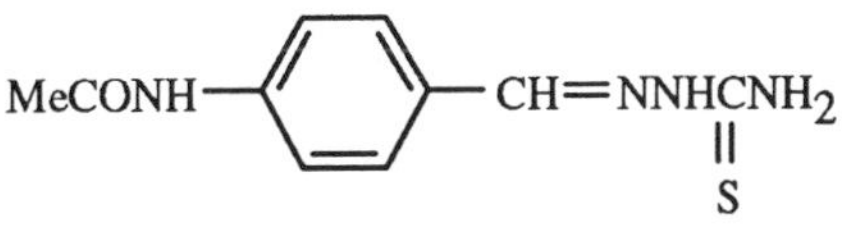

MeCONH—⟨benzene ring⟩—CH=NNHCNH$_2$ (C=S)

(**446**)

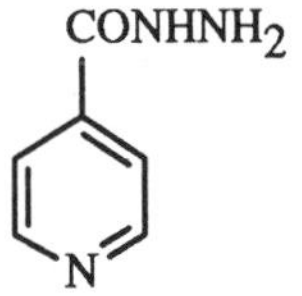

CONHNH$_2$ (pyridine, N)

(**447**)

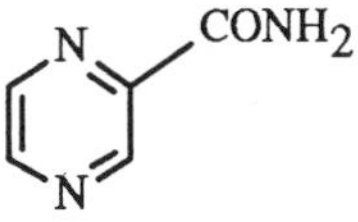

CONH$_2$ (pyrazine, N, N)

(**448**)

CH$_2$OH
|
CH$_2$NHCHEt
|
CH$_2$NHCHEt
|
CH$_2$OH

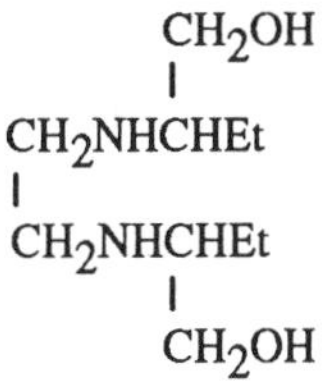

(**449**)

1.2.8.4 Anticancer Agents

Reference has been made in Section 1.2.7.3.7 to the introduction of antifolates in the treatment of leukaemia. These drugs belong to the group of therapeutic agents described as antimetabolites, so named because their close structural similarity to natural metabolites enables them to enter into competition with them at the active site of enzymes. Those which have proved effective in cancer chemotherapy are usually capable of binding extremely firmly to the active site of one of the key enzymes involved in purine or pyrimidine syntheses. Although many compounds are effective as antimetabolites *in vitro*, only a handful have fulfilled their promise in the clinic.

The antifolate activity of the drugs supplied to Farber for evaluation in leukaemic patients had been developed by screening them for their ability to inhibit the growth of *Lactobacillus casei*, an organism dependent on exogenous supplies of folic acid in the culture medium. Following the success of the antifolates, a range of purines were submitted to a similar screening procedure by Hitchings and his colleagues. This soon led to the clinical testing of diaminopurine (**450**) in leukaemic patients. Although remissions were produced in a few cases, the drug was clearly inferior to methotrexate. However, further screening revealed promising activity in a compound which had been prepared as a key intermediate for the synthesis of more aminopurines. This was mercaptopurine (**451**). When tested in the clinic, it had similar activity to methotrexate in children with acute leukaemia.[263] Hundreds of analogues of mercaptopurine were subsequently examined for antileukaemic activity. One of the inactive ones was shown to be a strong inhibitor of xanthine oxidase. It was tested successfully by Rundles[264] in the treatment of gout, where it blocked xanthine oxidase to interfere with the formation of uric acid. It was marketed for this purpose in 1966, with the approved name of allopurinol (**452**).

In 1961, azathioprine (**453**), an analogue of mercaptopurine, emerged from a screening programme as the most promising immunosuppressant examined. The screening was again conducted by Hitchings and his colleagues, involving an assessment of the ability of different drugs to interfere with the hemagglutinin reaction in mice challenged with foreign blood cells. The introduction of azathioprine as an immunosuppressant with superior activity to cortisone represented an important advance in human organ transplantation.

The success of methotrexate and mercaptopurine did much to stimulate investigations into other antimetabolites. Vast numbers of these have been synthesized, but only a few have been safe enough to permit their use even against cancer. Fluorouracil[265] (**454**) was reported to be active against transplanted tumours in 1957, and subsequently demonstrated clinical value against several types of tumours. Its analogue flucytosine (**455**) is effective in the treatment of systemic yeast infections. Idoxuridine (**456**), another halogenated pyrimidine, was introduced in 1959 for the topical treatment of herpes infections. It was too toxic for systemic use, unlike the purine antimetabolite acyclovir (**457**). This was introduced in 1977 by Elion and her colleagues[266] after they found that arabinosides of diaminopurine and of guanine were active against DNA viruses. Acyclic analogues showed the greatest activity. Arabinosides had previously been shown to behave as antimetabolites, with cytarabine (**458**) being introduced in the chemotherapy of leukaemia in the early 1960s.

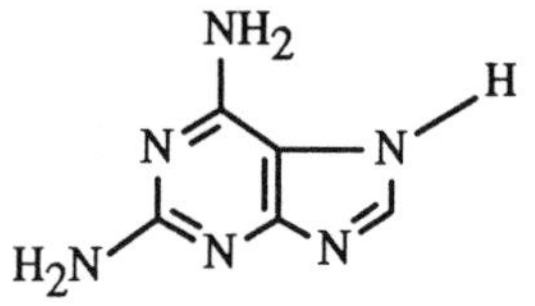

(450) 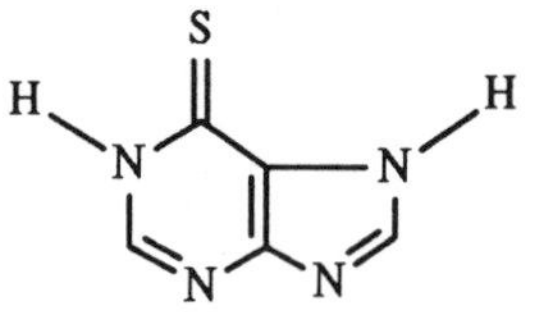 (451) 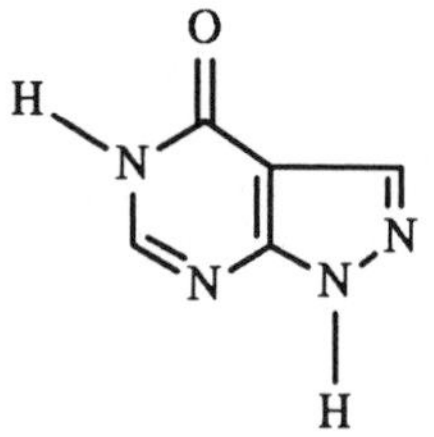 (452)

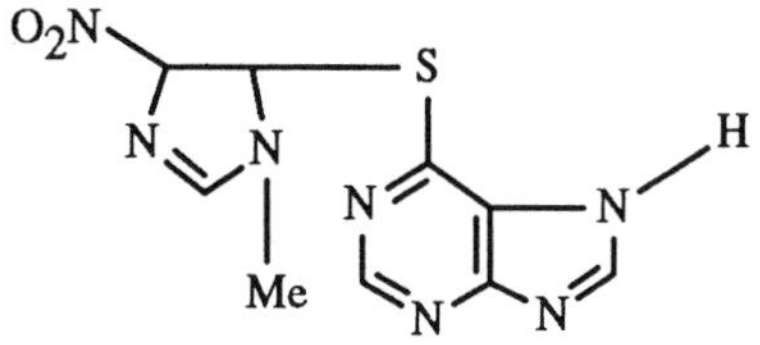

(453) 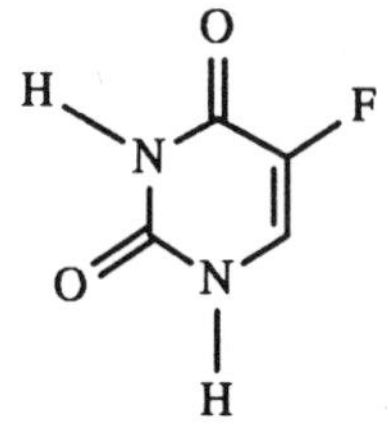 (454) 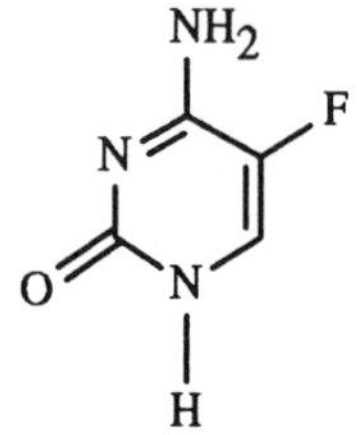 (455)

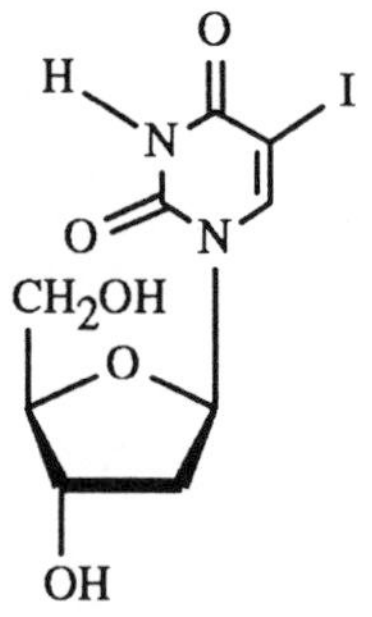

(456) 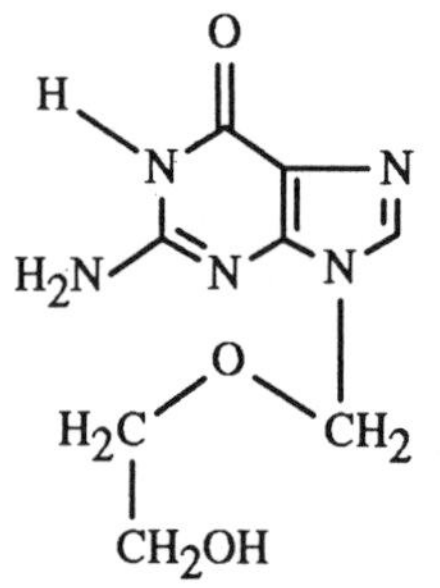 (457) 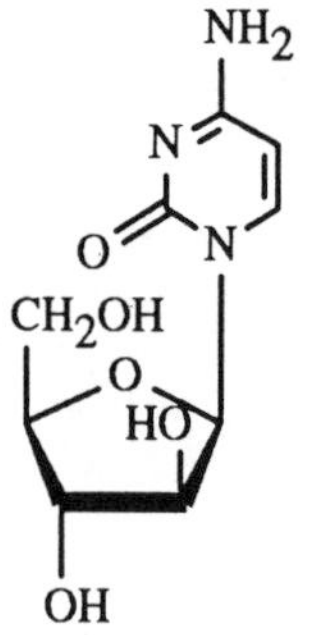 (458)

In 1955, the US Congress allocated millions of dollars to enable the Cancer Chemotherapy National Service Center to establish a major screening programme. Within three or four years, over 1000 compounds were being tested each month. Many of these were natural products, but it was an intermediate used in organic synthesis which was to provide a new drug prototype. The compound was 1-methyl-1-nitroso-3-nitroguanidine (**459**), which was found in 1959 to possess antileukaemic activity when tested in mice.[267] Its analogues carmustine (**460**) and lomustine (**461**) have proved to be valuable cancer chemotherapeutic agents. However, the programme to seek anticancer drugs of clinical value from plants was eventually abandoned after 20 years.

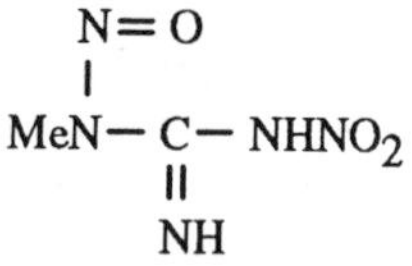

(459) 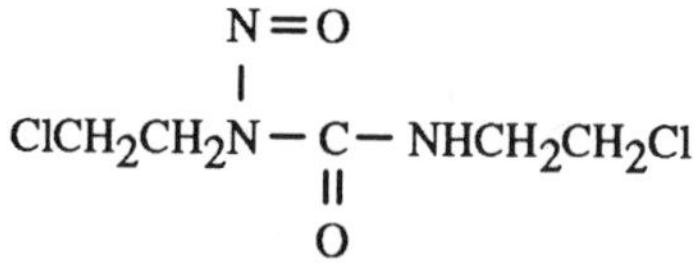 (460) 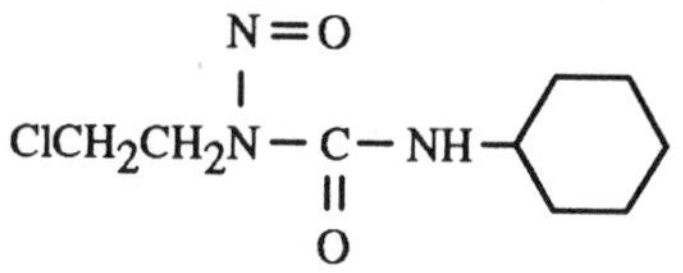 (461)

1.2.9 REFERENCES

1. W. Sneader, 'Drug Discovery: the Evolution of Modern Medicines', Wiley, Chichester, 1985.
2. N. Taylor, 'Plant Drugs that Changed the World', Dodd Mead, New York, 1965.
3. M. B. Kreig, 'Green Medicine', Harrap, London, 1964.
4. W. Sneader, 'Drug Development: from Laboratory to Clinic', Wiley, Chichester, 1986, p. 2.
5. J. Lind, 'Treatise on Scurvy', ed. C. P. Stewart and D. Guthrie, Edinburgh University Press, Edinburgh, 1953.
6. J. K. Aronson, 'An Account of the Foxglove and Its Medical Uses 1785–1985', Oxford University Press, Oxford, 1985.
7. J. Pringle, *Philos. Trans. R. Soc. London*, 1750, **46**, 480–488, 525–534, 550–558.
8. G. Sharp, *Med. Mag. (London)*, 1909, **18**, 568.
9. E. F. Hartung, *Ann. Rheum. Dis.*, 1954, **13**, 190.
10. W. Sneader, *Drug News Perspect.*, 1988, **1**, 185.

11. E. H. Ackernecht, 'Therapeutics from the Primitives to the 20th Century', Macmillan, London, 1973.
12. B. Ebbel, 'The Ebers Papyrus the Greatest Egyptian Medical Document', Levin and Munksgaard, Copenhagen, 1937.
13. C. D. Leake, 'The Old Egyptian Medical Papyri', University of Kansas Press, Lawrence, KS, 1952.
14. J. Stannard, *Bull. Hist. Med.*, 1961, **35**, 497.
15. R. T. Gunther, 'The Greek Herbal of Dioscorides' (translated J. Goodyar), Haffner, New York, 1968.
16. J. M. Riddle, in 'Dictionary of Scientific Biography', ed. C. C. Gillispie, Scribner, New York, 1971, vol. 4, p. 119.
17. C. Singer, *J. Hellenic Studies*, 1927, **47**, 1–52.
18. F. Kudlien and L. G. Wilson, in 'Dictionary of Scientific Biography', ed. C. C. Gillispie, Scribner, New York, 1972, vol. 5, p. 227.
19. S. Hamarneh, *Physis*, 1972, **14**, 5.
20. M. Levy, 'Early Arabic Pharmacology', Brill, Leiden, 1973.
21. W. Sneader, *Drug News Perspect.*, 1988, **1**, 317.
22. W. Pagel, 'Dictionary of Scientific Biography', ed. C. C. Gillispie, Scribner, New York, 1974, vol. 10, p. 304.
23. W. Sneader, *Drug News Perspect.*, 1989, **2**, 62.
24. G. Urdgang, 'The Apothecary Chemist Carl Wilhelm Scheele', American Institute for the History of Pharmacy, Madison, 1942.
25. J. Lesch, *Hist. Stud. Phys. Sci.*, 1981, **11**, 305.
26. H. M. Wuest, *Chem. Ind. (London)*, 1937, 1084.
27. A. D. Wright, *Med. Hist.*, 1968, **18**, 62.
28. H. Coenen, *Arch. Pharm. (Weinheim, Ger.)*, 1954, **287**, 165.
29. P. J. Hanzlik, *J. Am. Pharm. Assoc.*, 1929, **18**, 375.
30. G. Lockemann, *J. Chem. Educ.*, 1951, **28**, 277.
31. P. J. Robiquet, *Ann. Chim.*, 1832, **51**, 259.
32. H. How, *Quart. J. Chem. Soc.*, 1853, **6**, 125.
33. A. Crum Brown and T. R. Fraser, *Trans. R. Soc. Edinburgh.*, 1868, **25**, 151, 693.
34. C. R. A. Wright, *J. Chem. Soc.*, 1874, **27**, 1031.
35. H. Dreser, *Dtsch. Med. Wochenschr.*, 1898, **24**, 185.
36. M. Freund and E. Speyer, *J. Prakt. Chem.*, 1916, **94**, 135.
37. L. Small, H. M. Fitch and W. E. Smith, *J. Am. Chem. Soc.*, 1936, **58**, 1457.
38. J. M. Gulland and R. Robinson, *J. Chem. Soc.*, 1923, **123**, 980.
39. R. Grewe, *Angew. Chem.*, 1947, **59**, 194.
40. N. B. Eddy and E. L. May, in 'Narcotic Antagonists', ed. M. C. Braude *et al.*, Raven Press, New York, 1974, p. 9.
41. S. Archer, N. F. Albertson, L. S. Harris, A. K. Pierson and J. G. Bird, *J. Med. Chem.*, 1964, **7**, 123.
42. I. J. Pachter and Z. Matossian (Endo Labs.), *US Pat.* 3 393 197 (1968) (*Chem Abstr.*, 1968, **69**, 87 282q).
43. I. Monkovitch, T. T. Conway, H. Wong, Y. G. Perron, I. J. Pachter and B. Belleau, *J. Am. Chem. Soc.*, 1973, **95**, 7910.
44. K. W. Bentley, in 'The Alkaloids', ed. R. F. Manske, Academic Press, New York, 1971, vol. 13, p. 75.
45. D. I. Macht, *Ter. Arkh.*, 1916, **17**, 786.
46. C. Mannich and O. Walther, *Arch. Pharm. (Weinheim, Ger.)*, 1927, **265**, 1.
47. W. Buth, F. Kulz, and K. W. Rosenmund, *Chem. Ber.*, 1939, **72**, 19.
48. F. F. Blicke and E. Monroe, *J. Am. Chem. Soc.*, 1939, **61**, 91; F. F. Blicke and F. B. Zienty, *J. Am. Chem. Soc.*, 1939, **61**, 93.
49. A. Berman, in 'Dictionary of Scientific Biography', ed. C. C. Gillispie, Scribner, New York, 1974, vol. 10, p. 497.
50. M. D. Grmek, 'Dictionary of Scientific Biography', C. C. Gillispie, Scribner, New York, 1974, vol. 9, p. 6.
51. J. M. D. Olmsted, 'Francois Magendie', Schuman, New York, 1944.
52. B. Anft, *J. Chem. Educ.*, 1955, **32**, 566.
53. M. Delepine (translated R. E. Oesper), *J. Chem. Educ.*, 1951, **28**, 454.
54. M. L. Duran-Reynals, 'The Fever Bark Tree. The Pageant of Quinine', Doubleday, New York, 1946.
55. Henry and Delondre, *J. Pharm.*, 1833, **19**, 623.
56. W. Filehne, *Berlin Klin. Wochenschr.*, 1883, **20**, 77.
57. L. Knorr, *Chem. Ber.*, 1884, **17**, 2032.
58. W. Filehne, *Berlin Klin. Wochenschr.*, 1896, **33**, 1061.
59. R. Christison, *Monthly J. Med. Soc., London Edinburgh*, 1855, **20**, 193.
60. F. H. Rodin, *Am. J. Ophthalmol.*, 1947, **30**, 19.
61. J. A. Aeschlimann, *J. Soc. Chem. Ind.*, 1935, 135.
62. E. Hardy, *Bull. Soc. Chim. Fr.*, 1875, **24**, 497.
63. A. W. Gerrard, *Pharm. J.*, 1875, **5**, 865.
64. A. Ladenburg and L. Rugheimer, *Chem. Ber.*, 1880, **13**, 373.
65. J. von Braun, O. Braunsdorff and K. Rath, *Chem. Ber.*, 1922, **55**, 1666.
66. R. R. Burtner, 'Medicinal Chemistry', ed. C. M. Suter, Wiley, New York, 1951, vol. 1, p. 151.
67. O. Eisleb and O. Schaumann, *Dtsch. Med. Wochenschr.*, 1939, **63**, 967.
68. N. B. Eddy, *J. Am. Pharm. Assoc.*, 1947, **8**, 536.
69. A. Niemann, 'Ueber eine neue organische Base in den Cocablattern', Huth, Gottingen, 1860.
70. W. Filehne, *Berlin Klin. Wochenschr.*, 1887, **24**, 107.
71. G. Merling, *Ber. Dtsch. Pharm. Ges.*, 1896, **6**, 173.
72. A. Einhorn and R. Heinz, *Munch. Med. Wochenschr.*, 1897, **44**, 931.
73. E. Fourneau, 'Organic Medicaments and their Preparation', Churchill, London, 1925, p. 61.
74. A. Einhorn, *Liebigs Ann. Chem.*, 1910, **371**, 125.
75. F. J. Bove, 'The Story of Ergot', Karger, New York, 1970.
76. J. C. Moir, *Am. J. Obstet. Gynecol.*, 1974, **120**, 291.
77. A. Hofmann, *Agents Actions*, 1970, **1**, 148.
78. A. R. McIntyre, 'Curare, Its History, Nature, and Clinical Use', University of Chicago Press, Chicago, 1947.
79. P. Smith, 'Arrows of Mercy', Doubleday, New York, 1969.
80. B. Thomas, 'Curare—its History and Usage', Pitman Medical, London, 1960.
81. H. King, *J. Chem. Soc.*, 1935, 1381.

82. A. E. Bennett, *Anesth. Analg. (Cleveland)*, 1968, **47**, 484.
83. A. M. Betcher, *Anesth. Analg. (Cleveland)*, 1977, **56**, 305.
84. R. C. Gill, *Anesthesiology*, 1946, **7**, 14.
85. D. Bovet, F. Depierre, S. Courvoisier and Y. de Lestrange, *Arch. Int. Pharmacodyn.*, 1949, **80**, 172.
86. R. B. Barlow and H. R. Ing, *Br. J. Pharmacol.*, 1948, **3**, 298.
87. W. R. Buckett, C. L. Hewett, and D. S. Savage, *J. Med. Chem.*, 1973, **16**, 1116.
88. A. J. Everett, L. A. Lowe, and S. Wilkinson, *J. Chem. Soc., Chem. Commun.*, 1970, 1020.
89. J. B. Stenlake, R. D. Waigh, J. Urwin, G. Dewar and G. C. Coker, *Br. J. Anaesth.*, 1983, **55**, 3s.
90. G. Pines, H. Miller and G. Alles, *J. Am. Med. Assoc.*, 1930, **94**, 790.
91. R. K. Dikshit, *Trends Pharmacol. Sci.*, 1980, **1**(16), viii.
92. J. M. Muller, E. Schlitter and H. J. Bein, *Experientia*, 1952, **8**, 338.
93. I. S. Johnson, H. F. Wright and G. H. Svoboda, *J. Lab. Clin. Med.*, 1959, **54**, 830.
94. H. A. Kelly, *J. Am. Med. Assoc.*, 1901, **30**, 1083.
95. D. C. Schechter and H. Swan, *Surgery (St. Louis)*, 1961, **49**, 817.
96. R. Godlee, 'Lord Lister', Macmillan, London, 1917.
97. K. P. Link, *Circulation*, 1959, **19**, 97.
98. Paterson and Locock, *Appl. Ther.*, 1967, **9**, 60.
99. A. Stoll, 'The Cardiac Glycosides', The Pharmaceutical Press, London, 1937.
100. T. Maclagan, *Br. Med. J.*, 1876, **1**, 627.
101. R. Ellmer, *Lernen ± Leisten (Frankfurt/Main)*, 1978, **7**, 82.
102. G. P. Rodnan and T. G. Benedek, *Arthritis Rheum.*, 1970, **13**, 145.
103. J. Nicholson, in 'Chronicles of Drug Discovery', ed. J. S. Bindra and D. Lednicer, Wiley, Chichester, 1982, p. 149.
104. Y. Gaoni and R. Mechoulam, *J. Am. Chem. Soc.*, 1964, **86** 1646.
105. H. Stahelin, *Eur. J. Cancer*, 1973, **9**, 215.
106. A. F. W. Hughes, *J. Hist. Med. Allied Sci.*, 1977, 292.
107. A. Q. Maisel, 'The Hormone Quest', Random House, New York, 1965.
108. M. Borrell, *Bull. Hist. Med.*, 1976, **50**, 309.
109. G. R. Murray, *Br. Med. J.*, 1891, **2**, 796.
110. E. C. Kendall, *J. Am. Med. Assoc.*, 1915, **64**, 2042.
111. R. Pitt-Rivers, in 'Hormonal Proteins and Peptides', ed. C. H. Li, Academic Press, New York, 1978, vol. 6, p. 391.
112. G. Oliver and E. A. Schafer, *J. Physiol. (London)*, 1894, **16**, 1.
113. H. Konzett, *Arch. Exp. Pathol. Pharmakol.*, 1940, **197**, 27.
114. D. Hartley, D. Jack, L. H. C. Luntz and A. C. Ritchie, *Nature (London)*, 1968, **219**, 861.
115. J. W. Black and J. S. Stephenson, *Lancet*, 1962, **2**, 311.
116. R. G. Shanks, *Trends Pharmacol. Sci.*, 1984, **5**, 405.
117. C. R. Scholtz, *Ind. Eng. Chem.*, 1945, **37**, 120.
118. I. Murray, *Scot. Med. J.*, 1969, **14**, 286.
119. I. Murray, *J. Hist. Med. Allied Sci.*, 1971, **26**, 150.
120. V. du Vigneaud, *Science (Washington, D.C.)*, 1956, **123**, 967.
121. E. Allen (ed.), 'Sex and Internal Secretions', Williams and Wilkins, Baltimore, 1939.
122. E. Knauer, *Arch. Gynecol.*, 1900, **60**, 322.
123. E. Allen and E. Doisy, *J. Am. Med. Assoc.*, 1923, **81**, 819.
124. S. Ascheim and B. Zondek, *Klin. Wochenschr.*, 1927, **6**, 1322.
125. A. Butenandt, *Trends Biochem. Sci.*, 1979, **4**, 215.
126. E. C. Dodds, L. Goldberg, W. Lawson and R. Robinson, *Nature (London)*, 1938, **141**, 247.
127. M. J. K. Harper and A. L. Walpole, *Nature (London)*, 1966, **212**, 87.
128. L. Fraenkel, *Arch. Gynecol.*, 1903, **68**, 438.
129. W. M. Allen and G. W. Corner, *Proc. Soc. Exp. Biol. Med.*, 1930, **27**, 403.
130. V. Petrow, *Chem. Rev.*, 1970, **70**, 713.
131. G. Pincus, *Vitam. Horm. (N.Y.)*, 1959, **17**, 307.
132. F. C. Koch, C. R. Moore and T. F. Gallagher, *13th Int. Physiol. Congr.*, 1929, abstract 148.
133. E. Laqueur, K. David, E. Dingemanse, J. Freud and S. E. de Jongh, *Acta Brevia. Neerl. Physiol. Pharmacol. Microbiol.*, 1935, **4**, 5.
134. F. B. Colton, L. N. Nysted, B. Riegel and A. L. Raymond, *J. Am. Chem. Soc.*, 1957, **79**, 1123.
135. R. Hunt and R. Taveau, *Br. Med. J.*, 1906, **2**, 1788.
136. H. H. Dale, *J. Pharmacol. Exp. Ther.*, 1914, **6**, 147.
137. O. Loewi and E. Navratil, *Pfluegers. Arch. Gesamte. Physiol. Menschen Tiere*, 1926, **214**, 689.
138. A. Simonart, *J. Pharmacol. Exp. Ther.*, 1932, **46**, 157.
139. U. S. von Euler, *Acta Physiol. Scand.*, 1946, **12**, 73.
140. G. C. Cotzias, M. H. Van Woert and L. M. Schiffer, *New Engl. J. Med.*, 1967, **276**, 374.
141. T. Y. Shen, T. B. Windholz, A. Rosegay *et al.*, *J. Am. Chem. Soc.*, 1963, **85**, 488.
142. H. H. Dale and P. P. Laidlaw, *J. Physiol. (London)*, 1910, **41**, 318.
143. D. Bovet and A.-M. Staub, *Compt. Rend. Soc. Biol. (Paris)*, 1937, **124**, 547.
144. A.-M. Staub, *Ann. Inst. Pasteur (Paris)*, 1939, **63**, 400, 485.
145. B. N. Halpern and F. Walther, *Compt. Rend. Soc. Biol., (Paris)*, 1945, **139**, 402.
146. R. W. Brimblecombe, W. A. M. Duncan, G. J. Durant, C. R. Ganellin, M. E. Parsons and J. W. Black, *Br. J. Pharmacol.*, 1975, **53**, 435P.
147. U. S. von Euler, *Arch. Exp. Pathol. Pharmakol.*, 1934, **175**, 78.
148. H. Chick, *Prog. Food Nutr.*, 1975, **1**, 1.
149. S. A. Goldblith and M. A. Joslyn, 'Milestones in Nutrition', Avi Publishing, Westport, 1964.
150. E. V. McCollum, 'A History of Nutrition', Houghton Mifflin, Boston, 1957.
151. R. A. Morton, *Int. Z. Vitaminforsch.*, 1968, **38**, 5.
152. B. C. Jansen, *Nutr. Abstr. Rev.*, 1956, **26**, 1.

153. H. H. Mitchell, *J. Biol. Chem.*, 1919, **40**, 399.
154. R. Kuhn, P. Gyorgy and T. Wagner-Jauregg, *Chem. Ber.*, 1933, **66**, 317.
155. C. A. Elvehjem, R. J. Madden, F. M. Strong and D. W. Woolley, *J. Biol. Chem.*, 1938, **123**, 137.
156. S. Lepkovsky, *J. Biol. Chem.*, 1938, **124**, 125.
157. P. Gyorgy, *J. Nutr.*, 1967, **91**, Suppl. 1, 5.
158. G. W. Corner, 'George Hoyt Whipple and his Friends: the Life Story of a Nobel Prize Pathologist', Lippincott, Philadelphia, 1963.
159. T. H. Jukes, *Trends Biochem. Sci.*, 1980, **5**, 112.
160. W. A. Waugh, *J. Chem. Educ.*, 1934, **11**, 69.
161. T. H. Jukes, *Trends Biochem. Sci.*, 1980, **5**, 140.
162. P. E. Baldry, 'The Battle Against Bacteria, A Fresh Look', Cambridge University Press, London, 1976.
163. J. Brunel, *J. Hist. Med. Allied Sci.*, 1951, **6**, 287.
164. R. Dubos, *J. Exp. Med.*, 1939, **70**, 1.
165. H. W. Florey, E. B. Chain, N. G. Heatley, M. A. Jennings, A. G. Sanders, E. P. Abraham and M. E. Florey, 'Antibiotics', Oxford University Press, Oxford, 1949, vol. 1, p. 1.
166. H. W. Florey, E. B. Chain, N. G. Heatley, M. A. Jennings, A. G. Sanders, E. P. Abraham and M. E. Florey, 'Antibiotics', Oxford University Press, Oxford, 1949, vol. 2, p. 631.
167. S. Selwyn, 'The beta-Lactam Antibiotics: Penicillins and Cephalosporins in Perspective', Hodder and Stoughton, London, 1980, p. 1.
168. J. C. Sheehan, 'The Enchanted Ring. The Untold Story of Penicillin', MIT Press, London, 1982.
169. D. Wilson, 'Penicillin in Perspective', Faber and Faber, London, 1976.
170. R. Hare, 'The Birth of Penicillin', Allen and Unwin, London, 1970.
171. H. W. Florey and E. P. Abraham, *J. Hist. Med. Allied Sci.*, 1951, **6**, 302.
172. A. E. Elder, 'The History of Penicillin Production', American Institute of Chemical Engineers, New York, 1970.
173. F. R. Batchelor, F. P. Doyle, J. H. Nayler and G. N. Rollinson, *Nature (London)*, 1959, **183**, 257.
174. F. P. Doyle, J. H. Nayler, H. Smith and E. R. Stove, *Nature (London)*, 1961, **191**, 1091.
175. E. P. Abraham and P. B. Loder, 'Cephalosporins and Penicillins', ed. E. H. Flynn, Academic Press, New York, 1972.
176. J. C. Gentles, *Nature (London)*, 1958, **182**, 476.
177. J. F. Borel, *Agents Actions*, 1976, **6**, 468.
178. S. A. Waksman, *J. Hist. Med. Allied Sci.*, 1951, **6**, 318.
179. B. M. Duggar, *Ann. N.Y. Acad. Sci.*, 1948, **51**, 177.
180. J. Controulis, M. C. Rebstock and H. M. Crooks, Jr., *J. Am. Chem. Soc.*, 1949, **71**, 2463.
181. J. Dutcher, *Dis. Chest.*, 1968, **54**, Suppl. 1, 296.
182. P. Sensi, in 'Chronicles of Drug Discovery', ed. J. S. Bindra and D. Lednicer, Wiley, Chichester, 1982, p. 201.
183. S. A. Waksman (ed.), 'Actinomycin', Interscience, New York, 1968.
184. H. W. Thomas, *Br. Med. J.*, 1905, **1**, 1140.
185. E. Baumler, 'Paul Erlich, Scientist for Life', Holmes and Meier, New York, 1984.
186. P. Ehrlich, in 'The Collected Papers of Paul Ehrlich', ed. F. Himmelweit, Pergamon Press, London, 1960, vol. 3, p. 282.
187. I. Galdstone, 'Behind the Sulfa Drugs', Appleton-Century, New York, 1943.
188. J. Parascandola and R. Jasensky, *Bull. Hist. Med.*, 1974, **48**, 199.
189. E. Schmidt, *Angew. Chem.*, 1930, **43**, 963.
190. B. Rosenberg, L. Van Camp and T. Krigas, *Nature (London)*, 1965, **205**, 698.
191. H. D. Dakin, *Br. Med. J.*, 1915, **2**, 318.
192. T. S. Clouston, *J. Ment. Sci.*, 1868, **14**, 305.
193. J. F. C. Coindet, *Q. J. Sci. Lit., Arts*, 1821, **11**, 408.
194. D. Marine, *Medicine (Baltimore)*, 1924, **3**, 453.
195. B. M. Duncum, 'The Development of Inhalational Anaesthesia', Oxford University Press, London, 1947.
196. T. E. Keys, 'A History of Surgical Anaesthesia', Dover, New York, 1963.
197. F. L. Taylor, 'Crawford W. Long and the Discovery of Ether Anaesthesia', Hoeber, New York, 1928.
198. T. C. Butler, *Bull. Hist. Med.*, 1970, **44**, 168.
199. Sunday Times Insight Team, 'Suffer the Children. The Story of Thalidomide', Deutsch, London, 1979.
200. A. Hauptmann, *Munch. Med. Wochenschr.*, 1912, **54**, 1907.
201. T. J. Putnam, in 'Discoveries in Biological Psychiatry', ed. F. Ayd, Jr. and B. Blackwell, Lippincott, Philadelphia, 1970.
202. T. L. Brunton, *Lancet*, 1867, **2**, 97.
203. W. Murrell, *Lancet*, 1879, **1**, 80.
204. H. Erdtman and N. Lofgren, *Sven. Kem. Tidskr.*, 1937, **49**, 163.
205. N. Lofgren and B. Lundqvist, *Sven. Kem. Tidskr.*, 1946, **58**, 206.
206. J. H. Gaddum, *Chem. Ind. (London)*, 1954, 266.
207. W. Lange and G. von Kreuger, *Chem. Ber.*, 1932, **65**, 1598.
208. J. Morgenroth and R. Levy, *Berlin Klin. Wochenschr.*, 1911, **48**, 1560.
209. R. Slotta and R. Tschesche, *Chem. Ber.*, 1929, **62**, 1398.
210. S. L. Shapiro, V. A. Parrino and L. Freeman, *J. Am. Chem. Soc.*, 1959, **81**, 2220, 3728.
211. N. von Jancsó and H. von Jancsó, *Z. Immun. Forschr.*, 1935, **85**, 81.
212. E. M. Lourie and W. Yorke, *Ann. Trop. Med. Parasitol.*, 1939, **33**, 289.
213. J. F. J. Cade, in 'Discoveries in Biological Psychiatry', ed. F. J. Ayd, Jr. and B. Blackwell, Lippincott, Philadelphia, 1970, p. 218.
214. J. G. Kidd, *J. Exp. Med.*, 1953, **93**, 565.
215. J. D. Broome, *Nature (London)*, 1961, **191**, 1114.
216. J. K. Crellin, *J. Hist. Med. Allied Sci.*, 1981, **36**, 9.
217. A. Cahn and P. Hepp, *Centralbl. Klin. Med.*, 1886, **7**, 561.
218. A. Vogl, *Am. Heart J.*, 1959, **39**, 881.
219. E. M. Schultz, E. J. Cragoe, Jr., J. B. Bicking, W. A. Bolhofer and J. M. Sprague, *J. Med. Pharm. Chem.*, 1962, **5**, 660.
220. T. Mann and D. Keilin, *Nature (London)*, 1940, **146**, 164.

221. R. O. Roblin, Jr. and J. W. Clapp, *J. Am. Chem. Soc.*, 1950, **72**, 4890.
222. K. H. Beyer, *Perspect. Biol. Med.*, 1977, **20**, 410.
223. L. N. Gay and P. E. Carling, *Science (Washington, D.C.)*, 1949, **109**, 359.
224. S. Farber, L. K. Diamond, R. D. Mercer, R. F. Sylvester and J. A. Wolff, *New Eng. J. Med.*, 1948, **238**, 787.
225. G. H. Hitchings, *Cancer Res.*, 1969, **29**, 1895.
226. A. E. Caldwell, 'Origins of Psychopharmacology. From CPZ to LSD', Thomas, Springfield, 1970.
227. J. Ravn, in 'Discoveries in Biological Psychiatry', ed. F. J. Ayd, Jr. and B. Blackwell, Lippincott, Philadelphia, 1970, p. 180.
228. R. Kuhn, in 'Discoveries in Biological Psychiatry', ed. F. J. Ayd, Jr. and B. Blackwell, Lippincott, Philadelphia, 1970, p. 205.
229. H. Franke and J. Fuchs, *Dtsch. Med. Wochenschr.*, 1955, **80**, 1449.
230. N. S. Kline, in 'Discoveries in Biological Psychiatry', ed. F. J. Ayd, Jr. and B. Blackwell, Lippincott, Philadelphia, 1970, p. 194.
231. G. A. Stein, M. Sletzinger, H. Arnold, D. Reinhold, W. Gaines and K. Pfister, III, *J. Am. Chem. Soc.*, 1956, **78**, 1514.
232. J. A. Dates, L. Gillespie, S. Udenfriend and A. Szoerdsma, *Science (Washington, D.C.)*, 1960, **131**, 1890.
233. R. Cash, A. J. Brough, M. N. P. Cohen and P. S. Satoh, *J. Clin. Endocrinol. Metab.*, 1967, **27**, 1239.
234. M. L. Gujral, P. N. Saxena and R. S. Tiwari, *Ind. J. Med. Res.*, 1955, **43**, 637.
235. R. Charonnat, P. Lechat and J. Chareton, *Therapie*, 1957, **12**, 68.
236. M. J. Thullier and R. Domenjoz, *Anaesthesist*, 1957, **6**, 163.
237. P. A. J. Janssen, in 'Discoveries in Biological Psychiatry', ed. F. J. Ayd, Jr. and B. Blackwell, Lippincott, Philadelphia, 1970, p. 165.
238. L. H. Sternbach, *Agents Actions*, 1972, **2**, 193.
239. G. B. Kauffman, *Educ. Chem.*, 1982, **19**, 168.
240. W. D. M. Paton and E. J. Zaimis, *Br. J. Pharmacol.*, 1949, **4**, 381.
241. A. L. A. Boura and A. F. Green, *J. Auton. Pharmacol.*, 1981, **1**, 255.
242. G. D. Laubach, S. Y. P'An and H. Rudel, *Science (Washington, D.C.)*, 1955, **122**, 78.
243. J. A. Sutton, *Glaxo Volume*, 1972, **36**, 5.
244. F. M. Berger, in 'Discoveries in Biological Psychiatry', ed. F. J. Ayd, Jr. and B. Blackwell, Lippincott, Philadelphia, 1970, p. 115.
245. E. B. Astwood, *J. Am. Med. Assoc.*, 1943, **122**, 78.
246. G. W. Anderson, I. F. Halverstadt, W. H. Miller and R. O. Roblin, Jr., *J. Am. Chem. Soc.*, 1945, **67**, 2197.
247. L. S. Goodman, M. W. Winetrobe, M. T. McLennan, W. Dameshek, M. J. Goodman and A. Gilman, in 'Approaches to Tumour Chemotherapy', ed. F. R. Moulton, American Association for the Advancement of Science, Washington, 1947, p. 338.
248. J. L. Everett, J. J. Robertson and W. C. J. Ross, *J. Chem. Soc.*, 1953, 2386.
249. F. Bergel and J. A. Stock, *J. Chem. Soc.*, 1954, 2409.
250. H. Arnold, F. Bourseaux and N. Brock, *Nature (London)*, 1958, **181**, 931.
251. J. Dressel (translated R. E. Oesper), *J. Chem. Educ.*, 1961, **38**, 620.
252. W. Schulemann, *Proc. R. Soc. Med.*, 1932, **25**, 897.
253. A. Albert, 'The Acridines', Arnold, London, 1951.
254. C. H. Browning, *Scot. Med. J.*, 1967, **12**, 310.
255. C. H. Browning, R. Gulbransen, E. L. Kennaway and L. H. D. Thornton, *Br. Med. J.*, 1917, **1**, 73.
256. R. J. Schnitzer, *Ann. N.Y. Acad. Sci.*, 1954, **59**, 227.
257. R. J. Schnitzer, *Dtsch. Med. Wochenschr.*, 1929, **55**, 1888.
258. H. C. Richards, in 'Chronicles of Drug Discovery', ed. J. S. Bindra and D. Lednicer, Wiley, Chichester, 1982, p. 257.
259. H. Horlein, *Proc. R. Soc. Med.*, 1936, **29**, 313.
260. G. Domagk, *Dtsch. Med. Wochenschr.*, 1935, **61**, 250.
261. L. E. H. Whitby, *Lancet*, 1938, **1**, 1210.
262. G. Domagk, *Naturwissenschaften*, 1946, **33**, 315.
263. G. B. Elion, E. Burgi and G. B. Hitchings, *J. Am. Chem. Soc.*, 1952, **74**, 411.
264. R. W. Rundles, *Ann. Intern. Med.*, 1966, **64**, 229.
265. C. Heidelberger, N. K. Chaudhuri, P. Danneberg, D. Mooren, L. Griesbach, R. Duschinsky, R. J. Schnitzer, E. Pleven and J. Scheiner, *Nature (London)*, 1957, **179**, 663.
266. G. B. Elion, P. A. Furman, J. A. Fyfe, P. de Miranda, L. Beauchamp and H . J. Schaeffer, *Proc. Natl. Acad. Sci. USA*, 1977, **74**, 5716.
267. T. P. Johnston, G. S. McCaleb and J. A. Montgomery, *J. Med. Chem.*, 1963, **6**, 669.

1.3
Evolution of the Pharmaceutical Industry

JONATHAN LIEBENAU
London School of Economics, UK

1.3.1 INTRODUCTION: THE MANUFACTURE AND SUPPLY OF DRUGS

Therapeutic substances have always been a part of medical practice. Archaeological and anthropological evidence shows that medicines of various kinds played a central role in healing rituals and formed key props for healers. Although it would be easy to ignore medicines used before theories of specific action and pharmacodynamics showed how, in modern scientific terms, drugs work, nevertheless the place of medication within modern western health care systems, the relationships between the producers and consumers of medicines, and even some specific classes of materia medica are to a large extent derivatives of traditional medical practice.

While the placebo effect may be regarded as a major factor in the use of drugs, today as well as in ancient times, coherent bodies of theory and groups of substances associated with specific effects did contribute to the panoply of things associated with healing. There was increasing demand for large scale production, storage and supply as these substances, herbs such as garlic, spices such as cloves, metals such as mercury and iron, were increasingly used as standard materials. By late antiquity common mixtures and dosage forms stimulated demand. Previously, and continuing to some extent into this century, medicines had been prepared and dispensed by people whose social roles had included healing. Typically, medicines were prepared based on need at the time, but, increasingly, small batches for future use were produced.

The first major changes which brought larger scale production, warehousing and later the special cultivation and collection of basic materia medica probably came about as the result of the establishment of large standing armies. The classical Greeks and Egyptians may have had special storage area in their early hospitals such as Hippocrates' complex on the island of Cos, but it was not until the early pharmacies in Islamic North Africa, and later in Persia, Baghdad and elsewhere, that documented large scale production of standard medicines is known. It should be pointed out that other urbanizing peoples also had well-organized medicine preparation, but most, like the Chinese, developed complex traditions and practices of freshly mixing materia medica into medicines within pharmacies on demand.

North Europeans learned of the practice of storing large quantities of medicines in ceramic vessels from the Islamic urban pharmacies and drug storage facilities used for the military. Most notably in

Spain and then Italy and Southern France, drug pots and jars became important means of storing medicines systematically. The organization, as well as the administration of such stores, forced the systematization of at least large scale drug storage and distribution, even if not industrial style production.

The apothecary shop became a recognized business as it grew from two distinct but occasionally overlapping traditions, that of the dispensing physician and that of the general goods stores which sold medicines. As apothecary facilities became more specialized, expert dispensers began to distinguish themselves. Reputations, the forerunners of brands, were recognized beyond the confines of the local community, and trade began to include some prepared products in addition to raw materials and simple intermediates.

By the late 17th century and into the 18th century, the business of importing from diverse foreign sources, remixing and then exporting to traditional markets became an accepted business practice. The major western trading nations excelled at this sort of business. Typically this trade was mixed with other forms of business, commonly other chemicals and hardware, but, increasingly, specialists emerged who based their reputations on supplying high quality materia medica. Increasingly they were called upon to mix their own preparations, providing either the service of mixing difficult or expensive concoctions on behalf of busy, small scale pharmacy traders or, occasionally, offering some unique preparations in branded forms.

For the most part, these businesses were very small, employing no more than a handful of family members, apprentices and workers. They were not at all concerned with developing new preparations, and had only small incentives to consider improved production processes. The main business was one of balancing sources with markets. Reputations, while vulnerable to complaints about quality, were primarily based on trade concerns—credit facilities, promptness of delivery and completeness of stock.

Opportunities for trade improved only slowly through the 18th century, but markets changed more dramatically. By the early 19th century new urban markets had changed the character of retail business in general. Wholesalers, suppliers, importers and, increasingly, manufacturers became attuned to the opportunities afforded by more regular trade between major ports and the new demands of urban markets.

Changes in the medical professions in the late 18th and early 19th centuries also contributed to the new character of trade. With apothecaries defining a distinctive role apart from physicians in Germany, the low countries, Britain, and, to some extent, France, as well as elsewhere where urban cultures affected national professional development, these distinctions aided the growth of a large scale trade in refined materia medica and prepared medicines. When physicians ceased routinely to compound and sell all the drugs they prescribed and apothecaries increasingly began to rely upon physicians for diagnoses, the distinction between trade and practice was much clearer. From that point the course of development from drug store, apothecary or chemist's shop to manufacturing pharmacy was continuous.

1.3.2 FROM DRUG STORE TO MANUFACTURING PHARMACY

The growth of the modern pharmaceutical industry took place largely in western Europe and the United States. At the beginning of the 19th century little distinction could be found between, on the one hand, the American corner shop which sold drugs, the British chemist's shop or in Germany and elsewhere the apothecary, and, on the other hand, importers, manufacturers and wholesale drug traders. Drug exchanges operated in major trading ports, so that in Philadelphia and New York, London and Amsterdam, the trade had a self-governing body which helped to minimize bad business practices, especially overcharging by shippers. For the most part, however, trade consisted of representatives or proprietors of large pharmacies going to the ports when shipments came in and bargaining directly with shipping agents for bulk purchases. These purchases would typically be processed in a very preliminary fashion, cleaned, sorted and packaged, and then enter the wholesale trade as materia medica. Companies such as Allen & Hanbury, Jacob Bell, and Thomas Morson, all small London operators, established standard working arrangements where they would conduct retail trade from their city shops and maintain warehouses with compounding rooms from which they would serve as wholesalers to retailers in the southeast England region. Trade was well established by the early years of the century, with most suppliers offering complete ranges of medicines, but perhaps concentrating on larger scale processing of a few types of items. So when Thomas Morson returned from studying in Paris, where he came under the influence of Pelletier and others working on quinine products, he began to cultivate a reputation for processing quinine for the British market. Similarly, Allen & Hanbury began a company practice of processing oils as part of their efforts to supply their market with some lines less well served previously.

In France the distinction between planning and the professional practice of medicine was secured in the 18th century. At that time there were many hundreds of small pharmacists and only a couple of large scale producers. Antoine Baume, who invented a number of technical improvements for manufacturing, had a production laboratory which prepared 2400 items in 1775. Despite this early example, however, French pharmacy remained limited well into the 20th century and the coordination of a national body in the mid-19th century served to split the sector into one central cooperative wholesale and manufacturing enterprise plus numerous small concerns.

As pharmaceutical firms grew from apothecary shops to family companies and then to large corporations, their relationship with the medical community also changed. Pharmacists held a tenuous place in the medical world of the 19th century, although they provided important services to those members of the public who did not use physicians, and supplied those who did with medicines. Towards the end of the century drug makers distanced themselves from pharmacists and their informal practice of medicine. As the distinction grew between reputable manufacturers who sought the professional market—'ethical' pharmceutical producers—and the popular 'patent medicine' makers, the large companies looked for a compromise between the two areas. They hoped to establish reputations for quality but also to reap the profits from patent or proprietary medicines.

The early 19th century developments in the United States is instructive. The foundation for the pharmaceutical industry there was laid in Philadelphia between 1818 and 1822 with the establishment of half a dozen enduring fine chemical manufacturers. America had formerly been dependent on Britain for most of its medicine, as for other manufactured products. With the economic disorder created by the war of 1812 and its aftermath, this pattern of dependence was broken. Since importation was disrupted and high tariffs were levied on those goods which did get through, it became easier and more profitable to manufacture many products than to ship them from Britain.

The production of most common medicines was within the reach of many apothecaries. There was little need for high capital investment and only a general knowledge of pharmacy practice was necessary for them to produce such basic remedies as quinine sulfate, opium powders, or calomel. Demand was constant and universal, and profits were temptingly high. The laboratories of many pharmacies could be turned into manufacturing plants. A boiler, a drying room for plants and a suitable storage area were among the few necessities, along with a jacketed copper pan, a filter press and open furnaces. With little more investment small producers were able to compete on the local wholesale market with major manufacturers and importers, making a full range of opiate preparations, herbals and materia medica. One such company in Philadelphia had such success that by 1850 they had accumulated sufficient capital to move into a factory building and abandon the retail business. Their move gave them the opportunity to introduce steam-driven stirring and grinding apparatus, further increasing their capacity. Since the mechanical and chemical processes are basically similar for most preparations, they were able to manufacture a wide variety of products. This large scale production of medicines lay midway between the manufacture of chemicals and the practice of pharmacy. Most drugs at the time were made by such basic processes as crushing, drying, extracting and distilling. For example, cloves were dried, crushed and powdered to produce a popular anaesthetic; the juice from belladonna plants was extracted and then distilled to concentrate it so that it could be used as a heart stimulant. Analogous procedures were followed by chemical manufacturers to produce pigments, acids, sodas and other commonly used substances. Pharmacists not only borrowed from the chemists in their procedures, they also overlapped in the products, typically including calomel, vermilion, red precipitate, corrosive sublimate, cinnabar and some mercury preparations, as well as crystallized soda, aqua fortis, hydrochloric acid and other chemicals, thus supplying both industrial and medical needs. The simplicity of procedures allowed considerable diversification, which in turn permitted the development of wide markets or the strategic use of particular product lines according to demand.

Drug use has always been associated to some extent with therapeutic theory, although less directly than is often thought. In the early 19th century medicines were compounded and administered both by doctors and by pharmacists. Both patients and medical men identified the immediate and dramatic reactions caused by most drugs as part of the cure. Competing in a crowded marketplace for patients, as part of the cure doctors needed to demonstrate the efficacy of their treatments, and the remedies they prescribed were designed to cause easily perceptible changes in the physical state of the patient. Coupled with the mutual faith which physicians and patients had in these treatments, drastic remedies served to satisfy the demands of patients for effective cures.

The basis of this faith was a coherent system of therapeutic theory. Beneath it lay the notion that the body was in a constant dynamic relationship with its environment. Equilibrium was associated with health, imbalance with disease, and the purpose of therapeutics was to restore order. Humeral theory and localistic models of disease were reconciled by the principle that every part of the body was related inextricably to every other. The early 19th century conception of the body was as a system of intake and outgo. Physicians and patients could see and judge excretions or appetite; they therefore seemed an obvious monitor of health. Therapy, as a result, concentrated on diet and excretion, perspiration and ventilation as the aspects which could be controlled to produce a stable system. If a person had a wound, for example, he or she might be treated with a salve locally, but the holistic implications for the body might well necessitate the administration of a stimulant in addition. Enormous quantities of medicines were consumed in order to maintain or re-establish health. Special medicines were also administered in life crises, or at changes of season, when a patient's body was more liable to lose its healthy equilibrium. Cathartics, for example, were given in the spring and autumn to help the body adjust to these cyclical changes. Mercury treatments and bleeding were the most common severe therapies, both producing dramatic effects. This therapeutic framework was not necessarily interpreted by a physician: in the first half of the 19th century many people avoided doctors altogether and consulted pharmacists. Preference for self-dosing and the lesser expense of druggists stimulated trade in patent medicines. These remedies, often standard preparations of common drugs sold by brand name or unusual concoctions of stimulants, alcohol and flavouring, sold exceedingly well. Most of the drug companies which originated in pharmacies or from wholesalers depended upon patent preparations for the bulk of their profits.

The involvement of physicians created a special range of problems. Although a large number of physicians in the United States were involved in the sale of drugs, medicine makers had a long tradition of strained ties with mainstream practitioners. The United States differed from Great Britain, where apothecaries had early formed a guild and had been in direct, open competition with doctors. But even though the rift was never so wide or formalized in the United States, there was suspicion among some physicians about the extent to which pharmacists were actually diagnosing and prescribing.

The first part of the 19th century saw the establishment and steady growth of a number of small pharmaceutical firms. By mid-century a few of these firms were becoming particularly well established. In the decades between 1840 and 1870 regular physicians increased their emphasis on diet and regimen, with particular stress on the use of alcohol as a stimulant. Former practices were not abandoned, but they were used less routinely and, particularly under the influence of such new tenets as homoeopathy, smaller doses were administered. There was a major change in attitude from dispensing massive doses of basic medicines to the careful prescription of small quantities of specifically targeted, commercial preparations.

Large scale manufacturing in pharmacy grew during the second half of the 19th century for a number of reasons. Innovations in methods of drug preparation allowed factories to use machinery, and the growth in popularity of standardized preparations permitted new economies of scale. Changes in therapy and the influence of early research conducted in companies are other reasons for early expansion. The use of machinery, for example, enabled E. R. Squibb and Eli Lilly to secure large government contracts for basic drugs such as opium, quinine and ergot during the American Civil War. Fast-operating tableting machines, many of which were patented in the 1880s and 1890s, were examples of one aspect of large scale production. Innovations of this type were certainly advantageous for those few manufacturers who, like Squibb, could afford to invest in a large production plant and who felt that their market was sufficiently stable for them to predict consumption patterns. Collaboration with manufacturing chemists was also important in leading to the use of machinery, particularly in the United States.

Nevertheless, small scale producers were by no means defeated during the latter part of the 19th century by their lack of mass production capabilities. As late as 1890, a number of pharmacists were still increasing their production in order to enter the wholesale market. As long as they held at least one assured market they were able to survive competition from large manufacturers. Moreover, early efforts to develop new drugs and production processes even in the larger companies were not 'research', but rather the routine recombination of accepted remedies into new products. Investment in research or large scale manufacturing did not become a competitive advantage for most firms until well after the first decade of the 20th century.

Physicians increasingly relied on pharmacopoeia, a trend that encouraged the growth of a market for standard preparations and, conversely, a smaller one for apothecary-compounded drugs. Apothecary practice also changed because visiting salesmen began to deliver orders frequently and this, together with increasingly reliable transport, meant that pharmacies began to stock smaller quantities of drugs. Changes in therapy towards more specific drugs with measured doses also encouraged the introduction of the standard preparations produced by companies.

The British pharmaceutical industry in the 19th century was primarily composed of a number of small firms serving domestic consumers and exporting to traditional markets. They relied largely on raw materials imported from trading nations and tropical colonies or on standard chemicals and by-products. The typical range of products would include galenical preparations, alkaloids and numerous creams, infusions, dressings, and hundreds of miscellaneous medications. Leading firms such as Whiffen, Morson, Allen & Hanbury, May & Baker, and others in Britain all dealt in largely the same sort of products. By the 1830s some companies specialized in product areas such as alkaloids, as was the case with Morson's quinine trade in London and Macfarlain's opiates business in Edinburgh. For the most part, however, the products on offer overlapped among the competing firms. They also often listed a full range of medicines and acted as suppliers of products from associated companies where they could not cover the range adequately by manufacturing themselves.

All of these companies were small, usually employing fewer than 100 workers throughout the 19th century, and with the exception of a few scientifically minded proprietors such as William Allen or Thomas Morson, there was little commitment to pharmaceutical investigation or product development. Even those who did establish laboratories never integrated them into the normal course of business.

Adapting to developments in scientific medicine, by the 1890s many of the leading companies were projecting a new and avowedly scientific image: they began to employ medical men and maintain laboratories for quality control, standardization and, in some cases, product development. These small laboratories gradually evolved into significant research and development facilities which by the First World War were supplying new products to a medical world newly insistent on novel scientific therapies.

In Germany the pattern was slightly different. The German pharmaceutical industry was made up of a few old, well-established firms which had grown out of pharmacies and expanded with new production techniques in the mid-19th century. Companies such as Merck and Schering conducted a traditional business in galenicals, basic chemicals and other common medicines. Both these companies participated in the booming German manufacturing export trade of the late 19th century. Both established subsidiaries in the United States which were to become independent in the early 20th century and later to dwarf their parents. Nevertheless, within Germany their scope was limited to the drug business and they maintained traditional ties with pharmacists. This changed in the 1880s when the new dyestuffs manufacturers began to produce synthetic therapeutics. Although Britain and France held the dominant positions in coal-tar dye production in the 1860s, by the mid-1870s German production accounted for six times the value of British production, and overshadowed French output by over seven times. The German industry was also becoming composed mainly of big firms. Major changes in the structure and behaviour of the industry followed.

1.3.3 THREE DECADES OF MAJOR CHANGE: 1880–1910

In the mid-1880s two problems coincided to throw the German dye industry into crisis. The price and marketing convention governing the crucial red dyes broke up in 1885, forcing prices down to half their previous level within a year. At the same time the price of coal tar rose as new methods of producing coal gas reduced the quantity of by-product made. Laboratories helped to resolve the crisis with new dyestuffs and later by opening the new field of synthetic medicines.

At the Bayer company in Leverkusen there was no commitment to organized research until the late 1880s. The crisis in the dyestuffs trade of the early 1880s made them receptive to change. Research offered Bayer a strategy that might reduce vulnerability to the fluctuations of the business cycle. For chemists too, the depressed economy stimulated change. Their numbers had been growing steadily due to expanded opportunities for training in the 1870s. When that expansion slowed in the 1880s, academic careers were blocked and as mobility was reduced, chemists more readily accepted industrial employment. It was at that time that synthetic organic colours were first produced in academic laboratories, and that the connections were made between the chemical intermediates used for the production of dyestuffs and the manufacture of pharmaceutical products.

The Hoechst company near Frankfurt had started off small and late. In 1863 it consisted of the three founders, five workers, one chemist and an office girl. They established their centralized company laboratory in 1883 in a building peripheral to the main plant. It was in charge of a variety of tasks, including producing new colours and testing for constant quality. There was only a small staff of chemists, but they maintained very frequent contact with academic scientists. Paul Ehrlich, later to found modern chemotherapy, was one such young academic who came to the attention of the laboratory on account of his doctoral thesis on the bacteriological use of methylene blue. As attention became directed to synthetic medicinals, the main interest at Hoechst turned to the antipyretic effects of alkaloids. In the late 1880s one such antipyretic drug, pyramidon, became a major money-maker for Hoechst.

Another major influence on the decision to commit resources to research on pharmaceuticals was the widespread campaign to find a treatment for tuberculosis. Robert Koch, the leading medical scientist of the day, had contacted Hoechst around 1890 about producing his preparation, tuberculin, and this was seen as a marvellous opportunity to break into a huge new area of the drug business. Koch suggested that Emil Behring be contacted about a possible role in the development of tuberculin and other bacteriologically based medicines. Behring, who later was to win the first Nobel prize for medicine, continued to work in Berlin until 1892 when he was able to leave the armed forces, where he had been obliged to serve as a medical officer. When he moved to Frankfurt to work on Hoechst-sponsored research in to the immune response in animals, he found a staff of around a dozen chemists and pharmacists testing drugs and monitoring the production of tuberculin and pyramidon. The level of sophistication varied greatly within the laboratory. The director was an imaginative scientist and a very well-respected research director. His staff included a number of young graduates who conducted routine tasks using only the most basic equipment. Although the level of sophistication of the equipment was not particularly high, the application of chemical theory was considerable.

The German pharmaceuticals industry began with company-sponsored research and development in a haphazard, reactive way. No strategy for corporate laboratory development was discussed, and even the notion that there should be an adequate return on investment was missing. The people brought in were for the most part either chemists or general medical scientists, and rarely specialists in pharmaceutical chemistry. By the period immediately before the First World War this had changed, as it became apparent that laboratories capable of conducting academic-style research were necessary. For competitive advantage and to be able to keep up with the raised technical standards of both industry and the medical profession, laboratories had to be integrated into the newly expanding corporate structures of large fine chemicals and drug companies.

In Britain this began to change with the founding in 1894 of the Wellcome Physiological Research Laboratories. But although the Wellcome laboratories were founded in response to the bacteriological and immunological work which had led to the development and commercialization of diphtheria antitoxin in Germany and France, Henry Wellcome did not model the laboratory after the established company research and development facilities at Hoechst or Bayer.

Instead, by the end of the century the major British pharmaceutical firms were relying on cartel, convention and licensing agreements with German and Swiss companies to be able to offer new products. Upon the outbreak of the First World War the industry was, in business terms, reasonably stable but unable to supply the domestic market with many of the products which had so changed the industry abroad. There were no major industrial laboratories for product development, and British manufacturers, which had been complimented at numerous trade exhibitions for the quality of their standard products, seemed incapable of doing much else. This inability was recognized and much commented upon, but little changed. During the First World War they continued to obtain shipments from Continental suppliers, while publicly resolving to rectify their inadequacies. Major contracts to supply the armed forces had to be given to United States companies, and UK subsidiaries of US manufacturers such as Parke Davis & Co. did a booming business.

Between the years 1894 and 1926 products of the pharmaceutical industry came to be tested and certified by official bodies: first in Germany, followed shortly by the United States, much later in Britain. The form the laws took, the technical specifications they prescribed, the extent of their coverage, even the informal manner in which they were administered were broadly similar. But their genesis in each country was different. These differences reveal much about the character of governmental attitudes towards regulation, about the state of the pharmaceutical industry, and about the perception of the role of law within the respective medical communities.

Regulation of medicines was one of the clearest ways in which the established medical community could exercise their special control over therapeutics. In this way also those manufacturers who catered especially for physicians rather than lay consumers could be assured that their special—'ethical'—status was maintained. How then did the legal mechanisms come about in these three countries, and what role did drug manufacturers play?

All three countries had long had vague and unenforceable rules guarding against the sale of harmful agents. In Britain the sale of poisons was controlled and laws intended to curb opium abuse were tried at various times. But it was not all that clear in Britain whether selling a medicine which differed substantially from its *British Pharmacopoeia* description was illegal. In various test cases pharmacists were able to argue either that the customer had not specified that medicine exactly, or that the *Pharmacopoeia* did not have what was wanted, or that its description was inadequate. This vagueness prevailed in Germany also, and in the US state legislation varied widely, having in common only their weak provisions for enforcement.

The German government of the 1890s was the least exercised of the three about their lack of control over medicines. In the public health service and through military physicians they had a smoothly operating integration of official and medical functions. Their legislative apparatus was comparatively unencumbering and there seemed to be little dissent from practitioners.

As far as medicines were concerned, the full range of products was offered by the type of companies which existed in Britain and the United States. There were 'patent medicines' makers, suppliers of apothecaries, and a few firms which already by 1890 offered new synthetic remedies individually packaged and sold under brand names. These new medicines differed from other proprietary preparations by their instructions 'to be used only under the supervision of a physician'. It was this class of remedies which attracted much attention in the last decade of the century. Based on organic chemistry and bacteriology and embodying the appeal of scientific medicine, they held the promise of powerful new cures. But did they need to be controlled by the government? Clearly not, it seemed. The medical profession was always sharp and openly critical. With many important journals, a deceitful or worthless product would be exposed. That the system worked could be seen in numerous cases in the journals. Trials, assessments, commentary and correspondence often appeared which dealt with new therapeutics.

The change in official attitudes toward regulating medicines came with the development of diphtheria antitoxin in Germany and France. Soon after it became clear that antitoxin was to enter the market in a big way, the German health ministry seized the opportunity to apply regulations on it. Biologicals were seen to be particularly in greater need of standardization because they were introduced directly into the circulatory system rather than the digestive tract. They were normally administered in full doses and were shown to be most effective when given during the early stages of a disease. Therefore, if the first dose proved worthless or below strength, the loss of time could cost the life of the patient. This was particularly emotive since in Germany, during the first years, much of the antitoxin was being used in children's hospitals under government jurisdiction. Already, by November 1894, legislation was in place limiting availability only through apothecaries and with a physician's prescription. The criteria for approval of antitoxin included an assessment of its potency using a method of assay analyzing the amount necessary to neutralize a given volume of toxin. To facilitate this laboratory work, the Control Station for Diphtheria Antitoxin was established early in 1895. Already in Germany there were three major commercial producers of antitoxin—Hoechst, Merck and Schering. Hoechst and Schering cooperated and allowed all their antitoxin to be tested and standardized.

The need for technical standards was emphasized in 1896 by a report of a special commission established by the *Lancet*, a leading British medical journal, on 'Relative Strengths of Diphtheria Antitoxic Serums'. The *Lancet* tested antitoxin from three main German suppliers alongside three British suppliers and a French, a Belgian and a Swiss product. They found large discrepancies among the brands, and levels of potency widely off that claimed on the labels. Of the samples tested, those from Schering and Hoechst were evidently superior. None the less, Paul Ehrlich, a powerful and well-respected German medical scientist, remarked upon the *Lancet* results and cited them as proof of the need for an international standard which was easy to follow and clearly enforceable. He worked closely with the government and the Hoechst company to develop increasingly stringent standards to enforce.

Leading German dyestuffs companies tested many of their new products routinely using bacterial staining techniques similar to those used by Ehrlich. A combination of this work and changes in techniques of identifying chemical structures of simple naturally occurring drugs led to the developments of pyramidon, an antipyretic from Hoechst in 1884, aspirin, Bayer's famous antipyretic in 1889, vronal and urotropine a sedative and an internal antiseptic developed at Schering in 1900 and 1904, and salvarsan, the antisyphilitic developed at Hoechst by Ehrlich in 1909.

The structural implications for companies taking up the production of new synthetic medicines were profound. Both Bayer and Hoechst, principally dyestuffs companies, took advantage of the general research climate in Germany to develop many new products. Collaboration with universities was not something which could have been taken for granted. In Germany as in the USA a sea change overcame science-based industry in the late 19th century with the coincidence of the increased size of large companies, the growth of national markets and the availability of new scientific tools and staff. The climate created by these factors gradually forced both scientists and companies to reassess their traditionally separate realms.

The actions which brought together scientists and companies were diverse and spread over a long period of time. There were strong objections from scientific societies and long debates within departments over the propriety of professional scientists, to say nothing about university teachers, working even part time for pharmaceutical companies. The American Society for Pharmacology and Experimental Therapeutics banned any member who worked for drug companies until 1941, but some universities encouraged individuals and whole departments to take advantage of consultancy fees and company laboratory facilities in collaborative ventures. There were increasingly more cases where single academics would work in a continuing, exclusive consultancy with a single company, advising on a wide range of issues. There were specialist consultants who offered advice in relatively narrow areas of research or laboratory techniques and sold their services widely. And there were university departments which collaborated as teams with companies in large scale long term development projects.

By the early 1930s, despite continuing professional scepticism, university academics were eager to collaborate with companies. In the United States, the Eli Lilly Company collaborated with Toronto, Rochester and Harvard universities from the 1920s. There were drawbacks to this new trend. Misunderstandings between the University of Toronto researchers and the Lilly company over trademarks and patent protection for insulin led to serious acrimony. The informality of Lilly's later collaborative research efforts with Harvard and Rochester preserved the good feelings of the researchers, but they probably did not benefit the universities as much as the arrangements could have.

The advantages were less ambiguous in the cases where inividuals rather than university research teams were primarily involved. Companies such as Merck, Parke Davis, and Abbott Laboratories made mutually advantageous arrangements with individual scientists under a variety of terms. These frequently led to profitable new products and enriched, or at least supplemented the incomes of, university researchers. As both companies and universities slowly tested the appropriateness of collaboration, false starts and early failures probably sank from view, at great expense or potential loss. What it clearly led to, however, was a new attitude towards what the companies should do for themselves. The advantages of continuity and control of research staff based on these experiences hastened the trend towards the internalization of research.

In the United States the possibility of a federal agency which could work in collaboration with the leading manufacturers to regulate product standards was unknown in the 1890s. At the state and municipal levels there were some strong public health offices where extensive testing was conducted, often with the aid of local manufacturers, but the development of the federal Hygienic Laboratory stemmed from a different origin. The Hygienic Laboratory grew out of the predecessor to the Public Health Service, the Marine Hospital Service. Bacteriological investigations in the Service began at the Marine Hospital building in Staten Island, New York, where a 'laboratory of hygiene' was established in 1887, mainly to monitor the influx of immigrants. In 1895 it moved to Washington and began production of an American diphtheria antitoxin. By the end of the century the Hygienic Laboratory was equipped with facilities and had the experience in the staff to assess biological products.

There was an evident need for such assessment. Within months of the announcement in 1894 that diphtheria antitoxin was efficacious, the *Journal of the American Medical Association* cautioned physicians about the possibility of there being fake antitoxin on the market. Even to produce real antitoxins, 'any kind of a stable, a little technical skill, and a fair amount of nerve' were all that was needed to get into business. Furthermore, these fears and the model of German regulatory procedures were not enough. As with so much legislation, especially concerning drugs, it took some tragedies to mobilize popular, professional and commercial opinion. Two events in the autumn and winter of 1901 provided the final stimulus. In the St. Louis public health department a tetanus-infected horse had carelessly been used to produce diphtheria antitoxin which soon killed 13 children. There was an enquiry which aroused the interest and indignation of people throughout the country and which led to the dismissal of the laboratory head and his assistant amid general outcry against the dangers of serum therapy. At around the same time nearly a hundred cases of post-vaccination tetanus, and the subsequent death of nine children in Camden, New Jersey, led to a similar enquiry into the safety of commercially prepared smallpox vaccine.

With the stimulus of these two affairs and the model of cooperation between Hoechst and the regulators in Germany, the industry along with medical and public health professionals drafted a law to provide controls on the production and sale of biologicals. The act, passed in 1902, specified that manufacturers must be licensed through the Hygienic Laboratory, and each package of regulated viruses, sera, toxins and analogous products must be clearly labelled and dated. It also allowed for 'reasonable inspection' of company properties at any time and provided guidelines for enforcement. Regulation strengthened the ties between the government laboratory and the leading companies in the United States as it had in Germany. The two leading producers, Parke Davis and the H. K. Mulford Company, were in close contact with members of the staff of the laboratory.

The Federal Food and Drug Act of 1906, the basis of current American durg regulation, developed separately. The 1906 Act was used in a number of celebrated cases against patent medicine makers, but its effect as a regulatory device for manufacturers of ethical pharmaceuticals was not onerous. It came about as a result of extended lobbying for controls over adulterated and contaminated foods and was administered by the Department of Agriculture's Bureau of Chemistry. The Act made official the specifications of the *US Pharmacopoeia* and its supplement. Its overall impact as far as most established firms were concerned was to discourage a large but marginal group of the industry. It restricted the entry of small manufacturers and forced existing ones either to expand to the point where they could support a scientific staff, or to merge with larger companies. The significance of legislation and litigation concerning drugs finally hinged on the extent to which the pharmaceutical manufacturers themselves influenced the setting of standards and whether they could avoid prosecution. Leading firms maintained close contact with the Bureau and had some effect on the standards set. Their concern was to set criteria which could not easily be met by competitors.

In addition to standardization, a key issue which concerned those who were involved in research into therapeutic substances was the relationship between research organizations and the pharmaceutical industry. Nowhere was this conflict clearer than at the Pasteur Institute. In the early part of the 20th century the Pasteur Institute, like other national medical laboratories, had an ambiguous and ambivalent relationship with the pharmaceutical industry. There were a number of potential areas of conflict from the start. Funding was in any case a difficult matter, requiring constant attention from many of the leading figures in the Institute. The opportunities provided by commercial exploitation were in themselves tempting but the conflicts of interest they were seen to bring mitigated against exploitation. Efficient dissemination also proved to be double edged. Using companies was most effective for rapid dissemination, but control and potential income had to be forfeited. In some celebrated cases, such as with the production of diphtheria antitoxin, that meant that there was the potential for competition between the governmental laboratory and drugs producers and distributors. That conflict was faced in different ways in Germany and the United States, the two other significant producers of antitoxins. In those two countries national laboratories had well-defined functions in relation to producers, especially as regarded potential regulatory activities.

In Germany, where Paul Ehrlich presided over the national laboratory from 1897, there was well-coordinated cooperation between major pharmaceuticals producers and the researchers, as well as regulators, in government service. Conflicts of interest were not a serious concern, and the size and prominence of the national laboratory assumed its central role in many aspects of commercial activity.

In the United States a legal structure preceded the equipping of a major national laboratory which was, by virtue of being placed in Washington, separate from the commercial and academic centres of the medical industry. The lack of contact which this tended to create was overcome largely through close cooperation between government scientists and a couple of large manufacturers which manoeuvred into positions of influence, largely through superiority in technical work. In the content of the major scientific enterprises in these two countries, there were strong similarities between the leading national laboratories and the largest company laboratories.

In France, however, pharmaceutical companies had made little move to invest in research until late in the 20th century. Furthermore, the Pasteur Institute tended to work further afield of the mainstream of therapeutic research than did the German and American laboratories. The French pharmaceutical sector was thriving, but it depended on small family-dominated firms marketing 'specialties' as over the counter medicines, and until at least the late 1930s few companies sustained any significant research effort. Consequently, they exerted little pressure for aid. Poulenc (later Rhône-Poulenc) were the only exception, building up a small pharmaceutical research team from 1903 onwards.

In addition to this lower level commercial demand, Louis Pasteur's personal experiences of industry gave him reason to protect the distinction. Despite his personal interest in fermentation, for example, he refused the opportunity to become involved with a brewing research institute to be founded in the district near the Institute in case this caused the Institute to deviate from its 'purely and narrowly scientific aims'.

As a result of the success of the Institute's initial public appeal, it was awkward and rather embarrassing to return to the public with requests for further funds to pay for the ambitious programme of growth. However, following the research by Emile Roux and his coworkers on German-developed antidiphtheria serum, the Institute received much good publicity. This success created its own dilemmas. As Roux pointed out to the Minister of the Interior in 1897. 'Requests for serum were coming from all over. . . . A rising tide threatened to submerge the bacteriologists'. With support from the mothers of suffering children, reviving the spirit of the first rabies appeal, the institute tried to extend the coverage offered by the treatment. Eventually a grant was made by parliament to cover distribution, although in accordance with Roux's request, distribution was not of totally free serum, as this would have lead to seepage abroad, and the Institute would not then be able to cope with increased demand. Roux pointed out that any profit made on sale of serum would be recycled through the Institute, and eventually fund new research for the public good.

Dissemination was recognized from the start as a key issue of concern. A different code of practice, however, seems to have applied in discoveries where human health was not at issue. Pasteur and Roux devised a vaccine for veterinary use against anthrax in 1886. An associate of the Institute, M. de Ste Marie, was given the concession to exploit the discovery. This was done through a newly formed company, the Societé de Vulgarisation des Vaccins Carbonneux Pasteur, with the aim of maximizing revenues from the vaccine. This step was taken to move production away from the Institute and establish the business along commercial lines. Given the Institute's charitable status, customers often expected philanthropy to extend to cattle, and include free distribution of the vaccine. Also, the new company was more free to advertise and stimulate demand than the Institute was. The experience proved less than happy, however, when revenues from the anthrax vaccine did not grow as expected. Relations soured with the commercial partners but as the Institute tried to back away from the agreement, the businessmen involved insisted on exercising their prerogative and held them to the full contract of 30 years. In 1895 relations reached a nadir and a member of the Institute took the company to court for using his name on their letterheads, which he felt implied that he personally was making money out of the business.

Under Emile Roux's directorship (1904–1933) the mixed funding which he had championed from 1897 continued. The Institute's activities were determined by the receipts from the state for the provision of serum therapy, and the revenue from the gradually increasing capital fund. There were problems. Roux was aware that the distribution system for sera was open to abuses, particularly in the cases where sera were given freely. Also, it was estimated that state subsidies for the distribution of sera were insufficient to cover costs, and that between 1895 and 1911 the short-fall had amounted to around FF 1.5 million.

Of all the areas of research likely to provide commercially attractive results, chemotherapy seemed to offer the best prospects. The area became one of Roux's great enthusiasms and it was under his leadership, in 1911, that Ernest Fourneau was asked to establish a department of chemotherapy. Fourneau had joined the Etablissements Poulenc Frères in 1903 as head of their new research laboratory. His early work at Poulenc concentrated on the development of a nonaddictive cocaine substitute. Stovaine was the first result of this work and made Fourneau a celebrity. Stovaine was the first synthetic drug to be developed outside Germany. At this time Bayer tried to lure Fourneau away from Paris to work in their laboratories, but he is said to have resisted 'on patriotic grounds'. Fourneau's appointment at the Pasteur Institute was made explicitly conditional on his giving up all contact with his former employers. Nevertheless, the appointment was controversial since it was effectively a change of direction for the Institute, which was likely to prove expensive and stretch the Institute's resources.

The timing of the appointment and the intensity of Roux's support were probably due to the first reports from Ehrlich's laboratory of the results of the salvarsan experiments. Several of the Institute's biologists had been involved in the feld, and they felt ready to respond quickly. In this atmosphere it was possible for Fourneau to continue his own work on arsenicals once he had joined the Institute and with Roux's blessing he continued his technical collaboration with Poulenc. In at least one instance Poulenc supplied chemicals which otherwise Fourneau and his staff would have had to prepare themselves. It was largely as a result of these continuing informal contacts that Poulenc were in a position to market a salvarsan substitute shortly after the outbreak of the First World War.

The collaboration with Poulenc continued in the post-war period. His work at the Pasteur Institute again met with conspicuous success when in 1924 he managed to analyze the structure of Bayer 205, a nonmetallic urea compound found to be active in the treatment of trypanosomiasis (sleeping sickness). At that time the Germans were unable to defend themselves as French patent law did not allow protection for any medicinal products until 1959. At around the same time Fourneau was working on stovarsol, an antisyphilitic arsenical compound. An analogue of stovarsol, orsanine, was marketed in 1930 by Specia, the pharmaceutical business of the newly formed Rhône-Poulenc.

By the interwar period, then, the relationship between the Institute and companies such as Poulenc and Specia became more or less routine, with privileged access to new research at the Institute traded for small practical favours, such as cheap raw materials. It is possible that the collaboration was more formal than appears at first, and a specific contract of some form existed between the two, so that the company paid for the technical expertise it obtained from Fourneau, perhaps with Roux's consent but unknown to the rest of the governors of the Institute.

Another case where there was a need for commercial involvement came with the work of Calmette and Guerin on a vaccine for tuberculosis in new-born children, which resulted in the BCG vaccine. As with the expansion of serum production, the Institute was anxious only to cover its expenses and to be able to disseminate the popularized new techniques. The scale of the publicity effort required to assure the adoption of BCG, however, was not at first appreciated by the Institute administration. It soon became apparent that the Institute lacked the resources and the task was delegated to the Comité National de Defense contre la Tuberculose (CNDT). The CNDT had been set up with help from the Rockefeller Foundation immediately after the First World War and Calmette became its leading figure. Although the Institute retained control over the manufacturing of BCG vaccine, they had little control over the diffusion campaign. A new production laboratory was built in the early 1930s with funding split between the Institute's resources, FF 5 million, and a state grant of FF 2 million. The running costs of this laboratory required a further annual grant from the state of FF 4 million, a much larger sum than the FF 1.4 million given to subsidize serum production. Income from the sale of vaccines and viruses in 1932 totalled FF 1.3 million, compared to the pre-Stock Market crash income of FF 3.7 million. For the first time, then, the Pasteur Institute was forced to concede the need for constant commercial activity in order to remain solvent. Instead of trying to solve these problems through a direct arrangement with pharmaceuticals producers, the Institute struck out on its own. In contrast to foreign national laboratories and their relations with industry as awkward regulators, in France a new model of cooperation, including room for potential competition, was tried.

The problems which the Pasteur Institute had to face were as much related to issues associated with marketing as with industrial research. The peculiarities of marketing in the pharmaceutical industry have frequently been noted. The distanced relationship between producer and final consumer, the role of the physicians as mediator and the actions of the state as price fixer or limit setter complicate the relationship. The desire to maintain quality at the expense of marginal costs and the willingness to take the advice of experts during times of illness have all been seen as constituting the unusual criteria for marketing medicines. This characteristic of drug marketing developed slowly, taking on its recent form only in the late 19th century. Previously medicines had been marketed to the general public based on the understanding that common knowledge, or a 'commonplace book', would provide any necessary information.

The drug industry has been credited with two major innovations in marketing. One was the extensive use of newspaper advertisements to achieve national brand recognition. The other was the use of 'detail men': company representatives with substantial technical knowledge who visited physicians in their offices to explain the contents, use and value of new products. These two innovations developed into one of the most elaborate and expensive forms of marketing of any industry.

The market for medicines has long been distinguished from other consumer goods. With physicians acting as intermediaries between producer and ultimate consumer the responsibility was spread widely among those making, selling, prescribing and buying drugs. Occasionally there were demands for a remedy for a particular disease but more generally manufacturers offered what they had on hand. This allowed for a substantial amount of initiative on the part of the companies. They took it upon themselves not only to promote their products, but also to ensure that their market recognized and understood the significance of a new medicine. Drug sellers turned to ingenious advertising before other manufacturers out of necessity. Operating in conditions of almost unlimited potential demand with products easy to produce on a large scale, the pressures to pioneer marketing techniques were strong.

Although direct contact with persuasive salesmen was the best means of imparting technical information, pamphlets were the most common method used. The style of publications changed over the years, remaining 'serious' and 'scientific' in appearance but variously taking the form of sales leaflets, general information pamphlets, scientific reports (either reprinted from a journal or mimicking scholarly form), or even the style of review articles, summarizing the literature and debate over a therapy. Their content varied far less. Generally the drug would be introduced first, rather than the company, which was sometimes identified on the title page only as the pamphlet's publisher. A discussion of the therapy would follow. Explicit instructions would be given for dosage, administration and follow-through, with careful distinctions made for differing forms of the illness. Early pamphlets placed these instructions within the context of basic bacteriology and immunology, explaining, for example, the relationship between bacteria and toxin and the neutralizing effect of antitoxin.

The key to this growth and sophistication and parallel rise in costs was the increasingly technical character of new therapeutics and the consequent distance between the knowledge held by the company and necessary for the use of the drug, and the knowledge which every practitioner could be expected to have. As drug companies required increasingly technically advanced people for research and development, and as medical scientists were more available for industrial employment during the interwar years, the use of new science became central elements of competition.

Another interesting way to contrast the development of the pharmaceutical industry in Britain, Germany and the United States is to consider their activities with regard to cartels. It was in the German chemical industry in the late 19th century that techniques for building cartels and strategies for using cartels became highly developed. First, in association with each other, later extending internationally, these cartels, selling agreements and pricing conventions did more than manipulate national markets for single products: they strongly influenced the rate and direction of the industry's growth.

Based on the strength of the dye industry and the expansion of Germany's economic sphere of influence, German companies began to compete on a variety of fronts. To organize that competition, German chemical makers began to use marketing agreements amongst themselves, and soon to include their foreign competitors. Examples of agreements entered into between 1887 and 1910 included those covering iodine, camphor, salicine, bromine, strychnine, caffeine and codeine. For the most part these agreements specified prices and markets, defined geographically. Sometimes they specified further what proportion of raw material available generally could be allocated among signatories.

From the point of view of the British industry especially, the arrangements they entered into became a major aspect of their business. Regarded as 'remedial' and 'defensive', rather than monopolistic, these agreements were initially only oriented toward raising field prices in the aftermath of price wars. Combinations also allowed temporary alliances without requiring significant rationalization, increased productivity or expanded marketing. They also preserved individual autonomy and helped to maintain family control. Companies such as Whiffen and Howards were particularly deeply involved because of their traditional strengths in importing raw materials and relying on their shipping networks to supply markets.

Particularly good examples are the caffeine, iodine and camphor cartels. The first two are illustrations of the reaction of firms to the need for combinations, and the third is an example of the fragility of these agreements due to their inability to address the cause and not just the symptoms. These cartels were typical of arrangements used by German and British companies to try to stabilize trade, but they served the two national industries quite differently. To the Germans it opened markets overseas and took advantage of superior British supply routes. This facilitated their growth especially by opening up new areas. For the British, however, the agreements were usually drawn up with the intention of covering products which were already being imported, processed and sold or exported. Their advantage was primarily in price-fixing and it was a defensive strategy, which staved off competition. British interests were directed at imperial territories and the United States, the first to postpone German commercial encroachment and the second in the naive hope that they could regain their early 19th century position as major suppliers. What the British failed to understand was that by emulating the German industry, American manufacturers, with effective governmental aid, were being transformed fundamentally.

1.3.4 MASS PRODUCTION, WORLD WAR AND THE FOUNDATIONS OF A MODERN INDUSTRY

The First World War provided the opportunity for leading companies to consolidate their positions and to build the foundations for long term growth. At the same time the pressures of competition and the expectations of the consumers for 'advanced' or novel products forced many American manufacturers to institutionalize research and development and to solidify their links with the medical community. After the First World War the German industry responded to defeat by merging into I. G. Farben, while the British industry failed to capitalize on their short-felt resolve to compensate for their reative weakness. At the same time the Swiss industry emerged as a major force. First Hoffmann-La Roche grew from a medium-sized drugs manufacturer to a European-wide industrial giant, followed by Sandoz and Ciba, who benefited tremendously from the expanded markets during the war.

The rise of the Swiss pharmaceutical industry is one of the most spectacular in all of industrial history. From small roots as local pharmacies and later influential dye makers serving the textile industry of southeastern France and northwestern Switzerland, what are today three of the largest drugs companies emerged. Geigy was in the 18th century a trading company, which went into agricultural chemicals and medicines only in the early 20th century. Similarly, Ciba was in the 19th century a small silk factory, which moved from synthetic silks into synthetic dyes and then medicines. Merged only in 1969, they formed part of the prosperous Basel community which began to benefit initially from their neutrality during the world wars and capitalized on their trading advantages by cultivating strong links with universities to promote research and product development.

Sandoz was also a dye manufacturer until the end of the 19th century, when they tentatively followed their German competitors into snythetic medicines. With important developments in ergot-based products, they were able to secure a niche which propelled them into a major multinational corporation. Sandoz was set up as a partnership between a dyestuffs salesman, E. Sandoz, and a financier, A. Kern, in 1886. Within two years they were able to set up a small factory in Basel to produce synthetic colours. The business grew slowly and steadily until the First World War.The Swiss suffered both as a result of the small size of the Swiss domestic markets, particularly at a time when much of Europe was resorting to protectionism or cartel arrangements, and through Switzerland's lack of the necessary natural resources.

Sandoz, along with the other main Basel dyestuffs companies, Ciba and J. R. Geigy, were radically transformed by their experiences during the First World War. They had almost a complete monopoly of dyestuffs supply to the Allied powers and sales rose as fast as production could be increased. Raw materials continued to be a problem, and, with the company making record profits, directors began to look around for possible opportunities to diversify, anticipating a post-war slump in dyestuffs. One of the areas explored, as with the German companies, was pharmaceuticals. With the enthusiastic backing of a key member of the Board of Directors, Arthur Stoll, a young pharmacologist, was appointed to research this new area. Stoll's own research field was the standardization of ergot preparations, and the first pharmaceuticals launched by the new department built on his experience in this area.

Many of the first pharmaceutical products were sold at or near cost, particularly overseas, in order to get the company's name known. Different markets were charged differently, with exports to Germany sometimes priced at around one third the Swiss price, and close to a quarter the production cost.

In the immediate post-war period, the pharmaceutical department also undertook some basic preparations of alkaloids. Any profits made were used to subsidize the longer term research into synthetic pharmaceuticals. However, this business declined rapidly once the League of Nations initiative to restrict the trade in dangerous drugs and narcotics began to affect the pharmaceuticals industry. By that time, though, Sandoz's first proprietary products were coming to the market. With new opportunities and strong pressures from the small size of the Swiss home market, Sandoz decided in the early 1920s to market aggressively their new range of pharmaceuticals in several of the overseas markets where they already had an established presence in the dye business.

Only Hoffmann-La Roche among the Basel firms began as a pharmacy in the 1890s. They grew by exporting internationally and had subsidiaries or major agencies worldwide before the outbreak of the First World War. Hoffmann-La Roche were leaders in psychoactive drugs since they introduced new opium preparations in 1909. Their later work on analgesics, and in the 1960s on new psychopharmaceuticals, provided the foundation for growth which has made them one of the largest privately owned companies in the world.

There was a wide range of marketing practices used in the industry, differing internationally and coming to maturity only around the First World War and during the interwar period. The change from marketing medicines as a product similar to others to marketing medicines as a special item to be used with scientific supervision came at different times in different countries. The pharmaceutical industry in Germany and the US showed signs of change before the First World War. In Britain, the change in business was only fully appreciated by the middle of the interwar period, although it had been partly forced through by the experiences of the war. French pharmaceutical companies were also late to recognize the possibilities of new marketing methods and products.

Major new products of the interwar period bolstered the image of the industry. The sulfonamides and insulin in particular attracted much attention, and new chemotherapeutics promised cures for a wide range of illnesses. Vitamins also claimed miraculous benefits as nutritional supplements as well as therapeutic substances and were effectively exploited by firms such as Glaxo in Britain.

In contrast to Germany and America, Britain had virtually no regulation covering medicines until well after the First World War. Other than food and drug laws prohibiting gross acts of adulteration, and even these were difficult to define and enforce, manufacturers had virtually free hands. Nor were there many suggestions that British companies should be inspected as American firms were, or that their products should be tested and certified as German law prescribed. There was no outcry, that is, until the relative weakness of the British pharmaceutical industry was dramatically revealed by the outbreak of war.

It was the war and the apparent disadvantage which shocked the British industry into reassessing their position. In the case of one drug in particular, a new apparatus was devised to encourage its production utilizing the German patent and regulating its quality. That was the antisyphilitic salvarsan '606', an invention of Paul Ehrlich and a product of the Hoechst company. Salvarsan and its related arsenic-based chemotherapeutics were regulated in Frankfurt in the same way as the biologicals. This was partly because of Ehrlich's need to maintain control over the extended clinical trials, and partly because of the institutional structure in which it was introduced, where company ties and government responsibility played such a large part. Salvarsan was also potentially highly toxic. With a special need for Salvarsan identified because of the mobilization of soldiers and the well-known rise in incidence of venereal diseases in wartime, the newly formed Medical Research Committee (MRC) identified an active role for itself in regulating it. In collaboration with the Board of Trade, which coordinated the seizure of enemy-held properties, including patents, licences were granted to Wellcome and the French Poulenc Frères to produce salvarsan for the British market. The companies were held to unique provision, however, that all samples 'be submitted to biological tests' by the MRC. Official certificates were granted by the committee for each batch, but the cost of testing was charged to the manufacturers. The whole procedure was undertaken in a new laboratory, which maintained good contact with Wellcome in particular. This arrangement seemed to function well for all parties involved, and within a year the MRC was petitioning the government to extend regulations to cover sera, vaccines, other biologicals, chemotherapeutics and certain new medicines as they entered the market. Leading British medical scientists saw this as a move both to bring the UK into line with other countries whose products were regulated, and to put an end to a 'period of anarchy'.

During the interwar period in both the United States and Britain a wide range of scientific qualities, such as specific effect and level of toxicity, came to be associated with new drugs and were used in the promotion by the leading companies. A range of acceptable terminology was implicitly established. Drugs could be described, for example, as having been 'tested' or 'standardized'. Advertisements placed in the medical press and on pamphlets directed at physicians more often cited technical literature, a practice extended to include a wide range of English, French and German works. While the producers of biologicals were the first routinely to use extensive technical citations in advertisements for their medicines, the practice soon spread to other sectors.

This period also saw a new involvement by governments in the pharmaceutical industry, in America in the form of enforcement of the 1902 and 1906 Acts regulating drug production. The effects on marketing of regulation were widespread. Leading manufacturers saw their incentives to set criteria which could not easily be met by competitors. The companies insisted on a high standard of purity that could not be met by small operations. They could also rely upon the support of the medical community to testify to the efficacy of their scientifically advanced medicines. Even though the 'ethical' manufacturers could not hope to take over from patent medicine makers completely, they did gain in prestige as champions of scientific therapeutics by vilifying the so-called quacks.

With drug makers expanding into ever more highly theoretical areas like chemotherapeutics, physicians increasingly relied on them to provide information. The educational function which a growing number of companies assumed before and during the First World War became the acceptable and expected function of drug makers during the 1920s. Although physicians began to look more critically at the information they were receiving, they were also resigned to the fact that only scientists within companies and physicians who had conducted clinical trials for manufacturers had full information about a new product. Within a few months after the introduction of a drug, medical journals might begin to have independent results to publish, but this was not always the case and the results were often far less complete or conclusive than company-provided information.

Whatever the bewildering range of functions early company scientists actually performed, in every case their presence was recognized to be a distinction for the firm. It was a sign that they were manufacturing new and unusual products, medicines which required scientists for their manufacture and, presumably, comprehension. It was not far from this company role to the development of a broader image which leaders in the pharmaceutical industry adopted: the image of high technology companies. What this self-styled elite among manufacturers produced and sold was not simply a range of pharmacy products, but the very image of advanced medical science itself.

By the interwar period, then, when the potential market for many new remedies was opening up world-wide, national and international standards and regulations were in place in Britain, the United States and Germany. For Germany the initiative came from the community of Berlin medical scientists with the 'enlightened' patronage of the Minister of Education and Health. In collaboration with leading manufacturers, Hoechst in particular, standards were set and easily enforced. In the United States regulation was partly modelled after the German system, and spurred on by the tragedies in St. Louis and New Jersey. Regulation was quickly used by American producers to run out their competitors.

The case of the development of insulin and the response of the medical industry to it is particularly instructive of both official and commercial reactions to new products. Developed at the University of Toronto in 1921–1922, insulin immediately attracted attention in Canada and the United States. The Toronto researchers, Frederick G. Banting, Charles H. Best and John J. R. MacLeod, already had arranged with the Eli Lilly Company to manufacture and distribute in North America, and there was the possibility that the British Medical Research Council might have a comparable role in Europe.

Problems emerged near the start. In general the large scale manufacture of the extract, both at the University of Toronto and at Lilly in Indianapolis, was inadequate. The yield was below that obtained in small scale experiments and the potency was irregular. Furthermore, there lacked a simple test to measure critical characteristics. The patent claim seemed weak, vague and too general. Any attempt to specify the practices then in use for the purpose of a specific process patent would bring about easy circumvention. Already by the end of 1922, there were numerous rival patents being filed in America. Eli Lilly had registered their trade name Iletin and Fairchilds, a maker of ferment preparations, were advertising an extract of pancreas which they implied contained the active hormone, but which clearly had no insulin. Evidently even the most fundamental purposes of the patent were inadequately served. The Toronto inventors were unable to control the quality and price of the substance, and even the US government's Hygienic Laboratory had difficulty legally adding insulin to its list of regulated substances because of the limited definition provided by the patent.

Nevertheless, the medical community was anxious to begin using some remedy for diabetes and the early reports of insulin's successes stimulated great excitement. Lilly took careful advantage of their favoured status and foreign companies scrambled to get production or at least importing rights. Insulin also tested the ability of governments to respond quickly to the needs for regulation and control. Following on from the precedents of the early antitoxins and chemotherapeutic agents, government laboratories, especially in Britain and the United States, had to clarify their roles. In Britain especially, this challenge to the government threw into light the whole relationship between government and industry. Not only did the Medical Research Council have to decide what kind of links, through commercial practice as well as research, it would have, but companies had to decide whether they were willing and capable of taking on such research. This became an even more charged issue as the production of penicillin began to seem feasible.

A survey of the research capacity of five of the leading British companies was carried out in 1942. This showed a wide variance in size and emphasis. In the period 1936–1941, May & Baker held the lead in the group for the number of British patent applications (40), but their publication rate was slow, having produced only 11 scholarly articles out of a staff of 58 university graduates, 15 of whom held doctorates. Burroughs Wellcome had placed only six patent applications over the period, but their 66 degree-holding scientific staff, 24 of whom held doctorates, had published 220 journal articles in the period. Glaxo stood between these two research leaders with a less-well-qualified scientific staff (eight doctorates) but a publication record of 345 articles and 13 British patent applications. For the same period BDH had only five doctorates on the staff, but produced 32 publications and seven patents. Even so they were generally regarded as having little research potential. Boots was similarly regarded. They had a very large staff, amounting to almost 270 scientific workers, 24 of whom held doctorates. Nevertheless, they still only published 10 articles and applied for only 12 patents over the five year period.

It was in this context that efforts to capitalize on penicillin took place. Alexander Fleming made the initial observations of the antibacterial effect of the *P. notatum* mould in 1928. He noted its potential, but there was little more done with penicillin until 1939 when Ernst Chain looked to

Fleming's work for additional data on lysozyme, and turned Howard Florey's Oxford laboratory into a penicillin research centre. Not much was done to produce it, the traditional version of the story of the development of penicillin tells us, until Florey and his colleague Norman Heatley travelled to the United States in the summer of 1941 and brought penicillin to the attention of some American manufacturers, in particular Pfizer, Merck, Squibb and Lederle, along with United States federal government science administrators at the Office of Scientific Research and Development. From that time a number of changes took place, in particular in the strains used and the techniques of mass production by deep fermentation at the Peoria, Illinois, government laboratory.

1.3.5 ANTIBIOTICS AND POST-WAR GROWTH, 1940–1970

In Britain the bitterness of the penicillin episode stayed with the industry into the 1950s. Demoralized by the criticisms associated with penicillin and insecure of its place within the new National Health Service, the industry suffered an invasion of foreign takeovers of old firms, such as the acquisition of Morson's by Merck Sharp and Dohme, and numerous consolidations, such as the ones which built Glaxo into a major drugs manufacturer. There were of course a few major contributions by the British industry during the 1950s in areas such as antibiotics and with cortisone research, and Beecham did build itself by concentrating on penicillin production. The thalidomide tragedy, however, did unprecedented damage to the public image of the industry and forced major reassessments of how industry practice 'allowed it to happen'. The answer to that question, like the answer to the question of how British companies lost the lead in antibiotics, is the direct consequence of the history of the industry.

Two new types of products stimulated the growth of the industry during the period from the Second World War: antibiotics and psychoactive drugs. They provided massive new markets on which to base further corporate growth, and changed the role of research from an opportunistic to a necessary function. The success of the American cooperative effort to produce penicillin on a large scale is only part of the story. The less successful efforts of the Therapeutic Research Corporation, a consortium of British manufacturers and of German and French researchers, are important to understand in order to explain the subsequent distribution of strengths in the industry. After the war and the rise of Glaxo, Beecham and Boots in the UK, Pfizer and Lederle in the United States, and the reestablishment of Schering, Hoechst and Bayer in Germany, along with the continuing success of the three major Swiss manufacturers, the industry took on a marked new international and transnational character.

During the Second World War the drug industry was challenged to produce penicillin on a large scale. The use of antibiotics radically altered therapeutics; it also fundamentally affected the drug manufacturers. The production of antibiotics was ordinarily more complex than that of the biologicals or even chemotherapeutics, and hence even larger and more highly trained scientific staffs had to be assembled. Moreover, the subsequent search for other antibiotic drugs necessitated creative fundamental research and product development on a greater scale than ever before. Their proliferation was then matched by those of the pscychoactive drugs of the 1950s. But none of this was unprecedented. The company structure in which this took place, the laboratories and testing facilities, even the special relationships between marketing departments and physicians had all been established in the years preceding the Great Depression.

The character of the industry changed in the post-war years with the introduction of new antibiotics. Selman Waksman, working in the late 1940s at Rutgers University in New Jersey and in conjunction with the Merck company, developed a technique of screening soil samples to find new antibiotics and produced streptomycin, the first new antibiotic since penicillin.

These new antibiotics were also patentable, leading to a wide range of new products as a result of new research, as opposed to new production technology. The industry, responding to these new opportunities, began producing a wide range of antibiotics and knitting together licensing agreements, which changed the way companies gained entry into new product areas. This persisted until the broad spectrum antibiotics were systematically introduced. Starting with aureomycin, developed at Lederle Laboratories in 1948, chloromycetin at Parke-Davis in 1949, and terramycin at Pfizer Laboratories in 1950, all the leading firms began looking for their own antibiotics to market. Among the changes brought about by this boom in antibiotics was the eclipse of Pfizer, which had quickly jumped to the lead in the penicillin market.

Increasingly through the 1950s problems of patents and licences brought the pharmaceutical industry into the courts. The particular entanglements associated with the introduction of tetracycline brought into sharp relief the complex legal and structural condition which was emerging in the American industry.

The thalidomide tragedy brought the industry unprecedented shame and gave fuel to a range of opponents of drugs manufacturers who objected to their high profits, their collusive behaviour and their evidently inadequate checks on safety. A new relationship with government was supposed to have emerged, but the real effect in the 1960s was to put the industry under stringent scrutiny over prices, and to slow the process of drug approval by regulators. The 1960s saw the rise of the Japanese industry as part of the general boom in the Japanese economy, and the broadening of the base for multinationalization.

The 1960s also saw the rise of a new scrutiny of the pharmaceutical industry. The Kefauver Committee in the United States Senate focused on the industry in their antimonopoly investigations. Following these investigations, many discussions in the industry revolved around how to avoid abuses of monopoly-type power, while continuing to reward pharmaceutical companies for the financial risks which they took in developing new medicines which were beneficial to society. In Britain there was a sharp response to the findings of Kefauver, heightened by the impact of the thalidomde disaster. The Sainsbury Committee in 1967 considered the question of pharmaceutical pricing, and the Labour government toyed with the idea of nationalizing the whole pharmaceutical industry.

The size of the pharmaceuticals market has been growing tremendously in recent years. The world market was estimated to be around $10 billion (unadjusted) in the mid 1960s. In the mid 1970s the world market was estimated at about $36 billion. By the early 1980s the world pharmaceutical market was worth over $80 billion, most of it in sales to the developed market economies. The market is dominated by a small number of very large companies organized on a multinational basis and located in the United States, Germany, Switzerland, the United Kingdom and Japan. There is continuing regulation over the introduction of new products in all the largest countries, and most national governments regulate the prices which can be charged. The European market is divided along national and regional lines and accounts for just over 20% of the world market. European production, however, accounts for more than 25% of world output, concentrated in 33 firms which have a capacity for serious product innovation. Additionally, there are approximately 1500 other manufacturers, which produce primarily out of patent drugs, specialized products, or work under licence.

Drug consumption varies internationally according to cultural and historical patterns. These differences have a great deal to do with attitudes towards therapeutics and towards the role of physicians and pharmacists. Other variations of especial importance follow economic and demographic trends. Drugs for the treatment of the diseases of old age have come to dominate, as might be expected, in those countries with prosperous, generally healthy but aging populations.

As for drug marketing, we have seen the evolution of an established manufacturing sector into a high technology industry, and its adaptation of a scientific image which was most useful for marketing its products. The question of the control of information about drugs continues to plague the medical world and its regulators. Recent studies in Britain and the United States have shed some light on the continuing role of manufacturers as sources of information. Company representatives still supply most information about the existence and cost of drugs, and whereas scholarly periodicals may be of greater influence and regarded as more reliable sources about the efficacy of new medicines, they are less accessible.

Through the systematic use of trained representatives, the detail men, drug companies continue to maintain close personal contact with physicians. They frequently give doctors gifts and 'reminder items'. In the US they gave away almost 200 small gifts per doctor in 1973, just to establish a relationship and solidify the contact. Marketing and sociological studies have shown that doctors seeing detail men tend to prescribe drugs from the detail men's companies after the visits. There seems to be a direct relationship between the number of visits from a particular company and the propensity to prescribe a particular drug, and this closely matches the company's advertising expenses.

Throughout the industry drug marketing has evolved from small scale selling in competition with patent medicine makers to expensive and intensive marketing which reflects the sophistication of their products. While the industry in some countries, such as Britain, grew more slowly than the American, Swiss and German industry during the post-war years, and in particular relied on traditional markets and more conservative production methods, with the increasing internationalization of the industry, techniques for controlling information continued to concentrate in the industry.

1.3.6 REFERENCES

1. M. L. Burstall, J. H. Dunning and A. Lake, 'Multinational Enterprises, Governments and Technology; Pharmaceutical Industry', OECD, Paris, 1981.
2. M. L. Burstall, 'The Community's Pharmaceutical Industry', European Communities—Commission, Brussels, 1985.
3. W. Davis, 'The Pharmaceutical Industry, A Personal Study', Pergamon, Oxford, 1967.
4. E. Kremers and G. Urdang, 'History of Pharmacy', 4th edn. (revised by Glenn Sonnedecker), Lippincott, Philadelphia, 1976.
5. J. Liebenau, 'Medical Science and Medical Industry, The Formation of the American Pharmaceutical Industry', Johns Hopkins University Press, Baltimore, MD, 1987.
6. J. Liebenau, 'Industrial R&D in pharmaceutical Firms in the early twentieth century', *Business History*, 1984, **26**, 329.
7. J. Liebenau, 'The British success with penicillin', *Social Studies Sci.*, 1987, **17**, 69.
8. J. Liebenau, 'Ethical business: the formation of the pharmaceutical industry in Britain, Germany and the United States before 1914', in, 'The End of Insularity, Essays in Comparative Business History', ed. R. P. T. Davenport-Hines and G. Jones, Cass, London, 1988.
9. T. Mahoney, 'The Merchants of Life, An Account of the American Pharmaceutical Industry', Harper, New York, 1959.
10. M. Robson, 'The pharmaceutical industry in Britain and France, 1919–1939', PhD Thesis, University of London, 1989.
11. D. Schwartzman, 'Innovation in the Pharmaceutical Industry', Johns Hopkins University Press, Baltimore, MD, 1976.
12. M. Silverman and P. R. Lee, 'Pills, Profits and Politics', University of California Press, Berkeley, 1974.
13. P. Temin, 'Taking Your Medicine; Drug Regulation in the United States', Harvard University Press, Cambridge, MA, 1980.

1.4 Development of Medicinal Chemistry in China

HUANG LIANG AND ZHOU JIN
Chinese Academy of Medical Sciences, Beijing, People's Republic of China

1.4.1 INTRODUCTION

Medicinal chemistry development in China may be divided into three stages. The first stage, which depended upon traditional Chinese medicine (TCM), lasted several millennia and shifted to the second stage after the introduction of medicine originating from the 'West' to China in the early 19th century. During the second stage TCM and 'Western' medicine co-existed independently. In the third stage, three medical systems, TCM, 'Western' medicine, and one resulting from integration of Chinese traditional medicine and 'Western' medicine co-exist and are characteristic of the last 40 years of health care in China.

Since the 1950s more attention has been paid to drug research and production of 'Western' drugs along with traditional ones. The 1985 edition of the Chinese Pharmacopoeia lists 713 items comprising traditional drugs and selected set prescriptions in part I and 776 items in part II related to chemical pharmaceuticals, antibiotics and biologicals which are now in production in China. Through the effort of physicians and scientists in the medical field, contagious diseases such as plague, smallpox, cholera, typhus, relapsing fever, poliomyelitis, measles and leprosy, and the parasitoses like malaria, schistosomiasis, hookworm and filariasis are under control through treatment and preventive measures. The health condition of the people has been significantly improved in the last 30 years through a unique, close partnership of TCM and 'Western' medicine. As a result, the average life expectancy has reached 67.9 years, close to those found in the developed countries. The current main tasks of drug research are the search for effective agents against vascular diseases, cancer, hepatitis, ageing and for contraception.

The pharmaceutical industry in China has recently made great progress. In the early 1950s, the Chinese chemical–pharmaceutical industry started by adhering to the principle of 'bulk pharmaceuticals come first, within which the life-saving drugs come first'. Thus the production of the first six categories of principal drugs, *i.e.* antibiotics, sulfonamides, antipyretics, vitamins, drugs for epidemics and anti-TB drugs were high on the priority list. They were followed as the 1950s turned into the 1960s by the initiation of the production of steroids including contraceptives, anticancer drugs, cardiovascular drugs, drugs for colds and asthma, X-ray contrast agents and others. The production of bulk pharmaceutical chemicals in 1985 was about 57 600 tons of drugs in the above-mentioned 12 categories (about 1200 chemical identities) with antipyretics (21 600 tons), antibiotics (11 600 tons), sulfonamides (9 290 tons) and vitamins (4 430 tons) at the top of the list. Traditional drugs are procured at a rate of around 500 000 tons of crude drug per year from which 150 000 tons are converted into 3800 formulated remedies.

1.4.2 EVOLUTION OF TRADITIONAL CHINESE MEDICINE

The beginning of TCM is hard to trace but it certainly is of great antiquity. The first written medical text 'Huang Di Nei Jing' ('Emperor's Canon of Internal Medicine') which was compiled about 300 BC, provided the comprehensive theory for both diagnosis and treatment, and served as the basic philosophical concept for further development, elaboration and proliferation of the diverging schools of Chinese medicine during the following two millennia. Since then as many as 10 000 medical works have appeared. 'Shang Han Za Bing Lun' ('Treatise on Exogenous Febrile and Internal Diseases') of 219 AD by Zhang Zhongjing was the first systematic medical classic on 'bian zheng lun zhi' (determination of treatment based on the differentiation of symptoms and signs). 'Zhen Jiu Jia Yi Jing' ('A—B Classic of Acupuncture and Moxibustion') by Huang Fumi, 282 AD, 'Mai Jing' ('The Classic of Sphygmology') by Wang Xi, 280 AD, 'Zhi Bing Yuan Hou Lun' ('General Treatise on the Causes and Symptoms of Diseases') by Chao Yuangfang, 610 AD, and 'Wen Yi Lun' ('Treatise of Pestilence') by Wu Youxing, 1642 AD, represented the early classics in the fields of acupuncture, diagnosis, pathogenesis and pestilence respectively.

Traditional Chinese drug (TCD) usage was evolved along with clinical practice. The legendary Shen Nong (Holy Farmer), testing hundreds of herbs and being poisoned 70 times a day, is a fictitious description of how the Chinese *materia medica* originated from simple trial and error practices. The first written work on *materia medica*, which bore the title 'Shen Nong Ben Cao Jing' ('Shen Nong's Canon of *Materia Medica*'), appeared around the first centuries before and after Christ, and described the characteristics, processing, classification, and physiological and pharmaceutical effect of 365 entries (252 derived from plants, 67 from animals and 46 related to minerals) and also the theoretical bases of their prescriptions. In this work, drugs were classified into three ranks, upper, middle and low. Those in the upper rank such as renshen (*Panex ginseng*), gouqizi (fructus *Lycii chinense*), *etc*., are considered to be nontoxic and can be taken as tonics for long periods, while in the middle rank entries such as mahuang (herba *Ephedrae sinicae*) showed dose-related toxicity, whilst the more toxic members such as chuanwu (radix *Aconiti carmichaeli*) belong to the low rank. Afterwards, about 300 more texts of Chinese *materia medica* were compiled. 'Tang Ben Cao' (657–659 AD), the first official pharmacopoeia, was a cooperative work, in 53 volumes, of about 20 medical officers in the Tang Dynasty. The world renowed comprehensive classic *materia medica* written by Li Shizhen, 'Ben Cao Gang Mu' (1596), took the author 30 years to complete. It contained 1892 entries, 57.8% of which came from plant material, 23.5% the zoological domain and 14.5% the mineral field, and described 11 000 prescriptions. It has been translated into five languages—Latin, English, French, German and Russian.

The basic philosophy of the self-contained theories of TCM was developed through repeated autochthonous practices incorporating the concept of cosmology in the ancient period. The basic idea considered that the human body, like all objects both simple or complex, is composed of two opposite and complementary dynamic forces, Yin and Yang, which are inseparable in the unit but separable in their identity. Yin and Yang are symbolic for negative and positive, moon and sun, female and male, weak and strong, cold and hot, hypofunction and hyperfunction, etc. In the healthy normal state, Yin and Yang are in dynamic equilibrium. Imbalance of Yin and Yang, that is either Yin or Yang in excess or in deficiency, caused diseases. Excess of one means deficiency of the other. Conceptually, the logical medical treatment is to restore the equilibrium of Yin and Yang and bring them into harmony by drugs or physiotherapy. This means that TCM expresses disease concepts in the context of symptoms expressed by the whole body.

Accordingly drugs are classified into cold, hot, warm and cool. Cold- and cool-natured drugs such as zhimu (rhizoma *Anemarrhenae asphodeloidum*), huanglian (rhizoma *Coptidis chinensis*), huangbai (cortex *Phellodendri amurensis*), daqingye (folium *Isatidis tinctoriae*) have the effect of clearing away heat and toxic material, and relieving the damage of 'fire evil'. Thus they are used to treat patients with the symptoms of heat syndrome, such as thirst, hot flushing, with red cheeks and eyes, and strong but fast pulse. Drugs such as fuzi (subordinate radix *Aconiti carmichaeli*), ganjiang (rhizoma *Zingiberis officinalis*), rougui (cortex *Cinnamomi cassiae*), *etc*., are warm or hot natured and are used for cold syndromes with symptoms of cold feelings, cold extremities, pale looks and weak pulse.

TCM usually uses complex prescriptions, for which many principles were developed. An orthodox prescription consists of four parts, the 'monarch', the main ingredient against the disease; the 'minister', which is auxiliary to the monarch to increase the effect of the main ingredient; the 'assistant', which is present for relieving minor symptoms or as an antidote; and the 'guide', which transports the drug to the desired tissues or organs. Drug interactions have long been realized through use and are summarized under the term 'seven modes' of the nature of drugs in TCM classics which describe the influence of one drug over the other by effects such as promoting, decreasing, inhibiting or eliminating either its activity or toxicity.

1.4.3 STUDIES ON THE TRADITIONAL CHINESE MEDICINAL MATERIALS (TCMM) IN THE NEW ERA

TCMM as well as TCM were usually passed down and developed within individual families by imparting knowledge and experience from father to son and master to apprentice. The practical experiences evolved over many generations have produced formidable theories and principles which have been followed for years to serve the health of Chinese. However, they are limited by elements of social and historical background, especially its parochialism. Their theories and principles are naive and need to be verified, explained and improved by modern techniques and sciences. In recent years, various aspects of TCMM have been studied in China.

1.4.3.1 Botanical Studies

Up to 60% of TCMM are plant materials and their identification and grading were, in the old days, dependent upon the experience of individual herbalists. In order to standardize the biological effect of herbs, botanical identification and authentication have been carried out. The results have been summarized in many dictionaries and texts of TCMM. In these dictionaries each entry is usually described with a drawing, the identified botanical name, traditional names, the part of the plant for drug use, processing procedure, properties according to TCM aspect, biological effects, indications, chemical constituents being isolated, *etc.* Among them, 'Chinese *Materia Medica* Dictionary' (1978) which in two volumes listed 5767 entries; 'Flora of Chinese *Materia Medica*' (four volumes, 1959–61) and 'Chinese Traditional and Herb Medicine Compilation' (2202 drugs in two volumes, 1975–1978) are worth mentioning. A valuable comprehensive reference work in Chinese *materia medica*, to be published by The People's Medicinal Publishing House of China, is 'A Color Pictorial Book of Chinese Medicinal Herbs', which will be 25 volumes containing traditional and herb medicines with pictures and descriptions of their resources and applications. The first eight volumes have recently been published in Japanese. Herbariums of authentic specimens of TCMM have been set up in many institutes and colleges of pharmaceutical sciences.

China is endowed with abundant resources of medicinal plants. Through extensive surveys, 5136 plants have now been identified as being used in Chinese medicine. There are 281 species in Thallophytes, 38 in Bryophytes, 395 in Pteridophytes, 55 in Gymnosperms, 3690 in Dicotyledons and 676 in Monocotyledons. Among them about 1000 species are most frequently used.

TCMM were mostly growing wild. In order to keep pace with a growing population, their cultivation, transplantation and plant protection have been studied. These measures have increased the supply of medicinal plants. The achievement in cultivating valuable materials such as *Saussurea lappa*, *Gastrodia elata*, *Panax ginseng* and others has been most beneficial. *Panax ginseng* originally grew in the forests of the northern part of China. Since medicine uses the roots of plants which are several years old, ginseng has been rendered scarce and expensive. Now, however, the cultivated ginseng is in abundant supply.

Some fungi have been used as medicinal materials in China since ancient times. *Ganoderma lucidum*, *Poria cocos*, *Polyporus umbellatus*, *Omphalia lapidescens*, *Cordyceps sinensis*, *Cordyceps sobolifera*, *Batrytis bassiana*, *Calvatia gigantea*, *Tremella fuciformis* and *Lentinus edodes* have been described in 'Shen Nong Ben Cao Jing' and other medical classics as having various biological effects. Success in the cultivation of fungi has enabled previously scarce materials to be made readily available. For example, *Ganoderma lucidum*, whose traditional name, lingzhi, symbolizes spiritual grass, had been known as a legendary magic remedy able to rescue the dying. Now it is obtainable either in fruiting body form by cultivation or as the extract of its mycelia.

Plantlet regeneration from tissue culture has been successfully applied to ginseng (plantlet *via* anther culture), *Angelica sinensis*, *Codonopsis pilosula*, *Dendrobium candidum*, *Dendrobium tosaense*, *etc.* The biological effects of these cultivated substances are comparable to the natural products, as shown by the appropriate pharmacological tests.

1.4.3.2 Pharmaceutical Studies

Plant, animal and mineral materials (raw medicines) usually go through certain treatments before they are used as therapeutic agents. The treatment with the technical name of 'paozhi' (processing) concerns soaking, steaming, simmering or stewing with liquid such as water, salt solution, wine or vinegar, and roasting, burning and baking alone or with sand, bran or honey, wine, *etc.* Each individual procedure is supposed to have its own specific effect in modifying the biological action of the specified raw material. Studies on the chemical and pharmacological changes caused by processing and the standardization of proper processing procedures have recently been attempted. In cases where the main purpose seems evident from current knowledge they will be easy to work out. For the kernel of *Prunus armeniaca*, which is used as an anticoughing ingredient, processing procedures involving cutting off the tip of the skin of the soaked kernel, or steaming and boiling with water were listed in 'Ben Cao Gang Mu'. If it is assumed that the result of the processing is deactivation of the enzyme which hydrolyzes the active principle amygdalin (**1**; equation 1), then, by determination of the rate of deactivation of this particular enzyme[1] in various conditions, and the amount of HCN evolved by chemical hydrolysis of the processed material,[2] the best procedure could be suggested. The same rational basis has been applied to the heat processing of seeds of *Brassica juncea* and of *Sinapis alba*[3] for which the main purpose appears to be the deactivation of the enzyme which catalyzes the rearrangement of the active principles sinigrin (**2**) and sinalbin (**3**) to irritant isothiocyanate (equations 2 and 3).

$$C_6H_5-CH(CN)-O-\beta\text{-D-glucopyranosyl-6-}\beta\text{-D-glucose} \xrightarrow{\text{enzyme}} C_6H_5-CHO + HCN + \text{glucose} \quad (1)$$

(1)

$$\text{glucose}-SC(CH_2CH=CH_2)=NOSO_3K \xrightarrow{\text{enzyme}} CH_2=CHCH_2NCS \quad (2)$$

(2)

$$HO-C_6H_4-CH_2C(=NOSO_3\text{-sinapine})S-\text{glucose} \xrightarrow{\text{enzyme}} HO-C_6H_4-CH_2NCS \quad (3)$$

(3)

For certain cases the amount of known toxic or active constituents present before and after treatment can be compared, thus revealing the hidden effects of specific treatments. The quantity of toxic alkaloids, *i.e.* aconitine, mesaconitine and hypaconitine of unprocessed and processed radix *Aconiti kusnezoffii* were determined[4] by scanning thin layer chromatograph. The figures obtained for a high pressure steamed sample were 0.02, 0.0248 and 0.0277 for aconitine, its mes- and hyp-isomers respectively. Correspondingly, results of 0.0026, 0.0095 and 0.0100 were obtained from water-boiled samples, whilst in an untreated sample they were much greater, *i.e.* 0.0499, 0.0385 and 0.1974. The indication that the above procedure was a detoxifying process is what has been described in 'Ban Cao Gang Mu'.

Processing of radix *Rhei palmati* with wine or vinegar was recommended in TCM. It was believed that processing would alleviate the cold nature of this radix and that the wine treatment would be more effective in clearing away heat and toxic material than the vinegar processing which, however, is better in promoting blood circulation and removal of blood stasis. Recent pharmacological studies on raw and processed *Rheum* were undertaken.[5] Raw radix showed a strong laxative effect which was much milder in the processed samples. However, the degree of laxative effect is not in parallel with the content of rhein glucosides but is related to the amount of sennosides present. In rats, raw radix inhibited the secretion of gastric acid and the activity of gastroproteinase, while no such effect was shown by a wine-simmered sample. Meanwhile gastrointestinal side effects of raw radix such as vomiting, nausea and abdominal pain were not observed in wine-treated samples. The simmered radix also showed significantly less toxicity to rats in the acute and subacute toxicity tests. The antipyretic effect of both raw and treated radix in rats is comparable to aspirin. The inhibitory effects against *Staphylococcus aureus* and some bacillus *in vitro* also were demonstrated, but different treatments caused variation in the antibacterial activities. No correlation between this inhibitory activity and the amount of anthraquinones was observed as the decoction of raw radix freed from tannins and anthraquinones still retained its inhibitory action. All samples, whether treated or not, show antiinflammatory activity in rats but the wine-simmered is weaker. The haemostatic effect of raw radix was described in ancient texts. This effect was verified in experimental animals as well as in patients for both raw and processed forms, but the former was quicker in action. Nevertheless, the wine-treated form is more acceptable due to its low side effects.

Wan (bolus), san (powder), gao (plaster or concentrated extract) and dan (pellet) are the four main forms of traditional pharmaceutical preparations which are still in use. They have been studied and improved by modern concepts, such as freedom from contamination, dissolution, stability, *etc.* Many popular set prescriptions are now extracted and prepared in pharmaceutical plants instead of decocting the herbs in the patient's home. They are available in tablet form or in water soluble granular or powder forms which can be taken conveniently by patients.

1.4.3.3 Clinical and Pharmacological Studies

In order to further the development of TCM, verification and evaluation of the clinical efficacy of TCM and interpretation of the theory of TCM in line with modern biochemical and pharmacological concepts is necessary. The following are a few examples of preliminary results which are currently being reported.

The effects of the Yang tonifying drug (Yang drug) fuzi (*Aconitum carmichaeli*) and the Yin nourishing agent (Yin drug), Liu Wei Di Huang Fang (containing radix *Rehmanniae glutinosae*, fructus *Corni officinalis*, rhizoma *Dioscoreae oppositae*, rhizoma *Alismae orientalis*, cortex *Paeoniae suffruticosae*, and sclerotium *Poria cocos*) on renovascular hypertensive rats[6a–e] were studied. The Yang drug further increased the elevated blood pressure and level of urinary aldosterone excretion (UA) and high hydroxyproline level in the left ventricular wall whilst the Yin drug caused no change on the blood pressure and UA but decreased myocardial hydroxyproline. The lowered levels of M-enkephalin (MEK) and L-enkephalin (LEK) of brain stem, hypothalamus and striatum in the hypertensive rats were further decreased by the Yang drug, but the Yin drug restored LEK and MEK levels almost to normal. This means that in the renovascular hypertensive rats the Yang and Yin drugs have opposite effects with the Yang drug showing a deteriorative effect and the Yin drug a beneficial one. It also suggested that this particular model might serve as an experimental Yin deficiency model for studying TCM.

In contrast, in hypertensive rats induced by adrenal stimulation, the Yang drugs fuzi and rougui (*Cinnamomum cassia*) markedly reduced the blood pressure and urinary aldosterone. Rougui also significantly increased the lowered LEK in rat brain tissue. Lesions caused by hypertension in the endothelial cells and subendothelial layer of aortic intima of rats were improved by these two herbs. Thus it is considered that this hypertensive model belongs to a Yang deficiency model of TCM rather than a Yin deficiency one. These results may support the hypothesis of TCM that a Yin nourishing or a Yang tonifying drug may exert allopathic effects on the Yin and Yang deficiency.

The nature of 'cold' and 'heat' syndromes in TCM was studied[7] by determining the amount of catecholamine and cyclic nucleotides excreted by healthy persons (control) and patients diagnosed as having heat and cold syndromes. The levels of catecholamine and cyclic nucleotides in the urine of the heat syndrome patients were higher than those of the control group. In contrast, in the cold syndrome group, the levels of urine catecholamine and cAMP were decreased but cGMP increased, causing a marked lowering of cAMP/cGMP ratio. These changes can be considered as the characteristic biochemical features, indicating hyperfunctioning of the sympathetic adrenal medullary system in the heat syndrome in contrast to hypofunctioning of the sympathetic system and hyperfunctioning of the parasympathetic system in the cold syndrome. This coincided well with observations from animal studies.[8] Rats given cold-natured drugs developed hypofunction of the sympathetic adrenal medullary system, whilst animals treated with warm drugs increased the activity of this system. The levels of norepinephrine and dopamine in the rats of the warm drug group increased slowly but steadily and stayed elevated for a long period while administration of cold drugs significantly increased 5-HT.

Those patients with symptoms of feeling cold and suffering from dropsy and who were classified as having Yang deficiency syndrome in TCM, showed lower levels of T_3 (3,5,3′-triiodothyronine) and T_4 (thyroxine).[9a,b] The T_3 level was significantly raised on treating with Yang tonifying drug; T_4 was also increased but not significantly. In hyperthyroid rats, Yin nourishing agent such as *Rehmannia glutinosa* or *Platrum testudinis* restored the elevated plasma cAMP and maximum binding capacity of the β-adrenergic receptor of kidneys to normal while the Yang tonifying drugs rougui and fuzi caused further elevation.[10] The raised oxgen consumption rate in the hyperthyroid rat was lowered by the Yin drugs but the opposite effects were exerted by Yang drugs.

Many other types of animal models and biomedical indices from both experimental animals and humans were applied to study TCM. For instance huo xie hua yu yao (drugs for promoting blood circulation and removing stasis) which were recommended for coronary heart disease, collagen diseases and blood stasis syndrome, have been studied for their effects on platelet aggregation, β-thromboglobulin, platelet factor II, thromboxane B_2, TBA_2/PGI_2 balance, 6-ketoprostaglandin $F_{1\alpha}$, fibrin and haemorrheological parameters.

Whilst all the above attempts are preliminary, they do represent an encouraging start for studying TCM.

1.4.4 PRESENT TRENDS AND RECENT PROGRESS IN DRUG RESEARCH AND DEVELOPMENT

There are, in addition, three further lines of enquiry in drug research in China: (i) following leads from 'Western' basic medical ideas, known drugs, random screening of synthetic compounds and natural products; (ii) exploring, systematizing and developing TCMM in a more or less traditional way; and (iii) following leads from TCMM and studying them with modern pharmacological and chemical techniques and concepts.

1.4.4.1 Developing Drugs following 'Western' Trends

A few anticancer drugs which have been developed may serve as examples for drug development based on the leads provided by known drugs. Sodium purine-6-mercaptosulfonate (tisupurine) **(4)**[11] and glyciphosphoramide **(5)**[12a,b] are derivatives of 6-MP and cyclophosphamide respectively. Tisupurine has the advantage over 6-MP because of its aqueous solubility and is available in

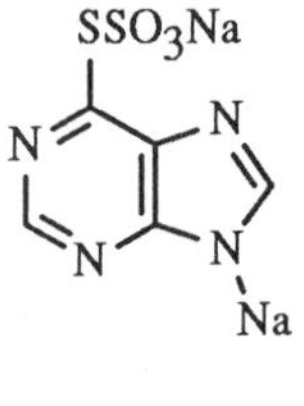

SSO_3Na … N … Na

(4)

$(ClCH_2CH_2)_2NP(O)(NHCH_2CO_2Et)_2$

(5)

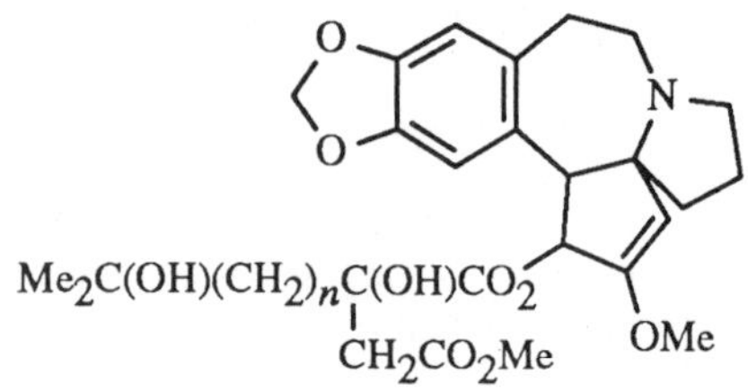

$Me_2C(OH)(CH_2)_nC(OH)(CH_2CO_2Me)CO_2$ … OMe

(6) $n = 2$

(7) $n = 3$

Et, OH

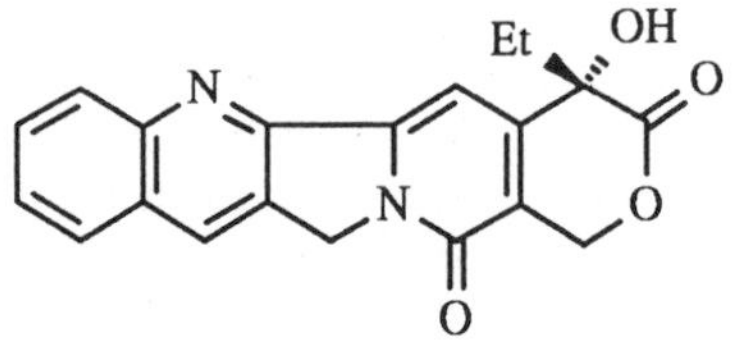

(8)

ampoules. Glyciphosphoramide was originally designed as a derivative of cytoxan with glycine as carrier and it turned out to be favourable for topical application in the management of cancerous ulceration. The antileukaemia agents, harringtonine (**6**) and homoharringtonine (**7**), from the Cephalotaxus plant which is indigenous to China, have been developed by following Powell's screening data.[13] They are now used in the treatment of nonlymphocytic leukaemia with satisfactory results.[14] Camptothecin (**8**) from *Camptotheca acuminata*, which is also indigenous to China, and whose positive screening results were reported by Wall,[15] is now used clinically in China for solid tumours.[16] The anticancer antibiotic Pingyangmycin[17a,b] was isolated from *Str. verticillus var. pingyangensis n. var.* Its main constituent is bleomycin A_5 instead of bleomycin A_2. Pingyangmycin is also now commercially available for cancer treatment.

Steroid contraceptives and prostaglandins are still actively studied for the purposes of inducing menstruation and interrupting early pregnancy.

1.4.4.2 Developing Drugs Based on Leads From TCMM

Chinese scientists, by following the indication of qinghao (*Artemisia annua*) as a remedy for malaria in 'Zhou Hou Bei Ji Fang' ('Prescriptions for Emergency') of 340 AD and 'Ben Cao Gang Mu', isolated qinghaosu (**9**; artemisinine) an antimalarial with a novel structure in 1972.[18] Along with artemisinine were obtained six other sesquiterpenes with closely related structures, particularly qinghaosu-III (**10**), but they lack the peroxide bridge and are devoid of antimalarial activity. Thus the peroxide bridge in the molecule is crucial for its antimalarial activity. Reduction of artemisinine by $NaBH_4$ gave the hemilactol, hydroartemisinine (**11**), a mixture of C_{12} α- and β-OH, which is twice as active as the parent compound. Hydroartemisinine, a derivative which retains the original peroxide skeleton, however, provides a hydroxy group for derivation. From this about 100 derivatives[19a–e] were prepared in the form of ethers, carboxylates and carbonates. Most of the derivatives are more active than artemisinine and the esters are more potent than the ethers. Their QSAR, which is given in the following two equations, illustrated the close relationship between lipophilicity and antimalarial activity. The electronic character of substituents also plays its role in ethers as the introduction of Taft's σ^* gives a better correlation coefficient.

$$\text{Ester:}\quad \log(1/c) = 1.202 - 0.1983(\log P)^2 + 1.549\log P - 0.6125I\alpha,\beta$$
$$n = 49,\; r = 0.9111,\; s = 0.153,\; \log P_0 = 2.91$$
$$\text{Ether:}\quad \log(1/c) = 0.9575 - 0.2231(\log P)^2 + 1.620\log P - 0.837\sigma^* - 0.1451I\alpha,\beta$$
$$n = 14,\; r = 0.906,\; s = 0.168,\; \log P_0 = 2.60,\; I\alpha,\beta = 1 \text{ for } C_{12}\text{-}\beta\text{-substitution and } 0 \text{ for the } \alpha\text{-epimer.}$$

Artemisinine, artemether (**12**) and sodium artesunate (**13**) are much less toxic than chloroquine, on the basis of their therapeutic indices against a chloroquine sensitive strain of *P. berghei* in mice. They are also effective against chloroquine resistant *P. berghei*. These three artemisinines are used in the clinic for the treatment of patients infected with *P. vivax* and chloroquine sensitive and resistant *P. falciparum*.[20]

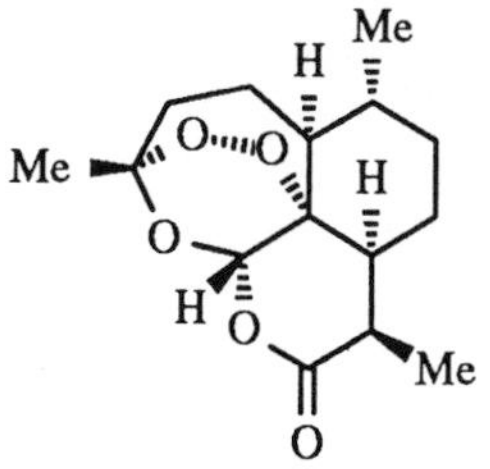

(**9**) 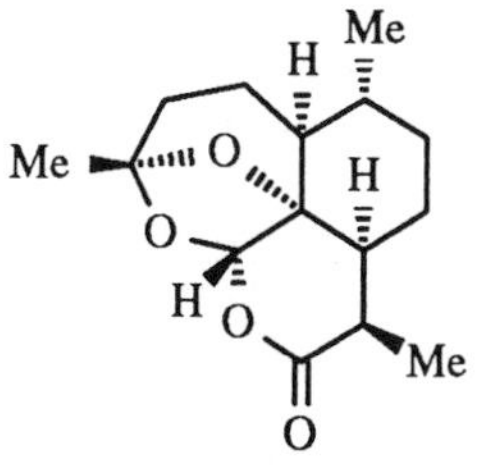 (**10**)

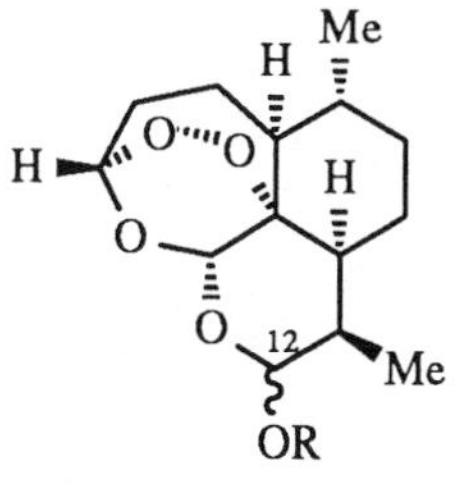

(**11**) R = H
(**12**) R = Me
(**13**) R = $COCH_2CH_2CO_2Na$

Another sesquiterpene antimalarial agent yingzhaosu A (**14**) isolated[21] from the herb *Artabotrys uncinatus* has a basic skeleton quite different from that of artemisinine, but it is most amazing that a peroxide bridge is also present in the molecule. The coincidence is indisputable evidence of the significance of a peroxide bridge. Such a unique structure for an antimalarial agent would hardly

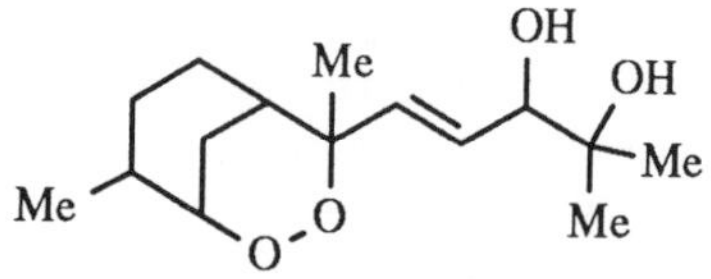

(14)

have been revealed if it were not for the traditional medicinal practices of many generations. Chemical studies on various species of *Artemisia* (including *A. annua*) have been conducted since 1936 outside China, but nothing was ever reported concerning its antimalarial activity before 1977.

Chronic myelocytic leukaemia (CML), for which incidently there is no such term in traditional medical classics, is treated by correlating patients' symptoms with TCM concepts. The patients have been classified as 'gan shi zheng' (liver sthenia syndrome). A readily available bolus, danggui luhui wan, with the effect of 'xie gan shi huo' (purge sthenic liver fire) was selected as the drug for treatment with the result that 16 cases responded positively to it. By subsequent deduction and investigation of the 11 ingredients in the bolus, qingdai (cold-natured drug with the effect of clearing away heat and toxic materials and cooling blood) was singled out as the active constituent. Qingdai is the floating solid formed after indican containing leaves of *Isatis tinctoria* have been soaked in water for two days and then treated with lime. The major organic component of qingdai is indigo and there are also many minor compounds present including indirubin. Further pharmacological and chemical studies have shown that indirubin (**15**) is mainly responsible for the antileukaemia activity. It inhibited leukaemia L1210 and Lewis lung carcinoma in mice and Walker 256 in rats. Unlike most antileukaemia agents, indirubin has no obvious effect on the haemotopoietic and immune systems in normal rats and mice. A response rate of 87% was observed in the clinic for CML patients but with the side effects of abdominal pain and diarrhoea.[22a,b] The profile of the amounts of uric acid (a metabolite of purine) and of 3-aminoisobutyric acid (a metabolite of thymidine) eliminated in the urine of treated patients, are different from those obtained with myleran, 6-MP and radiation therapy. Thus the mechanism of the action of indirubin might be different from that of these drugs. It inhibited the synthesis of DNA in Walker 250 ascitic cells. Indirubin has been a known compound for about 100 years and can be easily synthesized by condensation of indoxyl with isatin. But no antileukaemia action was observed until recently. However, isolation of indirubin from patients suffering CML and reduction of white blood count in guinea pigs being given indirubin intramuscularly were reported about 40 years ago. These observations serve as a support for the therapeutic action of indirubin against CML.

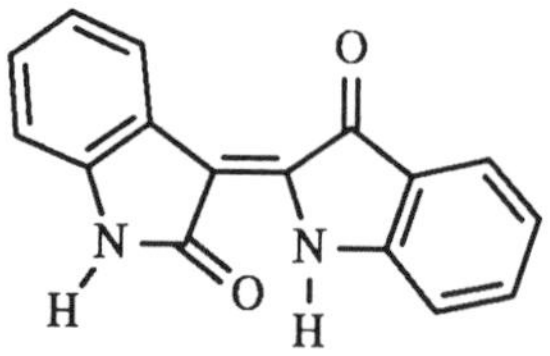

(15)

Indirubin, with its high melting point, is insoluble in water and most organic solvents. It was assumed that modification of its structure by introduction of substituents on the nitrogen atom, which would inhibit the intramolecular hydrogen bonding, might change the crystal lattice and thus its energy and solubility. Further, the structures of its less soluble isomers isoindigotin (**16**) and indigo were modified in the same way. Gratifyingly, it was found that the *N*-alkyl-substituted derivatives of the three isomers showed better solubility in lipophilic solvents and those with lower alkyl substituents showed higher activity against rodent tumours than the parent compounds.[23] The isoindigotin differs from indirubin in the position of the bridge between the two rings, *i.e.* 3,3′

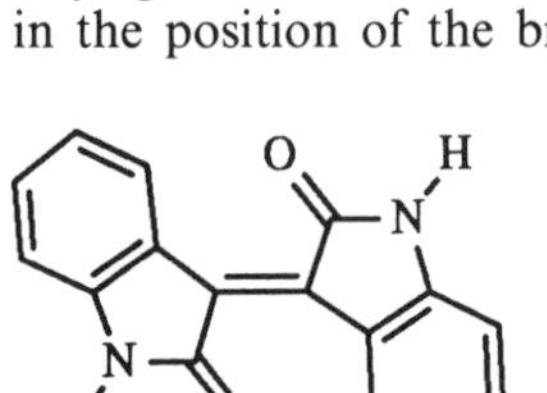

(**16**) R = H
(**17**) R = Me

instead of 2,3′. Its methyl derivative has been chosen for development and now is in clinical use. Whilst both indirubin and *N*-methylisoindigotin (**17**) are not new compounds, they are novel as anticancer agents.

In an attempt to relieve the stiff neck of hypertensive patients, which usually persists even after the blood pressure has been restored to normal, a search of TCMM classics disclosed that the seven-ingredient prescription 'gegen tang' was described in the medical classic 'Shang Han Lun' ('Treatise on Exogenous Febrile Diseases') 219 AD, to be effective in the treatment of stiff neck. Its major constituent, gegen (radix *Puerariae lobatae*), has been known to be nontoxic and is consumed as food in certain regions. Its aqueous decoction and alcoholic extract gave satisfactory results clinically in relieving the symptom of stiff neck. Chemical[24] and pharmacological studies on gegen indicated that the activity resided mostly in isoflavone mixture of alcoholic extracts. From the mixture of isoflavones, daidzein (**18**) as well as its three glucosides, daidzin (**19**), 7,4′-diglucoside (**20**) and puerarin (**21**) were isolated. The alcoholic extract, puerarin or daidzein are now used not only in treatment of stiff neck but also for angina pectoris and other vascular diseases, without having any noticeable adverse effects on patients.

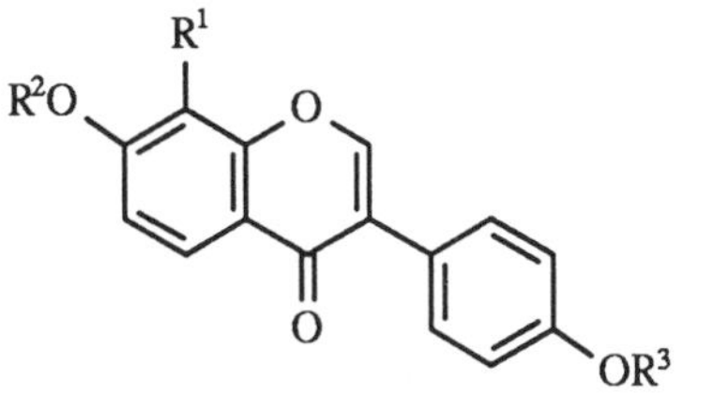

(**18**) $R^1 = R^2 = R^3 = H$
(**19**) $R^1 = R^3 = H$, R^2 = glucopyranosyl
(**20**) $R^1 = H$, $R^2 = R^3$ = glucopyranosyl
(**21**) $R^2 = R^3 = H$, R^1 = glucopyranosyl

Pharmacological studies[25] revealed the following effects of these drugs on the nervous and circulatory systems: (a) decreasing the activity of the sympathetic nervous system by lowering the elevated catecholamine level of hypertensive patients and patients suffering from severe angina pectoris; (b) increasing cerebral and coronary blood flow and decreasing the coronary and cerebral vascular resistance; (c) inhibiting the platelet aggregation and the release of 5-HT from platelets; and (d) improving myocardial ischaemia by increasing the oxygen supply and decreasing oxygen consumption and lactate production in the ischaemic myocardium. These physiological actions showed that a rational scientific basis lies behind TCM and further that the potential usage of TCMM might be explored.

Since the early 1970s, Chinese traditional doctors have used the honey bolus of the kernel of wuweizi (*Schizandra chinensis*), which was formally described as a tonic, an astringent, *etc.*, to treat hepatitis. The positive therapeutic effects which were obtained soon gave impetus to chemical and pharmacological studies on the herb. Seven isolated dibenzocyclooctadiene lignans demonstrated various effects on liver function.[26] Of these, five decreased the CCl_4, thioacetamide- and acetaminophen-induced hepatoxicity in mice and rats. They inhibited microsomal lipid peroxidation induced by CCl_4, stimulated liver drug metabolism and exhibited antioxidation activity. Clinically, the mixture of the lignans decreased the serum transaminase level of patients with chronic viral hepatitis B and restored the levels to normal in 70% of cases after three months treatment. In the synthesis of schizandrin C (**22**), an isomer (DDB) (**23**) of the intermediate displayed protective

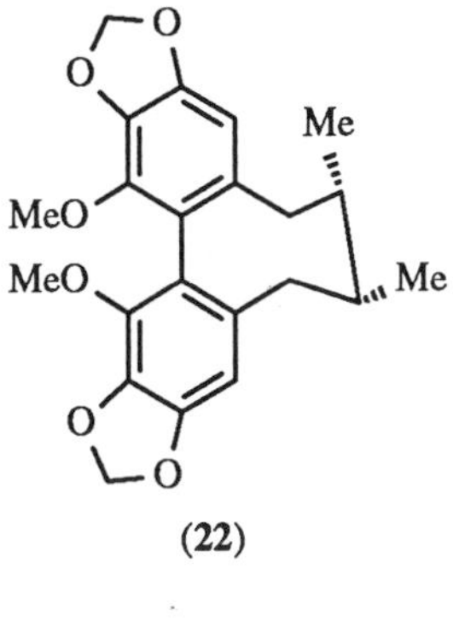

(**22**) 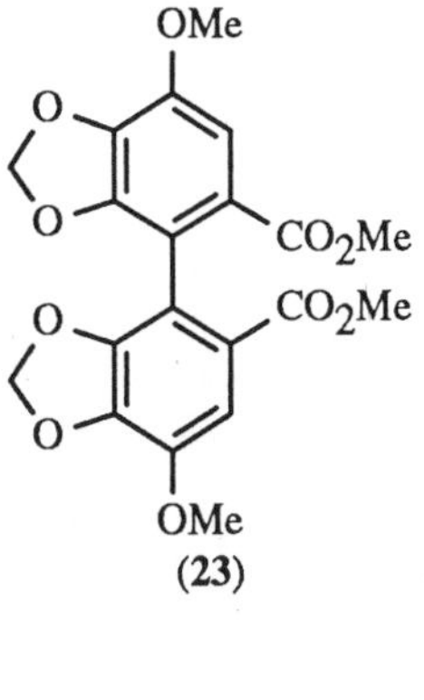 (**23**)

action against hepatotoxicity induced by CCl_4, *etc.* DDB increased the liver detoxification function and antagonized the mutagenicity of aflatoxin B. Furthermore DDB showed low levels of toxicity and no teratogenic and mutagenic action were observed. It has been in clinical use for the treatment of chronic viral hepatitis B since 1977. Elevated serum transaminase (SGPT) levels are reduced to normal in about 80% of the treated patients and symptoms are also relieved after a three-month treatment[27] period. DDB is now widely used to treat chronic viral hepatitis and drug-induced hepatitis in China.

In response to the necessity for family planning in China, the search for new, active contraceptives from TCM classics has been undertaken since the 1960s. In 'Ben Cao Gang Mu' there are listed 72 entries (42 herbs) as abortifacients, and 85 entries (41 herbs) were prohibited to pregnant women because of their potential ability to interrupt the pregnancy. Trichosanthine, which can be obtained from the juice of *Trichosanthes kirilowii*, has been described as inducing menstruation and expulsion of the afterbirth. Trichosanthine was isolated and proved to be a 234 amino acid residue protein with a molecular weight of 24 000 whose primary and three-dimensional structures have recently been determined.[28] The purified product has been used locally to interrupt both second trimester[29] and early pregnancies[30] with success rates of 99% and 88% respectively.

The known compound tetramethylpyrazine (TMP) (**24**) was one of the active principles isolated from chuanxiong (*Ligusticum wallichii*), a herb which, according to TCM, has the ability to improve blood circulation. It showed therapeutic value in the treatment of patients with coronary disease and this can be verified by modern pharmacological studies.[31a,b] TMP increased the blood flow in the coronary artery of anaesthetized dogs, relieved the myocardial anoxia in rabbits caused by pitressin, and improved microcirculation in the mesentery of rabbits. *In vitro* it not only inhibited platelet aggregation caused by ADP but also disaggregated aggregated platelets. On isolated rabbit aortic strips, TMP inhibited the contraction of the smooth muscle of blood vessels. Further, TMP could pass through the blood–brain barrier and be distributed widely in the brain stem. The preliminary results of pharmacodynamic and electrophysiological studies on cardiovascular tissue showed that the specific effects of TMP strikingly resemble those of verapamil. As a result it was suggested that TMP was a Ca^{2+} antagonist. The clinical value of tetramethylpyrazine in coronary disease and ischaemic cerebral vascular disease has thus been identified.

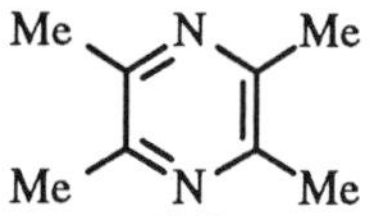

Me N Me
Me N Me
(**24**)

Tanshen (radix *Salviae miltiorrhizae*) has been reported to have the effect of promoting blood circulation, removing blood stasis and draining pus. The isolated tanshinones of diterpenequinone structure showed various antibacterial activities. The alcoholic extract of tanshen showed many beneficial effects on cardiovascular diseases in both pharmacological and clinical studies.[32] A water soluble derivative of tanshinone II-A, sodium tanshinone II-A sulfonate (**25**) showed protective action against anoxia, ischaemia and platelet aggregation in rats.[33] In coronary patients, it reduced anginal pain and the feeling of chest tightness and improved the ischaemic changes in the electrocardiogram.

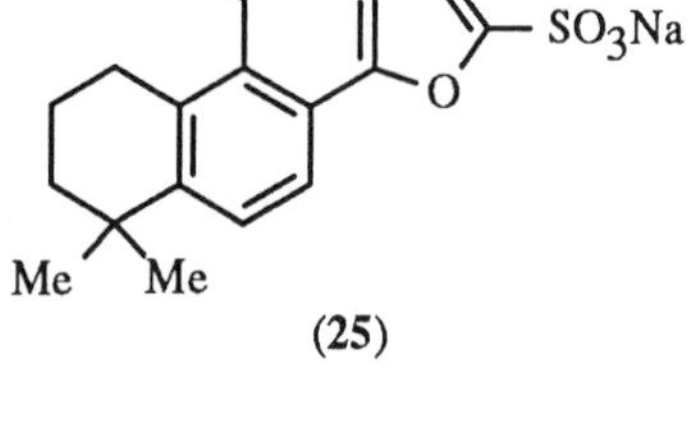

O Me
O
SO_3Na
O
Me Me
(**25**)

1.4.4.3 Experimental and Clinical Pharmacological Studies on Some Tonics and Antiageing Drugs of TCM

There is an important group of drugs, the tonics, which are indispensible in TCM. From a consideration of their modes of pharmacological action, they can be defined as biomodulators.[34]

They are usually prescribed to patients recovering from serious diseases, those debilitated from other causes and the aged. Biomodulators comprise a distinct class of TCM. They modulate the physiological and biological activities at various levels of body systems, such as the CNS, cardiovascular, digestive, haematopoietic, endocrine, renal, sexual and immune systems. Their administration depends upon the symptoms of the particular deficiency. They can be classified into four classes based upon TCM: bioenergetics, blood modulators, Yin modulators and Yang modulators. Bioenergetics are drugs which reinforce or invigorate the vital energy. Renshen (*Panax ginseng*) is the best known member of this class. Botanical, biological and chemical studies have been undertaken over many years not only in China but also in Korea, Japan, the Soviet Union and the United States. Its clinical and pharmacological effects can be summarized as: having antistress activity and antishock effects in circulatory failure; improving or facilitating memory and learning processes; modulating cardiovascular, neuroendocrine and hypothalamic–pituitary–adrenal–gonadal systems, cellular metabolic processes on carbohydrate, fat and protein metabolism and immune activities; promoting haematopoiesis and protecting against radiation and liver intoxification.

Renshen contains 4% triterpene glycosides, which have been considered to be the active principles in the herb. Therefore the levels of these triterpenes have been taken as an objective measure for quality control and grading of the clinical potential. However, some recent reports have indicated that the amount of saponins present is not completely parallel to their biological activities. Renshen also contains sapogenins, polysaccharides, amino acids, peptides, flavones, minerals, *etc.* The diverse activities combined with many different types of chemical constituents make it very difficult to sort out the major biological active constituents or to correlate the chemical constituents correctly with the biological actions. Therefore, the herb itself, or its extracts, rather than any of its constituents, is still the most popularly used product. Dangshen (radix *Codonopsis pilosulae*) and huangqi (radix *Astragali membranacei*) have clinical and pharmacological effects similar to renshen in many respects. Dangshen is regarded as an effective substituent for renshen, but is less potent in its stimulating effect on the CNS and CVS. However, it was recently shown to be effective in the treatment of gastrointestinal diseases, especially in duodenal ulcer. Huangqi is extensively prescribed to strengthen bioenergy. Many studies on its constituent polysaccharides have indicated its remarkable immunomodulatory effects on cell-medicated immunity, humoral immunity and on monocytic macrophages.

Danggui (radix *Angelicae sinensis*) may be taken as an example of blood modulatory drugs. It is a most valuable drug for women. It nourishes the blood, improves the rhythmicity and tonicity of uterine muscles. It modulates the cardiovascular activities to promote the relief of local circulatory stasis, shows an antithrombotic effect and reduces platelet aggregation. It also has haematopoietic, antiinflammatory, central sedative and analgesic effects. As an immunodepressant, it is used in certain types of renal glomerulus nephritis and allergic arthritis.

Yin modulators are drugs that replenish cellular constituents and promote anabolism and the feedback system in cellular activities. Gouqizi (fructus *Lycii chinense*) stimulates the reticulo endothelial system and protects against liver damage. It shows modulation of cardiovascular activities and gastrointestinal motility. The lycium polysaccharides increase the biosynthesis of DNA of thymus and induce a high degree of lymphocyte transformation both in the thymus and spleen cells.

Yang modulators are drugs which can activate the cellular metabolism, reinforce catabolism and promote the production of releasing factors in cellular activity. Yinyanghuo (*Epimedium brevicornum*) whose Chinese name suggests use as an aphrodisiac, especially in arousing the mating instinct of the goat, is used in the treatment of sexual impotence. It modulates neuroendocrine activities and cardiovascular activity and shows antihyperglycemic and antilipemic effects. Its polysaccharides stimulate both suppressor T cells and helper T cells, while the flavanoid glycoside, ticariin, enhances only the helper T cells but depresses the suppressor T cells. This suggests that the two agents occurring in the same herb work together to modulate the immune activity in complementary and contradictory ways.

Antiageing herbs or prescriptions have been emphasized in TCM classics. From the above discussion, it is obvious that many biological actions of the biomodulators would be most beneficial to the aged. Indeed in TCM, a combination of different biomodulators have been used to treat senile symptoms, to induce a better quality of life for the aged and perhaps also to improve longevity. Recently, those drugs traditionally used or regarded as antisenility agents, either as single herb or compound preparations, were studied by modern pharmacological methods. The results showed that they usually exhibit some of the properties which have been considered to be effective against ageing. Such activities are: (i) modulation of the endocrine systems, *i.e.* the hypothalamic–

pituitary–gonadal, hypothalamic–pituitary–adrenal and hypothalamic–pituitary– thyroid systems; (ii) modulation of the immune system; (iii) scavenging of free radicals; (iv) prolongation of the life of test animals; (v) modulation of the nervous system; and (vi) promotion of catabolism and anabolism. Clinically subjective and objective positive responses have also been observed in vital energy, immunity, memory, *etc.*

1.4.5 CONCLUSIONS

China's medicinal chemists and pharmacologists realize what great wealth they have inherited, and therefore much effort has been devoted to probe the secrets of Chinese Traditional Medicine using modern techniques and advanced biomedical theories and concepts. Yet to verify and evaluate the clinical value of the traditional *materia medica* and to bridge the gap between traditional theory and modern aspects of health care is not easy. The meagre experience thus far obtained, both successful and not, has indicated that TCM, with its accumulated clinical experience, will provide better chances for obtaining biologically active principles by use of directed pharmacological study than random screening. Moreover, sophisticated study of TCM with the integration of modern theory might provide invaluable leads for novel remedies and probably also new ideas, principles or even new theories in pharmacological and pharmaceutical sciences. This difficult and complicated but valuable task demands not only great effort and persistence over several generations but, being a multidisciplinary effort, also requires the cooperation of scientists in various fields including experienced Chinese traditional physicians and herbalists.

1.4.6 ADDENDUM

The first four of the six volumes of the second revised edition of 'Flora of Chinese Materia Medica', the second edition of 'Chinese Materia Medica Dictionary' (1986) and two of the ten volumes of 'Pictorial Record of Chinese Materia Medica' (Zhongguo Bencao Tu Lu) have been published recently. The last one is a joint work of The People's Medicinal Publishing House of China and the Commercial Press (Hong Kong) Ltd.

1.4.7 REFERENCES*

1. Q. Y. Yao, L. Zhang and H. Y. Xie, *Bull. Chin. Mater. Med.*, 1987, **12**, 464.
2. Y. Y. Tu and M. H. Chen, *Bull. Chin. Mater. Med.*, 1987, **12**, 407.
3. H. B. Shen, G. P. Peng and Z. P. Jie, *Bull. Chin. Mater. Med.*, 1987, **12**, 210.
4. C. J. Liu, Q. Zheng, G. Lin, D. G. Tang and C. J. Miao, *Bull. Chin. Mater. Med.*, 1987, **12**, 83.
5. W. J. Jiang, *Bull. Chin. Mater. Med.*, 1986, **11**, 707.
6. (a) A. K. Kuang, D. G. Gu, T. H. Gu, S. Y. Mao and H. Wang, *Chin. J. Integr. Med.*, 1984, **4**, 742. (b) D. G. Gu, A. K. Kuang, T. H. Gu, D. J. Song and X. M. Chen, *Chin. J. Integr. Med.*, 1985, **5**, 48. (c) D. G. Gu, A. K. Kuang, X. C. Qui, T. H. Gu and S. Y. Mao, *Chin. J. Integr. Med.*, 1985, **5**, 105. (d) A. K. Kuang, D. G. Gu, V. Z. Zhang, T. H. Gu and J. Q. Shen, *Chin. J. Integr. Med.*, 1985, **5**, 167. (e) A. K. Kuang, D. G. Gu, D. J. Song, X. C. Qui and S. Huang, *Chin. J. Integr. Med.*, 1986, **6**, 353.
7. Z. F. Xie, Z. J. Tang and H. Ma, *Chin. J. Integr. Med.*, 1986, **6**, 651.
8. Y. H. Liang, S. L. Niu, G. S. Liu, H. F. Li, J. Wang and Z. F. Xie, *Chin. J. Integr. Med.*, 1985, **5**, 82.
9. (a) Z. Y. Zhu, G. Q. Wang, S. G. Peng and Z. R. Yuan, *Chin. J. Integr. Med.*, 1986, **6**, 245. (b) B. G. Qiu, X. Y. Wang and X. Nin, *Chin. J. Integr. Med.*, 1985, **5**, 479.
10. G. P. Feng, Z. X. Rong, N. G. Yi, S. D. Zhang, W. M. Zhang and Z. Q. Xia, *Chin. J. Integr. Med.*, 1986, **6**, 606.
11. Anon., *Chung-Kuo K'o Hsueh (Chim. Ed.)*, 1977, 281.
12. (a) S. He, X. J. Ji, T. X. Zhu, N. G. Wang, X. L. Su, F. Z. Cao, Z. K. Pan, Z. R. Li, T. T. Bao and R. Han, *Acta Acad. Med. Sin.*, 1984, **6**, 334 (*Chem. Abstr.*, 1985, **102**, 160 101m). (b) Anon., *Acta Acad. Med. Sin.*, 1984, **6**, 273.
13. R. G. Powell, D. Weisleder and C. R. Smith, Jr., *J. Pharm. Sci.*, 1972, **61**, 1227.
14. Z. R. Li and R. Han, *Chin. Tradit. Herb. Drugs*, 1986, **17**, 135.
15. M. E. Wall, M. C. Wani, C. E. Cook, K. H. Palmer, A. T. McPhail and S. A. Sim, *J. Am. Chem. Soc.*, 1966, **88**, 3888.
16. Anon., *Zhonghua Yixue Zazhi (Beijing)*, 1975, 274.
17. (a) Y. S. Zhen, D. D. Li, F. T. Lin, Q. Li, P. Y. Tien, Y. C. Xue and X. P. Yang, *Acta Pharm. Sin.*, 1979, **14**, 83 (*Chem. Abstr.*, 1980, **92**, 34 284k). (b) Anon., *Chin. J. Oncol.*, 1979, **1**, 172.
18. Anon., *Kexue Tongbao (Engl. Transl.)*, 1977, **22**, 142.
19. (a) Y. Li, P. L. Yu, Y. X. Chen, L. Q. Li, Y. Z. Gai, D. S. Wang and Y. P. Zheng, *Kexue Tongbao (Engl. Transl.)*, 1979, **24**, 667. (b) Y. Li, P. L. Yu, Y. X. Chen, L. Q. Li, Y. Z. Gai, D. S. Wang and Y. P. Zheng, *Acta Pharm. Sin.*, 1981, **16**, 429

* It is recognised that some readers may have difficulty in obtaining those references which have not been abstracted by the Chemical Abstract Services. In this case they are invited to contact the authors directly for assistance in obtaining the article.

(*Chem. Abstr.*, 1982, **97,** 92 245n). (c) Anon., *Tradit. Chin. Med.*, 1982, **2**, 9. (d) Y. Li, P. L. Yu, Y. X. Chen and R. Y. Ji, *Acta Chim. Sin.*, 1982, **40**, 557 (*Chem. Abstr.*, 1983, **98**, 4420h). (e) P. L. Yu, Y. X. Chen, Y. Li and R. Y. Ji, *Acta Pharm. Sin.*, 1985, **20**, 375 (*Chem. Abstr.*, 1985, **103**, 160 711f).
20. Anon., *Tradit. Chin. Med.*, 1982, **2**, 45.
21. X. T. Liang, D. Q. Yu, W. L. Wu and H. C. Deng, *Acta Chim. Sin.*, 1979, **37**, 215 (*Chem. Abstr.*, 1980, **92**, 146 954k).
22. (a) Anon., *Zhonghua Neike Zazhi (Beijing)*, 1979, **18**, 83. (b) Anon. *Chin. J. Hematol.*, 1980, **1**, 132.
23. X. J. Ji and F. R. Zhang, *Acta Pharm. Sin.*, 1985, **20**, 137 (*Chem. Abstr.*, 1985, **103**, 98 313x).
24. Q. C. Fang, M. Ling, Q. M. Sun, X. M. Liu and W. Y. Lang, *Zhonghua Yixue Zazhi (Beijing)*, 1974, 271.
25. L. L. Fan, D. H. Zhao, M. Q. Zhao and G. Y. Zheng, *Acta Pharm. Sin.*, 1985, **20**, 647 (*Chem. Abstr.*, 1986, **104**, 45 507c).
26. T. T. Bao, G. T. Liu, Z. Y. Song, G. F. Xu and R. H. Sun, *Chin. Med. J.* (*Peking, Engl. Ed.*), 1980, **93**, 41.
27. G. T. Liu, *Proc. Chin. Acad. Med. Sci. Peking Union Med. Coll.* (*Engl. Ed.*), 1987, **2**, 228.
28. K. Z. Pan, Y. C. Dong and C. Z. Ni, *Chung-Kuo K'o Hsueh (Chim. Ed.)*, 1987, 257.
29. Y. C. Jin, C. W. Li and J. Y. Liu, *Reprod. Contracept.*, 1985, **5**(1), 15.
30. K. W. Liu, F. Y. Liu, Y. J. Li, G. Zhu and M. Z. Liu, *Reprod. Contracept.*, 1985, **5**(3), 55.
31. (a) Anon., *Chin. Med. J. (Peking, Engl. Ed.)*, 1978. **4**, 319. (b) Y. L. Wang and Y. K. Ba, *Chin. J. Integr. Med.*, 1985, **5**, 291.
32. W. Z. Chen, *Acta Pharm. Sin.*, 1984, **19**, 876 (*Chem. Abstr.*, 1985, **102**, 159 822r).
33. W. Z. Chen, Y. L. Dong, C. G. Wang and G. S. Ting, *Acta Pharm. Sin.*, 1979, **14**, 277 (*Chem. Abstr.*, 1980, **92**, 51 936s).
34. J. H. Zhou, Proceedings of International Symposium on Traditional Medicines and Modern Pharmacology, May, 1986, Beijing, p. 11–20.

1.5
Contribution of Ayurvedic Medicine to Medicinal Chemistry

NITYA ANAND

Central Drug Research Institute, Lucknow, India

1.5.1 INTRODUCTION

'Those alone are wise who act after investigation'—Charaka: Sutrasthana 10:5

Traditional systems[1] of medicine continue to be widely practised in India. It is estimated that 50 to 75% of the population use traditional drugs, some perforce because of the lack of easy access to drugs of the modern system, but many by deliberate choice on account of their faith in them. There is no escaping the fact that for a long time to come a large segment of the population in India and many other developing countries will continue to be largely dependent upon drugs from the traditional systems. Further, there is a strong belief that these systems possess drugs for certain conditions, particularly chronic conditions, for which the modern system offers inadequate or no remedies. The societal relevance and continued existence of these systems cannot, therefore, be ignored. In order to extend their application and usefulness it would help to understand their basic principles.

The major traditional systems of medicine in India are the Ayurveda, Siddha and Unani systems. Of these, Ayurveda and Siddha originated in India itself, while the Unani system came from Persia and found in India a very fertile soil for its growth. These systems have much in common, both in theory and practice. On account of the constraints of space, this chapter is mainly devoted to Ayurveda which occupies a preeminent position amongst the three systems in India. Ayurveda is also prevalent in some neighbouring countries, such as Nepal, Bhutan, Sri Lanka, Bangladesh and Pakistan, while the principles and practices of traditional systems of medicine of some other countries in this region, such as those of Tibet, Mongolia and Thailand, appear to be derived from Ayurveda.

The theories of Ayurveda are discussed first, in order to give an idea of its doctrinal base. This is followed by a presentation of its *materia medica* and of the present-day status of practice of Ayurveda and its contribution to the development of modern drugs.

1.5.2 TRADITIONAL SYSTEMS OF MEDICINE IN INDIA

1.5.2.1 Ayurvedic System

Ayurveda (ayur = life, veda = knowledge; science of life or longevity) is not merely a system of medicine, but rather a general philosophical approach to the maintenance of good health and long life and to the treatment of diseases. It prescribes drugs, diets and other regimens, which include codes of conduct conducive to the maintenance and promotion of positive health as well as the prevention and cure of disease. Ayurveda takes a 'holistic' view of man in the universe—its aim is not just the cure of a disease but the maintenance of a positive healthy state of body, mind and spirit, in a healthy environment and in harmony with the universe. The ancient Hindu sages attached great importance to maintaining bodily health, since they felt that disease would interfere with the spiritual development of an individual.

The origin of Ayurveda is lost in antiquity. As was the case with many branches of human knowledge in prehistoric times, Ayurveda developed in close association with religion and mythology. It can be considered as a stream of knowledge coming down from time immemorial which from time to time was reinterpreted and added to. Its characteristic concepts matured between 2000 and 500 BC in India. This was a period of intense intellectual activity which saw the development of many important schools of Indian philosophy. The growth of Ayurveda was a part of this general philosophical development and many of its concepts were derived from various schools of philosophy, particularly the Samkhya, Nyaya and Vaiseshika schools. Ayurveda is often considered as an upa-veda (subsidiary Veda) or even as a fifth Veda.[2]

The first mention of drugs and disease is found in the Rigveda and Yajurveda (both around 2000 BC). The earliest comprehensive description of the beginnings of Ayurveda is available in the Atharvaveda (1600–1000 BC), which contains descriptions of human anatomy and rudiments of the classification of diseases, and which refers to the practice of medicine by both itinerant physicians and those with formal training in the medical sciences (Vaidyas) and to herbal medicines, although treatment during the Vedic period was mainly by recitation of prayers and incantations (mantras).[3]

In the post-Vedic period, Ayurvedic knowledge was advanced further and was placed on a sound scientific basis. The available knowledge was classified, codified and interpreted in many Samhitas (Samhita = compilation) between 1000 and 200 BC. Charaka Samhita[4–6] and Sushruta Samhita,[7,8] which emphasize medicine and surgery respectively, are considered the most authoritative classical Ayurvedic texts. The texts give a description of how the Samhitas came to be composed. Charaka Samhita opens with a description of a meeting of sages in the Himalayas to discuss how to alleviate human suffering and ensure a long, healthy and satisfying life to all, and continues with a description of how the sages decided to take the necessary steps to acquire knowledge for this purpose. Later, one of them, Atreya Punarvasu, organized a series of symposia in different parts of the country and, after discussion, formulated the basic concepts of Ayurveda. He asked six of his disciples to compile his oral teachings in writing. Of them, Agnivesa documented the precepts of his teacher most faithfully in a compendium known as Agnivesa-Tantra. This was revised, enlarged and annotated by Charaka between 500 and 200 BC and came to be known as Charaka Samhita. This was redacted and enlarged later by other scholars, the most notable being Drdhbala, and the text of Charaka Samhita most commonly referred to is this redacted version. Principles and methodology of surgery were similarly formulated by the school of Dhanwantari and compiled by Sushruta during the same period, and later redacted by Nagarjuna. The Charaka and Sushruta Samhitas are thus the best-known compiled texts of Ayurveda and are still available more or less in their original complete forms.

These Samhitas are didactic texts in prose and verse in the form of questions and answers. In fact, the major portion of Charaka Samhita is presented in the form of questions and answers between the disciple Agnivesa and his teacher Atreya. The chapters that are entirely in verse form deal with the enumeration of symptoms or therapeutic prescriptions, and are the ones which it would be useful to know by heart for practising, whereas the passages in prose deal with theoretical principles.

In general the contents of the two Samhitas are similar, with the difference that Sushruta Samhita gives more emphasis to surgery than Charaka Samhita. Both of them present their knowledge in an extremely condensed form, but because of the abundance of material they are still voluminous. The Samhitas have tried to put the knowledge on a scientific and rational basis, basically excluding magic and mysticism, unlike the texts of Vedic medicine. References can be found to different schools of philosophy which had influence on the information and knowledge base. These references show that life was considered as a part of the total cosmological existence with common laws applicable to the whole universe. Experience, along with a keen sense of observation, had obviously played an important role in the development of therapeutics and general principles of good health. In this sense the Samhitas are an unusually rich record of medical and dietetic experience.

The Samhitas give a detailed classification of diseases and describe their etiology, pathophysiology, diagnosis and prognosis. Causes and symptoms of diseases and their treatment are dealt with. Embryology, obstetrics, anatomy, physiology, personal hygiene, sanitation, training and duties of physicians as well as other theoretical and practical aspects of medicine are dealt with in some detail. The origins of medical science, the fundamental causes of conception and birth and of physical deformities are discussed. Methods of preparation of drugs are described at some length. These Samhitas divide Ayurveda into eight branches which, expressed in modern terminology, are (1) internal medicine and therapeutics (kayachikitsa), (2) surgery (shalya), (3) diseases of the eye, ear, nose and throat (shalakya), (4) paediatrics, obstetrics and gynaecology (kaumarabhrtya), (5) toxicology (agadatantra), (6) psychiatry (bhutavidya), (7) rejuvenation, promotive therapy (rasayana) and (8) virility (vajikarana). Although the Samhitas do not deal with each of the above branches in separate chapters, one particular aspect of all branches is described in one chapter.

1.5.2.1.1 Charaka Samhita

The Charaka Samhita is considered to be an exhaustive compendium on internal medicine, incorporating all the knowledge in this branch known at that time, although it also contains chapters devoted to the other seven branches. The treatise describes the causes, symptoms and treatment of different diseases, and some of the 10 aspects of disease discussed are physiology, etiology, pathology, treatment, effect of age, sex and season, role of the physician, medicines

Table 1 The Scope of Charaka Samhita

Section number	*Section*	*Coverage*	*Number of chapters*[a]
1	Sutrasthana	General principles and philosophy	30
2	Nidhanasthana	Causes of disease	8
3	Vimanasthana	Nourishment and general pathology	8
4	Sharirasthana	Anatomy and embryology	8
5	Indryasthana	Diagnosis and prognosis	12
6	Kalpasthana	Pharmacy	12
7	Chikitsasthana	Treatment	30
8	Sidhisthana	Cure of disease	12

[a] 120 chapters in total.

and appliances, and the sequence of their application. It is divided into eight sections covering 120 chapters, and the broad subject matter of each section is given in Table 1.

Charaka Samhita classifies over 200 disease entities and about 150 pathological conditions, which include congenital defects and inherited diseases as well. The classification is based on the site of occurrence of the disease (topology), the symptoms and physical appearance of patients, and combinations of these, and, when expressed in terms of modern medicine, this classification would be as given in Table 2.

Table 2 Charaka's Classification of Disease, Expressed in Modern Medicinal Terms

(1) Infectious and parasitic diseases	(9) Diseases of the digestive system
(2) Neoplasms	(10) Diseases of the genito–urinary system
(3) Endocrine, metabolic, immunological and nutritional disorders	(11) Complications of pregnancy, child birth
(4) Diseases of blood and blood-forming organs	(12) Diseases of skin
(5) Mental disorders	(13) Diseases of the musculo–skeletal system and connective tissues
(6) Nervous system disorders	(14) Congenital anomalies
(7) Diseases of the circulatory system	(15) Perinatal diseases
(8) Diseases of the respiratory system	(16) Ill-defined conditions

Charaka describes at great length the method of examining a patient for diagnosing a disease, selecting the appropriate drugs and thereafter the procedure for treatment. In order to aid diagnosis, pathological examinations, *e.g.* of urine or faeces, are recommended.

1.5.2.1.2 Sushruta Samhita

This Samhita is similar to Charaka Samhita in format but its emphasis is on surgery, which is considered as the foremost speciality because of its ability to provide instant relief. The treatise consists of five sections corresponding to sections 1, 2, 4, 6 and 7 of Charaka Samhita and a sixth miscellaneous section. Various types of surgical problems and situations are described, which include intestinal obstruction, bladder stones, requirements for plastic surgery as in rhinoplasty, abdominal delivery of a foetus and amputation of limbs. Sushruta describes 32 surgical operations and lists 101 types of blunt instruments and 21 kinds of sharp instruments, which include scalpels, forceps, pincers, trocars, speculums, syringes, canulae, dilators and bone levers. Accessories include thread, caustic-coated thread (ksharsutra) for non-surgical treatment of anal fistula, twine for ligature, 14 types of splints, bandages and gauze.

From the vivid descriptions given of surgical situations, instruments and operations, surgery as a branch of Ayurveda seems to have been highly developed. And yet, in the current practice of Ayurveda, surgery is very rarely performed. It is not very clear at what period the practice of surgery declined. During the Buddhist period, as a consequence of the philosophy which discouraged vivisection, surgery was less favoured, and it is very likely that this fact coupled with the complications which can arise in surgery due to the non-availability of anaesthetics and aseptic conditions might have led to the decline of surgical practice.

1.5.2.1.3 Other classical Ayurvedic texts

Vagbhatta was another important compiler of Ayurvedic knowledge who lived around 700 AD. His work, Astanga Hridaya, although based mainly on the Charaka and Sushruta Samhitas, is even more precise and is considered unrivalled in its presentation of the basic principles and practice of Ayurvedic medicine.[9] It contains 7444 verses divided into six sections with 120 chapters. Charaka, Sushruta and Vagbhatta are considered the powerful Triad (Vrihat Traya) of Ayurveda.

Other important Ayurvedic texts include: Madhava Nidhana, written in the 12th century, dealing primarily with diagnosis; Sarangdhara Samhita, which contains a systematized *materia medica* and consists of three parts and 32 chapters; and Bhava Prakash, written around 1550 AD, which contains 10 831 verses, and deals primarily with treatment; in addition there are over 70 pharmacy lexicons (Nighantu Granthas), written mostly between the 7th and 16th centuries, which provide very valuable information about medicinal plants used in Ayurveda; of these, Raja Nighantu and Madanpala Nighantu are considered the most valuable.

1.5.2.1.4 Essential doctrines and fundamental principles of Ayurveda

As mentioned earlier, the period from 2000 BC to about the beginning of the Christian era was one of intense intellectual upsurge in the Indian subcontinent, and this was the period during which most of the concepts of Ayurveda were formulated. Ayurvedic thought was, therefore, a part of the general philosophical development of that period, as applied to medical sciences. Ayurveda adopted a number of concepts from various schools of philosophy, particularly from the Samkhya and Nyaya–Vaisheshika schools.[10]

The Samkhya school takes a view of life which corresponds almost to the position of logical positivists. The discussions and arguments in Samkhya philosophy are of a scientific character and have a bearing on the practical side of life.

According to Samkhya philosophy, all objects that we see in the universe are the result of transformation within one primal substance, prakriti (physical form would be its manifestation), the first cause in the universe of all existence excepting purusha (soul), which is uncaused. Purusha is considered as an independent entity, but without purusha, prakriti would not be active. The subject is 'purusha' and the object is 'prakriti' and the coming together of these two is the first step in evolution. Thus all objects in the universe are related. This was the Hindu concept of evolution and adopted by Ayurveda philosophy. Man is thus considered as an integral part of all that exists in the universe.

All material objects in the universe, whether animate or inanimate, are considered to be made up of the same five forms of matter, the panchbhootas (proto-elements). The elementary states of matter were not classified only as 'solid', 'liquid' or 'gas' but as five proto-elements, ether or space (akasa), fire or radiant energy (tejas), earth or solid (prithvi), water or liquid (ap) and air or gas (vayu) and each of these is perceived by one of the five senses, hearing, vision, smell, taste and touch respectively. This division of matter is similar to the four elements of Greek philosophy. Prithvi confers the property of hardness, water confers fluidity, tejas confers body heat, air denotes the vital breath and ether denotes the void body spaces. There is a sixth element, the spirit or self (atman) which pervades the animate world, and thereby distinguishes it from the inanimate. Living things in addition have two other traits, the intellect and ego, and have sense organs.[11–13]

These five elements are considered to combine in different proportions to form the seven basic body tissues, the saptadhatus, which are lymph or plasma (rasa), blood (rakta), flesh (mansa), fat (meda), bone (asthi), marrow (majja) and semen (shukra).

The three more concrete of the elements discussed above, namely air, fire and water, were conceived as playing a fundamental role in life, health and disease. This led to the tribhuta/tridosha theory, according to which body function is regulated by the three essential elements, vata, pitta and kapha (compare the four humours of Greek/Unani medicine), which can be literally translated as air, bile and phlegm, but which have a broader physiological connotation. Apart from a few exceptions, all human beings are considered to have a predominance of one or the other of these three humours from the very moment of conception, which imparts to them a constitutive character. These types can be distinguished by their physical, physiological and psychological characteristics. In normal health the three doshas are balanced and all pathological conditions of the body or mind and also their degree of severity are considered to be the direct outcome of an imbalance of humours (dosha also means defect or derangement).

Each dosha, working in unison with the other two, is considered to be responsible for a particular set of functions, which are listed in great detail and further categorized; vata represents the dynamic force regulating the psychic and nervous system and occupies a preeminent position, pitta controls the metabolic, digestive and enzymatic activities, while kapha regulates the state of energy, assimilation, attachment and lubrication.

Connected with this is the theory[14] of metabolism, digestion and excretion of waste products (malas). Tridoshas regulate the formation of body tissues (saptadhatus) from basic elemental constituents (panchbhutas) and in this process excretory products (malas) are formed, which for normal functioning must be excreted. The panchbhutas–saptadhatus–tridoshas–malas are basically physiological concepts formulated to explain the regulated functioning of the human body.

Each disease is characterized by a specific status of the three doshas. Nosology studies the symptoms of diseases and is therefore related to observation, and it tries to interpret these symptoms as a function of the status of the three dosha. This was aimed at giving a pathophysiological basis to the etiology of each disease. Diagnosing, therefore, consists of not only recognizing a disease but also of a judgement concerning the role exercised by the three doshas. The drug has to be such as would have an antagonistic effect on the doshas concerned.

These doctrines reflect the concern for having a comprehensive and integrated view of the universe, including both the microcosm and the macrocosm, and for evolving universally applicable concepts, theories and laws. Among Ayurvedic thinkers, as perhaps with intellectuals of many other schools of that period, there appears to have been a great desire, amounting almost to an obsession, to classify and categorize facts and knowledge and to explain a maximum number of phenomena with a minimum number of postulates; the brevity of presentation and expression of Ayurvedic texts is remarkable.

Considering the state of knowledge then existing of physiology and body functions, these theories were remarkable for their perception and power of correlation and generalization. However, modern developments in physiology, based as they are on the very sophisticated investigative methods now available, have by-passed the intuitive and perceptive power of the old scholars and there is a need to have a fresh look at these old doctrines, keeping at the centre the 'holistic' view of man in the universe which is one of the most important and abiding concepts of Ayurveda.

1.5.2.1.5 Method of acquiring knowledge in Ayurveda

Both Charaka Samhita and Sushruta Samhita lay great emphasis on the method of enquiry for acquiring knowledge and drawing conclusions, based on the concept of 'pramana' of the Nyaya and Samkhya schools of philosophy.[3,12] Pramana can be described as a means of correct cognition of truth (prama = to achieve a clear perception of the object). Pramana is dependent on trained and disciplined senses through which knowledge is obtained, and then analyzed and evaluated. According to the concept of pramana, the sources of valid knowledge are:

(i) Authoritative testimony (apt-upadesha). Whose testimony should be considered as authoritative has been stated, but it is clarified that even this knowledge must stand the test of experimental observation and only then accepted.

(ii) Direct observation (pratyaksha). It is emphasized that, while knowledge obtained through all the senses is valid, knowledge gained by direct visual perception is the most dependable.

(iii) Logical assumption (anumana). From observed knowledge by the processes of induction, deduction or analogy, and is of three types: *a priori*, *post priori* and commonly observed.

To these three means which are common with Samkhya philosophy, Ayurveda added the need for 'yukti' (reasoning) with a view to arrive at a correct judgement or conclusion as a method of determining the truth. 'Yukti' was considered a function of intellect (buddhi), which evaluates and correlates several causes with regard to the three aspects of time—past, present and future. Charaka illustrates the operation of this approach by examples of a number of phenomena. The Samhitas emphasize that, before drawing final conclusions, it is necessary to submit them to different kinds of experiments or tests. Conclusions emerging out of these procedures are then to be declared as having been established beyond doubt and are termed as 'sidhanta'. Charaka often uses the word 'parikhsha', examination, instead of pramana. These Samhitas have thus laid great emphasis on scientific methodology and the power of intellectual reasoning to arrive at a balanced scientific judgement and valid conclusions. This method, it is stated, should be applied both in the practice of medicine, *e.g.* for diagnosing and treating patients, and also for creating new knowledge, *e.g.* for discovering new drugs. Modern scientific methodology is not very different from Ayurvedic pramanas.

1.5.2.2 Unani System

The framework of the Unani system (Unani = Greek) is based on the teachings of the great physician Hippocrates (460–370 BC) and Galen (131–210 AD). It was developed into an elaborate medical system by the Arabs, noteworthy contributions being made by the Arab physicians Rhazes (850–925 AD) and Avicenna (Ibn Sina) (980–1037 AD).[15,16]

The system was introduced by the Muslims into India where it has interacted very intimately with the Ayurvedic system and flourished. Its practice is mainly confined to the Indian subcontinent. It is based on the humoral theory of Hippocrates, who proposed that there are four humours in the body, namely blood, phlegm, yellow bile and black bile (compare the three humours of Ayurveda, Section 1.5.2.1.4); each individual is constitutively characterized by the preponderance of one of the humours, which determines their temperament (termed 'mizaj'). If the humoral balance is disturbed, disease results. The body has a natural power of adjustment of this balance, and medicines help to regain this power and thereby restore the balance. There is also emphasis on correct diet to maintain this balance. The constitutive theory of five elements is an essential part of Unani medicine. Thus, on the theoretical plane, basic similarities exist between the Ayurvedic and Unani systems. The method of diagnosis is characterized by a heavy emphasis on examination of the pulse. Great stress is laid on treating the individual as a whole and not the disease only, taking into account the constitutive characteristics of the individual. Treatment is mainly with plant products; single drugs in crude form or as formulations are preferred, although some compound formulations are also used. Its *materia medica* has much in common with Ayurveda, and the two systems are known to have borrowed much from each other.

1.5.2.3 Siddha System

The Siddha system (sidh = to achieve self-realization and perfection) is practised mainly in the Tamil Nadu State of India. It is claimed to have originated in the pre-Vedic period and was practised in the Mohenjo-daro era (5000–3000 BC). The Siddha system is embodied in 96 principles described in the work of 18 Siddhas. They have composed treatises in which, besides medicine, other aspects of philosophy and culture are covered. The best-known work is Nadishastra. An important aspect of the Siddha system is that, along with herbal drugs, it makes liberal use of metals and minerals, particularly mercury, sulfur and salts. Among plants, *Azadirachta indica* (neem), lemon and garlic are highly valued in this system.

1.5.3 MEDICINAL SCIENCE IN AYURVEDA

1.5.3.1 Basics of Therapeutics

Charaka considered the physician, drug, nurse and patient as the four pillars for the treatment of disease, the success of which depended on a proper balance of their individual performances;[18] the expectations from each are listed.

The good physician was considered to be one who had theoretical knowledge, sound judgement and long experience. It is stated by Charaka that a good physician is one who knows the science of the administration of drugs with reference to the time and season, and who applies this knowledge only after examining each and every patient individually. Hence the skilful physician was one who had adequate experience in addition to theoretical knowledge. Detailed knowledge of the drug by the physician is also emphasized: 'a drug that is not understood well is comparable to poison'; 'the drug whose name, form and properties are not known, or though known, is not properly administered will cause disaster'. There is an elaborate description given of the way the detailed history of each patient is to be taken, based on which the diagnosis would be made. It is mentioned that diagnosing does not consist of only recognizing a disease but also of judging the role exercised by the three doshas. The method of deciding the diagnosis is based on 'pramanas', perception based on direct observation, induction and logic based on observed facts or based on authoritative testimony. The physician's judgement is expected to be in accordance with: (1) the testimony of accepted authorities, (2) direct perception, (3) inference and (4) analogy; these are the four pramanas of Ayurveda discussed earlier. This would require the physician to be familiar with the method of inductive and deductive logic.

Ayurveda has a well-developed system of classification of disease and of drugs, and of the basis of their development and usage. Charaka lists over 200 diseases and about 150 pathological conditions and congenital defects which look quite compatible with the modern classification of diseases.[19]

Diseases according to Ayurveda can be due to three causes: (1) intrinsic/internal, such as metabolic diseases and psychosomatic disorders; distinction is also made between genetic and congenital and acquired disorders; (2) external, which include infections, poisons, accidents, natural calamities, *etc.*; and (3) supernatural, effect of evil spirits, *etc.*

Parasites (visible and invisible) are mentioned by Charaka as important external agents causing disease. Parasites found in faeces, mucus, other external excreta and blood, skin and hair are described and classified according to their location. Their connection with unhygienic conditions—dirty water and infected foods, including fish and meat—is recognized. It is stated that some parasites present in blood are invisible and may cause diseases like leprosy—this is reminiscent of the germ theory of disease. Different treatments which include physical removal of the parasite, specific drugs, purgation, enemas and fumigation of the environment are described.

Both metabolic and infectious diseases are attributed to improper adjustment of the three doshas; there is immunity from disease as long as these three constituents are in equilibrium. Equilibrium of the tridoshas gives health, whilst lack of balance between them results in disease. Therapeutic intervention is dominated by the idea of a soothing or exciting action of all the measures, drugs, diet, *etc.* on the balance of these doshas. However, the theory is kept sufficiently flexible to be adjusted in the light of experience, as it is repeatedly emphasized by Charaka that knowledge of the action of remedies had arisen from their usage.

A disease can be treated by a judicious mix of medicines, diet and other practices prescribed jointly or severally; there is great stress on a suitable diet; not too much and not too little. Medicines according to Charaka are of two types; one kind is promotive of vigour in the healthy; the other destructive of disease· in the ailing. The special emphasis in Ayurveda on promotive health is noteworthy and is an important dimension of Ayurvedic therapeutics, which is discussed later. It is mentioned that drugs should be effective at a low dose, non-toxic, easy to use, have an acceptable taste and be properly formulated, 'a drug while inducing therapeutic effect should not evoke untoward effects'. Drugs were developed mainly on the basis of experience of long usage in patients, although there are some references to testing of drugs on pet animals and on worms.

Charaka classifies drugs into 50 groups on the basis of their pharmacological action, as shown in Table 3. In addition to this classification, drugs have also been divided into 37 groups according to their therapeutic uses; there is an overlap between these two types of classification. The properties of each drug have been discussed under five headings: (a) immediate action, such as taste or sensation (rasa); (b) overall class of the drug, which covers physical traits and action (guna); (c) systemic effect of the drug (veerya); (d) metabolism and effect on metabolism (vipaka); and (e) the specific action or potency or power of the drug (prabhava).

Table 3 Charaka's Classification of Drugs by Pharmacological Action

(1) Promoting longevity	(26) Enematics (oil based)
(2) Increasing fat	(27) Increasing nasal discharge
(3) Reducing obesity	(28) Antiemetics
(4) Promoting excretions	(29) Decreasing thirst
(5) Bone healing	(30) Relieving hiccup
(6) Digestive	(31) Laxatives
(7) Promoting strength	(32) Changing colour of stools
(8) Improving complexion	(33) Antidiuretics
(9) Improving voice	(34) Changing colour of urine
(10) Giving joy and happiness	(35) Diuretics
(11) Removing satiety	(36) Antitussives
(12) Destroying pills	(37) Antiasthma
(13) Destroying leprosy/skin diseases	(38) Antiinflammatory
(14) Destroying pruritis	(39) Antipyretic
(15) Anthelmintic	(40) Removing fatigue
(16) Antidote for poisons	(41) Removing burning sensation
(17) Galactagogue	(42) Removing cooling sensation
(18) Correcting mammary defects	(43) Removing urticaria
(19) Increasing semen	(44) Removing pain in limbs
(20) Correcting seminal disorders	(45) Removing stomach pain
(21) Emollients	(46) Styptics
(22) Diaphoretics	(47) Analgesics
(23) Emetics	(48) Antisterility
(24) Purgatives	(49) Restoring consciousness
(25) Enematics	(50) Preventing aging

Dose–response relationships and the need for a correct dose regimen are well recognized in Ayurvedic practice; the administration of more drug could be harmful and of less may be ineffective. It is mentioned that the dose may need to be adjusted in individual cases and some of the factors to be taken into consideration are: the weight and age of the patient (children and old people may need a lower dose); severity of the disease (severe disease may require a higher dose); digestive power, tolerance and vitality of the patient; time and season of drug administration was considered important, different diseases may require drugs to be given at different times of the day (morning, noon or at bed-time, or before meals, with or in the middle of meals or after meals), the dosage of drugs may need to be varied in different seasons; compatibility and incompatibility of drugs with each other and with the diet was emphasized.

1.5.3.2 Pharmaceutical Practices[20]

1.5.3.2.1 Sources of drugs

Almost 2000 drugs are listed in the Charaka and Sushruta Samhitas and the following sources are mentioned:

(1) Vegetable (living beings that do not move), which includes flower and fruit-bearing plants, creepers and grasses. Parts of plants which can be used include the leaf, stem, bark, root, tuber, rhizome, flower, fruit and exudate.

(2) Animal (living beings that move about), the products used being honey, secretions, bile, fat, bone-marrow, blood, flesh, excreta, urine, skin, semen, tendon, horn, feathers, nails, hoof, bristles and pigments.

(3) Mineral (lifeless), which includes preparations of metals, more specifically of gold, silver copper, lead, zinc, tin, iron, arsenic, mercury and antimony; sand, lime, diamond, emerald, pearl, common salts and chalk.

(4) The use of natural physical forces such as sunlight, heat, cold, wind, rain, *etc.* is considered as a fourth method of treatment by Sushruta.

1.5.3.2.2 Plant collection

More than 70% of the drugs are of herbal origin. In the collection of plants, importance is attached to its identification, the locality (habitat) from which a plant was collected, the soil in which it had grown, the season of collection, the part to be used, the method of collection, drying, storage and preservation and the period for which a plant should be stored; detailed descriptions on these points are given in various books. For example, it is stated that winter is the best time for gathering leafy parts of plants, and plants which fall in this category are named. There is stress on the use of fresh plants, but those which have to be stored should be kept in a well-ventilated room, free from pests, smoke, dampness or draught and hung or kept in earthen pots and used within one year of collection.

1.5.3.2.3 Mode of administration

Different modes of administration of drugs were practised which included: oral administration; external and local application; infusion through rectum, urethra or vagina as enemas, clysters or

Table 4 Drug Formulations in Ayurveda

Solid forms	*Liquids*
Powders (churna)	(a) Aqueous
Alkalies (kshar)	Fresh juice (swarasa)
Pills (vatika)	Extract (kashaya), infusion or decoction (kusatha)
Exudates (guggula)	Milk decoction (kshira-paka)
Collyria (netranjan) micronized	Perfumed waters (Sugandhita-Jal)
Snuff (nasya)	(b) Oily
Suppositories (phalavarti)	Medicated oil (taila)
Fumigants (dhaumpana)	Medicated butter (ghrita)
Ointments/viscous liquids	Emulsion (mantha)
Pastes (kalka)	(c) Vinegars (kanjali) and mineral acids (samkhadravaka)
Soft extracts (avaleha)	(d) Spirits
Confections (khandapaka)	Wines (sura)
Gruels (yavagu)	Tinctures (asava)
	Fermented materials (arishta)

suppositories; and inhalation, snuffing, smoking or fumigation. A number of pharmaceutical forms of drugs (delivery forms) are listed in Table 4, many of which are still used.

1.5.3.2.4 Pharmaceutical processes

Pharmaceutical processing was well developed and there are clear descriptions of the pharmaceutical processes and apparatus used. These processes include: expression (for making juice); extraction (hot and cold) and filtration through cloth; distillation and concentration (for making extracts, decoctions, infusions, syrups and pastes); maceration with liquid extract, especially of metals and minerals, till the solid is totally soaked; extraction with oils or ghee and decantation; fermentation in earthenware or wooden vats; purification of water (by heating it or dipping hot iron, clay or sand balls into it, or exposing it to the sun, clarifying by adding salts, precious stones or some plants, and by adding aromatic products); and roasting, particularly for making bhasmas (ash). Pills were made by mixing the drug extract with excipients, drying in the sun, and compressing by hand; pills were also made with the help of honey or syrup.

1.5.3.2.5 Toxicology and the use of poisons

Ayurvedic physicians attached great importance to poisons both for their treatment and their use as drugs; in fact toxicology was considered as one of the eight specialities of Ayurveda. Poisons, whether of animal or plant origin, though dangerous as such, could be used as drugs after proper processing and purification (shodhan kriya). The methods of processing included boiling or macerating them with milk, urine or plant extracts. Some of the poisonous plants used as drugs are *Aconite* spp., *Abrus precatorius, Calotropis gigantea, Datura metel, Gloriosa superba, Nerium odorum* and *Strychnos nuxvomica.*

The concern of Ayurvedic physicians with the safety of drugs is reflected in the emphasis on purification to make them fit for human use. In books on *materia medica*, descriptions are given of purification processes. There are reports that drugs and foods used to be checked for their safety by administration to pets (cats and dogs) or cattle.

1.5.3.2.6 Processing of metals

The use of minerals and metals is common in traditional medicine. Metals most commonly used are gold, silver, copper, lead, mercury, tin, iron, and zinc. Processes for the preparation of medicaments based on metals are described in great detail. These processes convert the metals into salts/compounds which are not toxic and are also absorbed. These include treating with juices and extracts of plants and acids, or calcining. One special process is called 'killing' of metals, which was particularly used for mercury and arsenic. There are many books devoted only to the making of metal preparations. One such publication is Rasaratuakara by Nagarajuna, written about 700 AD.

1.5.3.2.7 Hospitals and pharmacies

Although treatment in ancient times was mostly carried out at the residence of the physician or of the patient, there is also mention of hospitals, particularly from the Buddhist period, *i.e.* from around 500 BC. It is stated that a hospital should be constructed in a pleasing clean airy place, should be well ventilated but free from draughts, not exposed to glare but also not behind a big building. Detailed descriptions of pre-natal clinics, children's rooms and maternity homes, along with complete lists of equipment, are given in the Charaka Samhita. The dispensing of drugs seems to have been done mainly by the physician himself, although there is mention of pharmacies/drug stores. The need for the preparation of drugs under hygienic conditions and their proper storage is elaborately described. Drug manufacturing was carried out mainly by the physicians themselves.

1.5.4 PRACTICE OF AYURVEDIC MEDICINE TODAY

Currently, there are around 250 000 (April 1986) registered medical practitioners of the Ayurvedic system (the total number of practitioners of all traditional systems is around 291 000) as compared to about 700 000 practitioners of the modern system; in every state in India about one-third of the government medical posts are occupied by physicians belonging to the traditional systems.

Training of students is an essential part of the Ayurveda tradition. There are records of organized/regular training programmes for medical practitioners in the old Universities such as that of Taxila from 500 BC onwards. Much of the training, of course, was also imparted by tradition and heritage. At present, there is organized teaching and training in India for traditional systems more or less on the same pattern as for the modern allopathic system (Table 5).[21]

Table 5 Teaching Institutes/Hospitals/Dispensaries of Ayurvedic System (April 1988)

Graduate training colleges	97
Postgraduate training institutes/departments	23
Approximate annual outflow of students	2 400
Hospitals	1 460
Dispensaries	12 111
Registered practitioners	251 071

The extent of drug usage is hard to quantify for traditional systems as by tradition most of the practitioners manufacture and formulate their own prescriptions, although there are now about ten centralized traditional system drug manufacturers whose individual annual production is over $10 million, whilst the largest one has an annual production of about $112 million (June 1988). Their total present annual turnover would be about $350 million (as compared to $2320 million for modern drugs in 1987). The production by these large manufacturers, although carried out according to traditional prescriptions, is mechanized and automated, and conforms to the normal good manufacturing practices. The drugs marketed include both generic drugs and some branded products. In addition, many of the modern drug companies do market a few branded Ayurvedic products.

In 1959, the Government of India decided to recognize the various traditional systems of medicine as distinct entities. Since 1964, and as modified in 1982, the Drugs and Cosmetics Act of the Government of India has included special provisions for traditional system drugs. Each drug of the traditional systems has also got to be licensed and registered with the state Drug Control Authority. If any new drug to be introduced is manufactured strictly according to traditional system *materia medica*, the Drug Controller need not insist on safety and clinical trial data. There is a separate book of standards for Ayurvedic formulations.[22] All the Indian states are empowered to appoint independent Drug Controllers for traditional systems. In practice, however, in most Indian states the Drug Controller is common for all the systems of medicine. At the federal government level the Drug Advisory Board and the Drug Consultation Committee are also common for all systems of medicine. An Ayurvedic medicine pharmacopoeia is available. A Central Pharmacopoeal Laboratory of Indian Medicine has been established to monitor drug standards of Ayurvedic, Siddha and Unani system drugs. As most of the drugs are a mixture of plants, standardization is a difficult job, but one that has to be done. The Central Government has three separate councils, one each for Ayurveda (including Siddha), Unani and Homoeopathy, to advise on all matters pertaining to the practice of these systems of medicine.

It is estimated that a total of about 1000 Ayurvedic remedies are used at present, prepared from some 750 plants; about 430 remedies are included in the Ayurvedic Pharmacopoeia[22] brought out by the Indian government. The Central Council for Research in Indian Medicine and Homoeopathy has published a compendium of commonly used Ayurvedic remedies.[23]

1.5.4.1 Research on Ayurvedic Drugs

There has been considerable research on Ayurvedic drugs, both to put their usage on a modern scientific footing and also to obtain leads for the development of drugs for the modern system.

With the compilation of the Ayurvedic Pharmacopoeia, attention has been focussed on the need to ensure standards of quality and reproducibility in products. Most Ayurvedic drugs are a mixture of a number of ingredients, which no doubt present difficulty in quality control assays by conventional methods. However, the sophisticated spectroscopic and separation methods and the bioassay methods, both *in vitro* and *in vivo*, now available offer the possibility of developing suitable quality control criteria for these preparations. Further, the safety of these drugs cannot be taken for granted merely because they have the sanction of centuries of use. It is necessary to confirm their safety by toxicity studies. This will help to make traditional system remedies more acceptable in modern therapeutics. Clinical trials, particularly of drugs used for acute life-threatening situations, may also be necessary to place these drugs *vis-à-vis* the modern drugs in the correct community health perspective.

Traditional system drugs have been the starting point for the discovery of many important modern drugs. This fact has led to chemical and pharmacological investigations of these plants and to the undertaking of general biological screening programmes of plants not only in India but all over the world.[24–27] Some of the contributions of Ayurveda on this count are discussed later. Much of the work under these programmes has been centred round individual plants and often the emphasis has been on chemical investigation. It is necessary to first test the products as they are used in traditional medicine and, if the desired activity is confirmed, to then investigate individual plants, before going on to active fractions and pure constituents of each plant. While general screening programmes have their own place to help identify new and unexpected leads, the chances of finding activity are greater if drugs are tested for the activities already described in the classical texts; one such example, that of gugulipid as a hypolipidaemic agent, will be described later. Broad biological screening of plants and specific testing for an activity for which the drug is valued in Ayurvedic texts are thus complementary approaches to new drug development and are not mutually exclusive.

A question which is often asked is what are the areas in which Ayurvedic drugs are of special value. There are some diseases for which there is a need for new drugs as the modern system has either no or only inadequate drugs, while Ayurveda, because of its special traits, does seem to offer some remedies which could be fruitful areas of investigation. A few such areas will now be discussed.

1.5.4.1.1 Hepatoprotectors

The liver is the organ of metabolism and excretion in the body and provides protection against foreign substances by detoxifying and eliminating them. Any derangement of liver function caused by chemicals, infection or nutritional deficiencies can thus have serious repercussions on body functions. Liver disorders include acute or chronic hepatitis, hepatosis and cirrhosis. The drugs available in the modern system of medicine for providing protection against these disorders are most inadequate and at best provide only symptomatic relief. However, in the traditional systems of medicine of many countries a number of herbal drugs are claimed to provide protection against hepatitis and are widely used as cholagogues/choleretics. This subject has been adequately reviewed recently.[28, 29] In India, based on Ayurvedic texts, over 30 proprietary products are marketed for liver disorders; these represent a variety of combinations of about 50 plants.[28] Many of these preparations are widely used by modern physicians as well. The more commonly used plants in these preparations are given in Table 6.

Table 6 Some Plants Common to Ayurvedic Drugs for Liver Disorders

Achillea millifolium	*Phyllanthus niruri*
Andrographis paniculata	*Phyllanthus emblica*
Apium graveolens	*Picrorrhiza kurroa*
Aloe indica	*Piper nigrum*
Berberis aristata	*Solanum nigrum*
Boerhaavia diffusa	*Tamarix gallica*
Eclipta alba	*Terminalia arjuna*
Capparis spinosa	*Tinospora cordifolia*
Luffa echinata	

Extracts of a number of these plants have been tested individually, both in different experimental models of liver disorder and, in a few cases, clinically, but only *Andrographis paniculata* and *Picrorrhiza kurroa* have shown activity.

The major chemical constituent of *A. indica* is a diterpenoid andrographolide (**1**). It has been shown that CCl_4-induced increase of serum glutamic pyruvic transaminase (SGPT) and serum glutamic oxaloacetic transaminase (SGOT) in rats was prevented by pretreatment with a single dose of leaf extract (500 mg kg^{-1}) or of andrographolide (5 mg kg^{-1}) given orally.[30] The plant extract also caused an increase in biliary flow and liver weight and a decrease in the duration of hexobarbital sleeping time, indicating that it possibly causes induction of hepatic drug-metabolizing enzymes.[31] Clinical studies with the extract in infective hepatitis cases revealed significant symptomatic improvement, accompanied by improvement in liver function tests.[32]

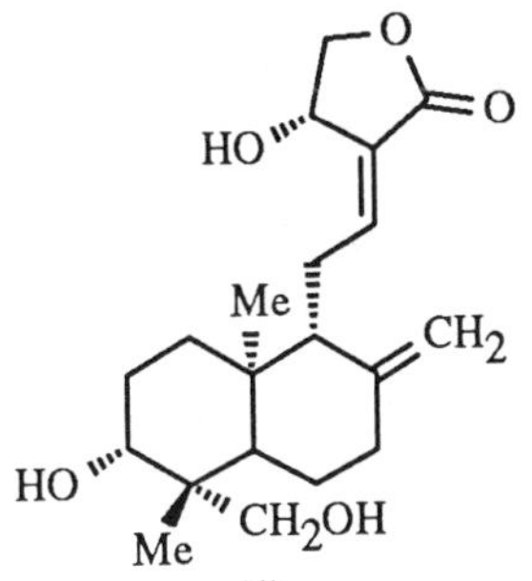

(1)

From *Picrorrhiza kurroa*, a mixture named kutkin, consisting of picroside I (**2**), kutkoside (**3**) and a few other minor glycosides, was isolated.[33] An extract of the plant was found to provide protection against CCl_4-induced damage to liver in rats, and to increase bile flow in dogs with experimental biliary fistulas.[34] Picroside I showed a protective effect against liver intoxication by CCl_4 and choleretic activity in rats.[35] Recently, it has been shown that kutkin at 6 mg kg^{-1} provided protection against CCl_4-induced and galactosamine-induced liver damage.[36]

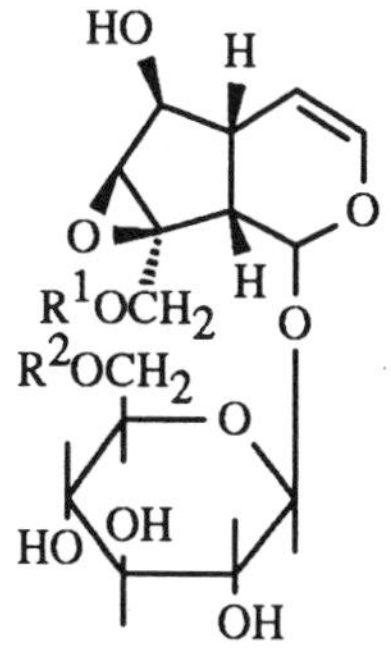

(**2**) R^1 = H, R^2 = cinnamoyl
(**3**) R^1 = vanilloyl, R^2 = H

Phyllanthus niruri is also a common household remedy for the treatment of jaundice. An extract of the plant has been shown to bind *in vitro* to HBsAg[37,38] and to inhibit endogenous DNA polymerase of hepatitis B virus and woodchuck hepatitis virus,[38] and also eliminate the surface antigen titre and DNA polymerase activity in infected woodchucks.[38] Preliminary open clinical trials with a drug containing *P. niruri* as the main constituent have shown encouraging results.[39,40]

There is thus significant presumptive evidence for the efficacy of a number of Ayurvedic drugs for liver disorders. In some cases where safety is assured, direct clinical trials may be in order to evaluate their efficacy. Although many individual plant constituents have been tested and found to exhibit activity in different experimental models, more comprehensive structure–activity relationship studies are needed to understand the different pathological situations in which different classes of compounds act. Some new experimental models may also have to be developed for these studies.

1.5.4.1.2 Health promoters in Ayurveda

According to Ayurveda, drugs which 'promote vigour in the healthy' are differentiated from curative drugs. They are termed 'rasayana' and presumably help to build up the non-specific body resistance to infections.[19] Ayurveda also had a developed concept of immunity as 'vyadhi kshamatva'. Similar concepts have been present in the medical systems of other cultures also, and the current concept of adaptogens or antistress drugs, tonics, and 'umstimmungs therapie' in German tradition[41] are reminiscent of similar ideas. The use of ginseng in Chinese tradition is one such drug. In Ayurveda, there has been a particular importance attached to this concept and a number of products are available which are widely used as rasayanas, and some of these are extremely common household remedies. A few of the important Ayurvedic drugs used for this purpose are chywanprash, *Withania somnifera*, *Altingia excelsa*, *Ocimum sanctum*, *Diospyros perigrina* and *Picrorrhiza kurroa*. Extracts of some of these plants and products have recently been screened by Singh and coworkers for their antistress activity in various experimental models.[42–44] Of the plants tested, the extract of *Ocimum sanctum* (tulsi) was found to be the most effective. It is interesting that this plant has been considered to be the most sacred in Hindu culture and almost worshipped in traditional homes.

Related to this in Ayurveda is the role of drugs in preventing the aging process (rejuvenation); in fact, rejuvenation and counteracting aging is considered as one of the eight specialities of Ayurveda. These rasayanas and rejuvenators are considered by Charaka to bring about longevity, restoration of intelligence and memory, disease-free state, youthful vigour, excellence of lustre, and complexion and improved voice. Apart from the drugs mentioned above, a number of plants are specially valued for old age, memory and learning. Two of the most important Ayurvedic plants in this category are *Bacopa moniera* and *Centella asiatica* (brahmi's). Recent research carried out at the Central Drug Research Institute has shown that two triterpenoid glycosides obtained from *B. monieri*, baccosides A and B, exhibited facilitatory effects on the mental retention capacity in experimental models.[45]

Other areas in which Ayurveda offers useful drugs are wound healing, arthritic conditions and urolithiasis, where modern drugs are inadequate, and it would be useful to investigate these drugs. The use of metal preparations including gold, mercury, iron and zinc in Ayurveda (including Siddha) for a variety of disease conditions and as tonics is also noteworthy and may have some leads to offer.

1.5.5 AYURVEDA AND NEW DRUG DISCOVERY

1.5.5.1 Rauwolfia Alkaloids in Hypertension

Rauwolfia serpentina (sarpgandha, pagal-ki-booti, pagal = insane) is a highly reputed drug in traditional medicine in India for the treatment of insanity and insomnia. This led to the chemical investigation of this plant by Siddiqui and associates in Delhi who in the early 1930s reported the isolation of five alkaloids from the roots,[46] which included ajmaline (**4**) and ajmalicine (**5**; raubasine) (named after Hakim Ajmal Khan, the doyen of Unani medicine of this century in India). About this time the hypotensive,[47–49] sedative and antipsychotic activities[49,51] of the crude

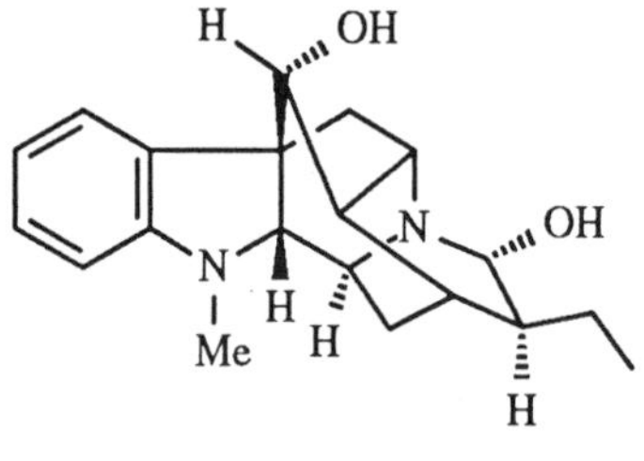

(4)

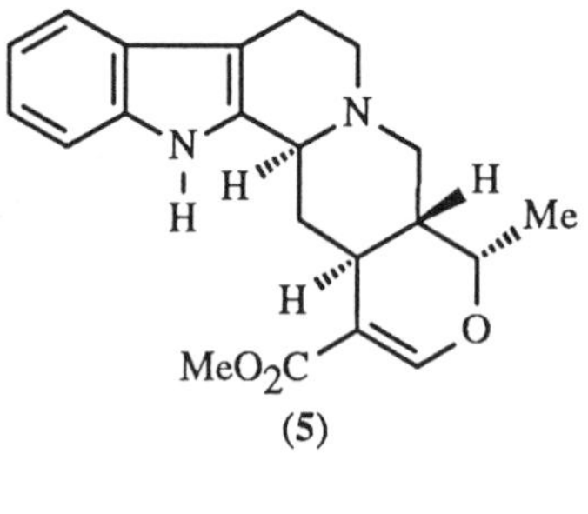

(5)

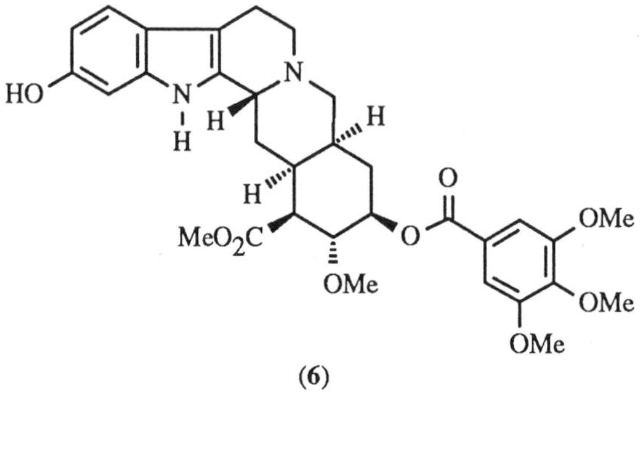

(6)

drug and of its extracts were also reported. Soon after knowledge of its clinical potential was brought to the attention of the world's scientific community by the report of Vakil,[52] the isolation of reserpine (**6**), which has very potent hypotensive and tranquillizing activities, was announced by the scientists of Ciba, Basel.[53,54] This was a big event in drug research, which sharply focussed world attention on plant products and traditional remedies as sources and leads for new drugs. Furthermore, this world interest, with occasional ups and downs, has been maintained ever since.

The discovery of reserpine led to extensive chemical investigation of other rauwolfia species and related plants and pharmacological evaluation of the alkaloids isolated therefrom.[55,56] The three alkaloids at present in clinical use are reserpine (for hypertension), to some extent ajmaline (as antiarrhythmic and antihypertensive) and ajmalicine (as a peripheral and central vasodilator).[57,58] Some physicians still prefer to use in place of reserpine a standardized powder or total extract of the root which they claim has a better therapeutic effect and safety margin.[59]

1.5.5.2 Psoralens in Vitiligo and Psoriasis

Psoralea corylifolia (babchi) seed powder is used in Ayurvedic medicine for vitiligo and other skin diseases. It is given orally and also applied locally in the form of a paste or ointment. Chemical and pharmacological investigations have shown that the active constituent is a furocoumarin, psoralen (**7**)[60] which stimulates/induces the formation of melanin pigment on exposure of the skin to UV/sunlight. Xanthotoxin (**8**; 9-methoxypsoralen), obtained from an Egyptian plant, *Ammi majus*, causes similar photostimulation of skin pigmentation. Psoralen's administration combined with UV light treatment (PUVA) is used to increase tolerance to sunlight and in the treatment of idiopathic vitiligo, but has no effect on the disease caused by infection or trauma.[62–64] Promising results have also been obtained in psoriasis following PUVA treatment;[62–64] excessive DNA synthesis in the lesions is reduced. However, there is some evidence of genetic damage which necessitates caution in treatment.[64,65]

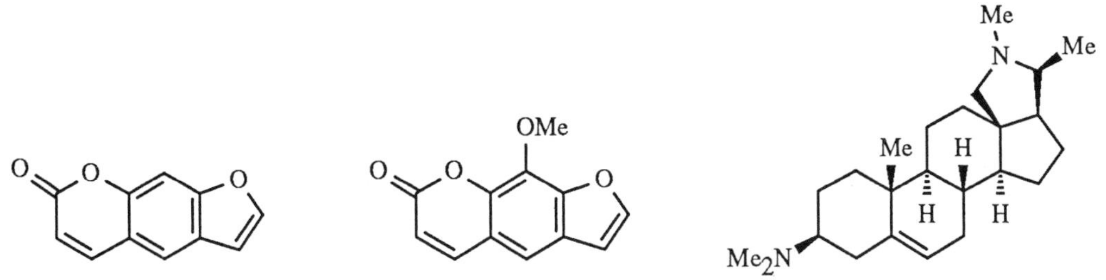

(**7**) (**8**) (**9**)

1.5.5.3 Holarrhena Alkaloids in Amoebiasis

Holarrhena antidysenterica (stem and root bark) is a reputed remedy of the traditional systems for amoebic dysentery and other intestinal ailments. Its major alkaloid conessine (**9**) was reported by Haines[66] as far back as 1858. In the early 1930s Siddiqui and his associates[67] reported the isolation of six more alkaloids and established their mutual relationship by interconversion. Conessine was found to be the most active and most abundant alkaloid in this plant. It has been used as a substitute for emetine, although it is less active, but is not much used now.

1.5.5.4 *Commiphora mukul* Steroids as Hypolipidaemic Agents

Gum guggul is the exudate of a plant, *Commiphora mukul*, which is an important drug of the Ayurvedic system. It has so far been used mainly for the treatment of various types of arthritis, although its beneficial effect in the treatment and correction of lipid disorders (coating and obstruction of channels by fats) and obesity is mentioned in Sushruta Samhita,[68] which also describes lucidly the etiopathogenesis of obesity and lipid disorders. This observation prompted Satyavati and Dwarkanath in the late 1960s to investigate experimentally its hypolipidaemic effect.[69] They showed that the crude gum not only lowered significantly serum cholesterol and phospholipids in hypercholesterolaemic rabbits but also protected these animals against cholesterol-induced atherosclerosis.[69] This was followed by exploratory clinical studies of the hypolipidaemic activity of the gum and some fractions thereof.[70, 71] These early results led to a detailed chemopharmacological examination of the gum by Dev and Nityanand and their associates.[72,73] It was found that the ethyl acetate-insoluble but water-soluble polysaccharide fraction, which constituted 55% of the gum, was toxic to rats and devoid of hypolipidaemic activity, while the ethyl acetate-soluble fraction exhibited marked hypolipidaemic activity.[72] It is interesting to note that in Ayurveda gum guggul is purified by boiling it with an aqueous extract of a mixture of three tannin-bearing plants (triphala), discarding the water-soluble fraction and using the insoluble fraction. The ethyl acetate-soluble fraction has been found to consist of a number of steroids, diterpenoids, triterpenoids, lignans, fatty alcohol (tetrol) esters, hydrocarbons, acids and bases, most of which have been identified;[74] tetrol esters and two diterpenes, mukulol and cumbrene, are the major constituents. The hypolipidaemic activity was found to be due mainly to two pregnadienediones, named (*Z*)- and (*E*)-guggulsterone (**10** and **11**), which were the most abundant of the steroids and which constituted about 2% of the gum (4.4% of the ethyl acetate-soluble fraction). The ratio of *Z* to *E* isomer on average was 4:1. The two guggulsterones had been synthesized earlier,[75] but their occurrence in nature was recorded for the first time in *C. mukul*. Both the isomers showed similar orders of activity.

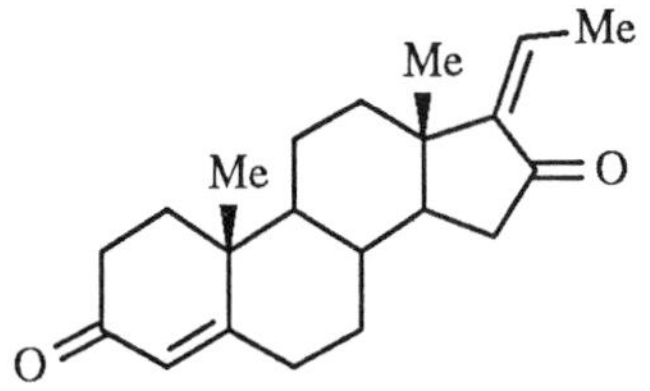

Me Me Me O O

(10)

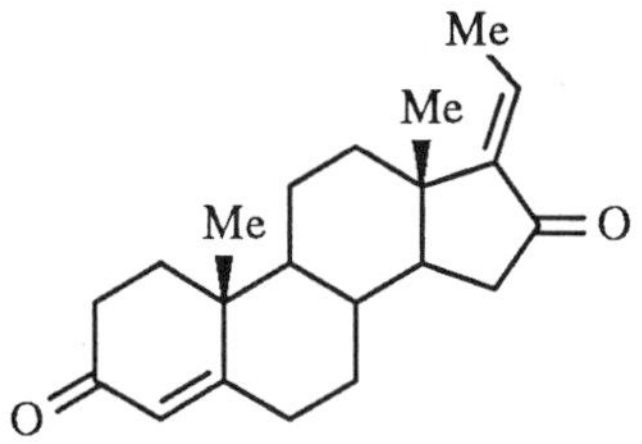

Me Me Me O O

(11)

In detailed studies it was found that the ethyl acetate-soluble fraction had almost the same order of activity as the mixture of pure guggulsterones, although the content of guggulsterones in the fraction was only about 4.4%. It would, therefore, appear that some of the other constituents of the extract, though inactive *per se*, contribute to enhancing the activity of the guggulsterones. It has not been possible so far to assign this adjuvant effect to any component. In view of the almost equal order of activity shown by this fraction and the mixture of guggulsterones contained in it, the total extract was taken up for preclinical development after standardization on the basis of its guggulsterone content. This standardized extract, named gugulipid, was obtained as a yellowish-brown viscous liquid and formulated into a free-flowing powder for oral administration. Gugulipid exhibited a dose-related lowering of serum cholesterol and triglycerides in normal and hypolipidaemic rats, rabbits and monkeys and the lowering was comparable dose for dose with that produced by clofibrate. It also caused regression of atheromatous lesions produced in rabbits by a high fat diet. In experimental animals it increased the excretion of bile acids and mobilized stored cholesterol and in *in vitro* experiments it inhibited cholesterol biosynthesis. It therefore seems to act by a number of mechanisms. In toxicological studies in rats, monkeys and dogs, gugulipid did not show any abnormal pathology and was not mutagenic in the Ames test. After statutory Phase I, II and III clinical studies,[76] gugulipid has been registered in India as a new hypolipidaemic agent and is being marketed under the brand name of Guglip.

A number of pregnane derivatives structurally related to guggulsterone have been tested for their hypolipidaemic activity, and one of these (Code No. 81/574), developed by the Central Drug Research Institute in Lucknow, is now in Phase I studies.

Gugulipid provides an interesting example of how an observation recorded in ancient medical texts,[68, 74] when investigated by modern scientific methods, led to the successful development of a new drug and also provided a lead compound for further modification. In the development of plant products as drugs, it may not always be necessary to insist on the use of single pure constituents so long as a therapeutically effective product can be properly formulated and its composition can be standardized.

1.5.5.5 Coleus Diterpenoids as Antihypertensive and Cardiotonic Agents

Coleus forskolii is an example of a plant which is not used in Ayurvedic medicine but in which promising biological activity was discovered under a programme of general pharmacological screening of plants. Two groups of investigators independently observed the root extract of this plant to exhibit antihypertensive activity.[77,78] Chemical investigation led to the isolation of a number of diterpenoids of which forskolin (**12**) (named coleonol by the second group) was found to be the most active with marked antihypertensive, inotropic and adenylate cyclase-stimulating activities.[79,80] Forskolin stimulates adenylate cyclase in membranes and in intact cells, even in the absence of hormone agonists, which suggests that it acts directly at the catalytic unit of adenylate cyclase. Because of its unique action and novel structure, forskolin has become a very important tool and lead prototype structure for development of new hypotensive/inotropic agents and conditions needing adenylate cyclase stimulation.[81]

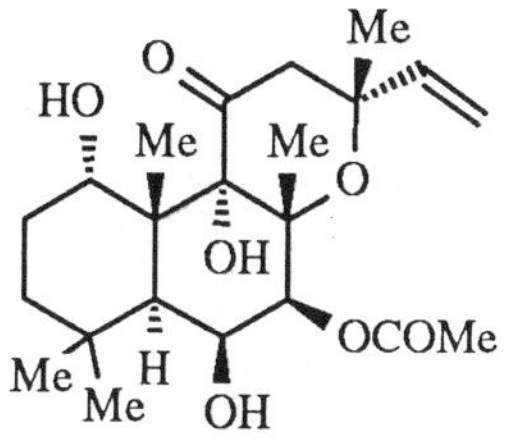

(12)

1.5.6 PERSPECTIVE

Ayurveda, in spite of its ancient origin, is of great relevance even today on account of its widespread use and the continued faith of the people in this system of medicine in the Indian subcontinent and neighbouring countries.

The classical texts of Ayurveda are a unique record of the long experience of many generations of physicians in the patient observation of symptoms of various diseases, their diagnosis and treatment by the use of medicaments in combination with appropriate dietetic and general health measures. It should be emphasized that experience played an important role in the choice of medicines and of the complete regimen prescribed. The doctrinal base of Ayurveda reflects a concern of the philosopher–physicians for a unified 'holistic' approach to the maintenance of positive health and not merely for the alleviation of disease symptoms. Much of the knowledge contained in the texts, reinterpreted in the light of present knowledge, can be applied to current problems of health. With the recent advances in biomedical sciences and investigative methods, a fresh look at the basic principles of Ayurveda is called for, keeping in view its essential emphasis on the concept of the common physiological disturbances underlying different disease processes.

The usefulness of Ayurveda can possibly be extended by the use of newer techniques of drug and medical research to uncover remedies for diseases for which modern drugs are inadequate and for which there is presumptive evidence of efficacy in Ayurveda. A few such examples have been mentioned in this chapter. The possibility of finding new drugs is much greater if investigations of Ayurvedic remedies are based on their specific mention in Ayurvedic texts. However, in order to make this possible, development of novel experimental models which would approximate better to the human situation may be necessary. In cases where the human situation is difficult to replicate in an animal model, direct clinical trial of selected drugs may be taken up. However, in spite of the sanction of long usage, a limited toxicity study may be necessary to ensure their safety.

A shortcoming of Ayurvedic drugs is the lack of suitable quality control standards, resulting sometimes in difficulties in ensuring uniformity of their composition and consequently efficacy of the final products. The fact that many drugs are composed of a number of ingredients does present difficulties in conventional quality control assays, but by using modern techniques, including 'fingerprinting' and bioassay of one or more active/major constituents of the drug, their standardization is possible on a scientific basis. Thus, in order to make traditional remedies more acceptable in modern therapeutics and promote the wider application of Ayurvedic knowledge and practice for human use, quality control and toxicological evaluation of Ayurvedic remedies would be essential.

ACKNOWLEDGEMENT

I would like to thank Dr B. N. Dhawan, Director, Central Drug Research Institute, Lucknow, Dr G. V. Satyavati, Deputy Director-General, Indian Council of Medical Research, New Delhi and Dr N. S. Bhatt, Zandu Pharmaceutical Works, Bombay for valuable discussions.

1.5.7 REFERENCES

1. The term 'traditional system' is used for systems indigenously developed and should be distinguished from folk medicine; the former are based on a body of organized knowledge having a theoretical basis and framework, while the latter are records of random and empirical observations without any theoretical basis; the term 'modern system' is used for the allopathic system introduced into India by the British.
2. The four Vedas (Veda literally means knowledge), Rigveda, Yajurveda, Samveda and Atharvaveda, are the oldest Hindu scriptural writings dating back to around 2000 BC, presented in the form of hymns, and based on knowledge of all spheres of life during that period.
3. S. Dasgupta, 'A History of Indian Philosophy', Cambridge University Press,Cambridge, 1932, vol. II.
4. 'Charaka Samhita', edited by the Shri Gulab Kunverba Society, with introduction, commentary and indices in English, Jamnagar, India, 1949, vols. I–VI.
5. P. Roy and H. Gupta, 'Charaka Samhita: A Scientific Synopsis', 2nd edn., Indian National Science Academy, New Delhi, 1980.
6. S. P. Sharma, 'Charaka Samhitas (Agnivesa treatise refined and annotated by Charaka and redacted by Drdhabala): Text with English Translation', Chaukamba Orientalia, Varanasi, India, 1981, vols. I–IV.
7. P. Roy, H. Gupta and M. Roy, 'Sushruta Samhita: A Scientific Hypothesis', Indian National Science Academy, New Delhi, 1980.
8. K. S. Bhisagratna, 'Sushruta Samhita', Chaukhamba Sanskrit Series, Varanasi, India, 1963.
9. Pt. B. Harishastri, 'Vagbhatta—Astanga Hridaya', Nirnaya Sagar Press, Bombay, 1939.
10. S. Dasgupta, 'A History of Indian Philosophy', Cambridge University Press, Cambridge, 1922, vol. I.
11. J. Filliozat, 'The Classical Doctrine of Indian Medicine', translated by D. R. Chananna, Munshiram Manoharlal Oriental Publishers, Delhi, 1964.
12. C. Dwarkanath, 'Fundamental Principles of Ayurveda', Bangalore Press, Mysore, India, 1952, vols. I–III.
13. H. V. Savnur, 'A Handbook of Ayurvedic Materia Medica', Jathar, Belgaum, Maharashtra, India, vol. 1.
14. C. Dwarkanath, 'Digestion and Metabolism in Ayurveda', Shri Baidyanath Ayurveda Bhawan, Calcutta, 1967.
15. G. V. Satyavati, *Indian J. Med. Res.*, 1982, **76** (suppl. 1), 1.
16. H. A. Razzak, 'Unani System of Medicine in India: A Profile', Central Council for Research in Unani Medicine, New Delhi, 1987.
17. Ref. 6, Sutrasthana section, p. 62, 3–9.
18. R. D. Lele, 'Ayurveda & Modern Medicine', Bharatiya Vidya Bhawan, Bombay, 1986, p. 66.
19. Ref. 6, Chikitsasthana section, p. 3, 3–14.
20. G. P. Srivastava, 'Abstracted from History of Indian Pharmacy', Pindar's, Calcutta, 1953.
21. G. V. Satyavati, paper presented at the 'International Symposium on Plants and Traditional Medicine in PHC' at University of Illinois, Chicago, USA, 1987.
22. 'Pharmacopoeial Standards for Ayurvedic Formulations', Central Council for Research in Ayurveda and Siddha, Government of India, New Delhi, 1987.
23. 'Handbook of Domestic Medicines & Common Ayurvedic Remedies', Central Council for Research in Indian Medicine and Homoeopathy, Ministry of Health and Family Welfare, Government of India, New Delhi, 1978.
24. R. N. Chopra, S. L. Nayar and I. C. Chopra, 'Glossary of Indian Medicinal Plants', Council of Scientific and Industrial Research, New Delhi, 1956.
25. R. N. Chopra, I. C. Chopra and B. S. Verma, 'Supplement to Glossary of Indian Medicinal Plants', Council of Scientific and Industrial Research, New Delhi, 1969.
26. G. V. Satyavati, M. K. Raina and M. Sharma, 'Medicinal Plants of India', Indian Council of Medical Research, New Delhi, 1976, vol. I; G. V. Satyavati, A. K. Gupta and N. Tandon, 'Medicinal Plants of India', Indian Council of Medical Research, New Delhi, 1987, vol. II.
27. Z. Abraham, D. S. Bhakuni, H. S. Garg, A. K. Goel, B. N. Mehrotra and G. K. Patnaik, *Indian J. Exp. Biol.*, 1986, **24**, 48, and earlier parts in the series.
28. S. S. Handa, A. Sharma and K. K. Chakraborti, *Fitoterapia*, 1986, **57** (5), 307.
29. A. Gajdos, T. M. Gajdos and R. Horn, in 'New Trends in the Therapy of Liver Diseases—Proceedings of an International Symposium, Tirrenia, 1974', ed. A. Bertelli, Karger, Basel, 1975, p. 114.
30. B. R. Choudhury and M. K. Poddar, *IRCS Med. Sci.: Libr. Compend.*, 1984, **12**, 466.
31. S. K. Chaudhuri, *Indian J. Exp. Biol.*, 1978, **16**, 830.
32. K. P. Singh, *J. Int. Inst. Ayurveda, Coimbatore, India*, 1983, **2**, 208.
33. B. Singh and R. P. Rastogi, *Indian J. Chem.*, 1972, **10**, 29.
34. V. N. Pandey and G. N. Chaturvedi, *Indian J. Med. Res.*, 1969, **57**, 503; V. N. Pandey and G. N. Chaturvedi, *J. Res. Indian Med.*, 1970, **5**, 11 (*Chem. Abstr.*, 1971, **75**, 3646t).
35. P. Kloss and W. Schwabe, *Ger. Pat.* 2 203 884 (1973) (*Chem. Abstr.*, 1973, **79**, P108 054r).
36. R. A. Ansari, B. S. Aswal, B. N. Dhawan, B. N. Garg, N. K. Kapoor, D. K. Kulshreshta, H. Mehdi, B. N. Mehrotra, G. K. Patnaik and S. K. Sharma, *Indian J. Med. Res.*, 1988, **87**, 401.
37. S. P. Thyagarajan, K. Thiruneelakantan, S. Subramanian and T. Sundaravelu, *Indian J. Med. Res.*, 1982, **76** (suppl), 124.
38. P. S. Venkateswaran, I. Millman and B. S. Blumberg, *Proc. Natl. Acad. Sci. U.S.A.*, 1987, **84**, 274.
39. S. P. Dixit and M. P. Achar, *J. Natl. Integr. Med. Assoc.*, 1983, **25** (8), 269.
40. R. Thyagarajan, R. Uma, C. P. Ramanathan and K. Ganapathiraman, *Journal of Research in Indian Medicine, Yoga and Homoeopathy*, 1977, **12** (2), 1.
41. H. Wagner, in 'Proceedings of the Fifth Asian Symposium on Medicinal Plants and Spices', ed. B. H. Han, D. S. Han, Y. N. Han and W. S. Woo, National Products Research Institute, Seoul, Korea, 1984, p. 33.
42. N. Singh, *Curr. Med. Pract.*, 1981, **25**, 50.
43. K. P. Bhargava and N. Singh, *J. Res. Educ. Indian Med.*, 1985, **4**, 27.
44. N. Singh, *Ann. Acad. Indian Med.*, 1986, **1**, 1.
45. H. K. Singh, R. P. Rastogi, R. C. Srimal and B. N. Dhawan, *Phytotherapy*, 1988, **2**, 70.
46. S. Siddiqui and R. Z. Siddiqui, *J. Indian Chem. Soc.*, 1931, **8**, 667.
47. B. B. Bhatia, *J. Indian Med. Assoc.*, 1942, **11**, 262.
48. A. S. Paranjpe, *Med. Bull. (New York)*, 1942, **10**, 135.

49. G. Sen and K. C. Bose, *Indian Med. World.*, 1931, **2**, 194.
50. J. C. Gupta and A. K. Deb, *Indian Med. Gaz.*, 1943, **78**, 547.
51. R. N. Chopra, J. C. Gupta, B. C. Bose and I. C. Chopra, *Indian. J. Med. Res.*, 1943, **31**, 71.
52. R. J. Vakil, *Br. Heart J.*, 1949, **11**, 350.
53. J. M. Muller, E. Schhttler and H. J. Bein, *Experientia*, 1952, **8**, 338.
54. L. Dorfman, A. Furlenmeier, C. F. Huebner, R. Lucas, H. B. MacPhillamy, J. M. Mueller, E. Schlitter, R. Schwyzer and A. F. St. Andre, *Helv. Chim. Acta*, 1954, **37**, 59.
55. E. Schlittler, in 'The Alkaloids, Chemistry and Physiology', ed. R. H. F. Manske, Academic Press, London, 1965, vol. 8, p. 287.
56. H. J. Bein, *Pharmacol. Rev.*, 1956, **8**, 435.
57. J. E. F. Reynolds (ed.), 'MARTINDALE, The Extra Pharmacopoeia', The Pharmaceutical Press, London, 1982, p. 1373.
58. J. E. F. Reynolds (ed.), 'MARTINDALE, The Extra Pharmacopoeia', The Pharmaceutical Press, London, 1982, p. 1632.
59. J. E. F. Reynolds (ed.), 'MARTINDALE, The Extra Pharmacopoeia', The Pharmaceutical Press, London, 1982, p. 162.
60. H. S. Jois, B. L. Manjunath and S. Venkatarao, *J. Indian Chem. Soc.*, 1933, **10**, 41.
61. S.Schonberg, *Nature (London)*, 1947, **160**, 468.
62. T. F. Anderson and J. J. Voorhees, *Annu. Rev. Pharmacol. Toxicol.*. 1980, **20**, 235.
63. A. Kornhauser, W. G. Warner and A. L. Giles, Jr., *Science (Washington, D. C.)*, 1982, **217**, 733.
64. J. E. F. Reynolds (ed.), 'MARTINDALE, The Extra Pharmacopoeia', The Pharmaceutical Press, London, 1982, p. 497.
65. M. J. Ashwood Smith and E. Grant, *Br. Med. J.*, 1976, **1**, 342.
66. R. Haines, *Trans. Med. Soc., Bombay*, 1858, **4**, 28.
67. S. Siddiqui and P. P. Pillay, *J. Indian Chem. Soc.*, 1932, **9**, 553.
68. G. V. Satyavati, *Indian Council of Medical Research Bulletin*, 1987, **17** (1), 1.
69. G. V. Satyavati, C. Dwarkanath and S. N. Tripathi, *Indian J. Med. Res.*, 1969, **57**, 1950.
70. S. N. Tripathi, V. V. S. Shastri and G. V. Satyavati, *J. Res. Indian Med.*, 1968, **2**, 10.
71. S. C. Malhotra and M. M. S. Ahuja, *Indian J. Med. Res.*, 1971, **59**, 1621.
72. S. Nityanand and N. K. Kapoor, *Indian J. Exp. Biol.*, 1971, **9**, 376; 1973, **11**, 395; *Indian J. Pharmacol.*, 1975, **71**, 60; *Indian J. Biochem. Biophys.*, 1978, **15**, 77.
73. V. P. Patil, U. R. Nayak and S. Dev, *Tetrahedron*, 1972, **28**, 2341.
74. S. Dev, *Proc. Indian Acad. Sci., Sect. A*, 1988, **54A**, 12.
75. W. R. Benn and R. M. Dodson, *J. Org. Chem.*, 1964, **29**, 1142.
76. R. C. Agarwal, S. P. Singh, R. K. Saran, S. K. Das, Nakul Sinha, O. P. Asthana, P. P. Gupta, S. Nityanand, B. N. Dhawan and S. S. Agarwal, *Indian J. Med. Res.*, 1986, **84**, 626.
77. M. P. Dubey, R. C. Srimal, G. K. Patnaik and B. N. Dhawan, *Indian J. Pharmacol.*, 1974, **6**, 15; J. S. Tandon, M. M. Dhar, S. Ramkumar and K. Venkatesan, *Indian J. Chem., Sect. B*, 1977, **15B**, 880.
78. S. V. Bhatt, B. S. Bajwa, H. Dornauer, N. J. deSouza and H. W. Fehlhaber, *Tetrahedron Lett.*, 1977, 1669; E. Lindner, A. N. Dohadwala and B. K. Bhattacharya, *Arzneim.-Forsch.*, 1978, **28**, 284.
79. N. J. deSouza, A. N. Dohadwala and A. J. Reden, *J. Med. Res. Rev.*, 1983, **3**, 201.
80. K. B. Seoman and J. W. Daly, in 'Advances in Cyclic Nucleotide and Protein Phosphorylation Research', ed. P. Greengard and C. A. Robison, Raven Press, New York, 1986, vol. 20, p. 1.
81. Y. Khandelwal, K. Rajeshwari, R. Rajagopalan, L. Swamy, A. N. Dohadwala, N. J. deSouza and R. H. Rupp, *J. Med. Chem.*, 1988, **31**, 1872.

2.1
Physiology of the Human Body

J. ROBERT McCLINTIC
California State University, Fresno, CA, USA

2.1.1 LEVELS OF BODY ORGANIZATION

The human body's units of structure and function are its cells. Cells that are similar in structure and function, together with associated intercellular material, constitute a tissue. Four tissue groups compose the body: epithelial tissues cover or line all surfaces of the body or of the organs within it; connective tissues form the supporting tissues of the body and include blood, bone, cartilage and adipose tissue among many types; muscular tissues are capable of contracting or shortening and create movement of the body as a whole and of materials through the body; nervous tissues serve as the transmission pathway for electrochemical disturbances called nerve impulses, and produce chemicals that control many body processes.

Two or more tissues organized to perform a particular task constitute an organ, and several organs, usually organized 'in series' to carry out a broader process, such as digestion, form a system. About a dozen interdependent systems form the human organism. The body systems and some of the most important organs composing each are presented in Table 1.

2.1.1.1 Descriptive Terms to Locate Body Organs

Internal body organs lie in body cavities that have no direct communication to the body surface. The cranial cavity contains the brain, the vertebral (spinal) canal contains the spinal cord. The thoracic (chest) cavity houses the heart and lungs. The diaphragm separates the thoracic cavity from the abdominopelvic cavity that houses digestive (stomach, intestines, liver, pancreas), excretory (kidneys, ureters, bladder), reproductive (ovaries, uterine tubes, uterus, prostate, seminal vesicles) and endocrine (adrenals, pancreatic islets, ovaries) organs.

To more accurately position organs within the abdominopelvic cavity, one may draw imaginary horizontal and vertical lines through the navel (umbilicus) that divide the subject's abdomen into four quadrants called right upper (RUQ), left upper (LUQ), right lower (RLQ) and left lower (LLQ). The quadrants and the major organs they contain are: RUQ—duodenum, liver, right kidney, right adrenal; LUQ—stomach, spleen, pancreas, left kidney, left adrenal; RLQ—small intestine, large intestine, right ovary; and LLQ—small intestine, large intestine, left ovary.

2.1.2 THE INTEGUMENTARY SYSTEM

This system includes the skin and organs derived from or associated with the skin such as hair, nails, sweat and sebaceous glands and a variety of sensory corpuscles for heat, cold, touch and pressure. Blood vessels in the skin provide a route for ridding the body, by radiation, of heat produced by metabolic activity.

2.1.2.1 Structure and Functions of Skin

An outer epidermis has many layers of keratin-filled cells forming a tough, nearly waterproof and microorganism-resistant layer protecting the body's internal environment. Keratin is a protein that can be 'dissolved' by keratolytics, such as aspirin, which is often included in salves to be rubbed on the skin for its analgesic (pain killing) property.

The lowest layer of the cells of the epidermis are dividing constantly to renew the epidermis. About three weeks are required to renew the epidermis (a 'new skin' every three weeks).

Beneath the epidermis is a layer of connective tissue called the dermis, in which lie the blood vessels, glands and sensory receptors. It is a compact layer containing elastic and collagenous (strong but not elastic) fibers. Intradermal injections are rather painful because only a small volume of fluid can be accommodated before stretching causes stimulation of pain fibers.

Table 1 Body Systems and Their Organs

System	*Organs*	*Comments*
Muscular	Named skeletal muscles	Move body through space
Skeletal	Bones and cartilages	Support, protect organs, store minerals
Nervous	Brain, spinal cord, nerves	Controls organ activity
Circulatory	Heart, blood vessels, blood	Set up to supply needs of cells
Respiratory	Nasal cavities, larynx, trachea, lungs	Supplies O_2, removes CO_2
Digestive	Mouth, stomach, intestines, liver, pancreas	Supplies nutrients to body
Urinary	Kidneys, ureters, bladder	Processes wastes, regulate blood composition
Reproductives	Testes, ovaries, uterus, vagina	Produce sex cells and hormones
Integumentary	Skin, hair, nails	Covers body
Immune	Certain blood cells, lymph nodes	Protects body from microorganisms
Endocrine	Glands of internal secretion: pituitary, thyroid, adrenals and others	Produce hormones that control body activity

A hypodermis (subcutaneous layer) lies beneath the dermis and typically contains much adipose tissue (fat). It is loosely arranged and can accommodate larger fluid volumes with less pain as might occur with a subcutaneous injection.

Glands in the skin include the eccrine sweat glands that produce a watery fluid, which, on evaporation, carries heat from the body to help control body temperature. Sodium chloride may be excreted in significant amounts during heavy sweating. Sebaceous or oil glands contribute cellular products that form an acidic surface antiseptic layer on breakdown.

Loss of the skin, as in burns, removes the protective functions of the skin and lays the body open to microorganismic invasion.

2.1.3 THE MUSCULOSKELETAL SYSTEM

Often considered together, these systems include the bones of the skeleton and the skeletal muscles attached to the bones, using them as points of origin and levers to achieve body movement. The skeleton also provides protection for vital body organs (brain in skull, heart and lungs in the thorax), serves as a reservoir for calcium and phosphate, and houses the red bone marrow that produces several types of blood cells.

There are three types of muscular tissue: skeletal (striated) muscle over which we have voluntary control and which attaches to and moves the skeleton; cardiac muscle found only in the heart, an involuntary type of muscle whose contraction circulates the blood; smooth (visceral) muscle, a slow-contracting involuntary type of muscle found in internal body organs such as the stomach, intestines, blood vessels, uterus and many other areas.

In all three types of muscle, the contractile mechanism is the same. The comments to follow describe the process as it would occur in skeletal and cardiac muscle and any differences in smooth muscle will be presented at the end of this section.

Four basic muscle proteins occur. (i) Myosin — a large double helix bearing on the ends enlarged 'heads'. The heads have binding sites on them for adenosine triphosphate (ATP) and for actin. At rest, a molecule of ATP is attached to the ATP site. (ii) Actin — a smaller double helix bearing binding sites for the myosin heads, which, at rest, are covered by tropomyosin. (iii) Tropomyosin — a fibrous protein occurring as two strands intertwined with the actin molecules, giving a four-stranded helix. It is physically attached to troponin. (iv) Troponin — a globular protein occurring at intervals along the four-stranded actin–tropomyosin complex. Troponin has a strong affinity for binding calcium ion (Ca^{2+}).

Calcium ion is the trigger for contraction and is sequestered in hollow cavities in the muscle fibers called the sarcoplasmic reticulum (SR). A stimulus causes the release of Ca^{2+} from the SR, which diffuses to and binds to the troponin molecules. Binding causes the troponin to undergo a change in shape that pulls the attached tropomyosin to the side of the actin molecules, uncovering the actin's myosin-binding sites. Ca^{2+} also occasions the splitting of the ATP, attached to the myosin heads, to adenosine diphosphate (ADP) and inorganic phosphate (P_i), these products remaining attached to the myosin heads. ATP splitting activates the myosin and a cross-bridge is formed between myosin and actin. The myosin head then moves, pulling on the actin molecules and drawing them toward the center of the muscle fiber, shortening it. The ADP and P_i are then released from the myosin head, a new ATP molecule binds to the myosin, and this breaks the cross-bridge and causes its movement back to its original position. New cross-bridges are formed and the shortening process is repeated.

Relaxation occurs when Ca^{2+} is returned to the SR, the troponin returns to its original shape permitting tropomyosin to recover the sites on the actin molecules and no cross-bridges can then be formed.

In smooth muscle, the proteins are more randomly arranged, there is no SR, the Ca^{2+} being stored in vacuole-like caveolae in the cells, and the muscle contraction is much slower than in the other types of muscle.

Calcium-blocking agents prevent the influx of Ca^{2+} through the muscle cell and thus the muscle cell is encouraged to relax. Such agents are most commonly used as antianginals or antihypertensives to relax vascular smooth muscle and increase blood flow. Among such agents are diltiazem hydrochloride, nifedipine and verapamil hydrochloride.

2.1.3.1 The Neuromuscular Junction

Skeletal muscle is normally caused to contract *via* nerves that confer voluntary control over its activity. An axon, the peripheral extension of a nerve cell located in the spinal cord, bears

Table 2 α and β Receptor Effects

Organ affected	*Receptor*	*Response*
Heart	β	Increased rate, stronger contraction
Blood vessels	α	Constriction
	β	Dilation

enlargements on its endings at a skeletal muscle fiber. In these enlargements are vesicles containing acetylcholine (Ach) that was synthesized in the enlargement itself. The enlargement is associated with a folded invagination of the muscle fiber membrane that bears Ach-binding sites and also attaches molecules of the enzyme cholinesterase that can cleave Ach. This whole assembly is termed the neuromuscular junction.

A nerve impulse traveling along the axon alters the membrane permeability at the enlargement to extracellular Ca^{2+} that moves into the nerve fiber. This entry causes movement of the vesicles to the membrane (exocytosis) and the Ach is released into the junction. It diffuses to the muscle fiber membrane and binds to the Ach receptors. This in turn causes an impulse that ultimately releases Ca^{2+} from the SR. Ach is then split by the enzyme so that it does not continue to act on the muscle membrane to cause, ultimately, muscle contraction.

Smooth muscle cells are believed to possess receptor proteins designated α receptors and β receptors that bind catecholamines, such as epinephrine and norepinephrine. α receptors promote such responses as vasoconstriction (narrowing) and relaxation of intestinal muscle. β receptors promote vasodilation and uterine muscle relaxation. While α effects are usually excitatory and β effects inhibitory, there are exceptions. Table 2 gives such examples. Thus, a 'β-blocking' agent will decrease heart action, lowering blood pressure. Such chemicals are most often used as anti-hypertensives (lower blood pressure). Some common β blockers are labetalol, metaprolol tartrate, pindolol and timolol.

2.1.4 THE HEART

2.1.4.1 General Structure

The heart is a four-chambered organ whose task is to create pressure to circulate the blood through the body's blood vessels. Two upper chambers, the atria, receive blood from the body generally (right atrium) and from the lungs (left atrium). The two lower chambers or ventricles receive blood from their respective atria and are the pumping chambers. The right ventricle sends blood to the lungs, while the left ventricle sends blood to the body generally. The circulation to and from the lungs is called the pulmonary circulation and that to and from the body generally is the systemic circulation. The left ventricle is thicker-walled than the right ventricle for it must create a pressure about four times that of the right ventricle to force blood through the many more resistances imposed by the many more organs in the systemic circulation. Figure 1 shows the general plan of the circulation.

2.1.4.2 Cardiac Muscle

The muscular tissue of the heart is arranged in the form of individual branching fibers (cells) separated from one another by double membranes called intercalated discs. The discs possess close junctions where the two membranes are 20 Å apart rather than the usual 150–200 Å. These junctions permit electrical impulses to pass from one cell to another as though there were no membranes present at all. Thus only a small area need be stimulated to excite the whole mass. If stimulated, the muscle will give a maximal contraction according to oxygen levels, acidity (pH), temperature and chemical factors in its environment. In this way, that is by controlling the chemical environment, the rate and strength of the the beat may be altered. Additionally, tension or stretch of the muscle will, to a point, result in a stronger contraction. This provides an automatic increase in blood volume pumped when more blood is returned to the heart as during exercise. The muscle cannot be tetanized or thrown into a held contraction. This permits continued filling and emptying.

2.1.4.3 Nodal Tissue

Cardiac muscle must be stimulated to contract and the nodal tissue inherent in the heart provides this stimulation. The tissue is disposed as several discrete masses or bundles within the heart.

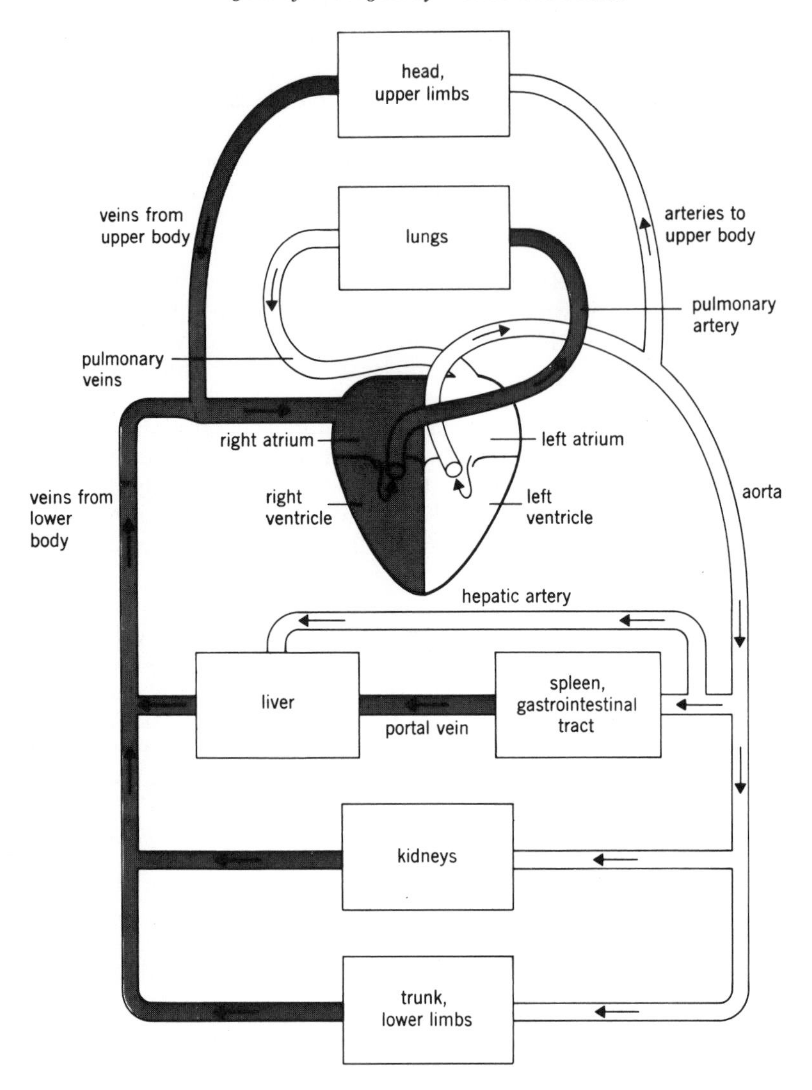

Figure 1 A schematic representation of the circulatory system

The sinoatrial (SA) node lies in the upper outer portion of the right atrium. It serves as the pacemaker for the heart beat at a basic rate of about 80 $\min^{-1}$. Impulses pass from the SA node through the atrial muscle, stimulating its contraction.

The impulses next arrive at the atrioventricular (AV) node located in the right atrium where the four chambers come together. This node slows the impulse to permit completion of atrial contraction and then passes the impulse to a rapidly conducting system that goes to the ventricular muscle.

The AV bundle lies in the upper part of the interventricular system.

The right and left bundle branches are derived from the AV bundle and pass through the septum to terminate as Purkinje fibers on the inner layer or two of ventricular muscle. The impulse then passes from cell to cell *via* the close junctions and the whole mass contracts nearly simultaneously.

2.1.4.4 Control of Nodal and Cardiac Muscle Activity

Rate and strength of the heartbeat is controlled by nervous and chemical factors. The heart receives catecholamine-secreting nerves from the sympathetic nervous system to both nodal tissue and muscle. These nerves cause increase in SA node activity (rate rises) and strength of contraction. Acetylcholine-secreting nerves pass to the heart from the parasympathetic nervous system and decrease SA node activity (rate falls) and strength of contraction. These nerves are derived from neurons located in the brainstem that are largely under reflex control from the periphery.

Electrolytes determine the excitability of the heart tissue and thus mainly the rate of beat. Among the more important electrolytes and their effects are: sodium, of which an increase causes fall of rate as tissue becomes more difficult to stimulate, whilst a decrease causes fall of rate as tissue takes a longer time to depolarize; potassium, of which an increase causes fall of rate as nodal tissue conduction is slowed, whilst a decrease causes fall of rate as tissue takes a longer time to depolarize; and calcium, of which an increase causes decrease of rate due to longer depolarization time, whilst a decrease causes increased rate, and decreased strength as less is available for contraction.

Other chemicals and their effects include: deslanoside, digitoxin, digoxin, which promote entry of Ca^{2+} into muscle cells (strengthens beat); atropine sulfate, which is anticholinergic and enhances nodal conduction and elevates rate; and quinidine, which depresses conduction rate, thus slowing beat.

2.1.4.5 Cardiac Output

Cardiac output (CO) refers to the volume of blood ejected per minute from one ventricle. Output averages 4.2 to 5.6 $\mathrm{L\,min^{-1}}$ at rest and may rise to 12 $\mathrm{L\,min^{-1}}$ during strenuous activity. Cardiac output is the product of two factors: stroke rate (SR) or beats per minute, and stroke volume (SV) or volume ejected per beat. To illustrate the calculation, assume a resting rate of 70 beats per minute and a volume of 70 mL per beat.

$$70 \text{ beats min}^{-1} \times 70 \text{ mL beat}^{-1} = 4900 \text{ mL or } 4.9 \text{ L min}^{-1}$$

With exercise, assume a rate increase to 100 beats per minute and a volume increase to 120 mL per beat.

$$100 \text{ beats min}^{-1} \times 120 \text{ mL beat}^{-1} = 12\,000 \text{ mL or } 12 \text{ L min}^{-1}$$

At rest, a ventricle does not completely empty its full contents so that the remaining blood may become available during times of stress.

2.1.4.6 Control of Cardiac Output

Cardiac output obviously can vary according to body requirements and is altered by changes in SR, SV or both. Several important factors are involved in altering these components of the output.

Venous inflow. The amount of blood returning to the right atrium from the body generally is the venous inflow. This volume of blood will enter the right ventricle during its relaxation (diastole) and this constitutes the end diastolic volume. To eject this blood, the ventricle must generate enough pressure (about 18 mmHg or 2.4 $\mathrm{kN\,m^{-2}}$) to open the pulmonary valve in the base of the pulmonary trunk, the artery leaving the right ventricle. Returning to the left atrium and then ventricle, a pressure of 80 mmHg (10.7 $\mathrm{kN\,m^{-2}}$) must be generated to open the aortic valve in the aorta, the artery leaving the left ventricle. During exercise, an increased volume of blood is returned to the right atrium by the contraction and relaxation of skeletal muscle (massaging action), this volume stretches the chambers, a stronger contraction results, and SV increases.

Peripheral resistance. Many arteries of the body have circularly disposed smooth muscle around their cavities. If the muscle contracts, the resistance to flow of blood through the vessel will increase. This is the peripheral resistance. Pressure between the vessels and the heart will increase, less blood will be ejected, and stroke volume will increase due to tension on the ventricular wall from the increased volume of blood.

Chemicals. Catecholamines increase both the rate (chronotropic effect) and the strength (inotropic effect) of the heart beat. Acetylcholine has the opposite effect.

Temperature. Acceleration of rate of SA nodal generation of impulses and conduction rate are increased by rise of temperature. About a 4 $\mathrm{beat\,min^{-1}}$ rise occurs for each degree rise of blood temperature.

Nervous factors. As mentioned in Section 2.1.4.4, the heart is supplied by parasympathetic and sympathetic nerves rising from cardiac centers in the brainstem. These centers include paired cardioinhibitory (CIC) and cardioaccelerator centers (CAC) that receive fibers from the periphery that signal the centers to elevate or depress heart rate. Table 3 presents several important cardiac reflexes.

Table 3 Cardiac Reflexes

Reflex name	*Source of input to centers*	*Stimulus triggering input*	*Effect on*		*Result*
			CIC	*CAC*	
Aortic depressor	Stretch receptors in aorta	Increased pressure in aorta	Stimulate	Inhibit	SR ↓ CO ↓ BP ↓
Carotid sinus	Stretch receptors in carotid artery	Increased pressure in artery	Stimulate	Inhibit	SR ↓ CO ↓ BP ↓
(Pain)	Thalamus, cerebrum	Painful stimulus to any nerve	Inhibit	Stimulate	SR ↑ CO ↑ BP ↑

2.1.4.7 The Cardiac Cycle

2.1.4.7.1 Timing

One cardiac cycle has occurred when a complete series of events has taken place in heart activity as from one SA node generation to the next. The length of a cycle varies inversely with heart rate. At a typical resting rate of beating of 75 min^{-1}, a cycle lasts 0.8 s. During this time, the atria are contracting 0.1 s and are relaxing 0.7 s. The ventricles are contracting 0.3 s and are relaxing 0.5 s. Filling of a chamber occurs during the relaxation phase, as does flow through the coronary vessels, since they are not collapsed shut by muscle contraction. With rapid heart rates (160–170 beats min^{-1}) resulting in insufficient time for filling, cardiac output will fall.

2.1.4.7.2 Pressure changes

Contraction of a chamber decreases the volume of the chamber and increases the pressure on the contained blood. Right and left atria serve as receiving chambers for blood and when they contract, a pressure elevation of 3 to 4 mmHg (400 to 530 $N\,m^{-2}$) occurs. Ventricular relaxation creates a pressure lower than that in the atria and 'draws' blood into the chamber rather than its being 'pushed' into the ventricle by atrial contraction. Right ventricular contraction produces a maximum resting pressure of about 30 mmHg (4 $kN\,m^{-2}$), while left ventricular pressure rises to about 130 mmHg (17 $kN\,m^{-2}$). During ventricular relaxation, blood remains under pressure in the arteries, a resistance the ventricles must overcome to again eject blood. The maximum pressure developed is the systolic pressure, the lower pressure is the diastolic pressure. As measured in a peripheral artery, the pressure is represented as systolic/diastolic or 120/80 in the normal individual.

2.1.4.7.3 The electrocardiogram

Recordable electrical disturbances are created as the nodal and muscle tissue creates or passes impulses. The record is termed an electrocardiogram (ECG). A series of upward or downward deflections occurs that are labeled by the letters P, Q, R, S and T as illustrated in Figure 2.

The P wave is associated with atrial excitation, the QRS wave with ventricular excitation, and the T wave with recovery of the tissues and a readiness to respond to another stimulus. The lengths between the waves represent time and the height of the waves voltage. Dead tissue, poor conduction and other conditions are reflected as deviations in the normal ECG, so medical personnel can determine the nature of a problem or follow recovery from heart disorders.

2.1.4.7.4 The heart sounds

Sounds may be heard as the heart works. There are four sounds that are heard with sensitive equipment. Two are easily heard with an ordinary stethoscope. These are designated as the first sound (S_1) and the second sound (S_2). S_1 occurs as ventricular contraction takes place and is

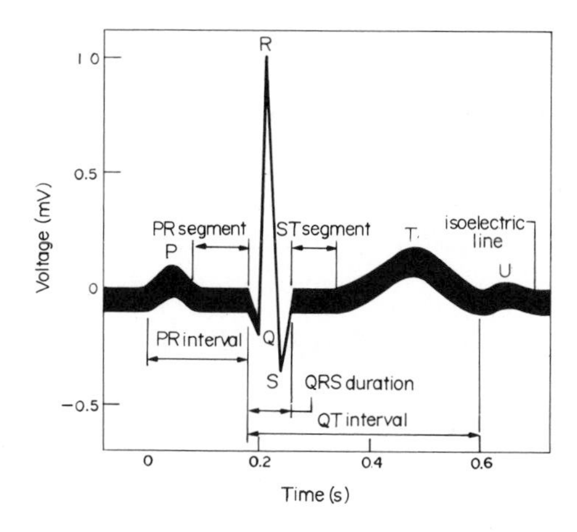

Figure 2 Normal electrocardiogram

attributed to the rush of blood against a closed atrioventricular valve, the sounds as the strands holding the valve shut vibrate, and to the sounds the muscle itself creates as it contracts. S_2 occurs as ejected blood tries to come back to the ventricle and shuts the pulmonary and aortic valves. Alterations in the normal quality of S_1 and S_2 usually reflect valves that do not open or close properly, and are called murmurs.

2.1.5 THE BLOOD VESSELS

2.1.5.1 Types of Vessels

The body's blood vessels distribute the blood to our tissues and organs and form a closed circulation that permits pressure to be developed on the blood. Vessels carrying blood away from the heart are called arteries. There are several types of arteries.

Elastic arteries, such as the aorta, have diameters of 20–25 mm and their walls are largely composed of stretchable elastic connective tissue. When the ventricles eject blood into such vessels, they expand to contain the additional blood volume, and then recoil when the ventricles relax to aid in pushing the blood onward.

Muscular arteries have circularly disposed smooth muscle around their cavities that can alter the diameter of the vessels. These vary in size from 10 mm to 20 μm as one progresses through the arterial system. Vessels thus decrease in size and increase in number through the arterial system.

Capillaries permeate the body tissues and organs. They are the most numerous and smallest (7–9 μm) of the blood vessels, and have only one cell layer in their walls. Because of their thinness, these vessels serve as the area of exchange of materials between bloodstream and body cells.

Veins carry blood to the heart. The smaller veins have some circular smooth muscle in their walls, but the larger ones serve to carry blood at low pressure and have little muscle. They become larger and less numerous as they approach the heart.

An important result of the vessel arrangement is the fact that total cross-sectional area of vessels increases from arteries to capillaries and decreases from capillaries to veins. This is an important determinant of pressure and flow in the vascular system.

2.1.5.2 Hemodynamics

Several factors determine pressure and blood flow in the blood vessels.

Cardiac output determines the volume of blood entering the arteries and thus the pressure in those vessels.

Peripheral resistance is determined by the diameter (total cross-sectional area) of the muscular vessels. Smaller vessels offer greater resistance to flow and raise pressure in the arteries between the heart and muscular vessels.

Elasticity of the large arteries determines the rise of systolic pressure and fall of diastolic pressure. A stiffened vessel results in a higher systolic pressure and lower diastolic pressure.

Volume of blood in the vessels of a given capacity determines a basic pressure. The capacity of the vessels may change but blood volume usually does not.

Viscosity of the blood determines the pressure required to circulate it, a 'thicker' fluid requiring a higher pressure.

Several principles govern the pressure, velocity and flow in the vascular system. Keep in mind that the vessels become more numerous and smaller toward capillary beds and fewer and larger from capillaries toward the heart. These principles may be stated as follows.

'Resistance to flow is directly proportional to length and inversely proportional to total cross-sectional area of the tubes'. This means that larger tubes or more tubes create a greater frictional surface that 'robs' the blood of its pressure. Pressure falls continually from arteries to capillaries to veins as there is no device to restore pressure on the blood until it returns to the heart.

'Flow is directly proportional to the fourth power of the radius of the tube'. Doubling the diameter results in a 16-fold increase in flow. Thus, small changes in size result in large alterations in flow.

'Pressure is directly proportional to cardiac output and inversely proportional to cross-sectional area'. Pressure is highest close to the heart and falls throughout the system.

'Velocity of flow is directly proportional to pressure and inversely proportional to cross-sectional area'. Speed is highest close to the heart, decreases to a minimum in capillaries where cross-sectional area is highest and where exchange occurs, and increases in veins.

Some actual values for these parameters are presented in Table 4.

Those vessels with smooth muscle in their walls can change size and this will influence pressure and flow in the vascular system. Of particular importance in this regard are the muscular arteries. These vessels are served by vasoconstrictor and vasodilator nerves rising from centers located in the brainstem. Vasoconstrictor nerves usually secrete norepinephrine at their endings and vasodilator nerves secrete acetylcholine. Activity of the brainstem centers is largely reflexively controlled through the same afferent pathways described for cardiac reflexes, and heart and vascular responses compliment one another. Thus, increased heart rate is accompanied by vasoconstriction to achieve a quicker response.

2.1.6 THE RESPIRATORY SYSTEM

2.1.6.1 General Structure and Activities

A respiratory system exists to provide a large surface area for diffusion of oxygen to the bloodstream and for diffusion of carbon dioxide in the opposite direction. The nasal cavities, pharynx, larynx, trachea and bronchi comprise the conducting division of the system, tubes too thick-walled to permit gas diffusion. Respiratory bronchioles, alveolar ducts and alveolar sacs contain thin-walled alveoli that do permit gas diffusion with surrounding capillary beds, and these organs constitute the respiratory division of the system. A surface area of 50–70 m^2 is provided by the alveoli. All parts of the system beyond the bronchi are contained within the lungs, located in the thorax.

The nasal cavities have the additional tasks of warming, moistening and cleansing inhaled air.

The physiology of the respiratory system may be considered under several headings: (i) pulmonary ventilation (breathing) brings air into or out of the lungs and is divided into inspiratory (intake) and expiratory (output) phases; (ii) gas exchange occurs between lungs and bloodstream (external respiration) and between bloodstream and cells (internal respiration); (iii) gases must be transported in the bloodstream; and (iv) control of rate and depth of breathing is provided by nervous and chemical mechanisms.

Table 4 Hemodynamic Parameters

Vessel(s)	*Total cross-sectional area* (cm^2)	*Pressure* (mmHg)	*Velocity* ($cm\ s^{-1}$)
Aorta	2.5	130 systolic 85 diastolic	100–140
Small arteries	60	40 av.	4–5
Capillaries	2500	28 av.	0.3 $mm\ s^{-1}$
Veins	325	10–0 at right atrium	60

2.1.6.2 Breathing

Inspiration of air provides a volume of gas of which about 30% remains in the conducting division, unavailable for exchange. This volume is called dead air, the tubes containing it are the dead space, and this is dead space ventilation. The remaining volume reaches the alveoli and constitutes alveolar ventilation, air available for exchange.

A basic principle involved in inspiration is that if the volume of a cavity is increased, the pressure in that cavity will fall. The chest is a closed cavity whose volume is increased by contraction of the diaphragm and the intercostal muscles attached to the ribs. These muscles increase the vertical and lateral dimensions of the chest. Suspended within the chest are the lungs. Their surfaces are in contact with the chest wall and are kept in contact by a fluid film that causes cohesion of the two surfaces. As the chest size is expanded, the lungs are increased in size, and the pressure within the lungs decreases. Since the lungs communicate with the atmosphere, air will enter the lungs to equalize the pressure. As the lungs are expanded, elastic tissue within is stretched, and on relaxation of the muscles of respiration and a decrease in chest volume, the recoil of this elastic tissue creates a higher pressure in the lungs than in the atmosphere, and air is driven from the lungs in expiration. Any condition that causes loss of lung elasticity will make the lungs harder to inflate and will remove the recoil that forces air out. In emphysema, fibrous non-elastic tissue replaces the elastic material and makes breathing a much harder task.

2.1.6.3 Surfactant

The alveolar surfaces are coated with a phospholipid called surfactant that lowers the surface tension between the lining and air. As alveoli are expanded, surface tension increases, and as they deflate, surface tension decreases due to the surfactant. Without the substance, surface tension would increase on deflation and the alveoli would have a tendency to collapse on deflation. Lack of the substance results in hyaline membrane disease in infants and the disorder is characterized by alveolar collapse and inability to reinflate them leading to oxygen deficiency in the blood.

2.1.6.4 Gas Exchange

The process of diffusion accounts for gas exchange between lungs, blood and cells. This process occurs only from a region of higher concentration of gas to a lower concentration. Concentrations must thus be maintained so as to ensure an 'inward' movement of oxygen and an 'outward' movement of carbon dioxide. Table 5 shows gas concentration in various areas of the body.

Oxygen levels are highest in the lungs, lowest in the tissues, and oxygen diffuses from lung to blood to tissue. Carbon dioxide levels are highest in the tissues, lowest in the lungs and it diffuses in the opposite direction to be breathed out.

2.1.6.5 Volumes of Exchange and Capacities

Tidal volume (TV) refers to the volume of gas inhaled or exhaled at any level of body activity. At rest, TV averages about 500 mL, of which 70% (350 mL) actually reaches the alveoli. Since air is about 20% oxygen, this results in about 70 mL of oxygen available per breath for exchange. Reserve inspiratory volume (RIV) is air that may be inhaled above tidal volume and it averages 3000 mL. Reserve expiratory volume (REV) is air that can be exhaled beyond tidal volume and it averages

Table 5 Gas Partial Pressures

Gas[a]	*Atmosphere*	*Alveoli*	*Arterial blood*	*Venous blood*	*Tissues*[b]
O_2	158.0	100.0	95.0	40.0	40.0
CO_2	0.3	40.0	40.0	46.0	46.0
Water vapor	5.7 av.	47.0	47.0	47.0	47.0

[a] Concentrations in mmHg partial pressure (% of gas x total atmospheric pressure). [b] Varies due to level of metabolic activity. Values represent at rest averages.

1100 mL. It should be obvious that a reserve of air is available beyond at-rest requirements to supply the needs of activity.

With even the most forcible expiration, air will remain in the lungs because they do not collapse on expiration. This air is called residual volume (RV) and amounts to some 1200 mL.

The sum of TV, RIV and REV is called the vital capacity and reflects the maximum exchangeable volume of gas. It gives a rough measure of lung size. The sum of REV and RV is called the functional residual capacity and serves to continuously oxygenate the blood in the face of intermittent renewal of gases by breathing.

We breath, at rest, about 16 times a minute. Multiplying this figure by TV gives a minute volume of respiration, a quantity akin to cardiac output of the heart. Sixteen, times 500 mL, gives 8000 mL (8 L) per minute, a typical figure for at-rest exchange. A maximum exchange volume would be obtained by multiplying vital capacity (4600 mL) by a maximum breathing rate (about 30 times a minute). This would be 128 000 mL (128 L) per minute, and gives an idea of the reserves available in the system.

2.1.6.6 Gas Transport

Oxygen diffuses into the bloodstream and combines with the respiratory pigment hemoglobin carried in the red blood cells (erythrocytes). Ninety-five percent of the oxygen is carried in this manner as what is called oxyhemoglobin. Five percent is carried dissolved in the blood plasma. Oxyhemoglobin releases its oxygen at the tissue level, again by diffusion.

Carbon dioxide is transported in three ways.

(i) As bicarbonate ion resulting from the reaction of water and carbon dioxide as shown in equation (1). This reaction occurs at the tissue level, and is reversed, aided by an enzyme in the lungs, so as to recover the CO_2 and breathe it out (equation 2). About 64% of the CO_2 is carried in this manner.

$$H_2O + CO_2 \longrightarrow \underset{\text{Carbonic acid}}{H_2CO_3} \longrightarrow \underset{\text{Bicarbonate ion}}{H^+ + HCO_3^-} \quad (1)$$

$$H^+ + HCO_3^- \longrightarrow H_2CO_3 \xrightarrow[\text{anhydrase}]{\text{Carbonic}} CO_2 + H_2O \quad (2)$$

(ii) Up to a maximum of 27% of the CO_2 is carried in combination with hemoglobin, but not on the same part of the molecule that carries oxygen. The term carbaminohemoglobin is given to the combination of CO_2 and hemoglobin.

(iii) About 9% of the CO_2 is carried dissolved in the plasma.

2.1.6.7 Control of Breathing

Central influences and peripheral influences determine rate and depth of breathing.

Located within the brainstem are several sets of respiratory centers that determine breathing patterns. These include the following: inspiratory centers provide the impetus for inspiration *via* nerves to the muscles of breathing; apneustic centers provide reinforcement to the inspiratory centers to ensure inspiration; expiratory centers are involved in partially inhibiting inspiration; pneumotaxic centers contribute to expiration; and vagal nerve nuclei may monitor cerebrospinal fluid acidity and contribute to expiration.

The initial stimulus for inspiration is thought to be provided by carbon dioxide in the blood or by the hydrogen ion that results from the reaction of water and carbon dioxide. In any event, the inspiratory centers are stimulated to generate impulses that are sent to the inspiratory muscles *via* nerves. Carbon dioxide may also stimulate the apneustic centers. As the inspiratory centers become active, impulses are sent that stimulate the pneumotaxic center that in turn sends inhibitory impulses back to the inspiratory center, partially shutting down its activity. As the lungs expand, stretch-sensitive receptors are stimulated, nerve impulses pass to stimulate the expiratory center and to further inhibit the inspiratory center. The expiratory center then provides a final inhibitory signal to the inspiratory center. Expiration is achieved by periodically interrupting continuous inspiratory activity.

Chemoreceptors in the aorta and carotid arteries continually monitor blood oxygen and carbon dioxide levels. A fall of oxygen levels or a rise of carbon dioxide levels stimulate the receptors to send impulses *via* nerves to stimulate rate and depth of breathing that renews oxygen supply and eliminates the excess carbon dioxide.

2.1.6.8 Protection in the Respiratory System

The alveolar surfaces are in direct contact with inhaled air that may contain dust, pollens and microorganisms that could pose a threat to the body. Several mechanisms operate to reduce this threat.

The nasal cavities and the respiratory system from larynx to smallest bronchi are lined with a ciliated, mucus-covered epithelium. The mucus provides a sticky surface to which particulates adhere, and the ciliary beat is toward the throat from both areas. The term mucociliary escalator is used to reflect the constant 'endless belt' nature of this process.

The alveoli contain macrophages that can engulf (phagocytose) particulates that escape the mucus. These cells then move to an area of ciliated epithelium and are swept out of the lungs.

A γ-globulin designated immunoglobulin A (IgA) is produced by glands in the walls of the system and secreted into their cavities. IgA is capable of causing the lysis of microorganisms as they contact the secretion on the epithelial surface.

Coughing and sneezing are reflex acts triggered by irritation in the system from larynx or below, or in the nasal cavities (respectively). The opening to the larynx (glottis) is sealed, a powerful contraction of the abdominal muscles is triggered, the glottis is opened suddenly and an explosive expulsion of air occurs that should remove the irritant.

2.1.6.9 Other Functions of the Lungs

Functions of the lungs unassociated with gas exchange include the following: surfactant secretion; production of kinins, which are vasodilating substances; inactivation of gastrin, which is a stimulant to gastric secretion; metabolizing serotonin, which is a vasoconstrictor; metabolizing insulin, which is a hormone regulating blood sugar level; and acid–base balance of the body; lungs play a role by their ability to rid the body of carbon dioxide and the hydrogen ion that would result from its reaction with body water. The lungs provide a rapid way to compensate for excess hydrogen ion in the body.

2.1.7 THE DIGESTIVE SYSTEM

2.1.7.1 General Organization and Functions

A digestive system (Figure 3) provides a route for intake of nutrients and processing those nutrients to make them available to body cells. The tubular organs of the system (mouth, pharynx, esophagus, stomach, small and large intestines) form the alimentary tract. Emptying into or contained within the walls of the tract are the accessory organs of digestion. These include the tongue and teeth, salivary glands, liver and pancreas. The functions served by the system include: ingestion, the intake of nutrients; digestion, the mechanical and chemical (enzymatic) reduction of large molecules to ones small enough to pass through cell membranes; absorption, the passage of end products of digestion into the bloodstream for distribution to cells; and egestion, the elimination of residues of the digestive process from the body.

2.1.7.2 Digestion and Absorption in the Mouth

In the mouth, ingested foods are chewed by the teeth to reduce their size to a point at which they can be easily swallowed. Also, smaller particles are more efficiently digested by enzymes as they present more surface for enzyme action. The tongue guides food between the teeth for chewing and forms a bolus for swallowing.

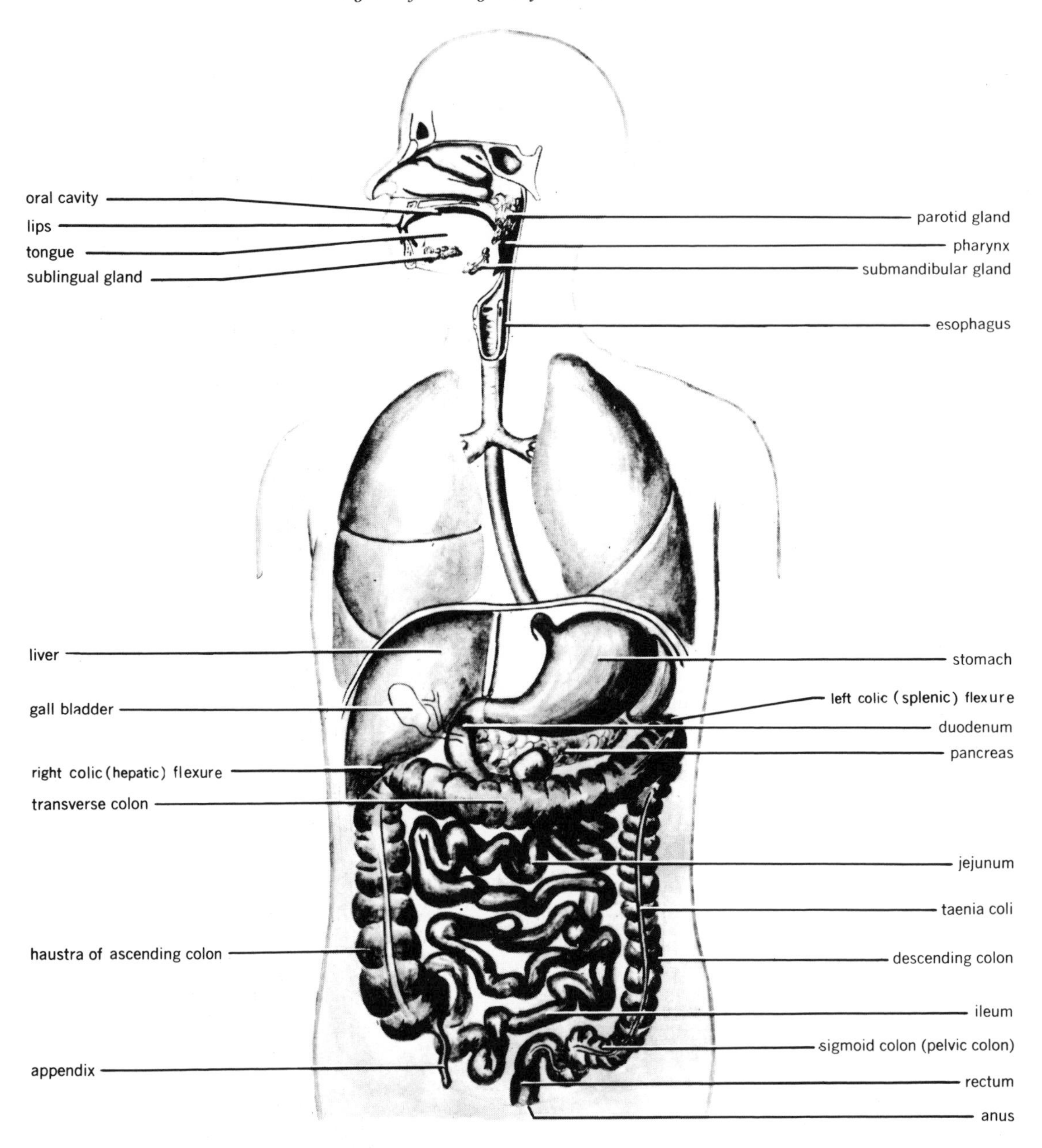

Figure 3 The organs of digestion

Three pairs of major salivary glands empty their secretion, saliva, into the mouth. The parotid, submandibular and sublingual glands produce a slightly acidic, buffered fluid containing a carbohydrate-digesting enzyme called salivary amylase (ptyalin). The enzyme breaks chemical bonds between simple sugar molecules in starches. Though capable of reducing starches to disaccharides (two simple sugar molecules) the foods remain so short a time in the mouth that only a small amount of disaccharides are produced (3–5%) while the bulk (95–97%) of the starches are reduced to dextrins, units containing many simple sugar molecules. After food is swallowed, amylase action may continue in the stomach until the gastric juice penetrates the digesting mass.

Because of the large size of the molecules in the mouth and the short time they spend here, absorption from the mouth is minimal. Medicines, as in lozenges, that are retained within the mouth may undergo significant absorption.

2.1.7.2.1 Control of salivary secretion

Production of saliva is continuous at low levels to keep the mouth moistened. When food is present in the mouth, a nervous reflex greatly increases salivary secretion. Molecules of digesting food enter taste buds located chiefly on the tongue. Nerve impulses travel over nerve fibers to the brainstem salivatory centers. From here, nerve impulses pass to the salivary glands whose secretion is increased. Saliva moistens food and mouth linings, softens and dissolves food, its mucus content lubricates food for swallowing and glues the bolus together.

2.1.7.3 Digestion and Absorption in the Stomach

Foods entering the stomach come in contact with gastric juice secreted by an estimated 35 million gastric glands in the stomach wall. Gastric juice is a watery solution of hydrochloric acid, with a pH of about 2, that contains a protein-digesting enzyme called pepsin and small quantities of electrolytes (Cl^-, Na^+, Ca^{2+}, Mg^{2+}). Mucus is also abundant in the juice in the stomach. It is derived from mucus-filled epithelial cells lining the stomach, and an unbroken layer of mucus protects the stomach itself from the action of the gastric juice. The hydrochloric acid is produced by oxyntic (parietal) cells of the gastric gland from CO_2, H_2O and NaCl (equation 3).

$$
\begin{aligned}
CO_2 + H_2O &\longrightarrow H_2CO_3 \longrightarrow HCO_3^- + H^+ \\
NaCl &\longrightarrow Na^+ + Cl^- \\
H^+ + Cl^- &\longrightarrow HCl
\end{aligned}
\qquad (3)
$$

Cells called zymogenic or chief cells produce pepsinogen, an inactive precursor of pepsin. Pepsinogen is converted to pepsin by hydrochloric acid. Pepsin begins the digestion of proteins by converting them to proteoses and peptones, units containing 4 to 12 amino acids.

Small quantities of pepsin B_5 and gastricsin are other protein-digesting enzymes produced by stomach cells. A gastric lipase is also produced that breaks down milk fat. Of no importance in adults, these three enzymes are of more importance in the infant stomach where full HCl and pepsin production have not yet been achieved.

The still large size of digesting food molecules in the stomach prevents much absorption of such products through the stomach wall. Some water, Na^+ and alcohol can pass through the organ's wall.

2.1.7.3.1 Control of gastric secretion

It is appropriate to have quantities of such a powerful juice in the stomach only when there is food present for it to work on. Three mechanisms determine gastric juice secretion.

(i) In the psychic or cephalic phase, foods in the mouth (taste buds) cause impulses to be sent over nerves to the nucleus solitarius of the brainstem. Fibers of the vagus nerve then pass to the stomach to cause the secretion of 50 mL or so of gastric juice. Swallowed food then is subjected to a small degree of digestion by this juice.

(ii) In the gastric phase the products (proteoses and peptones) of the psychic-controlled juice cause APUD (amine precursor uptake decarboxylase) cells in the gastric glands to produce a hormone called gastrin that appears to have similar activities to histamine (H_2). The gastrin is absorbed by the venous drainage of the stomach, enters the circulation and is distributed over the entire body including the stomach. (A blood-borne messenger is really the only way to contact thousands of cells in millions of glands.) On reaching the gastric glands, production of some 750 mL of gastric juice is stimulated. Cimetidine is a drug that binds to histamine H_2 receptors on the stomach cells but which does not cause gastric secretion and is used to treat ulcers from overproduction of gastric juice.

(iii) By the intestinal phase, additional small quantities of juice are secreted by the mechanism of acid entering the first part of the small intestine, causing 'intestinal gastrin' to be produced that follows the same course to the gastric glands as did gastrin.

2.1.7.4 Digestion and Absorption in the Small Intestine

As gastric digestion proceeds, small quantities of the mass are moved at a time into the first part of the small intestine (the duodenum). In the duodenum, the foods are subjected to the action of two secretions: pancreatic juice and bile from the liver.

Pancreatic juice is an alkaline fluid (pH about 8) that contains bicarbonate ion and enzymes that work on all three basic foodstuff groups (carbohydrates, lipids, proteins). The alkaline nature of the fluid neutralizes gastric acid and sets the proper environment for enzyme action. Pancreatic enzymes include the following.

Proteinases, enzymes acting on proteins, are secreted as inactive precursors and are converted to active principles in the duodenum.

Trypsinogen is converted to trypsin by enterokinase, an enzyme produced by duodenal epithelial cells. It breaks peptide bonds adjacent to the amino acids lysine and arginine. Chymotrypsinogen is converted to chymotrypsin by trypsin and breaks peptide bonds next to amino acids containing aromatic rings.

Procarboxypeptidase is converted to carboxypeptidase by trypsin and breaks peptide bonds on terminal amino acids (those at the ends of a large unit).

The net result of pancreatic enzyme action is to convert proteoses and peptones into dipeptides (units containing two amino acids) and some free amino acids.

Amylase — the same carbohydrate-digesting enzyme as was found in saliva, it converts dextrins to disaccharides.

Lipase — this acts to separate fatty acids from their attachment to glycerol in compounds called triacylglycerols (triglycerides).

2.1.7.4.1 Control of pancreatic secretion

Release of pancreatic juice requires two hormones, secretin and cholecystokinin, which are produced by APUD cells in the duodenal glands.

Secretin is a substance containing 27 amino acids that is released when acid (the most effective stimulus) enters the duodenum from the stomach. It enters veins and is ultimately distributed to the pancreas *via* the bloodstream. Secretin causes the pancreas to release a watery, alkaline, bicarbonate-buffered fluid deficient in enzymes. Its purpose is to neutralize gastric acid.

Cholecystokinin (CCK) contains 33 amino acids and is produced when protein hydrolysates or fatty acids enter the duodenum. It follows the same pathway as secretin, and, upon arriving at the pancreas, stimulates enzyme release.

2.1.7.4.2 Role of bile in digestion

Bile is a product continuously produced by the liver and is stored in the gall bladder until required in the digestive process. It contains no enzymes. Production of bile by the liver is stimulated by secretin, and gall bladder contraction is caused by CCK.

The emulsifying action of bile refers to its ability to lower the surface tension of lipid droplets and minimize their tendency to coalesce into larger masses that would be less efficiently digested by lipase. Its hydrotropic action refers to the ability of cholic acid in the bile to combine with fatty acids to form bile salts and render the fatty acids more water soluble for easier absorption.

2.1.7.5 Small Intestine Epithelial Cell Involvement in Digestion

Completion of carbohydrate and protein digestion occurs by enzymes located in the microvilli of the intestinal epithelial cells. Each epithelial cell contains an estimated 1700 microvilli that increase the total intestinal surface by 15 to 40 times what it would be without them. Entering the membranes, dipeptides are broken to individual amino acids by aminopeptidase and dipeptidase. Disaccharides encounter specific amylases that free their constituent simple sugars. Maltose is acted on by maltase and two glucose molecules result; sucrose is acted on by sucrase to free glucose and fructose; lactose is acted on by lactase to free glucose and galactose. All end products are now ready for absorption.

Absorption of end products of digestion and other substances (vitamins, electrolytes, water) occurs to different degrees in different parts of the small intestine.

Monosaccharides, vitamins A, B, C, D, E and K, salts and some amino acids are absorbed in the first one-third of the intestine; fats are absorbed in the middle third; bile salts and vitamin B_{12} in the last third. Larger lipids pass to a lymph vessel in the intestinal villi called a lacteal; all other products pass to blood vessels in the villus.

Absorption of monosaccharides, amino acids and electrolytes is by active processes (ones that require cellular activity), while small lipids appear to be passed by endocytosis. As these solutes cross cell membranes, water follows osmotically.

2.1.7.6 The Colon

The colon, most of the large intestine, has no digestive function. Some 300 to 400 mL of water are removed by the organ per day, along with Na^+, K^+, Cl^- and HCO_3^-. A normal microorganismic flora present in the organ produces nutritionally significant amounts of vitamins K, B_1, B_2 and B_{12}, which the host absorbs. The colon accumulates residues of digestion, dehydrates them and eliminates them as the feces.

2.1.7.7 The Liver

Other than its involvement in bile production, the liver must be considered a vital organ for its other functions: it acts as a 'blood reservoir' containing 0.5–1.0 L of blood that can be shunted to the circulation; phagocytic cells of the liver (macrophages) remove aged erythrocytes from the circulation and cleanse the blood flowing from the intestines of microorganisms entering from the gut; in the fetus, the liver produces blood cells; it detoxifies (renders harmless) many substances including alcohol, ammonia and drugs; and it stores iron, vitamins A and D and glycogen, and synthesizes proteins, triacylglycerols, glycogen and urea.

Liver damage is associated with elevated blood levels of several enzymes that are relatively unique to liver cells. Lactic dehydrogenase (LDH), glutamic-pyruvic transaminase (GPT), serum glutamic-oxaloacetic transaminase (SGOT) and isocitric dehydrogenase (ICD) are enzymes tested for when liver damage is suspected.

2.1.8 THE EXCRETORY SYSTEM

2.1.8.1 General Structure

The human excretory system consists of two kidneys, two ureters that carry urine to a urinary bladder, and the urethra that carries urine to the exterior.

The bean-shaped kidneys measure about 4 inches in length, 2 inches wide and one inch in depth ($10 \times 5 \times 2.5$ cm). They lie against the posterior abdominal wall about half covered by the last pair of ribs. Each kidney is supplied by a single renal artery derived from the aorta, a fact that ensures a high pressure to the organ. The renal artery branches repeatedly as it enters the kidney and forms a tiny afferent arteriole (a muscular vessel) that next forms a network of capillaries called a glomerulus. From the glomerulus rises an efferent arteriole that forms a second network of capillaries surrounding the tubules of the functional units of the kidney called nephrons. It is into this second capillary network that substances will be placed from the kidney tubules for return to the circulation.

2.1.8.2 Nephrons

The functional units of the kidney, nephrons (Figure 4), number about 1.5 million per kidney. Each microscopic unit begins with a double-walled cup surrounding a glomerulus. This cup is called the glomerular (Bowman's) capsule. The inner layer of the capsule applied to the glomerulus, forms a filtration membrane to govern the passage of substances from bloodstream to the cavity of the capsule. A proximal convoluted tubule, measuring about 14 mm in length by 60 μm in diameter, attaches to the capsule and lies coiled near the capsule it serves. A loop (of Henle) is derived from the proximal tubule and extends toward the medial aspect of the kidney and then curves back toward the periphery in a hairpin shape. A distal convoluted tubule, measuring about 5 mm in length by 20–50 μm in diameter, connects to a collecting tubule that becomes a papillary duct to empty fluid into cavities inside the kidney called calyces. The calyces form the renal pelvis from which the ureter rises.

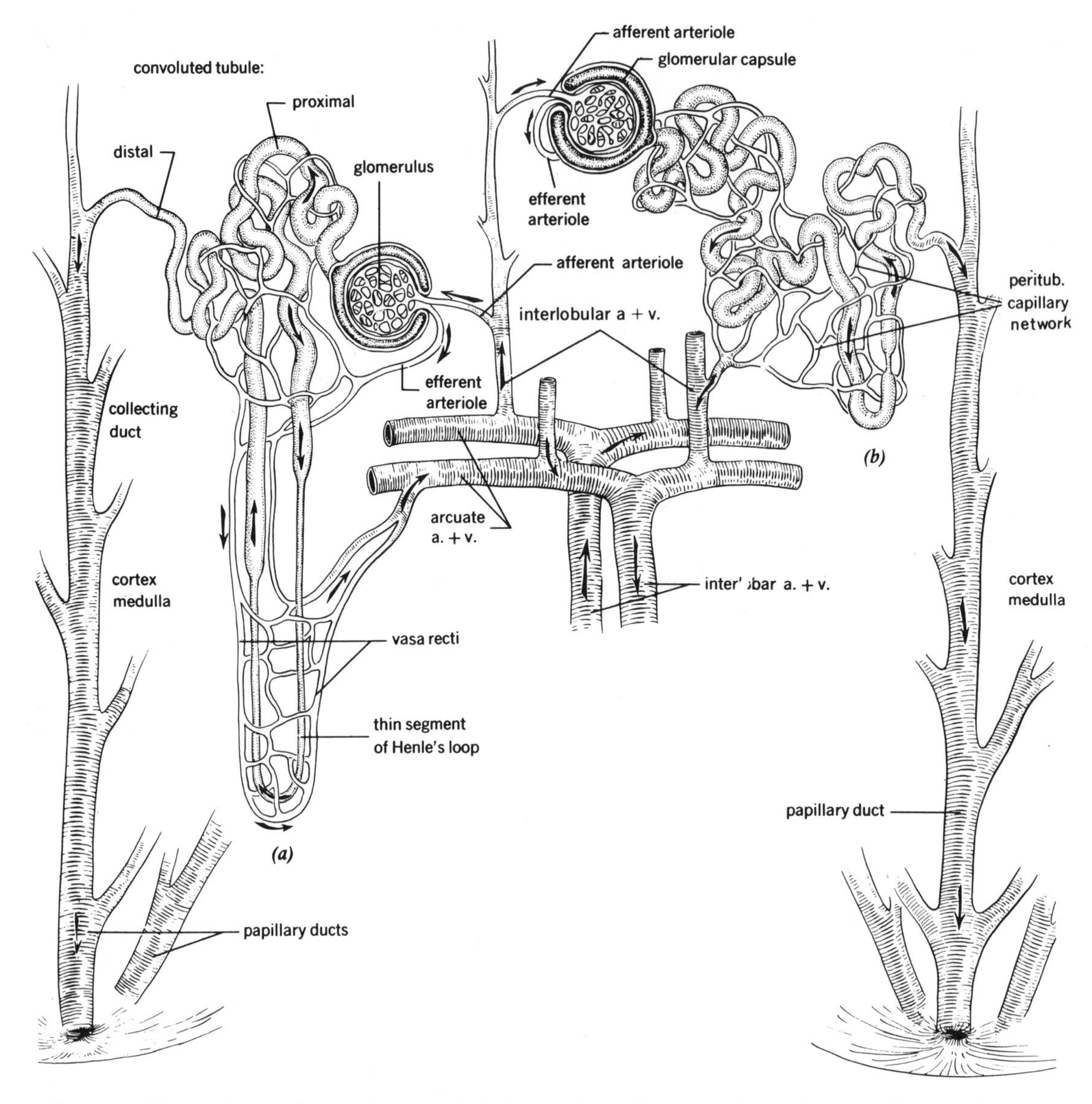

Figure 4 The nephrons of the kidney and their associated blood vessels: (a) juxtamedullary nephron; and (b) cortical nephron

2.1.8.3 Functions of the Kidneys

The kidneys have three basic functions: (i) to form urine; (ii) to regulate the blood composition of some three dozen substances including electrolytes, water and a variety of organic molecules; and (iii) to aid in controlling the pH (acid–base balance) of extracellular fluid.

In carrying out these functions, the nephrons utilize several processes: glomerular filtration; tubular transport; countercurrent multiplier and countercurrent exchanger; acidification; and concentration of the fluid.

2.1.8.3.1 Glomerular filtration

Blood arrives at the glomerulus at a pressure of about 75 mm Hg. This pressure forces (filters) substances through the capillary and inner capsular membranes according to their size (molecular weight). All substances with a molecular weight less than 10 000 pass quite freely through these membranes. The substances filtered include water, glucose, small lipids, amino acids, vitamins, salts, all substances of value to the body, as well as wastes such as urea. The fluid formed is called the filtrate. The blood pressure (P_B) is opposed by an osmotic pressure (P_o) created by the plasma proteins that do not filter, and this amounts to about 30 mmHg. Additional opposing pressures (P_{other}) amount to 20 mmHg. An effective filtration pressure (P_{eff}) is calculated by subtracting all opposing pressures from the blood pressure and it represents the actual force available to filter the blood. Expressed as a mathematical formula

$$P_{eff} = P_B - (P_o + P_{other})$$
$$25\text{ mmHg} = 75 - (30 + 20)$$

Changes in blood pressure will obviously influence rate of filtration and a pressure of less than 15 mmHg will not clear the blood of wastes.

2.1.8.3.2 Tubular transport

The filtrate moves to the proximal convoluted tubule where it is subjected to active transport that is called reabsorption if substances are moved out of the tubules into the surrounding capillaries, or secretion if substances are moved from cells into the filtrate. Active transport involves carrier molecules that are specific for the material being transported. A selectivity as to type and amount of substance transported is thus established and by this means the nephrons control blood composition. Electrolytes, amino acids and glucose are reabsorbed. Wastes, because they have no carriers, are not reabsorbed actively, although their concentrations may rise to where some return by diffusion occurs. Eighty to 90% of the electrolytes are reabsorbed and all of the glucose and amino acids, if the levels of the latter are less than the maximum ability of the cells to reabsorb them. As solutes are removed, water follows, so it also achieves an 80 to 90% reabsorption. The filtrate has been reduced in volume but has not changed osmolarity because solute and water pass one-for-one.

2.1.8.3.3 Countercurrent multiplier and exchanger

The filtrate next enters the loop of Henle that has a descending and an ascending limb. In the ascending limb there is an active transport of chloride ion from the filtrate into the fluids around the loop. As the negative ion is removed, sodium and potassium ions follow, and the concentration of solutes in the fluids is increased. The ascending limb is not permeable to water so that as the ions move there is not an equivalent osmotic movement of water. The filtrate becomes hypoosmotic, to about one-third the concentration of the original filtrate. Urea that accumulated in the proximal tubule passively diffuses into the fluids to add to the accumulation of solutes. The basic purpose of this multiplier is to increase solute concentration in the fluid around the nephrons.

To keep the solutes in the fluids and prevent their being carried away by the bloodstream as fast as they are transported from the filtrate, the exchanger comes into play. A looped blood vessel follows the loop of Henle. Solutes flow into the vessel and water out as it enters the area of increased solute concentration, and solutes flow out and water in as the vascular loop ascends. The solute is recirculated and largely kept within the kidney fluids.

2.1.8.3.4 Acidification

The distal convoluted tubule is an area wherein there is an active transport of H^+ from tubule cells into the filtrate. The source of the ion is the reaction of carbon dioxide and water within the cell. As the positive H ion is removed from the cell, a positive sodium ion is returned to the cell, where it combines with the bicarbonate ion from the CO_2–H_2O reaction and the sodium bicarbonate is returned to the bloodstream to maintain its alkalinity.

If this secretion is not sufficient to maintain acid–base balance, secretion of potassium ion along with hydrogen ion will result in two sodium ions being reabsorbed to maintain electrical neutrality. This doubles the bicarbonate returned to the bloodstream.

2.1.8.3.5 Concentration

Fluid exiting the loop of Henle is hypoosmotic, having lost solute without equivalent water loss. The now acidified but still hypoosmotic filtrate enters the collecting tubule that runs toward the renal pelvis through the area of high solute concentration created by the multiplier. There is a tendency for water to move osmotically from the tubule but this will not occur unless permitted by antidiuretic hormone (ADH). ADH secretion is directly proportional to blood osmolarity reaching the hypothalamus of the brain, the site of production of the hormone. It is *via* ADH that the fine tuning of blood water levels is achieved.

No further change in the composition of the fluid will now occur, and it is called urine.

2.1.8.3.6 Composition and characteristics of the urine

Normal urine is an amber to yellow transparent fluid with a pH of 5–7, and containing significant amounts of certain substances, as shown in Table 6.

Lacking in normal urine are glucose, amino acids (completely reabsorbed), protein (little filtered; what does is reabsorbed), and blood cells (too large to filter).

2.1.8.3.7 Other functions of the kidney

When the blood supply to a kidney is low in oxygen or pressure, the kidney produces an enzyme-like substance called renin. It acts on a plasma component called angiotensinogen (hypertensinogen) converting it to angiotensin I (hypertensin I). A plasma-converting enzyme changes angiotensin I to angiotensin II (hypertensin II), the most potent vasoconstrictor known. The generalized vasoconstriction that occurs with angiotensin II elevates blood pressure with the 'aim' of restoring filtration pressure to the nephrons. Aldosterone secretion by the adrenal cortex is increased by angiotensin II and the aldosterone increases sodium reabsorption with chloride and water following. This increases blood volume, also raising blood pressure.

Erythropoietin is also produced by a kidney receiving low-oxygenated blood. It stimulates production of red blood cells.

2.1.9 THE NERVOUS SYSTEM

The nervous system provides an ability to appreciate and react to stimuli arising from without or within our bodies. It also is the source of creativity, thoughts and intellect. The cells of the system that enable such activities to occur are known as neurons. Supportive and nutritive cells called glia are associated with neurons.

2.1.9.1 Neurons

Neurons are the structural and functional cells of the nervous system. They are highly excitable and conductile cells that originate and distribute nerve impulses for control of many body functions.

Table 6 Urine Characteristics

Substance/property	*Sources*	*Amount per day*
Water	Diet and metabolism	1.2–1.5 L
Urea	Deamination of amino acids	30.0 g
Uric acid	Metabolism of nucleic acids	0.7 g
Chloride (as NaCl)	Diet	12.0 g
Phosphate	Diet and metabolism of phosphate-containing compounds	3.0 g
Sulfate	Diet and metabolism of sulfate-containing compounds	2.5 g
Calcium	Diet, bones	200 mg
Osmolarity	—	800–900 mOs (plasma is 285 mOs)

2.1.9.1.1 Structure

Neurons are classified by size, shape and type of impulse carried.

Large-bodied neurons with cellular extensions (processes) that are very long (sometimes to > 1.0 m) are called Golgi I neurons. These cells relay impulses from the periphery to the cord and brain and in the opposite direction. Small-bodied and short-process (processes measure in mm) neurons are called Golgi II neurons. These are exemplified by the internuncial neurons set between other neurons in the brain and spinal cord that can amplify, subdue or create new pathways for nerve impulses.

By shape, neurons may be unipolar, having a single process from the cell body, bipolar with two processes, or multipolar, having three or more processes.

By function, neurons are described as transmitting sensory impulses, dealing with sensations, or motor impulses that cause a reaction.

2.1.9.1.2 Function

The excitable state, or a condition of the cell membrane enabling the formation of a nerve impulse in response to many types of stimuli, depends primarily on an imbalance of ions on the two sides of the membrane and a resulting electrical difference (potential) across the membrane. This separation is created by passive and active processes. If one examines the types and concentrations of ions outside and inside a neuron, the differences would appear as shown in Table 7.

Both sides of the membrane show a balance of positive and negative charges. A measured transmembrane potential of -70 to -90 mV to the exterior exists that is believed to be due to an active pumping of Na^+ to the exterior of the membrane with a coupled inward pumping of K^+. The membrane at rest may also be demonstrated to be more permeable to K^+ than to Na^+ and an outward leakage of K^+ accumulates an excess of positive charges outside the cell. This resting state is called a polarized state.

A stimulus capable of initiating a nerve impulse causes a local increase of 500-fold in membrane permeability to Na^+ and a 40-fold increase in that of K^+, creating a depolarized area. Sodium ion diffuses rapidly into the cell, causing a reversal of potential from the resting state. An electric current next flows from the polarized to the depolarized areas that are adjacent and this current is the nerve impulse. This current acts as an adequate stimulus to depolarize the next segment of the membrane and the disturbance is transmitted along the membrane—it is conducted. As the impulse moves away from the original point of stimulation, the Na^+ is pumped outward and the membrane becomes repolarized.

The nerve impulse, once generated, loses no strength as it is conducted along the membrane (decrementless conduction). A stimulus of sufficient strength to depolarize the membrane causes a complete response; one of insufficient strength brings no response. This is called the all-or-none law. When in the depolarized state, the membrane cannot form another impulse to another stimulus—it is refractory. The period is quite short, about 0.001 s for nerve cells, so that volleys of impulses can be carried along nerve membranes at high frequencies.

2.1.9.2 Glial Cells

Glial cells outnumber neurons by about five to one. They do not exhibit high degrees of excitability and do not conduct impulses. There are several types of glial cells.

Table 7 Ionic Distributions in Body Fluids

Ion	*Outside* (mM)	*Inside* (mM)
Na^+	145.0	12.0
K^+	4.0	155.0
$Other^+$	5.0	—
Cl^-	120.0	4.0
HCO_3^-	27.0	8.0
$Other^-$	7.0	155.0[a]

[a] Organic ions that are not diffusible.

Astrocytes are found in the brain and spinal cord (the central nervous system or CNS). They form supporting frameworks for neurons and their processes and transfer nutrients from brain capillaries to neurons. They are, in part, responsible for the 'blood–brain barrier' a concept that some substances are restricted in their passage from blood vessels to neurons in the brain.

Oligodendrocytes are also in the CNS and produce fatty coverings called myelin sheaths on nerve processes (axons) that speed the rate of conduction of nerve impulses.

Ependymal cells form an epithelium for the cavities of brain (ventricles) and spinal cord (central canal). These cavities originate from the formation of the CNS as a tube.

Microglia are CNS cells that remain quiet until there is injury or inflammation in the CNS. They then become actively ameboid and phagocytic and aid in cleanup and healing processes in the damaged region.

Schwann cells are in the part of the nervous system outside the brain and cord, called the peripheral nervous system (PNS). They produce myelin sheaths for peripheral nerve fibers.

Satellite cells form capsule-like coverings around neuron cell bodies in ganglia (grouping of peripheral neuron bodies) of the PNS. They probably act in supportive and nutritive fashions.

2.1.9.3 The Synapse

Neurons form pathways to carry nerve impulses from one part of the body to another. Such pathways contain at least two neurons, and a junction between the two neurons is called a synapse. The synapse is a functional but not an anatomical junction between two neurons. The endings of the processes of one neuron lie 200–300 Å from those of the next neuron, creating a synaptic gap. A synaptic transmitter is used to bridge the gap since a nerve impulse cannot. The presynaptic neuron (the one ahead of the synapse) contains in its process endings a chemical. The postsynaptic neuron (the one beyond the synapse) has binding sites for the transmitter on its processes. A nerve impulse alters the presynaptic terminal's permeability to Ca^{2+}, it enters the neuron, causes release of the transmitter that diffuses and binds to the postsynaptic membrane depolarizing it. Some device is present to remove or destroy the transmitter so that it does not remain to continually stimulate the postsynaptic cell.

There are many different substances known or suggested to act as synaptic transmitters, which includes those in Table 8.

2.1.9.3.1 Synaptic properties

A chemical mode of synaptic transmission confers on the synapse a set of properties different from those of the neurons composing it.

A synapse exhibits one-way conduction. The transmitter is only in the presynaptic neuron and so it is the only one capable of releasing the agent.

Transmitters may facilitate (enhance) or inhibit the passage of the impulse. γ-Aminobutyric acid (GABA) and glycine block transmission, while most other agents enhance transmission. A degree of control over impulse passage is established at the synaptic level.

Table 8 Synaptic Transmitters

Transmitter	*Method of removal*	*Body area where agent is in greatest concentration*
Acetylcholine	Cleavage by cholinesterase	Periphery, neuromuscular junctions
Norepinephrine	Oxidation by monamine oxidase (MAO)	Periphery
Serotonin (5-hydroxytryptamine)	Tryptophan hydroxylase	C N S
Dopamine	Tyrosine hydroxylase	C N S
Histamine	Histidine decarbonylase	C N S
Amino acids	Reuptake or decarboxylation	Brain
Enkephalins ('endogenous pain killers')	Cleavage by enkephalinase	C N S, gut
Angiotensin II	Enzymatic destruction	Blood vessels, adrenal cortex
Somatostatin	Metabolized	Pituitary gland

Synaptic delay occurs. This is a measurable time it takes for an impulse to cross the synapse because of the time required for release, diffusion and binding of the transmitter. It amounts to about 0.5 ms per synapse.

Sensitivity to environmental effects is exhibited at the synapse. Hypnotics and analgesics slow or block transmission. Strychnine blocks the effect of glycine by antagonizing its action on its receptor. Cocaine blocks reuptake of catecholamines, permitting the transmitter to continue stimulation of the postsynaptic neuron (the 'high') and depletion of the transmitter occurs (the 'low').

2.1.9.4 The Reflex Arc

The reflex arc forms a simple type of neural pathway used to automatically and involuntarily control body functions. A reflex arc always has five parts: (i) a receptor to detect change either from inside or outside the body; (ii) an afferent neuron carrying an impulse to the central nervous system; (iii) a synapse or junction within the central nervous system; (iv) an efferent neuron carrying the impulse away from the central nervous system; and (v) an effector (muscle, gland) that does something to maintain or alter body function.

The response to the arc's operation is termed the reflex act or the reflex. Reflexes have certain characteristics: they are involuntary, requiring no conscious involvement by the body; they are stereotyped, so that a given arc always gives the same response; they are predictable, a fact which depends on the stereotypical nature of the reflex and is used by medical personnel to test function of various parts of the central nervous system; and they are purposeful, serving to restore or maintain normal function.

2.1.9.5 The Brain

The portion of the central nervous system housed within the skull is the brain. Two cerebral hemispheres fill most of the skull, the cerebellum lies in the lower back portion of the skull, and the brainstem (medulla, pons, midbrain) and diencephalon lie centrally placed in the brain.

2.1.9.5.1 Cerebrum

Each cerebral hemisphere consists of an anterior frontal lobe, a central parietal lobe, a posterior occipital lobe, and a lateral temporal lobe. Covering the convoluted surface of each hemisphere is a thin (1.5–4 mm) layer of gray matter (mostly neuronal cell bodies) that constitutes the cerebral cortex. The cortex overlies a large mass of white matter (myelinated nerve fibers) that is called the medullary body. Deep within the medullary body are masses of neuron cell bodies that form the basal ganglia.

There are areas within the cortex of the lobes of the cerebrum that, if stimulated, result in specific responses. In the posterior part of the frontal lobe lies the primary voluntary motor area. Stimulation here causes contraction of skeletal muscles. Nerve fibers rising from cells in this area form the upper motor neurons that travel from cerebrum to spinal cord, there to synapse on neurons coursing from spinal cord to muscle (lower motor neuron). The body is represented upside-down in the area, with lower parts of the body on the upper part of the lobe and *vice versa*. The more complex the movement in a body area the larger the area given over to it in the lobe. The greater parts of the frontal lobe are given over to intellectual functions (creativity, memory, control of emotions, moral and social sense). In the lower portion of the left frontal lobe is a speech area that is closely associated with the head portion of the primary motor area and the auditory area of the temporal lobe. This area (speech) is involved in articulation of speech and in learning to speak (by hearing the spoken word and imitating it). In the anterior portion of the parietal lobes are the somesthetic or general sensory areas. These regions represent the ending of pathways carrying pain, heat, cold, touch and pressure (the 'general senses') from the periphery. As in the motor area, the body is represented upside-down with larger area given to body regions having more receptors for the sense.

Behind the somesthetic area (still in the parietal lobes) are association areas that permit appreciation of textures, shape, size, degrees of heat or cold or pain, and other facets of the general senses. The occipital lobe contains visual areas that mark the termination of pathways originating in the retinas. In front of these are areas that give color intensity, depth preception, hue discrimination and other aspects of vision. The temporal lobes contain the auditory areas where fibers from the cochlea (organ of hearing) terminate. The lobe also is an important memory storage area.

The basal ganglia consist of a number of nuclei that are concerned with modulating motor activities to prevent tremors, coordinate gross movements, and monitor cerebral motor activity. Parkinson's disease, with its tremors and disturbances of locomotion is associated with basal ganglia damage.

The medullary body contains three types of fibers. (i) Association fibers connect different parts of one cerebral hemisphere so that visual or auditory or general sensation can be correlated with motor or memory storage functions. (ii) Commissural fibers cross from one hemisphere to the other. A large band of commissural fibers called the corpus callosum contains an estimated 300×10^6 fibers. In epilepsy, the callosum is sometimes cut to prevent spread of the seizure from one hemisphere to the other. (iii) Projection fibers enter and/or leave the cerebrum from or to the periphery. The major motor, visual and auditory pathways are composed of such fibers.

2.1.9.5.2 Diencephalon

The diencephalon lies superior to the brainstem and in the midline between the cerebral hemispheres. It is composed of an epithalamus, a thalamus and a hypothalamus on each side.

The epithalamus is the most superior part and contains the pineal gland. The gland secretes melatonin, a hormone that retards testis and ovary development of sex cells. It is of little significance in humans.

The thalamus is the largest (20 g) portion of the diencephalon. The most important function of the thalamus is to act as a relay station for the general and special (sight, hearing, taste, smell) sensations. Synapses in these pathways occur in the thalamus which then sends the impulses to the appropriate cerebral region for interpretation and action.

The hypothalamus weighs about 4 g and lies beneath the thalamus. It is one of, if not the most important, brain regions for controlling activities necessary for organism survival.

The hypothalamus controls the pituitary gland, the so-called 'master gland' of the endocrine system. The anterior lobe of the gland is controlled by hormones called regulating factors that are produced in the hypothalamus and sent to the gland through a system of blood vessels. The posterior lobe of the gland stores hormones made in the hypothalamus and sent to the posterior lobe over nerve fibers. Thus a neuroendocrine link is established by which nervous activity can be translated into endocrine activity. Body temperature is controlled by 'heat gain' and 'heat loss' centers in the hypothalamus. Both centers send nerve fibers to skeletal muscle, sweat glands and cutaneous (skin) blood vessels. Elevated external temperature requires activation of the heat loss center that stimulates sweating (increased cooling), skin blood vessel dilation (increased radiation of heat) and decreased tone (degree of involuntary muscle contraction) that decreases heat production. The opposite series of responses occurs with a low external temperature and activation of the heat gain center.

Water balance is assured by hypothalamic neurons that constantly monitor blood osmotic pressure and adjust ADH (antidiuretic hormone) production. Increase in osmotic pressure results in increase of ADH secretion that signals the kidney to reabsorb increased amounts of water to dilute the blood. The opposite effect occurs if the blood has been diluted by excessive fluid intake. Urine volume increases.

Feeding and satiety refer to food intake, and cessation of feeding. Two hypothalamic centers control these activities. The blood sugar (glucose) levels may control these centers. Fall of blood glucose may trigger feeding; as blood glucose rises, feeding is stopped (satiation or 'fullness').

2.1.9.5.3 Cerebellum

The cerebellum operates to involuntarily coordinate motor activity.

Any movement involves cerebral signals to skeletal muscles that constitute the 'intent' of the movement. As the movement occurs, signals originate within the muscles, joints and skin that convey information about rate, force and direction of the movement. This may be called 'performance'. The task of the cerebellum is to compare intent and performance, and control and coordinate the movement. Four interrelated aspects of this control are described.

Error control. Constant adjustment of performance *versus* intent.

Prediction. Using visual information, determining when to slow and stop a motion.

Damping. Preventing the tendency for oscillations to occur at the end of movements.

Progression. Ensuring that muscles contract in proper order to achieve appropriate movement.

Damage in the organ results in incoordination, the severity depending on the extent of trauma and not on location within the organ.

2.1.9.5.4 Midbrain

The midbrain is the most superior part of the brainstem. Through it, and through the brainstem generally, run the motor and sensory pathways on their way from the cerebrum and to their connections in the thalamus and cerebellum. On the posterior aspect of the midbrain are two pairs of elevations called the superior and inferior colliculi. The superior pair provide a relay from visual input to the motor systems, while the inferior pair connect auditory input to the motor systems. Also in the midbrain are nuclei giving rise to several of the cranial nerves, specifically those for nerves 3 (oculomotor) and 4 (trochlear). These nuclei control visual reflexes (3) such as pupillary responses to light and five of the six extrinsic eye muscles on each eyeball (both 3 and 4).

2.1.9.5.5 Pons

The pons also transmits sensory and motor pathways and is distinguished by a large anterior bulge containing nerve fibers from cerebrum to cerebellum that convey ‘intent’ of the cerebrum for cerebellar activity and control of movement. The pons also contains a pneumotaxic center, part of the control mechanism for expiration, and the nerve cell bodies for cranial nerve 5–8 (trigeminal, abducent, facial, vestibulocochlear).

2.1.9.5.6 Medulla

Sensory and motor pathways pass through the medulla, and the area contains the nerve cell bodies for cranial nerves 9–12 (glossopharyngeal, vagus, accessory, hypoglossal). The medulla also contains ‘centers’, groups of neuronal cell bodies concerned with control of several essential body functions. These neurons constitute the so-called ‘vital centers’. These centers include: vasomotor centers that govern the diameter of muscular blood vessels and therefore the blood pressure; cardiac centers that control heart rate and cardiac output; and respiratory centers (inspiratory, apneustic, expiratory, vagal) that determine the rate and depth of breathing.

There are also ‘nonvital centers’ that are involved in controlling salivation, sneezing, coughing and vomiting located in the medulla.

2.1.9.6 Spinal Cord

The spinal cord attaches to the medulla and extends some 45 cm (18 inches) through the vertebral column to terminate in the middle lower back.

A cross-section of the spinal cord shows an inner area called gray matter that contains nerve cell bodies, and an outer region, called white matter, that contains myelin-covered nerve fibers of the spinal tracts conveying motor and sensory impulses down and up the cord.

The cord gives rise to 31 pairs of spinal nerves that supply muscles, skin and visceral organs.

2.1.9.6.1 Functions of the spinal cord

The white matter contains functional areas called tracts that convey specific types of sensory and motor impulses. These tracts transmit impulses to and from the brain as the first function of the cord. The major tracts of the cord may be listed as follows.

Gracile and cuneate tracts lie in the posterior white matter and carry the sensations of touch and pressure to the thalamus.

Spinocerebellar tracts lie in the lateral white matter and carry sensory impulses concerned with rate, force and direction of movements to the cerebellum.

Spinothalamic tracts, in lateral and anterior white matter, convey pain, heat and cold sensations to the thalamus.

Corticospinal tracts, located in the lateral and anterior white matter, carry the motor impulses for voluntary muscle activity.

Involuntary motor tracts, mainly in lateral and anterior white matter, control muscle tone and posture that maintains body equilibrium and balance.

The other function of the spinal cord deals with control of several types of reflex activity concerned with skeletal muscles and viscera.

Myotatic reflexes are triggered when skeletal muscles are stretched. The stretched muscles contract so as to maintain, mainly, body posture against the force of gravity.

Flexion reflexes originate in the skin in response to potentially harmful stimuli (pain, heat, cold). Skeletal muscles contract to withdraw the stimulated part from the stimulus.

Crossed-extension reflexes combine a flexion reflex with an opposite side muscular contraction that extends the body part to maintain posture, as in stepping on a painful object.

Visceral reflexes including urination and defecation that have a voluntary inhibitory component but which are entirely spinal.

2.1.9.7 The Meninges

The central nervous system is surrounded by three membranes collectively called the meninges (singular, meninx). The pia mater (the vascular meninx) is a delicate membrane lying directly on the brain and spinal cord that contains many blood vessels which nourish the organs themselves. A space filled with fluid (cerebrospinal fluid) called the subarachnoid space separates the pia mater from the arachnoid. The dura mater, a thick tough protective membrane lies against the arachnoid, separated from the latter by a potential subdural space. The cerebrospinal fluid 'floats' the brain and cord and acts as a shock-absorber when movements occur that could damage these organs.

2.1.9.8 Ventricles and Cerebrospinal Fluid

Within the brain are fluid-filled cavities and connecting channels that form the ventricular system. In each cerebral hemisphere is a lateral ventricle with horns extending into the frontal, occipital and temporal lobes. Interventricular foramina (holes) connect the lateral ventricles with a single slit-like third ventricle between the diencephalon halves. The cerebral aqueduct extends through the midbrain to connect to a fourth ventricle located in the pons and medulla. From the fourth ventricle the fluid exits to the subarachnoid space from which it is absorbed by arachnoid villi into the venous circulation of the brain. The CSF is produced by choroid plexuses, vascular structures, in the lateral, third and fourth ventricles. A circulation of cerebrospinal fluid (CSF) is established that, if disturbed, can lead to accumulation of fluid within the ventricles or in the subarachnoid space (hydrocephalus).

2.1.9.9 The Peripheral Nervous System

All nervous tissue outside the brain and spinal cord constitutes the peripheral nervous system. The spinal nerves, cranial nerves and their associated ganglia (areas of nerve cell bodies and synapses) form the peripheral nervous system.

The 31 pairs of spinal nerves are divided into eight pairs of cervical nerves, 12 pairs of thoracic nerves, five pairs of lumbar nerves, five pairs of sacral nerves and one coccygeal pair from the upper to lower regions of the spinal column. These nerves carry somatic efferents to skeletal muscles and skin glands, somatic afferents from skin and muscles to the CNS, and visceral efferents and afferents to and from body viscera.

The latter two types of fibers belong to the autonomic nervous system (ANS) that controls visceral activity.

The 12 pairs of cranial nerves rise from the brain. Their names and functions are presented in Table 9.

2.1.9.10 The Autonomic Nervous System

This portion of the nervous system operates reflexively to control blood vessel size, visceral organ operation and visceral gland secretion. One neuron is present on the sensory side of the system, two on the motor side. There are two divisions, the parasympathetic and sympathetic, to the system.

Table 9 The Cranial Nerves

Number and name	*Impulses*[a]	*General origin*	*General termination*	*Function*
1 Olfactory	S	Upper nasal cavities	Frontal lobes	Sense of smell
2 Optic	S	Retinas	Occipital lobes	Sense of sight
3 Oculomotor	M, S	M—midbrain	Eye muscles	Eye movement
		S—ocular muscles	Midbrain	Muscle sense
4 Trochlear	M, S	M—midbrain	Eye muscles	Eye movement
		S—ocular muscles	Midbrain	Muscle sense
5 Trigeminal	M, S	M—pons	Chewing muscles	Chewing control
		S—scalp, face	Pons, to parietal lobes	Sensation from origin
6 Abducent	M, S	M—pons	Eye muscle	Eye movement
		S—ocular muscle	Pons	Muscle sense
7 Facial	M, S	M—pons	Facial muscles	Facial expression
		S—ant 2/3$_s$ tongue	Pons	Taste
8 Vestibulocochlear	S	Inner ear	Pons	Hearing and equilibrium
9 Glossopharyngeal	M, S	M—Medulla	Throat muscles	Swallowing
		S—Post 1/3 tongue	Pons	Taste
10 Vagus	M, S	M—Medulla	Viscera	Visceral secretion, movement
		S—Viscera	Medulla	Visceral sensation
11 Accessory	M	Medulla	Throat, neck muscles	Swallowing, head movement
12 Hypoglossal	M	Medulla	Tongue, neck muscles	Speech, swallowing

[a] M = motor, S = sensory.

The parasympathetic division is composed of cranial nerves 3, 7, 9 and 10, and the sacral spinal nerves. Its effects conserve body resources, that is tend to hold activity to normal or resting levels.

The sympathetic division is composed of all thoracic and lumbar spinal nerves, and elevates body activity to resist stressful situations (the 'fight-or-flight' reaction).

Most body organs receive fibers from both divisions establishing a dual innervation that permits rapid and precise response. The effects of the two divisions on any given organ are typically opposite and are mediated by the production of acetylcholine, by parasympathetic fibers, and norepinephrine, by sympathetic fibers, at the organ itself.

Drugs that have the same effect as parasympathetic nerves are called parasympathomimetic; those that have the same effect as sympathetic fibers are called sympathomimetic.

2.1.9.11 Endorphins and Enkephalins

The brain produces substances that were first recognized as pain-killing agents. Endorphins are large molecules that contain several copies of smaller enkephalins. The latter are released from the endorphin by enzymatic action. The two most widely studied enkephalins are pentapeptides (five amino acids) called metenkephalin and leuenkephalin. In addition to reducing sensitivity to pain, these substances may act as synaptic transmission inhibitors. The enkephalins must bind to receptor sites on other cells to exert their effect, and receptor sites have been discovered in the hypothalamus, midbrain, pons, medulla, spinal cord, gall bladder, pancreas and alimentary tract. Most enkephalins reduce pain without involving addiction, and may become substitutes for such addictive pain-killers as morphine.

2.1.9.12 The Electroencephalogram

By placing electrodes on the scalp, it is possible to record continuous electrical activity from the brain. The record is termed the electroencephalogram (EEG), and is evidence of a living active brain. The nature of the 'waves' changes according to levels of consciousness, presence of seizures or damage to the brain. There are four basic wave patterns.

α waves have a frequency of 10–12 Hz and voltages of about 50 μV and are present when the subject is relaxed and awake but usually with the eyes closed. They are evidence of a brain that is not actively processing information.

β waves have a frequency of 13–25 Hz and voltage of 5–10 μV. They replace α waves when the eyes are opened or when there is mental activity.

δ waves have a frequency of 1–5 Hz and voltages of 20–200 μV. They are characteristic of sleep.
θ waves have a frequency of 5–8 Hz and have voltages of about 10 μV. They occur in newborns and in adults during stress or when there is damage in the brain.
Alterations in the normal EEG pattern can be used as evidence of abnormal activity from drug use and a 'flat' EEG (no activity) is a modern definition of death.

2.1.9.13 Wakefulness and Sleep

These states represent opposite ends of a conscious–unconscious scale. The waking state appears to depend on heightened activity in the reticular formation, several columns of neurons in the brainstem. Sensory input to the formation is relayed to the cerebrum generally to stimulate or arouse it. The reticular formation–cerebrum pathway is called the reticular activating system (RAS). The waking state permits thought, awareness of surroundings, perception and memory storage.
Sleep represents a temporary interruption of the waking state that seems to provide an essential need for the organism. Sleep may result from diminished sensory input to the RAS, or by production of a 'sleep chemical' such as serotonin, dihydroxy phenylalanine (DOPA) or γ-aminobutyric acid (GABA) all known as synaptic transmitters. Sleep centers have been proposed to exist in the pons, thalamus and medulla oblongata.
Sleep proceeds in several stages: stage 1 is characterized by a slowing and decrease in α EEG rhythm, accompanied by sensations of drowsiness and drifting; stage 2 sees the appearance of 'sleep spindles' in the EEG, bursts of 14–15 Hz waves lasting several seconds; stage 3 is characterized by the appearance of δ waves in the EEG, with breathing and pulse being slow; and stage 4 is deep sleep from which the sleeper is aroused only with difficulty.
Rapid eye movement (REM) sleep occurs every 90–120 minutes during stage 4 sleep. Eye movements, muscular activity and vocalization occur and this type of sleep is associated with dreaming. REM sleep may 'exercise' the brain or provide a release of tension.
Coma is a state of unconsciousness from which even the strongest stimuli cannot arouse the subject. It usually reflects brain damage from drugs, vascular accidents (strokes) or trauma, and seems to require involvement of the brainstem reticular formation.

2.1.9.14 The Limbic System

The limbic system is involved in expression of emotions and includes nuclei in the temporal lobes (amygdaloid nuclei), pathways to and from the hypothalamus (fornix and supracallosal striae) and the olfactory portions of the brain. Aggressive and passive behavior, and vascular and visceral changes accompanying emotions are governed by this system.

2.1.10 THE ENDOCRINE GLANDS

The ultimate chemical control of body activity resides in substances produced by glands of the endocrine system. All members of this system share certain characteristics: they secrete directly into the bloodstream; they are very vascular, deriving their building blocks from the blood and placing their products into the blood; and their products, called hormones, have well-defined effects on the body and excess or deficiency of a hormone produces characteristic symptom development.
The organs included as endocrines are the pituitary, hypothalamus, thyroid gland, parathyroid glands, pancreatic islets, adrenal glands, testes, ovaries, pineal gland and certain cells in the alimentary tract.

2.1.10.1 Pituitary Gland (Hypophysis)

The pituitary gland lies beneath and is attached to the hypothalamus. It has two major portions, the anterior lobe (pars distalis) and the posterior lobe (pars nervosa). The anterior lobe maintains a vascular connection with the hypothalamus called the pituitary portal system. The posterior lobe maintains hypothalamic connections by way of nerve fibers called the hypothalamic–pituitary tract.

2.1.10.1.1 Anterior lobe

With special staining techniques, six types of cells may be distinguished in the anterior lobe and six major hormones are produced here, each assigned to a specific cell.

Somatotroph cells produce somatotropin, also known as somatotropic hormone (STH), growth hormone (GH) or human growth hormone (HGH). A polypeptide containing 191 amino acids with a molecular weight of 22 000, this species-specific hormone controls growth of hard and soft body tissues. It also increases conversion of carbohydrates to amino acids, increases cellular uptake of amino acids, mobilizes fats from storage areas and increases blood glucose levels. The fat-mobilizing effect is separated by some endocrinologists and attributed to a separate hormone called β-lipotropin, with a molecular weight of 11 000. (It may be a portion or segment of somatotropin.) Growth hormone has been synthesized by genetic-engineering techniques and is used to increase growth in patients who suffer retarded growth from hormone deficiency.

Prolactin, also known as lactogenic hormone or luteotropic hormone (LTH) is produced by lactotroph cells and is a polypeptide of 205 amino acids and a molecular weight of 22 000. It causes milk release from a developed mammary gland as its major effect.

Adrenocorticotropic hormone (ACTH, corticotropin) is produced by corticotroph cells, and is a straight chain polypeptide containing 39 amino acids. Its molecular weight is 4567. ACTH controls all aspects of the activity of the two inner zones of the adrenal cortex, and mobilizes fats, produces hypoglycemia (low blood glucose levels) and increases muscle glycogen stores.

Thyroid-stimulating hormone (TSH, thyrotropin) is a glycoprotein with a molecular weight of about 28 000. Thyrotroph cells produce it, and it controls thyroid gland activity.

Luteinizing hormone (LH, luteotropin) is sometimes called interstitial-cell-stimulating hormone (ICSH) in the male. It is produced by luteotroph cells and is a glycoprotein with a molecular weight of about 28 000. It is essential for ovulation, corpus luteum formation, zygote implantation in the uterus and controls interstitial cell activity in the testes.

Follicle-stimulating hormone (FSH) is produced by folliculotroph cells and is a glycoprotein with a molecular weight about 33 000. In both sexes, FSH controls sex cell maturation.

Control of anterior lobe activity involves two mechanisms. The hypothalamus produces small molecules of 3 to 14 amino acids that qualify as hormones. These are placed in the pituitary portal system to control anterior lobe cells. These factors may stimulate or inhibit pituitary activity, and eight chemicals have been isolated. Since the hypothalamus is influenced by other parts of the central nervous system, this can be translated into hormone production or decrease. Once produced, anterior lobe hormones that control other endocrines have their levels controlled by a negative feedback mechanism from the target endocrine. This maintains both the pituitary and target organ hormone within narrow limits.

2.1.10.1.2 Posterior lobe

The posterior lobe produces no hormones, but hormones may be isolated from it. Neurons of the hypothalamus produce two hormones that are sent along axons of the hypothalamic-posterior lobe tract to be stored in and released from the posterior lobe. Vasopressin-antidiuretic hormone (V-ADH or ADH) is a nonapeptide (nine amino acids) that causes vascular smooth muscle to contract, and which also increases kidney water reabsorption (antidiuretic effect). Oxytocin, also a nonapeptide, stimulates pregnant uterine smooth muscle contraction and aids emptying of milk from a lactating mammary gland.

Control of ADH production is by the hypothalamic neurons monitoring blood osmotic pressure and adjusting production to keep osmotic (and fluid) levels nearly constant. Suckling of an infant and uterine cervix dilation at birth increase oxytocin production.

2.1.10.2 Thyroid Gland

The thyroid gland is a two-lobed organ located on the anterior aspect of the lower larynx and upper trachea in the neck. It weighs about 20 g and is composed of many small (0.3–2 mm diameter) hollow thyroid follicles. The follicles are lined with two types of cells that produce two different hormones.

Follicular cells produce thyroxin (T_4) a tetraiodinated amino acid with a molecular weight of 777. Cellular removal of one iodine atom produces triiodothyronine (T_3), a more potent hormone that appears to be the form that cells actually use. All body cells are affected by thyroxin. It accelerates degradative chemical reactions increasing heat production. It promotes growth, especially of the brain, increases protein catabolism and interferes with ATP production in excess. Control of thyroxin production is by pituitary TSH, with thyroxin exerting a negative feedback (inhibitory) action on TSH production.

Parafollicular cells produce calcitonin, a hormone that increases deposition of calcium and phosphate in bones. Control of calcitonin production is by blood calcium level. A rise of blood calcium increases calcitonin secretion.

2.1.10.3 Parathyroid Glands

Four parathyroid glands lie on the posterior aspect of the thyroid gland. Each measures about 5 mm in diameter and collective weight is about 120 mg. Principle (chief) cells produce parathyroid hormone (PTH), and oxyphil cells are reserve cells that produce no hormone. PTH is a polypeptide containing 74–80 amino acids. PTH increases gut absorption of calcium and phosphate, increases kidney reabsorption of calcium while decreasing that of phosphate, and demineralizes bone. Blood calcium level controls PTH production, a fall of blood calcium level increasing hormone release.

2.1.10.4 Pancreas

The pancreas contains many small islets of endocrine tissue among its digestive-enzyme-producing cells. The islets contain two major types of cells that produce two hormones.

α cells produce glucagon, a polypeptide hormone with a molecular weight of 3485. Glucagon promotes conversion of glycogen to glucose, raising blood sugar levels. A fall of blood sugar causes an increased release of glucagon.

β cells produce insulin, a polypeptide hormone with a molecular weight of about 6000. Insulin increases uptake of glucose by most body cells (neurons excepted) and increases storage of glucose as glycogen in liver and muscle. A rise of blood sugar level increases insulin release. Deficiency of insulin or refractory cells (which do not respond to insulin) results in diabetes mellitus, characterized primarily by high blood glucose levels, sugar in the urine, thirst and great hunger. If insulin deficiency is severe, insulin itself may be required to control the condition. Dietary control of carbohydrate intake and oral medications may suffice to control mild diabetes.

2.1.10.5 Adrenal Glands

Paired, hat-shaped adrenal glands lie above each kidney. An internally placed medulla secretes epinephrine (adrenalin), a sympathomimetic hormone. Controlled by nerves of the sympathetic nervous system, the medulla is not essential for life. Epinephrine reinforces sympathetic activity in 'fight-or-flight' response to acute stress. An outer cortex secretes steroid hormones and is divided into three zones or layers.

The external cortical layer is called the zona glomerulosa and secretes several mineralocorticoids including aldosterone. Aldosterone increases kidney reabsorption of sodium ion, bicarbonate and chloride ions and water follow sodium, and the net effect is to increase blood volume and blood pressure. Aldosterone secretion is controlled by angiotensin.

A middle zona fasciculata produces glucocorticoids including cortisol. Cortisol increases glucose formation from amino acids and fats, stimulates amino acid degradation, maintains muscle ATP levels, and exerts antiinflammatory effects.

An inner zona reticularis produces sex hormones, chiefly androgens (male type), whose physiological significance is not known.

The inner two zones of the adrenal cortex are controlled by pituitary ACTH, with cortisol exerting a negative feedback effect on ACTH production.

2.1.10.6 Ovaries

The ovaries have the dual role of ovum production and hormone production. As ova mature after puberty, they eventually form a vesicular (Graafian) follicle that produces steroid estrogens, including estradiol as the major one. Estradiol develops the female secondary sex characteristics (breast development, menstrual cycles, fat and hair distribution) and causes uterine growth. After ovulation, the tissues remaining in the ovary become a corpus luteum that produces progesterone (progestin). This hormone develops further the mammary glands (especially in a pregnant female), prepares the uterine lining for pregnancy and suppresses further ovum development *via* the hypothalamus.

Pituitary FSH and LH control the development of the follicles and their hormone production, with both estradiol and progesterone inhibiting FSH and LH secretion.

If pregnancy occurs and a placenta is formed, that organ produces estradiol, progesterone and human chorionic gonadotropin (HCG), the latter an LH-like hormone. HCG is the hormone tested for by early pregnancy tests to confirm or deny pregnancy as having occurred.

2.1.10.7 Testes

The testes, like the ovaries, produce sex cells (sperm) and a hormone. The steroid hormone testosterone is produced by interstitial cells lying between the sperm-producing tubules of the testes. Testosterone develops the male secondary sex characteristics (muscular development, hair pattern, voice changes) at puberty and causes growth of prostate gland, seminal vesicles and penis. Testosterone secretion is governed by pituitary LH (ICSH) with testosterone exerting a negative feedback effect on LH. 'Steroids' as used by sports figures are testosterone-like substances designed to increase muscle mass and strength.

2.1.10.8 Pineal Gland

The pineal gland located on the posterior aspect of the diencephalon produces melatonin, a hormone that exerts antigonadotropic effects. The hormone is of little importance in humans.

2.1.10.9 Prostaglandins

Prostaglandins (PGs) are cyclic unsaturated oxygenated derivatives of a fatty acid called prostanoic acid. They are subdivided according to location of the unsaturated bonds and structural characteristics into F, E, A and B types. Prostaglandins are widespread in the animal kingdom and are produced by many different human cells. The basic function of PGs appears to be to increase cellular levels of cyclic AMP that in turn stimulates a wide variety of cellular chemical reactions. PGs also induce destruction of the corpus luteum (abortifacient effect), inhibit gastric secretion, lower blood pressure and stimulate uterine muscle contraction (menstrual 'cramps').

2.1.10.10 Gastrointestinal Hormones

A variety of substances are produced by amine precursor uptake decarboxylase (APUD) cells in the stomach and intestines that control secretion or release of the stomach, pancreas, liver and intestines of their digestive fluids. Table 10 presents the currently known hormones and their effects.

2.1.11 BODY FLUIDS AND ACID–BASE BALANCE

Water is the single most abundant body constituent. On the average, water comprises 55% of the body weight and has a typical volume of 46 L. Water forms the medium for the body's chemical reactions, transports the heat of metabolic reactions to areas (skin) of elimination and dissolves or suspends many body chemicals.

Total body water is subdivided by membranes into two major compartments, one of which has several subdivisions.

(i) Extracellular fluid (ECF) is water outside body cells. It forms about 40% of total body water and has a volume of 17 L. It is subdivided into the following. (a) Plasma is the liquid portion of the blood, is contained within the cardiovascular system, averages 4% of the body weight and has a

Table 10 Gastrointestinal Hormones

Hormone	*Effect(s)*
Gastrin	Stimulates gastric secretion
Secretin	Stimulates pancreatic fluid secretion; stimulates liver bile production
Cholecystokinin (CCK)	Stimulates pancreatic enzyme release; causes gall bladder contraction
Glucagon	Raises blood glucose level
Gastric inhibitory polypeptide (GIP)	Inhibits gastric secretion
Vasoactive intestinal polypeptide (VIP)	General vasodilation
Bombesin	Stimulates gastric and pancreatic secretion and gall bladder contraction
Somatostatin	Inhibits pancreatic insulin secretion
Motilin	Increases stomach motility
Chymodenin	Stimulates pancreatic chymotrypsin release
Bulbogastrone	Inhibits gastric secretion
Entero-oxyntin	Stimulates gastric secretion
Pancreatic polypeptide	Stimulates gastric secretion

volume of 3 L. (b) Interstitial fluid (ISF) is water outside of blood vessels and outside of cells, it is the medium in which cells live. Forming about 28% of the body weight it has a volume of 14 L. (c) Transcellular fluids are the fluids in hollow body organs, eyes, cerebrospinal and joint fluids; this compartment varies greatly in volume. They average 1–3% of the body weight.

(ii) Intracellular fluid (ICF) is water within body cells. It constitutes 60% of the total body fluids and has a volume of 29 L.

2.1.11.1 Composition of the Compartments

Plasma is the only fluid compartment that undergoes a constant circulation. Its diffusible or filterable constituents are found in the same concentrations in ISF, while large protein and lipid molecules and cells tend to remain in the circulation, so that ISF is deficient in such constituents. ICF composition varies greatly depending on cell type. Table 11 compares plasma, ISF and ICF for important constituents.

It may be seen that ECF is rich in sodium, chloride and bicarbonate, while ICF is rich in potassium, magnesium, protein and organic phosphate. Electrolyte imbalances are usually due to active transport systems that keep these ions separated across cell membranes.

2.1.11.2 Exchanges Between Compartments

Plasma is the source of substances for all other compartments. Under the pressure created by heart action, substances small enough to pass through capillary membranes are filtered into the ISF. Large molecules and blood cells are retained in the bloodstream where they exert an osmotic force that tends to draw water back into the blood vessels by osmosis. Small constituents will be swept along with the fluid flow. Some 90% of the filtered constituents are put back into the bloodstream. The remaining 10% of the fluid goes into lymphatic vessels that return it to the blood vessels.

Cells, because of their high concentrations of solutes, tend to take on water by osmosis from the ISF. To offset this tendency, cells actively pump electrolytes, chiefly sodium, from the cell and water follows osmotically.

2.1.11.3 Sources and Routes of Loss of Fluids

Intake and output of fluids must balance on a day-to-day basis if body composition is to remain essentially constant. Table 12 summarizes routes of acquisition and loss of fluid.

Antidiuretic hormone is the main regulator of fluid loss. Its production and urine loss are adjustable, while little control is afforded over skin, lung and fecal loss. Loss *via* the urine is inversely proportional to loss by any other route. ADH thus controls osmolality of body fluids, while aldosterone mainly regulates fluid volume by controlling kidney sodium and therefore water reabsorption. Stretching of the right atrium causes the release of a natriuretic hormone that

Table 11 Plasma, Interstitial and Intracellular Fluid (Comparison of Constituents)

Constituent	Plasma	Value (meq L^{-1}) ISF	ICF
Na^+	140.0	145.0	7.0–30.0
K^+	4.0	4.1	133.0–166.0
Ca^{2+}	5.0	3.4	0.0–4.0
Mg^{2+}	1.6	1.3	6.0–35.0
Cl^-	120.0	118.0	4.0–6.0
HCO_3^-	25.0	28.0	12.0–18.0
Protein	15.0	0.0–1.0	30.0–55.0
Phosphate	2.2	2.3	4.0–40.0[a]
Other	6.0	5.5	10.0–90.0

[a] Organic phosphate.

Table 12 Routes of Intake and Loss of Body Water

Source	Volume (mL day^{-1})	Route of loss	Volume (mL day^{-1})
Drinking water, beverages	500–1600	Urine	600–1600
Water in food	800–1000	Lungs, skin	850–1200
Water produced by metabolism	200–400	Feces	50–200
Total	1500–3000	Total	1500–3000

increases sodium and therefore water elimination through the kidney. A fine hormonal control over blood volume is afforded.

2.1.11.4 Acid–Base Balance

Constant production of hydrogen ion (H^+) or, more specifically, the hydronium ion (H_3O^+) from the reaction of hydrogen ion and water poses a continual threat to the body's acid–base balance. ECF pH is normally maintained at 7.4 ± 0.04, a slightly alkaline solution. Deviation beyond these limits causes changes in enzyme shape and function with consequent disruption of cellular processes.

An acid is defined as a substance that donates H^+, a base as an H^+ acceptor. A strong acid donates much H^+, a weak acid little. Volatile acids may be excreted as gases *via* the lungs (carbonic acid containing CO_2), while fixed acids must be excreted from the body in a solution. Whatever the pH of a given solution is, it is reflective of a balance between acid and basic substances as on a scale. Change of pH can therefore occur by adding or removing acid or base on one side of the scale without touching the other side.

Sources of H^+ for the body fluids include the following: the reaction of carbon dioxide and water to produce the weak acid carbonic acid, that then dissociates to give hydrogen and bicarbonate ions; metabolic production of fixed acids that dissociate hydrogen ion; and ingestion of acidic substances as in aspirin or ammonium compounds.

The body must have devices to minimize the effects of H^+ production and to eliminate the acid from the body. Among these devices are the following.

Buffering. By providing a substance (conjugate base) that contributes an anion with which H^+ can react to form a weak acid, the H^+ can be 'soaked up' until it can be eliminated. Sodium bicarbonate forms the basic ECF base to react with H^+.

Lung elimination of carbon dioxide. Carbonic acid is enzymatically broken down in the lungs to liberate CO_2 that is breathed out. By preventing CO_2 buildup, H^+ release is lessened.

Kidney secretion of hydrogen ion. The kidney can eliminate H^+ while reabsorbing bicarbonate ion and sodium ion on a one-for-one basis.

Failure of these mechanisms can result in one of four basic types of acid–base disturbance.

Respiratory acidosis results from any condition that results in carbon dioxide retention (failure to rid it *via* the lungs). Airway obstruction, pneumonia, respiratory depression, muscle paralysis are some conditions causing retention. The kidney can compensate for some of the excess CO_2 by increasing H^+ secretion.

Metabolic acidosis results from excess H^+ from any other cause than one involving the lungs.

Respiratory alkalosis results from excessive CO_2 elimination by the lungs as in hysteric state or anything that causes hyperventilation.

Metabolic alkalosis results from any cause not involving the lung, as in vomiting, that causes loss of acid from stomach contents.

The two most important organs are the lungs and kidneys for regulating ECF pH.

2.1.12 REFERENCES

The references are grouped according to various bodily functions as:

Body organization	1–3	Digestive	30–33
Skin	4–8	Urinary	34–37
Musculoskeletal	9–15	Nervous system	38–49
Circulatory	16–22	Endocrines	50–64
Respiratory	23–29	Fluid and acid–base	65–69

1. J. R. McClintic, 'Basic Anatomy and Physiology of the Human Body', 2nd edn., Wiley, New York, 1980.
2. J. R. McClintic, 'Human Anatomy', Mosby, St. Louis, 1983.
3. J. R. McClintic, 'Physiology of the Human Body', 3rd edn., Wiley, New York, 1985.
4. R. L. Edelson and J. M. Fink, *Sci. Am.*, 1985 (June), **252**, 34.
5. T. B. Fitzpatrick and A. J. Sober, *N. Engl. J. Med.*, 1985, **313**, 818.
6. L. A. Goldsmith (ed.), 'Biochemistry and Physiology of the Skin', Oxford Univeristy Press, New York, 1983.
7. P. F. Millington and R. Wilkinson, 'Skin', Cambridge University Press, New York, 1983.
8. P. A. Nicoll and T. A. Cortese, Jr., *Annu. Rev. Physiol.*, 1972, **34**, 177.
9. G. H. Bourne, 'The Structure and Function of Muscles', Academic Press, New York, 1972.
10. G. H. Bourne (ed.), 'The Biochemistry and Physiology of Bone', Academic Press, New York, 1976.
11. D. B. Drachman, *N. Engl. J. Med.*, 1987, **316**, 743.
12. Y. E. Goldman, *Annu. Rev. Physiol.*, 1987, **49**, 637.
13. G. Hoyle, 'Muscles and Their Neural Control; Wiley, New York, 1983.
14. H. Huddart and S. Hunt, 'Visceral Muscle, Its Structure and Function', Halstead (Wiley), New York, 1975.
15. B. M. Twarog, R. J. C. Levine and M. M. Dewey (eds.), 'Basic Biology of Muscles', Raven Press, New York, 1982.
16. R. W. F. Campbell, *N. Engl. J. Med.*, 1987, **316**, 29.
17. M. Cantin and J. Genest, *Sci. Am.*, 1986 (February), **254**, 62.
18. W. J. Cliff, 'Blood Vessels', Cambridge University Press, New York, 1976.
19. R. Gilles (ed.), 'Circulation, Respiration and Metabolism', Springer-Verlag, New York, 1985.
20. A. M. Katz, 'Physiology of the Heart', Raven Press, New York, 1977.
21. C. M. Rodkiewicz (ed.), 'Arteries and Arterial Blood Flow: Biological and Physiological Aspects', Springer-Verlag, New York, 1983.
22. M. Simionescu and N. Simionescu, *Annu. Rev. Physiol.*, 1986, **48**, 279.
23. M. E. Avery, H. W. Taeush and J. Floros, *N. Engl. J. Med.*, 1986, **315**, 825.
24. A. J. Berger, R. A. Mitchell and J. W. Severinghaus, *N. Engl. J. Med.*, 1977, **297**, 92, 138, 194.
25. R. Grenville–Mathers, 'The Respiratory System', 2nd edn., Churchill–Livingston, New York, 1983.
26. M. P. Kalia, *Annu. Rev. Physiol.*, 1987, **49**, 595.
27. L. M. G. van Golde, *Annu. Rev. Physiol.*, 1985, **47**, 765.
28. T. P. Lim, 'Physiology of the Lung', Thomas, Springfield, IL, 1983.
29. M. Newhouse, J. Sanchis and J. Bienenstock, *N. Engl. J. Med.*, 1976, **295**, 990, 1045.
30. R. J. Bolt and P. E. Palmer, 'The Digestive System', Wiley, New York, 1983.
31. S. J. Henning, *Annu. Rev. Physiol.*, 1985, **47**, 231.
32. T. E. Machen and A. M. Paradiso, *Annu. Rev. Physiol.*, 1987, **49**, 19.
33. P. L. Rayford, T. A. Miller and J. C. Thompson, *N. Engl. J. Med.*, 1976, **294**, 1093, 1157.
34. A. Erslev, *N. Engl. J. Med.*, 1987, **316**, 101.
35. R. L. Jamison and R. H. Maffly, *N. Engl. J. Med.*, 1976, **295**, 1059.
36. J. A. Schafer and J. C. Williams, Jr., *Annu. Rev. Physiol.*, 1985, **47**, 103.
37. K. Schmidt–Nielsen, *Sci. Am.*, 1981 (May), **244**, 118.
38. O. Appenzeller, 'The Autonomic Nervous System', Elsevier, New York, 1983.
39. G. Austin, 'The Spinal Cord', 3rd edn., Igaku-Shoin, New York, 1983.
40. J. Axelrod, *Sci. Am.*, 1974 (June), **230**, 58.
41. H. F. Bradford, 'Chemical Neurobiology', Freeman, New York, 1985.
42. N. Dunant and M. Israël, *Sci. Am.*, 1985 (April), **252**, 40.
43. M. L. Karnovsky, *N. Engl. J. Med.*, 1986, **315**, 1026.
44. Readings in Scientific American, 'Progress in Neuroscience', Freeman, New York, 1985.
45. *Sci. Am.*, (Sept. issue entirely on Brain), 1979, **241**.
46. S. H. Snyder, *Sci. Amer.*, 1977 (March), **236**, 44.
47. S. P. Springer and G. Deutsch, 'Left Brain, Right Brain', Freeman, New York, 1985.
48. R. F. Thompson, 'The Brain: An Introduction to Neuroscience', Freeman, New York, 1985.
49. R. J. Wurtzman, *Sci. Am.*, 1982 (April), **246**, 42.

50. S. Brinkley, *Sci. Am.*, 1979 (April), **240**, 50.
51. S. W. Carmichael and H. Winkler, *Sci. Am.*, 1985 (August), **253**, 30.
52. L. Crapo, 'Hormones, The Messengers of Life', Freeman, New York, 1985.
53. M. F. Cuthbert (ed.), 'The Prostaglandins', Lippencott, Philadelphia, 1973.
54. J. H. Dussalt and J. Ruel, *Annu. Rev. Physiol.*, 1987, **49**, 321.
55. D. P. Figlewicz, F. Lacour, A. Sipols, D. Poste, Jr. and S. C. Woods, *Annu. Rev. Physiol.*, 1987, **49**, 383.
56. L. A. Frohman, *N. Engl. J. Med.*, 1972, **286**, 1391.
57. T. A. Howlett and L. H. Rees, *Annu. Rev. Physiol.*, 1986, **48**, 527.
58. H. Lagecrantz and T. A. Slotkin, *Sci. Am.*, 1986 (April), **254**, 92.
59. J. C. Marshall and R. P. Kelch. *N. Engl. J. Med.*, 1986, **315**, 1459.
60. M. M. Rechler, S. P. Nissley and J. Roth, *N. Engl. J. Med.*, 1987, **316**, 941.
61. R. Reiter (ed.), 'Pineal Research Reviews', A. R. Liss, New York, 1983.
62. M. Rosenblatt, *N. Engl. J. Med.*, 1986, **315**, 1004.
63. A. V. Schally, A. Arimura and A. J. Kastin, *Science (Washington, D.C.)*, 1973, **179**, 341.
64. T. R. Schwab, B. S. Edwards, W. C. DeVries, R. S. Zimmerman and J. C. Burnett, Jr., *N. Engl. J. Med.*, 1986, **315**, 1398.
65. J. Cohen and J. D. Kassirer, 'Acid–Base' Little, Brown, Boston, 1982.
66. J. L. Gamble, 'Acid–Base Physiology: A Direct Approach', Johns Hopkins, Baltimore, 1982.
67. O. W. Hand, 'Acid–Base Chemistry', Macmillan, New York, 1986.
68. J. L. Keyes, 'Fluid, Electrolyte, and Acid–Base Regulation', Wadsworth Health Sciences Division, Monterey, California, 1985.
69. J. H. Laragh, *N. Engl. J. Med.*, 1985, **313**, 1330.

2.2
The Architecture of the Cell

PETER D. KENNEWELL
Roussel Laboratories, Swindon, UK

2.2.1 INTRODUCTION

The cell is the fundamental building block of all independently viable forms of life. It is the essential constituent of the tissues, organs and body structures described in Chapter 2.1. All drugs act on or in cells and have to pass through membranes formed from cells on their path from the point of entry to the body to their point of action. An understanding of cellular architecture and biochemistry is therefore essential if a medicinal chemist is to understand how drugs act.

This chapter is written from the point of view of a chemist and is intended to present an overview of the subject. More detailed discussions of some elements are given in other chapters of this work. Thus, for instance, DNA replication, transcription and translation are dealt with in Chapter 3.7, cell membranes in Chapter 11.1 and cell walls in Chapter 9.1.

The structure of cells, their major constituents and their functioning will be described in this chapter. Most of the discussion will be concentrated on mammalian cells, but bacterial and yeast cells will also be described. This is partly in order to illustrate particular features but also to show differences between the cells. Drugs are only useful if they act selectively against an invading organism or against a receptor or enzyme whose malfunction causes a disease and have only minimal effects on the unaffected organs of the host. The theme of selectivity will be picked up again in Chapters 2.3, 2.4 and 2.6, but in a very real sense selectivity is the theme running through the whole of this treatise. Clearly an understanding of the different structures and functioning of parasitic cells compared with those of host cells is essential for the design of specific drugs.

2.2.2 GENERAL CONSIDERATIONS

Independently viable forms of life vary in size and complexity from unicellular bacteria with a diameter of 1 μm to Giant Sequoia trees which weigh over 1000 tons. However, all systems are classified according to how their genetic material is packaged within the cell. The simplest case is that of prokaryotes, which comprise only bacteria and blue-green algae, in which the genetic material is not contained within a membrane-bound organelle. All other systems are eukaryotes and their genetic information is found within one or more membrane-bound nuclei. Thus of the parasites which infect man bacteria are prokaryotic, whilst protozoa, fungi and yeasts are eukaryotes, as is also, of course, man. A discussion of the other infective agents, viruses, viroids and prions, which are less obviously 'living', is deferred to the end of this chapter.

A vast number of unicellular organisms thrive, and indeed thrive under conditions as varied as the Arctic permafrost and hot springs and oil wells. Thus in the appropriate environment an organism which can perform all the essential functions, ingestion of food, elimination of waste materials and reproduction, can prosper. However, these organisms are at the mercy of the environment, particularly its pH, temperature and ionic strength, over which they can have little control.

Grouping together individual cells to produce a multicellular organism has a number of advantages. It is no longer necessary for all cells to be able to perform all functions, thus leading to specialization and increasing efficiency. The internal environment can now be more easily controlled and maintained separately from the external environment. Specialization of function leads to increases in the variety of shapes and sizes of cells found in such organisms and this is certainly the case for those of the human body. Figure 1 shows representations of nerve, skeletal muscle, visceral muscle and ciliated columnar cells. The cells of the gut, kidney and liver all have deeply invaginated surfaces which greatly increase their surface areas. As one of their major functions is to absorb materials from the surrounding blood or extracellular fluids, this extra surface area will greatly aid this process. Muscle cells are constantly expanding and contracting and are therefore long, relatively thin and packed with fibrils to help their operation. Nerve cells control the activity of the body by the transmission of electric charges. Thus they consist of a central body and long axons leading to the muscle being controlled. The axons are covered by myelin, which is highly impervious to ions

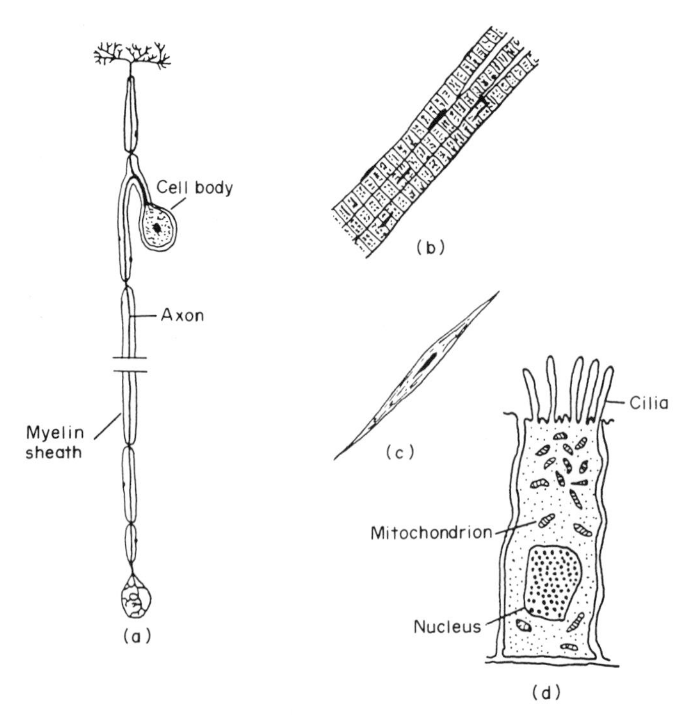

Figure 1 Representations of some mammalian cells

Table 1 Scale of Sizes in the Biological World

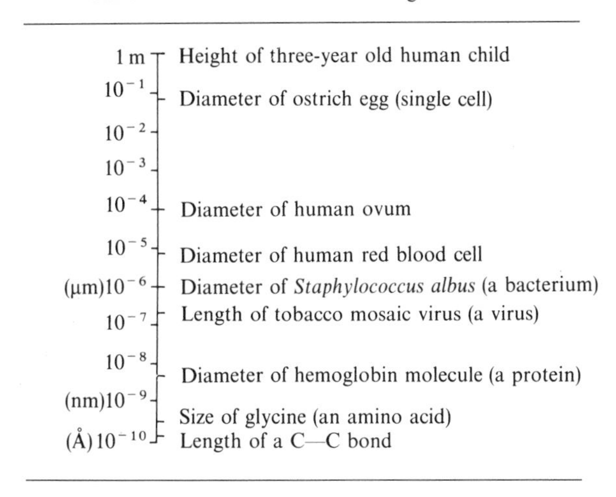

1 m	Height of three-year old human child
10^{-1}	Diameter of ostrich egg (single cell)
10^{-2}	
10^{-3}	
10^{-4}	Diameter of human ovum
10^{-5}	Diameter of human red blood cell
(µm)10^{-6}	Diameter of *Staphylococcus albus* (a bacterium)
10^{-7}	Length of tobacco mosaic virus (a virus)
10^{-8}	Diameter of hemoglobin molecule (a protein)
(nm)10^{-9}	Size of glycine (an amino acid)
(Å)10^{-10}	Length of a C—C bond

and other polar materials and which thus facilitates rapid electrical transmission. The arteries and veins of the body are made up of tough smooth muscle cells which facilitate blood circulation, while the walls of the capillaries are only a single cell thick in order to permit rapid transfer of gases, nutrients and drugs between the blood and the tissues.

2.2.3 SIZE

The lowest viable size for a cell is governed by the volume required to hold sufficient enzymes and structural elements necessary to maintain life. One of the smallest bacteria, *Dialester pneumosintes*, has the dimensions 0.5 × 0.5 × 1.0 µm and has been calculated to contain about 800 different proteins having an average molecular weight of 40 000 Da. Thus the maximum number of reactions available to the organism would be 800, although of course it would actually be less than this because some of the proteins must be structural rather than enzymatic. From other calculations, this would appear to be around the minimum necessary to maintain viability.

At the upper limit, size appears to be controlled by the organizational and cooperational difficulties involved when the structure becomes too large and by the fact that the ratio of surface area to volume of a sphere falls as the size increases. This limits the amount of material, nutrients and waste products, which can be exchanged by the cell with its environment. Specialized cells, such as nerve cells, overcome this problem by becoming long and thin or deeply invaginated.

Thus the 'ideal' size results from a cell having sufficient volume to contain all the essential elements, whilst still being able to exchange materials with its environment in an efficient manner. In the case of human cells, the ideal range appears to be between 1 and 20 µm. Table 1 is presented to give the organic chemist a feel for the relative scale of sizes in the biological world.

2.2.4 CELL STRUCTURES

Figures 2 and 3 show stylized diagrams of prokaryotic and eukaryotic cells, the major features of which will now be discussed.

2.2.4.1 The Cell Wall

Bacterial and plant cells possess a strong, rigid cell wall which is completely missing in mammalian cells and this provides one of the main points of action of specific antibacterial agents. The cells of yeasts and bacteria have high internal osmotic pressures (up to 20–25 atm for Gram-positive bacteria) caused by the presence of high concentrations of ions, which is retained by the wall. However, as the organism grows, the cell wall has to be added to by the synthesis of new material, so, if an agent inhibits this biosynthesis, the wall will be incomplete and the osmotic pressure will cause the cell to burst. This is indeed how the penicillin and cephalosporin antibiotics are believed to act. Since mammalian cells do not need to biosynthesize similar cell walls, they do not possess this biosynthetic machinery and a high degree of selectivity can be obtained.

Bacteria are subdivided according to whether or not they are able to retain the basic dye, crystal violet, following staining and an alcohol wash. The test is named after Christian Gram who

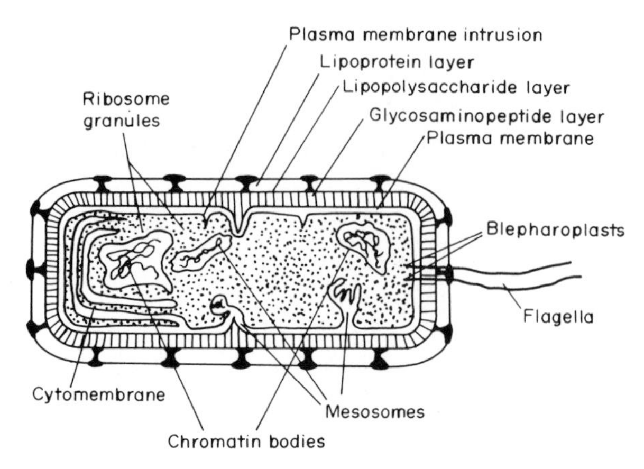

Figure 2 Representation of a prokaryotic cell (reproduced from P. L. Carpenter, 'Microbiology', 3rd edn., 1972, p. 87 by permission of W. B. Saunders, Philadelphia)

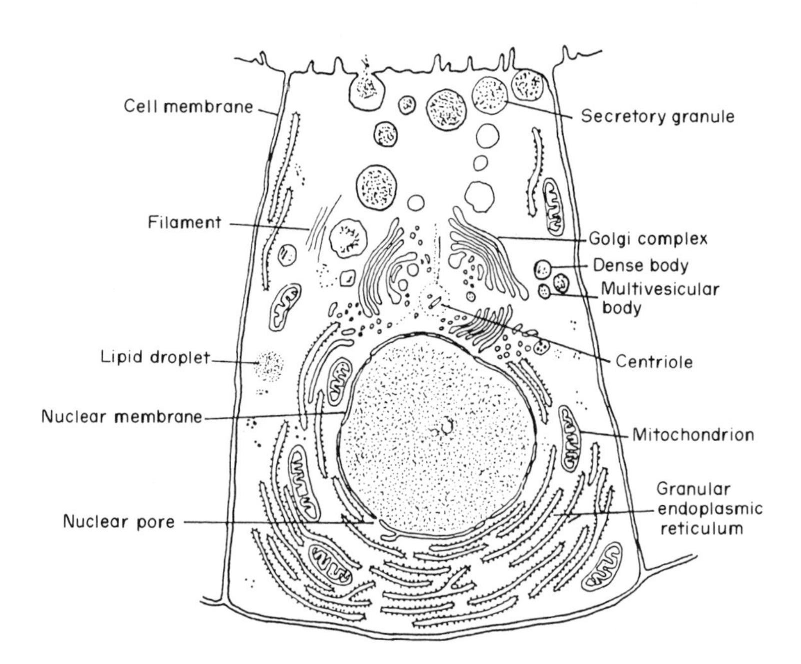

Figure 3 Representation of a eukaryotic cell: a diagram of a secretory cell as it would appear in an electron micrograph (reproduced from W. M. Copenhauer, R. P. Bunge and M. B. Bunge, in 'Bailey's Textbook of Histology', 16th edn., 1971, p. 15 by permission of Williams & Wilkins Co. Ltd., Baltimore)

introduced it in 1884 and Gram-positive bacteria retain the dye; Gram-negative ones do not. These differences are, as might be expected, due to differences in the chemical and physical nature of the walls. Gram-negative bacteria seem to have a more complex structure than that of Gram-positive ones and this is shown very diagrammatically in Figure 4. Gram-positive bacteria lack the outer lipoprotein–liposaccharide layers and the rigid, peptidoglycan layer is much thicker. This layer is the one of particular interest and the one which is not found in mammalian cells.

The wall as a whole comprises 20–25% of the dry weight of the cell and its principal components are shown in Table 2. The walls of Gram-negative bacteria have high lipid and low amino sugar compositions and carry the full range of amino acids found in proteins. On the contrary, those of Gram-positives have little or no lipid, a high content of amino sugars, but only a limited range of amino acids.

Peptidoglycan is a polysaccharide which comprises 50–80% of the walls of Gram-positives but only 1–10% of Gram-negatives. This unusual structure has a polysaccharide backbone made up of an alternating sequence of *N*-acetylglucosamine (**1**) and *N*-acetylmuramic acid (**2**) linked by β-1→4

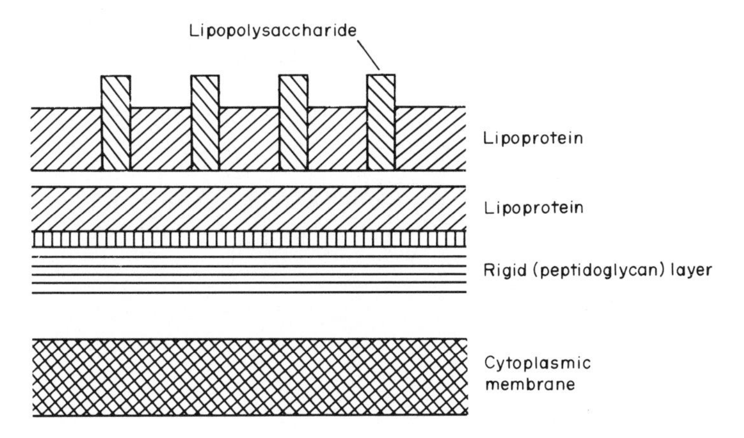

Figure 4 Schematic representation of the cell envelope of *Escherichia coli* (reproduced from J. Mandelstam and K. McQuillen (eds.) 'Biochemistry of Bacterial Cell Growth', 2nd edn., 1973, p. 72 by permission of Blackwell Scientific Publications, Oxford)

Table 2 Bacterial Cell Wall Components

	Gram-positives	*Gram-negatives*
Peptidoglycan	+	+
Polysaccharide	+	+
Protein	± (not all)	+
Teichoic acid and/or teichuronic acid	+	−
Lipid	−	+
Lipopolysaccharide	−	+
Lipoprotein	−	+

bonds. The acid group of (**2**) carries an unusual tetrapeptide; unusual in that it contains both D- and L-amino acids; unusual in that while three of the acids are L-Ala, D-Glu and D-Ala, the fourth can be L-homoserine, L-diaminobutyric acid, L-ornithine, L-lysine or *meso*- or LL-diaminopimelic acid (**3**); and unusual in that the glutamate is linked through the 4-acid group.

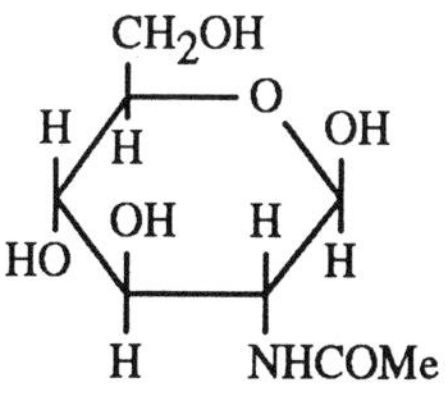

(**1**) 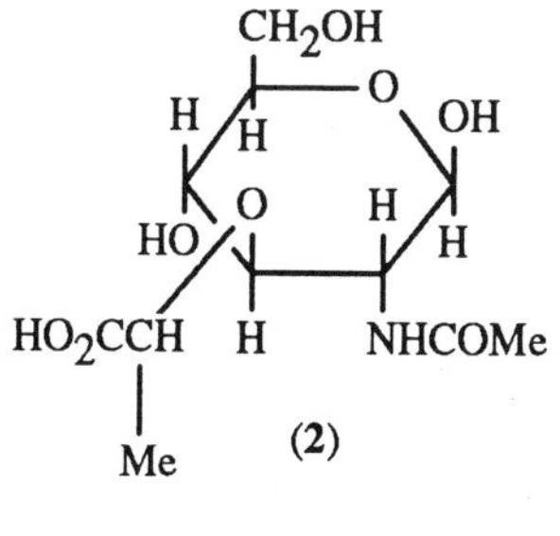 (**2**)

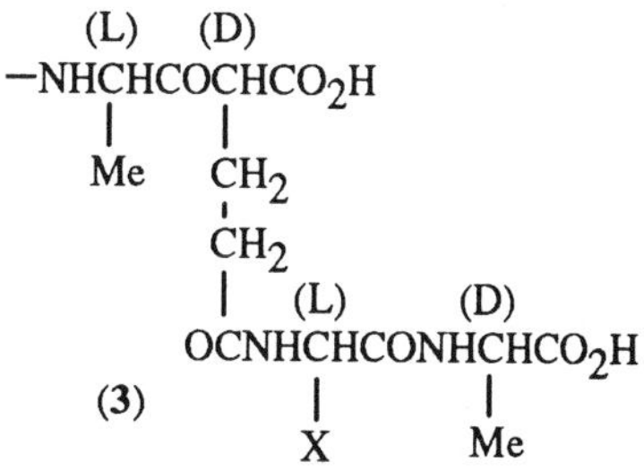

(L) (D)
$-NHCHCOCHCO_2H$ (Me on the first CH; CH_2CH_2 on the second)
(L) (D)
$OCNHCHCONHCHCO_2H$ (X on the first CH; Me on the second)

(**3**)

-NHCHCO- (with X) is L-homoserine, L-diaminobutyric acid, L-ornithine, L-lysine, LL- or *meso*-diaminopimelic acid

These tetrapeptide units are usually cross-linked to give the matrix greater rigidity. As might be expected, the nature of the cross-linking is variable but four main types have been identified. The simplest, which seems to be common to all Gram-negatives, is a direct linkage between D-alanine of one chain and the amino group on the D carbon atom of *meso*-diaminopimelic acid in another chain. A second type, more commonly found in Gram-positives, has a short peptide chain extending from the D-alanine to a free amino group of a diamino acid in another group. In the third type, the linking peptide unit comprises varying repeating units of the same tetrapeptide. Finally, when no diamino acid is present in the tetrapeptide, the link is between the D-alanine and the 2-carboxylic acid group of glutamic acid *via* a diamino acid. The potential complexity and variability of the resultant cross-linked polymer is obvious even from this superficial account and a medicinal chemist seeking to design drugs to interfere with a specific organism will clearly have to determine its precise constitution.

The role of the minor constituents is less clear, but that of teichoic acids, which are polymers of glycerol phosphate, may be to provide a negatively charged environment for enzymatic activity.

In the plant world, fungi and their single-cell equivalents, the yeasts, have somewhat simpler cell walls. That of fungi is a mosaic of various carbohydrates, the major one of which is chitin, poly-*N*-acetylglucosamine, with a little lipid and protein. Pentachloronitrobenzene, a commercial agricultural fungicide, causes the production of cell walls deficient in chitin and thus reduced viability.

Griseofulvin (**4**), a systemic fungicide used both in man and plants, also appears to act at least partly by inhibiting chitin production with the result that high internal osmotic pressures rupture the cell.

OMe O OMe Me O MeO O H Cl

(**4**)

Yeast cell walls contain little chitin but have two interlocking polysaccharide polymers, one formed by mannose polymers covalently linked to peptides and the other, glucan, a polyanhydride of glucose.

A much fuller account of bacterial cell walls and the targetting of drugs to interfere with their synthesis is given in Chapter 9.1.

2.2.4.2 The Cell Membrane

Since the functioning of the cell results from a complex interplay between all of its constituent parts, it would be ridiculous to single out any one part as being more important than any other. Nevertheless, the membranes of the cell are fascinating structures, and since drugs must react with them, or pass through them, they are of vital importance to the medicinal chemist.

The limits of a cell are defined by its outer, plasma membrane; most of the organelles which will be discussed below are surrounded by membranes; many vital biochemical processes take place on or in

Table 3 The Relative Amounts of Membrane Types in Two Eukaryotic Cells[a]

Membrane type	*Percent of total cell membrane*[b,c] *Liver hepatocyte*	*Pancreatic exocrine cell*
Plasma membrane	2	5
Rough ER membrane	35	60
Smooth ER membrane	16	< 1
Golgi membrane	7	10
Mitochondria		
outer membrane	7	4
inner membrane	32	17
Nucleus		
inner membrane	0.2	0.7
Secretory vesicle membrane	ND	3
Lysosome membrane	0.4	ND
Peroxisome membrane	0.4	ND

[a] From B. Alberts, D. Bray, J. Lewis, M. Raff, K. Roberts and J. D. Watson, 'The Molecular Biology of the Cell', 2nd edn., Garland Publishing, New York, 1989. [b] The pancreatic cell has approximately 5 × the volume of the hepatocyte and approximately 8 × its area. [c] ND not determined.

membranes and there is a constant traffic of material through intracellular membranes and exchange between the cell and its environment.

Naturally such an important structure has been much studied but a lot of the work has been performed on the membrane of the red blood cell, the erythrocyte. This is because it is readily available from many species, including man, and because it has no nucleus or internal organelles. Thus the only membrane present is the plasma membrane. Further, placing erythrocytes in a solution of low ionic strength causes them to burst with loss of internal constituents and the resultant fragments of membrane, known as 'ghosts', are readily obtained pure. This is in contrast to other eukaryotic cells shown in Table 3 where the plasma membrane accounts for less than 5% of the total mass of membrane in the cell. In the latter case it is of course much more difficult to produce pure, unaltered, plasma membranes free from contamination by the other intracellular membranes. However, it might be that the erythrocyte is not totally representative and some conclusions drawn from studies on it might not be relevant to other cells.

In constitution, membranes are complex mixtures of lipids, proteins and carbohydrates whose composition varies with the function of the cell. Thus, for example, myelin, the membrane which surrounds and insulates nerve cells, has less than 25% of its mass as protein, whilst in membranes involved in energy transfer, the proportion of protein can be as much as 75%. More usually, the relative proportion of lipid to protein is 50:50, which means that because of their smaller size there are about 50 lipid molecules for each protein molecule.

2.2.4.2.1 Membrane lipids

Three major types of membrane lipid are found in biological membranes. The most abundant of these are phospholipids and the structures of four representatives, phosphatidylcholine (**5**), sphingomyelin (**6**), phosphatidylserine (**7**) and phosphatidylethanolamine (**8**) are shown in Figure 5. It can clearly be seen that they consist of a long hydrophobic fatty acid chain connected to a polar hydrophilic head which makes them amphipathic. It should also be noted that only phosphatidylserine (**7**) carries a net negative charge and so might be expected to possess significantly different properties to the other phospholipids. The second lipid type is cholesterol (**9**) and the third comes from a group of carbohydrates with long fatty acid side chains known as glycolipids. The structure of a representative glycolipid, galactocerebroside (**10**), is shown on p. 177. Table 4 shows the variation in relative proportions of these lipids in different membranes.

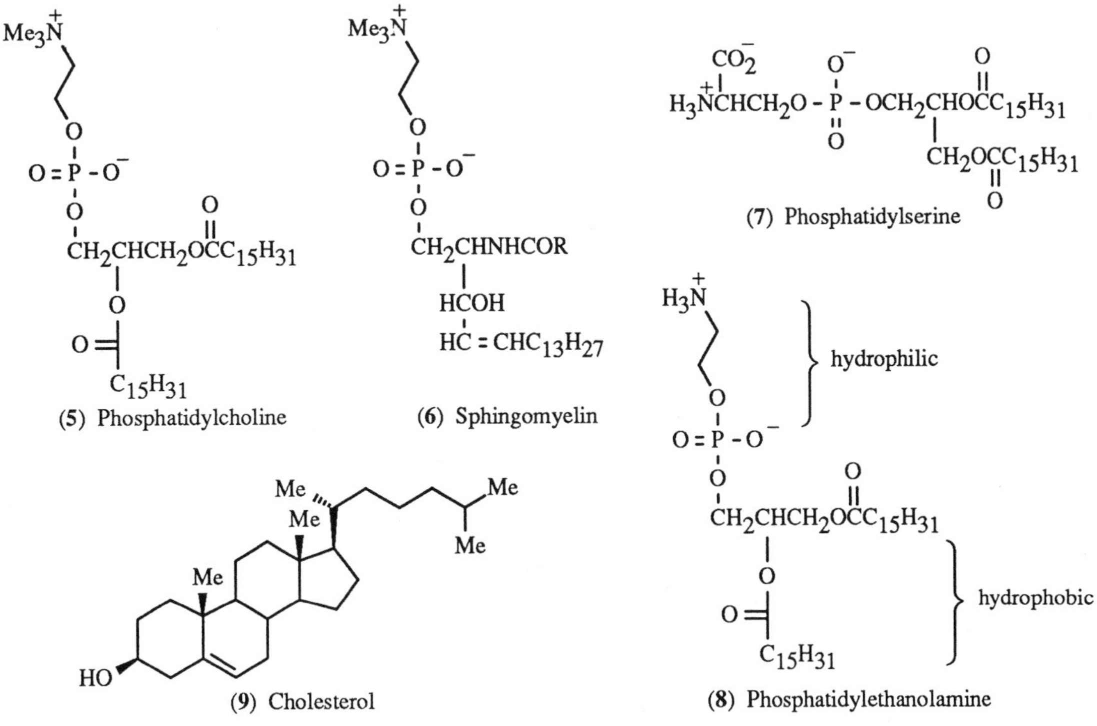

(**5**) Phosphatidylcholine

(**6**) Sphingomyelin

(**7**) Phosphatidylserine

hydrophilic

hydrophobic

(**9**) Cholesterol

(**8**) Phosphatidylethanolamine

Figure 5 Representative phospholipids

Table 4 Approximate Lipid Compositions of Different Cell Membranes[a]

	Percentage of total lipid by weight					
Lipid	*Liver plasma membrane*	*Erythrocyte plasma membrane*	*Myelin*	*Mitochondrion (inner and outer membranes)*	*Endoplasmic reticulum*	*E. coli*
Cholesterol	17	23	22	3	6	0
Phosphatidyl-ethanolamine	7	18	15	35	17	70
Phosphatidylserine	4	7	9	2	5	Trace
Phosphatidylcholine	24	17	10	39	40	0
Sphingomyelin	19	18	8	0	5	0
Glycolipids	7	3	28	Trace	Trace	0
Others	22	13	8	21	27	30

[a] From B. Alberts, D. Bray, J. Lewis, M. Raff, K. Roberts and J. D. Watson, 'The Molecular Biology of the Cell', 2nd edn., 1989, Garland Publishing, New York, p. 281.

Early experiments on erythrocyte ghosts established that the lipids must exist as a bilayer: a natural consequence of their amphipathic nature. Any phospholipid placed in water will form either a bilayer or a micelle (Figure 6), in which the polar heads aggregate together and interact with the water, while the hydrophobic chains interact with each other and not the polar solvent. What was not realized as quickly was that the two layers do not readily intermingle and that the membrane can be asymmetric. Lipid molecules in each layer can readily diffuse laterally with a diffusion coefficient of approximately $10^{-8}\ cm^2\ s^{-1}$. This would mean that an average lipid molecule would take about 1 s to diffuse along the length of a large bacterium. However, diffusion between the layers is much more difficult because the polar head of the lipid is forced to pass through the hydrophobic heart of the membrane. The uncatalyzed process, which has been given the colourful designation of 'flip-flop' (Figure 7), occurs less frequently than once a month. In cases where such translation is necessary, *e.g.* as the lipids are actively synthesized on the endoplasmic reticulum (see p. 182), specific enzymes known as phospholipid translocators are present.

Thus the picture emerges of a two-dimensional fluid membrane in which lipid molecules can readily diffuse. The fluidity of the system can be affected by a number of factors, particularly the nature of the fatty acid chain. The introduction of a single double bond in this chain can have a significant effect because the bend introduced into the chain increases its bulk and its fluidity.

The asymmetry of the bilayer can be quite marked. In the erythrocyte, the outer layer contains almost all of the phospholipid molecules which contain choline, while those like phosphatidylethanolamine and phosphatidylserine with primary amine groups are found in the inner layer. The functional necessity for the asymmetry is probably due to information flow into the cell. Associated with the inner cell membrane are inositol phospholipids and phosphatidylserine, the substrates for protein kinase C, which are heavily involved in the generation of 'second messengers'

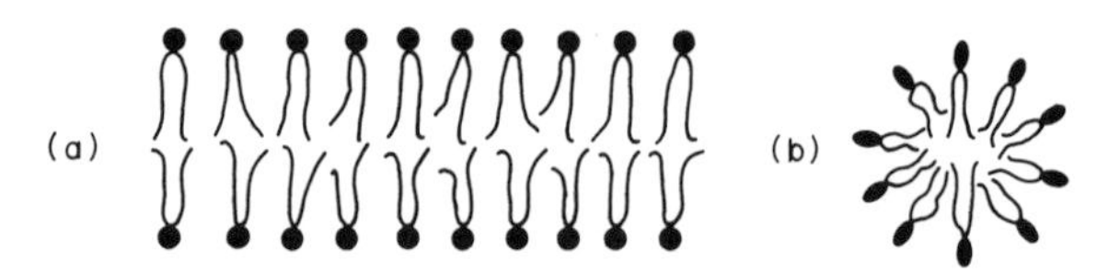

Figure 6 (a) Lipid bilayers and (b) micelles

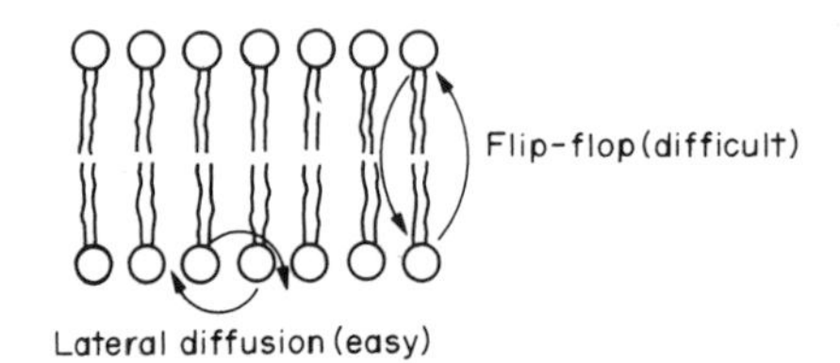

Figure 7 Phospholipid motion in bilayers

whose function is to conduct the effect of an outside stimulus on a cell into its interior. This is a highly significant area of biology and biochemistry which is discussed in more detail in Volume 3.

Table 4 shows that cholesterol is a major component of the plasma membrane of eukaryotic but not prokaryotic cells. Again it is amphiphilic with a polar hydroxyl group and a nonpolar steroid ring system. Accordingly, it is found in the membrane with the hydroxyl groups associated with the polar heads and the steroid nucleus with the fatty acid chains. However, the rigid steroid nucleus partly immobilizes the fatty acid chains close to their polar heads. This makes the layer less fluid, but also less liable to crystallize. Because of the small size of the hydroxyl group, cholesterol can readily flip-flop between the layers.

The final lipid constituents are the glycolipids and these are found exclusively in the outer layer where they comprise approximately 5% of the material. They are found in all cell types, but differ in precise structure markedly from species to species and tissue to tissue. In bacteria and plants, the glycolipids are derived from glycerol-based lipids but in animal cells they are solely derived from ceramide (**11**). The polar head groups in all cases are formed from sugars and can contain from 1 to 15 neutral sugar residues. An example is galactose as found in galactocerebroside (**10**), a major constituent of myelin. More complex structures, the gangliosides, are derived from sialic acid (**12**) and are found in the membranes of neurones.

HOH_2C, HO, H, OH, H, H, H, OH — O — $O{-}CH_2CH(NHCOC_{15}H_{31})CHOHCH{=}CHC_{13}H_{27}$

(**10**)

$HOCH_2CHNHCC_{15}H_{31}$ (C=O)
$\quad|$
$HCOH$
$\quad|$
$HC{=}CHC_{13}H_{27}$

(**11**)

MeCOHN, H, O, CHOH, CO_2^-, H, H, H, OH, HO, H
CHOH
$\quad|$
CH_2OH

(**12**)

Since the glycolipids are found only in the outer layers, with their sugars in the extracellular environment, it is presumed that they play a role in cell signalling. However, so far only one ganglioside has a defined role; as a cell surface receptor for cholera toxin and it seems highly probable that other, more useful roles, exist.

2.2.4.2.2 Membrane proteins

Most of the specific functions of the membrane are performed by proteins. Early models of membranes had the proteins lying as a further layer on top of the polar head groups of the lipids. This is now accepted as being incorrect; the proteins lie in the fluid bed of the lipids (Figure 8) either as transmembranal proteins or in association with only one of the layers (Figure 9). This arrangement has been called the 'fluid mosaic model'. As their name suggests, transmembrane proteins have part of their structures outside the cell, part embedded in the membrane and part inside the cell. They are profoundly asymmetric and the extra and intra portions perform different tasks. The protein passing through the membrane adopts an α-helical conformation in order to maximize the internal hydrogen bonds between the amino and carbonyl groups. Further, hydrophobic side chains are placed on the outside of the helix to maximize the bonding to the lipids. Some proteins have a single section in the membrane, others have more. Proteins associated with the outer layer only do so by being covalently bound, usually *via* oligosaccharides to the minor phospholipid, phosphatidylinositol, the fatty acid chains of which are embedded in this layer. Those associated with the cytoplasmic face can be either covalently bound to a fatty acid immersed in this layer, or non-covalently bound to a transmembranal protein.

Most transmembranal proteins are glycosylated on the extracellular portion and whilst thiol groups on this section are usually oxidized to disulfide bridges, those in the cytosol are present as free thiol groups, reflecting the reducing nature of the inside of the cell.

While the lipids in the bilayer are free to diffuse laterally, this does not also seem to be the case for the proteins. Thus, if we consider an epithelial cell, proteins required for the metabolism and ingestion of foodstuffs need to be found on the cilial portion of the cell, not generally spread over the

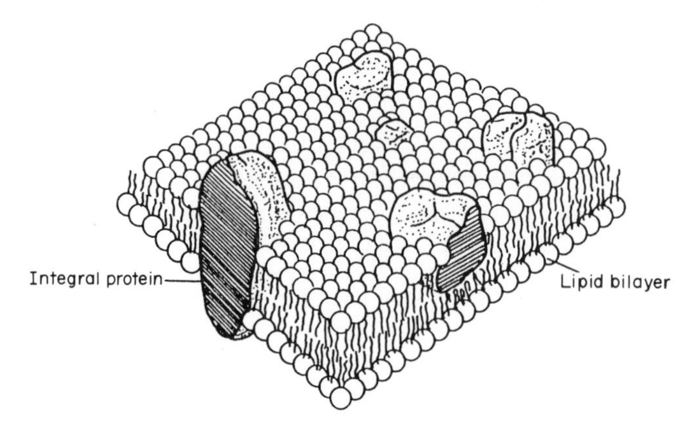

Figure 8 Fluid mosaic model of plasma membranes (reproduced from S. J. Singer and G. L. Nicolson, *Science (Washington, D.C.)*, 1972, **175**, 720 by permission of the American Association for the Advancement of Science)

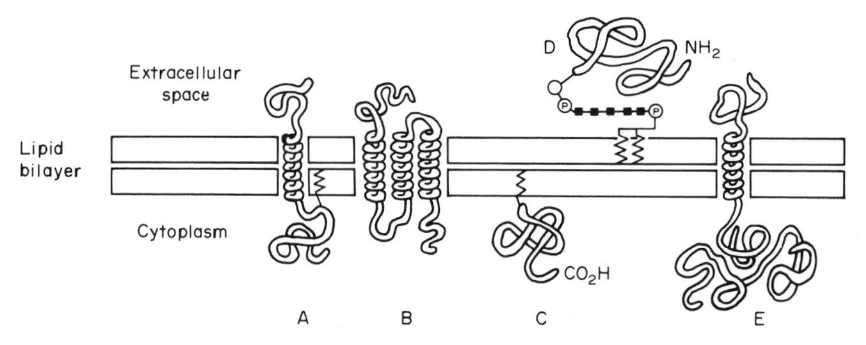

Figure 9 Representation of various membrane-associated proteins: (A) transmembrane protein with one transmembrane region; (B) transmembrane protein with multiple transmembrane regions; (C) cytoplasmic membrane anchored to membrane by fatty acid chains; (D) extracellular proteins anchored to membrane *via* oligosaccharide linked to fatty acid chains; and (E) cytoplasmic proteins associated noncovalently with transmembrane protein (reproduced from B. Alberts, D. Bray, J. Lewis, M. Raff, K. Roberts and J. D. Watson, 'The Molecular Biology of the Cell', 2nd edn., 1989, p. 284 by permission of Garland Publishing, New York)

other faces. The restraint can be induced by aggregation of proteins into massive complexes, which greatly reduces the rate of diffusion, by close association with extra- or intra-cellular structures or by means of tight junctions formed when the membranes of two adjacent cells fuse together. The mechanism of insertion of proteins into the cell membranes will be discussed later in the section on the endoplasmic reticulum.

2.2.4.2.3 Membrane carbohydrates

The total carbohydrate portion of the cell membrane represents between 2 and 5% of its total weight. Most of this is present either as glycolipid, discussed above, or as poly- or oligo-saccharides covalently bound to asparagine, serine or threonine residues of membrane proteins. Most expressed proteins are glycosylated; only one in ten lipid molecules are. This means of course that in absolute numbers more lipid molecules carry sugars than do proteins. However, a combination of greater polysaccharide chain lengths and multiple oligosaccharides found on proteins means that the greater carbohydrate mass is protein bound. The term 'cell coat' or 'glycocalyx' is given to the carbohydrate rich peripheral zone found on the outside of most eukaryotic cells. As with the glycolipids, the true function of the carbohydrate has not yet been determined, although it is hypothesized that it must play a role in cell–cell signalling and recognition processes.

The above has been written largely with reference to eukaryotes, but the membranes of prokaryotes appear to be very similar. Protein accounts for 60–80% of the mass of the membrane and it exists in the same ways in a fluid lipid bilayer. In Gram-positive bacteria the cytoplasmic membrane is found between the cell wall and the cytoplasm, whilst in Gram-negatives an additional

Table 5 Lipid Differences Between Gram-positive and Gram-negative Bacteria

	Phospholipids	*Fatty acids*
Gram-positives	Phosphatidylethanolamine, phosphatidic acid, phosphatidylglycerol, *etc.*	Branched chain; very little unsaturated
Gram-negatives	Mainly phosphatidylethanolamine	Saturated and unsaturated C_{16}–C_{18}; cyclopropanes

membrane is found on the outside of the cell wall. The outer layer of this membrane seems to be particularly rich in lipopolysaccharide. In both types, the predominant lipids are phospholipids with smaller amounts of glycolipids and, as mentioned above, no cholesterol. There are a number of differences in lipid composition between the two types and these are summarized in Table 5.

The ratio of saturated to unsaturated acids in Gram-negatives can be markedly dependent on the temperature of the growth medium. At low temperatures unsaturated chains predominate, while at higher temperatures saturated ones are favoured. This is presumably a response by the cell to keeping its membrane fluidity at an optimum level.

2.2.4.2.4 ***Passage through membranes***

The plasma membrane delineates the limits of the eukaryotic cell and also controls the passage of materials into and out of the cell. Such control results in highly significant differences between intra- and extra-cellular concentrations of a number of solutes. Thus, for example, the concentration of sodium ions is 10–30 times greater outside the cell than inside it; that of potassium is 20–30 times greater inside than out. The resultant concentration gradient is of major importance in transport processes across the membrane, as will be shown later, and also leads to an electrochemical imbalance with the inside of the cell being negatively charged compared to the outside.

It was shown above that much of the area of the membrane is a lipid bilayer, which could be readily predicted, and equally readily shown experimentally to be pervious to small uncharged molecules such as oxygen, carbon dioxide and urea but to be impervious to large polar molecules such as glucose and ions such as sodium and potassium. Thus the transport of these materials must involve the transmembranal proteins. It transpires that there are two processes: passive transport in which the material in question moves down a concentration gradient and an active process where the reverse is true. Of course, in the latter case energy must be supplied to the system to run the process. It also transpires that two types of protein are involved, carrier proteins and channel proteins, and both have more than one segment of their backbone passing through the membrane. Using proteins for this purpose of course also allows an element of selectivity to be introduced. Specific proteins can be used for specific substrates and thus add a high degree of control to the cell's exchange of materials.

Carrier proteins are envisaged to react with their substrate outside the cell membrane. The act of binding then induces a conformational change which allows the substrate to pass through the protein and out through a newly created opening into the cell. When the carrier protein operates solely on one substrate, it is called a uniport. However, there are many carrier proteins which must operate in conjunction with a second substrate. Such coupled transporters can be either symports, where both substrates go the same way, or antiports, where one substrate goes into the cell and the other, either simultaneously or sequentially, exits. Channel proteins on the other hand form a water-filled pore which when open allows the ion to pass through the pore and thus through the membrane. Such pores are usually closed by a 'gate' which opens in response to a stimulus to the membrane. Gated ion channels which open in response to a neurotransmitter reacting with its receptor are extremely important and will be dealt with in greater detail in chapters in Volume 3 of this series.

Channel proteins can only operate in the passive mode; the substrate flow can only be down a concentration gradient. Carrier proteins, on the other hand, can operate both passively and actively and can transport materials against a concentration gradient. There are two main sources of energy for this process: biochemical and a favourable concentration gradient of another substrate.

One of the most completely understood active transport systems is the sodium–potassium pump that maintains the imbalance of sodium and potassium ions across the membrane. The pump is tightly coupled to a sodium–potassium–ATPase enzyme which hydrolyzes ATP to ADP and phosphate and requires both sodium and potassium ions for activity. For every molecule of ATP hydrolyzed three sodium ions are pumped out of the cell and two potassium ions pumped in, both against concentration gradients. A similar ATPase-driven pump serves to pump calcium ions out of a cell against a thousandfold concentration gradient. Bacteria have similar ion-driven uptake mechanisms except that these tend to be due to protons rather than sodium ions.

The use of the electrochemical gradient provided by sodium ions to power the uptake of solutes is widely practised by a number of cells, particularly intestinal and kidney epithelial cells. Thus, for example, an intestinal epithelial cell can take up glucose using the electrochemical potential of sodium ions. The excess sodium ions are then pumped out of the cell *via* the Na^+–K^+–ATPase. However, the system is even more complicated than this. Epithelial cells extract nutrients from the fluids in the lumen of the gut and pass them on to extracellular fluids which bathe the other side of the cell. Direct contact between the gut contents and the extracellular fluids is, of course, prevented by the epithelial membrane. On the other side of the cell, another protein allows the diffusion out of the cell of the glucose whose cellular concentration is greater than that of the extracellular fluid. Figure 10 attempts to represent this process which is also a very interesting example of the asymmetric distribution of proteins in the plasmic membrane which was described earlier.

Clearly the presence of a hole through the membrane as provided by channel proteins could be detrimental to the cell if it allowed the transport of too many or the wrong kind of ions into or out of the cell. Ionophores are hydrophobic molecules which create such a pore and one, A 23187, which allows calcium ions to pass freely into a cell, is widely used to investigate the biochemical consequences of such an influx. An antibiotic, valinomycin, works by facilitating the loss of potassium ions from the cells whose membranes it invades.

The systems described above are only relevant to the processing of small molecules and ions. When proteins, polysaccharides or even larger particles are involved, different techniques are used. The particle passes through the membrane engulfed in a small spherical pocket known as a vesicle. The process of uptake is known as endocytosis, that of release, exocytosis. In exocytosis, a particle synthesized within the cell is surrounded by a vesicle, which then moves to the membrane. Coalescence of the two membranes with opening of the vesicle allows the particle to be expelled from the cell. Endocytosis is subdivided into two processes, pinocytosis and phagocytosis, ostensibly differentiated by the size of the vesicle but also differing in fundamental mechanisms. Pinocytosis (Figure 11) is applied to vesicles of less than 150 nm in diameter and is an ongoing process which is part of the cell's continual regeneration of its membrane. This begins in a region of the membrane

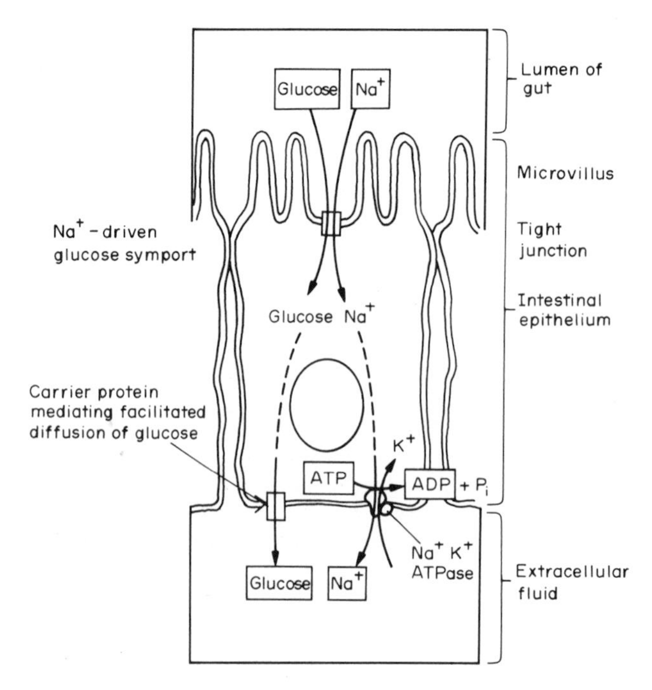

Figure 10 Schematic diagram showing facilitated transport of glucose through intestinal epithelial cells (reproduced from B. Alberts, D. Bray, J. Lewis, M. Raff, K. Roberts and J. D. Watson, 'The Molecular Biology of the Cell', 2nd edn., 1989, p. 312 by permission of Garland Publishing, New York)

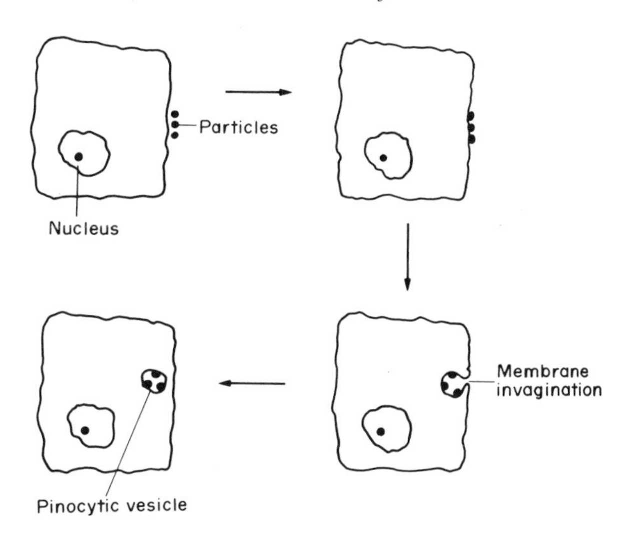

Figure 11 Schematic representation of the mechanism of pinocytosis. After the particles have been absorbed on the cell membrane, the membrane invaginates and the invaginated portion breaks away from the cell surface forming a pinocytic vesicle

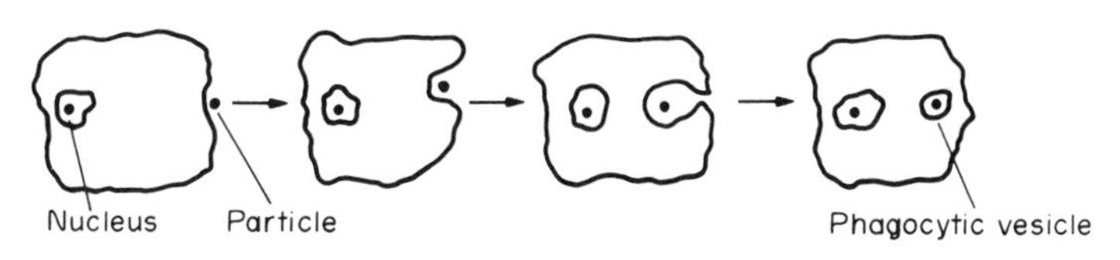

Figure 12 Schematic representation of the phagocytic process

called a 'coated pit', where it is covered with a network of polymers constructed from a number of protein complexes of which the best characterized is clathrin. How the subsequent invagination process is initiated, how the vesicle is pinched off from the membrane and how the clathrin returns to the membrane has yet to be determined. However, any particle trapped in a developing vesicle will be transported to the cytoplasm, there to be broken down by hydrolytic enzymes of the lysosome. It should be noted that this process is not entirely random. In most cells receptors for particular targets are found near the coated pits. After binding to the receptor, the target is absorbed. This process is used by many animal cells to absorb their supplies of cholesterol needed for the membranes. Cholesterol circulates in the blood in low density lipoproteins (LDL), a complex of lipids, proteins and its fatty acid esters. Specific receptors are present on the membrane which induce the pinocytosis of the LDLs. Subsequently enzymatic hydrolysis frees the cholesterol. The use of this type of specific recognition process gives the cell an additional degree of control over its constitution. When too much cholesterol is present, the cell stops production of the receptor and further uptake ceases; when too little is present, extra receptors are synthesized.

Phagocytosis (Figure 12) involves vesicles of a size greater than 250 nm and is largely a function of specific cells, particularly macrophages and neutrophils. These are the first lines of defence of the body against attack by microorganisms and they also act as scavengers, mopping up the detritus of old, damaged and dead cells. Phagocytosis is not a constitutive process and its initiation requires signal mechanisms from the cell membrane as it comes into contact with the target. The most developed trigger is that due to antibodies. When an antigen provokes antibody production, the combination of antibody and antigen presents the Fc fragment of the antibody to the macrophage, recognition of which initiates a process whereby the cell engulfs the complex by growing pseudopods, or cellular projections, around it. The target is then absorbed and ultimately degraded by the hydrolytic enzymes of the cell.

2.2.4.3 The Internal Structure of the Cell

Reference back to Figures 2 and 3 will show that whereas the interior of the prokaryotic cell is essentially unstructured, quite the reverse is true for eukaryotes, which have a host of identifiable components known as organelles. The space between the nucleus and the plasma membrane is known as the cytoplasm and the liquid which permeates the cytoplasm as the cytosol. The cytoplasm is highly structured with a honeycomb of membranes which support and separate the organelles and also provide a base on which many enzymatic reactions take place. The roles of these structures will now be examined.

2.2.4.3.1 The endoplasmic reticulum

The endoplasmic reticulum (Figure 13) is a complex, convoluted single membrane found in all eukaryotic cells except the erythrocyte. It extends throughout the cytoplasm to such an extent that it usually comprises 50–60% of the total membrane mass of the cell. It is difficult to be certain, but it does seem that it comprises a single sheet enclosing a single internal space called a lumen. A single membrane separates the lumen from the nucleus, but at least two membranes separate it from the interior of other organelles.

Two forms of endoplasmic reticulum have been identified, a rough and a smooth form, with the former being distinguished by the presence on its surface of knobbly protrusions absent in the latter. These lumps are ribosomes and give away its function. Simply, the endoplasmic reticulum is the site of much of the cell's protein synthesis and virtually all of its lipid synthesis.

As will be shown in greater detail in Chapter 3.7, proteins are synthesized on ribosomes. Ribosomes are large complexes of ribonucleic acid, RNA, and proteins, which, before initiation of protein synthesis, are found freely floating in the cytosol. Following initiation of protein synthesis, the ribosome–protein complex can either remain in the cytosol, or transfer and become attached to the endoplasmic reticulum. Which of these alternatives is taken depends on the final destination of the protein. If it is destined to remain in the cytosol, to be passed to the mitochondria, peroxisomes or the nucleus, then the whole synthetic process takes place in the cytosol. If, however, the protein is to be secreted from the cell, or is to be retained in the endoplasmic reticulum, the Golgi apparatus, the plasma membrane or the lysosomes, then the ribosome becomes attached to the endoplasmic reticulum.

This is achieved by the first few amino acids of the peptide chain emerging from the ribosome acting as a signal, which is then bound by a signal recognition particle. This binding slows or stops completely the synthesis to give the ribosome time to move to the reticulum. Here there is found a signal recognition particle receptor, or docking protein, which displaces the signal recognition particle and binds the ribosome. Synthesis restarts and the whole protein is produced in very close proximity to the membrane of the endoplasmic reticulum. Subsequently, it can either pass right through the membrane or become embedded in it. Soluble proteins destined for excretion from the cell pass into the lumen by first synthesizing a highly hydrophobic disposable N-terminus fragment, which becomes embedded in the membrane. This creates a hole big enough for the hydrophilic portion of the protein to pass through the membrane into the lumen. Subsequent cleavage of the disposable section releases the soluble protein into the lumen. In the case of a transmembranal protein, the process begins in the same way until the hydrophobic section which is embedded in the membrane is synthesized. This remains in the membrane and the rest of the protein is synthesized in the cytosol. Cleavage of the leader section leaves a transmembranal protein straddling the mem-

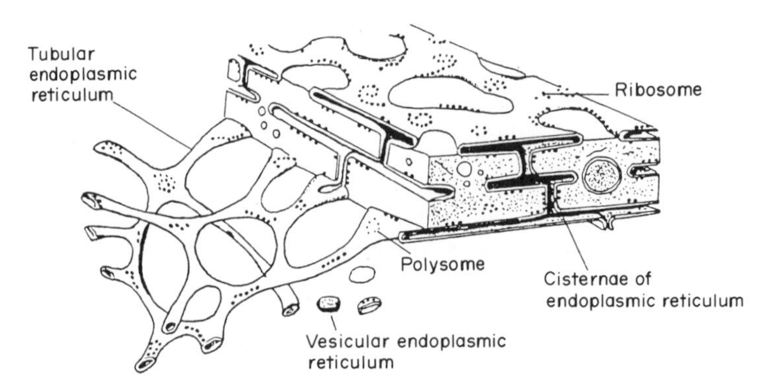

Figure 13 Schematic representation of the interconnected cisternae and tubules of granular endoplasmic reticulum (reproduced from W. M. Copenhauer, R. P. Bunge and M. B. Bunge, 'Bailey's Textbook of Histology', 16th edn., 1971, p. 23 by permission of Williams & Wilkins Co. Ltd., Baltimore)

brane. A slightly more complicated process along the same lines produces transmembranal proteins with multiple passes through the membrane.

A further refinement arises from the presence in the membrane of the endoplasmic reticulum of an enzyme with its active site on the lumen side of the membrane. This enzyme glycosylates proteins to produce glycoproteins, and as the same reaction does not occur in the cytosol, this differentiates the products of the two schemes.

As implied at the beginning of this section, prokaryotes do not possess an endoplasmic reticulum, and the ribosomes of these species are found in the cytosol in the form of long linear complexes closely associated with the complicated fibrillar structures of the cytoplasm. In fungi, the endoplasmic reticulum is sparse, much simpler than in mammalian cells, but some ribosomes are found associated with it.

The endoplasmic reticulum is also the site of synthesis of virtually all the lipids required by the cell for construction of its membranes. Synthesis takes place on the smooth endoplasmic reticulum and does not involve the ribosomes. It might therefore be expected, and is indeed the case, that those cells which produce more lipids, *e.g.* the hepatocytes which produce lipoprotein particles for excretion, have larger proportions of smooth endoplasmic reticulum. The synthetic process takes place in the cytosolic half of the membrane of the endoplasmic reticulum and an enzyme, 'flippase', more rapidly transfers phosphatidylcholine between the separate membrane layers than would be possible by the noncatalyzed flip-flop mechanism. This produces the asymmetry which is maintained in the final functioning membrane.

The smooth endoplasmic reticulum is also the site of the major metabolic and detoxification enzymes, including the cytoplasmic *P*-450 enzymes, and this further increases the amount of smooth endoplasmic reticulum found in hepatocytes.

2.2.4.3.2 The Golgi apparatus

When a protein has been synthesized on the endoplasmic reticulum, it is released in a transport vesicle and passes to the Golgi apparatus. This is a complex structure, unique to eukaryotic cells, whose function is to complete the post-synthetic processing of the protein and to direct it to its final destination.

In cross section, and the form usually seen in electronmicrographs, the Golgi apparatus has the appearance shown in Figure 14; more detailed investigations show the complex three-dimensional structure represented by Figure 15. The core of the structure is a stack of four to six flattened, membrane-bound cisternae, which look rather like plates. The membrane is smooth and peripheral tubules join the cisternae to one another. Closely associated with the structure are large numbers of small transport (Golgi) vesicles of about 50 nm diameter, and a smaller number of larger secretory vesicles.

The system has two faces: *cis*, which is the side to which the transport vesicles are delivered, and the *trans*, on which the secretory vesicles develop and then depart. The whole processing system works in a continuous flow with the protein arriving (in a vesicle) at the *cis* face, moving on to the medial (central) cisternae and then to the *trans* face. Each of the cisternae with which a protein comes into contact has its own specified set of enzymes responsible for specific transformations. Transport from cisternae to cisternae is again *via* vesicles and thus soluble or transmembranal proteins never cross a membrane directly. Vesicles fuse into membranes releasing their contents; processed proteins are packaged in vesicles within the lumen before passing out of the cisternae. Further, the asymmetry of the transmembranal proteins is maintained.

The main modifications to proteins in this system are further changes to the oligosaccharides attached to the protein in the endoplasmic reticulum and cleavage of peptide fragments from

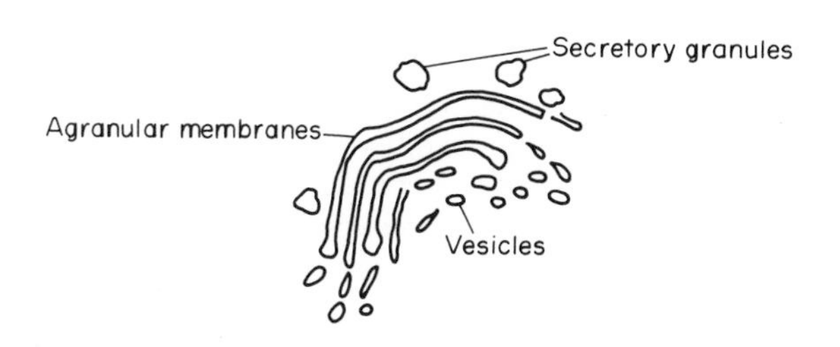

Figure 14 Schematic representation of the Golgi apparatus as seen in electron micrographs

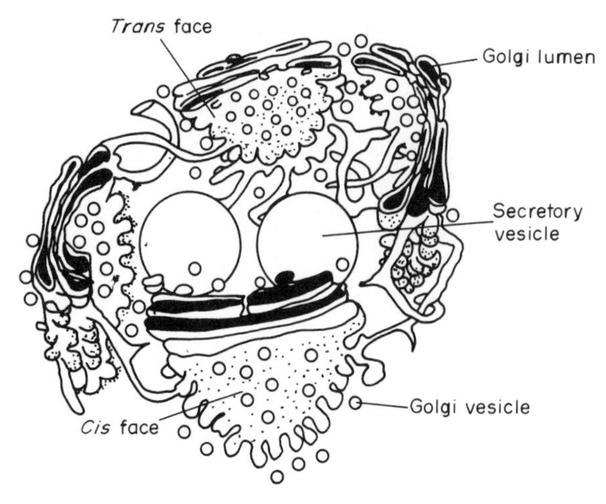

Figure 15 The Golgi apparatus (reproduced from B. Alberts, D. Bray, J. Lewis, M. Raff, K. Roberts and J. D. Watson, 'The Molecular Biology of the Cell', 2nd edn., 1989, p. 451 by permission of Garland Publishing, New York)

proteins, thus releasing the active entity. The most heavily glycosylated proteins of all, the proteoglycans, a major constituent of cartilage, are produced in the Golgi apparatus by extensive modification of proteoglycan core proteins synthesized in the endoplasmic reticulum. Sulfation of glycosaminoglycan chains and tyrosine residues in proteins is also performed in the Golgi apparatus.

The modified proteins are finally expelled from the *trans* walls in secretory vesicles and dispatched to the plasma membrane, lysosomes or expelled from the cell altogether. In view of the involvement of the Golgi apparatus in the secretory process, it is perhaps not surprising that it is more highly developed in secretory cells. Thus in the guinea pig pancreatic cell, it occupies 6–10% of the cell volume, whilst it is equally well developed in the mucosal cells which secrete digestive juices into the gastrointestinal tract. On the other hand, it is poorly developed in muscle cells.

2.2.4.3.3 The mitochondrion

Mitochondria are responsible in aerobic organisms for the majority of the extraction of energy from food, a role which quite properly qualifies them for the designation 'the powerhouse of the cell'. This organelle is usually depicted as a small hard capsule some 1–10 μm × 0.5 μm, but it is known that it can be remarkably flexible and mobile. However, it is also found fixed in areas requiring large amounts of energy, *e.g.* close to the myofibrils of muscle cells and the flagellae of sperm. The number of mitochondria per cell varies from around 50 to several thousands, depending on cell size and rate of expenditure of energy.

A schematic drawing of a representative mitochondrion is shown in Figure 16, which shows that the structure essentially consists of an outer membrane over an inner one which has numerous projections, called cristae, into the internal space. The result of the projections is of course to greatly enlarge the total surface areas of the membranes and reference back to Table 3 will show that the mitochondria contain between 20 and 40% of the total of cell membranes. The inner layer is of course considerably more extensive than the outer. Fitting in with the concept that the main function of the organelle is to supply energy, it has been found that the inner structures are more extensive in heart muscle cells, which have high respiratory rates, than in metabolically less active cells.

The two membranes also divide the volume into two: an intramembranal area and an inner area called the matrix. The two membranes have quite different properties so these two volumes will have different constitutions. The outer membrane is pierced by numerous pores caused by a transport protein called porin, which means that it is permeable to virtually all molecules with a molecular weight of less than 10 000 Da. The inner membrane is entirely different and is especially impermeable to ions. Because of their special functions, described below, the membranes have need of large numbers of protein molecules and this results in the ratio of protein to lipid for the membranes being about 70:30. The intramembrane space has a constitution very similar to the cytosol, whilst the

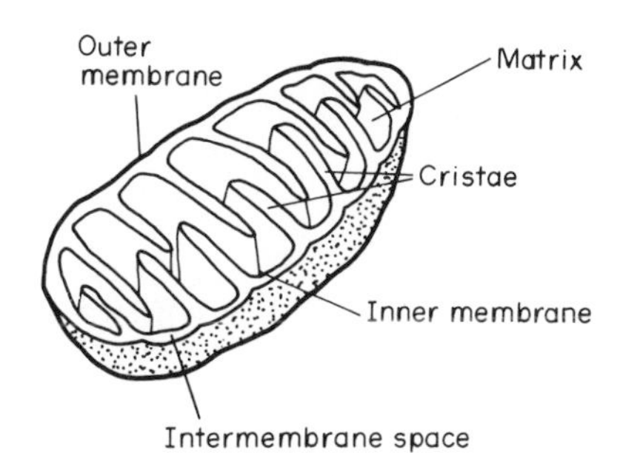

Figure 16 Schematic view of a mitochondrion showing proposed inner structure

matrix volume contains only oxygen and those elements for which active transport processes exist in the inner membrane.

Mitochondria are only found in eukaryotic cells. Prokaryotic bacteria have instead a rudimentary membrane system called a mesosome, which is continuous with the plasma membrane and which permeates the cytoplasm as tubular or vesicular invaginations.

In higher life forms, including man, the process of extracting the energy needed to maintain life starts in the gut, where the complex mixture of fats, carbohydrates and proteins in food is first broken down by the gastric juices. The constituents, glucose, fatty acids and peptides, enter the cytoplasm, where the first stage of metabolism takes place. This converts glucose by a nine-stage enzymatic process into pyruvic acid. This stage also yields a certain amount of energy, which is actually stored by the cell in the form of adenosine triphosphate (ATP). The hydrolysis of ATP to adenosine diphosphate (ADP), and inorganic phosphate is exothermic and the vast majority of the endothermic reactions of the body are driven by their being linked to the hydrolysis of ATP. All of the reactions in this part of the metabolic process do not require the involvement of oxygen and there is only a net gain of two molecules of ATP per molecule of glucose metabolized. The rest of this process, and the reason for discussing it here, occurs in the mitochondrion. Pyruvic acid freely enters the intermembranal space and then is actively transported into the matrix where it is first decarboxylated to acetyl-coenzyme A (CoASAc). This enters a complex cycle of reactions known as the citric acid or Krebs cycle, which has the overall effect of converting the acetyl group into carbon dioxide and water. In chemical terms the acetyl group is oxidized; oxygen is reduced. The energy released by this oxidation produces up to 30 molecules of ATP. The whole process is a stepwise one and this allows the recovery of the maximum amount of energy. It also can receive other substrates, for example from the metabolism of fatty acids, and can expel constituents of the cycle from the mitochondrion to the cytoplasm, where they can act as substrates for other biosynthetic reactions. All the enzymes of the Krebs cycle are found in the matrix of the mitochondrion. It should be realized that this part of the metabolic process generates vastly more usable energy than the other and that it is the only part to require oxygen. It follows that anaerobic processes are much less efficient than aerobic ones and that such systems have to devise other ways of dealing with the pyruvic acid. Thus, yeast fermentation generates alcohol and muscle cells produce lactic acid in times of oxygen debt.

While oxygen is the ultimate oxidizing agent in this cycle, it is only involved, *per se*, in the final step. The initial oxidation involves hydride ion transfer to the coenzyme NAD^+ (oxidized nicotinamide adenine dinucleotide) to generate NADH (reduced nicotinamide adenine dinucleotide). This looses a proton, to regenerate NAD^+, and two electrons to the 'respiratory chain', a complex of 15 different transmembranal proteins gathered in three major groups. These carriers are of sequentially reducing energy so that the electrons flow from one to the other until the final receptor is oxygen. Further, they are all found in the inner membrane of the mitochondrion and their presence, and that of all the other enzymes involved in the metabolic and transport processes, explains the high ratio of protein to lipid found for these membranes. The flow of electrons down the respiratory chain is used to pump protons out of the matrix through the inner membrane against a concentration gradient. This sets up a potential difference as the inside of the membrane will be negatively charged relative to the outside. Thus protons will now flow back into the matrix down this electrochemical proton gradient. This proton flow accomplishes two major tasks. It stimulates the enzyme, ATP-synthetase, which converts ADP into ATP, and it cotransports other substrates, notably pyruvate and phosphate, through the membrane in the manner described earlier. The importance of the ATP-synthetase is underlined by the fact that it comprises some 15% of the total membrane protein.

Thus the oxidation can be summarized as a chemical reaction initiating an electron flow, which stimulates proton pumping and consequent conversion of ADP to ATP. The name chemiosmosis has been given to this complicated process.

It is recognized that this has been a very superficial account of a very complicated process and the reader requiring further information is referred to any good modern biochemical textbook.

The mitochondrion has an unusual feature which perhaps betrays its origin: mitochondrial DNA. This is a single double helix circular form of DNA which in humans has 16 569 base pairs, *i.e.* about 10^{-5} times that of the nuclear DNA. This is sufficient to code for two ribosomal RNAs, 22 transfer RNAs and 13 different polypeptide chains, of which five are membrane proteins. This number is clearly insufficient to account for all the proteins found in the mitochondrion. The rest must be synthesized from nuclear DNA. Quite why the mitochondrion should retain the ability to produce such a small range of proteins has not yet been explained. The presence of this DNA is one line of evidence that supports the hypothesis that mitochondria are the descendants of specialized prokaryotic cells which were endocytosed by larger anaerobic organisms when oxygen began to appear in the earth's atmosphere in much greater quantities than had previously been the case some 1.5×10^9 years ago.

2.2.4.3.4 Lysosomes

All eukaryotic cells contain organelles whose function is to act as the major sites of intracellular digestion. These are the lysosomes, a group of vesicles of widely varying size and shape which contain a potentially lethal concoction of enzymes. To date, 40 enzymes, including proteases, phospholipases, sulfatases, phosphatases, lipases, glycosidases and nucleases, have been identified as constituents. All of these are acid hydrolases with a pH optimum of around 5. The membrane of the vesicle contains an ATP-driven proton pump which maintains this pH. Clearly, this combination of enzymes would be lethal to the cell itself if it was released and part of the cell's defence mechanism is to have the cytosol at a pH of approximately 7.2 at which they are very much less active. Nevertheless the lung disease silicosis is possibly caused because silica dust phagocytosed by lung macrophages disrupts the membrane of lysosomes, thus releasing these enzymes. The subsequent cell death causes further cellular changes, which leads to the characteristic signs of the disease.

The proteins of the lysosome, both enzymes and membrane bound, are synthesized as usual on the endoplasmic reticulum and passed, as described above, to the Golgi apparatus for packaging. It is one of the remaining puzzles of cell biology as to how the Golgi apparatus distinguishes between the proteins and packages them appropriately. In this case it appears that *N*-oligosaccharide chains on these proteins are further modified by the addition of mannose 6-phosphate groups. Mannose 6-phosphate recognition proteins on the *trans* face of the Golgi apparatus recognize these proteins and select them from the rest. These are then packaged in transport vesicles, which are dispatched to an intermediate structure, the endolysosome. Further modifications in the endolysosome, particularly lowering the pH to 5.0, mature this into the lysosome.

As stated above, the role of the lysosome is to digest unwanted material and there are three possible sources of this material: two from outside, one from inside the cell. Vesicles from pinocytosis form endosomes which then fuse with transport vesicles from the Golgi apparatus to form the endolysosome where hydrolysis commences. In phagocytic cells such as macrophages and neutrophils the phagosome, formed by the phagocytic process, fuses with the lysosome and the internal enzymes destroy the invading particle. In the third process, the lysosome destroys unwanted parts of the cell itself in a controlled way. The part to be destroyed is first surrounded by membrane from the endoplasmic reticulum. This is then fused with either the lysosome or the endolysosome to destroy it. The use of membrane fusion in all of these processes to prevent the lysosomal enzymes from being released into the cytoplasm should be noted.

2.2.4.3.5 Microbodies

Mitochondria are not the only organelles in eukaryotic cells involved in oxidation of food. Nor are they the only possible vestigular remains of a primitive prokaryotic organism which existed in a symbiotic relationship with early eukaryotic cells. All eukaryotic cells contain microbodies, small single membrane surrounded collections of oxidative enzymes. These are usually less than 0.5 μm in diameter and are found in various numbers. For instance, liver cells have between 400 and 800, although in certain circumstances, to be discussed below, they can be increased. In mammalian cells some 40 different enzymes have been detected in microbodies, but one of these is always catalase and this underlines one of the major differences between the microbodies and mitochondria. It will be recalled that the end product of the reduction of oxygen in the mitochondrion was water. In contrast, the end product in the microbody is hydrogen peroxide and this has led to the organelle being renamed the peroxisome. However, the hydrogen peroxide is potentially toxic to the cell and thus the microbody invariably contains the enzyme catalase, which decomposes it to water. The range of substrates oxidized by the other enzymes includes D- and L-amino acids, hydroxy acids, fatty acids, alcohols, amines and a purine. In fact, some 50% of ingested alcohol is oxidized by liver microbodies.

The second major difference to the mitochondria lies in the fact that peroxisomes have no phosphorylation system and therefore cannot produce ATP. Thus no useful energy is produced by this oxidation system. Finally, peroxisomes have no DNA and all the enzymes and membrane proteins are synthesized in the cytosol and imported into the organelle.

The levels of peroxisomes in cells are responsive to their environment. Thus yeast cells growing in a sugar-based medium have small peroxisomes. If the yeast is grown on methanol or fatty acids, the peroxisomes become very much larger and contain enzymes capable of oxidizing these substrates. Similarly, the lipid-lowering drugs such as clofibrate are found to cause a massive tenfold increase in peroxisomes in rat liver cells. The lipid-lowering effect then results from the increased fatty acid oxidation by these peroxisomes.

Two other morphologically related organelles are hydrogenosomes and glycosomes. The former have been found in trichomonatids, parasites which exist in the human genital tract and cause sexually transmitted diseases. These organisms are capable of reducing protons to hydrogen under anaerobic conditions. They can also reduce nitroimidazoles to species which are highly toxic to the cell and this is the basis for the selective use of these agents in treating this infection, since this reaction does not occur in the host's cells. Glycosomes contain the enzymes responsible for part of the glycolytic chain and are found in protozoan parasites which cause sleeping sickness, Chagas' disease and liechmaniasis. Unfortunately, it has not yet been possible to find a selective agent capable of exploiting this difference in the cellular architecture.

This completes the discussion of the organelles found in the cytoplasm. Attention will now be focused on the remaining constituents of the cell.

2.2.4.3.6 The cytosol and cytoskeleton

The cytosol is defined as that part of the cytoplasm occupying the space between the organelles and it commonly constitutes about 50% of the total cell volume. It is a highly organized, extremely productive volume, which is packed with enzymes, structural and nonstructural proteins and ribosomes on which most of the protein synthesis takes place. It is the site of much of the cell's intermediate metabolism and the major site of post-translational modification of proteins with more than 100 such processes having been identified. Many of these proteins remain in the cytosol; some are acylated with fatty acids which then anchor them in the membrane, while another group function only briefly before being degraded. These latter ones are often enzymes responsible for crucial steps in biosynthetic or degradation processes where turnover of the enzyme exercises fine control over the whole scheme.

The weight of proteins often accounts for about 20% of the mass of the cytosol and thus it is probably more accurately thought of as a gel rather than as a liquid. Nevertheless, some small molecules, including even small proteins, diffuse through it almost as fast as they do through pure water. Larger proteins and organelles diffuse very slowly but move along proteinaceous filaments by energy-requiring transport processes.

Embedded in the cytosol is a vast network of filaments of varying sizes, structures and complexities which give shape and form to the cell, transport materials, hold organelles in place, act as a template for enzymatic reactions and play vital roles in cell movement and cell division. This has been called the cytoskeleton but this may be misleading if it conjures up a picture of a rigid unchanging framework. In fact, the system is more dynamic than rigid and degradation–resynthesis of the framework is used in cell movement and division. Three, perhaps four, different major categories of filaments have been identified as making up the cytoskeleton. The largest of these are the microtubules which are made up from an array of tubulin molecules. Tubulin is a dimeric protein made up of two globular 450 amino acid proteins, α- and β-tubulin, of very similar structures. Linear chains of alternating α and β units form protofilaments, which are then joined in bundles of 13 to form an open tube 25 nm in external diameter and 15 nm internal diameter. All protofilaments are

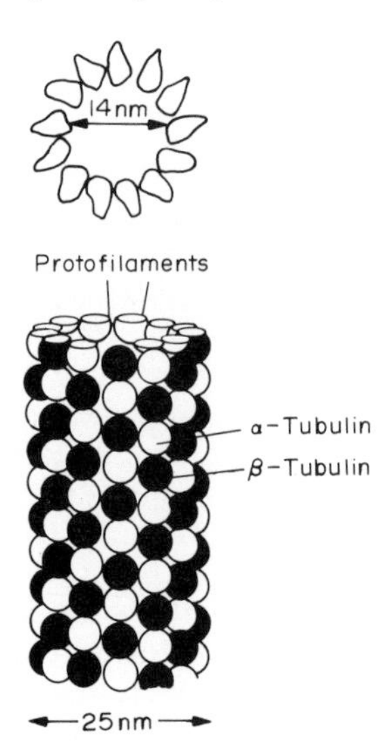

Figure 17 The microtubule in cross section and side view (reproduced from B. Alberts, D. Bray, J. Lewis, M. Raff, K. Roberts and J. D. Watson, 'The Molecular Biology of the Cell', 2nd edn., 1989, p. 646 by permission of Garland Publishing, New York)

aligned in the same direction as shown in Figure 17. Microtubules stretch throughout the cytoplasm and, in some cases, most notably that of the axon of nerve cells, this can be a considerable distance. However, it is not thought that these are continuous structures; it is more likely that there is an overlapping arrangement of smaller fragments with a maximum length of 10–25 μm. The importance of microtubules in nerve cells is perhaps indicated by the finding that tubulin comprises some 10–20% of soluble brain proteins.

Microtubules come together as groups of nine closely aligned parallel triplets forming a tube 0.2 μm wide by 0.4 μm long. Two of these units at right angles to one another form a body called a centriole (Figure 18). A pair of these centrioles form another body, a centrosome (or cell centre), which is found to one side of the nucleus and which plays a crucial role in cell division. The topic of cell division is too large to be dealt with here in any detail and the reader wanting more information is referred to more specialized texts. Suffice it to say that it is an ongoing process with even mature organisms needing to constantly replace worn out or damaged cells; for example, the human adult replaces millions of cells per second. Division begins when the cell has reached a sufficient size and involves sharing the organelles of the cytoplasm roughly equally between the two daughter cells, whilst the DNA of the nucleus is shared exactly equally. Two processes initiate the division: in the nucleus DNA is replicated and this is discussed fully in Chapter 3.7, while in the cytoplasm the centrosome acts as a centre from which a microtubule array radiates out. The assembly–disassembly–reassembly of the microtubules characterizes the process whereby the cell moves into and through division. During division, the centrioles and centrosome double and then divide. A number of drugs, including colchicine and vinblastin, can stop microtubule elongation and halt cell division. As cancer cells divide more rapidly than normal ones, such agents will have greater effects on cancer cells and thus can be used as a form of anticancer therapy.

A poorly defined amorphous cloud around the centrosome acts as an anchor point for one end of the microtubule, while the other end grows into the cytoplasm as a single filament. Microtubules are found in most eukaryotic cells but, as indicated above, are most developed in neurones. These cells have particular shapes (Figure 1a) with a cell body containing the nucleus, the endoplasmic reticulum and Golgi apparatus and a long axon ending in a synaptic cleft and a junction with

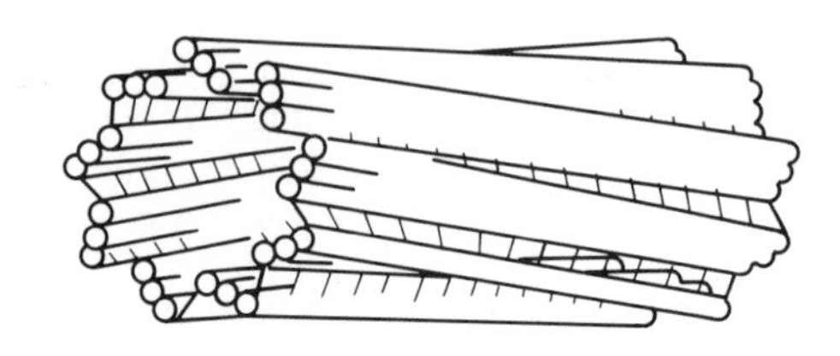

Figure 18 The centriole (reproduced from B. Alberts, D. Bray, J. Lewis, M. Raff, K. Roberts and J. D. Watson, 'The Molecular Biology of the Cell', 2nd edn., 1989, p. 650 by permission of Garland Publishing, New York)

another neurone or muscle fibre. The axons may be very long, up to 1 m, and the cell has the problem of transporting neurotransmitters synthesized in the cell body to the synaptic cleft. Clearly, random diffusion could not achieve this. What happens is that the vesicles containing the neurotransmitter are transported along the microtubule from the cell body to the synaptic cleft. Vesicles containing fragments of the cell membrane and endocytosed material destined for destruction in the lysosomes make the reverse journey. Remarkably, any single microtubule seems able to transport the vesicles both ways. Mitochondria are also shuffled along the microtubules backwards and forwards according to energy needs. Clearly, a mechanism which accommodates such transport is subtle, to say the least. One theory to account for this has small protein arms sticking out of the microtubule which can pass vesicles from one to another with conformational changes in the arms, driven by ATP hydrolysis, conveying movement along the filament.

As well as playing such a vital role in the cytoplasm of the cell, microtubules are a major constituent of cilia and flagella. Cilia are tiny hair-like appendages found on the surface of many endothelial cells whose function it is to move fluid over the surface of the cells. They are especially marked in epithelial cells in the human respiratory tract where they move mucus towards the mouth. Flagella are rather longer and stronger entities which propel sperm and some protozoa. The core structure of both is provided by a complex of nine double microtubules which form a circle around a further pair of single microtubules. The doublets and central pair are interconnected by several different protein links. One of the proteins, dynein, projects from the surface of the microtubules and causes the two components of the doublet to slide relative to one another. Again, this process is powered by ATP hydrolysis. Shear resistance produced by protein links between the microtubules in each pair causes the resultant bend.

Cilia and flagella with this core structure are restricted to eukaryotes. Prokaryotes use an entirely different mechanism. The flagella in this case is formed from a helical protein which is connected to a circular protein in the bacterial membrane. ATP hydrolysis drives the rotation of the membrane protein and consequent movement of the flagellum which propels the bacterium.

The intermediate filaments are smaller than the microtubules and have more varied constitutions. These tough, rope-like fibres are usually 8–10 nm in diameter and are constructed from six or seven different proteins. They are found in both the cytoplasm and the nucleus. In the nucleus a proteinaceous mesh some 10–20 nm thick lines the inside of the nuclear membrane, giving it strength and form. The cytoplasmic intermediate filaments form a basket around the nucleus from which the fibres gently curve out towards the plasma membrane.

The function of the cytoplasmic intermediate filaments is not entirely clear, as some eukaryotes, most notably the glial cells, are completely lacking this structure. It is probable that it serves to resist tension and stress in the cell.

Smaller still are the actin filaments. These have a diameter of around 8 nm and are made solely from actin, a globular protein of a molecular weight of about 42 000 Da. In most eukaryotic cells actin is the most abundant protein comprising about 5% of the cell's protein. Actin filaments are mainly distributed in the cell cortex, an especially dense network of filaments found just below the plasma membrane. There also appear to be covalent links between transmembrane proteins and the cell cortex, which presumably enhance its role in stabilizing the cell perimeter. Parts of the cortex are further stabilized by being cross-linked to another protein, filamin. As with the microtubules, the construction of the actin filaments is a dynamic one with a pool of monomeric actin available to be polymerized to the filament as required. It is believed that the rapid polymerization–degradation process allows ready changes to the cortex and thus to the shape of the cell. Such changes would be of great importance in cell movement and endocytosis.

Studies using the very high magnification available with high voltage electron microscopy have suggested the presence of yet another, thinner matrix of protein filaments linking mitochondria, microtubules, the actin filaments, ribosomes and other organelles. If this is truly present, it would complete a picture of a highly complex, interlinked network of filaments of a range of sizes which give a rigid, yet changing structure to the cell. It would be a matrix in which all the organelles of the cytoplasm could be held in the appropriate positions, relative to each other, for the most efficient functioning of the cell.

All of the above refers only to eukaryotes; similar structures do not seem to have been identified in prokaryotes.

2.2.4.3.7 ***The nucleus***

The last, but most definitely not the least, important structure to be considered is the nucleus. This contains DNA, the genetic material of the cell, and a significant part of the replication and

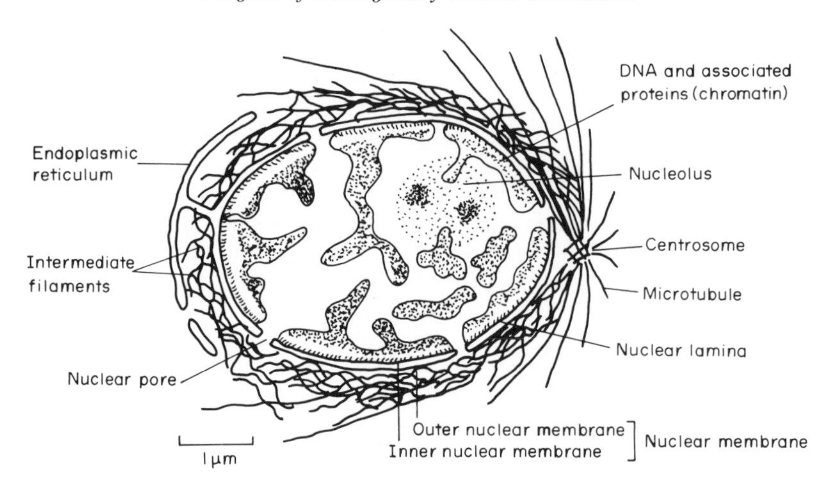

Figure 19 The nucleus (reproduced from B. Alberts, D. Bray, J. Lewis, M. Raff, K. Roberts and J. D. Watson, 'The Molecular Biology of the Cell', 2nd edn., 1989, p. 481 by permission of Garland Publishing, New York)

transcription machinery. By definition, only eukaryotic cells contain their DNA in a defined membrane-bound unit: no such structures are found in prokaryotes.

Figure 19 shows a diagrammatic representation of the structures seen in electron micrographs of the nucleus. It will be realized that this is considerably less structured than the cytoplasm with only the nuclear envelope, the nuclear lamina, chromatin and the nucleosome being distinguished. There is still confusion over whether or not the nucleus has a fine filamentous structure, but in view of the efficiency gains such a structure gives in keeping consecutive enzymes properly aligned, it seems likely that this does exist.

The nuclear envelope is a double membrane in which the inner membrane is connected *via* transmembrane proteins to the nuclear laminar matrix of intermediate filaments discussed in the previous section. The outer membrane is very similar to the endoplasmic reticulum with which it is continuous. It is studded with ribosomes and the proteins synthesized on these are released into the perinuclear space, *i.e.* the space between the membranes, which is continuous with the lumen of the endoplasmic reticulum. From there these proteins can be passed to the Golgi apparatus for further packaging and subsequent release. The outer membrane is stabilized by the cytoplasmic intermediate filaments which are found in close proximity to it.

Numerous water-filled pores are found in the nuclear membrane. These allow peptides of less than 5000 Da molecular weight to pass freely through them. Proteins of more than 60 000 Da do not diffuse through at all, while those of intermediate weights diffuse at intermediate rates. However, the nucleus contains many proteins with molecular weights much greater than 60 000 Da and since these are all synthesized in the cytosol, there must be mechanisms whereby these can pass through the pores. This is accomplished by the presence of specific receptor proteins set in the periphery of the pores, which recognize the required proteins and widen the pore to allow access.

Chromatin is a complex of DNA and many types of protein, the most abundant of which are histones. These are relatively small proteins containing large amounts of lysine and arginine to help them to bind to most forms of DNA. The total masses of DNA and histones are roughly equal. Chapter 3.7 gives a full and excellent account of DNA replication and transcription to RNA and so the process will not be described here except to say that these processes take place in the nucleus. Further, there is extensive modification of RNA after initial transcription before it is packaged as ribosomal, messenger or transfer RNAs. All these processes take place also in the nucleus but the final stage, that of protein synthesis, occurs in the cytosol. This is quite different from prokaryotes where RNA is being synthesized from its 3′ end at the same time as its 5′ end is controlling the synthesis of proteins on the ribosomes. Bacterial DNA forms a single thread lying in the cytoplasm and attached to the membrane at one point. It is unprotected by even histones and should therefore be more vulnerable to specific agents than eukaryotic DNA.

The nucleolus is a distinguishable but ill-defined region which is not membrane bound. It is the area in which the constituent parts of the ribosomes, the proteins and the ribosomal RNA, are assembled into two fragments, which are then released from the nucleus. In the cytosol these two fragments coalesce before being further activated to produce the functioning ribosome. It appears that the structure of the nucleolus is actually defined by the ribosomal protein fragments themselves.

These are the major structures seen in the nucleus, although as stated above it seems highly likely that others exist. The reason for eukaryotes having an elaborate nuclear membrane is not clear but may be because the very long DNA molecule is fragile and could be damaged by the considerable stresses found in the flexible cytoplasm. In prokaryotes with a rigid cell wall such stress is very much less and presents less of a danger to the DNA.

The structures described above are all found in cells which by the usual biological definitions can be regarded as 'living'. The last part of this chapter will deal with systems which do not fit the normal definitions of being alive, but nonetheless cause disease and suffering to mammals and plants and thus represent worthwhile targets for medicinal chemists. Such entities are viruses, viroids and prions.

2.2.5 VIRUSES

Viruses are responsible for a large number of human, animal and plant diseases. Virus diseases of animals include the common cold, influenza, smallpox, herpes, hepatitis, polio, rabies, the acquired immune deficiency syndrome (AIDS) and cancers such as adult T-cell and hairy cell leukemias. Plant viruses include the first virus to be fully characterized, the tobacco mosaic virus which is responsible for the distortion and blistering of leaves in a number of plants. One class of viruses, the bacteriophages even infect bacteria.

The basic form of a virus is a particle smaller than a bacterium, in which the central core of genetic material is surrounded by a protein shell known as a capsid. Figure 20 shows a simplified view of a fairly complex virus, the human immunodeficiency virus, the presumed cause of AIDS. The genetic material can be either DNA or RNA, but not both, and proteins are often found in close association with the chromosome. The capsid is made from one or more proteins, which are often found in several layers. Outside the capsid is often found a lipid membrane in which the transmembranal proteins are closely associated with the proteins of the capsid. The shape of the virus is determined

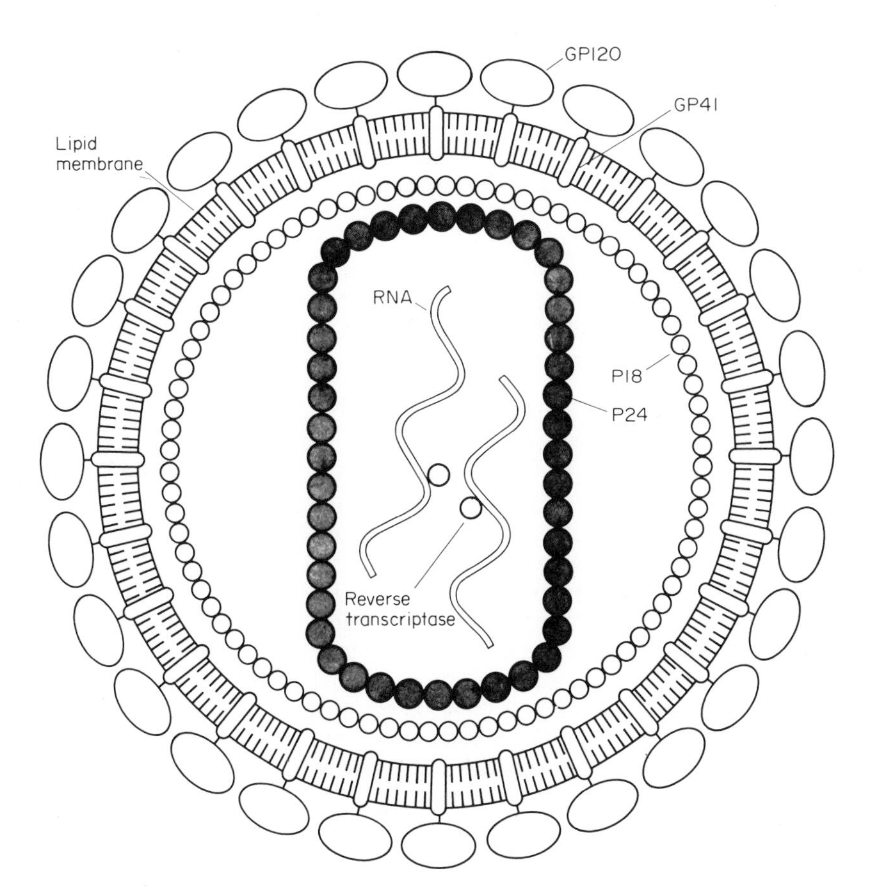

Figure 20 Schematic representation of the HIV virus (reproduced from R. C. Gallo, *Sci. Am.*, January 1987, **256** 39 by permission of the publisher)

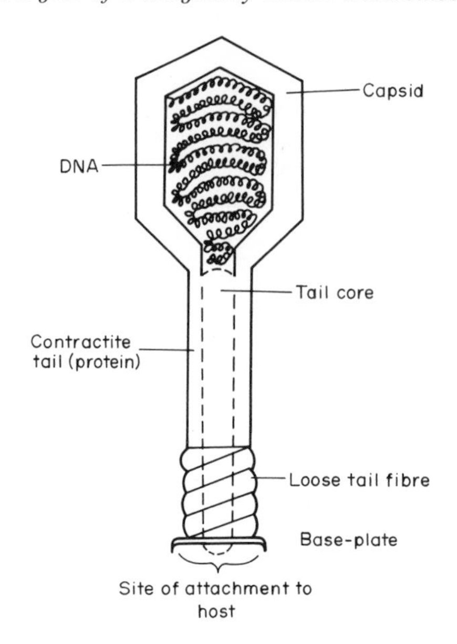

Figure 21 Representation of a bacteriophage (reproduced from A. Albert, 'Selective Toxicity', 5th edn., 1973, p. 129 by permission of Chapman & Hall, London)

by the structure of the capsid with perhaps the most common one being an icosohedron. However, the virus which causes measles is a spiral and the very large poxviruses are brick shaped. Bacteriophages often have the characteristic tailed shape shown in Figure 21.

Viruses cannot 'live' outside host cells. They contain no energy-producing elements and have no machinery for reproduction. Instead they invade a target cell, possibly by first binding to recognition sites on the membrane, and then penetrating the membrane or cell wall. The viral genetic material can then be released into the cell. Thus, for example, the bacteriophage T2 attacks *E. coli* by the long tail fibres aligning the spiked base plate onto the bacterial cell wall. A lysosome-like enzyme in the plate opens the cell wall allowing the genetic material to pass into the bacterium. Once inside the cell, the viral chromosome then hijacks the cell's reproductive machinery to produce, initially, multiple copies of the chromosome, which are then transcribed to produce all the proteins and enzymes required to produce copies of the invading virus. The new viruses burst out of the cell by lysis, bringing about the death of the cell. The death of the cell causes the characteristic sores of infections such as that caused by the herpes virus. Alternatively, the virus can leave by budding off vesicles from the cell membrane such that the cell is not lyzed and thus survives the attack. These viruses will be coated with the host cell's membrane and the process is of course greatly to the benefit of both the host and the parasite.

The genetic material of viruses can be either DNA or RNA with approximately equal quantities of each being responsible for human diseases. Thus typical DNA viruses are adenoviruses which cause upper respiratory tract infections, pox and herpes viruses. RNA viruses include myxoviruses, which are responsible for influenza, paramyxoviruses which cause mumps and measles, and those responsible for yellow fever and AIDS. The chromosome, whether RNA or DNA, is much smaller than that of man and comes in a remarkable variety of forms. Tobacco mosaic virus has a single-stranded RNA chain, reovirus a double-stranded RNA helix, parvoviruses a linear single-stranded DNA chain and M13 bacteriophage a circular single-stranded DNA chain. In addition, linear DNA double helices, circular DNA double helices and even more complex linear double helices have been found.

Whatever its shape, the chromosomal material is considerably shorter than that of the mammalian hosts and consequently codes for a much more restricted range of proteins. However, at the very least, it has to encode for proteins which make its continued replication favoured over that of the host chromosome. RNA viruses have the additional problem that the host does not copy RNA molecules so their genetic material has to code for RNA dependent nucleic acid polymerase enzymes in order to replicate.

In fact there are two types of RNA viruses. In the first type, the virus simply sidesteps the DNA part of the transcription process and makes RNA copies of itself. The second carries with it an enzyme, reverse transcriptase, which transcribes the viral RNA into complimentary DNA molecules.

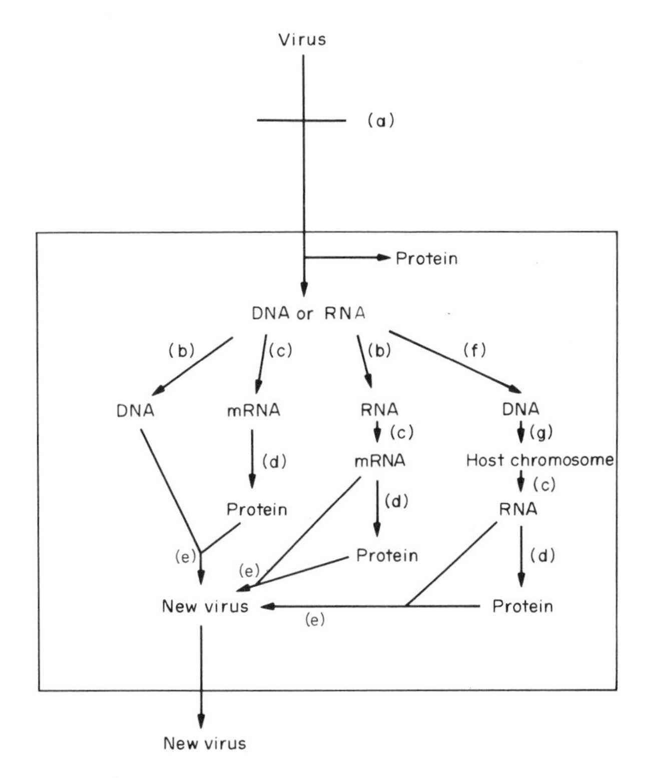

Figure 22 Possible sites of action of antiviral drugs

The DNA thus produced can be incorporated into the host's genome, resulting in permanent genetic change. In the meantime, the invading RNA is replicated, protein synthesis initiated and new viruses produced. Retroviruses are responsible for human cancers such as adult T-cell leukemia. HIV is also a retrovirus.

Selective attack on viruses has been difficult because of the way in which they reproduce only inside the cell. Nevertheless there are a number of points at which attack might take place (Figure 22) and an increasing number of agents are in clinical use. The first possibility is to prevent the virus from binding to the host cell (a). Inside the cell DNA viruses could be affected by inhibiting the replication (b) or transcription (c) processes, although it will be difficult to achieve selectivity if the host's own systems are used. Translation (d) and assembly of new viruses (e) are other targets. For RNA viruses replication (f), transcription (g) and, particularly, reverse transcription are attractive targets since they are not normal host cell processes.

Current antiviral treatments include 3′-azido-2′,3′-dideoxythymidine, AZT (**13**), which is widely used in AIDS treatment. This agent is a prodrug which in the cell is phosphorylated to AZT triphosphate, which acts in competition with the true substrate for DNA synthesis, namely thymidine triphosphate (**14**) on the reverse transcriptase enzyme. Selectivity over the host cell mechanisms arises because AZT has 100 times greater affinity for reverse transcriptase than for DNA polymerase. AZT lacks a 3′-hydroxy group and the DNA chain stops growing at this point. A similar mechanism applies to acyclovir (**15**), which acts on the herpes virus found in genital herpes simplex infections. Again, acyclovir has to be phosphorylated before acting on DNA polymerase;

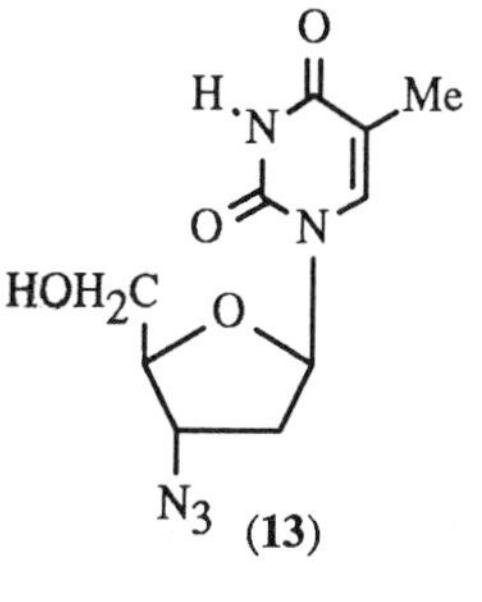

(**13**) 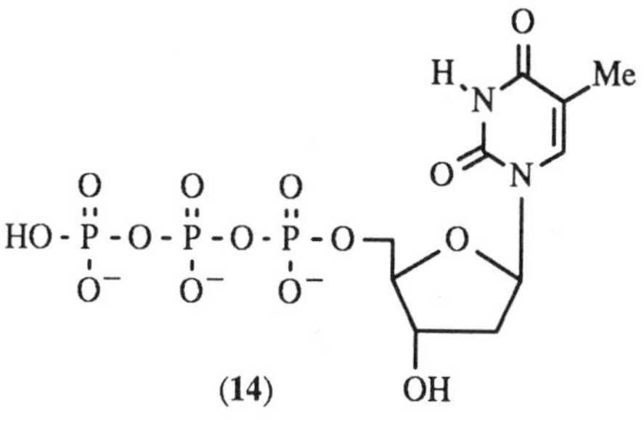 (**14**) 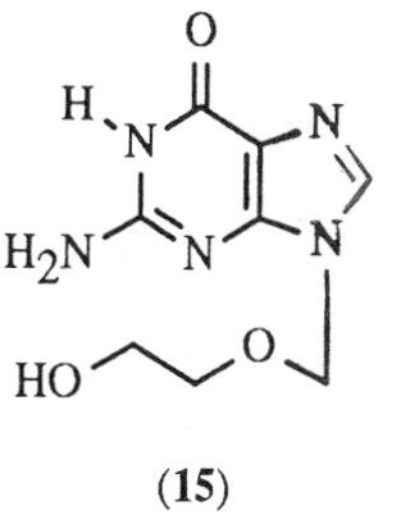 (**15**)

a phosphorylation reaction catalyzed by an enzyme, thymidine kinase, found only in infected cells. In noninfected cells other enzymes phosphorylate acyclovir to a very much lesser extent. Thus the agent is selectively activated where required and then acts as a nucleoside mimic which is taken up by the DNA polymerase of the herpes virus to a greater extent than the host polymerases. Finally the drug lacks a free 3′-OH group and polymerization ceases. Very much less is known about the mode of action of other antiviral agents, a situation which must become much clearer as a result of the ongoing work directed against the HIV virus.

2.2.5.1 Viroids

These are naked single strands of RNA with a molecular weight of less than 13 000 Da, which have been implicated in about 12 different diseases affecting higher plants. They have also been postulated as the agent involved in a number of human cerebral disorders. Quite how they can operate is very unclear as they do not seem to have sufficient genetic material to produce meaningful amounts of protein.

2.2.5.2 Prions

If the viroids are difficult to rationalize, then the prions are totally obscure. These appear to be short lengths of protein which have been implicated in the fatal human CNS disorder, Kreutzfeld–Jacob's disease, and scrapie, the fatal sheep disease. So far no DNA or RNA has been definitely associated with these particles, so how they could subvert the functioning of a host cell is difficult to imagine.

ACKNOWLEDGEMENT

The help of Dr. P. Miller is gratefully acknowledged.

2.2.6 SELECTED ADDITIONAL READING

1. B. Alberts, D. Bray, J. Lewis, M. Raff, K. Roberts and J. D. Watson, 'The Molecular Biology of the Cell', 2nd edn., Garland Publishing, New York, 1989.
2. A. Albert, 'Selective Toxicity', 5th edn., Chapman & Hall, London, 1973.
3. J. Mandelstam and K. McQuillen (eds.), 'Biochemistry of Bacterial Growth', 2nd edn., Blackwell, London, 1973.
4. P. L. Carpenter, 'Microbiology', 3rd edn., Saunders, Philadelphia, 1972.
5. L. E. Hawker and A. H. Linton (eds.), 'Micro-organisms', Arnold, London, 1974.
6. Scientific American articles on specific topics:
Golgi Apparatus: J. E. Rothman, September 1985, p. 84.
Mitochondrial DNA: L. A. Grivell, March 1983, p. 60.
Myelin: P. Morell and W. T. Norton, May 1980, p. 74.
Microbodies: C. de Duve, May 1983, p. 52.
Microtubules: P. Dustin, August 1980, p. 58.
Microtubules: R. D. Allen, February 1987, p. 26.
Microtubules: J. H. Schwartz, April 1980, p. 122.
Ribosomes: J. A. Lake, August 1981, p. 56.
Ribosomes: M. Nomura, January 1984, p. 72.
Antivirals: M. S. Hirsch and J. C. Kaplan, April 1987, p. 66.
Prions: S. B. Prusiner, October 1984, p. 48.
Viroids: T. O. Diener, January 1981, p. 58.
Ground Structure: K. R. Porter and J. B. Tucker, March 1981, p. 40.
Cell Movement: M. S. Bretscher, December 1987, p. 44.
Cell Movement: E. Lazarides and J. P. Revel, May 1979, p. 88.
AIDS: whole issue, October 1988.
The Nucleosome: R. D. Kornberg and A. Klug, February 1981, p. 52.
Reverse Transcription: H. Varmus, September 1987, p. 48.
7. S. J. Singer and G. L. Nicolson, Fluid mosaic model of cell membranes, *Science (Washington, D.C.)*, 1972, **175**, 720.
8. V. T. Marchesi, H. Furthmayr and M. Tomita, Red cell membranes, *Annu. Rev. Biochem.*, 1979, **45**, 667.
9. V. Bennett, The membrane skeleton of human erythrocytes and its implications for more complex cells, *Annu. Rev. Biochem.*, 1985, **54**, 273.
10. E. Lazarides, Intermediate filaments, *Annu. Rev. Biochem.*, 1982, **51**, 219.
11. H. G. Wittman, Architecture of prokaryotic ribosomes, *Annu. Rev. Biochem.*, 1983, **52**, 35.

2.3
Macromolecular Targets for Drug Action

TERRY P. KENAKIN
Glaxo Research Laboratories, Research Triangle Park, NC, USA

2.3.1 INTRODUCTION

'The most interesting feature of drug action is the extraordinary specificity of the action of drugs and the manner in which slight changes in chemical constitution alter their action . . . ' A. J. Clark (1937)

The rational design of a chemical structure to perturb a biological system is obviously a difficult task since so often the components of the biological cascade that is to be perturbed are unknown. Therefore the definition of a 'target' for the initial interaction between the drug and the biological system is of paramount importance to the success of a drug design program. Considering the number of chemicals involved in cellular processes, it is not surprising that pharmacologists, physiologists, medicinal chemists and biochemists have found numerous places in cellular pathways to interfere with normal or pathophysiological cellular function and alter physiological responses. The following is a brief overview of some cellular 'targets' for drug action and how they have been exploited for therapeutic advantage.

2.3.2 AGONISTS AND ANTAGONISTS

As a preface to the discussion of the interaction of drugs with macromolecules, a definition of agonism and antagonism would be useful. The occupation of a macromolecule by a drug that prevents the activation of the same macromolecule by a hormone, neurotransmitter, autacoid or other drug will be regarded as antagonism and the drug referred to as an antagonist. Similarly, a drug which activates the macromolecule to initiate a physiological response is an agonist. This latter category is further subdivided into full agonist for drugs which produce the maximal response that can be elicited from a tissue and partial agonist for drugs which produce a maximal response that is less than the tissue maximum (as measured by the effects of a full agonist). These latter differentiations can lead to ambiguity, since the ability of a drug to produce the maximal tissue response often varies from tissue to tissue depending on the amplification factors for receptor activation inherent in each tissue stimulus–response mechanism. Therefore, a drug such as the β adrenoceptor drug prenalterol can be a full agonist in rat atria, a partial agonist in cat atria and an antagonist in canine coronary artery. These differences are tissue dependent and do not relate to the macromolecule responsible for extracellular recognition (in this case, the β adrenoceptor). Such differences are confusing and make the classification of agonists by the magnitude of tissue response cumbersome; classification by relative intrinsic efficacy, a dimensionless proportionality constant (*vide infra*), is preferable when possible.[1] Other problems arise when multivariate systems are used to assess response. For example, an antagonist of the enzymatic hydrolysis of cAMP by phosphodiesterase produces an elevation of this second messenger and a subsequent increase in cardiac contractility. Thus, an enzyme antagonist becomes an agonist with respect to the tissue. Wherever possible, the classifications in this chapter will refer to the effects nearest to the initial interaction between the drug and the macromolecule. In the previous case, in spite of the observed increased inotropy, the drug will be classified as an antagonist of the enzyme phosphodiesterase. In general, the capricious nature of the properties of agonism by drug molecules with respect to the complex systems used to measure response sometimes makes drug classification and nomenclature contradictory.

2.3.3 MEMBRANE-BOUND RECEPTORS

Drugs can produce effects *via* a number of mechanisms by virtue of surfactant properties (amphotericin), their ability to denature proteins (astringents), acidic or basic properties (antacids, protamine), osmotic properties (laxatives, diuretics) and physicochemical interactions with membrane lipids (general and local anesthetics). A large proportion of drug effect is mediated by selective interaction of drugs with cognitive proteins on the cell surface referred to as receptors. Drug interaction with receptors is a major pathway of communication between molecules in the extracellular space and the cell. The concept of specific receptors for hormones, neurotransmitters, autacoids and other molecules emerged at the turn of the century from studies by Ehrlich (1854–1915) and Langley (1852–1926) and the ideas that describe the effects of drugs on receptors are referred to collectively as drug receptor theory.

2.3.3.1 Receptor Theory

The mathematical models used to describe the interactions of drugs with receptors are derived from those developed for the study of enzyme–substrate interactions. There are two important caveats to be recognized in the use of these models for this purpose; the first relates to the fact that, unlike well-mixed biochemical enzymatic reactions in structured organs, diffusion is often restricted and concentrations of drugs in the receptor compartment may have a temporal restraint. Since these equations are based on the assumption that drug concentrations rise to equilibrium levels instantaneously, miscalculation can result. Secondly, the observed effect of a drug–receptor interaction is often the product of a complex chain of enzymatic processes that amplify the effects of the initial drug–receptor binding. Therefore, response cannot be used directly to assess the extent of drug binding to receptors.

The interaction between a drug (A) and receptor (R) is characterized by a kinetic rate of onset (A binding to R, k_1) and rate of offset (A dissociating from R, k_2), as shown in equation (1).

$$\mathrm{A} + \mathrm{R} \underset{k_2}{\overset{k_1}{\rightleftharpoons}} \mathrm{A \cdot R} \tag{1}$$

At any instant, the fraction of receptor molecules bound by drug (A·R complex) is given by the Hill equation[2,3] as

$$\frac{[\mathrm{A \cdot R}]}{[\mathrm{R}]} = \frac{[\mathrm{A}]}{[\mathrm{A}] + K_{\mathrm{A}}} \tag{2}$$

where K_{A} refers to the equilibrium dissociation constant of the drug–receptor complex and is defined by $K_A = k_2/k_1$. The quantal stimulus received by a receptor upon interaction with a drug molecule is referred to as the intrinsic efficacy[4] and denoted ε. Thus, the stimulus[5] (S_R) per receptor is given by

$$S_R = \frac{\varepsilon \cdot [A]}{[A] + K_A} \tag{3}$$

The tissue receives a net signal that is the summation of these quantal stimuli for the receptor density $[R_t]$.

$$S = [R_t] = \frac{\varepsilon \cdot [R_t]}{1 + K_A/[A]} \tag{4}$$

The response (*i.e.* effect, E) is an unspecified function (f) of the receptor stimulus.[5]

$$E = f\left(\frac{\varepsilon \cdot [R_t]}{1 + K_A/[A]}\right) \tag{5}$$

It can be seen from equation (5) that the magnitude of the drug effect is dependent upon four factors. Two of these are related to the tissue, namely the receptor density $[R_t]$ and the amplification function f which comprises the cascade of cellular reactions that translate receptor stimulus into organ response. Two of these factors are strictly drug related, namely the equilibrium dissociation constant of the drug–receptor complex (K_A) and the intrinsic efficacy ε. These latter factors should be unique for every drug–receptor pair and thus should transcend organ type, species and function; as such, these parameters can be used for drug and drug–receptor classification. The equilibrium dissociation constant of the drug–receptor complex is a chemical term ($K_A = k_2/k_1$) and has meaning on a molecular level. The intrinsic efficacy is a proportionality constant with no molecular meaning. It will be seen in later sections that there are approaches to drug efficacy which define efficacy in molecular terms (*vide infra*). However, this term is useful from the point of view of drug classification since it quantifies the relative propensity of a group of drugs to produce receptor stimulus.

In terms of biological data of relative drug activity for use in structure–activity relationships and further medicinal chemical synthesis, equation (5) illustrates the possible confusion inherent in the reliance on biological effect as a read-out of activity. If the magnitude of response is used, then drug classification is dependent upon the tissue and species with all their variances since the factors f and $[R_t]$ are involved. On the other hand, if the factors K_A and relative ε can be obtained, these should be predictive of effects in all organs and species including man. Also, it can be seen that reliance upon K_A and ε should greatly diminish the amount of biological data a chemist must ingest in studies of structure and activity.

In general, the affinity of a drug for a receptor (often expressed as the reciprocal of the equilibrium dissociation constant of the drug–receptor complex; affinity = K_A^{-1}) describes the strength of the forces which keep the drug molecule within the force field of the receptor and the intrinsic efficacy describes what the drug does when it gets there. It is worth considering these separately.

2.3.3.2 Drug Affinity

Drugs bind to receptors because chemical forces create a position of minimum free energy (the equilibrium position) within the force field of that receptor. In general, most of the intermolecular forces involved depend upon distance by a general function[6]

$$u = cr^{-p} \tag{6}$$

where c is a constant and p a power of inverse relationship. The effect of distance on some intermolecular forces is given in Table 1. From this table it can be seen that different forces dominate at different distances. The forces which attract a drug to a receptor are counteracted by the universal repulsive force resulting from the invasion of van der Waals envelope of atoms for which $p = 9$–12. It can be seen from this high value of p that repulsive forces become insurmountable only when the drug molecule comes very near to the receptor. The combination of these effects creates a distance of minimum free energy at which the drug most probably will reside. Drugs escape from this position only when thermal agitation provides enough kinetic energy to overcome the attractive forces. Drug molecules move away from the receptor to a position where the acquired kinetic energy balances the

Table 1 Strengths of Electrostatic Bonds[7]

Type	*Energy of bond*[a] (kJ mol^{-1} per bond)	*Strength vs. distance Relationship*
Ion–ion	20–40	$1/d$
Ion–dipole	8–20	$1/d^2$
Dipole–dipole	3–15	$1/d^3$
Hydrogen	5–25	$1/d^4$
Induced dipoles	0.5–5	$1/d^5$–$1/d^8$

[a] At typical bond length.

increased potential energy of its new position and then will return to the equilibrium distance. The same occurs as the drug approaches toward the receptor beyond the position of minimum free energy.

In terms of drug design, there are examples of rational structure–activity relationships leading to valuable drugs. For example, two very important classes of drugs, the β blockers and histamine H_2-receptor antagonists, were designed on the principle that since the receptor recognizes the endogenous ligand (in this case epinephrine and histamine respectively) these should be the starting points for chemical design.[8–11] Thus, chemical modifications of epinephrine and histamine were made until intrinsic efficacy was eliminated but affinity was conserved. A sequence of the key structures leading to the prototype histamine H_2-receptor antagonist cimetidine is shown in Table 2.

There are many examples of drugs with a high affinity for receptors which structurally resemble the endogenous ligand but are larger more flexible molecules with added lipophilic groups,[12,13] and some examples of antagonists which bear little resemblance to the endogenous ligand. The study of the binding of these drugs to receptors has shown the importance of accessory binding sites on drug receptors distinct from the active site of endogenous ligand binding. Thus, the aromatic ring structures of benzylcholine may bind to such an accessory binding site in order to block acetylcholine response (see Figure 1a).[14] There are also experimental data which strongly suggest that the agonist acetylcholine and cholinergic antagonist α-bungarotoxin bind to separate sites on acetylcholine receptors.[15]

A thermodynamic rationale for this effect has been given in the 'Charniere theory' of antagonism,[16] where it is postulated that van der Waals forces tightly bind the large aromatic moieties of, for example, lachesine and leave the polar chain to compete for binding to the agonist recognition site of the receptor with acetylcholine (Figure 1b). Alternatively, binding of an antagonist to accessory sites separate from that of agonist binding may induce allosteric perturbation of the receptor thereby inducing a change in the tertiary structure of the agonist recognition site and preventing agonist binding. From the point of view of drug design, these effects essentially support a 'receptors unlimited' concept, in that a vast array of binding fields could be used to block pharmacologic response to endogenous agonists. The relevance of these sites to agonism, however, is more difficult to assess.

Such interactions raise questions about the value of biochemical binding studies of radioligands for drug discovery. For example, the histamine H_2-receptor antagonist ranitidine blocks histamine-induced tachycardia in isolated guinea pig atria with an equilibrium dissociation constant of 2.5×10^{-7} mol dm^{-3};[17] this corresponds to the concentration of ranitidine that occupies half of the receptor population. However, the concentration of ranitidine required to displace the radioactive histamine H_2-receptor blocker [^{3}H]cimetidine is 1900 times greater (4.8×10^{-4} mol dm^{-3}),[18] indicating that the two antagonists, cimetidine and ranitidine, bind to separate sites on the histamine H_2-receptor. Similarly, although nicotine displaces the radioligand [^{3}H]dihydro-β-erythroidine from nicotinic receptors in rat cortical membranes, the nicotinic antagonists mecamylamine, hexamethonium and pempidine were essentially inactive,[19] indicating separate binding sites for the agonist and antagonist. These data strongly indicate a need for the use of intact physiological systems in the drug development process as an essential complement to biochemical binding studies.

2.3.3.3 Intrinsic Efficacy

In general, there are two ways in which a drug can produce a change in a receptor that can lead to a biological response. The first is by conformational induction,[20] where the drug produces a

Table 2 Some Important Chemical Structures in the Discovery of Histamine H_2-receptor Blocking Agents *en route* from the Natural Endogenous Agonist Histamine to the Clinically Effective Blocking Drug Cimetidine[10]

Compound	*Structure*	*Antagonist in vitro*, K_B[a] (10^{-8} mol dm^{-3})	*Activity in vivo*, ID_{50}[b] (μg mol kg^{-1})
Histamine	4(5)-imidazolyl–$CH_2CH_2NH_2$		
α-Guanylhistamine (the 'lead', a weakly active partial agonist)	4(5)-imidazolyl–$CH_2CH_2NHC(=\overset{+}{N}H_2)NH_2$	130	800
SK&F 91486 (lengthening the side chain increases activity)	4(5)-imidazolyl–$CH_2CH_2CH_2NHC(=\overset{+}{N}H_2)NH_2$	22	100
SK&F 91581 (thiourea analogue is much less active as an antagonist, but is not an agonist)	4(5)-imidazolyl–$CH_2CH_2CH_2NHC(=S)NHMe$	115	c
Burimamide (lengthening the side chain again dramatically increases antagonist activity)	4(5)-imidazolyl–$CH_2CH_2CH_2CH_2NHC(=S)NHMe$	7.8	6.1
Metiamide (introducing —S— in the side chain and in the ring alters imidazole tautomerism and increases activity)	5-Me-4-imidazolyl–$CH_2SCH_2CH_2NHC(=S)NHMe$	0.92	1.6
Guanidine isostere (replacing C=S by C=CH gives a basic side chain and reduces activity)	5-Me-4-imidazolyl–$CH_2SCH_2CH_2NHC(=\overset{+}{N}H_2)NHMe$	16	12
Cimetidine (introducing a CN substituent reduces basicity and increases activity)	5-Me-4-imidazolyl–$CH_2SCH_2CH_2NHC(=N-CN)NHMe$	0.79	1.4

[a] Dissociation constant K_B determined *in vitro* on guinea pig right atrium against histamine stimulation. [b] Activity *in vivo* as an antagonist of near maximal histamine-stimulated gastric acid secretion in anesthetized rats using a lumen perfused preparation. The ID_{50} is the intravenous dose required to produce 50% inhibition. [c] No antagonism seen up to an intravenous dose of 256 μmol kg^{-1}.

conformational change in the tertiary structure of the receptor thus triggering the cascade toward cellular response. It would be supposed that the complementary binding of the drug molecule to the receptor would produce the structural change. A second general method to produce receptor stimulus is by conformational selection.[20] In this scheme, the receptor coexists in two interchangeable forms governed by an equilibrium constant (often referred to as an allosteric constant). Only one of the receptor forms leads to biological stimulus for response and, in the absence of an agonist, the equilibrium between the receptors lies heavily toward the inactive form. An agonist could produce response by selectively binding to the active form of the receptor, thereby producing a bias in the equilibrium and causing a net increase in the quantity of active form of the receptor.

It is known that drugs differ in their ability to produce response, *i.e.* they possess different magnitudes of intrinsic efficacy, but the actual mechanisms by which different drugs elicit different levels of response vary with the type of receptor. Thus drugs may differ in the quantity of structural

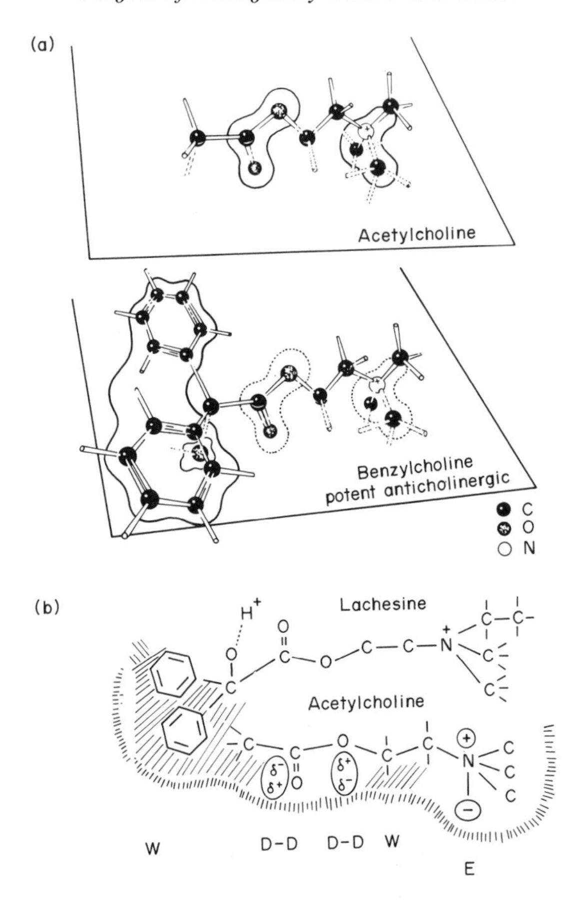

Figure 1 (a) Comparison of the receptor surfaces occupied by the potent agonist acetylcholine and the potent anticholinergic antagonist benzylcholine; note the large hydrophobic binding area for the aromatic rings of benzylcholine (reproduced from ref. 14 by permission of the New York Academy of Sciences). (b) The Charniere effect; the antagonism of acetylcholine receptors by lachesine. The large aromatic rings of lachesine may bind with van der Waals forces to an accessory site thus anchoring lachesine in place and allowing the polar chain to compete for acetylcholine binding at the receptor (reproduced from ref. 16 by permission of North-Holland Publishing Co.)

perturbation they impart to the receptor, the length of time they produce an activated state (for example, recent evidence shows that a range of nicotinic drugs produce an identical ion-channel opening but the channel remains open for differing amounts of time[21]) or how often they dissociate from and rebind to the receptor to induce an active state.

Although the mechanisms by which drugs produce agonist response are often not known, the operational use of the property of relative intrinsic efficacy can still be useful from the point of view of drug development. The use of relative intrinsic efficacy can simplify and organize biological data for use in structure–activity relationships. For example, Figure 2 shows dose–response curves for the α adrenoceptor agonists norepinephrine and oxymetazoline. In the rat anococcygeus muscle, oxymetazoline produces the maximal tissue response and is twice as potent as norepinephrine; it is more active than norepinephrine. In the rat vas deferens, oxymetazoline can be said to be less active by virtue of the fact that it produces only 20% of the maximal response of norepinephrine. Therefore a confusing tissue variable is introduced into the data, raising the question how can oxymetazoline be more active than norepinephrine in rat anococcygeus muscle and less active than norepinephrine in rat vas deferens? This problem can be circumvented by the calculation of the relative affinity and intrinsic efficacy of these drugs. In this case, it was found that oxymetazoline has six times the affinity but only one-third of the intrinsic efficacy of norepinephrine.[22] Since affinity determines potency, *i.e.* what concentration of receptors will be occupied by the drug, and intrinsic efficacy determines the power to produce response, the relative profile of these drugs would be predicted. The greater affinity of oxymetazoline for α adrenoceptors causes the dose–response curve to this drug to be shifted to the left of that for norepinephrine; the lower intrinsic efficacy dictates a lower response but, in the case of the rat anococcygeus muscle, a highly efficient receptor-coupling mechanism (*i.e.* a large receptor reserve) enables even a relatively weak drug such as oxymetazoline to produce the maximal tissue

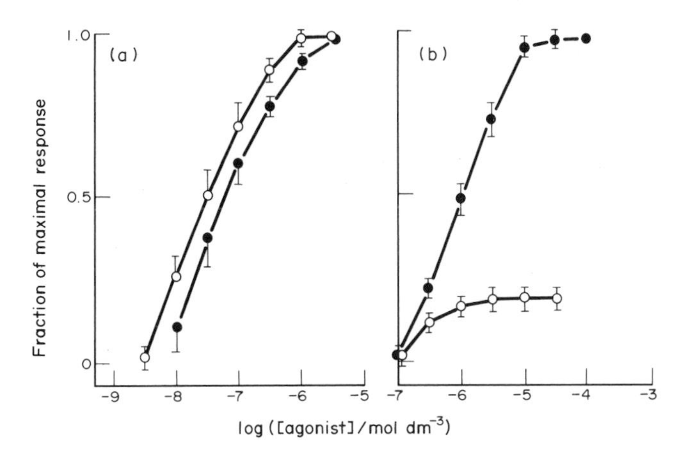

Figure 2 Responses of (a) rat anococcygeus muscle and (b) rat vas deferens to the α adrenoceptor agonists norepinephrine (●) and oxymetazoline (○). Norepinephrine has a lower affinity but higher efficacy for α adrenoceptors (reproduced from ref. 22 by permission of Macmillan Journals Ltd.)

response. This is not true in the poorly receptor-coupled rat vas deferens, where the low efficacy determines the low level of response.

One valuable aspect of determining affinity and efficacy is that these are receptor-specific quantities, the values of which should transcend species, organ and function. Therefore, providing a reference drug such as norepinephrine is included in the analysis, the relative activity of oxymetazoline can be predicted in all tissues possessing α-1 adrenoceptors. Therefore, the medicinal chemist need not be concerned with a plethora of biological screening data in a range of tissues and species but, assuming that the α adrenoceptors in the tissues are homogeneous, can use the relative affinity and intrinsic efficacy of drugs as indicators in structure–activity relationships.

There is a large body of evidence to suggest that the structure–activity relationships for affinity and efficacy are quite different. For example, Figure 3 shows the changes in affinity and efficacy of

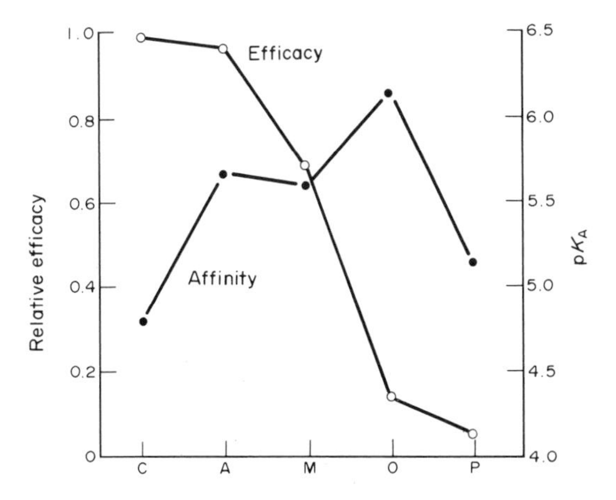

Figure 3 Structure–activity relationships for a series of cholinomimetics for muscarinic receptors; changes in relative efficacy (carbachol = 1) and affinity (expressed as the pK_A, *i.e.* minus the logarithm of the equilibrium dissociation constant of the drug–receptor complex). Data shown for carbachol (**1**; C), acetylcholine (**2**; A), methacholine (**3**; M), oxotremorine (**4**; O) and pilocarpine (**5**; P). Calculated from experimental data in refs. 23 and 24

(1) **(2)** **(3)**

(4) **(5)**

cholinomimetics for muscarinic receptors. Knowledge of relative affinities and efficacies can be very helpful in the prediction of drug response in pathological states and upon chronic treatment. For example, the β adrenoceptor partial agonist pirbuterol has been used as a cardiotonic for the treatment of congestive heart failure. However, it has only 0.01 times the intrinsic efficacy of isoproterenol for β adrenoceptors and thus is much more sensitive to changes in receptor number than is isoproterenol. Figure 4 shows the effects of β adrenoceptor down-regulation, by chronic exposure to isoproterenol, of pirbuterol and isoproterenol in rat left atria. Whereas the dose–response curve to isoproterenol is shifted to the right, the curve to pirbuterol is completely depressed, and this partial agonist functions as a β blocker in down-regulated tissue. This is because pirbuterol, being of lower intrinsic efficacy, must occupy a proportionally greater fraction of the receptors to elicit response and therefore is more sensitive to decreases in receptor number. This effect was observed clinically when, upon chronic treatment with pirbuterol for one month, the inotropic response to pirbuterol diminished to undetectable levels.[26]

The reliance of low efficacy agonists on larger numbers of receptors can be demonstrated acutely. Figure 5 shows the effects of irreversible receptor removal (by chemical alkylation) on guinea pig ileal responses to the high efficacy muscarinic agonist carbachol and the relatively lower efficacy agonist oxotremorine. With serial removal of the receptor population, the responses to the lower efficacy agonist oxotremorine are preferentially depressed over those to carbachol. This is because oxotremorine requires a relatively higher receptor density to produce a response. Note, however,

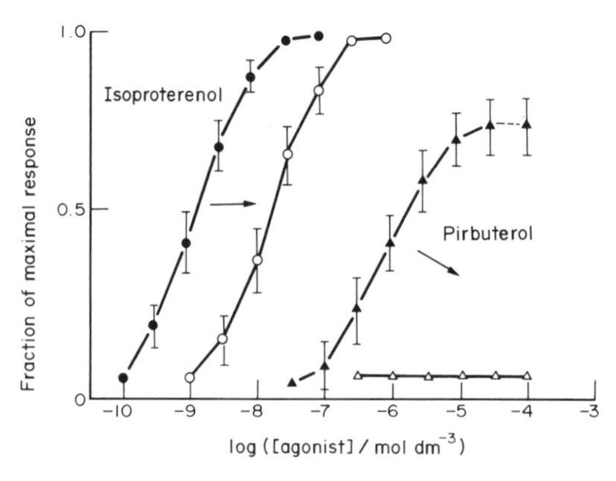

Figure 4 The effect of receptor down-regulation on rat atrial responses to isoproterenol and pirbuterol. Ordinate: increased inotropic response of rat left atria as a fraction of the maximal response to isoproterenol. Abscissa: logarithm of molar concentration of agonist. Responses to isoproterenol (●, $n = 5$) and prenalterol (▲, $n = 5$) in normal atria and those with down-regulated β adrenoceptors (by implantation of miniosmotic pumps delivering 400 $\mu g\ kg^{-1}\ h^{-1}$ isoproterenol for four days); isoproterenol (○, $n = 6$) and prenalterol (△, $n = 6$) (reproduced from ref. 25 by permission of Raven Press)

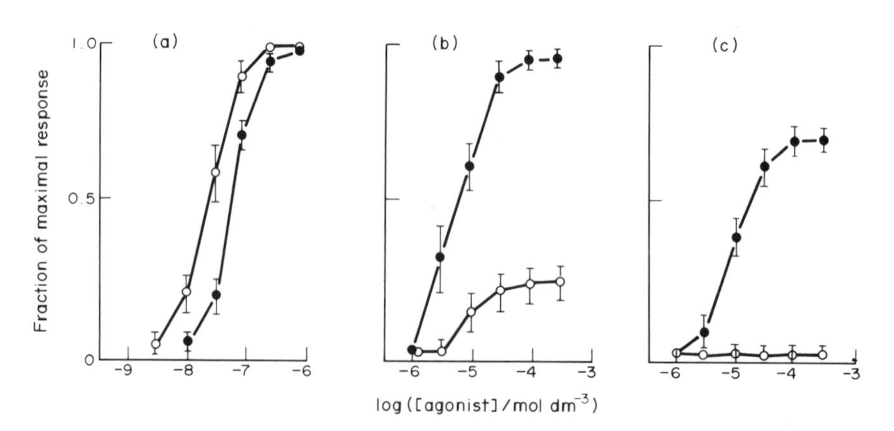

Figure 5 The effect of serial muscarinic receptor alkylation on guinea pig ileal responses to carbachol (●, $n = 8$) and oxotremorine (○, $n = 8$). Ordinates: fraction of maximal contraction to carbachol of guinea pig ileum. Abscissae: logarithm of molar concentration of agonist. Responses in normal ileum and those with increasingly reduced populations of muscarinic receptors (reduced by controlled exposure to the muscarinic receptor alkylating agent phenoxybenzamine, POB): (a) control; (b) 10^{-5} mol dm^{-3} POB (12 min); (c) 3×10^{-6} mol dm^{-3} POB (20 min) (reproduced from ref. 26 by permission of the Massachusetts Medical Society)

that before alkylation, the receptor density in the guinea pig ileum is sufficient to allow both agonists to produce the maximal tissue response.[27]

In general, the use of relative intrinsic efficacy as a proportionality constant to relate the relative fractional receptor occupancy for equiactive doses of agonist can be a useful method of quantifying the ability of drugs to produce an agonist response.

2.3.3.4 Stimulus–Response Coupling Mechanisms

A major difference between the study of drug binding to macromolecules or the production of a product from a substrate–enzyme reaction and the study of drug response in whole organs is the indirect nature of organ response with respect to the interaction of drugs with receptors. While the Langmuir absorption isotherm governs the initial binding of drug to receptor and the dose–response curve of an agonist in a tissue is usually sigmoid, there are many intervening biochemical reactions producing a cascade from the binding of the drug to the receptor to the final expression of organ response. Various tissues differ markedly in the 'efficiency' with which they process drug–receptor events into cellular responses, *i.e.* some tissues require relatively little initial receptor stimulus to produce total syncytial organ response, while others require more. Figure 6(a) shows a series of dose–response curves for isoproterenol in six isolated tissues; a seventyfold difference in potency is observed.[28] When this data is replotted as receptor occupancy–response curves, it can be seen that the six tissues possess differing powers of amplification of receptor stimulus.

The amplification of receptor stimulus into organ response usually occurs because some reactions in the sequence toward response are saturable at minimal levels of stimuli. For example, the production of glucose by β adrenoceptor stimulation may have an amplification factor of eight orders of magnitude, *i.e.* one occupied β adrenoceptor produces eight molecules of glucose.[29] This necessitates the use of null methods to determine drug–receptor parameters such as affinity and relative intrinsic efficacy. Therefore, equiactive drug concentrations are usually compared in order to cancel the effects of receptor-coupling mechanisms.

As knowledge about the intrinsic efficiency of receptor coupling is gained for various organs and in various disease states, drugs may be targeted for clinical use by virtue of specific magnitudes of intrinsic efficacy.

2.3.4 ENZYMES

Considering the multitude of cellular processes controlled by enzymes, it is not surprising that they should be frequent targets for drug intervention. One major advantage of targeting enzymes for therapeutic effect is that, in some cases, the initial systems used for discovery of activity are relatively simple, *i.e. in vitro* enzymatic assays. There are quantitative methods available for studying enzymatic reactions which are applicable to whole cell or organ systems and structure–activity

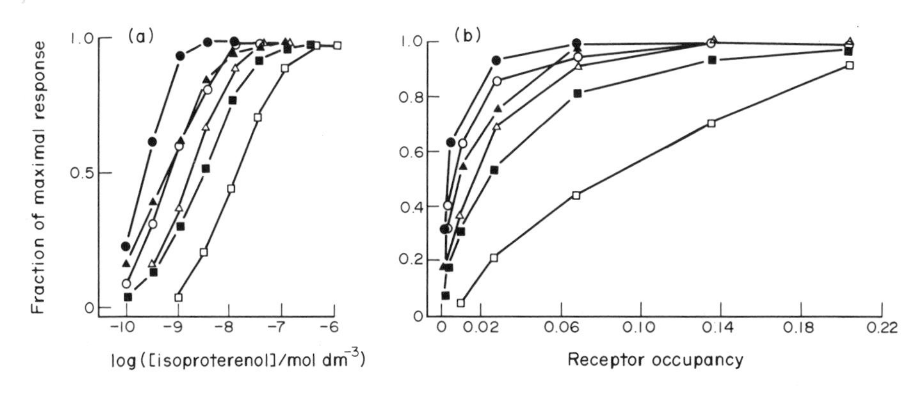

Figure 6 The efficiency of β adrenoceptor coupling. (a) Dose–response curves to isoproterenol in isolated tissues. Ordinate: fraction of the maximal response to isoproterenol. Abscissa: logarithm of molar concentration of isoproterenol. Responses in guinea pig trachea (●, $n = 6$), rat left atria (○, $n = 8$), cat left atria (▲, $n = 6$), cat papillary muscle (△, $n = 6$), guinea pig left atria (■, $n = 5$) and guinea pig extensor digitorum lognus muscle (□, $n = 6$). (b) Occupancy–response curves for data shown in (a). Ordinate: fraction of the maximal response to isoproterenol. Abscissa: receptor occupancy as calculated by the Hill equation (equation 2). Hyperbolic occupancy–response curves indicating the differences in the efficiency of receptor coupling in the various tissues (reproduced from ref. 28 by permission of Williams & Wilkins Co.)

relationships are usually easily determined. However, to a greater extent than for membrane-bound receptors, accessibility to compartments containing enzymes becomes a problem, *i.e.* drugs must usually pass through the cell membrane and enter the cell to be useful. Thus, a structure–activity relationship determined on the native enzyme may not be meaningful in whole cell systems if the drug cannot penetrate the cell.

A further problem can be encountered if substrate availability is saturating. For example, inhibitors of the enzyme carbonic anhydrase are useful for the reduction of aqueous humor production in the eye and thus are effective treatments for glaucoma.[30] However, one limitation of this approach is the fact that a 99% inhibition of the enzyme is required before a therapeutic effect is observed, presumably because of excess substrate.[31]

When specific enzymes are associated with pathological conditions they are obvious targets for drug inactivation. This approach has been exploited especially in the field of antibacterials and antivirals. For example, bacteria require a lattice structure to give rigidity in their cell walls. Penicillin acylates a transpeptidase to form a penicilloyl enzyme in these bacteria that is incapable of cross-linking the heteropolymeric components needed to form this cell wall structure, thereby destroying the integrity of the bacteria.[32] Trimethoprim prevents bacterial reduction of dihydrofolate to tetrahydrofolate by inhibiting the enzyme dihydrofolate reductase thus causing bacterial cell death.[33]

The antiviral drug acyclovir inhibits thymidine kinase and, by virtue of a two-hundredfold greater affinity for this enzyme in herpes virus (over mammalian cells), high levels of acyclo-GTP accumulate in the virus leading to the inhibition of DNA polymerase and the incorporation of acyclo-GTP into viral DNA. This, in turn, terminates the biosynthesis of the viral DNA strand.[34] Many such target enzymes are in a cascade of reactions which culminate in the replication of genetic material. For example, rifampin inhibits the DNA-dependent RNA polymerase of mycobacteria and other microorganisms to interfere with replication.[35]

There are enzymes which become targets for drugs only in pathological conditions. For example, allopurinol blocks xanthine oxidase to reduce the plasma concentration and renal excretion of uric acid in gout.[36] Aspirin acetylates a serine residue in the active site of cyclooxygenase in inflammation[37] and the enzyme aldose reductase may be of paramount importance in the treatment of diabetic complications.[38] In hypertension and congestive heart failure, the inhibition of the renin–angiotensin–aldosterone system leads to favorable hemodynamic changes. The angiotensin-converting enzyme inhibitors captopril and enalapril have now become among the most valuable drugs available for the treatment of these diseases. The fact that angiotensin-converting enzyme is plasma borne allows for correlation of enzyme inactivation and therapeutic effect. Figure 7 shows the fall in systolic blood pressure in spontaneously hypertensive rats after captopril 30 mg kg^{-1} p.o. and the concomitant inhibition of angiotensin converting enzyme.[39]

The inhibition of specific enzymes can be therapeutically useful in a normal physiological setting as well. For example, the reduction of serum cholesterol levels has probable value in prophylaxis

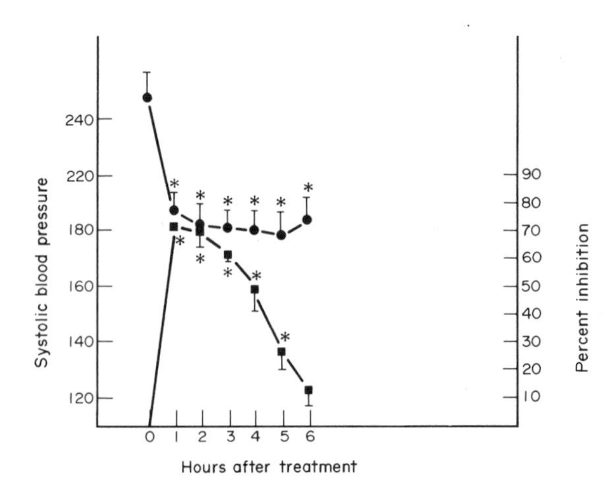

Figure 7 Effects of captopril (30 mg kg^{-1} p.o.; $n = 6$) on systolic blood pressure in rats (left-hand ordinate axis; ●) and percent inhibition of pressor response to angiotensin I (right-hand ordinate axis; ■) as a function of hours after administration of captopril. Asterisks represent statistically significant differences from control values. Percent inhibition of responses to angiotensin I represents inhibition of angiotensin converting enzyme (reproduced from ref. 39 by permission of North-Holland Publishing Co.)

against cardiovascular disease; in this case, the enzyme HMG CoA reductase has been targeted in the cholesterol pathway.[40]

In congestive heart failure, a fundamental change in the myocardium causes the heart to become weak and therefore unable to pump blood commensurate with the needs of organs. This, in turn, leads to incomplete emptying of the heart and myocardial swelling. The problem is exacerbated by the reflex vasoconstricting mechanisms which engage in response to the fall in perfusion pressure. In this situation, a drug that produces vasodilation (to reduce the load against which the heart must pump) and positive inotropy (to make the heart stronger for better emptying and organ perfusion) could be useful. Since it is known that the intracellular messenger cAMP produces vascular relaxation and positive inotropy, a logical approach is a drug intervention that increases intracellular cAMP. The levels of cAMP in cells available for physiological response are controlled by a family of degradatory enzymes known as the phosphodiesterases. Therefore, one method of treating congestive heart failure has been to use inhibitors of phosphodiesterase, the rationale being that inhibition of the degradatory enzyme should increase the intracellular levels of this second messenger.

An example of such a drug is fenoximone (MDL 17043), a selective inhibitor of Type IV phosphodiesterase in enzymatic assays conducted *in vitro* (Figure 8a).[41] This drug appears to enter the cell readily and block the enzyme in whole organ systems; Figure 8(b) shows that the pharmacologic effects of fenoximone in guinea pig left atria are greatly potentiated by prior treatment of the tissues with prenalterol, a low efficacy β adrenoceptor partial agonist. This has the effect of inducing a low level of activation of the enzyme that produces cAMP, adenyl cyclase, and thereby sensitizing the cell to inhibition of phosphodiesterase by fenoximone. These data indicate that, in the concentrations shown, fenoximone is an effective inhibitor of phosphodiesterase in a whole organ system. Figure 8(c) shows how this activity is translated *in vivo*; a positive inotropic response as well as vasodilation is observed.[43]

By far the greater number of drug interventions with respect to enzymes involve enzyme inhibition. However, there are examples where enzyme activation may be useful. For example, the diterpene forskolin is a potent activator of the adenyl cyclase catalytic subunit, thereby elevating intracellular levels of cAMP; this activity may have relevance to the treatment of a number of disease states.[44]

2.3.5 TRANSPORT PROCESSES

Another method of introducing bias into biological systems is to affect the active transport processes of biologically active endogenous molecules or ions across cell membranes. A classic example is the inhibition of the enzyme sodium–potassium ATPase which transports potassium into the cell (across the sarcolemma) and sodium out of the cell. Drugs such as digitalis inhibit this enzyme and cause an increase in intracellular sodium, which then is extruded by the myocardial cell by a sodium/calcium exchange mechanism. This in turn leads to an increase in intracellular 'trigger' calcium, which interacts with calcium stores in the sarcoplasmic reticulum and results in a net increase in cytosolic free calcium ion which then is available for contraction. The resulting myocardial inotropic response is useful in the treatment of congestive heart failure. Another ion pump with physiological relevance to gastric secretion is parietal cell hydrogen/potassium ATPase. Inhibitors of this pump such as omprazole[45] reduce gastric secretion and may have utility in the treatment of gastric ulcer.

Although not a continuous ion pump, calcium slow channels control the entry of calcium ion into the cytosol from the extracellular space. These channels are controlled either by autonomic receptors or membrane potential and their blockade by calcium channel inhibitors such as nifedipine, diltiazem and verapamil produces inhibition of excitation contract coupling in cardiac and smooth muscle. Such effects are useful in the treatment of angina, hypertension and various other disorders.[46]

Another setting for interference with a transport system is in failing neurotransmission. Thus in myasthenia gravis, a disease marked by weakness and fatigability of skeletal muscles, blockers of acetylcholinesterase such as physostigmine have been found to be useful.[47] The rationale for use of these drugs in this setting is that, since acetylcholinesterase degrades neuronally released acetylcholine from endplate terminal neurons, and since a deficiency of cholinergic neurotransmission is thought to cause the pathological condition, then potentiation of the remaining released acetylcholine by blockade of its degradation may strengthen the failing acetylcholine release. A similar mechanism may be operative in the use of acetylcholinesterase blockers such as physostig-

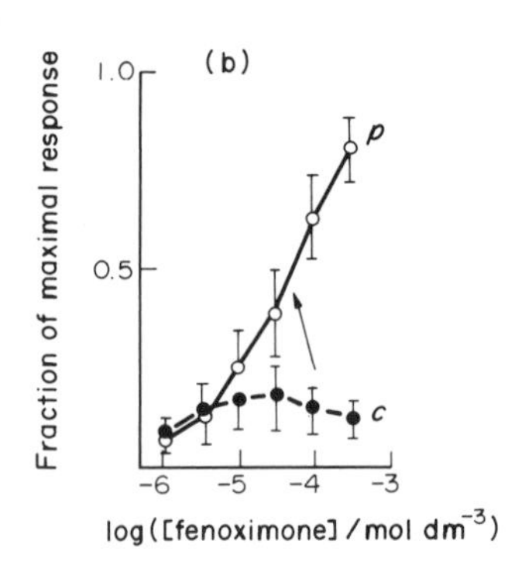

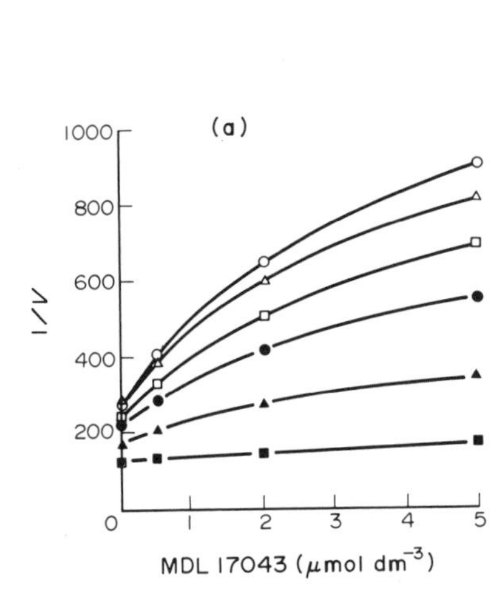

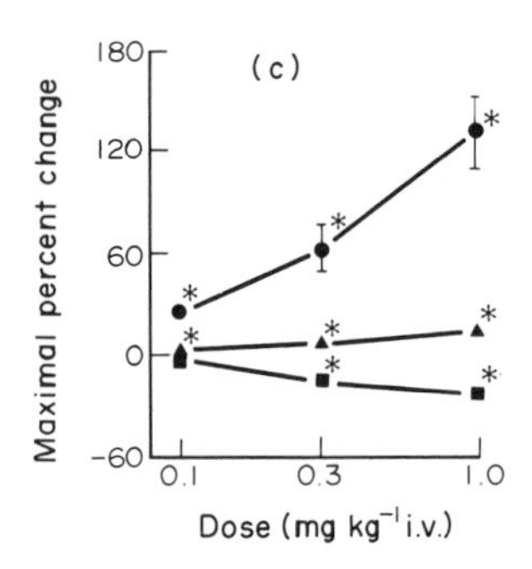

Figure 8 Phosphodiesterase (PDE) blocking and positive inotropic properties of fenoximone (**6**; MDL 17043). (a) Inhibition of dog cardiac cAMP PDE by MDL 17043. Ordinate: reciprocal of enzyme velocity for hydrolysis of cAMP ($V = \mu$mol substrate hydrolyzed min^{-1} (mg protein)$^{-1}$). Abscissa: molar concentration of MDL 17043. Concentration of cAMP (μmol dm^{-3}): ○, 0.25; △, 0.29; □, 0.33; ●, 0.5; ▲, 1.0; ■, 4.0 (reproduced from ref. 41 by permission of Raven Press). (b) Potentiation of inotropic responses of guinea pig left atria to fenoximone after prior incubation of tissues with the weak β adrenoceptor partial agonist prenalterol. Ordinate: increase in twitch contraction as fraction of the maximal contraction to isoproterenol. Abscissa: logarithm of molar concentration of fenoximone. Responses in the absence (●, $n = 4$) and presence (○, $n = 4$) of prenalterol. It should be noted that prenalterol had very little positive inotropic effect in these tissues but only served to provide a subthreshold activation of adenyl cyclase. The phosphodiesterase blocking property of fenoximone was made evident by the increased inotropic response in atria with slightly activated adenyl cyclase (reproduced from ref. 42 by permission of Raven Press). (c) Effects of MDL 17043 on cardiac contractile force (●), heart rate (▲) and mean blood pressure (■) in anesthetized dogs. Asterisks represent differences significant at the $p < 0.05$ level (reproduced from ref. 43 by permission of Raven Press)

O Me N-H MeS N H O

(6)

mine in Alzheimer's disease. In this case, the problem appears to be a deficiency of functional cholinergic neurons in the brain. Physostigmine has been shown to produce improvements in memory in the early stages of the disease.[48]

The interference with transport processes can be beneficial in normally functioning systems as well. For example, acetylcholinesterase inhibitors are useful in the treatment of glaucoma by potentiating the existing neuronal cholinergic tone in the eye and thereby reducing intraocular pressure.[49] The coronary vasodilators dipyridamole, dilazep and hexobendine owe some if not all of their vasodilating activity to a blockade of the uptake of extracellular adenosine.[50] This purine is released from ischemic heart tissue, produces coronary vasodilatation and is subsequently taken back up into cells by an active uptake process. These drugs inhibit the transport of adenosine into cells thus potentiating its action and producing a greater coronary vasodilatation.

An advantage of inhibiting physiological transport processes is the resulting regional selectivity. Thus, the selective stimulation of catecholaminergic receptors by inhibition of the neuronal uptake of norepinephrine by nerve terminals is quite different from the effects of stimulation of all norepinephrine receptors by large saturating concentrations of norepinephrine. The former situation results in a selective stimulation of innervated norepinephrine receptors; considering the complicated interactions of synapses in, for example, the central nervous system, it would be predicted that this would result in a different overall physiological effect than that of total receptor population stimulation by exogenous norepinephrine. This effect may play a role in the interference with norepinephrine neuronal uptake in the treatment of depression. Tricyclic drugs such as demethylimipramine are potent inhibitors of norepinephrine uptake and also are useful antidepressants. An interesting component in the action of these drugs is the need for chronic treatment. The antidepressant effect often coincides with regional receptor down-regulation of norepinephrine receptors, a response thought to be due to the chronically elevated concentrations of norepinephrine in the synaptic cleft.[51] The same effect is thought to be the cause of the antidepressant action of monoamine oxidase inhibitors.[51] In this case, the degradation of norepinephrine is blocked by the inhibition of the degradatory enzyme monoamine oxidase; this results in elevated concentrations of norepinephrine at the receptor much like that produced by the blockade of neuronal uptake. The interesting feature of this therapeutic effect is that, unlike the acute benefits obtained from the inhibition of other transport processes, antidepressant activity may be the result of adaptive changes in the receptor systems of the brain in response to chronically elevated concentrations of norepinephrine.

2.3.6 DNA AND GENETIC MATERIAL

There are a number of drugs which target DNA or other genetic material such as RNA or ribosomes for primary binding and expression of effect. This should be distinguished from those drugs which produce pharmacologic effects by binding to enzymes (*i.e.* DNA polymerase) involved in cell replication. Although the net effect is the same (*i.e.* prevention of cell division), the primary target for the initial drug–receptor binding process is different.

Drugs targeted to DNA or genetic material interfere with the normal process of cell replication and thus cause cell death. The most common setting for this effect is in cancer chemotherapy and the treatment of bacterial and viral infections.

Many drugs used for the treatment of neoplastic disease are known to function as antimetabolites, *i.e.* they are transformed enzymatically into structural analogues of nucleotides which then become incorporated into genetic material (DNA) to produce strands which cannot function normally and thus inhibit cell division. For example, 5-fluorouracil is converted by a number of enzymatic reactions to a product which is incorporated into RNA and DNA. Similarly, cytabarine is activated enzymatically and then incorporated into DNA with resulting inhibition of cell division.[52]

Another method of inhibition of cell replication is by alkylation of DNA to produce cross-linked complexes that cannot function normally. Thus mitomycin becomes active when the quinone ring is enzymatically reduced to produce a trifunctional alkylating agent that inhibits DNA synthesis, cross-links DNA and causes single-strand DNA breakage.[53] It is debatable if the primary targets for such drugs can be considered genetic material since they must first interact with an enzyme to be transformed into the active species. A drug which directly interacts with DNA is cisplatin, which produces platinum complexes that react with DNA to form intrastrand and interstrand cross-links and thus block cell replication.[54]

Other drugs which have DNA as their primary target are DNA intercalators such as dactinomycin, daunorubicin and doxorubicin. For example, the planar phenoxazone ring structure of dactinomycin is known to intercalate between adjacent guanine–cytosine base pairs of DNA to form a tight complex, and thus inhibit the function of RNA polymerase. This results in the inhibition of DNA transcription.[55]

Finally, many antibiotics including tetracyclines, chloramphenicol, erythromycin and clindamycin primarily target genetic material (*i.e.* ribosomes) to inhibit cell replication. These drugs and their mechanisms of action are dealt with more fully in subsequent chapters of this work.

2.3.7 CONCLUSIONS

This chapter attempts to generally introduce the types of primary targets that can be used for drug interventions for therapeutic advantage. As understanding of the biochemistry of cell function increases, more clearly defined targets will become obvious, hopefully bringing more selective therapy. Some of these targets may represent new directions as technology progresses; for example, a new frontier in the treatment of congestive heart failure may be to avoid the calcium overload in cells produced by drugs which increase calcium entry and function by sensitizing the contractile proteins of the myocardial cell to the existing intracellular calcium concentrations. This may be one mechanism of action of the cardiotonic sulmazole.[56,57]

2.3.8 REFERENCES

1. T. P. Kenakin, *J. Pharmacol. Methods*, 1985, **13**, 281.
2. J. N. Langley, *J. Physiol. (London)*, 1878, **1**, 339.
3. A. V. Hill, *J. Physiol. (London)*, 1909, **39**, 361.
4. R. F. Furchgott, *Adv. Drug Res.*, 1966, **3**, 21.
5. R. P. Stephenson, *Br. J. Pharmacol. Chemother.*, 1956, **11**, 379.
6. A. S. V. Burgen, *J. Pharm. Pharmacol.*, 1966, **18**, 137.
7. W. C. Bowman and M. J. Rand, in 'Textbook of Pharmacology', 2nd edn., ed. W. C. Bowman and M. J. Rand, Blackwell, Oxford, 1980.
8. J. W. Black and J. S. Stephenson, *Lancet*, 1962, **II**, 311.
9. J. W. Black, W. A. M. Duncan, C. J. Durant, C. R. Ganellin and E. M. Parsons, *Nature (London)*, 1972, **236**, 385.
10. C. R. Ganellin and C. J. Durant, in 'Burgers Medicinal Chemistry', 4th edn., ed. M. E. Wolf, Wiley, New York, 1981, part III, p. 489.
11. J. W. Black, in 'Catecholamines in the Non-Ischemic and Ischemic Myocardium', ed. R. Riemersma and W. Oliver, Elsevier/North-Holland Biomedical Press, New York, 1982, p. 3.
12. E. J. Ariens, A. J. Beld, J. F. R. de Miranda and A. M. Simonis, in 'The Receptors, a Comprehensive Treatis', ed. R. D. O'Brien, Plenum Press, New York, 1979, p. 33.
13. J. S. Morley, *Trends Pharmacol. Sci.*, 1983, **4**, 370.
14. E. J. Ariens and A. M. Simonis, *Ann. N. Y. Acad. Sci.* 1967, **144**, 842.
15. S. T. Carbonetto, D. M. Fambrough and K. J. Muller, *Proc. Natl. Acad. Sci. U.S.A.*, 1978, **75**, 1016.
16. M. Rocha e Silva, *Eur. J. Pharmacol.*, 1969, **6**, 294.
17. M. L. Torchiana, R. G. Pendleton, P. G. Cook, C. A. Hanson and B. V. Clineschmidt, *J. Pharmacol. Exp. Ther.*, 1983, **224**, 514.
18. D. R. Bristow, J. R. Hare, J. R. Hearn and L. E. Martin, *Br. J. Pharmacol.*, 1981, **72**, 547.
19. M. Williams and J. L. Robinson, *J. Neurosci.*, 1984, **4**, 2906.
20. A. S. V. Burgen, *Fed. Proc., Fed. Am. Soc. Exp. Biol.*, 1981, **40**, 2723.
21. P. Gardner, D. C. Ogden and D. Colquhoun, *Nature (London)*, 1984, **309**, 160.
22. T. P. Kenakin, *Br. J. Pharmacol.*, 1984, **81**, 131.
23. B. Ringdahl, *J. Pharmacol. Exp. Ther.*, 1984, **229**, 199.
24. R. F. Furchgott and P. Bursztyn, *Ann. N. Y. Acad. Sci.*, 1967, **144**, 882.
25. T. P. Kenakin and R. M. Ferris, *J. Cardiovasc. Pharmacol.*, 1983, **5**, 90.
26. W. S. Colucci, R. W. Alexander, G. H. Williams *et al.*, *N. Engl. J. Med.*, 1981, **305**, 185.
27. T. P. Kenakin, in 'The Pharmacologic Analysis of Drug Receptor Interaction', Raven Press, New York, 1987.
28. T. P. Kenakin and D. Beek, *J. Pharmacol. Exp. Ther.*, 1980, **213**, 406.
29. N. D. Goldberg, in 'Cell Membranes: Biochemistry, Cell Biology and Pathology', ed. G. Weissmann and R. Claiborne, H. P. Publishing Co., 1975, p. 1985.
30. B. R. Friedland and T. H. Maren, in 'Pharmacology of the Eye, Handbook of Experimental Pharmacology', ed. M. L. Sears, Springer-Verlag, Berlin, 1984, vol. 69, p. 279.
31. T. H. Maren, *J. Pharmacol. Exp. Ther.*, 1963, **139**, 140.
32. J. A. Kelly, P. C. Moews, J. R. Knox, J. Frere and J. Ghuysen, *Science (Washington, D.C.)*, 1982, **218**, 479.
33. G. H. Hitchings, *Trans. N. Y. Acad. Sci.*, 1961, **23**, 700.
34. G. B. Elion, *Am. J. Med.*, 1982, **73** (suppl. 1), 1.
35. W. Wehrli, *Rev. Infect. Dis.*, 1983, **5** (suppl. 3), S407.
36. G. B. Elion, in 'Uric Acid, Handbook of Experimental Pharmacology', ed. W. N. Kelley and I. M. Weiner, Springer-Verlag, Berlin, 1978, vol. 51, p. 485.
37. G. R. Roth and C. J. Siok, *J. Biol. Chem.*, 1978, **253**, 3782.
38. C. A. Lipinsky and N. J. Hutson, *Annu. Rep. Med. Chem.*, 1984, **19**, 169.
39. C. S. Sweet, P. T. Arbegast, S. L. Gaul, E. H. Blaine and D. M. Gross, *Eur. J. Pharmacol.*, 1981, **76**, 167.
40. A. Yamamoto, H. Sudo and A. Endo, *Atherosclerosis (Berlin)*, 1980, **35**, 259.
41. T. K. Kariya, L. J. Wille and R. C. Dage, *J. Cardiovasc. Pharmacol.*, 1982, **4**, 509.
42. T. P. Kenakin and D. L. Scott, *J. Cardiovasc. Pharmacol.*, 1987, **10**, 658.
43. R. C. Dage, L. E. Roebel, C. P. Hsieh, D. L. Weiner and J. K. Woodward, *J. Cardiovasc. Pharmacol.*, 1982, **4**, 500.
44. K. B. Seamon, *Annu. Rep. Med. Chem.*, 1984, **19**, 293.
45. E. Fellenius, B. Elander, B. Wallmark, H. F. Helander and T. Berglindh, *Am. J. Physiol.*, 1982, **243**, G505.
46. H. Myeer, S. Kazda and P. Bellemann, *Annu. Rep. Med. Chem.*, 1983, **18**, 79.
47. D. Grob, *Ann. N. Y. Acad. Sci.*, 1981, **377**, 1.
48. L. J. Thai, P. A. Fuld, D. M. Masur and N. S. Sharpless, *Ann. Neurol.*, 1983, **13**, 491.
49. P. L. Kaufman, T. Weidman and J. R. Robinson, in 'Pharmacology of the Eye, Handbook of Experimental Pharmacology', ed. M. L. Sears, Springer-Verlag, Berlin, 1984, vol. 69, p. 149.
50. S. J. Msutafa, *Biochem. Pharmacol.*, 1979, **28**, 2617.
51. W. J. Frazee, C. J. Ohnmacht and J. B. Malick, *Annu. Rep. Med. Chem.*, 1985, **20**, 31.
52. D. W. Kufe and P. P. Major, *Med. Pediatr. Oncol.*, 1982, suppl. 1, 49.
53. S. T. Crooke and W. T. Bradner, Cancer Treat. Rev., 1976, **3**, 121.
54. L. A. Zwelling and K. W. Kohn, in 'Pharmacologic Principles of Cancer Treatment', ed. B. A. Chabner, Saunders, Philadelphia, 1982, p. 309.
55. H. M. Sobell, *Prog. Nucleic Acid Res. Mol. Biol.*, 1973, **13**, 153.
56. J. W. Herzig, K. Feile and J. C. Ruegg, *Arzneim.-Forsch.*, 1981, **31**, 188.
57. R. J. Solaro and J. C. Ruegg, *Circ. Res.*, 1982, **51**, 290.

2.4
The Concept of Bioselectivity

TERRY P. KENAKIN
Glaxo Research Laboratories, Research Triangle Park, NC, USA

2.4.1 INTRODUCTION

'There is only a quantitative difference between a drug and a poison . . .' Walter Straub (1874–1944).

It is a common finding that drugs usually have more than one action and are selective rather than specific. For example, yohimbine has appreciable affinity for both subtypes of α-adrenoceptors, as well as receptors for serotonin (Figure 1a). Amitriptylene has affinity for histamine, muscarinic, α-adrenergic receptors and neuronal uptake sites for catecholamines (Figure 1b). Yet drugs usually are most useful when they produce selective organ responses, since the other activities often lead to unwanted side effects. For example, the catecholamine epinephrine produces a number of pharmacological responses. If these could be produced selectively, then epinephrine could alternately function as a useful drug treatment for conditions such as congestive heart failure, glaucoma or asthma (Table 1). Organ selectivity can be achieved by three general methods.

First, selectivity (with respect to affinity and/or intrinsic efficacy) for biological macromolecules (receptors, enzymes) will lead to bioselective drug response. A second method to attain bioselectivity for drugs with multiple activities is to control drug concentration in the receptor compartment. For example, 30 nM amitriptylene will produce a significant blockade of histamine H_1-receptors selectively, since this concentration is below the threshold for binding to other macromolecules (Figure 1b). A third method of achieving bioselectivity is with drugs possessing dual (or even multiple) activities that are expressed in a common concentration range (drug synergy). Each of these approaches will be discussed separately.

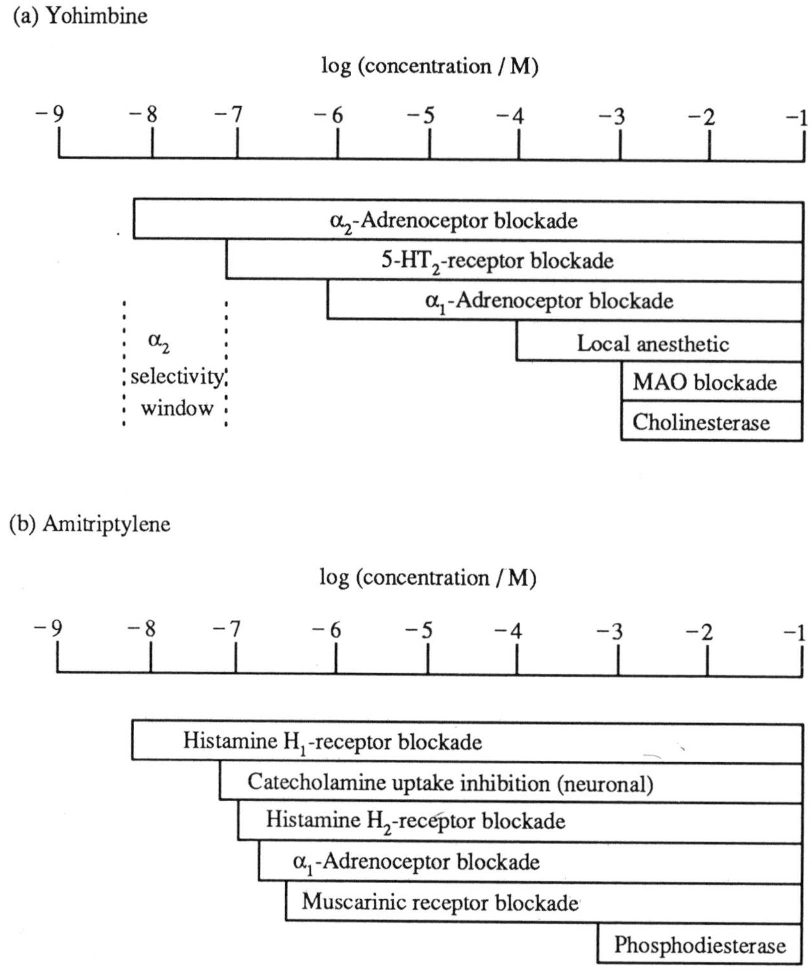

(a) Yohimbine

log (concentration / M)

−9 −8 −7 −6 −5 −4 −3 −2 −1

α_2-Adrenoceptor blockade
5-HT_2-receptor blockade
α_1-Adrenoceptor blockade
Local anesthetic
MAO blockade
Cholinesterase

α_2 selectivity window

(b) Amitriptylene

log (concentration / M)

−9 −8 −7 −6 −5 −4 −3 −2 −1

Histamine H_1-receptor blockade
Catecholamine uptake inhibition (neuronal)
Histamine H_2-receptor blockade
α_1-Adrenoceptor blockade
Muscarinic receptor blockade
Phosphodiesterase

Figure 1 Concentration ranges for yohimbine (a) and amitriptylene (b), where activity for a series of autonomic receptors and functions is expressed (reproduced from ref. 1 by permission of Raven Press)

Table 1 Some Pharmacologic Effects of Epinephrine

Response	*Disease*		
	Congestive heart failure[b]	*Glaucoma*[b]	*Asthma*[b]
Hypertension	−	−	−
Tachycardia	−	−	−
Digital tremor	−	−	−
CNS effects	−	−	−
(+)-Inotropy	+++	−	−
Decreased IOP[a]	0	+++	0
Bronchodilation	0	0	+++

[a] IOP = intraocular pressure. [b] +++ = desired response; − = undesired response; 0 = not specifically harmful to primary effect.

2.4.2 SELECTIVE RECEPTOR AFFINITY AND/OR EFFICACY

2.4.2.1 Interaction with Receptors

If a given drug has selective affinity for a cognitive component of a cell (whether that be a membrane-bound receptor, an enzyme, a transport site or some other cellular structure), then bioselectivity can be achieved with a dosage sufficient only to interact with that particular component. For example, the pK_B (−log of the equilibrium dissociation constant of the drug–receptor complex) of propranolol for β-adrenoceptors is 8.4, while that for α-adrenoceptors is 5.2.[2] Therefore, at a concentration of propranolol between 4 nM and 0.33 μM in a compartment containing both α- and β-adrenoceptors, little appreciable binding of propranolol to α-adrenoceptors will take place. Yet at a receptor compartment concentration of 0.33 μM, 98.8% of the β-adrenoceptors as opposed to only 5% of the α-adrenoceptors will be occupied by propranolol. This will be expressed as an 84-fold shift to the right of a dose–response curve of epinephrine on β-adrenoceptors and essentially no blockade (1.08-fold shift) of the α-adrenoceptor-mediated effects of any given concentration of the same hormone. Thus, providing the appropriate receptors are present in the target organ(s), bioselectivity can be achieved by selective affinity. In the case of agonists, an added dimension, namely intrinsic efficacy, must be considered since selective agonism can be achieved without selective affinity if the drug has a high intrinsic efficacy for a particular receptor type. For example, dobutamine has affinity and intrinsic efficacy for both α- and β-adrenoceptors. However, although the affinity of dobutamine for α-adrenoceptors is 224 times greater than the affinity for β-adrenoceptors,[3] predominantly β-adrenoceptor-mediated responses are observed *in vivo*.[4] This is because dobutamine has a higher intrinsic efficacy for β- as opposed to α-adrenoceptors and thus needs to occupy a smaller proportion of the receptor population to produce β-adrenoceptor-mediated responses. Therefore, wherever agonism is involved, intrinsic efficacy as well as affinity for receptors must be considered for bioselective response production.

Given a drug with a selective receptor affinity and/or intrinsic efficacy for a receptor, a response will be observed in those organs which possess that particular receptor. Wherever the cognitive target can be associated with a pathological state, then a particularly useful bioselectivity results. For example, bacteria, unlike the host cells, cannot utilize presynthesized folic acid but instead must synthesize it from *p*-aminobenzoic acid. Therefore, inhibition of this synthetic pathway (for example by sulfonamides) produces a selective destruction of bacteria with no deleterious effects to the host (the antimetabolite theory). Another application of this approach was utilized by Hitchings and coworkers[5] in the use of drugs to interfere with nucleic acid metabolism. In this case fraudulent nucleic acids are incorporated into genetic material and cell replication is prevented. Bioselectivity for quickly dividing cells is obtained since the effects of the fraudulent genes are expressed in those cells before those of the host. This approach has produced drugs useful for the treatment of lymphoblastic leukemia, chronic granulocytic leukemia, malaria, gout and organ transplantation.

2.4.2.2 Selectivity Due to Receptor-coupling Mechanisms

Cells generally amplify receptor signals. The interaction between drugs and receptors is usually the first step in a cascade of saturable reactions leading to cellular response. The fact that drug receptors in different organs are coupled to the stimulus–response machinery of cells with varying efficiency can lead to organ selectivity. The key element to organ selectivity by this means is the magnitude of the intrinsic efficacy of the drug, *i.e.* the inherent ability of the drug molecule to initiate a quantal stimulus to the receptor.

Cellular stimulus–response mechanisms can be considered power amplifiers of the signals given to receptors by agonist molecules. The stimulus given to a single receptor by a drug is measured by an abstract quantity called the intrinsic efficacy.[6] This quantity is unique to the chemical structure of the drug and the receptor to which it binds. In this sense, the summation of the quantal stimuli received by the organ by the total receptor population can be considered as the incoming signal to a radio antenna, the output being the volume. The cellular stimulus–response mechanisms which translate this stimulus into organ response are the power amplifiers of this signal. The fact that various organs have power amplifiers set at different volumes can lead to a bioselective organ response. The effects of these various levels of amplification are different for drugs of different intrinsic efficacy.

The actual magnitude of the intrinsic efficacy can be very important in terms of the maximal organ response produced by a drug. This can best be illustrated with a physical analogy. If the intrinsic efficacy of a drug is equated to a mass, assume then that the downward displacement of a lever when the mass is placed on it is equal to the intensity of receptor stimulus. The upward displacement of the opposite end of that lever is the observed response and the farther from the fulcrum this process is viewed, the greater is the stimulus amplification. Therefore, the vantage points along the lever represent organs of varying amplification capability. The model is depicted in Figure 2 for a drug of high efficacy (large mass) and low efficacy (small mass). For a large mass, the displacement is sufficient to exceed a defined limit, in this case the tissue maximal response, and irrespective of where along the lever this process is viewed (tissue I, II or III) this agonist will produce the tissue maximal response. For a drug with low intrinsic efficacy, the situation is quite different. In this case, the position of the observed displacement along the lever is critical to the magnitude of the response. Thus for a weak agonist of low intrinsic efficacy, the efficiency of stimulus–response coupling can determine whether it will be a full agonist, partial agonist or antagonist. Examples of both of these types of agonist are shown also in Figure 2. Isoproterenol is a full agonist in a range of tissues possessing β-adrenoceptors, whereas prenalterol, a drug with 0.004 times the intrinsic efficacy of isoproterenol,[8] is nearly a full agonist, partial agonist and antagonist in these same tissues (Figure 2). Thus, for prenalterol, the efficiency of stimulus–response coupling is much more important than for isoproterenol in terms of the production of tissue response. In general, low efficacy agonists are much more subject to the efficiency of stimulus–response mechanisms than are high efficacy agonists.[9] This can lead to bioselectivity if the intrinsic efficacy of the agonist is such that it produces responses only in the most efficiently coupled organs. For example, the rat left atrium is an efficiently coupled organ with respect to β-adrenoceptor responses and both isoproterenol and prenalterol produce a response (Figure 3a). However, in the canine coronary artery, a tissue with the same receptors but coupled less efficiently to stimulus–response mechanisms, isoproterenol produces a response but prenalterol does not (Figure 3b). Thus, complete bioselectivity with respect to responses in the rat atrium and canine coronary artery is obtained simply as the result of the low efficacy of prenalterol. In the case of prenalterol, this may lead to a useful selectivity *in vivo*; Table 2 shows the effects of isoproterenol and prenalterol in the anesthetized cat. Whereas isoproterenol produces β-adrenoceptor-mediated responses in the heart and blood vessels, prenalterol produces only cardiac inotropy. In this case, it is probable that prenalterol does not possess the intrinsic efficacy to activate arterial β-adrenoceptors and thus a useful positive inotropy results. However, the bioselectivity with respect to response should not be confused with a lack of affinity for receptors, for example, although prenalterol does not produce a response in the canine coronary artery, it still

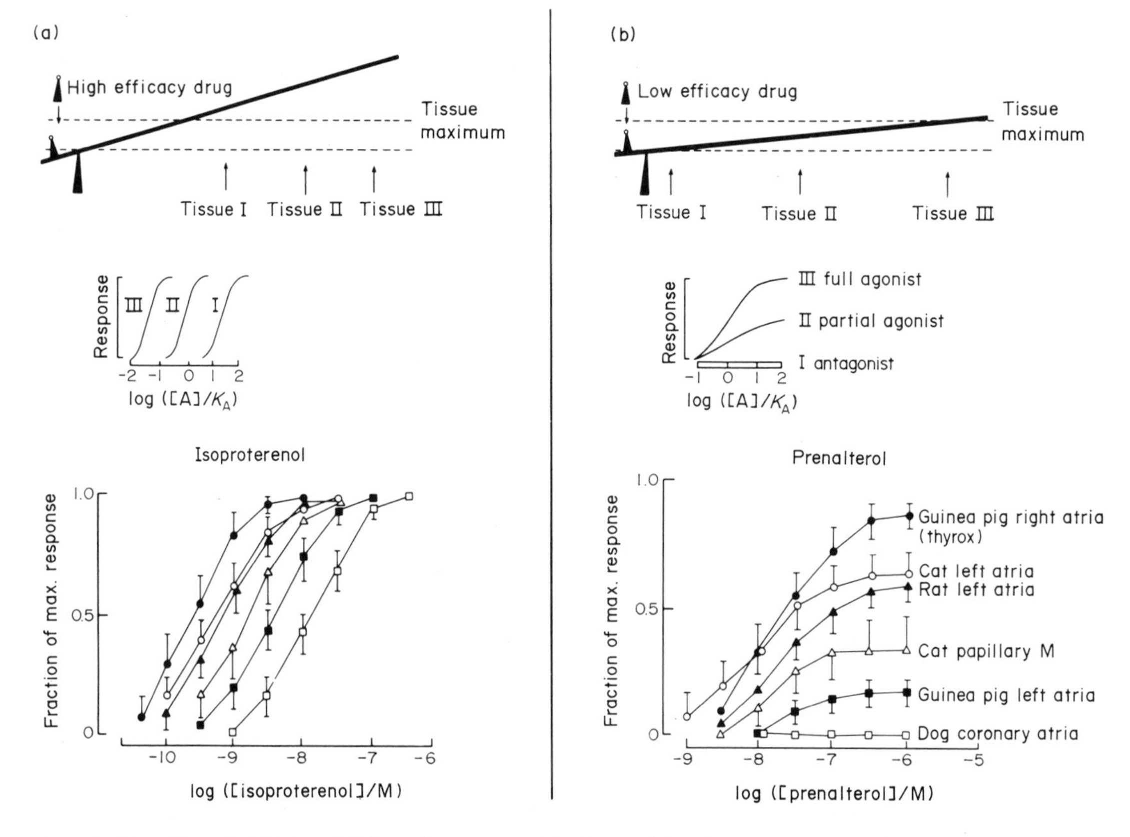

Figure 2 The effects of (a) high and (b) low efficacy agonists in tissues with different receptor-coupling efficiencies to cellular stimulus–response mechanisms. Intrinsic efficacy is equated to the weight placed on one end of the lever; the upward displacement represents the resulting response. Where along the lever the displacement is viewed represents different amplification factors, *i.e.* tissues with different efficiencies of receptor coupling. Whereas a powerful agonist can produce the maximal response in all tissues, a weak agonist may be a full agonist, partial agonist or antagonist in those same tissues. This is illustrated by data for isoproterenol, a high efficacy agonist for β-adrenoceptors and prenalterol, a low efficacy agonist (reproduced from ref. 7 by permission of Williams and Wilkins)

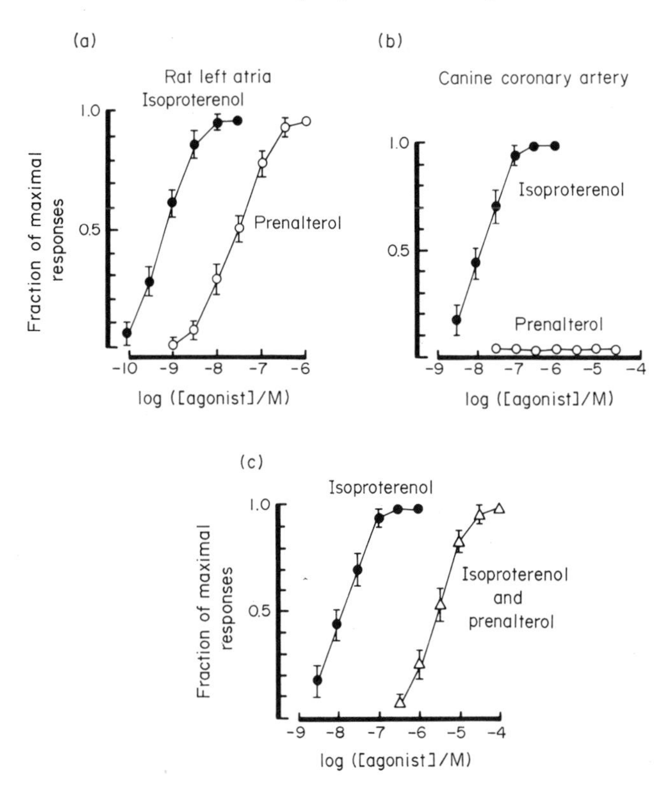

Figure 3 Effects of isoproterenol (●) and prenalterol (○) in rat left atria and canine coronary artery. Ordinates: responses as fractions of the maximal response to isoproterenol. Abscissae: logarithms of molar concentrations of agonists. (a) Effects in rat left atria. (b) Effects in canine coronary artery. (c) The effects of prenalterol (30 μM) on responses of canine coronary artery to isoproterenol (reproduced from ref. 10 by permission of Williams and Wilkins)

Table 2 Effects of β-Adrenoceptor Activation in Anesthetized Cats[a]

	ED_{50} (mol kg^{-1} i.v.) *Cardiac inotropy*	*Vasodilation*	*Ratio*
Isoproterenol	0.19	0.053	0.28
Prenalterol	9.7	> 15 100	> 1500

[a] Data from ref. 11.

binds to the receptors since they are indistinguishable from those found in the rat atrium. Thus, prenalterol functions as a competitive antagonist in this tissue (Figure 3c). The antagonism of receptors in poorly receptor-coupled organs by low efficacy agonists can also produce responses indirectly when it results in the blockade of intrinsic physiological tone *in vivo.*

In living organisms, organ function is controlled by basal levels of circulating hormones, and neuronal input and antagonists in the course of blocking the receptors for hormones and neurotransmitters can produce depressant responses. An added dimension, in terms of the resultant organ response, can be obtained with antagonists possessing intrinsic efficacy that is lower than that of the physiological transmitter (*i.e.* low efficacy partial agonists) since the depression is tempered with agonism due to the partial agonist. In these cases, the basal level of physiological tone is very important with respect to whether stimulation or depression is observed. Figure 4 shows the effects of the low efficacy β-adrenoceptor partial agonist pindolol in two types of preparations of anesthetized cat. In cats anesthetized with urethane/pentobarbital, resting basal heart rates are elevated because of high circulating levels of catecholamines, thus a 'high tone' preparation results (basal heart rate 169 ± 6 beats min^{-1}). Anesthesia with chloralose/pentobarbital produces a corresponding 'low tone' preparation (132 ± 5 beats min^{-1}). It can be seen from Figure 4(a) that pindolol

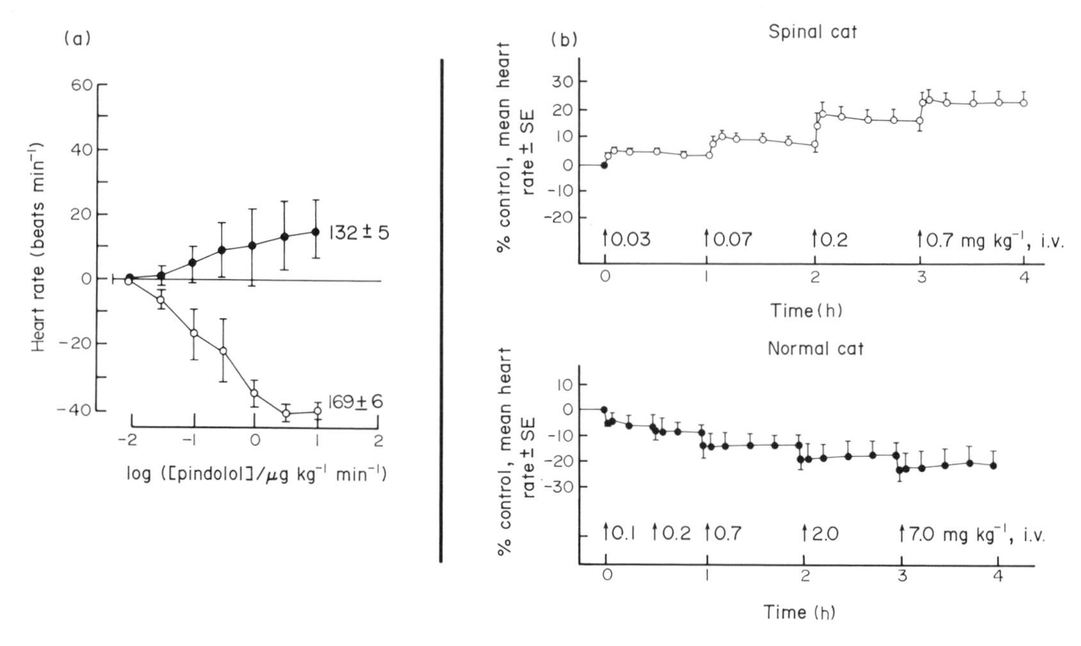

Figure 4 (a) The effects of pindolol in two preparations of anesthetized cat. Ordinates: changes in basal heart rate in beats min^{-1}. Abscissae: logarithms of intravenous doses (i.v.) of pindolol. Responses in anesthetized cats with 80 mg kg^{-1} chloralose/6 mg kg^{-1} sodium pentobarbital with low basal heart rates (●) and cats anesthetized with 800 mg kg^{-1} urethane/15 mg kg^{-1} sodium pentobarbital with high basal heart rates (○) (reproduced from ref. 12 by permission of Academic Press). (b) The effects of labetalol in anesthetized spinal cats (with no basal catecholaminergic tone) and normal anesthetized cats (reproduced from ref. 13 by permission of Raven Press)

produces substantial bradycardia in 'high tone' cats, presumably because pindolol blocks the stimulation of β-adrenoceptors by the high efficacy endogenous catecholamines norepinephrine and/or epinephrine. In 'low tone' cats where the receptors are not stimulated in the basal state, the low intrinsic efficacy of pindolol can be observed as a modest tachycardia (Figure 4a).

A similar situation is found for the low efficacy β-adrenoceptor partial agonist labetalol. In anesthetized spinal cats where there are no appreciable amounts of circulating catecholamines or neural tone, labetalol produces a β-adrenoceptor-mediated agonist response (tachycardia—see Figure 4b). However, in normal anesthetized cats with basal catecholaminergic tone, labetalol produces bradycardia due to blockade of existing catecholamines (Figure 4b).[13]

In vivo, the interplay of partial agonists with a range of organs and the various levels of basal physiological activation can lead to complex patterns of response; in these patterns, useful bioselectivity may result. Figure 5 shows the results of theoretical modeling of three organ systems *in vivo*. Tissue I has a low amplifying capability with respect to receptor stimulus. Tissue II has medium capability and tissue III produces high amplification of stimulus. The responses of these tissues to a full agonist, a full antagonist and a low efficacy partial agonist are shown under three different conditions of basal physiological tone: low, medium and high. The parameters and calculations used to construct the model are given in Appendix 1. These calculations illustrate some general predictions, namely that a full agonist will produce a response that will be additive to existing physiological tone, while an antagonist will block existing physiological tone and produce the opposite response. However, an interesting array of responses may result from the addition of a low efficacy partial agonist to these systems. In the presence of low physiological tone, the low intrinsic efficacy of the partial agonist causes the production of an agonist response, the magnitude of which is directly proportional to the efficiency of receptor coupling in each organ and inversely proportional to the magnitude of basal physiological tone. In preparations of higher physiological tone, the receptor occupancy by the low intrinsic efficacy partial agonist produces antagonism of physiological tone. The antagonism is expressed most importantly in organs where receptor coupling is least efficient (for example note Figure 3c). In general, it can be seen that, depending upon the efficiency of receptor coupling in the various organs, and the basal physiological tone, a full agonist produces added response or no effect, a full antagonist generally produces depression of the physiological tone and a low efficacy partial agonist produces a variety of responses from stimulation to depression.

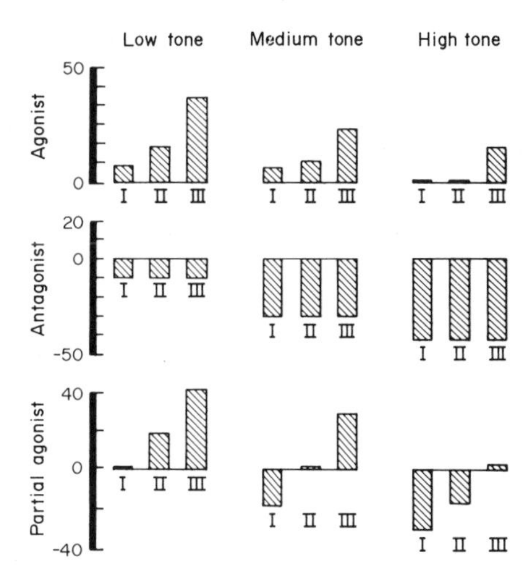

Figure 5 Theoretically calculated *in vivo* responses of preparations of varying physiological basal tone to a full agonist, antagonist and partial agonist. All changes calculated as described in Appendix 1. Ordinates: changes in basal responses as fractions of the maximal response to the endogenous agonist. Responses in organs with high (tissue III), medium (tissue II) and low (tissue I) amplifying properties for receptor stimulus

Depending upon the desired effects, low intrinsic efficacy in a drug molecule can be used to advantage either to produce a low level stimulation, as in the case of prenalterol for inotropy, or a cancellation of some of the depression produced by receptor antagonism. This latter effect is the rationale for the prediction of a greater therapeutic ratio for β-blockers with intrinsic sympathomimetic activity in asthmatics since it is postulated that a low level of β-adrenoceptor stimulation would prevent bronchospasm by producing a bronchodilatory effect.

2.4.3 SELECTIVE DRUG CONCENTRATIONS IN RECEPTOR COMPARTMENTS

Another general method to attain bioselectivity is to control the concentration of drug in the receptor compartment of various organs. There are three general methods by which this can be done.

2.4.3.1 Drug Distribution

One method to obtain bioselectivity is by controlling the route of drug entry into the receptor compartment. Some particularly accessible tissues in this regard are the skin (topical application), the lungs (aerosol delivery), the nasal passages (inhalation), the eyes (eyedrops) and the gastrointestinal tract (oral administration). By delivering drugs to these organs directly, a significant bias with respect to drug concentration can be introduced whereby the concentration in the receptor compartment of the particular organ is greater than in the surrounding tissues. For example, β-adrenoceptor agonists produce a useful bronchodilation for the treatment of asthma but taken orally they also produce a disturbing tachycardia and digital tremor. By inhalation, these side effects are greatly reduced and useful bronchodilation can readily be achieved. Likewise a reduction of intraocular pressure is produced by ocular application of the β-blocking drug timolol and by this route of administration some of the cardiovascular depression produced by systemically administered timolol is reduced. However, these methods usually only introduce bias insofar as a concentration gradient is produced between the target organ and the rest of the body and if no other mechanism removes the drug from the biophase, eventually, the drug will generally be distributed throughout the rest of the body. Thus, given enough β-adrenoceptor agonist, even by aerosol, digital tremor may result and bradycardia to intraocular timolol eventually may be observed.

There are examples where chemical modification of drug molecules limits the volume of drug distribution and in these cases steady-state differences in concentration can be achieved. For example, triprolidine (**1a**) is a potent antihistamine for H_1-receptors ($pK_B = 10.2$) and like most H_1-receptor-blocking agents, one limitation of its use is sedation. By introduction of an acrylic acid moiety to triprolidine to produce acrivastine (**1b**), a potent antihistamine ($pK_B = 8.6$) was created which, by virtue of the addition of that ligand, does not substantially cross the blood–brain barrier.[14] Thus, the distribution of acrivastine is limited to the periphery and no sedation is encountered upon oral administration.[15]

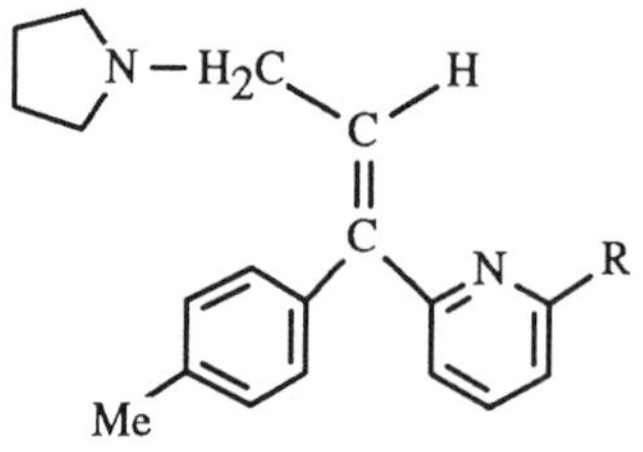

(**1a**) R = H
(**1b**) R = CH=CHCO$_2$H

The lipid solubility of a drug can affect its relative concentration in plasma and tissues. There are examples where lipid solubility may actually affect the steady-state concentrations of drugs in different receptor compartments to produce organ selectivity. For example, the relative potency of atropine as an antagonist of muscarinic mediation of gastric secretion *in vitro* is 14 times less than for other tissues (*i.e.* guinea pig trachea). This differential is most likely a result of the loss of atropine into the gastric secretion, an effect that produces a diminishing concentration gradient for atropine between the organ bath and the receptor compartment.[16] This is not observed with the considerably less lipid soluble antimuscarinic blockers *N*-methylatropine and pirenzepine.[16] This suggests that water solubility may predispose these drugs to selective antagonism of gastric secretion when taken by the oral route of absorption. This effect may carry over into the *in vivo* situation where lower plasma levels of pirenzepine, relative to atropine, are required to block acid secretion because pirenzepine, unlike atropine, is not lost through the oxyntic cells into the gastric juice.[17]

2.4.3.2 Active Drug Production (Prodrugs)

Another method of producing bioselectivity by controlling drug distribution is with the use of prodrugs (*i.e.* chemicals that are substrates for biological enzymes), which produce the active drug after enzymatic catalysis. Usually prodrugs are used to produce an increase in absorption or the extension of the duration of action of drugs. However, bioselectivity can be introduced when high levels of the enzyme are associated with the target organ; in essence, the target organ becomes a factory for the production of active drug. Thus, as with selective distribution, a concentration gradient is produced whereby the concentration of the active drugs is greatest at the target organ. For example, epinephrine produces a useful reduction of intraocular pressure for the treatment of glaucoma but this catecholamine does not penetrate the cornea readily and is unstable and short acting. The application of intraocular dipivalylepinephrine greatly increases the usefulness of epinephrine in this condition since dipivalylepinephrine is more easily absorbed through the cornea (17 times more effectively than is epinephrine)[18] and the active drug is produced at the site of action by hydrolysis with esterase enzymes (Figure 6a). Thus, epinephrine is produced where it is needed and then is degraded naturally before it can enter the general circulation and produce side effects.

There are methods to increase the passage of drugs through the blood–brain barrier, thus providing a means of concentrating drugs in the brain. One such method is drug latentiation, *i.e.* the conversion of hydrophilic drugs into lipid soluble drugs, usually by masking hydroxyl, carboxyl and primary amino groups.[19] A concentrating effect can be achieved if an enzymatic hydrolysis occurs within the brain to liberate a polar drug, which cannot then cross into the periphery. For example, the lipid soluble diacetyl derivative of morphine (heroin) crosses the blood–brain barrier at a rate 100-fold greater than morphine, thereby gaining entry into the CNS.[20] Once in the brain, pericapillary pseudocholinesterase produces deacetylation back to morphine[21] (Figure 6b). Another example of drug latentiation can be found in the selective delivery of γ-aminobutyric acid (GABA) into the central nervous system. Normally, GABA does not cross easily into the CNS but latentiation as the Schiff base amide progamide (Figure 6c), provides a useful vehicle which releases GABA after crossing the blood–brain barrier.[22,23] These approaches could be useful in the treatment of central nervous system disorders such as depression, anxiety, Alzheimer's disease, Parkinsonism and schizophrenia.

As with selective affinity, if a particular enzymatic conversion mechanism is associated with a specific organ, then bioselectivity may be achieved with a prodrug that is a substrate for that

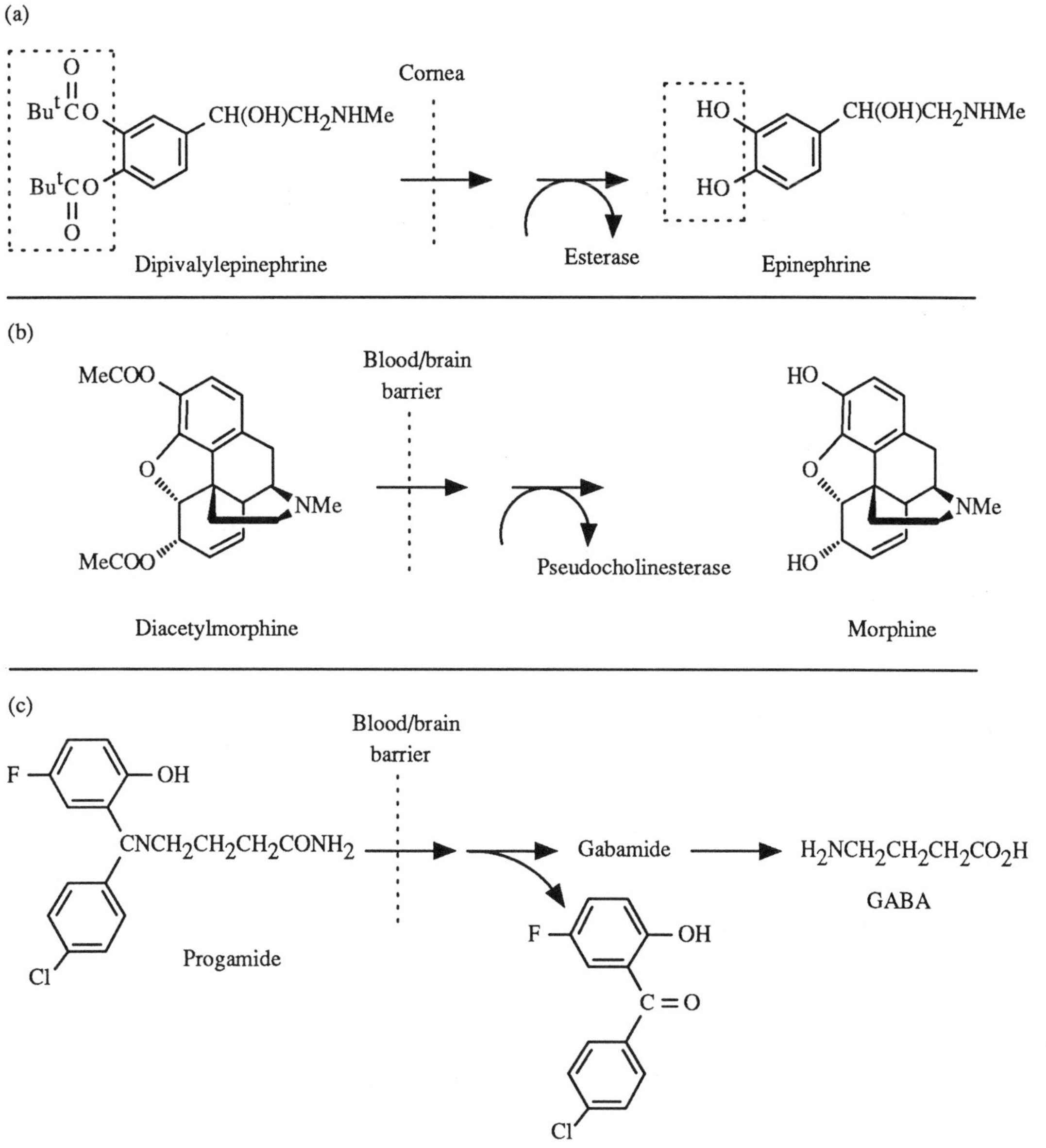

Figure 6 Prodrugs for bioselectivity. (a) Dipivalylepinephrine enters the eye readily *via* the cornea and then becomes hydrolyzed to the active antiglaucoma drug epinephrine. (b) The diacetyl derivative of morphine (heroin) crosses the blood–brain barrier at a rate 100 times greater than morphine and, once in the central nervous system, it is converted enzymatically to morphine. (c) The Schiff base amide, progamide, readily enters the CNS and releases the putative neurotransmitter γ-aminobutyric acid (GABA)

mechanism. For example, the kidney has a particularly high level of γ-glutamyltranspeptidase and aromatic L-amino acid decarboxylase activity. Thus, selective renal vasodilation can be obtained with the prodrug for dopamine, γ-glutamyldopa; the resulting levels of dopamine in the kidney with this drug are five times higher than those achieved after administration of L-dopa.[24] Since dopamine is relatively unstable in the periphery, a useful concentration gradient of dopamine in the kidney results.

2.4.3.3 Drug Degradation and/or Removal

Many organs possess degradatory or removal mechanisms for natural hormones and neurotransmitters. Physiologically, these allow for a delicate control of organ function and also are a method of conservation of specific chemicals needed for biosynthesis. In some cases, these removal mechanisms can be used to achieve bioselectivity. The interaction of receptors with drugs is a dynamic process and if a removal mechanism for the drug is present in the receptor compartment, the tissue responds to a flux of drug created by entry *via* diffusion and exit *via* the removal process. Therefore, the rate of entry of drug molecules becomes as important as the rate of removal with respect to the flux of drug concentration produced and the magnitude of steady-state response produced by the tissues. Drug removal mechanisms can be remarkably efficient in producing a deficit of drug concentration in the receptor compartment, especially if the drug must first penetrate the matrix of the removal mechanism before it can activate receptors (Figure 7). Figure 7 shows this schematically; a drug that is not removed by the removal mechanism (drug I in Figure 7b) does not experience a decreasing concentration gradient between the reservoir and the receptor compartment. However, the concentration of a drug that is a substrate for the removal mechanism can be significantly depleted *en route* to the receptor compartment (drug II in Figure 7b). Under these circumstances, there can be a considerable difference between the true potency of the drug in the receptor compartment and the observed potency when it is added to a reservoir outside the removal barrier.

Organs differ markedly in their capacity to remove hormones and neurotransmitters. Figure 8(a) shows the effects of acetylcholine in three different isolated tissues; the concentrations were adjusted such that they would have produced equal responses in the absence of acetylcholinesterase, a degradatory enzyme for acetylcholine present at neuromuscular junctions. It can be seen from this figure that whereas acetylcholinesterase does not significantly remove acetylcholine from the receptor compartment of guinea pig ileum, in rat anococcygeus muscle (RAM)—a considerable deficit of acetylcholine is produced by this enzyme. There can be many reasons for this variability. First, the capacity of the removal mechanisms such as nerves and degradatory enzymes can determine the degree of drug removal. For example, the innervation of guinea pig trachea is sparse and relatively little neuronal uptake of norepinephrine occurs (Figure 8b). In contrast, the rat anococcygeus muscle is highly innervated[26] and these nerves avidly remove norepinephrine from the receptor compartment. Figure 8(c) shows that the neuronal uptake produces a 400-fold decrease in the sensitivity of this tissue to norepinephrine.[27] Second, the proximity of the removal sites to the receptors can be critical. For example, in tissues such as the rat vas deferens where the nerves are close to the postsynaptic receptors (neuromuscular interval = 100–300 Å), neuronal uptake produces a large deficit of norepinephrine in the receptor compartment.[28] By contrast, in the pulmonary artery where the neuromuscular interval is considerably larger (4000 Å),[29] little neuronal removal of norepinephrine is observed. Finally, other factors such as the tortuosity factor[30] with respect to ease of diffusion of the drug within the tissue and the efficiency of muscle cell coupling to produce a syncytial response all contribute to the importance of removal mechanisms on the magnitude of drug effect in different tissues. These factors all combine in various organs to produce a range of capabilities for drug degradation and/or removal to affect potency. This can lead to bioselectivity if appropriate substrates for these mechanisms are used for response.

It is difficult to predict the effects of removal mechanisms on drug potency from *in vitro* experiments since the magnitude of response often is determined by the rate of drug entry and this differs from the presentation of drug *via* the bloodstream *in vivo*. However, there are examples *in vivo*

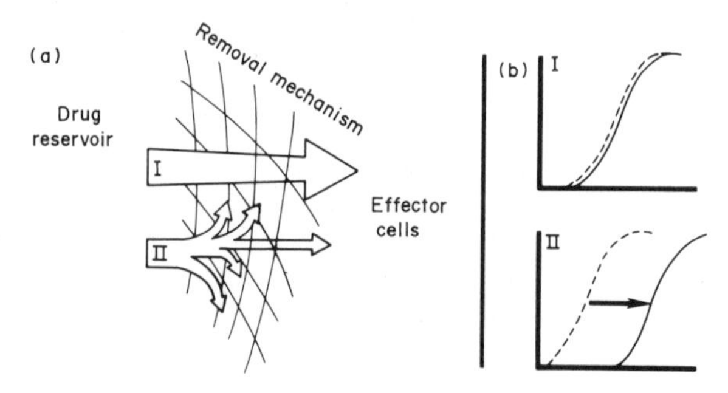

Figure 7 (a) If a drug II is a substrate for an active removal process and must pass through a barrier of active removal before reaching the receptor compartment, then a substantial concentration deficit between the pool of drug (*i.e.* the circulation) and the receptor compartment can result. This is not the case for drugs that are not substrates for the removal process (drug I). (b) The dose–response curves are for both drugs in tissues where the removal mechanism is blocked (broken lines) and operative (solid lines). Note the decrease in potency for drug II, the substrate for the removal process, in the presence of an operative removal process

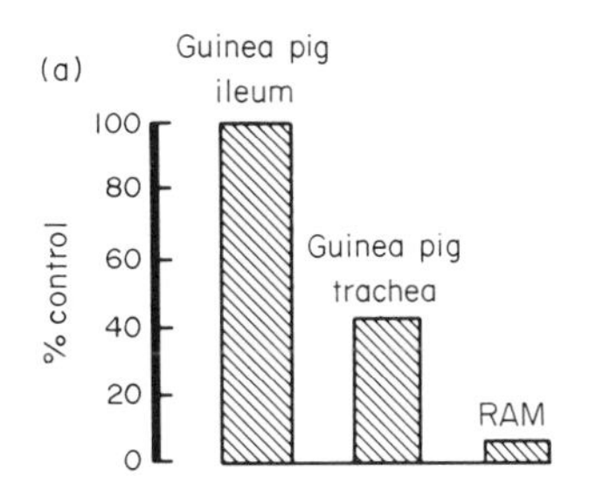

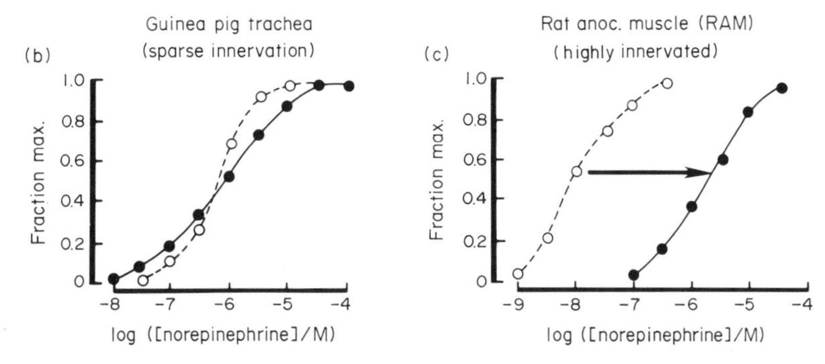

Figure 8 The effects of agonist removal mechanisms on the sensitivities of isolated tissues to acetylcholine (a) and norepinephrine (b). For (a), ordinates are responses to a given concentration of acetylcholine expressed as a fraction of the concentration that produces 85% of the maximal response when acetylcholinesterase is completely blocked. Data shown for guinea pig ileum, guinea pig trachea and rat anococcygeus muscle (RAM) (reproduced from ref. 25 by permission of Williams and Wilkins). (b) Concentration–response curves of guinea pig trachea to norepinephrine. Ordinates: relaxation of carbachol (10 μM) induced tone as a fraction of the maximal relaxation to norepinephrine. Abscissae: logarithms of molar concentrations of norepinephrine. Responses in the absence (●) and presence (○) of demethylimipramine, a blocker of the neuronal uptake of norepinephrine. (c) Concentration–response curves of rat anococcygeus muscles to norepinephrine. Ordinates: contractions as fractions of the maximal contraction to norepinephrine. Abscissae: logarithms of molar concentrations of norepinephrine. Responses to norepinephrine in normal tissues (●) and in tissues with neuronal uptake blocked by cocaine (○). The arrow shows the effects of neuronal uptake on the sensitivity of this tissue to norepinephrine (reproduced from ref. 27 by permission of Williams and Wilkins)

where drug uptake mechanisms significantly contribute to bioselective responses. For example, the sino-atrial node of the heart is richly innervated and with this innervation comes a substantial removal capability for catecholamines by active neuronal uptake in the region of sino-atrial node β-adrenoceptors. The β-adrenoceptor agonist isoproterenol, a drug not taken up by adrenergic nerve endings,[31] is considerably more potent in terms of producing tachycardia for a given increase in inotropy than the substrate catecholamine for neuronal uptake, norepinephrine. This disparity is abolished by blockade of neuronal uptake mechanisms by cocaine.[32] Another example of where a drug removal mechanism produces effective bioselectivity is the cerebral circulation of primates. In these blood vessels α-adrenoceptor-mediated vasoconstriction and β-adrenoceptor-mediated vasodilatation can be produced by catecholamines. However, the cerebral circulation is protected from the vasoconstricting effects of epinephrine by neuronal uptake mechanisms which prevent the attainment of receptor compartment concentrations of this catecholamine sufficient to activate α-adrenoceptors.[33] Another example of self-cancellation is the muscarinic antagonist and acetylcholinesterase blocker ambenonium. This drug reduces muscarinic receptor activation by competitively blocking muscarinic receptors but also elevates synaptic concentrations of acetylcholine by blocking the degradation of this muscarinic agonist by acetylcholinesterase.[25] The two effects of diminution and potentiation of muscarinic responses cancel each other and ambenonium is a weak muscarinic antagonist of acetylcholine responses in organs with an active acetylcholinesterase degradation as opposed to those organs where the enzyme is less important to the synaptic concentrations of acetylcholine. In effect, the acetylcholinesterase-blocking property of ambenonium restricts the subset of organs in which this drug will produce muscarinic receptor blockade from all those with muscarinic receptors to those which possess muscarinic receptors *and* which do not have active acetylcholinesterase (Figure 9). This can lead to a striking organ selectivity; for example, ambenonium is 13-fold more potent as an antimuscarinic blocking agent in guinea pig ileum than in

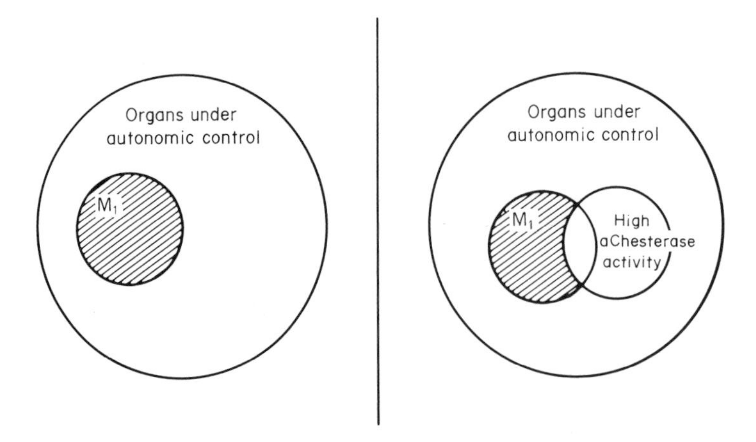

Figure 9 Schematic diagram of the subsets of organs affected by a standard antagonist of muscarinic M_1-receptors and an antagonist of muscarinic M_1-receptors and the degradatory enzyme for acetylcholine, acetylcholinesterase. This latter subset includes organs possessing muscarinic M_1-receptors as well as nicotinic receptors. The tissues which possess an active acetylcholinesterase *and* muscarinic M_1-receptors may not effectively demonstrate M_1-receptor blockade (see text) and thus will be excluded from the subset of organs possessing muscarinic M_1-receptors blocked by the drug

rat anococcygeus muscle because of self-cancellation due to blockade of acetylcholinesterase.[25] In effect, such opposing activities become a fine-tuning mechanism for bioselectivity due to selective receptor affinity.

2.4.4 DRUG SYNERGY

Another mode of drug bioselectivity is by the expression of two or more activities by one molecule in a common range of concentrations. In essence, this modality restricts the range of drug responses to a subset of organs which either have or do not have receptors for both activities. For example, if a drug has activity for two receptors which mediate physiologically opposing responses in tissues, then the drug would express one of the two activities in tissues which possess only one of the receptors but a diminished (or even negligible) response in organs having both. One drug with such self-cancelling properties is dobutamine, which produces vascular relaxation by activation of β-adrenoceptors and vascular contraction by activation of α-adrenoceptors. Figure 10 shows the responses of canine saphenous vein to dobutamine *in vitro*; in the absence of α- or β-adrenoceptor blockade, little

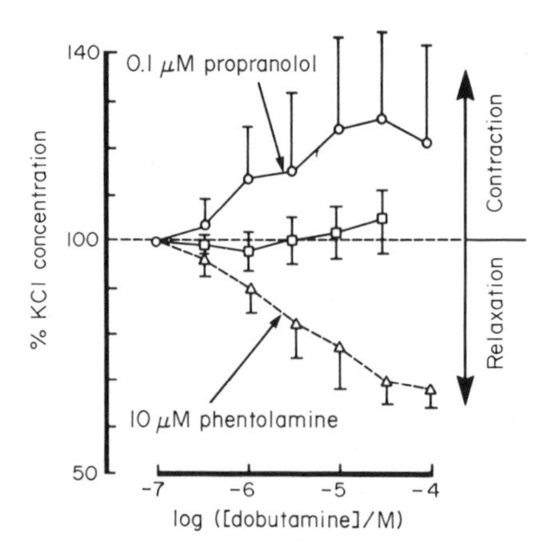

Figure 10 The effects of self-cancellation in organs to produce bioselectivity. The effects of dobutamine on canine saphenous vein *in vitro*. Ordinates: changes in tension of rings of canine saphenous vein as a percent of resting tension (relaxation) or maximal contraction to potassium chloride (contraction). Abscissae: logarithms of molar concentrations of dobutamine. Responses in normal veins (□; $n = 3$) and after blockade of relaxant β-adrenoceptors by propranolol (○; $n = 4$) and after blockade of contractile α-adrenoceptors by phentolamine (△; $n = 4$) (reproduced from ref. 3 by permission of Williams and Wilkins)

response to dobutamine is observed. This is a result of concomitant β-adrenoceptor relaxation and α-adrenoceptor contraction in this tissue (Figure 10).[3] Therefore, depending upon the relative importance of α- and β-adrenoceptor mechanisms in vascular smooth muscle throughout the body, dobutamine will produce a pattern of response much different from drugs with activity for only one of the receptors.

2.4.5 CONCLUSIONS

This chapter describes methods by which bias, with respect to drug effect, can be introduced into biological systems. In some cases the bias has a chemical basis in that it is the result of selective affinity and/or intrinsic efficacy for a cellular component. In other cases, bioselectivity results from a pharmacokinetic bias resulting in an unequal distribution of drug concentration *in vivo*. Lastly, a combination of drug properties may result in bioselective drug responses if these are expressed in the appropriate concentration range. It should be noted that in these latter modes of bioselectivity, relatively nonselective drugs may still be valuable in the therapy of disease. Thus the modification of existing drug molecules either to restrict biodistribution or to introduce another bioactivity should be considered as a viable avenue for pharmacological research.

2.4.6 APPENDIX 1

The total stimulus (S_T) produced by a full agonist (A) and a partial agonist (P) present in the receptor compartment concomitantly is defined by classical receptor theory as

$$S_T = \frac{\{([A]/K_A)[R_T]\varepsilon_A\} + \{([P]/K_P)[R_T]\varepsilon_P\}}{1 + ([A]/K_A) + ([P]/K_P)} \tag{1}$$

where ε_A and ε_P refer to the respective intrinsic efficacies of A and P, K_A and K_P the respective equilibrium dissociation constants of the receptor complexes with A and P, $[R_T]$ is the total receptor density. For this model, the relative intrinsic efficacy of $\varepsilon_A/\varepsilon_P$ is 220 (ε_A for the physiological transmitter = 3 and ε_P for the partial agonist = 0.0136).

The response is assumed to be a rectangular hyperbolic function of S_T given as

$$\text{Response} = S_T/(S_T + \beta) \tag{2}$$

where β is a fitting parameter. It should be noted that the nature of this function does not determine the outcome of the calculations but rather simply allows the modeling of stimulus–response

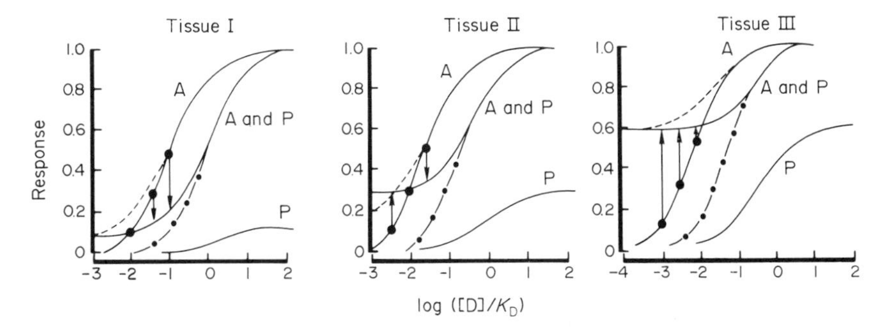

Figure 11 Theoretical effects of a full agonist (A) and partial agonist (P) in an *in vivo* physiological system. Ordinates: response calculated with equation (2) from total stimulus calculated with equation (1). Abscissae: logarithms of molar concentrations of drugs (D) as multiples of the equilibrium dissociation constants of the drug–receptor complexes (K_D). Tissues I to III are of varying efficiency of coupling from a relatively low amplification (tissue I, β for equation 2 = 0.3), to medium amplification (tissue II, β for equation 2 = 0.1) to high amplification (tissue III, β for equation 2 = 0.03). Dose–response curves for the full agonist (endogenous transmitter) and partial agonist alone denoted by A and P respectively. The dose–response curve to the full agonist in the presence of a defined ($[P]/K_P = 10$) concentration of partial agonist is denoted by A + P. The effects of a full antagonist at the same concentration are shown by the curve –·–·– and the additive effects of a full agonist of equal intrinsic efficacy ($[A']/K_{A'} = 0.01$) by the curve ––––. The heavy dots at response = 10%, 30% and 50% represent low, medium and high physiological tone preparations respectively. The arrows show the direction of the change in basal response to a concentration of partial agonist $[P]/K_P = 10$. The magnitudes of these changes are the responses shown in Figure 5

relationships of differing efficiencies of amplification.[8] For this simulation, three organs of differing receptor-coupling efficiency were used; tissue I, $\beta = 0.3$ = relatively poorly coupled; tissue II, $\beta = 0.1$ = medium coupling; tissue III, $\beta = 0.03$ = highly amplified coupling.

Three levels of basal physiological tone are used; low tone refers to a state where the basal tone due to the endogenous transmitter produces only 10% of the possible maximal response, medium tone = 30% of maximal response and high tone = 50% of the maximal response. The dose–response curves of the three tissues to the full agonist (A) alone, the partial agonist (P) alone and the full agonist in the presence of a given concentration ($[P]/K_P = 10$) of partial agonist are shown in Figure 11. This latter condition would correspond to the effects of administering that concentration of the partial agonist to a whole animal which was under a constant basal stimulation by a low level of high efficacy endogenous agonist. The arrows represent the change in observed response in a condition of low physiological tone (basal response to the endogenous agonist 10% of the maximum response), medium tone (30%) and high tone (50%). Also shown in broken lines on these figures are the effects of the agonist (A) in the presence of a given concentration ($[A']/K_A = 0.01$) of a full agonist of equal intrinsic efficacy and also a full antagonist ($[B]/K_B = 10$). The changes in observed basal response are calculated from these figures for the responses shown in Figure 5.

2.4.7 REFERENCES

1. T. P. Kenakin, in 'The Pharmacologic Analysis of Drug Receptor Interaction', Raven Press, New York, 1987.
2. O. D. Gulati, S. D. Gokhale, H. M. Parikh, B. P. Udwadia and V. S. R. Krishnamurty, *J. Pharmacol. Exp. Ther.*, 1969, **166**, 35.
3. T. P. Kenakin, *J. Pharmacol. Exp. Ther.*, 1981, **216**, 210.
4. R. R. Tuttle and J. Mills, *Circ. Res.*, 1975, **36**, 185.
5. G. H. Hitchings, in 'Design and Achievement in Chemotherapy—A Symposium in Honor of G. H. Hitchings', Science and Medical Publishing Co., New York, 1976.
6. R. F. Furchgott, *Adv. Drug Res.*, 1966, **3**, 21.
7. T. P. Kenakin, *Pharmacol. Rev.*, 1984, **36**, 165.
8. T. P. Kenakin and D. Beek, *J. Pharmacol. Exp. Ther.*, 1980, **213**, 406.
9. T. P. Kenakin, *J. Cardiovasc. Pharmacol.*, 1985, **7**, 208.
10. T. P. Kenakin and D. Beek, *J. Pharmacol. Exp. Ther.*, 1982, **220**, 77.
11. E. Carlsson, C. G. Dahloj, A. Hedberg, H. Persson and B. Tangstrand, *Naunyn Schmiedeberg's Arch. Pharmacol.*, 1977, **300**, 101.
12. T. P. Kenakin, *Adv. Drug Res.*, 1986, **15**, 71.
13. A. S. Tadepalli and P. J. Novak, *J. Cardiovasc. Pharmacol.*, 1986, **8**, 44.
14. H. J. Leighton, R. F. Butz and J. W. A. Findlay, *Pharmacologist*, 1982, **25**, 163
15. A. F. Cohen, M. J. Hamilton, S. H. Liao, J. W. Findlay and A. W. Peck, *Eur. J. Clin. Pharmacol.*, 1985, **28**, 197.
16. J. A. Angus and J. W. Black, *Br. J. Pharmacol.*, 1979, **67**, 59.
17. J. W. Black and N. P. Shankley, *Br. J. Pharmacol.*, 1985, **86**, 601.
18. A. I. Mandell, F. Stentz and A. E. Kitakchi, *Ophthalmology (Rochester, Minn.)*, 1978, **85**, 268.
19. W. M. Pardridge, *Annu. Rep. Med. Chem.*, 1985, **20**, 305.
20. W. H. Olendorf, S. Hyman, L. Braun and S. Z. Olendorf, *Science (Washington, D.C.)*, 1972, **178**, 984.
21. C. E. Inturrisi, M. B. Max, K. M. Foley, M. Schultz, S. U. Shin and R. W. Houde, *N. Engl. J. Med.*, 1984, **310**, 1213.
22. J. P. Kaplan, B. Raizon, M. Desarmenien, P. Feltz, P. M. Headley, P. Worms, K. G. Lloyd and G. Bartholini, *J. Med. Chem.*, 1980, **23**, 702.
23. P. Worms, H. Deportere, A. Durand, P. L. Morselli, K. G. Lloyd and G. Bartholini, *J. Pharmacol. Exp. Ther.*, 1982, **220**, 660.
24. S. Wilk, H. Mizoguchi and M. Orlowski, *J. Pharmacol. Exp. Ther.*, 1978, **206**, 227.
25. T. P. Kenakin and D. Beek, *J. Pharmacol. Exp. Ther.*, 1985, **232**, 732.
26. U. Trendelenburg, in 'Handbook of Experimental Pharmacology', ed. H. Blaschko and E. Muscholl, Springer-Verlag, Berlin, 1972, vol. 33, p. 726.
27. H. S. Leighton, *J. Pharmacol. Exp. Ther.*, 1982, **220**, 299.
28. M. A. Verity, *Physiol. Pharmacol. Vasc. Neuroeff. Syst., Proc. Symp. 1969*, 1971, 2.
29. M. A. Verity and J. A. Bevan, *J. Anat.*, 1968, **103**, 49.
30. E. Page and R. S. Bernstein, *J. Gen. Physiol.*, 1964, **47**, 1129.
31. A. S. V. Burgen and L. L. Iversen, *Br. J. Pharmacol.*, 1965, **25**, 34.
32. C. M. Furnival, R. J. Linden and H. M. Snow, *J. Physiol. (London)*, 1971, **214**, 15.
33. T. A. McCalden, B. H. Eidelman and A. D. Mendelow, *Am. J. Physiol.*, 1977, **233**, H458.

2.5
The Immune System

STEVEN C. GILMAN and ALAN J. LEWIS
Wyeth Laboratories, Philadelphia, PA, USA

and

PAUL H. WOOLEY
Ayerst Laboratories, Princeton, NJ, USA

2.5.1 INTRODUCTION

The immune system consists of a heterogeneous group of lymphoid cells whose collective function is to provide adequate protective mechanisms against potentially harmful foreign substances (*i.e.* bacteria, viruses, *etc.*) and prevent or limit the abnormal growth of host tissues (*i.e.* malignant cells). Normal and appropriate immune function is necessary for homeostasis; either an insufficient immune response (immune deficiency) or a misdirected immune response (autoimmunity, allergy) can result in severe and life-threatening disease.

2.5.2 ORGANIZATION AND ANATOMY OF THE IMMUNE SYSTEM

Scattered throughout the body are collections of lymphoid cells organized into various lymphoid tissues whose structural integrity varies dramatically in complexity.[1] The three central components of the lymphoid system are the stem cells, the central lymphoid organs and the peripheral lymphoid systems. Pluripotential hematopoietic stem cells originate from the bone-marrow and are the progenitors of all classes of lymphoid cells. The central lymphoid organs, the thymus and the bursa of Fabricius (or 'bursal equivalent' in mammals), are responsible for promoting the development of mature T and B cells, respectively (Figure 1). The peripheral lymphoid system represents those lymphocyte populations which have been processed by the central lymphoid organs which are directly responsible for the immune response observed upon antigen exposure, including the spleen, lymph nodes, gut-associated lymphoid tissues (GALT) and bronchial-associated lymphoid tissues (BALT). The organs of this system have specialized functions and structure and the cells therein can interact with each other *via* blood and lymphatic circulatory systems.

Cells of immunological importance include polymorphonuclear leukocytes (PMN), granulocytes (basophils, eosinophils, mast cells), monocyte/macrophages and lymphocytes. Substantial heterogeneity exists within each cell type. Different subsets of T and B lymphocytes can be identified based on the expression of certain surface antigens which are defined by monoclonal or polyclonal antibodies. For example, while all human peripheral T cells express the membrane antigen identified by a monoclonal antibody designated OKT3, subpopulations of these cells express either the OKT4 or OKT8 antigens.[2] These antigenic markers correlate with cell function, OKT4 being associated with 'helper' T cell function and OKT8 with 'suppressor/cytotoxic cell' function (see below). However, the relationship between surface antigen expression and T cell function is not absolute, as both $OKT8^+$ and $OKT4^+$ cells produce helper factors (cytokines) and both can function as cytotoxic cells in certain circumstances. B lymphocyte subpopulations express different B-cell-specific surface antigens and the display of surface immunoglobulin classes (primarily IgD, IgM and IgG) differs depending on the maturational state of the B cell.[3] Macrophage subpopulations have also been described differing in membrane surface antigen expression (HLA/DR/'Ia'), function and buoyant density.[4] No easily distinguishable subsets of PMNs have been described, but heterogeneity of mast cells has been reported. Clearly, immunity results from many complex interacting cell types and subtypes all of which have specialized functions and distinguishing characteristics.

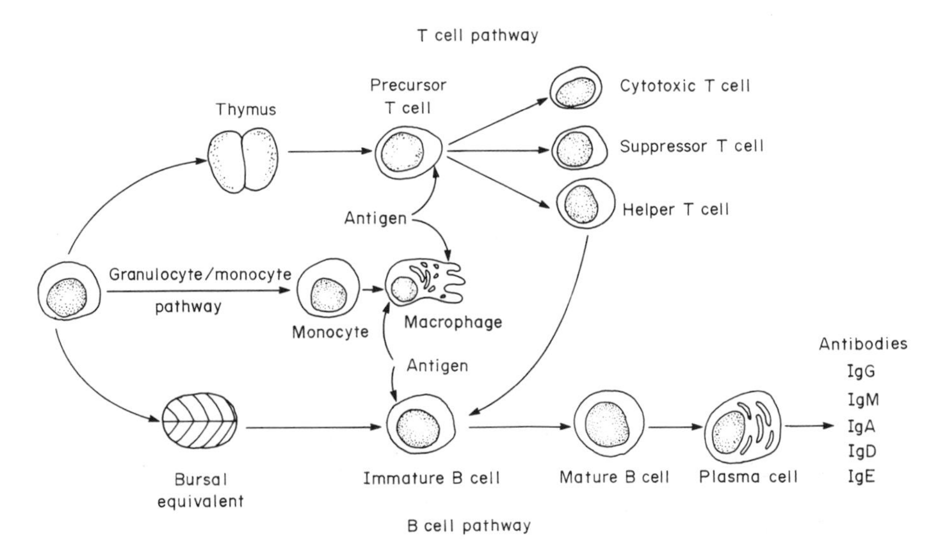

Figure 1 Schematic representation of the maturational development of lymphoid cells. Pluripotent stem cells, originating in the bone-marrow, migrate to the thymus or bursal equivalent where they receive differentiative signals and undergo several maturational events. The immunocompetent T and B cells then emigrate from these tissues and populate the peripheral lymphoid organs. Monocyte/macrophages are also derived from bone-marrow stem cells, but do not require thymic or bursal influence for their development

Two general types of cellular defense mechanisms can be distinguished depending on whether or not antigen stimulation is required for the particular defense mechanism to operate efficiently. Nonspecific or 'natural' cellular defense mechanisms are those which do not require prior antigen exposure in order to be activated. One principal mechanism is the phagocytosis and intracellular killing of infectious agents by PMNs, granulocytes and macrophages. Phagocytic efficiency can be enhanced by the presence of antibody (opsonization) or activated complement components. Natural killer (NK) cells are a subpopulation of T cells which are able to lyze a variety of tumor cells and virus-infected cells without prior activation, although their cytotoxic activity can be augmented by certain stimuli such as interferon.[5] Macrophages also express an antigen-nonspecific cytolytic activity.[6]

Antigen-specific cellular immune reactions are those in which immune cells specifically recognize antigenic determinants through specific membrane receptors and evoke immunological memory which is responsible for the more rapid and vigorous nature of secondary (anamnestic) responses to the same antigen. These cellular immune reactions involve numerous cell types which interact with each other in a complex but orderly sequence culminating in cellular activation and the generation of effector cells or molecules capable of eliminating the invading agent. The induction phase of T cell activation begins when antigen-presenting cells such as macrophages and dendritic cells initiate early events in T cell activation by ingesting and processing the antigen which is then presented in an immunogenic form to T lymphocytes bearing antigen receptors and certain membrane proteins (called 'Ia' or 'class II' antigens) which are encoded in the immune associated (Ia) region of the major histocompatibility gene complex (MHC).[7] Macrophages also provide the necessary T helper factors such as interleukin 1 (IL-1, formerly called lymphocyte-activating factor) to T cells.[8]

The regulatory phase of T cell activation is characterized by clonal expansion of subsets of antigen-reactive T cells and the generation of T helper or T suppressor cells which provides a system of checks and balances on the magnitude of the T cell response. Finally, other subsets of antigen-reactive T cells through a combined action of helper cells and soluble helper factors derived from these cells differentiate into antigen-specific effector cells such as cytotoxic T cells (CTLs) capable of lyzing virally infected cells, allogenic cells, tumor cells, *etc.*

B lymphocytes are the mediators of humoral immunity and are responsible for the production of antibody molecules (IgM, IgG, IgA, IgD, IgE and their subclasses). When exposed to antigen, mature B cells become activated and undergo proliferative and differentiative processes which are regulated by macrophages and T cells.

Circulating antibody can directly inactivate bacteria or viruses and, with the aid of a cascade of serum protein enzymes called the complement system, can cause tumor cell lysis. Alternatively, effector lymphoid cells including NK cells, macrophages, B cells and some T cells which express receptors for the Fc region of the immunoglobulin molecule (see Section 2.5.3.2) can be armed with specific antibody and lyze antigen-bearing cells in a process termed antibody-dependent cellular cytotoxicity (ADCC).

Tissue and skin mast cells are responsible for allergic and hypersensitivity reactions to specific antigens which in this case are termed 'allergens'. Antigen binding to IgE antibody which is itself bound to the IgE receptor on the cell membrane initiates a complex mast cell activation process culminating in the release of numerous biologically active mediators such as histamine, bradykinin, serotonin and leukotrienes. Under normal circumstances, these mediators are thought to regulate the microvascular tone, permeability and blood flow, but under abnormal conditions (*i.e.* hypersensitivity) they can cause swelling, pain, edema and eventual respiratory failure resulting in death.

2.5.3 HUMORAL IMMUNITY

2.5.3.1 Immunoglobulin Secretion

Immunoglobulin molecules are the products of antibody-secreting cells (plasma cells), which develop from the activation and differentiation of B lymphocytes.[3] Mature B lymphocytes are characterized by the presence of cell surface antibody molecules of a single specificity. As a consequence of the interaction of surface antibody with specific antigen, usually in combination with helper T cells and soluble factors, the B cell is activated and undergoes proliferation or clonal expansion. This results in an increased number of B cells with the same specific antibody receptor. After clonal expansion, a number of these B cells undergo differentiation into immunoglobulin-secreting plasma cells. This differentiation involves an increase in size, and cytoplasmic changes which include an increase in the number of mitochondria and a highly developed Golgi apparatus.

The surface-bound antibody disappears and the plasma cell produces high levels of secretory Ig molecules which have the same specificity as the original surface antibody. However, plasma cells are short-lived with a 2–3 day lifespan.

2.5.3.2 Immunoglobulin Structure and Subclasses

The structure of an immunoglobulin molecule is diagrammatically represented in Figure 2. The basic Ig unit is constructed from four polypeptide chains: two identical light chains with a molecular weight around 23 000 kDa and two identical heavy chains with a molecular weight around 58 000 kDa. The chains are then joined together by disulfide bonds, which link one light chain to one heavy chain, and join both of these subunits together at the hinge region. The light chains are divided into two domains: a variable region (V_L) and a constant region (C_L). These regions are juxtaposed to variable (V_H) and constant (C_H1) regions on the heavy chain. Three-dimensional folding of the molecule results in an interaction between the two variable regions to create an antigen-binding region. The heavy chain also contains several other domains, which are involved with the binding of complement (C_H2) and other biological functions of the individual Ig classes (C_H3).[9] Enzymatic cleavage allows the molecule to be divided with respect to function. Fab fragments consist of an intact light chain in association with the V_H and C_H1 regions of the heavy chain. Cleavage distal to the hinge region produces $F(ab)_2$ fragments. Both Fab and $F(ab)_2$ fragments retain the ability to bind to antigen, whereas the remaining portion of the heavy chain dimer (the Fc component) does not. There are two classes of light chains, kappa and lambda, which can be distinguished in a single species by variations in the polypeptide sequence of the C_L domains. There is no known preference of binding of kappa *vs.* lambda light chains to heavy chain partners, and the marked variation in the ratio of kappa:lambda chains seen between mice (95:5) and man (70:30) is attributed to the number of variable region genes which occur in the segments encoding for the respective regions. Major variations in the C_H domains have given rise to five major classes or isotypes of immunoglobulin molecules. IgG is the most frequently occurring isotype in serum and has been divided into four subclasses (IgG1, IgG2a, IgG2b and IgG3 in mice and IgG1, IgG2, IgG3 and IgG4 in humans). Most IgG subclasses fix serum complement *via* the interaction of the first component of the complement system (C1q) with the C_H2 domain. Membrane-bound Fc receptors found on the surface of macrophages and B lymphocytes appear to possess the highest capacity to interact with the Fc region (domains C_H2–C_H3) of IgG molecules.

IgA is the most frequently occurring isotype in secretions (tears, saliva, mucous, *etc.*). Secretory IgA occurs as two basic Ig molecules joined together by a J chain and associated with a glycoprotein known as the secretory component. IgM antibodies are the first immunoglobulin class to appear during a primary immune response, and are thus thought to be the earliest class from an evolutionary standpoint. Monomeric IgM is the major antibody found on the surface of mature B cells. Serum IgM occurs as a pentamer, with the five IgM units bound together by a J chain.

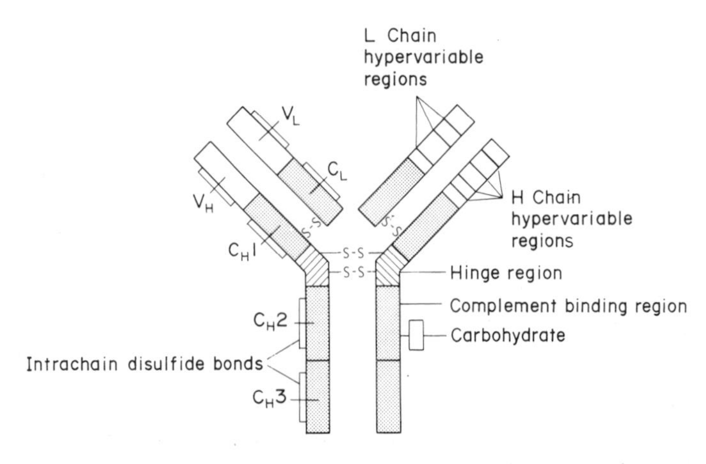

Figure 2 The immunoglobulin molecule. The Ig molecule is represented by the four subunits (two heavy chains and two light chains) interconnected by disulfide bonds as indicated. The juxtapositioning of the variable regions of light and heavy chains (V_H and V_L) provide the antigen-binding sites, while constant region (LC) domains provide for interaction with complement components and cell-bound Fc receptors

IgE antibodies mediate allergic responses. Mast cells and basophils express Fc receptors which are specific for the IgE molecule. If surface-bound IgE molecules interact with the appropriate antigen, a degranulation process is triggered which results in the release of vasoactive amines, arachidonic acid metabolites and other bioactive mediators. IgD molecules share with IgM the property of expression on the surface of mature B cells; however, the level of IgD in normal serum is extremely low and the function of this immunoglobulin class is, at present, unknown.

2.5.3.3 Idiotypes and Antiidiotypes

Since the antigen-binding site of the antibody molecule results from the association of the hypervariable regions of the heavy and light chains, there is an element of uniqueness in each specific set of antibodies. Due to the three-dimensional structure of the immunoglobulin molecule, the conformational variability in the antigen-binding pocket also gives rise to conformational changes exterior to this region. These small variations in the structure of antibody molecules give rise to restricted antigenic determinants, and antibodies raised to such determinants may be used to compare antibody sets with similar specificities. Single structural variations seen between antibodies due to their differences in specificities are known as idiotopes, and the sum of the different idiotopes on a given antibody molecule is referred to as the idiotype.[10] Only antibodies derived from a monoclonal source are completely identical in both common and individual antigenic determinants; however, polyclonal antibodies raised against a common antigen may share idiotypic markers. The potential for common idiotypes is raised if the antibody family under study is directed against a single antigen epitope or hapten.

Antiidiotype antisera have been used for several functional applications. The inheritance of a particular idiotype can be monitored to follow the familial transfer of a particular variable gene segment. However, this process is complicated if the idiotype is dependent upon the coexistence of both the V_L and V_H of the antibody in question, rather than a single variable domain alone. Recently, antiidiotype antisera have been proposed as a means of regulating the antibody response. This has direct application to the treatment of diseases where autoantibodies have been detected. Hahn *et al.*[11] have shown that a monoclonal antiidiotype antibody raised against antiDNA antibodies from lupus-prone mice can deplete the antiDNA antibodies carrying the idiotype recognized by the monoconal. This in turn provides significant amelioration of the symptoms of the spontaneous disease in MRL lpr/lpr mice.

2.5.4 CELLULAR IMMUNITY

2.5.4.1 Polymorphonuclear Leukocytes and Macrophages

Natural immunity may be considered as the inherent ability to mount a response to foreign bodies, in particular bacteria. Natural immunity does not involve the generation of specificity, beyond the recognition of the substance eliciting the response as foreign. In contrast, acquired immunity implies the generation of a specific response by the immune system to pathogens and the development of an immunological memory to that agent. From an evolutionary standpoint, the coexistence of two such mechanisms provide a distinct advantage; an immediate broad-based defense allows for the time needed to develop a specific controlled response appropriate in specificity and magnitude. The primary focus for natural immunity is usually centered on the role of polymorphonuclear leukocytes (PMNs) and macrophages. These cells possess the ability to engulf bacteria by phagocytosis, enveloping the pathogens in vacuoles and exposing them to enzymatic degradation. PMNs are more involved in the response to acute infection, whereas macrophages are more important in chronic infection.[12] This difference may be related to the mechanisms of intracellular processing between the two cell types. PMNs use several antimicrobial mechanisms to destroy organisms within the phagocytic vacuole. For example, myeloperoxidase is an oxygen-dependent enzyme present in high quantities within cytoplasmic granules, which in conjunction with hydrogen peroxide, halide cofactors and a low pH causes antibacterial, antifungal and antimycoplasma activity within the phagocytic vacuole. PMNs also have oxygen-independent systems, including the enzymes lactoferrin, lysozyme and elastase, which have marked antimicrobial properties. Macrophages do not possess myeloperoxidase beyond their maturation from early monocytes. They may employ catalase as an antimicrobial agent, through the generation of hydrogen peroxidase.[13]

Macrophages and other accessory cells also play the initial role in the development of acquired immunity, *via* the process of antigen presentation.[7] *In vitro* experiments show that macrophages express antigenic epitopes on their cell surfaces shortly after exposure to soluble antigens.[14] Antigen is believed to be internalized by the macrophage and subjected to a 'processing' or partial degradation. Physical contact with these antigen-bearing macrophages is necessary for the development of T cell help. There is no acquired immunity in the antigen-presenting process, since naive macrophages are equally efficient in presentation as macrophages from animals previously exposed to the antigen *in vivo*.[7]

2.5.4.2 T Lymphocytes

T lymphocytes are derived from immature circulating stem cells which have undergone a differentiation and maturation within the thymus. Thymic maturation results in the expression of different cell surface glycoproteins, which have permitted the division of T cells into subsets. The various subsets of the T cells are responsible for the regulation of the immune response, as well as several effector functions.[15]

2.5.4.2.1 T cell subsets

T cells may be characterized by glycoprotein antigens on their cell membranes, and these surface markers also correlate with various immune functions. For example, all mature circulating human T cells express the OKT3 antigen, whereas subpopulations of OKT3$^+$ cells bear either OKT4 or OKT8 antigens, which are associated with helper and cytotoxic/suppressor functions, respectively.[16] There are three broad T cell populations in mice, based on the Lyt antigens. T cells may be Lyt123$^+$, Lyt1$^+$ or Lyt23$^+$.[17] The majority (50%) of the T cells in peripheral blood are Lyt123$^+$. These cells appear to be in an intermediate state of differentiation and can subsequently mature into either T helper or T suppressor cells. Lyt1$^+$ cells comprise the T helper subset, and these cells interact with B cells to promote their clonal expansion and differentiation into antibody-secreting cells. Lyt23$^+$ cells are T suppressor cells, and are responsible for the down-regulation of the immune response. The same cell surface phenotype is also expressed by the cytotoxic T cell population, which is capable of causing the lysis of cells bearing appropriate antigens. Similar cell surface markers have been identified on rat T cell subpopulations.[18]

2.5.4.2.2 Cytotoxic T lymphocytes

Certain antigens are capable of invoking the development of a cytotoxic T cell response. These antigens are usually expressed on the surface of other cells: examples are virally induced cell surface antigens, foreign MHC antigens and tumor-associated cell surface markers. These antigens are recognized in association with 'self' MHC class I antigens on the antigen-bearing cell surface, giving rise to the concept known as MHC restriction of antigen recognition in CTL killing. It is not known whether CTLs recognize antigen and self MHC simultaneously by one receptor ('altered self') or possess two receptors, one specific for antigen and the other recognizing the coexpression of self class I MHC antigens. CTLs effect lysis of the target population by direct contact, and this process is independent of serum factors such as complement. The precise mechanism by which lysis is produced is poorly understood, beyond the requirement for a viable CTL bearing the appropriate receptor for the specific antigen. Lysis occurs independently of cell division and protein, RNA and DNA synthesis. A single CTL is capable of killing several target cells in sequence in a process called CTL recycling.[19]

2.5.4.2.3 Delayed-type hypersensitivity (DTH)

The DTH response is a T-cell-dependent *in vivo* immune phenomenon manifested by an inflammatory response with a duration of greater than 24 hours at the site of antigen challenge in an immune animal.[20] The measurement of DTH responses has provided valuable information concerning the specific function of various T cell populations in the immune response and its regulation. It is envisaged that primed T cells react to the local antigen deposit with subsequent release of soluble factors and lymphokines, particularly migration inhibition factor (MIF). These products recruit further lymphocytes and inflammatory cells into the local area, giving rise to the localized inflammatory response. The cells which mediate the DTH reaction are UV-radiation-sensitive T cells which express the $Lyt1^+$ phenotype. In a similar manner to the generation of immune help, the T cells involved in the generation of both priming and activation of the DTH response require the relevant antigen to be presented in context with class II MHC determinants on the surface of an antigen-presenting cell. The $Lyt1^+$ effector cell is down-regulated by interaction with $Lyt23^+$ T suppressor cells. The range of antigens that can elicit a DTH reaction include protein antigens (usually immunization with complete Freund's adjuvant is required to provoke a sustained sensitivity), contact-sensitization reagents such as dinitrofluorobenzene, bacteria and bacterial products, and foreign cell surface antigens such as allogenic grafts and viral or tumor-induced antigens.

2.5.4.2.4 T cell activation

As previously described, the activation of the T cell system is dependent upon the uptake, processing and presentation of antigen in context with class II MHC antigens (the Ia antigens of mice and HLA D-family antigens of man). Only very few 'T-independent' antigens are able to bypass this mechanism and generate an immune reaction from the B cells directly. Antigen presentation is usually a function of cells of the macrophage lineage; however, under some circumstances, other Ia positive cells such as B lymphocytes and stimulated endothelial cells are capable of presenting antigen to T cells.

The activation of T cells is a receptor-mediated event, although the precise structure and nature of the T cell receptor(s) involved are currently unknown. Antigen in context with Ia is recognized by membrane-bound glycoprotein receptors on the T cell surface. T cells may also be activated by mitogens, particularly the plant lectins phytohemagglutinin (PHA) and concanavalin A (Con A). Mitogens polyclonally stimulate T cells by binding to the carbohydrate moieties associated with the membrane glycoproteins. Although most experimental evidence has been derived from mitogen work, the activation process of T cells is believed to be essentially similar for both antigen- and mitogen-induced T cell activation. Provided sufficient cell surface binding takes place to effect receptor cross-linking, signals which commence cell activation and division are released intracellularly. Cell activation is accompanied by rapid changes in the flux of intracellular calcium and potassium ions. These cation fluxes result in a depolarization of the plasma membrane, the significance of which is not fully understood, but may be important for cell-to-cell signalling events during the immune response.[21] Membrane ion events are also accompanied by an increase in phospholipid methylation and activation of phospholipase A_2(PLA_2). PLA_2 acts on phospholipids to produce lysophospholipids and release free fatty acids (mainly arachidonic acid). Phospholipids such as lysolecithin enhance guanylcyclase activity and suppress adenylcyclase activity, with a subsequent increase in GMP levels. This is thought to result in the phosphorylation and activation of protein kinases essential to increased mRNA activity and protein synthesis required for cell division.[22]

2.5.4.2.5 T cell regulation of the immune response

The T cell populations act to direct, increase and decrease the immune response to specific antigens, and also modulate immune activation in a nonspecific (polyclonal) manner. The level of reactivity against a specific antigen is under immunogenetic regulation, and this may well be effected through the role of the T helper cell in the recognition of antigen in context of Ia on the surface of antigen-presenting cells. It is well established that the differentiation of B cells into antibody-secreting plasma cells requires T cell help; the precise mode of action by which T cells interact is not fully understood. Several mechanisms may interplay in T cell help, including the concentration of antigen on the T cell surface with subsequent interaction of antigen with the B cells surface Ig receptor. This antigen bridge may also trigger cell-to-cell signalling, which is important for activation and differentiation of the B cell. The intercellular signals may be mediated by soluble factors, particularly lymphokines such as interleukin 2 (IL-2), or involve the direct interaction of cell surface receptors, or a combination of both. For many antigens, a recognition by the T helper cell of surface Ia antigens on the B cell is a requisite for activation. However, certain soluble factors such as B cell growth factor (IL-4) may be released by activated T cells and act on B cells in a nonspecific manner, resulting in polyclonal B cell activation.[23] T cell recognition of the B cell surface Ig may also be involved, since T helper cells are able to regulate the production by B cells of specific allotypes and idiotypes, in addition to antibody specificity.[24]

The complex mechanism by which T cells suppress an immune response is also poorly understood, although antigen-specific and -nonspecific suppressors, and idiotypic-specific suppressor mechanisms have all been described. $Lyt2^{+}3^{+}$ T suppressors are primarily induced by antigen, rather than the products of the immune response itself, through an interaction with inducer T cells, which are phenotypically $Lyt1^{+}$, $I\text{-}J^{+}$. This interaction may be mediated *via* antigen-specific suppressor factors. T suppressor factors (TsF) bind antigen and possess determinants that map to the MHC I-J region in mice, but are not MHC restricted.[25] The association of TsF with antigen may trigger a response from $Lyt2^{+}3^{+}$ T cells to release a second factor (T suppressor effector factor) which suppresses the T helper ($Lyt1^{+}$) cell. The second factor is antigen specific and MHC restricted, although the restriction is not mediated through I-J.[26] In addition, the $Lyt2^{+}3^{+}$ T cells may release a polyclonal or antigen-nonspecific mediator of suppression, soluble immune response suppressor (SIRS), which was identified by *in vitro* polyclonal mitogen stimulation.[27]

A simplistic representation of the regulatory network is shown in Figure 3. Several common features are employed throughout the immune response to effectively retain control of the specificity and response magnitude. Three of these features have particular significance in the control of the immune response. First, cells of the same basic lineage mediate most of the functions through a series of differentiation processes that are reflected by alterations in the expression of specific cell membrane glycoproteins. Second, antigen recognition and the degree of help and suppression which is generated are all mediated by a system that recognizes self in context with antigen. Third, both immune effector mechanisms and regulatory control mechanisms utilize soluble proteins (cytokines) with both antigen-specific and polyclonal effects. These similarities may reflect conservation in a complex system which has evolved to result in the flexibility and specificity characteristic of the immune system.

2.5.4.3 Role of Cytokines in Immunity

'Cytokine' is a collective term used to describe soluble factors derived from cells which mediate biological effects on other cell types. Cytokines can be produced by monocyte/macrophages (monokines) or by lymphocytes (lymphokines). Over 100 cytokines have been described in the literature based upon activity of supernatant fluids from activated lymphoid cells in one or another bioassay procedure.[28]

Interleukin 1 (IL-1) is a 17 kDa protein which is produced by monocyte/macrophages, as well as a variety of other cell types. The peptide plays a critical role in T cell activation as it provides a signal necessary for T cells to produce another glycoprotein cytokine called IL-2 (molecular weight (MW) = 15 kDa in humans, 30 kDa in mice).[29,30] IL-2 then acts as a direct growth factor to stimulate T cell proliferation and differentiation, completing the IL-1/IL-2 cascade system in T cell activation.[31] IL-1 also enhances the cytolytic activity of NK cells.[29]

Originally, it was believed that IL-1 and IL-2 were involved only in T cell activation, but recently it has become clear that these cytokines play a role in the activation of B cells as well.[30,32] Moreover,

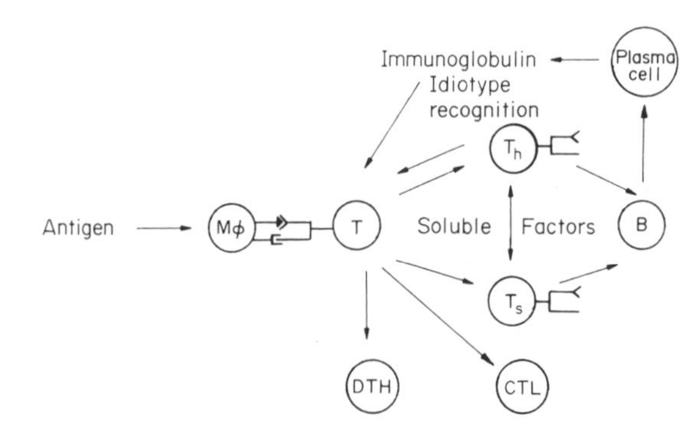

Figure 3 The immune network. Antigen is processed by macrophages ($M\phi$) and presented in context with self class II MHC antigens to T cells. T cell differentiation procedes, giving rise to appropriate functional cell types and secretion of soluble factors (IL-2, BSF-1, interferons, *etc.*). Mature T cells mediate help (T_H), suppression (T_s), cytotoxicity (CTL), delayed-type hypersensitivity (DTH) and other effector functions. B cells differentiate into antibody-secreting plasma cells under the influence of regulatory cells. The idiotypes expressed by both the cells and immunoglobulin provide a feedback mechanism to regulate immune responses *via* idiotype recognition

IL-1 has been shown to have numerous biological effects distinct from those involved in lymphocyte activation.[29,30] For example, IL-1 mediates fever and acute phase reactant synthesis and induces chondrocytes and synovial fibroblasts to produce prostaglandin E_2 and a variety of proteases. For these reasons, IL-1 is thought to be an important pathological mediator of certain immunoinflammatory disorders such as rheumatoid arthritis.[33,34] In support of this, IL-1 has recently been shown to directly induce articular synovitis when injected into the joint capsule.[35,36]

Two different IL-1 peptides, derived from distinct IL-1 genes, are now recognized and are termed IL-1α and IL-1β.[37] Both are 17 kDa molecular weight proteins which have isoelectric points of pI = 5.5 and 7.0, respectively.[37] The biological activity of IL-1α and IL-1β are very similar[38,39] and both bind to the same receptor on IL-1-responsive cells.[40]

Tumor necrosis factor (TNF) is another macrophage-derived 17 kDa peptide mediator of biological importance.[41] As its name implies, this peptide causes necrosis and regression of certain tumors.[42] However, TNF production is thought to be responsible for the wasting disease (cachexia) observed during chronic infection and, for this reason, was originally named cachectin.[43] Cachexia induced by TNF presumably results from the inhibition by TNF of lipoprotein lipase. It has been recently recognized that TNF produces a number of biological effects in addition to cachexia and tumor necrosis including fever, induction of acute phase protein synthesis and activation of arachidonic acid metabolism in certain cell types.[39,43] Although TNF and IL-1 have many features in common, they are distinct proteins derived from different genes which share only limited sequence homology. Furthermore, IL-1 and TNF do not bind to the same cellular receptor.[43]

Interferons (IFN) are another biologically important class of peptide cytokines. Three families of interferons exist—α (leukocyte), β (fibroblast) and γ (immune) and at least 17 distinct interferon genes have been identified.[44] Both α and γ interferons have immune effects including the ability to enhance NK cell cytotoxicity and increase the cytotoxic function of T cells, NK cells and macrophages.[45,46] γ-IFN is also believed to be a critical cofactor in the production and utilization of IL-2 by T cells. The peptide augments the expression of class II histocompatibility antigens (HLA/DR in humans, Ia in mice) on several cell types including macrophages, T cells and endothelial cells, thereby promoting the antigen-presentation function of these cells.[44] Treatment of macrophages with γ-IFN activates their ability to ingest and destroy intracellular pathogens and enhances their cytolytic activity.[43,44] Finally, several studies have shown that nearly all types of IFNs have direct antiproliferative effects on certain tumor cells.[43–45].

Colony-stimulating factors (CSFs) are glycoprotein cytokines which induce the proliferation and differentiation of bone-marrow cells and thereby regulate hematopoiesis. Some CSFs act on granulocyte/macrophage precursors (GM-CSF), while others act on granulocyte, erythroid, lymphoid, eosinophil or mast cell precursors.[47,48] A pluripotential murine CSF derived from WEHI cells has been termed IL-3, although the use of IL-3 in this context has not received wide acceptance. CSFs are produced by macrophages and T cells during activation processes and some CSFs, for example GM-CSF, can alter the function of mature macrophages and granulocytes.[49]

B cell stimulatory factor 1 (BSF-1) is a 20 kDa peptide cytokine produced by activated T cells which augments B cell immunoglobulin secretion.[50] Recently, BSF-1 has been shown to also stimulate T cell growth and have a variety of effects on several other hematopoietic cells. BSF-1 has recently been named interleukin 4.[50]

2.5.5 IMMUNOLOGICAL ABNORMALITIES IN HUMAN DISEASE

Several human diseases result from or are associated with abnormal immune response. Some of the more important diseases associated with immunological abnormalities are shown in Table 1. Operationally, we have listed these diseases in distinct categories based primarily on the clinical manifestations and our current understanding of the underlying immunological mechanisms involved. However, it is important to note that these categories, while useful, are not rigid and many overlaps exist; for example, a significant number of autoimmune reactions occur in patients with immunodeficiency disease. Moreover, individual disorders within a given category vary widely with respect to their clinical symptomology, immune abnormalities, pathogenesis and laboratory findings.

2.5.5.1 Immunodeficiency Diseases

Immunodeficiency disease represents a spectrum of immune defects which occur either spontaneously (primary immunodeficiency) due to an underlying inherited trait (metabolic disorders,

Table 1 Potential Therapeutic Targets for Immunomodulatory Agents

Category	*Example*	*Primary cellular defect*[a]
Cancer	Leukemia, lymphoma melanomas, *etc.*	Mϕ, T, NK
Autoimmune diseases	Rheumatoid arthritis	Mϕ, PMN, T, B
	Systemic lupus erythematosis	T, B
	Type 1 diabetes	T
	Phemphigus	B
Immunodeficiency disease		
Primary	SCID (several combined immunodeficiency disease)	T
	Purine nucleoside phosphorylase (PNP) deficiency	T
	Adenosine deaminase deficiency	T
	Ataxia telangectasia	T
	DiGeorges' syndrome	T
	X-linked agammaglobulinemia	B
	IgA deficiency	B
	Common variable immune deficiency	B
Acquired	Due to tumor (see above)	Mϕ, NK, T
	Viral infection	Mϕ, NK, T
	Bacterial infection	Mϕ, T, PMN
	Fungal infection	Mϕ, T
	Miscellaneous	
	post-surgical	T
	burn patients	T
Allergy	Asthma	Mast, B
	Dermatitis	Mast, T
	Insect sting allergy	Mast
	Systemic anaphylaxis	Mast, B

[a] Mϕ = macrophage.

enzyme deficiencies, *etc.*) or occur secondarily due to cancer or infection (acquired immune deficiency).[51–53]

2.5.5.1.1 ***Primary immune deficiency diseases***

Primary immunodeficiencies are those disorders which occur naturally and have primarily a genetic rather than an infectious basis. A spectrum of primary immunodeficiency disorders has been documented in the clinical literature and they are characterized by an inability to manifest normal cell mediated and/or humoral immunity. In the general population, the incidence of primary immune deficiency is quite low and individual cases show marked heterogeneity in inheritance pattern and both laboratory and clinical findings. The most common clinical features of primary immune deficiencies include a failure to thrive, recurrent and severe bacterial and viral infection and an early mortality.

Some primary immunodeficiencies such as severe combined immunodeficiency disease (SCID) affect the function of both T and B cell lineages. Many SCID patients are hypogammaglobulinemic and have absent or low peripheral blood lymphocyte (PBL) responses to the T cell mitogen, PHA. On the other hand, some primary disorders predominantly affect either the B or T lineage. Thus, several humoral immune deficiencies such as X-linked agammaglobulinemia have been described in which the primary defect is in the B cell lineage and, conversely, T-cell-specific malfunctions have also been noted (an example is the genetic absence of purine nucleoside phosphorylase, PNP).[54] In addition, congenital disorders primarily affecting phagocyte (PMN and macrophage) function, for example Chediak–Higashi syndrome, are also known.

Clinical and laboratory studies have demonstrated at least two mechanisms through which primary immune deficiency can be manifested. The first is a genetic enzyme deficiency, as typified by adenosine deaminase deficiency (seen in many patients with SCID), purine nucleoside phosphorylase deficiency (resulting in cellular immune deficiency) and absence of transcobalamin II (which results in agammaglobulinemia).[54] The second possible mechanism is aberrant lymphoid cell maturation in which lymphoid cell precursors fail to differentiate appropriately into mature functional lymphocytes due to intrinsic stem cell defects or reduced thymus or bursal-equivalent function.[55]

Based on these considerations, it is apparent that drugs capable of enhancing the reduced immune function or promoting lymphocyte differentiation provide a rational therapeutic approach to these disorders, especially when given in conjunction with standard antibiotic and antiviral therapy and/or immune globulin injections. However, the type of immunostimulatory agent must be carefully selected to suit the particular immune disorder. For example, agents which stimulate B cell differentiation and antibody secretion would be appropriate for patients with hypo- or agammaglobulinemia but not for patients with T cell immunodeficiency, *etc.*

*2.5.5.1.2 **Acquired immune deficiency diseases***

(i) Cancer

The use of immunomodulatory drugs in cancer therapy has been a major driving force for development of these agents.[56] Indeed, it was the limitations of conventional cancer treatments that provided the impetus for utilizing the immunostimulatory properties of a variety of bacterial products and other simple chemicals in the treatment of malignancy and this led to the application of this method of immune enhancement to autoimmune diseases and immunodeficiency. While immunostimulation in cancer therapy is not a recent concept, interest in this approach has been rejuvenated following the development of the biological response modifier (BRM) program under the auspices of the National Institutes of Health. The primary goal of this program is to identify, characterize and ultimately evaluate clinically the antitumor effectiveness of natural and synthetic agents which modulate immune reactivity.[56]

The role of nonspecific defense mechanisms appear to be very important in controlling tumor growth and reducing the probability of metastasis. Thus, immunomodulators which have the ability to augment the cytotoxic activity of NK cells and macrophages (*e.g.* interferons) have the strongest rational basis for development in this area. However, even though interest in the role of specific T-cell-mediated antitumor immunity and antitumor antibody in controlling tumor growth has waned somewhat, in part due to the difficulty in demonstrating tumor-specific antigens in several human tumors, agents which stimulate T and B cell reactivity may also be useful in tumor therapy.

Immunomodulatory drugs which stimulate antitumor immunity will not in general be effective first-line therapy in cancer patients. Evidence to date indicates that this type of drug will be most useful as an adjunct to cytoreductive therapy (surgical resection, irradiation, cytotoxic drug therapy). Thus, in animal tumor model systems, these agents rarely influence tumor growth or survival if the tumor burden is large, but do slow or stop tumor growth and reduce metastatic spread following cytoreductive therapy.[57]

(ii) Infectious diseases

The use of immunomodulatory drugs in the treatment of viral, bacterial and fungal infections is a promising therapeutic arena. This is particularly true if these agents are designed to be used as adjunctive therapy in combination with antiviral, antibiotic and antifungal agents. In principle, the antiviral/antibiotic would directly reduce the magnitude of the infection, while the immunomodulatory agents used would stimulate the functional activity of those particular lymphoid cells which attack the infectious agent in question, effecting a more rapid and complete elimination of the infectious agent and reducing the possibility of recurrence. The immunomodulatory agents must be selectively used depending upon the type of infection. As macrophages and PMNs are important antibacterial effector cells, agents which stimulate their functions (*i.e.* phagocytosis, intracellular killing, mobility, *etc.*) would be of primary interest in bacterial infections. On the other hand, T cell or NK cell stimulators would be more appropriate for viral infections. The immunomodulatory therapy in infectious diseases would be particularly valuable in patients who are immunocompromised due to either the infection itself or due to surgical trauma, irradiation, cancer, severe burns, *etc.*[57]

The idea that immunomodulatory agents, primarily those with immunopotentiating effects, may be useful in antiinfective therapy is the result of numerous studies in man and experimental animals which clearly show that a transient blunting of immune responsiveness accompanies many viral, bacterial and fungal infections.

Perhaps the most poignant example of a viral infection of lymphoid cells resulting in immunosuppression is the acquired immune deficiency syndrome (AIDS).[58] The principle etiological factor in this disease appears to be infection of helper ($OKT4^+$) T cells by the human T cell leukemia virus type III (HTLV-III, also known as lymphodenopathy associated virus, LAV). Both the absolute number and functional activity of helper T cells are dramatically reduced in AIDS patients and this leads to multiple clinical manifestations, the most prevalent and important of which are lymphodenopathy, Kaposi's sarcoma, *Pneumocystis carinii* pneumonia and numerous other secondary bacterial and viral infections.[58]

(iii) Autoimmune diseases

The treatment of autoimmune disorders is another major focus for the use of immunomodulatory agents. This includes diseases such as rheumatoid arthritis (RA), systemic lupus erythematosis (SLE), type 1 (juvenile) diabetes and autoimmune thyroiditis. While the primary pathological features of autoimmune disease are due to an overactive immune reaction to 'self' antigens, the hypothesis that these diseases are due to abnormal immunoregulation has become quite attractive.[53] Thus, in SLE and RA, where hyperactive B cell function and elevated autoantibody secretion are characteristic features, regulatory T cell functions such as lymphokine synthesis and suppressor T cell activity are diminished. This suggests that both immunosuppressive and immunostimulatory agents could be therapeutically useful, the former by directly inhibiting those lymphocytes reacting against host components and the latter by stimulating regulatory cell function (*e.g.* suppressor cells). Indeed, both immunosuppressive agents such as cyclophosphamide as well as immunostimulators such as levamisole and thymic peptides do show clinical utility in RA patients.[53]

(iv) Allergic disease

Immediate hypersensitivity is the immune response that has become synonymous with the term allergy, in part because it is the most common of all allergic reactions. However, there are actually four major types of allergic reactions which vary considerably. Patients with allergic diseases often possess a constellation of immunological defects. For example, bronchial asthmatics and allergic rhinitics have demonstrated the involvement of both immediate (manifested by positive skin tests, antiallergen-specific IgE antibody and increased levels of serum IgE) and delayed hypersensitivity (usually depressed cell-mediated responses). Patients with atopic dermatitis have elevated serum IgE but also depressed cell-mediated immunity and abnormal leukocyte chemotaxis.

IgE production by antigens (*i.e.* allergens) is pivotal to immediate hypersensitivity reactions. This antibody binds to surfaces of mast cells, which are abundant in the respiratory and gastrointestinal tracts as well as the skin, and basophils, which are circulating leukocytes. When the antigen reacts with surface-bound IgE on the mast cells or basophils, these cells release their mediators which include histamine, bradykinin, serotonin, leukotrienes, prostaglandins, platelet-activating factor (PAF), eosinophil chemotactic factor of anaphylaxis (ECF-A) and neutrophil chemotactic factor (NCF). These mediators work in concert to produce the symptoms of allergy: bronchoconstriction, enhanced secretions from eyes, lungs, and nose, sneezing, itching, wheal-flare, abdominal pain and diarrhea.

2.5.6 IMMUNOTHERAPY

2.5.6.1 General Considerations

Immunotherapy can best be described as the process of altering the immune system in such a way as to effect a therapeutic benefit. For example, this can be effected by administering natural or synthetic substances (immunomodulatory agents) which act directly or indirectly on immune cells to modify their functional capacity. There are numerous potential mechanisms through which immunomodulation can be achieved; for example, increased T cell reactivity could result from either a specific enhancement of T helper cell activity or an inhibition of T suppressor cell functions. In fact, however, most immunopharmacological agents to date are not truly specific in this sense and are capable of enhancing or depressing immune responses depending upon the experimental conditions.[59] Indeed, probably more than any other class of drugs, the dose, frequency of treatment and the time and route of administration greatly affect the response of the host both qualitatively and quantitatively. Moreover, several host characteristics also influence the nature and magnitude of drug response including genetic factors, whether or not these are linked to the major histocompatibility complex, the animal's physiological status (sex, age, diet, neuroendocrine function, *etc.*) and overall health (*e.g.* intercurrent infections, spontaneous autoimmune disease, neoplasias, organ dysfunctions). Further, the characteristics of the antigen under study are also capable of modifying the end result. The dose, immunogenicity and size of the antigen are critical. For example, it may be necessary to reduce the tumor size by surgery, X-ray irradiation or chemotherapy, or pathogen number by concomitant antimicrobial therapy in order to be successful with immunomodulatory therapy.

2.5.6.2 Characterization of Immunomodulatory Agents

In view of the diverse mechanisms associated with immunomodulation, it is unlikely that a single *in vitro* or *in vivo* model system will be satisfactory for screening all types of immunomodulators. Although no consensus on specific assays has emerged, the use of a battery of appropriate tests is recommended, including infectious models, autoimmune models and neoplastic models (Table 2). Unfortunately, a comparative evaluation of the sensitivity, predictiveness and limitation of such tests employing a diverse group of immunopharmacological agents is lacking.[60]

The pharmacological approach to evaluation of a new immunomodulator often begins with an investigation of its effects *in vitro* on various cell populations that constitute the immune system which are isolated from either normal animals or normal subjects.[61,62] This allows the range of concentrations which modulate cell reactivity to be identified. However, these observations may not reflect events induced *in vivo* by the compound, effects on 'abnormal' cells or even effects on cells of the same lineage but obtained from different anatomical sites.

Since most immunomodulators restore rather than potentiate immune responses, *in vivo* models involving animals with suboptimal resistance or tests involving suboptimal derangement are more

Table 2 Commonly Used Assays to Identify and Characterize Immunomodulatory Agents

	Immune function (in vitro/in vivo)	*Disease models*
T cells	T cell cytotoxicity	Organ transplantation skin graft survival
	Blastogenesis (Con A, PHA, alloantigen)	graft *vs.* host disease (popliteal lymph node assay)
	Lymphokine production (IL-2)	
	Th/Ts ratios	Autoimmune diseases experimental allergic encephalomyelitis
	DTH (sheep red blood cells (SRBC), methylated bovine serum albumin (MBSA), oxalone)	immune lupus (MRL/1, NZB)
		murine immune complex nephritis (DBA/2 into C57B1/6 × DBA/2F) Inflammation rat carrageenan edema/pleurisy rat adjuvant arthritis rat/mouse collagen arthritis
B cells	Blastogenesis [pokeweed mitogen (PWM), lipopolysaccharide (LPS)]	Infection
	Plaque-forming cells (PFC) to SRBC	viral prophylaxis (herpes simplex) bacterial and fungal therapeutic (*Salmonella typhimurium*, *E. coli*, *Candida albicans*, *Listeria monocytogenes*)
Nonspecific	NK cytotoxicity	Tumor primary and metastatic (Lewis lung, lymphoma, Madison lung M109, mammary AC, fibrosarcoma *etc.*)
	Macrophages (phagocytosis; pinocytosis; chemotaxis, cytotoxicity; release of O_2 radicals, lysosomal enzymes, arachidonate metabolites and monokines; Ia and C3 expression)	
	Antibody-dependent cellular cytotoxicity (ADCC)	
	Neutrophils (chemotaxis; release of O_2 radicals, lysosomal enzymes)	

appropriate than those involving optimal response conditions. Excessive challenge doses of pathogens and tumor cells should also be avoided otherwise the result is overwhelming and may not be modifiable by this class of drug.

The choice of animal is often critical. For example, some strains of mice are very susceptible to certain diseases (*e.g.* DBA/2, Balb/c and C_3H) whereas others (*e.g.* C57B1/6-related strains) are highly resistant. The choice of resistance model should also depend largely on the suspected primary mechanism of action of the immunomodulator. For example, immunomodulators in general will exert little or no effect in bacterial models of resistance unless macrophages are the primary targets of their activity. The basis for the selection of tumor models should include an evaluation of the immunogenicity of the tumor. Highly immunogenic tumors may not be optimal due to the strong immunoprotection in the absence of the immunomodulator, whereas a relatively nonimmunogenic tumor may also be counterproductive. Tumor cells should also be sensitive to killing by the effector cells stimulated by the immunomodulator under test. For example, tumor variants exist that are resistant to specific CTL- or NK-cell-mediated cytotoxicity and may thus be inappropriate.[57]

A problem that commonly arises in the testing of immunomodulators with stimulatory activity is the bell-shaped dose–response curve. Increasing the dose will thus result in an increased response followed by a peak or plateau and finally a reversal of the effect. The precise interval between the maximally effective dose and the paradoxical effect varies with the agent tested. The highest tolerated dose is, therefore, not always the most effective.

Some immunomodulators may exert a degree of selectivity towards specific cell types and exert uniform *in vivo* responses; however, absolute selectivity does not yet exist and the possibility of finely targeted immunomodulation remains a future goal. As a consequence, all of the current immunomodulators have inherent limitations for the treatment of disease. The leap from laboratory to clinic is also complicated by our ignorance of the etiological, immunological and physiological bases of many human diseases and so the relevance of animal models frequently cannot be properly validated. In some diseases it may also be necessary to combine immunomodulator administration with more established therapy in order to demonstrate maximum benefit. An example is the combination of immunomodulators with macrophage-activating, T-cell-restoring and/or NK-cell-enhancing activity with cancer cytoreductive therapy (chemotherapy, surgery or irradiation).

2.5.6.3 Immunomodulatory Agents Currently in Use or Under Development

Table 3 lists some of the more established biologically and synthetically derived immunomodulators with clinical efficacy or under clinical investigation. The biologically derived immunomodulators are largely isolated from microorganisms such as bacteria and fungi. The synthetic immunomodulators have been divided into stimulators and suppressors according to the primary direction of modulation. Although not listed here, agents capable of modulating the immune response exist in a number of other drug categories including calmodulin antagonists, cannabinoids, bromocriptine, diazepam, insulin, thyroxine and tricyclic antidepressants. The immunological activities of such a diverse group of agents is more clearly understood when one considers the variety of surface receptors, including β adrenergic, dopamine, cholinergic H_2-histamine, opiate and benzodiazepine receptors, that have been observed on immune cells.

2.5.6.4 New Approaches to Immunotherapy

2.5.6.4.1 Monoclonal antibodies

Immunotherapy with monoclonal antibodies has been proposed as a viable means of detecting or treating cancer, autoimmune disease and other disorders.[63,64] Monoclonal antibodies have already led to rapid advances in the understanding of immune regulation and how the normal immunoregulatory balance is perturbed in patients with autoimmune immunodeficiency disorders, cancer, *etc.* These antibodies can be used alone, or can first be coupled to a toxin (*i.e.* ricin A) or other drug to 'target' such chemicals directly to the desired tissue (*e.g.* tumor site, rheumatoid pannus), thereby reducing the total dose of drug and diminishing the probability and/or severity of side-effects. Attaching immunostimulants or immunosuppressants could be used to target such drugs to specific lymphoid cell populations. In fact, some monoclonal antibodies, for example antibody to the T3 molecule on human T cells, directly stimulate T cell proliferation (at least *in vitro*) and could be used essentially as immunostimulatory 'drugs' themselves. Similarly, antibodies which suppress lympho-

Table 3 Examples of Some Immunomodulatory Drugs in Current Use or Under Development

Biologically derived	*Synthetic*
	(a) Stimulators
BCG and extracts	Azimexon
Bestatin	Diethyl dithiocarbamate
Brucella abortus	Isoprinosine
Corynebacterium parvum	Levamisole
Glucans	Lipoidelamines
Interferons (α, β, γ)	Maleic anhydride/divinyl ether copolymer (MVE)
Krestin (PSK)	
Lentinan	MDP analogs
Lymphokines (IL-1, IL-2, CSF, TNF)	NED-137
Muramyldipeptide (MDP)	NPT-15392
OK432 (Picibanil)	Polyinosinic–polycytidylic–poly(L-lysine) (Poly IC:LC)
Thymic factors	
thymosins (α_1, α_5, α_7, β_3, β_4)	Pyrimidinones
thymopentin (thymopoietin)	Tilomisole (Wy-18,251)
thymostimulin	
thymic humoral factor	
thymulin (factor thymique serique)	(b) Suppressors
Trehalose dimycolate (cord factor)	Corticosteroids
Tuftsin	Cyclophosphamide
	Cyclosporin A
	6-Mercaptopurine and azathioprine
	Methotrexate

cyte activation, such as antibodies to class II major histocompatibility antigen (Ia or HLA/DR), could be used directly as immunosuppressive agents. While the application of monoclonal antibodies has shown good promise, to date, the viability of this approach remains to be proven in a controlled clinical trial.[65]

2.5.6.4.2 *Cytokines*

As discussed earlier, lymphokines are important mediators of a variety of effector functions of immune cells. Lymphokines which stimulate T cells (IL-1, IL-2), macrophages (macrophage-activating factor, interferon), NK cells (IL-1, IL-2, interferon) and B cells [IL-1, IL-2, T-cell-replacing factor (TRP)] as well as lymphokines which suppress lymphoid cell reactivity have all been described.[28] In addition, cytolytic and tumoricytotoxic lymphokines such as lymphotoxin and tumor necrosis factor (TNF) are known. *In vitro* data utilizing these and other lymphokines suggest that they could be used therapeutically to augment or suppress immune function. Only interferons, primarily α and γ, have been extensively assessed in the clinic so far, with disappointing results overall (with the exception of some specific tumor types).[66] Early clinical trials with IL-2 and TNF have shown promise, although significant toxicity has been observed.[67,68] Nonetheless, further information on the biology, pharmacokinetics and mode of action of lymphokines may provide solutions for the successful therapeutic use of these agents.

2.5.7 SUMMARY

The immune system is a highly ordered, complex and tightly regulated biological defense mechanism. Numerous cell types, membrane molecules and soluble products work in concert to generate a diverse array of effector mechanisms capable of protecting the host from harmful insult, be it an infectious agent, opportunistic pathogen or malignant tumor cells. An improperly functioning immune system can result in a number of serious diseases. The use of agents which modify immune responsiveness may have broad therapeutic applications.

2.5.8 REFERENCES

1. N. E. Kay, S. K. Ackerman and S. D. Douglas, *Semin. Hematol.*, 1979, **16**, 252.
2. L. Chess and S. F. Schlossman, *Adv. Immunol.*, 1977, **25**, 213.
3. P. W. Kincade and R. A. Phillips, *Fed. Proc., Fed. Am. Soc. Exp. Biol.*, 1985, **44**, 2874.

4. G. J. Dougherty and W. H. McBride, *J. Clin. Lab. Immunol.*, 1984, **14**, 1.
5. R. B. Herberman and H. T. Holden, *Adv. Cancer Res.*, 1978, **27**, 305.
6. R. Gallily and H. Eliahu, *Cell. Immunol.*, 1976, **25**, 245.
7. E. R. Unanue and P. M. Allen, *Science (Washington, D. C.)*, 1987, **236**, 551.
8. C. A. Dinarello, *Rev. Infect. Dis.*, 1984, **6**, 51.
9. J. B. Natvig and H. G. Kunkel, *Adv. Immunol.*, 1973, **16**, 1.
10. M. Potter, M. Pawlita, E. Mushinski and R. J. Feldmann, in 'Immunoglobulin Idiotypes', ed. C. Janeway, Academic Press, New York, 1981, p. 1.
11. B. H. Hahn and F. M. Ebling, *J. Immunol.*, 1984, **132**, 187.
12. A. J. Lewis, R. P. Carlson and J. Chang, in 'Handbook of Inflammation', ed. I. L. Bonta, M. A. Bray and M. J. Parnham, Elsevier, Amsterdam, 1985, vol. 5, p. 371.
13. C. F. Nathan, *J. Clin. Invest.*, 1986, **79**, 319.
14. J. J. Ellner, P. E. Lipsky and A. S. Rosenthal, *J. Immunol.*, 1977, **118**, 2053.
15. E. L. Reinherz, P. C. Kung, G. Goldstein and S. F. Schlossman, *Proc. Natl. Acad. Sci., USA*, 1979, **76**, 4061.
16. E. L. Reinherz and S. F. Schlossman, *Cell*, 1980, **19**, 821.
17. J. A. Ledbetter, R. V. Rouse, H. S. Micklem and L. A. Herzenberg, *J. Exp. Med.*, 1980, **152**, 280.
18. D. W. Mason, R. P. Arthur, M. J. Dallman, J. R. Green, G. P. Spickett and M. L. Thomas, *Immunol. Rev.*, 1983, **74**, 57.
19. M. Cohn, *Cell*, 1983, **33**, 657.
20. M. I. Greene, S. Schatten and J. S. Bromberg, in 'Fundamental Immunology', ed. W. E. Paul, Raven Press, New York, 1984, p. 685.
21. J. G. Kaplan and T. Owens, *Ann. N. Y. Acad. Sci.*, 1980, **339**, 191.
22. R. F. Ashman, *Immunol. Today*, 1982, **3**, 349.
23. K. Yoshizaki, T. Nakagawa, T. Kaieda, A. Muraguchi, Y. Yamamura and T. Kishimoto, *J. Immunol.*, 1982, **128**, 1296.
24. E. E. Sercarz, R. L. Yowell, D. Turkin, A. Miller, B. A. Araneo and A. Adorini, *Immunol. Rev.*, 1978, **39**, 108.
25. J. Theze, C. Waltenbaugh, M. E. Dorf and B. Benacerraf, *J. Exp. Med.*, 1977, **146**, 287.
26. P. Flood, K. Yamauchi and R. K. Gershon, *J. Exp. Med.*, 1982, **156**, 361.
27. T. Tadakuma and C. W. Pierce, *J. Immunol.*, 1978, **120**, 481.
28. J. M. Hanson, V. M. Rumjanek and J. Morley, *Pharmacol. Ther.*, 1982, **17**, 165.
29. J. J. Oppenheim, E. J. Kovacs, K. Matsushima and S. K. Durum, *Immunol. Today*, 1986, **7**, 45.
30. W. R. Benjamin, P. T. Lomedico and P. L. Killian, *Annu. Rep. Med. Chem.*, 1985, **20**, 173.
31. R. J. Robb, *Immunol. Today*, 1984, **5**, 203.
32. R. J. M. Falkoff, J. L. Butler, C. A. Dinarello and A. S. Fauci, *J. Immunol.*, 1984, **133**, 692.
33. J. Chang, S. C. Gilman and A. J. Lewis, *J. Immunol.*, 1986, **136**, 1283.
34. J. T. Dingle, D. P. P. Thomas, B. King and D. R. Bard, *Ann. Rheum. Dis.*, 1987, **46**, 527.
35. E. R. Pettipher, G. A. Higgs and B. Henderson, *Proc. Natl. Acad. Sci. USA*, 1986, **83**, 8749.
36. S. C. Gilman, T. Hodge and J. Chang, *Arthritis Rheum.*, 1987, **30**, S29.
37. C. J. March, B. Mosley, A. Larsen, D. P. Cerretti, G. Braedt, V. Price, S. Gillis, C. S. Henney, S. R. Kronheim, K. Grabstein, P. J. Conlon, T. P. Hopp and D. Cosman, *Nature (London)*, 1985, **315**, 641.
38. E. A. Rupp, P. M. Cameron, C. S. Ranawat, J. A. Schmidt and E. K. Bayne, *J. Clin. Invest.*, 1986, **78**, 836.
39. S. C. Gilman, *J. Rheumatol.*, 1987, **14**, 1002.
40. S. K. Dower and D. L. Urdal, *Immunol. Today*, 1987, **8**, 46.
41. D. Pennica, G. E. Nedwin, J. S. Hayflick, P. H. Seeburg, R. Derynck, M. A. Palladino, W. J. Kohr, B. B. Aggarwal and D. V. Goeddel, *Nature (London)*, 1984, **312**, 724.
42. A. A. Creasey, L. V. Doyle, M. T. Reynolds, T. Jung, L. S. Lin and C. R. Vitt, *Cancer Res.*, 1987, **47**, 145.
43. B. Beutler and A. Cerami, *Nature (London)*, 1986, **320**, 584.
44. G. L. Mannering and C. B. Deloria, *Annu. Rev. Pharmacol. Toxicol.*, 1986, **26**, 455.
45. E. M. Bonnem and R. K. Oldham, *J. Biol. Response Modif.*, 1987, **6**, 275.
46. G. Trinchieri and B. Perussia, *Immunol. Today*, 1985, **6**, 131.
47. J. D. Watson and R. L. Prestidge, *Immunol. Today*, 1983, **4**, 278.
48. A. W. Burgess and D. Metcalf, *Blood*, 1980, **56**, 947.
49. L. S. Park, D. Friend, S. Gillis and D. L. Urdal, *J. Exp. Med.*, 1986, **164**, 251.
50. W. E. Paul and J. Ohara, *Annu. Rev. Immunol.*, 1987, **5**, 429.
51. F. Auiti, F. Rosen and M. D. Cooper, *Proc. Serono Symp.*, 1985, **28**, 1.
52. G. W. Siskind, G. L. Christian and S. D. Litwin, 'Immune Depression and Cancer', Gruen and Stratton, New York, 1975, p. 1.
53. S. C. Gilman and A. J. Lewis, in 'Antiinflammatory and Antirheumatic Drugs', ed. K. D. Rainsford, CRC Press, Boca Raton, FL, 1985, vol. 3, p. 127.
54. R. Hirschorn, *Birth Defects, Orig. Artic. Ser.*, 1983, **19**, 73.
55. J.-F. Bach, *Birth Defects, Orig. Artic. Ser.*, 1983, **19**, 245.
56. J. E. Talmadge and R. B. Herberman, *Cancer Treat. Rep.*, 1986, **70**, 171.
57. J. E. Talmadge, R. K. Oldham and I. J. Fidler, *J. Biol. Response Modif.*, 1984, **3**, 88.
58. H. C. Lan and A. S. Fauci, *Annu. Rev. Immunol.*, 1985, **3**, 477.
59. J. E. Talmadge, K. L. Benedict, K. A. Uithoven and B. F. Lenz, *Immunopharmacology*, 1984, **7**, 17.
60. G. Renoux, *Med. Find. Exp. Clin. Pharmacol.*, 1986, **8**, 45.
61. S. C. Gilman, R. T. Maguire and A. J. Lewis, *Handb. Exp. Pharmacol.*, 1988, **85**, 345.
62. S. J. Urbaniak, A. G. White, G. R. Barclay, S. M. Wood and A. B. Kay, *Handb. Exp. Immunol. (3rd Ed.)*, 1979, **3**, 47.
63. J. D. Rodwell, V. L. Alvarez, C. Lee, A. D. Lopes, J. W. F. Goers, H. Dalton King, H. J. Powsner and T. J. McKern, *Proc. Natl. Acad. Sci. USA*, 1986, **83**, 2632.
64. L. Olsson, *Allergy (Copenhagen)*, 1983, **38**, 145.
65. P. C. L. Beverly and G. Rethmuller, *Immunol. Today*, 1987, **4**, 101.
66. R. V. Smalley and E. C. Borden, *Springer Semin. Immunopathol.*, 1986, **9**, 73.
67. M. Blick, S. A. Sherwin, M. Rosenblum and J. Gutterman, *Cancer Res.*, 1987, **47**, 2986.
68. E. Platzer, M. Gramatski, M. Rollinghoff and J. R. Kalden, *Immunol. Today*, 1986, **7**, 185.

2.6
Selectivity

ROBERT L. LIPNICK

Environmental Protection Agency, Washington, DC, USA

'The present author is of the opinion that there are three main principles by which a biologically active agent can exert selectivity. Either it can be accumulated principally by the *un*economic species, *or*, utilizing comparative biochemistry, it may injure a chemical system important for the *un*economic (but not for the economic) species, *or*, it may react exclusively with a cytological feature that exists only in the *un*economic species.'[3]

2.6.1 INTRODUCTION

It is important to distinguish between activity and selectivity. The rules governing activity are discussed elsewhere in this work (most notably in Volume 4) but these do not help us to achieve selectivity. This is the ability of a compound to act preferentially or even uniquely on the target cells rather than on others. The means by which this selectivity might be achieved are discussed below.

Selectivity plays a pivotal role in the design of new drugs and pesticides. Ideally, a drug or other biological agent should elicit only the desired biological or pharmacological activity from the organism or cellular target site, yet produce no effects on other organisms or pharmacological sites. In practice, selectivity is frequently measured by an index of the therapeutic dose to that producing an adverse effect.[1]

In 1951, Albert, Professor of Medicinal Chemistry, Australian National University and Lecturer in the Department of Biochemistry, University College, London, published a monograph entitled 'Selective Toxicity with Special Reference to Chemotherapy', which he based upon a course of public lectures he had delivered in the Department of Biochemistry, University College, London.[2]

In the first edition of this book, Albert defined selective toxicity as 'the practice of injuring one species of living matter without harming another species with which the first is in intimate contact.' He defined the '. . . species to be injured as the uneconomic species and the species to be preserved as the economic species'. Albert also applied the term selective toxicity to pharmacology in which the '. . . uneconomic cells are part of the organism of the economic species'. He noted that 'in pharmacology, selectively toxic agents are often required to have a temporary rather than a permanent action.' He identified two factors governing the selectivity of chemicals, comparative accumulation and comparative biochemistry.

'Selective Toxicity' has now progressed to a seventh edition.[3] By the fifth edition,[4] published in 1973, separate chapters were devoted to each of three principles governing selectivity, namely, distribution, comparative biochemistry and cytological differences. The first edition consisting of 228 pages has grown to 750 pages in the most recent seventh edition, issued in 1985.

This chapter covers only a fraction of what is discussed in Albert's book. This author has given particular emphasis to the contribution of selective accumulation by passive diffusion because of the broad diversity of chemicals which can achieve selectivity on this basis alone. Many of the concepts fundamental to the factor of selective accumulation can be traced to the seminal work done at the turn of the century by Overton whose articles and monograph 'Studien über die Narkose' have had a profound effect on subsequent scientific work.[5,6]

What constitutes acceptable selectivity in a commercial product can change dramatically with the availability of new data. For example, in the first edition of 'Selective Toxicity', DDT, now severely restricted in use because of its persistence and environmental effects, was offered as an illustration of a selective pesticide due to its low toxicity to man. In general, there is increasing interest in these approaches to the design of safer chemicals.[7]

2.6.2 CELL PERMEABILITY AND SELECTIVITY

2.6.2.1 Overton's Plasmolysis Studies

Selectivity is chiefly of interest in the design of drugs or pesticides, but may be regarded in a more general sense as a measure of differences in biological response of any two species. This question is frequently raised in the field of ecotoxicology in reference to an attempt to determine the most sensitive organism for a chemical that may be released into the environment. Overton was one of the first scientists to investigate selectivity from this more general standpoint.

Overton's well-known contributions to the lipoid theory of narcosis and his use of partition coefficients evolved from his studies of cell permeability. He performed several thousand experiments between 1890 and 1896 to investigate the role of differences in cell permeabilities on the selectivity (Wahlwirkung) of poisons and drugs.[6,8] He considered the selectivity of drugs to be one of the two general findings in pharmacology within the prior 50 or 60 years. For example, curare was known to act selectively on the motor nerves without affecting other tissues, digitalis on the transverse muscles of the heart, strychnine on the spinal cord and chloroform on the cerebrum.

In addition to differences in cell permeability, Overton speculated that differential movement into the cell by other than passive diffusion and differences in the degree of change taking place within different cells could contribute to selectivity. For example, the urine of diabetics was known to contain up to 12% glucose, while their blood glucose never exceeded 10%, indicating a movement of glucose from a region of lower to higher concentration.

Overton chose plant cells for his initial permeability studies, since in general they could be studied in more detail and their osmotic properties examined more systematically than those of animal cells. The minimum concentration was determined that produced plasmolysis or separation of the cell protoplast from the cellulose membrane using sugar or another nonelectrolyte to which the cell was impermeable. A second nonelectrolyte tested at this same concentration also produced plasmolysis which disappeared with time. The rate of disappearance of plasmolysis with time was used as a measure of the rate of uptake of the chemical into the plant cell. While these rates varied in a systematic fashion with chemical structure, they were essentially independent of the type of plant cell used. Thus, for these chemicals and the route of administration studied, Overton observed a complete *lack of selectivity*. This provides a useful measure of comparison for other observations.

2.6.2.2 Selectivity Within the Cell: Tannin Precipitation Studies

Experiments involving tannin precipitation in algal cells demonstrated to Overton that selectivity could arise based upon differential binding affinities at the same concentration within a cell.

Spirogyra algal cells immersed in a caffeine solution of 1 part per 2000 were found to produce a tannin precipitate within the cell, which increased with doubling of the caffeine concentration. Transfer of the algal cells to solutions of decreasingly lower concentration caused a decrease in the precipitate until it disappeared at about 1 part in 20 000, consistent with the formation of a tannin–caffeine complex precipitate in equilibrium with dissolved complex and with the dissociated caffeine and tannin. These early observations presaged many later developments in the study of receptors and in the design of selective drugs based upon specific receptor binding.

Most alkaloids tested as the free bases were found to readily penetrate the spirogyra cell membrane and produce a tannin precipitate at low concentrations. These alkaloids were found to be toxic to plant cells within just a few hours to days, even at great dilutions. In contrast to simple nonelectrolytes, the toxic concentration varied enormously from one plant species to another, being attributed to a similar reaction in the cell protoplasts between protein and alkaloid. Since all studied plant and animal cells were very readily permeated by a wide variety of toxicants and drugs, including most of the known general anesthetics, hypnotics and antipyretics, the selective properties of these compounds were ascribed to differences in the concentration required within the cell to produce the observed biological effect, *i.e.* complexes formed in various types of cells have differing degrees of solubility and tendencies to dissociate.

2.6.3 SELECTIVE ACCUMULATION

2.6.3.1 Selectivity and Narcosis Mechanism

The behavior of nonelectrolyte organic chemicals acting by a narcosis mechanism provides an opportunity to demonstrate the most fundamental mechanisms leading to selectivity or lack of it. For such chemicals, there is little if any selectivity if accumulation takes place by pseudo steady-state partitioning between the donor phase at the site of administration and the hydrophobic site of action. Overton employed the production of narcosis in organisms as a probe to monitor uptake of toxicants from aqueous solution, and made the remarkable discovery that the lowest concentration producing narcosis in tadpoles and other aquatic organisms decreased with increasing olive oil/water partition coefficient. Not only did permeability of simple nonelectrolytes vary little among cells, but also there was little or no variation in the concentration in the blood of organisms from mammals to tadpoles at which anesthetic agents such as chloroform and ether produced narcosis independent of the route of administration. The concentrations in blood from a fatal overdose of eight drugs were correlated with their *n*-octanol/water partition coefficients,[9] consistent with Overton's observations.

2.6.3.2 Solubility Cutoff

Overton was able to identify other factors leading to selectivity by control of the distribution between the site of administration and the amount reaching the intracellular fluid. At a given partition coefficient, water solubility decreases with increasing melting point (equation 1)[10]

$$\log P = 6.5 - 0.89 \log S - 0.015\mathrm{MP} \qquad (1)$$

$$n = 27, \quad r = 0.96$$

where P is the *n*-octanol/water partition coefficient, S is the water solubility in $\mu\mathrm{mol\ dm^{-3}}$ and MP is the melting point in °C (a nominal value of 25 is used for liquid solutes).

Although Overton may not have appreciated the strong influence of increasing melting temperature (due to the contribution from the enthalpy of fusion), he indicated that solutes in which the narcotic concentration to tadpoles is just within water solubility can provide a dramatic example of selectivity between algae and aquatic animals.[5] Algae and other plants contain no ganglia cells and require aqueous concentrations six to ten times those needed for tadpoles to produce narcosis. Sulfonal has a water solubility of $2000\ \mathrm{mg\ dm^{-3}}$, the same as the minimum narcosis-inducing concentration in tadpoles. Algae are unaffected, regardless of the amount of sulfonal added. For a given partition coefficient, once equilibrium has been established the concentration of toxicant reaching the site of action within the algae is limited by the toxicant solubility. An incomplete but real contribution of the sulfonal toxicity to algae can be demonstrated by the reduced quantity of a second toxicant required to produce narcosis in the algae. Such additivity to an aquatic organism has been found to apply with as many as 50 different nonelectrolytes.[11]

2.6.3.3 Route of Administration

Inhalation of vapor by mammals and other organisms provides a second route by which equilibrium partitioning can be attained between the site of administration and the site of action. In this case, the concentration in blood is governed by the water/air partition coefficient and the lipoid concentration by the oil/gas partition coefficient.[5] This is illustrated in Figure 1, showing a correlation between the oil/gas partition coefficient and anesthesia in the mouse for a range of saturated monohydric alcohols and monoketones.[12] By contrast, $\log P$, where P is the octanol/water partition coefficient, yields no correlation. The signs and symptoms of anesthesia in fish exposed *via* aqueous solution directly parallel those of mammals exposed *via* gaseous inhalation.[13] Since the molecular properties governing partitioning from air are not the same as those governing partitioning from water, the relative toxicities of a series of chemicals will vary depending upon the route of administration and corresponding selective accumulation.

2.6.3.4 Variation with Temperature and Salinity of Test Water

The relationship between the concentration of toxicant in water and that at the lipoid site of action for aquatic organisms varies with temperature in proportion to the variation of the lipid/water partition coefficient.[5] The tadpole, a cold-blooded animal, assumes a body temperature the same as that of the water in which it is immersed. Meyer[14] reported that the variation in partition coefficients with water temperature directly parallels the temperature dependence of their toxicity to tadpoles.

The apparent sensitivity also varies for organisms living in salt water, in which the solubility of such toxicants is less than in fresh water, resulting in a higher oil/water partition coefficient.[5] Most invertebrates living in salt water have a salt concentration in their blood plasma equal to that of the sea water and, at equilibrium, equal concentrations of toxicant will be present in the sea water and the plasma of the invertebrate. Since the partition coefficient with respect to sea water is higher, such organisms will appear to be more sensitive. The net result is the same for fish, in which the salt concentration in the blood is lower than the salt concentration in the water, resulting in a higher concentration of toxicant in the blood than in the sea water, but the same concentration of toxicant partitioned to the lipoid site of action.

2.6.3.5 Pharmacokinetic Influence on Selectivity

Overton noted that in the literature prior to his own work there existed the erroneous belief that the relative sensitivity of different organisms could be evaluated by the length of time required to produce narcosis at a given concentration in the air.[5] It is true that a bird, mouse and frog exposed to ether vapors are anesthetized at different rates: the bird in four minutes, the mouse in ten minutes and then the frog. However, these rates reflect differences in both the pharmacokinetics of uptake and in lipid/air partition coefficient as a function of the body temperature, but not in the intrinsic sensitivities of these organisms at the site of action.

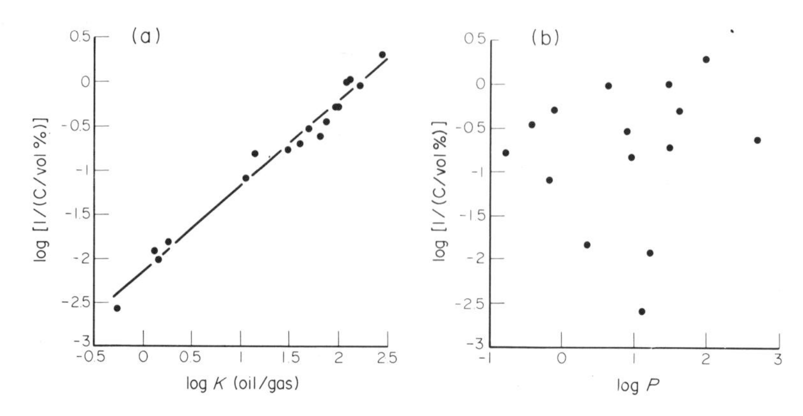

Figure 1 Correlation of mouse inhalation anesthesia of a series of nonelectrolytes with oil/gas partition coefficient (a) but not with octanol/water (b) (reproduced from ref. 12 by permission of Elsevier Scientific Publishing Co.)

Differences in the rate of uptake from water by aquatic organisms will also give rise to a corresponding pharmacokinetically based selectivity. The rate of transfer to the organism is governed not by the permeability of individual cells but by the physiology of whole tissues and organs, *i.e.* the surface area of the gills and the rate of blood flow adjacent to them.[15] Konemann[16] found that nonelectrolytes having log P values up to 5.8 achieve steady-state partitioning, *i.e.* exhibit a linear log–log relationship between partition coefficient and toxicity to the guppy within a 14-day test duration. By contrast, toxicity was found to be pharmacokinetically limited at a log P of 4.5 in a 96-hour test[17] and at a log P of 3 in a 24-hour test.[18] The goldfish has traditionally been considered a less sensitive test organism than other fresh water fishes such as the rainbow trout. This selectivity is for the most part a reflection of the slower rate of uptake of chemicals by the goldfish relative to that of these other fish species.

In the extreme case, pharmacokinetics of uptake can be the major factor in governing selectivity. In routes of administration other than from water or air, steady state cannot normally be attained and the amount of toxicant reaching the site of action is strongly influenced by competition between the rate of uptake and the rate of excretion. Figure 2 shows a plot of the relationship between log P and toxicity to mice of alcohols and ketones that have been administered intravenously. The bilinear relationship[19] can be interpreted in terms of a pharmacokinetically controlled distribution of toxicant above log $P=3$, reflecting a complex combination of transport rates to excretion as the intact molecule or as more hydrophilic metabolites.

Uptake *via* the gastrointestinal tract becomes increasingly less efficient beyond about a six-carbon alcohol due to the pharmacokinetic limit imposed by the limited transit time within the gut and competition between the rate of uptake and excretion. This phenomenon seems to have been first discovered in France by Cros,[20] who in 1862 indicated that the toxicity to mammals of these alcohols increased with increasing molecular weight and decreasing water solubility until a cutoff was reached due to limiting water solubility.[21] Hansch and co-workers[22] found that data on the hypnotic activity of many classes of organic compounds were fitted to a parabolic model (equation 2)

$$\log(1/C) = a\log P - b(\log P)^2 + c \tag{2}$$

where the optimum partition coefficient or log P_0 was found to be about two. The bilinear model, proposed by Kubinyi[19] to model the transport of drugs under nonequilibrium conditions, has now replaced the use of the parabolic model in many cases, as it provides a better fit to the experimental data. Data on rat oral LD_{50}, mouse oral LD_{50}, mouse intravenous LD_{50}, and other LD_{50} and ED_{50} data for saturated monohydric alcohols were fitted to a bilinear model.[12,23] These models were developed to serve as a measure of baseline toxicity to mammals for various routes of administration. An analysis of log P values of hypnotics, sedatives and other drugs showed that most have log P values close to two.[24] The authors concluded that if drugs acting at other sites were designed with log P values outside of this range, entrance to the brain and undesirable side effects could be avoided. The above findings for narcosis toxicity, which is considered to reflect a common molar concentration at the biophase site of action, do not support this as a feasible approach, unless the required drug uptake takes place by other than a passive diffusion mechanism.

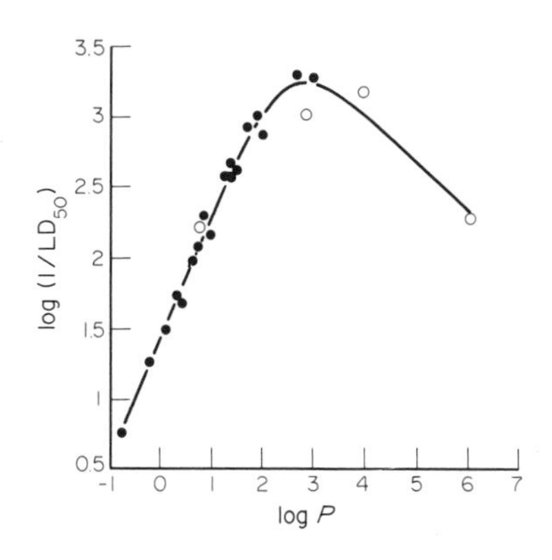

Figure 2 Bilinear relationship between log P and mouse intravenous LD_{50} of saturated monohydric alcohols (●) and saturated monoketones (○) (reproduced from ref. 12 by permission of Elsevier Scientific Publishing Co.)

2.6.3.6 The Influence of pH on Selectivity

Distribution is strongly influenced by the degree of ionization of a chemical at the site of uptake relative to its state at the site to which it has been distributed. The transport process can be modeled using the distribution coefficient defined as the ratio of the concentration of the nonionized solute in the lipophilic phase, *e.g.* *n*-octanol, with respect to the concentration of the nonionized and ionized solute in the aqueous phase.[25]

$$\log D = \log P + pK_a - pH \quad \text{for acids} \tag{3}$$

$$\log D = \log P - pK_a + pH \quad \text{for bases} \tag{4}$$

Ionized organic compounds are selectively transported from the stomach or the intestine into the bloodstream according to the amount ionized at the pH at each site. The distribution will be independent of the direction of transport if equilibrium can be achieved. Thus, Schanker[26] found that basic drugs administered intravenously to dogs appeared in the gastric juice at concentrations up to 40 times that found in the blood plasma. On the other hand, the opposite was found for acidic drugs, which were not found in the stomach or at no greater than 0.6 that of the plasma concentration, the ratios generally in close agreement with those calculated.

The distribution coefficient also governs the degree to which metabolic transformation assists excretion through the production of more hydrophilic substances.[27] For example, the oxidation of an aliphatic C—H bond to an aliphatic C—OH moiety produces a decrease in log *P* of 1.99, while the equivalent aromatic transformation affords a decrease of only 0.67 log units. The ability to estimate log *P* values directly from chemical structure using the CLOGP computer program[28] and knowledge of metabolic differences between the 'economic' and 'uneconomic' cells or species could be useful in designing more selective safer drugs.

2.6.4 COMPARATIVE BIOCHEMISTRY AND CYTOLOGICAL DIFFERENCES

By introducing the Chemotherapeutic Index (ratio of minimal curative dose to maximal tolerated dose) Ehrlich in 1910 placed the concept of selectivity on a numerical basis for the first time. The whole of pharmacology was to greatly benefit from this in the decades that followed.

Albert invoked two additional principles governing selectivity, differences in biochemistry and cytology. Although these factors are distinct, in some cases selectivity results from two or even all three of these principles acting in concert. Classic studies by Albert of the bacteriostatic activity of aminoacridines may have served as the basis for his subsequent interest in the subject of selective toxicity. Penicillin, trimethoprim and tetracycline provide useful examples of widely used current drugs in which selectivity has been attributed to cytological differences, biochemical differences and differences in accumulation between the economic and uneconomic species, respectively.

2.6.4.1 Bacteriostatic Activity of Aminoacridines, Selective Binding of Cations and Differences in Biochemistry and Accumulation

Proflavine, or 3,6-diaminoacridine, was reported by Browning in 1913 as antibacterial against Gram-positive and -negative bacteria but was noted to be of low toxicity to human tissues. Albert and co-workers synthesized 101 different acridine derivatives and examined their effects on 22 species of bacteria.[3,29] Albert hypothesized that the biological activity was related to the cationic properties of these compounds and made many measurements of dissociation constants, at a time when few data of this type were available. It is likely that the preparation of Albert's other books[30–32] and his systematic approach to heterocyclic chemistry[33] can be traced to the inspiration of this study.

He determined, following a systematic study of many derivatives and related compounds, that two properties, delocalized cationic charge and a minimal flat area, were required for activity. This requirement is related to a molecular mechanism of action involving intercalation into the DNA helix in which both electrostatic attraction between the cationic group and DNA phosphate backbone as well as van der Waals forces dependent upon a minimal flat area of the drug are required for DNA binding and bacteriostatic action.

The selectivity of the aminoacridines appears to be due to two factors, the location of the solitary bacterial chromosome in an exposed environment outside of the nuclear membrane of the cell (selective accumulation) and the preference for intercalation by the circular bacterial DNA (differences in biochemistry). Mammalian cells, including those from the brain, accumulate aminoacridines in the nucleic acids of living vertebrate cells without adverse effect. No accumulation and staining (observable under a fluorescence microscope) take place on proteins. In bacteria, part of the cellular DNA is located where the ring-shaped chromosome penetrates the cytoplasmic membrane, requiring no membrane penetration to reach the molecular site of action. The addition of lipophilic substituents on the acridine ring produced a decrease rather than an increase in bacteriostatic activity, the opposite of what would have been expected if the site were located within the bacteria.

2.6.4.2 Penicillin: Selectivity from Cytological Differences

The penicillins are the most widely used of the antibacterial drugs. Their high degree of selectivity to bacteria depends upon a special cytological feature of bacterial cells not found in mammalian cells.[3] Bacteria are under very high osmotic pressure and require a strong cell wall to maintain this condition. The structural integrity of the cell wall is provided by a polysaccharide–polypeptide polymer known as murein or peptidoglycan, consisting of acetylmuramic acid (**1**) linked to polypeptide chains such as (**2**). The penicillins act by inhibiting the synthesis of this cell wall at a key transpeptidation reaction. When cell growth takes place, new cell wall cannot be synthesized and the cells burst from the extreme osmotic pressure. However, when such penicillin-treated bacteria are placed in a 0.3 $mol\,dm^{-3}$ sucrose solution bursting does not take place, since the high osmotic pressure of the solution balances the extreme internal pressure. All penicillins contain a lactam group in a strained four-membered ring (**3**) which can act as an acylating agent in the inhibition process. The structural similarity of penicillins to D-alanyl-D-alanine has been proposed as a key feature governing the selectivity of penicillins for this site.[3,29]

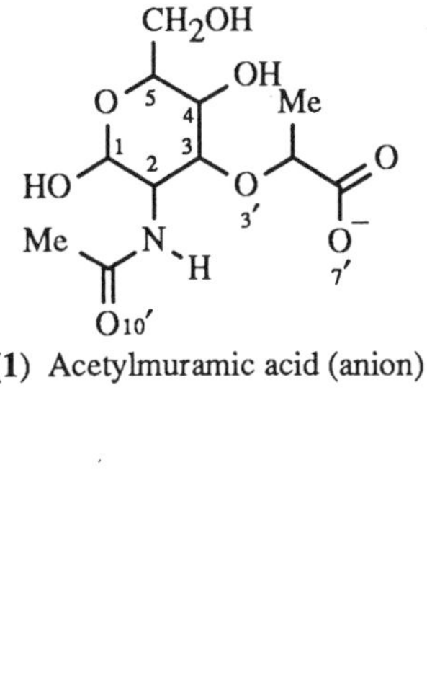

(**1**) Acetylmuramic acid (anion)

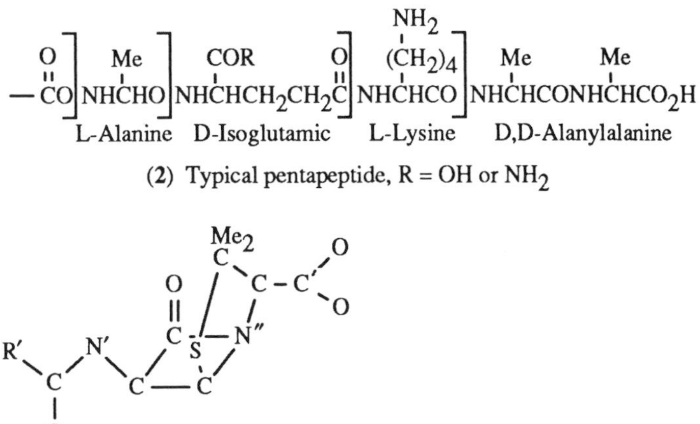

—CO[NHCH(Me)CO][NHCH(COR)CH$_2$CH$_2$CO][NHCH((CH$_2$)$_4$NH$_2$)CO][NHCH(Me)CONHCH(Me)CO$_2$H]

L-Alanine D-Isoglutamic L-Lysine D,D-Alanylalanine

(**2**) Typical pentapeptide, R = OH or NH_2

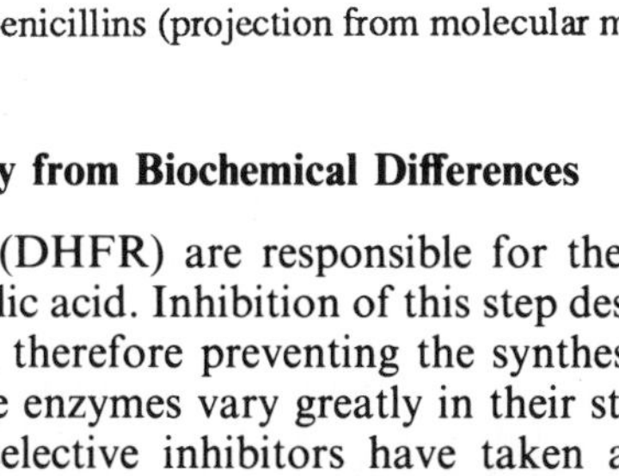

(**3**) The penicillins (projection from molecular models)

2.6.4.3 Trimethoprim: Selectivity from Biochemical Differences

The dihydrofolate reductases (DHFR) are responsible for the biochemical transformation of dihydrofolic acid to tetrahydrofolic acid. Inhibition of this step destroys the ability of the organism to convert uridine to thymidine, therefore preventing the synthesis of DNA and subsequent cell growth and multiplication. These enzymes vary greatly in their steric and electronic requirements for binding and many highly selective inhibitors have taken advantage of these biochemical differences in the form of antibacterial agents, antitumor agents and antimalarial drugs. A number of quantitative structure–activity relationship (QSAR) studies have been performed on various types of dihydrofolate inhibitors. More recently, the availability of X-ray crystallographic studies of these enzymes from various sources have provided a valuable target for molecular modeling studies to investigate the basis for such selectivity.[34,35] Trimethoprim (**4**), discovered in 1962 as a highly selective inhibitor of DHFR in both Gram-positive and -negative bacteria, shows considerably less activity to mammalian cells.[3,29]

(4) Trimethoprim

2.6.4.4 Tetracycline: Selectivity from Differences in Accumulation

Chlortetracycline ('Aureomycin') was isolated from a soil sample in 1945 by Dugar at Lederle Laboratories. A second tetracycline derivative, oxytetracycline ('Terramycin'), was isolated several years later by a group at Pfizer, which subsequently found that removal of the chlorine from chlortetracycline produced the parent tetracycline (**5**; 'Tetracyn') an antibiotic with fewer gastrointestinal side effects.[36] Tetracycline and its derivatives have been shown to owe their selectivity to bacteria to selective accumulation by these organisms. Franklin reported in 1971 that all bacteria accumulate tetracyclines but that very little accumulation takes place in mammalian cells, owing to differences in the cytoplasmic membranes, even though the tetracyclines were found to repress protein synthesis in both bacteria and mammalian cells by having similar inhibitory action at their ribosomes. It has been proposed that the greater accumulation in bacteria is due to the removal of a lipophilic magnesium complex from the bacterial membrane.[3,29]

(**5**) Tetracycline

2.6.5 CONCLUSIONS

Three factors are responsible for the selectivity of drugs and other bioactive agents. These consist of differences in accumulation, biochemistry and cytology. In some cases, two or even all three of these factors together contribute to the observed selectivity. A search for the basis of selectivity in a particular case can proceed by gaining an insight into the relationship between activity and chemical structure. As a first step, Albert[2] has recommended listing the physical, chemical and biochemical properties of active substances. This critical step, unfortunately, is often performed with insufficient insight and imagination. Each of these listed properties can then be made the *limiting factor* in testing each hypothesis, as was elegantly demonstrated by Albert in the case of the aminoacridine and oxine studies. The determination in this fashion of the mechanism of action to the target species or site will then provide insight into its specific selectivity.

2.6.6 REFERENCES

1. A. G. Gilman, S. E. Mayer and K. L. Melman, in 'The Pharmacological Basis of Therapeutics', 6th edn., ed. A. G. Gilman, L. S. Goodman and A. Gilman, Macmillan, New York, 1980, p. 38.
2. A. Albert, 'Selective Toxicity: with Special Reference to Chemotherapy', Methuen, London, 1951.
3. A. Albert, 'Selective Toxicity: The Physicochemical Basis of Therapy', 7th edn., Chapman and Hall, London, 1985.
4. A. Albert, 'Selective Toxicity: the Physico-Chemical Basis of Therapy', 5th edn., Chapman and Hall, London, 1973.
5. E. Overton, 'Studien über die Narkose, zugleich ein Beitrag zur allgemeiner Pharmakologie', Fischer, Jena, 1901.
6. R. L. Lipnick, *Trends Pharmacol. Sci.*, 1986, **5**, 161.
7. E. J. Ariens, 'Drug Design', Academic Press, New York, 1980, vol. IX, p.1.
8. E. Overton, *Z. Phys. Chem., Stoechiom. Verwandschaftsl.*, 1896, **22**, 189.
9. S. L. Cassidy, P. A. Lymphany and J. A. Henry, *J. Pharm. Pharmacol.*, 1988, **40**, 130.
10. S. Banerjee, S. H. Yalkowsky and S. S. Valvani, *Environ. Sci. Technol.*, 1980, **14**, 1227.
11. H. Konemann, *Toxicology*, 1981, **19**, 229.

12. R. L. Lipnick, C. S. Pritzker and D. L. Bentley, in 'QSAR in Drug Design and Toxicology', ed. D. Hadzi and B. Jerman-Blazic, Elsevier, Amsterdam, 1987, p. 301.
13. W. N. McFarland, *Publ. Inst. Mar. Sci., Univ. Tex.*, 1959, **6**, 23.
14. H. Meyer, *Arch. Exp. Pathol. Pharmakol.*, 1901, **46**, 338.
15. R. L. Lipnick, in 'QSAR in Toxicology and Xenobiochemistry', ed. M. Tichy, Elsevier, Amsterdam, 1985, p. 39.
16. H. Konemann, *Toxicology*, 1981, **19**, 209.
17. G. D. Veith, D. J. Call and L. T. Brooke, *Can. J. Fish. Aquat. Sci.*, 1983, **40**, 743.
18. R. L. Lipnick, K. R. Watson and A. K. Strausz, *Xenobiotica*, 1987, **17**, 1011.
19. H. Kubinyi, *J. Med. Chem.*, 1977, **20**, 625.
20. A. F. A. Cros, Thesis, Faculte De Medecine de Strasbourg, 1863.
21. R. L. Lipnick, *Environ. Toxicol. Chem.*, 1989, **8**, 1.
22. C. Hansch, A. R. Steward, S. M. Anderson and D. L. Bentley, *J. Med. Chem.*, 1967, **11**, 1.
23. R. L. Lipnick, C. S. Pritzker and D. L. Bentley, in 'QSAR and Strategies in the Design of Bioactive Compounds', ed. J. K. Seydel, VCH, Weinheim, 1985, p. 420.
24. C. Hansch, J. P. Bjorkroth and A. Leo, *J. Pharm. Sci.*, 1987, **76**, 663.
25. R. A. Scherrer, *ACS Symp. Ser.*, 1984, **255**, 225.
26. L. S. Schanker, *Annu. Rev. Pharmacol.*, 1961, **1**, 29.
27. C. N. Manners, D. W. Payling and D. A. Smith, *Xenobiotica*, 1988, **18**, 331.
28. A. Leo and D. Weininger, 'Medchem Software Release 3.33', Medicinal Chemistry Project, Pomona College, Claremont, CA, 1985.
29. A. Albert, 'Xenobiosis: Food, Drugs and Poisons in the Human Body', Chapman and Hall, London, 1987.
30. A. Albert, 'The Acridines: their Preparation, Properties, and Uses', 2nd edn., Arnold, London, 1966.
31. A. Albert and E. P. Serjeant, 'The Determination of Ionization Constants', 3rd edn., Chapman and Hall, London, 1984.
32. A. Albert, 'Heterocyclic Chemistry: An Introduction', 2nd edn., The Athlone Press, London, 1968.
33. E. Campaigne, *J. Chem. Educ.*, 1986, **63**, 860.
34. J. M. Blaney, C. Hansch, C. Silipo and A. Vittoria, *Chem. Rev.*, 1984, **84**, 333.
35. C. Hansch and T. E. Klein, *Acc. Chem. Res.*, 1986, **19**, 392.
36. W. Sneader, 'Drug Discovery: The Evolution of Modern Medicines', Wiley, Chichester, 1985.

3.1
Classification of Drugs

ALFRED BURGER
University of Virginia, Charlottesville, VA, USA

3.1.1 INTRODUCTION

The classification of drugs is an arbitrary procedure, governed by conventions and depending on who chooses to do it, for whom it is performed, and on the point of view of the respective persons. In Volume 6 of this series it is shown that there are approximately 5500 chemical entities which in suitable formulations can be used to treat diseases in humans. Faced with such a collection of items it will be a characteristic human response to classify them into groups in order to facilitate comprehension and assimilation. The difficulty in arriving at a generally acceptable definition arises from the very nature of all drugs. Drugs are regarded as biologically active chemical compounds, mostly but not necessarily with a therapeutic purpose. Not a single drug has ever been encountered which exhibits only one biological activity. Indeed, side effects are associated with all drugs and have formed the greatest stumbling block in searches for a specific agent that is to achieve a specific biological purpose. In many cases the side effects are so pronounced, or even overpowering, that the original biological or therapeutic mission of such a drug is dimmed. Numerous examples abound where a given agent has been reclassified after a better preclinical appreciation of its side effects which may become main effects in an alternate course of therapy.

3.1.2 CHEMICAL CLASSIFICATION OF DRUGS

A chemist will think of a drug in terms of its chemical structure. Even though a research assignment may specify drugs with a given therapeutic mission, too many diverse structural types will usually come into consideration in one biological area. Structural designations are of value only if they are used by experts. Physicians who quite routinely have forgotten their early college course in organic chemistry will call any polycyclic antidepressant 'tricyclic' because the early standards (imipramine, amitriptyline) satisfied this description. Second-generation antidepressants with four or five rings and sometimes divergent structural features are still called 'tricyclic' as long as their biochemical role is the same as or similar to that of the historic prototypes.[1] An excuse may be offered for such a chemical confusion. If the terms tricyclics, tetracyclics and pentacyclics were to be used side by side, might that not imply that they have a different biochemical activity, at least to the chemically naive? Therefore it would be better to classify such drugs as inhibitors of synaptic reuptake of biogenic amines but psychiatrists, accustomed to analytical practices, may well find that cumbersome. The short though inadequate catchword 'tricyclics' at least sets such reuptake inhibitors apart from monoamine oxidase inhibitors, not to mention from neuroleptics and antianxiety agents which so often are lumped together as 'tranquilizers'.

In some cases structural classifications come close to being the best for the purpose. We think of organo-arsenicals, -antimonials and -bismuthials as classical chemotherapeutic agents for spirochetal, protozoan and similar infections. Chemists, pharmacologists and physicians can agree on this, and the organic chemical structures of such drugs pale before the presence of heavy metal atoms in their midst. Thus when we speak of arsenical drugs we relegate their many toxic effects to the back of our minds, even if some pronounced side effect may foreshadow some utility in some chemotherapeutically or pharmacologically unrelated field.

One of the most common activities of medicinal chemists is to synthesize 'me-too' drugs that might rival or improve the properties of a successful prototype agent. In the same way many, mostly industrial, pharmacologists test such compounds, usually by the same methods that had characterized the activity profile of the earlier drugs in such a series. The implication of these procedures for drug classification is that similar chemical structures should yield similar biological activities. On the whole that holds true but one has to account for the frequent black sheep in such families of compounds that stand out by their lack of activity in the tests used, their irregularly narrow margin of safety, or even their different type of activity. Thus one has to classify structurally closely analogous compounds in the same overall activity series with some caution and not commit oneself too broadly to such a classification. The physical reasons for such exceptions to an overall pattern will be discussed in chapters dealing with the individual drug types. They are often traceable to steric conditions, receptor fit and lipophilicity.

Chemists who supply untold numbers of compounds for tests in screening or crash programs have one biological purpose in mind and fervently hope that some of their structurally assorted substances will respond positively in such undertakings. Compounds arising from such trials may or may not have structural relationships to earlier prototypes and become new 'lead' compounds. Also chemists who isolate and study natural products must keep an open mind about the biological activity of such materials, even though an often questionable history of folklore may point to a given preference of testing. Again, natural products will thereby become novel 'leads' for molecular modification. Modern methods of biochemistry will often elucidate the mechanisms of action of even structurally unorthodox compounds and thereby facilitate their rational classification.

Structural analogy need not be restricted to visibly close formal similarity in order to make tentative assignments for drug classification. Bioisosteric analogs from different series can be included in such predictions. Thus ester-type local anesthetics can be extended to hydrolytically more stable amide types or to ethers such as dimethisoquin even though a prediction of local anesthetic activity for the latter would have been tentative at best.

Medicinal chemists who work in a structurally limited series for some time will inevitably become experts in the literature, methodology and structure–activity relationships of their subject (leucotrienes, steroid lactones, benzodiazepines, *etc.*). For their purposes, a chemical structural classification of drugs is of real advantage. This classification used to be so prevalent before 1950 that several textbooks of organic chemistry for pharmacy students taught therapeutic agents according to their structure, disregarding biochemical and biological relationships. With the growing understanding of such relationships the classification of drugs has shifted increasingly to biochemical and therapeutic headings, subdivided, to be sure, into structural types. It would be foolish nowadays to attempt to state that all amino alcohols have a given biological activity, all arylalkanoic acids are antiinflammatory, *etc.* For these reasons, the classification of drugs on the basis of therapeutic purpose has become generally accepted by scientists, the public, government officials and entrepreneurs.

3.1.3 BIOLOGICAL CLASSIFICATION OF DRUGS

The medical profession and the public seeking pharmacotherapeutic and chemotherapeutic aid in disease states have resorted to disease-oriented classifications of medicinal agents which are, overall, descriptive but not very accurate on scientific grounds. Examples are the terms pain killer, heart drug, antiarthritis medicine, diuretic, local anesthetic, chemotherapeutic drug (here the public thinks of anticancer drugs although chemotherapy stood only for antimicrobial therapy for seven decades), worm medicine, *etc.* In each of these categories one will find compounds with entirely different chemical structures and, more importantly, different biochemical modes of action. For scientific purposes, much more accurate classifications are demanded. They will be dictated by meaningful biochemical and biological testing methods and the mechanisms of interaction with physiological targets these tests reveal. The 'Annual Reports in Medicinal Chemistry' have proposed a broad classification of drugs based on gross overall biological effects. An outline is shown in Table 1.

Table 1 Classification of Drugs Based on Their Gross Biological Effect

(1) *CNS agents or psychopharmaceuticals*	
Antipsychotic agents	Anxiolytics
Anticonvulsants	Sedative–hypnotics
Analgesics	Drugs for cognitive disorders
Antiparkinsonian agents	Stimulants
Antidepressants	
(2) *Pharmacodynamic agents*	
Antihypertensive agents	Pulmonary and antiallergic agents
Calcium modulators	Antiarrhythmics
Antieschemic (antistroke) drugs	Antiglaucoma agents
Antiulcer—gastrointestinal agents	Vasodilators
Antithrombotics	Antianginals
(3) *Chemotherapeutic agents*	
Antimicrobials	Anticancer, antineoplastics
Antifungals	Antiparasitics
(4) *Metabolic diseases and endocrine function*	
Contraceptives	Antiatherosclerotics
Antiandrogens	Antirheumatics and autoimmune disease drugs
Antiinflammatories	
Dermatologicals	Antiobesity drugs

Table 2 lists established drug classifications and suitable subclassifications and cross-references of many drugs. It is evident from this tabulation that the classification of drugs on the basis of their principal activity is convenient but neither adequately specific nor broadly correct scientifically. Even noticing one or more important secondary activities does not satisfy a comprehensive classification. The rate of biochemical advances over relatively short periods of time of months or a few years demands a more explicit subdivision and ultimately a mechanism-based arrangement wherever possible.

A few examples, by no means all-inclusive, illustrate these dilemmas. Chloroquine is still the most important antimalarial, perhaps one should say the most important plasmodial suppressant for infection with the four types of human malaria as well as several experimental plasmodiases of a variety of animals. Clearly, the most common classification of chloroquine is as an antimalarial, and, if you will, a 4-dialkylaminoalkylquinoline antimalarial. But chloroquine is also a clinically reasonably effective antiinflammatory drug. No causative connection between the malarias and arthritis has been suggested. In addition, chloroquine is also of clinical value in amebiasis including amebic hepatitis. To a bioscientist, an overriding classification of chloroquine in any of these three categories by themselves remains unsatisfactory. Perhaps future researchers will establish an inhibitory activity by chloroquine of one or several enzyme systems which play a role in the life cycles of plasmodia and entamebae and perhaps even in the etiology of arthritis. Until such a day arrives, a side-by-side listing of the various activities of chloroquine will remain in force.

Another antimalarial agent, the cinchona alkaloid quinidine, a diastereomer of quinine, has antiarrhythmic activity. This cardiac action is more useful in medicine than the antiplasmodial properties of the alkaloid. Classification under two headings is demanded.

A physician is accustomed to treat hypertension with one drug after another, choosing from a battery of agents available from the pharmaceutical industry. Among the many widely prescribed classes of these drugs are the thiazide agents (a convenient chemical structural classification) and the β-adrenergic receptor inhibitors. Both of these types of drugs had a history that illustrates the complications of classification even if fundamental mechanisms of action are taken into account. The thiazides are cyclic sulfonamides; they were conceived and developed as inhibitors of carbonic anhydrase based on the corresponding properties of other heterocyclic sulfonamides such as acetazolamide. The thiazides revealed their antihypertensive activities only after they had been introduced as diuretics.

The β-adrenergic receptor antagonists, spearheaded by their clinically useful serial member, propranolol, were developed as pharmacological tools to regulate and inhibit cardiac arrhythmias.

Table 2 An Alphabetical List of Major Categories of Drugs, Some Subclassifications and Alternate Cross-listings

Principal drug classifications	*Subclassifications*	*Cross-listings*
Adrenergics	Catecholamines Dopamine receptor agonists Norepinephrine receptor agonists Vasoconstrictors	Amphetamine Ephedrine Phenylpropanolamine Soterenol
Adrenergic receptor antagonists	Antiarrhythmics Antihypertensives Vasodilators	Antiocular pressure agents
Amebicides	Metalloorganic drugs	Antimalarials Antitrichomonal agents
Anabolic agents	Steroids	
Analeptics		Cardiac stimulants CNS stimulants
Analgesics, antiinflammatory	Nonsteroidal antiinflammatory agents	Acetaminophen Aryl- and heteroaryl-alkanoic acids Salicylates
Analgesics, potent	Opioides	Enkephalins Pethidine Methadone
Androgens	Nonaromatic steroids	
Anesthetics, general	Oral anesthetics Parenteral anesthetics Volatile anesthetics	
Anesthetics, local		Cocaine Amide, ester, ether-type anesthetics
Anorexigenics	Amphetamine analogs	Antiappetite agents
Anthelmintics	Drugs for cestode, nematode, trematode infections	
Antiaging drugs	Free radical scavengers Immunosuppressants	Hormones, pituitary
Antiallergenic drugs	Antihistaminics Histidine decarboxylase inhibitors Bradykinin antagonists	Anticholinergics Serotonin antagonists Slow-reacting substance of anaphylaxis
Antianxiety agents	Benzodiazepines	Carbamates CNS Depressants

Antibiotics	Aminoglycosides Antibacterials β-Lactam antibiotics Macrolides Polyenes	Antimycobacterial drugs Antiprotozoal drugs Antitumor drugs Antiviral drugs
Anticholinergics	Antiacetylcholine drugs Antispasmodics Antiulcer drugs Mydriatics	Antihistaminics Antiserotoninergics Atropine Scopolamine
Anticoagulants	Antithrombotics Clot-dissolving proteins	Coumarins Heparin Indanediones Snake venoms
Anticonvulsants	Barbiturates Benzodiazepines Hydantoins and related systems Valproate	
Antidepressants	Inhibitors of catecholamine and serotonin reuptake Monoamine oxidase inhibitors	'Tricyclic' antidepressants
Antifungal agents	Chlorinated compounds Phenolic substances	Antibiotics
Antihyperglycemic drugs		Peptide hormones (insulin) Sulfonylureas
Antihyperlipidemic agents	Clofibrate and analogs: inhibitors of lipid absorption	Sterol-binding agents
Antihypertensive drugs	Inhibitors of angiotensin-converting enzyme	β-Adrenergic receptor antagonists Clonidine Diuretics Ganglionic-blocking agents Methyldopa Vasodilators
Antiinflammatory drugs	Steroidal Nonsteroidal	Analgesic–antiinflammatory agents
Antimalarials	Antiplasmodials Aminoalkylaminoquinolines Pyrimidines Quinoline amino alcohols Terpene peroxides	Amebicides Antiinflammatory drugs, nonsteroidal Dihydrofolate reductase inhibitors

Table 2 *Continued*

Principal drug classifications	*Subclassifications*	*Cross-listings*
Antimycobacterials	Antibiotics Bis(aminoaryl) sulfones Hydrazines Rifamycins	Antileprotics Antituberculous drugs Atypical antimycobacterials
Antiparkinsonism drugs	Dopa	Anticholinergics
Antiseptics	Disinfectants Phenols, halogenated Quaternary ammonium ions	Antibacterials Antifungals Antivirals
Antithyroid drugs	Thyroidal hormone inhibitors Goitrogens	
Antitumor agents	Bleomycin antibiotics Colchicin (mitosis inhibitor) Doxorubicin	Alkylating agents Antibiotics Antimetabolites Nitroso compounds Vinca alkaloids
Antitussives	Centrally acting Peripherally acting	Demulcents Expectorants Mucolytics
Antiviral agents	Adamantane derivatives Antibiotics Interferon and other proteins Thiosemicarbazones	Metabolite analogs
Calcium channel regulators	Calcitonin Parathyroid hormones Vitamin D	Antianginal agents Calmodulin Verapamil
Cardiac drugs	Antiarrhythmic agents Antilipemic drugs	β-Adrenergic receptor antagonists Antianginal agents Cardiac glycosides
Cathartics	Intestinal lubricants Phthaleins	Anthraglycosides Inorganic laxatives
Central nervous system depressants	Antianxiety agents Anticonvulsants Neuroleptics Sedative–hypnotics	Anesthetics, general Barbiturates Benzodiazepines Bromide ion
Cholinergics		Anticholinesterases Curaremimetics
Coagulants		Vitamin K Snake venoms

Constipants (antiperistaltics)	Opium (morphine) Diphenoxylate (and difenoxin)	Antacids Antidiarrheals, inorganic Astringents
Diuretics	Carbonic anhydrase inhibitors Thiazides and related compounds	Antihypertensives Uricosurics
Estrogens	Ring A aromatic steroids	Contraceptives 1,2-Diphenylethylenes Progesterone
Ganglionic-blocking agents	Quaternary onium ions Sterically hindered amines	Antihypertensives Curaremimetics
Histamine H_2-receptor antagonists		Antiulcer drugs Imidazoles Nitrofuran derivatives
Hypnotics–sedatives	Barbiturates and analogs Benzodiazepines	Antihistaminics Carbamates Chloral hydrate Ureas, halogenated
Immunostimulants and immunosuppressants	Alkylating agents Metabolite analogs	Alkaloids (colchicum, vinca) Antibiotics Cyclosporin Interferons Steroids
β-Lactam antibiotics	Penicillins Cephalosporins Cephamycins	Clavulanic acid Nocardicins Thienamycins
Muscle relaxants	Depressants, polysynaptic	Benzodiazepines Carbamates
Neuroleptics	Antipsychotic agents Dopamine receptor antagonists	Butyrophenones Dialkylaminoalkylphenothiazines
Neuromuscular-blocking agents	Curare	Bisquaternary ions
Peptide and protein agents	Hormones: antihyperglycemic, hypothalamic, intestinal, pancreatic, parathyroid, pituitary	Immunosuppressants Peptide antibiotics
Prostaglandins	Cyclic nucleotide-controlling agents Hormones, natriuretics, renal Inflammatory agents	Abortifacients Gastrointestinal drugs

Table 2 *Continued*

Principal drug classifications	*Sub-classifications*	*Cross-listings*
Psychosomimetics	Glycolate esters Lysergic acid derivatives Tetrahydrocannabidiol Tryptamine derivatives	Cocaine Heroin Methaqualone and other synthetics Mescaline
Radiation protective agents	Radiosensitizers Radical scavengers	
Radioopaques	Diagnostic agents	Iodinated compounds
Steroids	Adrenocortical hormonal agents Anabolic agents Androgens Anticancer drugs Antiinflammatory steroids	Cardiac glycosides Contraceptives Corticosteroids Estrogens Glucocorticoids Mineralocorticoids Progestational agents Vitamin D
Sulfonamides and sulfones	Aminoaryl sulfones Antibacterials Carbonic anhydrase inhibitors	Sulfanilamides Diuretics
Sulfonylureas		Antihyperglycemics
Thyromimetics	Thyroid hormones	Antithyroid drugs
Trypanocides	Antiamebics Antileishmania drugs Antiplasmodials Antitrypanosomal agents	Antibiotics Nitroimidazoles Sulfonamides Triazines and related structural types

After the US Food and Drug Administration had reluctantly consented to admit propranolol for this purpose, propranolol faced the same administrative hurdle again several years later when its pronounced antihypertensive effect by a central nervous system mechanism was discovered. Cardiologists learned that coadministration of a thiazide 'diuretic' and a 'β-blocker' resulted in a particularly gratifying reduction of the blood pressure in hypertensive patients. How should we classify these two types of drugs? A cross-reference between diuretic and antihypertensive entries is a minimal requirement for a thoughtful printed entry and would be facilitated by computer memory for quick recall. Other β-adrenergic receptor antagonists need further classification; timolol is primarily though not exclusively effective in reducing ocular pressure in glaucoma. Also, there are now increasingly more selective inhibitors of β_1- and β_2-adrenoceptors. Perhaps all these properties should indeed be stored in a computer rather than in a printed chapter such as this. In cases of more extensive biological multiactivity this may well become the choice of recalling the necessary classifications.

Ibuprofen is generally listed as a nonsteroidal antiinflammatory agent. Like aspirin, another carboxylic acid but otherwise structurally divergent, it inhibits arachidonic acid cyclooxygenase and thereby interrupts the biosynthesis of prostaglandins.[2] It also inhibits other enzyme systems, among them lipoxygenase. Products from the reactions catalyzed by this enzyme such as chemotactic factor contribute to ischemia-induced myocardial damage, and inhibitors of their formation should alleviate the injury.[3] Ibuprofen counteracts polymorphonuclear leukocyte-induced edema formation *in vivo*.[4,5,6] If at least in the scientific literature an attempt at biochemical accuracy were made, these variant activities would deserve classification. However, even here revisions would have to be made as further experiments advance our detailed understanding of these processes.

Another example concerns antihistaminic drugs. They are among the most poorly classified agents with alternate and overlapping activities in several other areas. It is safe to state that no specific antihistaminic drug has been found in the 50 years since the introduction of these drugs. Central nervous system effects are most pronounced. A number of antihistaminics are available as over-the-counter sedative–hypnotics, and CNS depression is a common side effect of all of them. Related to this property is the use of several antihistaminics to suppress motion sickness. Others are used as antispasmodics; in fact, a medicinal chemist would be hard put to guess whether a given traditional antihistaminic will also be anticholinergic. All this does not take into account biochemical considerations, whether a given antihistaminic effect is achieved by protecting an H_1-receptor from histamine occupancy, or preventing histidine decarboxylase from aiding in the biosynthesis of histamine. The effects of H_2-histamine antagonists are not included in this account nor are antihistaminic actions of phenothiazine neuroleptics considered here.

Simple classification becomes virtually impossible for antimicrobial agents, there is too much overlap of effects on different organisms. β-Lactamase inhibitors started out as agents for Gram-positive bacterial infections but long since, several curative agents for Gram-negative infections by different types of organisms have been found among them. Successful antibacterials now include dihydrofolate reductase inhibitors but such drugs also inhibit isozymes in protozoan cells such as plasmodia. Quinolinecarboxylic acids and other structures are effective antiprotozoan agents, and metronidazole is a true broad-spectrum drug in this field. For a hint about additional multiactive chemotherapeutic drugs see Table 1.

Hundreds of thousands of compounds from commercial sources, from the collection of academic chemists and from every type of natural and synthetic substances have been submitted to a battery of tests against experimental cancers. More than 100 drugs have become available in the last 30 years for the treatment of clinical malignancies. They include antimetabolites, alkylating agents, drugs that intercalate between sections of polynucleotides, nitroso compounds, alkaloids, terpenes, steroids and every kind of odd structure, and not yet understood biochemical mechanisms. To lump them together as antitumor agents would be to admit scientific ignorance. The medical profession classifies such drugs in broad categories but the general public calls their action 'chemotherapy'. It will be years before a consensus on rational identification will be worked out.

3.1.4 CLASSIFICATION OF DRUGS BY THE LAY PUBLIC

'Drugs' are, in the public's view, chemicals that alter the mind, lead to habituation, dependence liability, and cause dependent individuals to lose control over legitimate or legal actions. In other words, the public equates most, if not all, drugs with central nervous system active agents, natural or synthetic.

Table 3 Characteristics of Common Routes of Drug Administration

Route	*Absorption pattern*	*Special usefulness*	*Limitations*
(a) Via *gastrointestinal tract*			
Oral	Variable	Most convenient, safe and economical	Requires patient cooperation. Absorption potentially erratic and incomplete for poorly soluble drugs. Can cause gastric irritation.
Rectal	Good	Useful for medications which cause gastric distress; avoids first pass effect	Poor patient-acceptability
Sublingual	Good	Rapid onset; especially useful for materials absorbed erratically from gastrointestinal tract. Avoids first pass effect	Taste. Requires lipid soluble agents
(b) Parenteral			
Intravenous injection	Absorption circumvented	Potentially immediate effects; valuable for emergency use; allows titration of dose; can be used with large volumes and dilute solutions of irritating substances	Increased risk of adverse effects; slow injections required; needs trained personnel, not suitable for oily or insoluble substances.
Intramuscular (i.m.) injections	Prompt from aqueous solution, slow and sustained from depot preparations	Suitable for moderate volumes, oily vehicles and some irritating substances	Often causes irritation around injection site; needs trained personnel
Subcutaneous injection	As for i.m.	Suitable for some insoluble suspensions and implantation of solid pellets	Not suitable for large volumes; possible slough from irritating substances. Can be done by untrained personnel
(c) Lung			
Aerosol or spinhaler	Good	Drug acting on lung or bronchorespiratory tract	Systemic absorption erratic. Spinhaler may be difficult for small children to use
(d) Topical			
Cream and ointment	Erratic into systemic circulation	Applying high levels of drug to required site	

As in politics and other human preoccupations, the memory of the average person does not reach back very far. Forgotten are the times when no medications for infections were available, when diabetes was a terminal illness, and when the pharmacological armamentarium for dozens of disease states contained only ill-concocted placebos. Barely any newspaper reader has even the slightest conception how medicines are created, how they act, how much it costs to put one on the market, and what happens to a drug after it has been administered.[7,8] A stockbroker asked recently why it would take so long to develop a drug for acquired immune deficiency syndrome (AIDS) and why a company that had expressed instant hope of producing such a drug (and have the price of its shares rise) did not fulfill this promise.

These attitudes have a direct bearing on the classification of drugs. The few broad public classifications (pain reliever, tranquilizer, cancer drug, antihemorrhoidal medicines, diet control pills, sleep inducers, 'penicillin', 'Novocaine', *etc.*) are, of course, scientifically inadequate. The weekly assurance of the news media that Doctor so and so has found a new medicine for some incurable disease creates the impression that new medicines arise out of the mind of physicians. Medicinal chemists and experimental biologists have before them a vast educational task to inform at least the educated lay person, that is, the informable member of society, about all aspects of the chemistry, biological action and fate of medicinal agents.

3.1.5 CLASSIFICATION OF DRUGS ACCORDING TO COMMERCIAL CONSIDERATIONS

The manufacturers and distributors of therapeutic agents have to amortize their operational expenses, their research investment, and they have to make a reasonable profit so that they can finance new ventures in the future. This limits the types of drugs an industry can consider for development. An adequate number of prospective users and patients is necessary to induce a pharmaceutical firm to undertake work in a given area.

Medicines for relatively rare diseases are called orphan drugs. Such drugs often lack patent protection, production costs are high and demand is limited. The US Congress has permitted companies to write off as tax credits up to 73% of the costs of clinical testing of orphan drugs. In 1984 the term orphan drugs was redefined to designate agents for a patient population under 200 000. Drugs and vaccines for tropical diseases are also classified as orphan drugs because the patients suffering from tropical diseases, although numbering in the tens of millions, are too poor to pay the price for their medications. For further information on orphan drugs, see Chapter 4.11.

In a few special cases the term orphan drug may have to be redefined further. Until 1985 human growth hormone was extracted from the pituitary glands of cadavers, but now the 191-amino-acid protein can be produced commercially by recombinant DNA techniques. The patient population of approximately 7100 children in the USA which may use the new hormone (somatropin) is not large but this number must be multiplied by the cost of $11 000 per year per patient. In spite of this profitable market the hormone is still classified as an orphan drug.

Recently, tissue-type plasminogen activator (t-PA), which dissolves blood clots has been manufactured by genetic engineering. Here the number of potential users has been estimated at several million worldwide. Some enzymes, streptokinase and urokinase, have similar though less potent clot-dissolving properties. The classification of these proteins in medicinal terms is still pending.[9]

Another classification depends on the route of administration of a drug. Drugs are given orally, parenterally by several methods, by inhalation, sublingually, rectally, *etc.* On the whole, orally active drugs are preferred by physicians and patients alike. The oral activity may also be an important commercial consideration as was the case with cephalosporins. Such drugs must be reasonably resistant to digestive influences and enzymatic attack in the gastrointestinal tract but this requirement can be supported to some extent by encapsulation, time-release methods and other pharmaceutical procedures. For rapid, life-saving therapy or for the administration of many peptide and protein drugs, parenteral injection is preferred. Volatile anesthetics are, of course, administered by inhalation. Table 3 summarizes such characteristics of common routes of drug administration.

3.1.6 REFERENCES

1. C. Kaiser and P. E. Setler, in 'Burger's Medicinal Chemistry', 4th edn., ed. M. E. Wolff, Wiley, New York, 1981, vol. 3, p. 997.
2. T. Y. Shen, in 'Burger's Medicinal Chemistry', 4th edn., ed. M. E. Wolff, Wiley, New York, 1981, vol. 3, p. 1205.

3. K. M. Mullane, J. A. Salmon and R. Kraemer, *Fed. Proc., Fed. Am. Soc. Exp. Biol.*, 1987, **46**, 2422.
4. P. J. Flynn, W. K. Becker, G. M. Vercellotti, D. J. Weisdorf, P. R. Craddock, D. E. Hammerschmidt, R. C. Lillehei and H. S. Jacob, *Inflammation (NY)*, 1984, **8**, 33.
5. M. Rampart and T. J. Williams, *Biochem. Pharmacol.*, 1986, **35**, 581.
6. I. Rivkin, J. Rosenblatt and E. L. Becker, *J. Immunol.*, 1975, **115**, 1126.
7. A. Burger, 'Drugs and People', The University Press of Virginia, Charlottesville, VA, 1986.
8. F. Gross, 'Decision Making in Drug Research', Raven Press, New York, 1983.
9. R. M. Baum, *Chem. Eng. News*, 20 July, 1987, 11.

3.2
Lead Structure Discovery and Development

GEORGE deSTEVENS

Drew University, Madison, NJ, USA

3.2.1 INTRODUCTION

It has often been said that the medicinal chemist is like a riverboat gambler: against overwhelming odds he believes that the next chemical he synthesizes and submits for biological evaluation will be the long-sought breakthrough drug. This hope, this dream, this stimulus to his imagination is the driving force behind his quest for the discovery of a useful drug which will be of value in restoring health to literally millions of people. This goal is what keeps the chemist going year after year, somehow convinced that amidst the many combinations of atoms which are the building blocks of organic chemistry lies one molecule, the product of his conception, which will take its rightful place amongst other useful medicines. How the creative forces within an individual come together to give rise to a new chemical entity with a unique biological activity is a very complicated process and requires the juxtaposition of a number of scientists with their own particular expertise. It is the purpose of this chapter to analyze the various organizational structures and to suggest those features most suited for optimizing lead discovery and the subsequent development to a marketed drug. Several examples will be given to support and enforce the point of view under consideration. In addition, the selection criteria for leads will be discussed with an emphasis on the crucial significance of this process.

3.2.2 ORGANIZATIONAL STRUCTURE

In order for scientists to work productively and creatively it is essential that a certain attitude exists within a corporate entity. It is conceded that all the necessary accoutrements for doing research will be available, that is state of the art laboratories which are well equipped with the most advanced instrumentation and supporting services. It is also of paramount importance that management articulate clearly to the research and development department the objectives of the corporation in their broadest sense, sometimes referred to as *corporate philosophy*. In addition, management must be clear in outlining specific areas of interest and markets to penetrate, to expand one's franchise or to avoid others. These decisions are taken with the input of all operating divisions. In Figure 1 is outlined a scheme whereby this can be achieved.

Three factors which influence the management's decision to enter a field of research are: (i) medical need; (ii) commercial potential; and (iii) scientific resources. Each of these is important in its own right and one really does not have priority over the others. For example, there may be a vital medical need for a severe disease which afflicts a very small number of people (*e.g.* less than 5000 people). Obviously, the commercial potential is limited and it would not be of interest for management to designate this as a target area (see Chapter 4.9). On the other hand, a large number of people may be afflicted with a disease for which there is no treatment. The commercial potential is evident but it may be the case that scientific knowledge in this area is limited, thus requiring notable expansion of the research effort to ensure one's chance of success. With full input from marketing, research and planning a decision is taken by management which will define the course of action and research commitment for the corporation for years to come. There may be alterations in the plan at intervals to evaluate the progress being made but again the courses of action to be taken will be determined by the medical need, changes in the marketplace and in the evolution of scientific knowledge in the project under consideration.

The most difficult parameter for management to assess is the chance of success of the scientific program. In this regard, it is not necessary (although certainly desirable) that a corporate president understand the details of the research program. However, it is important for management to know that the discovery process is at times slow, somewhat tedious, always requiring patience, tenacity, objectivity and intellectual integrity. Scientists, to be innovative, must work in a corporate environment in which the management not only recognizes these factors but makes every effort to let their importance be known to the scientists. In addition, the periodic evaluation of a scientific program should not be made with an emphasis on return on investment or how to minimize risk.

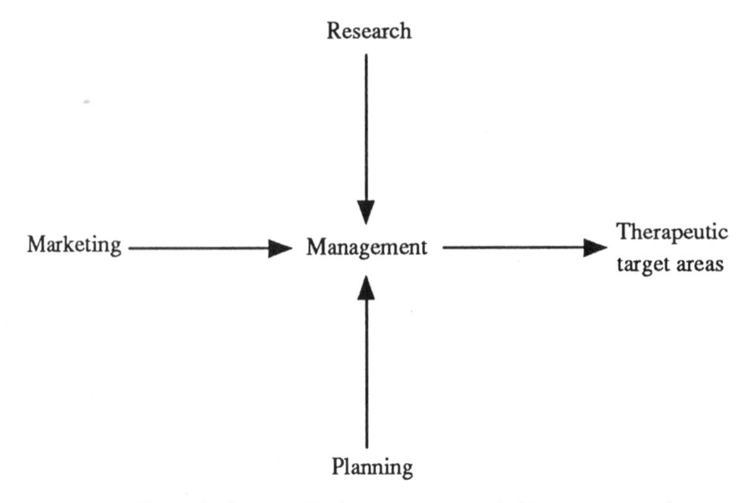

Figure 1 Inputs affecting management decision process

Research *is* risky and its cost is necessary for the company's survival. After all, the name of the game is not to avoid failure, but to give success a chance.

Thus, assuming that the company has agreed on reasonable target areas of research and has in place highly competent scientists with knowledgeable and experienced research leaders, it is of interest to explore the most efficient and effective organizational structure for research and development in a pharmaceutical company.

There are several structures which have been used with varying degrees of success. The principal divisions involved in the process are:

Exploratory research	*Development*
Chemistry	Drug metabolism
Biology	Toxicology
Physical chemistry	Pharmacy
Biochemistry	Analytical
Supporting technical services	Chemical development
Medical	*Drug regulatory affairs*
Clinical pharmacology	Preclinical
Phase 1	Clinical
Phase 2	Marketed drugs
Clinical research	Quality surveillance
Phase 3	
Clinical support services	
Statistics	
Clinical research associates	

Within some companies each of the above divisions functions as a separate department, each headed by a vice-president reporting to the president. Another model is the combination of exploratory research and development separated from medical, which may or may not incorporate drug regulatory affairs (DRA). Still another possibility is to separate exploratory research from a single other department made up of development, medical and drug regulatory affairs. The above paradigms suffer from a number of deficiencies due to fragmentation or bifurcation of the process.

The author's preference is for exploratory research, development, medical and drug regulatory affairs to be separate divisions within a single research and development department. Consequently, the whole process from chemical synthesis and lead discovery to new drug application (NDA) is under the overall direction of one research management. Under these circumstances the responsibility and accountability of all (from line scientists and subdivision directors to divisional vice-presidents) are clear and unequivocal. Decisions concerning lead structure selection, development, and clinical evaluation and registration are taken by one research management, thus minimizing interdepartmental bias.

3.2.2.1 Drug Discovery Organization

A previous paper[1] gives a detailed account of different approaches to the drug discovery process. As noted, this can be accomplished by means of serendipity or structured research.

Serendipity by definition indicates that in the course of solving a particular problem an unanticipated finding results which leads to the discovery of a new and *unexpected* phenomenon. In the case of medicinal research such a chance finding can result in the discovery of new fundamental knowledge which becomes critical in the finding of a new drug, or the unexpected result leads directly to the discovery of a new drug. This approach is inherently more difficult and certainly unpredictable in drug discovery; however, it is a method which can lead to major innovative breakthroughs in drug therapy. The risks of failure are great, but the benefits, if successful, are immense. This type of discovery cannot be planned or foreseen within an organization. It arises quite unexpectedly, especially in a research environment in which scientists are allowed to explore unusual ideas, to challenge dogma and to think beyond the ordinary experiment. Sometimes in those circumstances 'luck comes to the prepared mind' and a bizarre experimental finding is turned into a major discovery.

The second approach, structured research, can be divided into three parts: (i) *ad hoc* group; (ii) project team; and (iii) matrix.

The *ad hoc* group structure refers to an informal association of a chemist and a biologist working together on an idea or group of ideas directed towards the discovery of a drug in a particular therapeutic area.

The project team approach is to define a particular therapeutic goal and to assign a group of chemists and biologists (project team) to discover a drug in that area. When the task is completed, the project team is disbanded and the scientists are in turn assigned to other projects. Thus, the hallmark of this approach is *flexibility*.

The matrix method for drug discovery is the epitome of structured research. According to this approach, scientists from various disciplines, *e.g.* chemists, biologists, clinicians, toxicologists, pharmacists, as well as business managers, market research analysts and cost accountants, *etc.*, are assigned on a *permanent* basis to be members of a therapeutic area, the purpose of which is to discover, develop and bring to the market place a drug entity. Each therapeutic area is, in a sense, a minicompany within a company and each has its own director who is responsible for the program. Each therapeutic area group competes with other therapeutic groups for funds to maintain or increase the level of support for their programs. The specific group committee approves not only all funding within the program but, as a consequence, approves the chemical research program, including types and numbers of compounds to be synthesized and the cost to prepare the same. The matrix system suffers from several deficiencies: (i) it stifles innovation and creativity; (ii) it insulates the scientist of one therapeutic area from others; (iii) it assigns chemists to a specific area for an interminable period of time; (iv) it lacks flexibility; (v) its year to year progress is governed by return on investment; and (vi) it encourages 'me-tooism' in lead structure discovery.

The above drawbacks should give management a reason for serious pause when considering the matrix system for drug discovery. It leaves much to be desired.

On the other hand, flexible project teams based on therapeutic modalities are much more promising. Due to the explosive growth in cell biochemistry, receptor structure and receptor-mediated intracellular events, it seems more realistic to establish within the biology line function several therapeutic teams, each with a specific expertise and knowledge for a target area. For example, the cardiovascular–renal team looking for new antihypertensive agents would comprise cardiovascular pharmacologists, renal physiologists, biochemists well versed in ligand–receptor technology, enzyme inhibition and Ca^{2+} mediation in cell function. To this group would then be added several organic chemists. This combination of scientists would make up the project team for this mechanism-driven target area. It would be headed by a project leader (chemist or biologist) who would periodically report scientific progress to the research and development management. The chemists and biologists of the team would have full responsibility for the day to day program. However, the final decision for selection of a compound for preclinical and clinical development would be made by research management (director of research, vice presidents and functional line directors) with strong input from the therapeutic team scientists. At this point, some of the chemists may stay within that therapeutic team or be transferred to another one. In addition, the therapeutic team may emphasize another target area within the cardiovascular–renal area leading to a change in chemist assignments. This *modus operandi* ensures a dynamic flexibility in the system, thus offering new challenges to all the scientists.

The members of each project team are associated administratively with their line subdivisions (chemistry, biology) and their support service needs are supplied by the same. However, during the course of the program, they are scientifically assigned to the project team. In the case of the biologists, there is little if any organizational (administrative or scientific) change. However, over a period of years chemists can be members of several different project teams. As one project is completed, some chemists can be transferred to another. This flexible system is vital to maintaining creativity, productivity and a high level of morale for chemists.

Generally speaking, a well-balanced exploratory research division should have a chemist to biologist ratio of 1:3. However, this ratio in turn will vary with each therapeutic area. If the biological and biochemical knowledge-base is limited for a therapeutic area, then a more than usual number of biologically oriented personnel will be required. Since the scientists will have established a unique expertise, this level of biology support will probably be maintained throughout the life of the program. Should this expertise serve as the foundation for an innovative lead structure discovery, then it would be in the best interest of the company to strengthen its scientific know-how therein.

The occasion may also arise in which the ratio of chemist to biologist should be increased. For example, an involved or difficult synthesis problem may be encountered in the lead discovery process. Additional chemists could be thrown into the breach to expedite its solution. Such concentration of effort would certainly be more critical in a competitive situation *vis-à-vis* another company.

The development, medical and DRA divisions should be of sufficient size to support the drug discovery program. These three divisions *in toto* usually have approximately twice the number of scientists assigned to exploratory research. Their specific function will be discussed in detail in other chapters of this work (see Chapters 4.7 and 4.8). However, some commentary on their significance is worthy of note.

Toxicology, drug metabolism and pharmaceutics should be brought into the lead structure evaluation as soon as possible. The clinical evaluation of a potential drug is, after lead structure selection, the most critical area in the whole process. The poor design of clinical protocols and the lack of good control of clinical studies (wrong patient selection, violation of clinical protocol, compromised patient record forms, *etc.*) result in protracted delays and possible damaging, inconsistent and/or inconclusive clinical data, all of which can raise questions of efficacy and risk *versus* benefit of the drug. Such problems in turn seriously affect the drug's marketing and commercial potential.

3.2.3 LEAD STRUCTURE DISCOVERY

As we begin this section, a few important and fundamental assumptions must be made. Firstly, management with input from marketing, research and planning has clearly outlined target areas of research. Assuming that the necessary scientific personnel are in place, the drug innovation sought for the therapeutic areas must meet minimum criteria which will show some improvement over marketed products.

Another assumption is that a project team is established to pursue the lead discovery program in the therapeutic areas under study. The biology component of this team will consist of biologists and biochemists, *etc.*, who have recognized expertise in the area. A variety of *in vitro* and *in vivo* test systems will be available. These test systems will have been standardized *versus* well-established drugs. All secondary screens also will have been rigorously developed. It is also imperative that the biologists on the team devote some time to the development of new methodology. Such research can lead to unique drug evaluation which will permit the demonstration of marked advantages over existing products. This type of research can also enhance the lead selection process.

The chemistry members of the project team interface closely with the biologists in initiating the chemistry program. Biochemical principles play a vital part of the process. In fact, it is essential that the chemist fully understands the basic biochemical reactions (if known) in the therapeutic area. An understanding of the *in vitro* and *in vivo* tests in addition gives the chemist an appreciation for the significance of the test data. The chemist searches for a starting point in the program either from internal leads (pharmacologically or biochemically inspired) or from external leads reported in the scientific literature. The information scientist can be of great assistance to the team in presenting up-to-date relevant reports from the latest journals. Neither the chemist nor the biologist should discount any unusual or unexpected result in their research. Serendipity may be most useful in lead discovery. Furthermore, to optimize lead structure discovery, representative compounds from each of the project teams should be tested in all therapeutic areas.

3.2.3.1 What is a Lead?

During the course of a testing program, many compounds show some degree of activity in a primary screen. These are not all leads. A lead is more than just a compound active in the primary screen. A lead is a compound which meets specific criteria, the consequence of which is that the project team decides to implement a major chemical program to explore its activity with the objective of finding a clinical candidate.

The basic criteria for lead selection do not vary significantly from one therapeutic area to another. The primary screen must be well established for ascertaining a certain biological activity. Validation of the screen must be made *versus* standard known drugs and the statistical significance of the test must be well defined. All secondary screens must be equally validated. The screens must be validated statistically so that their accuracy and reproducibility are known. In some cases the primary screen is an *in vitro* test. It is then necessary to establish whether or not the compound is active *in vivo* and to what extent.

With this information in hand one can now ascertain the pharmacological potential of the compound. This should incorporate its strong points as well as its shortcomings.

3.2.3.2 The Lead Structure Selection Process

At this juncture it is the purpose of the project team to initiate a broad chemical program to minimize the shortcomings and to enhance the pharmacological profile of the lead compound. The lead selection process is a crucial step. The selection of the wrong lead compound can lead to months, and sometimes years, of fruitless effort. In fact, it is axiomatic that it is better to have *no* lead structure than to have the wrong one.

With the selection of the lead compound, the chemists and biologists embark on an extensive program to improve its potency, the specificity of biological effect with concomitant reduction in toxicity, oral absorption, duration of action, metabolic pattern and pharmacokinetic profile. This will involve extensive structure–activity relationship (SAR) studies. Needless to say, the lead structure series must be patentable.

The stage is now set for the interdigitation of chemist and biologist to bring forth the compound with properties which closely meet the criteria established by the project team for study in humans.

A wide variety of structures are usually evaluated in the primary screen. But usually only about 10% of these compounds will be of sufficient interest to go into the secondary screen. The selection process really begins to take form at this point. The structures under consideration may be of the same or different chemical classes. Additional SAR studies, with the help of computer graphics, will further assist in the lead structure selection. Depending on the shortcomings of the compound(s) evaluated in the secondary screen, the extent of the chemical program is variable. The major influence is the maturity of the therapeutic area and the criteria established for clinical candidate selection. For example, the search for a *pure* β_1-adrenergic receptor antagonist with α_1-antagonist activity would require much more work than the search for a drug with one of these activities. Both fields have been extensively explored and the type of selectivity sought becomes no mean task.

In a virgin field, the lead structure selection also can be formidable. The cimetidine story is a case in point. According to Ganellin:[2] 'In the first four years some 200 compounds were synthesized and tested, without providing a blocking drug. Toward the end of this time many doubts were expressed about whether it would really prove possible to block the action of histamine on gastric acid secretion and, indeed, there was considerable pressure within the Company to abandon this approach. The scientists involved in the project were, however, firmly convinced to continue, and during this period the test system was refined and chemical ideas began to crystallize. A most important aspect of research is to conduct the work in such a way as to learn from negative results . . . Some 700 compounds were synthesized on the road to the discovery of cimetidine.'

The types of the tests incorporated in the total screening program are of considerable importance. The *in vitro* tests are valuable in suggesting the mechanism of action of a class of compounds. However, a follow-up test (*in vivo*) in animals is imperative since it will establish potency, oral absorption, selectivity, duration of action and metabolic pattern of the compound. In the intact animal other pharmacological effects can be noted, and these may or may not be associated with the metabolism, pharmacokinetics and toxicity.

After lead structures are put through the battery of secondary and tertiary screens, it is of interest to determine the metabolic pattern of the primary lead compound in the two species of animals in which the subchronic toxicity studies will be carried out. This information will assist in establishing the toxicity dose levels. At the completion of the toxicological studies the lead structure which has the desired profile of action is ready for clinical trials.[3]

The time-frame from chemist discovery program to first clinical studies is shown in Table 1.[4]

The whole process can take from four to five years. The lead structure discovery phase by the chemists might be improved, depending on lead ideas and their realization; but this is too speculative.

Table 1 Time Frame for Preclinical Phase (months)

Discovery of substance by chemists	12–24	Toxicology (90-day studies)	12
Biological profile studies	12	Preparation of IND	3

Going from other animals to humans is a quantum jump. The activity may be evident but at a lower potency level. This is usually associated with more rapid metabolism in humans. The absorption, distribution and excretion may also be different. These may cause unanticipated adverse reactions. Thus, early clinical studies are still part of the lead structure selection process. Several examples will serve to support this.

3.2.3.3 Clinical Feedback in the Lead Structure Discovery Process

3.2.3.3.1 Pronethalol and propranolol

Black, Stephenson[5, 6] and colleagues at ICI had established that on the basis of the β-adrenergic agonist/antagonist activity of isoprenaline (DCI), this structure served as a starting point for the synthesis of a true antagonist. Pronethalol (**1**) was eventually synthesized and was shown to reduce the effects of sympathetic stimulation upon the heart. Clinical trials showed that the exercise tolerance of patients suffering from angina pectoris was improved and in addition it was found to elicit an antihypertensive effect. However, long-term rat toxicity studies revealed that pronethalol was carcinogenic at high dose levels and consequently further clinical studies were terminated.

Nevertheless, the clinical information confirmed that β-adrenergic antagonists could be used to treat angina and hypertension. Stephenson and coworkers embarked on a broad SAR program recognizing that the aromatic ring and the aminoethanol group —$CH(OH)CH_2NHR$ with R equal to isopropyl were essential for activity. After much experimentation it was found that insertion of an oxymethylene group between the aromatic ring and the aminoethanol group resulted in compounds with high potency which were devoid of intrinsic sympathomimetic activity. Propranolol (**2**) was introduced into clinical medicine in 1968 and today is the standard against which all other β-adrenergic blockers are measured.

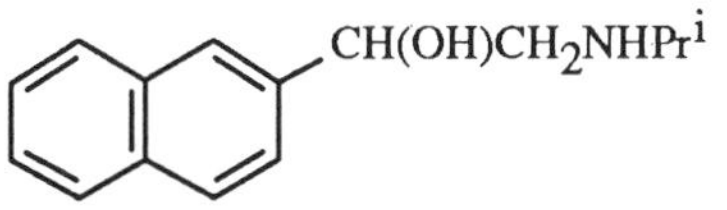

$CH(OH)CH_2NHPr^i$

(**1**) Pronethalol

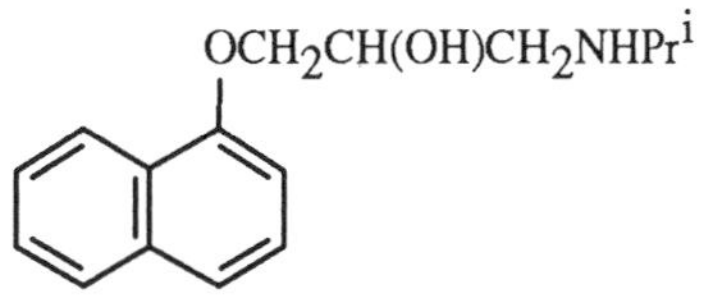

$OCH_2CH(OH)CH_2NHPr^i$

(**2**) Propranolol

3.2.3.3.2 Su-4029 and guanethidine

Su-4029 (**3**) was synthesized by Mull[7] and found by Maxwell to be antihypertensive in renal hypertensive rats and dogs. In clinical trials Page and Dustan[8] noted the marked reduction in blood pressure caused by Su-4029 but also observed that the substance caused fever. The antihypertensive effect in patients on Su-4029 was of sufficient interest to encourage Mull and Maxwell to expand the lead structure discovery program. The consequence was the discovery of guanethidine (**4**), a potent antihypertensive drug useful in the treatment of severe hypertension.

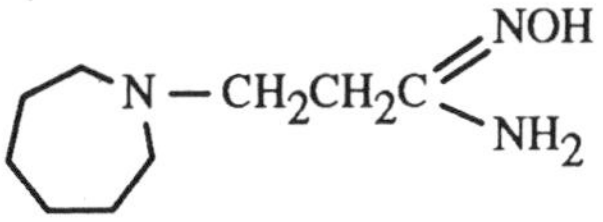

N—CH_2CH_2C(=NOH)NH_2

(**3**) Su-4029

N—CH_2CH_2NHC(=NH)NH_2

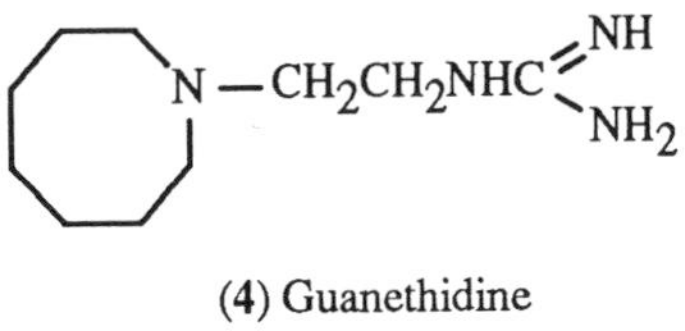

(**4**) Guanethidine

3.2.3.3.3 Burimamide, metiamide and cimetidine

The story of the development of cimetidine (**7**) has been well documented by Ganellin.[2] Briefly, after the synthesis of several hundred compounds Black *et al.* were finally able to report on a burimamide (**5**) which was a pure H_2-antagonist. It had all the criteria set by the project team for a clinical candidate except that it was not sufficiently active when given orally. This led Ganellin and coworkers to make further structural changes which resulted in metiamide (**6**). Metiamide was indeed a potent H_2-inhibitor without agonist activity and it was orally active. In clinical trials metiamide was very effective in healing duodenal ulcers. However, a small number of the 700 patients treated suffered from granulocytopenia which, although reversible, put an end to further clinical work. Ganellin and fellow chemists went back to the laboratory feeling quite confident that a clinically useful drug could be obtained. It was postulated that the granulocytopenia associated with metiamide could be caused by the thiourea group. Without going into detail concerning the replacement of guanidine by thiourea and the consequences of diminished potency due to equilibria between the guanidinium cation and its three conjugate bases, the group finally settled on incorporation of the cyanoguanidine group, which gave rise to cimetidine (7), one of the most successful prescription drugs to be used in medicine.

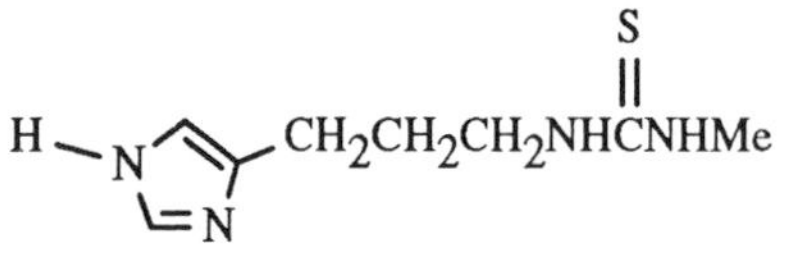

(**5**) Burimamide

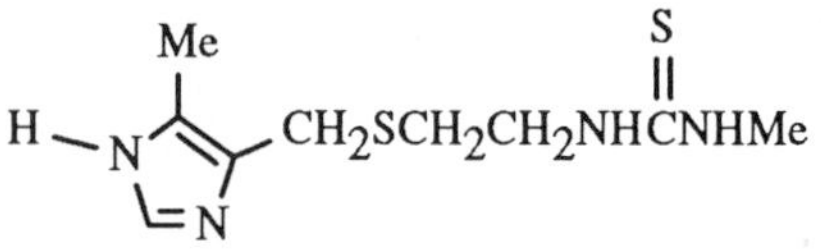

(**6**) Metiamide

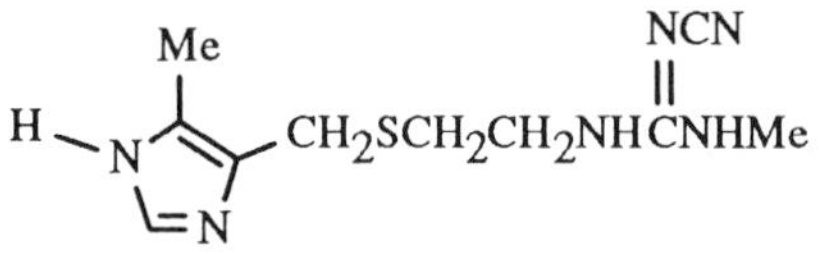

(**7**) Cimetidine

3.2.3.3.4 The antiinflammatory oxicams

The Pfizer project team, which was initially made up of chemists Kadin and Lombardino and pharmacologist Wiseman,[9] began a program of looking for antiinflammatory lead structures different from nonsteroidal isopropionic acids. The initial compounds, isoquinolinecarboxanilides showed antiinflammatory potency similar to phenylbutazone. The working premise was that the enhanced acid properties of these cyclic β-diketones were responsible in part for their biological activity. After much structure–activity studies tesicam (**8**) was selected for clinical trials. Metabolism studies in animals had shown that a chlorine substituted in the 4-position of the phenyl ring extended the half-life and the duration of action. In clinical trials in patients with rheumatoid arthritis and osteoarthritis, tesicam (250 mg b.i.d; bio in die; twice daily) improved clinical symptoms of the arthritis. Thus, the long action noted in animals was observed in man, but potency was not believed to be fully sufficient to satisfy the criteria initially aimed for.

This led Lombardino[10] to synthesize the unknown 2*H*-1,2-benzothiazine-3(4*H*)-one 1,1-dioxides as bioisosteres of tesicam. This series did not prove fruitful. Consequently, the isomeric carboxamides of the 4-hydroxy-2*H*-1,2-benzothiazine 1,1-dioxides were prepared. Extensive SAR studies finally gave rise to sudoxicam (**9**) as a clinical candidate. This compound was more potent than tesicam and had a longer duration of action. The metabolic studies in man revealed that the metabolic plasma concentrations were markedly above those of the parent drug at daily doses of 200 mg. These clinical data encouraged further synthetic research, which eventually gave rise to piroxicam (**10**), which is active at 2 mg per day with a half-life of 45 h and this half-life does not vary with plasma concentration, as was the case with sudoxicam. Along with other parameters, the biotransformation studies in man were vital to the preclinical scientists in the planning of the synthetic program and in the lead structure selection process.

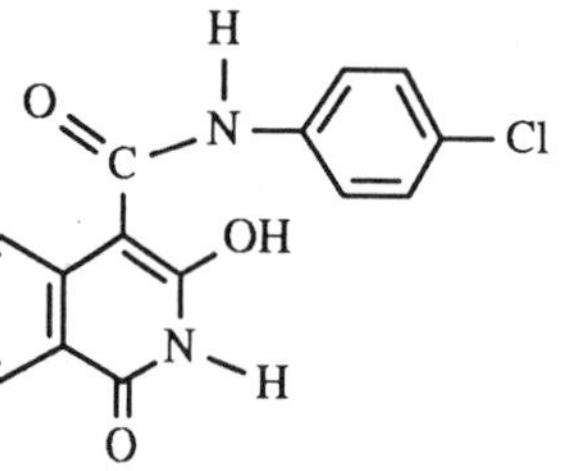

(**8**) Tesicam

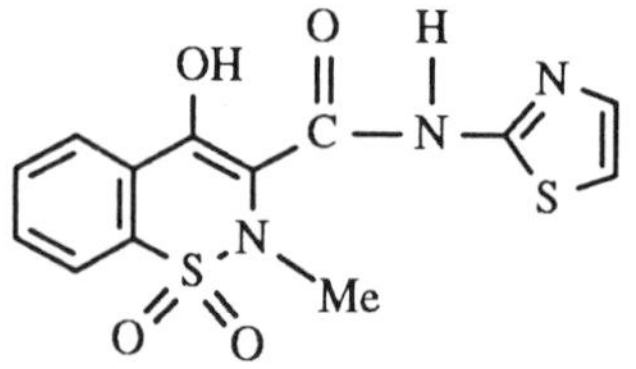

(**9**) Sudoxicam

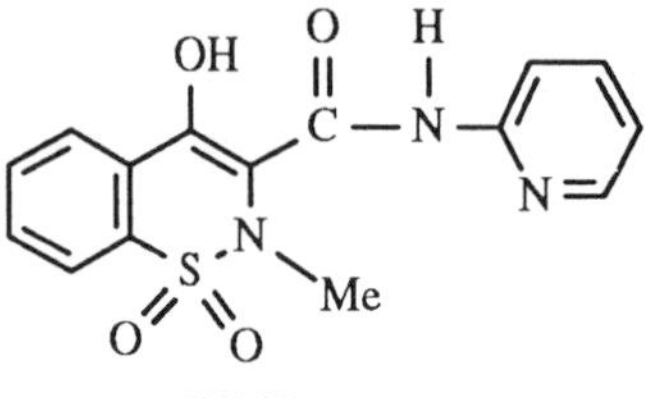

(**10**) Piroxicam

The four examples cited above have several common features, all of which may seem obvious, but are often neglected in part or wholly. For this reason they are worth emphasizing: (i) the project teams (chemists and biologists) worked as a tightly knit unit with a single-minded purpose; (ii) SAR studies had established certain criteria for optimizing leads; (iii) SAR studies gave rise to back-up compounds, a necessary-part of lead structure discovery; and (iv) both negative and positive information from toxicology, pharmaceutics, drug metabolism and clinical pharmacology were used to select the optimum compound for lead development.

3.2.4 LEAD STRUCTURE DEVELOPMENT

The selection of a substance for development for clinical trials sets in motion the activities of several disciplines.

Drug metabolism is an essential factor. The metabolite pattern must be determined in two or three animal species: the animal(s) in which the principal biological activity has been determined as well as the rodents and dogs (possibly monkeys) in which the subchronic and chronic toxicity studies will be carried out. The half-life of the drug as well as a principal metabolite (if present) must be determined. The metabolites must be chemically identified and synthesized if necessary.

A radioactively labeled lead structure must be synthesized and studied in animals to determine absorption, distribution and excretion. The labeled compound also facilitates studies on the mode and rate of metabolism.

These studies are important in establishing oral absorption of the drug, as well as its distribution, its half-life and its excretion in addition to the same information for its metabolites.

Chemical Development is a vital part of the R and D process. The most efficient synthesis of a lead structure at the early stages of the discovery process is not, as a rule, of great importance to the medicinal chemist. He/she is primarily interested in sending 2–3 g of pure substance to the biologist for testing. However, should the substance exhibit activity of interest and larger quantities of material be desired, then the synthetic procedure will definitely be modified and improved if necessary. Sometimes new methodology is developed which expands the lead structure discovery program to compounds which were not originally in the plan.

When a substance has been selected for development, large quantities (kg) of material are required for toxicological studies and eventually also for clinical trials. Chemical development chemists and engineers working in the pilot plant refine the synthesis further by working with solvents which do not represent environmental hazards and by improving the synthesis and yields at each step so that the final drug product is economically viable. Standards of purity must also be established since these will be part of the manufacturing master file and to which each batch of drug product must conform (see Chapter 3.6).

Pharmaceutical and *analytical studies* should be well underway before subchronic toxicity is commenced. The lead structure and back-up members should be stable to air, light and hydrolysis. Stability in acid solution is particularly important since the stomach solutions are strongly acidic. In order to facilitate absorption it is preferable to have a water-soluble lead structure wherever possible. The selection of a stable salt is also mandatory. Stability studies are carried out at room temperature and at elevated temperatures (accelerated) to establish long-term stability. The appropriate pharmaceutical composition must also be developed for each drug.

Toxicological studies represent the last and decisive step before the substance can be considered for clinical trials (see Chapter 4.5). This subject has been comprehensively covered in a landmark paper by Traina.[11] For the purposes of this discussion only a few pertinent points will be referred to. After the preliminary acute range-finding studies and therapeutic ratio have been established on a substance scheduled for further development a decision to go into clinical trials is made. This will necessitate the initiation of subchronic toxicity studies which involve 90-day studies in at least two species of laboratory animals. The rat is often selected as the rodent model, while the dog or monkey serves as the nonrodent species. Most toxicologists administer daily doses that are multiples of the anticipated human clinical dose or multiples of the effective pharmacological doses in animal models. Reproductive studies of segments 1, 2 and 3 are also carried out. The subchronic studies usually can eliminate about 50% of lead structures from further development. Pending the favorable outcome of these matters, the developmental compound is ready for submission of an investigational new drug (IND) application. The DRA division scientists work closely with the project team and the development group to prepare the IND.

3.2.5 CLINICAL RESEARCH PROCESS

Figure 2 outlines the various stages of taking a clinical candidate from IND submission to the new drug application (NDA) approval. In Phase 1 the clinical candidate is tested for safety and the determination of metabolic disposition. The Phase 2 studies are usually divided into two parts: (i) early efficacy studies to establish whether or not the substance shows the efficacy in humans as noted in animals; dose–response data are also accumulated; (ii) additional efficacy studies are carried out under controlled conditions, usually *versus* placebo and *versus* a standard drug. These studies should suggest whether or not a potentially useful drug is on hand.

It should be noted that during the Phase 2 trials, long-term toxicity testing in animals, including a one-year chronic study in nonrodents and 18-month or two-year carcinogenicity studies in rodents, are initiated to support the expanded clinical program.

The Phase 3 clinical trials are designed to expand both the safety and efficacy profile of the compound. Several controlled trials are initiated, the length of which is determined by the nature of the indication studied. Usually several hundred patients are incorporated in these trials. For some therapeutic areas specific studies must be carried out. For example, in the development of an antihypertensive agent the Food and Drug Administration (FDA) mandates that at least 100 patients on a drug must be followed for one year to evaluate safety under treatment conditions. Considering the time of entry of patients into the study, the Phase 3 time-span for an antihypertensive drug could be at least 36 months.

A meeting with FDA at the end of each phase is desirable and sometimes necessary. Certainly at the end of Phase 3 a meeting with FDA could be most instructive as feedback for preparation of the NDA.

A review of Figure 2 suggests that the whole process from IND submission to NDA approval could extend from eight to ten years. Even with NDA approval, postmarketing studies are needed to expand the competitive position of the drug and to develop other indications.

These considerations emphasize the huge effort by multiprofessional scientific talent to take a lead structure discovery in research through many complicated steps in the research and development process to a viable drug product. The whole endeavor along with the many lead compounds that drop out along the way in the total program can take as much as 12 years at a cost of \$80–100 000 000 per drug.

3.2.6 PRODUCT DISCOVERY ASSESSMENT

Several years ago the 'product discovery' statistics from four companies (Ciba-Geigy, Pfizer, Sandoz and Merrill Dow)[12] were compiled to establish a survival rate of compounds taken through the research and development process (see Table 2). These are the results of a 10-year experience.

It is worthy of note to analyze these statistics, and each company may want to check their data against these. On the basis of the dearth of new products, it would appear that these figures may be representative of the industry.

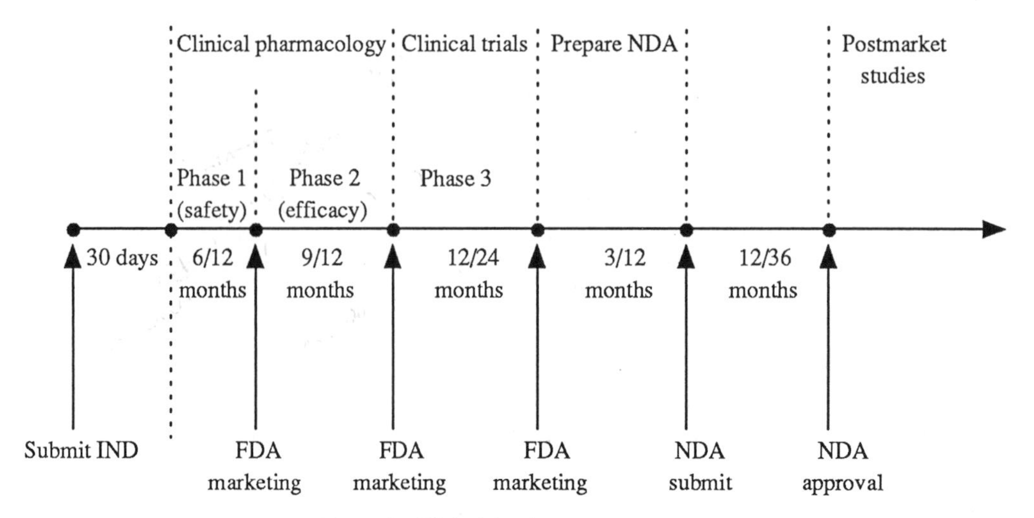

Figure 2 Clinical development process

Table 2 Industry 'Product Discovery' Statistics

Number of compounds	*Probability of survival*	*Research hurdle*
20	50%	Subchronic toxicity
10	40–50%	Phase 1/phase 2
4–5	40–50%	Phase 3
2	50%	NDA/FDA
<1	Reach market as variable product	
	Overall chance of survival is 5%	

This table tells us that of 20 compounds selected for subchronic toxicity, approximately 50% do not survive due to undesirable toxicity. Of the surviving 10 new chemical entities (not derivatives of known drugs) entering Phase 1/Phase 2, 40–50% survive this hurdle. Half of the clinical candidates entering Phase 3 make it to the NDA stage. At this point FDA approval should be assured, but for a variety of reasons this is not the reality. Thus the statistics suggest that of the 20 new chemical entities entering subchronic toxicity studies approximately *one* substance reaches the market as a viable product (the overall chance of survival is a dismal 5%).

The survival rate has the best chance for improvement at the subchronic toxicity level. Greater or more extensive input from cell biochemistry and drug metabolism may indicate potential toxicity. This information will assist in selecting the compound with the best therapeutic index.

In a previous section several examples were given of how clinical research assists in lead structure discovery. However, there are many compounds which have not survived clinical trials in spite of the fact that they appeared to have real therapeutic value.

In 1970, Su-13, 437 (**11**), a hypolipidemic discovered by Ciba, was in world-wide Phase 3 clinical trials.[13] The efficacy results were excellent and a promising new drug appeared to be on the horizon. However, long-term toxicity studies in rats revealed that at high doses it caused carcinogenicity of the kidney with metastasis to the lungs. All clinical studies were terminated. One of the metabolites, albeit found in small amounts, was compound (**12**), which could be considered the causative agent.

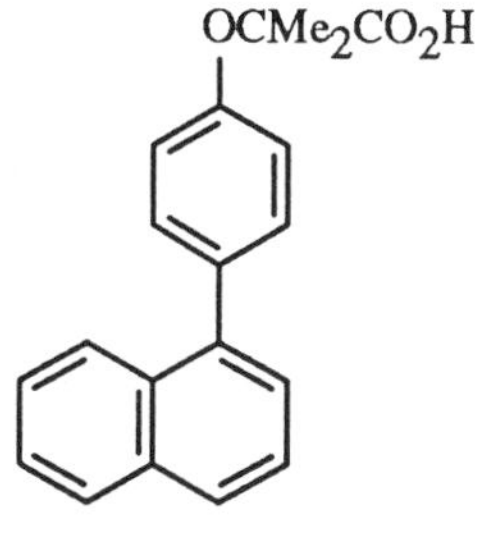

$OCMe_2CO_2H$ $OCMe_2CO_2H$

(**11**) Su-13, 437 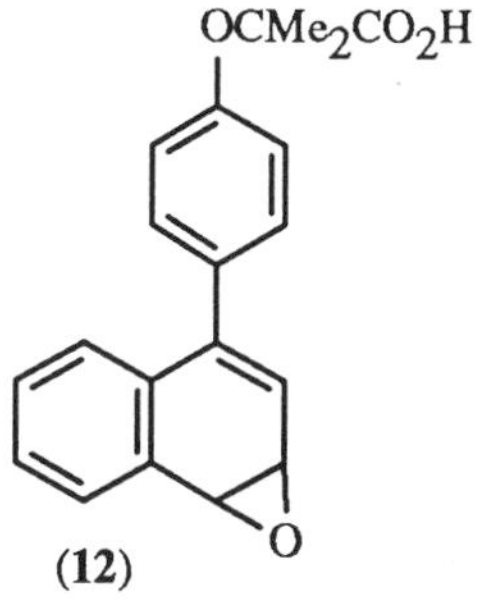 (**12**)

The Wander Research Institute[14] in 1969–1970 introduced into clinical trials clozapine (**13**), a neuroleptic with unique properties. The drug proved to be antipsychotic and practically free of extrapyramidal side effects. It was marketed in several European countries and its acceptance was excellent. However, in 1975 soon after its introduction in Finland a high incidence of agranulocytosis was reported. This clinically adverse reaction has limited the use of this drug. Extensive research has been underway in a number of laboratories to find a substance of this class which is devoid of the side effect.

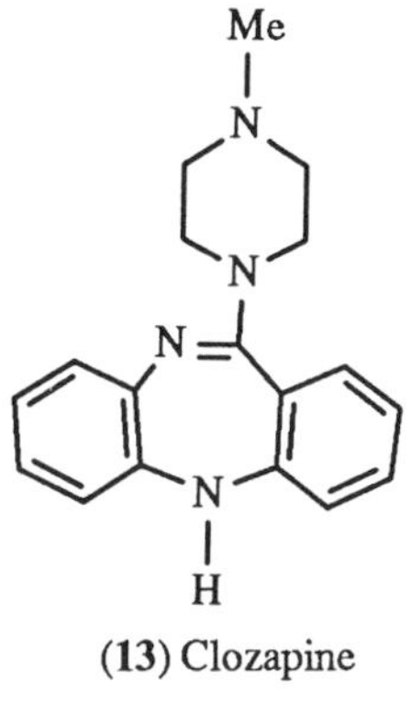

(**13**) Clozapine

Several years ago Schering Laboratories[15] discovered and developed for clinical trials a potent broad-spectrum penem antibiotic (**14**). Although the drug was most effective, it had to be withdrawn from Phase 3 trials because in its metabolic breakdown ethanethiol was formed. Obviously its offensive odor was manifest in the breath and urine of patients.

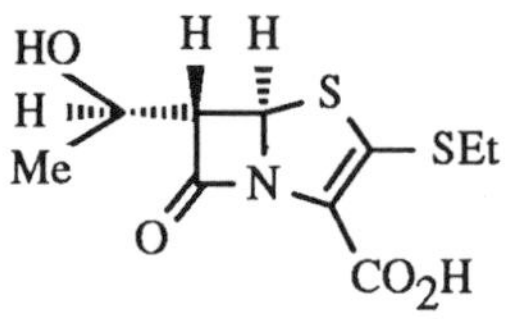

(**14**) Sch-29, 482

Zimelidine (**15**) was found by Astra[16] to be an excellent antidepressant by acting as a serotonin uptake inhibitor. Its selective action made it a useful drug for certain kinds of endogenous depression. However, its clinical trials in the US were terminated because of liver toxicity.

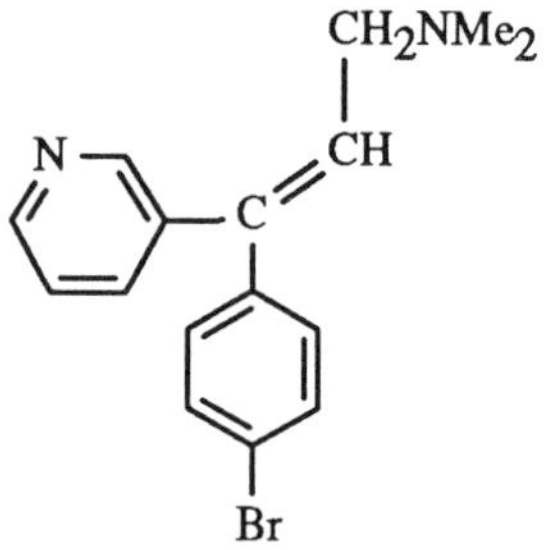

(**15**) Zimelidine

On the other hand, some substances do survive some apparent preclinical and clinical problems and go on to become useful drugs.

Amiloride (**16**) is a case in point. Cragoe[17] and coworkers at Merck, Sharp and Dohme discovered this potassium-sparing diuretic. After several years of successful clinical trials, the NDA was submitted to FDA but it was not approved for use in the US because of what appeared to be a narrow therapeutic range between potassium sparing and tendency to cause hyperkalemia. However, amiloride was introduced in several European and Asian countries. Its excellent tolerance and usefulness, especially when used in combination with hydrochlorothiazide, led finally to its approval by FDA in 1981.

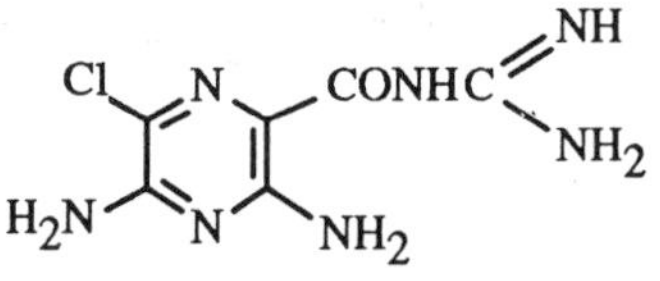

(**16**) Amiloride

Omeprazole (**17**), an antiulcer compound which acts by inhibiting the H^+-ATPase system in the parietal cell has been under development for several years by Astra and Merck, Sharp and Dohme.[18] In 1985 all clinical trials were stopped for what appeared to be carcinogenic effects in long-term toxicity studies in rats. However, after careful study of the data by a group of experts the carcinogenic effects were not found to be statistically significant. Consequently, the health authorities have permitted a recommencement of world-wide clinical trials.

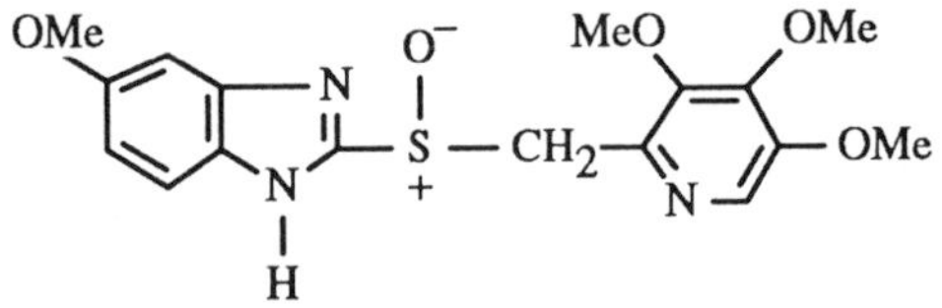

(**17**) Omeprazole

Considerable attention has been devoted here to the importance of clinical investigation in drug discovery. A highly competent medical department is a *conditio sine qua non* in the process of bringing a new drug to the medical profession. The clinical pharmacologists working in the Phase 1/2 area must maintain close contact with the exploratory research project teams. The input from pharmacology, biochemistry, drug metabolism, pharmacokinetics and toxicology will be of immense value in preparing clinical protocols for early safety and efficacy studies.

The Phase 3 clinical studies must be well thought out and planned. The physicians in charge of Phase 3 studies must be well organized so that the clinical plan is executed efficiently. All patient record forms must be properly processed and checked for pertinent information. The physician working with clinical research associates must be sure that the clinical protocol is not compromised. Any medical incompetence at the Phase 3 level must be rooted out, otherwise years of work can be lost. By and large, the health authorities will approve or disapprove an NDA on the quality of the efficacy data obtained in Phase 3 clinical trials. Its importance cannot be overemphasized.

3.2.7 DRUG REGISTRATION

The objective of the lead structure discovery and development program is to assemble all the data accumulated in the process and to submit the NDA to FDA (for the US) and to the health authorities in countries outside the US for marketing approval. The various regulations regarding drug registration in the major markets have the effect of producing internal company rules, regulations and standards concerning how a compound will be selected, developed, evaluated in the clinic and subsequently registered and sold.

A multinational company usually will be carrying out clinical trials on a world-wide scale. Thus, the registration and introduction of the drug is not a simultaneous event all over the globe but is staggered. Again the registration requirements in each country are the governing factors. As a rule, for reasons which are too involved and numerous to discuss in this chapter, the NDA approval time in the US is the most protracted. Whereas the drug in question may be in clinical use in many other countries, the US may be one of the last countries to make it available to its medical profession. Table 3 lists a group of cardiovascular–renal drugs which were approved by FDA between 1981 and 1983. These drugs were already available for several years in other countries. By and large, an innovative drug which has been properly studied and for which substantially good preclinical and clinical data have been accumulated appears to be approvable within 18–24 months of submission of the NDA.

3.2.8 TIMING OF DRUG DISCOVERY

The introduction of a drug in the marketplace, to be successful, must be timely. If it is unique, innovative and represents a breakthrough in therapy, then the need and demand will be great and its success is assured. Usually the prototype drug sets in motion a great flurry of activity in other laboratories to find a derivative with improved activity and less toxicity or adverse effects. This type of molecular alteration can be most productive and in many cases has led to improved therapy.

Table 3 Selected NME Approvals from Cardiorenal (HFN-110)

Name of drug	*Firm*	*Approval date*	*Approximate time* (months)
Oxprenolol (1-C)	Ciba-Geigy	December 29, 1983	72
Indapamide (1-C)	USV	July 6, 1983	34
Ranitidine (1-C)	Glaxo	June 9, 1983	18
Bumetanide (1-C)	Roche	February 28, 1983	27
Diltiazem (1-C)	Marion	November 5, 1982	21
Pindolol (1-C)	Sandoz	September 3, 1982	41
Nifedipine (1-B)	Pfizer	December 31, 1981	21
Sucralfate (1-B)	Marion	October 31, 1981	28
Amiloride (1-C)	Merck	October 5, 1981	32
Verapamil (1-B)	Knoll	August 21, 1981	16
Atenolol (1-C)	Stuart-ICI	August 19, 1981	32
Captopril (1-B)	Squibb	April 6, 1981	19

However, timeliness is vitally important since many laboratories will be engaged in this process. Under these circumstances the lead structure discovery and development process must be highly focused, concentrated and driven with single-minded purpose to get the clinical candidate to market as soon as possible. In other words, 'me-tooism' will be worthwhile provided a company is on the market with the number *two* entity of a new drug group. Several examples serve to illustrate the point.

3.2.8.1 Antidepressants

Prior to 1958, endogenous depression was a serious mental disease, the treatment of which was primarily by electroshock. The success rate was low and the incidence of suicide was indeed high. It was at about this time that Schindler[19] of the Geigy Company synthesized the tricyclic imipramine (**18**), which was sent to Kuhn[20] at Munsterlingen, Switzerland, to evaluate its psychotherapeutic effects. Kuhn discovered imipramine's antidepressant action and the results were confirmed in many other laboratories. Over the next few years, several other laboratories, recognizing the breakthrough nature of this clinical finding, embarked on programs to discover improved antidepressants. Wendler[21] and coworkers at Merck, Sharp and Dohme decided to incorporate isosteric substitutions in the molecule and thereby prepared amitryptalline (**19**). Now, with quite a few depressed patients, there is an underlying anxiety component. Imipramine has little effect on controlling the anxiety; on the other hand, amitryptalline not only is an effective antidepressant, but it also controls the anxiety. In other words, it elicits a bipolar effect.

Maprotoline (**20**), discovered by Schmidt and Wilhelm[22] of Ciba, also shows a bipolar effect, but in clinical trials in the US it was noted that it caused a higher than normal incidence of seizures at higher doses.

Although imipramine has been considered the prototype discovery in this area, amitryptalline through the years continues to be the drug of choice for the treatment of endogenous depression.

$CH_2CH_2CH_2NMe_2$ $CHCH_2CH_2NMe_2$ $CH_2CH_2CH_2NHMe$

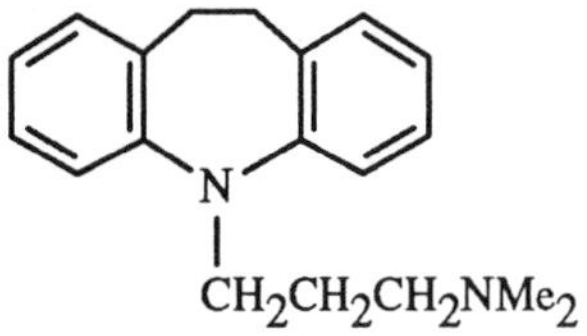

(**18**) Imipramine 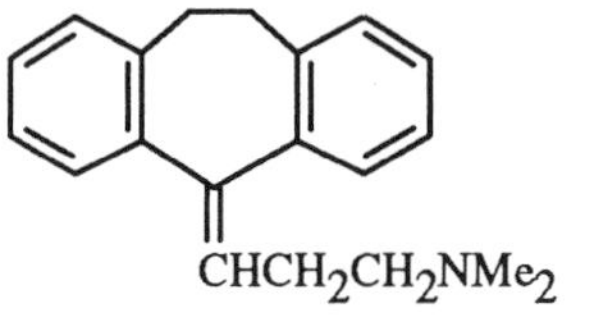 (**19**) Amitryptalline 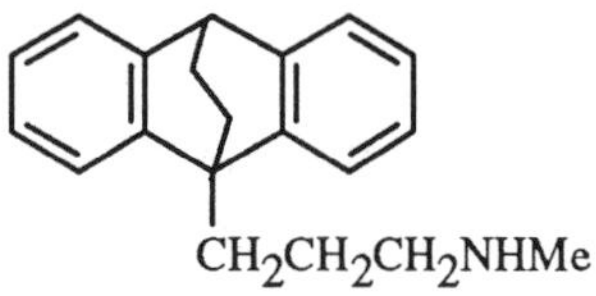 (**20**) Maprotoline

3.2.8.2 Diuretics

Novello and Sprague[23] in 1957 at Merck, Sharp and Dohme synthesized chlorothiazide (**21**), the first nonmercurial orally active diuretic/saluretic showing significantly low carbonic anhydrous inhibition (as distinguished from acetazolamide). Chlorothiazide was clinically effective at 250 mg to 500 mg once or twice per day. Its significance in the treatment of edema and hypertension was immediately obvious. Shortly thereafter (1958), deStevens, Werner[24] and colleagues from Ciba prepared hydrochlorothiazide (**22**), which was 10 times more active than chlorothiazide. Hydrochlorothiazide became and still is the diuretic of choice, both alone and in combination in the treatment of hypertension. Many other hydrothiazides were prepared and introduced but none showed any advantage over hydrochlorothiazide.

Further modifications of the disulfonamide compounds by Hoechst[25] gave rise to furosemide (**23**), the first of the high-ceiling diuretics which have been found to be very useful for the treatment of severe edema and congestive heart failure.

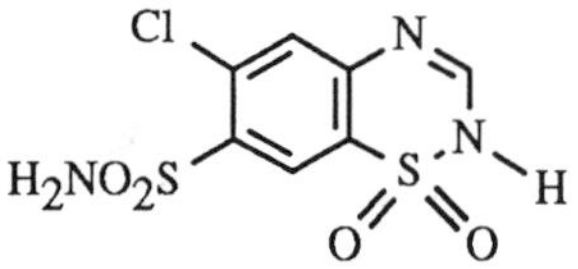

(**21**) Chlorothiazide 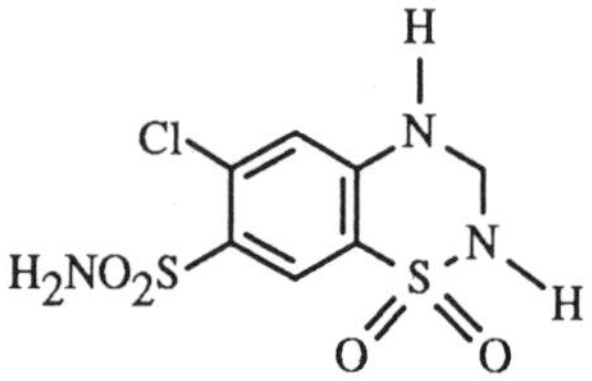 (**22**) Hydrochlorothiazide

(**23**) Furosemide

3.2.8.3 β-Adrenergic Antagonists

Propranolol has already been discussed in a previous section of this chapter. This drug was found to antagonize not only the β_1-receptors of the heart but also the β_2-receptors of the lung. The latter effect is undesirable in hypertensive patients with bronchial conditions. Therefore, the need for a β_1-blocker was clearly evident.

Haessle[26] and Ciba-Geigy jointly developed metoprolol (**24**), a relatively selective β_1-blocker. This was the second β-blocker to be introduced in the US market. It was effective at 100 mg b.i.d. and it became an immediate success. Within a year ICI[27] introduced atenolol (**25**) whose cardioselectivity was similar to that of metoprolol. However, atenolol had a longer half-life and thus could be administered once daily, facilitating compliance. In addition, it was also found to be less hydrophilic and did not cross the blood–brain barrier diminishing the possible central side effects. Although atenolol was the third β-adrenergic antagonist to be introduced in the US, it has surpassed metoprolol in the clinical treatment of hypertension.

(**24**) Metoprolol

(**25**) Atenolol

3.2.8.4 Histamine H_2-inhibitors

After the tremendous medical and commercial success of cimetidine was realized, many laboratories became involved in searching for improved H_2-inhibitors. The immediate drawbacks of cimetidine appeared to be dosage (400 mg) administration of three to four times per day and there was some evidence that in some patients it inhibited the cytochrome *P*-450 system in the liver, the consequence being that oxidative metabolism would be affected. Within three years of the cimetidine introduction, Glaxo Laboratories[28] introduced ranitidine (**26**), a derivative which appears to overcome those shortcomings. Ranitidine is prescribed at 300 mg once daily and does not affect the cytochrome *P*-450 system. It has gained wide acceptance becoming the most-prescribed drug in medicine today. Yamanouchi[29] and Merck, Sharp and Dohme have jointly developed famotodine (**27**) for the US market. Its advantages over ranitidine have not been clearly defined, thus its impact is too early to be determined.

(**26**) Ranitidine

(**27**) Famotodine

3.2.8.5 Angiotensin-converting Enzyme (ACE) Inhibitors

One of the most important discoveries in cardiovascular medicine in the past 20 years has been the ACE inhibitors.

The classic work of Laragh[30] and others on the renin–angiotensin system and its effect on the control of blood pressure offered excellent biochemical insight to this program. The dogma is that as angiotensin II is formed in greater than metabolically accepted amounts, then vasoconstriction and increased sodium ion reabsorption occurs, resulting in hypertension. Ondetti[31] and his group at Squibb reasoned that an ACE inhibitor would correct this imbalance and thus lower blood pressure. The consequence of this research is captopril (**28**), the first ACE inhibitor. It cannot be overemphasized that this is a landmark discovery in antihypertensive research. Captopril has been widely accepted by the medical profession. However, a small number of patients have been reported to develop proteinurea, which has been considered to be due to the free —SH group in captopril.

Based on Ondetti's model of the ACE receptor site, Patchett[32] and collaborators at Merck, Sharp and Dohme prepared enalapril. Enalapril (**29**) is deesterified *in situ* to the parent acid enalaprilat. The biochemical rationale in this case was twofold: (i) eliminate the —SH group in captopril to reduce the side effects; and (ii) introduce in its place a phenylalanine-like moiety, which would fit better into the active site.

Me
$HSCH_2CHCO$—N
CO_2H

(**28**) Captopril

EtO_2C Me CO_2H
$PhCH_2CH_2CHNHCHCO$ — N

(**29**) Enalapril

In Figure 3 is outlined the fit of captopril onto its active site with binding of —SH to the zinc ion.

Figure 4 shows the proposed molecular fit of enalapril on the surface of angiotensin-converting enzyme. According to Patchett this type of bonding has resulted in a more potent antihypertensive drug with fewer side effects.

Captopril has captured a significant share of the antihypertensive market but enalapril is also expanding its franchise significantly and promises to be a widely used drug.

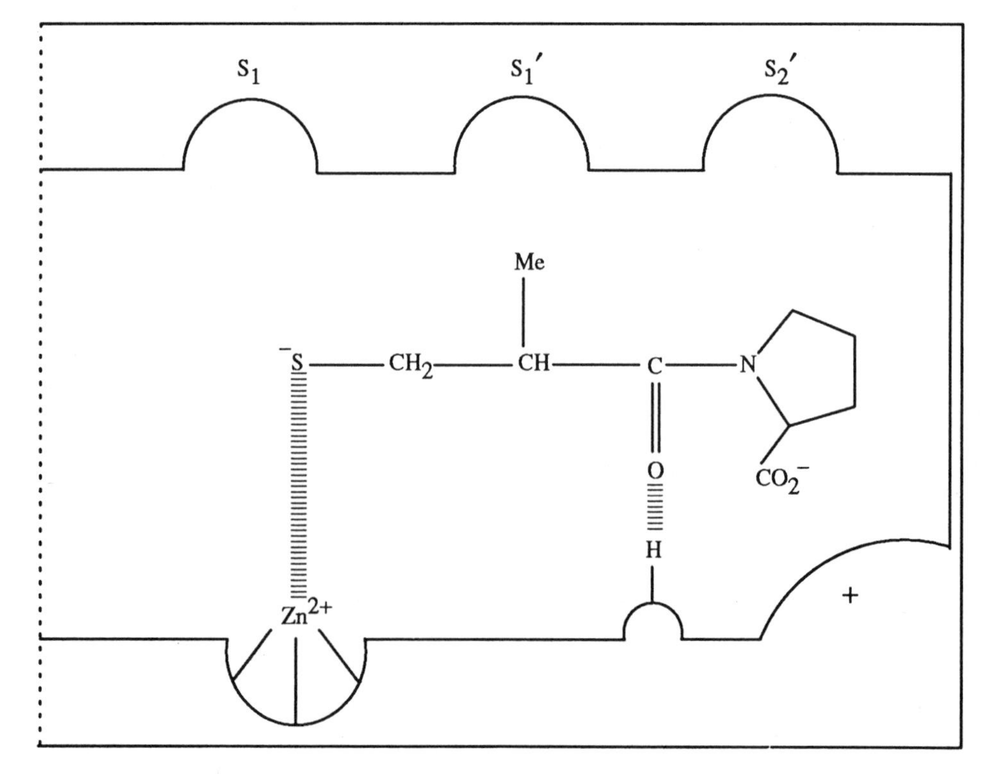

Figure 3 The binding interactions of captopril with the active site of ACE as hypothesized by Ondetti

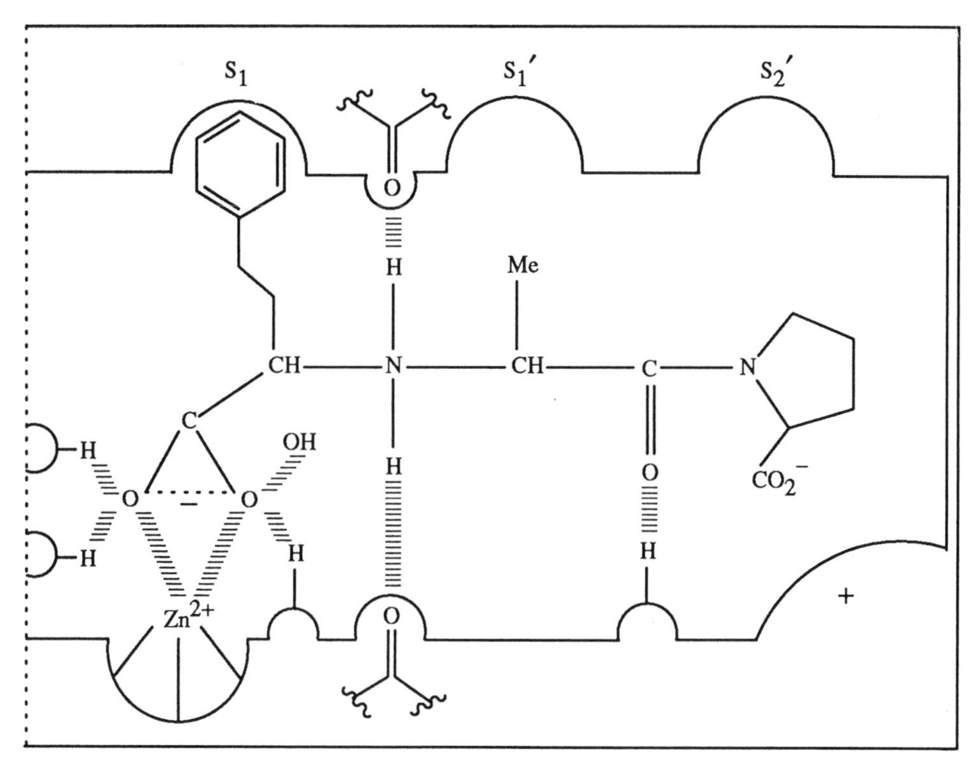

Figure 4 Proposed binding interactions between enalaprilat (MK-422) and angiotensin-converting enzyme

Presently, it is estimated that there are 35 ACE inhibitors in clinical trials in the US. Many of these are variations of captopril and enalapril. It is doubtful that they will make much of an impact unless they show major advantages over the established products.

Each of the above examples illustrates the importance of timing in the discovery, development and introduction of a product in a new therapeutic area. Delays of weeks or a few months can spell the difference between success and failure.

3.2.9 LEAD STRUCTURE DISCOVERY OF THE FUTURE

The foregoing sections have emphasized certain factors in the past which have been most influential in the lead structure discovery process. A variety of examples have been given to make the point. However, within the past 20 years there has been a marked emphasis on the rational or mechanistic approach to the design of drugs. The single most important influence affecting this change has been the virtual explosion in the science of molecular biology. DNA and its influence on life processes through protein synthesis and function are at the core of all cell action. As a consequence, the three-dimensional structure of receptors is being elucidated. Detailed information on enzyme structure and function has also been clarified and elaborated. This knowledge has given the medicinal scientist a clearer insight to physiological and pathological processes. The medicinal chemist is now approaching the design of drugs according to the biochemical process to be enhanced or inhibited. Of necessity, receptor site structures will have a profound effect on how the medicinal chemist plans the design of a potential drug. Computer terminals in every laboratory are now commonplace since computer graphics have become a powerful tool. Today's practising medicinal chemist must be well versed in all these disciplines in order to be in the position to discover the important new drugs of tomorrow.

3.2.10 REFERENCES

1. G. deStevens, *Prog. Drug. Res.*, 1986, **30**, 189.
2. C. R. Ganellin, 'Chronicles in Drug Discovery', ed. J. Bindra and D. Lednicer, Wiley, New York, 1982, vol. 1, p. 1.
3. N. Finch, *Med. Res. Rev.*, 1981, **1**, 337.

4. G. deStevens, *Prog. Drug Res.*, 1976, **20**, 181.
5. J. W. Black and J. S. Stephenson, *Lancet*, 1962, **2**, 311.
6. J. W. Black, A. F. Crowther, R. G. Sharks, L. H. Smith and A. C. Dornhorst, *Lancet*, 1964, **1**, 1080.
7. R. P. Mull, R. A. Maxwell and A. J. Plummer, *Nature (London)*, 1957, **180**, 1200.
8. I. M. Page and H. P. Dustan, *JAMA, J. Am. Med. Assoc.*, 1959, **170**, 1265.
9. J. G. Lombardino in 'Anti-inflammatory Agents, Chemistry and Pharmacology', ed. R. A. Scherrer and M. W. Whitehouse, Academic Press, New York, 1974, p. 129.
10. J. G. Lombardino and E. H. Wiseman, *Med. Res. Rev.*, 1982, **2**, 127.
11. V. M. Traina, *Med. Res. Rev.*, 1983, **3**, 43.
12. This information was obtained from Mr. George Ohye, R. W. Johnson Pharmaceuticals Research Institute, and Charles Pesterfield, formerly of Pharmaceuticals Division, Ciba-Geigy Corp.
13. W. Bencze, R. Hess and G. deStevens, *Prog. Drug Res.*, 1969, **13**, 217.
14. J. Schmutz and E. Eichenberger, 'Chronicles in Drug Discovery', ed. J. Bindra and D. Lednicer, Wiley, New York, 1982, vol. 1, p. 39.
15. A. Afonso, F. Hon, J. Weinstein, A. K. Ganguly and A. T. MacPhail, *J. Am. Chem. Soc.*, 1982, **104**, 6138.
16. T. Hogberg, B. Ulff, A. L. Renyi and S. B. Ross, *J. Med. Chem.*, 1981, **24**, 1499.
17. E. J. Cragoe, Jr., O. W. Woltersdorf, Jr., J. B. Bicking, S. F. Kwong and J. H. Jones, *J. Med. Chem.*, 1967, **10**, 66.
18. H. Larsson, E. Carlsson, U. Junggren, L. Olbe, S. E. Sjostrand, I. Skanberg and G. Sundell, *Gastroenterology*, 1983, **85**, 900.
19. W. Schindler and F. Hafliger, *Helv. Chim. Acta*, 1954, **37**, 472.
20. R. Kuhn, *Schweiz. Med. Wochenschr.*, 1957, **87**, 1135.
21. R. D. Hoffsommer, D. Taub and N. L. Wendler, *J. Org. Chem.*, 1962, **27**, 4134.
22. M. Wilhelm and P. Schmidt, *Helv. Chim. Acta*, 1969, **52**, 1385.
23. F. C. Novello and J. M. Sprague, *J. Am. Chem. Soc.*, 1957, **79**, 2028.
24. G. deStevens, L. H. Werner, A. Halamandaris and S. Ricca, Jr., *Experientia*, 1958, **14**, 463.
25. K. Sturm, W. Siedel, R. Weyer and H. Ruschig, *Chem. Ber.*, 1966, **99**, 328.
26. B. Ablad, E. Carlsson and L. Ek, *Life Sci I.*, 1973, **12**, 107.
27. D. J. Le Count, 'Chronicles of Drug Discovery', ed. J. Bindra and D. Lednicer, Wiley, New York, 1982, vol. 1, p. 113.
28. A. M. Creighton and S. Turner (ed.), 'The Chemical Regulations of Biological Mechanisms', Special Publication No. 42, The Royal Society of Chemistry, London, 1982, p. 1.
29. M. Takada, T. Takagi, Y. Yashima and H. Maeno, *Arzneim.-Forsch.*, 1982, **32**, 734.
30. J. H. Laragh, *John Hopkins Med. J.*, 1975, **137**, 184.
31. M. A. Ondetti and D. W. Cushman, *Annu. Rev. Biochem.*, 1982, **51**, 283.
32. M. J. Wyvratt and A. A. Patchett, *Med. Res. Rev.*, 1985, **5**, 483.

3.3

Computer-aided Selection for Large-scale Screening

LOUIS HODES

National Cancer Institute, Bethesda, MD, USA

3.3.1 INTRODUCTION

This chapter describes how computer programs for estimating not only activity but also novelty were used to aid in the selection process. Novelty is important because we wish to test a wide range of substances to find new leads. The first section shows the need for primary and secondary screening in the search for new leads from a large body of available compounds. When the capacity of the primary screen is more restricted than the availability of the compounds, then some selection becomes necessary.

3.3.2 PRIMARY AND SECONDARY SCREENING

The search for new lead compounds for treatment becomes almost trivial if one has a good biological model for the disease. A model will give a response to drugs that ideally corresponds to treatment of the disease itself. If the model is reasonably fast, not too costly, and also provides a good indication as to whether a compound will be clinically useful, then one could find new leads by simply testing everything available.

The trouble is that there seldom exists such an ideal biological model. If one has a model that is accurate for the clinic it will usually be slow and expensive. If, in addition, one has a fast, cheap model which is not terribly accurate, one can put both models together by using the latter as a primary screen.

That has been the logic behind the search for new leads at the NCI for the past 10 years.[1] The primary screen was P388 mouse leukemia,[2] which turned out to have a yield of about 5% active compounds. These actives have diverse structures and include over 90% of the known anticancer agents. The 5% yield is relatively high for an animal model. The high yield and the response to diverse structures show it has low selectivity. Thus P388 makes a good primary screen. The secondary screen was a panel of mouse tumors and xenografts of human tumors on athymic mice.

The large capacity of the primary screen was nevertheless limited both by its own cost and the restricted capacity of the secondary screen. We were soon faced with more available input than the 10 000 compound per year capacity of the primary screen.

The problem was resolved by refining the available input so as to achieve a guided random selection. A purely random selection would take, say, every third compound from the available input. We feel there is no special virtue in randomness for this purpose. A better selection can be obtained by systematically pursuing the two criteria of novelty and potential activity.

Computer programs to estimate P388 activity and novelty were developed based on chemical structure. The computer scores were used as an aid in selection but could be overruled by the chemist. In this manner more than 50 000 compounds were selected for P388 screening.

The computer program for P388 activity can be considered a nonbiological model since it is fed data on compounds and produces measures of activity. In the following account we often use the term 'model' as a replacement for 'program for activity'.

The NCI anticancer screening program is undergoing a revision from mouse models to cell cultures taken from varieties of human cancers. As results become available from testing many diverse compounds against these disease-oriented screens, comparable computer models will be employed to aid selection.

3.3.3 THE COMPUTER PROGRAM FOR ANTICANCER ACTIVITY

The computer model depends on a training set. This is a previously tested set of diverse compounds with definitive biological results. Therefore, the training set contains an active and an inactive portion. By the nature of a primary screen the large majority of the compounds in the training set will be inactive.

The computer model uses molecular fragments[3] to evaluate compounds. These fragments should be of sufficient size to represent functional features, yet small enough so that many of them have substantial incidence in the training set. By exhaustive generation of all fragments of a standard size, we find a large range of incidence. Thus we can expect that any new compound will contain fragments that also occur in the training set among those of high and moderate incidence. It is worth noting that low incidence fragments are especially useful for picking up new leads.

For each fragment occurring in the training set, an activity weight is computed by the following method.[4] Suppose that the fragment is not relevant to activity. Let us represent the compounds of the training set as balls in an urn: a black ball represents a compound with the fragment and a white ball represents a compound without the fragment. Since the active compounds are assumed to be unrelated to the fragment, the active compounds can be represented by drawing a like number of balls from the urn.

Thus, the expected incidence of the fragment in the active compounds is in the same proportion as in the entire training set. The actual incidence of the fragment in the active compounds, which we also know, can be considered as evidence of a relationship between the fragment and activity, to the extent that it differs from the expected incidence. The statistical significance of this difference between the actual and expected incidence can be expressed as a number of standard deviations derived from the urn model.

In this manner each fragment obtains a weight for activity. This weight, expressed as a number of standard deviations, is equal to the difference between its incidence in the active portion of the training set and its expected incidence. For two or three occurrences of the same fragment, the weights are computed according to conditional probabilities. More than three occurrences are treated as three occurrences.

To evaluate a new compound we add the activity weights for all of its fragments which have appeared in the training set. This sum is taken as the activity score for the compound. As a check on the model, 20% of the training set is run through the program as a test set. Table 1 shows the distribution of active compounds at selected percentile levels in the ranking of the activity scores from a typical test run of the model. The scores at the decile levels have been used to rate the new compounds into 10 categories of activity score. By maintaining a record of the rating of each selected compound, we have been monitoring actual performance of the model as the screening results are received.

Table 1 Cumulative Percent Active Compounds in Computed Activity Ranking at Selected Percentile Levels

Percentile	*Highly active*[a]	*Moderately active*[a]
99	27	10
98	39	16
95	71	30
90	88	44
80	92	59
70	97	70
50	100	85
30		92
10		99
0		100

[a] Active compounds were recorded at two activity levels depending on the length of increased life span of the treated animals *versus* the controls.

3.3.4 THE COMPUTER PROGRAM FOR NOVELTY

To obtain new leads we are always looking for novel compounds. Here 'novel' means compounds unlike those in the training set.

An objection to the computer program for activity was that it would select only compounds that are like those that have already been tested. A partial answer to that objection was the removal of well-known active compounds and their analogs from the training set.[5,6]

A simple novelty measure was devised, based on structure fragments and their incidence in the training set. A compound containing a fragment which has never occurred in the training set is flagged as unique. Otherwise, the incidence of the fragment with the lowest incidence is taken as an inverse measure of novelty. The higher the incidence, the lower the novelty. A cutoff threshold was chosen such that all compounds with higher values become candidates for rejection. For example, when the threshold was 50, a compound in which all fragments occurred more than 50 times in the training set was likely to be rejected.

3.3.5 VALIDATION OF COMPUTER-AIDED SELECTION[7]

Considerable emphasis was placed on validating the predictions of the model. Validation was performed in two different ways. First the training set was separated into disjoint training and test sets by a series of 80%/20% cuts. The earlier work was highly biased since the test sets were subsets of the training sets. The disjoint runs also showed an enrichment of actives among the high scoring compounds. For example, there were almost 10 times as many active compounds in the highest decile as there were in the lowest decile.

The second, more interesting, validation study was a direct comparison of the performance of the computer with that of a chemist thoroughly familiar with the design of anticancer compounds. A sample of 988 previously untested compounds was processed by the computer and by the chemist independently of each other. Then all 988 compounds were sent to P388 testing. In this study, the computer proved to be at least equal to the chemist in predicting activity, although the choice of compounds differed. This was the first indication to the author of the difficulty of predicting which compounds are likely to be active, even for a well-informed chemist.

Table 2 shows some results from this test. As the top line indicates, there were only 26 confirmed actives among the 988 compounds at evaluation time. We also counted compounds with incomplete testing as follows. Compounds which met the P388 activity criteria are presumed active or 'presumptives' until a confirmation test. If they fail the second test, they become 'pass/fails' and are given one more chance to confirm. Also, toxic tests are repeated at lower doses. The table bottom line includes a summary formula which estimates the ultimate yield of confirmed actives, extrapolating the partial results according to the previous history of P388 testing.

This study showed that a good enrichment of active compounds could be obtained by simply using the computer-generated scores. The novelty measure helped, since for some unknown reason the compounds rated novel tended to be more active than those rated adequately represented. It became clear that the computer can complement the chemist in selection.

Table 2 Prediction Performance of Chemist *versus* Computer on Active Compounds including Extrapolation of Incomplete Testing

	Chemist[a]	*Computer 312*[b]	*Agree*[c]	*Computer 494*[d]	*All 988*
Confirmed actives (C)	11	13	8	20	26
Presumptives (P)	8	10	4	18	33
Pass/fails (PF)	5	4	2	6	10
Toxics (T)	19	14	12	18	27
C + 0.6P + 0.2PF + 0.1T	18.7	21.2	12.0	33.8	50.5

[a] Actual P 388 results on 312 compounds selected by chemist. [b] Results on highest 312 compounds by computer activity score. [c] Agreement, or overlap, between chemist and computer. [d] Results on highest 494 compounds (50%) by activity score.

3.3.6 EXPERIENCE AT THE NATIONAL CANCER INSTITUTE

During the years 1980–85 more than 50 000 compounds were tested in P388. The great majority of these compounds were selected by the chemist after evaluating the computer scores of about three times that many potential acquisitions. Those compounds which got low activity scores or were found to be adequately represented were therefore considered to be candidates for rejection.

Thus, we have collected some statistics on the P388 outcomes of the about 50 000 selected compounds. This is a biased sample since the rejected compounds were not tested. Still, some compounds were selected despite low ratings. These did rather poorly in P388.

There was a general pattern that persisted through the years, continuing under yearly updates of the training set. The compounds scoring in the top decile of the training set yielded about 8% actives. Those in the top half yielded about 4% actives and those in the lower half yielded about 2% actives. Beginning in 1983 records were kept for compounds in the top 5% of the training set range. These yielded about 20% actives, but the number of compounds in this range was proportionately less, perhaps 1% of the input.

3.3.7 VARIATIONS AND EXTENSIONS

A series of variations and extensions of the computer methods were performed. First was the application of the programs to literature surveillance. Next, by the appropriate choice of training sets, the activity prediction program was used to predict toxicity and also therapeutic index. Extensions to other test systems were incorporated when those systems acquired enough testing results to achieve substantial training sets. Finally, a method for adjoining the octanol/water partition coefficient to the structure features was developed. Each of these variations required some alteration but was highly compatible with the original method.

3.3.7.1 Literature Surveillance

An early hope for these methods was the ability to survey the chemical literature by running all the structures registered by Chemical Abstracts Service through the programs for determining activity and novelty. Since these compounds numbered roughly half a million per year, only the top few percent in the score would be reviewed to decide which compounds should be further examined.

One of the requirements of such a literature surveillance project is a sufficiently fast fragment-generating program. At about five compounds per second, the original NCI fragment-generating program was considered to be too slow. A new fragment-generating program was developed starting from a CAS program originally designed to extract functional groups. The new fragments differed somewhat from the earlier ones. They were bond centered rather than atom centered but were otherwise of similar size. However, they were generated at a rate of over 50 compounds per second. Comparison runs on the P388 data showed almost equal performance.

A great deal of experimentation on a subset of the 1977 data led to an optimized formula for combining the scores for novelty and activity to produce a single priority score. The automation circle was closed by automatically generating letters requesting a sample from authors. Although the entire 1977 data were eventually run, this project was abandoned mostly due to the earlier mentioned shift of the screening program from *in vivo* to *in vitro* tests.

3.3.7.2 Toxicity and Therapeutic Index

The standard P388 protocol requires testing at three dose levels: 200, 100 and 50 mg kg^{-1}. If the two lower levels result in toxic tests, then the overall test outcome is considered toxic and further testing is required at lower levels. About one sixth of P388 testing exhibits this first toxic outcome.

The application to toxicity was undertaken because an accurate prediction of toxicity would permit initial testing of presumed toxic compounds at lower levels. This is especially important when there is not enough original material for further testing.

A special model was created for toxicity in P388 by the use of appropriate training sets: a set of 'actives' that were toxic in P388 at 50 mg kg^{-1} and all higher doses and a set of 'inactives' that were nontoxic at 200 mg kg^{-1} and all lower doses. With these criteria, a search of the file yielded about 4000 toxic compounds and 30 000 nontoxic compounds.

Results from this model showed that toxicity could be predicted at least as well as antitumor activity. Of the compounds scoring in the highest decile of the training set range, one third yielded toxic outcomes. This was twice the normal result but still did not justify initial testing of compounds rated highly toxic at lower doses.

A model for high therapeutic index was prepared using a different approach. The P388 active training set was restricted to those compounds which met the activity criterion over three successive dose levels. This cut the number of active compounds from about 6000 to about 1000. This model, however, was not as useful in predicting actives as the original version.

3.3.7.3 Other Test Systems

Two of our most influential test systems, L1210 mouse leukemia and B16 mouse melanoma, were included in the panel of secondary screens. The higher yield of actives under secondary screening combined with the earlier compounds that had been tested. By 1983 there were substantial training sets available for these two test systems. Models were generated and a new composite model combined the P388, L1210 and B16 scores in equal proportion.

The composite model was not expected to predict as well for P388, but compounds which did pass P388 were destined to do better in secondary screening. Therefore, recommendations for selection were subsequently based on the composite score. The score for P388 was also carried along for comparison.

3.3.7.4 Log *P* as a Separate Component[8]

It is clear that physicochemical parameters such as the octanol/water partition coefficient could not realistically be treated simply as another variable in addition to the thousands of structure fragment variables. Indeed, the activity weight of a fragment should vary with the partition coefficient. This was accomplished by dividing the training set into disjoint subsets, according to ranges of partition coefficient.

Separation of the training set into ranges of log *P* was accomplished by using measured log *P* data from the Pomona file[9] and forming models for the desired ranges of log *P*. The log *P* range of a new compound can then be estimated by running it through the log *P* models and then assigning it to the range with the highest score. This was performed for every compound in the entire training set, creating a subset training set for each chosen range of log *P*.

When a compound is evaluated it is first run through the same log *P* models to estimate its range of log *P*. Then the fragment–weight table for the training subset of the corresponding log *P* range is used in computing its activity score.

The use of log *P* did lead to improvement in performance over the original method, mostly at the high percentile levels of activity score. Improvement would probably have been more substantial had there been more data on measured log *P*. Data on about 4000 compounds were used to classify about 100 000 more diverse compounds. Data were especially lacking for low log *P* compounds and those subsets yielded the worst performance.

3.3.8 REFERENCES

1. R. I. Geran, 'Anticancer and Interferon Agents', in 'Drugs and the Pharmaceutical Sciences', ed. R. M. Ottenbrite and G. B. Butler, Dekker, New York, 1984, vol. 24, p. 25.

2. '*In Vivo* Cancer Models', NIH Publication No. 84-2635, US Government Printing Office, Washington, DC, 1984.
3. L. Hodes, *J. Chem. Inf. Comput. Sci.*, 1981, **21**, 132
4. L. Hodes, G. F. Hazard, R. I. Geran and S. Richman, *J. Med. Chem.*, 1977, **20**, 469.
5. R. D. Cramer, III, G. Redl and C. E. Berkoff, *J. Med. Chem.*, 1974, **17**, 533.
6. L. Hodes, *ACS Symp. Ser.*, 1979, **112**, 583.
7. L. Hodes, *J. Chem. Inf. Comput. Sci.*, 1981, **21**, 128.
8. L. Hodes, *J. Med. Chem.*, 1986, **29**, 2207.
9. Pomona College Medicinal Chemistry Project Data Base, Issue 23, July 1983.

3.4
Isolation of Bioactive Materials and Assay Methods

WILFRID R. BUTT
Birmingham & Midland Hospital for Women, UK

3.4.1 INTRODUCTION

It has not always been easy to identify the origin of biologically active material. Research into the glandular source of hormones has been complicated by the numerous interactions between the secretions of different glands. Thus, for a long time the pituitary was regarded as an organ without any known function,[1] and when its importance was finally realized its interrelationships with other glands caused much confusion. For the experimentalist, the removal of the gland was a difficult operation to perform without damage to surrounding organs. On the other hand, the administration by mouth of preparations from pituitaries not unexpectedly gave confusing results. From very early times, human organ preparations have been recommended for therapeutic use, often with fairly obvious associations, *e.g.* brains for headache, genitals for infertility, kidneys for renal complaints, *etc.*, while primitive people drank the blood of their enemies and cannibal tribes ate the brains of dead relatives.

When advances had been made and there were basic scientific grounds for using preparations from tissues therapeutically, animal extracts were often used as they were relatively easy to obtain and sometimes the work-up of thousands of glands was required. When the active principals were small molecules, treatments were sometimes successful, but with large molecules, such as proteins, problems arose because of species differences and it became clear that it was essential to obtain human material.

Because of the vast number of biologically active substances isolated from natural sources, it is only possible in the space available to present examples as illustrations of general methodology and these examples will be confined to the hormones. Details of some of the hormones discussed in this chapter, with abbreviations, are listed in Table 1. Further details and other hormones are included in Chapter 3.5 (Tables 1 and 2 and Figure 1).

3.4.2 SOURCES

The extraction of steroids, thyroid hormones, insulin and other hormones from the glands of pigs, sheep and cows, dating from more than 60 years ago, proved useful for numerous studies in endocrinology.[2] These hormones were widely used as standards, reagents for assays and for biological and clinical studies and in many cases the animal glands remained the most important sources until chemical syntheses had been achieved. Protein and glycoprotein hormones, such as adrenocorticotrophin (ACTH), growth hormone (GH) and the gonadotrophins, were also obtained in large quantities from animal pituitary glands and this was the means by which their biological properties were discovered. When required therapeutically, however, initial successes were not maintained and it became clear that immunological reactions were produced because of species differences. It was therefore necessary to organize collections of large numbers of human pituitaries for extraction of human hormones. These glands were available in sufficient numbers only from post mortems and since about 1955 several important national collections have been organized.

3.4.2.1 Hypothalamus

The hormonally active substances discovered in the hypothalamus are peptides or polypeptides and species differences have not proved to be a problem. It has therefore been possible to use tissues from farm animals as a source for preparation of active substances. This is fortunate because, in order to isolate sufficient material to characterize activities, it has been necessary to work on tens of thousands of glands: over 250 000 porcine hypothalami were required in order to obtain a mere 7.5 mg thyrotrophin-releasing hormone (TRH)[3] and 500 000 ovine hypothalami to obtain 8.5 mg somatostatin;[4] in the initial studies even more glands were required.

In general, a crude extract is described as containing releasing (or release-inhibiting) activity, which is then termed releasing factor when better purified, and finally releasing (or regulatory) hormone when the structure is known.[5] At the time of writing, releasing and inhibiting hormones have been identified for each of the major hormones of the anterior pituitary and there is some overlap of their respective actions, *e.g.* TRH releases both thyroid-stimulating hormone (TSH) and prolactin, while somatostatin, besides inhibiting the release of GH, has actions elsewhere and also inhibits the release of insulin and glucagon.

The methods used for isolating these hormones were adapted with modifications from those that had been used earlier for the neurohypophysial hormones, vasopressin and oxytocin.[6,7] The tissues are extracted, usually in acidified solvents such as hydrochloric or acetic acids, the extract is defatted and the factors separated by various chromatographic procedures, lately including HPLC, *etc.* Successful isolation of individual hormones is dependent upon sensitive and specific monitoring of the separation processes. Difficulties arise because the activities of one factor sometimes overlap the activities of another and there may be both releasing and inhibiting factors in the same extracts.

Synthetic preparations are now available for most of the better-known regulating hormones. One of the most recent to be isolated was corticotrophin-releasing factor (CRF) and the work on this gives some good examples of the difficulties encountered. Substances releasing ACTH were recognized at least 30 years ago but the structure of CRF itself was only identified as a 41-residue peptide in 1981.[8,9] The biological assay was the first problem; *in vivo* methods lacked sensitivity and specificity and advances were not made until *in vitro* assays using pituitary cell cultures were developed to provide sensitive quantitative assays. Secondly, when crude hypothalamic extracts were split by chromatography, biological activity was distributed into at least three fractions and 70% of the ACTH-releasing activity was lost.[10] The peptide is hydrophobic and there is a tendency for its single methionine residue to oxidize with loss of activity, and originally vasopressin was considered to be the CRF. There is some species specificity and the sheep and rat peptides differ by seven amino acids. The human structure was deduced from recombinant DNA techniques and was found to be identical to that of the rat.[11]

Table 1 The Major Source, Abbreviation and Action of some Hormones Described in this Chapter

Major source	*Hormone*	*Abbreviation*[a]	*Main actions*
Hypothalamus[b]	Corticotrophin-releasing hormone	CRF	Stimulates synthesis and release of ACTH from anterior pituitary
	Gonadotrophin-releasing hormone (also known as luteinizing-hormone-releasing hormone)	GnRH (LHRH)	Stimulates synthesis and release of FSH and LH from anterior pituitary
	Gonadotrophin-releasing-hormone associated protein	GAP	Inhibits release of prolactin; stimulates release of FSH
	Somatostatin (growth hormone release-inhibiting factor)	—	Inhibits release of hGH and other hormones[a]
	Thyrotrophin-releasing hormone	TRH	Stimulates synthesis and release of TSH and also of prolactin
Anterior pituitary[c]	Adrenocorticotrophin (corticotrophin)	ACTH	Promotes adrenal steroidogenesis and maintains adrenal gland
	Precursor molecule = pro-opiomelanocortin, giving rise to:		
	Melanophore-stimulating hormone	MSH	Maintains melanophores and may influence adaptive behaviour in man
	β-Lipotrophin	β-LPH	Mobilizes fatty acids but function in man unclear
	β-Endorphin	—	Morphine-like action, inhibits release of LH
	Gonadotrophins		
	follicle-stimulating hormone	FSH	Stimulates sperm production in testis and growth of primordial follicles in ovary
	luteinizing hormone	LH	Stimulates steroidogenesis in Leydig cells of testis, induces ovulation and maintains corpus luteum
	human menopausal gonadotrophin (extracted from urine)	hMG	Exhibits both FSH and LH activity
	Growth hormone	GH	Promotes skeletal and visceral growth
	Prolactin	—	Stimulates lactogenesis and has weak GH activity
	Thyrotrophin (thyroid-stimulating hormone)	TSH	Stimulates thyroid gland to secrete thyroid hormones
Gonads	Inhibin	—	Inhibits FSH release
	Activin	—	Stimulates FSH release
	FSH-releasing protein	FRP	Stimulates FSH release
Placenta	Human chorionic gonadotrophin (extracted from urine)	hCG	Maintains corpus luteum of pregnancy, action similar to LH
	Human placental lactogen	hPL	Weak lactogenic and growth-stimulating properties
Liver	Insulin-like growth factors (somatomedins)	IGF-I, *etc.*	Intermediate in growth-promoting effects of GH
Kidney	Erythropoietin (extracted from blood)	—	Regulates erythropoiesis
Gastrointestinal tract[d]	Pancreastatin	—	Inhibits insulin and somatostatin secretion

[a] Prefix 'h' indicates hormone of human origin. [b] See also Chapter 3.5, Table 1. [c] See also Chapter 3.5, Figure 1. [d] See also Chapter 3.5, Table 2.

In the original preparation of the 41-residue peptide, almost 500 000 ovine hypothalamic fragments were employed.[8,9] Extraction was with ethanol/acetic acid followed by defatting with chloroform and then partitioning into aqueous and organic phases. The organic phases could be used for the preparation of gonadotrophin-releasing hormone (GnRH) and somatostatin. Ultrafiltration of the aqueous phase retained the active fractions and two zones were separated by gel filtration on Sephadex G-50. The lower molecular weight zone contained ACTH-releasing activity and was similar to vasopressin; the larger molecular weight fraction was further purified by gel filtration on SP-Sephadex, Bio-Gel-P10 and ultimately by HPLC. Columns with large pore size (300–330 Å), small particle size and monolayered end-capped octadecyl silica enabled final purification to be accomplished in two steps. Eventually 90 μg CRF was obtained with a purity of above 80%. The primary structure was determined by Edman degradation and the molecule was found to be a 41-residue straight-chain C-terminally amidated peptide. Synthesis was achieved by solid phase methodology[8] and radioimmunoassays were soon developed.

The peptide stimulated the release of ACTH *in vivo* and *in vitro* and was more potent in this respect than vasopressin. The whole structure is needed for full activity.

Another recent advance in the biochemistry of the hypothalamus concerns the gonadotrophin- and prolactin-regulating hormones. The existence of a release-inhibiting factor for prolactin other than dopamine had long been suspected. Work on the precursor protein of GnRH has cast some light on this.[12] Complementary DNA clones containing coding sequences for a GnRH precursor protein ($M = 10$ kDa) were isolated from libraries derived from human placental and human and rat hypothalamic mRNA. The precursor protein comprised the GnRH decapeptide preceded by a signal sequence of 23 amino acids and followed by a Gly-Lys-Arg sequence necessary for enzymatic processing and C-terminal amidation of GnRH (Figure 1). A sequence of 56 amino acid residues occupies the C-terminal region of the precursor and constitutes the GnRH-associated peptide (GAP), which has prolactin-release-inhibiting properties.

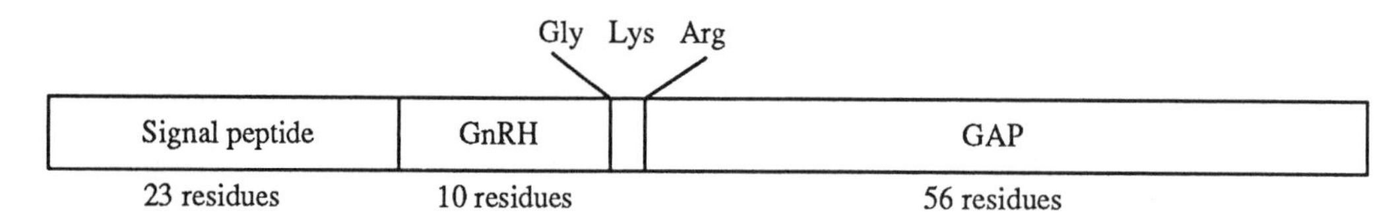

Figure 1 Precursor protein for GnRH. A signal peptide of 23 residues precedes the decapeptide, GnRH. This is separated from the 56-residue GnRH-associated peptide GAP by the 3 amino acids, Gly, Lys and Arg[12]

The purification of GAP was achieved by a combination of gel filtration and reverse-phase HPLC. The 56-residue peptide was released from the precursor by treatment with cyanogen bromide and, after reverse-phase HPLC, gel filtration was used to separate GAP from side products. It was finally purified by reverse-phase HPLC, the product showing a single immunoreactive band on polyacrylamide gel electrophoresis.

For the *in vitro* bioassay, anterior pituitaries from rats were dispersed by enzymatic treatment and the cells were kept in culture plates in medium supplemented with 10% foetal bovine serum. Incubations with GAP were performed on the fourth and fifth day of culture in serum-free medium. Pituitary hormones released over 4 h were then determined by double-antibody radioimmunoassay. GAP was used at a concentration of 10^{-8} M, which for most of the other regulating hormones elicits maximal response from anterior pituitary cell cultures. The release of both gonadotrophins, *i.e.* follicle-stimulating hormone (FSH) and luteinizing hormone (LH), was stimulated, while the basal secretion of prolactin was reduced by about 50%, comparable to inhibition levels reported for dopamine.

3.4.2.2 Pituitary

The human anterior pituitary is rich in protein and glycoprotein hormones which are valuable as reagents, standards and for therapy. These hormones are species specific and therefore for clinical studies it has been necessary to organize collections of human glands taken at autopsy. National collections were first organized some 30 years ago with the primary object of producing hGH for therapy.[13] For glands to be included in the collection, a number of precautions must be taken. The body must have been placed at a temperature not exceeding 8 °C within 24 h of death and the gland must be removed as soon as possible and not more than 96 h after death. Glands should not be included from patients dying of septicaemia, hepatitis, viral infections of the central nervous system (CNS) or dementia, *etc.* hGH was prepared and used successfully for the treatment of hypopituitary dwarfism until 1985 when four deaths were reported in patients who had been treated with hGH several years earlier.[14] The clinical histories indicated similarities to the Kreutzfeldt–Jacob syndrome, suggesting that the slow virus associated with this disease had probably contaminated the batches of glands from which the hGH had been prepared. Because of the possibility of this occurring again due to the difficulty of detecting the virus and possibly others related to it, many national collections, including that in the UK, were stopped. Patients are now treated with GH with the human amino acid sequence synthesized by recombinant DNA procedures.[15,16] Human glands are still required, however, as a source of reagents for immunoassays and as standards, and will continue to be essential until synthetic preparations become available.

Two methods for the preservation of the glands have been widely used. In the first, glands are placed in not less than 20 mL acetone per gland and stored at 4 °C. After a week the glands are removed and transferred to fresh acetone (1–2 mL per gland) after which they can be stored indefinitely at 4 °C. In the second method, glands are immediately frozen at −20 °C (−70 °C if possible) usually in batches of 25 until autopsy reports have eliminated the risk of contamination from viruses, *etc.* The advantage of using this second method is that yields of polypeptide hormones such as prolactin and ACTH are much better than from the acetone-preserved glands.

Subsequent fractionation of the glands is designed to preserve the activities of as many hormones as possible because of the limited availability of human pituitaries. Well-established fractionation procedures have been used which include chromatography on DEAE- and CM-celluloses, hydroxyapatite, gel filtration, electrophoretic procedures and lately HPLC. A useful preliminary fractionation of glycoproteins from proteins is by extraction in a mixture of ammonium acetate and ethanol; a scheme for the subsequent separation of hormones which has proved satisfactory for several years is given in Figure 2.[17,18]

One of the benefits of monoclonal antibodies is the possibility of constructing specific probes for each hormone with no cross-reaction so that purification can be achieved using these probes in immunosorbent columns. Gentle elution methods are possible so that biological activity can be retained, as demonstrated for hTSH.[19] Columns were prepared of cyanogen bromide-activated Sepharose 4B coupled to monoclonal antibody to hTSH with an affinity of 5×10^{-7} M L^{-1} using 4 mg antibody per mL of gel. Crude glycoprotein from human pituitaries was applied to the column in 50 mM borate/0.5 M NaCl (pH = 8.5) and the TSH antigen bound to the antibody with a capacity of 0.5 μg TSH per mL of gel. Elution was at pH = 3.5 (50 mM glycine/HCl containing 0.5 M NaCl). The immobilized antibody was stable and could be used repeatedly for up to 12 months. The hormone eluted in a biologically active form with <0.5% by weight of contaminating glycoprotein. Similar methods were described for FSH and LH[20] using automated systems. Sterile buffers are obligatory for such systems.

Final purification of glycoprotein hormones must usually be by gel filtration so that any dissociated subunits or polymers are removed. These glycoproteins each consist of two subunits, the α subunit, which is common to each, and a β subunit, which is hormone specific. The subunits can be reversibly dissociated by reagents such as 4 M guanidine for the gonadotrophins and 1 M propionic acid for TSH. The subunits themselves are biologically inert but they may react in immunoassays. Because of the similarities in structure it has been essential to use only highly purified preparations for raising antibodies of sufficient specificity for assay purposes.

During purification procedures, rapid radioimmunoassays are the most suitable methods for identifying fractions of interest. High sensitivity is not required, so that the methods can be designed using higher concentrations of antibody than those found necessary for clinical assays; incubation times can therefore be reduced and the process is speeded up.

Hormones prepared for therapeutic applications, *e.g.* hGH and gonadotrophins, must be assayed by bioassays. The assay based on the measurement of tibial width in hypophysectomized rats (rats whose pituitaries have been surgically removed) is most commonly used for hGH.[21] Over a limited range of doses the width is linearly related to the logarithm of the dose. Discrepancies between hGH potencies of crude extracts or plasma assessed by this assay and by radioimmunoassay have been noted and are probably related to the presence of other growth-promoting factors such as the insulin-like growth factors (IGF). There are a number of direct *in vitro* effects of GH, however, including stimulation of RNA, DNA and protein synthesis in a cell line of human leukaemic T lymphoblasts. Assays based on such effects are sensitive and correlate with radioimmunoassays.[22] Radioreceptor assays have also been described with sensitivities higher than *in vitro* bioassays

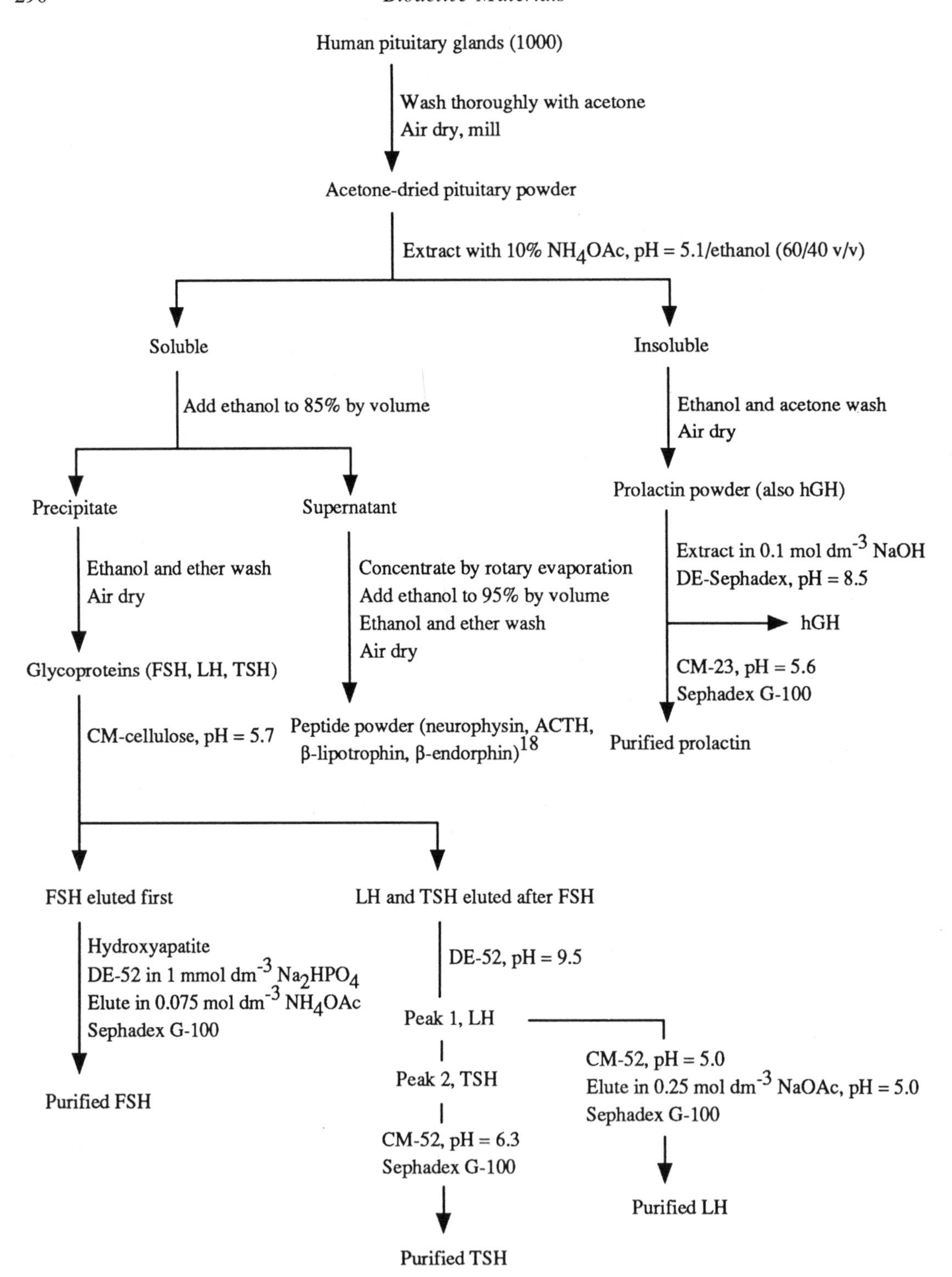

Figure 2 Scheme for the fractionation of human pituitary glands and the purification of individual hormones[17]

although lower than radioimmunoassays. Specific binding sites have been recognized in numerous tissues including the liver, kidney and human monocytes.

The first specific *in vivo* bioassays for gonadotrophins used hypophysectomized rats. For FSH, the assay depended on the stimulation of ovarian growth in the hypophysectomized animals and for LH it depended on the repair of interstitial tissues of the ovaries or on the increase in weight of the ventral prostate in hypophysectomized male rats. These are very laborious assays and simpler ones are now available which are also highly specific. The effect of FSH on the ovarian weight of immature rats is augmented by a large constant dose of LH (in the assay in the form of hCG) and is unaffected by other pituitary hormones (ovarian augmentation assay).[23] For LH, a highly specific assay depends on the depletion of ovarian ascorbic acid in intact immature rats made pseudopregnant by pretreatment with pregnant mare serum gonadotrophin and hCG (ovarian ascorbic assay depletion assay).[24] This is an acute assay in which the ovaries are dissected for ascorbic acid determinations 4 h after injection of the LH.

Gonadal target sites have formed the basis of sensitive *in vitro* assays for the gonadotrophins. Homogenates of rat testes contain binding sites suitable for the assay of both LH and FSH.[25] There are also assays depending on the steroidogenic response of gonadal tissue stimulated by gonadotrophins such as the production of testosterone from interstitial cells of the testes for LH and estradiol production from cultured Sertoli cells for FSH.[26] A cytochemical method for LH has also been described based on the principle of the ovarian ascorbic acid depletion method. The reducing activity of ovarian slices treated with LH was estimated by microdensitometry after staining with Prussian blue.[27]

3.4.2.3 Gonads

The gonads and the adrenal cortex were important sources of steroid hormones until these became available in synthetic form. Other secretions from these glands are now proving a fruitful source of several important peptide and glycoprotein hormones which modulate the control of gonadal function. Much of this work has been done on the follicular fluid, usually from the pig as litre quantities are required to produce enough bioactivity to study.

Several non-steroidal regulators in follicular fluid which control oocyte and follicular maturation have been described[28] but the one into which most progress has been made is inhibin. The concept that the gonads produced a substance which specifically regulated the secretion of FSH from the pituitary gland has existed for many years, but the purification of this substance has proved very difficult. There were reports of substances with inhibin-like activity ranging in molecular weight from 10–100 kDa, and the problem has only recently been resolved. It appears that the molecule contains not only an interlinked α and β subunit which possesses the expected FSH-inhibiting activity but that dimers of the β subunits are potent stimulators of FSH synthesis and secretion which raises the possibility of dual control of FSH from within the gonad (Figure 3).[29]

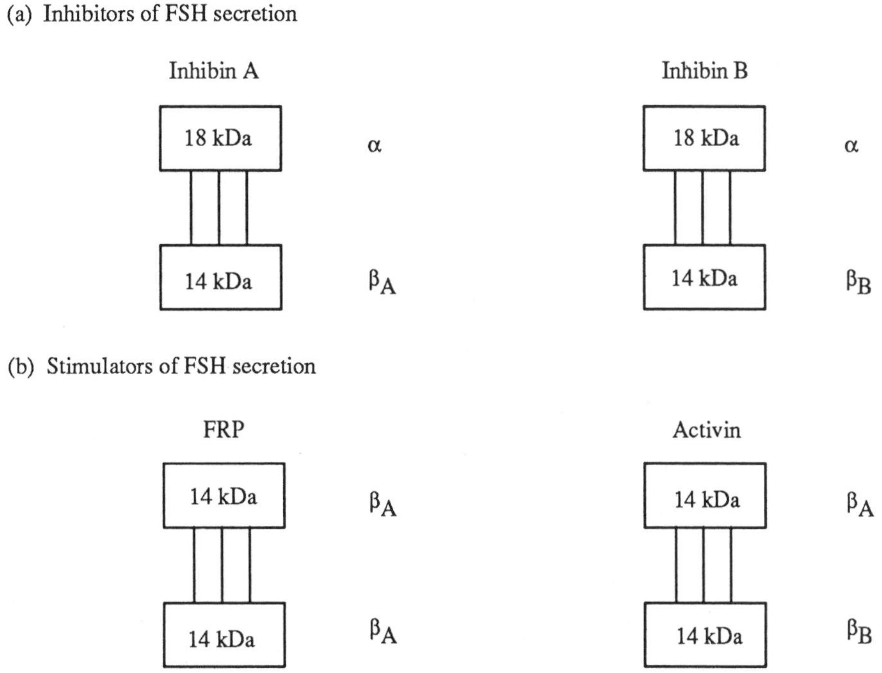

Figure 3 Diagrams of subunit structures of inhibin and inhibin-related factors which are stimulators of FSH release isolated from porcine and bovine follicular fluid.[29,31,33] The α subunit has a molecular weight of 18 kDa and the two β subunits, β_A and β_B, have molecular weights of 14 kDa and differ at the N terminus

Two forms of inhibin (A and B) with molecular weights of 32 kDa have been purified from porcine follicular fluid.[30] Heparin–Sepharose affinity chromatography, gel filtration on Sephacryl S-200 and four reverse-phase HPLC steps[31] were used to isolate a substance with both inhibin activity and FSH-releasing activity which was called activin (M = 24 kDa). The activities were monitored by an *in vitro* bioassay using a monolayer culture of dissociated rat anterior pituitary cells.[32]

In another method, preparative HPLC was used to isolate both an FSH-releasing protein (FRP) and an FSH-release-inhibiting zone (inhibin) from a 50% ammonium sulfate precipitation of 6 L of porcine follicular fluid.[33] By using gel filtration in a strongly dissociating buffer containing 6 M guanidine/HCl, cation exchange chromatography fast protein liquid chromatography (FPLC) and reverse-phase HPLC, highly purified FRP was isolated. It was found to consist of two inhibin β_A chains linked by disulfide bonds with a molecular weight of 28 kDa.

The course of purification was followed by monitoring by an *in vitro* bioassay.[34] Rat anterior pituitary glands were dissociated enzymatically and the cells were plated and washed twice. Test material was added and allowed to remain in contact for 48–72 h after which the fluids were removed and tested for FSH by radioimmunoassay. After ammonium sulfate precipitation and the first preparative HPLC column, 2800 ng protein per mL of medium was required to elicit a half-maximal response in the bioassay. At the final stage between only 0.5–1.0 ng protein per mL was required which represents a purification of more than five-thousandfold.

Two forms of inhibin with molecular weights of 65 and 30 kDa have been isolated from ovine follicular fluid[35] using a combination of gel filtration, reverse-phase HPLC and preparative polyacrylamide gel electrophoresis. The *in vitro* bioassay employed was based on the specific suppression of FSH cell content of rat pituitary cells in culture.[36a] On reduction of the 30 kDa form, four components were recognized, 20–21 and 16 kDa fragments being similar to the corresponding inhibin subunits isolated from porcine and bovine follicular fluids.

To avoid confusion, recommendations have been made about the definition and nomenclature of inhibin and related substances.[36b] Inhibin is considered to consist of two dissimilar disulfide-linked subunits termed α and β (not A and B). Inhibin of molecular weight 31–32 kDa has been purified and cloned from porcine, bovine, ovine and human sources; the higher molecular weight forms of 55–65 kDa should be distinguished by their molecular weight, *e.g.* α-44. Inhibin-like substances not of gonadal origin are designated α-inhibin (from seminal vesicles) and β-inhibin (from prostate). β dimers stimulating FSH secretion should be termed 'activin', the β_A dimer being termed activin A and the β_A-β_B dimer activin B.

3.4.2.4 Placenta

The developing human foetus and its placenta form an important partnership during pregnancy (the foetoplacental unit). Several important hormones are produced: the glycoprotein human chorionic gonadotrophin (hCG), the polypeptide human placental lactogen (hPL) and neuropeptides which were originally discovered in the hypothalamus, such as GnRH, TRH, somatostatin and ACTH and related peptides including β-endorphin.[37] Steroid hormones are also produced and metabolized by the foetus, placenta and mother. The large-scale extraction and purification of these hormones, however, has usually been from other glands and, in the case of hCG, from the urine of pregnant women. hPL, however, although showing some of the activity of hGH, is different and the placenta is the best source of this hormone.

In early work it was demonstrated that hPL could be prepared from placentae by extraction in aqueous media and purified by standard techniques involving fractional precipitation using salts, DEAE-chromatography, polyacrylamide gel electrophoresis and gel filtration. It became clear that high pH should be avoided otherwise deamidation and aggregation tended to occur with loss of biological activity. In a method yielding hPL of high purity, frozen placentae were extracted in ammonium bicarbonate at pH = 7.8 and then the hPL was adsorbed with DEAE-cellulose from 0.1 M bicarbonate and eluted in 0.4 M bicarbonate.[38] Purification was by precipitation from ammonium sulfate, gel filtration on Sephadex G-100 and polyacrylamide gel electrophoresis. Monoclinic crystals were produced from solution in 0.1 M sodium phosphate (pH = 7.8) and PEG 1000.

hPL shares a number of biological activities with hGH and prolactin. In most bioassays, however, it is less potent than the pituitary hormones and for monitoring purification procedures and for measuring the hormone in blood radioimmunoassays are available.

3.4.2.5 Other Glands

Corticosteroids from the adrenal cortex, thyroid hormones from the thyroid and insulin from the pancreas were originally extracted from natural sources but are now available by synthesis. More recently, the pancreas, stomach and other regions of the gastrointestinal system have been much studied and numerous active substances have been isolated.[39] Most of the hormones isolated have been from porcine tissue because of its ready availability. Stimulation of the secretion of gastrointestinal (GI) hormones arises from within the gut by food passing through, and from the CNS in anticipation of food. The cells producing the hormones are distributed among the target cells, scattered so that they respond as a stimulus is applied and so allowing an integrated response to food stimulus over a wide area. This is a reason why characterization of the hormones has been slow and painstaking.[40]

Much use was made of ethanol fractionation, ion exchange celluloses and gel filtration in the purification of these compounds. Active fractions were initially identified by bioassays and it was not until a few mg of purified material was available that more rapid and sensitive radioimmunoassays could be developed. Many of the GI peptides and analogues have now been synthesized.

The GI hormone pancreastatin is a 49-residue peptide isolated from porcine pancreas.[41] It was purified by ethanol fractionation, gel filtration and CM-cellulose with a final HPLC step. The hormone inhibits insulin and somatostatin secretion. Isolated rat pancreas perfused with glucose induces a biphasic insulin release; pancreastatin decreases the early release and has a less pronounced but significant effect on the late-phase secretion.

The structure is somewhat similar to that of bovine chromogranin,[42] which is possibly a prohormone for pancreastatin. This substance is one of a family of acidic glycoproteins which are abundant in the neuroendocrine system occurring in the brain, pituitary, retina, thymus and chromoffin cells of the adrenal medulla.[43] A C-terminal amide structure occurs only in neuroactive or hormonally active peptides and a chemical detection method depending on the enzymatic release of amide and conversion into a fluorescent dansyl derivative was used to further the purification.[44a]

Prohormones of gastrin and cholecystokinin are expressed from pituitary cells, but the pituitary levels of these peptides and the corresponding mRNA concentrations are several-fold lower than those of the classical pituitary hormones and the number of these pituitary cells is small compared with the number of gastrointestinal G- and I-cells. Recent advances in the sequences and syntheses of the prohormones for these peptides have been reviewed.[44b] The extremely low concentrations in tissue have necessitated the use of highly sensitive immunotechnology for their detection. Specificity is essential so that measurements must be based on several radioimmunoassays with monospecific antisera to precisely defined sequences and residue specificity. Chromatographic separations should be monitored by immunoassays and molecular characterization made by enzymography with immunochemical measurements before and after cleavage with enzymes directed against different processing sites.

3.4.2.6 Urine

Human urine is an important source of two gonadotrophins used therapeutically: the urine from women in the menopause, which is rich in human menopausal gonadotrophin (hMG) and contains both FSH and LH activities, and the urine of women in early pregnancy, which contains hCG, an LH-like hormone.

After the menopause the ovary no longer secretes estrogen in a cyclical manner as in the reproductive years. The negative feedback normally exerted by estrogen on gonadotrophin release is therefore no longer operative and consequently the amount of pituitary gonadotrophin released increases. Although the urine of menopausal women is rich in gonadotrophin, large volumes are required so that enough hMG can be obtained to satisfy the demands for treatment of infertility.

Usually urine is stored at 4 °C until sufficient is available for extraction. The pH is brought to 4.0 with hydrochloric acid and hMG is adsorbed on to acid-washed kaolin. Biologically active material can be eluted in alkali (ammonium hydroxide at pH = 11.0) and precipitated from five volumes of acetone or ethanol at neutral pH.[45] Further purification can be as for pituitary hormones or by immunoaffinity chromatography.[46]

The identification of active regions during chromatography is best monitored by radioimmunoassays. The potency of therapeutic material is determined by bioassay, the ovarian augmentation assay for FSH and the ovarian ascorbic acid depletion assay for LH. It is unnecessary for the therapeutic material to be highly purified and commonly the hMG preparation contains an equal unitage of FSH and LH.

Urine from women in the first trimester of pregnancy contains very high concentrations of hCG which is produced from the syncytiotrophoblast. Extraction and purification is similar to the methods used for hMG.[47–49] In early work, adsorption of the hormone to benzoic acid was used as the first stage, the benzoic acid precipitate being dissolved in acetone or ethanol, while simultaneously the adsorbed proteins were precipitated. Later, kaolin adsorption has been more usually used.

Considerable heterogeneity has been reported in the molecule, primarily because of variations in the carbohydrate structure, particularly in the number of terminal sialic acid residues. Since loss of these sugars leads to loss of activity, it is important that the residues are protected as much as possible, avoiding methods which lead to their removal, *e.g.* low pH. Conventional ion exchange and gel filtration methods have been used to produce hCG of high potency.[49]

Radioimmunoassay can be used for identifying active fractions during purification. The ovarian ascorbic acid depletion method is suitable for the determination of biological potency, the activity being similar to that of LH. hCG has a longer biological half-life than LH, however, and there may be differences in potency estimates for the two hormones when assays such as the ovarian ascorbic acid depletion, in which the test material circulates for a short time, are compared with ventral prostate assays which extend over several days.

Immunoassays for hCG form the basis of tests for early pregnancy and when these are used to identify the hormone soon after conception high sensitivity is required. It is important that sensitive assays are not affected by LH, and most antibodies to hCG also recognize the chemically similar LH. Frequently, individual antibodies are employed which have been raised to the specific β subunit of hCG to avoid this cross-reactivity. Caution is required with such assays, however, as the system may then recognize the β subunit of hCG but not the intact hormone.[50] It is uncommon for a polyclonal antibody raised to the intact hCG not to cross-react with LH, but when a specific antibody is found it is an extremely valuable reagent.

3.4.2.7 Blood

Hormones circulate in the blood in concentrations too low to make their extraction worthwhile on a large scale. The assay of these hormones, however, gives valuable information for diagnostic purposes.

There has been much interest in recent years in growth factors known as somatomedins and also as insulin-like growth factors (IGF) and discarded fractions from blood used for transfusions have proved to be useful sources of these substances.

For the preparation of IGF-I (also termed somatomedin C), the best known of these growth factors, Cohn Fraction IV (3.5 kg from 130 L plasma) was homogenized in 70 L acetone (2 mmol L^{-1}) containing NaCl (75 mmol L^{-1}) for 30 min and then centrifuged.[51] The material was purified twice on SP-Sephadex-C25 using gradient elution and concentrated on an Amicon hollow fibre ultrafiltration unit (HIP2 fibre, 2000 molecular weight cut off). The concentrate was desalted into acetic acid (2 mol L^{-1}) by filtration through a 5 × 90 cm column of Bio-Gel P2 (200–400 mesh) or desalted directly and then freeze-dried. There followed columns of Sephadex G-50F, polybuffer exchanger 94, hydroxyl apatite, Bio-Gel P2 and Sephadex G-50F. Finally, reverse-phase HPLC was used to obtain a product giving a single band of protein on polyacrylamide gel electrophoresis.

The study of this family of growth factors has been hampered by the limited availability of purified material for use as standards and reagents in immunoassays. There are several types of biological activity by which the factors can be recognized. Sulfation activity can be measured by the stimulation of the uptake of [^{35}S]sulfate by small discs of preadolescent porcine costal cartilage.[52] Mitogenetic activity is indicated by the stimulation of [^{3}H]thymidine uptake into growth-arrested mouse embryo fibroblasts 3T3 cells[53] and insulin-like activity by the measurement of the stimulation of lipogenesis in rat epididymal fat pad cells.[54]

3.4.3 EXTRACTION AND PURIFICATION

The general methods have been referred to under individual hormones. In the original work on a hormone, before its structure is known, classical methods of separation such as precipitation from salts, ion exchange adsorption, gel filtration, electrophoresis, *etc.*[55] have followed extraction in acid or alkaline aqueous media for protein hormones or solvents for steroids. Many of the long-established methods are being replaced by the newer techniques of HPLC[56] and, when purified material has been isolated and specific antibodies are available, by immunoaffinity chromatography.

Physicochemical methods of monitoring separations[57] are rapid and sensitive but not specific. Thus, elution patterns of proteins and peptides can be followed by UV absorption or colour reactions, but bioassays initially and immunoassays later are required in order to identify the active materials.

3.4.4 ASSAYS

3.4.4.1 Biological

3.4.4.1.1 **In vivo** *assays*

The molecular structure responsible for the biological activity of a hormone may not be identical to that which is recognized by non-biological methods, and since a hormone is defined by its biological activity this type of assay is necessary when the hormone is to be used therapeutically. The complete activity can only be assessed by assays of the *in vivo* type in order to take into account all the various components of biological activity, *i.e.* the transport of the hormone to the target site, the recognition of the target site and the binding and action at this site. Only some of these factors are recognized by *in vitro* type assays.

Unfortunately, *in vivo* assays tend to be inconvenient in terms of cost, time and expertise and are generally less precise and sensitive than non-biological assays. Frequently, the end-point is the assay of another hormone produced at the target site so that errors are magnified. It is also possible that bioassays are affected by non-specific factors in biological fluids and extraction and purification processes have to be included before the assay.

In order to attain specificity, particularly when dealing with closely related hormones such as the gonadotrophins, hypophysectomized animals have been frequently employed. There have been many attempts to avoid this difficult and costly procedure and in the example of the gonadotrophins suitable alternatives were found. Steelman and Pohley[23] needed a simple specific assay for their attempts to purify FSH and decided to make use of the interaction between FSH and LH on ovarian weight. It was shown that by administering a large excess of LH in the form of hCG to immature rats any LH contamination in the sample would not affect the ovarian weight response to FSH. The relatively simple assay which they developed proved to be highly specific and it explains the rapid progress that was then made in the purification of FSH. Until radioimmunoassays became available the method was the simplest, most sensitive and specific of all those available.

Assays are performed for various purposes including the estimation of the concentration of a substance in a test specimen, the comparison of the effects of two substances or treatments and the assessment of the response of two or more populations, individuals or tissues to a common treatment.[58] An essential component of assays is the standard; for an International Standard of a complex system such as a protein hormone, the preparation should consist of the natural and unaltered preparation that shows the highest potency in a 'classical' *in vivo* biological assay system. Other substances such as steroids may be regarded as pure chemicals, the whole exact structure of which can be determined by physical and chemical methods.

Descriptions of practical and well-proved 'classical' *in vivo* bioassays have been collected together with statistical procedures in pharmacopoeas,[59] and guidelines for the design of assays, the improvement of performance and the proper use of standards have been reviewed.[58]

3.4.4.1.2 **In vitro** *assays*

In vitro assays are easier to perform and each assay can include larger numbers of samples than the corresponding *in vivo* assays. They are therefore more suitable for monitoring fractions for activity during extraction and purification procedures.

There are many different types of *in vitro* assay; the biological tissue may, for instance, be tissue segments or thin sections, dissociated intact cells or fragments of cells or a solution of hormone receptors. The end-point may take many forms. The 'ligand' assay[60] depends on non-covalent binding to the target cell receptor, the binding action having an affinity constant of about 10^{-6} to 10^{-8} M L^{-1}. The binding can be measured by using a radioisotopically labelled hormone as in immunoassays and in protein-binding assays. The latter make use of binding proteins such as thyroid-binding globulin for thyroxine[61,62] or cortisol-binding globulin for cortisol.[63] In this type of 'saturation analysis' a tracer amount of labelled hormone is added to the hormone under test in the presence of a limited amount of receptor. The tracer is distributed between the binding protein (bound fraction) and the medium (free fraction), the amount that is bound being dependent upon the amount of test sample.

This binding type of assay is not so much a functionally directed assay as those which depend on the recognition and binding to the target followed by an interaction with target cells such as the activation of adenylate cyclase and the release of cAMP. The final product can be measured by any appropriate method such as radioimmunoassay to measure steroids produced from adrenal or gonadal tissues or pituitary hormones from pituitary tissue. A particularly sensitive type of assay uses a cytochemical end-point as already described for LH.[27] Similarly the depletion of ascorbic acid in adrenal segments from the guinea pig, with quantitative measurement by microdensitometry of a Prussian blue reaction, gives one of the most sensitive assays for ACTH.[64a]

The 'eluted stain assay' (ESTA) is derived from the cytochemical assay but is not so demanding technically.[64b] The end-point utilizes the reduction of a tetrazolium salt to a formazan by intracellular dihydrogenases but differs from the cytochemical assay by using uniform microcultures of target cells maintained in microtitre plates rather than tissue segments or sections. The cytochemical stain is eluted from the microcultures and measured in a microtitre plate reader. Systems have been described for hGH, prolactin, TSH and hCG, and although at present these are not as ultrasensitive as cytochemical bioassays, their sensitivity is adequate, the precision is good and no expensive microspectrophotometric apparatus is required.

A typical hormone bioassay may involve one or more sites on the hormone and one or more complementary combining sites on the target cell. Furthermore, the hormone may act on several different target tissues. Another problem is illustrated by the glycoprotein hormones, which include the gonadotrophins and TSH. The *in vivo* activity of these hormones is markedly affected by minor molecular modifications, particularly by loss of terminal sialic acid moieties.[65] Galactose residues are then exposed and there is increased liver uptake and degradation by peptidases. In consequence, the desialylated products possess lowered potencies compared to the intact hormones and in *in vivo* bioassay systems fewer molecules reach and interact with target cells. In contrast, in *in vitro* systems the hormone is incubated directly with target cells and the potencies of the desialylated and intact hormones may be similar, since sialic acid residues do not appear to be directly involved in the specific hormonal interaction with the receptors. The *in vitro* type of assay, therefore, cannot be assumed to have the same specificity as *in vivo* bioassays. However, it must be recognized that the latter can also be influenced by hormone metabolism, feedback mechanisms and many other interacting factors.

3.4.4.2 Immunoassays

Assays based on the use of antibodies have widespread applications in the monitoring of extraction and purification procedures and in diagnosis.[66] There are a number of basic requirements: a purified preparation of the substance to be analyzed, *i.e.* the antigen, the antibody and a system for detecting the reaction between antigen and antibody.

Purified antigen serves as immunogen for the raising of the antibody which for assay purposes has until recently been a polyclonal antibody. Small molecules such as steroids, thyroid hormones and peptides are available in pure form, but in order to render them immunogenic they must be linked to a larger molecule. Bovine serum albumin is most commonly used, but others include ovalbumin, thyroglobulin and keyhole limpet hemocyanin. The position of the linkage to the carrier is important since the biologically important groups in the antigen molecule should be left accessible for antibody response. Antibody specificity is directed primarily at that part of the antigen (hapten) molecule furthest removed from the linkage to the carrier. The best results for most steroids, therefore, are obtained when the linkage is at positions 6 or 11 since important groups in both rings A and D are available for antibody reactions. The methods of linking steroids have been reviewed;[67] commonly, hemisuccinates or *O*-carboxymethyl oximes are linked to protein by the mixed anhydride reaction.

For raising (specific) antibodies to protein and glycoprotein hormones, the best results are obtained by using highly purified preparations.[68] There are many different immunization schemes;[69] for glycoproteins the multisite intradermal method has proved efficient.[70] The immunogen is dissolved in saline and mixed with two volumes of Freund's complete adjuvant. This is administered (1.5 mL) to between 30 and 70 sites on an area about 20 × 30 cm on the backs of rabbits, the maximum amount of antigen being about 100 μg per rabbit. Booster injections are given by the same route but only at 10–20% of the amount used initially. The titre of antibody is tested at intervals and best results are obtained when high titres are produced by 14–20 weeks. Booster injections are given if necessary after the titre has fallen by at least 50% of the maximum reached, after which high acceptable titres should be reached within 2–3 weeks.

The screening and characterization of antisera includes assessment of their energy reactions (*i.e.* their sensitivity potential) and the determination of their specificity. The strength of binding of antigen to polyclonal antibody in typical assay systems is high, with affinity constants typically in the range of 10^{-8} to 10^{-11} M L^{-1}.[60]

In processes where high sensitivity is not a problem a labelled reagent is not necessary. The assay then depends upon a precipitin line which forms in a gel support when the antigen comes into contact with the antibody. To improve sensitivity and convenience the antigen can be coated on to red cells or latex particles. These particles agglutinate in the presence of antibody and if there are limited amounts of antigen-coated particles and of antibody there is competition for binding sites when the test sample containing antigen is added and this forms the basis of an inhibition test. These methods provide visual end-points and with latex particles give yes/no answers within minutes, as for instance in pregnancy tests for hCG.

Instead of coating particles, antigens can be labelled with a radioactive tracer. This forms the basis of radioimmunoassay, the most commonly used immunoassay procedure. The advantages are high sensitivity, potentially high specificity, convenience, speed, reproducibility if properly controlled and relatively low cost. The disadvantages are that the immunological sites recognized in the assay may not be those concerned with biological activity, the radioactive hazards and the relatively short effective lives of the labelled compounds. For many applications the advantages of the method far outweigh the disadvantages.

In the usual form of the assay the principle of saturation analysis is used. A fixed amount of antibody (Ab) is reacted with a fixed quantity of labelled antigen (Ag*) and a variable and unknown amount of antigen (Ag) in the sample or the standard (Figure 4). The reaction is allowed to equilibrate over a fixed time and the binding sites on the Ab become saturated with Ag* and Ag in proportion to their relative concentrations in the assay tubes. The Ab-bound and unbound (free) forms of Ag are then separated and the amount of label bound is counted and read off a standard calibration curve plotted as bound counts against standard dose of Ag suitably transformed if necessary.

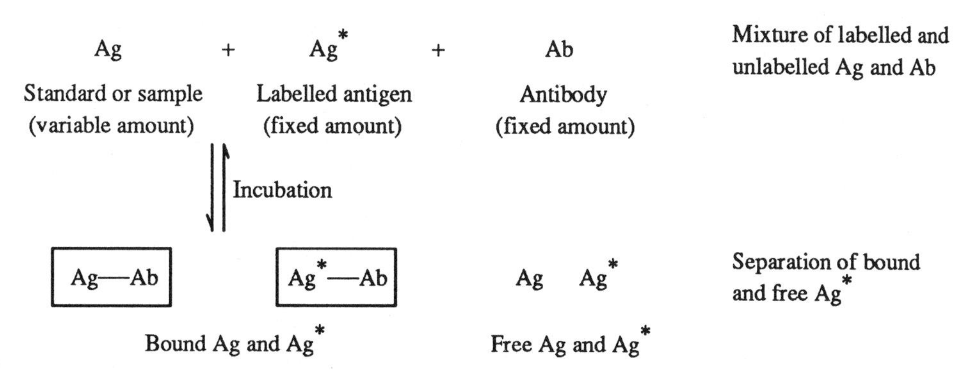

Figure 4 Diagram of a radioimmunoassay (saturation analysis). The standard or sample (Ag) is incubated with a fixed amount of Ag* and Ab. The bound Ag* is then separated from the free Ag* and is counted. The amount of Ag* bound is limited by the number of binding sites and rises inversely with the amount of Ag in the sample or standard

Methods for the separation of bound and free fractions vary considerably.[66] Solid phase methods using Sepharose or cellulose to which the antibody may be attached, or dextran-coated charcoal for the separation of small molecules such as steroids, have been widely used. One of the best methods, however, is the double-antibody method. Here a second antibody raised in a different species of animal to that in which the primary antibody was raised is used to precipitate the bound fraction. As with other methods this requires careful optimization with regard to titre, time of incubation, *etc.* There are many variations of the method, among which is the use of poly(ethylene glycol) to accelerate the precipitation and the use of solid phases for the second antibody. Magnetizable particles to which are coupled the second antibody provide a method of separation which does not need centrifugation and this method is now widely used.

There has been enormous interest in alternative non-isotopic assay methods and some of these have comparable performance characteristics to radioimmunoassay. Among the many labels proposed[71] three types have emerged as the most likely alternatives: enzymes,[72] luminescent compounds[73] and fluorescent labels.[74,75]

The main potential advantage of using an enzyme as a label is the amplification of signal effected by the enzyme acting on several substrate molecules. The enzyme-multiplied immunoassay technique (EMIT) is based on the principle of saturation analysis. The separation of bound and free fractions is avoided by selecting a system in which the binding of specific Ab to the enzyme label results in inhibition or enhancement of the enzyme–substrate reaction. The enzyme-linked immunosorbent assay (ELISA) uses the principle of two-site immunometric analysis of one Ab coated to the wells of a microtitre plate. Ag binds to this Ab and a second Ab labelled with enzyme binds to a different epitope on the Ag to form a sandwich. The unbound label is removed by aspiration and washing, and colour-developed by adding substrate which can then be read by spectrophotometric scanning of the plate. Another variation is based on the principle of enzyme amplification. One Ab is labelled with alkaline phosphatase, which at the end of the assay is used to catalyze a chain of reactions with signal amplification at each stage.

Fluorescence immunoassay has been improved greatly by the introduction of time-resolved fluorescence. Europium chelates, used as the fluorophore in the dissociation-enhanced lanthanide fluoroimmunoassay (DELFIA), have a large Stokes' shift and a relatively long-lived fluorescence. Fluorimeters have been designed which delay the measurement of the emitted light by, for instance, 400 μs during which time non-specific background fluorescence will have largely disappeared. The cycle can be repeated and measurements taken again, an advantage that this system has over some of the other non-isotopic methods. This type of assay is at least as sensitive as the best radioimmunoassays and the counting time is <1 s.

Chemiluminescent immunoassay is another important non-isotopic method. The emission of photons from a chemical reaction is less affected by background interference than are other techniques which depend upon photometry. A disadvantage is that the reaction can only take place once so that recounting is not possible. The reaction is oxidative and luminescent compounds such as luminol and related compounds can be linked directly to proteins and haptens, but the light output from the oxidation of luminol is present for only a few seconds. Instead, the catalytic component of the chemiluminescent reaction such as horseradish peroxidase can be used as a tracer molecule (Figure 5). Again, such reactions have low efficiency and much better results have been obtained by the discovery of compounds which enhance the light output from the oxidation of luminol and give prolonged output of light of high intensity.[76]

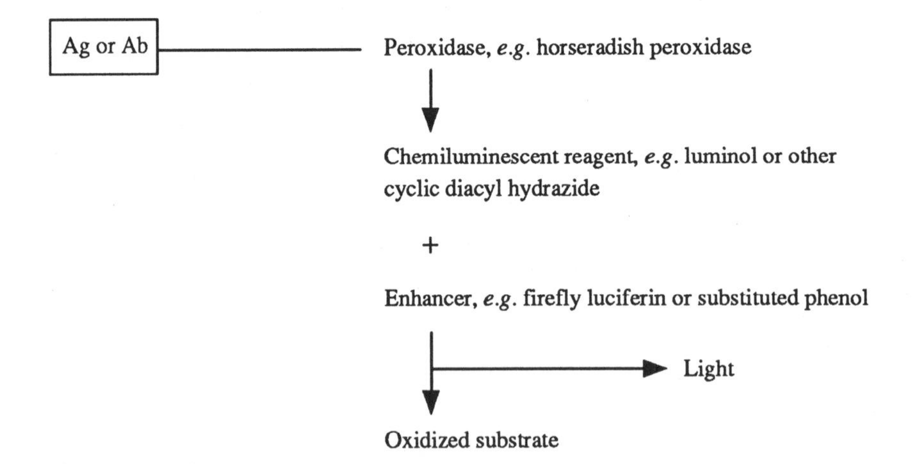

Figure 5 Example of enhanced luminescence immunoassay

Firefly luciferin was found to enhance the chemiluminescence of luminol several-fold, while actually depressing the interference from the reagent blank. Later, other enhancers were found such as dehydroluciferin and 6-hydroxybenzothiazole, which were even more efficient. The reaction can be optimized to give a stable output of light within 2 min of adding the reagent and this output only decreases slowly over the next 20–30 min so that repeated measurements are possible.

Chemiluminescent-labelled compounds are stable, free of hazard and the reagents are relatively cheap. The assays can be even more sensitive than radioimmunoassay and of comparable precision and they clearly may be considered as alternatives to radioimmunoassay.

The introduction of monoclonal antibodies, as well as assisting in extraction methods by the use of immunoaffinity chromatography, is changing immunoassay itself. This hybridoma technique[77] allows the polyclonal response to be dissected into its monoclonal compartments. This is achieved by fusing antibody-secreting cells to cells of a plasmacytoma line. The monoclonal antibody (MAb) will be specific for a single epitope.

The first stage is similar to the polyclonal technique, except that mice are used for immunization with the antigen of interest.[78] Spleen cells rather than serum are removed from the animal when antibody titre is satisfactory and they are fused with myeloma cells in the presence of poly(ethylene glycol). The hybridomas are selected for their ability to grow in a medium containing aminopterin, which poisons unfused myeloma parent cells, and for the hybridoma's ability to produce MAb of desired characteristics. The advantages of this technique over that for polyclonal Abs are that the immunogen need not be so highly purified, the immunoglobulin fraction of the MAb is homogeneous, the MAbs being produced by a colony of cells derived from a single hybridoma, and the cross-reactions and non-specific reactions are eliminated by the selection procedure.

MAbs have been applied as replacements for polyclonal Abs in most of the classical immunoassay procedures but they are particularly useful in labelled Ab and immunoradiometric (IRMA) methods (Figure 6).[79] Polyclonal Abs in these assays require affinity purification of the label and this may result in loss or partial destruction of high and low affinity Ab species. In the two-site IRMA, Ag reacts with solid phase Ab adsorbent. Unbound Ag is removed by washing and labelled Ab then reacts with the bound Ag and the label associated with the solid phase is directly proportional to the amount of Ag originally bound in the sample.

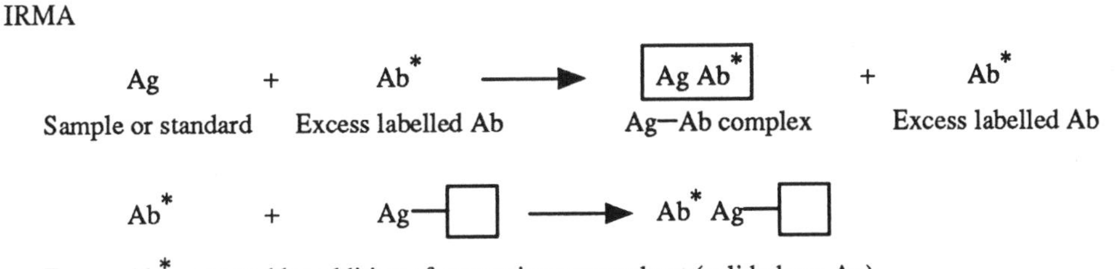

Two-site procedure

Ag + Ab + Ab* → Ab* Ag Ab + Ab*

Sample or standard | Excess solid phase Ab | Excess labelled Ab | Ag removed by solid phase Ab and linked to Ab* | Excess free Ab*

Figure 6 Immunoradiometric assay (IRMA) procedures

In the 'reverse two-step' method Ag reacts with labelled MAb in solution and, after an appropriate time for incubation, solid phase MAb is added. The amount of Ag originally present is related to the quantity of label associated with the solid phase.

The 'simultaneous' procedure is simple, as all the reagents and specimen are mixed together, *i.e.* solid phase MAb, Ag and labelled MAb. Again, the amount of label bound will be proportional to the amount of Ag.

Some of the advantages of IRMA over radioimmunoassay are that the Ag is measured directly as opposed to estimations based on the Ag's ability to compete with labelled Ag for the combining site. Secondly, the precision relates essentially only to the addition of the specimen, other reactants being in excess, and, thirdly, the Ag does not have to be labelled, a source of many technical problems related to damage to the antigenic site and to the relative instability of some radiolabelled Ags.

3.4.5 REFERENCES

1. V. C. Medvei, 'A History of Endocrinology', MTP Press, Lancaster, 1982, p. 58.
2. W. R. Butt, in 'Hormone Chemistry', 2nd edn., Horwood, Chichester, 1975, vol. 1; 1976, vol. 2.
3. A. V. Schally, T. W. Redding, C. Y. Bowers and J. F. Barrett, *J. Biol. Chem.*, 1969, **244**, 4077.
4. P. Brazeau, W. Vale, R. Burgus, N. Ling, M. Butcher, J. Rivier and R. Guillemin, *Science (Washington, D.C.)*, 1973, **179**, 77.
5. A. V. Schally, A. Arimura, C. Y. Bowers, A. J. Kastin, S. Sawano and T. W. Redding, *Rec. Prog. Horm. Res.*, 1968, **24**, 497.
6. A. H. Livermore and V. du Vigneaud, *J. Biol. Chem.*, 1949, **180**, 365.
7. R. A. Turner, J. G. Pierce and V. du Vigneaud, *J. Biol. Chem.*, 1951, **191**, 21.
8. W. Vale, J. Spiess, C. Rivier and J. Rivier, *Science (Washington, D.C.)*, 1981, **213**, 1394.
9. W. Vale, C. Rivier, M. R. Brown, J. Spiess, G. Koob, L. Swanson, L. Bilezikjian, F. Bloom and J. Rivier, *Rec. Prog. Horm. Res.*, 1983, **39**, 245.
10. G. E. Gillies and P. J. Lowry, *Nature (London)*, 1979, **278**, 463.
11. S. Shibabara, Y. Morimoto, Y. Furutani, M. Notake, H. Takehashi, S. Shimizu, S. Horikawa and S. Numa, *EMBO J.*, 1983, **2**, 775.
12. K. Nikolics, A. J. Mason, E. Szonyi, J. Ramachandran and P. H. Seeburg, *Nature (London)*, 1985, **316**, 511.
13. R. D. G. Milner, T. Russell-Fraser, C. G. D. Brook, P. M. Cotes, J. W. Farquhar, J. M. Parkin, M. A. Preece, G. J. A. I. Snodgrass, A. Stuart Mason, J. M. Tanner and F. P. Vince, *Clin. Endocrinol. (Oxford)*, 1979, **11**, 15.
14. J. Powell-Jackson, R. O. Weller, P. Kennedy, M. A. Preece, E. M. Whitcombe and J. Newsom-Davis, *Lancet*, 1985, **2**, 244.
15. D. V. Goeddel, H. C. Heyneker, T. Hozumi, R. Arentzen, K. Itakura, D. G. Yansura, M. J. Ross, G. Moizzari, R. Crea and P. H. Seeburg, *Nature (London)*, 1979, **281**, 544.
16. K. C. Olson, J. Fenno, N. Lin, R. N. Harkins, C. Snider, W. H. Kohr, M. J. Ross, D. Fodge, G. Prender and N. Stebbing, *Nature (London)*, 1981, **293**, 408.
17. S. S. Lynch, M. Bluck, P. Reay and W. R. Butt, *Acta Endocrinol. (Copenhagen)*, 1988, **119**, Suppl. 288, 12.
18. P. J. Lowry, R. E. Silman, J. Hope and A. P. Scott, *Ann. N. Y. Acad. Sci.*, 1977, **297**, 49.
19. G. W. Jack and R. Blazek, *J. Chem. Technol. Biotechnol.*, 1987, **39**, 1.
20. G. W. Jack, R. Blazek, K. James, J. E. Boyd and L. R. Micklem, *J. Chem. Technol. Biotechnol.*, 1987, **39**, 45.
21. F. S. Greenspan, C. H. Li, M. E. Simpson and H. M. Evans, *Endocrinology (Baltimore)*, 1949, **45**, 455.
22. C. S. Cockram, P. H. Sonksen and T. E. T. West, in 'Hormones in Blood', 3rd edn., ed. C. H. Gray and V. H. T. James, Academic Press, London, 1983, vol. 4, p. 65.
23. S. L. Steelman and F. M. Pohley, *Endocrinology (Baltimore)*, 1953, **53**, 604.
24. A. F. Parlow, in 'Human Pituitary Gonadotrophins', ed. A. Albert, Thomas, Springfield, IL, 1961, p. 300.
25. W. R. Butt, in 'Hormones in Blood', 3rd edn., ed. C. H. Gray and V. H. T. James, Academic Press, London, 1979, vol. 1, p. 411.
26. M.-P. Van Damme, D. M. Robertson, R. Marana, E. M. Ritzen and E. Diczfalusy, *Acta Endocrinol. (Copenhagen)*, 1979, **91**, 224.
27. R. M. Kramer, I. M. Holdaway, L. H. Rees, A. S. McNeilly and T. Chard, *Clin. Endocrinol. (Oxford)*, 1974, **3**, 375.
28. C. P. Channing, L. D. Anderson, D. J. Hoover, J. Kolena, K. G. Osteen, S. H. Pomerantz and K. Tanabe, *Rec. Prog. Horm. Res.*, 1982, **38**, 331.
29. C. G. Tsonis and R. M. Sharpe, *Nature (London)*, 1986, **321**, 724.
30. N. Ling, S.-Y. Ying and N. Ueno, *Proc. Natl. Acad. Sci. U.S.A.*, 1985, **82**, 7217.
31. N. Ling, S.-Y. Ying, N. Ueno, S. Shimasaki, F. Esch, M. Hotta and R. Guillemin, *Nature (London)*, 1986, **321**, 779.
32. N. Ling, S.-Y. Ying, N. Ueno, F. Esch, L. Deneroy and R. Guillemin, *Proc. Natl. Acad. Sci. U.S.A.*, 1985, **82**, 7217.
33. W. Vale, C. Rivier, A. Hsueh, C. Campen, H. Meunier, T. Bicsak, J. Vaughn, A. Corrigan, W. Bardin, P. Sawchenko, F. Petraglia, J. Yu, P. Plotsky, J. Spiess and J. Rivier, *Rec. Prog. Horm. Res.*, 1988, **44**, 1.
34. J. Rivier, J. Spiess, R. McClintock, J. Vaughan and W. Vale, *Biochem. Biophys. Res. Commun.*, 1985, **133**, 120.
35. L. J. Leversha, D. M. Robertson, F. L. de Vos., F. J. Morgan, M. T. W. Hearn, R. E. M. Wettenhall, J. K. Findlay, H. G. Burger and D. M. de Kretser, *J. Endocrinol.*, 1987, **113**, 213.
36. (a) R. S. Scott, H. G. Burger and H. Quigg, *Endocrinology (Baltimore)*, 1980, **107**, 1536; (b) H. G. Burger and M. Igarashi, *J. Clin. Endocrinol. Metab.*, 1988, **66**, 85.
37. T. Chard, 'An Introduction to Radioimmunoassay and Related Techniques', Elsevier, Amsterdam, 1982.
38. R. E. Hunt, K. Moffatt and D. W. Golde, *J. Biol. Chem.*, 1981, **256**, 7042.
39. J. D. Gardner and R. T. Jensen, *Rec. Prog. Horm. Res.*, 1983, **39**, 211.
40. H. Gregory and P. Scholes, *Top. Horm. Chem.*, 1978, **1**, 48.
41. K. Tatemoto, S. Efendic, V. Mutt, G. Makk, G. J. Feistner and J. D. Barchas, *Nature (London)*, 1986, **324**, 476.
42. A. Iacangelo, H.-U. Affolter, L. E. Eiden, E. Herbert and M. Grimes, *Nature (London)*, 1986, **323**, 82.
43. L. E. Eiden, *Nature (London)*, 1987, **325**, 301.
44. (a) K. Tatemoko and V. Mutt, *Proc. Natl. Acad. Sci. U.S.A.*, 1978, **75**, 4115; (b) J. F. Rehfeld, *J. Mol. Endocrinol.*, 1988, **1**, 87.
45. P. Donini, D. Puzzuoli and R. Montezemolo, *Acta Endocrinol. (Copenhagen)*, 1964, **45**, 321.
46. P. Roos, *Acta Endocrinol. (Copenhagen)*, 1968, **59**, Suppl., 131, 1.
47. J. J. Bell, R. E. Canfield and J. J. Sciarra, *Endocrinology (Baltimore)*, 1969, **84**, 298.
48. H. Van Hell, in 'Gonadotropins and Gonadal Function', ed. M. R. Moudgal, Academic Press, New York, 1974, p. 66.
49. R. E. Canfield, F. J. Morgan, S. Kammerman, J. J. Bell and G. M. Agosto, *Rec. Proc. Horm. Res.*, 1971, **27**, 121.
50. J. L. Vaitukaitis, *Clin. Chem. (Winston-Salem, N. C.)*, 1985, **31**, 1749.
51. D. J. Morrell, K. P. Ray, A. T. Holder, A. M. Taylor, J. A. Blows, D. J. Hill, M. Wallis and M. A Preece, *J. Endocrinol.*, 1986, **110**, 151.
52. G. S. G. Spencer and A. M. Taylor, *J. Endocrinol.*, 1978, **78**, 83.
53. D. J. Hill, R. Watson and R. D. G. Milner, *J. Clin. Endocrinol. Metab.*, 1984, **59**, 231.
54. A. J. Moody, M. A. Stan, M. Stan and J. Glieman, *Horm. Metab. Res.*, 1974, **6**, 12.
55. A. Braithwaite and F. J. Smith, 'Chromatographic Methods', 4th edn., Chapman and Hill, London, 1986.

56. W. S. Hancock, 'Handbook of HPLC for the Separation of Amino Acids, Peptides and Proteins', CRC Press, Boca Raton, FL, 1984, vols. I and II.
57. E. S.Yueng, 'Detectors for Liquid Chromatography', Wiley, Chichester, 1986.
58. D. R. Bangham, in 'Hormones in Blood', 3rd edn., ed. C. H. Gray and V. H. T. James, Academic Press, London, 1985, vol. 5, p. 256.
59. 'British Pharmacopoeia', HMSO, London, 1980, vols. I and II.
60. R. P. Ekins, in 'Hormone Assays and their Clinical Applications', 4th edn., ed. J. Loraine and T. Bell, Churchill Livingstone, Edinburgh, 1976, p. 1.
61. R. P. Ekins, *Clin. Chim. Acta*, 1960, **5**, 453.
62. B. E. P. Murphy and C. J. Pattee. *J. Clin. Endorcrinol. Metab.*, 1964, **24**, 187.
63. B. E. P. Murphy, *J. Clin. Endocrinol. Metab.*, 1967, **27**, 973.
64. (a) J. Chayen, J. R. Daly, N. Loveridge and L. Bitensky, *Rec. Prog. Horm. Res.*, 1976, **32**, 33; (b) P. A. Ealey, M. E. Yateman, S. J. Holt and N. J. Marshall, *J. Mol. Endocrinol.*, 1988, **1**, R1.
65. W. R. Moyle, O. P. Bahl and L. Marz, *J. Biol. Chem.*, 1975, **250**, 9163.
66. W. R. Butt (ed.), 'Practical Immunoassay', Dekker, New York, 1984.
67. B. F. Erlanger, *Methods Enzymol.*, 1980, **70**, 85.
68. S. S. Lynch and A. Shirley, *J. Endocrinol.*, 1975, **65**, 127.
69. B. A. L. Hurn and S. M. Chantler, *Methods Enzymol.*, 1980, **70**, 104.
70. J. Vaitukaitis, J. B. Robbins, E. Nieschlag and G. T. Ross, *J. Clin. Endocrinol. Metab.*, 1971, **33**, 988.
71. R. F. Schall, Jr. and H. J. Tenoso, *Clin. Chem. (Winston-Salem, N. C.)*, 1981, **27**, 1157.
72. M. J. O'Sullivan, in 'Practical Immunoassay', ed. W. R. Butt, Dekker, New York, 1984, p. 37.
73. L. J. Kricka and T. J. N. Carter (eds.), 'Clinical and Biochemical Luminescence', Dekker, New York, 1982.
74. S. Dakubu, R. P. Ekins, T. Jackson and N. J. Marshall, in 'Practical Immunoassay', ed. W. R. Butt, Dekker, New York, 1984, p. 71.
75. I. Hemmilä, *Clin. Chem. (Winston-Salem, N. C.)*, 1985, **31**, 359.
76. T. P. Whitehead, G. H. G. Thorpe, T. J. N. Carter, C. Groucutt and L. J. Kricka, *Nature (London)*, 1983, **305**, 158.
77. G. Kohler and C. Milstein, *Nature (London)*, 1975, **256**, 495.
78. E. D. Sevier, G. S. David, J. Martinis, W. J. Desmond, R. M. Bartholomew and R. Wang, *Clin. Chem. (Winston-Salem, N. C.)*, 1981, **27**, 1797.
79. L. E. M. Miles, in 'Handbook of Radioimmunoassay', ed. G. E. Abraham, Dekker, New York, 1977, p. 131.

3.5
Biomaterials from Mammalian Sources

WILFRID R. BUTT
Birmingham & Midland Hospital for Women, UK

3.5.1 INTRODUCTION

Many materials with important therapeutic applications which were originally obtained from mammalian sources have now been synthesized. These include such hormones as steroids, thyroid hormones, catecholamines, corticotrophin and recently human growth hormone. This chapter is mainly concerned with some of those hormones which are still only obtained from biological tissues and fluids. Some of these have well-established diagnostic and therapeutic applications. Others have been demonstrated to possess certain biological activities and others may have important applications in the future.

3.5.2 HYPOTHALAMUS

The hypothalamus contains a number of peptides which regulate the synthesis and release of pituitary hormones (Table 1). By 1980 nearly all the expected hypothalamic hormones had been

Table 1 Some Regulating Hormones Isolated from the Hypothalamus

Name	*Abbreviation*	*Number of residues*	*Remarks*
Releasing			
Thyrotrophin-releasing hormone	TRH	3	Also releases prolactin
Gonadotrophin-releasing hormone	GnRH	10	Releases both FSH and LH
Corticotrophin-releasing factor	CRF	41	
Growth-hormone-releasing factor	GRF	Biological activity residues in 27 residues	Forms of GRF with 37–44 residues isolated from tumours
Inhibiting			
Growth hormone release-inhibiting hormone (somatostatin)	GRIF	14	Inhibiting effect not specific for GH
Gonadotrophin-releasing hormone associated protein	GAP	56	Inhibits release of prolactin and stimulates gonadotrophin release

identified except releasing hormones for growth hormone and corticotrophin and the inhibiting factor for prolactin.[1,2] Many of these peptides have now been synthesized. Although originally isolated from hypothalamic tissue, the peptides are widely distributed in others areas, including nervous tissue, the gastrointestinal tract, gonads and placenta.

3.5.2.1 Gonadotrophin-releasing Hormone (GnRH)

GnRH is a single chain decapeptide which has diagnostic and therapeutic applications. Although originally termed luteinizing-hormone-releasing hormone, this name is misleading since, when administered, it releases both luteinizing hormone (LH) and follicle-stimulating hormone (FSH).[3]

There are probably two pathways involved in the mode of action of GnRH: within minutes of GnRH-receptor activation intracellular free calcium concentration rises and also polyphosphoinositides are broken down to generate diacylglycerol and inositol triphosphate. These compounds are involved in the action of protein kinase C and in calcium mobilization respectively. It was originally considered that adenylate cyclase was coupled to the GnRH receptor but what appears to happen is that activation of adenylate cyclase occurs secondarily to the rise in calcium and possibly cAMP plays a role in later trophic actions of GnRH.

The peptide has been used diagnostically in tests of pituitary function.[4,5] Usually 100 μg is given intravenously and serum gonadotrophins are measured immediately before and at intervals (*e.g.* 30 and 60 min) after the injection. In normal subjects serum LH and to a lesser extent FSH rises rapidly, reaching maximum levels at 20–30 min and falling significantly by 60 min. The response varies considerably during the menstrual cycle, being lowest in the early follicular phase when circulating estrogens are low and highest at mid cycle when estrogens are high also.[6] Experimentally it can be shown that estrogen sensitizes the pituitary to the action of GnRH.[7]

Caution is required in interpreting this test as normal responsiveness to GnRH depends on the presence of an intact hypothalamus and pituitary and a defect at either site impairs the response.[5] GnRH is involved in synthesis as well as release and if hypothalamic deficiency is of long standing, the pituitary will be depleted of gonadotrophin.

GnRH is normally released from the hypothalamus in a pulsatile manner, pulses occurring at about 90 min intervals in the follicular phase of the cycle.[8,9] In hypothalamic rather than pituitary failure, administration of GnRH in pulses allows pituitary gonadotrophin reserves to build up and the response gradually increases. This helps to distinguish hypothalamic from pituitary failure, in which condition there would be no response.

Therapeutically GnRH offers an important method for the treatment of anovulation related to a hypothalamic defect.[10–14] The hormone must be administered in a physiological, *i.e.* pulsatile manner. This can be achieved by the use of a battery-operated pump adjusted to administer doses of GnRH, usually in the range 10–20 μg subcutaneously, at 90 min intervals. Gonadotrophins (FSH and LH) are released and follicular growth should commence. Ultrasonic scanning or the determination of estrogens in blood or urine can be used to monitor follicular growth and, as in a normal cycle by 12 to 14 days, the follicle is between 17 and 25 mm diameter and is secreting maximum quantities of estrogen. The pituitary at this time is more sensitive to GnRH and a surge of LH with

FSH occurs as in the normal cycle inducing ovulation. In some centres treatment is continued throughout the luteal phase but good results are obtained by giving human chorionic gonadotrophin (hCG) (5000 i.u. intramuscularly) to augment the LH surge and then the pump is removed after a further 48 h.[15]

This method of treatment is extremely effective when used for patients with hypothalamic amenorrhoea and is without any serious hazards such as hyperstimulation and the increased risk of multiple pregnancies, which are problems with gonadotrophin therapy.

Chronic pulsatile GnRH treatment can also be used to promote spermatogenesis in men with hypogonadism related to hypothalamic/pituitary defects. Treatment has to be maintained for more than six months however, and is not successful in more than about 50% of patients.[16]

There is now good evidence that GnRH is capable of producing inhibitory effects on gonadal function in both males and females, although GnRH of hypothalamic origin is never produced *in vivo* in sufficient concentrations to produce these effects. In the ovary it is considered that the action of gonadotrophins on steroid biosynthesis is blocked by the binding of GnRH to granulosa cell and probably theca cell receptors. Leading on from this came the important clinical observation that GnRH and, even more so, synthetic agonists of GnRH, exert suppressive effects on certain neoplastic tissues. Perhaps the most important medical applications of these long-acting analogues which are being investigated are in the treatment of advanced prostatic cancer[17] and breast cancer.[18]

3.5.2.2 Thyrotrophin-releasing Hormone (TRH)

This tripeptide, pyro-Glu-His-Pro-NH_2, originally isolated from hypothalamic tissue is now known to be widely distributed in other areas of the brain, in nervous tissue such as the retina, the spinal cord and the gastrointestinal tract. In addition to its role of releasing thyroid-stimulating hormone (TSH) from the pituitary it also stimulates the release of prolactin both *in vivo* and *in vitro*.

These actions follow the binding of the peptide to specific receptors in the plasma membranes of specific pituitary cells. It is believed that the mode of action is through hydrolysis of phosphatidylinositol 4,5-diphosphate mediated by phospholipase C: the inositol triphosphate generated then acts as an intracellular messenger to enhance the levels of cytosolic calcium and subsequently to increase calcium efflux and hormone release.[19]

TRH was the first of the hypothalamic releasing hormones to be recognized as having neuronal functions outside the pituitary.[20] The most carefully studied function has been as a possible antidepressant. There was transient benefit from intravenous injections and other methods of administration, such as intramuscular injections, have been tried. A blunting of the TSH response to TRH has been noted in some patients with classical depression[21] but the effectiveness of TRH in treating psychiatric disorders has so far proved disappointing.

TRH may be useful, however, in the treatment of shock and spinal cord injury[22] and the possible induction of motor neurone regeneration has led to studies of the treatment of the weakness and spasticity in patients with amyotrophic lateral sclerosis.[23] TRH is believed to act as a neurotrophic agent from evidence that it increases choline acetyltransferase activity and neurite outgrowth, reverses neuronal damage[24] and stimulates myelin lipid synthesis in chick neural cultures.[25] It also affects heart rate, arterial blood pressure and respiration rate, possibly related to the increase produced in circulating adrenalin and noradrenalin.

The chief clinical use of TRH has been in the diagnosis of pituitary function.[5] A standard dose is 200 μg given intravenously after which concentrations of TSH in blood should begin to rise after 5 min, reaching a peak between 10 and 30 min and falling over the next hour or two. Specimens of blood are therefore taken before and at 20 or 30 min and 60 min after TRH injection. The test may give variable results in pituitary–hypothalamic disease and measurements of thyroid hormones will also be needed. Although an absent or impaired TSH response to TRH is found in hypothyroidism secondary to pituitary disease,[26] this is not diagnostic of thyroid failure since some patients may nevertheless be clinically euthyroid.[27] Low basal TSH and delayed peak response to TRH is often seen in patients with hypothalamic disease as TRH is responsible for both synthesis and release of TSH from the pituitary. In hyperthyroidism the TSH response to TRH is usually absent because of the feedback effects of the elevated thyroid hormones.

The TRH test has limited clinical usefulness and should be used in conjunction with thyroid hormone assessments. The main diagnostic use is in the investigation of borderline primary hypothyroidism and hyperthyroidism.

3.5.2.3 Growth Factors

A hypothalamic releasing factor for growth hormone (GRF) is the only hypothalamic hormone which was first isolated from an extrahypothalamic source—a pancreatic islet cell tumour. In the hypothalamus it occurs in the arcuate–ventromedial nucleus and the pituitary stalk: it is also found in the gastrointestinal tract, adrenals and placenta. Any function of GRF other than the stimulation of the release of growth hormone, however, is unknown.[28]

A major problem in the isolation of this hormone was the minute amount present in the hypothalamus and the high concentration of a growth hormone release-inhibiting hormone or somatostatin. GRF peptides containing between 37 and 44 residues have now been isolated from tumours and from the hypothalamus: biological activity, however, resides in the first 27 amino acids.

GRF shows a specific effect on human growth hormone (hGH) when administered to man, no other pituitary hormones being affected.[29] It can be administered intravenously with other hypothalamic hormones in tests for anterior pituitary function. When given at a dose of 1 $\mu g\,kg^{-1}$ body weight, there is a prompt release of hGH and when higher doses are given, there is a more prolonged elevation of hGH with a second increase 2 h after injection.

The peptide has potential clinical value in the treatment of children with idiopathic GH deficiency. Preliminary trials[29a] have indicated that pulsatile administration of GRF leads to the release of the growth factor somatomedin (or insulin-like growth factor, IGF-I) from the liver and peripheral tissues. This intermediate in the action of growth hormone is released by the growth hormone stimulated by GRF and is not a direct action of the hypothalamic peptide. Pulses of 1 $\mu g\,kg^{-1}$ body weight of GRF, given three hourly, promoted growth in five of seven children in a recent study.[29b] The growth rate was 4.4–7.5 cm by 1 year: only one of the two non-responders gained linear growth when given hGH itself. For long term therapy intravenous administration is hardly practical: the hormone is active in releasing GH when given intranasally but a 300-fold higher dose is required. Although this is encouraging, the quantities required would be so great that the method will only be economically feasible if less expensive analogues of GRF are developed.

Somatostatin, a 14-amino acid peptide, inhibits the release of several hormones, including GH. As well as occurring in the hypothalamus it is also found in the pancreas, gastrointestinal tract, placenta, thyroid and peripheral nerves. It has many actions in the gastrointestinal tract, such as reducing gastric acid and gastrin secretion, probably by blocking the stimulating action of histamine or acetylcholine on this process.[30] In the intestine it reduces motility, probably by reducing the secretion of motilin and cholecystokinin. Possibly it reduces the release of hormones, including GH, by affecting calcium mobilization.[31]

With its many effects it might be a useful therapeutic agent. Unfortunately it has a short half-life of only 3 min and is active only after intravenous administration. It also causes a rebound hypersecretion of hormones. For this reason long-acting analogues have been synthesized which have potential roles in the treatment of acromegaly, gastrointestinal function, diabetes mellitus, central nervous system disturbances and oncology.[32]

3.5.2.4 Corticotrophin-releasing Factor (CRF)

This 41-amino acid peptide is the most potent of several hypothalamic factors that release corticotrophin (ACTH). The more abundant hypothalamic peptide, vasopressin, was known to have this action long before the isolation of CRF.[33] Vasopressin and other peptides, while possessing intrinsic ACTH-releasing activity, appear to amplify the action of CRF. It appears that vasopressin and oxytocin act in a separate intracellular mechanism, which couples with that activated by CRF.[31] CRF enhances adenylate cyclase activity, while vasopressin activates phosphatidylinositol degradation, leading to inositol triphosphate production, which activates intracellular calcium mobilization. In this way the two processes act synergistically to release ACTH.

There have been both *in vitro* and *in vivo* studies to demonstrate release of ACTH, β- and γ-lipotrophin and β-endorphin following stimulation by CRF.[34] This peptide hormone clearly has a function as a diagnostic agent in the investigation of hypothalamic–pituitary function: an example of the ACTH stimulation by CRF is the diagnosis of pituitary-dependent Cushing's disease.[35]

CRF has been recognized outside the hypothalamus in other parts of the brain, stomach, duodenum, pancreas, adrenal and placenta.[36]

3.5.2.5 Prolactin-inhibiting Factors

For long it has been recognized that prolactin secretion is under the control of an inhibiting factor and dopamine has been considered to fill this role. Work on a GnRH prohormone in the human placenta led to the isolation of a GnRH-associated peptide (GAP) which, as well as stimulating gonadotrophins, inhibited prolactin secretion from cultured pituitary cells.[37] This same pro-hormone has now been detected also in the hypothalamus.[38]

The gonadotrophin-releasing properties were investigated in rat anterior pituitary cells and showed that GAP has a preference for releasing FSH over LH compared with the release from GnRH itself. The inhibitory action of GAP on the secretion of prolactin from pituitary lactotrophs *in vitro* showed maximal inhibition to 40 to 45% of the basal secretion by 3–4 h and the dose required was 10 times less than that required for gonadotrophin release. As a prolactin inhibitor, GAP was much more potent than dopamine, the half-maximal inhibitory dose being 2.5×10^{-11} M compared with 10^{-7} M for dopamine. In view of the great success of dopamine agonists such as bromocriptine in the treatment of hyperprolactinaemia, *in vivo* studies on GAP and of more potent analogues will be of great interest.

3.5.3 PITUITARY

The hormones of the human pituitary are species specific so that when they are required for the preparation of reagents or for therapy, glands from animals cannot be substituted. The risk of contamination from viruses, such as the slow virus of the Kreutzfeldt–Jakob syndrome, has prompted renewed efforts to produce the hormones by recombinant DNA techniques. Synthetic hGH is now available and no doubt other pituitary hormones will be synthesized eventually.

3.5.3.1 Growth Hormone

For more than 30 years the collection of human pituitaries has been organized for the extraction of hGH used in the treatment of children who do not grow because of the failure to produce growth hormone.[39] The hormone consists of a single peptide chain of 191 amino acids with two intrachain disulfide bridges.[40] Biologically active hGH has been produced from *Escherichia coli* using recombinant DNA methodology and is now used clinically in some countries.[41,42]

The normal pituitary contains large stores of hGH but the daily secretion rate is quite low.[43] There are discrete secretory episodes associated with sleep, exercise, stress and high protein meals. In children some 50% of the daily secretion occurs in sleep. Secretion from the pituitary gland is regulated by GRF and somatostatin, and influenced by many other agents from the hypothalamus.

The hormone has anabolic action related to the promotion of skeletal and visceral growth and metabolic action affecting fat and carbohydrate metabolism. The effects of hGH on skeletal tissue are not direct but are mediated by small growth factors (somatomedins) largely of hepatic origin, serum concentrations of which are dependent on growth hormone. Failure to grow therefore may originate from a primary deficiency of hGH itself or be secondary to a hypothalamic defect of the GRF or to a failure of somatomedin. The dwarfism described by Laron[44] was in patients with normal or raised growth hormone levels but with low somatomedin levels which did not increase after treatment with hGH. There is, therefore, either a receptor abnormality or a failure in the hepatic cellular generation of somatomedin. In the African pigmy, however, growth hormone level is normal and somatomedin increases after hGH administration. The retarded growth and absence of response to hGH in these pigmies therefore indicates some intrinsic somatomedin abnormality or a defective cell receptor for somatomedin or some other growth factor.[45] There is some evidence for an isolated deficiency of one of the somatomedins, somatomedin C or IGF-I.

The long-term use of hGH in pituitary deficiency dates from 1958 and the natural source was the only one for about 25 years until the biosynthetic material became available. The amount of hormone available was restricted and careful selection of patients was necessary, only those with established growth hormone deficiency being accepted for treatment. Basal serum determinations of hGH are of little diagnostic value because normal levels are so low: provocative tests are therefore required. The hormone is released during sleep and by exercise and in the commonly used insulin-induced hypoglycaemia test. Although this latter test is potentially dangerous it has been the most reliable: a satisfactory hypoglycaemic stimulus with a blood glucose of < 2.2 mM L^{-1} or a decrease of 50% of the basal blood glucose level should result in a rise in serum hGH to at least 15 u L^{-1} within 2 h. In the normal subject hyperglycaemia induced by oral glucose results in suppression of hGH. This suppression, however, is followed within about 5 h by a compensatory rebound and this test has also been used to assess hGH secretion in short children.

There has been considerable research and experience in the treatment of short children with hGH. Commonly a dose of four units is given three times per week by intramuscular injections. The first treatment should produce a dramatic increase in growth velocity, approximately three times the pretreatment velocity, with smaller increases in succeeding years. The use of growth hormone from human glands is essential for this treatment, antibodies being produced if growth hormone from animal species is used because of structural differences. Even with hGH the method of purification affects the product and if chemical changes occur during processing, antibodies are produced. Growth inhibition or a reduction in response to its high affinity antibodies has been noted in about 3% of patients.[38]

3.5.3.2 Gonadotrophins

The two pituitary gonadotrophins, FSH and LH, are glycoproteins each of molecular weight approximately 30 kDa. They contain a common α subunit which is shared with TSH and a hormone-specific β subunit. They are essential in the processes of reproduction in both male and female. In the testis the Leydig cells in the interstitial compartment secrete steroids, notably androgens, and LH plays a major role in stimulating this process.[46] The site of action is at the stage of conversion of cholesterol to pregnenolone: cytochrome *P*-450 is involved with cAMP as the second messenger, LH specifically activating Leydig cell adenylate cyclase.[47] FSH acts predominantly on the tubular compartment of the testis where the seminiferous tubules produce the sperm. The Sertoli cell is a target for FSH where the hormone stimulates adenylate cyclase activity, increases cAMP levels and cAMP-dependent protein kinase activity.[48] There is important feedback regulation from the testis, androgens, chiefly testosterone, controlling the release of LH, and the peptide inhibin from the tubules, controlling FSH.

In the ovary, primordial follicles are stimulated to grow by FSH in combination with some LH. As the follicles mature they secrete estrogens which first exert a negative feedback on gonadotrophin release and later, when the follicle has matured, a positive feedback in response to which there is a surge of FSH and LH to rupture the follicle and release the ovum.[49] The follicular remnant becomes the corpus luteum which is maintained by LH and stimulated to produce progesterone which with estrogen prepares the endometrium for implantation. The action of the gonadotrophins in the ovary is initiated by binding to high affinity receptors on the outer surface of the cell membrane: FSH receptors appear first and LH receptors increase in number with follicular maturity. Adenylate cyclase is activated and the second messenger cAMP is involved in both fast and slow responses to the gonadotrophin stimulation. The fast response is mediated by cAMP binding to kinase regulating the ovarian cells' enzymatic and structural proteins. The slow response leads to RNA and protein synthesis, cell growth, differentiation and division. Only a fraction of the receptors need be occupied to obtain maximal biological response.

The feedback mechanisms in both male and female act at the pituitary level and at the hypothalamus from whence the GnRH is secreted to regulate both the synthesis and the pulsatile release of the gonadotrophins.

Most of the early clinical trials on the treatment of male and female infertility made use of FSH and LH preparations from human pituitary glands.[50,51] The biological activities of these hormones are similar to those of human menopausal gonadotrophin (hMG) extracted from urine, and because of the risk of viral contamination from pituitary glands and the ready availability of commercially prepared hMG, the pituitary preparations are not much used nowadays. Furthermore, LH, which could be used to induce ovulation, resembles human chorionic gonadotrophin (hCG) obtained from the urine of women in early pregnancy, and this latter hormone is more readily available than the pituitary variety.

3.5.3.3 Other Pituitary Hormones

The other hormones of the pituitary are required as reagents and standards for assays but their therapeutic value is for the most part unexplored and now limited because of the risk of viral contamination.

3.5.3.3.1 TSH

Human TSH is a glycoprotein of molecular weight about 30 kDa consisting of an α subunit common to the α subunits of the gonadotrophins and a specific β subunit.[52] The principal effect of the hormone is to stimulate the rate at which the thyroid produces its hormones. TSH binding to specific receptors activates adenylate cyclase, cAMP and protein kinase, so leading on to the stimulation of iodine metabolism.

The release of TSH is under the control of the hypothalamic releasing hormone TRH, which is modulated by negative feedback from the thyroid hormones.[53] The assay of TSH and of thyroid hormones in blood is therefore informative in the investigation of pituitary function and in the differential diagnosis of thyroid disease.[54] Serum TSH concentrations are almost always elevated in untreated patients with primary hypothyroidism because lack of thyroid hormones leads to the absence of negative feedback. The assay forms the basis of screening tests for this disease in the newborn, important to diagnose because of the benefits of early treatment.

Human TSH has potential use as a therapeutic agent in the treatment of thyroid cancer as well as being a diagnostic reagent in the assessment of thyroid function. Unlike gonadotrophins, TSH cannot be obtained from urine so the human pituitary is at present the only source. Bovine TSH has been used as a substitute but has a lower potency in man than the human hormone and commonly produces allergic side reactions.

3.5.3.3.2 Prolactin

Human prolactin is a peptide hormone of 198 amino acids and is secreted by the lactotrophic cells of the anterior pituitary. The assay of the hormone is useful in the diagnosis of hyperprolactinaemia, a common cause of infertility in women with amenorrhoea.[55] The inappropriate secretion may arise from a pituitary tumour producing prolactin or from a non-secreting pituitary or hypothalamic tumour which blocks the flow of prolactin-inhibiting hormone from the hypothalamus to the pituitary. High prolactin levels reduce the frequency of pulsatile gonadotrophin secretion[56,57] and suppress the normal preovulatory gonadotrophin surge in response to high estradiol levels.[58] This results in either complete amenorrhoea, anovulatory cycles or ovulation with defective luteal function. The condition is treated by the administration of a dopamine agonist such as bromocriptine, an ergot alkaloid, dopamine itself being one of the inhibiting factors for prolactin release. There is also evidence that the presence of some prolactin is required for normal steroid synthesis by the corpus luteum.[59,60]

The lactogenic effect of the hormone is the one by which it was first recognized: high levels sometimes lead to galactorrhoea in anovular women and concentrations above basal are recognized during pregnancy[61] and although basal concentrations soon return to normal after delivery, prolactin secretion increases in response to suckling.[62] So far, however, there appears to be no clinical need for the hormone to be used therapeutically.

3.5.3.3.3 Corticotrophin

The synthesis of corticotrophin (ACTH) proceeds from a large glycosylated precursor, pro-opiomelanocortin of 239 residues.[63] Processing of this precursor yields both ACTH and β-lipotrophin (β-LPH), a peptide whose function in the human is unclear. A part of β-LPH, however, consists of β- and γ-endorphins, which are endogenous opioids (Figure 1). There are three melanophore-stimulating hormone (MSH) sequences, α-MSH at the N-terminal of ACTH, β-MSH in the middle of β-LPH and γ-MSH in the N-terminal portion of the prohormone (pro-γ-MSH).

There is a circadian rhythm in the secretion of ACTH, the lowest levels being usually about midnight with an increase at 03.00 to 04.00 h, reaching maximum concentrations by about 09.00 h. The secretion is also markedly pulsatile.[64] ACTH secretion increases in stress and is controlled by a negative feedback exerted by cortisol. The hormone is well known for its action on the adrenal cortex, promoting steroidogenesis and maintaining adrenal size. The significance of the other peptides has been discussed a great deal: they may complement the actions of ACTH, β-LPH mobilizing fatty acids and β-endorphin playing a part in stress-induced analgesia. Petersen, Brownie and Ling[65] found that pro-γ-MSH potentiated the steroidogenic action of ACTH if it was partially trypsinized before use: it also affects adrenal growth.[66] Pro-γ-MSH itself had no mitogenic effect but when trypsinized there was a significant increase in DNA synthesis in dexamethasone-suppressed rats.

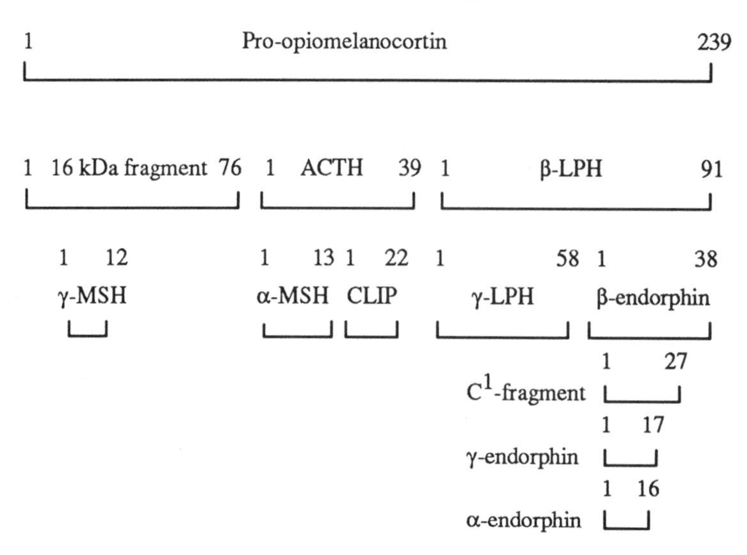

Figure 1 Some biosynthetic products from pro-opiomelanocortin: the figures represent the number of residues (CLIP = corticotrophin-like intermediate peptide)

3.5.3.3.4 Endogenous opioids

There are three known families of endogenous opioids of which β-endorphin is an example.[67] Each is derived by cleavage of a precursor polypeptide coded by a separate gene. The precursors are produced in separate neural systems and all are found in the hypothalamus. Pro-opiomelanocortin neurones producing β-endorphin are almost entirely in the arcuate nucleus/tuberal region. Proenkephalins are widely distributed in the brain:[68] they produce methionine- and leucine-enkephalins. The third group, dynorphins and neodynorphins, arise from prodynorphin which is also distributed in several brain areas.

These endogenous opioids may have effects on pain perception, appetite, thirst and mood[69] and they also affect the release of anterior pituitary hormones, inhibiting release of LH.[70] This last effect has been considered to explain the failure of ovulation in some women who are, for instance, professional athletes, ballet dancers or participants in other occupations entailing intensive exercise which is accompanied by release of excessive amounts of pro-opiomelanocortin, producing ACTH and β-endorphin.

It has been hypothesized that a disturbance of endogenous opioid production may explain some of the forms of premenstrual tension commonly encountered by women in the reproductive years.[69,71,72,73] β-Endorphin secretion increases in pregnancy and falls after delivery: some of the symptoms of postpartum depression resemble those of premenstrual tension and these include depressed mood, lethargy, anxiety and tearfulness, and the mechanisms may be related. These are typical of the 'withdrawal' effects of opiates, which are related to a decreased neurotransmission of biogenic amines: excessive activity also occurs and this is linked to irritability and aggression also common in premenstrual tension. Opioids also affect certain neuronal activities and the release of dopamine, noradrenaline and serotonin may be altered. Decreased serotonin levels have been related to increased depression, irritability, pain sensitivity and changes in libido.

Little is known about the role of the MSH molecules in the human. They may influence adaptive behaviour through an ability to increase awareness and enhance arousal and attention.[74] They may also influence the release of anterior pituitary hormones such as hGH and gonadotrophins.

3.5.4 GONADS

The importance of the steroid hormones of the gonads, estrogens, androgens and progesterone is well known: these hormones have been synthesized and have wide clinical applications. Recent interest has centred on a number of peptides and glycopeptides present in both ovaries and testes which have functions related to follicular development and feedback control to the hypothalamic–pituitary–gonadal axis.

One of the best known of these hormones is the glycoprotein inhibin. Bovine follicular fluid has been a source for the preparation of this hormone: it was isolated with a molecular weight of 58 kDa composed of two subunits of 43 kDa and 14 kDa bound by disulfide bonds.[75] More recently two 32 kDa varieties composed of two subunits were isolated from porcine follicular fluid:[76,77] each form is composed of two cross-linked subunits, a common α subunit, 18 kDa and two smaller distinctive β subunits of 14 kDa, designated β_A and β_B, which differ at the amino terminus. These two forms of inhibin, inhibin A containing the β_A subunits and inhibin B with the β_B subunit inhibit the secretion of pituitary FSH but not of LH or other pituitary hormones both *in vivo* and *in vitro*. An interesting reversal of activity occurs when two β_A subunits (FSH-releasing protein, FRP) or one β_A and one β_B subunit (activin) are held together by disulfide bonds. Both β dimers are highly potent stimulators of both the biosynthesis and secretion of FSH but have no effect on LH or prolactin. These dimers therefore differ from GnRH in their biological action and they also act more slowly, by greater than 4 h rather than within minutes. GnRH antagonists which completely block the GnRH-stimulated release of both FSH and LH *in vivo* have no effect on the secretion of FSH mediated by the β dimers. Presumably there are separate receptors for the β dimers and GnRH. The gonadal feedback role of FSH secretion is therefore more complicated than used to be thought: the development of reagents for the assay of these various forms of inhibin should help in further studies of the mechanisms involved.

There are several other factors controlling ovarian function which may eventually prove to be as important as inhibin.[78] These include an oocyte maturation inhibitor and a luteinization inhibitor and a stimulator.

Another glycoprotein, the anti-Müllerian hormone (AMH) has been isolated from foetal testes, where it is produced by the Sertoli cells[79] and it has also been found in ovarian follicular fluid produced from granulosa cells. The hormone is responsible for the regression of Müllerian ducts in the male foetus. The duct is sensitive to AMH only during a short and early period of development which precedes the first signs of Müllerian regression in the male. The persistence of the ducts in human intersexual states is usually associated with severe cryptorchidism and it has been suggested that AMH might play a role in the descent of the testes.[80] Possibly also it is involved in the failure of male germ cells to enter meiotic prophase.

There is some evidence that testicular proteins, including AMH, have effects on cancers of the female reproductive tract. Testicular proteins enriched with AMH delayed the growth of a human ovarian cancer cell line[81] and another preparation affected the growth of mouse endometrial cancer.[82] Now that highly purified preparations of AMH are available, further studies may indicate eventual pharmacological applications.

3.5.5 PLACENTA

The placenta is extremely active in the synthesis of proteins and other molecules. Some of these, such as the respiratory enzymes, are not specific for the placenta and even the enzymes responsible for steroid synthesis and interconversion do not differ from counterparts in the other steroid-producing tissues. There are also low concentrations of ACTH-related peptides and hypothalamic-releasing hormones which do not differ significantly from those of the hypothalamus and pituitary. Hormones such as human placental lactogen (hPL) and human chorionic gonadotrophin (hCG) have been considered specific hormones of the placenta but even these are now known to be produced elsewhere, such as in seminal plasma, in follicular fluid, and low concentrations of hCG have been recognized in many normal tissues and in the urine of postmenopausal women.[83]

hPL is a product of the syncytiotrophoblast and is detectable in maternal blood soon after implantation. Levels then rise progressively to reach a plateau by about the 35th week of pregnancy. Many biological activities have been proposed for the hormone, including growth promotion, lactogenesis, effects on carbohydrate and lipid metabolism, stimulation of the corpus luteum, erythropoiesis, inhibition of fibrinolysis and immunosuppression. In some of these activities it therefore resembles hGH and prolactin and there are similarities in the amino acid sequences. hPL is a polypeptide of molecular weight 21 kDa with a single chain of 191 amino acids and two intrachain disulfide bonds.[83] There is 86% sequence homology with hGH and 30% with prolactin, although this hormone has 198 residues and three intrachain disulfide bonds.[84] hPL has only weak somatotrophic activity compared with hGH although growth-promoting activity of hPL may be greater than that of hGH in some foetal tissues.[85] In the same way the hPL appears to be less potent than prolactin in most bioassays.

The physiological role of hPL in pregnancy has been much discussed and it is not clear that it is essential for any of the many processes in which it has been involved.[86] Other hormones of the placenta, such as estrogens and progesterone, and hormones from elsewhere in the body could well maintain such functions as the preparation of the breast for lactation, carbohydrate and lipid metabolism and growth of the foetus. For this reason there is at present no therapeutic application for hPL. The assay of the hormone, however, is used as a test of placental–foetal function in conjunction with other observations such as ultrasonic scanning.[87]

hCG is also secreted by the syncytiotrophoblast and can be detected in blood within a few hours of implantation. The concentration rises to reach a peak at 60–80 days of pregnancy and then falls to relatively low levels with a small secondary rise towards term.[82]

The hormone has biological activity similar to that of pituitary LH but has a longer biological half-life: there appears to be a component with a half-life of about 5.5 h and a slow component with half-life of about 24 h following intravenous injection.[88] It maintains the secretion of progesterone by the corpus luteum of pregnancy which is necessary in early pregnancy for the preparation of the endometrium for implantation. The assay of the hormone in early pregnancy is the basis of pregnancy tests and many kits are available which are highly specific and can differentiate between hCG and pituitary LH. It has therapeutic applications for the induction of ovulation and for the preparation of birth control vaccines. The material used, however, is normally prepared from urine rather than from placentae (see Section 3.5.6.2).

3.5.6 URINE

Human urine is the source of some important gonadotrophins used therapeutically, human menopausal (hMG) and human chorionic gonadotrophin (hCG). Urine of postmenopausal women is rich in both follicle-stimulating and luteinizing activities and hMG as usually prepared contains approximately equal unitage of FSH and LH. hCG is obtained in high concentrations in the urine of women in the first trimester of pregnancy and it is predominantly luteinizing in action.

3.5.6.1 hMG

The most important clinical application of hMG is the treatment of infertility in women who have primary pituitary failure.[89] It acts as a follicle stimulator and is administered intramuscularly for the 10 days or more needed for follicular maturation to occur, then hCG is given to induce ovulation. The sensitivity of individuals to hMG varies considerably[90,91] and because the difference between ineffective and effective doses is very small, treatment commences with low doses, which are gradually increased until a response occurs. Overstimulation is dangerous as gross ovarian enlargement with abdominal pain may result and in the more severe cases, fluid changes with ascites and pleural effusions and possibly changes in clotting factors leading to thrombosis may develop.[92,93] In addition, if conception occurs, the risk of multiple pregnancy is high.[94]

Some LH is required during the stage of follicular growth and the usual preparations of hMG with a 1:1 ratio of LH to FSH are satisfactory. In special cases such as the polycystic ovary syndrome, where basal LH may be elevated,[95] premature luteinization could occur. Theoretically FSH without LH should be more suitable and the preparations of FSH from hMG free of LH are available for therapy.[96] A further application of hMG is in the *in vitro* methods of fertilization. hMG is given with the anti-estrogen clomiphene to stimulate the growth of multiple follicles needed for successful egg recovery and *in vitro* fertilization.[97]

hMG has also been used in the treatment of the male with defective spermatogenesis.[98] The treatment is prolonged and is combined with hCG but results have been rather disappointing when compared with treatment of the female. The ovarian cycle is relatively short, follicular ripening requiring about 15 days, whereas the spermatogenic cycle is approximately 70 days.

3.5.6.2 hCG

hCG is more readily available than pituitary LH and therefore is generally substituted for LH in therapy. It has a longer circulating half-life than LH and chemically differs in a higher sialic acid content and an extended amino acid sequence at the C-terminus of the β subunit.[99]

The C-terminus extension is unique amongst the glycoprotein hormones and has been investigated as a reagent to produce a birth control vaccine.[100] The function of hCG in early pregnancy is to maintain progesterone production from the corpus luteum of pregnancy and if secretion fails the foetus is not maintained. Antibodies to hCG usually cross-react with LH so that ovulation would be interrupted if such an antibody was used for birth control. When the C-terminus of β-hCG is used, however, the peptide being linked to a larger protein to increase the immunogenicity, LH secretion is not affected. The first clinical trials are now proceeding.

3.5.6.3 Erythropoietin

Another glycosylated protein excreted in urine is erythropoietin.[101] The main source is the kidney and, in the foetus, the liver. It has been extracted from the urine of anaemic humans, there being no satisfactory alternative natural source.

Erythropoietin contains 166 amino acids with a 27 amino acid leader peptide.[102] The secreted glycoprotein is heavily glycosylated and has a molecular weight of 34 kDa. The hormone regulates erythropoiesis and when administered to animals it induces an increase in circulating red cell mass. Serum concentrations do not vary with age or sex but they increase during pregnancy.[103] The assay of erythropoietin may be helpful in the differential diagnosis of polycythemic states: in some instances raised levels arise from ectopic sources and then removal of a tumour or other localized lesion is accompanied by regression of the polycythaemia and a fall in serum levels.

Serum concentrations are inversely related to hemoglobin concentrations:[104] if anaemia is associated with renal failure, however, there is a parallel decrease of erythropoietin production and the reverse relationship with hemoglobin no longer holds. There is good evidence that deficiency of erythropoietin is the dominant cause of anaemia in chronic renal failure.

Treatment with erythropoietin-rich plasma cures anaemia in uraemic sheep and clinical trials are now being undertaken using recombinant material.[105,106] It reverses the anaemia of renal failure as demonstrated by 16 out of 35 patients dependent on transfusions no longer needing them. The shortened red cell survival and blood loss on dialysis were corrected and the increase in hemoglobin was accompanied by improved well being and exercise tolerance and there were few side effects. The therapeutic response, however, may be limited by factors such as iron deficiency and aluminum intoxication.

3.5.7 GASTROINTESTINAL TRACT

Around the turn of the century the first hormone, secretin, was discovered, when it was observed that the intravenous injection of an aqueous extract of jejunal mucosa stimulated pancreatic secretion. Since then many other hormones arising from the gastrointestinal (GI) tract have been identified, some of which are listed in Table 2. Some of these peptides, including secretin, clearly function as hormones, *i.e.* they are secreted by cells at one site and are carried *via* the blood stream to a target cell in a distant organ. Others, for example vasoactive intestinal peptide (VIP) function as neurotransmitters: these have some characteristics of a hormone in that they are secreted by one cell and carry a message to a target cell, but they do not travel through the blood stream. This, however, is not always a clear division, as catecholamines, such as noradrenaline, can function either as a hormone or as a neurotransmitter. Yet others, such as somatostatin, are paracrine messengers, regulating the activities of neighbouring cells.[107] In addition to the GI peptides there are local transmitters, such as acetylcholine, histamine and possibly prostaglandins, involved in the function of the GI tract. Furthermore, the GI peptides are not confined to the GI tract but are also found in the neurones of the central nervous system (CNS) and other tissues.[108] Much information on their clinical effects has come from a study of the hypersecretion of these peptides from tumours, tumours which derive from cells with 'amine precursor uptake and decarboxylation (APUD)', characteristics found most commonly in the pancreas.[109] The GI peptides exist in multiple forms:[110] precursors or prohormones of high molecular weight may be found in blood and hormones, such as gastrin, exist in forms with a sulfated and an unsulfated tyrosine residue.

The best known of the GI hormones is undoubtedly insulin, so widely used in the treatment of diabetes. It is synthesized in the β cells of the Islets of Langerhans from larger precursors.[111] The control and release of the hormone is complicated and speculative:[112] it is affected not only by glucose but by ions such as calcium and potassium, somatostatin, gastrointestinal peptide (GIP), by neural influences, prostaglandins, monoamine oxidase inhibitors and by vitamin D.

The action of insulin *in vitro* includes stimulation of glucose transport and oxidation, lipid synthesis, glycogen synthesis, amino acid transport and protein synthesis.[113] *In vivo*, however, the major hypoglycaemic action of insulin is the inhibition of hepatic glucose production: up to the concentrations reached in response to 100 g oral glucose this is the only demonstrable action. Only

Table 2 Some Hormones of the Gastrointestinal Tract

	Abbreviations	*Number of amino acids*	*Remarks*	
Glucagon		29	Glycogenolytic	
Secretin		27	Releases bicarbonate from pancreas	Secretin group: inhibit gastric acid secretion
Vasoactive intestinal peptide	VIP	28	Vasodilator, glycogenolytic, releases bicarbonate	
Gastrointestinal peptide	GIP	43	Potentiates insulin release	
Gastrin		34, 17, 13	Active site: 4 amino acids. 2nd messenger not found: may affect parietal cells directly or indirectly by releasing histamine	Gastrin family
Cholecystokinin–pancreozymin	CCK	33	Active site: 8 amino acids. Releases intracellular calcium which binds to calmodulin	
Pancreatic polypeptide	PP	36	Produced in some pancreatic tumours but hypersecretion has not been found to produce any specific clinical features	
Pancreastatin		49	May derive from chromogranin A. Inhibits insulin (and somatostatin) release	
Motilin		22	Causes increased motility in stomach	
Somatostatin		14	Multiple forms exist, some of which may be biosynthetic precursors Inhibits release of GH and TSH as well as several GI hormones	

at higher concentrations of insulin is stimulation of peripheral uptake of glucose observed. Insulin deficiency leads to increased glycogenolysis, proteolysis, lipolysis and ketogenesis.[114]

Insulin concentrations in blood are difficult to measure reproducibly at the low levels seen in the fasting state. The assay is important in the diagnosis of insulinomas[109] but otherwise is of limited value. Many standard assays do not distinguish proinsulin from insulin: proinsulin is converted to insulin and the C-terminal extension, C-peptide, one molecule of C-peptide always being released at the time of insulin secretion. The C-peptide is immunologically distinct from insulin and in diabetics treated with insulin antibodies are formed which interfere with the radioimmunoassay of insulin so that C-peptide measurements provide a valuable index of β cell activity.

Glucagon is a single-chain polypeptide of 29 residues, the porcine, bovine and human sequences being identical.[115] The best known source of the hormone is the α cells of the Islets of Langerhans. It acts on the liver to increase glucose production by a mechanism involving adenylate cyclase, cAMP, protein kinase and subsequent phosphorylation of glycogen synthetase, inhibition of glycogen synthesis and stimulation of glycogenolysis.[116] It also stimulates hydrolysis of adipose tissue triglycerides into glycerols and free fatty acids, *i.e.* lipolysis, an effect which is inhibited by the presence of insulin.[117] In common with other peptides of the secretin group (secretin itself, VIP and GIP), glucagon also has a direct stimulating effect on insulin secretion by the β cells.

The clinical applications of glucagon[118] include investigation of glycogen storage diseases and of phaeochromocytoma[119] and the inhibition of gastrointestinal motility.[120]

Just as the GI peptides are numerous, so their actions are many and various. Those acting on pancreatic acinar cells have been adequately reviewed:[107] these include cholecystokinin (CCK), gastrin, caerulein, substance P, secretin and VIP.

There may be actions elsewhere: VIP for instance stimulates biosynthetic activity of granulosa cells.[121] VIP, isolated from the duodenum, acts on smooth muscle and is known to act also on other cells, even in the CNS. It is believed to be involved in the release of prolactin and nerve fibres containing it or present in the uterus and ovarian stroma. It stimulates granulosa cells *in vitro* to synthesize progesterone, 20α-hydroxyprogesterone and estrogens, but in contrast to the action of FSH it does not increase the number of LH receptors in these cells.

Calcitonin is well known for its role in long-term skeletal maintenance:[122] it also has well-established actions on the GI tract. It increases intestinal secretions of Na, K, Cl and H_2O but inhibits gastric emptying, gastric acid secretion and the secretion of gastrin, insulin, pancreatic glucagon, motilin and pancreatic polypeptide.[123] It may have a neurotransmitter role in the gut and it is thought that in evolutionary terms it was originally a neuropeptide.

Pharmacological doses of gastrin, CCK and glucagon affect calcitonin secretion and the magnitude of the inhibiting effects approaches that of somatostatin.[124] It could possibly have a role in the management of tumours secreting gut hormones, as distinctive effects have been noted in patients with pancreatic tumours secreting gastrin, insulin, glucagon or pancreatic polypeptide given somatostatin.

3.5.8 BLOOD

Several growth factors have been extracted and purified from blood. They have been referred to as somatomedin A and C [C being identical to insulin-like growth factor I (IGF-I)], IGF-II and multiplication stimulating activity (MSA). Another somatomedin, B, originally described, is not now considered as in the same family as it lacks sulfation-factor activity and its mitogenic activity is probably attributable to epidermal growth factor (EGF).[125a] EGF is a polypeptide of molecular weight 6 kDa. It is a potent mitogen that hastens wound healing, inhibits gastric acid secretion and regulates spermatogenesis.[125b] It has been found to reduce the production of inhibin by granulosa cells *in vitro*[126a] and to stimulate the release of LH,[126b] hGH and prolactin and the hypothalamic hormone CRF.[126c] EGF is contained in blood platelets and is released after trauma.

Somatostatins are mediators of the skeletal growth-promoting effects of growth hormone *in vivo*. The various *in vitro* effects on cartilage muscle tissue and other tissues act through somatomedin receptors and on adipose tissue through insulin receptors.[127]

The somatomedins are transported bound to a carrier protein which is itself responsive to growth hormone. Binding to the carrier protein diminishes the biological activity and, as with steroids and other hormones, it is the unbound fraction which represents the active fraction.

The assay of somatomedins may be useful in the study of growth hormone defects to assist in the classification of short stature syndromes.[128] When done in parallel with assays for hGH the levels of both will be low in growth hormone deficiency. Stimulated growth hormone levels are also low but hGH treatment should lead to raised somatomedin levels. Applications in the diagnosis of Laron dwarfism[44] and in investigations of the African pigmy have already been described. In patients who have had surgery to correct for a craniopharyngioma hGH levels are very low but somatomedin is normal and there may be spontaneous growth without therapy.

3.5.9 OTHER TISSUES

The number of bioactive materials isolated from various other tissues is so great that they clearly cannot all be covered here. Reference can only be made to the widely distributed prostaglandins,[129] the kidney hormones[130] and the parathyroid glands.[131] The pineal gland has for long been suspected to be an important source of hormones, particularly in animal species: it is known to be a source of the methoxyindole, melatonin, which is secreted in a regular circadian rhythm with peak levels in the dark phase of the day. In animals whose breeding cycles are related to changes in the length of the day manipulation of melatonin levels either by pinealectomy or by the administration of the hormone results in changes in gonadal growth and the timing of the breeding cycle.[132] It has been suggested that if melatonin can affect the light–dark cycle it may have therapeutic use in the treatment of disorders such as jet lag.[133]

When administered to man, melatonin induces somnolence and sleep and stimulates hGH secretion and long-term treatment may depress gonadotrophins. Treatment with low doses over one month with doses adjusted to give peak levels earlier than the normal physiological peak had apparently no effect on cortisol or hGH profiles but the prolactin nocturnal peak was shortened.[134] There was also some suggestion that the secretion of endogenous melatonin was advanced by 1–3 h in the presence of exogenous melatonin. However, the potential therapeutic usefulness of the hormone as a hypnotic or in the treatment of jet lag is unlikely to be complicated by undesirable endocrine effects.

Finally, research on endometrial function and implantation has led to the identification of several proteins in the human endometrium. At present nomenclature is confusing and possibly several of the factors described are the same or are closely related to one another.[135] The endometrial protein 14 (EP14), α_1-pregnancy-associated endometrial protein, (α_1-PEG) and placental protein 12 (PP12) are similar and the N-terminal sequence of PP12 is identical to a 34 kDa somatomedin-binding protein of human amniotic fluid. Another group of six proteins including placental protein 14 (PP14) which has an N-terminal sequence identical to β-lactoglobulin are probably the same and these are not specific for the endometrium and occur elsewhere as in human milk.[136] Although endometrial tissue concentrations of the proteins rise after ovulation in line with increased progesterone production and in the endometrium of women treated with gonadotrophins, the physiological importance of the proteins for implantation is not yet clear.

3.5.10 REFERENCES

1. P. E. Cooper and J. B. Martin, *Ann. Neurol.*, 1980, **8**, 551.
2. T. Hokfelt, O. Johansson, R. Elde, M. Goldstein, S. L. Jeffcoate and N. White, in 'Neuroactive Drugs in Endocrinology', ed. E. E. Muller, Elsevier/North-Holland Biomedical Press, Amsterdam, 1980, p. 19.
3. R. N. Clayton, *Clin. Endocrinol. (Oxford)*, 1987, **26**, 361.
4. J. C. Marshall, *Clin. Endocrinol. Metab.*, 1975, **4**, 545.
5. J. C. Davis and L. J. Hipkin, in 'Management of Pituitary Disease', ed. P. E. Belchetz, Chapman and Hall Medical, London, 1984, p. 161.
6. R. W. Shaw, W. R. Butt, D. R. London and J. C. Marshall, *J. Obstet. Gynaecol. Br. Commonw.*, 1974, **81**, 632.
7. F. R. Kandeel, W. R. Butt, D. R. London, S. S. Lynch, R. Logan Edwards and B. T. Rudd, *Clin. Endocrinol. (Oxford)*, 1978, **9**, 429.
8. E. Knobil, *Rec. Prog. Horm. Res.*, 1980, **36**, 53.
9. S. S. C. Yen and A. Lein, in 'Marshall's Physiology of Reproduction', ed. G. E. Lamming, Churchill Linvingstone, Edinburgh, 1984, p. 713.
10. R. L. Reid, G. R. Leopold and S. S. C. Yen, *Fertil. Steril.*, 1981, **36**, 553.
11. G. Leyendeker and L. Wildt, *J. Reprod. Fertil.*, 1983, **69**, 397.
12. N. Santoro, M. Filicori and W. F. Crowley, *Endocrinol. Rev.*, 1986, **7**, 11.
13. G. Skarin, S. J. Nillius and L. Wide, *Fertil. Steril.*, 1983, **40**, 454.
14. H. S. Jacobs, in 'Therapeutic Applications of LHRH', ed. S. R. Bloom and H. S. Jacobs, Royal Society of Medical Services, London, 1986, p. 99.
15. V. Menon, W. R. Butt, R. N. Clayton, R. Logan Edwards and S. S. Lynch, *Clin. Endocrinol. (Oxford)*, 1984, **21**, 223.
16. D. V. Morris, R. Adeniyi-Jones, M. Wheeler, P. Sönksen and H. S. Jacobs, *Clin. Endocrinol. (Oxford)*, 1984, **21**, 189.
17. F. Labrie, A. Dupont, A. Belanger, R. St. Arnaud, M. Giguere, Y. Lacourciere, J. Emond and G. Monfette, *Endocrinol. Rev.*, 1986, **7**, 67.

18. A. Manni, R. Santen, H. Harvey, A. Lipton and D. Max, *Endocrinol. Rev.*, 1986, **7**, 89.
19. M. C. Gershengorn, in 'Prolactin: Basic and Clinical Correlates', ed. R. M. MacLeod, U. Scapagini and M. O. Thorn, 1985, Springer Verlag, Berlin, p. 155.
20. S. Reichlin, *Acta Endocrinol. (Copenhagen)*, 1986, **112**, Suppl. 276, 21.
21. P. T. Loosen and A. J. Prange, *Am. J. Psychiatry.*, 1982, **139**, 405.
22. A. I. Faden, T. P. Jacobs and J. W. Holaday, *N. Engl. J. Med.*, 1981, **305**, 1063.
23. W. K. Engel, T. S. Siddique and J. T. Nicoloff, *Lancet*, 1983, **2**, 73.
24. J. Freedman, T. Hökfelt, G. Jonsson and C. Post, *Exp. Brain Res.*, 1986, **62**, 175.
25. Y. Kamamoto, H. Karashima, S. U. Kim and Y. Eto, *Brain Res.*, 1986, **371**, 201.
26. R. Hall, B. J. Ormston, G. M. Besser, R. J. Cryer and M. McKendrick, *Lancet*, 1972, **1**, 759.
27. W. M. G. Tunbridge, R. A. Jackson, M. Inignez and T. Russell Fraser, *Proc. R. Soc. Med.*, 1973, **66**, 187.
28. M. O. Thorner, M. L. Vance, W. S. Evans, R. M. Blizzard, A. D. Rogol, K. Ho, D. A. Leong, J. L. C. Borges, M. J. Cronin, R. M. MacLeod, K. Kovacs, S. Asa, E. Horvath, L. Frohman, R. Furlanetto, C. J. Klingensmith, C. Brook, P. Smith, S. Reichlin, J. Rivier and W. Vale, *Rec. Prog. Horm. Res.*, 1986, **42**, 589.
29. (a) M. O. Thorner, M. L. Vance, W. S. Evans, K. Ho, A. D. Rogol, R. M. Blizzard, R. Furlanetto, J. Rivier and W. Vale, *Acta Endocrinol. (Copenhagen)*, 1986, **112**, Suppl. 276, 34; (b) L. C. K. Low, C. Wang, P. T. Cheung, P. Ho, K. S. L. Lam, R. T. T. Young, C. Y. Yeung and N. Ling, *J. Clin. Endocrinol. Metab.*, 1988, **66**, 611.
30. G. L. Pittenger, A. I. Vinik, A. A. Heldsinger and S. Seino, *Adv. Exp. Med. Biol.*, 1985, **188**, 447.
31. R. M. MacLeod, A. M. Judd, W. D. Jarvis and I. S. Login, *Acta Endocrinol. (Copenhagen)*, 1986, **112**, Suppl. 276, 9.
32. S. W. J. Lamberts, *Acta Endocrinol. (Copenhagen)*, 1986, **112**, Suppl. 276, 41.
33. W. Vale, J. Spiess, C. Rivier and J. Rivier, *Science (Washington, D.C.)*, 1981, **213**, 1394.
34. G. Gillies, S. Ratter, A. Grossman, R. Gaillard, P. J. Lowry, G. M. Besser and L. H. Rees, *Horm. Res.*, 1980, **13**, 280.
35. A. M. Landolt, A. Valvanis, J. Girard and A. N. Eberle, *Clin. Endocrinol. (Oxford)*, 1986, **25**, 687.
36. P. J. Lowry, F. E. Estivariz, G. E. Gillies, A. C. N. Kruseman and E. A. Linton, *Acta Endocrinol. (Copenhagen)*, 1986, **112**, Suppl. 276, 56.
37. K. Nikolics, A. J. Mason, E. Szönyi, J. Ramachandran and P. H. Seeburg, *Nature (London)*, 1985, **316**, 511.
38. H. S. Phillips, K. Nikolics, D. Branton and P. H. Seeburg, *Nature (London)*, 1985, **316**, 542.
39. R. D. G. Milner, T. Russell Fraser, C. G. D. Brook, P. M. Cotes, J. W. Farquhar, J. M. Parkin, M. A. Preece, G. J. A. I. Snodgrass, A. Stuart Mason, J. M. Tanner and F. P. Vince, *Clin. Endocrinol. (Oxford)*, 1979, **11**, 15.
40. U. J. Lewis, R. N. P. Singh, G. F. Tukwiler, M. B. Sigel, E. F. Vanderlaan and W. P. Vanderlaan, *Rec. Prog. Horm. Res.*, 1980, **36**, 477.
41. D. V. Goeddel, H. L. Heyneker, T. Hozumi, R. Arentzer, K. Itakura, D. G. Yansura, M. J. Ross, G. Moizzari, R. Crear and P. G. Seeburg, *Nature (London)*, 1979, **281**, 544.
42. K. C. Olson, J. Fenno, N. Lin, R. N. Harkins, C. Snider, W. H. Kohr, M. J. Ross, D. Fodge, G. Prender and N. Stebbing, *Nature (London)*, 1981, **293**, 408.
43. L. Lazarus, in 'Endocrine Disorders: A Guide to Diagnosis', ed. R. A. Donald, Dekker, New York, 1983, p. 273.
44. Z. Laron, *Isr. J. Med. Sci.*, 1974, **10**, 1247.
45. C. S. Smith, in 'Management of Pituitary Disease', ed. P. E. Belchetz, Chapman and Hall Medical, London, 1984, p. 461.
46. S. L. Jeffcoate, *Clin. Endocrinol. Metab.*, 1975, **4**, 521.
47. B. A. Cook, C. J. Dix and R. Magee-Brown, *Biochem. Soc. Trans.*, 1980, **9**, 40.
48. A. R. Means, J. R. Dedman, J. S. Tash, D. J. Tindall, M. van Sickle and M. J. Welsh, *Annu. Rev. Physiol.*, 1980, **42**, 59.
49. S. S. C. Yen and A. Lein, in 'Marshall's Physiology of Reproduction', ed. G. E. Lamming, Churchill Livingstone, Edinburgh, 1984, p. 713.
50. C. A. Gemzell, E. Diczfalusy and G. Tillinger, *J. Clin. Endocrinol. Metab.*, 1958, **18**, 1333.
51. A. C. Crooke, in 'Recent Research on Gonadotrophic Hormones', ed. E. T. Bell and J. A. Loraine, Livingstone, Edinburgh, 1967, p. 278.
52. P. G. Condliffe and B. D. Weintraub, in 'Hormones in Blood', ed. C. H. Gray and V. H. T. James, Academic Press, London, 1979, vol. 1, p. 499.
53. P. R. Larsen, *N. Engl. J. Med.*, 1982, **306**, 23.
54. H. K. Ibbertson, in 'Endocrine Disorders: A Guide to Diagnosis', ed. R. A. Donald, Dekker, New York, 1984, p. 417.
55. P. J. A. Moult, in 'Management of Pituitary Disease', ed. P. E. Belchetz, Chapman and Hall Medical, London, 1984, p. 116.
56. H. G. Bohnet, H. G. Dahlen, W. Wuttke and H. P. G. Schneider, *J. Clin. Endocrinol. Metab.*, 1976, **42**, 132.
57. P. J. A. Moult, L. H. Rees and G. M. Besser, *Clin. Endocrinol. (Oxford)*, 1982, **16**, 153.
58. M. R. Glass, R. W. Shaw, W. R. Butt, R. Logan Edwards and D. R. London, *Br. Med. J.*, 1975, **iii**, 274.
59. K. P. McNatty, *Fertil. Steril.*, 1979, **32**, 433.
60. A. Kauppila, P. Chatelain, P. Kirkinen, S. Kivinen and A. Ruokonen, *J. Clin. Endocrinol. Metab.*, 1987, **64**, 309.
61. S. Franks, in 'Hormones in Blood', ed. C. H. Gray and V. H. T. James, Academic Press, London, 1979, 3rd edn., vol. 1, p. 299.
62. J. E. Tyson, M. Kohjandi, J. Huth and B. Andreassen, *J. Clin. Endocrinol. Metab.*, 1975, **40**, 764.
63. R. E. Mains, B. A. Eipper and N. Ling, *Proc. Natl. Acad. Sci. USA*, 1977, **74**, 3014.
64. M. C. Quigley and S. S. C. Yen, *J. Clin. Endocrinol. Metab.*, 1979, **49**, 945.
65. R. C. Pedersen, A. C. Brownie and N. Ling, *Science (Washington, D. C.)*, 1980, **208**, 1044.
66. P. J. Lowry, *Biosci. Rep.*, 1984, **4**, 467.
67. R. J. Bicknell, *J. Endocrinol.*, 1985, **107**, 437.
68. V. Clement-Jones and L. H. Rees, in 'Hormones in Blood', ed. C. H. Gray and V. H. T. James, Academic Press, London, 1983, 3rd edn., vol. 4, p. 267.
69. U. Halbreich and J. Endicott, *Med. Hypoth.*, 1981, **7**, 1045.
70. M. E. Quigley and S. S. C. Yen, *J. Clin. Endocrinol. Metab.*, 1980, **51**, 179.
71. R. L. Reid and S. S. C. Yen, *Am. J. Obstet. Gynecol.*, 1981, **139**, 85.
72. J. Ff. Watts, W. R. Butt and R. Logan Edwards, *Br. J. Obstet. Gynaecol.*, 1985, **92**, 247.
73. W. R. Butt, in 'Functional Disorders of the Menstrual Cycle', ed., M. Brush and E. Goudsmit, Wiley, Chichester, 1988, p. 117.

74. L. H. Miller, A. J. Kastin, C. A. Sandman, M. Finle and W. J. Vanveen, *Pharmacol., Biochem. Behav.*, 1974, **2**, 663.
75. D. M. Robertson, L. M. Foulds, L. Leversha, F. J. Morgan, M. T. W. Hearn, H. G. Burger, R. E. H. Wettenhall and D. M. de Kretser, *Biochem. Biophys. Res. Commun.*, 1985, **126**, 220.
76. W. Vale, J. Rivier, J. Vaughan, R. McClintock, A. Corrigan, W. Woo, D. Kerr and J. Spiess, *Nature (London)*, 1986, **321**, 776.
77. N. Ling, S.-Y. Ying, N. Ueno, S. Shimasaki, F. Esch, M. Hotta and R. Guillemin, *Nature (London)*, 1986, **321**, 779.
78. N. Josso, *Clin. Endocrinol. (Oxford)*, 1986, **25**, 331.
79. C. P. Channing, L. D. Anderson, D. J. Hoover, J. Kolena, K. G. Osteen, S. H. Pomerantz and K. Tanabe, *Rec. Prog. Horm. Res.*, 1982, **38**, 331.
80. J. M. Hutson, *Lancet*, 1985, **2**, 419.
81. A. F. Fuller, G. P. Budzik, I. M. Krane and P. K. Donahoe, *Gynecol. Oncol.*, 1984, **17**, 124.
82. P. K. Donahoe, G. P. Budzik, R. Trelstad, M. Mudgett-Hunter, A. Fuller, Jr., J. M. Hutson, H. Ikawa, A. Hayashi and D. Machaughlin, *Rec. Prog. Horm. Res.*, 1982, **38**, 279.
83. W. R. Butt and T. Chard, in 'Marshall's Physiology of Reproduction', ed. G. E. Lamming, Churchill Livingstone, Edinburgh, 1989, vol. 3, in press.
84. H. D. Niall, M. L. Hogan, G. W. Tregear, G. V. Seagre, P. Hwang and H. G. Friesen, *Rec. Prog. Horm. Res.*, 1973, **29**, 387.
85. D. J. Hill, C. J. Crace and R. D. G. Milner, *J. Cell. Physiol.*, 1985, **125**, 337.
86. Y. B. Gordon and T. Chard, in 'Placental Proteins', ed. A. Klopper and T. Chard, Springer Verlag, New York, 1979, p. 1.
87. T. Chard, in 'Antenatal and Neonatal Screening', ed. N. J. Wald, Oxford University Press, 1984, p. 510.
88. K. D. Bagshawe, F. Searle and M. Wass, in 'Hormones in Blood', ed. C. A. Gray and V. H. T. James, Academic Press, London, 1979, 3rd edn., vol. 1, p. 363.
89. G. Bettendorf and V. Insler, 'Clinical Application of Human Gonadotropin', G. T. Verlag, Stuttgart, 1970.
90. A. C. Crooke, W. R. Butt and P. V. Bertrand, *Acta Endocrinol. (Copenhagen)*, 1966, **53**, Suppl. 111, 3.
91. C. Gemzell and E. D. B. Johansson, in 'Control of Human Fertility', ed. E. Diczfalusy and W. Borell, Wiley, New York, 1971, p. 241.
92. B. J. Lunenfeld, *J. Int. Fed. Gynaecol. Obstet.*, 1963, **1**, 153.
93. R. P. Shearman, *Am. J. Obstet. Gynecol.*, 1969, **103**, 444.
94. C. R. Thompson and L. M. Hansen, *Fertil. Steril.*, 1970, **21**, 844.
95. S. S. S. Yen, *Clin. Endocrinol. (Oxford)*, 1980, **12**, 177.
96. E. Drapier-Faure, *Rev. Fr. Gynecol. Obstet.*, 1986, **81**, 179.
97. J. L. Yovich, J. D. Stanger, A. I. Tuvik and J. M. Yovich, *Med. J. Aust.*, 1984, **40**, 645.
98. B. Lunenfeld and R. Weissenberg, in 'Modern Trends in Endocrinology', ed. F. T.-G. Prunty and H. Gardiner-Hill, Butterworth, London, 1972, vol. 4, p. 157.
99. F. J. Morgan, S. Birken and R. E. Canfield, *J. Biol. Chem.*, 1975, **250**, 5247.
100. V. C. Stevens, J. E. Powell, A. C. Lee and D. Griffin, *Fertil. Steril.*, 1981, **36**, 98.
101. P. M. Cotes, in 'Hormones in Blood', ed. C. H. Gray and V. H. T. James, Academic Press, London, 1983, 3rd edn., vol. 4, p. 1951.
102. S. Lee-Huang, *Proc. Natl. Acad. Sci. USA.*, 1984, **81**, 2708.
103. P. M. Cotes, C. E. Canning and T. Lind, *Br. J. Obst. Gynaecol.*, 1983, **90**, 304.
104. P. M. Cotes, *Br. J. Haematol.*, 1982, **50**, 427.
105. C. G. Winearls, D. O. Oliver, M. J. Pippard, C. Reid, M. R. Downing and P. M. Cotes, *Lancet*, 1986, **2**, 1175.
106. J. W. Eschbach, J. C. Egrie, M. R. Downing, J. K. Browne and J. W. Adamson, *N. Engl. J. Med.*, 1987, **316**, 73.
107. J. D. Gardner and R. T. Jensen, *Rec. Prog. Horm. Res.*, 1983, **39**, 211.
108. S. R. Bloom, *J. R. Coll. Physicians*, 1980, **14**, 51.
109. S. M. Wood and S. R. Bloom, in 'Endocrine Disorders: A Guide to Diagnosis', ed. R. A. Donald, Dekker, New York, 1984, p. 533.
110. H. Gregory and P. Scholes, in 'Topics in Hormone Chemistry', ed. W. R. Butt, Ellis Horwood, Chichester, 1978, p. 48.
111. D. F. Steiner, *Diabetes*, 1977, **26**, 322.
112. U. A. Parman, 'Hormones in Blood', ed. C. H. Gray and V. H. T. James, Academic Press, London, 1983, 3rd edn., vol. 4, p. 27.
113. P. H. Sönksen, in 'Endocrine Disorders: A Guide to Diagnosis', ed. R. A. Donald, Dekker, New York, 1984, p. 507.
114. P. H. Sönksen and P. M. Brown, in 'Topics in Therapeutics', ed. D. W. Vere, Pitman Medical, Marshfield, Mass., 1978, vol. 4, p. 176.
115. P. J. Lefebvre and A. S. Luyckx, in 'Hormones in Blood', ed. C. H. Gray and V. H. T. James, Academic Press, London, 1979, 3rd edn., vol. 1, p. 171.
116. H. G. Hers, *Annu. Rev. Biochem.*, 1976, **45**, 167.
117. J. E. Gerich, M. Lorenzi, D. M. Bier, E. Tsalikian, V. Schneider, J. H. Karam and P. H. Forsham, *J. Clin. Invest.*, 1976, **57**, 875.
118. J. A. Galloway, in 'Glucagon, Molecular Physiology, Clinical and Therapeutic Implications', ed. P. J. Lefebvre and R. H. Unger, Pergamon Press, Oxford, 1972, p. 299.
119. E. F. Sebel, R. D. Hull, M. Kleerekoper and G. S. Stokes, *Am. J. Med. Sci.*, 1974, **267**, 337.
120. R. E. Miller, S. M. Chernish, R. L. Brunelle and B. D. Rosenak, *Radiology (Easton, Pa.)*, 1978, **127**, 55.
121. J. B. Davoren and A. J. W. Hsueh, *Biol. Reprod.*, 1985, **33**, 37.
122. J. C. Stevenson, G. Abeyasekera, C. J. Hillyard, K. G. Phang and I. MacIntyre, *Lancet*, 1981, **1**, 693.
123. S. I. Girgis, D. W. R. Macdonald, J. C. Stevenson, P. J. Bevis, C. Lynch, S. J. Wimalawansa, C. H. Self, H. R. Morris and J. MacIntyre, *Lancet*, 1985, **2**, 14.
124. T. E. Adrian, A. J. Barnes, R. G. Long, D. J. O'Shaughnessy, M. R. Brown, J. Rivier, W. Vale, A. M. Blackburn and S. R. Bloom, *J. Clin. Endocrinol. Metab.*, 1981, **53**, 675.
125. (a) M. A. Preece, in 'Hormones in Blood', ed. C. H. Gray and V. H. T. James, Academic Press, London, 1983, 3rd edn., vol. 4, p. 87. (b) O. Tsutsumi, H. Kurachi and T. Oka, *Science (Washington, D.C.)*, 1986, **233**, 975.
126. (a) P. Franchimont, M. T. Hazee-Hagelstein, C. H. Charlet-Renard and J. M. Jaspar, *Acta Endocrinol. (Copenhagen)*, 1986, **111**, 122. (b) A. Miyoke, K. Tasaka, S. Otsuka, H. Kohmura, T. Wakimoto and T. Aono, *Acta Endocrinol.*

(Copenhagen), 1985, **108**, 175. (c) A. Luger, A. E. Calogero, K. Kalogeras, W. T. Gallucci, P. W. Gold, D. L. Loriaux and G. P. Chrousos, *J. Clin. Endocrinol. Metab.*, 1988, **66**, 334.
127. M. A. Preece and A. T. Holder, in 'Recent Advances in Endocrinology and Metabolism', ed. J. M. L. O'Riordan, Churchill Livingstone, Edinburgh, 1982, vol. 2, p. 47.
128. L. Lazarus, in 'Endocrine Disorders: A Guide to Diagnosis', ed. R. A. Donald, Dekker, New York, 1984, p. 273.
129. J. M. Bailey (ed.), 'Prostaglandins, Leukotrienes and Liposeins. Biochemistry, Mechanism of Action and Clinical Application', Plenum Press, New York, 1985.
130. J. W. Fisher (ed.), 'Kidney Hormones', Academic Press, London, 1986, vol. 3.
131. H. M. Kronenberg, T. Igarashi, M. W. Freeman, T. Okazaki, S. J. Brand, K. M. Wiren and J. T. Potts, Jr, *Rec. Prog. Horm. Res.*, 1986, **42**, 641.
132. L. Tamarkin, C. J. Baird and O. F. X. Almeida, *Science (Washington, D.C.)*, 1985, **227**, 714.
133. J. Arendt and V. Marks, *Br. Med. J.*, 1983, **287**, 426.
134. J. Wright, M. Aldhous, C. Franey, J. English and J. Arendt, *Clin. Endocrinol. (Oxford)*, 1986, **24**, 375.
135. S. C. Bell, *Hum. Reprod.*, 1986, **1**, 313.
136. M. Julkunen, R. S. Raikar, S. G. Joshi, H. Bohn and M. Seppälä, *Hum. Reprod.*, 1986, **1**, 7.

3.6
Development and Scale-up of Processes for the Manufacture of New Pharmaceuticals

TREVOR LAIRD
Formerly of Smith Kline & French Research Ltd, Tonbridge, UK

3.6.1 INTRODUCTION

The conversion of a synthetic route used for making gram quantities of a chemical by medicinal chemists to a process for manufacturing tonnage quantities of drug substance is a topic about which a great deal is known, but little published in the literature;[1-5] the 'tricks of the trade' are handed down within individual company research and development organizations and there is little shared experience between chemists in different companies. The result is a gap in the literature resulting in a lack of awareness (outside the industry itself) of what is involved in chemical development. The techniques used are similar to those used in chemical or process development units in other fine chemical industries (*e.g.* dyestuffs, agrochemical or photographic) but there are additional problems peculiar to the pharmaceutical industry which make the area more complex, but probably more interesting to the organic chemist. It is these aspects which this chapter will attempt to address—to discuss the ways in which a process is transformed in its development from bench to plant, and the issues which need to be resolved '*en route*'. Of course the drug development process varies from company to company and the generalizations below reflect the author's experience.

The scope of the discussion will begin from the point at which the target molecule has been selected from a group of probably closely related compounds having similar pharmacological profile to proceed further into development and hopefully, in five to ten years time, on to the market. The development chemist's role over these five to ten years may vary considerably from the almost academic organic chemistry involved in synthetic route selection to the control of manufacture on large plant-scale immediately prior to launch of the new product. With the increasing demands of regulatory requirements, resulting in additional studies in toxicology or the clinic, the quantities of chemical required prior to marketing over the five to ten year development period may often be in the 1–1.5 t range.[6,7] For a ten-stage synthetic route, with yields at each stage of 90%, 3–4.5 t of the first-stage intermediate will be needed, and possibly 16–25 t of intermediates may be manufactured prior to product launch.

At the stage at which the chemical development team in research and development takes over responsibility for the compound the following information is probably known. (i) A few grams of drug substance will have been synthesized so that at least one synthetic route will have been demonstrated. This may, however, be totally unsuitable for scale-up. It is possible that some alternative routes will have been examined (and possibly rejected) in an attempt to supply compound for initial toxicology. (ii) Preliminary analytical information on the drug substance will be available and some data on impurities may have been generated. (iii) A specification for the drug substance will not usually have been set. (iv) Enantiomers of the drug substance may have been separated and their relative activity assessed. A decision on whether to proceed into development with a racemate or single enantiomer is crucial for the development chemist's strategy but, for other reasons, it may be sensible to keep the options open until later in the development of the drug.

It is important at this stage that the development chemist is not inhibited by what has gone before. An apparent easy option at this early stage in development always is to carry out minor modifications to the synthetic route used to make the first few grams, scale-up this process and make supplies by this route. However, the reasons why this synthetic route was used by the medicinal chemist are probably different from those factors which the development chemist feels are significant. It is better that the development chemist takes a long-term viewpoint, *i.e.* to begin again to evaluate synthetic routes which are likely to be short, easy to scale-up and, probably, therefore produce a drug at minimal cost. It is more difficult to change synthetic routes halfway through the development process, and these early studies provide a chance to 'get it right' at the start. A radical approach to synthetic route selection is likely to produce long-term benefits when the drug reaches the market, it will give stronger patent coverage for the process which may extend the patent life of the drug, and should lead to cheaper manufacturing methods which result in cheaper drugs.

3.6.2 FACTORS WHICH AFFECT SYNTHETIC ROUTE SELECTION

In this section, the factors which may affect the choice of synthetic route are discussed. Of course the most important factor is whether the chemistry leads to the desired product and gives material of adequate purity, in the same way as for any synthetic chemical target. However, after a number of possible syntheses have been evaluated, other factors such as costs of material supply, of more interest to the development chemist, will become increasingly important and these factors are also discussed in this section. Possibly the medicinal chemist will feel these sections a little out of place but their relevance will become apparent as Section 3.6.3 develops.

3.6.2.1 Synthetic Route Analysis and Selection

Given the structure of the target drug substance, retrosynthetic analysis can be performed in a way analogous to any other synthetic target, either manually using a disconnection approach[8] and incorporating ideas generated in brain-storming sessions, or using computerized retrosynthetic analytical methods such as LHASA.[9] These methods generate vast amounts of information and the difficult task for the organic chemist, faced with producing the best synthesis in a limited amount of time, is to choose which options to examine—a daunting task. For example, the synthetic routes to make the antiulcer drug (H_2-antagonist) cimetidine[10] (Smith Kline & French's 'Tagamet') are delineated below. The product (**1**) can be broken down in a number of ways, leading to precursor molecules which can be further fragmented (Scheme 1). Each synthetic route leading to cimetidine from its immediate precursor will lead to different by-products, and therefore the 'impurity profile' for cimetidine prepared from each precursor will be different—this is very important for regulatory issues, where the toxicity of possible by-products is considered and in patent issues, when the impurity profile may become a 'fingerprint' to determine which synthetic route has been used and therefore whether a patent has been infringed.

As can be seen from Scheme 1 a large number of precursors to a drug substance are possible (each of these delineated in Scheme 1 has been patented by Smith Kline & French or its competitors) and of course each precursor can have many synthetic routes to it. The 'synthetic tree' which can be constructed will be extremely complex. The task of the development chemist is to select for study those routes which are likely to yield a cheap and simple manufacturing process.

After a preliminary study, it soon became clear that there were a variety of ways in which the 2,3-disubstituted imidazole unit could be constructed, that cysteamine itself was readily available (though rather costly) and that cyanoguanidines were readily available by nucleophilic addition to compounds of the type $X_2C{=}N{-}CN$, where X is a leaving group. The main decisions related to the order of construction of the bonds and the choice of leaving groups X to produce high-yielding syntheses capable of being scaled-up.

Synthetic route selection is very labour intensive and even when the field has been narrowed down it may not be practicable, within the time-scale required, to examine all the possible alternatives. The time-scale is limited by:

(i) Regulatory considerations. It may be difficult to change synthetic routes, which may lead to a different impurity profile in the final drug substance, once the compound has proceeded beyond phase II clinical studies. The requirement is that toxicological data is available, prior to clinical trials, on a drug substance having a similar impurity profile (see Section 3.6.4.4). Synthetic routes are often changed after some toxicological tests have been repeated, but this may be costly.

(ii) Scale-up considerations and the need to make supplies. As the synthetic route selection is proceeding, there will be demands from toxicologists and pharmacists for supplies of drug for testing. These quantities may be small, but with a long and poorly developed process, the lead time for synthesis may be extensive. If scale-up to pilot plant is required, it may be that the synthetic route has to be proven several months prior to the material supplies being required.

So in practice, only a few main synthetic pathways can be examined in detail. Some of the factors which influence the choice for laboratory study are:

(i) *Likelihood of success.* Does the chemistry use reactions which are well exemplified in the literature, which have been demonstrated on similar substrates, or does the chemistry use speculative reactions?

(ii) *Raw materials.* Are the raw materials readily available (in bulk?) and are they relatively cheap?

(iii) *Number of steps.* A short synthetic route, even if it involves speculative chemistry, will be most appreciated by the site of manufacture and by accountants (requires less plant, less capital

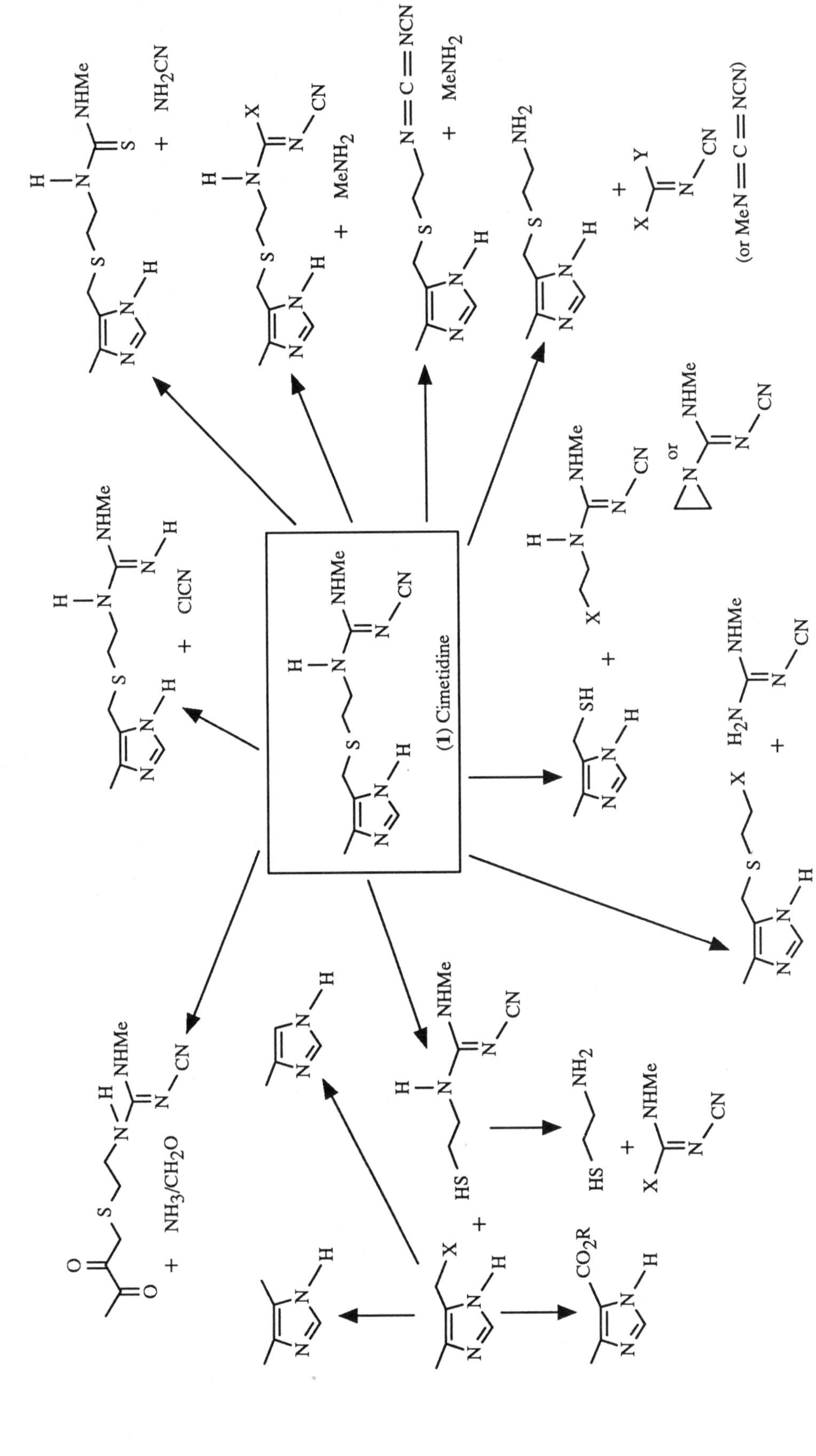

(1) Cimetidine

Scheme 1

investment and has smaller overheads). It should also produce cheaper drugs, since overall yields should be higher and overheads less.

(iv) *Ease of scale-up.* This factor may be outweighed by the need to use a short synthesis or the availability of a very cheap raw material which may dictate the technology to be used.

(v) *Selectivity.* Separations, particularly those involving chromatography are time consuming and costly, and selective processes are preferred. However, the selectivity of a chemical reaction can often be vastly improved during development by careful choice of reaction parameters.

(vi) *Environmental.* Increasingly effluent and by-product considerations have to be discussed at an early stage and in extreme cases, may affect the choice of route.[6] However, a properly managed effluent policy which involves chemical destruction of toxic by-products or recycling can make other factors outweigh the environmental issues. It is likely that environmental costs will significantly increase in the future, however, and the importance of this may correspondingly increase.

(vii) *Safety.* The use of highly toxic raw materials or reagents, or the likelihood of explosion hazards on scale-up, may make a synthetic route unattractive.

Although these factors may be taken into account in a paper study, it is difficult to quantify the relative importance of each. Decisions however, are often based on the experience (and often the prejudices) of the chemists involved and two chemists may come out with different solutions to the same problem. One may prefer a short speculative route to a longer, safer, more conventional approach, but both may end up with a high quality drug at a similar price. As always a balance has to be found.

The effect of different factors on synthetic route selection is shown in the following example of a development compound from the author's experience in Smith Kline & French. The H_1-antagonist SK&F 93944 (temelastine) could be made by the two routes shown in Schemes 2 and 3. The shorter route shown in Scheme 3 had tremendous advantages, being only two stages from the cheap 2,3-lutidine[11] to the pivotal intermediate (**2**). However, scale-up of the two steps proved difficult; in the first step, selective alkylation of the methyl group could be achieved, but the product was always contaminated with the other isomer and with dialkylated products. Bromination of the pyridine would only occur readily in oleum but the workup difficulties (quench into concentrated sulfuric acid, then into water, followed by neutralization and extraction) and resultant large aqueous effluent problems made the process no cheaper than the longer route (Scheme 2). Isomeric bromopyridines and dibromopyridines were also present so that on conversion to temelastine, a less-pure product was obtained. The longer route shown in Scheme 2, however, scaled-up well, and batches of final product in the 100 kg range were made before the project was completed.

3.6.2.2 Costing of Synthetic Routes

Costing is important at various stages of the development process for a new drug substance. Although the cost of goods may not be crucial in every drug's marketability, it may be important in its profitability. For some drugs in a competitive market situation, it is vital that a cost-effective process for manufacture of the active ingredient is available. As health authorities become more cost conscious, drug costs are likely to become an increasingly important issue; the chemist has a role to play in this, even though the cost of the active component usually only accounts for a minor component of the price of the marketed drug. However, a cheaper drug may allow a wider marketing strategy particularly for drugs aimed at the Third World, and will be significant once the patent life of the compound has expired, when stiff competition from low-cost generics can be expected. It is important, therefore, that costs are considered early on in the development process and that an understanding of the costing methods is widespread, since it is possible that decisions, affecting synthetic route selection, may be taken on this basis.

The cost of a chemical process will depend on the number of synthetic steps and, for each step, the raw material costs and the overheads (processing time, labour, effluent costs, *etc.*). The raw material costs will be affected by the following factors: (i) the chemical yield at each stage—when expensive raw materials are used, this is the most important factor in cost reduction (other than the number of steps in the synthesis); (ii) the reagents used and their costs (per mole), for example high molecular weight oxidants such as cerium(IV) ammonium nitrate can contribute high costs to an oxidation step; (iii) solvent costs—often significant in inefficient processes. In chemical manufacture, solvent recovery is important for the economics of the process and for reducing effluent streams; and (iv) the volume efficiency of the process, defined as the amount of product produced from unit volume of reaction medium, in unit time. In the context of raw material costs, the concentration of the raw

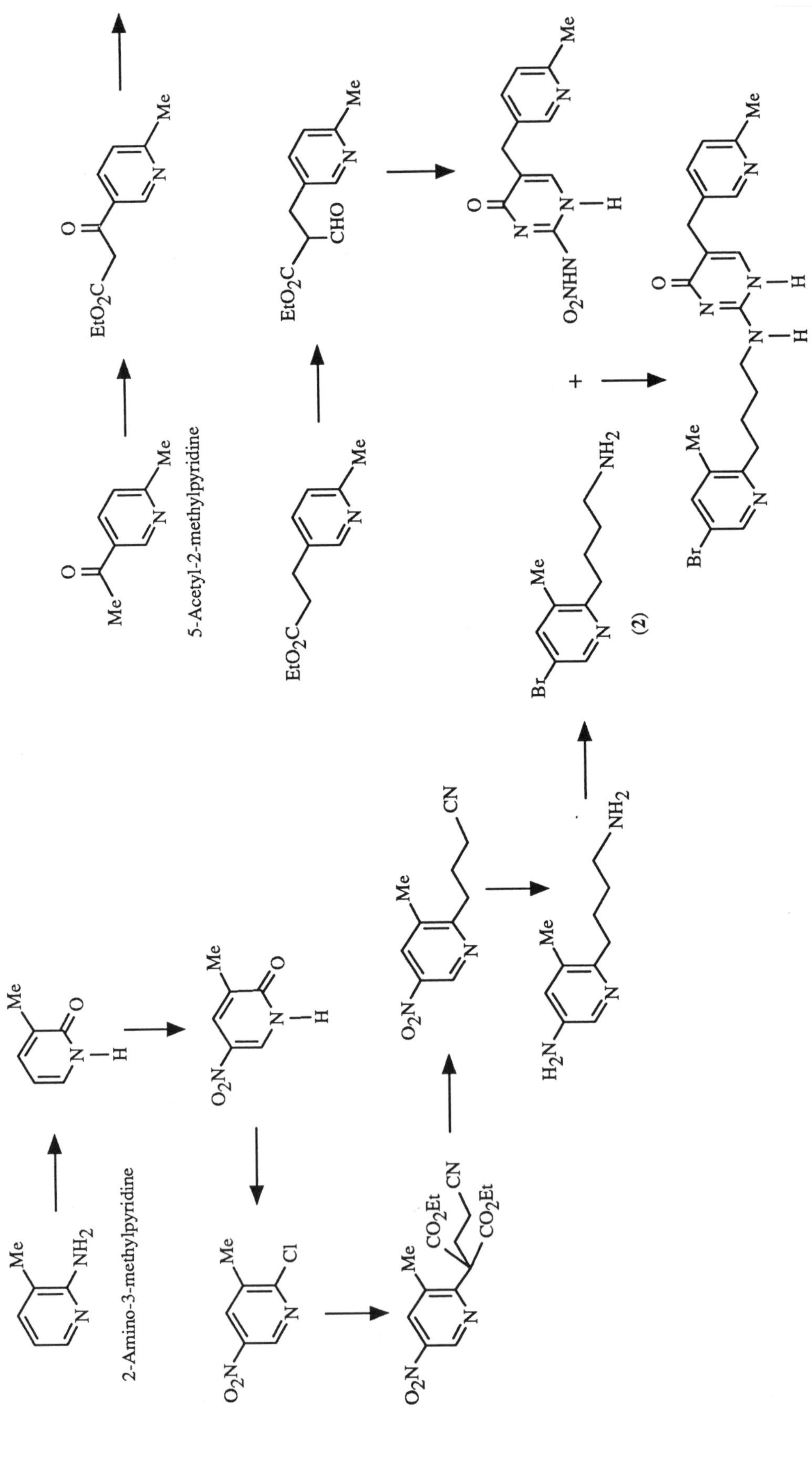

2-Amino-3-methylpyridine

5-Acetyl-2-methylpyridine

(2)

SK&F 93944

Scheme 2

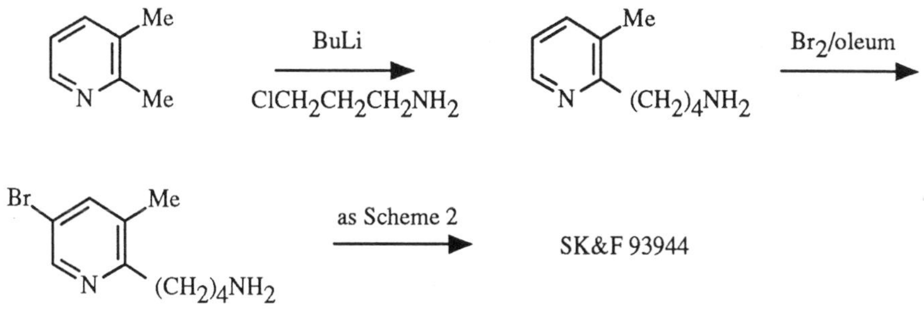

Me
Me
BuLi
$ClCH_2CH_2CH_2NH_2$
Me
$(CH_2)_4NH_2$
Br_2/oleum

Br
Me
$(CH_2)_4NH_2$
(2)
as Scheme 2
SK&F 93944

Scheme 3

material or intermediate in the reaction is therefore significant, but work up volumes must also be minimized (and, if possible, designed to allow easy solvent recovery) in efficient processes.

Overhead costs are more difficult to cost reliably, their method of calculation varying from company to company, and even from site to site within the same company. Overheads are generally a function of: (i) process time for each step; (ii) scale of manufacture—as the scale increases the proportion of raw material costs to overhead costs increases; (iii) labour costs, which also vary with (iv) site of manufacture; and (v) effluent costs, which are also geographically variable depending on a country's attitude to waste disposal.

As a general rule, chemical processes which use cheaper materials will be more dependent on overheads, and therefore on manufacturing costs. Scale of likely manufacture is also important; for example, for highly active drugs (*e.g.* prostaglandins), where ultimate manufacture may only be on the kg scale, but where the number of synthetic steps is large and the raw materials relatively expensive, conventional economics may not apply. During the development phase of a new drug substance, the cost of the drug may fall by a factor of 5–10 owing to process optimization and larger-scale manufacture, but also to cheaper raw materials, since they may be made on a larger scale, and purchased on tonnage quantities at a reduced rate.

When the synthetic chemist compares synthetic routes at an early stage of drug development, much of this information is unavailable. To the experienced development chemist, who has had previous expertise in scale-up, reasonable projections on the basis of future manufacture can be made. In some cases it may be possible to reject a synthetic route on the grounds that, even if 100% yield on each stage were obtained, the route would not compare to other routes.

A note of caution is necessary, however. Costing is a notoriously ambiguous area in which the inexperienced chemist can get into difficulties. Quoting costs of final drug substance based on projections from a laboratory synthesis may mislead senior personnel and marketing executives into a false position, which may be detrimental. It is vital that an understanding of the final figure is maintained, and that the figure is based on *bulk* raw material prices (not that of specialist suppliers!) which are, however, not easy to obtain. Exchange rate variations can also have a profound effect on cost projections!

3.6.2.3 Selection of Raw Materials

The choice of basic raw materials for the synthesis of complex molecules is difficult. It can often be left until later in the development process since the chemistry of the earlier stages of the synthesis can usually be changed without affecting the quality of the final product, whereas the later steps control the impurity profile of the drug substance. But raw material choice at any early stage can influence the development process markedly. It depends on the chemist having a degree of intuition about synthetic routes and a wide knowledge of what is available on the fine chemicals market—the latter is difficult to assess. Whilst it is relatively easy for the synthetic chemist to track down sources of laboratory chemicals, either through the well-known catalogues or computer search facilities,[12] it is comparatively difficult to identify the bulk (and usually primary) sources of fine chemicals. Compendia are available[13–17] but often, by the time the data are processed and the compendium published, the listings of chemicals for each supplier have been superseded. There is a definite need for a regularly updated, computerized listing of primary fine chemical sources, with facilities for searching by substructure but as yet this does not exist. Location of sources is often, therefore, carried out *via* a 'chemical grapevine'. It is also important to know not only what *is* available, but also what *can be made* available in bulk. Often the only reason for a chemical not being offered is the lack of demand, and fine chemical manufacturers are only too willing to be approached for possible new business.

From a retrosynthetic analysis and preliminary laboratory investigation it may be possible to identify attractive synthetic routes from chemicals which are unavailable in bulk. This is often true for heterocyclic compounds, where a limited range of substitution patterns and substituents on even the more common ring systems are available. Often, suppliers can be persuaded to develop processes for the manufacture of these raw materials, and a close link between the supplier and pharmaceutical company builds up as the drug proceeds into development.

Examples of this include the development of processes by Lonza for the manufacture of thiazole intermediates such as **(3)** and **(4)** used in the synthesis of the antibiotics cefotaxime **(5)** and ceftazidime **(6)**.

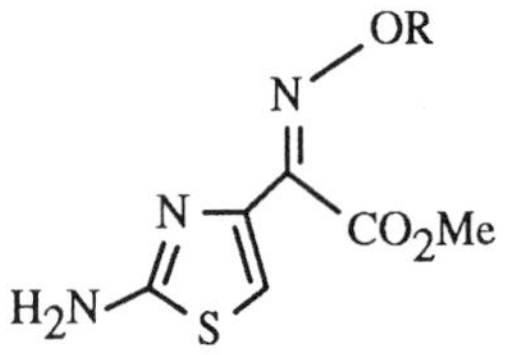

(3) R = Me
(4) R = $C(Me)_2CO_2H$

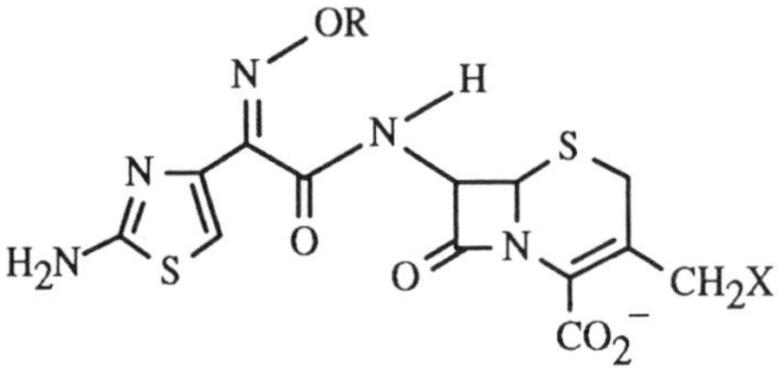

(5) R = Me; X = OAc
(6) R = $C(Me)_2CO_2H$; X = pyridinium

Close links between fine chemical manufacturers and the pharmaceutical industry are vital; the fine chemical company can develop processes to either raw materials or to intermediates in the synthetic route alongside the pharmaceutical manufacturer. Initially laboratory supplies are purchased and checked out for quality and potential price. As the requirements for the drug substance increase, the orders for raw materials and intermediates will increase sufficiently to finance the costs of scale-up by the fine chemical company. As the process develops during scale-up, unit costs and prices begin to fall. The fine chemical companies have to gamble on the success of the new drug, since their own development effort may be wasted if the drug fails in development. They recover some costs by charging high prices initially for the early kilogramme lots. However, in the event of termination of the drug development programme, often the back-up or second generation drug may have similar chemical fragments, or a 'me-too' drug from another company may also require a similar raw material or intermediate. The suppliers can then survive these difficulties, in the expectation of an occasional bonanza when a drug does reach the market and is highly successful. A further area of close collaboration between chemical companies and the pharmaceutical industry is in custom synthesis and toll manufacture,[18] where a synthetic method (sometimes only a laboratory method) for the manufacture of a raw material or intermediate is supplied by the pharmaceutical company, and supplies are made to order by the chemical supplier. Often this occurs when the contractor (usually small) has a special expertise, which allows handling of the more toxic or potentially explosive intermediates. As the drug requirements increase, the contractor can expand his facilities to cope with the increased demand.

A particular example of this collaboration has been in the development of intermediates for the synthesis of the H_2-antagonists, cimetidine **(1)** and ranitidine (**7**), which require supplies of the intermediates **(8)** and **(9)**, both made from carbon disulfide (Scheme 4). The extreme flammability and low flash point of carbon disulfide and the toxicity of dimethyl sulfate require a special plant and one or two companies (*e.g.* Fine Organics, now part of the Laporte group) have expanded tremendously on the basis of supply of these compounds to Smith Kline & French and Glaxo.

3.6.2.4 Enantioselective Synthesis or Resolution?

One of the most difficult decisions affecting drug companies in the last two decades has been, and will continue to be, whether to proceed with the development of a candidate drug, containing at least one chiral centre, as a single enantiomer or as a racemate. Of course if the drug contains more than one chiral centre, the choice may be more difficult, but, in general, few drugs have been marketed as a mixture of diastereoisomers. Traditionally, however, synthetic drugs containing chiral centres have

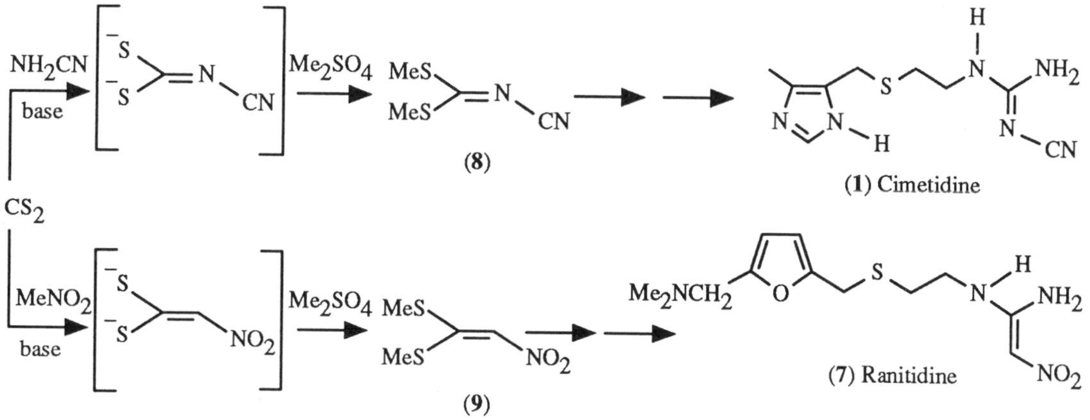

(8)

(1) Cimetidine

(9)

(7) Ranitidine

Scheme 4

been marketed as racemates, particularly in the β-blocker field, despite the fact that the biological activity of the two enantiomers is markedly different.[19] More recent cardiovascular drugs, such as the ACE inhibitor captopril (**10**), are sold as a single enantiomer. Products derived from natural products or *via* biosynthetic pathways, *e.g.* antibiotics such as aminoglycosides, erythromycin, tetracycline, penicillins and cephalosporins, polypeptides and steroidal products, however, have generally been marketed as single enantiomers.

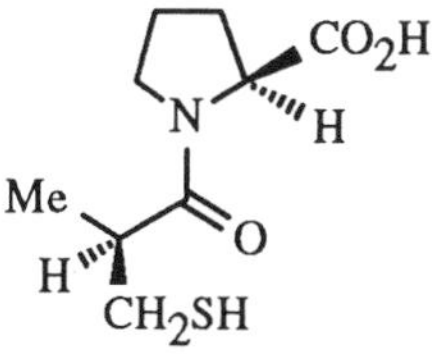

(**10**)

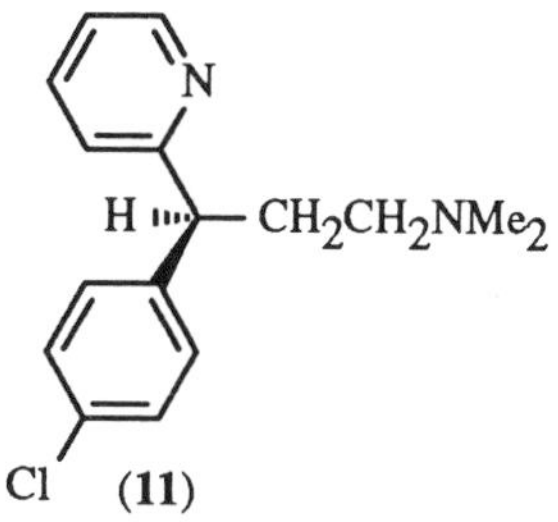

(**11**)

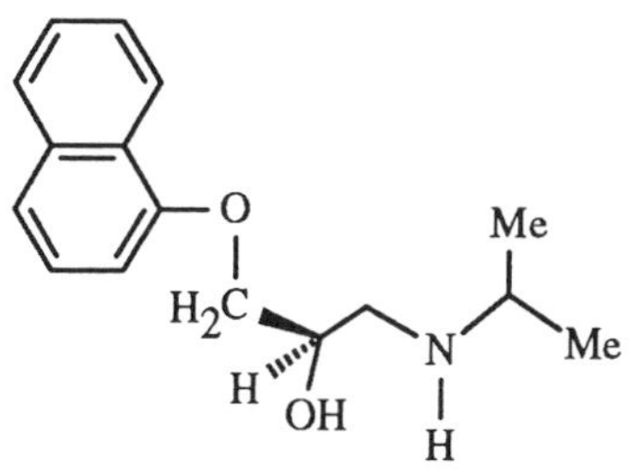

(**12**)

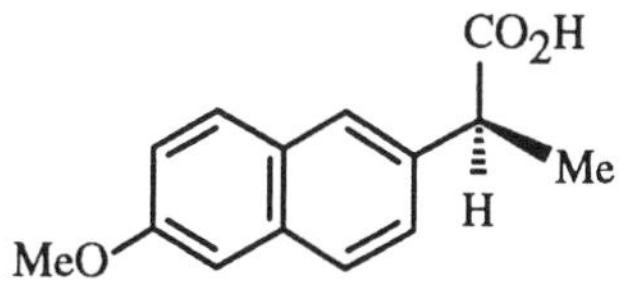

(**13**)

For example, the antihistamine (*S*)-chloropheniramine (**11**) is 100 times more potent than the (*R*) isomer, the β-blocker propranolol (**12**) shows a similar ratio compared to its enantiomer and for α-methylarylacetic acids, which are used as nonsteroidal antiinflammatory agents (*e.g.* naproxen; **13**), the (*S*) enantiomer exhibits all of the activity. Pressure is mounting from regulatory authorities, particularly in the US, to regard the presence of the less active isomer in the same way as an inactive (or toxic) impurity, and it may not be too long before it will be necessary for all new drugs to be marketed as mainly one enantiomer, with less than 5% of the less active enantiomer present.

It is likely that the overriding decision about whether to proceed with drug development with a single enantiomer or a racemate will be taken outside the realm of synthetic chemistry, but important factors which need to be considered at an early stage will be:

(i) Can the drug be synthesized easily in a chiral form (either by resolution or chiral synthesis)?[20] Can a similar synthesis be used for racemate and chiral compound? How efficient is resolution likely to be?

(ii) What effect will the chirality have on drug costs? The activity of a chiral drug should in general be twice that of the racemate but the cost of synthesis may be more than twice that of the racemate, particularly if resolution at the last step is not a feasible option. However, recycling of the unwanted isomer is often possible.

(iii) There are only a limited number of cheap chiral raw materials[21] available in bulk, which may make the chiral synthetic route long and expensive.

(iv) A chiral synthesis will probably be more difficult to scale-up, particularly if the chiral centre is prone to racemization in acidic or basic conditions. Better process control will be required during the reaction to ensure retention of chirality, but work up times on scale-up are often extended, and long contact with acids or bases during, say, extractions may be a problem. Extra recrystallizations during the synthesis may need to be introduced to improve the chiral purity of intermediates or final product.

As a general rule if a simple resolution at a late stage can be incorporated in the synthesis (particularly if a kinetic resolution can be achieved) a cheaper product *via* a simpler manufacturing method will be obtained. However, if short synthetic routes from readily available raw materials such as carbohydrates and amino acids can be achieved, the chiral route may be economic.

3.6.2.5 Choice of 'Final' Synthetic Route

One of the most difficult decisions in drug development is when to finalize the synthetic route. The only limitation on changing routes is that the final drug substance must have a similar impurity profile as previous batches otherwise some toxicology may have to be repeated. The chemist usually defines a change in synthetic route as when one of the intermediates used to synthesize the molecule changes. Changes in the reagents, solvents, stoichiometry, procedures, *etc.* are not normally classified as 'route changes' by the development chemist, although they can profoundly affect the impurity profile. Obviously changes to the chemistry early in the synthesis are likely to have lower implications than later stages, but during the drug development, the chemistry is likely to change significantly in all steps. The problem is more acute if scale-up difficulties in one route mean that a change to another is necessary, particularly after the drug has entered clinical trials. The development chemist is often faced with a race against time.

Ideally, the drug development timetable will give the chemist a rough deadline to meet, such that early clinical trials in the main use batches of chemical of comparable quality to those used in previous animal studies and in toxicological tests. For these animal studies, particularly if large animals such as dogs are used, large quantities of drug substance in the 10–50 kg range may be required which, in a 6–10 step synthesis, may take at least six months to manufacture; the process to make the material will have had to be proved in the laboratory a few months earlier than this, so the decision on which route to scale-up may well have to be taken a year earlier than the date the animal studies take place. Factors which are important in deciding on synthetic route changes include:

(i) *Where the change takes place.* If the last two or three synthetic steps remain the same, the chances of impurity profile changes are low. If the penultimate or last step changes, significant analytical effort will need to be made to prove to some regulatory authorities that significant new by-products are absent by a variety of analytical techniques.

(ii) *The quality of drug substance used in early toxicological tests.* If 99.9% pure drug substance is used in early toxicological tests, it may be very difficult to produce material on a large scale of comparable purity for the clinic. From the chemist's viewpoint, it is generally preferable, therefore, to test drug substances of 97–98% purity at early stages (possibly limiting single impurities to ~0.5% or below) to allow freedom to change the chemistry at a later stage. Otherwise the chemist may get locked in to a route which becomes difficult to scale-up or which produces a drug at an unacceptable cost.

(iii) *The use for which the drug is intended.* The specification for short-term therapy drugs (*e.g.* antibiotics, antimalarials) is generally wider than for maintenance therapeutics and this may allow some flexibility in synthetic route changes.

(iv) Cost of repeating toxicology on material of significantly different impurity profile, as measured against the potential cost savings of a cheaper or better process.

(v) *Patent issues.* When several companies are working on the same area and intermediates in the synthesis are patented by a competitor, a change in synthetic routes to circumvent the patent may be necessary.

An example which demonstrates these issues arose during the synthesis of the potential second generation H_2-antagonist SK&F 93479 (**14**),[22] where the intermediate (**15**) used to synthesize the drug during the earlier stages of development could not be used owing to a Glaxo patent. A simple but effective method of circumventing this difficulty was to change the order of the synthetic steps, carrying out the Mannich reaction, as the last step on the precursor (**16**), giving a method which turned out to be superior to the previous route. A change in the synthesis at such a late stage, however, gave rise to new impurities in the final drug substance and, despite extensive development work, small quantities of these impurities remained in the SK&F 93479. There was, therefore, no option but to repeat some toxicological tests on material made by the new route.

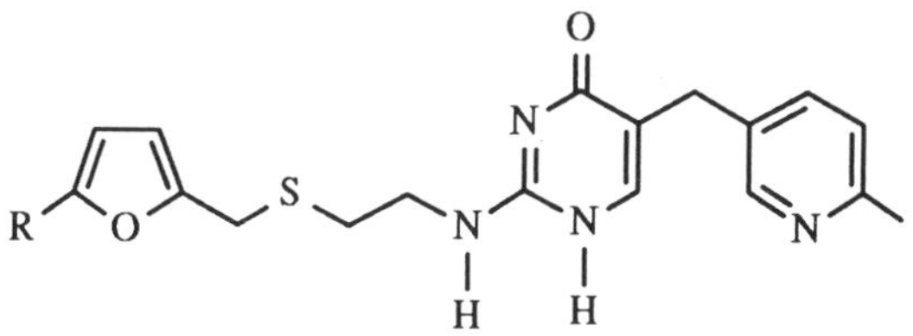

(**14**) R = Me_2NCH_2
(**16**) R = H

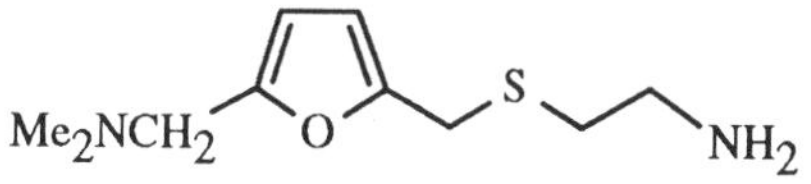

(**15**)

3.6.3 DEVELOPMENT AND SCALE-UP

3.6.3.1 Introduction

Development chemistry involves not only synthetic route selection, but also the optimization, scale-up and further improvement of the synthetic method until a routine and efficient process, suitable for manufacture by operators, who are skilled but have little chemical knowledge, is obtained. The development chemist's task is to convert a laboratory synthetic route, which uses complex and expensive reagents, to a robust manufacturing method, which is efficient and uses cheap raw materials. High yielding chemical reactions producing high quality intermediates and drug substances in tonnage quantities are the development chemist's target.

Often time is of the essence—nowadays the drug development process is so long that the development chemist must try not to be the rate-determining step in getting a new drug on to the market. This may mean compromise—the chemist must be prepared to scale-up the process before it is fully optimized (provided it is safe to do so!) so that batches of drug substance for vital studies are made available on time. However, this can bring benefits; the effect of scale-up on the chemical reaction is demonstrated at an early stage and gives ample opportunity to rethink and to carry out a more detailed laboratory development before further scale-up may be required.

In this author's opinion (and some bias is admitted) this is one of the more demanding but also the most satisfying areas of chemistry. The extra restraints of tight time-scales, increased safety considerations, and rigid analytical specifications place additional demands on the development chemist. The rapid development over the last 20–25 years of analytical methods (particularly HPLC) for quality control of drugs has increased the complexity of the drug development process. Prior to the routine use of HPLC in quality control, drug substance purity was assessed by assay methods such as titrimetric analysis and UV spectroscopy (which may assay impurities of similar structure to the drug substance) with impurity profiles being monitored by TLC. Whilst TLC is an excellent method for quick and reproducible qualitative analysis, it has only been useful for quantitative analysis in recent times. The advent of HPLC has provided the analyst with reproducible and quantitative methods of checking assay and impurity profiles so that, with TLC as a back-up, impurity profiles and drug specifications can be more strictly controlled. Regulatory authorities are now requiring tighter specifications for new chemical entities, with specifications being controlled by detailed HPLC analysis of impurities, and compounds present in $>0.1\%$ in the drug substance should ideally be isolated and characterized. New peaks at levels above the 0.1% level are generally considered to be inadmissible. As a result, to ensure that a new manufacturing method gives material which passes these tighter specifications, the chemist has to be able to produce high purity drugs routinely in the 99–99.5% range as standard, and this has implications for the control of processes in plant and the instrumentation and methodology used (see Section 3.6.3.8). The advent of HPLC has had its bonuses too. Quality control of intermediates is easier, and it is now possible to detect likely problems, at an earlier stage in the synthesis.

3.6.3.2 The Development Process—a Multidisciplinary Effort

Process development and scale-up requires multidisciplinary cooperation between organic chemists, analytical chemists involved in quality control and process engineers involved in plant trials, plant modifications or new plant construction.[23] Development and scale-up in the pharmaceutical industry requires even further collaboration with pharmacists, pharmacologists, toxicologists, clinicians and marketing specialists to ensure that the quality and quantity of drug substances are produced efficiently to an agreed programme over a number of years. Factors which often cause communication barriers in the drug development programme are: (i) the long lead times necessary to make even a few kilogrammes of new drug substance and the difficulty, particularly at an early stage when a synthetic route suitable for scale-up may not be available, of predicting when quantities of drug will be available; (ii) the difficulties in long-term prediction of likely quantities of drug substance, so that the chemist can plan ahead with his supplies campaign and (iii) quality differences between batches within the same specification.

It is therefore important that clear objectives and provisional time targets are set well in advance and that project planning involves periodic review of these targets.

In order to meet these targets, the development chemist must compromise between the need to optimize each of the synthetic steps prior to scale-up and the needs of the project timetable, which requires materials for toxicological or carcinogenicity testing or for clinical trials. Inevitably this means that the chemist has to proceed with scale-up into the pilot plant with a partially optimized process, *i.e.* with the best method which is safe to scale-up. Further optimization can then be carried out later.

The first 10–20 kg of drug substance are often the most difficult to obtain. If a new synthetic route is not required, or is required but not yet available, the method used earlier to make the first few grams must be optimized and scaled-up, even though this is not necessarily the method of choice in the long term. More typically, a new synthetic route will be evaluated but a conflict between the time taken to prove the route and the need to scale-up quickly to make the 10–20 kg drug substance will need to be resolved.

In the early stages of development, it may be difficult to guarantee the quality of the first batches of final drug substance and often multiple recrystallizations are needed to give material of adequate quality.

3.6.3.3 Optimization of Synthetic Routes

Optimization of a synthetic route used to make a new drug involves not only maximizing the yield (and quality) at each synthetic step, but more importantly in producing a drug of acceptable quality at the minimum cost, as measured at the manufacturing site. The importance of process costing (see Section 3.6.2.2) in optimization is therefore emphasized, and economic factors may outweigh chemical reasoning. For example, a chemical reaction which gives a 90% yield of product in DMF as solvent but involves a tedious aqueous drown out and extraction in the work up may be superseded by a process which uses toluene as the solvent, and gives only 75% yield but the product crystallizes from the reaction mixture. Thus simplification of the process is often important.

When the material costs in the process are high, the 'residence time' of that process in the manufacturing plant may not significantly affect the overall cost of the drug, but may impinge on the plant's ability to meet market demand, and to produce a drug to tight time schedules. Plant throughput must therefore be a long-term consideration, and is important in all manufacturing processes (often called debottle-necking!). When material costs are low, minimization of plant occupation (*i.e.* reduction in plant overheads) is probably the most important factor in optimization. Process costing in the early stages of development can often highlight the areas in the process where attention needs to be directed. Often the chemistry is good, leading to high solution yields, but the work-up and isolation lead to wastage.

Before beginning optimization, an overview of the synthetic route is advisable, particularly if rapid scale-up is required and safety considerations may be of vital importance. This overview will enable decisions on where the finite resources available are to be placed.

Some questions which have to be asked are: (i) What needs to be changed to make the process safe to scale-up? (ii) Are potentially hazardous raw materials, reagents and intermediates involved? Can they be substituted? Do they need to be isolated? (iii) Is the order of steps most appropriate for the synthetic route; can some steps be eliminated? (iv) Can steps be easily combined? (v) Are the raw materials available on the scale required or do they need synthesizing? (vi) Is the route likely to lead to the presence of highly toxic impurities (or heavy metals) in the final product? (vii) Are the by-products likely to lead to effluent problems?

Although all these questions need to be asked, and they may delay scale-up, it is unlikely that they will result in the process being abandoned. Almost any process can be scaled-up in properly engineered equipment if procedures are strictly adhered to, but the cost implications may be prohibitive. As a result, many pharmaceutical processes involve so-called hazardous reagents such as boron tribromide, substituted diazomethanes, thallium reagents, sodium–liquid ammonia reductions, phosgene, alkyllithiums, hydride reagents and ozonolysis, which, with good engineering, can be handled safely on the tonnage scale. Even if expertise is not available in-house, contract companies having particular expertise in a particular area may need to be involved.

In some cases, problems such as likely effluent difficulties may lead to alternative routes being evaluated. For example, in the process development of synthetic routes to ICI's H_2-antagonist, tiotidine (**17**), which was abandoned in 1980 owing to toxicological problems, the initial process involved release of 1 kg of methanethiol for every 2 kg of pure product in the reaction in Scheme 5 and was a major factor in the selection of an alternative reagent (**18**).[6]

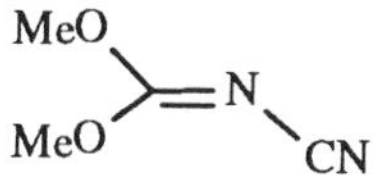

MeO, MeO, N, CN

(**18**)

Often, however, process optimization concentrates on yield improvement and streamlining the process.

3.6.3.3.1 *Yield improvement*

Yield improvement often depends on the choice of reaction conditions and on accurate (fine) control of parameters, *i.e.* attention to detail. In order to improve yields it is vital to have an accurate

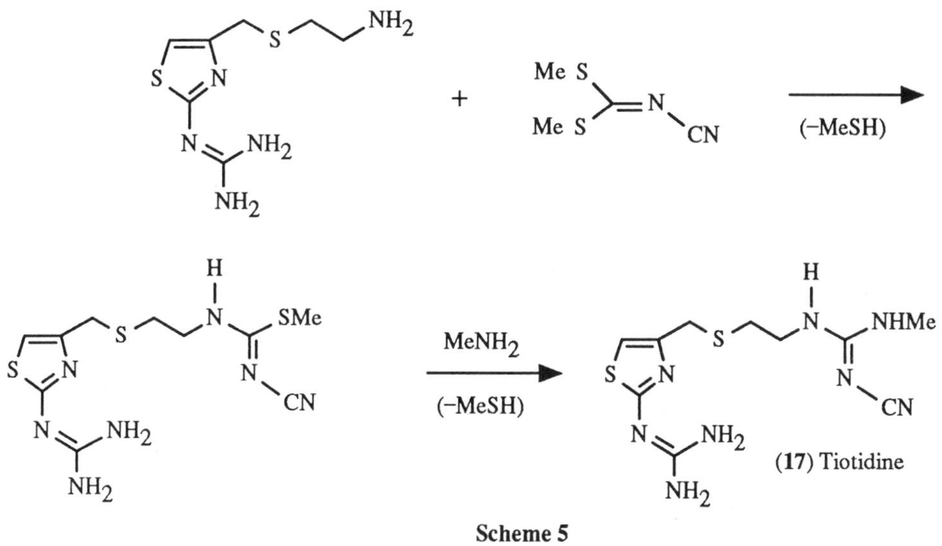

S, N, S, NH_2, N, NH_2, NH_2 + Me S, Me S, N, CN → (–MeSH)

H, N, S, SMe, S, N, N, CN, N, NH_2, NH_2 → $MeNH_2$ (–MeSH) → H, N, S, NHMe, S, N, N, CN, N, NH_2, NH_2 (**17**) Tiotidine

Scheme 5

assessment of the yield in each reaction. Therefore a prime requirement is an analytical assay method and a reference standard; the latter may be an arbitrary laboratory sample against which all others are referenced but preferably is a highly purified sample of the intermediate or final product. So often, yields quoted on an isolated weight basis can be misleading owing to the presence of inorganics or from the HPLC purity not being measured against a standard, *i.e.* long-running impurities may not have been eluted.

Yield improvement can be carried out either using an investigative approach or empirical methods such as simplex and factoral design (see Section 3.6.3.3.4) based on achieving the maximum yield in the minimum number of experiments. The skill in the latter methods lies in the choice of parameters to optimize.

3.6.3.3.2 The investigative approach

Firstly, for each stage, it is important to find out where the yield was 'lost' *i.e.* to obtain a material balance.

The following questions should be addressed:

(i) Has the reaction gone to completion? Is there any starting material left? If a reaction has not gone to completion (assuming adequate time has been given), it may be that not enough reagent is present or that reagent is being consumed in a side reaction or on further reaction with product. These factors may be different if the order of addition is changed or if the conditions are changed. Adventitious water either in reagents or solvents can also lead to decomposition of reagent so that it is important that all reagents/solvents are analyzed for purity.

(ii) Was the product formed but further reacted to give a by-product? Further reactions can often be detected by examining the effect of extended reaction time on impurity levels. If secondary products are a problem to separate, it may be advantageous to carry out the reaction with a slight deficiency of reagent, particularly if separation of starting material and product is relatively straightforward.

(iii) What by-products are formed and how can they be minimized? Isolation and characterization of by-products is one of the most valuable exercises in chemical development. They are best isolated by chromatography of samples, enriched in by-product, obtained by recrystallization of the crude product followed by evaporation of mother liquors. The structure of by-products can give valuable mechanistic information about the course of the reaction and allow a choice of reaction conditions so that by-product formation can be minimized. Occasionally, by-products are formed by addition of solvent (*e.g.* exchange of esters with alcoholic solvents) or by impurities in the solvent (for example, dimethylamine in DMF or acetone in isopropyl alcohol reacting with carbonyl compounds).

Impurity characterization is also important in deciding on the criteria for intermediate purity. Often, during the early stages of development, the chemist does not know how pure the intermediates need to be to give a final drug substance of adequate purity. Initially, therefore, a cautious approach is adopted and intermediates are upgraded by recrystallization at strategic points in the synthetic route. Once the structures of impurities are known, however, it may be realized that they can be left in the intermediate, since they may not react at the next stage.

Characterization of impurities allows the work-up to be designed specifically to remove that impurity. On occasions careful pH control of the aqueous phase during an extraction (based on the pK_a of the impurity) will allow a separation method to be devised.[22] Additionally choice of recrystallization solvent is aided by a knowledge of the solubility properties and the structure of impurities.

(iv) Did the reaction go to completion but the product was lost during work-up? Product isolation is one of the most vital—and most underrated—areas in development. Work up of a process needs careful design and should consider safety factors and ease of scale-up. Often, unstable products can be hydrolyzed during aqueous work-up, the problem being accentuated during scale-up. It is important to establish, therefore, a material balance to determine how development of the process should be carried out. The solution to a low yield problem may be as simple as a change in recrystallization solvent or a change in extraction conditions (*e.g.* change of pH, salting out, relative ratios of solvent and aqueous layers, reducing emulsions). More often, however, complex factors involving degradation of product by high temperature (during stripping of solvents for example), by aqueous acid or base (during extractions) or by reaction with solvent (*e.g.* reaction of dichloromethane with amines).[22] Occasionally, loss of yield due to adsorption of the product on a solid reagent or by-product such as manganese dioxide (from $KMnO_4$ oxidations) or aluminum salts (from $LiAlH_4$ reductions or Friedel–Crafts reactions) can be circumvented by a change in work-up.

Examples in the literature which illustrate these points include the development of the drugs SK&F 93479 at SK&F,[22] the cephalosporins at Glaxo[24] and Merck,[25] and thromboxane antagonists at ICI.[26]

3.6.3.3.3 Streamlining the process

Further development of a synthetic route, even when the yields in each step are high, can reduce the cost of drug substance substantially and impact greatly on the introduction of the process into routine production. Many of the factors below may help to improve the yield, but not all will do so; they may, however, improve throughput in the plant by improving 'volume efficiency' and by process simplification. Factors which may need to be varied (possibly using a statistical approach—see Section 3.6.3.3.4) are discussed below, and a rational approach to variation of conditions is an essential part of optimization.

(i) Change of reagent or catalyst

At the early stages of development, the chemist may wish to change reagent to improve yield or increase selectivity; towards the end of the development, cost reduction or effluent control may be the prime motive. An example of the importance of these factors in a commercial process was in the scale-up of the manufacture of the Merck drug, cefoxitin (**19**), where the change from a complex catalyst (*N*-silyltrifluoroacetamide) to a cheaper and simpler material (powdered molecular sieves) allowed not only the required acid-protecting groups to be removed, but also meant that a simpler amine-protecting group (Ts instead of CCl_3CH_2OCO) could be used in the sequence (Scheme 6).[25]

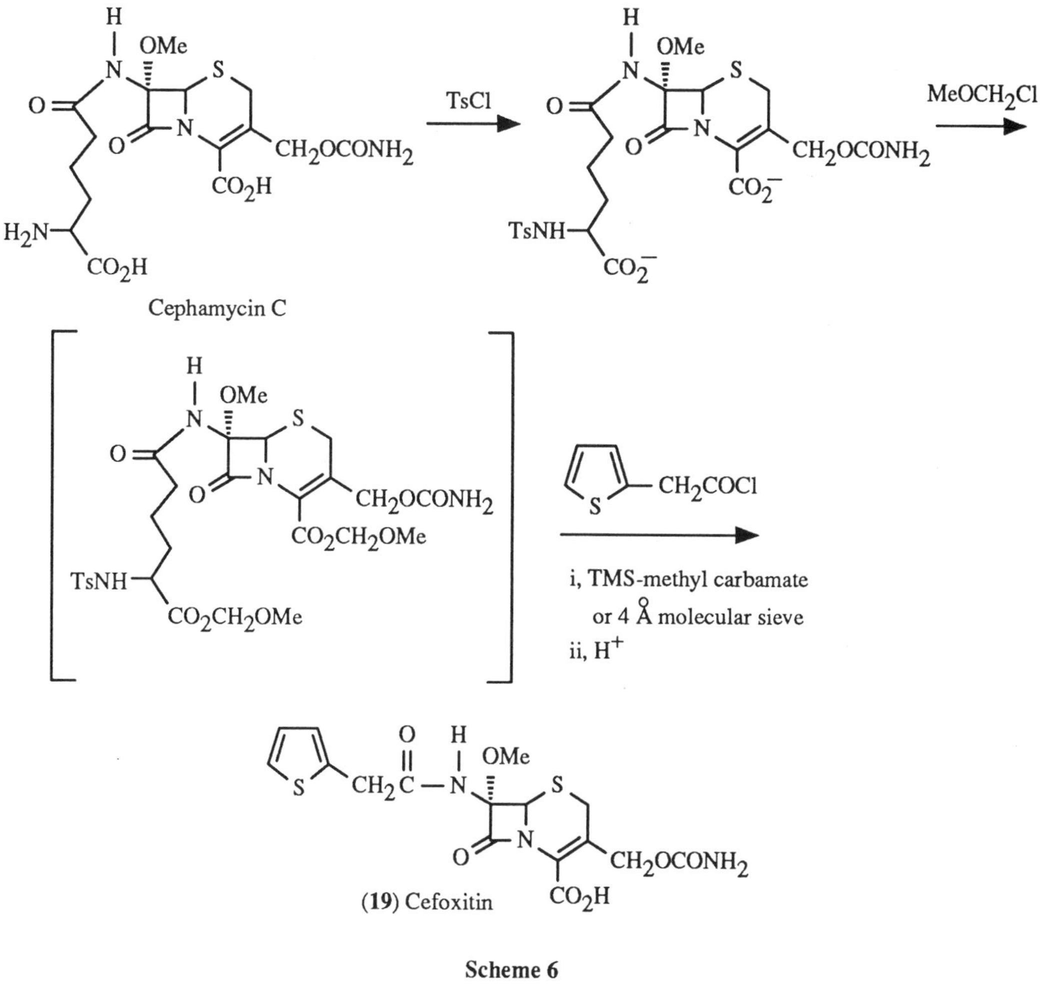

Scheme 6

However, further scale-up indicated that batch to batch differences in the molecular sieves exert an unacceptable variation in the process and the ultimate reagent chosen was the soluble catalyst trimethylsilylmethyl carbamate.

(ii) Minor change of intermediate

Although the synthetic route may remain substantially the same, it is possible to effect significant improvements by a minor change in the intermediate, *i.e.* by varying one of the following: (a) change in protecting groups (see Scheme 6); (b) change in the ester to increase/decrease rate of reaction and improve selectivity; (c) change in salt form of the intermediate, possibly to improve ease of isolation and (d) change in leaving group to increase rate of reaction or to circumvent a potential effluent problem (see Scheme 5).

(iii) Solvents[27]

Solvents are usually varied to increase the rate of reaction; increase selectivity (usually by decreasing the relative rate of reaction); to increase the solubility of raw materials or decrease the solubility of the product; to improve or simplify work-up and product isolation; and occasionally to change the crystal form of the product (see Section 3.6.4.5). Changing the solvent may not uniformly affect the rate of reaction and, occasionally, what are generally thought of as minor changes in solvent can profoundly affect the result. For example, in the scale-up of a Glaxo process for the imidazotriazinone (**20**), no reaction was observed with isopropyl alcohol as solvent in the cyclization of (**21**) to (**22**) (Scheme 7); ethanol led to by-product formation from reaction of the aminoguanidine bicarbonate with released diethyl oxalate; but methanol was satisfactory and aqueous methanol (1:99) gave improved product colour.[28]

Solvent effects are hard to predict and rationalize, even when a detailed knowledge of the kinetics and mechanism of the process is understood, and an empirical approach to the choice of solvent sometimes has to be taken.

(iv) Stoichiometry

Changing the relative ratios of reagent (and occasionally so-called catalysts) to raw material usually has a profound effect on the course of reaction and may allow a reaction to be driven to completion. Excess of reagents, however, may hinder work-up, (*e.g.* excess aluminum chloride in Friedel–Crafts reactions usually drives the reaction to completion by complexing with the product, but work-up is often difficult) and a balance has to be found. Occasionally excess of reagent may be detrimental, for example where it reacts with the required product (*e.g.* in oxidation of sulfides to sulfoxides), and this over-reaction may need to be carefully controlled. An essential part of this control would therefore be accurate assay methods for both reacting components.

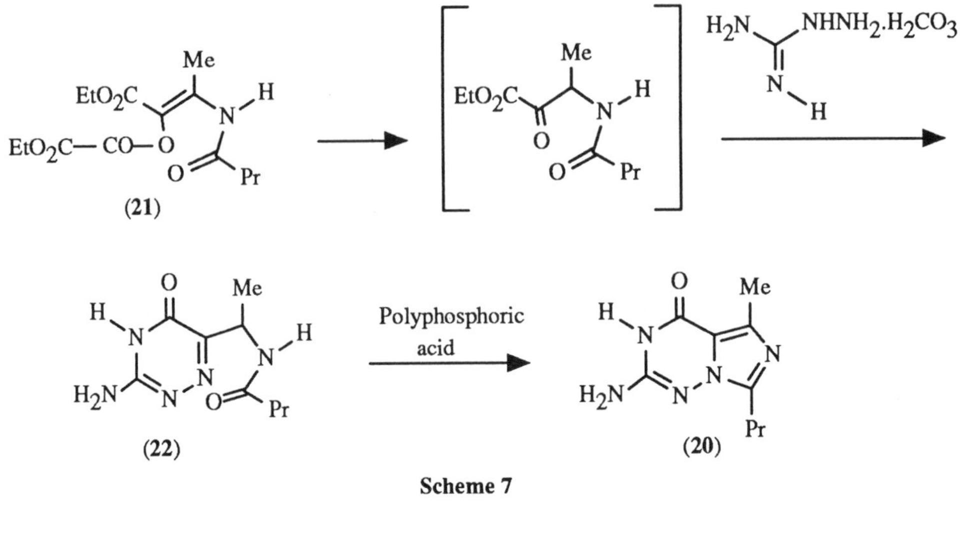

Me
EtO_2C
N H
EtO_2C—CO—O
O Pr
(**21**)
Me
EtO_2C
N H
O
O Pr
H_2N $NHNH_2.H_2CO_3$
N H
O Me
H N
N H
H_2N N N O Pr
(**22**)
Polyphosphoric acid
O Me
H N
N
H_2N N N
Pr
(**20**)

Scheme 7

(v) Rate and order of addition of reagent or catalyst

The rate and order of addition of reagents is often changed by development chemists to assist in the ease of handling on the plant, or to aid the control of exotherms, but may have a fundamental effect on the course of some reactions (*e.g.* Friedel–Crafts reactions). It is essential that the effect of these factors on yield and by-product formation is examined prior to scale-up.

(vi) Temperature

Increasing temperature obviously will increase the rate of a reaction but often selectivity is reduced. Conversely, reactions are usually carried out at low temperature to improve selectivity. Very low temperatures are rarely used in scale-up, and many reactions carried out in the literature at −78 °C (*e.g.* butyllithium reactions) can be carried out with similar results at −30 °C, this being within the capability of normal batch-processing equipment.

Accurate and reproducible control of temperature, particularly during exothermic reactions is a critical part of scale-up and the effect of temperature on reactions should be examined for hazard evaluation even if no yield or throughput improvement is envisaged—occasionally surprising results are found. For example, the well-known formylation of ketones and esters with sodium hydride and ethyl formate proceeds rapidly at 0–10 °C and gives high yields of α-formylcarbonyl compounds.[22] At 50 °C, however, it was found that no product was obtained, because ethyl formate is decomposed at this temperature by the hydride to give sodium ethoxide and carbon monoxide. This study does suggest, however, that the reverse process may be valuable, and that carbon monoxide in the presence of a strong base may provide useful formylating conditions.

(vii) Pressure

Pressure is rarely considered by chemists as a variable except when reactions such as catalytic hydrogenation are being carried out. Only very high pressures (10–20 kbar; 1 bar = 10^5 Pa) will affect solution-phase reactions, but recent interest in the subject,[29] and the availability of cheaper specialized equipment may mean that in the future, the development chemist may need to use pressure besides temperature as a means of improving selectivity (for example in cycloaddition reactions).

(viii) Time

In plant terms, time costs money, so that any factor which can decrease the plant-occupation time will result in cost reduction. Inevitably this means studying temperature and concentration variability. In this context it is important to have available an accurate method (usually TLC or HPLC) for checking completion of reaction, thus allowing work-up to take place as soon as possible. Simplification of work-up is also usually the best method of minimizing plant-occupation time.

(ix) Concentration

Studying the effect of concentration on yield and quality of product is vital to improve volume efficiency and minimize costs. The volume efficiency of a process is defined as the amount of product which can be produced in unit volume of reactor in unit time. It is thus a measure of plant throughput and is important at all stages of chemical development. At early preclinical stages when the process is first scaled-up and time is short for production of early supplies for vital toxicological tests, a doubling of volume efficiency may allow tight time schedules to be met. Obviously the main concern is to reduce solvent to reactant ratios, and to reduce work-up volumes in order to improve process efficiency. It should also be remembered, however, that increasing the concentration should have the benefit of increased rate of reaction, together with the disadvantage of increased thermal hazard; the lower the solvent volume, the greater the heat evolution per unit volume and the greater the cooling required. Once again a balance has to be found. Sometimes short cuts taken to try to improve volume efficiency can have additional benefits. For example, in phase-transfer reactions it is often possible to eliminate the use of water and to get even better results. The condensation of aromatic aldehydes with active methylene groups will often proceed with solid sodium hydroxide or sodium carbonate and a phase-transfer catalyst in the absence of water.

The above discussion has centred on changing process variables to improve usually yield and often product quality. One factor not discussed above is the effect of raw material (or intermediate) purity on yield, since small amounts of impurities can often have a significant effect on reactions. It goes without saying that a programme of optimization should, as far as possible, use raw materials of comparable quality (preferably taken from one large batch) but periodic checks over a number of months for changes in quality of raw materials (has the manufacturer changed or has the specification changed?) or changes in home-produced intermediates (does the intermediate deteriorate with time during storage?) are an essential part of any development programme. The importance of analytical methods for quality control of reagents, solvents and intermediates is once again stressed. Use-tests to check the quality of new batches of essential raw materials or intermediates are therefore important, particularly when the supplier is developing his process specifically for that drug. The supplier will be constantly improving his process but the quality may vary; occasionally higher quality raw materials give worse results—the trace of acid impurity, for example, can be vital!

Of similar importance is that reactants, catalysts and solvents used in process optimization are of similar grade and quality to those to be used in the plant, so that a direct correlation can be made. Of course, it is perfectly acceptable for optimization to take place initially with laboratory quality reagents, provided that the reagents to be used on the plant are adequately use-tested beforehand.

3.6.3.3.4 Process optimization using statistical methods

The previous discussion has centred on the variables which can be manipulated to gain a fundamental understanding of the chemistry and hence improve the process. However, this is a rather haphazard approach, varying in procedure from case to case. None of the variables discussed is likely to be independent of each other, and if a large number of process variables are shown by initial experiment to be important, changing each variable on a stepwise approach (keeping the others constant) will take a vast number of experiments and may never reach the optimum parameters.

Increasingly, therefore, statistical methods, which were developed at ICI in the 1950s and 1960s, have assumed more importance in chemical development.[30, 31] The aim of these methods, namely factorial design, simplex and, to a lesser extent, evolutionary operation, is to achieve the optimum yield (or purity or other measurable criteria) in the minimum number of experiments. These methods can also be used in drug design[32] and in the optimization of HPLC conditions.[33]

Statistical methods for optimization of yield or product quality were devised to try to study large numbers of variables using as few experiments as possible, and to take into account interacting variables. For example, in Figure 1, where the effect of temperature on reaction yield at two different pH values is shown, the two variables pH and temperature are said to interact. Obviously for large numbers of variables the situation is very complex. In these methods, the response (in this case chemical yield) is regarded in terms of a surface of n dimensions (where n is the number of variables) with contours of equal response level. The aim of each method is to reach the maximum response by the 'path of steepest ascent' (*i.e.* the minimum number of experiments). Two methods, factorial design and simplex useful in chemical development are described below, whereas evolutionary operation, more useful in a full production plant is briefly touched on later.

(i) Factorial designs[30, 31, 34, 35]

The simplest factorial designs are based on carrying out a prescribed set of experiments, varying each parameter between upper and lower values. Thus to complete the series varying five parameters, $2^5 = 32$ experiments are required. Algorithms exist so that the less meaningful experiments can be eliminated and possibly 16 experiments may be all that is needed to assess the relative importance of the five variables. Calculation of the path of steepest ascent (*i.e.* the road to the highest yield) may then afford the optimum in a further 5–10 experiments.

The brief description can only touch the surface (no pun intended) of this whole area of design of experiments and readers are encouraged to explore this area further, particularly through the text of Davies,[30] which was first published by the ICI team in the early 1950s. It was written with chemical experimentation in mind and is still relevant today.

(ii) Simplex methods[36]

Simplex, or self-directing optimization, is a rapid technique for calculating optimum reaction conditions using the minimum number of experiments by varying the level of each parameter after each experiment. It was first introduced by Spendley[36] (again of ICI!) in 1962, but if evidence from

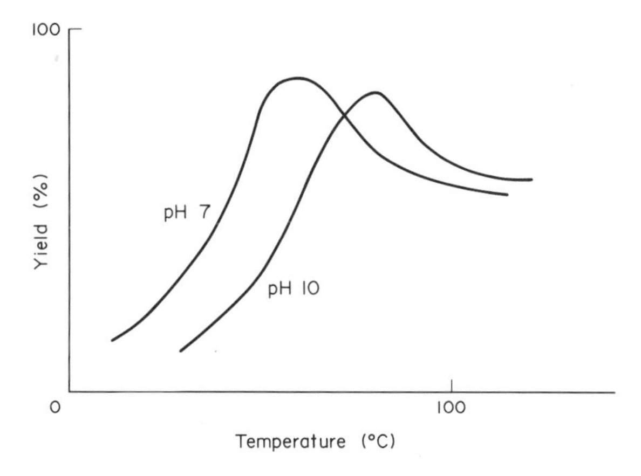

Figure 1 The effect of temperature on reaction yield at two different pH values

the published literature is a guide, it has been little used by organic chemists.[37] It is, however, ideal for process optimization and has been in use in some pharmaceutical companies for a number of years.

A simplex is defined as an n-dimensional figure with $n+1$ vertices; thus in two dimensions it is a triangle; in three dimensions a tetrahedron. Each dimension is used to represent a reaction variable on a contour map showing regions of equal response (in this case, yield). In the simple case of two dimensions (Figure 2), two variables (*e.g.* reaction time and temperature) would be represented by the contour map and three initial experiments represented by the points A, B and C would be carried out. The simplex then allows the parameters (temperature, time) to be calculated for the next experiment D by reflection in the axis BC. After experiment D has been completed, the results are used to calculate the parameters for experiment E, always eliminating the point which gives the lowest response (yield). As the simplex proceeds, the response increases as shown in Figure 3.

The value of the method is that for complex reactions, a large number of variables can be altered in the same set of experiments. The method is empirical, needing no knowledge of mechanistic organic chemistry except in the choice of initial conditions (which could well be a published procedure), in the choice of reaction variables and in the incremental changes in the variables. The latter determine the size of the simplex (the triangle ABC in Figure 2) and therefore the number of experiments which will be required to reach the optimum. For a published procedure, giving yields

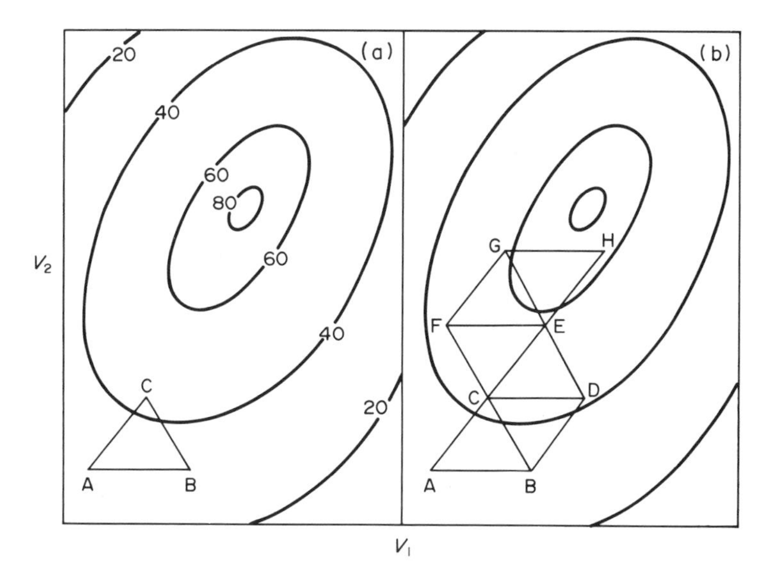

Figure 2 (a) A two-dimensional simplex superimposed on a contour map of isoresponse lines (percentage yields shown); (b) movement of the simplex toward optimum: replacing A by D, B by E, D by F, C by G, F by H

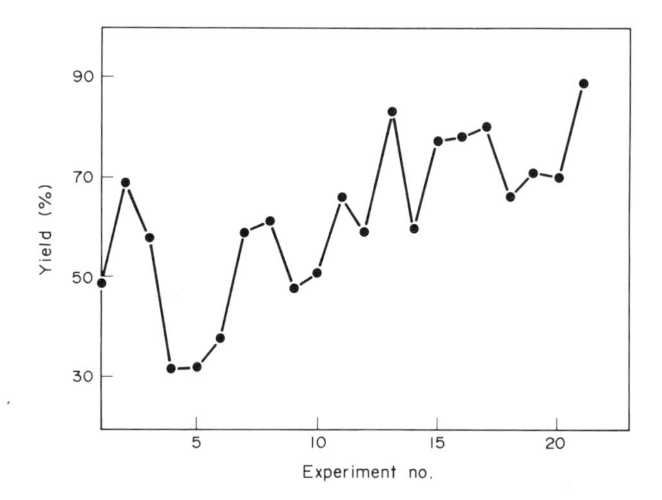

Figure 3 Progress of the sequential simplex toward optimum yield for the six-variable Bucherer–Bergs reaction of cyclohexanone

of 20–30%, larger increments may yield quick results; for fine tuning a process in which the yield is already 80%, small increments may allow the optimum to be reached. The method is illustrated by the following example[38] which used the method to optimize the yield in the Bucherer–Bergs method for the synthesis of hydantoins, important precursors of amino acids. The Bucherer–Bergs reaction is essentially the reaction of hydrogen cyanide, carbon dioxide (or carbon disulfide or carbon oxysulfide) and ammonia with ketones, giving hydantoins (or thiohydantoins) in moderate yields; hindered ketones often give only 20–25% yields. Scheme 8 shows a possible reaction mechanism. The initial parameters chosen to vary were concentrations of ammonia, carbon oxysulfide and hydrogen cyanide; temperature; time and per cent ethanol in the solvent. The pH was controlled only by the other six parameters. The results are shown in Table 1 for the first seven reactions, while Figure 3 shows the progress toward the optimum yield of 88% after 21 experiments.

Although the simplex method is empirical, it can lead to mechanistic understanding of a process. In the example in Scheme 8, the results show that the ratio of cyanide to ketone must be close to theoretical; the instinct of the organic chemist would be to increase the cyanide concentration which actually lowers the yield. Similarly, although unreacted intermediates are present in the reaction mixture, lengthening reaction time leads to a lower yield, again contrary to the instincts of the organic chemist.

(iii) Evolutionary operation[39]

In the 1950s Box proposed the technique of evolutionary operation for the optimization of processes on a plant. A systematic cycle of slight variants of the current process is repeatedly explored until a sufficiently strong indication of the desired changes emerges. The process is then modified and a new cycle of changes is initiated. It is really only applicable where large numbers of plant batches (at least 50) are being processed so that statistical significance is achieved. Thus it is best applied to a full production plant rather than during development. It was, however, from Box's original concept of evolutionary operation that simplex optimization was developed.

Table 1 Reaction Yields in the Bucherer–Bergs Reaction

Experiment no.	*Solvent (% EtOH)*	*[NH_3]*	*[COS]*	*[HCN]*	*Temperature (°C)*	*Time (h)*	*Yield (%)*
1	50	4	3	3	53	4	49
2	75	4	2	2	53	4	69
3	50	6	2	2	53	4	57
4	50	4	1	2	53	4	31
5	50	4	2	4	53	4	32
6	50	4	2	2	43	4	36
7	50	4	2	2	53	2	59

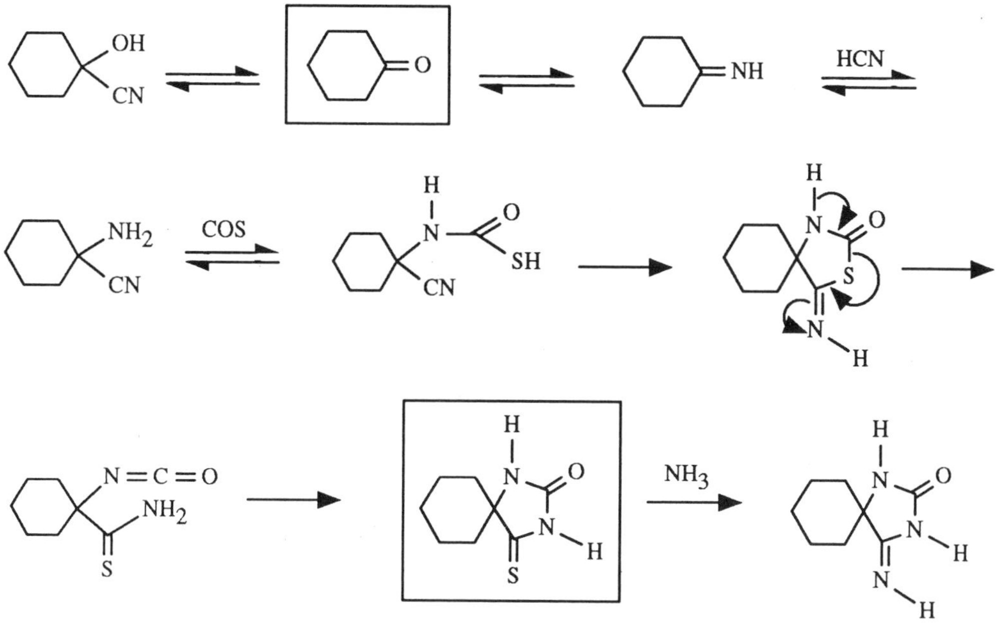

Scheme 8

3.6.3.4 Computer Control of Laboratory Processes

Automated computer-controlled reactors are now available for the fine tuning of processes during development work and for safety testing, but at the moment they are costly (of the order of £70 000). Many industrial laboratories, including Smith Kline & French,[40] Roussel-Uclaf,[41] Ciba-Geigy[42] and ICI,[43] have tried to devise their own systems for automating laboratory chemistry, when a series of experiments using the same reagents has to be repeated under varying conditions (as in a factorial design for process optimization). In general, the best systems have been those designed by chemists for use by chemists and of the in-house systems the one used by Roussel-Uclaf has received the most publicity. However this has been, until recently, difficult to obtain commercially in the UK.

The system developed in the Sandoz laboratories (built and marketed by Contraves)[44] and at Ciba-Geigy (marketed by Mettler)[45] are, however, readily available and are widely used throughout Europe although very few are in operation in the UK. The systems are user-friendly, requiring no computing knowledge and the 'recipes' are keyed into the computer using a question and answer dialogue which builds up an experimental programme quickly.

A basic system is shown in Figure 4. The reactor assembly comprises a 1 L jacketed reaction vessel, a reflux condenser and a stirrer. The reactor jacket is connected to a heating fluid used for all heating and cooling, and the fluid pump is controlled by external heat exchangers. Additions are metered in from reservoirs located on the balance, and can be programmed to be added in unit time, or at such a rate as to control the temperature in the reaction at a preset value. Measurements which can be made include reaction temperature, jacket temperature, pH, pressure and weight on balance; these are measured against time and can be plotted on the printer. Control of parameters is much better than a chemist could achieve (*e.g.* temperature change of 0.1 °C); rates of addition are very finely controlled (and recorded) so that subtle differences in process variables can be investigated. Measurement of jacket temperature allows (with the correct software) a heat balance and heat of reaction to be obtained and is thus of value in safety-testing processes. Not only is a measurement of heat of reaction obtained during a typical reaction, but a plot of heat evolved *versus* time will show exactly at which point the exotherm took place, and the rate of heat evolution. This will allow the chemist to calculate whether cooling capacity on plant is sufficient to allow further scale-up of the process.

Other models of the calorimeter are now coming on to the market and these have some similarities to the above system. This is a likely area for future expansion, particularly as detailed safety testing of processes becomes more important in the future.

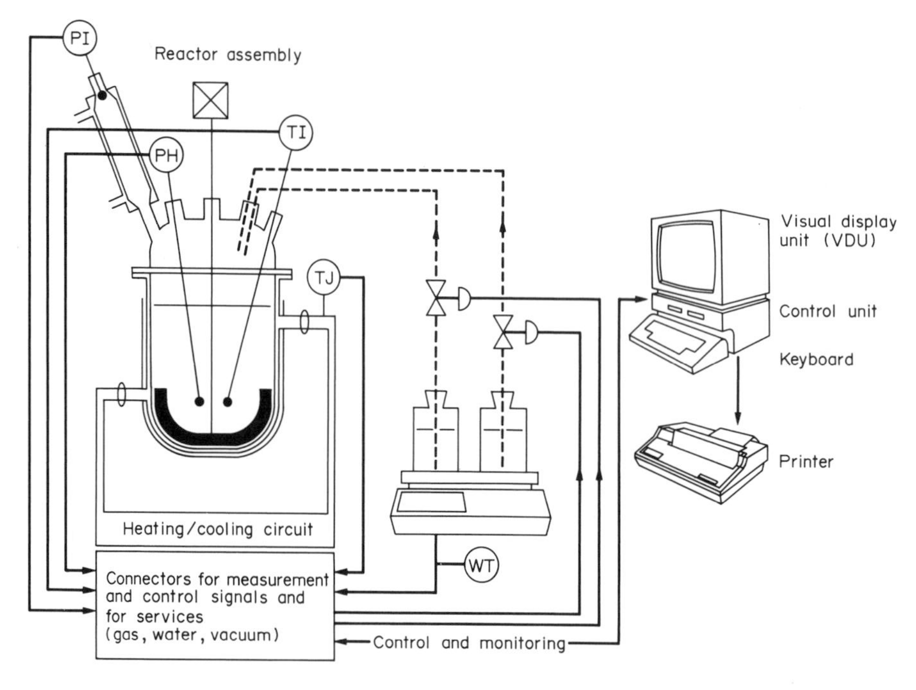

Figure 4 Computer-controlled laboratory reactor. Measurements which can be made include: temperature inside reaction vessel (TI); jacket temperature (TJ); internal pressure in reaction vessel (PI); pH value (PH) and weight on balance (WT)

3.6.3.5 Factors Important in Scale-up to Pilot Plant[46, 47]

Scale-up of organic reactions usually takes place initially in large glassware — up to 20 L capacity — in which the characteristics of small reactions are generally maintained. Physical handling of such quantities of material — flammable solvents, toxic materials — usually in large 'walk-in' fume cupboards, is difficult, and, in the author's opinion, one of the most dangerous operations involved in scale-up, particularly if standard laboratories are used. The consequences of a spillage of 10–20 L of highly flammable solvent in an area not fitted with flameproof electrics could be a disaster. Similarly the release of say 3–5 mol of toxic gaseous by-product may also cause handling problems within the laboratory. For this reason, scale-up to small glass or steel reactors in a purpose-built, flameproof pilot plant, where adequate scrubbing of waste gases can be controlled, is generally preferable. The definition of a pilot plant varies from the viewpoint of the chemical engineer involved in bulk chemical production, who envisages a pilot plant as a purpose-built plant designed to test a process, to the development chemist, who sees the plant as a logical extension of a typical laboratory set up of a three-necked flask with an addition funnel, stirrer and condenser. The latter view (anathema to the chemical engineer) will prevail in the following discussion, although some simple chemical-engineering concepts will be introduced; the cooperation of chemists and chemical engineers is vital to the success of scale-up, particularly the later stages of introduction of a process to full plant-scale.

The purpose of scale-up to pilot plant[46] is twofold. (i) *Information gathering* to determine the best way to: scale the reaction; handle reactants, intermediates, waste streams and off-gases; extract/isolate the product; filter and dry the product; distil/purify the product and to give additional safety and engineering data which may impact on further scale-up. (ii) *Production of supplies* for toxicological, carcinogenicity and clinical studies.

A typical multipurpose pilot plant is generally geared to batch processing in stainless steel or glass-lined jacketed reactors with internal capacities in the 20–1000 L range (usually at the lower end of this range) processing batches in the 1–30 kg range with full manufacture usually in the 1500–10 000 L region. The plant has flameproof electrics, with equipment rated for handling highly flammable liquids, and gases, but not usually group IIC materials such as carbon disulfide, acetylene and hydrogen, which require specialist zones. The plant is designed for flexibility, to allow as many different types of processes to operate in the equipment, bearing in mind that requirements may change on a weekly (occasionally daily) basis. Usually the plant is two-storey (occasionally, higher) with reactions and work-up taking place at the upper level; using gravity flow, filtration can then take place at the lower level.

The plant must be capable of handling all types of processes; coping with a wide variety of materials in different physical forms, with widely differing handling requirements and corrosion properties; protecting staff from toxic effects of materials; being flexible yet safe to operate routinely; being easy to clean, thus avoiding cross-contamination between batches and conforming to good manufacturing practice[48] and being updated to incorporate new ideas, techniques and technologies, with space for experimental rigs if necessary.

Whilst versatile pilot plants can be adapted to carry out most operations, some standard laboratory procedures can lead to plant problems and are best avoided if at all possible. These are shown in Table 2.

3.6.3.6 Pilot Plant Equipment

3.6.3.6.1 Reactor vessels

Batch process chemical reactors (Figure 5) are usually constructed of jacketed stainless steel or glass-lined (enamelled) mild steel, although recently fluorinated polymer (Fluoroshield) lining has become available for general purpose use.[49] Stainless steel reactors have the advantage of better heat transfer characteristics and usually better agitation but suffer from the obvious disadvantage of corrosion with aqueous acid conditions (acetic acid is usually satisfactory with type 316 stainless steel). Hastelloy C reactors have similar advantages and are relatively corrosion resistant but are expensive; all metal reactors, however, may interact with certain organics which can complex with metals, leaving traces of metal ions in products, or causing discolouration (for example with phenolic compounds). As a general rule, therefore, glass-lined reactors are preferred initially for most processes. However, their limitations are on temperature (usually $\simeq -30\,^{\circ}C$ to $150\,^{\circ}C$), extreme alkaline conditions, which can lead to dissolution unless special glass (*e.g.* Nucerite) is used, and reactions containing fluorides or fluoroborates, which may lead to etching—reactors coated with Fluoroshield polymer,[49] for example—are then of value.

Low temperature reactions at $-78\,^{\circ}C$ (such as alkyllithium processes)[50] and processes requiring very good heat transfer (*e.g.* cooling a Grignard reaction) are best carried out in stainless steel or Hastelloy vessels. High temperature processes and distillations also generally use stainless steel equipment although there may be exceptions if corrosion is a problem (*e.g.* Friedel–Crafts alkylations at high temperatures).

Vessel sizes should aim to give a wide flexibility, with small 25–100 L equipment for initial scale-up when vital intermediates may be in short supply, or when concentrated reactions (or distillations) may be required. Larger equipment (100–1000 L) will be useful not only for further scale-up, but also for processes with poor volume efficiency, exothermic processes carried out under dilute conditions to minimize the hazard, and for drown outs (*e.g.* $POCl_3$, $SOCl_2$ reactions) and separations.

Each reactor vessel will usually have an associated header tank, which may be of glass for simple, easy to view addition of 50–100 L of liquid or solution reagents, or may be a glassed steel tank in the 100–1000 L range, useful additionally for separations. Reflux and/or distillation facilities are provided with the condenser or other form of heat exchange in the return line (see Figure 5). The chemist's preference will be for all lines and condensers to be made of glass so that he can observe the process, but other dictates may require steel or lined steel pipework or carbon block heat exchangers (particularly if water reactives such as hydrides are to be used).

Low pressure reactions could be carried out in these reactors, but the limitation is usually the pressure rating of the associated glassware, which varies with the size of equipment.

3.6.3.6.2 Reactor services

The jackets of general purpose vessels are connected to steam (usually reaching temperatures to $160\,^{\circ}C$), hot water, cold water, refrigerant (either brine, methanol or glycol giving temperatures down to $-20\,^{\circ}C$ and occasionally below) and nitrogen to 'blow back' when service fluid is changed. Cold water is usually connected to the condenser and refrigerant is also advantageous if low boiling solvents such as dichloromethane are being distilled under vacuum.

A more modern method of control of temperature is to use a secondary heat-transfer fluid circulating around the reactor jacket. This avoids the cross-contamination of aqueous and refrigerant services and the requirement for nitrogen purging when changing services, and minimizes vessel corrosion.

Table 2 Laboratory Procedures Which Lead to Plant Problems

Laboratory method	*Scale-up problem(s)*	*Possible solutions/alternatives*
1. Evaporation to dryness to produce an oil. The oil may crystallize or be induced to crystallize by trituration	1. (a) Excessive stripping time in plant area on a large-scale rotary evaporator means lengthy exposure to heat and decomposition (b) It is difficult to remove the product from the vessel without redissolving	1. (a) Use solution of product in the next stage (b) Replace the solvent to be evaporated by the solvent in the next stage (if of higher boiling point) by adding second solvent prior to partial evaporation
2. Charge all reactants then heat	Difficult to control exotherms on large scale	Charge one component when required temperature has been reached
3. Adding catalyst last	Difficult to control exotherm	Change order of addition
4. Chromatography	Tedious on plant-scale—wasteful on solvents which are difficult to recover when mixed eluents are used	Avoid by developing fractional crystallization or distillation procedures
5. Use of drying agents	Involves filtration of agent (handling losses) and loss of product absorbed on drying agent	Azeotope to dry if necessary by addition of second component
6. Reaction in a dipolar aprotic solvent (*e.g.* DMF, DMSO) followed by work up by aqueous quench and extraction into a water immiscible solvent	Volume increases during process by as much as tenfold. Often interface problems during separations. Nonaqueous phase retains water, yet backwashing to remove dipolar solvent is sometimes problematical. Large volumes of effluent produced	Precipitate product directly from solvent using nonpolar solvent. Remove any inorganics by recrystallization Use toluene–DMF (or DMSO) (10:1) as solvent; often there is little diminution in rate of reaction

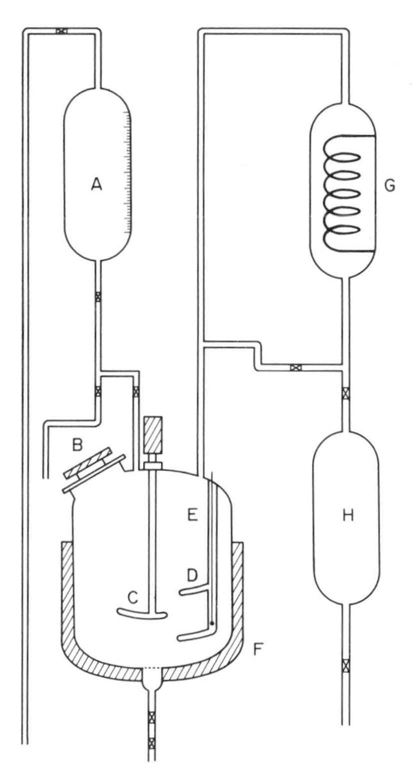

Figure 5 Typical layout for a pilot plant vessel. Header tank (A); charge hole (B); agitator (C); baffle (D); thermowell (containing temperature probe) (E); jacket (for supply of all services including steam, cold water, refrigerant *etc.*) (F); condenser (or other form of heat exchanger) (G) and receiver vessel (H)

Gaseous by-products are usually removed by fan-assisted wet scrubbing using packed towers or Venturi scrubbers.[51] Generally aqueous sodium hydroxide solution (for acidic vapour) and sodium hypochlorite (for sulfurous and other noxious vapours) are the common fluids, but on occasions water or ammonia may need to be used. Some companies use carbon absorption as a back-up for final removal of trace quantities of, say, thiols.

A vacuum is also connected to each vessel for general purpose distillation and solvent removal, whilst nitrogen should be available for blanketing as an inert atmosphere.

Specialist vessels which require high temperatures (to 300 °C) for distillation or special reactions will use hot oil (such as Dowtherm or Santotherm) as a heat-transfer medium in the jacket; the choice of which type of oil will depend on the temperature range required. Low temperature processes will normally use a solid CO_2–isopropyl alcohol or acetone fluid, probably cooled through a secondary fluid such as glycol (CO_2–solvent is difficult to pump round without getting vapour locks). Recently, liquid nitrogen, fed directly on to the reaction mixture has been used to cool reactions to below -100 °C in stainless steel equipment.

Split jacket vessels which only allow for heating on the lower or on both jackets may be useful if baking of solid product on the sides of the vessel is likely to be a problem (with decomposition to by-products or for safety reasons).

3.6.3.6.3 Agitators and baffles

Agitation is important not only for mixing (or in chemical engineers' terminology, mass transfer) but also to assist in heat transfer,[52] particularly with viscous mixtures or slurries. Turbine agitators are used for high speed, low viscosity applications, whereas anchor impellers are used when agitation close to the vessel wall is needed. Baffles are essential in stirred batch reactors to provide good mixing patterns and to increase turbulence, thereby improving heat transfer.

3.6.3.6.4 Equipment for reagent addition

Liquid or solution reagents can be added from header tanks or pumped from drums or other vessels, whereas solids (if not dissolved in a solvent) can be added through the charge hole or *via* a hopper arrangement. Controlled addition of reagents is important in controlling exothermic processes and in ensuring reproducibility of batches, and therefore the fine control offered by metered liquid addition is preferable — in this case the principles are the same as in laboratory reactions.

3.6.3.6.5 Filters and centrifuges[53]

The usual methods for separating solid products from liquid waste streams are filtration and centrifugation. Vacuum or pressure filtration, originally rather similar to laboratory filtration, has now been updated, and modern Nutche filters allow not only good filtration and washing of the filter cake, but also agitation during this process, often in an enclosed environment. Filters with heating jackets are also available, allowing their use as a form of crude drier. The perforated basket centrifuge, although more expensive than other filtration equipment, is of wide applicability, particularly if lined with fluorocarbon coatings such as Halar. Nitrogen-purged centrifuges allow safe filtration and cake washing using highly flammable solvents to be carried out. Plate or cartridge filters are used to remove unwanted solids (*e.g.* charcoal, drying agents, *etc.*) from liquids. Ideally filtration equipment should be segregated to minimize contamination.

3.6.3.6.6 Filter-driers and reactor filter-driers

Combined filter-driers such as the Rosenmund filter are important in production processes and occasionally justify their expense in a pilot plant. Reactor filter-driers, for example the Nutrex, are important for the processing of toxic solids, since the whole process can be carried out in one unit with no handling problems. Alternatively they can be used to handle sensitive materials where transfer from a reactor to a filter then a drier may cause difficulties. These combination units are expensive, and are available in stainless steel (or, even more expensively, Hastelloy), but may avoid the need for separate clean room facilities.

3.6.3.6.7 Driers[54]

A wide variety of drying equipment is available on the market. The most versatile, but not necessarily the most efficient, are steam-heated tray driers (either air or vacuum) but other equipment such as fluid bed driers, rotating cone vacuum driers, vacuum pan driers and paddle driers are used. Ease of cleaning is an important factor on a pilot plant, and it is preferable that each drier is located in a separate cubicle containing facilities for washing the equipment — this reduces the chance of dusty product contaminating other processes.

3.6.3.6.8 Distillation and evaporation equipment

Short-path distillation units[55] or thin-film evaporators are important for the concentration of labile solutions or for the purification of heat-sensitive liquids. Large rotary evaporators are also available for concentrating solutions as in the laboratory.

Fractionation-equipment to separate mixtures of products or purify compounds is important for intermediates rather than final products since few final drugs these days are liquids.

Purification of small batches can also now be carried out with preparative HPLC, or GLC, an expensive but important addition to a modern pilot plant.[56]

3.6.3.6.9 Miscellaneous equipment

Liquid–liquid extraction is usually carried out in the standard agitated vessels although more modern equipment such as reciprocating plate columns give more efficient separation.

Transfer of fluids and slurries between vessels and other equipment is best carried out using flexible hoses with quick release couplings.

3.6.3.7 Important Differences Between Laboratory and Plant which Affect Scale-up

3.6.3.7.1 Heat transfer

In the laboratory, heating and cooling is very quick, but as the vessel size increases, the surface area (proportional to the square of the diameter) to volume (proportional to the cube of the diameter) ratio changes. The effect is that as the batch size increases, heat up and cool down times will be lengthened, and exotherms will be more difficult to control. The shape of the vessels in plant often means that heating and cooling only takes place at the sides, not at the bottom of the vessel, further reducing the surface available for heat transfer. Thus if only a small amount of liquid is present in the vessel at the start of a reaction (*e.g.* Grignard reaction), temperature control may be difficult.

There will be a temperature gradient between the vessel wall and the centre of the vessel close to the agitator, which may be accentuated in viscous solutions or if solid becomes caked on the sides of the vessel. Good mixing is vital for good heat transfer, and baffling in the vessel will ensure turbulence so that material does not become 'locked' in a vortex close to the agitator. Heat transfer is better in metal vessels, whereas glass lining adds another surface through which heat has to be transferred.

3.6.3.7.2 Mass transfer

Mass transfer, which will determine the rate of reaction, rate of extraction, rate of crystallization, *etc.*, will vary with the shape and size of the vessel, type of agitator and baffle and agitator speed, the quantity of material in the vessel, and with the viscosity of the liquid or solution. The shape of reactor vessels and the presence of a bottom outlet mean that coarse solids often drop into the well and may form a plug which prevents vessel emptying. Oily products which crystallize in the bottom run-off can cause a similar problem. For this reason, some companies use conical-shaped vessels for many reactions.

Mass transfer is better in stainless steel vessels, where the agitator can be close to the vessel wall; in glass-lined vessels tolerances are such that the 'minimum stirred volume' is higher than in stainless steel vessels.

Gas additions usually take place *via* dip pipes, or sparge tubes and the shape and size of the vessel and type of agitation will determine how good the gas absorption is.

3.6.3.7.3 Separations

Liquid–liquid separation are carried out by stirring, not shaking.

3.6.3.7.4 Visibility

Chemists are used to observing processes and monitoring parameters such as colour, gas evolution, dissolution of solids *etc.* In the plant, it is difficult to see clearly and much more reliance must be placed on instrumentation. Alternatively, more frequent sampling may be required to determine the progress of the reaction.

It is also difficult to tell whether a reactor is clean, and a standard TLC check for cleanliness after washing the vessel with a solvent should be carried out.

3.6.3.8 Control and Instrumentation in the Pilot Plant[56,57]

The following parameters are usually measured in plant equipment:

(i) *Temperature*: measured by a probe in a well or a probe inserted in the bottom run-off. The position of the probe will be important in determining the accuracy of measurement. There is usually a time-lag before an accurate reading is obtained, particularly in large vessels and this must be taken into account during reagent additions.

(ii) *Pressure*: usually measured inside the vessel and controlled by pressure relief valves or bursting discs, with lines containing these safety devices vented to air or, more satisfactorily, to a 'dump tank'. Thus if overpressurization takes place, the contents of the vessel will be ejected safely, rather than leading to further pressure build up and explosion.

(iii) *Agitation speed.*

(iv) *Volume (and sometimes weight) of reactants*: volume can be measured using ultrasonic or other instruments, whereas batch weight needs a vessel on load cells. This is common in automated full plant but rare in pilot plants.

(v) *pH control*: pH probes can be fitted to the vessel but are prone to corrosion problems when used continuously. As a result, pH may need to be measured externally.

Because of the need for flexibility pilot plants have traditionally been manually operated. Many pilot plants are operated by skilled chemists who prefer to retain the freedom to control processes as in the laboratory. The trend, however, is towards some computerization in the pilot plant to assist the chemist in controlling the parameters more carefully. Computer control in fine chemicals manufacturing is commonplace[57] and multistep syntheses can now be controlled so that intermediate quality does not need to be checked. Computerization is ideal for dedicated processes but the flexibility required in pilot plant operation has meant that chemists have been reluctant to automate.

In particular, the following improvements may result from computerization:[58] (i) better batch to batch reproducibility, since rates of addition of reagent and rate of heating (ramping) can be more carefully controlled. Control of rate of cooling is important for crystallizations and control of particle size (see Section 3.6.4.5); (ii) improved control of safety; and (iii) better information gathering and information storage.

It is likely that computer control of pilot plant processes will follow the trend set in the control of laboratory reactions (see Section 3.6.3.4) and be an important factor in process control in the 1990s.

3.6.3.9 Thermal Hazard Testing of Processes and Intermediates

Although the fine chemical and pharmaceutical industries have, in general, a satisfactory safety record, there have been occasional incidents in which reactions have gone out of control, resulting in fire or explosion which has damaged plant, caused pollution of the environment or, in severe cases, loss of life. Such incidents result in loss of production capacity but perhaps, more importantly, alienate public opinion, and it is therefore vital that such happenings are minimized by thorough screening of processes prior to scale-up to pilot plant and production. A recent example which resulted in the death of an employee and complete destruction of a chemical plant concerned the oxidative treatment of a waste stream from manufacture of intermediate (**9**), a precursor of the H_2-antagonist ranitidine (Scheme 4).

The development of a new process is more prone to this type of hazard since little information on materials and reaction exotherms is available in the literature. The different grades of raw materials used in plant, compared to laboratory purified grades, and the presence of other contaminants (*e.g.* rust from plant corrosion) which may catalyze side reactions, are factors which have led to uncontrolled exotherms in the past. Loss of temperature control can often occur after equipment failure, such as loss of agitation or cooling water interruption, and equipment must be engineered to be fail safe.

Even relatively minor changes in raw materials, reaction solvent (or its water content) and construction materials of the vessel can be potentially hazardous. Change of scale, which affects agitation and mass transfer, cooling capacity and temperature gradients, is one of the most significant factors. The other main cause of problems is equipment failure not only as a direct cause but indirectly. When equipment breaks down, a time-delay occurs during which a batch may be held for an extended period under abnormal conditions (elevated temperature, high pH) which may lead to exotherms when reprocessing begins.[22]

For these reasons, all changes of process should be adequately investigated in the laboratory prior to scale-up and new sources of materials should be use-tested prior to pilot plant usage. Examples where changes in raw material specification can cause significant changes in rate are: (i) the amount of sodium present in lithium metal or hydrides;[59] (ii) the particle size and grade of aluminum chloride in Friedel–Crafts reactions; (iii) catalysts such as Raney nickel and palladium-on-charcoal for hydrogenations or catalytic-transfer hydrogenation; and (iv) water content of solvents used for Grignard reactions, which may prevent initiation. The temptation is then to add too much alkyl halide so that when reaction does begin, an uncontrollable exotherm takes place.

3.6.3.9.1 Principles of hazard testing[60,61,62]

Each stage of a chemical process should be tested so that confidence in the ability to scale-up safely is achieved; that potential hazards can be dealt with; and that equipment failure will not lead to disaster, *i.e.* the process should be fail safe. The most appropriate time to consider safety testing is when scale-up to pilot plant is being carried out, since the scale-up factor from laboratory runs may be hundredfold, whereas further scale-up from pilot plant to full manufacture is often only a further tenfold increase in scale.

Ideally the study should include an evaluation of raw materials, solvent, intermediates (even if not normally isolated), final product, significant by-products and distillation residues and waste streams. The effect of changes in circumstances, which may be caused by human error, should also be considered. These include: (i) a change in the order of addition of reagents; (ii) the effect of omitting one component; (iii) under- or over-charging of a reagent; (iv) incorrect pH; (v) overheating the batch; and (vi) extended time of reaction.

These factors may influence the type of equipment used, the amount of technical supervision employed, or the safety devices (*e.g.* automatic cut outs) used.

The greatest potential exothermic hazard is a change of scale. The total heat generated in a process is proportional to the total volume (and obviously the concentration of the solution), which in turn is proportional to the cube of the vessel diameter. The rate of heat removal is dependent on the surface area available for cooling, proportional to the square of the diameter of the vessel, to the heat transfer coefficients of the surfaces (metal being better than glass) and the cooling capacity of the coolant, usually water. The larger batches are more liable to get out of control, and the consequences are potentially disastrous, since an increase in temperature will increase the rate of reaction (or decomposition) so that temperature may rise exponentially after the onset of an exotherm.

Furthermore, in larger batches, the increased distance through which heat has to be transferred from the centre of the reactor to the heating or cooling surface may mean that temperature gradients occur, *i.e.* that parts of the batch may be at a different temperature from that measured. If agitation fails during a critical phase of an exotherm, there is therefore a greater risk of loss of control.

These factors must be taken into account during scale-up. If the scale-up of a reaction increases tenfold, the relative rate of heat dissipation must be reduced by a factor of at least 2.15. To maintain the same margin of safety, the rate of reaction may need to be halved, *i.e.* the temperature may need to be reduced by $\sim 10\,^{\circ}C$. By extrapolating this argument, safe operating procedures can be devised when exotherms are detected in the hazard-testing procedure. Thus if the safety tests detect onset of an exotherm on 100 mL scale at 150 °C, on a 10 000 L reactor (10^5 scale-up), 100 °C is likely to be the maximum temperature for safe operation. In general, operating conditions should always be 50 °C below the temperature of onset of an exotherm which is likely to lead to loss of control. Often exotherms can be controlled at a set temperature by suitable choice of refluxing solvent; the additional cooling capacity of the condenser or heat exchanger gives an added safety margin. It should be pointed out, however, that onset of an exotherm may be rather imprecise, varying from batch to batch of the same compound; different crystalline modifications and impurity levels may give widely differing results.

3.6.3.9.2 Methods of thermal-hazard testing[61,62]

(i) Desk screening

Valuable information can be obtained prior to any experimental safety evaluation and it is important that 'desk screening' is carried out, since safety testing itself may present hazards with some compounds. There are examples where safety testing has led to expensive calorimeters being damaged (*e.g.* with the monosulfoxide of **9**). Physical data for each component in the reaction mixture should be gleaned from the literature[63] and data obtained for new compounds (see Section 3.6.3.10). The following, however, may give some indication of potential hazards: (a) presence of unstable groups such as acetylene, nitro, azide, perchlorate, diazonium; (b) comparison with known explosives or potentially explosive materials; (c) calculation of oxygen balance (see Table 3); and (d) thermodynamic data calculation (for example using the CHETAH computer program).[61] From the molecular formula of each component the heats of formation and reaction can be calculated to give an order of magnitude accuracy. Positive or small negative heats of formation indicate instability, whereas large negative heats of reaction indicate a high output; the rate at which this heat is generated, however, is unknown.

Table 3 Oxygen Balance

Hazard potential	*Oxygen balance*
High	-80 to $+120$
Medium	$+240$ to $+120$
	-160 to -80
Low	$>+240$
	<-160

Oxygen balance $= -1600[2x+(Y/2)-z]/M$ where M is the molecular weight; x is the number of carbon atoms; y is the number of hydrogen atoms and z is the number of oxygen atoms (other heteroatoms are ignored for the purposes of this calculation).

(ii) Laboratory evaluation of new compounds

Methods for the laboratory screening of new compounds have increased in sophistication over the last 15 years. Differential scanning calorimetry (DSC) and thermogravimetric analysis (TGA) will give an indication of exothermic decomposition in the solid state but may give an overoptimistic view of safety margins. Adiabatic calorimetry gives most valuable information and commercially available instruments such as Columbia's accelerating rate calorimeter (ARC)[64] or Systag's sikarex and sedex[65] are now used routinely to generate self-heating parameters, kinetic data for decomposition, and pressure and temperature maxima. Potential runaway reactions are quickly identified and their hazard potential assessed.

An example[66] from the author's experience demonstrating the value of these techniques was in the scale-up of a process to make an H_2-antagonist, oxmetidine (**23**), from the nitroamino compound (**24**; Scheme 9). DSC–TGA data indicated that an exotherm took place at 190–200 °C on dry heating generating 500 J g^{-1} energy. Data from an isoperibolic screen on a Sikarex calorimeter, however, revealed that maintaining the compound at 110 °C for 2 h (as in a drier for example) induced a self-heating which above 130 °C escalated to destruction. ARC data confirmed this conclusion and allowed a heat of reaction of 800 kJ mol^{-1} to be measured and a likely pressure build up of 2000 psi (1 psi = 0.175 N m^{-2}) was indicated. Safe operating conditions could thus be determined from this data. The data generated by adiabatic calorimeters on small samples of intermediates is invaluable but the instruments are expensive and have some limitation. Reaction mixtures can be handled, but adequate agitation was not available on earlier instruments. Simpler equipment, such as a Dewar flask arrangement,[61] is a good approximation to an adiabatic calorimeter and more closely mimics actual operating conditions. Dewar flask methods are recommended by the ABPI[61] in the UK for all hazard evaluation of reaction mixtures—it has been estimated by one pharmaceutical company that one in four processes tested by this method will require further investigation before safe scale-up can be carried out. The computer-controlled reactors[44,45] described in Section 3.6.3.4 for use in chemical development are invaluable for safety assessment, the major advantage of their data handling being that the region in the reaction where the exotherm takes place is pinpointed, and the rate of heat evolved is measured accurately. This provides excellent data for the chemist and engineer, involved in scale-up, to calculate required cooling capacity and to predetermine whether the chosen vessel will be able to adequately control the exotherm.

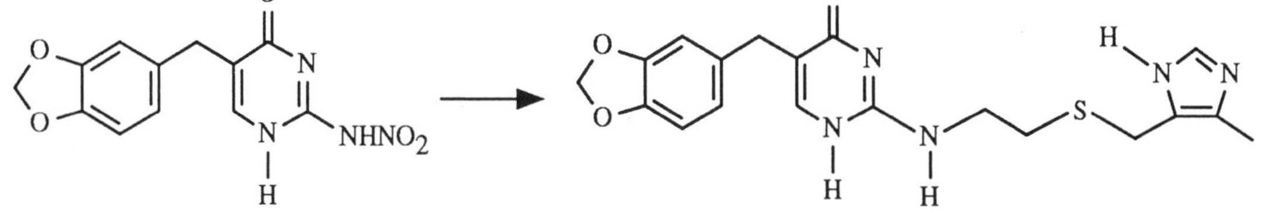

(**24**) (**23**) Oxmetidine

Scheme 9

3.6.3.10 Toxicity Considerations in Scale-up

Handling toxic reagents, solvents, intermediates and by-products in the laboratory causes little difficulty if adequate fumehood extraction is available. In the plant, however, the vessel itself may be scrubbed, and to some extent air local to the vessel will be extracted, but in general chemists and pilot plant operators will be exposed to large quantities of toxic materials unless adequate safety precautions are taken. It is therefore important to identify likely hazards prior to scale-up so that satisfactory procedures can be incorporated in the pilot plant working directions.

An essential part of this procedure should be the preparation of safety data sheets for each raw material, solvent, intermediate and product with which personnel are likely to come into contact. Raw material and solvent data sheets can be compiled from manufacturer's data, or from standard reference texts.[67–75] Safety data sheets for intermediates may need compilation from literature data, with some best guess attempts to correlate data with compounds of similar structure, erring always on the side of caution. For example, a new α-haloketone could be predicted to be lachrymatory and possibly vesicant, and handling precautions can therefore easily be assessed, since no contact at all should be allowed—*i.e.* full protective clothing (airhood and suit) should be worn. In some chemical and pharmaceutical companies, preliminary toxicity testing of new intermediates may be initiated, particularly if the synthetic route has been fixed. Contract toxicology houses will carry out LD_{50} and skin irritancy tests on animals, using approximately 30 g intermediate, with a minimum lead time of 4–8 weeks. It is arguable, however, how useful this information is, since oral poisoning is unlikely to be an issue, inhalation of vapour or dust being the most likely route. Intermediates should be always be treated with caution, since their structural similarity to the final drug substance is likely to render them pharmacologically active. The greatest danger with intermediates is from a splash from solution, or when drying, through dust inhalation.

Probably the greatest potential toxicity hazard, however, lies in the exposure to volatile reagents and solvents for which adequate data exist. Manufacturers have a duty under current legislation to provide data sheets to customers; many produce detailed booklets, illustrating not only valuable safety data, but also useful technical engineering information such as compatibility with materials of construction, vapour pressures, or viscosities at various temperatures, and storage conditions. Usually, if contacted, manufacturers are only too willing to assist and advise.

A typical safety data sheet prepared prior to scale-up to pilot plant, is illustrated in Figure 6. Preparation at this stage is a useful exercise for the development chemist, but effluent disposal routes should also be considered prior to this stage. However, the safety data sheet is often a useful trigger to encourage more detailed consideration.

3.6.3.11 Effluent Disposal

Methods of disposal of all waste streams are required once scale-up has taken place so it is important to consider this aspect early in the development phases. Off-gases during a reaction such as acids, ammonia, nitrous fumes or thiols can be adequately scrubbed, but scrubber liquors still require a disposal route; in the case of thiol vapours trapped in sodium hydroxide, this can be a difficult problem on a large-scale. On large plants, off-gases can, however, be incinerated and the resultant species (*e.g.* SO_2) scrubbed.

Organic solvent streams miscible with water are usually biodegradable in a biological treatment plant but traces of other components may kill the 'bugs' and early laboratory trials are recommended. Organic solvents immiscible with water (*e.g.* toluene) are often recoverable, but may be sold as fuel for incinerators, if not highly contaminated with halogenic or sulfurous materials. Chlorinated solvents can be recovered, and there is a market for recycling used chlorinated solvents in paint strippers. The presence of chlorinated solvents in other waste streams, however, causes problems.

Heavy metals which can be precipitated from solution may have some market value for recycling; nickel (in the UK) and, especially, precious metal catalysts are valuable. Mostly, however, effluent streams containing metals such as copper, chromium, manganese, lead, *etc.*, will require special contract disposal which will be expensive. Therefore minimal use of these reagents is necessary to control environmental pollution.

Aqueous streams can often be treated in-house by special biological treatment, with the water emanating from the treatment plant being controlled for purity by the local water authority. Chemical plants close to the sea have this as an alternative disposal route for treated aqueous streams. However, many streams are toxic to the standard organisms used in treatment plants,

IDENTITY

NTP PREFERRED NAME: Sulfacetamide

Synonyms: p-Aminobenzenesulfonoacetamide

N-Acetylsulfanilamide

CAS Registry Number: 144-80-9

NIOSH Registry Number: AC8450000

Formula: $C_8H_{10}N_2O_3S$

Molecular Weight: 214.24

WLN: ZR DSWMV1

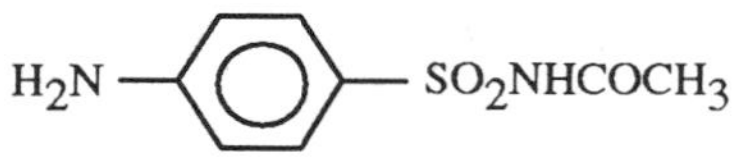

$H_2N-C_6H_4-SO_2NHCOCH_3$

PHYSICAL PROPERTIES

Physical Description: Colorless powder

Melting Point: 182-184 °C

Boiling Point: Not available

Density: Not available

Specific Gravity: Not available

Flammability: Not available

Stability: Stable under normal laboratory storage conditions.

Flash Point: Not available

Reactivity: Not available

Solubiliity In: **Water:** 6 mg/mL at 20 °C; **Acetone:** 140 mg/mL at 20 °C; **DMSO:** Not available; **Ether:** Insoluble; **Ethanol:** 60 mg/mL at 20 °C; **Benzene:** Not available

Other Physical Data: Aqueous solution is acidic to litmus paper.

SHIPPING

D.O.T. Shipping Name: Hazardous Substance, Solid, N. O. S.

D.O.T. Identification Number: NA9188

D.O.T. Hazard Classification: ORM-E

Other Shipping Regulations: None; no limit with passenger or cargo aircraft.

Exceptions: None. Specific Requirements, 173.1300 in Hazardous Materials Regulations of the Department of Transportation (1981).

NTP PREFERRED NAME: Sulfacetamide

HEALTH HAZARDS

Acute Hazards: Moderate toxicity

Symptoms: Unknown

Exposure Limits: Not regulated

FIRST AID

Skin Contact: Flood all areas of body that have contacted the substance with water. Don't wait to remove contaminated clothing; do it under the water stream. Use soap to help assure removal. Isolate contaminated clothing when removed to prevent contact by others.

Eye Contact: Remove any contact lenses at once. Flush eyes well with copious quantities of water or normal saline for at least 20-30 minutes. Seek medical attention.

Inhalation: Leave contaminated area immediately; breathe fresh air. Proper respiratory protection must be supplied to any rescuers. If coughing, difficult breathing or any other symptoms develop, seek medical attention at once, even if symptoms develop many hours after exposure.

Ingestion: If convulsions are not present, give a glass or two of water or milk to dilute the substance. Assure that the person's airway is unobstructed and contact a hospital or poison center immediately for advice on whether or not to induce vomiting.

ADDITIONAL INFORMATION

Storage Precautions: Store in a refrigerator or in a cool, dry place.

Spills and Leakage: Dampen spilled material with alcohol to avoid dust, then transfer material to a suitable container. Use absorbent paper dampened with alcohol to pick up remaining material. Wash surfaces well with soap and water. Seal all wastes in vapor-tight plastic bags for eventual disposal.

Suggested Gloves: Not available

Uses: Antimicrobial

Additional Reference Sources:

Merck Index. M. Windholz et al, 9th Ed. , p. 14 (1976), Merck.

Figure 6 A typical safety data sheet

although modern biotechnology is now producing organisms which will degrade chlorohydrocarbons, for example. Many water authorities control the final effluent very tightly, with low limits on even inorganic ions such as sulfate and chloride ions. Increasingly, therefore, disposal of aqueous effluent by landfill or dumping at sea is necessary. Landfill sites are, however, becoming scarce, and it is likely that effluent costs will increase dramatically in the future, causing development chemists to think earlier about minimizing the amount of waste streams from a process.

3.6.3.12 Further Scale-up

The quantity of drug substance required for toxicity, carcinogenicity and human testing prior to marketing is often in the 500–1000 kg range and occasionally more than 1 t of final product may be required.[6,7] In a multistage synthesis, *e.g.* tiotidine,[6] of say 8–10 steps with an average yield of 80% (i.e. overall yields of 11–17%), the quantities of early stage intermediates required to be manufactured may well be of the order of 5–6 t per stage. Often work will be carried out on full production scale in 1000–10 000 L equipment, either within the pharmaceutical company, or if capacity is short or specialized equipment is required, by subcontract to a fine organics manufacturer.

Carrying out processes on this scale often means the process may have to be run by skilled operators with very little technical supervision, since the production site may be geographically far removed from the research and development/pilot plant site. Therefore, process optimization to ensure the 'robustness' of the process must be carried out. Furthermore, since data gathered from these large-scale runs can be used to cost the process for eventual marketing considerations, it is important that the most economic process is operated, *i.e.* that yields and volume efficiencies have been improved as far as possible. Computer modelling and simulation techniques may be useful.[76]

Safety considerations may also play a role—a process which may be satisfactorily run at 250 L may require modification if carried out on a 10 000 L scale, particularly if cooling capacity is likely to be a problem on a multipurpose production plant, where vessels on different processes may use a common refrigerant. It is essential that on this scale a reevaluation of safety data and hazard analysis is carried out, and that the process is revised to take account of the change in scale.

For example, in increasing the scale of a batch by twentyfold, it is unlikely that the larger equipment could cope with a twentyfold increase in gas evolution, particularly if the gas had to be scrubbed; thus rates of addition may need to be modified. Heat transfer in a larger vessel will be less efficient, particularly for viscous reaction mixtures, and some baking (caused by overheating) on the sides of the vessel may occur; this may lead to product of lower quality or darker colour. Extended reaction times, inevitable on a larger scale, may also lead to reaction intermediates decomposing, resulting in lower yields or poorer quality product. The importance of close monitoring of the early batches, following reactions using some analytical technique such as HPLC, GLC or TLC, cannot be stressed too highly.

3.6.4 ANALYTICAL ISSUES IN DEVELOPMENT AND SCALE-UP

This section will examine the analytical issues which are important to the development chemist's scaling-up processes.

3.6.4.1 Sampling[77]

An analytical result is only valid if the sample submitted is representative of the whole batch. Whilst this is not normally a problem for small laboratory batches, as scale-up proceeds and large batches are obtained, accurate sampling becomes crucial and variations may occur owing to: (i) the position in the filter, centrifuge or drier, leading to different amounts of solvent; (ii) inadequate agitation in vessels causing nonuniform reaction; (iii) differences in heating (*e.g.* baking on the sides of the reactor) may cause variations in the level of impurities; (iv) physical contamination; and (v) nonuniform particle size.

Sampling of undried intermediates is notoriously unreliable and results generated from these samples should be treated cautiously. Usually dried batches can be sampled rigorously using standard methodology, and final products should be sieved to ensure uniformity before a sample is taken.

3.6.4.2 Analytical Methodology[78]

Analytical methods for the first batches of a drug substance are mainly directed towards proving the structure and giving a semiquantitative guide of overall purity. As the drug development programme proceeds, the analytical methods must develop alongside the synthetic methodology, so that, on the one hand, the development chemist is aware of the effect changes in chemistry can have on impurity profile, and on the other, the analyst is aware of how changes in the chemistry should affect the development of good assay and impurity profile methods. Of vital importance to the analyst is the early availability of reference materials, which should be in the form of a working reference (*i.e.* a typical batch of intermediate or final product against which all other batches are measured) or a primary reference. The latter is a highly purified sample prepared by multiple recrystallization to constant purity or by a chromatographic technique.

Analytical chemistry involvement in assessment of raw material and intermediate purity is important for the development chemist, who needs to know the tolerance levels for each stage of the synthetic route, *i.e.* whether impurities are carried through to the next stage. In addition, good analytical methodology may allow significant mechanistic information, which can often affect process optimization, to be obtained.

As the drug proceeds further into development, and scale-up takes place giving kilogramme batches of drug substance, the analyst will need to determine in more detail the impurity profile of the individual batches. The analyst requires a sensitive method, usually HPLC or HPTLC these days, which separates all known impurities from the main peak. Impurities present in greater than 0.1% are often isolated, possibly by preparative HPLC of enriched samples, characterized and synthesized independently. Standard samples of these impurities can then be used to develop accurate assay methods for the compounds most likely to arise in full-scale batches or which are mentioned in the specification.

Knowledge of the structure of impurities may give information about where they arose, and how they could be eliminated. Only when the structures are known, and accurate assays are available can the importance of a particular impurity be assessed. Many impurities will have structures similar to the drug substance and may be active, or possibly toxic. In some cases toxicological information may need to be obtained.

Quality control of repeat batches of final drug substance needs constant vigilance. Not only the synthetic route may change, but also the synthetic methods within each route will change as the development chemist constantly strives for a better and cheaper process. All these changes have to be monitored yet at the same time the analytical method will also be changing as it, too, is improved. Close contact with the development chemist will ensure that the analyst knows what to look for, and so maintain a check on the development process.

The above discussion has concentrated on impurity profile, but often it is for apparently more trivial reasons that batches of drug substance are rejected. These include sulfated ash (or residue-on-ignition; inorganic substances, usually); water and residual solvent contents; correct salt form (*i.e.* presence of cations or anions other than the required form), or correct polymorph and solvate. For chiral compounds, control of optical purity may be an issue.

3.6.4.3 Specifications

The issue of when and how tightly to set a specification for the drug substance will vary from one company to another, but ultimately the aim will be to satisfy regulatory authorities that the purity of the drug is as stated and that the analytical methods have been adequate to control the purity.

For early toxicological work, some companies prefer to use impure batches (say 97% pure) with a good spread of impurities present, so that an impurity likely to arise in future manufacture has been present in the early toxicology studies and, that later, purer batches will be used for human studies. An alternative viewpoint is, however, that the presence of a large number of impurities, some of which may be highly toxic, or the presence of some impurities in the 0.5–1.0% region may jeopardize the drug's development if the resultant toxicological studies prove unacceptable. It is then difficult to know whether the drug substance or the impurity has caused the problem. For this reason, high purity batches may be used in early toxicology. However, material used for the clinic must also then match this purity level.

On statistical grounds, the likelihood of a small amount of an impurity causing a large effect in toxicology is small but examples are known.[79] The consequences of setting too tight a specification early on in the development of the drug may also be unpalatable; the manufacturing method on a large-scale may never be able to meet the specification, or may meet it only with costly purification procedures, which may make the drug more expensive than it needs to be. Thus it is advisable during early development that specifications reflect a balance between what the chemist can achieve and what the toxicologist would prefer to have, *i.e.* some scope for further tightening is available. Specifications should initially have: (i) an appearance test, possibly incorporating a colour determination; (ii) an identity test, usually IR spectroscopy; (iii) an assay, either by UV or titrimetric methods or by HPLC, with a limit set at 97.0 or 98.0%; (iv) an impurity profile method, usually HPTLC, HPLC or GLC, with impurity limits set at perhaps a maximum of 0.5%, with a total impurity content of probably 2.0% (not including sulfated ash or solvents); (v) a test for sulfated ash (usual maximum 0.3%); (vi) a test for water and/or residual solvent, which may be a simple loss on drying (LOD) or a more sophisticated test for solvents by GLC; and (vii) a test for the presence of heavy metals (usually with a limit of 20 p.p.m.).

In addition, further tests may probably be carried out, but not mentioned in the specification. These may include: (i) NMR and/or mass spectral verification of impurities or solvent levels; (ii) DSC–TGA confirmation of solvent levels, and to check for presence of polymorphs and solvent adducts; (iii) particle size determination; (iv) optical purity check; and (v) microanalysis as a check on overall purity.

As the drug moves further into the clinical phases of development, and the manufacturing methods become clear, the specification can be tightened to reflect the quality of large batches in the 50–100 kg range, which should typify the quality likely in future manufacture. Often these batches are much purer than those used in earlier toxicology, owing to improvements instituted by the development chemist and the more controlled crystallization methods used on a large-scale. It may well be that a specification of 99% purity can be easily achieved, with no single impurity at greater than 0.3%.

3.6.4.4 Regulatory Issues

Provision of data for submission to regulatory bodies to ensure the timely approval of new drugs is often a task assigned to the development chemist. A rudimentary knowledge of the basic processes of drug registration, and what is required to surmount the 'hurdles' is therefore necessary. The regulatory bodies, however, do not provide specific information on what is required in the way of chemical information, only guidelines which can have different interpretations.

In general, the UK, Australia and Canada currently have the most comprehensive guidelines, with the US likely to require a similar level in the future—at present only draft guidelines are available. Many countries (*e.g.* Austria, Belgium, France, Italy, Portugal and Spain) issue no guidelines and requirements have to be addressed on past experiences. West Germany requires no chemistry data to gain approval for clinical trials.

In the UK the regulatory process is controlled by the Department of Health and Social Security (DHSS), which issues a Product Licence for new drugs.[80] Clinical trials are controlled by a clinical trial certificate (CTC) or an exemption (CTX) application;[81] the latter has a shorter assessment time and was introduced to accelerate the process by which new drugs can be tested. The UK requires in the application a flow sheet indicating the sequence of reactions and usually an outline process description, giving reaction conditions and relative molar quantities, and an indication of the scale of manufacture. Alternative procedures should not be used without a clear indication of the circumstances under which the change will be implemented. The DHSS must be informed of changes in the synthetic route where such a change will affect the range or level of impurities. Minor changes in the synthetic process can be made during the work of clinical trials provided that the quality of the drug substance does not alter *significantly* and remains within the existing specification. If the specification is wide, then it is prudent to inform the DHSS of any major changes, together with any supporting data.

The DHSS usually requires a specification, brief analytical methodology and batch analysis (to show the consistency of the method of synthesis) for a CTX application but more comprehensive data are required for the CTC. Impurity profile data are particularly important in the UK and is one of the most common reasons for rejection of CTX applications.

In the US, the FDA (Food and Drug Administration) requires an IND (Investigational New Drug Application) to be applied for before human trials and an NDA (New Drug Application) prior to marketing. Control of manufacturing methods after approval is controlled by a Drug Master File (unfortunately for chemists, abbreviated to DMF). The recent FDA draft guidelines[82] indicate that for phase I and II trials in the US, the synthesis, isolation, extraction and purification methods for the drug substance are required—usually a flow sheet and process description for each synthetic step is sufficient. It is possible to change the synthetic route without prior approval by the FDA but the IND should be updated when this occurs. As the synthetic process becomes more defined, and the drug proceeds into phase III clinical trials then the following will be required: (i) a flow sheet containing structures of reactants, molecular weights, stereochemistry, intermediates (isolated or *in situ*), solvents, catalysts, conditions (pH, temperature, time), reagents and important by-products; (ii) a description of each synthetic step including equipment used, conditions, work-up and isolation methods, *etc.* together with any other critical data required to carry out the synthesis; yields must be quoted; (iii) purification methods for the drug substance, especially detailed recrystallization data and rework methods; and (iv) detailed quality control methods, not only for the final product, but also for intermediates and raw materials.

3.6.4.5 Crystallization and Polymorphism

The performance of many drug substances depends on the ability to manufacture reproducibly a particular crystal form, *i.e.* a certain crystal habit, a solvate or a polymorph. One crystalline modification may show 10 times the solubility of another form of the same compound and this will affect the bioavailability. Since it is now recognized that most organic compounds, when studied carefully, exist in more than one crystal form, the subject becomes of great importance for the development chemist interested in scale-up. Not only the final drug substance but also the intermediates may be polymorphic and this may affect reactivity, solubility and particularly solids-handling characteristics. The latter will be important in product isolation, when the ability to remove impurities in the mother liquors will depend to some extent on how well the solid retains the liquors; this is dependent on crystal form. Development chemists should therefore be aware of likely problems during scale-up. It has been stated[83] that 'scale-up of a crystallization process is probably more difficult than any other unit operation in chemical engineering', owing to the fine control required; computer control must surely be of value in the future in this respect.

3.6.4.5.1 Crystal habit[84]

The crystal habit determines the external shape of the crystal, and is affected by variations in crystallization conditions such as temperature, level of supersaturation, rate of cooling, rate of agitation, solvent polarity (especially water content), the nature of impurities and their concentration, and viscosity; the latter changes as crystallization proceeds. These factors affect both the initiation of crystallization and the rate of crystal growth but not necessarily in the same ways. Careful control of process parameters is crucial to guarantee production of a reproducible product, which always performs the same during formulation.

3.6.4.5.2 Solvation[85]

Solvates are molecular complexes which have incorporated the crystallization solvent into the lattice. Solvates are different from polymorphs but the two areas are related; removal of solvent from a solvate on drying often results in a new polymorphic form being produced. Solvates and hydrates are formed by many drug substances; for example, oestradiol forms solvates with all 30 solvents studied. Adducts have different solubilities and dissolution rates from the anhydrous compounds and this affects the bioavailability. Solvates are often more soluble, but hydrates usually are less soluble than the anhydrous form and are likely to be absorbed more slowly. Solvates can be characterized by DSC–TGA techniques, and by IR, GLC and NMR methods.

3.6.4.5.3 Polymorphism[86,87]

Polymorphism is defined as the ability of the compound to exist in more than one crystal form. This means that the arrangement of the atoms within the lattice is different in different polymorphs. Polymorphism arises out of a molecule's ability to change its conformation, affecting the balance of intramolecular and intermolecular hydrogen bonds. Thus lattice energies of polymorphs are different, and differences in heats of fusion, heats of solution, dissolution rate, solubility, bulk density and melting point are to be expected. Solid state properties, particularly IR spectra and X-ray diffraction patterns may be different. Polymorphism is exhibited by a wide range of substances including barbiturates (phenobarbital has 15 polymorphs and two hydrates),[88] antibiotics (ampicillin,[89] erythromycin[88]), corticosteroids (methylprednisolone),[90] H_2-antagonists (cimetidine,[91] tiotidine[6]). It may affect intermediates as well as final products and it is an important topic which development chemists need to be aware of, particularly in the following areas.

(i) *Solids handling.* Polymorphs often have different bulk densities and abilities to retain solvent and require different isolation strategies. Chunky crystals filter well and shed mother liquors easily and are easy to dry, whereas needles form felted masses which retain mother liquors (and therefore impurities) tenaciously and are difficult to wash and dry.

(ii) *Drying.* Products which are difficult to dry often lead to surface deterioration, producing off-coloured product, nonuniform products (lumps) which need sieving or milling, and are wasteful of energy.

(iii) *Reactivity and stability.* The different dissolution rate of polymorphs may affect reactivity,[92] whereas the ability to absorb solvent may produce thick reaction mixtures which are difficult to agitate. Some polymorphs are more light sensitive and stable to heat than others.

It cannot be stressed too highly that polymorphism needs to be investigated early in the lifetime of a drug substance. There are examples in the pharmaceutical and explosives industries in which a product has been manufactured in one crystal form for many years.[89] Later, a second form has appeared which is thermodynamically more stable, and, in some instances, even after that the first form cannot be obtained, even in laboratories many miles away! Ampicillin exists as a trihydrate and two polymorphic anhydrous forms, but an earlier monohydrate has not been seen since the trihydrate was formed. Xylitol, first prepared as an oil in 1891, was crystallized in 1942 with m.p. 61 °C and the experiments were repeated many times by others. Later a new form (m.p. 94 °C) arose, and the earlier polymorph has not been obtained since![89] The consequences of this phenomenon occurring during drug development are horrific and it is the duty of the chemist to always try to prepare stable polymorphs; often improvements in processes result.

Polymorphs can be monotropic or enantiotropic, depending on the phase diagram. For monotropic systems, only one crystal form is stable below the melting point. If a metastable form is obtained during manufacture, however, reversion to the stable form may occur during tablet manufacture or in a suspension on storage, especially in the presence of solvents.

For enantiotropic polymorphs, one form is stable above the transition temperature, the other below, and since transition temperatures are often in the 20–100 °C range, conversion from one form to another can occur during routine chemical manufacture. An understanding of the relationship between polymorphism, transition temperatures and stability can thus aid in choosing conditions for reproducible manufacture.

3.6.4.5.4 Study of a compound to determine whether polymorphism exists

Changing the solvent of recrystallization, particularly from a nonpolar to a polar solvent, will often lead to a different polymorphic form. Changing the rate of cooling may also be effective, but 'crash' cooling produces metastable forms, which may be difficult to reproduce on a large-scale. Use of wet solvents may give hydrates which, on careful drying, will dehydrate to new polymorphs. Drying should always be carried out at as low a temperature as possible to avoid further solid–solid transitions taking place.

For example, crystallization of the drug tienilic acid[93] from acetone or ethanol gave the stable polymorph A, which could also be obtained by slurrying in water. Recrystallization from toluene or xylene, however, gave metastable polymorph B. Rapid cooling of hot solutions in chloroform or dichloroethane gave form B, but slow recrystallization gave A. Interconversion from B to A could not be effected by heating dry samples at 115 °C for 48 h, but heating in water at 100 °C for 1 h was sufficient to complete the transformation.

One of the simplest ways to induce solid–solid transformations is to suspend the compound in a solvent in which the compound is sparingly soluble and maintain the temperature as near as possible to the melting point of the compound for several hours. The solid must then be quickly isolated and characterized. Use of a solvent which boils at the required temperature simplifies the procedure. The inability to produce polymorphs by this method after different solvents have been tried would give some confidence that the stable form had been isolated initially.

3.6.4.5.5 Characterization of polymorphs

Polymorphs often have slightly different melting points, and DSC studies can be used to characterize the polymorphs in this way; mixtures of polymorphs will give the melting endotherm of each, unless interconversion takes place during the determination.[94] Samples of pure polymorphs may show differences in IR, solid state NMR and X-ray diffractograms. In some cases ^{14}N-quadrupole spectroscopy has been useful to prove that the polymorphic form was unchanged after tabletting (pressure can sometimes cause polymorphic transitions[6]).[95]

A study of all isolable polymorphs, their transition temperatures and methods of interconversion will allow the chemist to choose methods of manufacture which will reproducibly give the polymorph the pharmacist desires.

3.6.5 REFERENCES

1. H. L. White, 'Introduction to Industrial Chemistry', Wiley, Chichester, 1986.
2. C. A. Heaton, 'The Chemical Industry', Blackie, Glasgow, 1985.
3. L. H. Werner and E. Donaghue, in 'Riegel's Handbook of Industrial Chemistry', ed. J. A. Kent, 8th edn., Van Nostrand, New York, 1983, p. 718.
4. G. T. Austin in 'Shreve's Chemical Process Industries', 5th edn., McGraw-Hill, London, 1984, p. 795.
5. D. G. Jordan, Chemical Process Development', Interscience, New York, 1968; C. A. Clausen, III and G. C. Mattson, 'Principles of Industrial Chemistry', Wiley-Interscience, New York, 1978.
6. G. E. Robinson, *Chem. Ind. (London)*, 1983, 349.
7. C. B. Rosas, *Chem. Ind. (London)*, 1987, 238.
8. S. G. Warren, 'Designing Organic Syntheses', Wiley, Chichester, 1978.
9. W. T. Wipke and W. J. Howe, *ACS Symp. Ser.* 1977; **61**; A. Long, S. D. Rubenstein and L. J. Joncas, *Chem. Eng. News.*, 1983, May 9th, 22.
10. C. R. Ganellin, in 'Medicinal Chemistry, The Role of Organic Chemistry in Drug Research', ed. S. M. Roberts and B. J. Price, Academic Press, London, 1986, p. 93.
11. C. G. M. van der Moesdijk, *Chem. Ind. (London)*, 1986, 129.
12. Chemquest, available from Pergabase, 12 Vandy Street, London, EC2A 2DE.
13. 'Directory of Chemical Producers—Western Europe', SRI International, Menlo Park, 1978, vols. 1 and 2.
14. A. F. Plant, 'Chemcyclopedia', American Chemical Society, Washington, DC, 1988.
15. 'Chemical Sources USA', Directories Publishing Company Inc., Orlando, FL, 1987 (now also available as an updated on-line searching facility).
16. 'Directory of World Chemical Producers', Chemical Information Services Ltd., Oceanside, NY, 1984.
17. 'Chemical Industry Directory', Benn, Tonbridge, UK, 1986.
18. 'Custom Chemical Synthesis Services in the UK', IAL Consultants Ltd., London, 1986; similar compendia are available for France, Germany and Western Europe from the same source.
19. D. T. Witiak and M. N. Inbasekaran, in 'Kirk-Othmer Encyclopedia of Chemical Technology', 3rd edn., ed. M. Grayson and D. Eckroth, Wiley-Interscience, New York, 1982, vol. 17, p. 311.
20. J. D. Morrison, 'Asymmetric Synthesis', vols. 1–5, Academic Press, Orlando, FL, 1985.
21. J. D. Morrison and J. W. Scott (eds.), in 'Asymmetric Synthesis', Academic Press, Orlando, FL, 1984, vol. 4, p. 1.
22. T. Laird, *Chem. Ind. (London)*, 1986, 134.
23. A. M. Brandeis, *Chem. Ind. (London)*, 1986, 90.
24. E. M. Wilson, *Chem. Ind. (London)*, 1984, 217.
25. L. M. Weinstock, *Chem. Ind. (London)*, 1986, 86.
26. S. Lee, *Chem. Ind. (London)*, 1987, 223.
27. C. Reichardt, 'Solvents and Solvent Effects in Organic Chemistry', 2nd edn., VCH, Weinheim, 1988.
28. D. R. Marshall, *Chem. Ind. (London)*, 1983, 331.
29. N. S. Isaacs and N. V. George, *Chem. Br.*, 1987, 47.
30. O. L. Davies, 'The Design and Analysis of Industrial Experiments', Longman, London, 1978.
31. G. E. P. Box, W. G. Hunter and J. S. Hunter, 'Statistics for Experimenters—An Introduction to Design, Data Analysis and Model Building', Wiley, New York, 1978.
32. R. D. Gilliom, W. P. Purcell and T. R. Bosin, *Eur. J. Med. Chem.—Chim. Ther.*, 1977, **12**, 187.
33. S. N. Deming, J. G. Bower and K. D. Bower, *Adv. Chromatogr. (NY)*, 1984, **24**, 35; S. N. Deming and S. L. Morgan, *Anal. Chem.*, 1973, **45**, 278A; D. E. Long, *Anal. Chim. Acta*, 1967, **46**, 193.
34. W. E. Biles and J. J. Swain, 'Optimisation and Industrial Experimentation', Wiley-Interscience, New York, 1980; J. F. Scuotto, D. Mathieu, R. Gallo, R. Phan-Tan-Luu, J. Metzger and M. Desbois, *Bull. Soc. Chim. Belg.*, 1985, **94**, 897; L. Beu, I. Farcasanu, J. Russu, D. Breazu and G. Bora, *Rev. Chim. (Bucharest)*, 1985, **36**, 496 (*Chem. Abstr.*, 1985, **104**, 129 761).
35. R. Carlson, T. Lundstedt, R. Phan-Tan-Luu and D. Mathieu, *Nouv. J. Chim.*, 1983, **7**, 315.
36. W. Spendley, G. R. Hext and F. R. Himsworth, *Technometrics*, 1962, **4**, 441; W. K. Dean, K. J. Heald and S. N. Deming, *Science (Washington, D.C.)*, 1975, **189**, 805.
37. L. Rigal and A. Gaset, *Biomass*, 1985, **8**, 267 (*Chem. Abstr.*, 1985, **104**, 110 049); T. Lundstedt, P. Thoren and R. Carlson, *Acta Chem. Scand., Ser. B*, 1984, **38**, 717; R. Carlson, A. Nilsson and M. Stromqvist, *Acta Chem. Scand., Ser. B*, 1983, **37**, 7; P. Victory, R. Nomen, M. Garriga, X. Thomas and L. G. Sabate, *Afinidad*, 1984, **41**, 241 (*Chem. Abstr.*, 1984, **101**, 151 724); R. Lazaro, P. Bouchet and R. Jacquier, *Bull. Soc. Chim. Fr.*, 1977, 1171.
38. F. L. Chubb, J. T. Edward and S. C. Wong, *J. Org. Chem.*, 1980, **45**, 2315.

39. G. E. P. Box and N. R. Draper, 'Evolutionary Operation', Wiley, New York, 1969; G. E. P. Box, *Appl. Stat.*, 1957, **6**, 3; *J. R. Stat. Soc., Ser. B*, 1951, **13**, 1; G. J. Hahn, *Chem. Technol.*, 1975, **5**, 496; G. J. Hahn, *Chem. Technol.*, 1975, **5**, 561; *Chem. Tech. (Leipzig)*, 1976, 142; B. H. Carpenter and H. C. Sweeney, *Chem. Eng. (Rugby, Engl.)*, 1965, **72** (14), 117, 126.
40. H. Winicov, J. Schainbaum, J. Buckley, G. Longino, J. Hill and C. E. Berkoff, *Anal. Chim. Acta*, 1978, **103**, 469; D. F. Chodosh, F. E. Wdzieckowski, J. Schainbaum and C. E. Berkoff, *J. Autom. Chem.*, 1983, **5**, 99; D. F. Chodosh, S. H. Levinson, J. L. Weber, K. Kamholz and C. E. Berkoff, *J. Autom. Chem.*, 1983, **5**, 103.
41. M. Legrand and P. Bolla, *J. Autom. Chem.*, 1985, **7**, 31; M. Legrand and A. Foucard, *J. Chem. Educ.*, 1978, **55**, 767. This system (Logilap) is marketed by Orso Electronique, 15 Passage de la Main D'Or, 75011 Paris, France.
42. F. Brogli, G. Giger, H. Randegger and W. Reganass, *Int. Chem. Eng. Symp. Ser.*, 1981, **68**, 3/M: 2.
43. M. G. Kemp, *Chem. Ind. (London)*, 1983, 335.
44. Available in the UK from Contraves Industrial Products Ltd., Times House, Station Approach, Ruislip, Middlesex, HA4 8LH, UK.
45. For example, the Mettler RC1, Reaction Calorimeter System available in the UK from MSE Scientific Instruments, Manor Royal, Crawley, Sussex, RH10 2QQ, UK.
46. P. Dawson, *Chem. Ind. (London)*, 1986, 99.
47. J. C. Mecklenburgh, 'Process Plant Layout', Wiley, New York, 1985.
48. 'Guide to Good Pharmaceutical Manufacturing Practice (The Orange Guide)', HMSO, London, 1983.
49. Fluoroshield is the registered trade mark of N. L. Gore and Associates Inc. Corbond is Fluoroshield-lined process equipment available from Corning Process Systems.
50. R. Anderson, *Chem. Ind. (London)*, 1984, 205.
51. M. Lambert, *Process Equip. News*, April 1987, 26.
52. P. Fletcher, *Chem. Eng. (Rugby, Engl.)*, April 1987, 33.
53. D. B. Purchas, 'Solid–Liquid Separation Equipment Scale up', Uplands Press, London, 1986.
54. G. Nonhebel and A. A. H. Moss, 'Drying of Solids in the Chemical Industry', Butterworths, Sevenoaks, 1971.
55. K. J. Erdweg, *Chem. Ind. (London)*, 1983, 342.
56. J. L. Dwyer, *Kem.-Kemi*, 1985, **12** (3), 247.
57. I. G. Gostelow, *Chem. Ind. (London)*, 1983, 339.
58. J. Love, *Chem. Eng. (Rugby, Engl.)*, June 1987, 34.
59. D. Service, *Chem. Br.*, 1987, **23**, 27.
60. For important reviews see papers from the conference on 'Control and Prevention of Runaway Chemical Reaction Hazards' held in Amsterdam, November 1986; available from IBC Technical Services Ltd., Byfleet, Surrey.
61. 'Guidance Notes on Chemical Reaction Hazard Analysis', ABPI, London, 1981.
62. C. F. Coates and W. Riddell, *Chem. Ind. (London)*, 1981, 84.
63. L. Bretherick, 'Handbook of Reactive Chemical Hazards', 3rd edn., Butterworths, Sevenoaks, 1985.
64. Available from Columbia Scientific Industries, 101 Garamonde Drive, Wymbush, Milton Keynes, MK8 8DD, UK.
65. Available from Systag, CH-8803, Ruschlikon, Switzerland.
66. R. L. Webb and J. T. Buckley, Smith Kline & French Ltd., unpublished results.
67. 'Handling Chemicals Safely', 2nd edn., Dutch Chemical Industry, The Hague, The Netherlands, 1980.
68. N. I. Sax, 'Dangerous Properties of Industrial Materials', 7th edn., Van Nostrand-Reinhold, New York, 1988.
69. L. Bretherick, 'Hazards in the Chemical Laboratory', 4th edn., Royal Society of Chemistry, London, 1986.
70. N. V. Steere, 'Handbook of Laboratory Safety', 2nd edn., CRC Press, Boca Raton, FL, 1970.
71. G. D. Clayton and F. E. Clayton, (ed.) 'Patty's Industrial Hygiene and Toxicology', 3rd edn., Wiley-Interscience, New York, 1978.
72. Royal Society of Chemistry Laboratory Hazard Data Sheets, published monthly in 'Laboratory Hazards Bulletin' by Royal Society of Chemistry, London.
73. 'NIOSH: Registry of Toxic Effects of Chemical Substances', 1985; US Dept. of Commerce: available from Micrinfo Ltd., Alton, Hants, UK, 1986.
74. 'Kirk-Othmer Encyclopedia of Chemical Technology', 3rd edn., Wiley-Interscience, New York, 1984, vols. 1–24.
75. L. H. Keith and D. B. Walters, 'Compendium of Safety Data Sheets for Research and Industrial Chemicals', VCH, Weinheim, 1985.
76. A. R. Wright, *Chem. Eng. Res. Des.*, 1984, **62**, 391; A. Husain, 'Chemical Process Simulation', Wiley, New York, 1986.
77. B. W. Woodget and D. Cooper, 'Samples and Standards', Wiley, Chichester, 1987.
78. K. Florey, 'Analytical Profiles of Drug Substances', Academic Press, Orlando, FL, 1984, vols. 1–14.
79. *Wall Street Journal*, Feb. 19th 1987, 6.
80. 'Medicines Act 1968; Guidance Notes on Applications for Product Licences', HMSO, London, June 1984.
81. 'Medicines Act 1968; Guidance Notes on Applications for Clinical Trial Certificates and Clinical Trial Exemptions', HMSO, London, June 1984.
82. 'Draft Guidelines for Submitting Supporting Documentation in Drug Applications', FDA, Rockville, MD, April 1985.
83. J. W. Mullin, 'Crystallisation', 2nd edn., Butterworths, London, 1978.
84. J. K. Haleblian, *J. Pharm. Sci.*, 1975, **64**, 1269.
85. M. Kuhnert-Brandstatter and P. Gasser, *Microchem. J.*, 1971, **16**, 590.
86. R. P. Bouche and M. Draguet-Brughmans, *J. Pharm. Belg.*, 1977, **32**, 23.
87. N. K. Jain and M. N. Mohammedi, *Indian Drugs*, 1986, **23**, 315.
88. N. Stanley-Wood and G. Riley, *Pharm. Acta Helv.*, 1972, **47**, 58.
89. G. D. Woodard and W. C. McCrone, *J. Appl. Crystallogr.*, 1975, **8**, 342.
90. J. K. Haleblian and W. C. McCrone, *J. Pharm. Sci.*, 1969, **58**, 911.
91. B. Prodic-Kojic, F. Kajfez, B. Belin, R. Toso and V. Sunjic, *Gazz. Chim. Ital.*, 1979, **109**, 535.
92. A. Gavezzotti and M. Simonetta, *Chem. Rev.*, 1982, **82**, 1.
93. Unpublished results obtained at Smith Kline & French Ltd.
94. M. Kuhnert-Brandstatter and R. Vollenkle, *Fresenius' Z. Anal. Chem.*, 1985, **322**, 164.
95. P. M. G. Bavin, D. Stephenson and J. A. S. Smith, *Z. Naturforsch., Teil A*, 1986, **41**, 195.

3.7

Genetic Engineering: The Gene

KARL DRLICA

Public Health Research Institute, New York, NY, USA

3.7.1 INTRODUCTION

The development of recombinant DNA technology has led to an explosion of knowledge in molecular genetics. Genes are being purified, their information content is being determined, and rare regulatory proteins are being produced in large quantities. These procedures allow detailed examination of interactions between proteins and DNA. It is now possible to address questions about gene function by chemically altering purified genes, placing them into living cells, and evaluating the effects of the altered genes. Recombinant DNA technology is also beginning to deal directly with human disorders. The medical applications currently fall into two general areas, diagnosis of genetic diseases and large scale production of rare human proteins by bacteria. Some of these applications have already become controversial. For example, the techniques used in the screening of fetuses for genetic disorders also reveal the sex of the fetus; thus, parents could use selective abortion to choose the sex of their children. As other human characteristics become equated with specific nucleotide sequences, these too may be evaluated by genetic screening procedures—it has even been proposed that eligibility for certain types of employment or insurance coverage require one to pass genetic screening tests. Thus genetic engineering is likely to have political as well as scientific ramifications.

The goal of this chapter is to introduce the terminology and concepts underlying genetic engineering and recombinant DNA technology by describing how genetic information is stored, organized and utilized. It is hoped that with this background the reader will be able to quickly assimilate technical details encountered in chapters describing applications of the technology. The general strategy is first to provide an overview of how genetic information is stored in DNA and how the information is made available for directing the chemistry of cells. This is done through a discussion of the central dogma of molecular genetics, a formulation of the fundamental concepts of information flow in genetic systems. Then four aspects of DNA biology are described in more detail to introduce the tools used in genetic engineering, to develop a framework for interpreting future findings in genetics, and to provide an appreciation for the depth of knowledge existing about informational macromolecules. The detailed discussion begins with DNA and chromosome structure, for DNA, with its associated proteins, is the substrate upon which many regulatory proteins act. We then focus on initiation of DNA replication. Knowledge of this process is accumulating rapidly, and understanding initiation of replication may make it possible to selectively control the growth of cells and thus control many types of disease. Mobile genetic elements are the third topic; this subject is already important for some types of leukemia and AIDS, and it will become increasingly important to the medical community when considering genetic engineering in higher organisms. Next the principles of gene control are outlined, first employing bacterial examples and then using examples taken from higher organisms. The diversity of molecular mechanisms involved in switching genes on and off will become apparent. The chapter concludes with a brief laboratory-level description of one strategy for cloning genes.

Much of our understanding of molecular genetics has been obtained from studies with bacteria, and it is likely that many of the principles developed from bacterial studies will prove to have validity with all organisms. However, there are processes taking place in higher organisms that have not been found to occur in bacteria. Consequently, our discussion of higher organisms focuses on phenomena that are not common to bcterial systems. By higher organisms we usually mean yeast and fruitflies. The biochemistry of humans is likely to be very similar to that of other higher organisms, but human examples are rarely the best to illustrate a point because they are usually not as well characterized experimentally.

It is important to caution against uncritical application of knowledge from bacterial models to animal or human systems. Bacteria lack a true nucleus, and this leads us to expect differences in DNA packaging between bacteria and higher organisms (organisms without a true nucleus are called prokaryotes while those with a true nucleus are called eukaryotes). We might also expect bacterial genetic systems to be less complicated than those of higher organisms, for bacteria do not develop highly specialized tissues. In general, these expectations are borne out, and some of the more interesting questions in molecular biology focus on how and why prokaryotes differ from eukaryotes.

Readers interested in a more elementary description of genes are referred to Drlica;[9] those interested in broad advanced treatments are referred to Watson *et al.*[37]

3.7.2 THE CENTRAL DOGMA

The central dogma of molecular genetics states that genetic information flows unidirectionally from the long, linear polymer called DNA through a similar polymer called RNA and finally to a chemically unrelated polymer called protein. A schematic representation is shown in Figure 1. The dogma can also be stated in the following way: information is stored in DNA, selectively copied by production of RNA using a mechanism called transcription, and transferred from its RNA form into proteins by a process called translation. Proteins then direct the set of biochemical reactions that we call life, including the reproduction (replication) of DNA itself. Although a few exceptions to the central dogma have been uncovered, the concept has survived for 30 years as the major paradigm guiding modern biology. Almost all of the concepts and terminology needed to understand developments derived from genetic engineering can be introduced through a description of the three central processes, transcription, translation and replication. The rudiments of each process are outlined below.

DNA (deoxyribonucleic acid) is a long polymer comprised of millions, and sometimes billions, of subunits (nucleotides) arranged in a linear array (many naturally occurring DNA molecules are actually circles, but this point is irrelevant for the present discussion). There are four types of nucleotide, and genetic information is stored in DNA through the precise order of the four types of subunit. Genes are small regions of the DNA composed of thousands of nucleotides; thus, a single DNA molecule comprised of millions of nucleotides can be organized into thousands of genes, much as the information in a motion picture film is arranged as a series of many scenes. In this analogy each nucleotide would correspond to a frame in the film. All of the information required for life is contained in the sequence of nucleotides in DNA, and it is the orderly expression of the information that allows complicated processes such as development to occur. The information for controlling when a specific gene is expressed into a protein form is also contained in specific nucleotide sequences. Two aspects of gene expression, transcription and translation, are discussed below.

3.7.2.1 Transcription: Selective Copying of Genetic Information

Transcription is the process whereby the information from a gene, *i.e.* the specific order of nucleotides within a precise region of DNA, is transferred to an RNA (ribonucleic acid) molecule. This occurs by the synthesis of an RNA molecule using the information contained in one strand of DNA as a template. Thus an RNA molecule from a particular gene will have its subunits ordered in a way that is dictated by the order of nucleotides in the DNA of the gene.

Three points are central to understanding transcription. First, DNA is a two-stranded polymer in which each subunit (nucleotide) of one strand interacts noncovalently with a corresponding nucleotide in the other strand. The chemical structure of DNA is shown in Figure 2. Second, the interaction between nucleotides in opposite strands obeys the complementary base-pairing rule. There are four nucleotides in DNA, adenine (A), thymine (T), guanosine (G) and cytosine (C). According to the complementary base-pairing rule an **A** in one strand of DNA always pairs with a **T** in the opposite strand; likewise, **C** always pairs with **G** (for more details see Section 3.7.3.2). Strict conformity to this rule allows a strand of DNA to act as a template for the formation of a new strand of DNA or RNA (see Figure 3). The third point is that an enzyme called RNA polymerase can travel along one strand of DNA and polymerize nucleotides free in solution to generate an RNA molecule. During RNA synthesis the complementary base-pairing rule is obeyed; thus, the nucleotide sequence in the RNA is complementary to that in the DNA strand used as a template. Of course the other

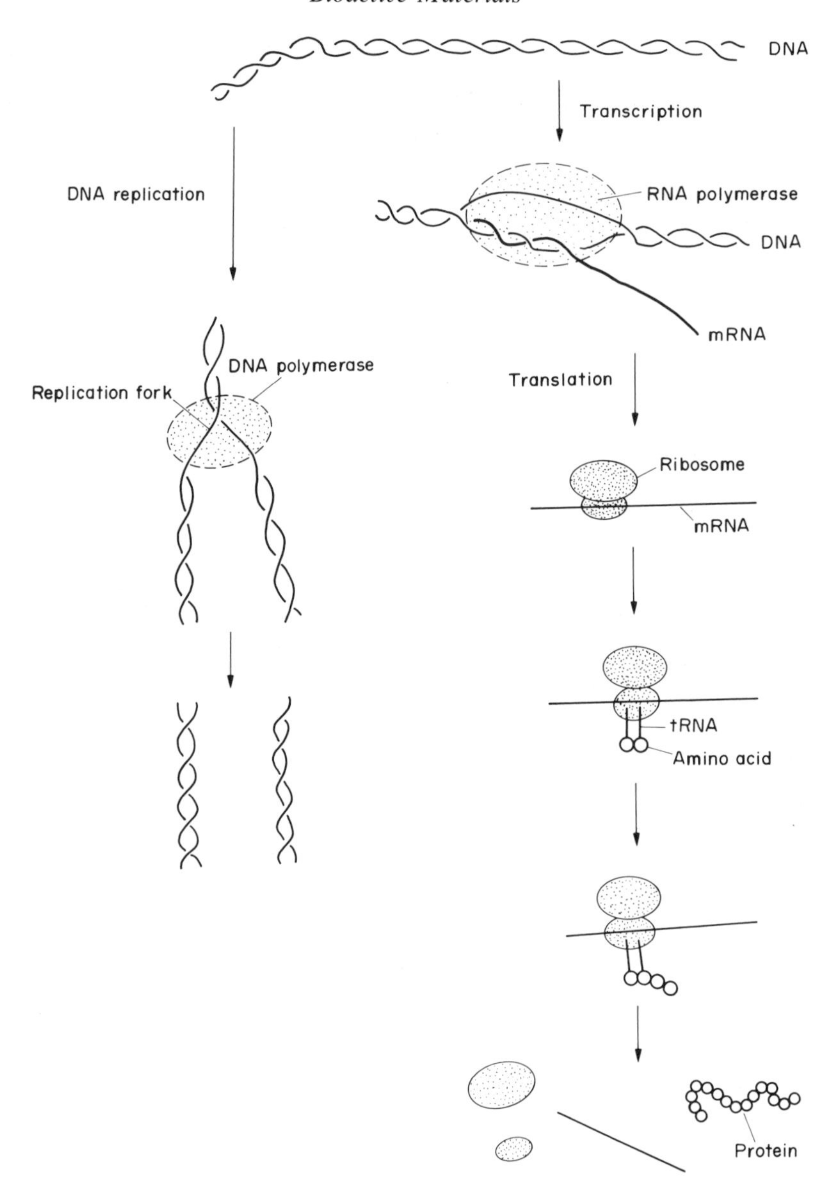

Figure 1 *The central dogma.* The information in DNA is transcribed into a messenger RNA (mRNA) molecule by RNA polymerase. The mRNA then attaches to a ribosome, and transfer RNA (tRNA) molecules, each bearing an amino acid, bind to the ribosome–mRNA complex. The amino acids join to form a protein. This process of protein synthesis is called translation. The two DNA strands separate and are copied by an enzymatic process called DNA replication.

DNA strand is also complementary to the template DNA strand, so the information in the RNA will be identical to that in one strand of DNA and complementary to that in the other.

RNA differs from DNA in several ways. First, the sugar moiety of each nucleotide contains a hydroxyl group where only a hydrogen is found in DNA. Second, the base called uracil is found in RNA rather than the one called thymine. Thus, opposite to an **A** in the template strand of DNA, a **U** will be inserted into the growing RNA strand. A third difference is size. RNA molecules, since they contain information from at most a few genes, are much shorter polymers than the corresponding DNA molecules from which they are made.

Much of the current effort by molecular geneticists focuses on understanding how transcription is controlled, for understanding this process is central to understanding how and when cells regulate the expression of their genetic information. Our current understanding of several examples of gene control is described in Sections 3.7.6 and 3.7.7.

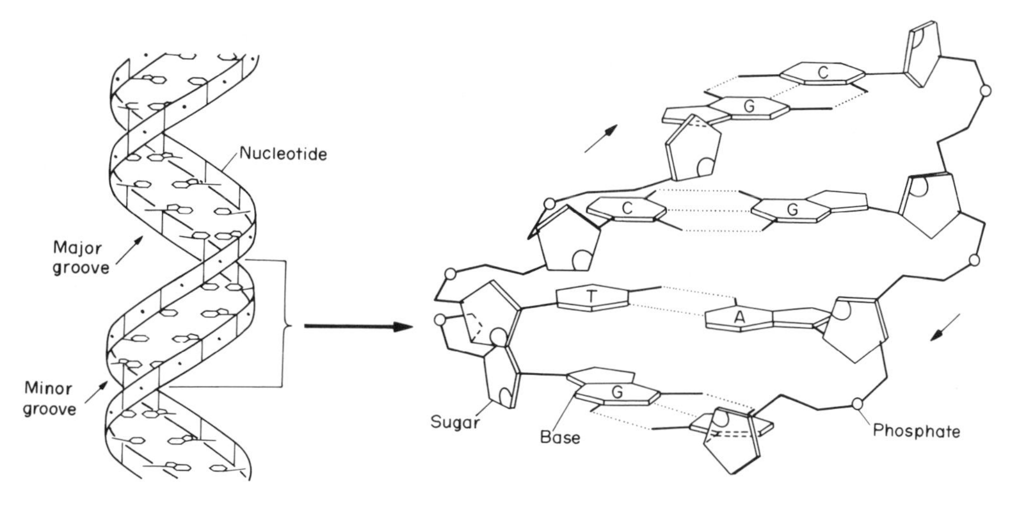

Figure 2 *Schematic diagram of a short section of a DNA double helix.* Each DNA strand is composed of three types of chemical structure: bases, sugars and phosphates. A nucleotide is a unit composed of one base, one sugar and one phosphate. The sugars and phosphates connect to form the backbone of each strand with a base attached to each sugar. The four different bases are represented by the letters A, T, G and C. The bases of one strand point inward and toward those of the other. Attractive forces called hydrogen bonds (represented by dotted lines) exist between the bases of opposite strands and contribute to holding the two strands together. The two strands run in opposite directions (see Figure 9) (reproduced from ref. 9 by permission of John Wiley & Sons, Inc.)

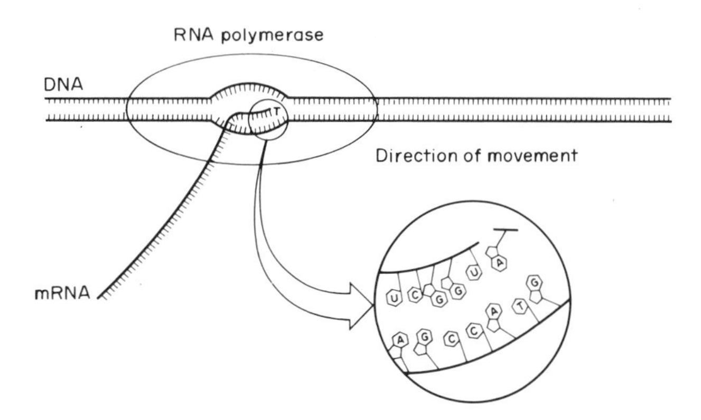

Figure 3 *Transcription.* The enzyme complex called RNA polymerase causes the DNA strands to separate within a short region (10 to 20 base pairs). The polymerase moves along the DNA, and as it does, it forms an RNA chain using free nucleotides. The order of the nucleotides in RNA is determined by the order of nucleotides in one of the DNA strands by the complementary base-pairing rule. In the example shown, the nucleotide sequence of the RNA is complementary to that of the lower DNA strand. For simplicity, the DNA strands have not been drawn as an interwound helix (reproduced from ref. 9 by permission of John Wiley & Sons, Inc.)

3.7.2.2 Translation: Making New Proteins

The synthesis of proteins is called translation because genetic information encoded by only four different nucleotides, the subunits of DNA and RNA, is converted into the 'protein' language in which 20 different types of subunits (amino acids) are used to specify protein structure and function. The thousand or so nucleotides in an RNA molecule are read three at a time during translation to produce a protein containing several hundred amino acids. Each nucleotide triplet that specifies a particular amino acid is called a codon. Four different nucleotides, taken as triplets, can be arranged in 64 combinations, *i.e.* there are 64 possible codons. The correspondence of specific triplets, codons, with each of the 20 different amino acids has been called the genetic code. Some amino acids are specified by more than one nucleotide triplet, but not *vice versa.* The RNA conveying sequence information for making proteins is called messenger RNA to distinguish it from two other types called ribosomal RNA and transfer RNA. The role of each of these types of RNA is discussed below.

After the messenger RNA has been made (or while it is being made in the case of bacteria), a site near one end binds to a large ball-like structure called a ribosome. It is on ribosomes that the process of assembling the amino acid subunits into proteins occurs. Ribosomes are composed of several specialized RNA molecules (ribosomal RNA) and about 50 different types of protein. Part of the function of these protein–RNA complexes is to properly orient the messenger RNA for translation. A particular set of three nucleotides, called the start codon, is exposed for reading; the information in these three nucleotides specifies which amino acid will be the first in the new protein (see Figure 4).

The reading process occurs through the interaction of very small RNA molecules (transfer RNA) with nucleotides in the messenger in such a way that a series of amino acids, attached singly to the transfer RNA molecules, are brought close enough together to be covalently joined into a protein chain. There are many types of transfer RNA, at least one type for each of the 20 types of amino acid (amino acids that are specified by more than a single codon also have more than a single cognate transfer RNA). Each transfer RNA is covalently bonded to its cognate amino acid through an enzymatic reaction in which the transfer RNA and the amino acid are specifically recognized by an enzyme called an amino acyl transfer RNA synthetase. As expected, there is at least one type of synthetase for each type of transfer RNA and amino acid.

Translation is a reiterated series of binding reactions between transfer RNAs and messenger RNA (see Figure 5). It begins when a particular type of transfer RNA, attached to a particular amino acid, recognizes the start codon on the messenger RNA and binds to it (Figure 4). The recognition is governed by the complementary base-pairing rule. A specific set of three nucleotides (the anticodon) on the transfer RNA base-pair with the start codon of the messenger RNA. A similar reaction occurs between the second codon on the messenger RNA and the appropriate transfer RNA. This places the first two amino acids close together, and they are enzymatically joined by a hydrolytic reaction that forms a peptide bond. Peptide bond formation leads to the release of the transfer RNA bound to the first amino acid; thus the growing protein chain is linked only to the second transfer RNA. The ribosome then shifts one position relative to the messenger RNA, exposing the third codon of the gene. A third transfer RNA with its cognate amino acid binds to the messenger RNA at the third anticodon. The third amino acid is then covalently joined to the second, the second transfer RNA is released from the complex, and the growing protein chain is now three amino acids long and is attached to the third transfer RNA. The process is repeated until a special stop codon is reached. Stop codons do not specify amino acids but instead signal the ribosome to halt translation and release the new protein from the ribosome. Proper folding of the protein is spontaneous and is dictated by the order of the amino acids.

3.7.2.3 DNA Replication: Making New DNA

Each time a cell divides to form two daughter cells, the genetic information in the DNA must be duplicated so each daughter cell can receive a copy. Duplication of DNA is called DNA replication. It is an enzymatic process in which the two strands of DNA are unwound, pulled apart, and two new strands are synthesized using each old strand as a template. By complying with the complementary base-pairing rule, DNA replication produces two daughter DNA molecules which have exactly the same information as the parental molecule (see Figure 6).

Replication of DNA, which has been most thoroughly studied in bacterial systems, requires the activity of several enzymes.[19] DNA helicases appear to supply the activity necessary for unwinding the DNA strands, and topoisomerases probably provide the necessary swivels. The points where the strands separate are called replication forks. The enzyme that synthesizes new DNA strands using the old strands as templates is called DNA polymerase (see Figure 7). DNA polymerase moves along a template strand in only one direction, from the 3′ end of the template toward its 5′ end. Since the two strands of DNA have opposite polarity (see Figure 2), both strands cannot be replicated by the same mechanism at a single replication fork. This enzymatic idiosyncracy requires a complicated solution. One DNA polymerase moves forward in an uninterrupted manner on one DNA strand, incorporating nucleotides into a new strand as it moves. This process is called leading strand synthesis. Another DNA polymerase on the other template strand moves in the same direction as the first in a chemical sense, but since the two DNA strands have opposite polarity, this polymerase must physically move in a direction opposite to that of the other polymerase. In a sense, the second polymerase moves backward, away from the replication fork. A short tract of DNA is polymerized, and then the polymerase skips 'forward' toward the replication fork. It then synthesizes another short stretch of DNA as before, in the direction opposite to fork movement. This process is called

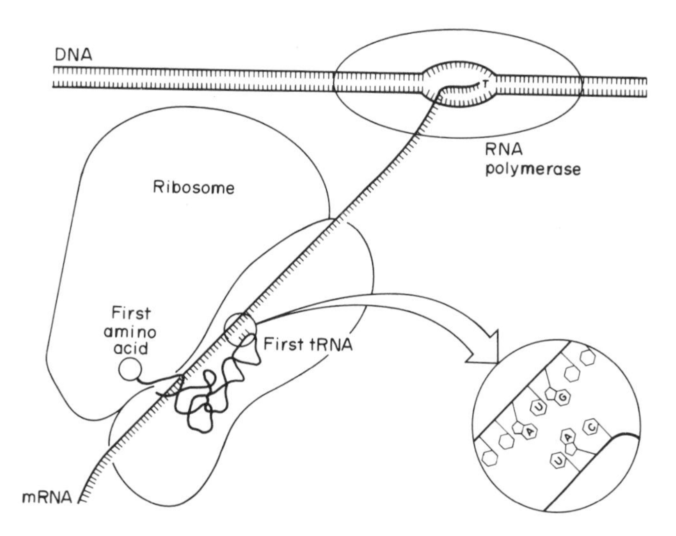

Figure 4 *Recognition of codons by anticodons.* The start codon (A-U-G) on the messenger RNA and the anticodon (C-A-U) of the first transfer RNA bind on the ribosome. The amino acid destined to be first in the new protein chain is already attached to the first transfer RNA. RNA and DNA have distinct left and right ends, and by convention nucleotide sequences are always written left to right; however, when base pairing occurs, one strand runs left to right and the other runs right to left. Consequently, if the codon is written as A-U-G, the anticodon must be inverted during base pairing and appears as U-A-C in the drawing (reproduced from ref. 9 by permission of John Wiley & Sons, Inc.)

lagging strand synthesis. Thus both DNA polymerases always synthesize DNA in the same direction in a chemical sense but in different directions relative to the moving replication fork (see Figure 7).

DNA polymerase has two additional complicating features. First, it cannot join short fragments of DNA together. Thus on the lagging strand other enzymes must fill in gaps with nucleotides and seal together the fragments made by DNA polymerase. In bacteria the filling enzyme is a different DNA polymerase, and the sealing enzyme is called DNA ligase. Both of these enzymes have become valuable tools in genetic engineering. The other feature is that DNA polymerases, unlike RNA polymerases, cannot begin synthesis of a new nucleic acid chain *de novo*; instead, they add onto the end of a preexisting nucleic acid chain called a primer (see Figure 8). Consequently, an enzyme called a primase is associated with the replication fork to synthesize primers on the lagging strand. In summary, DNA synthesis in bacteria requires the activity of a helicase, a topoisomerase, a primase, a ligase and at least two different DNA polymerases.

Occasionally errors occur during replication. If these errors are passed from one generation to the next, they may be observed as mutations. Several types of mutation can occur. A single nucleotide may be incorrectly copied, and this might result in a change of a codon in the messenger RNA. An incorrect amino acid might then be incorporated into the protein made from the information in the mutant gene. The incorrect amino acid may or may not affect the biological activity of the protein. In other cases information is deleted from DNA. Since genetic information is read (translated) as triplets with only start and stop signals as punctuation to orient the translation machinery, a deletion can have several effects. If the deletion is in multiples of three, the protein will simply lack the amino acids specified by the deleted material, and the protein will be shorter by the amount deleted. If, however, the deletion is not a multiple of three, then the reading frame will be shifted so all subsequent codons will be out of phase, and they may *all* specify incorrect amino acids. Occasionally the information in the DNA will be altered to create a stop codon in the middle of the gene. This produces an abnormally short protein. Likewise, the normal stop codon can be altered to generate an abnormally long protein.

3.7.3 DNA AND CHROMOSOME STRUCTURE

Understanding DNA structure is important for two aspects of the utilization of genetic information. In one, the primary structure of DNA, *i.e.* the specific arrangement of nucleotides, determines the order of amino acids in proteins and thereby determines the structure of all proteins; abnormal arrangements of information often lead to abnormal proteins and thus to genetic diseases. The other aspect involves how nucleotide sequences control the production of proteins; in this case abnormal

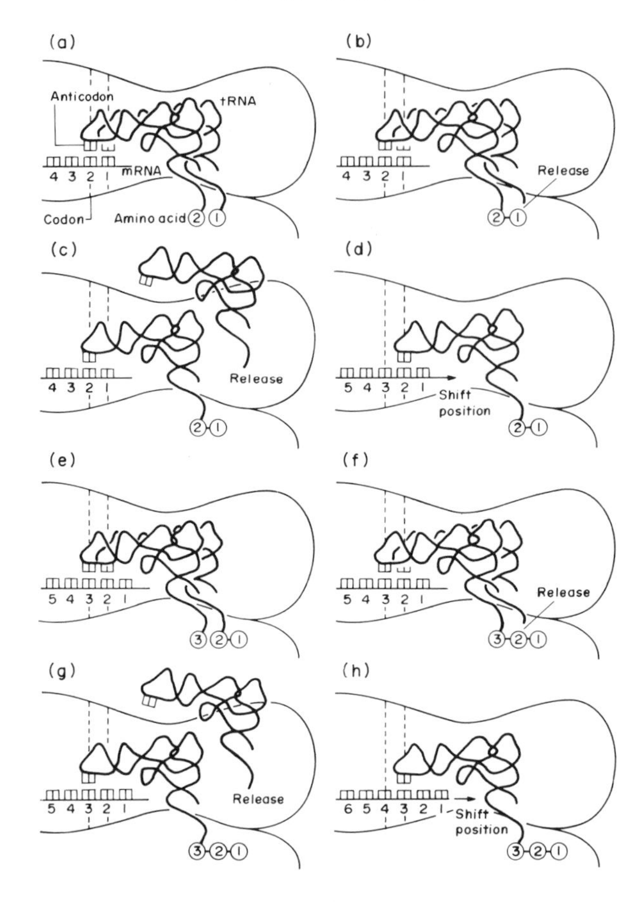

Figure 5 *Ordering amino acids during protein synthesis.* (a) After the messenger RNA (mRNA), initiator aminoacyl–tRNA, and ribosome have formed a complex (Figure 4), a second tRNA, with its attached amino acid, is ordered on the ribosome when its anticodon region base-pairs with the second codon of the mRNA. Dashed lines are reference lines to show positioning of codons on the ribosome. (b) Amino acids 1 and 2 are bonded covalently; amino acid 1 is released from tRNA 1 (note break). (c) tRNA 1 is released from the ribosome. (d) mRNA and tRNA 2, now attached to two amino acids, are translocated (shifted one position on the ribosome). This brings codon 3 into position on the ribosome. (e) Aminoacyl–tRNA 3 attaches to the ribosome and forms base pairs with codon 3. (f) Amino acid 3 binds to amino acid 2, repeating step b. (g) tRNA 2 is released from the ribosome, repeating step c. (h) tRNA 3 and the growing protein are translocated, repeating step d. This process continues until a stop codon is reached. At this point the protein chain is released from the last tRNA (reproduced from ref. 9 by permission of John Wiley & Sons, Inc.)

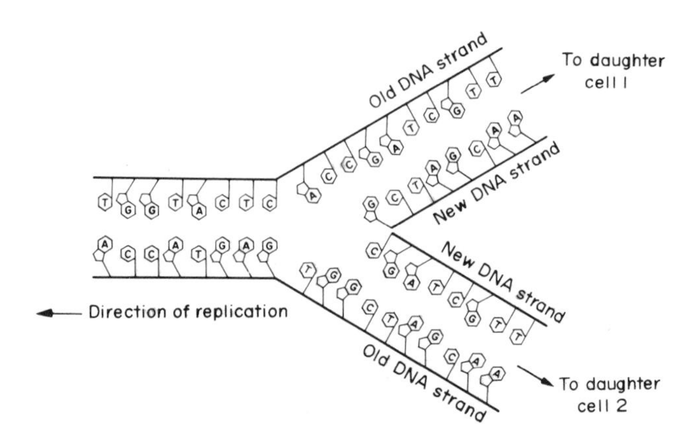

Figure 6 *Two DNA molecules arise from one.* Base-pairing complementary allows information to be copied exactly. Notice that the set of information transmitted to the two daughter cells is identical. For clarity, the strands have not been drawn as an interwound helix (reproduced from ref. 9 by permission of John Wiley & Sons, Inc.)

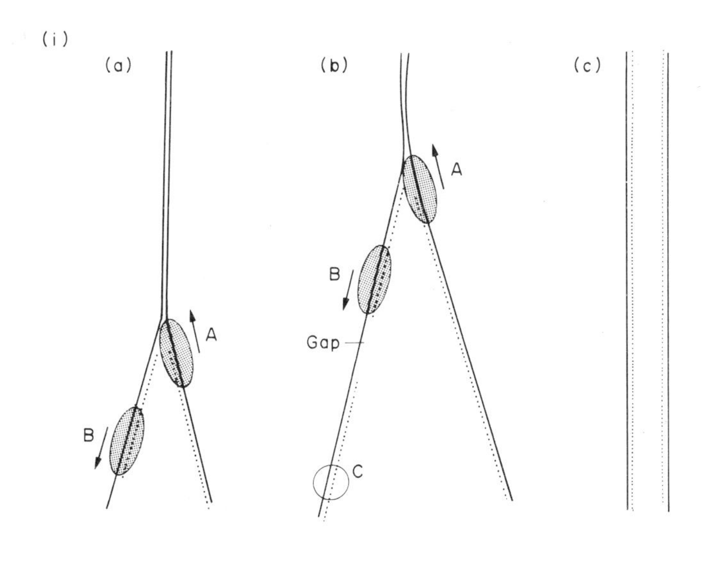

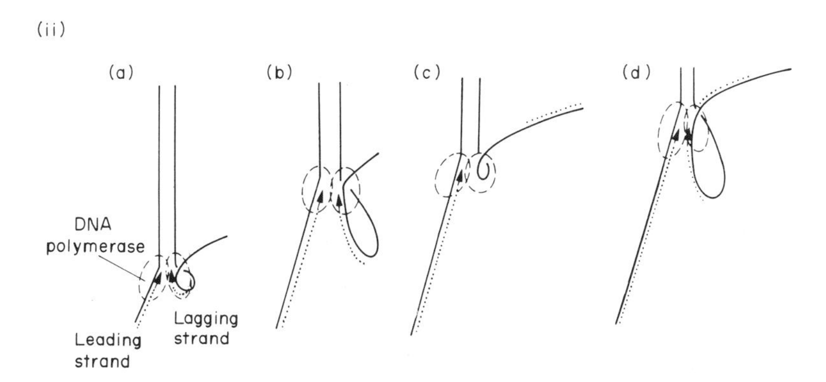

Figure 7 *Discontinuous DNA synthesis.* (i) Separate DNA polymerase molecules: (a) one DNA polymerase complex (A) moves continuously along an old single strand in the direction of replication fork movement while the other (B) moves in the opposite direction; (b) polymerase B synthesizes short patches of DNA with small gaps being filled in by another type of DNA polymerase, and the DNA fragments being spliced together by DNA ligase (C); (c) the result is two double-stranded DNA molecules. (ii) Two attached DNA polymerase molecules: (a) biochemical evidence suggests that the DNA polymerase molecules on the leading and lagging strands physically interact so that both move in the same direction as the replication fork (arrows indicate the direction of DNA synthesis); (b) in this model the lagging strand polymerase pulls a loop of DNA backwards for a short distance; and then (c) releases it; (d) another loop is pulled as another short stretch of DNA is synthesized. Dotted lines represent newly synthesized DNA

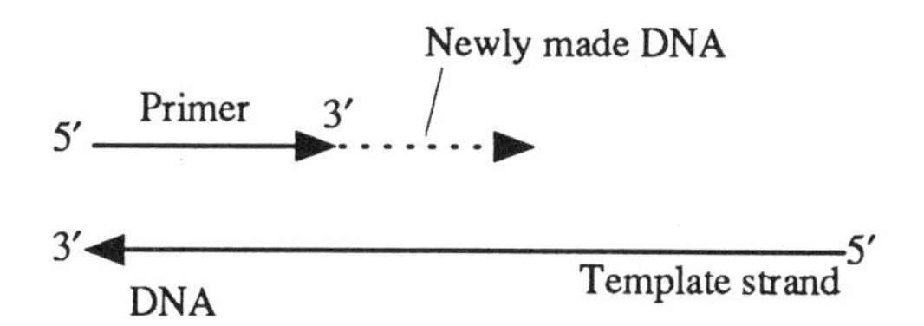

Figure 8 *DNA synthesis from a primer.* Two complementary DNA strands are shown in which one is longer than the other. One end of the shorter chain can serve as a start point, a primer, for DNA polymerase. In this case DNA synthesis would extend the top strand, using the bottom strand as an information template

information can create abnormal amounts of specific proteins. It is this second aspect that has captured the attention of most molecular biologists, and Sections 3.7.6 and 3.7.7 focus on how gene expression is controlled. In one of the major regulatory strategies, effector molecules (usually proteins) bind transiently to specific regions of DNA to control production of specific messenger

RNA molecules and thus specific proteins. Precisely where the effectors bind is determined by nucleotide sequences in the DNA. When they bind is usually determined by the intra- and extra-cellular environment.

Inside cells DNA is complexed with proteins that help compact it to fit within cells often less than 1/1000 the length of the DNA. The protein–DNA complexes are called chromosomes, and the precise structure of the chromosome in the vicinity of a particular gene may be important in determining whether a regulatory protein will bind to its target DNA sequence. Precisely how chromosome structure affects gene expression is still poorly understood; however, there is no doubt that it does. In this section our first point is that DNA is a dynamic molecule whose precise shape is readily perturbed by environmental factors. Our second is that torsional strain can be introduced into DNA and that this strain probably plays an important role in the energetic activation of DNA for processes involving DNA strand separation and the wrapping of DNA around proteins. Finally, we briefly consider DNA packaging, for the location of DNA packaging proteins may affect gene expression.

3.7.3.1 B-form DNA: Probably the Dominant Form in Solution

During the early studies of DNA that established the helical nature of the molecule, it was discovered that DNA could assume different structures depending upon the environment. B-form DNA seemed to be the most likely structure for DNA in solution because it was the form present at relative humidities approaching aqueous conditions. Textbooks began describing DNA as a right-handed double helix with the bases of the nucleotides stacked perpendicular to the helix axis such that the centers of their planes are separated by a spacing of 3.4 Å (see Figure 2). In this structure the sugar–phosphate backbones follow helical paths at the outer edge of the molecule while the bases are in the central core of an imaginary cylinder. The bases are arranged so that a base in one strand is always paired with a base in the other. As mentioned earlier, according to the Watson–Crick complementary base-pairing rule, adenine (A) always pairs with thymine (T) and guanine (G) always pairs with cytosine (C). Since a base pair always contains one double-ringed purine (A or G) and one single-ringed pyrimidine (T or C), the length of each base pair, *i.e.* the diameter of the cylinder, is nearly the same at all points in the DNA. Both bases in a base pair lie in the same plane (perpendicular to the helix axis), and in B-form DNA the pairs are rotated by 36° relative to each adjacent pair.

DNA is not a smooth cylinder; it contains two grooves (Figure 2). The larger groove, called the major groove, is deep and wide while the smaller, the minor groove, is narrow and shallow. It is currently thought that many regulatory proteins bind to one face of the DNA and interact by forming contacts in adjacent major grooves. In a sense these proteins dock on one side of the DNA.

As mentioned earlier, nucleic acids have polarity; the ends are chemically distinct, and this feature provides directionality to the molecule (Figure 2). The nomenclature of the ends is taken from the nomenclature of the sugar moiety at either end of the nucleic acid. The terminal carbon of the sugars is a 5′ carbon at one end and a 3′ carbon at the other (Figure 9a). In the case of double-stranded DNA, the two strands are antiparallel, *i.e.* they run in opposite directions (Figure 2 and Figure 9b). DNA and RNA polymerases recognize this polarity and only synthesize new nucleic acids in the 5′ to 3′ direction (relative to the newly synthesized molecule, see Figure 9b).

3.7.3.2 Base Stacking, Hydrogen Bonding and Phosphate Repulsion: Factors Affecting Double Helix Stability

Many of the biological processes that DNA undergoes involve separation of the two strands. Thus, it is important to understand the forces involved in stabilizing the double helix, for modulation of these forces will control the biology of a particular DNA molecule. Three major forces are usually discussed when considering the dynamics of DNA structure: base stacking, base-pair hydrogen bonding and backbone phosphate charge repulsion.

In both double- and single-stranded DNA, adjacent bases tend to interact through van der Waals forces, stacking one on top of the other. These stacking interactions provide the major stabilizing force acting on the double helix. Hydrogen bonding between bases in opposite strands provides base-pairing specificity (see Figure 2). Formation of the double helix requires that hydrogen bonds between the bases and water be broken. In complementary helices these hydrogen bonds between bases and water are replaced by ones associated with base-pairing, and there is little change in net

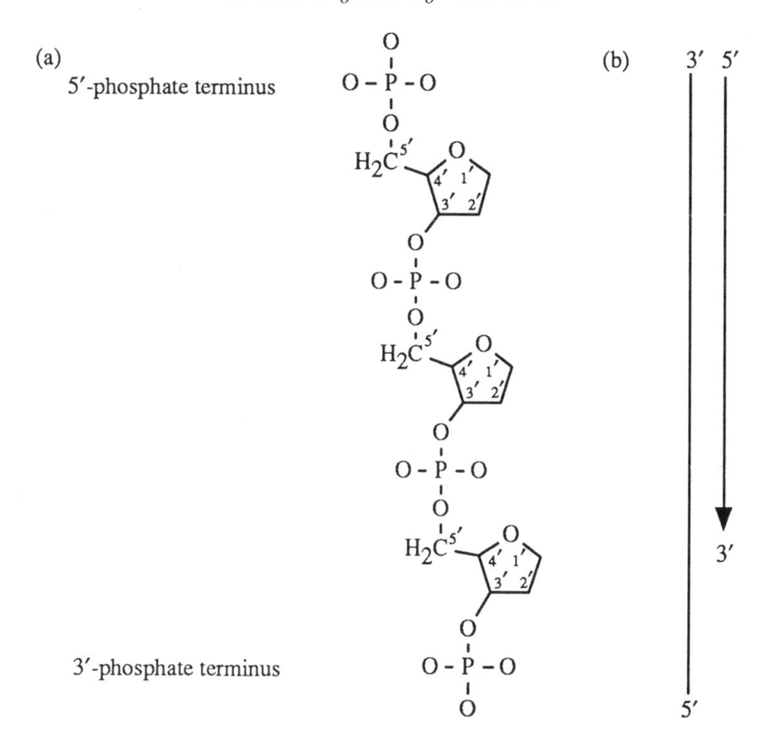

Figure 9 *Directionality of DNA.* (a) A portion of a single-stranded DNA is shown in which both the 5′ and 3′ ends are linked to phosphates. For simplicity not all of the carbon and hydrogen atoms in the deoxyribose moieties are shown. (b) A schematic drawing of partially double-stranded DNA. The arrow indicates the direction of DNA synthesis

bonding energy. Non-complementary base pairing is energetically unfavorable because not all of the hydrogen bonds between the bases and water are replaced by base–base hydrogen bonds. These considerations suggest that base pairing itself provides little stabilization energy to the double helix, but it is sufficient to insure that the complementary base-pairing rule is obeyed. A major destabilizing force in the double helix appears to be the electrostatic repulsion existing between the charged phosphate backbones. As expected, counterions, either as simple salts, polyamines or proteins, dramatically increase helix stability in solution.

The two strands of a DNA molecule can be separated by heating or by extremes in pH. The separation process has been called denaturation, and at moderate salt concentration temperatures approaching 100°C are required to completely separate the two strands of DNA. Upon cooling or neutralization, the strands will reassociate. The ability of two strands to associate and form base pairs depends on the two having complementary, or nearly complementary, nucleotide sequences. Thus the relatedness of two strands can be assessed by their ability to associate (hybridize). One popular method for measuring hybridization is to denature one DNA species, immobilize it by attachment to a membrane filter, and then incubate the filter-attached DNA in a solution containing a second type of DNA (or RNA) which has been radioactively labeled. Following incubation and washing of the filter, the amount of radioactivity associated with the filter-bound DNA is a measure of relatedness. The molecular basis for this measurement is sketched in Figure 10. Nucleic acid hybridization is the basis for important techniques used in cloning genes, in measuring the synthesis of specific RNA molecules, and in genetic screening.

The phosphate groups on DNA and RNA give nucleic acids a net negative charge. Consequently, they move in an electric field, and short nucleic acids are frequently fractionated by electrophoretic methods. In gels composed of acrylamide or agarose the electrophoretic mobility of a nucleic acid is proportional to its length. This technique is sensitive enough to distinguish single nucleotide differences in the lengths of molecules hundreds of nucleotides long.

3.7.3.3 Other Forms of DNA: DNA as a Dynamic Structure

Very early in the study of DNA, it became clear that DNA molecules are structurally dynamic. By varying the environmental conditions used for X-ray diffraction study of DNA fibers, it was possible

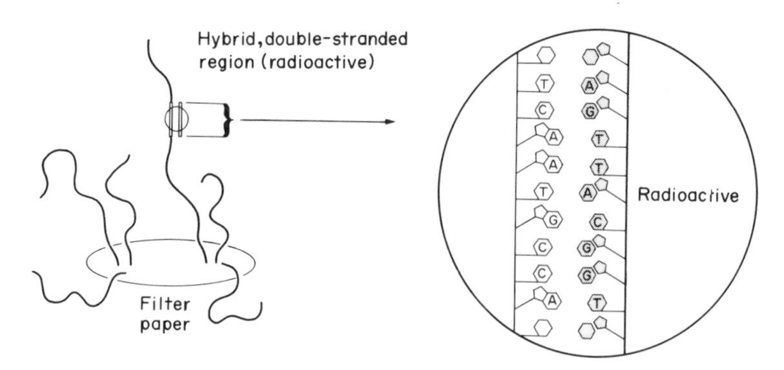

Figure 10 *Nucleic acid hybridization.* Under the appropriate conditions two complementary single-stranded nucleic acids will spontaneously form base pairs and become double-stranded. If single-stranded, nonradioactive DNA is fixed tightly to a filter and then incubated in a solution containing single-stranded, radioactive DNA, double-stranded regions will form where the two types of DNA have complementary nucleotide sequences; the radioactive DNA will become indirectly bound to the filter through its attachment to a specific region of nonradioactive DNA (open). The amount of radioactivity bound to the filter depends upon the relatedness of the two DNAs, which may thereby be assessed (reproduced from ref. 9 by permission of John Wiley & Sons, Inc.)

to observe alternate forms of DNA (A- and C-forms), and transitions between the forms could be obtained by changing the relative humidity and salt concentration. More recently, conditions have been found in which right-handed B-form DNA flips into a left-handed Z-form. It is likely that the precise structure for any given region of DNA depends on the local environment as well as on the nucleotide sequence, raising the intriguing possibility that regulatory proteins recognize these structural differences along a DNA molecule.

The finding that regions of double-stranded DNA frequently open to become single-stranded reinforced the idea that alternate conformations[40] could exist. This transient opening phenomenon, called breathing, was recognized by the observation that formaldehyde very slowly denatures DNA. Since formaldehyde reacts with the amino groups of the bases of single-stranded DNA and eliminates their ability to base pair, the effect of formaldehyde was most easily explained if regions of the DNA were constantly fluctuating between double- and single-strandedness.

When the DNA double helix is destabilized by changing the solution environment, denatured (single-stranded) regions appear in the DNA. Regions rich in AT base pairs denature first. Consequently, it has been popular to think that AT-rich regions might serve as entry sites for proteins known to act by unwinding DNA. Indeed, inspection of nucleotide sequences called promoters, locations where RNA polymerase binds and unwinds DNA, does reveal a high AT content. However, as discussed below, promoter strength is a complicated phenomenon and does not relate simply to the AT content of the region.

We expect that DNA conformations having regulatory significance will depend on specific nucleotide sequences. Z-form DNA may represent a striking example. Z-form DNA is a left-handed polymer that gains its name from the zigzag path that the sugar–phosphate backbone takes along the helix in this conformation.[28] Z-form DNA tends to form most readily in polymers that have an alternating purine–pyrimidine sequence, particularly alternating G's and C's. A stretch of only 30 alternating G's and C's is sufficient to generate Z-form DNA. Plasmids, small autonomously replicating DNA molecules (see Section 3.7.8.3), have been constructed in which a short region of left-handed Z-form DNA can exist in a DNA molecule otherwise in the right-handed B-form (see Figure 11a). RNA polymerase movement is blocked at a B–Z junction, and it has been speculated, but not demonstrated, that Z-DNA might serve as a control signal influencing gene expression.

Another sequence-dependent structure is bent DNA.[38] Short runs of 5 or 6 A's appropriately spaced cause DNA fragments to migrate abnormally slowly during electrophoresis in gels when the A's are in the middle of the fragment but not when they are near the ends (see Figure 11b). This anomalous electrophoretic mobility has been explained in terms of formation of a bend in the DNA. A number of nucleotide sequences that serve as recognition sites for specific proteins contain DNA bends. One idea is that during recognition of the DNA the proteins coil DNA and that the bend aids in the coiling. Also important in the recognition process are DNA supercoiling (Section 3.7.3.4) and accessory proteins that can themselves coil DNA.

Still another nucleotide sequence-dependent structure is a cruciform. Such a structure can arise when self-complementary, inverted repeat sequences are located close to each other on the same DNA strand (Figure 11c). To form a cruciform, the DNA must be partially denatured to overcome

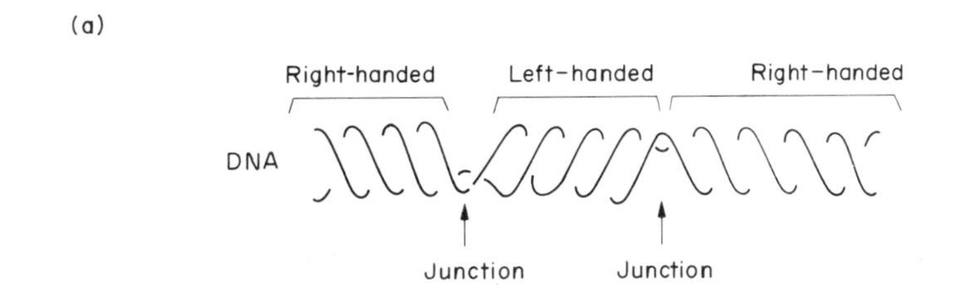

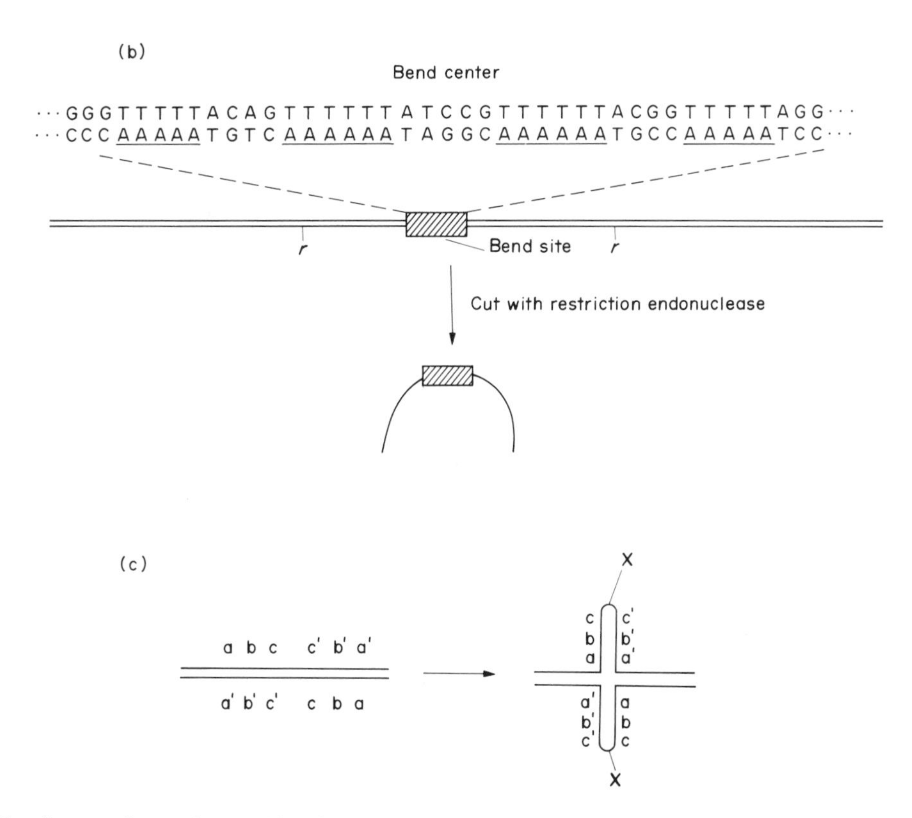

Figure 11 *Alternate forms of DNA.* (a) B-form DNA containing a short region of Z-form DNA. Under appropriate environmental conditions some nucleotide sequences will flip from right-handed B-form DNA into left-handed Z-form. A junction between B- and Z-form DNA acts as a barrier to RNA polymerase movement. (b) Bent DNA. Some DNA molecules contain runs of A's spaced such that they create a bend site. When DNA is cut with a restriction endonuclease so the bend site is in the middle of the resulting fragment, the fragment is bent and migrates abnormally slowly during gel electrophoresis. (c) Cruciforms. Inverted repeat sequences can convert from a linear duplex to a cruciform in which all of the bases form a double helix except those at the tips of the arms (arrows labeled X). Here the primed and unprimed letters indicate complementary base pairs

an activation energy barrier. Cruciforms have been detected in DNA isolated from natural sources, and examples of nucleotide sequences capable of forming cruciforms have been found in regions thought to contain control signals.

It is important to point out that although the alternate structures of DNA have been detected in stretches of nucleotides known to be control regions, purified DNA was used for these studies, and it has been difficult to firmly establish that any of the alternate structures contribute to the structure and biology of DNA inside living cells.

3.7.3.4 Negative Supercoiling: Activation of DNA Molecules

DNA molecules inside many types of cell are circular. For example, the chromosome of the common intestinal bacterium *Escherichia coli* is a single, circular DNA molecule. The absence of ends blocks free rotation of the strands and allows torsional strain to be introduced into DNA. This strain is called supercoiling or superhelical tension. Supercoiling in the negative sense (see below) is thought to be biologically important because it can lower the activation barrier for all processes involving strand separation, for conversion of B-form DNA to Z-form, and for the formation of cruciforms. At least in bacteria, supercoiling is known to be an important parameter in determining the structure of DNA inside living cells.[8, 36]

The phenomenon of supercoiling has been most extensively studied with small, covalently closed, circular DNA molecules called plasmids (Section 3.7.8.3). Supercoiling is generally described in terms of a parameter called linking number, the number of times one strand crosses the other when the molecule is conceptually constrained to lie in a plane. Circular DNA molecules isolated from natural systems have lower linking numbers (fewer strand crossings) than expected from a comparable circular DNA molecule in which one strand is broken so the DNA strands are free to rotate. Thus when constrained to lie in a plane, a negatively supercoiled molecule would not have the full number of helical turns and could not be fully base-paired (Figure 12a). In solution there are forces which drive the DNA toward a fully base-paired conformation; thus, in a DNA lacking the full number of helical turns, these forces cause the molecule to writhe (supercoil, Figure 12a). The writhe (W) is related to linking number (L, as defined above) and twist (T, the number of times one strand passes over the other when the DNA is in solution rather than lying on a plane) by the simple algebraic expression $W = L - T$. When T is greater than L, W is negative, the situation occurring with most circular DNA isolated from living cells.

A supercoiled DNA molecule is under torsional tension, and a break in one strand, which acts as a swivel and allows strand rotation, spontaneously relieves superhelical strain (Figure 12b). Thus superhelical strain is energetically unfavorable, and any process that removes strain will be favored in a superhelical DNA relative to a relaxed one. Processes that unwind DNA relieve strain (Figure 13); consequently, superhelical energy will tend to drive these reactions. In a sense, superhelical strain is a way in which circular DNA molecules can store energy to drive strand separation processes. For the present discussion, one of the most important reactions that could be driven by superhelical energy is the opening of the DNA helix by RNA polymerase during transcription. Studies with bacterial systems, both *in vitro* and *in vivo*, suggest that superhelical tension may have an important influence on gene expression. Another important process that involves supercoiling is DNA replication. A critical feature of replication is the separation of the two parental DNA strands, a process that requires DNA unwinding and is favored by negative supercoiling.

DNA inside living bacteria has been shown to be under negative superhelical tension by dye-binding studies and by topological measurements. Whether some of the DNA in higher cells is under superhelical tension is not yet known.

3.7.3.5 DNA Topoisomerases: Regulation of Supercoiling

The importance of DNA supercoiling became clear when enzymes called DNA topoisomerases were discovered. These enzymes change the topology of DNA by making transient breaks in the DNA strands. During the breakage reaction, one (or both) strands of DNA are passed through the break (Figure 14). This reaction is equivalent to the enzyme acting as a swivel to change the linking number. In bacteria, there are at least two enzymes involved in maintaining the level of supercoiling. The enzyme called DNA gyrase introduces supercoils utilizing hydrolysis of ATP as an energy source. The competing activity is supplied by an enzyme called DNA topoisomerase I which relaxes negative supercoils without requiring an external energy source (Figure 15). Mutations in the gene encoding topoisomerase I lead to higher-than-normal levels of supercoiling and mutations in the genes encoding gyrase lead to lower-than-normal levels. Eukaryotic cells also contain two types of topoisomerase; both types relax DNA *in vitro*, but one requires ATP hydrolysis for activity. The ATP-requiring enzyme may turn out to have a gyrase-like activity when proper cofactors are found.

Since DNA topoisomerases can pass DNA strands through transient breaks in DNA, they are capable of tying and untying DNA knots and interlinking or unlinking DNA circles (see Figures 16 and 17). The latter property is probably an important aspect of DNA replication since the product of replication of circular DNA is two interlinked daughter DNA molecules. Topoisomerase mutations in bacteria and in yeast appear to prevent duplicated daughter DNA molecules from fully separating.

Biochemical studies have identified potent inhibitors of topoisomerases that are beginning to play a role in medicinal chemistry. The quinolones, of which nalidixic acid is the best known, appear to trap bacterial gyrase on DNA in a way that leads to cell death. The quinolones have little effect on

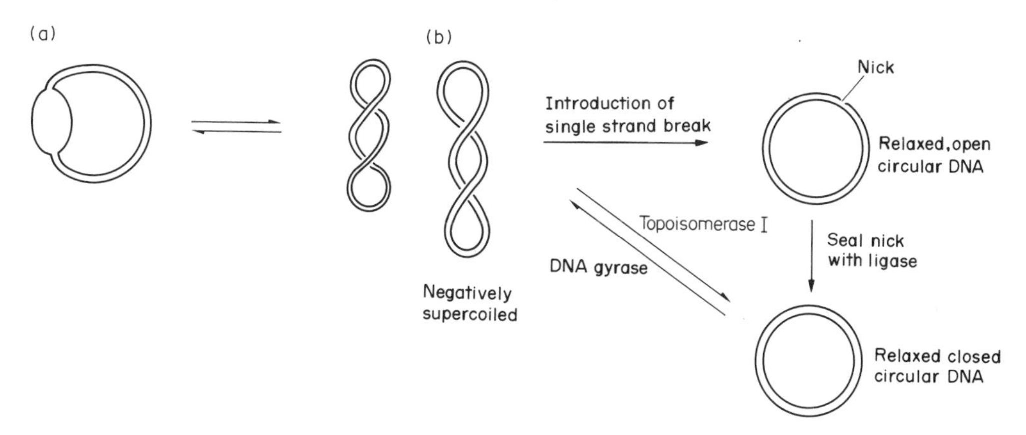

Figure 12 *Negative supercoiling.* (a) Negatively supercoiled DNA has a deficiency of duplex turns, so when the molecule is constrained to lie in a plane, not all of the DNA can form a double helix (left). In solution all of the bases will pair forming a complete double helix. This places torsional strain on the molecule, and it will twist (right). (b) Supercoils dissipate spontaneously if one or both of the DNA strands are broken. Nicked DNA is still circular, and the nick can be sealed by ligase to produce a covalently closed, relaxed, circular DNA. DNA gyrase can introduce supercoils into closed circular DNA. All topoisomerases can remove supercoils from DNA

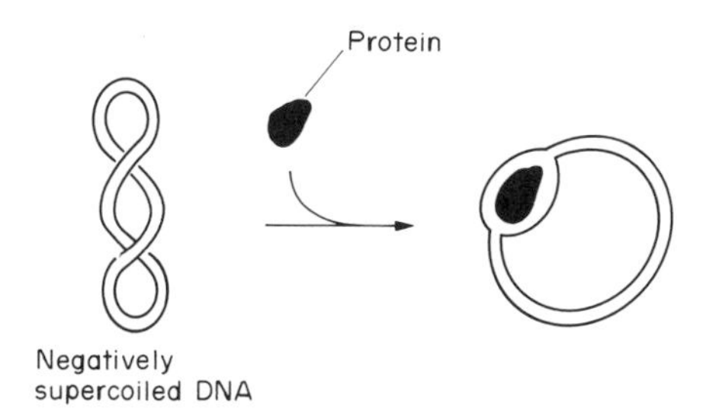

Figure 13 *Strand separation relieves superhelical strain.* Some proteins, notably RNA polymerase, unwind DNA when they bind. The unwinding occurring during binding to negatively supercoiled DNA leads to DNA relaxation, an energetically favored process

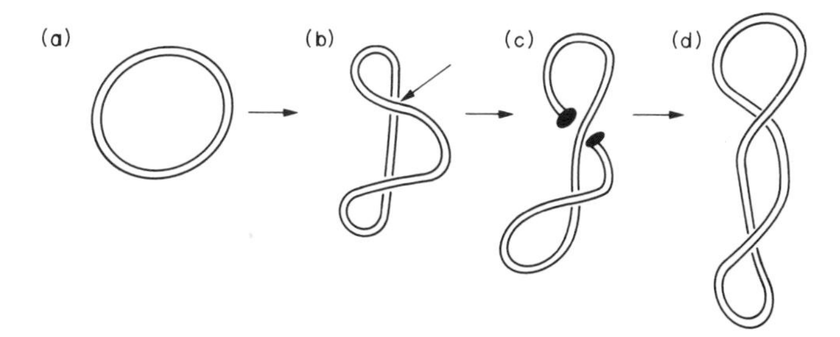

Figure 14 *The topology of topoisomerase action.* All topoisomerases change DNA topology by breaking DNA strands, passing strands through the break, and then rejoining the broken strands. Gyrase is shown introducing two negative supercoils into relaxed DNA (a) by breaking both strands at the arrow in (b), passing the lower strand through the break (c), and then sealing the break (d)

topoisomerases in higher organisms, so these compounds are receiving considerable attention as antibacterial agents. The eukaryotic type II topoisomerases are affected in a similar way by a number of anticancer drugs among which are the epipodophyllotoxins. In this case the drugs appear to be effective against tumors because rapidly growing cells are particularly susceptible.

3.7.3.6 DNA Loops and Folds: Creation of Topological Domains

Chromosomal DNA in both prokaryotic and eukaryotic cells is constrained into large loops containing between 50 000 and 100 000 base pairs. It is likely that the loops partition the DNA into a

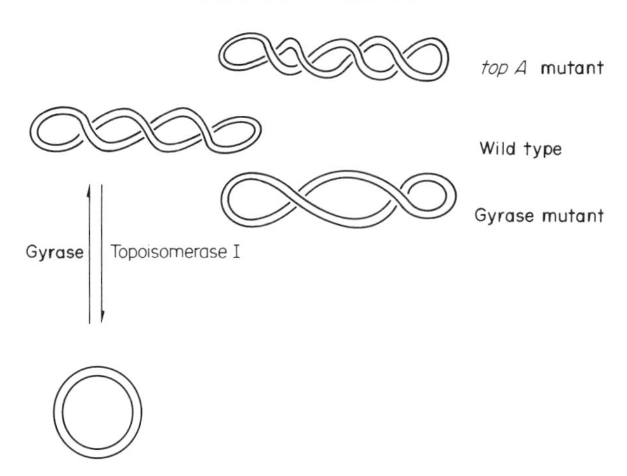

Figure 15 *Control of supercoiling in bacteria.* Gyrase introduces supercoils and topoisomerase I removes them. Mutations in the genes encoding gyrase can lead to lower-than-normal levels of supercoiling, and mutations in *topA*, the gene encoding topoisomerase I, can lead to higher-than-normal levels

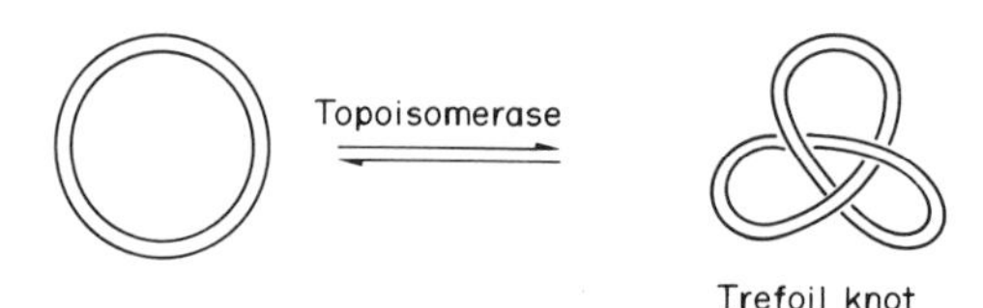

Figure 16 *Knotting.* Topoisomerases can tie and untie knots. A simple trefoil knot is shown, but more complicated knots have also been observed

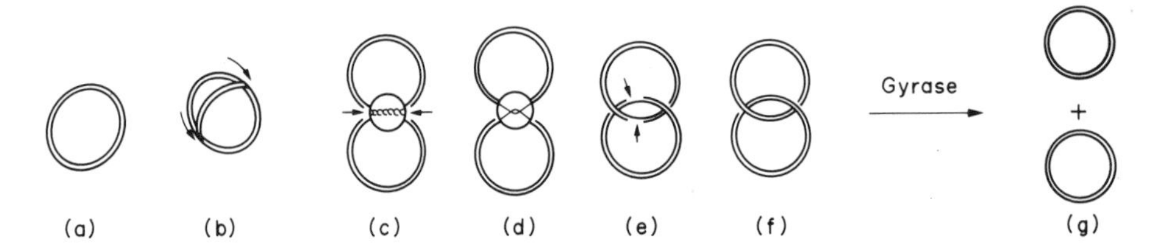

Figure 17 *Catenanes.* A circular DNA molecule (a) initiates replication in two directions (b). The direction of replication fork movement is shown by arrows. In (c) the central enlargement shows helical overlaps in parental DNA. If the replication forks halt while two single-strand overlaps remain (enlargement in d), then the two circles remain linked (catenated) as shown in (e) where gaps (arrows) remain in each daughter chromosome. The gaps can be enzymatically filled and sealed, producing a covalently closed, circular catenane (f). If replication forks stop before (d), catenanes will be multiply intertwined. In bacteria gyrase appears to be involved in decatenating the interlinked circles so the daughter chromosomes can separate (g) during cell division

number of topologically independent domains, for multiple nicks are required to relax the supercoils in chromosomal DNA. The partitioning makes it possible for the majority of the DNA to remain under superhelical tension even though nicks or gaps occur transiently in the chromosome as a result of DNA replication and repair. As pointed out above, maintaining the proper level of superhelical tension in DNA is important for processes involving DNA strand separation and for proteins to recognize the specific nucleotide sequences involved in site-specific recombination and initiation of DNA replication.

3.7.3.7 Nucleosomes: The Fundamental Unit of Chromatin Structure

During most of the cell cycle in higher organisms, chromosomal DNA is diffuse, and discrete chromosomes are not microscopically observable in the nuclei. This period in the cycle is called interphase. When interphase nuclei from eukaryotic cells are suspended in solutions of low ionic strength, they rupture and release fibers called chromatin. If stretched out, a chromatin fiber resembles a string of beads. The beads are protein-containing particles, called nucleosomes, in which the DNA is wrapped around the protein. The thread connecting the beads is double-stranded DNA. By cutting the DNA between the nucleosomes it is possible to release individual nucleosomes from the chromatin. Physical and chemical analyses of isolated nucleosomes show that they are remarkably similar from one organism to another. Each nuclesome contains about 160 base pairs (bp) of DNA and two each of the four highly basic proteins called histones (H2A, H2B, H3 and H4). The DNA is wrapped in slightly less than two superhelical coils around the histones (see Figure 18).

The beads-on-a-string appearance of chromatin occurs only when the chromatin has been stretched. At low ionic strength, unstretched chromatin, containing the additional protein called histone H1, is a continuous filament of tightly packed nucleosomes having a diameter of about 10 nm.[33] Under more physiological conditions, the 10 nm filament appears to be folded to form what has been called a 30 nm fiber. One attractive model maintains that the 10 nm filament is coiled into a 30 nm solenoid-like structure partially through interactions among the H1 histones (see Figure 19). Still more complex foldings have been postulated to account for electron microscopic observations of chromosomes, but so far they have not led to a universally accepted view of higher order chromatin structure.

In bacteria it is not known whether some of the DNA is packaged into nucleosome-like particles, for nucleosomes comparable to those described above have not been isolated from bacteria. Nevertheless, histone-like proteins have been isolated, and it is possible to form nucleosome-like structures by adding the histone-like bacterial protein called HU to DNA *in vitro*. Why chromatin cannot be easily extracted from bacteria is still a mystery.[10]

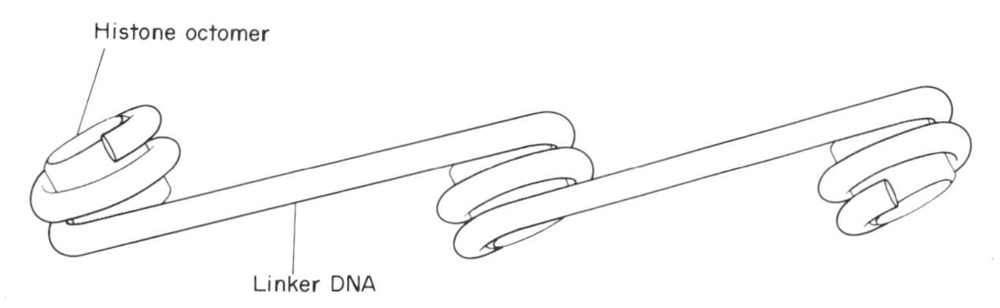

Figure 18 *Nucleosome structure.* In eukaryotic organisms, DNA (about 145 base pairs) is wrapped around an octomer of eight histones (H2A, H2B, H3 and H4) into a structure called a nucleosome. DNA between the nucleosomes is called linker DNA; its length depends on the species of organism and is generally about 50 base pairs

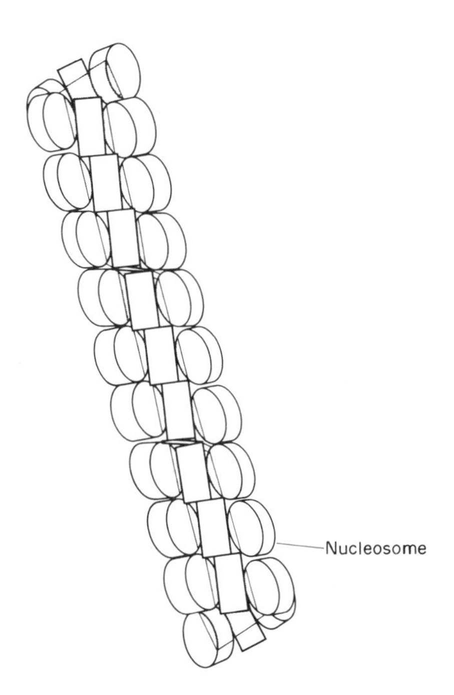

Figure 19 *Packed nucleosomes.* Nucleosomes are shown condensed into a solenoid-like structure, presumably by the action of histone H1. Fibers similar to the one shown may in turn be condensed into still more complex structures (redrawn from ref. 33)

3.7.3.8 Chromosomes: Units for DNA Segregation

Eukaryotic cells pass through a stage called mitosis, a brief period in the cell cycle when the individual DNA molecules condense up to 10 000 fold (5 to 10 times more condensed than DNA in chromatin). In this form the DNA is able to segregate to daughter cells as an intact unit. Genetically, individual chromosomes behave as discrete linkage groups, consistent with each chromosome containing a single piece of DNA.

Dyes are available which bind to DNA, some preferentially to certain bases; consequently, stained chromosomes take on banding patterns that are distinctive of a particular chromosome. Microscopic examination of stained chromosomes has become a valuable tool for detecting chromosomal abnormalities, particularly breaks, rearrangements and extra chromosomes.

3.7.3.9 Heterochromatin: DNA Condensation Associated with Gene Inactivation

As described above, DNA inside eukaryotic cells is wrapped by histones into nucleosomes. Microscopic examination of nuclei has revealed regions that are more highly condensed than others, and it is generally thought that genes in these highly condensed regions are transcriptionally inactive and not easily activated. Most of the genes in less condensed regions are also inactive, but they are thought to be capable of activation.

Several types of microscopically visible chromosomal structures fall into the category of highly condensed chromatin containing inactive DNA. Mitotic chromosomes are one example in which the DNA becomes transcriptionally inactive during the brief period in the cell cycle when DNA condensation facilitates proper chromosome segregation to daughter cells. Other examples occur during interphase. Although at this stage of the cell cycle most of the chromatin appears to be diffuse, some highly condensed regions, called heterochromatin, persist and are correlated with inactive genes.

The most striking example of heterochromatin is the inactive X chromosome of female mammals. In placental mammals, females contain two X chromosomes while males contain only one. Early in embryonic development, one of the female X chromosomes is inactivated, resulting in the dosage of active genes on the X chromosome in females being the same as that in males. Inactivation of X chromosomes is random and clonal; once it has occurred, it is fixed in the cell's somatic heredity. The inactive X chromosome is highly condensed and has a distinctive cytological appearance which in human cells is called a Barr body. As with other forms of heterochromatin, the inactivated X replicates late in the cell cycle relative to the rest of the DNA. Inactivation of X chromosomes is not restricted to somatic cells, for it also occurs in the germ line. In germ cells inactivation is reversed when the cells enter meiosis, the process that produces ova having only one copy of each chromosome.

While many models have been proposed to explain X inactivation, the process is not well understood at the molecular level. A satisfactory model must account for a number of remarkable features. First, there must be a counting mechanism, for only one X is maintained in an active state. Even cells containing five X chromosomes have only one active X chromosome. Second, inactivation can distinguish between two identical DNA molecules, and once initiated, the region of inactivation spreads only along the chromosome to be inactivated. Somehow this state is maintained for hundreds of cell divisions. Third, as pointed out above, inactivation is reversible.

A potentially productive approach, a hint toward understanding inactivation, involves DNA methylation[11] (see Figure 20 for methylated nucleotides). The key observation is that a transient treatment of cells with 5-azocytidine, an inhibitor of DNA methylases, converts an inactive X chromosome into an active state. This new state is stably inherited. Thus, the presence of methyl groups in certain nucleotide sequences may be one of the signals for inactivation of the chromosome. Although many other explanations could be invoked for the effect of 5-azocytidine, blocking methylation is an attractive hypothesis because expression of a number of genes in other eukaryotic systems is known to be affected by methylation (see below). Moreover, methylation patterns can be easily inherited (see Figure 21). While site-specific methylation may serve as a signal for inactivation, how this signal spreads along the chromosome and how the chromatin condenses are still unknown.

3.7.3.10 Euchromatin: Activatable Genes

Active genes tend to be found in the diffuse-appearing chromatin called euchromatin.[17] As with heterochromatin, euchromatin contains DNA packaged into nucleosomal particles. It has been

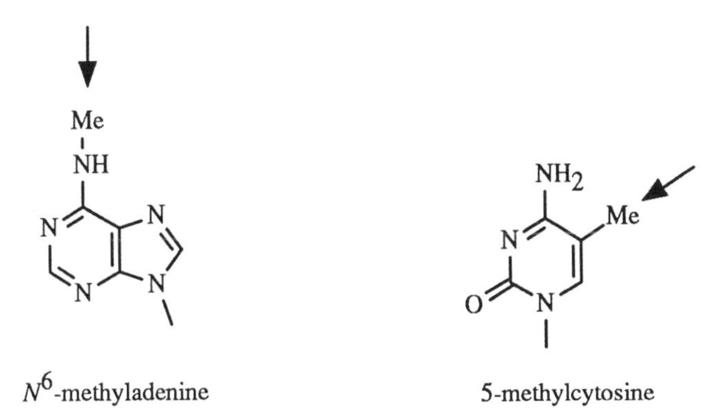

Figure 20 *Methylation of nucleotides.* Two nucleotides, adenine and cytosine, are occasionally found to be methylated as indicated by arrows

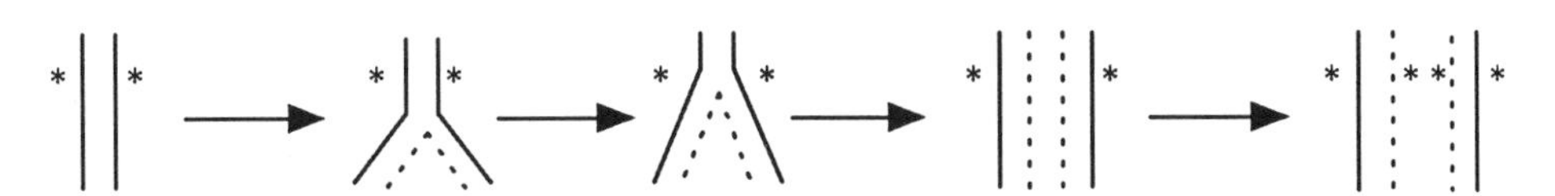

Figure 21 *Inheritance of methylation.* Prior to replication both strands are methylated at sites indicated with asterisks. After the replication fork passes through the region, a new, unmethylated copy of the DNA is made. Eventually the new copy is methylated. If the methylase enzyme only methylates DNA in which one strand is already methylated, then methylation at particular sites will be transmitted from one generation to the next

postulated that nucleosome structure of active genes may differ from that of inactive genes. In a special case, that of transcription of DNA encoding ribosomal RNA genes, electron microscopic observations suggest that nucleosomes are absent during transcription. Similar measurements cannot be made with more widely scattered and less actively transcribing genes. Nevertheless, active genes do appear to be more loosely packaged because they are degraded more readily by a nuclease than are inactive ones. Nuclease sensitivity often extends over thousands of base pairs, frequently beyond the nucleotide sequences of the gene contained in the region. It is important to note that nuclease sensitivity has been found for a variety of genes, some of which are infrequently transcribed. Thus it appears that this enzymatic probe is defining domains of the chromosome that are capable of being transcribed, *i.e.* regions that are activatable but not necessarily undergoing transcription at that time.

Another structural correlation found with activatable genes is the extent of methylation. A lower level of methylation of cytosines (Figure 20) appears to be associated with activatable genes. For example, cloned genes introduced into animal cells show more expression if they are unmethylated when introduced. In another example, a developmental correlation between activity and methylation has been observed with β-globin genes: β-globin genes are unmethylated in embryonic tissue where they are expressed, and many of the potential sites are methylated in adult erythroid tissues where these genes are never expressed. While these correlations are interesting and will stimulate additional research, there is still doubt that methylation can be a general method for gene control since some organisms such as fruit flies exhibit very little DNA methylation.

Once a gene is in the activatable state, it must then be activated, presumably by mechanisms similar to those described below in Sections 3.7.6 and 3.7.7. One feature of some active eukaryotic genes is the presence of sites that become hypersensitive to digestion by DNase I when the genes are being actively transcribed. The interesting sites tend to be upstream from the gene, *i.e.* on the 5′ side. Such hypersensitive sites have been found on the 5′ side of active β-globin genes in embryonic cells but not near the same genes when inactive in adult erythroid cells. In another example a mutant of *Drosophila* has been found in which a deletion removed the two upstream hypersensitive sites but left intact the gene and more than 250 base pairs of upstream DNA. This mutant does not express in the gene. In these and other cases it appears that a structural change in chromatin occurs in the region of the hypersensitive site since a number of enzymes can preferentially cleave DNA there. Precisely how these putative structural changes affect gene expression is not yet known.

3.7.4 INITIATION OF DNA REPLICATION

DNA replication must occur before cells can divide, and this process has become the target of a variety of therapeutic agents. The key to therapeutic success is specificity—the growth of invading microorganisms or proliferating cancer cells must be preferentially halted. The point in replication having the greatest specificity, and thus vulnerability, is initiation of replication. Viruses, plasmids, bacteria and human cells have unique locations in their DNA where replication begins, and for every system examined there are specific proteins that recognize these sites and control when a round of replication is to begin. The rudiments of DNA replication were presented earlier in Section 3.7.2.3; here emphasis is placed on initiation of replication. Several aspects of bacterial DNA replication are described below to provide a framework for thinking about the development of replication-specific antibacterial and antiviral agents.

3.7.4.1 Bacterial DNA Replication: An Overview

DNA replication begins at specific locations on DNA molecules called origins. There the two parental strands begin to separate, forming two replication forks (see Figure 1) at the points of separation (see Figure 17). The forks progress around the circular chromosome until they meet at a place called the terminus; there replication stops. This type of DNA replication is called bidirectional. In bacteria it takes about 40 minutes for the two forks to completely replicate a circular chromosome of 4 million base pairs. Thus each fork travels at a rate of 50 000 base pairs per minute, unwinding DNA at about 5000 revolutions per minute (rpm). At the end of the replication process, gyrase appears to be necessary to unlink the two daughter chromosomes: in a gyrase mutant, chromosome doublets can be isolated, and the addition of gyrase resolves the doublets into singlets.

The situation is slightly more complex in higher cells[3] since they often contain a thousand times more DNA than bacteria. To replicate their DNA higher cells have multiple origins scattered over the genome; thus many regions are simultaneously synthesizing DNA. As with bacteria, the daughter chromosomes appear to be interlinked, and a topoisomerase is required to separate them.

In the simplest terms, initiation of replication involves generating a primer at a specific place on the DNA. DNA polymerase can then synthesize new DNA by extending the primer (see Figure 8). One strategy would be for an initiation protein to specifically bind to the origin of replication, cut the DNA to form a 3′ primer, and then allow DNA polymerase to replicate the DNA (see Figure 22). Some plasmids and bacteriophages do use this strategy for beginning replication from one of the DNA strands. Initiation of chromosomal replication has turned out to be more complex, perhaps because the timing of replication may be more crucial for the chromosome.

Almost 20 years ago a search was made for bacterial mutants which were defective in DNA replication when the cells were incubated at moderately high temperatures. Several mutants were obtained in which DNA replication stopped slowly when the cells were shifted to the higher temperatures (nonpermissive for growth), and in these cells the amount of residual DNA synthesis corresponded to that expected if ongoing replication could continue but no new cycles could initiate. The genes in which these mutations were found were assigned the names *dnaA* and *dnaC*. Later it was found that mutations in *dnaB* and the genes encoding gyrase could also block initiation of replication. Thus genetic studies identified four proteins involved in the initiation process.

3.7.4.2 Anatomy of *oriC*: Sites for DnaA Binding and Base Modification

With the development of gene cloning technologies it became possible to remove the chromosomal origin of replication from the DNA of the enteric bacterium *Escherichia coli* and place it into a

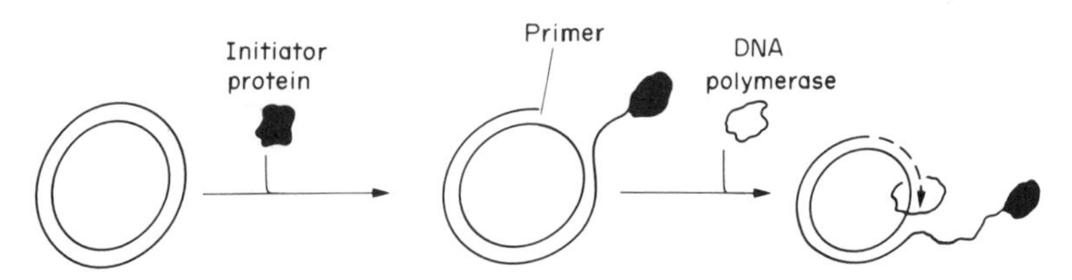

Figure 22 *Initiation by primer extension.* In some forms of viral and plasmid DNA replication an initiator protein binds to circular DNA at the origin of replication and nicks one strand. This leaves a primer suitable for extension by DNA polymerase

plasmid such that replication of the plasmid depends on the integrity of the chromosomal origin. The chromosomal origin is called *oriC*, and analyses of *oriC*-containing plasmids has led to the conclusion that the minimum size of an active *oriC* is about 250 nucleotides. Within this region, there are four binding sites for the DnaA protein (see Figure 23). The role of this protein is described in more detail below. The *oriC* region also contains about a dozen sites which are potential targets of an enzyme that adds a methyl group to adenine. If this region of the DNA is not methylated, the DNA does not serve as a origin for DNA replication. Methylation may be part of the timing mechanism—newly made DNA is unmethylated, and a new round of replication may not be able to begin until *oriC* becomes fully methylated (see Figure 24).

3.7.4.3 Replication *in vitro*: Sequential Action of Several Proteins

Replication of plasmid DNA molecules containing *oriC* has been achieved outside living cells using about a dozen purified proteins.[1] One of the earliest steps appears to be the binding of DnaA to *oriC* on a circular, supercoiled DNA substrate. The *oriC* region appears to be wrapped around the complex of DnaA subunits, and then three more proteins, DnaB, DnaC and HU, interact to form a slightly bigger complex. The DnaB protein is a helicase capable of separating the two DNA strands. DnaA and DnaC appear to be responsible for directing DnaB to *oriC* where the DnaB protein then begins to unwind the double helix. Once DnaB opens a small bubble in the DNA, the single-stranded regions created at *oriC* are coated with a small protein called Ssb. Ssb stabilizes the single-stranded conformation, preventing reformation of double-stranded regions. Since the substrate is a circular DNA molecule, further unwinding and separation of the strands encounters a topological barrier, and gyrase must be added to reaction mixtures to relieve the strain. Then the DnaB helicase is able to easily unwind most of the DNA. The stimulatory role of HU has not been defined; it probably assists in wrapping DNA around the DnaA protein complex and forming the correct structures for DnaB activity.

Once the double helix has been opened, an enzyme called primase, the product of the *dnaG* gene, binds to the DNA within *oriC* and synthesizes short runs of RNA on each template DNA strand. These short RNA pieces serve as primers for DNA polymerase III, the major replication enzyme; DNA polymerase III attaches deoxyribonucleotides to the end of the primers and uses the template strand to order the nucleotides. Leading and lagging strand synthesis occurs as described in Section 3.7.2.3. Removal of RNA primers and filling of gaps with deoxyribonucleotides is accomplished by another polymerase called DNA polymerase I. This second polymerase cannot join the DNA fragments, so still another enzyme must be called into play: DNA ligase seals all of the nicks so the final product is a covalently closed, circular, double-stranded DNA molecule. The two circular daughter molecules are interlinked, but if gyrase is present in the reaction mixture, it separates the interlinked circles. Gyrase also introduces superhelical tension to activate the DNA for another round of replication.

3.7.4.4 Initiation of Bacteriophage λ Replication: Variation on a Theme

The origin of replication of bacteriophage λ DNA is recognized by the product of the O gene of the phage; in a sense the O protein is an analogue of the host protein DnaA. Once O binds to the λ origin, the product of the phage P gene, along with the DnaB protein of the host, binds to the origin. The P protein appears to be an analogue of DnaC. DnaB, after the action of two more host proteins, DnaJ and DnaK, opens the double helix with its helicase action, and host DNA polymerases then replicate the phage DNA as they would any DNA. Thus the general strategy for initiation is similar for the virus and the host, but proteins of different specificity are used to recognize the appropriate origins. By controlling these specificity proteins one can selectively control the replication of DNA molecules—in general, specificity proteins should be good targets for antiviral agents.

3.7.5 MOBILE GENETIC ELEMENTS

According to classical genetics, the genomes of both prokaryotic and eukaryotic organisms are relatively static arrangements of nucleotide sequences. Maps positioning the genes relative to each other can be constructed, and these maps are remarkably similar among related organisms, even organisms that have been evolutionarily separated for a long period of time. But within the

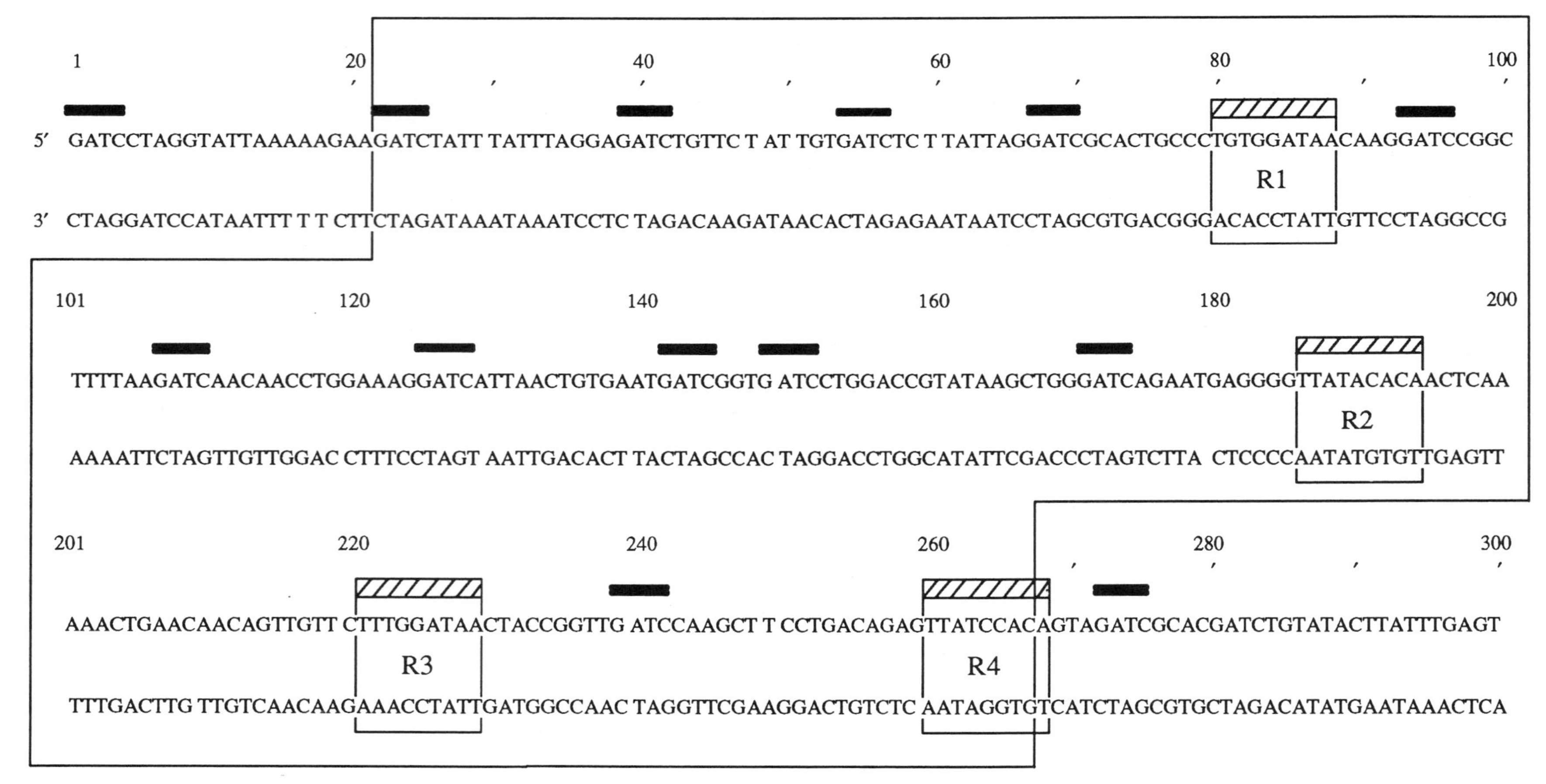

Figure 23 *Anatomy of* oriC. The nucleotide sequence of *oriC* from *E. coli* is shown. The large box encloses the minimal nucleotide sequence required for function. Solid bars denote GATC sequences which are potential methylation sites. Boxes labelled R1, R2, R3 and R4 are DnaA protein binding sites

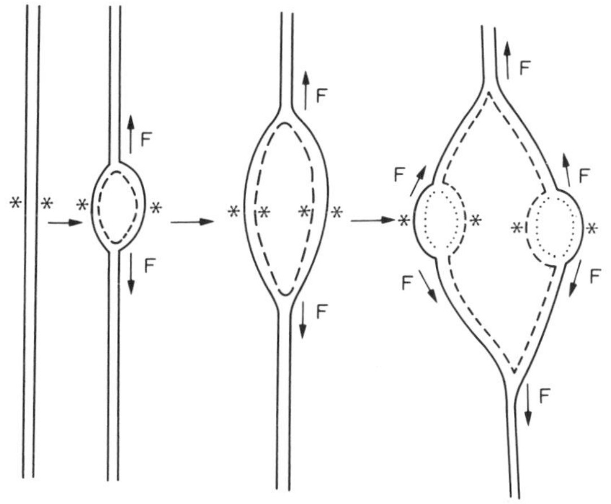

Figure 24 *Methylation as a timing device.* Sites of methylation are depicted by asterisks. The first round of replication produces two replication forks (F), and the newly made DNA (---) remains unmethylated for a short time. Once both strands have been methylated, a second round of replication can begin, this time producing four forks. According to this model the time required to methylate the new DNA dictates the spacing of the rounds of replication

background of fixed sequences exist short regions that can move. These mobile genetic elements are collectively called transposons, and the process of moving is called transposition. Each transposon contains genes required for its own relocation; transposition also utilizes the products of genes located outside the transposon.

Combinations of genetic manipulations and nucleotide sequence analyses have led to a detailed understanding of bacterial transposons. These genetic elements were initially discovered when they inactivated a gene through insertion of their nucleotide sequences into the gene. The resulting negative mutations could be reverted only by excision of the inserted DNA. Many transposons carry drug resistance genes, and an easy way to measure transposition is through the transfer of the drug resistance factors from one type of plasmid to another (see Figure 25). Both donor and recipient plasmids involved in transposition have been isolated, and examination of nucleotide sequence changes has led to several general statements. First, transposons are discrete genetic units that contain at least one gene encoding a protein required for transposition. Second, at each end transposons contain identical, or nearly identical, nucleotide sequences, usually in an inverted orientation (see Figure 26). Third, for many transposons, insertion into the target DNA can occur at a wide variety of nucleotide sequences. Fourth, a short nucleotide sequence (5 to 12 base pairs) in the target DNA is found directly repeated at each end of the transposable element following transposition. While the nucleotide sequence of the target varies from one transposition event to another, the length of the repeat is invariant for each particular transposable element. Figure 27 illustrates how these sequence duplications might arise. Fifth, transposition often (but not always) involves duplication of genetic information so that both donor DNA and recipient DNA contain a copy of the transposon following the transposition event.[7]

3.7.5.1 Insertion Sequences: Independent Transposable Units

Insertion sequences (IS elements) were the first transposons discovered in bacteria, and they appear to be the least complicated. IS elements tend to be about 1000 base pairs long, with each end terminating in a short (15 to 40 base pairs) inverted repeat sequence. The terminal repeats, both of which are required for transposition, are usually similar but not identical. The repeats are *cis*-acting. Operationally, *cis*-acting means that mutations in the region affect only the DNA molecule in which they occur. It is generally thought that the inverted repeats are binding sites for specific proteins involved in the transposition event. Although each IS element DNA contains several open reading frames (regions capable of encoding a protein), the gene products and their involvement in transposition are still poorly understood.

Examples have been found in which two IS elements are separated by a drug-resistance gene, and all three transpose as a unit called a composite transposon. The transposon called Tn10 serves as a good example of this class. The structural organization of Tn10 is outlined in Figure 28; in this case, the two IS elements, IS10L and IS10R, flank a gene encoding tetracycline resistance. As is common

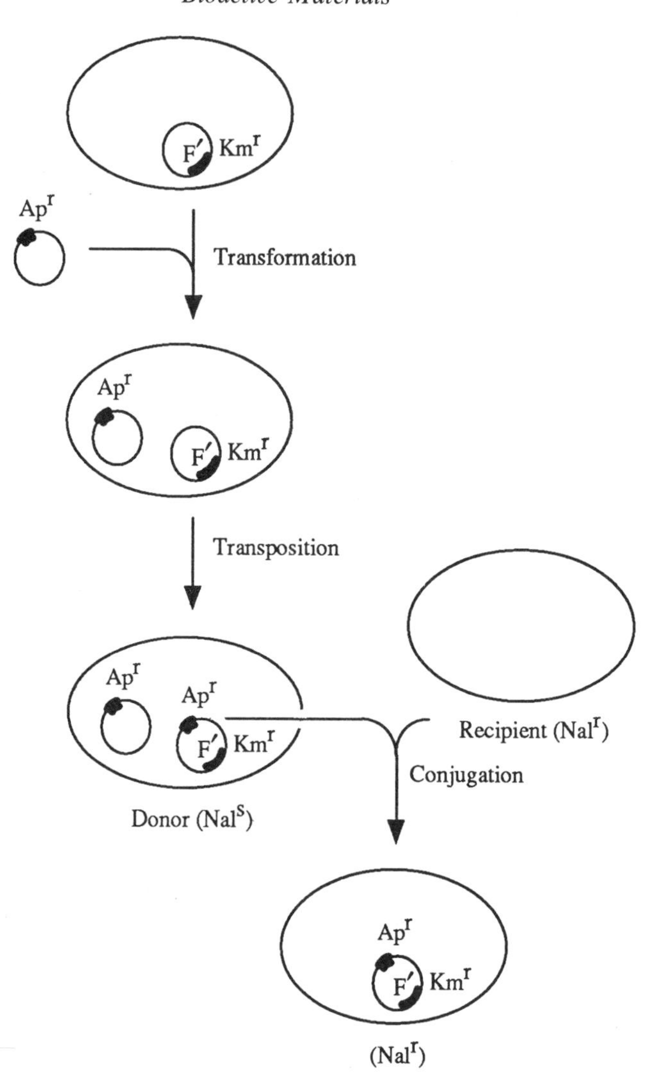

Figure 25 *Measurement of transposition.* (i) A small, nonconjugative plasmid carrying an ampicillin-resistance transposon is introduced by transformation into an F′-containing *E. coli* cell (F′ is a large plasmid capable of direct transfer from one *E. coli* cell to another by a process called conjugation). The F′ carries a gene for resistance to kanamycin. Transformants are selected by their ability to grow on ampicillin- and kanamycin-containing plates. (ii) Transposition occurs. (iii) The F′ is transferred by conjugation into a recipient cell which carries a chromosomal drug marker, in this case resistance to nalidixic acid (nal^r). The number of transposition events (movement of Ap^r into the F′) is proportional to the number of $Ap^rKm^rNal^r$ colonies recovered

with many composite transposons, only one of the IS elements, IS10R, can transpose by itself. This may simply reflect an evolutionary loss of activity from the second, and therefore, unneeded, copy. IS10R encodes a protein of 402 amino acids that is required for transposition of Tn10. Unlike transposition by Tn3 (see below), transposition by Tn10 appears to be a conservative process: transposition results in the excision of Tn10 from its donor site and insertion of that DNA into a target site in the absence of extensive replication of the element. Thus the element hops from one location to another.

As with other IS elements, transposition by Tn10 is tightly regulated (unregulated transposition would quickly lead to the accumulation of mutations in the host's chromosome). One regulatory feature is that the transposase from IS10 is not readily diffusable. Transposition is also regulated at the transcriptional level by the transposase promoter (for discussion of promoters see Section 3.7.6.2). This promoter is inactivated if it becomes methylated; consequently, transcription occurs only in the short time following DNA replication before the DNA becomes fully methylated (see Figure 24). Control is also exerted at the translational level. Small RNA molecules complementary to the 5′, untranslated region of the messenger RNA are thought to bind to the

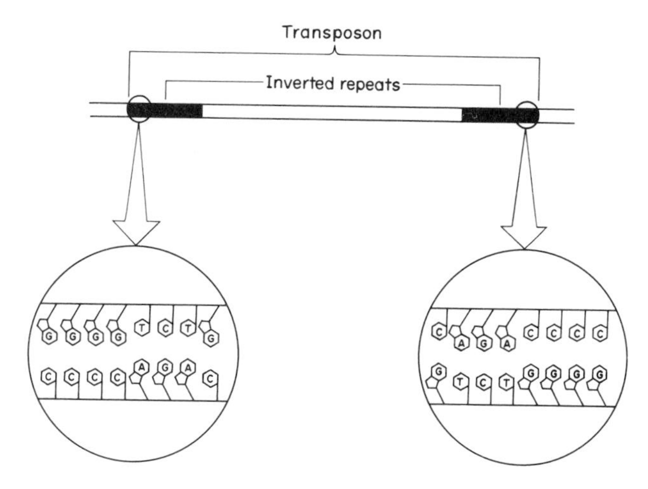

Figure 26 *General structure of a transposon.* Transposons contain repeated nucleotide sequences at each end. The repeated sequences are generally in an inverted orientation. Genes involved in transposon movement lie between or within the repeats (reproduced from ref. 9 by permission of John Wiley & Sons, Inc.)

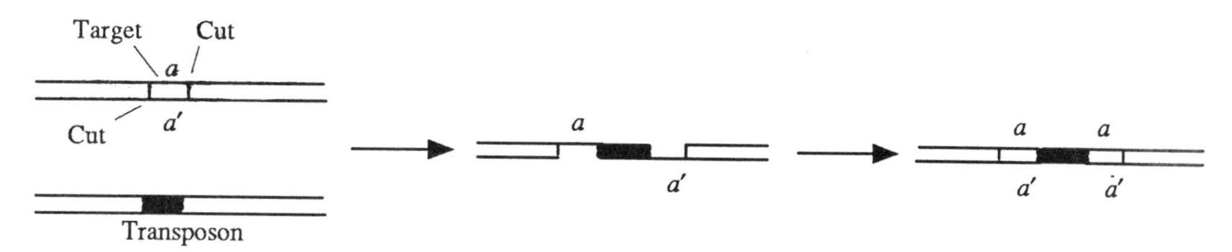

Figure 27 *Duplication of target sequences.* A short target nucleotide sequence (generally 4–12 base pairs long) has been labeled *a* and *a'*. Repair synthesis duplicates *a* and *a'* so the transposon is now bounded by a direct repeat

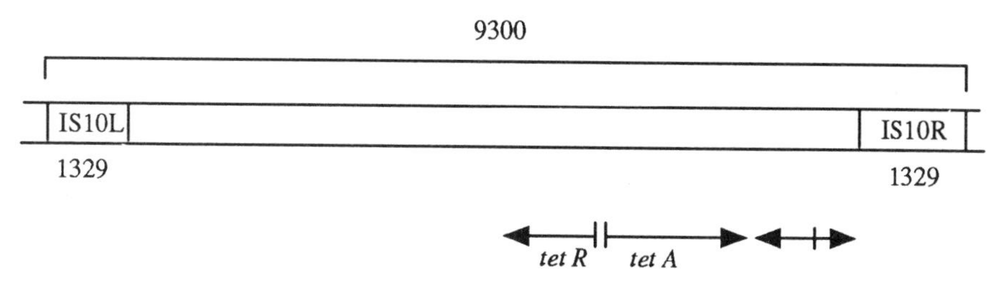

Figure 28 *Anatomy of Tn10.* Tn10 is composed of two IS sequences, IS10L and IS10R, which flank genes responsible for tetracycline resistance. Gene *A* is involved in the efflux of the drug from the cell and gene *R* encodes a repressor of gene *A*. Arrows indicate the direction of transcription of these, and two other genes of unknown function which reside between the IS sequences. The numbers indicate distances in nucleotide pairs (for details see ref. 30)

messenger and interfere with translation. The net result of these controls is that transposition by Tn10 is inefficient.

3.7.5.2 Tn3: Transposition Involving Cointegrates

The Tn3 family consists of large DNA elements (about 5000 base pairs long) which contain two separate gene products involved in the process of transposition.[13] In addition, the elements contain genes conferring drug resistance which add a selective advantage to cells that carry the transposon. The anatomy of Tn3, the prototype member of this class, is outlined in Figure 29. The *tnpA* gene encodes a 1021 amino acid protein that functions as a transposase. The *tnpR* gene encodes a 185 amino acid protein that has two functions. First, the *tnpR* gene product represses transcription of both *tnpA* and *tnpR*. Second, the *tnpR* gene product provides a resolvase function as described

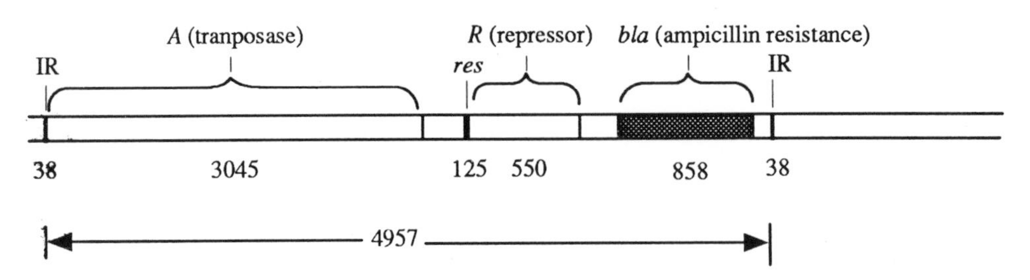

Figure 29 *Arrangement of genes in Tn3.* Tn3 contains three genes. *A*, *R* and *bla*, located between the 38 base pair inverted repeats (IR). The repressor protein binds to Tn3 DNA at region *res*. Numbers indicate nucleotide pairs in each region (reproduced from ref. 9 by permission of John Wiley & Sons, Inc.)

below. The *bla* gene encodes β-lactamase which confers resistance to ampicillin. The 38 base pair inverted repeats may be binding sites for proteins involved in transposition.

Transposition by members of the Tn3 family appears to be a two step process involving formation of a structure called a cointegrate (Figure 30). Since the cointegrate contains two copies of Tn3, replication of the Tn3 must occur during transposition. Resolution of this intermediate requires the *tnpR* gene product: mutations in the *tnpR* gene prevent separation of the cointegrate into two circles. To form the two circles TnpR catalyses a strand exchange reaction that involves the breaking and re-joining of DNA strands (see Figure 31).[14] This process is called recombination, and in this case the breaking and re-joining occurs at the site called *res*.

3.7.5.3 Retroviruses: A Viral Model for Transposition

Retroviruses are single-stranded RNA viruses that appear to violate the central dogma by converting their genetic information from RNA into a DNA form as a normal step in their life cycle.[2] These viruses infect animals, and in the process the DNA copy of the viral genetic information integrates (inserts) into animal cell chromosomes where it then serves as a permanent template for viral RNA and protein production. In a variety of animals viral DNA is passed from one generation to the next, and sometimes integration of viral DNA leads to malignant transformation of cells.[18] The viral integration process shares many properties with bacterial transposition, and it has become the paradigm for transposition in higher organisms.[24] The integration process has also received attention because it is required for establishment of a productive infection by these viruses, the agents that cause many types of leukemia in mammals and AIDS in humans.[6,35]

The viral genome, which is about 10 000 base pairs long, is organized into three 'genes' encoding polyproteins which are subsequently cleaved by proteases to form about half a dozen proteins. The *gag* gene is at the left side of the genome, and it gives rise to proteins that package the RNA in the core of the virion. The middle gene is called *pol*. It encodes a protease that processes polyproteins, a DNA polymerase called reverse transcriptase that synthesizes viral DNA from RNA, and a third protein required for integration. The gene at the right side is called *env*, and it encodes proteins that, along with components of the cell membrane, form the envelope of the virus particle.

During synthesis of viral DNA from viral RNA, long terminal repeats (LTR's) are generated (see Figure 32). The LTR's range in size from 250 to 1400 base pairs, depending upon the particular virus. These ends can be joined by a ligase to form a circle in which the two ends generate a pair of tandemly repeated LTR's. Each end of an LTR contains a short inverted repeat, and as with insertion elements in bacteria, mutations in these short inverted repeats block integration (transposition). During integration either four base pairs from the junction between the tandem repeats or two from each end of the linear DNA are lost (it is not known whether the intermediate is circular or linear). As with many types of bacterial transposition, the target sequence of the host is duplicated to form a direct repeat at either end of the integrated viral DNA. The number of nucleotides in this repeat is 4, 5 or 6, depending on the particular virus. Integration is seemingly random with respect to the target host sequences, and generally 1 to 10 copies of the viral DNA integrates into each infected host cell.

Searches for antiviral agents generally focus on processes that occur in the viral life cycle but are absent in normal host metabolism. The best characterized is the reverse transcription step. Another is the integration of viral DNA into the host chromosome. Both are required for productive infection to occur, and interruption of either blocks infection.

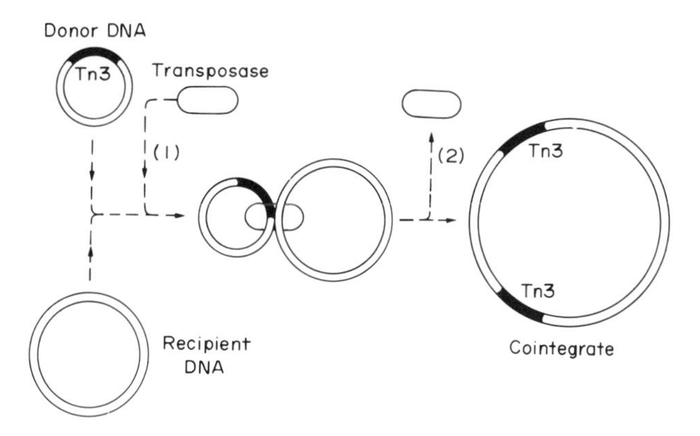

Figure 30 *Scheme for formation of cointegrates by Tn3.* (1) Transposase mediates the joining of a DNA molecule containing Tn3 (donor DNA) with a DNA lacking Tn3 (recipient DNA). (2) Tn3 is replicated (duplicated by DNA polymerase), and in the process the donor and recipient DNAs are joined to form a larger circle (cointegrate). The transposase leaves the DNA and is presumably free to initiate a new round of transposition (reproduced from ref. 9 by permission of John Wiley & Sons, Inc.)

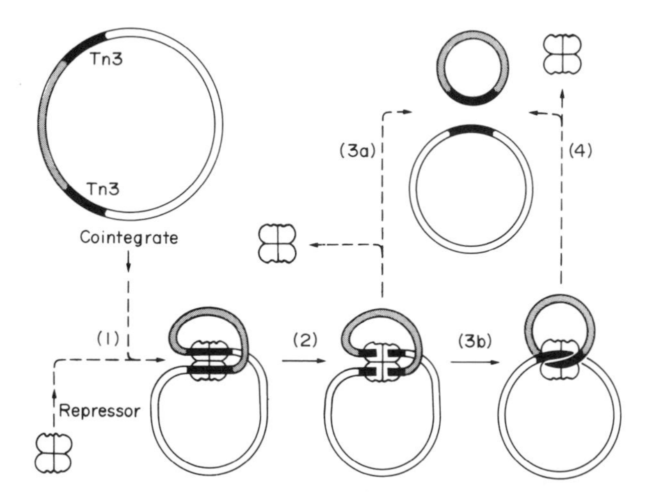

Figure 31 *Recombination between two Tn3 transposons in a cointegrate.* (1) The repressor protein (also called resolvase) binds to the cointegrate, probably twisting the DNA molecule so that the two copies of Tn3 (solid) align. (2) A break occurs within the *res* site (see Figure 29) of each Tn3. (3) The two broken ends of the DNA realign, and the breaks are sealed. This produces two rings, one of which is the donor (shaded) and the other is the recipient (open) as defined in Figure 30. Both contain a copy of Tn3. Recombination has two possible outcomes: two separate rings are produced (3a) or two interlocked rings arise (3b). In both cases the repressor dissociates from the rings (3a, 4). (4) A topoisomerase will separate the interlocked rings (reproduced from ref. 9 by permission of John Wiley & Sons, Inc.)

3.7.5.4 Ty and Copia: Other Eukaryotic Transposons

Eukaryotic organisms play host to other mobile genetic elements that are structurally related to retroviruses and bacterial transposons. The elements are small, have terminal repeats, are bounded by short, direct target sequence repeats, and are found at a variety of locations in DNA from different individuals of the same species. As in bacteria, these elements have been found through a combination of genetic studies and nucleotide sequencing analyses.

Yeast contain an element called Ty which resembles composite transposons of bacteria. Each Ty element is about 6300 base pairs long, and common laboratory strains of yeast carry about 30 copies in their genomes. At each end of Ty are 330 base pair direct repeats called δ. The δ element is also found alone (solo δ), occurring in about 100 different places in the yeast genome. A target sequence of 5 base pairs is found as a direct repeat at each end of a Ty element. Movement of Ty elements can be detected by selecting mutations arising from the insertion event. The Ty element is transcribed into two RNA molecules, both starting 95 base pairs inside the right boundary of the element. One transcript terminates after about 5000 base pairs while the other extends to within 40 base pairs of

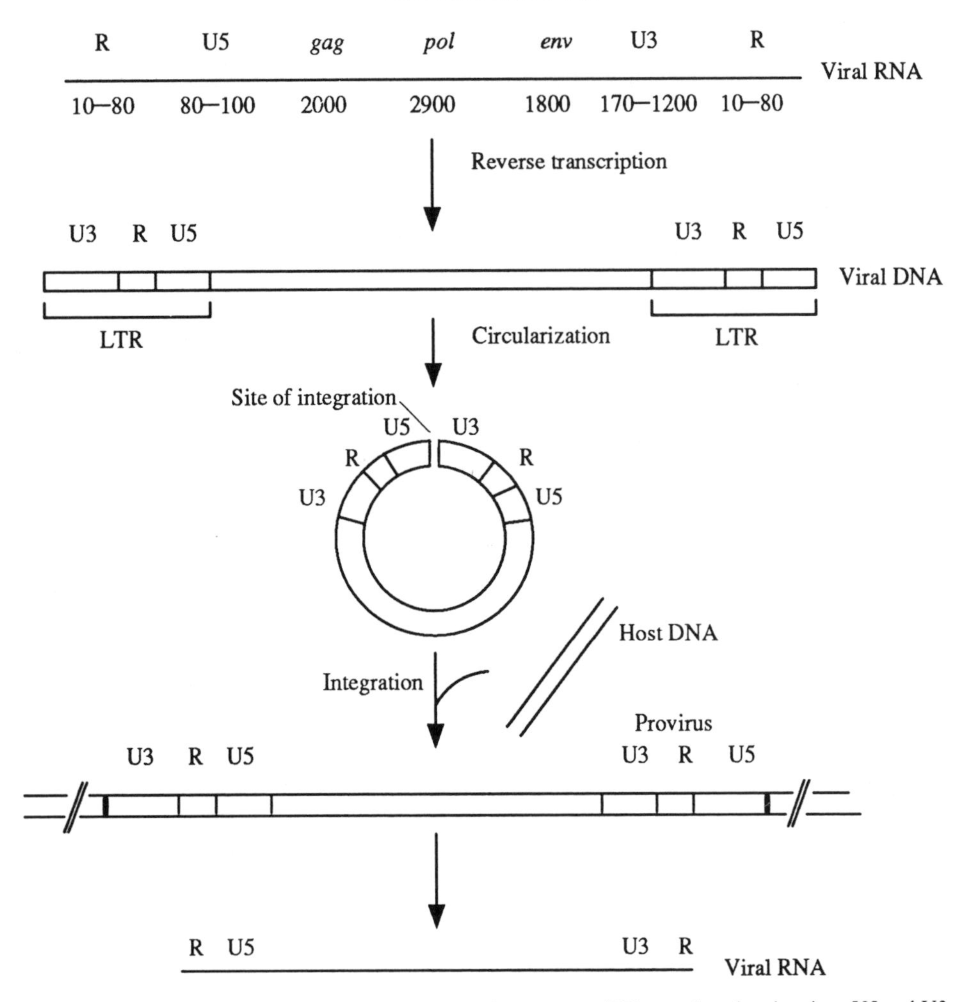

Figure 32 *Integration of a retrovirus.* Viral RNA contains direct repeats (RT) at each end and regions U5 and U3 unique to each end. Through the polymerizing action of reverse transcriptase viral RNA is converted into linear viral DNA. In the course of reverse transcription regions U3 and U5 are duplicated to generate long terminal repeats (LTRs). If the viral DNA circularizes, the two LTR's become tandem, adjacent repeats. The junction between them may be the site for integrative recombination. During transposition into the host chromosome four base pairs at the junction or two base pairs from each end are deleted from the viral DNA. Copies of viral RNA are transcribed from the provirus for assembly into new virus particles

the left end. This is similar to the transcription pattern found in some bacterial transposons. Transposition of Ty occurs through formation of an RNA intermediate, suggesting that retroviruses may serve as a good model for understanding Ty.

The fruit fly *Drosophila* contains several types of transposon, the best studied of which is a family called copia. Copia and copia-like elements are very abundant, and together they account for 1% of the genomic DNA. Copia is about 5000 base pairs long with identical direct repeats of 276 base pairs at each end, and like the LTR's of retroviruses, each repeat is bounded by related inverted repeats. But unlike δ elements in yeast and IS elements of bacteria, the terminal repeats of copia are not found by themselves in the genome. However, copia elements are occasionally found as free circular DNA molecules and as RNA molecules in virus-like particles. As with Ty, the retrovirus model appears to be a good one for understanding copia.

3.7.6 GENE CONTROL IN BACTERIA

As pointed out in our discussion of the central dogma, not all of the information stored in DNA is being used at any given time by cells. It is selective utilization that allows cells to adapt to new environments and to develop specialized functions as parts of multicellular organisms. Selective utilization of genetic information involves elaborate mechanisms whose details are only beginning to be uncovered. The best understood are those in bacteria. After presenting additional features of transcription, two examples of gene control are detailed to provide a sense of how degradative and biosynthetic pathways are controlled. We then briefly consider two additional examples: control by genomic rearrangement and control at promoter-distal sites.

3.7.6.1 RNA Polymerase: The Transcription Enzyme

RNA polymerase, purified from the common intestinal bacterium *E. coli*, has been studied for more than 20 years. It is a large enzyme composed of five different types of subunit that together bind to specific nucleotide sequences on DNA called promoters. After RNA polymerase binds at a promoter, it moves along one strand of the DNA and polymerizes ribonucleotides to form an RNA molecule having a nucleotide sequence complementary to that of the DNA strand. During transcription in bacteria RNA polymerase separates the two DNA strands over a stretch of about 17 base pairs, and this 'bubble', plus the enzyme, travel along the DNA. As messenger RNA is being made, it attaches to ribosomes and is translated into proteins.

The five polypeptide complex ($\alpha_2\beta\beta'\sigma$) of RNA polymerase is called the holoenzyme, and it is this complex that recognizes the specific promoter nucleotide sequences. Soon after RNA synthesis has initiated (six to nine nucleotides into the growing chain), the σ subunit is released from the enzyme complex. The remaining core complex ($\alpha_2\beta\beta'$) then continues the elongation phase of RNA synthesis. The σ subunit of RNA polymerase appears to be involved in specific promoter recognition. This concept is supported experimentally by studies with purified enzyme and nucleic acid templates. The core RNA polymerase without the σ subunit will initiate transcription on a DNA template, but initiation takes place at random sites. When the σ subunit is added, initiation of transcription now occurs at specific sites that had been previously identified as promoters. Thus it is the σ subunit that controls the specific place where RNA polymerase starts transcription. The specificity of the σ subunit for the promoter presents a powerful way for the cell to control gene expression. By synthesizing a new type of σ subunit that will recognize a different set of promoter sequences, a cell could switch transcription from one set of genes to another. Such an event appears to occur when the soil bacterium *Bacillus subtilis* switches from vegetative growth to sporulation. The process of sporulation is a series of timed stages beginning with the shutdown of vegetative growth and ending with the production of a dormant structure resistant to many environmental extremes. Many different stages of the sporulation cycle can be correlated with the appearance of different σ subunits.

3.7.6.2 Promoters: Recognition Sites for RNA Polymerase

As pointed out above, the segment of DNA that captures RNA polymerase and positions the enzyme to initiate transcription is called a promoter. A number of statements can be made about the anatomy of promoters found in *E. coli.*[21] First, upstream from genes are two highly conserved six base pair nucleotide sequences centered on positions -10 and -35 (position $+1$ represents the first nucleotide of the nascent RNA molecule; there is no position designated 0; see Figure 33). Mutations in these two DNA stretches alter the ability of the region to act as a promoter for RNA synthesis. Promoters are asymmetric, a feature required to give directionality to the RNA polymerase and to align the enzyme on the correct (sense) strand of DNA. The distance between the -10 and the -35 regions is quite uniform ($17+/-2$ base pairs between the most conserved positions of the -10 and -35 regions), and the polymerase covers a rather large region of DNA (about 75 base pairs from -55 to $+20$). Particularly tight contacts appear to be on the same side of the DNA in the -10 and -35 regions. The RNA polymerase unwinds the double helix between bases -9 and $+3$. Although these statements are generally true, it is important to stress that promoters from different genes are not identical. The precise structure probably plays an important role in determining how often RNA polymerase will bind to a particular promoter and thus in determining the frequency of transcription of a particular gene. Rates of initiation of transcription range from once per second for some ribosomal RNA genes to as low as once per cell generation for some repressor genes.

A popular hypothesis is that a promoter is a specifically oriented matrix of hydrogen bond donors and acceptors and that this matrix is recognized by a complementary matrix on the RNA polymerase. The hydrogen bonding pattern provides for the specificity of the interaction between polymerase and DNA. The stability of the polymerase–DNA interaction probably comes from two sources: (i) electrostatic attractions between DNA phosphates and basic residues in the protein; and

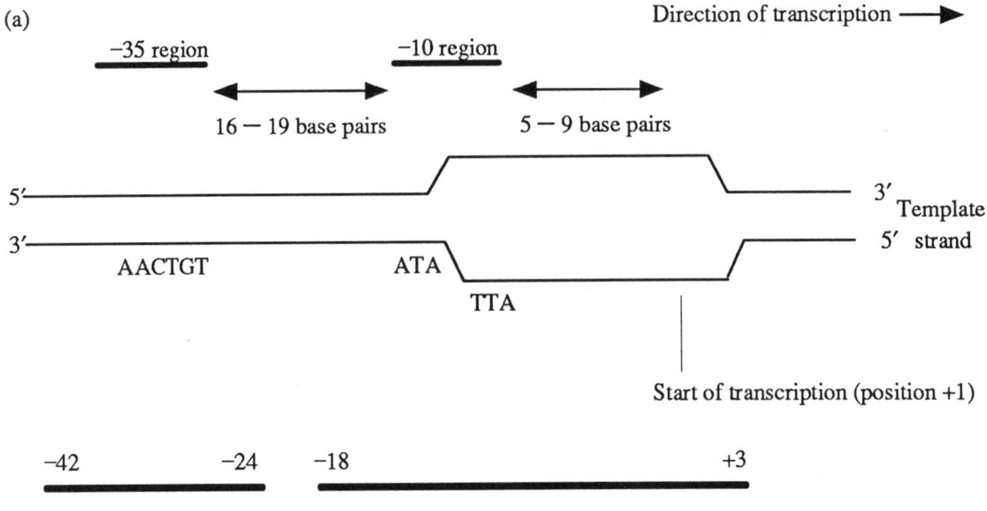

(b)

−35 region	16 − 19 base pairs	−10 region
$A_{85}A_{83}C_{81}T_{61}G_{69}T_{52}$	. . .	$A_{89}T_{89}A_{50}T_{65}T_{65}A_{100}$

Figure 33 *Anatomy of bacterial promoters.* (a) The nucleotide sequences in many bacterial promoters have been compared, revealing highly conserved sequences in two regions centered at −35 and −10. RNA polymerase opens the helix between +3 and −11. (b) A consensus sequence for *E. coli* promoters is shown in which the subscripts represent the percentage of the time that the indicated nucleotide is found at a particular position

(ii) hydrophobic interactions between hydrophobic residues in the protein and methyl groups of thymine protruding into the major groove of DNA. This hypothesis predicts that promoters for different genes will share nucleotide sequence homologies, and indeed many such homologies have been found. The homologies of more than 50 promoter sequences have been distilled into what is called a consensus sequence (see Figure 33).

We envision that initiation of transcription occurs in the following way. First, RNA polymerase holoenzyme binds nonspecifically to the chromosome. Such binding is diffusion limited, and the chromosome provides a large target. Then the polymerase slides along the double-stranded DNA. Studies with the *lac* repressor suggest that DNA binding proteins can slide along DNA at a rate of about a thousand base pairs per second *in vitro*. In the compact, ball-like chromosome the polymerase also may jump from one region of the DNA to another during this scanning stage. When the polymerase encounters a promoter, the appropriate matrix of hydrogen bond donors and acceptors, the enzyme forms what is called a closed complex. An isomerization then occurs in which the DNA double helix opens over a short region, and in the presence of nucleoside triphosphates, RNA synthesis begins.

Often there are nucleotide sequences adjacent to promoters which allow the binding of accessory proteins. These accessory proteins in turn modulate the binding of RNA polymerase. In one type of interaction a protein competes with RNA polymerase for binding to the promoter region. In another, a protein binds near a promoter and increases the probability of formation of a closed complex between RNA polymerase and promoter. Examples of both types of system are discussed in Section 3.7.6.5.

3.7.6.3 Termination: Stopping at the End of a Gene

Two mechanisms have been found which terminate synthesis of an RNA polymer.[16, 26] One type utilizes what is called an independent terminator. This is a nucleotide sequence which requires no additional proteins or factors to stop RNA synthesis catalyzed by purified RNA polymerase *in vitro*. Many terminators of this type have been identified, and analysis of nucleotide sequences has led to the formulation of a consensus sequence containing two parts (see Figure 34). The first is a short stretch of nucleotides having many G and C residues and diad symmetry capable of forming a stem–loop structure in the RNA transcript. The postulated stem–loop structures range from 7 to 10 base pairs and are centered 16 to 20 nucleotides upstream from the 3′ terminus of the RNA. The second part is a short (4 to 8 nucleotides) run of uridine residues in the transcript sequence. The transcript terminates in the oligo-U sequence or shortly beyond it. Both parts appear to be important for termination because the strength of a particular terminator can be increased by increasing the stability of the stem–loop structure or by increasing the length of consecutive uridines. The stem-loop structure appears to cause the RNA polymerase to pause at a point where the 5′ end of the RNA:DNA hybrid is composed largely rU:dA base pairs; these particular base pairs are weak and decrease the ability of the messenger RNA to remain base-paired with the template. Dissociation of the hybrid occurs and transcription stops.

The second type of termination requires the presence of a protein factor called ρ and is termed ρ-dependent termination. Rho is a 46 kDa protein which appears to bind to RNA as a hexamer, covering 72 to 84 nucleotides. No strong nucleotide sequence homologies have been found among the different ρ-dependent terminators, and it is currently thought that ρ binds to regions of RNA lacking secondary structure. When RNA polymerase pauses at particular sites, ρ, in a still undefined way, is able to cause a transcription complex to dissociate from the DNA. The location and strength of the pause site determines the precise 3′ terminus of the RNA.

3.7.6.4 Degradative and Biosynthetic Operons: General Control Strategies

Bacteria experience many fluctuations in their environment, and as a result they must be able to respond quickly to change. At the same time they must be efficient. A general rule has emerged from the study of bacterial systems: in teleogical terms, a gene is expressed at high levels when it is needed and at low levels (sometimes transcribed only one or two times per cell per generation) when it is not needed. The most efficient way to obey this rule is to control the production of messenger RNA, and several examples are described below.

Bacterial genes tend to be clustered on the chromosome so that those involved in the same pathway are adjacent. This makes possible coordinate regulation through production of a single messenger RNA made from all of the genes of the pathway, *i.e.* production of all the enzymes of a pathway can be controlled by a single event. Genes coordinately regulated in this way are said to comprise an operon.[22] Eukaryotic organisms seem to employ the same principle, but not at the level

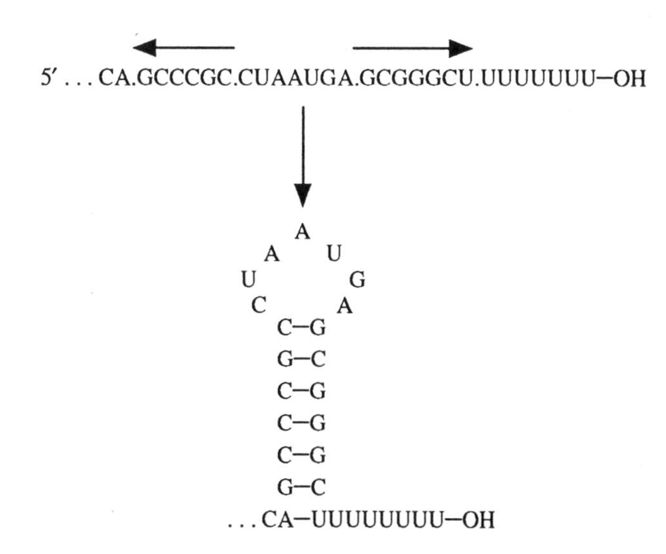

Figure 34 *Rho-independent terminator.* The nucleotide sequence at the 3′ end of an RNA is shown in which there is a diad symmetry (arrows) and a run of uridine residues. Such a structure could fold as indicated, weakening any base-pairing interaction the RNA might have with DNA during transcription. The many uridines might further destabilize RNA–DNA interactions, causing release of RNA from the template

of messenger RNA synthesis—in higher organisms messenger RNA tends to be made from only one gene at a time. Instead, a single gene sometimes encodes a polyprotein which is subsequently processed by proteases to form a number of smaller proteins.

Some bacterial genes are also regulated at the level of translation and/or at the level of messenger RNA stability. Although such mechanisms are less efficient than transcriptional control, they allow a cell to respond more rapidly to environmental factors. Important examples of these post-transcriptional control systems have been found with genes that encode proteins involved in making cells resistant to antibiotics.

Biochemical pathways can be grouped into two general types, degradative and biosynthetic. The degradative pathways tend to be regulated by the concentration of the substrate of an enzyme which acts early in the pathway: a high substrate concentration leads to high levels of degradative enzyme production. If regulation is of the negative type, an inhibitor usually blocks transcription from the set of genes involved in the pathway. The presence of the substrate of the first enzyme may then act as an anti-inhibitor, *i.e.* an inducer. Positive regulation occurs if the substrate of the first enzyme interacts with an activator to induce expression. Some degradative systems have both types of control. In biosynthetic pathways, high levels of the product of the pathway often inhibit transcription from the operon. Thus the product controls its own production. Two operons are described below. The lactose operon is a degradative system involving both positive and negative regulators; this is the classic paradigm for inducers and repressors. The tryptophan operon is a biosynthetic operon which has an additional fine tuning system called attenuation.

3.7.6.5 The Lactose Operon: A Degradative System

The lactose system is composed of four genes, three of which are adjacent and one is nearby (see Figure 35a). The products of the adjacent genes are involved in the transport, metabolic modification and degradation of the sugar lactose. The *lacA* gene encodes β-galactoside permease, a membrane protein which mediates the transport of lactose into the cell. The *lacY* gene encodes the β-galactoside transacetylase, an enzyme which transfers an acetyl group from acetyl-CoA to β-galactosides. Its biological function is unlcear, but it may be involved in detoxification of certain non-metabolizable analogues of β-galactosides. The *lacZ* gene encodes β-galactosidase, the enzyme which hydrolyses lactose. The nearby gene, *lacI* encodes a negative regulator called the *lac* repressor. The three genes *lacZ*, *lacY* and *lacA* comprise the *lac* operon and are transcribed into a single messenger RNA so that they are all expressed in unison. The *lacI* gene is transcribed into its own messenger RNA.

Genetic study of the *lac* operon led to the discovery of the transcriptional control mechanism called repression. In this system, a protein, the repressor, binds to a region of DNA near the start of the gene and blocks transcription by RNA polymerase. The region where the repressor binds is called the operator. In the *lac* system, cells growing in the absence of lactose do not 'need' the *lac* enzymes and thus their synthesis is repressed. In this state the repressor is bound to the operator, RNA polymerase cannot bind to the promoter, and the operon cannot be transcribed efficiently. In the presence of lactose, an inducer is produced which binds to the repressor and causes a conformational change in the repressor so it no longer binds to the operator. In this situation, RNA polymerase can initiate transcription that leads to production of the enzymes involved in lactose utilization.

The existence of most of the functional components of the *lac* operon was deduced solely by genetic analysis. The system was defined by obtaining mutations in the various genes. Structural gene mutations in *lacZ* or *lacY* simply result in cells unable to utilize lactose as a nutrient source. Regulatory mutations are more complicated. For example, mutations in *lacI*, the gene encoding the repressor, cause cells to produce high levels of messenger RNA from the *lac* operon whether the inducer is present or absent. When a wild type (normal) *lacI* gene is present on a plasmid in a cell in which the chromosomal gene is defective, the cell behaves as if it is wild type—the plasmid gene product is able to diffuse to the chromosome and act in the way a repressor normally acts. It is said to act in *trans*, *i.e.* on any DNA molecule in the cell regardless of the DNA responsible for its synthesis. In this situation the wild type repressor is dominant over the mutant repressor. Another class of mutation is located close to the beginning of the operon, just 5′ (upstream) from the *lacZ* gene. As with *lacI* mutations, these mutations cause the operon to be expressed independently of lactose concentration. However, a plasmid carrying the homologous region from a wild type cell has no effect on expression of the *lac* operon when placed in a cell containing this type of *lac* mutation. In this case a diffusible substance is unable to overcome the mutation. Such mutations are said to

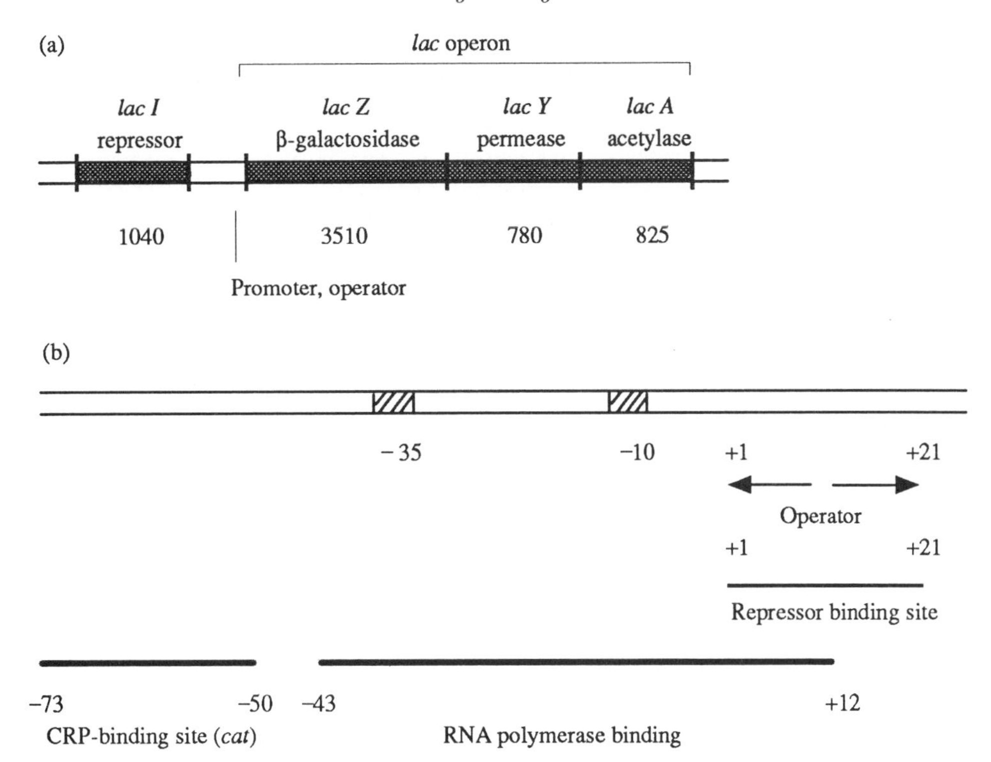

Figure 35 *The lactose operon.* (a) Three of the genes in the system (*lac Z*, *Y* and *A*) are organized into a single transcription unit controlled by a single promoter. Expression from the promoter is controlled by an upstream gene that encodes a repressor. Numbers indicate the length of each region in base pairs. (b) The control region of *lac* has a repressor binding site (the operator) which overlaps with the RNA polymerase binding site. Upstream is the *CRP* binding site (*cat*) where a positive regulator binds

operate in *cis*, and in this case they define the operator, the region of the DNA where the repressor binds. A third class of mutant cannot be induced by lactose. Some members of this class reside in the promoter region, making RNA polymerase unable to bind; others are in *lacI*, and they result in a mutant repressor which is unable to bind to lactose and thus always remains bound to the operator, blocking messenger RNA synthesis.

Natural induction occurs by a convoluted pathway. Normally a small amount of permease is made from *lacA*, allowing some lactose to enter the cell. Enough β-galactosidase is made constitutively from *lacZ* to break down lactose into glucose and galactose. Galactose is then transferred to acceptor molecules (glucose) which, through a rearrangement, is converted to allolactose, a very strong inducer of the *lac* operon.

The lactose system is also regulated in a positive sense by cyclic AMP and cyclic AMP regulatory protein (CRP). cAMP and CRP always act together in wild type cells. CRP is a dimeric protein in which each subunit binds one molecule of cAMP. In the presence of cAMP, CRP binds to DNA at a site called *cat*, which is about 25 nucleotides long (see Figure 35b). Mutations have been found in the gene encoding CRP, and some of these mutations allow CRP to act as a positive regulator without cAMP. Thus cAMP may be an allosteric effector; the mutant protein may be in the active state even without cAMP. In the lactose operon CRP seems to cover and bridge an impotent promoter, increasing the probability that RNA polymerase will bind to the active promoter. It is important to note that cAMP–CRP complexes do not act the same way in all operons. For example, the complex has two different effects in the galactose operon, another operon involved in sugar utilization. This operon is transcripted from two nearby promoters, P1 and P2. Near *gal* P1, the cAMP–CRP complex seems to interact with the RNA polymerase to stimulate expression from the operon. In *gal* P2, cAMP–CRP may compete with RNA polymerase for the promoter, lowering initiation of transcription from this promoter.

3.7.6.6 The Tryptophan Operon: A Biosynthetic System

The *trp* operon is one of the best-studied biosynthetic operons. It is composed of a cluster of genes encoding enzymes that constitute a pathway for making the amino acid tryptophan. There are five genes in the cluster, *trpE*, *D*, *C*, *B* and *A*, arranged adjacent to one another in the order in which they participate in the biosynthetic pathway. These five genes are transcribed into a single messenger RNA beginning with *trpE* and ending with *trpA*. As with the lactose operon, the tryptophan system is controlled by a repressor. The promoter and operator for the operon are upstream from the first gene, *trpE*. A gene called *trpR*, located far away on the bacterial chromosome, encodes a repressor which binds to the *trp* operator and blocks transcription. But unlike the *lac* repressor, the *trp* repressor does not block transcription by itself; it must be bound to tryptophan. Thus tryptophan, the product of the pathway, participates in inhibiting transcription from the genes responsible for making it.

The repressor–operator system in the *trp* operon does not act as a total on–off switch. Instead it lowers the system from fully on to an intermediate level of expression by only partially preventing messenger RNA synthesis from the operon. The finding that most of these RNA molecules terminate before reaching the first structural gene (*trpE*) led to the discovery of attenuation, a second control system.

An attenuator is a transcription terminator (see Section 3.7.6.3), which in the case of many biosynthetic operons, is located at the end of a 100–200 nucleotide leader RNA immediately preceding the first structural gene of the operon (*trpE* in the *trp* operon). In the attenuator region there are two different stem–loop structures that can form in the RNA; the important element in this regulatory system is that the two stem–loops are mutually exclusive. In one RNA conformation transcription termination occurs; in the other transcription continues. The position of ribosomes on the translated leader RNA sequence determines which RNA stem–loop structure forms. The leader RNA contains tandem tryptophan codons located where the presence of a ribosome would inhibit formation of the terminator stem–loop structure. When the ribosome reaches this position, its progress depends upon the concentration of charged tryptophanyl transfer RNA (transfer RNA to which tryptophan is covalently bound). The concentrations are high when the tryptophan concentration is high, and the ribosome proceeds through the region. Then the terminator stem–loop structure forms, and RNA synthesis stops. At low tryptophan concentrations, the ribosome stalls at the two tryptophan codons because there is little charged tryptophanyl transfer RNA in the cell. Then the alternate stem–loop structure forms, allowing RNA synthesis to continue through the operon. Deletion of the attenuator region increases tryptophan synthesis by a factor of about six. Attenuators are found in many biosynthetic operons, and in each case they are factor-independent terminators of transcription similar to that shown in Figure 33.

3.7.6.7 Genomic Rearrangement: Control by Inversion of a DNA Segment

In *Salmonella typhimurium*, a bacterium related to *E. coli*, a case has been found in which a gene changes its physical orientation as a way of controlling its expression. The *Salmonella* genome contains two genes (H1 and H2) which each encode a protein that is a component of the flagella (whip-like appendages responsible for motility). Only one of these genes is expressed at any given time; thus there are two types of flagella, the H1 type and the H2 type. The H2 gene is in an operon containing a gene for a repressor that acts only on the gene encoding the other flagellar gene, H1 (see Figure 36). Thus when H2 is being expressed, so is the repressor that prevents transcription from H1. When H2 and repressor are not expressed, then H1 is. The promoter for the H2 operon lies within a 993 base pair invertable segment of DNA immediately upstream from the H2 gene. In one orientation the promoter functions for H2 and repressor expression. In the other, it does not. About once every 1000 bacterial generations, the invertable segment switches orientation. The inversion either places the promoter for H2 and repressor next to the H2 gene, pointing into the gene, or places the promoter so it points away from the H2 gene, preventing expression of H2 and the repressor. Thus, a system has evolved in *Salmonella* in which a surface protein regularly changes, and presumably this increases the ability of this bacterium to survive the immune response of mammalian hosts.

The invertable segment resembles transposable elements in that it has short inverted repeat sequences at each end. Moreover, within the invertable segment is a gene called *hin*, whose protein product facilitates the inversion of the segment (mutations in the *hin* gene reduce inversion by a factor of 10 000). It is thought that the inversion occurs by a site-specific recombination event much like the integration of bacteriophage or resolution of transposon Tn3 cointegrates. Similar examples of genomic rearrangements have been found in higher organisms, and three of them are described in Sections 3.7.7.5, 3.7.7.6 and 3.7.7.7.

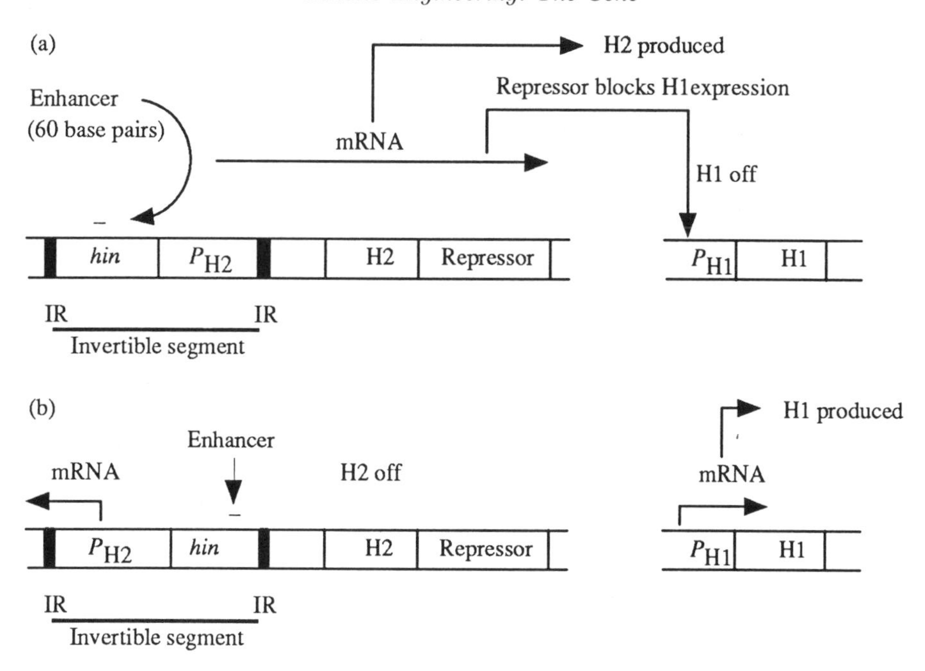

Figure 36 *Control of gene expression by inversion.* (a) In one orientation, mRNA originating at P_{H2} leads to production of H2 protein and the repressor of H1. Located within the *hin* gene is a 60 base pair enhancer which, if deleted, drastically lowers inversion. Regions labeled IR are 26 base pair inverted repeats. (b) In the other orientation expression from P_{H2} is directed away from H2, so that protein is not made. Instead, H1 is synthesized

3.7.6.8 Antitermination and Retroregulation: Control at Promoter-distal Sites

The gene product of the N gene of bacteriophage λ stimulates antitermination at both independent and ρ-dependent terminators (Section 3.7.6.3) in the phage DNA. Its action requires the presence of two nucleotide sequence elements. One element has a diad symmetry consisting predominantly of GC residues which could presumably form a stem–loop structure. The second is a sequence called box A which is 8–14 base pairs upstream from the diad. Mutational analyses have implicated a number of host genes in the antitermination process. Precisely how antitermination works is not yet understood. One idea is that the N protein binds at Box A-*nut* sites, and when RNA polymerase transcribes through that region, N modifies the polymerase so that specific terminators, sometimes hundreds of nucleotides farther downstream, are bypassed.

Retroregulation is a term used to describe selective degradation of a specific region of messenger RNA. The best-studied example is found in the bacteriophage λ *int* gene whose transcription is normally halted at an independent terminator. A second transcript of this gene has been found that bypasses this termination signal (antitermination). The longer RNA folds back to form a specific secondary structure that creates a cleavage site for RNAase III. RNA cleaved by the nuclease lacks the hairpin structure associated with the normal termination site, and this message is rapidly degraded following the first cleavage by RNAase III. In the λ life cycle, antitermination and the *int* gene product are not present at the same time, and this mechanism assures that to be the case. But a more general principle emerges—the hairpin structure created by the normal termination site at the 3′ end of the messenger RNA may be crucial for messenger stability.

3.7.7 GENE CONTROL IN HIGHER ORGANISMS

Although bacterial gene regulation serves as a model for thinking about gene control in higher organisms, one might expect differences in regulatory circuits because prokaryotic and eukaryotic cells experience such different environmental situations. In prokaryotic systems, the cells are free living and are repeatedly exposed to environmental changes. Their regulatory systems seem to be designed to provide maximum growth rate for a particular environment. Maximizing energy utilization appears to be very important, even to the extent that the same gene can code for two different proteins. Eukaryotic cells, except those that are free-living such as yeasts, protozoa and algae, face a different set of problems. As an organism develops into a multicellular form, cells differentiate—they change their morphology and biochemistry. By and large, eukaryotic cells are not subjected to such environmental extremes as prokaryotic ones. Also, in the adult organism, cell division often halts, and a cell need only maintain itself rather than grow.

At the molecular level there are differences between eukaryotic and prokaryotic cells that influence genetic information storage and utilization. One of these differences is the amount of DNA. Eukaryotic cells often contain a thousand times more DNA than bacterial cells. Eukaryotic DNA also appears to be packaged differently. DNA in eukaryotes is organized into a structure called a nucleus which is bounded by a membrane, and the DNA is wrapped around histones to form nucleosomes. In bacteria no membrane separates the DNA from the rest of the cell. Whether a nucleosome-like packaging of DNA occurs in bacteria is not known. Another difference is that eukaryotes contain a few base sequences that are highly repeated; the prokaryotic genome is largely a unique nucleotide sequence. Moreover, much of a eukaryotic genome is never translated, and genes and gene products are not colinear in eukaryotes as they are in bacteria. Some of these differences are described in more detail below to provide an appreciation for the complexities involved.[20]

3.7.7.1 Eukaryotic Transcription: Three Types of RNA Polymerase

Our inability to easily obtain mutations in higher organisms has made it difficult to study functional aspects of transcription. We expect the principles to be similar to those found in bacteria but with added complexity. One example is that eukaryotic nuclei contain three different RNA synthesizing systems. Each has its own type of RNA polymerase, its own type of promoter, and its own type of RNA processing strategy. The three are generally categorized according to their polymerase type. Polymerase I accounts for 50–70% of the total cellular transcription, but it synthesizes only a single type of product, the ribosomal RNA transcript. Polymerase II is responsible for transcription of genes that encode proteins. This enzyme is distinguished from polymerase I by its sensitivity to α-amanitin. Polymerase III is a minor activity involved in the transcription of small RNAs, 5S ribosomal RNA, and transfer RNAs. These transcripts are usually short (less than 300 nucleotides), and they tend to play roles in protein synthesis.

3.7.7.2 Eukaryotic Promoters: Three Different Motifs

As in bacteria, eukaryotic cells have regions in their DNA that have been identified as promoters. Since polymerase I is responsible for transcription only of ribosomal RNA genes, it was first thought that consensus sequences for promoters would be found easily through nucleotide sequence comparisons among related organisms. Obvious promoter sequences did not emerge. Mutational studies do establish, however, that promoters exist between nucleotides −45 and +10 relative to the start of transcription. It appears that many transcription factors (proteins) are involved in controlling ribosomal RNA synthesis. Evolutionary divergence of these factors, along with co-evolution of their DNA targets (rRNA promoters), could explain the lack of promoter similarity from one organism to another.

Polymerase II is responsible for transcription of very diverse genes, and it must respond to a variety of control signals. Thus, we expect polymerase II promoters to exhibit a variety of configurations. In the many cases examined, the control signals are located upstream from the beginning of a gene (see Figure 37). At about −25 there is often a seven base pair sequence called the TATA or Hogness box, and it resembles the −10 region found in bacterial promoters. The TATA box is not absolutely essential for transcription, but the accuracy of initiation decreases if the TATA box is removed. Removal of the TATA box has little effect on the frequency of initiation. Farther upstream (to about −100) are other conserved sequences that are called upstream promoter elements. They are probably involved in the frequency of transcription initiation. Often several blocks of conserved sequences are present, and it is likely that they are binding sites for specific regulatory proteins.

A very different situation is observed with promoters for polymerase III. They occur within the gene. In the case of the 5S ribosomal RNA gene, the promoter lies between +55 and +80. Deletion into this region markedly reduces the frequency of transcription. This region of DNA binds to a 5S-specific transcription factor, presumably needed to properly position polymerase III. In the case

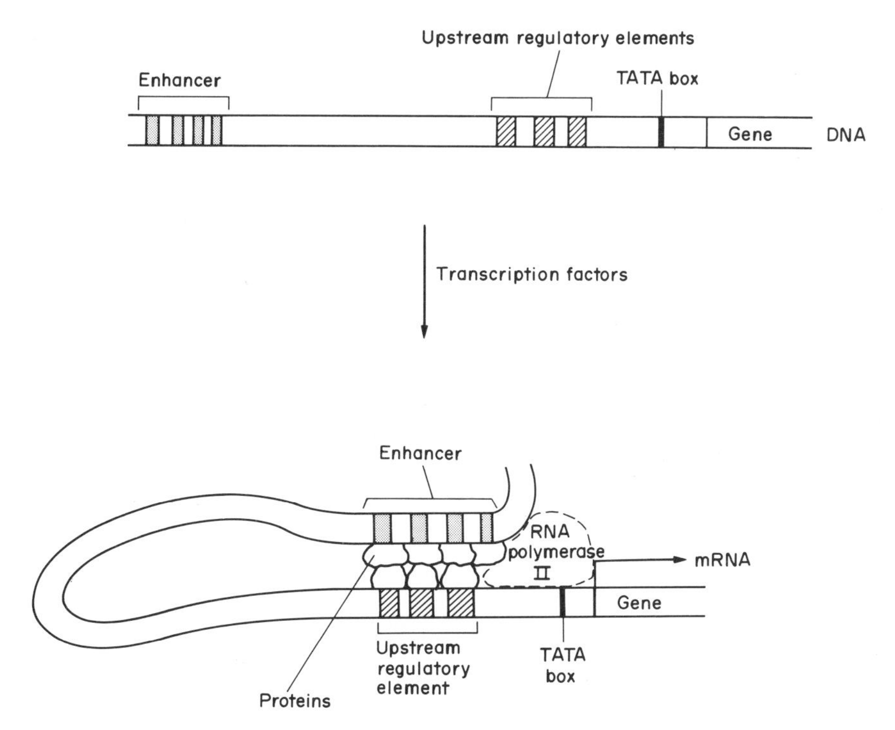

Figure 37 *Model for enhancer action.* Tissue-specific proteins (transcription factors) are postulated to bind specifically to enhancer elements and to upstream promoter sequences. These sets of proteins may interact to fold the DNA into a local nucleoprotein structure that serves as an attractive site for RNA polymerase II binding

of the methionyl transfer RNA gene, the promoter is split in two, extending from +8 to +19 and from +52 to +62. Both sections must be intact for initiation of transcription. Very different DNA sequences can be inserted between the promoter regions with little effect.

3.7.7.3 Enhancers: Nucleotide Sequences that Alter Transcription Frequency from a Distance

Enhancers are modular elements, sometimes including several hundred base pairs, that influence transcription frequency by acting at a distance, often up to 5000 base pairs away from the promoter.[27] Enhancers function in either orientation, and they are position-independent, *i.e.* they work on the 5′ or 3′ side of genes or even within genes. It now appears that enhancers are binding sites for specific proteins, and that may explain why particular enhancers are specific for certain cell types. One attractive idea, which explains why enhancers exhibit different activities with different promoters, is that they physically interact with upstream promoter elements[15] (see Figure 37). The folding shown in Figure 37 might generate a preferred site for polymerase II binding.

Enhancers have been found in many viral systems, and when viral enhancers are present in host DNA, they can elevate transcription from some host genes. An important observation is that enhancers are often found in the LTR regions of retroviruses (see Figure 32). LTR regions are known to stimulate transcription from adjacent host genes, and since there is such a plethora of integrated retroviruses, this may be an important natural regulatory factor. It may also be important in the development of certain types of cancer since integration of the virus could markedly affect expression of adjacent host genes. Viral genes must compete with those of the host; thus, it is likely that viral enhancers are more powerful than host ones. An interesting possibility is that highly differentiated tissues may have enhancers associated with genes that are highly expressed in that tissue. Viral enhancers are generally found in regions of DNA hypersensitive to DNAse, so they may affect chromatin structure (see Section 3.7.3.10).

3.7.7.4 Interrupted Genes and RNA Splicing: Multiple Usage of Information

A number of eukaryotic genes are split, *i.e.* they contain internal nucleotide sequences that do not appear in the mature messenger RNA. Following transcription, the extra information, the intron (for intervening sequence), is removed, and the remaining fragments of the exons (for expressed sequence) are spliced together to form mature messenger RNA. The introns may be very long: a case has been found in which the gene is 31 000 base pairs long but the mature messenger RNA, after splicing, is only 1600 nucleotides long.

All interrupted genes share several common features. First, the order of the information in an interrupted gene is the same in the DNA as it is in the mature messenger RNA (see Figure 38). Second, introns (regions removed by splicing) of nuclear genes generally have termination codons in all three reading frames so they cannot be translated. Third, the presence of an intervening sequence is an invariant feature of a gene; it is present in the gene in germ line and in somatic tissue, and its presence is not related to whether the gene is expressed. Fourth, all classes of eukaryotic genes have been found to contain examples of interrupted genes (but not interrupted in every species): nuclear genes encoding proteins, nuclear genes encoding ribosomal RNA and transfer RNA, and mitochondrial and chloroplast genes. However, intervening sequences are rare in bacterial genes, having been found only in bacteriophage T4.[29]

The rules governing RNA splicing are not well understood. In the cases that have been extensively studied, there is a preference for removing certain introns before others, but this preference is not absolute. Splicing does not begin at one end of the gene and work toward the other. Examination of nucleotide sequences near the splice junctions has revealed several interesting features. First, there are distinct left and right junctions which share no complementarity, thus base pairing between the junctions probably does not play a role in splicing. Second, junctions from a large number of organisms (yeast to mammals) have very similar sequences. A consensus sequence can be written for nuclear genes of eukaryotes (the consensus does not extend to yeast transfer RNA, mitochondrial or chloroplast genes, suggesting that they may employ different splicing mechanisms). Third, any left junction appears capable of being spliced to any right junction. Host genes that are placed in viral DNAs by recombinant DNA technologies form RNAs that are spliced at the normal places even though the precursor RNA is largely of viral origin. Moreover, tissue-specific factors are not required for proper splicing, since the splicing described above can take place in a tissue where the gene is normally not expressed.

The splicing process is mandatory for translation of split genes, and several human diseases have been traced to nucleotide sequence alterations within the consensus sequences involved in splicing. In an α-thalassemia (a defect involving the blood protein hemoglobin), messenger RNA production from a globin gene is blocked by deletion of a 5′ nucleotide sequence from the left junction of intron # 1. In a β-thalassemia, there is no β messenger RNA, and the DNA shows a point mutation at the left junction of intron # 2. In still another case, a mutation inside an intron creates a right splice junction in the wrong place, causing splicing to occur about 20 base pairs upstream from the normal site. Thus the consensus sequences and proper splicing appear to be essential for proper expression of genes containing introns.

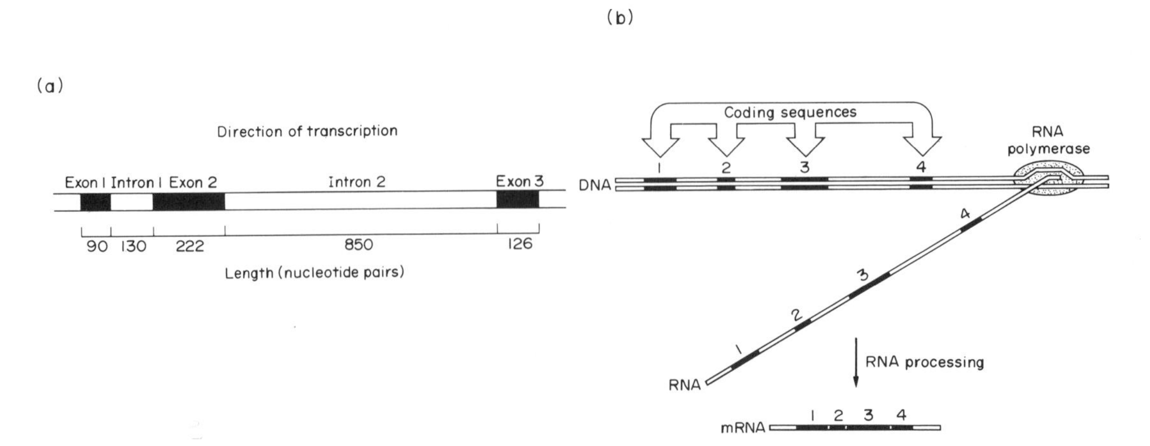

Figure 38 *RNA splicing.* (a) Intervening sequences in the β-globin gene of humans. The exons code for the amino acids in the protein. (b) Arrangement of sequences in a gene from a higher organism. The regions of DNA coding for amino acids in the protein product are interspersed with noncoding regions, which are processed out before mature messenger RNA is formed. Coding sequences 1 through 4 are part of a single gene. Note that in higher organisms ribosomes do not bind to messenger RNA before it is released from the DNA (reproduced from ref. 9 by permission of John Wiley & Sons, Inc.)

Biochemical examination of splicing reactions has been carried out in several *in vitro* systems. One of the more interesting findings has emerged from examination of splicing of the ribosomal RNA precursor from the protozoan *Tetrahymena*.[4,5] Addition of GTP, GDP, GMP or guanosine to purified RNA leads to RNA cleavage, circularization of the single intron of the gene and splicing. No proteins are required for the reaction to occur—this RNA splices itself!

Nuclei contain small RNA molecules having properties which suggest that they may participate in splicing. These highly abundant molecules (10^5 to 10^6 copies per cell) are called snurps. They are 100–300 nucleotides long and are found bound to specific proteins. A particular snurp called U1 has at its 5′ terminus a nucleotide sequence which is complementary to the 5′ consensus splice junctions. Antibodies against U1 block splicing but not transcription when added to nuclei from adenovirus-infected cells. Moreover, digesting away the 5′ end of U1 prevents splicing *in vitro*. Thus it appears that a snurp is an important component in splicing. Precisely how it functions is not known.

A general rule defining the function of interrupted genes and RNA splicing has not yet emerged, but it is likely that differential splicing allows cells to utilize the same genetic information in several different ways. One example can be seen with immunoglobulin genes. These genes differentially splice at their 3′ end to vary antibody localization in the body.

It is instructive to trace the progress of RNA once it has been transcribed. Soon after polymerization begins, the 5′ end of the messenger is enzymatically modified. This modification (Figure 39) is called a cap[32] and probably protects the RNA from degradation by 5′ exonucleases. It may also serve as a binding site for proteins that help properly position the messenger on ribosomes for translation. The primary transcript containing introns (formerly called heterogeneous nuclear RNA or hnRNA) is often four to five times longer than the average messenger RNA found in the cytoplasm. Shortly after the transcript is synthesized, the 3′ end is polyadenylated by the enzyme poly(A) polymerase using ATP as the substrate. At this point the 5′ and 3′ ends of the RNA are clearly defined. Splicing then occurs, presumably under the influence of tissue-specific proteins and snurps. Mature messenger RNA passes from the nucleus to the cytoplasm packaged as ribonucleoprotein particles. Since polyadenylation can be specifically blocked by cordycepin, its relationship to splicing and RNA transport can be assessed. Polyadenylation is not required for splicing, but it is required for RNA to exit from the nucleus. The introns that are removed from the primary transcript are rapidly degraded, perhaps because they lack a 5′ cap. Since so much of the primary transcript is intron material, it is not surprising that most of the RNA made in the nucleus never reaches the cytoplasm. However, there seems to be an additional control circuit during embryonic development. In the sea urchin, many genes are transcribed, but only a small fraction appear as mature messenger RNA. As development progresses, the fraction of genes reaching the cytoplasm changes. Thus it appears that one type of gene control is operating at the level of RNA processing. Our ignorance of these processes is emphasized by the sense that they must be energetically very costly, and it is not obvious how the benefits outweigh the costs.

Before closing this section it is important to point out that not all eukaryotic genes undergo splicing nor do all of them become polyadenylated. It will be interesting to discover how these genes differ from those that do undergo the elaborate processing described above; perhaps then we can begin to understand how splicing originated.[31]

3.7.7.5 Yeast Mating Type: Control by Genomic Rearrangement

Earlier we pointed out that DNA sequences can invert in bacterial cells to regulate gene expression (Section 3.7.6.7). Similar processes occur in eukaryotic cells. In one type, a gene may be moved to a new location to elicit transcription from the gene. The mating type locus of yeast is an example of this type of control.[23]

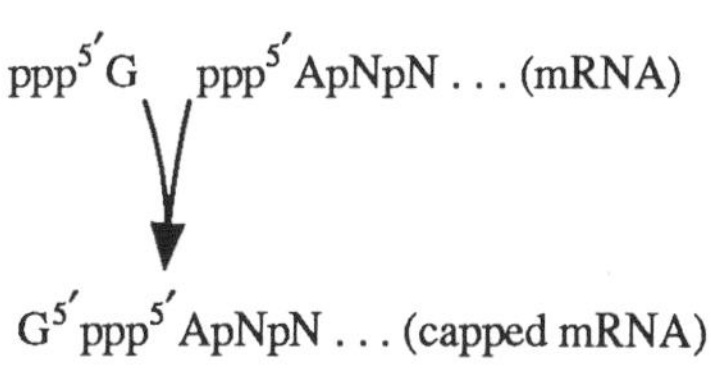

$$\mathrm{ppp}^{5'}\mathrm{G} \quad \mathrm{ppp}^{5'}\mathrm{ApNpN}\ldots\ (\mathrm{mRNA})$$

$$\downarrow$$

$$\mathrm{G}^{5'}\mathrm{ppp}^{5'}\mathrm{ApNpN}\ldots\ (\text{capped mRNA})$$

Figure 39 *Capping of messenger RNA*. After synthesis by RNA polymerase, many eukaryotic messenger RNA molecules react with guanosine triphosphate to form an unusual structure called a cap. The caps are then decorated with methyl groups at several locations not indicated in the figure. The letter *N* indicates any nucleotide (A, T, G or C) and the letter *p* indicates a phosphate moiety

The common bakers' yeast exists in two mating types called *a* and α. An unusual feature of this yeast is that it can switch between the two mating types, often as frequently as once per generation. This is much more rapidly than would be expected by mutation. Figure 40 depicts the arrangement of some of the genes involved in the mating type switch. On yeast chromosome 3 there is a short region specifying mating type α (HML) and another short region specifying mating type *a* (HMR). Neither α nor *a* is expressed from these two regions: HML and HMR are simply genetic storage sites. Between the two is an active region called MAT. MAT contains a copy of either the α or *a* region. During switching, the information from either HML-α or HMR-*a* replicates and replaces the information at the active MAT site. The exchange is not reciprocal, and the information being replaced is lost. Generally the switch is heterologous, *i.e.* in an α type cell the α gene at the active MAT site will be replaced by the *a* gene from HMR and *vice versa.*

Several other genes are also involved in this system. A set of genes called SIR keeps the HMR-*a* and HML-α regions silent. Mutation in SIR leads to expression of the silent regions and also to switching from them. The target of the SIR gene products is a region called *E*. *E* regions are located 1500 base pairs away from HMR and HML, and in a sense *E* acts as a negative enhancer (Section 3.7.7.3): it functions in *cis* and in either orientation. There is no *E* region near MAT, thus allowing the α or the *a* gene present to be expressed. A dominant gene, HO, is probably needed to initiate the switch, perhaps by encoding a nuclease that recognizes a site within the MAT locus (see Figure 40). One current model maintains that switching is initiated in the MAT locus; once the nuclease-induced cut is made in the DNA of that region, pairing occurs with either HMR or HML, as appropriate, and degradation of MAT and replication of either HMR or HML then occur. The site at which the HO nuclease cuts is also present in HML and HMR, but the SIR gene products protect those areas from being cut.

3.7.7.6 Antigenic Variation in Trypanosomes: Rearrangements that Protect a Parasite

Trypanosoma brucci, the microbial agent causing sleeping sickness in humans, is covered by a compact, regular surface coat composed of a single protein species called the variant surface

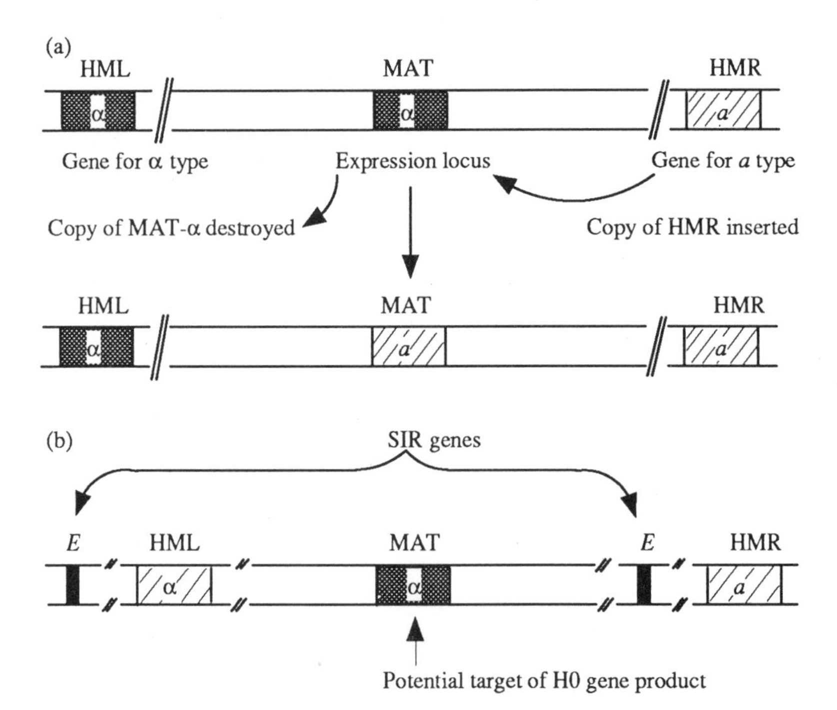

Figure 40 *Yeast mating type switching.* (a) Information for the two mating types α and *a* is stored in silent regions called HML and HMR. A copy of either HML or HMR is present at a third site called MAT. In the example a copy of HMR replaces the α-region present in MAT, and from the *a*-region newly inserted into MAT the cell produces a set of proteins that give it the *a* mating type. (b) Products of the SIR genes keep the HML and HMR regions genetically silent, probably by binding to distant regions called *E*. The H0 gene product, a site-specific endonuclease, initiates switching and degradation of the existing MAT allele by cutting the DNA within MAT

glycoprotein (VSG). The VSG is the only antigenic structure on the surface of the living organism. The host immune system actively responds to the VSG, and most of the trypanosome population is rapidly destroyed. However, every 7 to 10 days a new population of trypanosome having an antigenically distinct surface arises. The trypanosome genome contains 300 to 1000 genes encoding different VSG's, and switching transcription from one VSG gene to another allows the parasite to evade the immune response of the host.[12, 25]

One mode for gene switching appears to be generated by duplication of one of the VSG genes and its transposition into an expression site. Thus the general strategy is similar to that observed with the switching that occurs with the yeast mating type locus. Nucleotide sequences have been determined for the active and inactive copies of VSG genes, and some of the anatomical features are sketched in Figure 41. At the 5′ end of the gene is a series of 76 base pair repeats that are probably involved in transposition. A different set of homologies occur at the 3′ end. The 3′ break site for transposition occurs inside the gene, thus only part of each different VSG gene is placed in the expression site and all VSG messenger RNA's contain identical 3′ ends.

Another feature of trypansome gene expression is discontinuous transcription.[34] Messenger RNAs are formed by the splicing of two exons from different genes, genes which sometimes reside on different chromosomes. The 5′ mini-exon is short (35 nucleotides) and does not code for protein. It is probably involved in proper translation of the messenger since addition of RNA complementary to the 5′ mini-exon blocks translation of all trypanosome messengers *in vitro*. Thus, in this case RNA splicing eliminates the need to have a copy of the mini-exon in front of every gene.

3.7.7.7 Antibody Formation: Increasing Options by Rearrangements and Splicing

Genetic rearrangements that occur to produce mature antibodies represent another type of control mechanism. Antibodies are proteins that recognize and bind to antigens, that is, to substances not normally found in our bodies. As such, antibodies form part of our immune system, the elaborate network of molecules and cells that protect us from many types of disease. Millions of antigens can be recognized by antibodies. Since each different antigen is recognized by a different antibody, our bodies must be able to produce antibodies of millions of different types.

Each antibody is composed of four protein chains, two identical heavy chains and two identical light chains. The chains are folded and connected to form a 'T' as shown in Figure 42. Comparison of amino acid sequences from many different antibodies has revealed several interesting features. First, antibodies can be grouped into classes based on the amino acid sequences and properties of the heavy chains. Second, within a class there are sections of the protein chains that are identical from one antibody to the next. These sections are called constant regions, and they determine the behavior of the antibody in our bodies. For example, heavy chain antibodies with one type of constant region circulate in the blood, those with another type attach to the surface of the cell that

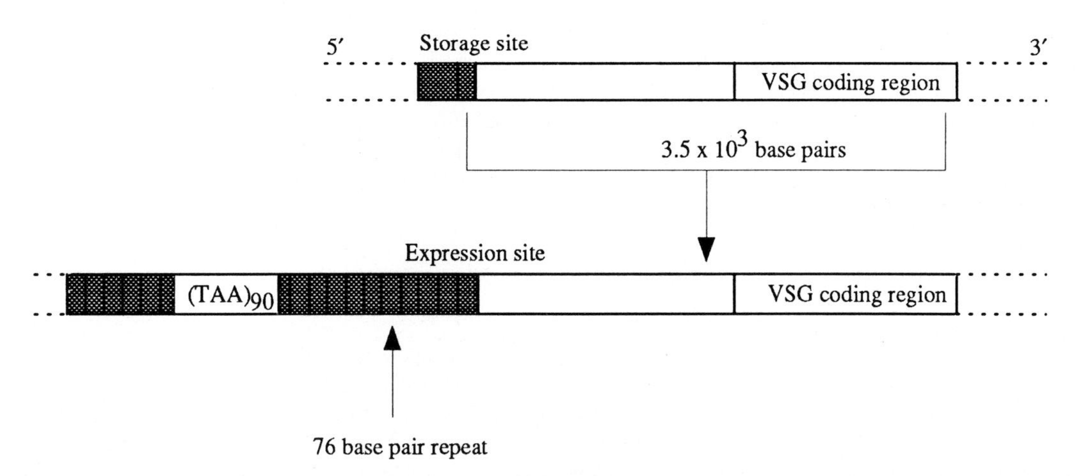

Figure 41 *Trypanosome surface antigen switching.* The DNA of some trypanosomes contains hundreds of genes for a surface protein called VSG. These genes are normally silent, *i.e.* no messenger RNA is made from them. Periodically a copy of one gene is moved from its storage site into an expression site, and the information from that gene is then used to make the protein. At the 5′ end of the transposed region are 76 base pair repeats. In the expression site there is a run of 90 repeats of the triplet TAA

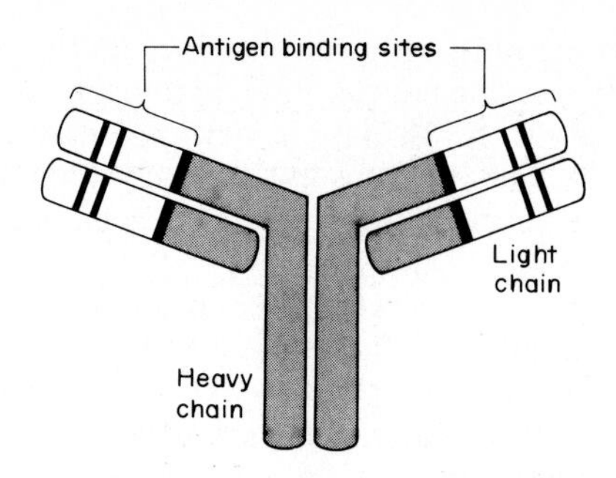

Figure 42 *Schematic diagram of an antibody molecule.* Two heavy chains pair with each other and with two light chains to form the active antibody. The amino acid sequences are divided into constant regions (shaded), variable regions (open) and hypervariable regions (solid). Two antigen binding sites are present, one in the variable region of each arm (reproduced from ref. 9 by permission of John Wiley & Sons, Inc.)

produced them, and still others bind to specific cells that release histamines. The third point is that each light chain and each heavy chain have regions of amino acids that are unique to that antibody. These regions are called variable regions, and it is this part of the antibody that binds to foreign substances such as viruses and bacteria. Since the shape and structure of a protein are dramatically affected by small changes in the sequence of amino acids, the slight differences in amino acids found in the variable regions result in millions of different antibodies, each able to recognize a particular antigen.

DNA sequencing studies revealed that most of our cells do not have a complete set of antibody genes. Instead, they have bits and pieces that can be combined in many different ways, thus producing millions of distinct antibodies from a small amount of genetic information. The rearrangements occur inside blood cells called B lymphocytes, the cells responsible for making antibodies.

By comparing the nucleotide sequences in DNA from embryonic cells with those in DNA from antibody-producing cells, it has been possible to develop a general idea about how gene shuffling results in antibody chains. In the case of light chains (Figure 43), the embryo contains several hundred variable regions genes (V) widely separated from five short, joining genes (J). DNA breakage and rejoining occurs so that one of the V genes is placed next to one of the J genes. RNA polymerase transcribes this region and continues until it also transcribes a constant region gene (C). This long RNA molecule is then spliced to remove the sequence between the V/J region and the C region, producing mature messenger RNA. The messenger RNA is then translated into an antibody light chain. Since any one of perhaps 150 V genes can join to any of 5 J genes, roughly 750 combinations (150×5) can occur. Moreover, the joining sites are not precisely located; thus, the actual number of possible combinations is probably closer to 7500.

The same principles apply for heavy chain formation. However, more elements are included in creating heavy chain diversity (Figure 44). In humans there are about 80 V (variable) genes, 50 D (diversity) genes, and 6 J (joining) genes. Thus there are about 24 000 combinations ($80 \times 50 \times 6$) that can form. Flexibility in the V/D and the D/J junctions probably adds 100 more ways to combine the genes, so the total number of heavy chain combinations is about 2.4 million ($24\,000 \times 100$). The total number of antibody combinations is the product of the light chain and heavy chain combinations, or 18 billion (7500×2.4 million). Thus enormous diversity can be produced by about 300 embryonic DNA segments.

3.7.8 CLONING A GENE

The goal of this final section is to provide an introduction to the logic of gene cloning at the experimental level. A number of steps are involved. First, DNA containing the gene of interest is cut into large, discrete fragments by restriction endonucleases. The fragments are next spliced into small DNA molecules capable of invading bacterial cells and replicating. Then the bacterial cells that contain the cloned gene must be identified and separated from cells that lack it. Finally, the cloned gene must be recovered from the bacteria in a purified form for detailed studies. The principle features of these steps are described below.

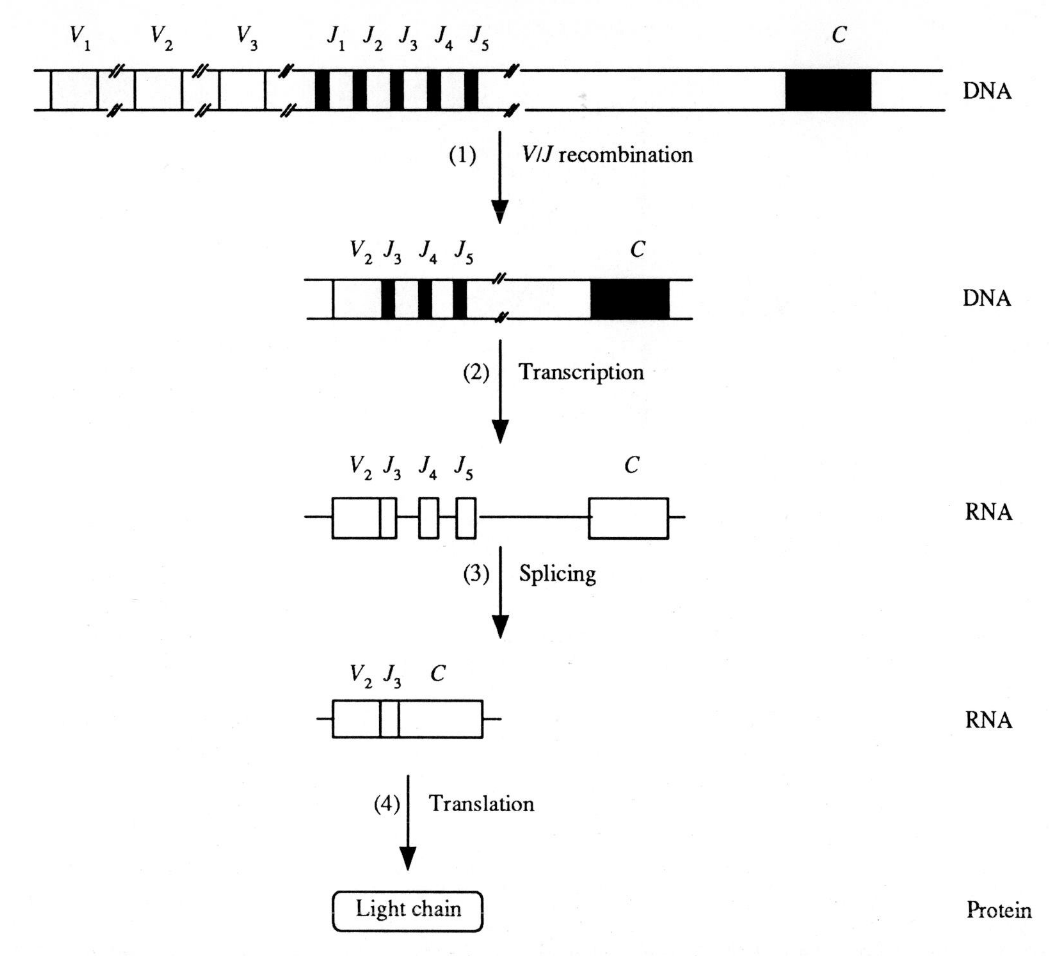

Figure 43 *Schematic representation of the formation of an antibody light chain.* (1) One of the approximately 150 variable genes V recombines with one of the five joining genes J. In the example V_2 is moved so it becomes adjacent to J_3. (2) RNA is synthesized from this DNA to produce a primary transcript. (3) Splicing occurs to remove all the RNA between J_3 and the constant gene C, producing mature messenger RNA. (4) This messenger RNA is translated into the antibody light chain. Discontinuities in the DNA indicate large distances between the genes (reproduced from ref. 9 by permission of John Wiley & Sons, Inc.)

3.7.8.1 Restriction Endonucleases: Enzymes that Cut DNA

One of the ways in which bacterial cells protect themselves from invasion by foreign DNA is through a system called restriction–modification.[39] Bacterial cells contain enzymes called restriction endonucleases which recognize specific nucleotide sequences and cut double-stranded DNA. When foreign DNA bearing the recognition sequence for the restriction endonuclease enters a cell, it will be cut and destroyed by the endonuclease. The cell protects its own DNA by modifying (methylating) nucleotides in the recognition sequence so the endonuclease cannot cut. This is accomplished by cellular methylases that specifically recognize the restriction site. Usually a foreign DNA molecule will contain many sites that are recognized by the endonuclease and the methylase, and when it enters the cell, a race begins between the two enzymes. Since only one cut is required to inactivate the foreign DNA, the nuclease usually wins. In the case of an entering DNA the size of bacteriophage DNA, only one in about 10 000 molecules will become methylated at all potential sites and survive to replicate.

As a part of this self-recognition system the restriction endonucleases from different bacteria recognize different nucleotide sequences. A large collection of these enzymes is now commercially available, providing many different cutting options. The recognition sequences are often four or six base pairs long, and most nucleases cut both strands of the DNA within the recognition site. In some cases the two DNA strands are not cut opposite each other, rather, the cuts are staggered. Thus, once the cuts have been made, only a few base pairs hold the two DNA strands together. When DNA cut

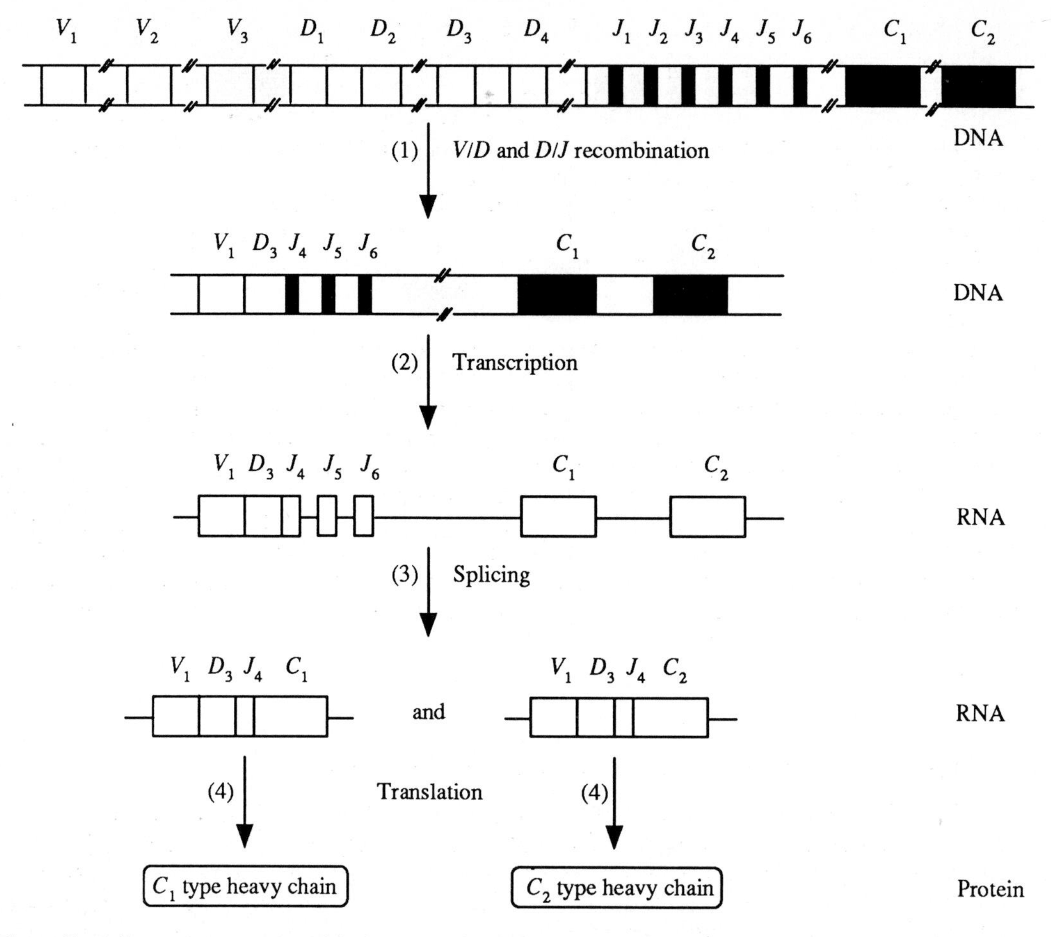

Figure 44 *Schematic representation of the formation of antibody heavy chains.* (1) One of 80 V regions joins with one of about 50 D regions and one of 6 J regions to form a recombined DNA molecule in a cell called a B lymphocyte. (2) A primary transcript is made that contains two different C regions. (3) By differential splicing, two types of heavy chain messenger RNA can be made. (4) When the messenger RNA's are translated, they produce two types of heavy chain protein. Since the $V_1/D_3/J_4$ regions are the same for both, the two heavy chain proteins will have identical antigen binding sites. Discontinuities in the DNA indicate large distances between the genes (reproduced from ref. 9 by permission of John Wiley & Sons, Inc.)

in this manner is gently warmed, the base pairs between the cuts will break apart, and the DNA molecule will separate into fragments.

3.7.8.2 Ligation: Splicing Two DNA Molecules

As mentioned above, some restriction endonucleases generate staggered cuts. The few nucleotides in the single-stranded ends of the DNA molecules are complementary to the ends of other molecules generated by cutting with the same restriction endonuclease. When two DNA molecules having complementary ends collide, the single-stranded ends form base pairs, and the two molecules tend to stick together. Thus the ends are called sticky ends. As mentioned in Section 3.7.2.3, DNA ligase is an enzyme that performs the essential function of joining DNA molecules together after DNA replication. If this enzyme is present when two DNA molecules having sticky ends come together, it will repair the breaks that had been introduced by the restriction endonuclease. Not all ligases require sticky ends. A bacteriophage DNA ligase has been discovered that will join blunt ends; thus both types of end can be spliced.

3.7.8.3 Plasmids: Vehicles for Cloning

Plasmids are dispensable, circular, double-stranded DNA molecules that occur naturally in bacteria. Many plasmids are small, occur as multiple copies per cell, and are easily isolated (see

Section 3.7.8.7). As with all natural DNA molecules, plasmids contain an origin of replication that serves as a start signal for DNA polymerase and ensures that the plasmid DNA molecule will be replicated by the host cell. Thus, if a gene of interest is spliced into a plasmid, the plasmid can then be added to a bacterial culture in such a way that the plasmids invade individual cells, take up residency, and naturally multiply. This makes it possible to obtain large amounts of particular genes by growing large cultures of the engineered bacteria.

Many kinds of plasmid have been discovered that differ in length and in genetic organization. Some of the smaller plasmids, which are popular in gene cloning, have about 5000 nucleotide pairs, enough DNA to code for about five average-sized proteins. In comparison, the bacterium *E. coli* contains slightly more than four million nucleotide pairs in its DNA, and humans have about four billion nucleotide pairs in theirs.

An important aspect of plasmid DNA molecules is that they often contain genes that make their host bacterial cell resistant to antibiotics. Such resistance turns out to be extremely useful in genetic engineering. For example, in a popular cloning procedure DNA fragments are spliced into a plasmid DNA that also contains a gene conferring resistance to tetracycline. Then the plasmid DNA is added to a culture of bacteria normally killed by tetracycline. Under the proper experimental conditions, the plasmid DNA enters the cell and multiplies along with the bacterial cell. If the bacteria are spread on an agar plate containing tetracycline and incubated, most of the bacteria are killed. However, the few cells that acquire a plasmid become tetracycline-resistant, and they grow into colonies on the agar. Thus, when testing colonies to determine which took up a plasmid, antibiotics are used to avoid examining the millions of bacterial colonies that fail to be transformed by a plasmid.

3.7.8.4 Cloning with Plasmids: A Way to Enrich for Specific Genes

Generally DNA containing a gene of interest will be purified and cut into many thousands of different pieces. These pieces will be randomly spliced into a large population of plasmid DNA molecules, and then plasmids will be incubated with bacteria. The bacteria take up the DNA by a process called transformation. At this stage the problem is to locate the few bacterial cells that contain the cloned genes. Only a tiny fraction of these will contain the gene of interest. Following the initial transformation, the bacterial cells can be divided into four classes: (i) Cells that failed to take up any plasmid DNA; (ii) Cells that took up plasmid DNA without any other type of DNA spliced in; (iii) Cells that took up plasmid DNA with another type of DNA spliced in but not the particular gene being sought; and (iv) Cells that took up plasmid DNA into which the desired gene was spliced. The fourth category is the important one, and members of this category are generally very rare, perhaps one in a billion.

Two procedures are used to increase the probability for finding plasmids with the gene of interest. First, the plasmid chosen as a cloning vehicle contains a gene for resistance to tetracycline. Thus, all the cells can be spread onto an agar plate that contains tetracycline, and only the cells containing a plasmid having a gene for tetracycline resistance will grow and form colonies. This eliminates category 1.

The second procedure distinguishes cells containing plasmids spliced with another type of DNA (categories 3 and 4) from those that have only plasmid DNA (category 2). A plasmid is used which contains two genes for antibiotic resistance. Often one gene is for tetracycline resistance (tet^R) and the other is for ampicillin (penicillin) resistance (amp^R). Since restriction endonucleases cut in very specific locations, an endonuclease can be found that cuts the plasmid DNA only inside the ampicillin-resistance gene. Consequently, whenever another type of DNA is inserted into this plasmid, it will be spliced into the middle of the ampicillin-resistance gene, inactivating the gene. Cells containing plasmids with inserts in the ampicillin-resistance gene will be resistant only to tetracycline. On the other hand, cells containing a plasmid lacking inserted DNA will be resistant to both drugs. Thus all of the colonies that formed on the tetracycline-containing agar plate must be tested to find ones that fail to grow on ampicillin-containing agar.

A piece of sterile velvet is carefully placed on the surface of the tetracycline-containing agar plate so it touches the bacterial colonies. Some of the cells from each colony will stick to the velvet. The velvet is then removed and set onto a clean ampicillin-containing agar plate. Cells from each colony will come off the velvet and stick to the agar of the clean plate. There the cells will grow into colonies if they are resistant to ampicillin. This technique is called replica plating, and it distributes cells onto the second plate in a pattern identical to the distribution of the colonies on the first plate. Only colonies that grow on tetracycline but not on ampicillin would be saved for further examination. This eliminates category 2 in the list above.

At this stage one still doesn't know which gene or genes are contained in any particular bacterial colony. In fact, the chance of obtaining a colony that has the gene of interest is very low. Consequently, the plating procedure must be repeated many times to produce a large collection of colonies. The replica plating technique makes it possible to quickly screen thousands of colonies for growth in different drugs by simply looking for differences in the distribution patterns of colonies on agar plates. Once several thousand have been collected, one can proceed to the next procedure—identifying bacterial clones that have the particular gene sought.

3.7.8.5 Radioactive Probes: Nucleic Acids Complementary to the Gene Being Sought

The most popular way to locate the colony of interest is by nucleic acid hybridization. In this procedure a radioactive nucleic acid is obtained which is complementary to all or part of the gene being sought (see Section 3.7.3.2). This nucleic acid is then used as a probe to find the appropriate bacterial colony.

If the protein product of the gene can be isolated and its amino acid sequence determined, knowledge of the genetic code allows the appropriate nucleotide sequence of the gene to be deduced (there is redundancy in the code, so the exact nucleotide sequence for the entire gene cannot be obtained this way). From this information a radioactively labeled oligonucleotide can be synthesized which has enough complementarity with the gene to allow the oligonucleotide and DNA from the gene to hybridize.

Another method utilizes antibodies raised against the protein product of the gene of interest. The antibodies will selectively precipitate ribosomes that are synthesizing the protein of interest. These ribosomes are attached to the messenger RNA from the gene, and the messenger RNA can be purified. It is then converted into a highly radioactive DNA form using radioactive nucleotides and reverse transcriptase obtained from retroviruses: this complementary DNA (cDNA) is then used to hybridize with DNA from the gene of interest.

Occasionally the gene of interest is expressed so actively in a certain type of cell that it is the dominant messenger RNA of the cell. By first purifying the particular cells, it is possible to greatly enrich for the RNA of a specific gene. As above, radioactive complementary DNA is synthesized to use in identifying the gene by hybridization.

3.7.8.6 Colony Hybridization: Finding the Cloned Gene

A small sample of cells from each colony found to contain cloned DNA by replica plating on agar containing antibiotics is spotted on another agar plate to produce a grid-like arrangement. These colonies are representatives of categories 3 and 4 in the list described above in Section 3.7.8.4. After the cells grow into colonies on the agar, a piece of nitrocellulose filter paper is placed on the agar and then removed. Bacterial cells stick to the paper, forming the same grid pattern they had held on the agar. The paper is then placed in a dilute sodium hydroxide solution to lyse the cells. Some of the cell debris plus the cellular DNA stick tightly to the paper. The sodium hydroxide also causes the DNA to become single stranded. The sodium hydroxide is then neutralized, and the paper is incubated with the radioactive probe, prepared by strategies similar to those described above in Section 3.7.8.5.

Both the radioactive probe and the cellular DNA attached to the paper are single stranded. They will form base pairs with any complementary nucleotide sequences they contact. If the paper-bound DNA contains the gene being sought, the radioactive probe will form base pairs with the paper-bound DNA (see Figure 10). The radioactive complementary DNA will then become indirectly bound to the paper, and the location of the radioactivity on the paper will identify the bacterial colony which contains the gene being sought.

To determine where the radioactivity is located, the paper is removed from the hybridization mixture and washed thoroughly to remove any radioactive probe that is not base-paired with paper-bound DNA. X-Ray film is then placed next to the paper. Wherever radioactive probe is base-paired to paper-bound DNA, it exposes the film. The exposed regions are aligned with the bacterial colonies on the agar plate to identify the ones containing the gene of interest, thus separating members of category 4 from those of category 3 in the list outlined in Section 3.7.8.4.

3.7.8.7 Isolation of Plasmid DNA: Recovery of the Cloned Gene

Once the colony containing the desired DNA is identified, the cloned DNA fragment must be retrieved by purifying the plasmid containing it. The first step in obtaining plasmid DNA is to prepare a liquid bacterial culture containing billions of cells harboring plasmids. Each of these bacterial cells contains the cloned DNA fragment previously identified, and the fragment is still part of the circular plasmid. Next the DNA molecules must be removed from living cells. Enzymes and detergents are added to a concentrated bacterial suspension to dissolve the cell walls of the bacteria, releasing both bacterial DNA and plasmid DNA molecules from the cells.

Once the DNA molecules have been released from the cells, the plasmid DNA must be physically separated from the bacterial DNA. The two types of DNA differ in length as well as nucleotide sequence. Depending on the particular plasmid, the bacterial DNA may be up to a thousand times longer. Consequently, one step in the purification procedure is to separate the DNA species by velocity centrifugation. Under the appropriate conditions the large bacterial DNA will form a pellet in the bottom of a centrifuge tube while the much smaller plasmid DNA will stay in the supernatant fluid. Unfortunately, there is usually so much more bacterial DNA than plasmid DNA that this centrifugation procedure does not completely separate the two kinds of DNA molecules. The great length of bacterial DNA, however, makes it possible to carry out an additional type of separation. Bacterial DNA is easily sheared by pipetting. This procedure does not break the smaller plasmid DNA, so it remains circular. In the absence of ends, DNA rotation is restricted. Thus intercalating dyes such as ethidium bromide, dyes which unwind DNA upon binding, bind more to linear DNA at high dye concentration. Dye binding alters the buoyant density of the DNA, so linear and circular molecules can be easily separated by isopycnic density-gradient centrifugation in cesium chloride (see Figure 45).

3.7.9 CONCLUDING REMARKS

Recombinant DNA technologies have been responsible for an enormous improvement in the precision of biological experiments. By combining the knowledge of the nucleotide sequence of a gene with specific changes made by mutation, it has been possible to correlate precise alterations in genetic information with physiological effects. The methods for determining nucleotide sequences are highly reproducible, and since the information exists as a linear array, it is easily computerized. Many laboratories now contribute to the database, and almost everyone has electronic access to it. The database provides a wealth of information for studies ranging from molecular evolution to forensic medicine. However, there is still a large gap remaining between knowing the amino acid sequence of a given protein and having precise knowledge about how particular amino acids contribute to protein function. Determining three-dimensional structures of proteins is still very laborious, so mutants have not been studied extensively at the protein level. Thus one of the next challenges for medicinal chemists is to develop rapid methods for determining protein structure.

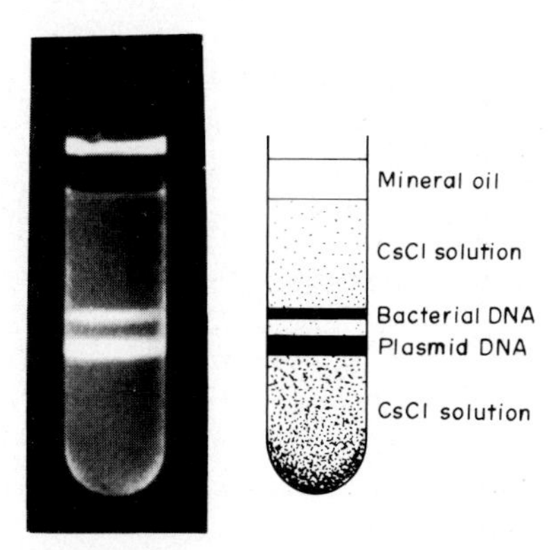

Figure 45 *Separation of plasmid and bacterial DNAs by dye buoyant density centrifugation.* Plasmid DNA, bacterial DNA, ethidium bromide and cesium chloride were mixed and centrifuged for two days at 35 000 rpm to form a density gradient. The DNA species formed distinct bands. Before centrifugation, mineral oil was added to fill the plastic centrifuge tube to prevent its collapse from the force of the centrifugal field. After centrifugation, the tube was illuminated with UV light, and the DNA can be seen as bright orange bands due to fluorescence of ethidium intercalated between the base pairs (reproduced from ref. 9 by permission of John Wiley & Sons, Inc.)

These could then be combined with the DNA strategies to develop rational methods for designing the proper chemicals to combat disease.

ACKNOWLEDGEMENTS

The following are thanked for critical comments on the manuscript: J. Bargonetti, R. Burger, M. Gennaro, J. Komblum, B. Kreiswirth, E. Murphy, S. Projan and D. Turner. The author's work on bacterial chromosome structure has been supported by grants from the American Cancer Society, the National Institutes of Health, the European Molecular Biology Organization, and the National Science Foundation.

3.7.10 REFERENCES

1. T. Baker, K. Sekimizu, B. Funnell and A. Kornberg, *Cell*, 1986, **45**, 53.
2. J. Bishop, *Annu. Rev. Biochem.*, 1983, **52**, 301.
3. J. Campbell, *Annu. Rev. Biochem.*, 1986, **55**, 733.
4. T. R. Cech, *Cell*, 1986, **44**, 207.
5. T. Cech and B. Bass, *Annu. Rev. Biochem.*, 1986, **55**, 599.
6. N. T. Chang, J. Huang, J. Ghrayeb, S. McKinney, P. K. Chanda, T. W. Chang, S. Putney, M. G. Sarngadharan, F. Wong-Staal and R. C. Gallo, *Nature (London)*, 1985, **315**, 151.
7. K. M. Derbyshire and N. D. F. Grindley, *Cell*, 1986, **47**, 325.
8. K. Drlica, *Microbiol. Rev.*, 1984, **48**, 273.
9. K. Drlica, 'Understanding DNA and Gene Cloning: A Guide for the Curious', Wiley, New York, 1984, p. 205.
10. K. Drlica and J. Rouviere-Yaniv, *Microbiol. Rev.*, 1987, **51**, 301.
11. W. Doerfler, *Annu. Rev. Biochem.*, 1983, **52**, 93.
12. P. Englund, S. Hajduk and J. Marini, *Annu. Rev. Biochem.*, 1982, **51**, 695.
13. N. D. F. Grindley, *Cell*, 1983, **32**, 3.
14. N. Grindley and R. Reed, *Annu. Rev. Biochem.*, 1985, **54**, 863.
15. L. Guarente, *Cell*, 1984, **36**, 799.
16. W. M. Holmes, T. Platt and M. Rosenberg, *Cell*, 1983, **32**, 1029.
17. T. Igo-Kemenes, W. Horz and H. Zachau, *Annu. Rev. Biochem.*, 1982, **51**, 89.
18. R. Jaenisch, *Cell*, 1983, **32**, 5.
19. A. Kornberg, 'DNA Replication', W. H. Freeman, San Francisco, 1980, p. 724.
20. S. Leff, M. Rosenfeld and R. Evans, *Annu. Rev. Biochem.*, 1986, **55**, 1091.
21. W. McClure, *Annu. Rev. Biochem.*, 1985, **54**, 171.
22. J. Miller and W. Reznikoff, 'The Operon', Cold Spring Harbor, NY, 1978, pp. 469.
23. K. Nasmyth, *Annu. Rev. Genet.*, 1982, **16**, 439.
24. A. T. Panganiban, *Cell*, 1985, **42**, 5.
25. M. Parsons, R. Nelson and N. Agabian, *Immunol. Today*, 1984, **5**, 43.
26. T. Platt, *Annu. Rev. Biochem.*, 1986, **55**, 339.
27. R. H. Reeder, *Cell*, 1984, **38**, 349.
28. A. Rich, A. Nordheim and H. Wang, *Annu. Rev. Biochem.*, 1984, **53**, 791.
29. F. J. Schmidt, *Cell*, 1985, **41**, 339.
30. K. Schollmeier and W. Hillen, *J. Bacteriol.*, 1984, **160**, 499.
31. P. A. Sharp, *Cell*, 1985, **42**, 397.
32. A. J. Shatkin, *Cell*, 1985, **40**, 223.
33. F. Thoma, T. Koller and A. Klug, *J. Cell Biol.*, 1979, **83**, 403.
34. L. H. T. Van der Ploeg, *Cell*, 1986, **47**, 479.
35. H. E. Varmus, *Science (Washington, D.C.)*, 1982, **216**, 812.
36. J. Wang, *Annu. Rev. Biochem.*, 1985, **54**, 665.
37. J. Watson, N. Hopkins, J. Roberts, J. Steitz and A. Weiner, 'The Molecular Biology of the Gene', Benjamin Cummings, Menlo Park, CA, 1987, p. 744.
38. H. Wu and D. M. Crothers, *Nature (London)*, 1984, **308**, 509.
39. R. Yuan, *Annu. Rev. Biochem.*, 1981, **50**, 285.
40. S. Zimmerman, *Annu. Rev. Biochem.*, 1982, **51**, 395.

3.8

Genetic Engineering: Applications to Biological Research

STEVEN P. ADAMS
Monsanto Company, St Louis, MO, USA

3.8.1 INTRODUCTION

In the last decade, scientific advances on a number of fronts have combined to synergize the developments of molecular and cellular biology, and biotechnology has become a powerful force in modern science. It has already transformed the methodology and the means by which fundamental knowledge of living systems is acquired. In the discipline of medicinal chemistry, biotechnology promises to aid in establishing a rational basis for drug design by significantly enhancing the development of assays, by providing powerful tools for understanding disease mechanisms at the molecular level, and by providing molecular probes and reagents to test molecular hypotheses. Moreover, biotechnology possesses the tools for producing large quantities of biopolymers such as proteins, nucleic acids and oligosaccharides that may be useful therapeutics.

This chapter will focus on applications of molecular biology to medicinal chemistry. Obviously, with tens of thousands of published papers to select from, the treatment will necessarily be selective. In Chapter 3.7, the fundamental principles of molecular biology and genetic engineering have been discussed; in this chapter, the basic elements of genetic expression vectors will be examined with specific examples. In order to exemplify the principles underlying current applications, case studies from the literature have been selected with emphasis on medically relevant proteins. Where basic studies are necessary to exemplify a point, an effort has been made to select examples that apply to real therapeutic problems or, at least, may have a bearing on drug design.

3.8.2 BACTERIAL EXPRESSION SYSTEMS

3.8.2.1 Vectors

In an expression system, a vector is the vehicle that is used to introduce DNA into a cell. Two broad classes of agents have been extensively employed as vectors: plasmids and viruses.

3.8.2.1.1 Plasmids

Plasmids are small circular DNA molecules usually composed of a few thousand nucleotides. Plasmids are observed widely in bacteria and are characterized by their ability to exist and replicate autonomously in the cell, independent of the replication of the chromosome. They frequently carry genes that encode antibiotic resistance and are capable of rendering a bacterium tolerant to antibiotics. Furthermore, multiple copies of a plasmid, as many as 50–100, are able to reside in a single cell, which offers a potentially significant amplification of any gene that may be resident on the plasmid. In practice, plasmids are manipulated by chemical and enzymatic means to introduce genes and sequences necessary for expression of the genes. The plasmid DNA is then introduced into the cell by incubating the cells with plasmid DNA following a treatment with calcium phosphate that makes the cells permeable and able to take up DNA.

3.8.2.1.2 Viruses

Viruses are infective agents composed of genetic material (DNA or RNA) that is enclosed in a protective protein coat. Viruses infect a cell by first attaching to the surface of the cell through their protein coat and by then being taken up by endocytosis. Once inside the cell, the coat proteins are degraded, exposing the genetic material; the viral chromosome ultimately inserts itself into the genome of the cell. Once integrated into the chromosomal DNA in the cell, the virus utilizes the cell's biosynthetic machinery to make viral proteins and to replicate. In its latent state a virus is relatively innocuous; however, when it replicates and produces new infective virus particles, the cell is killed. While viruses can be used in their natural form, it has been possible to alter some viruses and eliminate their cytotoxic activities while retaining their ability to infect and direct protein synthesis. Viral promoters, as will be noted below, have been utilized extensively in plasmid cloning vectors to direct the synthesis of heterologous proteins, proteins that are ordinarily not found in the expressing cell.

3.8.2.1.3 Selectable markers

An essential component of the cloning vector is a 'selectable marker', a gene that confers a unique characteristic on the cell into which a plasmid or virus has been introduced. The selectable marker distinguishes the cell containing the plasmid from the parent cell (see Figure 1). It is important to have a selectable marker because the introduction of a plasmid or virus into a cell is a relatively rare event; indeed, most of the cells do not take up the vector and identifying the desired transformed cell would otherwise be very difficult. The most widely used selectable marker is antibiotic resistance. The resistance determinant is typically a gene that encodes a protein that degrades or modifies the antibiotic in some way to render it inactive. If a cell that is ordinarily sensitive to an antibiotic such as ampicillin is transformed by a plasmid containing the ampicillin-resistance marker, the cell will grow and divide in the presence of antibiotic while all untransformed cells that did not take up the vector will die. In an expression system, the selectable marker serves initially to select for the transformed cells but it additionally insures that cells growing in culture retain the plasmid, because cells have a tendency to eliminate plasmids in the absence of selective pressure. In other words, if a

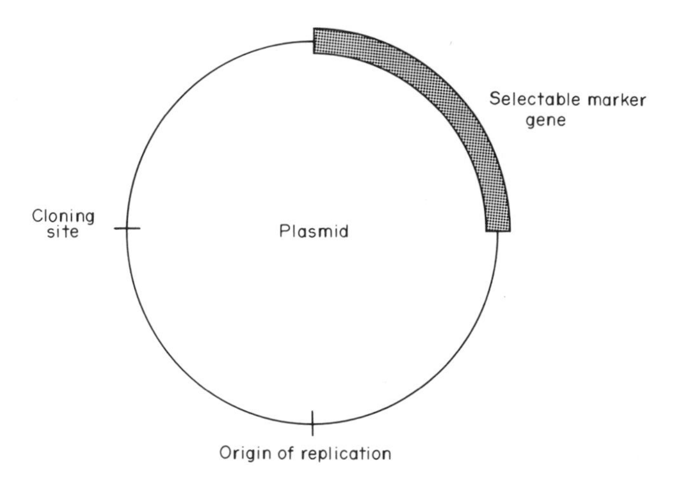

Figure 1 A simple plasmid cloning vector. The plasmid contains an origin of replication required for maintenance and replication of the vector in a cell, a selectable marker gene that confers a unique property on the recipient cell such as antibiotic resistance, and a cloning site, a unique restriction site into which a gene may be spliced

bacterial culture is maintained in a medium containing antibiotic, only those bacteria can survive that retain a plasmid bearing an antibiotic-resistant gene. An expression system incorporates the gene being expressed for protein production on the same plasmid that contains the selectable marker, therefore the selectable marker insures that all cells retain the capacity to produce the protein of interest.

Expression vectors also employ two other important elements: an origin of replication and a cloning site (Figure 1). The origin of replication is a sequence on the vector that directs its replication. The origin is essential for maintenance of the vector in the cell and, in the case of plasmids, for achieving high copy number. The cloning site is a unique restriction site into which DNA—such as a gene, a promoter and attendant regulatory DNA sequences—may be inserted. This site must be positioned such that it does not interfere with the selectable marker or any other functionally important sequences.

3.8.2.2 Genes

Bacterial expression vectors are composed of elements as outlined in Figure 2. Generally, the fundamental objective of an expression system is to produce a protein that is desired for further study, hence the central component of any expression system is the gene (the DNA sequence that is ultimately translated by the ribosome into a specific protein). Genes are typically obtained from the messenger RNA (mRNA) encoding a protein by the technique of complementary DNA (cDNA) cloning. In some instances it is not possible to obtain a cDNA copy of a gene, particularly in the cloning of eukaryotic genes, as will be discussed below; it may then be necessary to employ genomic cloning techniques. Alternatively, a gene can be assembled by enzymatically linking together chemically synthesized DNA fragments. Techniques for cDNA and genomic cloning have become highly refined and numerous laboratories around the world have isolated and cloned many different genes.

3.8.2.2.1 cDNA cloning

Complementary DNA cloning of a gene requires a source for its mRNA, therefore it is necessary to identify a cell or a tissue where the gene is expressed in order to obtain mRNA for cloning. For many genes, isolating mRNA for a gene of interest is relatively easy and cDNA cloning may proceed smoothly. However, the requirement for mRNA is occasionally problematical; if a ready source of the protein, and hence its mRNA, cannot be identified or if a message is present in extremely low concentration, limitations may be imposed on the technique. Furthermore, if an mRNA is very large it can be difficult, if not impossible, to achieve full length second-strand DNA synthesis using reverse transcriptase. This results in a truncated cDNA lacking 5′ sequences that correspond to the N terminus of the protein.

3.8.2.2.2 Synthetic genes

Chemically synthesized DNA sequences can be assembled into a gene by enzymatic ligation of the synthetic fragments. Recent developments in the methodology for chemical synthesis and purifica-

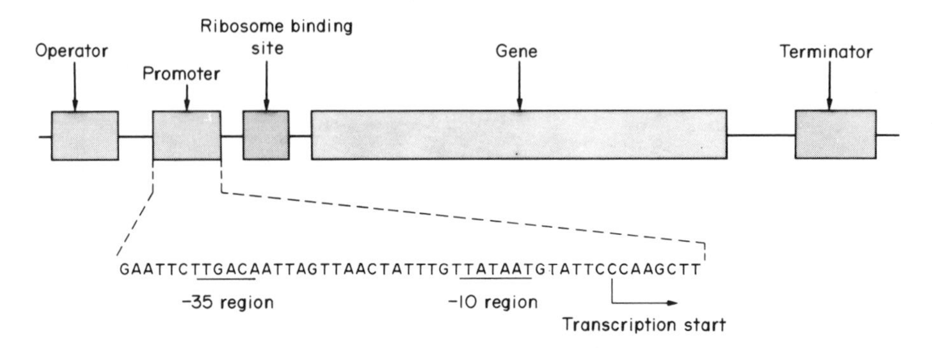

Figure 2 The genetic elements generally contained in an expression system. Each component is discussed in detail in Section 3.8.2. The concensus promoter sequence of Rossi *et al.*[7] is typical of bacterial promoters and exemplifies the conserved −35 and −10 regions

tion now allow for the preparation of DNA sequences 80–100 nucleotides in length of sufficient purity for gene assembly. The total chemical synthesis of genes has proven to be a powerful tool for obtaining genes to incorporate into protein expression vectors. However, gene synthesis requires that one know the gene sequence directly. This implies that the gene has already been isolated and cloned or that the protein sequence has been determined from which a gene sequence may be deduced. With recent advances in high performance techniques for protein purification and protein microsequencing technology, it is not uncommon for a protein to be entirely sequenced prior to gene cloning. Nevertheless, protein sequencing remains a challenging proposition, particularly for large proteins.

The major limitation of chemical gene synthesis is the size of the gene. The construction of a very large gene is very costly in time and resources, but if the protein is of sufficient interest, as has been the case recently with the interferons, growth factors and tissue plasminogen activator (a protein of 527 amino acids), the technology is sufficient to allow for the construction of large genes.

As will be discussed in the examples below, cDNA cloning, genomic cloning and chemical synthesis are powerful methods for obtaining a gene. Considered individually, each technique has advantages and limitations, but, most importantly, the techniques complement each other and frequently the scenario that unfolds for preparing an expression vector, and for optimizing its performance, involves a combination of all approaches.

3.8.2.2.3 Codon usage

Gene synthesis provides for great flexibility in the design of a gene sequence, allowing for the strategic placement of restriction sites that can significantly assist in manipulating and expressing the gene. Furthermore, synthesis also allows one to select a pattern of codon usage that is most appropriate for the host organism being employed in the expression system. In bacteria there exists a correlation between the efficiency of mRNA translation and the codons present in the mRNA.[1] This correlation extends to the relative levels of acceptor tRNA species in the cell.[2] Thus highly expressed genes utilize codons for which corresponding acceptor tRNA levels are high. This correlation also extends to yeast[3] and, by inference, because mammalian genes favor very specific codon usage, the correlation probably extends to higher organisms.[4] Moreover, a recent examination of glycogen phosphorylase isoenzymes from liver and from muscle reveal a striking difference in codon usage with the muscle enzyme using $>80\%$ guanosine (G) or cytidine (C) in codon third positions.[5] This trend appears to be universal for muscle and may represent an important tissue-specific property.

Experiments in *E. coli* were designed to test directly the consequences on gene expression of adjacent low use codons.[6] Two synthetic DNA linkers were inserted into the EcoRI site of the chloramphenicol acetyltransferase (CAT) gene (an antibiotic-resistance gene) which introduced the amino acid sequence Phe-Arg-Arg-Arg-Arg-Lys into the protein. In one construct, the linker utilized high preference arginine codons, while, in the other construct, low usage codons were chosen. The gene was expressed under control of the tryptophan promoter; protein levels were measured and the expression level monitored by gel electrophoresis of [^{35}S]methionine-labeled protein. The tryptophan promoter can be induced by the addition of indoleacrylic acid (IAA), therefore it was possible to examine low levels of expression without added IAA and to observe the effects of high expression levels after the addition of IAA to the fermentation medium. At low expression levels no difference in production of CAT was observed for either of the constructs. However, at high levels of expression a striking attenuation in the production of CAT was observed for the construct that utilized low preference arginine codons. Clearly, in this case as few as four adjacent unfavorable codons had a significant effect on protein translation in a highly expressing plasmid vector. These results suggest that, where possible, a judicious choice of codon usage should accompany the design of a synthetic gene.

3.8.2.3 Promoters

Gene expression in biological systems is regulated by a number of complex mechanisms,[15] most of which involve interactions between proteins and specific sites on nucleic acids. In bacteria, many of these control regions are encoded in relatively short stretches of DNA that are located immediately adjacent to the gene. The DNA sequence information required to promote transcription, the synthesis of mRNA, is called a promoter (Figure 2). Promoters are usually contained in about 40 base pairs of DNA immediately upstream from the gene. A large number of promoters from *E. coli*, the common laboratory organism used by molecular biologists, have been sequenced and considerable sequence variation is observed. However, there are two regions of sequence that are highly conserved in promoters that are known to support a high level of transcription. It is believed that these regions are involved in the obligatory initial step of RNA polymerase binding to the DNA template. One conserved sequence, designated the −35 region and having the conserved sequence TTGACA (T = thymidine, A = adenosine), is 35 base pairs upstream from the transcription start site, and the other sequence, TATAAT, is 10 base pairs upstream from the start site (−10 region). By comparing the sequences of the known *E. coli* promoters and by incorporating the conserved −35 and −10 region sequences, a consensus promoter sequence has been established. The consensus promoter has been synthesized and cloned in front of the CAT gene and the β-galactosidase gene,[7] each of which allowed for quantitation of mRNA transcribed from the promoter and for protein translated from the mRNA. The consensus promoter was functional in the expression of both genes; however, it was not particularly strong in promoting transcription of either of the genes in bacteria, even though RNA polymerase exhibited high affinity for the promoter. A single nucleotide change in the conserved −35 region or an alteration of the spacing between the −35 and the −10 region by two nucleotides resulted in significantly decreased activity. The results suggest that important promoter elements have been identified but that a refined molecular model for RNA polymerase interaction with a promoter, and subsequent transcriptional efficiency, is not yet possible; indeed, the gene itself and other DNA sequences that flank the promoter and the gene may influence promoter activity. While any good promoter will probably afford some level of expression of different genes, a single promoter construct that has been optimized for the expression of a given protein is not likely to be optimal for other proteins. There is no generic promoter system that affords optimal high level expression for every gene. However, sufficient information is available for selecting a strong promoter to serve as a starting point for an expression system, and the efficiency of the promoter may be optimized by small sequence alterations in the promoter and flanking sequences in the context of the gene to be expressed.

Several naturally occurring promoters known to give high levels of transcription in bacteria have been synthesized and employed in the high level expression of a variety of proteins, which will be discussed below.

3.8.2.4 Transcription Terminators

In addition to having promoter sequences necessary to initiate transcription, genes also carry sequences downstream from the gene that are required for termination of transcription and release of the newly synthesized mRNA from its DNA template.[16] The termination sequence for the tryptophan attenuator gene (trp a) presented in Figure 3 identifies two essential features: a region of self-complementary sequence having dyad symmetry that is rich in guanosine (G) and cytidine (C) followed immediately by a stretch of uridine (U) residues. It is believed that the self-complementary sequence hydrogen bonds to itself and folds to adopt a hairpin secondary structure. This hairpin arrests elongation of the mRNA by sterically causing RNA polymerase to pause. Because the RNA–DNA hybrid formed between the uridine residues in the mRNA and the adenosines on the DNA template is inherently unstable, the mRNA dissociates from the template and is released from the transcription complex. In other words, if RNA polymerase does not transcribe through a stretch of uridine residues quickly, dissociation of the mRNA and termination are likely to occur. Evidence in favor of this model includes the fact that alterations that affect the stability of the hairpin have a direct influence on termination. If hairpin stability is enhanced by increasing the length of the 'stem'

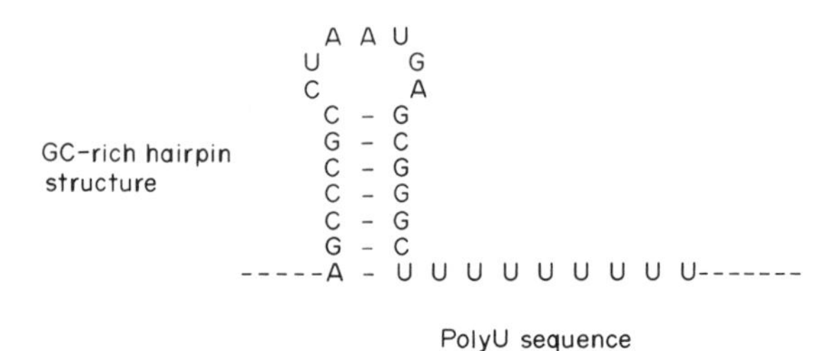

Figure 3 Sequence and proposed hairpin structure of the transcription terminator of the trp a gene. The sequence is for mRNA

of the hairpin structure or by increasing the GC content, polymerase pausing is increased and termination is more efficient.[8] On the other hand, decreasing the hairpin length and/or the GC content decreases termination. Additionally, the length of the uridine-rich sequence also has a significant impact, with termination increasing as the length of the uridine sequence increases. The mechanism for achieving efficient transcription termination described above has been effectively utilized in expression vectors. It is typically possible to simply use the natural terminator that accompanies a gene obtained in the cDNA cloning process. Alternatively, a synthetic terminator may be incorporated into an expression vector.

3.8.2.5 Ribosome Binding Sites

Another sequence associated with the promoter that has significant effects on gene expression in bacteria is the site where ribosomes attach to mRNA. The ribosome binding site on mRNA generally consists of a purine-rich region followed by the translation initiation codon, AUG. The spacing between these elements can vary from about four to ten nucleotides. It is believed that the ribosome binding site on mRNA pairs with a complementary sequence at the 3′ end of the 16S RNA molecule present in ribosomes. This pairing assists in the initial formation and stabilization of the mRNA–30S ribosome complex and it aligns the translation initiator codon AUG to begin protein synthesis.[9,10] Figure 4 demonstrates the homology between 16S ribosomal RNA and the ribosome binding site preceding the β-galactosidase gene. Sequences between ribosome binding sites vary, with many sites exhibiting even greater complementarity than β-galactosidase up to the AUG. As is the case with promoters, a good ribosome binding site is generally chosen for initial expression vector constructions; expression levels are then further optimized by altering the sequence in the ribosome binding site and flanking sequence, particularly the short sequence between the ribosome binding site and the initiator AUG where guanosine seems to decrease translation efficiency.[11]

3.8.2.6 Operators

Many promoters and genes are accompanied by DNA sequences of about 15 to 40 nucleotides that serve as specific binding sites for accessory proteins that control transcription. A classic example of this control principle that is frequently encountered in bacteria is the repressor. A repressor protein binds a region of DNA, termed the operator or repressor binding site, that is adjacent to or overlying the promoter site so that RNA polymerase is sterically prevented from binding to the DNA and initiating transcription. Various conditions in the cell reflected in the levels of intracellular mediators and metabolic products can effectively regulate the expression of a gene or a whole 'operon' of coordinately expressed genes by controlling the activity of the repressor.

Repressor systems are widely represented in nature and repressor activities may be regulated in a number of clever ways. The expression of a repressor molecule and its concentration in the cell may be controlled by another repressor or by the presence or absence of a small regulatory molecule or metabolite. In other cases repressors may bind to their DNA binding sites only when an effector molecule is bound to the repressor. Alternatively, repressor binding may occur only in the absence of an effector. In yet other instances, the repressor may bind only after it has been activated by a protease, or the repressor activity may be eliminated and gene expression induced by proteolytic cleavage of the repressor.

The promoter that controls lactose utilization in bacteria is a good example of a system that is controlled by repression in the absence of an effector (Figure 5).[12] The lac promoter is responsible for expression of several genes involved in the transport and metabolism of the disaccharide lactose which is composed of β-linked galactose and glucose. One of these genes encodes β-galactosidase (β-gal), an enzyme required to release β-linked galactose (β-galactoside) from oligosaccharides and

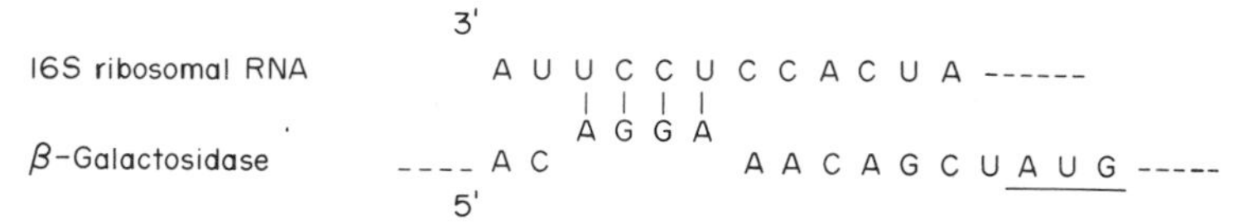

```
                            3'
16S ribosomal RNA            A U U C C U C C A C U A ------
                                 | | | |
                                 A G G A
β-Galactosidase        ---- A C          A A C A G C U A U G -----
                            5'
```

Figure 4 Sequence of the 3′ end of 16S ribosomal RNA and the ribosome binding site of the β-galactosidase gene showing complementary sequence and the translation initiator codon, AUG

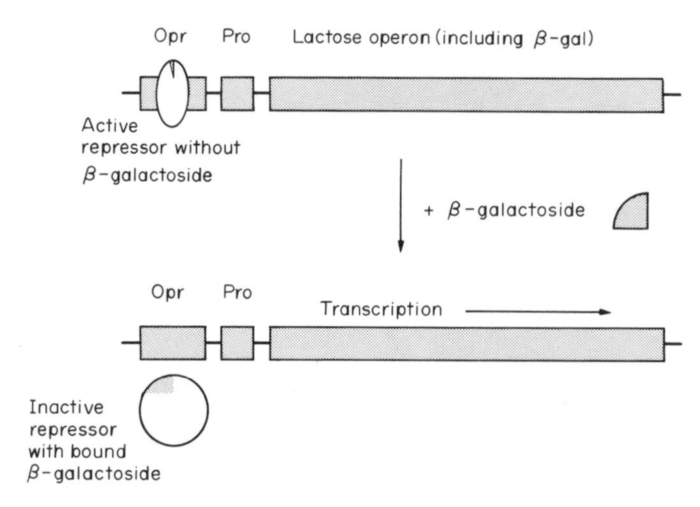

Figure 5 The lactose promoter–operator system. Repressor is bound to the operator preventing transcription. When β-galactoside is available, it binds to and inactivates the repressor so that transcription can begin (Opr = operator sequence, Pro = promoter sequence)

complex sugars. β-gal is a particularly useful 'reporter' molecule for detecting and monitoring expression of the lactose utilization genes. The cell expresses β-gal in high levels when glucose, the primary energy source for the cell, is unavailable and when a source of galactose, such as lactose, is present and can be utilized as an energy source. The expression of β-gal is consequently under control of a repressor that is sensitive to the levels of available β-galactoside. In the absence of β-galactoside, the lac repressor binds tightly to the lactose operator preventing transcription of the lac genes, including β-gal. This strategy ensures efficiency in the cell by preventing the expression of genes when there is no substrate for the protein products to act upon. On the other hand, when β-galactoside is present, it binds to the repressor and prevents it from binding to the lactose operator. Then in the absence of an active repressor the genes for lactose utilization can be expressed.

The tryptophan promoter (trp) that drives the transcription of genes involved in tryptophan biosynthesis is controlled in part by a repressor that is *stimulated* by an effector (Figure 6).[13] In this case, tryptophan, the end product of the trp genes, binds to the repressor and the resulting complex is then activated so it can bind to the trp operator region and prevent transcription. Thus, as tryptophan (the end product of the biosynthetic genes) increases in concentration in the cell, it 'feeds back' to control its synthesis by repressing transcription of the biosynthetic genes. If tryptophan levels fall in the cell, tryptophan is not available to bind repressor, which in turn is unable to bind the operator and prevent transcription. Tryptophan biosynthesis can then ensue until its level is once again sufficient to cause repression. As in the case of the lac repressor, the trp repressor assures efficient usage of metabolic resources in the cell.

The SOS system in *E. coli* is composed of a group of genes that are induced upon injury to the cell (Figure 7).[14] The genes are believed to be involved in repairing damage to DNA, thereby increasing survivability of the cell. This system exemplifies repression that is controlled by proteolysis. The SOS genes are coordinately regulated by a repressor called lex A that binds to an operator region in the promoters of the various SOS genes. When the cell is damaged an as yet unidentified signal activates the proteolytic function of a protein called rec A which is itself a product of one of the SOS genes. Upon activation of its proteolytic function, rec A cleaves lex A eliminating its ability to bind its operator and repress transcription. Because rec A is an SOS gene product, it is also controlled by lex A; therefore, as rec A inactivates increasing amounts of repressor, more rec A is expressed and the system is amplified, providing for a very rapid induction of the entire system. The resulting induction of the SOS genes repairs the cellular damage that elicited the original signal. As repair mechanisms progress, the signal that originally activated rec A disappears, active lex A levels build and the SOS functions are repressed.

In the cell, the regulation of lactose utilization, tryptophan biosynthesis and SOS functions are more complicated than described here—many more factors are involved in the cell that impact their regulation. However, effective bacterial expression systems have been constructed that utilize these promoters and repressors. In practice, when a gene is placed under control of the lac promoter in bacteria that contain a lac repressor, the gene is unexpressed until a β-galactoside derivative is added

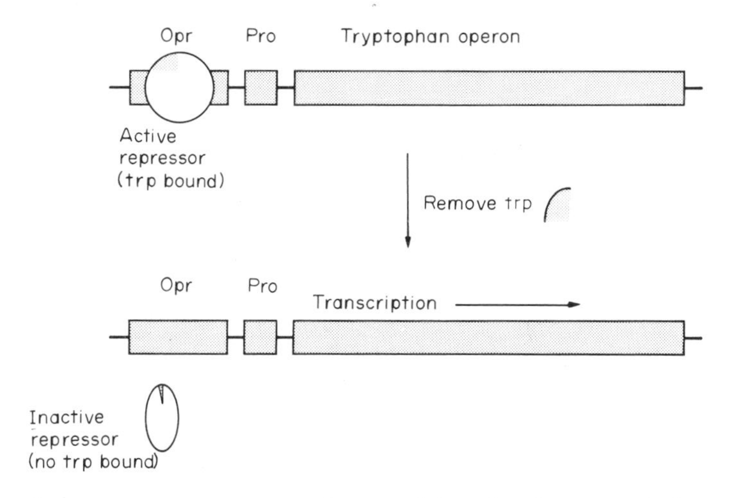

Figure 6 The tryptophan promoter–operator system. Repressor is bound to the operator in the presence of tryptophan, when its synthesis is not required. When tryptophan is absent, the repressor is inactivated and transcription is initiated (Opr = operator, Pro = promoter)

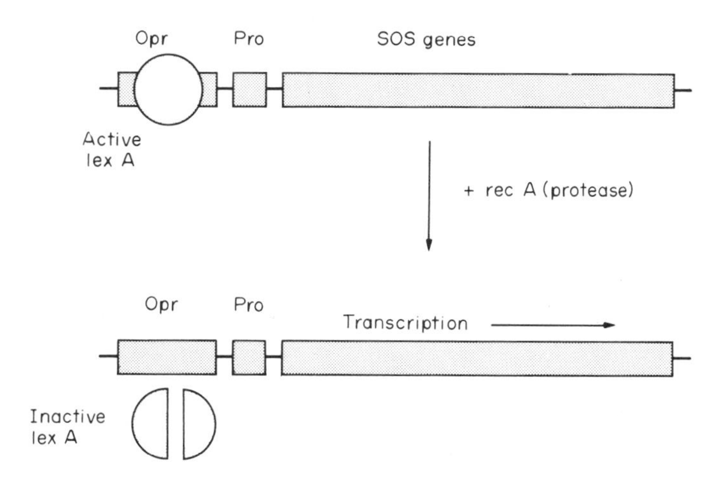

Figure 7 The promoter–operator system for the SOS genes. Lex A, the SOS repressor, binds to the operator until the proteolytic activity of rec A is induced, whereupon lex A is inactivated by proteolysis and transcription ensues

to the culture medium. This gives the molecular biologist control over the timing and extent of expression, a strategy that may be particularly useful in those instances where a gene product may be toxic or in some way deleterious to the cell. The tryptophan promoter is effectively induced by analogs of tryptophan that are able to bind the repressor and thus compete with tryptophan but where the resulting complex is incapable of binding to the operator. Indoleacrylic acid is such a tryptophan analog. Likewise, the rec A promoter can be induced by a number of stimuli that reflect damage to the cell, but the chemical agent, nalidixic acid, is easiest to control. Each repressor system has been effectively employed to allow for controlled expression of genes.

3.8.2.7 Periplasmic Secretion Leaders

While *E. coli* does not excrete proteins from the cell *per se*, a small number of proteins are transported to the periplasmic space, the compartment in *E. coli* between the inner membrane and the cell wall. One essential feature of proteins that are secreted into the periplasm is a signal peptide of about 20 hydrophobic amino acids on the N terminus of the protein. As will be discussed in detail in Section 3.8.3.2.2, this 'signal sequence' is involved in membrane association and translocation of the protein. Upon secretion into the periplasm the signal peptide is removed proteolytically by 'signal peptidase' to generate the mature periplasmic protein. If bacterial cells are submitted to an osmotic shock by treating the cells with a buffered 20% sucrose solution, a subsequent water wash causes proteins to leak out of the periplasmic space into the fermentation medium. Most intracellular proteins remain in the cell under these conditions, and this represents a potentially useful purification step for a protein localized in the periplasm. Considerable effort has focused on understanding the requirements for periplasmic transport in order to utilize the phenomenon to express heterologous proteins.[17]

The protein alkaline phosphatase is localized in *E. coli* in the periplasm and the responsible signal sequence has been investigated in the expression of human epidermal growth factor (hEGF), as will be discussed in detail below. Similarly, signal sequences derived from other periplasmic proteins have been examined in secretion vectors. It appears that eukaryotic and prokaryotic signal sequences can be interchanged and, in some cases, correct processing occurs with heterologous combinations. For example, proinsulin is efficiently secreted and processed by *E. coli* when directed by either the bacterial penicillinase leader or by the natural proinsulin secretion signal. Likewise, human growth hormone is secreted using either the bacterial alkaline phosphatase leader or the natural secretion sequence. It is important to note that some heterologous proteins are not directed to the periplasm by signal sequences but remain in the cytoplasm and, in other cases, while transported, the protein is not properly processed. It is clear that a signal sequence is a necessary condition for *efficient* secretion in *E. coli*. However, a leader sequence is not sufficient to assure secretion or that properly processed protein will be obtained; other factors associated with the protein may determine these points. Practical application of secretion constructs in E. coli is an intriguing possibility for protein expression; however, much work needs to be done to understand the mechanisms involved, and the ultimate outcome of a secretion construct remains an empirical proposition.

3.8.3 EUKARYOTIC EXPRESSION SYSTEMS

Bacterial expression systems typified by *E. coli* are the workhorses for the biotechnologist in the expression of heterologous proteins. However, in eukaryotic organisms many proteins must be modified post-translationally in order to exert their biological activity and these post-translational modifications are generally not observed in bacteria. Important examples of post-translational modifications include glycosylation, acetylation, fatty acylation and phosphorylation, to name a few. Additionally, because eukaryotic cells have mechanisms for secreting proteins from the cell, proteins traverse intracellular compartments that are significantly different from the environment inside the bacterial cell and it is common for proper folding of a protein to occur more readily while traversing the secretion pathway. Consequently, in order to obtain properly modified and properly folded proteins exhibiting desired biological activity, it is frequently necessary to express genes in eukaryotic expression systems.

Saccharomyces cerevisiae, the common laboratory yeast, is the prototypical eukaryotic microorganism used for expression. Yeast genetics are relatively well understood and effective transformation systems have been developed that make genetic engineering of yeast feasible. Additionally, yeast fermentation technology has developed to a level of sophistication that allows for very high levels of production of the organism at high density. Consequently, if the intrinsic expression level of a protein in yeast is less than in *E. coli*, the difference may be compensated for by enhanced efficiency available through yeast fermentation.

Many different mammalian cell lines that are capable of providing proteins of virtually any character have also found utility as expression systems in the laboratory. As will be noted below, microorganisms, including yeast, do not in general possess all of the biochemical and cellular machinery for a full range of protein modifications. For example, even though yeast is capable of glycosylating and secreting proteins, the glycosylation is limited to high-mannose oligosaccharides. Where complex or hybrid oligosaccharides may be important in a protein one must rely on mammalian cell culture. Mammalian cell expression, in general, is less efficient for protein production—mammalian cells have specialized requirements for growth, they grow more slowly, protein synthesis rates are lower than in microorganisms, they frequently must be anchored on a surface in order to grow and they tend to be more fragile than microorganisms. Furthermore, mammalian cell culture technology is not yet as mature as fermentation technology. Nevertheless, the importance of mammalian cell culture for protein production has motivated the development of a number of sophisticated large-scale culture technologies that have begun to revolutionize the field, and important new developments should continue to improve the efficiency and cost-effectiveness of cell culture.

3.8.3.1 Vectors

Eukaryotic expression systems use the same conceptual components as bacterial systems and the plasmid and viral vectors that have found application are similar in principle to prokaryotic vectors. However, eukaryotic cells are characterized by the presence of a nucleus, the repository of genetic information, and of an endoplasmic reticulum and golgi apparatus, the compartments of the cell most significantly involved in post-translational modification and maturation of proteins as well as in the intracellular movement and secretion of proteins. As a consequence these added complexities in the components of a eukaryotic expression system must be understood in order to develop and implement an effective expression vehicle.

3.8.3.2 Genes

3.8.3.2.1 Intervening sequences

The DNA that encodes a gene in bacteria is an uninterrupted linear sequence of nucleotides, thus the mRNA resulting from transcription from the DNA template is also an uninterrupted sequence. In eukaryotes, many (if not most) gene sequences are interrupted at various positions in the gene by DNA sequences that are not ultimately translated into protein. These untranslated regions are referred to as intervening sequences or 'introns', and the sequences that are translated are referred to as 'exons' (Figure 8). In the eukaryotic cell, the initial mRNA product transcribed from a gene that contains introns must undergo a maturation process where the introns are removed and the exons are spliced into a contiguous linear sequence. cDNA cloning procedures for eukaryotic genes that contain introns start with mature mRNA where the introns have usually been removed, and the resulting cDNA is usually compatible with bacterial expression. Occasionally, the molecular biologist is unable to specifically identify a cell or tissue where a protein is synthesized or, alternatively, the level of synthesis in the tissue of origin may be extremely low. In the face of such obstacles it can be difficult or even impossible to obtain mRNA for the gene to be cloned by the complementary DNA procedure. It is necessary, in such an event, to resort to genomic cloning to obtain the desired gene. However, the genomic clone obtained from a eukaryotic organism usually still contains introns. As bacteria are incapable of performing intron–exon splicing, it is necessary to employ a eukaryotic expression system whenever a genomic copy of a gene is being expressed. In those instances where genomic cloning is the method implemented for obtaining a gene, expression of that gene is usually a requisite intermediate step in characterizing the gene and in proving that it encodes the desired protein. Once a genomic clone has been obtained, it is possible to express the genomic construct in cell culture to prepare mature mRNA that is enriched in the desired gene. Possessing an enriched population of mRNA greatly simplifies the preparation of a cDNA copy of the gene that can then be employed for efficient protein expression in cell culture.

The best engineered mammalian expression systems typically utilize cDNA copies of genes as they are more readily accommodated in multicopy cloning vectors; however, it is of significance that some important mammalian proteins are obtained by simply growing cells in culture for which the protein is a 'natural' product. Obviously, expression is obtained from a genomic gene. Hybridoma production of monoclonal antibodies in cell culture is an important example of this type of 'genomic expression' (see Section 3.8.9.2).

3.8.3.2.2 Secretion leaders

A hallmark of eukaryotic cells is their ability to transport certain proteins to the outside of the cell. Cells secrete a relatively small number of their proteins; therefore, in cell culture, the process of secretion potentially offers a significant purification step to the molecular biologist, because a protein secreted into the culture medium can easily be separated from cells and from the large number of intracellular proteins. The protein product can then be readily purified from the relatively small number of other proteins, nutrient molecules and metabolites in the culture medium.

The initial event required for a protein to be secreted is transport of the protein into the endoplasmic reticulum (ER), which is the proximal compartment of the secretion pathway. The transport of the protein requires a 'secretion leader' or 'signal peptide' composed of a sequence of 10–20 hydrophobic amino acids, appended on the N terminus of the protein. The character of the eukaryotic secretion sequence is similar to the secretion sequences in bacterial periplasmic proteins, suggesting that the intrinsic mechanism for membrane transport is similar.

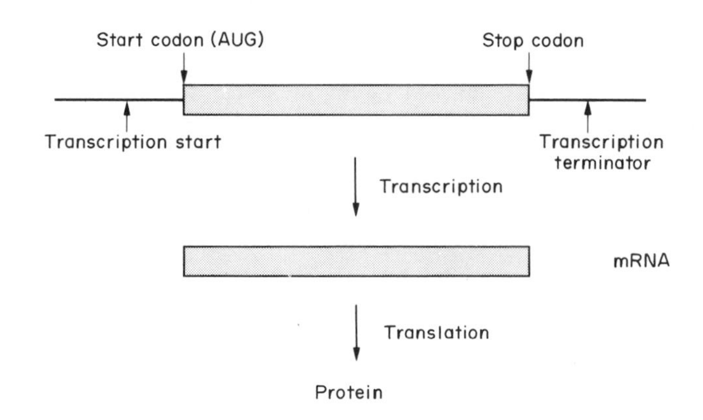

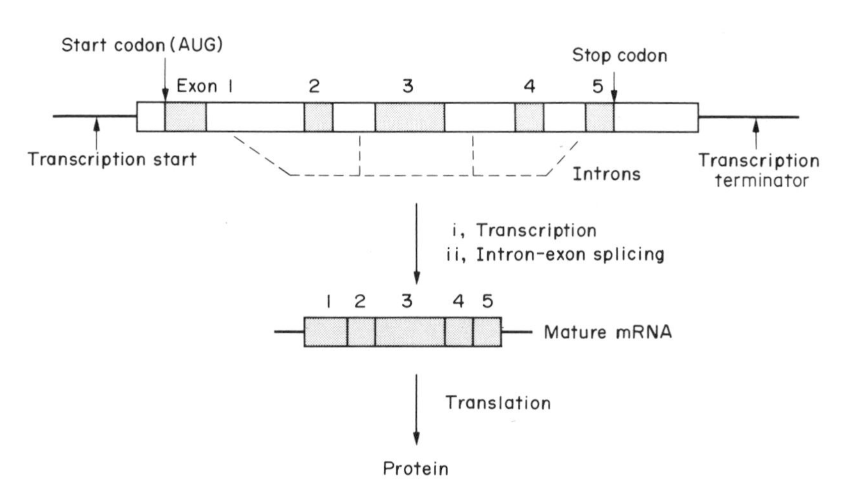

Figure 8 Prokaryotic and eukaryotic gene expression. In bacteria, genes are uninterrupted linear sequences, while in eukaryotes, genes are interrupted by intervening untranslated sequences (introns). Following transcription, introns are removed and the gene segments encoding protein (exons) are spliced together into a mature uninterrupted mRNA sequence that is then translated into protein

Following transcription and transport out of the nucleus, mRNA binds to ribosomes and translation of the message begins. Upon translation of the secretion leader, the nascent protein, the ribosome and accessory proteins, with the mRNA, bind to the ER membrane and, as protein synthesis continues, the signal peptide directs the protein across the membrane and into the lumen of the ER. Once translocated into the ER, the signal peptide is removed by a protease, termed 'signal peptidase', bound on the luminal face of the ER membrane.[17]

The structural features of a secretion leader are quite permissive as long as the general theme of a linear hydrophobic sequence is maintained.[18] Indeed, any of the hydrophobic amino acids in any order are tolerated. However, charged and hydrophilic amino acids in the signal sequence inhibit transport across the ER membrane and secretion of the protein is significantly decreased or even eliminated entirely.

Secretion leaders pose some practical considerations in the design of expression systems. When an intracellular bacterial expression system is devised for a secreted eukaryotic protein, tailoring of the cDNA construct is required in order to remove the portion of the gene encoding the secretion signal. Otherwise the resulting protein might retain the unwanted secretion leader due to the inability of the bacterium to remove the signal sequence. In those instances where attempts are made to secrete a protein in cell culture, the gene must have a sequence coding for a secretion leader introduced at the N terminus of the protein. Typically, a gene for a secreted eukaryotic protein obtained by cDNA cloning will already have a leader sequence; however, it is common to attempt secretion of proteins that ordinarily are not secreted and a signal sequence must be introduced, regardless of the source or the character of the gene.

3.8.3.2.3 *Secretion in yeast*[19]

A particularly useful and effective expression system for yeast utilizes the secretion signals of the yeast α-mating factor.[20] Yeast cells of the α-mating type secrete a 13 amino acid pheromone into the culture medium that attracts yeast of the opposite A-mating type, and it promotes conjugation between the two. α factor is synthesized as a precursor consisting of an 83 amino acid leader sequence followed by four copies of the pheromone separated by short spacers that are composed of variations of the sequence Lys-Arg-(Glu/Asp-Ala)$_{2,3}$ (Figure 9). The leader sequence directs the precursor into the secretion pathway where processing takes place to release the four copies of α factor from the precursor and to package the product for secretion.[21] The processing mechanism involves three proteolytic events, a trypsin-like cleavage on the C-terminal side of the dibasic site followed by removal of the dibasic residues by a carboxypeptidase-B-like activity and a dipeptidyl aminopeptidase cleavage to remove N-terminal Glu/Asp-Ala. In some cases, these proteases are the limiting activities in highly expressing constructs and incompletely processed products can be observed.[22,23] Consequently, when genes are spliced to the α-factor leader, the DNA sequences coding for the Glu/Asp-Ala sequences may be deleted, leaving the basic dipeptide. This alteration obviates the need for the limiting dipeptidyl aminopeptidase and the trypsin-like activity still functions adequately when the Lys-Arg sequence is immediately adjacent to the product. As will be discussed below, the α-factor secretion system has been used extensively to achieve secretion in yeast of small- to medium-size peptides and proteins.

3.8.3.2.4 *Glycosylation signals*

After a protein has entered the ER, it traverses various compartments of the golgi apparatus to be packaged ultimately into membrane-bounded vesicles and secreted from the cell. If the protein has the requisite glycosylation signals, Asn-X-Ser/Thr, presented in an appropriate conformation, glycosylation through the amide moiety of Asn may occur in the ER. The initial oligosaccharide is identical on every glycosylated protein, but as the protein travels through the ER and the golgi, the oligosaccharide is extensively processed and remodeled in a fashion that is believed to be dependent on the cell type and on the structure of the protein. Subsequent post-translational modifications also occur in the golgi that frequently confer important properties and activities on proteins.

The factors that determine the modifications of a protein and its ultimate destination are beginning to be understood. Phosphorylation of the mannose residues on glycoproteins in the golgi appears to be essential for localization of the protein in lysosomes. Moreover, short peptide sequences have recently been identified in proteins that target the nucleus, and other sequences cause proteins to be retained in the ER rather than proceeding through the secretion pathway. Much remains to be learned about principles governing post-translational modifications; however, they certainly depend on protein structure and on the nature and the degree of specialization of the cell in which expression occurs. For the molecular biologist this means that extensive experimentation with cell types and with culture conditions may be required to successfully obtain a useful expression system.

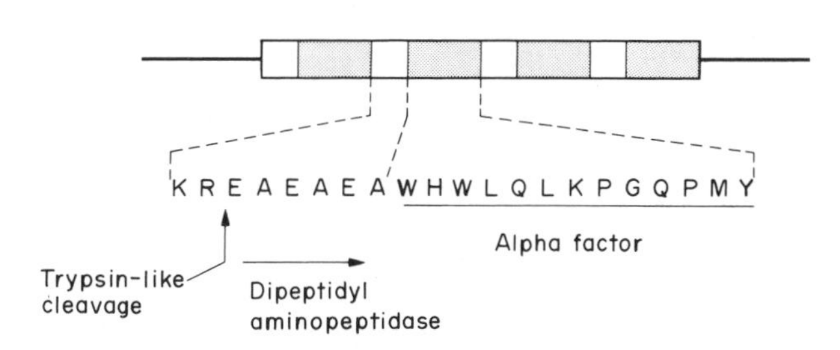

Figure 9 Structure of the alhpa-mating factor precursor containing four alpha-factor cassettes separated by processing sites. A trypsin-like cleavage exposes the (Glu/Asp-Ala)$_{2,3}$ sequences to a dipeptidylamino peptidase, and the Lys-Arg dipeptide is removed from each alpha-factor by a carboxypeptidase-B-like activity

3.8.3.3 Promoters

Eukaryotic transcription initiation sequences are similar to those in *E. coli*, with an essential feature being the stretch of nucleotide sequence rich in adenosine (A) and thymidine (T), usually TATAAA or AATAAA, located 25–30 residues upstream from the transcription start site. Likewise, transcription termination sequences retain the same general characteristics except that some termination signals lack the strict dyad symmetry characteristic of bacterial terminators. RNA polymerase II is the enzyme in eukaryotic cells responsible for synthesizing RNA that will ultimately be translated into protein and, while much more complicated and more highly regulated, it retains many of the features of its bacterial counterparts. Even so, significant differences in specificity exist between the two systems and it is essential to use eukaryotic promoters and terminators to construct functional eukaryotic expression vectors.

DNA transcription in eukaryotic cells takes place predominantly in the nucleus. Prior to transit out of the nucleus, the mRNA is covalently modified at the 5′ end, corresponding to the transcription start site, with a 'cap' structure consisting of a 7-methylguanosine derivative. Furthermore, eukaryotic cells contain a specific RNA polymerase that is responsible for polyadenylation, which is the addition to the 3′ end of the mRNA, following the transcription terminator, of 100–200 adenosine residues. These modifications are believed to be important for transport of the mRNA out of the nucleus. While prokaryotic mRNA contains ribosome binding sites important for initiating translation, eukaryotic mRNAs lack specific sequences that interact with ribosomes to initiate translation. However, there is evidence that the 'cap' structure promotes association with ribosomes and aligns the mRNA so that translation can begin at the first translation initiator AUG.

3.8.3.4 Operators

Eukaryotic cells employ virtually all of the gene regulatory mechanisms exemplified by bacteria, including repressor systems. More frequently, however, they have highly regulated positively acting controls in which activator proteins bind promoter operator regions in consort with RNA polymerase II and turn on transcription. Like a repressor, the activity of an activator protein can be modulated in many different ways.

Genes associated with the late phase of infection by adenovirus are transcribed from a strong promoter (adenovirus major late promoter) that has been very useful in expression studies and exemplifies a positively controlled system. This promoter and its attendant control regions represent a system that is highly regulated by activator proteins and stimulating factors. Elegant *in vitro* studies[24] have demonstrated that transcription from the promoter by RNA polymerase II requires the involvement of several transcription factors that bind control regions on the DNA in the vicinity of the TATA sequence (Figure 10). While the initiation factors are sufficient for transcription, an additional factor, upstream stimulating factor (USF), has been shown to bind a region of dyad symmetry located 60 base pairs upstream from the initiation site and immediately adjacent to the binding site of one of the transcription factors. USF acts as a transcription stimulator and enhances transcription twentyfold.

The metallothionein promoter is another good example of a regulated eukaryotic promoter that has found utility in expression systems (Figure 10).[25] Metallothionein (MT) is a protein in mammalian cells that binds and detoxifies heavy metals; consequently, its expression is induced by the presence of heavy metals such as zinc and cadmium. The protein is also induced by the presence of dexamethasone, a glucocorticoid, and thus also serves as a model for the many glucocorticoid-regulated genes in mammalian cells. In order to investigate the nature of induction of the MT gene, about 800 nucleotides of DNA sequence containing the MT promoter and operator sequences, including a region of dyad symmetry common to all known MT regulatory sequences, were fused to the thymidine kinase (TK) gene from herpes simplex virus. The construct was then transfected into a rat cell line lacking TK, and the plasmid-borne TK served as a selection for transfected cells and also as a reporter molecule for the MT promoter. In a number of rat cell transfectants, TK mRNA was induced by as much as fortyfold by cadmium and by dexamethasone. Induction of the TK gene was not significantly affected by the position of the MT operator region relative to the promoter and to the structural portion of the gene. While much remains to be learned, it is clear that heavy metals and dexamethasone interact directly or perhaps in association with activator proteins with the operator of the MT gene. This information has been used to advantage in the construction of a number of high level eukaryotic expression systems.

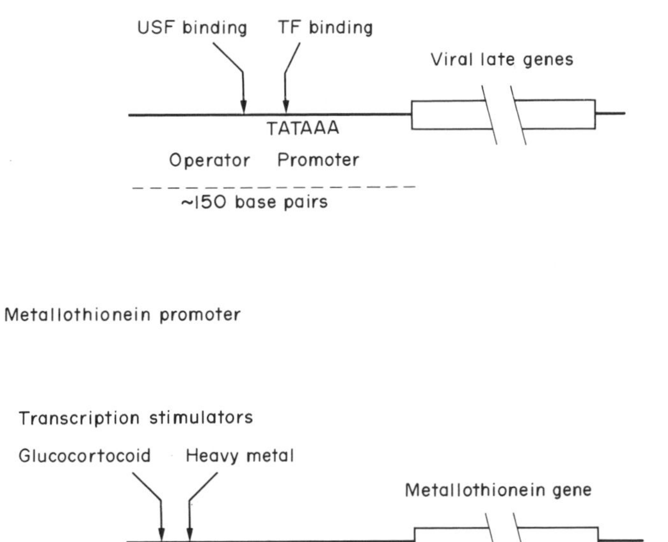

Figure 10 Positively controlled eukaryotic promoters. Adenovirus major late promoter has a binding site for transcription initiation factors (TF) near the promoter and a site for upstream stimulatory factor (USF) that binds TF and the operator in a cooperative manner to stimulate transcription. The metallothionein gene contains 800 bp of operator and promoter sequence that is positively regulated by glucocorticoids and by heavy metals

3.8.4 EXPRESSION STUDIES

3.8.4.1 Peptides

In the realm of peptide chemistry, bacterial expression systems encounter some important challenges. Classical methods of solution and solid-phase peptide synthesis have achieved a high degree of sophistication and any new method must compete on these terms. The level of difficulty, and cost, in classical peptide synthesis generally increases with increasing sequence length. Also, the technical limit to the size of a peptide that can be synthesized and purified is about 70 amino acids, depending on the experience of the laboratory. On the other hand, because small peptides are generally unstable in bacterial cells and because secretion efficiencies are rather low, it is more difficult to make small peptides by expression in bacteria. Are the two approaches actually complementary? The answer obviously lies in the specific peptide under consideration—every new peptide is a unique problem—however, a few conclusions are beginning to emerge. Classical peptide chemistry can provide small (milligram to gram) quantities of peptides very readily and is well suited to discovery projects where many peptides are needed quickly. Large-scale peptide chemistry at the kilogram level and above encounters high raw material costs, while development and capital costs are modest. Also, there are no significant savings in increases of scale in production. On the other hand, expression systems for producing peptides (and proteins for that matter) have relatively low raw material costs but have longer and more costly development requirements. Furthermore, production processes based on expression systems experience significant efficiencies in scale. Therefore, once a peptide candidate has been identified and targeted for development, the medicinal chemist does well to consider implementing an expression system for production of interim development quantities and ultimately for manufacturing the product. This is particularly the case if the peptide is very large or difficult to synthesize. While improved chemical methods for peptide synthesis continue to increase the achievable length, it appears that for multikilogram levels of production an expression system is most likely to win out. For the medicinal chemist involved in discovery, there is still no general approach for producing large numbers of pure small peptides by bacterial expression and synthesis is still clearly the method of choice. If activities focus on large peptides that exceed about 50–60 residues, it may very well be necessary to employ an expression system to achieve discovery objectives. Obviously, bacterial expression systems are not yet able to incorporate amino acid analogs in a useful way, so peptide analogs remain the realm of the synthetic chemist. It is important to note, however, that enzymatic coupling of amino acids and peptides has become a reality in the last few years and it is now possible to utilize a combined approach of synthesis to prepare analogs and enzymatic coupling to assemble the final product.

In this section, efforts to produce peptides by expression in microorganisms will be discussed.

3.8.4.1.1 Somatostatin

Somatostatin is an important peptide hormone that is present in the central nervous system and in peripheral tissues. Somatostatin inhibits the secretion of a number of hormones, including growth hormone, insulin and glucagon, that are important in anabolic processes. Aside from its intrinsic interest and potential therapeutic utility, Itakura *et al.*[26] chose somatostatin as a model system to explore the expression of small peptides and proteins in bacteria. When this work was published in 1977, early in the history of recombinant DNA research, it represented the first example of a biologically active human peptide expressed in bacteria. The principles established by the work continue today to guide concepts in peptide expression.

To study the expression of somatostatin, a 14 amino acid peptide, a synthetic gene was constructed from DNA fragments that had been chemically synthesized and enzymatically ligated together (Figure 11). The gene sequence was deduced from the known peptide sequence and included a methionine residue at the N terminus of the peptide and two stop codons at the end of the gene sequence. The synthetic gene included EcoRI and BamHI restriction sites that allowed it to be spliced into two different expression vectors that both utilized the lac promoter, operator and ribosome binding site. The first expression vector (pSOM1) incorporated the somatostatin gene immediately downstream from the lac promoter. The small peptide resulting from expression of the gene would be a hybrid composed of 10 amino acids followed by a methionine and somatostatin. In the second expression vector (pSOM11-3) the β-galactosidase gene was placed under control of the lac promoter and the somatostatin gene was inserted into an EcoRI restriction site located near the C terminus of the protein that maintained the somatostatin gene in the same reading frame with

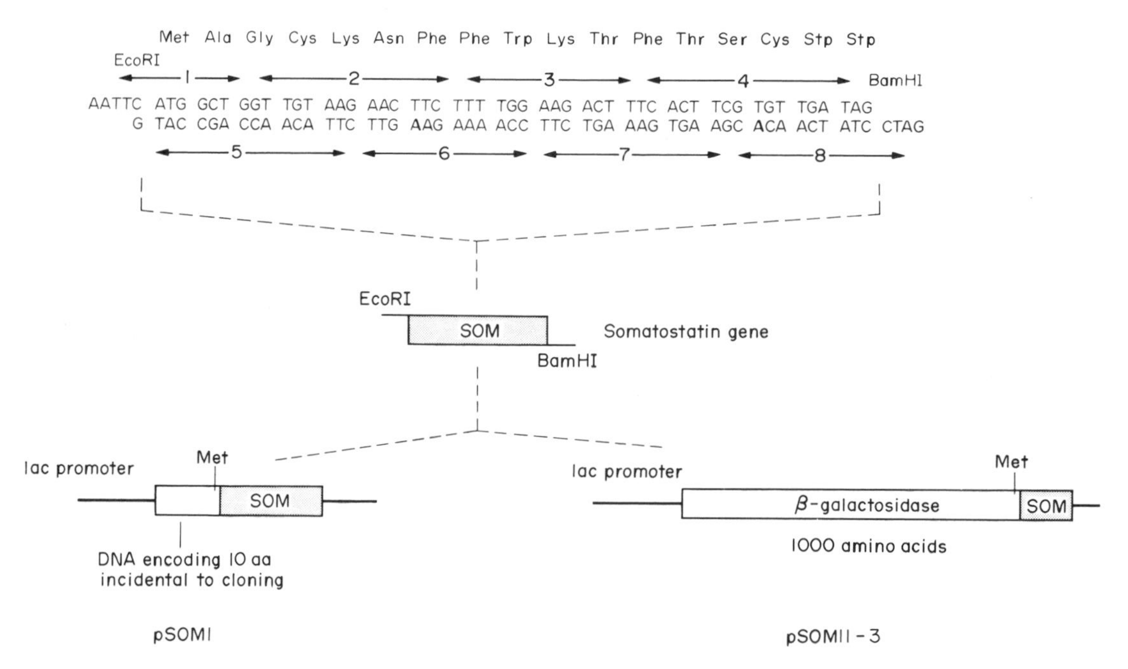

Figure 11 Expression constructs for somatostatin. Eight DNA fragments were synthesized and enzymatically ligated to prepare a somatostatin gene flanked by EcoRI and BamHI restriction sites. In pSOM1, the gene was spliced into an EcoRI/BamHI site downstream of the lac promoter. In pSOM11-3, the gene was spliced into an EcoRI site at the 3′ end of the β-galactosidase gene affording a C-terminal fusion

β-galactosidase. The resulting construct would produce a large fusion-protein between β-galactosidase and somatostatin. With the two expression vectors, it was possible to ask if a small peptide could be expressed directly in bacteria or if it would need to be attached to a larger more stable carrier protein. In both cases, somatostatin was to be cleaved from adjacent N-terminal sequences by treatment with acidic cyanogen bromide, a reagent that selectively cleaves peptides on the C-terminal side of methionine, and then detected by a sensitive radioimmunoassay (RIA).

When *E. coli* bearing multiple copies of the plasmid vector pSOM1 was grown in culture under various conditions and harvested, no somatostatin could be detected following cyanogen bromide treatment. Subsequently it was shown that when somatostatin prepared by chemical synthesis was added back to *E. coli* extracts it was rapidly degraded.

The fusion-protein construct pSOM11-3 was expressed and the extract was treated with cyanogen bromide; significant levels of somatostatin could be detected. Furthermore, the addition of β-galactoside further induced the promoter and increased the level of fusion protein and somatostatin by nearly tenfold. Even by today's standard, the production of fusion protein was quite respectable—about 3% of total cellular protein. However, because somatostatin was relatively small and because recovery was only 10%, recoverable somatostatin represented only about 0.01% of cellular protein.

Many studies in the intervening time have demonstrated that small peptides are almost always rapidly proteolyzed in *E. coli* and can be expressed directly only in very special cases. It is generally still necessary to prepare expression vectors in which a small- to medium-size peptide is fused to a 'carrier' protein to insure stability and acceptable levels of production.

3.8.4.1.2 Insulin

The great therapeutic importance of insulin in treating diabetes stimulated studies directed toward the production of insulin. The concepts established by the work with somatostatin were nicely extended to the expression and assembly of biologically active human insulin. However, insulin posed some unique challenges. The hormone is synthesized *in vivo* as a single chain prohormone that is proteolytically processed to the mature two-chain form where the two chains are joined by two interstrand disulfide bonds. The two chains, being different, required the construction and expression of two different constructs.

As in the case of somatostatin, insulin A- and B-chain genes were prepared from chemically synthesized DNA.[27] Each gene had EcoRI and BamHI restriction ends and a methionine located at the N terminus to allow for cyanogen bromide cleavage from the carrier protein. Two expression vectors were constructed[28] in which the A-chain gene and the B-chain gene were ligated into the EcoRI restriction site at the C terminus of the β-galactosidase gene (Figure 12). The respective plasmid constructs (pIA1 and pIB1) were each expressed and high levels (20% of total cellular protein) of the respective β-galactosidase hybrid proteins were obtained. When the fusion proteins were treated with cyanogen bromide, the insulin A- and B-chains were released from the β-galactosidase carrier and the cysteine residues were oxidized to *S*-sulfonates. Mixing of the two chains under conditions that promoted disulfide interchange afforded active insulin that could be readily purified from the mixture. Recovery of product was reported to be ten times better than somatostatin.

A complementary approach to the production of mature human insulin implemented a gene for the prohormone.[29] Normal intracellular processing of proinsulin removes a 35 amino acid peptide fragment, referred to as C-peptide, situated between the B chain and the A chain. In this study, the prohormone gene containing an N-terminal methionine was spliced into a gene (trpE) under control of the tryptophan promoter affording a fusion protein. After expression, cleavage with cyanogen bromide and cysteine *S*-sulfonation, the product was purified and submitted to disulfide interchange. Correct folding ensued, resulting in mature native proinsulin. Subsequent controlled proteolysis with trypsin and carboxypeptidase B afforded pure human insulin.

Yeast expression of insulin gene constructs has been quite successful in circumventing some of the problems inherent in the production of insulin by bacterial expression. Thim *et al.*[30] constructed a high copy plasmid vector for yeast that utilized the promoter and control sequences for yeast triose phosphate isomerase, one of the highly expressed genes involved in the glycolytic pathway. A cDNA copy of the human proinsulin gene from which the mammalian secretion leader had been removed was incorporated next to the secretion leader of the yeast α-mating factor. Several variant genes of proinsulin were also constructed, using oligonucleotide-directed mutagenesis techniques, where the C-peptide was deleted and short peptide spacers containing dibasic amino acids (Arg-Arg and

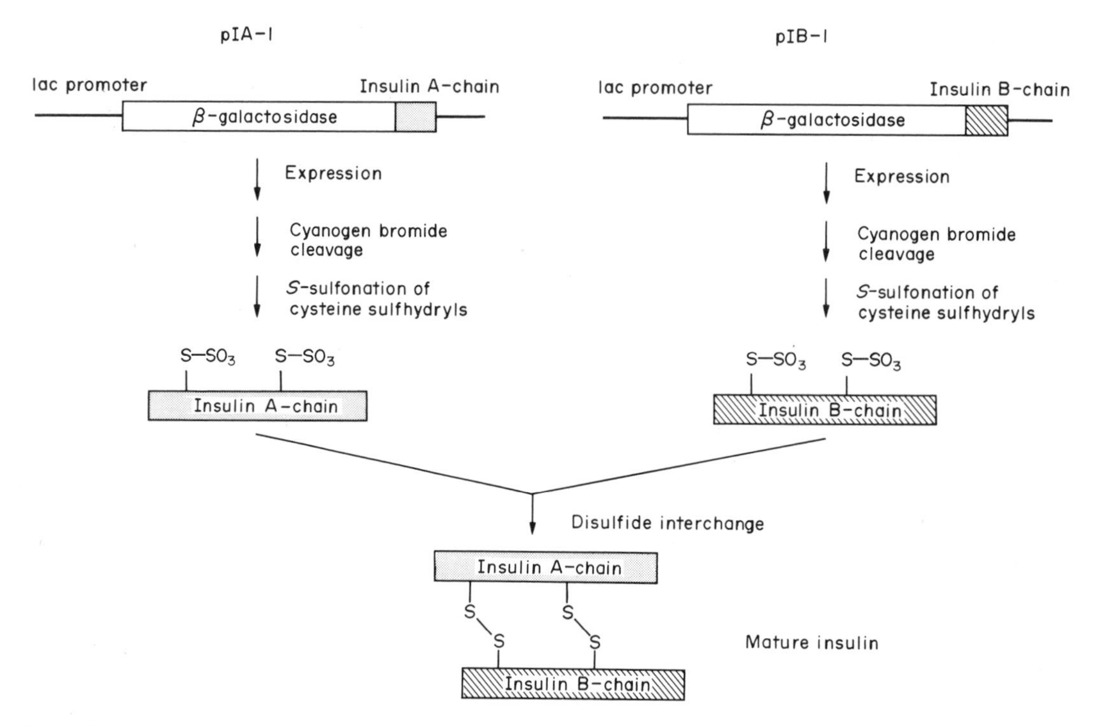

Figure 12 Expression of insulin A- and B-chains and preparation of mature insulin. Synthetic genes for the A and B chains were spliced into the EcoRI site at the 3′ end of the β-galactosidase gene. After expression, the fusion proteins were cleaved with cyanogen bromide and insulin chains were sulfonated. Subsequent mixing under disulfide interchange conditions afforded active insulin

Lys-Arg) were inserted between the B chain and the A chain. When these constructs were expressed in yeast, immunoreactive insulin was detected in the culture medium, indicating that properly folded insulin or insulin analogs had been secreted. Expression efficiency was excellent for yeast—as much as 10 mg of insulin per liter of culture. Upon closer examination, it was shown that the α-secretion leader had been properly removed but that processing of the spacers between the B and the A chains was incomplete. While direct expression of a properly processed product was hoped for, a nice solution to the problem was achieved by simply treating the product with trypsin followed by carboxypeptidase B. Enzymatic conversion to pure mature product were uniformly high and specific.

In this study, it was quite remarkable that by making an artificial prohormone by joining the two insulin chains with an unnatural peptide spacer, as short as two amino acids, correct folding and disulfide bond formation could be achieved. This observation attests to the value of the secretory pathway in offering a milieu conducive to proper folding, in addition to achieving secretion from the cell. It is apparent as well that the conformational constraints imposed on the artificial 'proinsulin' molecules by the short spacers were still compatible with correct alignment of the cysteine sulfhydryl groups required for correct folding. While expression levels for insulin were lower in yeast than can be achieved in well-tailored bacterial expression systems, the final yield of product was excellent by comparison, because direct expression in yeast circumvented the need for multiple processing steps required to obtain native material from bacterial expression systems.

3.8.4.1.3 Epidermal growth factor

Epidermal growth factor (EGF) has received considerable attention as a potential therapeutic agent. It possesses potent gastric antisecretory activity and also stimulates the proliferation of epithelial cells from a variety of tissues. Consequently, it has been viewed as a potential antiulcer agent and may have utility in accelerating wound-healing. EGF is a compact highly folded molecule composed of 53 amino acids containing six cysteine sulfhydryl groups that form intramolecular disulfide bonds. EGF is an excellent model for testing the limits of an expression system's ability to produce a relatively small molecule that requires extensive folding and disulfide bond formation.

Smith *et al.*[31] implemented an *E. coli* expression system designed around a synthetic gene driven by the tryptophan promoter. The gene sequence was deduced from the known protein sequence of EGF and codons were incorporated for two lysine residues at the N terminus of the protein (Figure 13). The construct did not place the EGF gene directly adjacent to the initiator codon but left a short stretch of nucleotides from the trpE gene and some linking sequence between the initiator AUG and the beginning of the EGF gene. It had been previously observed that EGF was stable to trypsin so the lysine residues were incorporated to allow for trypsin cleavage of the protein from the extraneous sequence. Upon expression in *E. coli*, active EGF was obtained at levels of 2–3 mg per liter of culture.

The same group devised a clever approach for enhancing the purification of recombinant proteins and implemented it in the context of hEGF.[32] Their idea was a variation on the theme of affinity purification where the molecule is endowed with a unique property that distinguishes it from other molecules in the expression milieu. Most proteins in *E. coli* are negatively charged and do not bind to a negatively charged affinity matrix (a cation exchanger). Therefore, a variant of EGF was designed that incorporated a sequence of five arginine residues on the C terminus of the protein (Figure 13). The construct was similar to the one implemented for intracellular expression; a synthetic gene containing five arginine codons on the 3′ end of the gene was incorporated into a plasmid expression vector containing the tryptophan promoter and operator. As in the earlier case, the construct also included a short sequence of the trpE gene next to the 5′ end of the EGF gene which resulted in the addition of a short peptide to the N terminus of EGF after expression. The expression of the polyarginine-tailed EGF was induced by the addition of indoleacrylic acid (a tryptophan analog) and intracellular accumulation of the product was observed by polyacrylamide gel electrophoresis performed on extracts of the cells. By comparison with the original intracellular construct, the level of accumulation of the product was considerably higher, presumably because the polyarginine sequence increased the stability of the protein in the cell. Furthermore, the properties of the hybrid protein product reflected high positive charge density—the p*I* was greater than 9.5 and the protein bound strongly to a cation exchange resin and was eluted with a relatively high salt concentration.[33] The resulting product was treated with carboxypeptidase B, which cleaves arginine from the carboxy terminus of proteins, and ion-exchange chromatography gave a product that was >95% pure. Mature EGF could then be obtained by digestion with trypsin. EGF ordinarily contains an arginine at its C terminus, so carboxypeptidase B actually removed the terminal arginine; however, this product had the same biological activity as the full length molecule. The affinity 'handle' employed in this study is generally applicable and could potentially be incorporated into any recombinant protein to enhance purification and recovery. Moreover, it was interesting and potentially important that the polyarginine fusion was more stable in the cell. Perhaps an analogous strategy would be appropriate for stabilizing intracellularly expressed heterologous proteins that are rapidly degraded.

The protein alkaline phosphatase is localized in *E. coli* in the periplasm and the signal sequence responsible for that localization has been investigated in the expression of EGF.[34] A gene was constructed using synthetic DNA that incorporated the *E. coli* alkaline phosphatase signal sequence at the N terminus of the protein separated by a signal protease cleavage site (Figure 14). The gene was expressed under control of the alkaline phosphatase promoter in the presence of low levels of phosphate which regulates the promoter, and EGF was secreted into the periplasm. Osmotic shock

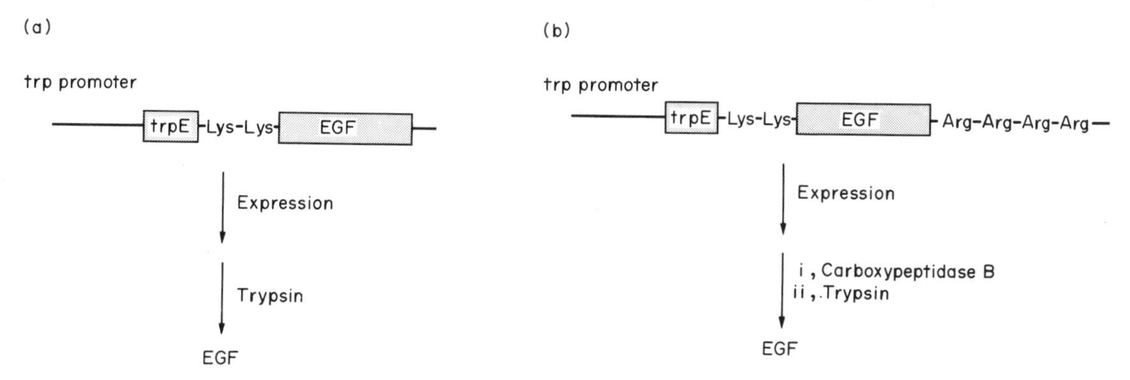

Figure 13 *E. coli* expression systems for epidermal growth factor (EGF). (a) The synthetic gene cloned in the trpE gene under control of the trp promoter. The TrpE sequences and EGF are separated by two lysine residues to allow for trypsin cleavage. (b) EGF gene with a polyarginine tail to facilitate purification. Removal was effected enzymatically under conditions where EGF was stable

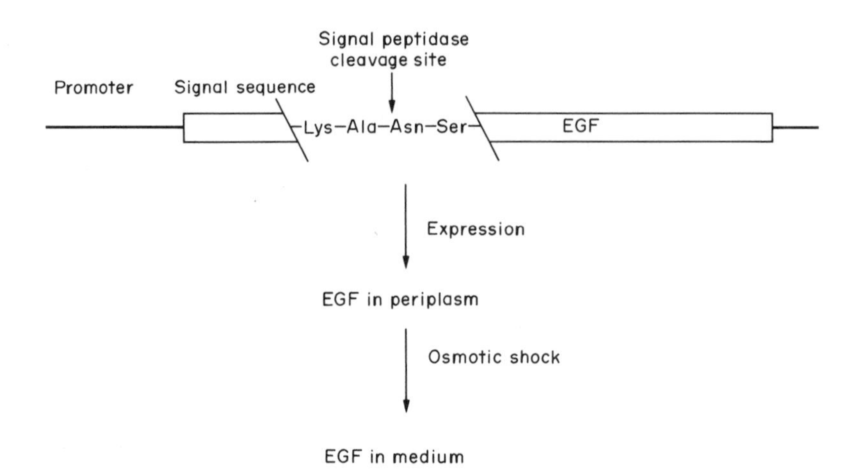

Figure 14 Secretion of EGF in *E. coli.* The EGF gene was fused to the signal sequence and promoter for *E. coli* alkaline phosphatase. Upon expression, EGF was secreted into the periplasm, and it was released into the medium by osmotic shocking of the cells

released most of the EGF from the cells and about 1 mg per liter of pure active protein could be recovered from the medium.

Expression studies on EGF have also been extended to yeast. Once again, the most facile entry to the gene for the relatively small protein was assembly of synthetic DNA fragments. In the first study,[35] the EGF gene was placed under control of the glyceraldehyde-3-phosphate dehydrogenase (GAPDH) promoter and alcohol dehydrogenase terminator sequences. GAPDH is a highly expressed protein that is part of the glycolytic pathway in yeast and it was expected to give high level expression of heterologous genes as well. GAPDH is a cytoplasmic protein and consequently it carries no secretion leader. The construct was expressed in yeast; however, production levels were very low, amounting to only about 30 μg per liter of culture. Apparently, expression levels were low or the protein may have been unstable in the cytoplasm of the yeast cells. Another series of constructs for yeast expression were examined where the EGF gene was cloned into the α-factor gene.[22] This construct included the α-factor promoter, the secretion leader, the Lys-Arg-(Glu-Ala)$_3$ cleavage sequence and terminator sequences. In one vector the natural α-factor cleavage site was retained, and in another the (Glu-Ala)$_3$ sequence was deleted while retaining the Lys-Arg. The latter construct was made to preclude the need for the dipeptidyl aminopeptidase activity that is required for complete processing to the desired product but is limiting in some highly expressing yeast cells. In distinct contrast to the disappointing results obtained with the intracellular yeast construct, both α-factor secretion vectors gave high levels (5–10 mg L^{-1}) of EGF secreted into the medium. As observed in other studies, the product requiring the limiting dipeptidyl aminopeptidase for correct processing still contained the (Glu-Ala)$_3$ sequence on the N terminus of the protein. On the other hand, the product that lacked the Glu-Ala repeat sequence, requiring only the trypsin-like cleavage at Lys-Arg for correct processing, afforded mature product.

3.8.4.2 Proteins

It is in the expression of large proteins that the techniques of recombinant DNA have truly revolutionized science. It did not take long, following the discovery of the tools and reagents of molecular biology, for important applications to appear. Early studies with proteins isolated from natural sources such as human and bovine growth hormone have clearly demonstrated the therapeutic importance of protein products, and small companies and large pharmaceutical concerns have aggressively sought to develop the technology and to bring new protein products to the marketplace. Concurrent with product development has been an equally committed effort in applying the tools of the biotechnology revolution to understanding the pathophysiology of disease and to implementing highly specific rational treatment modalities. This section will explore some of the applications of gene expression in the production of important proteins.

3.8.4.2.1 Growth hormone

Growth hormone (GH) is a 191 amino acid protein synthesized and secreted by the pituitary. It stimulates the growth of an organism and is responsible for mobilizing a vast array of biological responses necessary for growth. A deficiency of GH in humans results in dwarfism and stunted growth, and GH has been used successfully for a number of years in treating deficiency states in humans. Near normal growth patterns can be restored in deficient children if GH is administered while long bones are still lengthening prior to the attachment of the epiphyses to the bone shaft. Some proteins exhibit interspecies activity and treatment of human disease can be accomplished using material isolated from animals. This is the case with insulin where porcine insulin has been used for decades to treat diabetes. However, GH is species specific and, in the past, material for treating disease had to be isolated from cadaver pituitaries. This is a costly procedure, but of much greater concern has been the recent observation of a troublesome number of patients who have developed Kreutzfeldt–Jakob disease, a fatal neurodegenerative disease, after receiving pituitary hGH isolated from cadavers. The situation is believed to have resulted from contamination of the pituitary hGH preparations with a retrovirus. Consequently, in order to provide more readily available pure protein, hGH was identified as a highly desirable target for cloning and bacterial expression.

Goeddel *et al.*[36] implemented a range of techniques in the construction and refinement of an hGH expression system that with further optimization and considerable development has resulted in an important therapeutic and commercial product. An hGH gene was isolated and cloned in *E. coli* using cDNA cloning techniques. hGH is a secreted protein and the cDNA carried nucleotide sequences coding for a secretion leader that had to be removed, otherwise expression in bacteria would have afforded the prohormone that would probably not have been properly processed. The cDNA copy of the gene contained a unique HaeIII restriction site located at the position in the gene

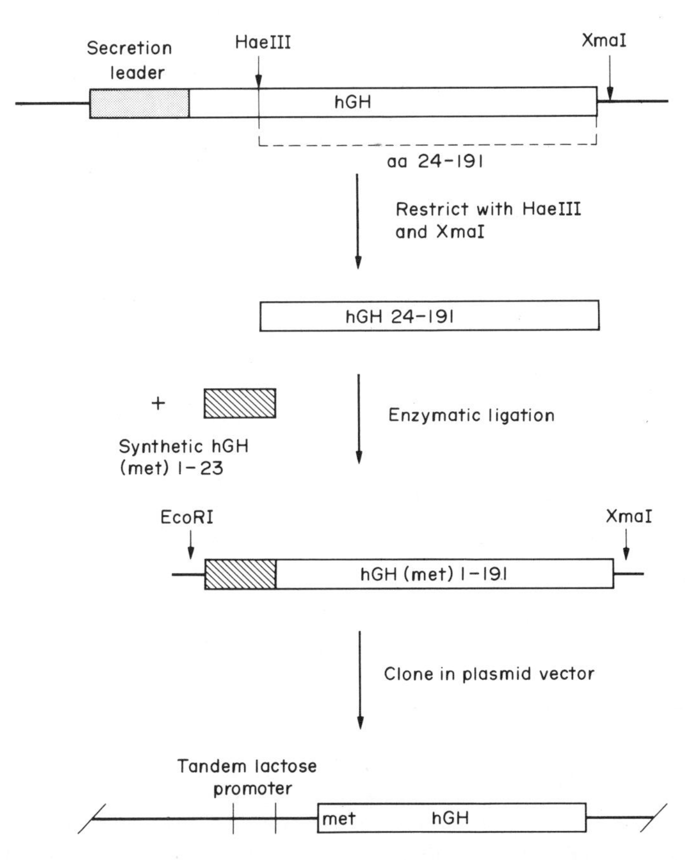

Figure 15 *E. coli* expression construct for human growth hormone (hGH). An HaeIII/XmaI restriction fragment of the hGH gene consisting of nucleotides encoding aa 24–191 was fused to a synthetic DNA fragment that reconstructed sequences for aa 1–23, including an initiator ATG. The resulting full length gene, as an EcoRI/XmaI fragment, was cloned in an expression plasmid under control of a tandem lactose promoter

corresponding to amino acid 24 and an XmaI site located in the untranslated sequences at the 3′ end of the gene (Figure 15). The cDNA was treated with HaeIII and with XmaI to provide a 512 base pair (bp) gene coding for amino acids 24–191 of hGH. Concurrently, an 84 bp DNA sequence containing an EcoRI 'sticky end' that reconstructed hGH 1–23 and included an ATG translation initiator codon was prepared by chemical synthesis. The two DNA fragments were enzymatically ligated to create the complete 1–191 sequence and the gene was inserted into a plasmid expression vector that contained a tandem lac promoter. The resulting construct was expressed in *E. coli* and mature product was detected in cell lysates to the extent of 1–3 mg dm^{-3}.

The hGH construct initially employed by Goeddel placed the translation initiator 11 bp downstream from the ribosome binding site. In the natural lac promoter system, the ribosome binding site is seven bp from the initiation codon and so, in order to increase expression, refinement of the construct was attempted by removing four bp from the DNA sequence to allow for the same spacing between the ribosome binding site and the AUG as observed in the natural promoter. In fact, the alteration actually decreased the expression level by half. This study and many others clearly indicate that a multitude of factors together determine a given expression level and what works in one instance may not be extensible to a different construct. Many principles about vector construction can be generalized but, in the end, final refinements require empirical testing.

In the context of bovine growth hormone (bGH), Matteucci and Heyneker prepared an expression vector containing a bGH gene and a tryptophan promoter in which an XbaI restriction site was placed between the ribosome binding site (RBS) and the initiation codon.[37] By inserting different sequences between the RBS and the translation initiation site, this construct allowed for testing the effect of sequence composition on expression efficiency while keeping the length of the sequence the same. Following restriction with XbaI, enzymatic manipulation with S1 nuclease and the Klenow fragment of DNA polymerase, an asymmetric site was generated into which a synthetic DNA duplex, 9 bp in length, was inserted. The synthetic fragment contained a wild-type sequence, the starting point, and a mixture of all possible combinations of the four nucleotides in the nine positions. The synthesis was conducted in a fashion that insured that only one or two changes would occur in each sequence inserted into the modified XbaI site. The resulting mixture of plasmids, differing from each other by variant sequences in the spacer region between the RBS and the AUG codon, was used to transform *E. coli*. The transformants were preselected for plasmids that contained one or two alterations from the wild-type sequence of the spacer region, and extracts of each of these transformants were screened for bGH activity using a sensitive radioimmunoassay. The general conclusion from this study indicated that introduction of guanosine residues in the spacer region significantly decreased expression, while increased amounts of A and T gave modest increases in translation efficiency. It was noted that the construct already gave high translation levels; indeed, there were no G residues in the wild-type spacer region, so positive mutations were modest in effect. The principle exemplified in this study, 'saturation mutagenesis' of a targeted sequence utilizing synthetic DNA fragments, is a powerful application to designing expression vectors. Obviously, any 'random' approach that generates large numbers of products requires a high throughput assay in order to screen for desired results. In a well-designed system, such as that exemplified by the bGH study, rationally targeting a specific domain of a vector or a gene allows for refinement of an expression construct and asks specific mechanistic questions. The principles clarified in the process can be quite useful in the specific instance, and they may be applicable to other systems as well.

Ikehara *et al.*[38] employed a synthetic hGH gene in constructing an efficient expression vector. By so doing, they were able to incorporate a number of strategic restriction sites in the gene that aided cloning and expression. Furthermore, they elected to use *E. coli* favored codons in the gene, anticipating that translation might be enhanced. This possibility is unique to synthetic gene construction because cDNA and genomic cloning afford genes that reflect the codon usage of the source organism. The construct utilized a tryptophan promoter and the trpE ribosome binding site that was contained on a unique HpaI–ClaI fragment of 32 bp, and the spacer region between the RBS and the initiator codon could be altered by switching the unique HpaI–ClaI fragment with synthetic fragments. In this fashion, vectors were constructed that had spacers between the RBS and the initiator codon of 8, 9, 11 and 13 nucleotides. Expression of the parental construct, containing an eight residue spacer, with induction by indoleacrylic acid was very good, affording 100 mg dm^{-3} of product. Consistent with the results of Goeddel was the finding that increased spacer length enhanced translation efficiency with a spacer of 11 nucleotides being optimal (50% enhancement) and a length of 13 being about the same as the parental construct.

Periplasmic secretion vectors have been designed that have allowed for successful secretion of hGH in *E. coli*.[39] Constructs including a tandem lac–trp promoter, a hybrid lac promoter and the

E. coli alkaline phosphatase promoter have been examined. Interestıngly, vectors incorporating either a bacterial secretion leader from alkaline phosphatase or the natural human signal sequence afforded comparable amounts of correctly processed secreted product. Moreover, the secretion leader responsible for transporting *E. coli* outer membrane protein (OmpA) to the periplasm, one of the most highly expressed proteins in *E. coli*, gave excellent results when fused to the hGH sequence.[40] Ultimate yields of secreted hGH were ten- to twenty-fold higher than in the previous secretion vectors.

3.8.4.2.2 ***Interferon***

In the last decade, the interferons have generated an enormous amount of scientific interest and have gained attention as potential therapeutic agents. These relatively small proteins (146–166 amino acids) are potent antiviral materials and they have been shown to be immune modulators and inhibitors of cellular proliferation. Furthermore, they have a direct inhibitory effect on tumor cell growth and may be effective in controlling malignancy. Interferons (IFs) fall into three general classes depending on their original cellular source. Interferon α (leukocyte) and interferon β (fibroblast) are each actually families of closely related molecules, while interferon γ (immune) appears to be a single unique molecule. The immunomodulatory and antitumor potency of IF-γ is substantially greater than the other molecules and it has been explored extensively as a candidate for clinical anticancer studies. Interferons are highly potent molecules and they are present in cells and in the circulation of higher organisms in extremely low concentration. Consequently, gene cloning and bacterial expression have been indispensable for characterizing the molecules and for producing sufficient quantities of material to allow for basic study and clinical implementation.

Early genetic studies on interferons focused on the cDNA cloning of genes from each of the classes. A cDNA copy of the gene for IF-α was incorporated into an *E. coli* expression vector composed of a tryptophan promoter, operator and ribosome binding site.[41] Because IF-α is a secreted protein, the cDNA included a secretion leader and expression of the intact cDNA in *E. coli* afforded significant levels of the protein as pre-IF-α that retained the secretion leader. A Sau3a restriction site was located near the beginning of the gene corresponding to amino acid 2 and it was possible to delete the portion of the gene coding for the secretion leader. The gene was reconstructed using synthetic DNA so that the final gene for the mature protein, including a translation initiator ATG, was fused directly to the promoter sequences. Expression of this construct in *E. coli* gave mature IF-α, albeit in relatively low amounts. A similar strategy was employed for the construction of an expression vector for IF-β,[42] and results similar to IF-α expression were obtained. Chemical gene synthesis has also been implemented for preparing IF expression systems and at the time they were prepared the synthetic genes, composed of nearly 600 bp, were the largest genes ever prepared by total synthesis.[43–45]

In the early stages of development, the expression systems for IF were inefficient; indeed, in many cases it was not possible to definitively show the presence of the desired protein by polyacrylamide gel electrophoresis in whole cell extracts, even though biological activity could be detected by sensitive IF assays. Subsequent refinements have significantly improved the expression systems. In the expression of IF-γ, Jay *et al.*[46] placed a synthetic IF-γ gene under the control of a synthetic promoter from bacteriophage T5 that had been shown to produce high levels of constitutive (continuous) transcription. Furthermore, they employed an efficient consensus ribosome binding site to enhance translation efficiency. The resulting expression system produced IF-γ at the level of 15–20% of total cellular protein.

The interferons have also served as an intriguing departure point for protein engineering. It was observed in early studies that the various IF species exhibited different relative responses in different cell lines and against different viruses. With a growing pool of genes for the various IF molecules, all of which were closely related and shared common restriction sites, attempts were made to mix and match the genes to create entirely new hybrid interferon molecules. The results have been quite interesting and the study exemplifies well the principles involved in the construction of hybrid molecules. In two separate studies by the Genentech group,[47,48] the domains of three different IF-α molecules, designated A, D and I, were interchanged (Figure 16). IF-D and IF-I are composed of 166 amino acids, while IF-A has 165 due to deletion of amino acid (aa) 44. Each of the genes share common BglII restriction sites, one at the position corresponding to aa 62 and the other at aa 151, as well as a PvuII site at aa 92. Restriction cleavage of the parental genes and reconstruction by interchanging the corresponding restriction fragments afforded, after expression, the hybrid molecules depicted in Figure 16. The hybrid molecules were tested for their antiviral activity against

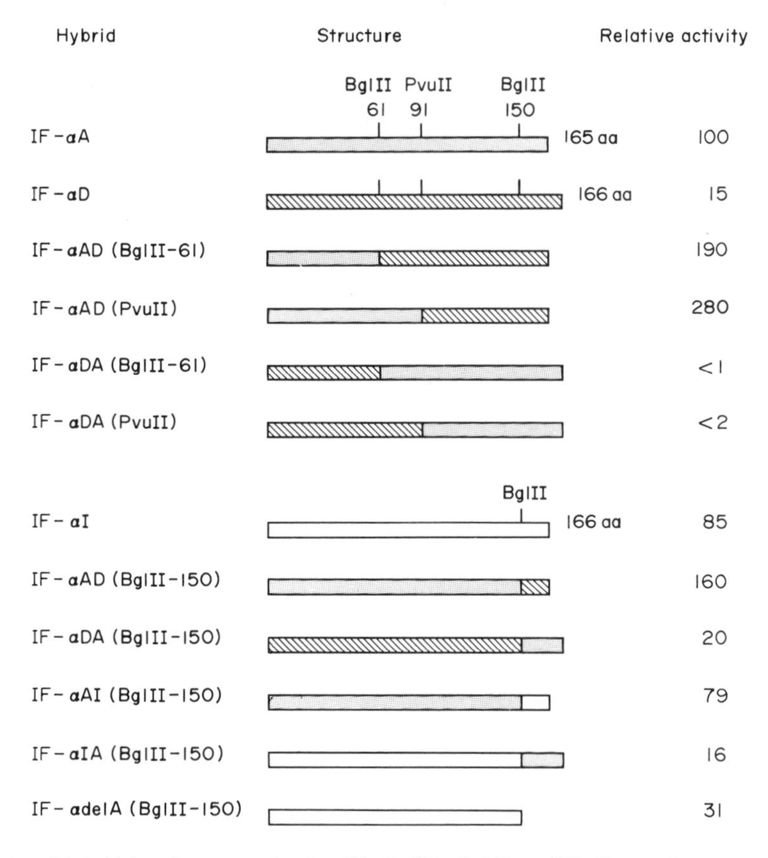

Figure 16 Activity of hybrid interferon α molecules. IF-αD (IF-α1), IF-αA (IF-α2) gene fragments were interchanged at common restriction sites to produce hybrid interferon genes. The products of expression were assayed for their ability to inhibit vaccinia-virus-induced plaque formation in a human amnion cell line (WISH cells). Activities are normalized to native IF-αA = 100

several viruses in a wide range of cell lines. Of particular interest was the result in a human amnion cell line (WISH) challenged by vaccinia virus (VS). The parental molecules IF-A and IF-I exhibited similar activity, while the activity of IF-D was tenfold lower. Hybrid molecules that retained an N-terminal fragment corresponding to A and that had C-terminal D or I fragments exhibited enhanced activity over any of the parental proteins. Furthermore, fusions composed of the N terminus of D and the C terminus of A were at least a hundredfold lower in activity. A comparable study has also been conducted on IF-β with similar results.[49] These studies did not allow for a detailed dissection of the factors contributing to activity; however, the hybrid molecules identified domains of activity that contributed to differential activity in the family of IF species.

The interferons have also been the object of yeast expression and excellent results have been obtained. Zsebo *et al.*[23] have examined the expression of a hybrid IF-α having a consensus sequence (IF-conα) derived from the various IF-α species whose genes have been cloned. A synthetic gene (the only possible source for a consensus sequence) was spliced into the yeast α-mating factor gene. In one construct, the wild-type cleavage sequence was retained and, in another, the Glu-Ala repeat sequence was removed to place the protein next to the dibasic cleavage site (see Section 3.10.3.2.3). Both constructs were effective in producing very high levels of product secreted into the periplasm—100 mg dm^{-3} or about 20% of all secreted protein. The wild-type construct afforded mature processed IF-conα, while the construct producing the protein with only a dibasic cleavage site was not processed. Small peptides that are expressed under control of the α-factor secretion system are efficiently secreted into the fermentation medium, a point that has been well demonstrated in the case of EGF (Section 3.8.4.1.3), calcitonin, endorphin and atrial natriuretic peptide.[50] It was significant, therefore, that the secreted IF-conα was retained in the periplasm and not subsequently released into the medium. It appears that larger proteins can be secreted into the periplasm but that only small peptides can traverse the cell wall barrier into the medium. These issues need to be further clarified. A second important point can be made from the study. IF-conα has two cysteine disulfide bridges that result from proper folding and are important for activity. Many expression studies on

IFs in *E. coli* are complicated by misfolded and aggregated products that result from polymerization through the cysteine sulfhydryl groups. In contrast, proteins containing disulfides that are expressed in yeast secretion vectors generally fold without difficulty and without yield losses due to polymerization, apparently because the yeast secretion pathway includes a disulfide isomerase that assures proper folding. This is also a considerable advantage in the production of small peptides, because chemical synthetic methods require an extra oxidation step to obtain active products and this frequently results in extra processing difficulties and expense as well as yield loss.

3.8.4.2.3 Superoxide dismutase

Normal metabolic processes in aerobic cells and organisms are responsible for the generation of reduced oxygen species such as superoxide and hydroperoxide. These materials are potent oxidants and are capable of inducing oxidative damage in the cell that can compromise function and viability. Consequently, under normal circumstances, they are rapidly and specifically degraded. The enzymes principally responsible for the degradation are superoxide dismutase (SOD) and catalase, respectively. It has been shown that when tissues are deprived of oxygen, as in the case of ischemia or anoxia, and are then reperfused, a large increase in superoxide concentration results. Furthermore, other studies have shown that when allopurinol or SOD are administered at the time of reperfusion, a dramatic decrease in damage to the ischemic tissue is observed. Moreover, superoxide is a product of phagocytic cells and appears to contribute to damage in immunoinflammatory conditions, such as osteoarthritis, where bovine SOD has been used clinically.

It is believed that the biochemical mechanism[51] for superoxide generation in reperfused anoxic tissue involves xanthine oxidase. In normal tissues and cells, hypoxanthine and xanthine (intermediates in purine metabolism to uric acid) are oxidized by xanthine dehydrogenase, which uses NAD as the reducing equivalent; however, when a tissue becomes anoxic, xanthine dehydrogenase is converted irreversibly to xanthine oxidase which utilizes molecular oxygen as the reducing equivalent, generating superoxide in the process. Upon reperfusion and reoxygenation of ischemic tissue, xanthine oxidase produces superoxide.

The potential utility of SOD in preventing post-ischemic damage induced by superoxide has stimulated the development of expression systems for the enzyme. Normally, cytoplasmic SOD is acetylated, a post-translational modification characteristic of eukaryotic cells. An *E. coli* expression system for human SOD (hSOD) was implemented that, with refinements of the ribosome binding site sequence and the sequence in the first few codons of the gene, was very effective in producing high levels of an active protein.[52] However, it was not acetylated and therefore was not identical to the native human enzyme. A yeast expression system was then designed that sought to produce intracellular product where acetylation might take place.[53] A hSOD cDNA was inserted into a plasmid containing the promoter for the GAPDH gene, a constitutive promoter that produces high cytoplasmic levels of the glycolytic enzyme. In stationary cultures of yeast transformed with the plasmid construct, extraordinary levels of hSOD were observed, amounting to as much as 70% of the cellular protein. hSOD is a very stable protein and may have accumulated as a consequence—its levels during exponential growth were much less. When the protein was isolated and purified, it was shown definitively that the N-terminal alanine residue was acetylated. This observation is quite important as it suggests that the yeast 'acetylase' is reasonably homologous with the human enzyme, and yeast may be an appropriate system for expressing human proteins that are normally acetylated.

3.8.4.3 Glycoproteins

There are a wide variety of mammalian proteins that require glycosylation for biological activity. In most cases it is not entirely clear what role glycosylation plays at the molecular level. There is evidence to suggest that glycosylation aids in protein folding or is involved in subunit assembly. Glycosylation is also strongly implicated as a targeting mechanism, both for trafficking of proteins inside the cell and also upon appearing on the cell surface. Once secreted, the oligosaccharides on a glycoprotein may determine its tissue localization and its lifetime in the circulation. Alternatively, if glycosylation served an intracellular function, once the protein is secreted the oligosaccharide becomes superfluous. In such instances, an unglycosylated product obtained from bacterial expression may be entirely acceptable.

Glycosylation is a post-translational modification characteristic of eukaryotic cells and, in principle, expression of human proteins that require glycosylation might be expressed in other eukaryotes such as yeast. However, glycosylation in yeast is restricted to the high mannose class of oligosaccharides. Very little is known at this point about whether or not yeast will glycosylate mammalian proteins at the same positions or, indeed, if high mannose forms of glycoproteins will exhibit the same activity as proteins with complex oligosaccharides obtained from mammalian cells.

3.8.4.3.1 **Renin**

The renin–angiotensin system is intimately involved in blood pressure homeostasis. When blood pressure falls or when salt concentration in the blood decreases, the kidneys release the proteolytic enzyme renin that acts on angiotensinogen present in the circulation to produce the peptide angiotensin I (AT I). This inactive peptide is then further processed by angiotensin converting enzyme (ACE) to produce angiotensin II (AT II), an exceedingly potent vasoconstricting agent. AT II directly constricts the vasculature to effectively decrease the vascular volume, and it also acts on the kidney to retain water and sodium which increases blood volume. Both actions of angiotensin serve to increase blood pressure. It has been well demonstrated using receptor antagonists of AT II and inhibitors of ACE that hypertension can be effectively controlled by inhibiting the production or the action of AT II. Renin is the actual control point for the renin–angiotensin system; angiotensinogen circulates freely in non-limiting concentration and, once formed, AT I is rapidly converted by ACE to AT II. Consequently, there is great hope that inhibitors of renin may be ideal antihypertensive drugs.

Many renin inhibitors prepared to date have shown significant differences when tested against renin isolated from different species, and it is now apparent that any work on renin inhibitors must be developed in the context of the human enzyme. There is, however, one caveat—circulating renin concentrations are low and it is difficult to obtain development quantities of the enzyme from natural sources. Moreover, every medicinal chemist ultimately hopes for a crystal structure to aid in drug design, a requirement that demands milligram quantities of protein. To meet these needs, renin has been the target of expression studies to produce enough enzyme to assist in developing renin inhibitors as antihypertensives.

Initial expression studies in *E. coli* succeeded admirably in the production of renin protein. Expression vectors were designed around a renin cDNA clone that encoded preprorenin and the tryptophan promoter and operator sequences.[54,55] Intracellular expression afforded high levels of aggregated protein as inclusion bodies; the excellent expression results were to no avail, however, because exhaustive attempts to refold the renin to obtain active enzyme were completely unsuccessful.[56] This folding problem is occasionally encountered in bacterial expression systems, especially with cysteine-containing proteins that must be oxidized to disulfides to achieve an active conformation. It appears that when proteins are aggregated in inclusion bodies, they exist in a reduced form and correct folding and oxidation required for activity must be preformed *in vitro*. It is frequently possible to successfully refold a protein from inclusion bodies, as in the case of the growth hormones which contain two disulfides (see Section 3.8.4.2.1), but with renin correct refolding was not possible.

Natural human renin and prorenin are secreted glycoproteins, so the next attempt at expression focused on mammalian cell production. Fritz *et al.*[57] incorporated a renin cDNA into a plasmid vector containing the human metallothionein promoter. This plasmid was cotransfected with a second plasmid containing an antibiotic-resistance marker into Chinese hamster ovary (CHO) cells. Following selection and expansion of the transformed cells, maximal expression was achieved by adding zinc to induce the promoter and human prorenin was secreted into the medium. By treating the prorenin with trypsin, mature human renin was readily obtained that was biologically equivalent with natural human renin. Another study by Evans *et al.*[58] used a similar construct with a mouse metallothionein promoter and transcription termination sequences from the SV40 virus. The construct included the bovine papillomavirus genome that exists in the cell as a multicopy episome just like a plasmid, thereby amplifying the renin gene sequences. Excellent results were obtained from mouse C-127 cells that could be maintained for months in culture while secreting constitutive levels of 4 $mg\,dm^{-3}\,day^{-1}$ of human prorenin. Using this expression system, the authors were able to obtain 12 mg of prorenin in 40 days. The prorenin was stable for extended time periods and could be activated by mild trypsinolysis to fully active product whose sequence was proven by protein sequencing to be identical with natural human renin.

These studies on renin exemplify the great utility that mammalian cell expression systems offer in the production of properly folded fully active glycoproteins. It is clear that the cellular milieu offered to secreted proteins by mammalian cells has oxidative potential that ensures proper disulfide formation.

3.8.4.3.2 Tissue plasminogen activator

Tissue-type plasminogen activator (tPA) is a serum glycoprotein that exhibits serine protease activity in specifically converting plasminogen to plasmin. Plasmin in turn cleaves fibrin (and fibrinogen) to small soluble peptide fragments. This proteolytic activation of plasminogen by tPA is significantly increased when tPA is bound to fibrin. Because blood clots are composed, in part, fibrin, tPA is endowed with an important property—it is able to activate plasminogen in the matrix of a blood clot which leads to fibrin degradation and clot lysis. Hence, tPA has the potential of inducing thrombolysis under conditions where systemic activation and attendant bleeding problems do not occur. This potential has been in large measure demonstrated by clinical experimentation and tPA represents a very attractive therapeutic product for the treatment of thrombotic disorders.

Cloning studies by Pennica *et al.*[59] succeeded in providing a full length cDNA copy of the tPA gene. This gene was initially altered to remove the secretion leader and was expressed at low levels in *E. coli* from a plasmid vector containing the trp promoter sequences. The resulting product was reported to exhibit fibrin-mediated plasminogen activation and fibrin degradation.

tPA was also incorporated into a yeast expression vector under control of the phosphate-inducible alkaline phosphatase promoter.[60] Active glycosylated product was observed at relatively low levels in the cell membrane fraction, but secreted tPA was not observed regardless of whether the alkaline phosphatase or the natural tPA signal sequence was present.

tPA is a glycoprotein and some questions remained concerning the role of carbohydrate in its activity. It appeared from the *E. coli* expression study that tPA retained its fibrin-mediated activity in the absence of attached oligosaccharide. This conclusion was borne out by growing tPA-producing mammalian cells in the presence of tunicamycin to inhibit glycosylation and, in a companion experiment, by treating tPA isolated from tissue culture with endoglycosidase H, an enzyme that removes N-linked oligosaccharides from proteins.[61] Both products retained fibrin-mediated plasminogen activation. However, the crucial question remains: what is the difference between fibrin-stimulated activity and basal activity in the absence of fibrin? It is clear that the difference is large for natural tPA; however, in tPA variants, if the basal unstimulated activity of tPA is the same as it is in the presence of fibrin, there is no real advantage—systemic plasminogen activation and hemorrhagic problems would be expected. There is some evidence that carbohydrate modulates the activity of tPA.[63,75] The specific activity of tPA containing three oligosaccharides is about 30% less than tPA containing two oligosaccharides. For this reason and others, tPA glycosylation may be important for optimizing the performance of tPA as a thrombolytic agent and it cannot be overlooked; hence, mammalian expression systems are generally favored for tPA production.

Another interesting principle that has been effectively explored in the context of protein expression is the idea that some proteins are a mosaic of autonomous structural domains that impart independent functions on a protein. The genes for many mammalian proteins are composed of separate exons that are spliced together to afford a mature mRNA that is translated into protein and, interestingly, autonomous functional domains are frequently organized at the gene level on distinct exons. Furthermore, families of proteins that have similar functional activity frequently have similar gene organization including exon/intron structure. It is interesting to speculate that different functional proteins each encoded by a single contiguous DNA sequence may evolve independently and that a new multifunctional protein may result from mixing and matching, at the gene level, of the separate proteins. Fine tuning of any protein activity to suit the selective environment of an organism might then provide for mutations away from the initial gene/protein structure. If this were the case, one might expect to observe strong sequence homologies between the exons and the functional domains of proteins exhibiting similar activity. tPA is a multifunctional protein—it has serine protease activity and it binds to fibrin. Does tPA exhibit similar exon sequences and are the corresponding protein domains functionally homologous to other serine protease and fibrin-binding molecules? Indeed, tPA shares striking sequence homology with trypsin-like serine proteases and its 'finger'and 'kringle' domains are homologous with corresponding domains in the fibrin-binding proteins prothrombin and fibronectin. Furthermore, tPA has another domain that is homologous with epidermal growth factor. van Zonneveld *et al.*[62] have made tPA variants that contain deletions of the various domains and they have expressed the genes in a mammalian expression vector. As shown in Figure 17, deletion of the second kringle domain that is homologous to the prothrombin kringle and deletion of the fibronectin-like finger domain eliminate fibrin-promoted plasminogen activation and only basal proteolysis is observed. Elimination of the finger domain alone does not affect the activity, while elimination of only the second kringle gives a product that is weakly

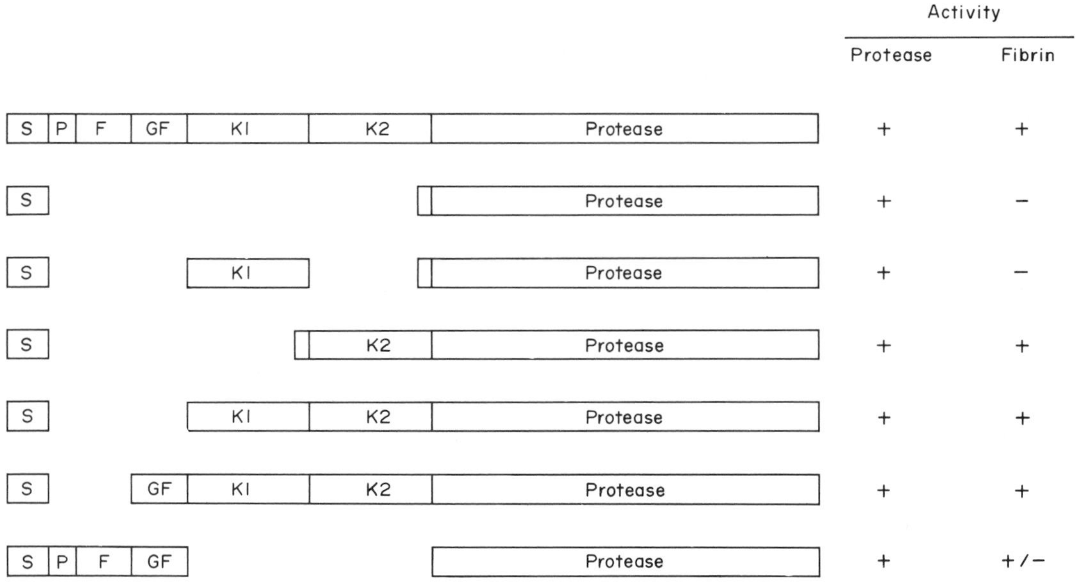

Domains							Activity: Protease	Activity: Fibrin
S	P	F	GF	K1	K2	Protease	+	+
S						Protease	+	−
S				K1		Protease	+	−
S					K2	Protease	+	+
S				K1	K2	Protease	+	+
S			GF	K1	K2	Protease	+	+
S	P	F	GF			Protease	+	+/−

Figure 17 Deletion variants of tPA. Activity of the variants was determined with a chromogenic substrate for plasmin in the absence of fibrin to determine basal proteolytic activity, and in the presence of fibrin to determine fibrin-stimulated activity. S = signal sequence, P = propeptide, F = finger domain, GF = growth-factor-like domain, K1 and K2 = kringle domains homologous with prothrombin kringle, protease = serine protease domain

activated by fibrin. One can therefore conclude that both the finger and the second kringle domains may be involved in fibrin binding. Within the limits of the assays employed, deletion of the growth factor domain and the first kringle had no significant effect on the activity.

3.8.4.3.3 α1-Antitrypsin

Neutrophils are involved in the production of elastase, the proteolytic enzyme involved in the degradation of elastin, collagen and various structural elements of lung tissue. This activity is believed to be important for tissue remodeling. Under normal circumstances, elastase activity is controlled by the specific protease inhibitor α1-antitrypsin (α1AT) that forms a strong bimolecular complex with elastase. When elastase activity is uncontrolled due to genetic deficiency or to underproduction or inactivation of α1AT, elastase is responsible for a number of serious pathophysiological conditions. Excessive elastase activity is characterized by degradation and destruction of connective tissue in the lung leading in turn to chronic obstructive pulmonary disease and emphysema. Supplementation therapy with α1AT has been shown to relieve the symptomology of emphysema and it represents an attractive course of action for individuals suffering from this condition.

It was possible to isolate a cDNA copy of the human α1AT gene and it was expressed at high levels in *E. coli*[64] using a plasmid vector bearing the pL promoter and operator sequences from bacteriophage λ. The *E. coli* host cell line carried a temperature-sensitive λ repressor that is inactivated at higher temperature. By growing the cells to high density at 28°C where the repressor was functional and then raising the temperature to 42°C where the repressor was inactivated, the pL promoter was induced and high level transcription of the α1AT gene proceeded. The actual product was a fusion between mature α1At and 17 N-terminal amino acids from the expression vector; nevertheless, the product exhibited similar antielastase activity compared with native material.

Yeast expression vectors employing a cDNA were also constructed that were successful in producing α1AT.[65] When expression was attempted using a gene that retained the natural secretion leader, it was not possible to obtain active product—the yeast cells did not process the precursor. However, by removing the DNA sequences encoding the secretion leader and expressing the mature protein directly, it was possible to obtain active α1AT. Examination of the activity and stability of bacterial and yeast-produced α1AT demonstrated that the unglycosylated protein had the same activity as the natural glycoprotein but it was significantly less stable to heat and to oxidation. Moreover, the rate of clearance from the circulation of unglycosylated α1AT was significantly faster than the natural protein. In another study, it was shown that α1AT could be obtained from yeast in glycosylated form when the secretion signal was retained in the gene; however, activity of the product was not reported.[66]

Garver *et al.*[67] have utilized a retroviral vector containing an α1AT cDNA and have succeeded in getting stable integration of the gene into the genome of mammalian cells. The vector consisted of sequences from Moloney murine leukemia virus (MMLV) and the simian virus 40 (SV40) early promoter with an antibiotic-resistance marker. Using this construct it was possible to obtain secretion of native glycosylated α1AT that was functionally indistinguishable from the natural molecule. While the expression efficiency was relatively low—indeed, one might expect refinements of the expression system to significantly improve production levels—the whole study portends the possibility of eventual gene 'transplants'. The lung is normally populated by alveolar macrophages that develop from bone-marrow-derived precursor cells, and it has been shown that these cells can be transformed with heterologous genes; therefore, incorporation of a wild-type α1AT gene into precursor cells of an individual lacking the α1AT gene might give rise to mature alveolar macrophages that could produce active α1AT in the lung and prevent elastase-induced pathology. Such are the dreams of molecular biology—induce cells to produce their own medicine, at the site where needed! And, as will be discussed in Section 3.8.8, the dream is not far from reality.

3.8.4.3.4 Coagulation Factor IX

The blood coagulation cascade is a complex series of proteases and interacting protein factors that orchestrate signals leading to blood clotting. Factor IX is a serine protease that is a central component in the cascade and exists in zymogen form. It is activated through proteolysis by Factor XI and it then acts in consort with Factor VIII, calcium ions and phospholipid to cleave Factor X. The absence of Factor IX in the circulation causes the clotting disorder hemophilia.

Factor IX is an extremely challenging protein and it is a prototypical model for complex post-translational modifications. Factor IX is synthesized in the liver and it contains 12 γ-carboxyglutamate residues that result from a vitamin-K-dependent enzymatic process in the ER, one β-hydroxyaspartate and three oligosaccharides, all of which are required for biological activity. Moreover, the protein is secreted and contains numerous disulfides, implying the need for extensive folding to achieve an active conformation. Obviously, mammalian cell expression is the only system likely to provide the product.

Several laboratories have prepared mammalian cell expression vectors for producing Factor IX in cell culture. The successful attempts incorporated a cDNA for Factor IX into plasmid and virus vectors that gave stable integration of the gene, and active Factor IX was secreted into the culture medium.[68,69] Expression levels were very low, so Kaufman *et al.*[70] transfected CHO cells deficient in the enzyme dihydrofolate reductase (DHFR) with a plasmid containing a copy of the DHFR gene in addition to a Factor IX gene driven by the adenovirus late promoter. By growing the transfected cells on increasing concentrations of methotrexate, an inhibitor of DHFR, the copy number of the DHFR gene and the Factor IX gene was significantly amplified providing cell lines that produced very high levels of glycosylated Factor IX protein. Levels of active protein could be increased by supplementing the culture medium with vitamin K; however, the active product was only a small fraction of the total Factor IX protein present. The problem resulted from inadequate γ-carboxylation, the crucial modification required for activity, apparently because the high level of protein synthesis exceeded the capacity for γ-carboxylation. Further refinements of the system will be required to make Factor IX in commercial quantities, perhaps by selecting or developing a cell line that has increased γ-carboxylation capacity.

3.8.4.3.5 Coagulation Factor VIII

If Factor IX is representative of the complexity of post-translational modifications that may be required of an expression system, FactorVIII exemplifies the challenges offered by large proteins. Factor VIII is a gigantic glycoprotein; it is composed of 2351 amino acids and has a molecular mass greater than 330 000, its gene consists of 26 exons distributed over nearly 200 000 bp of DNA, and its mRNA is 9000 bp in length. The cloning of the gene, accomplished by two groups at Genentech[71–73] and Genetics Institute,[74] required virtually every technique and trick in the molecular biologists armamentarium. It was not possible to obtain a full length cDNA and so a combination of genomic cloning and primed cDNA synthesis were required to obtain the full length gene.

Expression was then accomplished by transfecting mammalian cells with plasmids that incorporated the cDNA clones under the control of viral promoters and polyadenylation signals and active product was obtained from cell culture.

3.8.5 RECEPTOR STUDIES

Cell surface receptors are critical cellular components that are required for cells to detect external conditions and to respond in an appropriate manner. In some instances, receptors cause specific cells to adhere to each other by serving as sites of physical interaction. This activity is extremely important in a wide variety of developmental functions and in the organization of tissues and organs. Other receptors exist to select nutrient molecules from the extracellular medium and transport them into the cell. In a related sense, other cell surface receptors operate by clearing unneeded or undesirable molecules from the medium by binding the molecules, internalizing them and then transporting them to an intracellular compartment for degradation. Yet other receptors bind very specific messenger molecules, such as hormones and growth factors, and thereby monitor the external chemical environment of the cell. The receptor then generates signals within the cell that induce appropriate intracellular responses. Receptors are involved in one way or another with virtually every biochemical pathway in the cell. Because they play such an important role in mediating biological activities and because they are highly specific, receptors represent important sites for therapeutic intervention. Moreover, because receptors have extracellular domains and are accessible to the circulation they are attractive drug design targets.

Receptors are generally large complex proteins that are usually glycosylated and they may have other post-translational modifications. They are also frequently modular proteins composed of independent structural units that operate cooperatively to orchestrate a relatively complex sequence of events.

3.8.5.1 EGF Receptor

The epidermal growth factor (EGF) receptor, a prototypical receptor, is depicted schematically in Figure 18. The receptor is composed of a cysteine-rich extracellular domain, a hydrophobic transmembrane sequence and an intracellular tryosine kinase domain. The receptor functions by specifically binding EGF outside of the cell and transduces a signal to the intracellular tyrosine kinase domain that stimulates functions inside the cell by phosphorylating tryosine residues on itself (autophosphorylation) and on crucial intracellular proteins. The mechanism by which EGF occupancy of the binding domain is sensed by the tyrosine kinase domain is poorly understood. EGF may induce a conformational change that activates the kinase or it may involve the crosslinking and clustering of separate receptors through their cysteine-rich regions that bring separate tyrosine

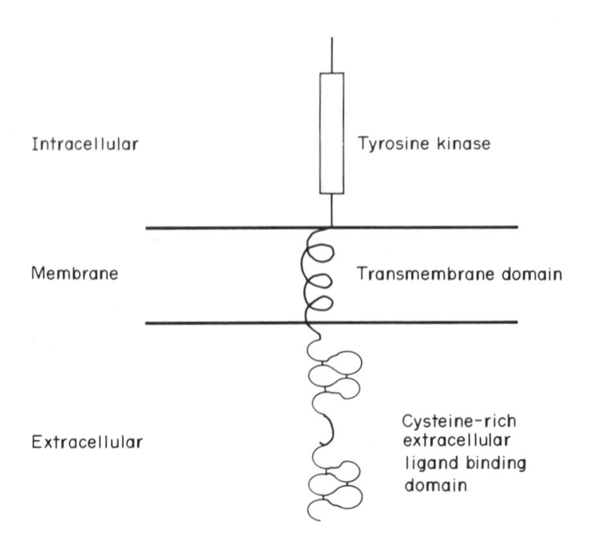

Figure 18 Schematic representation of the epidermal growth factor (EGF) receptor

kinase domains together in a multimeric assembly which induces phosphorylation. The EGF receptor gene has been cloned and sequenced in several laboratories,[76] as have an increasing number of other receptors, including, for example, the receptors for platelet-derived growth factor (PDGF),[77] insulin-like growth factor 1 (IGF-1)[78] and interleukin 2.[79] The sequence information and genetic probes thus obtained are intrinsically useful. By comparing their sequences, it has been possible to group receptors into families based upon sequence homology and inferred structural similarity. As will be discussed below, these kinds of comparative studies have been particularly important in cancer research by correlating the structure of oncogenes to growth factor receptors and their ligands and to molecules involved in growth factor signaling in the cell. In addition, it is now feasible to produce quantities of receptor in mammalian cell expression systems that make it possible to consider structural and biological studies including crystallization that would otherwise be very difficult.

3.8.5.2 Chimaeric Receptors

The modular nature of receptors and the interaction of separate domains to achieve vectorial signal transduction was nicely demonstrated by Riedel *et al.*[80] in the construction of a chimaeric receptor. In this study, the extracellular domains were interchanged between two different human cell surface receptors, the EGF receptor and the insulin receptor, both of which are tyrosine kinases. The chimaeric receptor, composed of the insulin receptor signal sequence and extracellular domain and the EGF receptor transmembrane sequence and tyrosine kinase domain, was constructed from the respective cDNAs and successfully expressed in COS cells, a monkey cell line. The chimaeric receptor was correctly processed and bound insulin to a similar extent as the native insulin receptor. Furthermore, insulin binding induced receptor autophosphorylation at ATP concentrations characteristic of the native EGF receptor. Thus, the insulin binding domain was able to communicate with the intracellular tyrosine kinase domain of the EGF receptor even though the extracellular domains of the two receptors are very different. In other words, insulin was able to transduce an EGF signal to the inside of cells bearing the chimaeric receptor.

3.8.5.3 Oncogenes

Molecular cloning techniques provided early evidence demonstrating an important relationship between protein factors that transform cells and growth factors and their receptors. It has been demonstrated in numerous cases that many oncogenes encode products that induce proliferative growth and tumorigenesis by interacting with growth factor signaling mechanisms and by deranging the control systems of growth in the cell.[81] Oncogenes are frequently encoded by viruses and are introduced into a cell upon viral infection. Alternatively, an oncogene may be generated from a normal cellular gene by mutation or by chromosomal rearrangement that causes aberrant expression of the gene or that produces a protein that circumvents growth control mechanisms in the cell.

Many human tumors have been shown to express different oncogenes and this observation has stimulated a great deal of work on the subject. An understanding of the molecular basis of oncogene-induced cellular transformation provides a conceptual framework for designing selective anticancer therapies; early experimental results suggest that inhibition of oncogene activity achieved by various mechanistic strategies may be therapeutically useful.

3.8.5.3.1 **neu** *oncogene*

The *neu* oncogene is an illustrative example of how molecular biology has begun to define the molecular basis of a neoplastic disease and how potential therapies may be designed.[82] The *neu* oncogene was identified in rat neuroblastomas that had been induced by treatment with ethylnitrosourea, a chemical carcinogen. Hybridization studies were conducted on DNA from cell lines derived from the neuroblastomas using DNA probes from the *erb B* oncogene, which is an EGF receptor homolog. It was shown that the *neu* oncogene produces a 185 kDa cell surface protein (p185) that is homologous with *erb B* and with the EGF receptor. Monoclonal antibodies against the EGF receptor were unable to recognize the *neu* oncogene product, indicating that p185 and the EGF receptor are distinct. It is possible that p185 acts like an EGF receptor that has escaped the normal controls on EGF and induces uncontrolled growth factor signals in the cell.

The fact that the *neu* oncogene encodes a cell-surface-receptor-like protein suggests that its extracellular domain might be selectively targeted by a monoclonal antibody. If so, the antibody could activate the immune system to destroy the transformed cell. In fact, a monoclonal antibody against p185 was prepared and it was successfully employed to significantly retard the progression of *neu*-oncogene-induced tumors in nude mice and in immunocompetent rats.[83] It could also prevent metastases from the primary tumor site. While not curative—the antibody lengthened the lifespan of the tumor-bearing animals but tumors eventually progressed to kill all animals—the results represent an exciting development.

3.8.6 PROTEIN ENGINEERING

An exciting application of recombinant DNA technology and protein expression has developed in the context of protein chemistry. For many years, chemical modification of proteins and chemical synthesis of peptides have been utilized to elucidate structure/function relationships in proteins. These techniques have delivered considerable information that has aided in understanding proteins. From these data together with structural information obtained by physical methods it has been possible to develop molecular models for protein function. The advent of sophisticated recombinant DNA techniques that allow for specific modification of genes at the single nucleotide level and the development of efficient expression systems for the production of proteins make it possible to introduce virtually any variation into a protein with single amino acid specificity. Furthermore, it is also possible to produce enough pure material to do extensive protein chemistry and to attempt crystallization. This approach has revolutionized the study of protein structure and function. These studies help elucidate the nature of binding in active sites and the processes involved in catalysis; consequently, they have been extremely useful in designing small molecules such as enzyme inhibitors and transition state analogs. And at the other end of the spectrum, it has now become possible to alter proteins to introduce improved properties and activities—to 'engineer' desirable properties into proteins.[84] Moreover, we can even begin to think about designing new proteins *de novo.*

Engineering a protein is fundamentally no different than designing a drug. Rational drug design begins with a lead compound that is tested in a relevant assay and a structural model is developed. The model is tested by designing new molecules that incorporate features predicted by the model. The new molecules are synthesized, evaluated in the assay system and the model is further refined. After any number of cycles through the process (preferably as few as possible!) a drug candidate is obtained. As outlined in Figure 19, protein engineering follows a similar rationale. Once a protein has been identified and its activity has been evaluated, it is highly desirable to develop a molecular model of the protein. Ideally, one would like to have a crystal structure; however, recent developments of a variety of physical techniques, notably two-dimensional nuclear magnetic resonance (2-D NMR) spectroscopy, make it possible to obtain structural information in the absence of a crystal structure. Computational techniques and computer modeling have also proven very useful in the development of structural models for proteins. With a model for the structure of the protein variations are then designed into the protein, perhaps as few as a single amino acid change at a

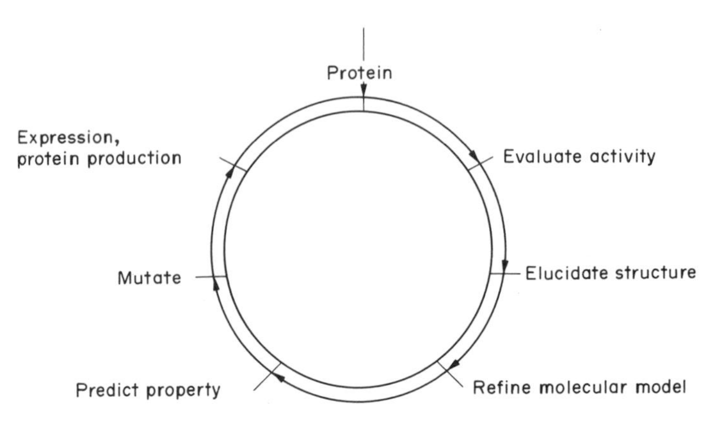

Figure 19 The protein engineering cycle

specific location, that test the model or that attempt to incorporate a predictable feature into the molecule. Once a variant protein has been designed, site-specific mutation of the gene is performed to incorporate the desired change. The resulting gene is then expressed and the variant is purified and submitted to the same biological and structural evaluations as the parental protein to evaluate the effect of the alteration and to correlate the result with the original prediction. The prospect of engineering desirable predictable properties into proteins is an immensely complex problem. Nevertheless, interesting and useful examples of protein engineering have begun to appear in the literature and one may expect that continued refinement of the knowledge base for protein structure and activity will enhance the ability to predictably alter proteins.

3.8.6.1 Site-specific Mutagenesis

The protein engineering cycle described in Figure 19 is relatively straightforward. However, before discussing specific examples of protein engineering it will be useful to examine the technique of oligonucleotide-directed mutagenesis, which is the most popular and powerful method for generating specific protein variants. The technique has had a tremendous impact on the specificity and relative ease with which mutations can be introduced into a gene.

The general approach to oligonucleotide-directed mutagenesis is outlined in Figure 20. The procedure utilizes a synthetic oligonucleotide that is hybridized to a single-stranded DNA template, usually from bacteriophage M-13, at the site where a mutation is desired. The oligonucleotide primer is then extended enzymatically from its 3′ end using the Klenow fragment of DNA polymerase, which lacks exonuclease activity, and the resulting strand is ligated to provide a circular complementary copy of the template DNA. The priming oligonucleotide is designed with an appropriate nucleotide sequence to introduce mutations at the desired location in the complementary DNA sequence. Because the oligonucleotide sequence is different from the template sequence in the region of the mutation, mismatches result when it is hybridized to the template. As long as adequate complementary sequence is maintained on either side of the mismatched base pairs, usually 8–14 nucleotides, the hybrid is stable enough to allow for the enzymatic extension and ligation reactions to take place. The resulting double-stranded DNA can then be used to transfect bacterial cells. When the transfected cells replicate the DNA and package it in bacteriophage particles some of the transfectants replicate the wild-type strand while others package the mutant DNA strand. Following selection of bacterial colonies harboring the mutated DNA, the cloned

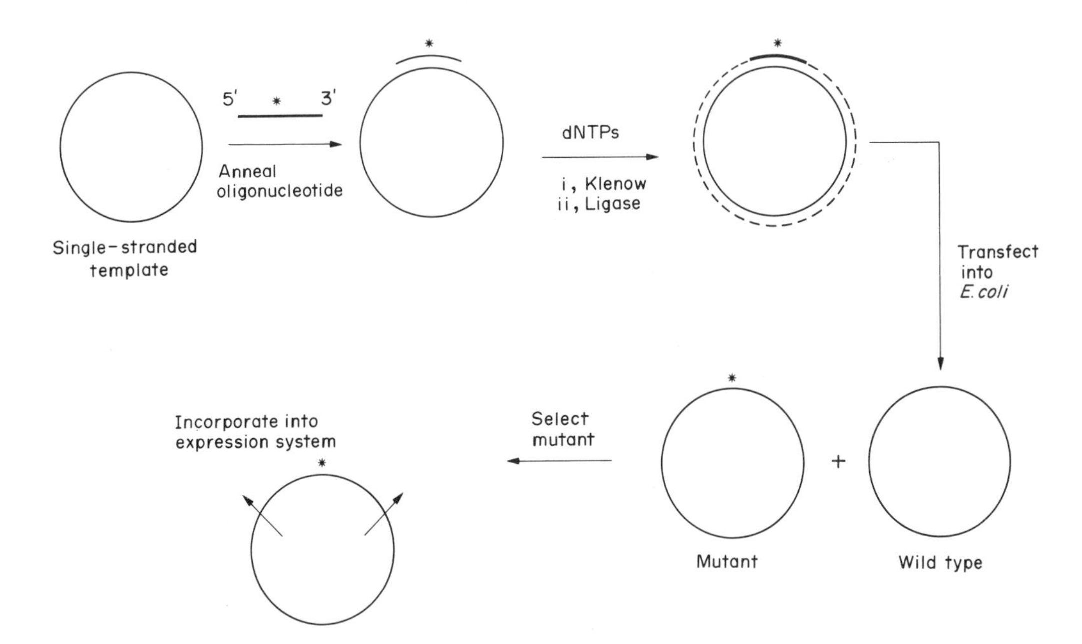

Figure 20 Oligonucleotide-directed mutagenesis. A synthetic DNA fragment is annealed to a single-strand template and second-strand synthesis is accomplished from the synthetic primer with the Klenow fragment of DNA polymerase. Following ligation of the second strand containing the desired mutation, the double-stranded vector is transfected into *E. coli*. Strand separation occurs and the plasmid is replicated in a certain number of transformants which are selected and amplified

DNA can be isolated and incorporated into an expression system to produce the variant protein. In principle, it is feasible to prepare many different types of mutations including transitions (purine to purine or pyrimidine to pyrimidine), transversions (pyrimidine to purine or purine to pyrimidine), insertions and deletions and even combinations of the various possibilities. Many modifications of the general procedure described here have been developed that enhance the efficiency and streamline the process.

3.8.6.2 Tyrosyl-tRNA Synthetase

Aminoacyl-tRNA synthetases are an important class of enzymes that are responsible for charging tRNA molecules with their appropriate amino acids for use in protein synthesis. The tyrosyl-tRNA synthetase from *Bacillus stearothermophilus* has been utilized with great success to study the mechanism of the reaction. Interesting modifications of the protein have been made utilizing oligonucleotide-directed mutagenesis that provide useful information about how to engineer a protein. A high resolution crystal structure of the protein is available, the gene has been cloned and an efficient expression system utilizing the bacteriophage M-13 is available. Though complex, the enzyme kinetics for the reaction are well understood, allowing for a sensitive measure of the interaction energies between substrates and the mutant enzymes. Taken together, these factors make tyrosyl-tRNA synthetase a very attractive model system for protein engineering, designed to elucidate the molecular mechanism of the reaction.

The enzyme catalyzes two distinct reactions (equations 1 and 2). The first reaction involves the nucleophilic attack of the tyrosine carboxyl group on the α-phosphate group of ATP and proceeds through a pentacoordinate transition state resulting in stereochemical inversion at the phosphate. The crystal structure of the enzyme containing bound tyrosine adenylate (Tyr-AMP) suggested that Thr-40 and His-45 are responsible for binding to the γ-phosphate residue of ATP. The binding appears to induce a conformation at the α-phosphate resembling the pentacoordinate transition state, and it properly aligns the pyrophosphate leaving group for S_N2 attack by the tyrosyl carboxylate. To test this model, mutant enzymes were prepared by oligonucleotide-directed mutagenesis that replaced Thr-40 with alanine and His-45 with glycine.[85] Both mutant enzymes had essentially identical binding constants for substrates and for the product; however, the rate of the reaction was dramatically reduced, 240-fold for His-45→Gly-45 and 6900-fold for Thr-40→Ala-40. When the enzyme was prepared that included both mutations, substrates and product again bound with similar affinity and the reaction rate was now reduced 300 000-fold. This study demonstrates in dramatic fashion that catalytic activity can be achieved by constraining the reactants in conformations that resemble the transition state of the reaction, using binding energy to overcome the decrease in entropy.

$$\mathrm{E + Tyr + ATP \longrightarrow E \circ Tyr\text{-}AMP + PP_i} \quad (1)$$

$$\mathrm{E \circ Tyr\text{-}AMP + tRNA \longrightarrow tRNA\text{-}Tyr + E + AMP} \quad (2)$$

In another study, Wilkinson *et al.*[86] predicted from the crystal structure for tyrosyl-tRNA synthetase that the sulfhydryl group of Cys-35 hydrogen bonds to the 3′-hydroxyl group of ribose in ATP. Furthermore, Thr-51 appears to provide a long hydrogen bond to the oxygen in the ribose ring. Mutations at these residues confirmed the predictions in an interesting way. Substitution of Cys-35 with Gly or Ser reduced the affinity of the enzyme for ATP; and, interestingly, when Thr-51 was substituted by Ala or Pro, an enhancement in ATP binding, reflected in a decreased K_m, was observed. These results suggested that, in an engineered protein, the removal of 'bad' hydrogen bonds that are too long or distored may contribute to enhanced binding.

Tyrosyl-tRNA synthetase in its active form exists as a dimer of identical subunits. An interesting study on the subunit interactions has been conducted that has significance for the engineering of intermolecular protein interactions.[87,88] At the interface between subunits, Phe-164 on one subunit interacts with the same residue on the second subunit. Using oligonucleotide-directed mutagenesis, Phe-164 was changed to Asp in one variant and to Lys in a second variant. In the Lys-164 variant, the C-terminal domain of the protein responsible for tRNA binding (aa 321–419) was deleted. This abolished tRNA binding and the second half of the reaction, but the binding properties of Tyr and ATP and catalytic formation of tyrosine adenylate in the first half of the reaction were essentially unaffected. When the Asp-164 variant was submitted to neutral pH conditions where aspartate is ionized, the protein existed as a monomer and was inactive; as the pH was lowered and Asp-164 became protonated, the subunits assembled properly, restoring activity. Likewise, the Lys-164 variant remained monomeric at low pH and began to associate and exhibit increased activity when the pH was raised, where lysine began to deprotonate. Interestingly, though the two variants tested alone were essentially monomeric and inactive at pH = 7.8, when they were mixed in equimolar concentration at pH = 7.8 (where both aspartate and lysine are predominantly charged) heterodimers were formed restoring wild-type binding and activity. From this study, it is clear that as little as a single amino acid alteration that introduces charge repulsion or an electrostatic interaction can control the assembly and stabilization of protein subunits. These conclusions may be very useful in the design and engineering of useful properties in proteins.

3.8.6.3 Engineering Protein Stability

The introduction of a covalent crosslink into a protein decreases the entropy of the unfolded protein, compared to the non-crosslinked molecule, by restricting the conformational degrees of freedom available to the protein. Consequently, the relative free energy difference of denaturation is less for a crosslinked protein, and this energy difference can be manifest in the protein as increased thermal stability. This assumes, of course, that any strain or conformational energy introduced by the crosslink does not offset the stabilization energy.

Subtilisin BPN′ from *Bacillus amyloliquefaciens* is a serine protease that has an alkaline pH optimum. Motivated by potential industrial applications of the enzyme, subtilisin has received considerable attention in protein engineering studies and attempts have been made to increase its stability.[89] Modeling from the crystal structure, disulfide bridges were designed into subtilisin (the wild-type enzyme contains no cysteine) that appeared to be energetically and conformationally acceptable, and site-specific mutagenesis was employed to introduce the requisite cysteine residues. One variant with a disulfide between amino acids 22 and 87 gave interesting results. When the variant protein was purified from the expression system, its chromatographic properties indicated that the disulfide had formed. Furthermore, the crosslinked variant protein exhibited increased thermal stability and its rate of inactivation was twofold slower than the rate exhibited by the reduced variant and by the wild-type enzyme. A crystal structure of the variant[90] revealed that the conformation of the disulfide had a relatively high energy, otherwise the effect may have been greater. Nevertheless, the result is encouraging for the protein engineer where stability is an issue.

One of the predominant mechanisms for irreversible thermal inactivation of an enzyme is the deamidation of asparagine residues. This has been shown for a number of enzymes including yeast triosephosphate isomerase (TIM), a glycolytic enzyme. In its active form TIM exists as a dimer and when asparagine residues 14 and 78 located at the interface between the subunits are deamidated, the resulting aspartic acid residues destabilize the dimer leading to dissociation and loss of activity. By replacing Asn-14 and Asn-78 with isoleucine or threonine, which are sterically and conformationally conservative, the enzyme could be stabilized against thermally induced deamidation and inactivation.[91]

Interleukin 2 (IL-2) plays an essential role in the stimulation of T lymphocytes and normal killer cells, and it has received attention as a possible therapeutic agent for stimulating the immune system. Like many proteins, IL-2 contains an odd number of cysteine residues and the free cysteine that is not involved in disulfide formation is a nuisance during expression and purification; the cysteine catalyzes improperly folded species and participates in intermolecular disulfide bond formation that generates inactive polymers and aggregates. Wang *et al.*[92] have overcome many of these problems by replacing the unpaired cysteine with serine. The resulting analog had the same activity as the wild-type protein, indicating that the original cysteine was not essential; moreover, the variant failed to form the problematical polymers seen with native IL-2 and purification of the variant from an *E. coli* expression system was greatly facilitated.[93]

With an understanding of the protein chemistry involved in the denaturation and inactivation of a protein, site-specific mutagenesis allows sophisticated changes to be made to engineer increased stability into the protein. Initial studies in this area are encouraging and increased effort should provide a predictable database to allow for further design.

3.8.6.4 Engineering Protein Binding Specificity

The hallmark of proteins is the specificity with which they interact with other biopolymers and with substrates and ligands. This specificity is achieved by a complex interplay between protein and ligand that involves electrostatic and hydrophobic interactions, hydrogen bonds, steric effects and solvation. In principle, it is possible to engineer a protein to alter binding specificity and to create entirely new interactions.

The protease subtilisin has an extended binding cleft that is capable of binding many different peptides and endows the protein with broad substrate specificity. Amino acid 166 in the protein can be seen in the crystal structure to interact closely with the sidechain of the P1 site of the substrate (the amino acid on the amino-terminal side of the cleavage site). Variant proteins were designed that incorporated amino acids with progressively larger sidechains at position 166.[94] The effect was dramatic. Increased sidechain size at amino acid 166 significantly reduced the catalytic efficiency of the enzyme toward substrates bearing a large sidechain in the P1 position. While the correlation could not be related strictly to a steric parameter, other factors appeared also to be operative, a trend based on sidechain size was clear. It was thus possible to design into subtilisin a high degree of specificity (as much as 5000-fold) that effectively excluded substrates with a large P1 sidechain.

Studies on alcohol dehydrogenase (ADH) have extended design concepts to small organic substrates.[95] While only sequence information is available for yeast ADH and a crystal structure has not yet been obtained, comparative kinetic studies on ADH from different species suggest that Trp-93 and Thr-48 are involved in the definition of the binding site and give the enzyme specificity for ethanol. A variant with conservative but less sterically demanding alterations, Phe-93 and Ser-48, was constructed and tested in the oxidation of the linear primary alcohols ethanol through octanol. A modest increase in activity toward the larger alcohols was observed compared with the wild-type enzyme.

The subtilisin system described above was used to effectively explore the role of electrostatic interactions in the binding site.[96] The P1 sidechain of the substrate is capable of interacting with sidechains on amino acids 156 and 165. When either of these residues were positively charged (lysine), a significant specificity for a substrate with glutamate at P1 was observed. Likewise, when the binding site was negatively charged (glutamate or aspartate), preference was exhibited for substrate with lysine in the P1 position.

The physiological significance of α1-antitrypsin has been discussed in Section 3.8.4.3.3. Many circulating protease inhibitors achieve specificity for their proteases by utilizing an active site that is similar to the normal substrate for the protease. For example, arginine-specific proteases cleave peptides and proteins on the C-terminal side of arginine (arginine is thus in the P1 position); in the naturally occurring inhibitors of these proteases, an arginine occupies the same site. In the case of α1-antitrypsin,[97] a methionine at amino acid 358, corresponding to the P1 position, makes the inhibitor selective for elastase. When the methionine is oxidized to methionine sulfoxide, the protein loses its ability to bind elastase. In an effort to increase the chemical stability of the inhibitor and to alter its specificity, site-specific mutagenesis was performed on a cDNA copy of the human gene to introduce a variety of different amino acids at position 358 and the resulting variants were produced in an *E. coli* expression system. It had already been shown that a Met-358→Arg-358 variant, a mutation observed in the population, changed the specificity of the variant from elastase to proteases specific for arginine and that the variant was a potent inhibitor of thrombin. Alterations that incorporated hydrophobic residues provided an interesting activity profile. Native α1-antitrypsin inhibits both elastase and cathepsin G, another autolytic protease, and the Leu-358 variant exhibited a similar profile. On the other hand, the Phe-358 analog had only weak activity against elastase but was a potent inhibitor of cathepsin G, exceeding the activity of the natural inhibitor. Any mutation away from methionine was effective in stabilizing the protein against oxidative conditions that ordinarily destroyed native α1-antitrypsin.

The engineering of alterations and refinements into protein interactions and into enzyme activities is an exciting prospect offered by the techniques of site-specific mutagenesis. Efficient expression systems allow for the preparation of material needed to conduct the extensive physical and biological studies required to develop structural models. Furthermore, the products thus obtained may be useful therapeutic agents.

3.8.7 VACCINES

One of the most significant developments in human medicine has been the implementation of vaccines to prime the immune system against invading pathogens and protect against disease. Effective protection against a variety of diseases has been achieved by immunizing with vaccines composed of the killed or attenuated organism or virus. Unfortunately, due to the inability to culture some pathogens and to the unavailability of sufficient material for widespread application, many diseases cannot be countered with vaccines produced by classical means. Moreover, it can be difficult to insure that a particular organism or virus is completely disabled and will not actually cause the disease in a vaccinated individual. This requirement precludes the development of vaccines for some of the most troublesome infections.

Molecular biology, as in other fields, has revolutionized the development of vaccines against diseases that have not yielded to classical techniques. While the results and the benefits are just becoming apparent, the promise of safe effective vaccines based on recombinant DNA technology is great.

Immunity against a pathogenic organism or a virus in general is achieved either by antibodies that recognize the pathogen and target it for destruction by components of the immune system or by cellular immunity mediated by direct T-lymphocyte interactions. In both cases, the immune system interacts with discrete molecular structures associated with the pathogen referred to as antigenic determinants or epitopes. These antigenic determinants are usually relatively small regions of 5–20 amino acids in proteins that are specific to the organism and they are frequently surface proteins or viral coat proteins. Furthermore, epitopes frequently exist in a specific conformation induced by the protein in which they reside. In order to protect against infection, the antibody or T cell must not only recognize an epitope but must induce a response by the immune system that is potent enough to destroy the invader.

Antigenic determinants on pathogenic organisms can be identified by determining which proteins associated with the organism react with antibodies or with T cells that have been isolated from infected individuals. Alternatively, serum from animals can be examined that have been challenged with the infective agent. Once identified, genes for the proteins can be cloned and sequenced and, by a combination of techniques, the epitopes can be elucidated. Once a dominant antigenic determinant has been characterized, the gene for the respective protein or a fragment thereof can be incorporated into an appropriate expression system and the protein can be produced in large quantities and purified. This approach represents a powerful tool for producing a safe vaccine incapable of causing infection and free of other potentially toxic proteins. A related approach utilizes small chemically synthesized peptides that correspond to an epitope.[98] Generally, peptides do not induce a potent immune response by themselves and are conjugated with other proteins or adjuvants to increase their effectiveness.

3.8.7.1 Hepatitis B

Individuals infected with the hepatitis B virus (HBV) have in their circulation particles consisting of multimers of the viral coat protein, referred to as surface antigen (HBsAg). These highly immunogenic particles have been isolated from carriers of the hepatitis virus and have been utilized as a vaccine because the virus cannot be grown in culture to produce the protein. Needless to say, even though the vaccine is effective, there are serious limitations with a vaccine obtained from the blood of HBV carriers and extensive efforts have focused on producing HBsAg by recombinant means. HBsAg is a 25 kDa protein and its gene has been cloned in yeast[99–101] and in mammalian cells[102] to provide a ready source of the protein. Monomeric HBsAg is not very immunogenic; however, the protein assembles *in vitro* to form particles that are chemically, morphologically and immunologically identical with HBsAg isolated from individuals infected with the virus. The recombinantly derived protein is effective at inducing neutralizing antibodies in a variety of species including man, and it has been shown to be effective in preventing HBV infection in chimpanzees.

Synthetic peptides corresponding to various regions of HBsAg and to its precursor, which is also present in viral particles, have been used to identify antigenic determinants on the intact proteins.[103,104] Rabbits were immunized with peptides conjugated to a carrier protein and the resulting peptide specific antisera were examined for their ability to react with the native protein. Those peptide antisera that did crossreact with the protein revealed epitopes of potential significance for vaccine development. Two different peptide sequences thus determined, one from the surface antigen precursor (pre-S2) and the other from the structural domain of HBsAg, were then used as immunogens to generate immune responses in chimpanzees such that when challenged with HBV partial or complete protection against infection was achieved.

3.8.7.2 Polio Virus

As noted for the recombinant hepatitis B vaccine, proteins carrying antigenic determinants are frequently more potent immunogens when they are polymerized or assembled into particles. The ability of tobacco mosaic virus (TMV) coat protein to assemble into high molecular weight particles has been effectively utilized to prepare a polio vaccine. Structural studies on the TMV coat protein indicate that the C terminus is exposed on the surface of the intact viral particle. Consequently, Haynes *et al.*[105] have prepared by recombinant DNA techniques a hybrid TMV coat protein in which an eight amino acid epitope from the polio virus viral coat protein was fused to the C terminus of the TMV coat protein. The fusion protein produced in *E. coli* assembled into rod-like particles at acid pH that were similar to the native TMV coat protein particles. Furthermore, the hybrid protein was a potent immunogen in rats and mounted an effective response against polio virus antigen.

3.8.7.3 Respiratory Syncytial Virus Vaccine from a Transformed Vaccinia Virus

The vaccinia virus, the native form of which is responsible for cowpox, has been utilized with great effect in vaccine regimens. When introduced intradermally, vaccinia causes a mild localized infection that is effective in generating high antibody titer against the viral surface proteins. Recombinant vaccinia viruses have been prepared that have been transformed with genes encoding coat proteins from other viruses. When expressed, these foreign proteins are incorporated into vaccinia virus particles and are capable of inducing a brisk immune response. A recombinant vaccinia virus has been prepared in which the gene for respiratory syncytial virus (RSV) G glycoprotein, a major antigenic determinant of RSV, has been incorporated into the vaccinia genome.[106] Immunization of rats with the recombinant virus resulted in the production of neutralizing antibodies against RSV that protected the animals against lower respiratory tract infection induced by RSV. A similar approach has been implemented for production of antibodies against the AIDS virus envelope glycoprotein,[107] and further studies will be needed to determine if it can serve as a protective vaccine. In a similar study, glycoprotein D from herpes simplex virus type 1 was incorporated into vaccinia.[108] When mice were infected with the transformed virus, they were protected against lethal doses of HSV-1 or HSV-2. Furthermore, the vaccine also protected the mice against latent HSV-1 infection.

3.8.8 GENE TRANSPLANTATION

3.8.8.1 Transgenic Mice

Applications of recombinant DNA research have reached a particularly elegant stage in the production of transgenic animals where a foreign gene has been introduced into the genome of experimental animals. The concepts learned from these applications have direct bearing on the possibility of performing ‘gene transplants’; that is, the introduction of a normal gene into somatic cells or even into the germ line of an organism that lacks the gene. Since many human diseases result from recessive mutations of a single gene, it is possible to envision a time not too distant when it will be possible to correct these genetic diseases by introducing a corrected copy of the defective gene into the individual.

In an elegant series of experiments, Palmiter *et al.*[109] introduced a ‘foreign’ gene into the germ line of mice such that the gene was expressed and was passed on to the progeny. By microinjecting DNA into the pronucleus of fertilized eggs *in vitro* and then introducing the ova back into a receptive pseudopregnant female, as many as 25% of the mice that developed carried the foreign DNA. In this study, a human growth hormone gene was fused to the mouse metallothionein promoter and transcriptional signals in a vector that was designed to integrate randomly into the mouse genome. The transgenic mice obtained were shown to carry the human gene and, furthermore, the gene was expressed to produce significant circulating levels of human growth hormone that induced the animals to grow to twice their normal size. Moreover, the gene was inherited by progeny mice in Mendelian fashion.

With transgenic mice it has been possible to examine the tissue specificity and developmental regulation of genes in a very refined manner. The strategy entails the preparation of a DNA construct composed of a foreign ‘reporter’ gene that is fused to the promoter and transcription control sequences for the gene under study. In general, a heterologous gene is selected whose protein product is easily assayed and is absent in wild-type mice. In an interesting example of this approach, the temporal regulation and tissue-specific expression of the gene for the ocular lens protein γ2-crystallin was studied by fusing the *E. coli* β-galactosidase gene (β-gal) to an 800 base pair DNA fragment containing the promoter and transcription control signals for the mouse γ2-crystallin gene.[110] It could be anticipated that β-gal would be expressed as if it were in fact γ2-crystallin. The

DNA was microinjected into 256 fertilized mouse eggs that yielded 69 live-born animals of which five carried the transgene. Offspring of these transgenic mice were examined for the expression of the β-gal gene and it was shown that β-gal was expressed exclusively in the nuclear fiber cells of the lens in a pattern consistent with the localization of γ2-crystallin. β-gal activity was absent in all other tissues in the animals. In this study, it was possible to conclude that the tissue-specific and temporal control of gene expression resides in the transcription control signals of a gene. This conclusion has been confirmed now in a number of studies involving other different genes.

The ability to construct transgenic mice promises to revolutionize the way biologists investigate the mechanisms of disease. Indeed, with transgenic animals it is possible to create highly specific disease models in which experimental treatments may be undertaken that bear on human disease. This principle is eloquently demonstrated in studies on 'shiverer mice'.[111,112] The shiverer phenotype results from a recessive mutation that is characterized by the absence of myelin basic protein (MBP), a protein that is an essential component of the myelin sheath that covers nerve fibers. In homozygous shiverer animals, there is significant hypomyelination of nerve fibers in the central nervous system that results initially in hind limb tremors and progresses to convulsions and early death. In order to study the role of MBP in shiverer mice, a copy of the wild-type gene was introduced into animals homozygous for the shiverer trait. Attempts to introduce an active MBP gene directly into shiverer mice were unsuccessful; consequently, a normal copy of the MBP gene containing transcription control sequences and termination signals was introduced into a normal mouse, producing a transgenic animal that carried two distinct copies of the MBP gene. This transgenic animal was used by a series of backcrosses and sibling matings with 'shiver mice' to produce transgenic animals that were homozygous for both genotypes, that is, animals that contained both the defective MBP gene and the normal MBP transgene. These animals no longer exhibited the shiverer phenotype and lived a normal lifespan. These studies confirm that the absence of MBP is responsible for the pathogenesis of the shiverer phenotype and that appropriate manipulation of the production of MBP can correct the defect.

A progressive demyelination characterizes the human malady multiple sclerosis (MS), and the shiverer mouse has consequently gained some attention as a model for MS. A normal MBP gene is observed in MS patients and the human disease has a different etiology that probably includes an autoimmune component, therefore the shiverer mouse is limited as a model. Nevertheless, MBP involvement in MS is clear and insights gained from the study of the shiverer model may help to define the human disease more clearly. Significant insights into disease mechanisms will certainly continue to unfold through the use of transgenic animals.

3.8.8.2 Transformation of Hematopoietic Cells

At the present time, medical science in consort with the molecular biologist faces the very exciting prospect of being able to perform gene transplants. In essence, the technology is being refined to deliver a corrected copy of a defective gene into people such that the gene can be expressed to provide the protein that is defective or deficient in the individual. Several crucial preliminary experiments demonstrate the feasibility of the concept.

There are a significant number of human diseases resulting from defects in genes that are expressed in bone-marrow-derived cells. This system has yielded some exciting preliminary results because bone-marrow stem cells can be removed from an animal, manipulated *in vitro* and returned to the animal in a viable state. Using retroviral vectors containing an antibiotic-resistance gene as a marker, it was possible to introduce the marker into bone-marrow stem cells.[113] The cells were grown on antibiotic-containing medium which selected those stem cells that expressed the foreign gene. The transformed cells were then injected back into lethally irradiated mice and the mice were examined 11–17 weeks later to find that the transformed stem cells were present in the marrow and that these precursor cells had differentiated to produce the normal peripheral hemopoietic lymphoid and myeloid cell types. Attempts to correct an actual defect in hemopoietic cells with a corrected gene have succeeded in introducing the gene; however, expression sufficient to correct the deficiency has not yet been achieved.

3.8.8.3 Transplantation of Transformed Fibroblasts

Another approach to gene transplantation described by Selden *et al.*[114] involves the introduction of an expression system into fibroblasts that can then be transplanted into an animal. Mouse L cells (a fibroblast line) were transformed with a construct containing the human growth hormone gene under control of the mouse metallothionein promoter and the thymidine kinase gene as a marker to allow selection on medium containing methotrexate. Following selection, the transformed cells were implanted in various tissues of the same strain of mice from which the L cells were obtained. Under a variety of conditions, significant levels of hGH could be detected in the serum of the animals for as long as three months.

3.8.8.4 Gaucher Disease

Initial experiments by Sorge *et al.*[115] have established the groundwork for extension of gene transplantation to humans suffering from Gaucher disease. This disease is characterized by a deficiency of glucocerebrosidase (GC), a lysosomal enzyme required for the catabolism of glycolipids, that results, in its most severe form, in neurological damage and death. Fibroblasts and lymphoid cells from a patient with Gaucher disease were transformed with a retroviral vector containing a correct copy of the GC gene and normal levels of GC were restored in these cells.

From these studies it is apparent that the principles underlying the production of proteins in a bacterial or mammalian cell expression system may be extended directly to eventual gene therapy. While a successful result will be far more complicated to achieve *in vivo*, we may expect to enventually see gene transplants in animals and humans that correct genetic defects.

3.8.9 ANTIBODIES

The immune system of animals is composed of a large number of interacting cellular and molecular species whose responsibility is to detect, identify and destroy foreign material such as infective organisms and their associated macromolecular constituents. B lymphocytes, white blood cells that originate in the bone marrow and are widely distributed throughout the body, are essential elements of the immune system that are responsible for synthesizing antibodies against foreign antigens. When an animal is presented with a foreign antigen, B cells are stimulated to synthesize antibodies that interact with different portions of the antigen, a process that identifies the antigen as foreign and causes it to be eliminated. Any given B cell makes only a single structurally unique antibody; therefore, when an antigen elicits an immune response that is directed against multiple determinants on the antigen, multiple B cells must be involved in making the various antibodies. One B cell is committed to make only one type of antibody. This is an important point that has significant bearing on the production of monoclonal antibodies that will be discussed below.

Antibodies are extremely valuable tools in biomedical research. Because they recognize very specific structural elements of the molecules that they are generated against (the antigen), they may serve as very specific reagents for detecting the antigen. Moreover, the detection can be accomplished for very small amounts of the antigen. This approach is a powerful diagnostic tool for detecting disease and it has become a mainstay for virtually every aspect of molecular and cell biology. Because they are specific for the antigen and generally exhibit high binding affinity, antibodies can also be used for affinity purification. It has become commonplace to prepare antibodies against proteins for purification purposes. In the development of a recombinant-DNA-based expression system, it is extremely useful to prepare antibodies to aid in the development of the expression vector and to facilitate product purification. Furthermore, because antibodies can be made that recognize a single specific protein, they have also been explored as vehicles for specific drug targeting.

3.8.9.1 Polyclonal Antibodies

In general, many applications that require antibodies are well served by making a 'polyclonal antibody'. In this procedure, an antigen is injected into an animal in the presence of an adjuvant containing bacterial lipopolysaccharides that stimulate the immune system. The animal generally mounts an immune response and within a few weeks antibodies can be detected in blood samples drawn from the animal. Usually, serum is prepared from the blood and upon dilution it can be used directly in many applications. The serum typically contains several different classes of antibodies that are directed against different domains in the antigen molecule and each of these antibodies originates from a distinct B cell, hence the serum is referred to as a polyclonal antiserum. Polyclonal antisera are usually easy to prepare either from isolated intact antigens such as proteins or from synthetic peptides derived from the protein sequence. In the latter case, the antiserum is directed against a specific small portion of the protein. The region-specific antisera derived from small synthetic peptides are similar in some ways to monoclonal antibodies and they are much easier to prepare. Additionally, antibodies can be generated against small organic molecules, such as drugs and drug metabolites, and the antibodies can be very useful in assaying for the drug and in studying metabolism and pharmacokinetics.

3.8.9.2 Monoclonal Antibodies

One of the most important developments in antibody research has been the implementation of techniques for producing monoclonal antibodies, the essential elements of which are outlined in Figure 21. A mouse is immunized with an antigen against which antibodies are desired. After an immune response is detected, in that antibodies against the antigen are observed in the mouse serum, the animal is sacrificed and the spleen is removed. Spleen cells containing high concentrations of B cells, usually including the B cells making antibodies against the immunizing antigen, are prepared and placed in culture with myeloma cells which are immortal bone-marrow-derived tumor cells. The cells are induced to fuse by the addition of poly(ethylene glycol). When a spleen cell and a myeloma

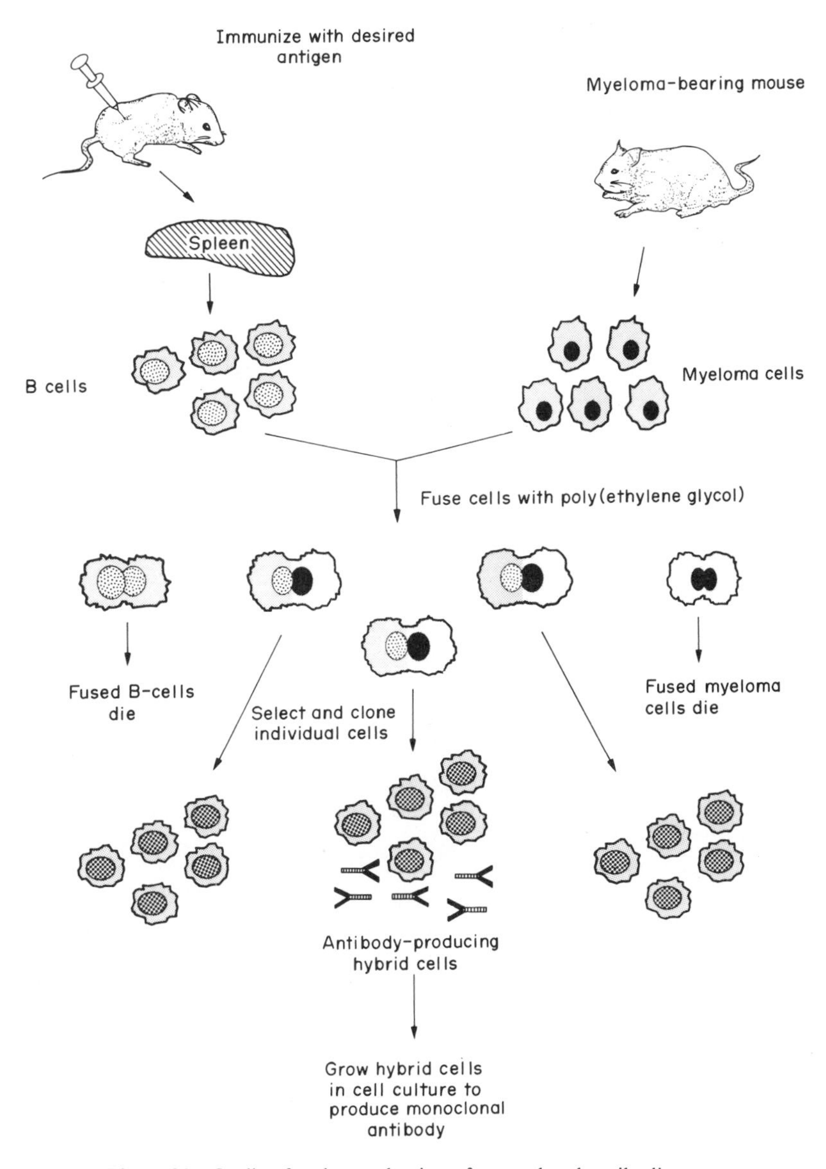

Figure 21 Outline for the production of monoclonal antibodies

cell fuse the nuclei also fuse and genetic material from each cell is combined. The resulting hybrid cells are thus endowed with genes from each of the progenitor cells. Under ideal circumstances, cells will result that contain the myeloma genes responsible for immortalization and also the B-cell genes that encode and express the desired antibody. By selecting and purifying separate antibody-producing cell lines, pure clonal populations of cells can be obtained that grow indefinitely in culture to produce virtually unlimited quantities of a single monoclonal antibody.

The application of antibodies in biomedical research, both polyclonal and monoclonal, are limited only by the imagination of the investigator. The literature is replete with clever applications, and therapeutic implementation is already a reality.

3.8.9.3 Therapeutic Applications

3.8.9.3.1 Metastatic disease

Monoclonal antibodies have revolutionized concepts in the early detection of cancer. In principle, because cancer cells frequently have different surface antigens than normal cells and because those surface antigens are frequently shed into the circulation, it is possible to make monoclonal antibodies against the unique tumor antigens that can detect the presence of the cancer cell. This concept has been nicely reduced to practice in the case of colon carcinoma.[116] Hybridomas producing mouse monoclonal antibodies generated against a human colorectal carcinoma (CRC) cell line were screened against CRC surface antigens and against normal cells and an antibody was identified that specifically recognized a unique tumor antigen. The antigen was subsequently identified as a monosialoganglioside.[117] The antibody was useful diagnostically in determining the presence of the antigen in the circulation of patients with CRC. Moreover, when the antibody was administered to nude mice that were concurrently injected with CRC cells, subsequent tumor growth was profoundly suppressed.[118]

The same monoclonal antibody was also investigated as a vehicle to selectively deliver ricin toxin or diptheria toxin to CRC cells.[119] The two toxins are composed of a protein that carries the biological activity (A chain) and a protein responsible for binding of the toxin to cells (B chain). The B-chain binding is quite non-selective, consequently ricin and diptheria toxins are toxic to a broad spectrum of cells. However, in the absence of the B chain (the cell binding domain), the A chain does not bind or gain entry into cells and is not toxic. By crosslinking the toxin A-chains to the CRC selective antibody, it was possible to produce a highly toxic agent that was specific for colon carcinoma cells and had no effect on normal fibroblasts or other tumor cells.

While the results with easily obtainable mouse monoclonal antibodies are very exciting, treatment of cancer and metastatic disorders in humans with mouse monoclonal antibodies presents serious problems due to side effects that result from the patient developing an immune response against the mouse protein. These side effects limit the utility of mouse monoclonal antibodies in treating human disease. However, several recently developed strategies may eventually make it possible to produce large quantities of human monoclonal antibodies that will effectively eliminate the problems with side effects and extend the usefulness of antibody-based immunotherapy.

Irie and Morton succeeded in producing a human monoclonal antibody against ganglioside GD2, which is a surface marker characteristic of many human melanomas.[120] A human monoclonal antibody against GD2 was produced by an immortalized lymphoblastoid cell line generated by Epstein–Barr virus transformation of peripheral blood B-lymphocytes that had been obtained from a melanoma patient. Preliminary studies in mice indicated that the antibody was effective in binding specifically to GD2 on human melanomas that had been implanted in the mice. The antibody was also able to fix complement, and it could achieve an antitumor effect by targeting the tumor for destruction by the mouse's immune system. The anti-GD2 monoclonal was then investigated in cases of disseminated cutaneous melanoma by injecting the lesions with the antibody. In a number of cases, a complete remission of the injected lesions was obtained; in those cases where only a partial remission or no effect was noted, the lack of resolution could be tied to the absence of GD2 in the lesion. These remarkable results were even more encouraging in light of the fact that virtually no side effects were observed in any of the patients.

3.8.9.3.2 Cardiovascular disease

As the molecular mechanism for a disease becomes known, it is frequently possible to devise strategies for therapeutic intervention that capitalize on the mechanistic understanding. Where an interaction occurs between a ligand and a receptor that leads to an undesirable consequence, it can be hypothesized that an antagonist of the receptor–ligand interaction will achieve a desirable effect. Indeed, many successful drugs are enzyme inhibitors or receptor antagonists. However, in order to validate the concept, it is necessary to have an antagonist, which means that the drug must already be discovered to prove that the mechanistic assumptions are correct—a nice case of circular reasoning and a real problem for the scientist attempting to justify a foray into a new field of study! Because antibodies can be generated against virtually any protein, it is possible, in principle, to prepare an antibody against a receptor that blocks interaction with its ligand—an antibody antagonist. Studies with functional antagonists of this type can be very useful in validating a therapeutic concept and in generating support for drug discovery.

In platelet biology, several lines of evidence point to the importance of the platelet gp IIb/IIIa receptor in the binding of fibrinogen and subsequent aggregation of platelets. This receptor-mediated process is important for normal hemostasis and it is also important in a number of serious thrombotic disorders. Coller *et al.*[121] have generated a monoclonal antibody against the platelet gp IIb/IIIa receptor that has been extremely useful in clarifying the function of the receptor and in validating the concept that antagonism of this receptor may lead to useful antithrombotic activity. When the antibody was incubated with platelets *in vitro*, normal activating agents such as ADP and thrombin were ineffective at inducing platelet aggregation. Moreover, antibody $F(ab)_2$ fragments were also active in *ex vivo* platelet aggregation and in *in vivo* models of thrombotic occlusion in dogs.[122] ($F(ab)_2$ is a pepsin fragment of any IgG antibody; it contains the antigen binding domain of the antibody but is unable to fix complement and generate an immune response.) While the antibody may be a useful therapeutic agent *per se*, the results obtained with the antibody clearly give credence to the idea that gp IIb/IIIa receptor blockade may be a useful therapeutic concept justifying a medicinal chemical approach.

3.8.9.4 Catalytic Antibodies

In a fascinating series of studies, monoclonal antibodies have been generated that possess catalytic activity. In these studies, the specific high affinity binding activity of an antibody was exploited to bind an organic ester molecule in a conformation resembling the transition state for ester hydrolysis and a significant enhancement in the rate of hydrolysis was observed in the presence of catalytic antibody.

The phosphorus atom in a phosphonate ester is tetracoordinate and roughly tetrahedral, and it serves as an excellent analog for the tetrahedral transition state encounted in carboxylic ester hydrolysis. Tramontano *et al.*[123,124] prepared a monoclonal antibody against an organic phosphonate ester and they theorized that, because the combining site of the antibody recognized and bound the tetrahedral phosphorus, it would also bind a homologous carboxylic ester in a tetrahedral conformation and drive the ester toward the transition state for hydrolysis. In fact, experiment confirmed the idea; the monoclonal antibody, specific for phosphonate ester, catalyzed the hydrolysis of a series of carboxylic ester homologs with rate enhancements of 100–1000. A similar study[125] using a cyclic phosphonate ester as the immunogen generated a monoclonal antibody that was effective in catalyzing the formation of the cyclic δ-lactone from the phenyl ester of 5-hydroxypentanoate with a 267-fold rate enhancement. Of even greater interest was the observation that a prochiral hydroxy ester was stereospecifically cyclized to a chiral lactone with an enantiomeric excess of 94%. It is apparent that the chirality of the antibody binding site can impose steric constraints on bond-forming reactions that results in chiral induction.

3.8.10 CONCLUSION

The dramatic advances of molecular biology in the past 10–15 years have ushered in the biotechnology revolution, a revolution that will be as important and as pervasive as any technological revolution witnessed by man. The potential for good is inestimable. The benefits of biotechnology to biology and to medical research are already apparent, and applications to medicinal chemistry are coming forth at a prodigious rate that we may expect will transform the way we design, prepare and test new therapeutic agents.

3.8.11 REFERENCES

1. M. Gouy and C. Gautier, *Nucleic Acids Res.*, 1982, **10**, 7055.
2. T. Ikemura, *J. Mol. Biol.*, 1981, **146**, 1.
3. T. Ikemura, *J. Mol. Biol.*, 1982, **158**, 573.
4. R. Grantham, C. Gautier and M. Guoy, *Nucleic Acids Res.*, 1980, **8**, 1893.
5. C. B. Newgard, K. Nakano, P. K. Hwang and R. J. Fletterick, *Proc. Natl. Acad. Sci. U.S.A.*, 1986, **83**, 8132.
6. M. Robinson, R. Lilley, S. Little, J. S. Emtage, G. Yarranton, P. Stephens, A. Millican, M. Eaton and G. Humphreys, *Nucleic Acids Res.*, 1984, **12**, 6663.
7. J. J. Rossi, X. Soberon, Y. Marumoto, J. McMahon and K. Itakura, *Proc. Natl. Acad. Sci. U.S.A.*, 1983, **80**, 3203.
8. G. E. Christie, P. J. Farnham and T. Platt, *Proc. Natl. Acad. Sci. U.S.A.*, 1981, **78**, 4180.
9. J. Shine and L. Dalgarno, *Proc. Natl. Acad. Sci. U.S.A.*, 1974, **71**, 1342.
10. W. F. Jacob, M. Santer and A. E. Dahlberg, *Proc. Natl. Acad. Sci. U.S.A.*, 1987, **84**, 4757.
11. M. D. Matteucci and H. L. Heyneker, *Nucleic Acids Res.*, 1983, **11**, 3113.
12. J. Beckwith, in 'Escherichia Coli and Salmonella Typhimurium Cellular and Molecular Biology', ed. F. C. Neidhardt, American Society for Microbiology, Washington, DC, 1987, vol. 2, p. 1444.
13. C. Yanofsky and I. P. Crawford, in 'Escherichia Coli and Salmonella Typhimurium Cellular and Molecular Biology', ed. F. C. Neidhardt, American Society for Microbiology, Washington, DC, 1987, vol. 2, p. 1453.
14. G. C. Walker, in 'Escherichia Coli and Salmonella Typhimurium Cellular and Molecular Biology', ed. F. C. Neidhardt, American Society for Microbiology, Washington, DC, 1987, vol. 2, p. 1346.
15. B. C. Hoopes and W. R. McClure, in 'Escherichia Coli and Salmonella Typhimurium Cellular and Molecular Biology', ed. F. C. Neidhardt, American Society for Microbiology, Washington, DC, 1987, vol. 2, p. 1231.
16. T. D. Yager and P. H. von Hippel, in 'Escherichia Coli and Salmonella Typhimurium Cellular and Molecular Biology', ed. F. C. Neidhardt, American Society for Microbiology, Washington, DC, 1987, vol. 2, p. 1241.
17. D. B. Oliver, in 'Escherichia Coli and Salmonella Typhimurium Cellular and Molecular Biology', ed. F. C. Neidhardt, American Society for Microbiology, Washington, DC, 1987, vol. 1, p. 56.
18. J. Garnier, P. Gaye, J.-C. Mercier and B. Robson, *Biochimie*, 1980, **62**, 231.
19. R. C. Das and J. L. Schultz, *Biotechnol. Prog.*, 1987, **3**, 43.
20. J. Kurjan and I. Herskowitz, *Cell*, 1982, **30**, 933.
21. D. Julius, L. Blair, A. Brake, G. Sprague and J. Thorner, *Cell*, 1983, **32**, 839.
22. A. J. Brake, J. P. Merryweather, D. G. Coit, U. A. Heberlein, F. R. Masiarz, G. T. Mullenbach, M. S. Urdea, P. Valenzuela and P. J. Barr, *Proc. Natl. Acad. Sci. U.S.A.*, 1984, **81**, 4642.
23. K. M. Zsebo, H.-S. Lu, J. D. Fieschko, L. Goldstein, J. Davis, K. Duker, S. V. Suggs, P.-H. Lai and G. A. Bitter, *J. Biol. Chem.*, 1986, **261**, 5858.
24. M. Sawadogo and R. G. Roeder, *Cell*, 1985, **43**, 165.
25. M. Karin, A. Haslinger, H. Holtgreve, G. Cathala, E. Slater and J. D. Baxter, *Cell*, 1984, **36**, 371.
26. K. Itakura, T. Hirose, R. Crea, A. D. Riggs, H. L. Heyneker, F. Bolivar and H. W. Boyer, *Science (Washington, D.C.)*, 1977, **198**, 1056.
27. R. Crea, A. Kraszewski, T. Hirose and K. Itakura, *Proc. Natl. Acad. Sci. U.S.A.*, 1978, **75**, 5765.
28. D. V. Goeddel, D. G. Kleid, F. Bolivar, H. L. Heyneker, D. G. Yansura, R. Crea, T. Hirose, A. Kraszewski, K. Itakura and A. D. Riggs, *Proc. Natl. Acad. Sci. U.S.A.*, 1979, **76**, 106.
29. B. H. Frank, J. M. Pettee, R. E. Zimmerman and P. J. Burck, in 'Peptides—Synthesis, Structure, Function. Proceedings of the 7th American Peptide Symposium', ed. D. H. Rich and E. Gross, Pierce Chemical Co., Rockford, IL, 1981, p. 729.
30. L. Thim, M. T. Hansen, K. Norris, I. Hoegh, E. Boel, J. Forstrom, G. Ammerer and N. P. Fill, *Proc. Natl. Acad. Sci. U.S.A.*, 1986, **83**, 6766.
31. J. Smith, E. Cook, I. Fotheringham, S. Pheby, R. Derbyshire, M. A. W. Eaton, M. Doel, D. M. J. Lilley, J. F. Pardon, T. Patel, H. Lewis and L. D. Bell, *Nucleic Acids Res.*, 1982, **10**, 4467.
32. J. C. Smith, R. B. Derbyshire, E. Cook, L. Dunthorne, J. Viney, S. J. Brewer, H. M. Sassenfeld and L. D. Bell, *Gene*, 1984, **32**, 321.
33. S. J. Brewer and H. M. Sassenfeld, *Trends Biotechnol.*, 1985, **3**, 119.
34. T. Oka, S. Sakamoto, K. Miyoshi, T. Fuwa, K. Yoda, M. Yamasaki, G. Tamura and T. Miyake, *Proc. Natl. Acad. Sci. U.S.A.*, 1985, **82**, 7212.
35. M. S. Urdea, J. P. Merryweather, G. T. Mullenbach, D. Coit, U. Heberlein, P. Valenzuela and P. J. Barr, *Proc. Natl. Acad. Sci. U.S.A.*, 1983, **80**, 7461.
36. D. V. Goeddel, H. L. Heyneker, T. Hozumi, R. Arentzen, K. Itakura, D. G. Yansura, M. J. Ross, G. Miozzari, R. Crea and P. H. Seeburg, *Nature (London)*, 1979, **281**, 544.
37. M. D. Matteucci and H. L. Heyneker, *Nucleic Acids Res.*, 1983, **11**, 3113.
38. M. Ikehara, E. Ohtsuka, T. Tokunaga, Y. Taniyama, S. Iwai, K. Kitano, S. Miyamoto, T. Ohgi, Y. Sakuragawa, K. Fujiyama, T. Ikari, M. Kobayashi, T. Miyake, S. Shibahara, A. Ono, T. Ueda, T. Tanaka, H. Baba, T. Miki, A. Sakurai, T. Oishi, O. Chisaka and K. Matsubara, *Proc. Natl. Acad. Sci. U.S.A.*, 1984, **81**, 5956.
39. G. L. Gray, J. S. Baldridge, K. S. McKeown, H. L. Heyneker and C. N. Chang, *Gene*, 1985, **39**, 247.
40. H. M. Hsiung, N. G. Mayne and B. W. Becker, *Bio/Technology*, 1986, **4**, 991.
41. D. V. Goeddel, E. Yelverton, A. Ullrich, H. L. Heyneker, G. Miozzari, W. Holmes, P. H. Seeburg, T. Dull, L. May, N. Stebbing, R. Crea, S. Maeda, R. McCandliss, A. Sloma, J. M. Tabor, M. Gross, P. C. Familletti and S. Pestka, *Nature (London)*, 1980, **287**, 411.
42. D. V. Goeddel, H. M. Shepard, E. Yelverton, D. Leung, R. Crea, A. Sloma and S. Petska, *Nucleic Acids Res.*, 1980, **8**, 4057.
43. M. D. Edge, A. R. Greene, G. R. Heathcliffe, P. A. Meacock, W. Schuch, D. B. Scanlon, T. C. Atkinson, C. R. Newton and A. F. Markham, *Nature (London)*, 1981, **292**, 756.
44. M. D. Edge, A. R. Greene, C. R. Heathcliffe, V. E. Moore, N. J. Faulkner, R. Camble, N. N. Petter, P. Trueman, W. Schuch, J. Hennam, T. C. Atkinson, C. R. Newton and A. F. Markham, *Nucleic Acids Res.*, 1983, **11**, 6419.
45. S. Tanaka, T. Oshima, K. Ohsuye, T. Ono, A. Mizono, A. Ueno, H. Nakazato, M. Tsujimoto, N. Higashi and T. Noguchi, *Nucleic Acids Res.*, 1983, **11**, 1707.

46. E. Jay, J. Rommens, L. Pomeroy-Cloney, D. MacKnight, C. Lutze-Wallace, P. Wishart, D. Harrison, W.-Y. Lui, V. Asundi, M. Dawood and F. Jay, *Proc. Natl. Acad. Sci. U.S.A.*, 1984, **81**, 2290.
47. P. K. Weck, S. Apperson, N. Stebbing, P. W. Gray, D. Leung, H. M. Shepard and D. V. Goeddel, *Nucleic Acids Res.*, 1981, **9**, 6153.
48. A. E. Franke, H. M. Shepard, C. M. Houck, D. W. Leung, D. V. Goeddel and R. M. Lawn, *DNA*, 1982, **1**, 223.
49. A. G. Porter, L. D. Bell, J. Adair, G. H. Catlin, J. Clarke, J. A. Davies, K. Dawson, R. Derbyshire, S. M. Doel, L. Dunthorne, M. Finlay, J. Hall, M. Houghton, C. Hynes, I. Lindley, M. Nugent, G. J. O'neill, J. C. Smith, A. Stewart, W. Tacon, J. Viney, N. Warburton, P. G. Boseley and K. G. McCullagh, *DNA*, 1986, **5**, 137.
50. G. P. Vlasuk, G. H. Bencen, R. M. Scarborough, P.-K. Tsai, J. L. Whang, T. Maack, M. J. F. Camargo, S. W. Kirsher and J. A. Abraham, *J. Biol. Chem.*, 1986, **261**, 4789.
51. J. M. McCord, *N. Engl. J. Med.*, 1985, **312**, 159.
52. R. A. Hallewell, F. R. Masiarz, R. C. Najarian, J. P. Puma, M. R. Quiroga, A. Randolph, R. Sanchez-Pescador, C. J. Scandella, B. Smith, K. S. Stemer and G. T. Mullenbach, *Nucleic Acids Res.*, 1985, **13**, 2017.
53. R. A. Hallewell, R. Mills, P. Tekamp-Olson, R. Blacher, S. Rosenberg, F. Oetting, F. R. Masiarz and C. J. Scandella, *Bio/Technology*, 1987, **5**, 363.
54. P. S. Kaytes, N. Y. Theriault, R. A. Poorman, K. Murakami and C.-S. C. Tomich, *J. Biotechnol.*, 1986, **4**, 205.
55. T. Imai, T. Cho, H. Takamatsu, H. Hori, M. Saito, T. Masuda, S. Hirose and K. Murakami, *J. Biochem. (Tokyo)*, 1986, **100**, 425.
56. S. K. Sharma, D. B. Evans, C.-S. C. Tomich, J. C. Cornette and R. G. Ulrich, *Biotechnol. Appl. Biochem.*, 1987, **9**, 181.
57. L. C. Fritz, A. E. Arfsten, V. J. Dzau, S. A. Atlas, J. D. Baxter, J. C. Fiddes, J. Shine, C. L. Cofer, P. Kushner and P. A. Ponte, *Proc. Natl. Acad. Sci. U.S.A.*, 1986, **83**, 4114.
58. D. B. Evans, T. F. Weighous, J. C. Cornette, W. G. Tarpley and S. K. Sharma, *Bio/Technology*, 1987, **5**, 705.
59. D. Pennica, W. E. Holmes, W. J. Kohr, R. N. Harkins, G. A. Vehar, C. A. Ward, W. F. Bennet, E. Yelverton, P. H. Seeburg, H. L. Heyneker and D. V. Goeddel, *Nature (London)*, 1983, **301**, 214.
60. J. F. Lemontt, C.-M. Wei and W. R. Dackowski, *DNA*, 1985, **4**, 419.
61. S. P. Little, N. U. Bang, C. S. Harms, C. A. Marks and L. E. Mattler, *Biochemistry*, 1984, **23**, 6191.
62. A.-J. van Zonneveld, H. Veerman and H. Pannekoek, *Proc. Natl. Acad. Sci. U.S.A.*, 1986, **83**, 4670.
63. G. Opdenakker, J. van Damme, F. Bosman, A. Billiau and P. de Somer, *Proc. Soc. Exp. Biol. Med.*, 1986, **182**, 248.
64. M. Courtney, A. Buchwalder, L.-H. Tessier, M. Jaye, A. Benavente, A. Balland, V. Kohli, R. Lathe, P. Tolstoshev and J.-P. Lecocq, *Proc. Natl. Acad. Sci. U.S.A.*, 1984, **81**, 669.
65. J. Travis, M. Owen, P. George, R. Carrell, S. Rosenberg, R. A. Hallewell and P. J. Barr, *J. Biol. Chem.*, 1985, **260**, 4384.
66. A. van der Straten, J.-C. Falque, R. Loriau, A. Bollen and T. Cabezon, *DNA*, 1986, **5**, 129.
67. R. I. Garver, Jr., A. Chytil, S. Karlsson, G. A. Fells, M. L. Brantly, M. Courtney, P. W. Kantoff, A. W. Nienhuis, W. F. Anderson and R. G. Crystal, *Proc. Natl. Acad. Sci. U.S.A.*, 1987, **84**, 1050.
68. H. de la Salle, W. Altenburger, R. Elkaim, K. Dott, A. Dieterle, R. Drillien, J.-P. Cazenave, P. Tolstoshev and J.-P. Lecocq, *Nature (London)*, 1985, **316**, 268.
69. S. Busby, A. Kumar, M. Joseph, L. Halfpap, M. Insley, K. Berkner, K. Kurachi and R. Woodbury, *Nature (London)*, 1985, **316**, 271.
70. R. J. Kaufman, L. C. Wasley, B. C. Furie, B. Furie and C. B. Shoemaker, *J. Biol. Chem.*, 1986, **261**, 9622.
71. J. Gitschier, W. I. Wood, T. M. Goralka, K. L. Wion, E. Y. Chen, D. H. Eaton, G. A. Vehar, D. J. Capon and R. M. Lawn, *Nature (London)*, 1984, **312**, 326.
72. W. I. Wood, D. J. Capon, C. C. Simonsen, D. L. Eaton, J. Gitschier, B. Keyt, P. H. Seeburg, D. H. Smith, P. Hollingshead, K. L. Wion, E. Delwart, E. G. D. Tuddenham, G. A. Vehar and R. M. Lawn, *Nature (London)*, 1984, **312**, 330.
73. G. A. Vehar, B. Keyt, D. Eaton, H. Rodriguez, D. P. O'Brien, F. Rotblat, H. Oppermann, R. Keck, W. I. Wood, R. N. Harkins, E. G. D. Tuddenham, R. M. Lawn and D. J. Capon, *Nature (London)*, 1984, **312**, 337.
74. J. J. Toole, J. L. Knopf, J. M. Wozney, L. A. Sultzman, J. L. Buecker, D. D. Pittman, R. J. Kaufman, E. Brown, C. Shoemaker, E. C. Orr, G. W. Amphlett, W. B. Foster, M. L. Coe, G. J. Knutson, D. N. Fass and R. M. Hewick, *Nature (London)*, 1984, **312**, 342.
75. S. C. Howard, A. J. Wittwer, N. K. Harakas and J. Feder, *Fed. Proc. Fed. Am. Soc. Exp. Biol.*, 1987, **46**, 2007.
76. A. Ullrich, L. Coussens, J. S. Hayflick, T. J. Dull, A. Gray, A. W. Tam, J. Lee, Y. Yarden, T. A. Libermann, J. Schlessinger, J. Downward, E. L. V. Mayes, N. Whittle, M. D. Waterfield and P. H. Seeburg, *Nature (London)*, 1984, **309**, 418.
77. Y. Yarden, J. A. Escobedo, W.-J. Kuang, T. L. Yang-Feng, T. O. Daniel, P. M. Tremble, E. Y. Chen, M. E. Ando, R. N. Harkins, U. Francke, V. A. Fried, A. Ullrich and L. T. Williams, *Nature (London)*, 1986, **323**, 226.
78. A. Ullrich, A. Gray, A. W. Tam, T. Yang-Feng, M. Tsubokawa, C. Collins, W. Henzel, T. Le Bon, S. Kathuria, E. Chen, S. Jacobs, U. Francke, J. Ramachandran and Y. Fujita-Yamaguchi, *EMBO J.*, 1986, **5**, 2503.
79. D. Cosman, D. P. Cerretti, A. Larsen, L. Park, C. March, S. Dower, S. Gillis and D. Urdal, *Nature (London)*, 1984, **312**, 768.
80. H. Riedel, T. J. Dull, J. Schlessinger and A. Ullrich, *Nature (London)*, 1986, **324**, 68.
81. R. A. Weinberg, *Science (Washington, D.C.)*, 1985, **230**, 770.
82. A. L. Schechter, D. F. Stern, L. Vaidyanathan, S. J. Decker, J. A. Drebin, M. I. Greene and R. A. Weinberg, *Nature (London)*, 1984, **312**, 513.
83. J. A. Drebin, V. C. Link, R. A. Weinberg and M. I. Greene, *Proc. Natl. Acad. Sci. U.S.A.*, 1986, **83**, 9129.
84. R. Wetzel, *Protein Eng.*, 1986, **1**, 5.
85. R. J. Leatherbarrow, A. R. Fersht and G. Winter, *Proc. Natl. Acad. Sci. U.S.A.*, 1985, **82**, 7840.
86. A. J. Wilkinson, A. R. Fersht, D. M. Blow, P. Carter and G. Winter, *Nature (London)*, 1984, **307**, 187.
87. D. H. Jones, A. J. McMillan, A. R. Fersht and G. Winter, *Biochemistry*, 1985, **24**, 5852.
88. W. H. J. Ward, D. H. Jones and A. R. Fersht, *J. Biol. Chem.*, 1986, **261**, 9576.
89. M. W. Pantoliano, R. C. Ladner, P. N. Bryan, M. L. Rollence, J. F. Wood and T. L. Poulos, *Biochemistry*, 1987, **26**, 2077.
90. B. A. Katz and A. Kossiakoff, *J. Biol. Chem.*, 1986, **261**, 15480.
91. T. J. Ahern, J. I. Casal, G. A. Petsko and A. M. Klibanov, *Proc. Natl. Acad. Sci. U.S.A.*, 1987, **84**, 675.
92. A. Wang, S.-D. Lu and D. F. Mark, *Science (Washington, D.C.)*, 1984, **224**, 1431.
93. T. Arakawa, T. Boone, J. M. Davis and W. C. Kenney, *Biochemistry*, 1986, **25**, 8274.

94. D. A. Estell, T. P. Graycar, J. V. Miller, D. B. Powers, J. P. Burnier, P. G. Ng and J. A. Wells, *Science (Washington, D.C.)*, 1986, **233**, 659.
95. C. Murali and E. H. Creaser, *Protein Eng.*, 1986, **1**, 55.
96. J. A. Wells, D. B. Powers, R. R. Bott, T. P. Graycar and D. A. Estell, *Proc. Natl. Acad. Sci. U.S.A.*, 1987, **84**, 1219.
97. S. Jallat, D. Carvallo, L. H. Tessier, D. Roecklin, C. Roitsch, F. Ogushi, R. G. Crystal and M. Courtney, *Protein Eng.*, 1986, **1**, 29.
98. R. Arnon, *Trends Biochem. Sci. (Pers. Ed.)*, 1986, **11**, 521.
99. W. J. McAleer, E. B. Buynak, R. Z. Maigetter, D. E. Wampler, W. J. Miller and M. R. Hilleman, *Nature (London)*, 1984, **307**, 178.
100. D. E. Wampler, E. D. Lehman, J. Boger, W. J. McAleer and E. M. Scolnick, *Proc. Natl. Acad. Sci. U.S.A.*, 1985, **82**, 6830.
101. P. J. Kniskern, A. Hagpian, D. L. Montgomery, P. Burke, N. R. Dunn, K. J. Hofmann, W. J. Miller and R. W. Ellis, *Gene*, 1986, **46**, 135.
102. E. J. Patzer, G. R. Nakamura, R. D. Hershberg, T. J. Gregory, C. Crowley, A. D. Levinson and J. W. Eichberg, *Bio/Technology*, 1986, **4**, 630.
103. J. L. Gerin, N. Alexander, J. W. Shih, R. H. Purcell, G. Dapolito, R. Engle, N. Green, J. G. Sutcliffe, T. M. Shinnick and R. A. Lerner, *Proc. Natl. Acad. Sci. U.S.A.*, 1983, **80**, 2365.
104. Y. Itoh, E. Takai, H. Ohnuma, K. Kitajima, F. Tsuda, A. Machida, S. Mishiro, T. Nakamura, Y. Miyakawa and M. Mayumi, *Proc. Natl. Acad. Sci. U.S.A.*, 1986, **83**, 9174.
105. J. R. Haynes, J. Cunningham, A. von Seefried, M. Lennick, R. T. Garvin and S.-H. Shen, *Bio/Technology*, 1986, **4**, 637.
106. N. Elango, G. A. Prince, B. R. Murphy, S. Venkatesan, R. M. Chanock and B. Moss, *Proc. Natl. Acad. Sci. U.S.A.*, 1986, **83**, 1906.
107. S.-L. Hu, S. G. Kosowski and J. M. Dalrymple, *Nature (London)*, 1986, **320**, 537.
108. K. J. Cremer, M. Mackett, C. Wohlenberg, A. L. Notkins and B. Moss, *Science (Washington, D.C.)*, 1985, **228**, 737.
109. R. D. Palmiter, G. Norstedt, R. E. Gelinas, R. E. Hammer and R. L. Brinster, *Science (Washington, D.C.)*, 1983, **222**, 809.
110. D. R. Goring, J. Rossant, S. Clapoff, M. L. Breitman and L.-C. Tsui, *Science (Washington, D.C.)*, 1987, **235**, 456.
111. C. Readhead, B. Popko, N. Takahashi, H. D. Shine, R. A. Saavedra, R. L. Sidman and L. Hood, *Cell*, 1987, **48**, 703.
112. B. Popko, C. Puckett, E. Lai, H. D. Shine, C. Readhead, N. Takahashi, S. W. Hunt, III, R. L. Sidman and L. Hood, *Cell*, 1987, **48**, 713.
113. J. E. Dick, M. C. Magli, D. Huszar, R. A. Phillips and A. Bernstein, *Cell*, 1985, **42**, 71.
114. R. F. Selden, M. J. Skoskiewicz, K. B. Howie, P. S. Russell and H. M. Goodman, *Science (Washington, D.C.)*, 1987, **236**, 714.
115. J. Sorge, W. Kuhl, C. West and E. Beutler, *Proc. Natl. Acad. Sci. U.S.A.*, 1987, **84**, 906.
116. H. Koprowski, M. Herlyn, Z. Steplewski and H. F. Sears, *Science (Washington, D.C.)*, 1981, **212**, 53.
117. J. L. Magnani, M. Brockhaus, D. F. Smith, V. Ginsburg, M. Blaszczyk, K. F. Mitchell, Z. Steplewski and H. Koprowski, *Science (Washington, D.C.)*, 1981, **212**, 55.
118. D. M. Herlyn, Z. Steplewski, M. F. Herlyn and H. Koprowski, *Cancer Res.*, 1980, **40**, 717.
119. D. G. Gilliland, Z. Steplewski, R. J. Collier, K. F. Mitchell, T. H. Chang and H. Koprowski, *Proc. Natl. Acad. Sci. U.S.A.*, 1980, **77**, 4539.
120. R. F. Irie and D. L. Morton, *Proc. Natl. Acad. Sci. U.S.A.*, 1986, **83**, 8694.
121. B. S. Coller and L. E. Scudder, *Blood*, 1985, **66**, 1456.
122. B. S. Coller, J. D. Folts, L. E. Scudder and S. R. Smith, *Blood*, 1986, **68**, 783.
123. A. Tramontano, K. D. Janda and R. A. Lerner, *Proc. Natl. Acad. Sci. U.S.A.*, 1986, **83**, 6736.
124. A. Tramontano, K. D. Janda and R. A. Lerner, *Science (Washington, D.C.)*, 1986, **234**, 1566.
125. A. D. Napper, S. J. Benkovic, A. Tramontano and R. Lerner, *Science (Washington, D.C.)*, 1987, **237**, 1041.

3.9
Genetic Engineering: Commercial Applications

PIERRE BOST and GUY BOURAT
Rhône-Poulenc Santé, Antony, France
and
ANNE-CATHERINE JOUANNEAU
Genex International, Paris, France

3.9.1 INTRODUCTION

Market studies have been conducted concerning the commercial impacts of genetic engineering in the field of medicinal chemistry following the first laboratory successes obtained in California by Boyer, Cohen and Berg in 1972–1973. Companies specializing in this type of prediction have regularly proposed, since that time, numerous forecasts which all tend to be particularly optimistic for the future of recombinant product pharmacy. So far, the delay of 10 years, classical in industrial pharmacy, between the laboratory discovery and the product launch, continues to apply, as the first drug derived from genetic engineering (insulin) was released in 1982. However, it should be noted that this case did not correspond to a new drug in the strict sense of the term, but to a new production technique for an already known drug. It is therefore probable that longer delays will be required for truly new drugs, which will have to be *proven in human clinical trials*, and that many predictions from the 1980s will undoubtedly have to be revised in terms of the prolongation of the calendar of release of these products. In any case, the first commercial successes have already been achieved, but before listing them, it is important to define the advantages and disadvantages of *in vitro* recombination for drug design and production.

The essential feature of genetic engineering is the possibility of obtaining, by fermentation techniques: (i) either proteins or glycoproteins, strictly conforming to human products, whether or not they are already used in therapeutics; or (ii) new proteins derived from natural proteins but with improved pharmacological and pharmacokinetic characteristics due to the possibilities of *in vitro* mutagenesis.

It should be noted that, unlike drugs derived from traditional organic chemistry, those obtained by genetic engineering (with the exception of secondary metabolites of microorganisms) can, in the majority of cases, only be administered parenterally. This is a minor disadvantage for single or subchronic administration drugs, but becomes prohibitive for the long-term drugs. Thus a whole range of drugs would appear to be unsuitable for *in vitro* recombination techniques, in the absence of a convenient generalization of routes accessible to polypeptides or proteins other than injection, such as the transnasal, percutaneous and transmucosal routes.

Up until now, cloning and expression of human genes have been performed in prokaryotes (*e.g. Escherichia coli, Bacillus subtilis*), in lower eukaryotes (*Saccharomyces*) or in higher eukaryotes (*i.e.* mammalian cells). Other hosts such as actinomycetes, filamentous fungi, protozoans and insects, for which the vectors are still known, are being specifically investigated but have not yet reached the industrial stage. As might be expected, each of the principal hosts listed above has both advantages and disadvantages. For prokaryotes the disadvantages include the impossibility of obtaining glycosylated products, protease degradation of intracellularly stored heterologous proteins in *E. coli* and, particularly for proteins rich in disulfide bonds, the risk of producing erroneous tertiary configurations. This latter problem requires the use of subsequent chemical denaturation–renaturation cycles which may have low yields. In addition, there is always the potential for the existence of toxic or allergenic residues which may be difficult to remove. On the positive side there is the possibility of high levels of production of heterologous proteins, of growing *E. coli* in very high densities in classical fermenters and of the excretion of the protein in *B. subtilis*. For lower eukaryotes the native structure of the heterologous protein is generally respected, although there is sometimes imperfect glycosylation in yeast. In addition it is possible to use cultures with very high biomass density in classical fermenters and there are few problems due to toxin residues in yeast.

Whilst the use of higher eukaryotes is the only approach possible for highly glycosylated proteins whose biological activity requires the presence of the correct carbohydrate structure, there is often only a low weight production of cloned protein. Other problems may arise from the presence of latent viruses in cells of human origin and the difficulty of maintaining the stability of cell lines apart from cancer lines, from which the genome residues must be eliminated.

In any case, regardless of the host adopted, the collection of the pure protein requires very intense purification using the most sophisticated separation techniques, particularly various forms of chromatography, which are generally restricted to the laboratory. In this case, these techniques must be scaled-up to the industrial level and, now that *in vitro* recombination techniques have become routine, it is very often the purification phase which becomes the bottle-neck to the development of the industrial manufacturing process.

About 100 proteins and glycoproteins have been defined as drugs suitable for genetic engineering techniques. Of course, they are not all of equal importance in terms of the advancement of industrial development or in terms of their real or supposed therapeutic value.

This chapter will be essentially concerned with recombinant products which are already on the market or which will be released in the near future. Section 3.9.12 will deal with proteins which present various problems, while Section 3.9.13 will be devoted to recombinant vaccines. Section 3.9.14 will show that genetic engineering is not confined to the production of human proteins, but that it is already operational in the improvement of the production of secondary metabolites for medical use (principally antibiotics) by industrial fermentation. Lastly, the legal aspects of manufacture and marketing of recombinant products as well as the specific problems of industrial protection will be discussed in Sections 3.9.15, 3.9.16 and 3.9.17.

3.9.2 INSULIN AND RELATED PRODUCTS

3.9.2.1 Introduction

Insulin is a polypeptide secreted by the cells of the islets of Langerhans of the pancreas, and its principal physiological activity is to stimulate the absorption of glucose by the tissues. Insulin secreted into the blood is the result of biochemical maturation consisting of three steps. Firstly the 108-amino acid preproinsulin is produced on the ribosomes. This then undergoes post-translational enzymatic cleavage in the endoplasmic reticulum of the first 23 hydrophobic amino acids of the N-terminal to produce proinsulin. Finally proinsulin is cleaved enzymatically in the Golgi vesicles to give insulin, a 34-amino acid connecting peptide and two dipeptides, Arg-Arg and Arg-Lys. In its final hormonal form, insulin has a molecular weight of 6000 Da and consists of two polypeptide chains A and B, containing 21 amino acids and 30 amino acids respectively, linked together by two disulfide bridges at positions 7-7 and 20-19, with a third disulfide bridge in position 6-11 in the A-chain. Note the absence of methionine in the molecule and the existence, *in vivo*, of forms of polymeric storage, particularly hexamers linked by a zinc atom.

Therapeutically, insulin is used to treat patients producing insufficient amounts of this hormone, suffering from so-called insulin-dependent diabetes. It should be noted that the aetiology of the diabetic phenomenon is complex and involves hormonal systems other than that of insulin. About 5 million patients out of the total of 230 million diabetics require insulin treatment, which is administered subcutaneously or intravenously at an average dose of 40 international units per day (24 IU = 1 mg of chemically pure insulin). The percentage of diabetic patients receiving insulin therapy is increasing by 5% per annum in industrialized countries.

3.9.2.2 Production and Market

The worldwide consumption of insulin is about 3000 kg, *i.e.* 70 billion IU. Its market increased from $125 million in 1977 to $270 million in 1981 and $632 million in 1986, and is distributed by country as shown in Table 1.

Two main companies dominate the world market. Eli Lilly (USA) has 54% of the world market and Novo (Denmark) 25%. The rest of the market is supplied by Hoechst-Behring, Organon Nordisk and Choay.

Up until 1982, all of the insulin was prepared by extraction from the pancreas of cattle or pigs, with a production yield of about 100 g of crystalline product per tonne of organ. The industrial product is extremely pure for a polypeptide, as it contains less than one part per thousand of impurities, which makes allergic reactions extremely rare. Porcine insulin differs from human insulin by the presence of an alanine C-terminal in the B-chain in place of threonine. This difference, which does not affect the biological activity of the hormone, is sufficient to induce the progressive appearance of an immunogenic reaction over the course of long-term treatment of diabetic patients.

3.9.2.3 Nonnatural Insulins

In order to palliate this disadvantage, insulin producers have tried, over the last 10 years, to prepare an insulin with a rigorously analogous structure to that of the human hormone.

Three approaches have been investigated.

3.9.2.3.1 Synthetic hormones

The synthesis of insulin in the laboratory, after elucidation of its sequence (1955), was simultaneously achieved in about 1965 by three teams of chemists in the FRG, the People's Republic of China and the USA. By using the technique of solid-phase synthesis developed by Merrifield, Hoechst and Ciba-Geigy tried to scale-up this process, which consists of 170 phases. Up until now, this approach has not proved to be economically feasible.

3.9.2.3.2 Semisynthetic hormones

The difference of a single amino acid at the end of the B-chain between porcine insulin and human insulin allows porcine insulin to be humanized by chemical transpeptidation after enzymatic cleavage of the terminal octapeptide. This approach was successfully achieved by Novo in 1980 with the final hormone being over 99% human insulin.

Semisynthetic insulin is now available in 40 countries and is supplied in various forms with variable durations of action under the names of Novoline, Actrapid and Monotard. Nordisk, in collaboration with Yamanouchi, have also developed a semisynthetic insulin.

Table 1 Insulin Market By Country

USA	46%	France	4%
FRG	15%	The Netherlands	4%
UK	12%	Spain	3%
Japan	6%	Italy	2.5%

3.9.2.3.3 Biosynthetic hormones

Together with growth hormone, insulin is the first example of the industrial success of *in vitro* DNA recombination techniques. Two Californian scientists, Itakura and Riggs from the City of Hope, Duarte, California, directed the development of the method from 1976. It should be noted that the first patents described in detail all of the techniques of cloning, amplification and expression of the heterologous genes in *E. coli* according to a general methodology which is difficult to circumvent. Their real or claimed preeminence in terms of industrial patent rights is the basis for disputes which have not yet been completely resolved.

Overall, the sequence of nucleotide bases in insulin (77 pairs for the A-chain and 104 pairs for the B-chain) was deduced from the knowledge of the amino acid sequence and the synthetic genes were prepared by decameric or pentadecameric fragments obtained by chemical synthesis according to the phosphotriesters method. After assembly of the HPLC-purified fragments, the genes carrying adequate restriction sites at the extremities were amplified by one half or by one third by cloning in *E. coli*, by means of pBr 322, and the clones carrying the heterologous sequences were detected by differential resistance to ampicillin and tetracycline. After ligation of the amplified fragments, the genes for the A- and B-chains were cloned separately, preceded by the methionine codon behind the gene for β-galactosidase controlled by the lactose operon of *E. coli*. Each chain was expressed in the state of insoluble hybrid protein constituting 20% of the cellular proteins. The pure chain was recovered after chemical cleavage at the level of methionine by cyanogen bromide (hence the importance of the absence of this amino acid in the chain). The A- and B-chains were then combined *in vitro* to ensure correct positioning of the disulfide bonds.

An improved recombination method in *E. coli* uses the tryptophan promoter–operator system from which the attenuator region has been deleted. A very large quantity of biomass is produced while the promoter is repressed by external tryptophan. This latter is subsequently eliminated, the repression is no longer effective and the expression of the heterologous protein is maximal. This insulin has been shown to be less immunogenic than porcine insulin.

Biosynthetic insulin can also be obtained by cloning in yeast; the Chiron and Nordisk companies plan to develop this method under licence from the University of California and Eli Lilly. Although Novo is the leader for semisynthetic insulin, Eli Lilly is the promoter of biosynthetic insulin as a result of agreements made with Genentech. Eli Lilly has invested a large amount of research in this field and the process has been licensed out to Shionogi and Kabivitrum. Biobras is also preparing to invest $300 million in a similar technology.

The insulin obtained by recombination in *E. coli* is currently in full development. It was launched in the UK in 1982 by Eli Lilly under the name of 'Humulin' and received FDA approval in October 1982. Its price is currently about $3700 g^{-1} compared with $250 g^{-1} for porcine insulin and $530 g^{-1} for semisynthetic insulin. The Humulin market was $85 million in 1986, *i.e.* 14% of the total market.

3.9.2.4 Related Products

3.9.2.4.1 Proinsulin

Although 20% of the total insulin 'potential' excreted by the pancreatic cells is in the form of proinsulin, this molecule was thought for a long time to be biologically inactive. We now know that this is not the case and that proinsulin not only represents a longacting form, but also possesses a specific action on hepatic glycogenolysis, enabling it to play a role in the treatment of noninsulin-dependent diabetes. Proinsulin, which cannot be obtained by extraction, can, on the other hand, be produced by recombinant DNA. For this reason, all of the companies involved in the insulin market are trying to develop proinsulin. The Genentech technique has been patented, while Eli Lilly is currently conducting phase III clinical trials in the USA and phase II trials in Europe, after awarding the licence to Shionogi. The annual market is estimated to be $50–100 million. Chiron and Nordisk propose to prepare proinsulin, like insulin, by cloning in yeast.

3.9.2.4.2 Insulin-like derivatives

Computerized protein-modelling techniques combined with the possibilities of directed mutagenesis and enzymology allow the creation of new polypeptides derived from insulin and proinsulin, which are more active than these natural proteins on the global regulation of blood glucose. Eli Lilly and Novo are very active in this field, but so far no products have appeared.

3.9.3 HUMAN GROWTH HORMONE

The hypophyseal origin of growth hormones was demonstrated in 1920, after being suspected for a long time. From 1944 onwards, these hormones were recognized as being specific to a given species and, in 1956, the human growth hormone (HGH) was isolated from cadaver hypophyses. HGH has a molecular weight of 22 000 Da and contains 191 amino acids and two disulfide bridges, but it is preceded *in vivo* by a signal-sequence of 26 amino acids which is eliminated at the time of secretion.

The history of the industrial production of HGH is exemplary in terms of the intrinsic advantages of genetic engineering techniques. In fact, the treatment of hypophyseal dwarfism with extracts of the human gland, which commenced in the 1960s, gave rise to complications which were only recognized in 1985 as being due to the presence of latent viruses in some of the donor cadavers. These slow viruses were responsible for a serious incurable nervous disease called Kreutzfeld–Jakob disease and a single viral-infected hypophysis was sufficient to contaminate the whole batch of HGH; hence the value of obtaining a recombinant hormone exempt, by definition, from the virus was very evident.

3.9.3.1 Recombinant HGH

The gene for HGH was cloned for the first time in *E. coli* by the teams of Genentech in 1979. When the gene does not include the signal-sequence, the HGH produced possesses a methionine molecule before the N-terminal phenylalanine (Met-HGH) derived from the initiation codon necessary for the expression of any protein in prokaryotes. In the particular case of HGH, however, the presence of this supernumerary methionine does not affect the biological properties of the hormone. Furthermore, as for all proteins cloned in Gram-negative bacteria, the protein expressed is not excreted and the bacteria have to be lyzed in order to recover the product. This operation requires difficult and expensive purification steps in order to eliminate the presence of impurities likely to induce immunogenic and/or allergenic phenomena in the final product. On the other hand, when the signal-sequence is cloned, the heterologous protein spontaneously possesses the correct conformational structure when it leaves the ribosome and, most importantly, it is excreted into the periplasm after excision of the signal-peptide without having to be subjected to the action of proteases. In this situation, it is possible to recover the periplasmic HGH by simple osmotic shock (Sanofi technique), which simplifies the industrial operations of purification. In the case of Met-HGH, the supernumerary methionine can be eliminated *in vitro* by enzymatic cleavage (Biotechnology General, Nordisk). The 191-amino acid recombinant HGH is absolutely identical to the natural hormone extracted from the hypophysis in terms of all of its physicochemical and biological characteristics.

Trials of cloning in eukaryote cells have been very successful (Serono, Celltrol and Calbiotech), but the production costs appear to be higher than those of recombination in *E. coli*.

Several companies are currently marketing either methioninated or nonmethioninated recombinant HGH. Thus Kabivitrum, Sumitomo and Boehringer Ing. use the licensed Genentech process ('Protropin') to produce Met-HGH, whilst Eli Lilly produces HGH ('Humatrope'). In addition Genentech, Hoechst, Kabivitrum, Sanofi, Biotechnology General and Nordisk all have drugs at the preregistration stage.

The recombinant HGH market was estimated at $120 million in 1987 and no marked growth is expected unless therapeutic indications other than dwarfism are discovered.

3.9.4 INTERFERONS

3.9.4.1 General Characteristics[5]

The term interferons refers to several groups of water-soluble multimeric glycoproteins produced by the cells of all vertebrates in response to any form of viral aggression. These agents, demonstrated for the first time in 1957 by Isaacs and Lindenmann, were rapidly found to be sufficiently heterogeneous to warrant a phenomenological rather than a physicochemical definition. Thus they have a number of characteristics. They are inactivated by proteases but not by nucleases, are specific for a given species and are active against a large number of viruses. However this activity does not involve direct molecular interaction with viruses, but rather involves differential inhibition of the mechanisms of transcription and translation of the host cell and the infecting virus. They are generally produced by numerous cell types for a given species and in several types by a given species. The chemical structures are very variable in terms of molecular weight, which ranges from 20 000 to 120 000 Da, and in terms of amino acid composition. However, it appears that the interferons of different species present marked analogies at the level of the monomeric subunits. All of the interferons are nondialyzable, fairly stable in acid pH and are nonsedimented by 4 h at 100 000 *g*. Apart from viruses, many organic substances obtained either by extraction from living organisms (endotoxins, polysaccharides, phytohemagglutinin, *etc.*) or by organic synthesis (polynucleotides, polyanions, dyes, fluorenone derivatives) are capable of inducing interferon production in a number of cell types.

Like their structures, the biological properties of interferons are very varied. Apart from their principal antiviral properties, they act on: (i) phagocytosis; (ii) cytotoxicity of T-lymphocytes; (iii) antibody and lymphokine production by lymphocytes; (iv) expression of cellular antigens; (v) antineoplastic properties of macrophages; and (vi) regulation of cellular synthesis of macromolecules and cell growth.

A number of these properties are membrane dependent and prostaglandins and cyclic nucleotides therefore play a role in their activation.

In relation to the antiviral action it has been shown that interferons act locally on the initial infection by intracellular protection of the first cells infected, then peripherally by intercellular or cellular diffusion.

The broad spectrum of biological activities of interferons in laboratory tests has led to the use of the increased quantity of material made available by genetic engineering in various clinical trials. To date the principal indications for interferon use are antiviral and anticancer therapies.

3.9.4.2 Natural Human Interferons

At the present time, four types of human interferon, namely α-interferon of leukocytic and lymphoblastoid origin, β-interferon of fibroblastic origin and γ-interferon (immune interferon) have been identified. Their biological activities are generally measured *in vitro* by inhibition of viral cultures of the vesicular stomatitis virus in various human cell lines.

3.9.4.2.1 α-Interferons

About 15 subvarieties have been identified. Cantell developed their preparation from blood leukocytes from which any erythrocytes were rigorously separated prior to infection and infected with Sendai virus. After precipitation of the supernatant and dialyses at different pH, an interferon was obtained with a titre of 10^6 U mg^{-1} of protein by means of gel or monoclonal antibody affinity chromatography.

The limitation of blood leukocytes (buffy coats) justified the investigation of a source of transformed lymphocytes. One line, called Namalva, was found to produce large amounts of α-interferon.

By means of a purification protocol analogous to Cantell's process, interferons are obtained with analogous specific activities, *i.e.* 10^6 U mg^{-1} of protein for a product suitable for clinical use and up to 10^8 U mg^{-1} of protein for an ultrapurified product.

3.9.4.2.2 β-Interferons

Using continuous cell lines of human fibroblasts induced by Newcastle viruses inactivated by UV radiation and superinduced by poly-I-poly-C, it is possible to produce large quantities of β-interferons (β_1 and β_2), for which the purification process by chromatography has been studied in particular detail, resulting in final purities of 10^6 U mg^{-1} to 10^8 U mg^{-1}, depending on the process.

3.9.4.2.3 γ-Interferon or 'immune' interferon

Leukocytic interferons induced by viruses are described as type I. On the other hand, when a pool of immunocompetent cells is activated *in vitro* or *in vivo* by various, but nonviral, inducers (antilymphocyte serum, lectins and various mitogens), an immune interferon-like substance is obtained, called type II. In view of the methods of production and the producing cells, it is obvious that type I and type II interferons may sometimes be secreted simultaneously. The distinction between the two types consists of a different stability in acid medium, very different antigenic properties towards antiinterferon sera and a selective affinity for certain strains. In terms of biological activities, type I interferons are involved in problems of cellular regulation, while type II interferon is involved in immune regulations and both types are endowed with antiviral properties.

3.9.4.3 The Therapeutic Properties of Interferons

3.9.4.3.1 Antiviral properties

The supplementation of endogenous interferon, naturally produced in response to a viral infection by means of exogenous interferon, is essentially justified by the possibility of obtaining much higher concentrations. This peripheral excess of exogenous interferon induces, *ipso facto*, side effects resembling the viral infection itself.

The clinical applications of interferon principally concern prophylaxis (transplantation of organs frequently carrying cytomegalovirus) and treatment: (i) topical applications (skin, eye, throat, bronchi) at a dose of 10^7 U per patient; and (ii) systemic applications: (a) herpes zoster at a dose of 10^8 U per patient, (b) hepatitis B at a dose of 10^9 U per patient.

More recently, new antiviral indications have been proposed: genital and labial herpes, encephalitis, juvenile laryngeal papilloma and the common cold. In the latter case, the opinions of the experts diverge concerning the benefit/risk ratio of such treatment.

3.9.4.3.2 Antineoplastic properties

The inhibitory effect of interferons on the growth of normal or neoplastic human cells justified, at a very early stage, clinical trials of anticancer treatment with intramuscular doses of up to 5×10^7 U per week for one year. The first successes were achieved with α- and β-interferons in cases of osteosarcoma, Hodgkin's disease, multiple myeloma, certain forms of leukaemia and malignant melanoma. Exogenous interferon has always been found to be superior to endogenous interferon obtained by means of various inducers. A much wider range of clinical trials has obviously been conducted since the availability of large quantities of recombinant interferons. All primary or metastatic solid cancers of the kidney, uterus, ovary, breast, bladder, lung, colon, throat and Kaposi's sarcoma respond in varying degrees to treatment with interferons. The best clinically documented indications at the present time are multiple myeloma, hairy-cell leukaemia and Kaposi's sarcoma.

3.9.4.3.3 Other indications

Clinical studies have been conducted with interferons in diseases affecting the immune system: rheumatoid arthritis, multiple sclerosis and in the treatment of AIDS with Kaposi's sarcoma, which has a simultaneous viral, cancerous and immunological aetiology. Combinations with tumour necrosis factor and with IL-2 have also been proposed.

3.9.4.3.4 Side effects

Benign or severe side effects of treatment with interferons were observed right from the start of clinical trials: influenza symptoms, arthralgia, aesthenia, anorexia, confusional states and even heart attacks. These effects were thought to be due to impurities in the interferons used. However, even with highly purified products, many of these side effects still persist, although not requiring discontinuation of treatment in cases of life-threatening viral or neoplastic diseases.

3.9.4.4 Recombinant Interferons

Although insulin was the first recombinant product to be released on to the market, it was closely followed by α- and β-interferons, as the possibilities offered by genetic engineering make interferon production a textbook case, illustrating the advantages and disadvantages of these methods.

The essential advantage concerns the number of units produced per culture volume: (i) 10^8 U L^{-1} by cell culture; (ii) 10^9 U L^{-1} by fermentation of recombinant yeast and (iii) 10^{10} U L^{-1} by fermentation of recombinant *E. coli.*

The cost of interferons was therefore divided 'by 100' and the quantities available no longer constituted the limiting factor for therapeutic trials. Furthermore, the presence of latent or adventist viruses, always possible with eukaryotic cells, is no longer a risk with cloned microorganisms. On the other hand, interferon cloned in *E. coli* is not glycosylated and, most importantly, each clone produces a single type of molecule, while eukaryote cells produce a 'cocktail' of interferons with a superior overall biological activity for an equal number of units.

Recombination techniques all involve the phase of cDNA in *E. coli* obtained from mRNA of the producing cells. The DNA is then cloned either in yeast or in a eukaryote cell, when pseudohuman or true human glycosylation is required. The unquestionable success of recombinant interferons possessing natural sequences has led producing companies to look for second generation interferons by means of genetic-engineering techniques (hybridization of DNA sequences, guided or random mutagenesis). In particular, Amgen, Cetus, Kyowa, Takeda and Wellcome are studying either mutated interferons with a supernumerary disulfide bridge or hybrid interferons possessing, on a single molecule, sequences derived from various α and/or β subvarieties and even IL-2.

This field may well renew interest in interferons, for which the therapeutic indications have finally proved to be more limited than initially expected.

3.9.4.4.1 Recombinant interferons available on the market

For the moment, the following α-interferons are available on the market: (i) α-2A Genentech process produced under licence by Roche and Takeda ('Roferon'); (ii) α-2B Biogen process produced under licence by Schering-Plough and Yamanouchi ('Bioferon'); and (iii) α-2C Boehringer Ing. process ('Berofor α 2').

3.9.4.4.2 Recombinant interferons at the stage of phase I to phase III studies

Table 2 lists the principal companies throughout the world currently developing interferons.

Table 2 Principal Companies Currently Developing Interferons

Type of interferon	*Cloning*	*Companies*	*Phase of clinical trials*
α-Con 1 (hybrid)	*E. coli*	Amgen	I
	Yeast	Coll. Research	I
α-D	Yeast and *E. coli*	Genentech	I
α	*E. coli*	Kabivitrum	II
β mutated (betaseron)	*E. coli*	Cetus	II
β	*E. coli*	Kyowa Hakko	II
γ	*E. coli*	Viragen	I
γ	*E. coli*	Amgen	II
γ	*E. coli*	Searle (agreement with Meiji Seika)	II
γ	*E. coli*	Suntory (agreement with Schering)	II
γ	*E. coli*	Transgene (agreement with Roussel-Uclaf)	II
γ	*E. coli*	Takeda (agreement with Roche)	III
γ (immunomax)	*E. coli*	Shionogi	III

3.9.4.4.3 The recombinant interferon market

Many companies throughout the world, including a number of Japanese companies, are interested in r-interferons. The 1987 market amounted to $50 million for a selling price of about $15 million kg^{-1}. Worldwide consumption of 50 kg is forecast but this figure is probably optimistic, unless hybrid interferons prove as valuable as is hoped. It should be noted that nonrecombinant interferons already are on the market.

3.9.5 PLASMINOGEN ACTIVATORS

3.9.5.1 Introduction

The regulation of haemostasis depends on a dynamic enzymatic equilibrium between coagulation and fibrinolysis. In the case of a major lesion of the vascular endothelium, the fibrinolytic side may be saturated, resulting in the appearance of pathological clots capable of totally obstructing the blood circulation. These very severe complications, which arise in the form of myocardial infarction, pulmonary embolism or cerebral stroke affect a large proportion of the population in industrialized countries, where they constitute the major cause of mortality and morbidity.

Once the three-dimensional fibrin network has been formed, only direct or indirect enzymatic treatment can dissolve it. This exogenous fibrinolysis is directly inspired by the endogenous mechanism in the case of plasminogen activators.

Plasminogen activators constitute a group of enzymes with a similar structure, of the serine-protease type, with the essential physiological role of transforming a circulating proenzyme, plasminogen, into plasmin, itself a proteolytic enzyme capable of lysing fibrin clots. The normal mechanism possesses a remarkable specificity of action in that fibrinolysis only occurs *locally*. However, the protease properties of plasmin are such that when it is generated not locally but systemically, they can induce marked breakdown of other serum proteins such as fibrinogen, prothrombin and factors V and VIII, resulting in the development of haemorrhagic side effects.

The therapeutic future of the various known plasminogen activators will depend on the distinction between local and systemic action.

3.9.5.2 Structures and Mechanisms of Action

Human serine-protease type plasminogen activators are classified into two groups: urokinase type activators (uPA) and tissue activators (tPA). These two classes of molecules possess a common ancestral origin, which is reflected by major structural homologies, but they can be distinguished by their tissue origin, their immunological reactivity, their affinity for fibrin and the mechanism of their fibrinolytic activity.

The activators of each class contain a catalytic domain (serine-protease domain) and several other structural domains implicated in the biological specificity of the molecule. In each class, the cleavage (by plasmin or by another 'trypsin-like' molecule) of a single peptide bond between the serine-protease domain and the other domains changes the molecule from its 'single-chain' form to its 'double-chain' form. The two fragments formed remain linked by the disulfide bond (between Cys-148 and Cys-279 for uPa and between Cys-264 and Cys-395 for tPA). The fundamental difference between uPA and tPA is that tPA retains its specificity for local fibrinolysis after cleavage, while the activity of uPA becomes systemic. Thus, only the 'single-chain' form of uPA (prourokinase) and not the double-chain form (urokinase) satisfies the criteria for a locally acting fibrinolytic agent.

Despite the similarities of their molecular structures, tPA and prourokinase activate plasminogen *via* different mechanisms. tPA has a high affinity for fibrin, due to the F and K_2 sequences. Fibrin also greatly increases its enzymatic activity and plays the role of a local cofactor.

3.9.5.3 Production of Plasminogen Activators

Four plasminogen activators can be used therapeutically and three of them can be produced by genetic engineering (Table 3).

Table 3 Plasminogen Activators

	Natural	*Recombinant*
Fibrin dependent		
KPA	+	+
Prourokinase	+	+
tPA	+	+
Streptokinase	+	
Urokinase	+	+

3.9.5.3.1 Natural products

The oldest natural product is streptokinase produced from cultures of β-haemolytic streptococci. When injected intravenously or intraarterially, streptokinase has the disadvantage of acting systemically and of being immunogenic. It is used at doses of 10^6 to 3×10^6 units over 24 h, costing \$100 to \$300 per treatment. The principal producers are Hoechst (Streptase) and Kabivitrum (Kabikinase). The worldwide market totals \$10 million.

Of the newer products, urokinase (UK) is used massively in Japan due to its supposed action in the treatment of metastatic cancer. It is extracted from urine and is marketed by ten or so companies, the most important being Green Cross. The Japanese market is worth \$150 million per year. In the USA, urokinase is produced by Abbott (Abbokinase) by renal cell culture, but the market is very modest, only about \$2 million.

In the case of KPA and prourokinase, Collaborative Research Inc. are conducting the phase II development of KPA production from a human renal cell line, while the Weizmann Institute uses human fibroblasts.

Finally for natural tPA, a large number of companies are currently at the stage of preclinical studies of tPA production by human melanoma, epithelial or muscle cells. They include: Bioresponse, Celltech, Ciba, Kanegafuchi, Meiji, Mochida, Monsanto, Porton and Yeda.

3.9.5.3.2 Recombinant products

In view of the mode of action described above, the companies interested in recombinant plasminogen activators are planning several successive 'generations' of products: (i) double-chain (systemic); (ii) then single-chain (nonsystemic); and (iii) then gene-deleted for the Kringle sequences (longer half-life).

The cloning of *E. coli* produces nonglycosylated but nevertheless active products, while cloning in eukaryotic cells (myeloma, Chinese hamster ovary cells, *etc.*) produces an activator analogous to the natural product. The third generation plasminogen activators obtained by rearrangement of the gene are currently being intensively studied, together with the possible synergies between tPA–KPA or tPA-protein–C-protein S, as these last two products are endowed with anticoagulant properties (Biogen, Eli Lilly, Genentech, Chiron, Monsanto, Upjohn).

The most advanced company on the road to marketing of tPA is Genentech, which has just obtained FDA approval (activase) and has licenced to Kyowa, Mitsubishi and Boehringer. However, a large number of companies are working on recombinant tPA, KPA and prourokinase in *E. coli* or in eukaryotic cells, from the preclinical phase up to phase II. They include: Asahi, Beecham, Biogen, Celltech, Chiron, Ciba, Damon, Genetics Institute, Kabivitrum and Lilly. Integrated Genetics and Green Cross are currently running phase II trials.

The recombinant or nonrecombinant plasminogen activator market has been the subject of frequently divergent, but always optimistic estimations, since the figures proposed range between \$150 to \$900 million per year. It is certain that even if only the indications of myocardial infarction and pulmonary embolism were accepted and if the success of the preliminary trials obtained by Genentech were confirmed, the lower limit mentioned above would be achieved and even greatly exceeded. On the basis of an annual consumption of 100 kg and a price of \$3 million kg^{-1}, the market could be worth \$300 million per year by about 1995. But streptokinase will probably remain a strong competitor.

3.9.6 LYMPHOKINES

Since 1966, it has been known that circulating white blood cells exchange information by means of soluble polypeptide mediators endowed with a variety of functions. The wealth of discoveries concerning these mediators over the last 20 years has created a number of confusions of terminology. It is useful to distinguish, under the general heading of lymphokines: (i) *interleukins*, which ensure interactions between leukocytes; and (ii) *monokines*, mediators secreted by monocytes and macrophages. These polypeptides act as growth factors stimulating specific cell proliferations by binding to specialized membrane receptors.

Genetic engineering, by separate cloning of the corresponding genes, has elucidated the very complex mechanisms of cellular interactions; this complexity is essentially due to the multiplicity of the messengers secreted by the same cell species or by species which are difficult to separate and/or to culture (see Chapter 2.5 for a fuller discussion of these complexities).

3.9.6.1 Interleukins (IL)

3.9.6.1.1 IL-1

This interleukin is associated with the following: lymphocyte-activating factor, thymocyte-activating factor, thymocyte-proliferation factor, mononuclear cell-derived factor, proteolysis-inducing factor, B-cell-activating factor, leukocyte endogenous mediator and epidermal T-cell-activating factor.

The two categories α and β of IL-1 are essentially produced by macrophages activated by various mitogens, endotoxins, immune complexes and other lymphokines.

Perhaps the most important functions of IL-1 defined so far are: (i) initial activation of T-lymphocytes by macrophages; (ii) intervention in the processes of inflammation and healing; (iii) degradation of cartilage; (iv) direct or indirect toxic action on cancer cells; and (v) modulation of hepatocyte receptors.

The two genes which code for a propolypeptide containing 270 amino acids have been cloned in a yeast and even in a higher eukaryote cell (Immunex, Syntex, Otsuka) and in *E. coli* (Dainippon, Hoffmann La Roche), but studies on the product expressed are still at the pharmacological stage. The treatments envisaged concern cancer, promotion of the immune response after vaccination and aid to healing. The treatment of arthritis and arthrosis by IL-1 inhibitors have been described (uromodulin). The potential uses for this mediator include cancer treatment and promotion of the immune response after vaccination.

3.9.6.1.2 IL-2 (T-cell growth factor)

This interleukin, which is secreted by helper T-lymphocytes, is probably the most important and is certainly the most extensively studied. Amongst a range of activities are the ability to act as a growth factor for natural killer (NK) cells, as a stimulant of the proliferation and differentiation of activated B-lymphocytes, as a stimulus for the secretion of other lymphokines by helper T-cells themselves and as an inducer of killer cells. From these properties it is not surprising that the potential therapeutic applications include immunotherapy of various cancers, *e.g.* Kaposi's sarcoma, osteosarcoma, chronic leukaemia, all types of solid cancers and the restoration of immunity in AIDS patients. It should also be of value in treatment *in vitro* after cytapheresis.

Some 30 companies have announced their interest in the production of IL-2. The most advanced are believed to be Cetus and Immunex-Roche who have initiated phase III studies whilst Collaborative Research, Interleukin Inc., Roussel and Takeda are at the phase II stage. Five different routes of preparation of IL-2 have been used. The culture of normal cells is used by Immuno-modulator, Interleukin, Imreg, Biogen, Collaborative Research, Ajinomoto and Shionogi, whilst mutated cells are used by Cetus. Cloning in eukaryotic cells (Roussel, Amgen, Collaborative Research, Dnax), yeast (Suntory:Chiron) and *E. coli* (Takeda, Biotechnology) is also used.

The worldwide market for IL-2 in 1990 is estimated at $500 million. This figure is multiplied by a factor of two or three, depending on the degree of optimism of the estimations. This optimism probably needs to be tempered, as in the case of interferon, especially as the stage of preregistration has not yet been reached in the USA.

3.9.6.1.3 *IL-3*

This interleukin is also referred to as the following: haematopoietic cell growth factor, mast cell growth factor, haematopoietin-2, multicolony-stimulating factor. It is isolated particularly in the mouse, is secreted by activated T-lymphocytes and allows the continuous culture of macrophages and mast cells and promotes the growth of pluripotent bone-marrow stem cells of the bone marrow.

The human gene was cloned in 1984 by Genetic Institute and by Dnax–Schering Plough. Acting prior to erythropoietin, IL-3 can be used in numerous forms of anaemia and leukaemia.

3.9.6.1.4 *IL-4 (B-cell-stimulating factor) and IL-5 (B-cell growth factor)*

These B-cell stimulation factors are a recent discovery and are still at the stage of laboratory studies, although the gene for IL-4 was cloned in 1986. No valid prospects can be proposed for the future.

3.9.6.1.5 *TNF (tumour necrosis factor)*

Two types of TNF with a similar composition but of different origins were discovered in 1975: (i) TNF α (cachectin) secreted by macrophages; and (ii) TNF β (lymphotoxin) secreted by helper T-lymphocytes. Both types of TNF are cytotoxic *in vitro* and *in vivo* against numerous tumour cell lines by binding to specific membrane receptors.

As for many anticancer products, Japan was the first to study TNF and clinical trials commenced in 1984. However initial phase II trials showed that injection of TNF was accompanied by major side effects, *e.g.* hypotension, fever and intravascular coagulation. TNF has been obtained from cultured human cells (Yamanouchi, Otsuku, Suntory) or by recombination techniques either in *E. coli* (Dainippon, Genentech, Asahi, Biogen) or mutated cells (Cetus).

3.9.6.1.6 *CSF (human colony-stimulating factors)*

Three types of CSF are distinguished: (i) granulocyte-CSF (G-CSF); (ii) macrophage-CSF (M-CSF); and (iii) granulocyte–macrophage-CSF (GM-CSF). These are glycoproteins which are still active in the nonglycosylated state and responsible for the differentiation of bone-marrow stem cells and the activation of differentiated cells. They are secreted by several cell types, but essentially by T-lymphocytes.

GM-CSF is a stimulator of granulocytes and macrophages; it promotes the destruction of bacteria and parasites by neutrophils, while increasing the number of white blood cells in immunodeficient patients. The potential indications concern the treatment of certain cancers and AIDS and bone-marrow transplant follow-up treatment. Schering Plough, Genetic Institute with Sandoz and Immunex with Hoechst are currently conducting phase I and II clinical trials.

G-CSF is an activator of neutrophils which can be used in bone-marrow regeneration, cancer chemotherapy follow-up, treatment of myeloid leukaemia and AIDS. The gene was cloned in 1986 and Amgen is currently developing phase II clinical trials.

M-CSF is a polypeptide which is capable of very specifically stimulating monocytes and macrophages and is potentially useful in the treatment of lung cancers. Although the gene was cloned in 1985, the product is not yet at the stage of clinical trials. The Cetus company is developing this product.

The CSF market has given rise to various estimates. The cost of a course of treatment would be of the order of $1000 to $3000 and, with the various indications envisaged, the worldwide market for CSF plus IL-3 is estimated at $600 million to $1000 million.

3.9.7 α_1-ANTITRYPSIN

Respiratory failure is a frequent cause of morbidity and mortality. Chronic bronchopulmonary diseases are due to cumulative damage to lung tissue by elastase, a leukocytic endopeptidase which normally combats the local presence of pollutants and bacteria. However, in excess, elastase hydrolyzes elastin, a fibrous scleroprotein responsible for the elasticity and resistance of the lung. The harmful proteolytic effects of elastase are normally counteracted by a serum protein which inhibits serine-proteases, α_1-antitrypsin. Deficiency of the gene, frequent in European populations, is responsible for insufficient production of this protein with the development of pulmonary emphysema, an inflammatory phenomenon aggravated by ingested toxins, in particular tobacco smoke. Acute and chronic bronchopulmonary diseases affect more than 6% of the white populations of industrialized countries. It is envisaged that treatment could entail the use of recombinantly produced α_1-antitrypsin along with pulmonary surfactant proteins. Treatment would be essentially based on recombinant proteins: α_1-antitrypsin together with related products and pulmonary surfactant proteins.

The complete sequence of α_1-antitrypsin has been identified. It is a glycosylated protein with 394 amino acids (51 000 Da), in which the active centre is situated on methionine-358 and which is, in some respects, analogous to antithrombin III and ovalbumin.

Directed mutagenesis techniques have produced more active nonglycosylated derivatives which are more resistant to oxidation.

Table 4 shows the companies developing these agents by genetic engineering, whilst Cutter Laboratories are producing them by nonrecombinant cell culture.

The potential annual worldwide market for α_1-antitrypsin and its variants is estimated to be $300 million, to which must be added $200 million for pulmonary surfactant proteins, also of recombinant origin.

3.9.8 VARIOUS PLASMA PROTEINS

3.9.8.1 Antithrombin III

Antithrombin III is a natural serum inhibitor of thrombin which acts on the latter either directly or as a cofactor of heparin. It inhibits the formation of fibrin from fibrinogen and consequently limits the formation of blood clots. Antithrombin III, normally extracted from plasma (Green Cross, Hoechst), has been cloned in yeast (Delta Biotechnology) and in *E. coli* (Genentech). The commercial future of the cloned product must take account of the fact that several tripeptides (accessible by chemical synthesis) have an analogous biological activity to that of antithrombin III.

3.9.8.2 Factor IX

Also known as autoprothrombin II, Christmas factor, antihaemophilic factor B, plasma thromboplastin, Factor IX is a vitamin K dependent, single-chain glycoprotein with a molecular weight of 56 000 Da involved in the clotting cascade. Deficiency of the corresponding gene, situated on the X-chromosome, is responsible for haemophilia B (Christmas disease). The gene has been cloned in *E. coli* by Genentech, in human hepatoma or mouse fibroblast by vaccinia virus (Transgene) and in the hamster renal cell (Zymogenetics).

3.9.8.3 Factor VIII: C

This is a high molecular weight (2330 amino acids) serum glycoprotein, the deficiency of which is responsible for type A haemophilia. There are approximately 100 000 cases of this disease. The factor is normally produced by extraction from blood plasma (Baxter, Porton, Hoechst, Rover) at the same time as other serum proteins. However, this origin was responsible for contamination of poly-

Table 4 Production of α_1-Antitrypsin

Company	*Method*	*Company*	*Method*
Chiron	Yeast	Transgene	*E. coli*
Smith Kline	Yeast	Genentech	*E. coli*
Zymogenetics	Yeast	Cal Biotech	Recombined eukaryote cell

transfused haemophiliacs by HIV virus present in the blood of certain donors. Demand for extraction product market is therefore likely to shift towards the recombinant product. The recombinant product must be glycosylated and can therefore not be produced in *E. coli*. The gene has been cloned and expressed in various mammalian cells (Chiron for Nordisk, Transgene, Green Cross), hamster kidney cells (Genentech for Cutter) and monkey kidney cells (Genetics Institute for Baxter). It should be noted that Biogen has shown, for Kabivitrum and Teijin, that fragments of Factor VIII may have both a superior biological activity and a longer half-life than those of the natural product. The worldwide market for factors VIII and IX represents $300 million for a total quantity of less than 1 kg.

3.9.8.4 Albumin

Human albumin, a nonglycosylated, single-chain serum protein with a molecular weight of 66 500 Da, present at a concentration of 4%, is currently produced by extraction from blood plasma (Green Cross, Hyland, Cutter, Armour CNTS) or from placental blood (Institut Merieux).

The worldwide market total of $680 million for a production of 200 t, is intended for the reconstitution of blood volume during transfusions. Production is expected to double before the market reaches saturation.

The gene for albumin has been cloned and expressed in *E. coli* and in yeast (Genentech for Mitsubishi, Delta Biotech, Genex for Green Cross). Human albumin represents a typical challenge for genetic engineering as it is a relatively easy product to obtain by *in vitro* recombination, but the cost price for mass production must be low. The current price of recombinant albumin appears to be twice that of extraction albumin. This market is therefore sensitive to psychological factors (fear of HIV virus, although it can be inactivated by heating) or commercial policy (competition with official human blood-collecting organizations).

3.9.9 ERYTHROPOIETIN

Erythropoietin is a glycoprotein with 166 amino acids, essentially produced by the kidney and the Kupfer cells, which acts on the differentiation of the bone-marrow stem cells into erythrocytes. Initially isolated in the urine of anaemic patients, its purification was long and difficult. Through the use of monoclonal antibodies by inverse affinity (elimination of impurities), a specific activity of 82 000 IU mg^{-1} has been obtained. The potential therapeutic indications essentially concern anaemia of renal failure patients on haemodialysis, *i.e.* 180 000 patients throughout the world. The preliminary clinical trials demonstrated a very marked action on the increase in haematocrit, but secondary complications such as thromboses and hypertension complicate the dose-ranging studies at the present time. It is possible that the use of IL-3 may prove to be more favourable.

The worldwide annual consumption could be of the order of 100 g, corresponding to a market of $350 million for all indications combined. Several companies are developing the production of recombinant erythropoietin: (i) Amgen for Kirin Brewery and Johnson and Johnson currently being released on to the market; (ii) Genetics Institute for Chugai; (iii) Integrated Genetics for Behring and Toyobo; and (iv) California Biotechnology, Biogen, Sumitomo Chemicals.

The glycosylation of erythropoietin is imperative. Cloning in mammalian cells was therefore selected by the manufacturing companies.

3.9.10 CELLULAR GROWTH FACTORS

About 30 cellular growth factors have been identified, of which epidermal growth factor (EGF) appears to have the most accessible and most valuable therapeutic applications at the present time.

3.9.10.1 Epidermal Growth Factor (EGF)

This factor is capable of accelerating the healing process of the skin and cornea; its indication in the healing of ulcers of various origins (particularly diabetic) appears to have a very promising future. The gene has been expressed in yeast by Chiron and in *E. coli* by Amgen, Creative Biomolecules and Sumitomo.

The potential markets, if the above indications are confirmed, would be $100 million per year in ophthalmology and $800 million per year in skin healing.

3.9.10.2 Fibroblast Growth Factor (FGF)

This factor stimulates fibroblasts and capillary angiogenesis and can be used in all situations of skin healing following burns, wounds and ulcers. Calbio, Synergen, Merck and Chiron are interested in the production of rFGF by cloning in *E. coli* or in mammalian cells.

3.9.10.3 Platelet-derived Growth Factor (PDGF)

PDGF is a glycoprotein which stimulates division of muscle cells and the collagenase activity of fibroblasts. The potential therapeutic indications concern healing. The gene has been expressed in yeast (Zymogenetics) and in *E. coli* (Amgen and Creative Biomolecules).

3.9.10.4 Transforming Growth Factor (TGFα and β)

These are molecules similar to EGF.

3.9.10.5 Insulin-like Growth Factor (IGF-1)

This is a factor inducing general stimulation of bodily growth, and which may promote the regeneration of bone and cartilaginous tissues. IGF has been expressed in *E. coli* by Amgen, in yeast by Chiron and in mouse fibroblasts by Merck.

In vitro recombination techniques may enter in competition, with *in vitro* activation of platelets, which provides a 'releasate' liquid containing a mixture of the majority of the above growth factors, for use in all healing and angiogenesis processes. This has not prevented specialized companies estimating the American market, with the exception of EGF, at $500 million per year (including $275 million for surgical operation, $200 million for ulcerated wounds and $25 million for burns).

3.9.11 SUPEROXIDE DISMUTASE (SOD)

SOD is an enzyme with a molecular weight of 32 000 Da, containing Cu and Zn. It dismutes the superoxide O_2^- anion into oxygen and hydrogen peroxide, which is subsequently broken down by catalase.

The potential therapeutic value of SOD is the prevention of acute tissue toxicity reactions due to the essentially extracellular presence of the superoxide anion. This excess can occur, in particular, after reperfusion of an ischaemic zone (infarction, major cardiac surgery, organ transplantation). The intervention of SOD in the treatment of inflammatory processes, arthritis, ulcers, nephropathy and, more generally, in the treatment of lipid membrane peroxidations, is also possible.

The gene for SOD was expressed in *E. coli* and in yeast by Chiron (for Gruenenthal). However, *in vitro* recombination is not the only source of SOD and one of the major problems with this drug is to increase the half-life of this enzyme, which is very short.

A market worth $300 million has been forecast, but this figure is probably excessive.

3.9.12 MISCELLANEOUS POLYPEPTIDES

Some 95 proteins and polypeptides of potential therapeutic value have been identified. *A priori*, all lend themselves to recombinant DNA techniques. The main fields of application are: (i) the cardiovascular and blood coagulation system; (ii) the immune system; (iii) the central and peripheral hormonal systems; (iv) the osteoarticular system; and (v) the cell differentiation and dedifferentiation system.

Without going into details of the polypeptides involved, the following may be mentioned: angiogenesis factors, endorphins, collagenase inhibitors, hormone-releasing factors, hirudin, proteins A, C, G and S, renin inhibitors, atrial natriuretic peptide, anti-AIDS CD4, *etc.* The fact that, for certain peptides, organic synthesis may compete with genetic engineering needs to be stressed. Considerable progress has recently been made in solid- or liquid-phase polypeptide synthesis processes and these may well become the preferred route for molecules of up to 30 amino acids. In addition, the real therapeutic value of many of these products is still very much open to doubt, and clinical trials alone will enable the initial hypotheses to be proven or refuted. Given these conditions, even the most approximate estimate of future sales would be illusory.

3.9.13 VACCINES

Until the development of genetic engineering, human and veterinary vaccines were based on three techniques: (i) the use of live immunogenic organisms made nonpathogenic; (ii) the use of killed immunogenic organisms; and (iii) the use of isolated chemical antigens, anatoxins or bacterial extracts. Despite the generally satisfactory results obtained with these techniques, several scientific and industrial disadvantages persist. These include the impossibility of multiplying certain viruses in *in vitro* cell cultures (*e.g.* hepatitis A and B) which leads to the need to use serum from chronic antigen carriers. This of course opens the possibility of contamination from other pathogenic viruses, especially and most fearfully HIV, against which precautions must be taken. Further when using live viruses it is difficult to obtain an equilibrium between sufficient and excessive attenuation as there is the possibility of partial recovery by mutation of the deactivated virus. Finally, there has been to date little success in the development of antiparasitic vaccines either because of the lack of an antigen, or frequent spontaneous mutations of the wall antigens or because of biochemical inhibition of the host immune response.

The use of genetic engineering palliates some of these disadvantages, particularly those related to live vaccines. Two approaches are possible. (i) modify the genome of the pathogenic organism *in vitro* by amputation of the genes expressing the pathogenicity or by limitation of the survival of latent viral forms. This technique is currently being developed for antiherpes, antipoliomyelitis, anticholera and antityphoid vaccines. (ii) integrate, into the genome of the nonpathogenic vaccinia virus, nuclear sequences coding for the antigenic epitopes of one or several pathogenic organisms. This technique is currently being studied for vaccination against influenza, herpes 1 and 2, hepatitis B, sylvatic rabies, foot and mouth disease and cholera. Regulatory constraints may restrict the generalization of this approach.

The industrial and commercial production of recombinant vaccines currently concerns hepatitis B, for which there are 200 million chronic carriers in the world.

The HB antigen can be produced by cloning in yeast (5 mL of culture medium per vaccinating dose) or in Chinese hamster ovarian cells by means of the SV 40 virus (0.5 mL of culture medium per dose).

A number of companies are already marketing or planning on marketing this vaccine: (i) Chiron for Merck and Shionogi: 'Recombivax HB'; (ii) Smith Kline for Yu Han and Pasteur Vaccins: 'Energix B'; and (iii) Biogen for Green Cross and Wellcome, Genentech for Mitsubishi, Rhône-Poulenc Merieux, Pasteur Vaccins, Amgen for Johnson and Johnson, Integrated Genetics for Connaught, Research Development Corp., Takeda and Wyeth.

It should be noted that, apart from *in vitro* recombination techniques, various synthetic antihepatitis B vaccines or vaccines obtained *via* a classical procedure from plasma HBs, are now available on the market: Cheil Sugar (hepaccine B), Green Cross (HB vaccine-1), Daichi Seryaku (HB vaccine). A Korean company, Korea Green Cross, produces a vaccine costing only $1 a dose. The worldwide market for antihepatitis B vaccine reached $50 million in 1987 and should double over the next five years.

Recombinant antiherpes simplex vaccines are currently the subject of clinical trials (Porton International) or preclinical trials: Rhône-Poulenc Merieux, American Cyanamid, Bristol-Myers Inc., Genentech, Roche. The expected worldwide market for the 1990s would be $50 million. A large number of companies are obviously involved in the development of a recombinant antiHIV vaccine, but it is too early to predict a date for commercial success.

Amongst the antiparasitic vaccines, the most necessary is that designed to prevent *Plasmodium falciparum* malaria. This disease affects 200 million people. The clinical results are relatively disappointing at the present time, although the feasibility of vaccination has been demonstrated. In the event of success in the next decade, the expected worldwide market would represent $200 million.

Vaccination against bilharziasis, due to schistosoma, is probably a more distant achievement, although promising results have been obtained by the Pasteur Institute. Recombinant DNA techniques will therefore be responsible for new generations of vaccines. However, we will probably not observe a systematic displacement of classical vaccines, but rather a coexistence, depending on the specific case, with other new vaccination techniques such as chemical vaccines with synthetic epitopes or antiidiotypic vaccines.

3.9.14 NATURAL PRODUCTS OF MICROBIAL FERMENTATION

Industrial fermentation of microorganisms for medicinal purposes has been used for almost half a century to produce either primary metabolites (amino acids, vitamins, nucleotides, *etc.*) or secondary metabolites (antibiotics). In contrast with recombinant products derived from heterologous cloning of a single gene (particularly human proteins), the improvement in the production of natural metabolites of microorganisms depends very generally on the homologous cloning of several genes: the biosynthetic pathways are generally complex and subject to multiple regulations. Under these conditions, the concept of a gene limiting a chain of biosynthesis must be taken into account in view of the fact that release of the limitation of a gene by cloning is immediately followed by the appearance of another limiting gene and so on until improvement of the complete enzymatic chain. This chain, when it is known, may depend on several dozen genes (about 30 in the case of vitamin B_{12}, for example). This explains why, particularly in the case of antibiotics, the improvement of producing strains has involved, up until now, mutation–selection or *in vivo* recombination techniques much more than *in vitro* recombination techniques. However, *in vitro* recombination techniques are proving to be increasingly useful, as it was recently realized that the majority of genes involved in the final specific steps of biosynthesis are generally arranged in clusters either on the principal chromosome or on a plasmid. This therefore allows easy manipulation in a single operation of several limiting genes as well as access to new products of unknown nature, by interspecies cloning.

These operations are facilitated by the fact that very effective plasmids are now available for *in vitro* recombinations in streptomycetes and, more generally, in actinomycetes, responsible for the great majority of antibiotics. This genetic technique, which owes a great deal to the British microbiologist Hopwood, is now being actively developed by industrial antibiotic manufacturers. However, from the commercial point of view, which concerns us here, we can only observe and predict the reinforcement of this tendency to use genetic engineering, keeping in mind that, in the majority of cases, however, industrial secrecy will cover the increases in productivity specifically related to the use of these techniques.

3.9.15 LEGISLATION CONCERNING MARKETING RECOMBINANT DRUGS

The reader is asked to note that the information presented below is derived, for the moment, from unofficial documents discussed at various national and international levels. It should not be considered to represent quotations from official legislative texts.[1]

The development of *in vitro* recombinant DNA technologies enables, at the present time, the preparation of a type of therapeutic agent that was unknown until very recently. In particular, the fermentation production of polypeptides analogous to human products raises specific problems of toxicity, purity and quality control, which are subject to extensive studies by the majority of national drug registration bodies. Traditionally, drug-marketing authorizations are awarded on the basis of drug files satisfying specifications and technical requirements which frequently vary from one country to another and these differences naturally apply in the case of products derived from methods of *in vitro* DNA recombination. However, a consensus concerning certain general rules for the compilation of drug registration files seems to have been established in all of the countries with a strong tradition of human pharmacy.

There are three such rules: (i) the novelty of the technology means that, *ipso facto*, any product derived from these techniques is considered to be a *new product* even if its chemical structure has been known for a long time; (ii) to evaluate its acceptability in human therapeutics, it is necessary to take into account the *entire* manufacturing process, as it is possible that the criteria of purity of the finished product are unable, in the current state of the art, to detect any difference introduced by chance, error or intent into this process; and (iii) the great variability of the known parameters and the always possible existence of unknown underlying parameters inherent to any biological process mean that, in addition to general specifications based on scientific knowledge, experience and common sense, each file will be evaluated in detail *case by case*. This rule is obviously applicable to recombinant vaccines, for which a complete description in terms of the chemical and biochemical structures constituting the final therapeutic agent is virtually impossible.

Any *in vitro* recombination manufacturing process resulting in the production of a polypeptide, homoprotein or heteroprotein is evaluated at the following stages: (i) preparation of a nucleotide sequence coding for the desired product; (ii) *in vitro* construction of an expression vector adapted to the host organism; (iii) expression of the sequence in the host; (iv) multiplication of the host by microbial fermentation or cell culture; (v) isolation and purification of the final product from the above cultures; and (vi) quality control of the process and the final product.

The criteria to be met for each of these stages include the following.

(i) The coding nucleotide sequence, composed of structural genes and possibly regulatory genes must be defined as precisely as possible, in particular by total sequencing or by restriction maps. The microorganism, eukaryotic cell or tissue from which the sequence is extracted must be specified and identified unambiguously. The strategy and the procedures for obtaining the gene or genes must be described: isolation of genome DNA by probes, isolation of mRNA, preparation of cDNA, synthetic sequences, *etc.*

(ii) The description of the genome vector must include: (a) the origin of the vector; (b) the components of the expression system: promotors, enhancers, coding or noncoding flanking DNA sequences; (c) restriction analysis of the vector; (d) physical state within the host organism: extrachromosomal or integrated into the genome; and (e) mechanism of translation: direct transcription and translation of the coding DNA, transcription and translation as a fusion protein.

(iii) The stability of the expression vector in the host organism must be confirmed over several generations together with the fidelity of the nucleotide sequence cloned in the master cell bank (see below). This requirement may be difficult to satisfy in the case of a sequence integrated into the host genome. The methods of transfer of the vector into the host cell must be specified (transformation, transfection, transduction, infection, microinjection, cell fusion) together with the methods of clone selection. The nature and the properties of the prokaryotic or eukaryotic host cells or vertebral cell line (primary, secondary, continuous lines) must be defined: (a) microorganisms: origin, phenotypic markers, growth characteristics, genetic stability; (b) higher eukaryotic cells: origin and history of the cell line, phenotypic (morphology, isoenzyme pattern) and immunological markers, karyotype, growth characteristics, carcinogenicity, presence or absence of relevant viruses and mycoplasma; (c) the protocols of cell transformation by virus or by physicochemical processes must also be described; (d) all of the above criteria, valid for the host cell, must also be evaluated after cloning of the heterologous sequences and (e) an appropriate clone of the recombined host organism will be used as the master cell bank (MCB), composed of a stock of ampoules stored under identical conditions: method of storage, characteristics of storage and stability of the cells after successive freezing and thawing should be defined.

A subculture derived from the master cell bank will be prepared for the purposes of production and will constitute the manufacturer's working cell bank (MWCB).

(iv) The conditions of culture of the microorganisms or the eukaryotic host cells must be described in detail: (a) exact composition of the medium of the liquid phase, of the gas phase (O_2, CO_2, *etc.*) and possibly the solid phase (microcarrier, *etc.*) of the bioreactor; and (b) yield and reproducibility of growth of the inoculum.

The outcome of the use of eukaryotic cells raises more difficult problems. For instance, the addition of antibiotics such as β-lactams or aminoglycosides classically used in the laboratory to prevent bacterial contamination is not recommended in industrial production. Further, control of the presence of viruses in the culture is essential, whether the virus is unwanted, *i.e.* due to adventist viruses or latent retroviruses present in the host cell, or deliberately introduced as a requirement for the cloning of the coding sequence. The possible presence of adventist viruses can be detected in a number of ways: (a) a lysate of the host cell, obtained by serial freezing–thawing, is deposited on a monolayer of sensitive cells. These cells could comprise the host cell itself, normal human embryonic or amniotic cells, monkey or rabbit kidney cells or HeLa cells. After four and twelve days of incubation, the cell suspension is tested in terms of its haemadsorption property on guinea pig or chicken red blood cells; (b) successive subcultures of the cells are examined by transmission electron microscopy, to look for the presence of viruses, or by scanning electron microscopy to look for the presence of mycoplasma, fungi or bacteria; (c) the cells are inoculated into mice *via* the intraperitoneal and intracerebral routes, and into guinea pigs *via* the subcutaneous or intradermal routes. The resultant diseased or dead animals can be autopsied to check for the presence of virus particles; and (d) the cells are cultured in embryonated hens eggs with detection of the presence of haemagglutinin.

(v) The purification and isolation of the cloned heterologous protein is a crucial step in the process. It determines not only the technical and economic feasibility of the industrial process, but also the degree of complexity of the intermediate and final analytical control procedures required by the drug registration authorities. Since the process consists of separating a protein molecule from numerous other molecules, frequently with a similar structure, several sequences of the arsenal of separation methods generally need to be applied: centrifugation, ultrafiltration, ion exchange chromatography, gel exclusion chromatography, affinity chromatography, high pressure chromatography, precipitation-crystallization, solvent extraction, dialysis, adsorption, electrophoresis, *etc.*

In the context of legislation, which is what concerns us here, all of these individual processes must be *validated* in terms of the absence of physicochemical modification of the protein, the non-introduction of new contaminants (toxic solvents, carrier residues, chromatographic residues *etc.*) and, as far as possible, the increasing elimination of undesirable impurities with each step.

(vi) At the end of the process, the recombinant protein will be characterized according to a range of several possible analytical criteria, some of which may be made obligatory by official bodies: (a) overall amino acid composition; (b) amino acid sequence; (c) two-dimensional and high performance liquid chromatography peptide mapping; (d) Western blot (following sodium dodecyl sulfate–polyacrylamide gel electrophoresis); (e) Western blot (following isoelectric focusing); (f) gel permeation; (g) reverse phase high performance liquid chromatography; (h) nuclear magnetic resonance; (i) circular dichroism; and (j) X-ray structure analysis.

All of these techniques should demonstrate: (a) the fidelity of transcription and translation in the host cell; (b) the involuntary presence or absence of terminal amino acids or oligopeptides; and (c) the correct arrangement of the disulfide bonds and the three-dimensional structure of the protein.

Apart from the criterion of physicochemical purity defined above, the verification of the actual identity of the heterologous protein and its biological activity by *in vitro* and *in vivo* trials is essential. In particular, it is possible to verify *in vitro*, by means of monoclonal or polyclonal antibodies, the presence of the correct number and affinity of the epitope sites. *In vivo* studies may be particularly difficult to perform, sometimes requiring the use of several hundred animals. However, if the product is very pure according to chemical criteria, it is possible to very significantly reduce, in statistical terms, the *in vivo* tests.

The absolute purity of a recombinant protein can obviously not be achieved; this means that the nature and biological properties of the impurities that cannot be eliminated must be assessed. It is obvious that the acceptable limit of these impurities depends on their toxicities and on the frequency of administration of the therapeutic protein. The impurities may be directly related to the product like, for example, polypeptide chains truncated at the C- or N-terminal. Their content may sometimes be as much as 5% of the final product and their major disadvantage is their frequent antigenic nature. Foreign impurities not related to the product are detected by immunoassay. A placebo clone, identical to the producing clone, but exempt of the heterologous sequence, is grown, then an antiserum directed against the cell proteins of the placebo is prepared. This serum is used to detect any proteins foreign to the manifestation of the heterologous gene. These impurities generally vary between 1 and 10 p.p.m., depending on whether or not the host cell is able to excrete the desired protein.

The presence of residual nucleic acid should also be monitored, especially when the host cell is a higher eukaryote. The levels detected are generally of the order of 1 p.p.m. for nonexcreting cells, 1 p.p.b. for excreting cells and 0.1 p.p.b. for DNA derived from the coding sequence. The possibility of oncogenic properties of residual DNA has led the World Health Organization to recommend that the quantity injected be less than 0.1 ng per dose. In view of the levels indicated above and the usual therapeutic doses, the repeated injection of 1000 doses would still be well below the threshold of oncogenicity.

Another type of impurity to be considered is that of growth factors coded by the oncogenes. Assuming that a dose of 0.3 p.p.m. is physiologically significant and taking into account the possible presence of 1 p.p.m. of transforming protein in the protein injected, the potential danger of these impurities appears to be very low. The oncogenic impurities/dose ratios indicated above show that even with the maximal levels encountered experimentally, the dangers associated with residual DNA are negligible. However, as an additional safety measure, the control authorities generally refuse registration of products obtained by cell cultures containing a virus known to be useful for transfection and demand the rejection of any batches contaminated with such a virus.

The presence of unknown viruses, although very unlikely, is still possible and, in this case, the risk must be accepted in relation to the therapeutic benefit of treatment.

3.9.16 LEGISLATION CONCERNING GENETIC ENGINEERING RISKS

The evaluation of the potential risks of the use of microorganisms for scientific and industrial purposes is based on the concept of the possible pathogenicity of these organisms for man, domestic animals and livestock, cultivated plants and the environment. The pathogenic nature of a strain is independent of the natural or recombinant nature of the microorganism. The generally accepted factors of pathogenicity are as follows: (a) pathogen–host relationships and possible production of toxic metabolites; (b) transmissibility and range of possible hosts; (c) infecting dose and route of infection; (d) biological survival and stability in the environment; (e) resistance to antiinfective and antiseptic antibiotics; (f) allergenic property; and (g) availability of adequate prophylaxis or treatments.

The principal criterion of pathogenicity is actually complex, as it depends on the genetic and physiological states not only of the pathogenic agent, but also of the host, as well as the dose and route of infection. The concept of nonpathogenicity is typically illustrated by the example of *E. coli* due to the existence of the strain K.12, the host selected for the first experiments in genetic engineering. Strain K.12 has lost a number of the phenotypic characteristics of the wild strain coded by five separate genes on the principal chromosome: surface antigen K; fimbriae allowing adherence to intestinal epithelial cells; resistance to lysis by complement; and resistance to phagocytosis by macrophages.

The probability of observing a natural revertant mutant for some of these potentially dangerous characteristics is therefore virtually nil. The same type of situation applies to other microorganisms used, in particular in *in vitro* recombination, where the evaluation of the absence of pathogenic risk is based on both the genetic information of the time and the absence of accidents observed over long historical periods of use. All of these criteria of more or less intense pathogenicity and of the possibility of prophylaxis have led to a classification according to the risk assessment of microorganisms proposed by the European Federation of Biotechnology.[2] The classification comprises the following four classes: (i) class 1: microorganisms which have never been responsible for disease in man, animals or plants; (ii) class 2: microorganisms which are potentially pathogenic for investigators, but without any possibility of dissemination in the environment and for which effective prophylactic methods are available; (iii) class 3: microorganisms with a high pathogenic potential for investigators and a minor risk for populations normally not exposed and for which effective prophylaxis is available; and (iv) class 4: microorganisms causing serious diseases in man and for which no known effective prophylaxis is available.

These classes have corresponding potential dangers (E2, E3, E4) for plants, animals and the environment. Apart from differences in terminology, the very great majority of the OECD countries consider this classification to be valid. Thus France, the UK, Japan, West Germany and the USA have published corresponding lists of microorganisms.

At the time of the first experiments in genetic engineering in the USA, the initiators of these new techniques, due to the entirely new nature of the recombined species, spontaneously decided to take extreme precautions (Asilomar Conference, 1975) and even to suspend the majority of the manipulations for a certain period of time. Fifteen years of experience have now demonstrated that these fears were totally unfounded and that *in vitro* recombinations in microorganisms do not introduce any additional dangers to those observed with natural microorganisms. Nevertheless, the majority of industrial countries and international bodies concerned have decided to define five new criteria of risk to be considered in these manipulations on the basis of the model proposed by the OTA:[3] *formation*: deliberate or accidental creation of genetically modified microorganisms; *liberation*: deliberate or accidental liberation of these microorganisms at the work site and/or in the environment; *proliferation*: multiplication, genetic reconstruction, growth, transport, modification and disappearance of these microorganisms in the environment and possible transfer of genetic material to other microorganisms; *installation*: installation of these microorganisms in an ecological niche and possible colonization of human beings or other living organisms; and *effect*: subsequent appearance of effects on human beings or on the ecosystem due to interactions between the organism and a host or an environmental factor.

The EEC countries, Japan and the USA have a very similar approach to the operations of genetic engineering in terms of the above five criteria and the common position summarized by the OECD has led to the important concept of GILSP (good industrial large-scale practice), corresponding to the conditions of manipulation of class 1 microorganisms. The host organism must: (a) be nonpathogenic; (b) not contain any incident pathogenic agent; and (c) be the subject of long experience of safe industrial use or possess optimal growth under industrial conditions combined with limited survival in the environment without any harmful consequences. The vector agent of the fragment inserted must: (a) be well defined and without any known harmful sequences; (b) have dimensions limited to a single desired function; (c) be minimally mobilizable; and (d) have no stabilizing effect on the survival of the host in the environment. The organism with recombined DNA must: (a) be nonpathogenic; and (b) be as harmless as the host organism under industrial conditions combined with limited survival in the environment without any harmful consequences.

It would appear that, in the majority of industrialized countries, the GILSP classification or another equivalent classification should allow nonregulated industrial use, even if this concept is not yet officially recognized in some of these countries.

The analysis of the problem in the USA is presented in Sections 3.9.16.1 and 3.9.16.2 as an example.

3.9.16.1 Classification of Microorganisms

(i) Modified microorganisms for which industrial use is not regulated. *(a) 'GILSP' microorganisms*: manipulations corresponding to the following criteria are included in this category: a nonpathogenic receptor organism, a vector for which the DNA fragment inserted is well defined and known not to have any dangerous consequences, and a nonpathogenic modified organism. *(b) Microorganisms similar to GILSP* microorganisms containing noncoding heterologous sequences, such as operators, terminators, ribosome attachment sites. The sequence inserted should be known and must not code for a protein, a peptide or a functioning RNA molecule. It can only exert a control function on the activity of the other coding sectors or as a recognition site for the initiation of nucleic acid or protein synthesis. *(c) Nonpathogenic microorganisms*: these microorganisms are considered to be nonpathogenic by experts in the field, to be a result of the transfer from a noncoding regulatory region of a pathogenic donor, and to be commensal or mutualist microorganisms.

(ii) Modified microorganisms for which the industrial use is regulated. (a) Recombined microorganisms containing genetic material derived from another species. The regulations are determined for each individual case according to the intended uses. (b) Virus or microorganism pathogenic for other living beings (man, animals, plants, microorganisms). (c) Microorganisms containing genetic material derived from an organism recognized to be pathogenic by a federal agency.

3.9.16.2 Classification of the Manipulations Affecting Health

The guidelines of the classification are defined by the National Institute of Health (NIH)[4] and take account of the recommendation of the Recombinant DNA Advisory Committee (RAC). The local application of these guidelines is controlled by an Institutional Biosafety Committee (IBC), at least two of the members of which are foreign to the establishment at which the manipulations are performed. Several cases are defined by decreasing order of requirements. (i) Experiments that require specific RAC review and NIH and IBC approval: (a) recombinant DNA genes for biosynthesis of molecules lethal for vertebrates at an LD_{50} of less than 0.1 p.p.b.; and (b) transfer of a drug resistance trait to microorganisms that are not known to acquire it naturally. (ii) Experiments that require IBC approval: (a) experiments using human or animal pathogens (class 2 to 5) as host–vector systems—containment equipment is specified for each class; (b) experiments in which DNA from human or animal pathogens (class 2 to 5) is enclosed in nonpathogenic prokaryotic or lower eukaryotic host–vector systems; (c) experiments involving the use of infectious animal or plant viruses (defective or not) in the presence of helper virus in tissue culture systems; and (d) recombinant DNA experiments involving whole animals or plants. (iii) Experiments that require only IBC notice: experiments in which all components derive from nonpathogenic prokaryotes or nonpathogenic lower eukaryotes. (iv) Experiments exempt from the guidelines: (a) use of recombinant DNA molecules not in organisms or viruses; (b) use of DNA segments from a single nonchromosomal or viral DNA source; and (c) use of DNA from a prokaryotic host including its indigenous plasmids or viruses, or from a eukaryotic host (excluding viruses), when propagated in that host or in a closely related strain; and (d) use of recombinant DNA molecules specified in a list by the NIH.

Recommendations of containment of microorganisms for large-scale applications not defined by the GILSP have been proposed by the OECD and are summarized below. Three categories of increasing requirements are defined (Table 5).

While recognizing that no scientific reason justifies the adoption of specific legislation to control the use of organisms with recombinant DNA, the OECD recommends that member countries: (i) share information concerning the principles of national legislation and risk assessment in order to facilitate harmonization of regulations; (ii) design mechanisms for monitoring and appraising of manipulations so as not to interfere with technical progress, referring to the opinions of international bodies on the subject; (iii) ensure that, as far as possible, only organisms with a low intrinsic risk are used in large-scale industrial applications of recombinant microorganisms, in accordance with GILSP; (iv) ensure that all measures of effective containment are taken when GILSP good practices cannot be applied; and (v) encourage research designed to improve methods of monitoring and managing of unintentional release of recombinant DNA organisms.

3.9.17 PATENTS AND GENETIC ENGINEERING

The general conditions required by the legislation of most countries for the protection of inventions are based on four fundamental requirements.

(i) The invention must be new, *i.e.* it must never have been presented publicly either orally or in writing and must not be used publicly (absolute novelty). In order to evaluate novelty, reference is made to the 'state of the technique' at the time of submission, which is defined as all of the knowledge available to the public. Several countries have introduced into their legislation a 'grace period', which corresponds to the period between publication of an invention and the submission of an application for the corresponding patent by the inventor. This period is two years in Canada, one year in the USA and six months in Australia and Japan. The great majority of the other countries do not accept this possibility and any publication automatically places the invention in the 'public domain', preventing, at the same time, the possibility of obtaining a valid patent.

(ii) The invention must imply an inventive activity; it must not be an obvious result of the state of the technique for someone trained in the field.

(iii) The invention must be 'useful' and must possess 'possibilities of industrial application'.

(iv) The invention must be sufficiently well described that someone trained in the field would be able to reproduce it. In the field of microbiology, this last criterion has given rise to a number of debates which are reflected in the legislation of various countries, as we shall see below.

Although the patentability of products for pharmaceutical or other uses derived from genetic engineering processes does not pose any particular problems, that of the processes and methods used for their production raises a number of questions, sometimes difficult to resolve, concerning the interpretation of positive law and even ethical problems. Frequently contradictory doctrinal positions exist and the rules applicable in this domain, which may vary from one country to another, can only be defined by jurisprudence.

Table 5 Recommendations for Containment of Microorganisms in Large-scale Applications not Defined by GILSP

	Categories		
	1	*2*	*3*
Manipulation of viable organisms in a production enclosure which physically separates the process from the environment	Yes	Yes	Yes
Processing of gases to minimize or totally suppress dissemination of microorganisms outside of the production enclosure	Minimize	Suppress	Suppress
Design of devices for sampling, addition or removal of substances in the enclosure to minimize or totally suppress dissemination of microorganisms outside of the enclosure	Minimize	Suppress	Suppress
Inactivation of culture media before removal from the enclosure	Yes	Yes	Yes
Airlock device situated in the controlled enclosures	Optional	Optional	Yes
Personnel showers and washroom with waste water collection and deactivation	No	Optional	Yes
Very effective filtering of air entering and leaving the zone	No	Optional	Yes
Controlled enclosures allowing containment of the entire contents of the enclosure	No	Optional	Yes
Disinfection device for controlled zones	No	Optional	Yes

In the field of microbiology, the principal difficulty is that of the patentability of microorganisms, whether they are natural or modified.

The particular case of plasmids, vectors of genetic recombination operations in prokaryotes or lower eukaryotes, appears to be resolved, as the majority of countries accept their protection in the same way as a 'classical' chemical product, provided that they satisfy the criteria of patentability.

The essential question was to determine whether microorganisms, which belong to the realm of living organisms, could be considered in the category of patentable inventions. The answer given by the US Supreme Court in 1980 in the Diamond–Chakrabarty case was positive. Although this precedent is only valid in the USA, it strongly influenced the policy adopted in other countries. Assuming that they can be included in the category of patentable inventions, can microorganisms *actually* be patented? A distinction must be made between natural microorganisms and modified microorganisms and between what constitutes a 'discovery' and what constitutes an 'invention'.

The isolation of a microorganism present and preexisting in nature was for a long time considered to be a 'discovery' which, consequently, is excluded from the category of patentable inventions. However, it now appears to be accepted that the pure culture of a natural microorganism may qualify for patentability, since obtention of such a culture requires human intervention and is necessary in order for the microorganism to be able to secrete the metabolite subsequently used in the industrial application.

The preparation of a modified microorganism by genetic recombination techniques, *i.e.* by means of a microbiological procedure, constitutes an 'invention' and the modified microorganism must be considered to be patentable inasmuch as it satisfies the general conditions of patentability. Consequently, the inventor should be able to obtain protection for the results of his or her research, which is frequently long and costly, and should be able to exercise his or her rights resulting from the protection obtained.

Whatever the technical field considered, the inventor, in return for the monopoly granted, must divulge his or her invention which must be described in sufficient detail to allow a third person to reproduce and verify the technique. In the field of microbiology, the invention can only be achieved if the microorganism used is accessible, as the taxonomic description of the microorganism alone is not sufficient to allow reproduction of the invention described. This requirement means that, prior to submitting a patent application, the inventor must submit a sample of the microorganism to an officially recognized collection and must authorize unreserved access to this microorganism by third parties as soon as the patent application has been published, *i.e.* 18 months after submission of the application or the priority date (in the majority of countries) or to patent approval (USA). In order to avoid submission of microorganisms in every country in which protection is required, the Treaty of Budapest (1977) instituted recognition of a single submission to an approved collection which is valid for all of the countries having ratified the treaty.

The most delicate problem is that of free access to the submitted microorganism. As the microorganism is a living organism capable of multiplying, it is important to establish rules designed to prevent its dissemination and to protect the inventor's rights. Although this problem has not yet been satisfactorily resolved, various solutions have been proposed such as, for example, Rule 28 of the Executive Regulation of the European Patent Convention (1978). An additional complication arises in the form of the natural or deliberately induced mutations from the submitted strain, raising problems of evidence in an eventual breach of patent law suit. In view of these difficulties, for which no satisfactory solution has yet been provided, the manufacturer's policy with regard to patents is essentially based on a choice between protection by patent, which implies making the new microorganism available, together with the risks involved, and maintaining secrecy. This choice may be influenced by the nature of the protection required.

In the presence of a new metabolite produced by a new microorganism, application for patent protection would appear to be essential as the metabolite will be protected regardless of the method of production. In the presence of a new natural or modified microorganism, producing a known metabolite, the decision is more delicate as it must take into account the problems involved with exercising the patent-holder's rights.

3.9.18 CONCLUSION

The pharmaceutical market for recombinant products is currently in a phase of active growth: over the last five years, since the release of the first of these products into the market, the most readily accessible products in terms of technique and those with the most obvious therapeutic value are now in the phase of clinical development and recent or imminent release on to the market. The expected turnovers predicted by marketing specialists for the next few decades have been indicated for each of the preceding monographs. It is always possible to calculate the sum of these figures to obtain a global percentage of the value of genetic engineering in the pharmaceutical field, provided that it is kept in mind that the figures obtained are highly subjective and very imprecise.

Some market studies estimate the total market for recombinant products to be worth $35 billion (base 1985) by the year 2000, which would represent about 18% of the global pharmaceutical market at that time. The share attributed to recombinant products would therefore be equal to the largest therapeutic class (cardiovascular system) for all types of drugs combined. In view of the very severe competition between major drug companies to launch the *same* recombinant products on to the market, it is unlikely that we will see this type of revolutionary situation by the year 2000. This obviously does not exclude the possibility of very profitable niches for companies arriving first in a particular therapeutic field.

3.9.19 REFERENCES

1. See especially: Y. H. Chiu, *et al.*, 'Interscience Colloquium', Paris, September 10–13th, 1987.
2. M. T. Kuenzi, *Eur. J. Appl. Microbiol. Biotechnol.*, 1985, **21**, 1.
3. 'Impact of Applied Genetics', Office of Technology Assessment, Washington, DC, 1981.
4. *Federal Register*, 7th May 1986.
5. D. A. Stringfellow, in 'Interferon and Interferon Inducers', Dekker, New York, 1980.

4.1
Industrial Factors and Government Controls

GEORGE E. POWDERHAM
Roussel Laboratories Ltd., Uxbridge, UK

4.1.1 INTRODUCTION: THE SHAPE OF THE MULTINATIONAL INDUSTRY

The international pharmaceutical industry is today dominated by 20 to 30 large multinational companies, all of them based in the USA, Switzerland, West Germany, the UK and France. These might be said to constitute the 'first division'.

There is then a range of 'second division' companies, based in Japan, Sweden, Italy and India. These companies, for the most part, have some international presence but their domestic markets predominate.

Reference to the accompanying Tables 1 and 2 illustrates a number of points. (i) The top 20 companies are all in the 'first division' category. (ii) The only non-American/European companies in the top 30 are those of Japan. (iii) While one or two companies like Merck Sharp & Dohme, American Home Products and Hoechst maintain a constant position within the top four or five places, the fortunes of others are much more volatile. For example, Glaxo 15 to 8, Roche 1 to 14, Smith Kline Beckman 22 to 5, Abbott 34 to 10 to 19, *etc.* (iv) The relatively massive US market provides the base, from which 13 of the top 20 companies are American.

Table 1 Leading Pharmaceutical Companies Pharmaceutical Specialities

		Market share 1986 (%)
Leading 10		
1. Merck & Co.	US	3.6
2. Ciba-Geigy	Swiss	3.0
3. American Home Products	US	2.7
4. Hoechst (including Roussel)	Franco/German	2.7
5. Smith Kline Beckman	US	2.5
6. Pfizer	US	2.5
7. Johnson & Johnson	US	2.3
8. Glaxo	British	2.2
9. Eli Lilly	US	2.2
10. Bayer	German	2.0
		25.7
Leading 20		
11. Sandoz	Swiss	2.0
12. Bristol-Myers	US	1.9
13. Roche	Swiss	1.9
14. Boehringer Ingelheim	German	1.8
15. Upjohn	US	1.7
16. Warner-Lambert	US	1.6
17. Schering-Plough	US	1.6
18. Squibb	US	1.5
19. Abbott	US	1.4
20. Cyanamid	US	1.4
		42.5

[a] Calculations based on pharmacy and hospital sales in 35 countries representing 90% of pharmacy and hospital market in the western world.

Both the leading and secondary elements of the industry are, to an ever-increasing extent, involved beyond ethical prescription branded drugs, with generics, the over-the-counter self-medication market and allied spheres such as nutritional foods and dietary 'fashions'.

In spite of opinions that have persisted for 20 years that the industry would steadily reduce in the number of major participants by merger and acquisition, such a trend has not yet accelerated. Such a view was (and is) supported by the argument that with the ever-increasing cost of research and drug development, fewer companies would be able to afford the pace.

That this is not happening might well be attributed to the very substantial shift in the total market to the detriment of the branded drugs in favour of the rapid growth of generics and self-medication. In this sector the cost of entry is less and one's share is primarily dependent on good commercial flair rather than scientific skill and invention.

However, whether one of the majors or part of the secondary group, the problems which face the industry in terms of investment and profits, on a stage increasingly subject to regulatory controls and exposed to public assessment, are the same and are reviewed in the sections which follow.

4.1.2 INTERNATIONAL INVESTMENT STRATEGIES

4.1.2.1 Investment

There are three areas of substantial investment for any comprehensive pharmaceutical operation: (i) fundamental research; (ii) chemical production; and (iii) pharmaceutical formulation production.

Table 2 Participation by Companies in the Pharmaceutical Market

Company	*Nationality*	*1970 sales* (£m)	*1970 ranking*	*1977 sales* (£m)	*1977 ranking*	*1984 sales* (£m)	*1984 ranking*
American Home Products	American	200	5	875	2	2 251	1
Merck Sharp & Dohme	American	280	2	831	4	2 178	2
Pfizer	American	174	8	634	9	1 778	3
Warner-Lambert	American	171	9	831	5	1 719	4
Hoechst (including Roussel)	W German	285	[a]	888	1	1 601	5
Smith Kline	American	90	19	325	22	1 581	6
Eli Lilly	American	176	6	515	11	1 514	7
Bayer (including Miles)	W German	151	—	818	3	1 477	8
Ciba-Geigy	Swiss	206	4	673	7	1 452	9
Abbott	American	77	24	206	34	1 273	10
Bristol-Myers	American	110	14	677	6	1 184	11
Glaxo	British	109	15	388	16	1 159	12
Upjohn	American	143	11	414	15	1 081	13
Roche/Sapac	Swiss	251	1	670	8	995	14
Sandoz	Swiss	145	10	546	10	991	15
Johnson & Johnson	American	50	29	298	24	967	16
Takeda	Japanese	87	22	419	14	918	17
Rhône-Poulenc	French	108	16	441	12	877	18
Boehringer Ingelheim	W German	88	20	429	13	833	19
Squibb	American	130	12	384	17	813	20
Cyanamid	American	102	18	278	25	789	21
Schering Plough	American	107	17	348	19	788	22
Wellcome	British	86	23	342	20	788	23
Sterling-Winthrop	American	175	7	362	18	770	24
Sankyo	Japanese	50	30	241	30	725	25
ICI	British	42	33	250	28	697	26
Beecham	British	55	26	310	23	612	27
Fujisawa	Japanese	—	—	244	29	610	28
Sanofi	French	(New grouping)		122	—	574	29
Shionogi	Japanese	—	—	241	31	551	30
Dow (including Merrell)	American	—	—	326	21	527	31
Syntex	Panamanian	28	36	191	—	507	32
Revlon	American	—	—	—	—	490	33
Schering AG	W German	88	21	267	26	477	34
Robins	American	—	—	—	—	472	35

[a] 1970 sales by Hoechst £210 m (ranked 3) and Roussel-Uclaf £77 m (ranked 24) (Reproduced from a Report, 'A new focus on pharmaceuticals', by NEDO: Pharmaceuticals Economic Development Committee, published late 1986, with original sources being 'companies' annual reports, industry and NEDO estimates)

Other areas of investment in, say, distribution or office automation are of minor consequence in comparison.

4.1.2.2 Research

Research needs are by no means the highest in terms of the initial and ongoing capital investment but demand a very high annual outlay in terms of maintaining, over a long period of time, a team of highly specialized scientists and the consumables of that operation. At 1988 levels, a capital outlay of around £100 000 per scientist would be an approximate scale for a research establishment.

4.1.2.3 Chemical Production

The investment in chemical production will, by comparison, be very much greater. It is difficult to give an accurate idea of the amount of investment necessary in order to set up a chemical production plant because so much depends on its location, the chemical processes involved and whether the plant is entirely new or a modification of an existing one. The relevant features of the chemical processes include the number of stages, the nature and hazards of the chemicals involved and whether the process runs continuously or in batches. It is a moot point, and one that is very dependent on particular circumstances, as to whether it is cheaper to construct a new plant or to modify an existing one, especially if extensive modifications are needed to bring the plant in line with modern pollution control requirements.

The effect of these factors means that even a very modest multipurpose chemical plant with an output of an unsophisticated product would involve a cost of £5/£10 million, whilst, at the other end of the spectrum, for a major complex involving, say, a sophisticated antibiotic with many stages of synthesis, one would be talking about 10 times that level of cost.

In addition, one must also bear in mind a number of other factors. Firstly, it will be necessary to thoroughly examine the process before deciding to use it industrially by conducting studies on the synthetic process in a pilot plant. This may also involve investment but the plant could also ensure provisional production before the main one was commissioned. Secondly, chemical production plants, unlike other complexes (*e.g.* research centres, pharmaceutical production units) are often specific and can only be transformed with substantial secondary investment. Thirdly, it may be necessary, even with a well-planned set-up to extend the facilities in future as a result of an unanticipated demand for the product. Fourthly, it is becoming increasingly necessary to take into consideration the demands of 'good laboratory practice' with its requirements for easy-to-clean utensils and surfaces, separate preparation and finishing rooms and the organized flow of materials. Finally, future set-ups will rely increasingly on computer-controlled processes to improve productivity, reliability and security.

In contrast to research, the manning of such an installation would be small numerically and the annual expense greatest in terms of energy costs.

4.1.2.4 Pharmaceutical Production

Pharmaceutical production for formulations varies enormously in terms of capital costs. The relatively least costly might be the mass production of tablets and capsules through various other forms such as ointments, suspensions, *etc.*, to the most expensive, highly sophisticated, sterile manufacture of injections.

Common to all forms of primary production would be an automated packaging operation. In today's highly regulated scene, the packaging operation is very sophisticated with electronic equipment for on-line counting, weighing, printing, labelling, inserts, *etc.* All elements of such production have to be supported by a substantial structure of quality assurance, control, engineering, *etc.* (see Chapter 4.6).

Estimates of capital cost would very much depend on the infrastructure of volumes produced in the various categories but, as a generality, for a multiproduct type output, figures would be to the order of: (i) for a limited range of tablets and ointments, one would be estimating in the region of £0.75 million per million units of production per annum; whilst (ii) for a comprehensive range of tablets, ointments, liquids and with sterile production facilities, £1 million per million units of production per annum.

Such broad estimates would presume a minimum size. It would not literally be possible to spend one million pounds and achieve anything practical. A minimum outlay, for the purposes of an example, should be considered as £10 million.

Clearly, there would also be many other factors qualifying such figures in terms of location and site costs, *etc.*

4.1.3 LOCATION

Given such facts, it is not surprising that the strategy of the industry is, inevitably, based on the fewest number of units possible. Strictly from a financial investment point of view, one major complex of production or research to serve all international markets would be the ideal.

4.1.3.1 Research

Perhaps there is no such thing as an 'ideal' size (or location) for a research establishment. Since research is to such a large extent a result of individual creativity, it follows that the individual researchers are more important than the structure.

Nevertheless, it is true that whilst an ideal environment will not turn second-rate researchers into world-beaters, the wrong establishment could have the reverse effect. An inordinately large research centre would seem to be counterproductive. The physical separation of facilities, the difficulties of knowing well all the fellow researchers with whom interaction is essential and the sheer bureaucracy required to keep track of projects, products, *etc.*, lead to inefficiencies.

On the other hand, a unit which is too small will not be able to employ a sufficiently wide range of specialists to investigate any disease in the depth necessary today; it will not be able to synthesize in a reasonable time enough compounds to properly evaluate candidate molecules and the cost of providing essential back-up services, *e.g.* computing, modern spectrometers, will become disproportionate to the overall investment.

This analysis suggests that a large corporation will have a number of research laboratories of manageable size in different locations, the size being chosen to maximize the interactions between scientists of different disciplines with a resultant high productivity level for the investment.

It is, of course, possible to envisage an operation which has a number of satellite establishments with a central one having available the most expensive facilities which are accessible for use by the other groups. With a multinational operation, then, it is possible to envisage the establishment of research centres virtually anywhere in the world. The factors which influence the choice of country can be broadly classified into scientific, cultural and economic.

4.1.3.1.1 Scientific

Clearly, there needs to be a good supply of trained scientists in many disciplines but, in particular, chemistry, pharmacology, biochemistry, biology, pharmacy and computer studies. By training, in this context, is meant at least graduate level but preferably research degree standard. This implies that the country has a well-established education system, leading to universities, polytechnics and research centres containing functioning research departments. Since it is essential that scientists continually upgrade and update their skills and knowledge, there should be opportunities for postgraduate refresher courses. Also, if the centre is not large enough to justify employing all possible specialists, the opportunity to interact with academic experts in those fields becomes essential.

4.1.3.1.2 Cultural

All research is based to a greater or lesser degree on the advances of the past and experience shows that countries with a research record over a number of years are best able to maintain that effort. Table 2 shows that different countries have different records in the introduction of new chemical entities and these differences cannot be ascribed simply to the level of research activities in those countries.

Language is becoming an increasing problem for the non-English-speaking part of the world as the sheer dominance of American workers is rapidly resulting in English becoming the dominant language of science. Scientists in these countries are unfortunately going to have to do a great amount of their reading and communicating in English. Since much of the science is communicated at conferences and symposia, the opportunity to participate in these meetings is essential.

In these days, a major influence on pharmaceutical research is the attitude of society in general towards the essential animal experimentation involved in our work. Chapter 4.10 discusses this in more detail but it is clear that too restrictive legislation in this area, or in clinical pharmacology with human volunteers, could have major influences on the siting of research centres.

4.1.3.1.3 Economic

Capital construction and equipment costs will vary from country to country, especially if major items have to be imported. However, as stated earlier, the salary costs are likely to be the most important current costs for any research centre and such salaries vary enormously around the world. To a large extent, these are historical and reflect the worth that a particular society is prepared to place on any profession. Salaries do, however, have to be large enough, or perceived status be large enough, to attract the relatively small proportion of individuals with the necessary intellectual skills to take up research as a career in preference to many other possible professions. It seems true to state that one of the reasons for American companies establishing research facilities in the UK has been the much lower salaries in comparison with those of the US.

However, salaries are only one of the equations; such salaries are provided by the profit of the organization. Whilst it is true that drug development costs today are so large that all products must be sold in as many countries as possible to recover these costs, the company must sell and be profitable in any country in order to support its research.

It must also be sure of being able to protect its investment by patents; a fact most clearly shown by the example of Canada, where a policy of forced licensing of products has essentially closed down basic pharmaceutical research. Other countries encourage research which carries kudos and the hope of balance of payment bonuses by allowing greater or lesser amounts of research costs against tax. The situation in the UK is described later in the section on PPRS. The Indian Government has encouraged the setting up of research institutes by the tight controls on profits allowed to be returned from the country to a foreign-based multinational. This latter policy has had the happy result of encouraging research into the supply of drugs from the many indigenous plants of the country.

4.1.3.2 Pharmaceutical Formulating

The containment of pressures demanding a multiple number of formulation units is one of the industry's major political problems. There was a period during the fastest growth of the industry in the 1950s and 1960s when many of the large multinationals set up local production units in many countries of the world only to find, in recent years, that these have become economic burdens.

This, because of the fast and continuous growth of the earlier years, has not been maintained. Production technology has improved enormously but always favouring high volume throughput and very expensive sophistications not repeatable in small scale units around the world.

Many governments continue to seek the investments of high technology industries and often impose tariffs or currency restrictions as a means of 'encouragement'. The Pharmaceutical Price Control Scheme in the UK has, as its main element, a system of granting rates of profits based on *capital employed*, *i.e.* mainly industrial investment.

India represents an interesting historical case where government regulations set up to control the massive imbalance of trade subsequent to independence have led to a very rapid development of a national pharmaceutical industry.

Perhaps only countries of the size of India could provide a domestic market capable of supporting an industry. Suffice to say that in spite of the onerous obligations, tight import restrictions, obligatory chemical as well as pharmaceutical capital investment, very restricted export of 'hard currency' in the form of royalties or dividends, the international pharmaceutical industry has invested. Such investment has, by law, had to be shared with Indian partners and on the back of such experience and shared technology has grown the indigenous Indian industry now supplying in excess of 50% of the domestic demand.

Today, the overall impression is that in the western world—in spite of such pressures—the number of production units is not increasing and is probably decreasing. The area of current expansion would be in South East Asia with the emergence of the 'Oriental pharmaceutical industry'.

Again, as with research investment, the industry strategies as to such new investments as must be made would take into account, apart from the political necessities, the availability of sufficient cost effective, skilled labour and the local tax treatment of capital investments including availability of substantial grants for location in areas of unemployment.

4.1.4 RESEARCH TARGETING

It is improbable that any pharmaceutical company today is trying to cover the whole span of problem diseases. What has happened is that companies have established reputations in given areas (*e.g.* Fisons with antiasthmatics, Roche with tranquillizers, Smith Kline & French with gastrointestinal products) and then build on this. The company has a cadre of chemists who are familiar with the synthetic routes and structure–activity relationships of groups of molecules active in 'their' area; the biologists may have invested many years of effort in understanding the intricacies of the disease process and, indeed, may in fact know more than anyone else about it; the clinical pharmacologists and clinical trials specialists will fully understand how to properly evaluate new drugs in the area both for their efficacy and for their potential side effects; the marketing department will know how to promote these products and lastly but by no means least, prescribing clinicians and doctors will associate that company with those activities. It thus becomes difficult, although not impossible, for any company to redirect this momentum into a new therapeutic area or for another company to move into it.

When selecting a research area, influences are felt from research, marketing and clinical departments.

(i) Research pressures

Before embarking on a project one needs to be certain that there are good chemical and biological reasons for expecting that there is a reasonable expectation of success. This usually means that there are biological theories of the underlying malfunction in the disease, that there are animal models which appear to be reasonably predictive of the clinical manifestations of the disease and that the chemists have some concepts around which they can design molecules. The latter can be lead structures already existing from clinical or pharmacological experience or an enzyme of known structure for which an inhibitor can be designed by the application of modern computer graphics.

(ii) Commercial

This is really not as great a pressure as might be thought at first sight. This is simply because the long timescale for development of a drug (at least 10 years) means that there are very few people who can confidently predict what the market demand will be that far in the future. Clearly, antibiotics, antivirals, anticancer and antipsychotic agents will continue to sell well for many years yet.

The example of cimetidine is perhaps illuminating. This drug was a genuine breakthrough: the first agent capable of controlling gastric ulcers without surgery and for which there simply was no existing market. In fact, the drug rapidly became the leading drug in the world in value of sales.

The problems and opportunities involved in research in orphan drugs are discussed in more detail in Chapter 4.11.

(iii) Medical

Clearly, there has to be a medical need—some failure of current therapy which needs improvement. Further, since the effort involved in developing a drug is the same whether no one or millions eventually use it, it is clearly preferable to provide agents that bring benefit to the greatest number of people. In the developed countries, this means the ailments of affluence—heart disease, geriatry, infection, cancer—receive much attention, whilst the demands of the underdeveloped world are more basic and more directed towards immediate survival. For both communities, the efficient control of fertility remains an important goal.

When a programme has been initiated, a series of chemical and biological projects will be initiated. Ideally in chemistry, there will be a mixture of those closely based on lead structures and those most speculatively based. The problem with those too closely related to existing agents is that whilst the prospect for success may be high, the probability is that the final drug will not offer much advantage over existing ones. The more distal project may be less likely to be successful but the rewards in terms of a novel agent may be that much greater.

Likewise in biology, the mixture of projects will include tests already well established to be predictive of clinical activity and more novel ones. The advantage of the latter is that the outcome may be an agent with a new profile of clinical activity. The disadvantage may be that it is difficult to persuade a clinical pharmacologist of the relevance of the test and to initiate studies in volunteers.

The subsequent development process is described in other chapters in this and other volumes in this work.

4.1.5 PROMOTION

4.1.5.1 General

The two areas of greatest controversy surrounding the pharmaceutical industry are product promotion and prices/profits.

The purist view within the industry is that all promotion is education of the medical profession and other users of its products, and the most severe critics say it is hard commercial selling bordering on corruption.

Such a state of affairs arises because of the vast complexity of the problem. The permutation of sickness and ailments in the world's population is (taking into account the range of different circumstances of geography, climatic conditions, ethnic variations) the spectrum of millions living in a situation of almost total deprivation and other millions excessively provided for and indulgent.

The resulting effects, both psychological and real, on the health of those many millions of people, is capable of a mathematically infinite number of possible diagnoses. Over many years the pharmaceutical industry has designed an enormous number and complexity of drugs to treat such problems of society—each drug with a multiplicity of dosages and strengths calculated to serve every need.

Between these two very large logistical masses is the medical profession, relatively very small in number but controlling the diagnosis and prescription of the treatment.

It is an interesting, as well as a complicating, factor that distribution of pharmaceuticals is predominantly from manufacturer to wholesaler to pharmacy or from manufacturer to hospital. The manufacturer 'never sees' the consumer. On the other hand, the physician who prescribes the product is not a buyer. Nor, for that matter, is the patient always a buyer. Indeed, in most cases, the patient pays little or nothing and the buyer is yet another party—the national welfare system, *i.e.* government!

The industry case is that no member of the medical profession can possibly expect to know at any time the range of drugs available and certainly not to keep up with the new innovations. Therefore, constant education, reminders and promotion are necessary.

The critics say that such is the competition within the industry over a range of comparable medicines (none of which are unique or exclusive) that promotion becomes hard selling and often gets to a level of buying sales and the corruption of the medical profession.

There is clearly a substance of truth in both views, as a result of which has evolved a complex of regulations on the industry to acknowledge the need for promotion but to control the techniques, content and volume.

4.1.5.2 Promotion Techniques

The promotion practices of the industry fall into three broad categories: (i) direct face to face conversations with the medical profession, primarily by the use of medical representatives; (ii) direct advertising either in the medical journals or through direct mailing, and (iii) group seminars in various forms—workshops, symposiums, international conferences, *etc.*, *i.e.* large scale 'education'.

Today such forms of contact with the medical profession utilize all the modern visual aids, audio visual, computer technology equipment, *etc.* Such techniques were originally used for small groups but are now equally capable of use in audiences of thousands or on a one-on-one basis.

The problem of the industry in promoting its products is the wide range of attitudes prevailing in the medical professions to the reception of such promotion. Both within general practice and in hospitals and specialist clinics there are the following. (i) Physicians that will never see medical representatives or attend seminars. It can also be presumed that they would not be avid readers of advertising material. (ii) There are those who will see representatives but only on a limited basis of frequency. (iii) Most physicians will tell you that they never read advertisements in journals and 'have no time' for direct mailings. Nevertheless, the industry persists with a substance of both—presumably for some purpose? (iv) Many physicians are receptive to seminars and workshops since they prefer the 'group style' and the opportunity it offers to mix with the fraternity. (v) Others—probably the minority—work enthusiastically with the industry and can be expected to be receptive to all forms of promotion.

In such circumstances, it is inevitable that the industry uses all forms of contact on a considerably larger scale and at greater cost than it would choose. Equally inevitably, such a form of response to the problem brings accusations of excess.

It is not at issue that the industry spends many millions of pounds on promotion every year. 'Who pays?' is equally not in doubt; ultimately, the consumer pays. The industry case is classical to business economics, *i.e.* that the legitimate promotion of its products is essential to the economic success and profitability of the industry—in turn, financing new research to the ultimate benefit of the consumer . . . . It might be said cynically that the ultimate benefit of the consumer is also coincident with the ultimate self-perpetuating growth and success of the pharmaceutical industry.

This point perhaps brings us to the current view of society in general, and governments in particular, that, from an ethical point of view, there must be a point at which promotion becomes excessive and can be said to be in the greater interest of the company rather than the patient. On such a premise is constructed the regulatory pattern attempting to restrain the advertising of the industry to what might be considered reasonable.

There are as many systems of control as there are health and social services which pay. The one common theme that is present in virtually all of them is a code of conduct on promotional practices.

4.1.5.3 Codes of Conduct

A comprehensive review of such codes is beyond the scope of this chapter but the main recommendations of the International Federation of Pharmaceutical Manufacturers Associations are as follows.

(i) Information on pharmaceutical products should be accurate, fair and objective, and presented in such a way as to conform not only to legal requirements but also to ethical standards of good taste.

(ii) Information should be based on an up to date evaluation of all the available scientific evidence and should reflect this evidence clearly.

(iii) Statements in promotional communications should be based upon substantial scientific evidence or other responsible medical opinion. Claims should not be stronger than such evidence warrants. Every effort should be made to avoid ambiguity.

(iv) Particular care should be taken that essential information as to pharmaceutical products' safety, contra-indications and side effects or toxic hazards is appropriately and consistently communicated subject to the legal, regulatory and medical practices of each nation. The word 'safe' must not be used without qualification.

(v) Promotional communications should have medical clearance or, where appropriate, clearance by the responsible pharmacist before their release.

(vi) Medical representatives must be adequately trained and possess sufficient medical and technical knowledge to present information on their company's products in an accurate and responsible manner.

(vii) Symposia, congresses and the like are indispensable for the dissemination of knowledge and experience. Scientific objectives should be the principal focus in arranging such meetings and entertainment and other hospitality shall not be inconsistent with such objectives.

(viii) Scientific and technical information shall fully disclose the properties of the pharmaceutical product as approved in the country in question based on current scientific knowledge including: the active ingredients using the approved names where such names exist; at least one approved indication for use together with the dosage and method of use; and a succinct statement of the side effects, precautions and contra-indications.

4.1.5.4 Control of Promotion

The general application of control varies from one country to another. For example, the UK code is 'self-administered' by the ABPI and complaints of infringements are considered by a Code of Practice Committee; whereas, in France, proposed literature and advertising copy was until the end of 1987 subject to prior approval by a 'Publicity Control Commission' which reported to the Minister of Health. Subsequently, that regulation has been relaxed and the French have reverted to a self-regulatory system as in the UK.

The UK Pharmaceutical Price Regulation Scheme (PPRS) includes a formula which limits each company's total spend on promotion to a percentage of turnover and involves punitive monetary penalties in respect of any excess spending.

It can, therefore, be observed that the controls (whether voluntary or mandatory) respond to the increasing concerns manifested by both the medical profession and the public interest. That is to say, the advertising content must be qualitatively 'ethical' in the sense of balanced, accurate, reasonable, *etc.* and that there should be some limit on the quantity in the sense of the sheer weight and intensity of pressure on the medical profession such as otherwise might justify the criticism of excessive.

It will be readily appreciated that such controls create a somewhat artificial circumstance for the pharmaceutical industry compared with the free marketing of consumer products in general. In a qualitative sense, no advertising 'licence' is permissible which might give a slanted impression to the doctor. Superlatives are either banned or discouraged and one cannot claim to be 'the best', 'the least' or 'the most'. Equally, the disadvantages (side effects, adverse reactions) of your product *must* be recorded.

Within the UK, monetary limitations on total spending put the small companies at a major disadvantage. Even if such a company develops a new compound with major advantages over all existing products, it could not spend up to the commercially justifiable level except at relatively high expense when including the penalties involved.

None of this would offer the professional marketer an attractive scenario. Nevertheless, as far as the qualitative rules are concerned, the industry now accepts that such standards are very necessary and have, since their inception, been of overall benefit to the industry in curbing all of the old excesses and improving the industry's responsible reputation.

In spite of these restraints and perhaps because of the difficulties thereby posed, the industry is always recruiting and developing well-educated candidates for 'marketing'. It is virtually mandatory with most companies to expect that candidates for strategic marketing posts have had direct selling experience. Recruitment for medical representatives is, therefore, usually directed at the life science graduates or those with higher level education who have had initial medical experience in hospital or laboratory work. Following a few years of selling experience, progress within sales management or product management has considerable potential.

4.1.6 PRICES AND PROFITS

Prices and profits are, inevitably, interrelated and the industry is forced, most of the time, to consider them as a related problem. The price question varies from country to country and depends on: (i) the welfare system of total, partial or non-reimbursement which, in turn, dictates whether the patient, the insurance company or the state, or some combination of all three, pays; (ii) the system of price negotiation which varies from a totally free market to one of total price control; (iii) the national price of competitive branded product competitors or generic copies; (iv) increasingly, as in the EEC, the international price comparators; and (v) some combination of all of those and where the overall rate of profitability of the company is also brought into account, as in the UK.

4.1.6.1 United Kingdom

The UK PPRS is worth special mention since it is perhaps one of the more sophisticated (and complicated) schemes evolved between the industry association and a government body—in this case the Department of Health.

The fundamental point of interest is that the DOH is charged by the government with the responsibility for administering the whole of the health care budget which, therefore, includes all spending on drugs. However, it is also the sponsoring Ministry for the pharmaceutical industry and thus has need to encourage investment, employment and exports in the same manner as, say, the Department of Energy exercises responsibility for the coal industry. The intellectual gymnastics required to engage the combination of stick and carrot result in considerable complexities in designing and administering the rules.

The way it works is that based on the submission by the company of its budget for the year ahead, the DOH will consider a request for a profit increase (*i.e.* price increases) if the company can show that its profits are not expected to be 'up to standard'. That standard is a percentage level of profit to capital employed *in respect of UK sales* of ethical pharmaceutical products. The percentage allowed by the DOH is set by them and is based in a given range of 16% to 22% depending upon whether the company has research, development, manufacturing, clinical research, *etc.* in the UK, with associated levels of employment, *etc.* That is to say, is the company a good citizen contributing to the generation of wealth in the UK economy?

The theory is that if such budget discussions with the DOH show that the company is likely to be profitable beyond the percentage agreed, they are expected to *reduce* prices. If, on the other hand, they are short of the profit level allowed, they will be given a lump sum of extra profit which they must spread in percentage price increases across their range of products at rates calculated to equate to the lump sum agreed. Subsequently, the company's, audited financial return (*i.e.* accounts presented in a standard DOH format) have to be submitted and agreed. If those accounts show a profit at odds with the original budget expectations then the company will be due to repay any excess to the DOH or can reflect a shortfall when submitting their next budget proposal.

Where the whole exercise gets more complicated and is subject to negotiation is in the analysis of the cause and effect of the various component parts of those budgets and accounts. For example, in the procedure just outlined, if the company shows that its profits were considerably better than budgeted after it got a price increase, due to its improved efficiency, higher productivity, *etc.*, then it will be allowed to retain that profit. This is referred to as the 'grey area'. That is to say a range of profitability over and above the standard set which the company can aspire to achieving by dint of its own efficiency or productivity and which it may then retain. The point of the grey area is clearly to encourage companies to be efficient. Without it, on a purely cost plus basis, there would be no incentive for efficiency and companies 'up to a point' could simply expect to make profits on higher and higher capital investment and with excessive costs.

Within the area of negotiation and required agreement, annually, on the basis of the Annual Financial Return (AFR) are other potentially difficult problems.

The DOH will require a foreign-owned company to produce evidence that imported raw materials from its parent or associated company are priced at 'arms length', *i.e.* consistent with a price that would be charged to an independent third party. The whole question of *transfer prices* is one of the major points of controversy in the international scene; every government being anxious to minimize imports *and* ensure that 'profits' are not being transferred abroad by means of excessive transfer prices for goods or services.

Research spending by the company must also be reasonable. Generally, this is interpreted as being at a rate in the UK consistent with the rate for the group internationally. Here, the DOH is concerned to support research in the UK but not to subsidize 'excessive' research at a cost to the UK taxpayer.

All other cost centres are judged against the average experience of the department. Having access to returns from the whole of the UK industry, the DOH is in a very good position to judge the relative economy or inefficiency of all the cost headings. Marketing and sales are particularly sensitive spend areas and subject to absolute limitations within a complex of rules but, generally, with an upper limit of around 8%/9% of UK sales.

It is worth re-emphasizing that this whole exercise concerns *only* the UK prescription (ethical) business. A company which has a mixture of ethical, generic, self-medication or export sales must divide its accounts between the ethical and all others. Clearly, this can also pose problems in terms of a fair apportionment of all the cost centres involved. Some, such as distribution or royalties, may be relatively simple but consider others like finance or research. One might consider, especially in a company with a large domestic market and modest exports, that the mainstay of such activities as finance and research were 100% for the UK activity and only of marginal benefit to other sales areas.

The major part of this problem is overcome by the fact that the Annual Financial Returns to the DOH must be certified as a "true and fair view' by the company's auditors. The whole system is, obviously, somewhat bureaucratic but in straight forward circumstances, agreement can usually be reached in a matter of a few weeks but, in other circumstances, it is known that disputes on AFRs have taken two or three years to clear!

4.1.6.2 France

Outside of the UK, there is a wide range of price control systems in operation. For example, in France each new reimbursable product, once the product licence is given, must be the subject of price submission from the company to the 'Commission de la Transparence' representing the Health, Social Security and Social Affairs Ministries. In effect, therefore, there are four parties to the negotiation for an agreed price.

The periodic review of the prices of all established products on the market is similarly initiated by a 'Commission Interministerielle'. Over the years the French have adopted a very pragmatic variety of policies—at one time giving increases to those companies who were increasing investment, employment or exports—and more recently 'across the board' increases of 1% or 2% but, at the same time, allowing companies to increase some products by up to 5% while decreasing others by up to 10%. While supposedly a zero immediate cost to the budget, the longer term effect would be to decrease the price of volume falling products while increasing the prices of growth products.

4.1.6.3 Canada

In Canada there was a free market as far as the GP (general practitioner) prescription market was concerned. Theoretically, companies could establish any price subject, of course, to the normal commercial regard for competitive products.

Recently, the industry and the government have concluded a national 'deal', commonly referred to as C.22 (having been the number of the Bill in Parliament). In simple terms, the government has given the industry guaranteed patent protection for up to 10 years from the *date of marketing* and discontinued the previous 'licence of right' rules enabling generic companies to copy patented products. In exchange, the industry has undertaken to increase national research spending over a period of five years up to a minimum of 8% of sales by 1991 and to 10% of sales by 1995. Secondly, the industry has accepted that it can be asked to justify its prices or proposed price increases which must be reasonable and cost-comparable to alternative medicaments.

However, in so far as hospital buying is concerned, this is regulated by the Provincial Health Authorities which place substantial periodic contracts. The Provincial Health Authorities are clearly in a position to negotiate very competitive prices.

The provinces have a system of formularies, *i.e.* products accepted for reimbursement and products which are overpriced compared to similar competitive products would not get on to the formularies—a considerable influence on the industry to keep prices down to competitive levels.

4.1.6.4 Future Prospects

From these examples, it can be seen that there is, as yet, a wide diversity of systems involved. Already, however, there is the prospect of greater uniformity being planned within the EEC.

Currently, under the fundamental rule of the Treaty of Rome, which provides for the free movement of goods, there has been considerable disruption of the markets in pharmaceuticals caused by parallel importing. This can happen where, due to the pricing systems used by different countries in the EEC and/or because of the relative strengths or weaknesses of the currencies involved, an identical product can be very much cheaper in one country than another. Given such a situation, the wholesalers or retail pharmacists have been quick to buy products from the cheapest source and 'parallel import' them into a higher price market.

A general realignment of all the EEC pharmaceutical regulations and the wider adoption of *original pack dispensing** will ensure that this is a relatively short term headache for the industry in Europe.

4.1.7 PHARMACEUTICAL INDUSTRY COOPERATION

There is a very well-developed structure representative of the international industry and although not authoritative in a hierarchical sense, it is indicative of the geographical spread.

Overall, the IFPMA (International Federation of Pharmaceutical Manufacturers' Associations), then, the EFPMA (European Federation of Pharmaceutical Manufacturers' Associations); then, on a national basis the US Pharmaceutical Manufacturers Association (PMA) and the Canadian Pharmaceutical Manufacturers Association (CPMA); whilst in Europe the Association of the British Pharmaceutical Industry (ABPI), the French Syndicat National Industrie Pharmaceutique (SNIP), the German Bundesverband der Pharmazeutischen Industrie EV (BPI), *etc.*

All these associations at their various levels of national interest cover the whole range of pharmaceutical activities; *e.g.* relationships with world health organization, government and public relations and consumer associations; prices and imports, exports—credit guarantees, *etc.*, good manufacturing and good laboratory practice and original pack dispensing; relationships with the medical profession on many subjects, *e.g.* post-marketing surveillance; marketing codes of practice and international trade; legal affairs, *e.g.* patent laws. This is just a sample of the many areas of interest and activity.

4.1.8 GOVERNMENT CONTROLS

The pharmaceutical industry is among relatively few others in every day life which are subject to massive government controls on virtually every activity. Other such 'high profile' industries would be aerospace, aircraft manufacturing and airlines, nuclear research and nuclear power.

All of these have in common the fact that from the initiation of research right through every activity to the selling prices and conditions of sale, and even to a point of after-sales responsibility, one government agency or another is involved to a greater or lesser extent.

---

* In many countries OPD has been the rule rather than the exception for many years but in the UK it has been the practice to provide both hospitals and pharmacies with bulk packs of 100's or 1000's of tablets or capsules from which the pharmacist dispenses in small quantities; production being cheaper for the industry and profitability greater for the 'dispensing pharmacist'. OPD, therefore, involves the industry producing every pack of product as a dispensing pack in terms of the normal prescribing by doctors for 7, 28 or a 90 day course of treatment.

In the pharmaceutical industry, all the following activities have elements of control.

(i) Research

(a) Laboratories must be licensed for the use of animals in research. (b) Named scientists must be nominated as responsible. (c) Controls and licensing on the use of dangerous or toxic chemicals.

(ii) Development

As for research.

(iii) Clinical Pharmacology

The testing of drugs on human volunteers is controlled by ethics committees which must include qualified and lay personnel.

(iv) Clinical Research

Subject to licensing by the Committee on Safety of Medicines by means of CTC (Clinical Trial Certificate). The CSM is an expert body appointed by the Department of Health. Clinical research must be conducted in accordance with approved protocols. The intensity, complexity and duration of such trials is dictated by the needs to get the product licensed and, therefore, have to be in accordance with the demands and standards of the CSM for each therapeutic class.

(v) Manufacturing

Every manufacturing establishment has to be licensed by the DOH. Every manufacturing operation is inspected on a regular basis by inspectors appointed by the DOH to ensure that all required standards of technique, physical control, record control, cleanliness and hygiene are satisfactory. For exporting companies, such inspections and licences are also required by many foreign authorities, *e.g.* both the US and Canadian Federal Authorities must approve UK factories making pharmaceutical products for North American markets.

All the ingredients used for any product and all primary and secondary packaging are incorporated in the original product licence. The source of primary chemicals has to be registered. Such sources are subject to approval by the Medicines Inspectorate. Any change to any such elements of production must be approved by the Registration Authority.

Product packaging must carry defined information as to content but also include expiry dates and batch codes such as would facilitate a recall of the product from the distribution chain, should the need arise.

(vi) Marketing

No product can be marketed without a product licence which will be granted by the CSM if they are satisfied that the clinical research establishes efficacy, tolerance, safety, *etc.* All old products already on the market are subject to investigation by the CRM (Committee for Reviewing Medicines) and their product licences must be reviewed from time to time. There is increasing demand for post-marketing surveillance on the basis of 'adverse reaction' information from the GPs to the DOH.

As previously rehearsed, all promotion activities are subject to the 'voluntary' code of practice. Promotional spending is limited by the rules of the PPRS.

Every product must have a data sheet or monograph approved by the licensing authority. Such data includes: full technical description of the product, ingredients, *etc.*; full list of all approved indications for use; dosage regimes; adverse reactions and/or potential side effects and general warnings of inappropriate use; required conditions of storage—shelf life, *etc.*; any and every change to the data sheet must be approved by the authority.

(vii) Distribution

The storage and distribution of pharmaceutical products are subject to Home Office approval specific to the storage of hazardous chemicals and those categories of drugs regarded as dangerous which come under Schedules 3 and 4 of the Medicines Act 1968.

It will be appreciated by the reader that these elements are specific to the activity and products of the industry and are additional to the requirements of company law and the civil and criminal law affecting all corporate activity.

4.1.9 CONCLUSION

This chapter has tried to give the reader an overall view of the substantive strategies and problems that influence the pharmaceutical industry today. As was referred to in the introduction, the industry is increasingly subject to government controls and regulations and increasingly exposed to public assessment. Both government and public are motivated by conflicting emotions and objectives.

In the ultimate, the public want medicines that are effective, safe and cheap. In response to that public pressure, governments have activated increasing regulations and controls in scientific terms for the former two and in financial terms for the latter. At the same time, both the public and governments would like to achieve the ideal of having a pharmaceutical industry that is sufficiently profitable to maintain research and medical progress but not so great as to have it said that the industry is *excessively* profitable.

It is a Utopia that is unlikely to be achieved. There are many pharmaceutical products that are very effective and offer complete cures for certain ailments. It is also the case that many medicines are at best palliatives and others have an action more psychological than scientific. Very few medicines are absolutely safe. The freak allergic reaction can occur, the problems of achieving patient compliance and the risk of addiction are ever present. The industry is often wrongly criticized for such occurrences.

The industry is clearly far too complex to work to the refined balance required by all parties.

Scientifically, one only has to consider the increasingly high economic risks of chemical and clinical research. The huge complexity of regulations in these spheres which, while necessary, increase the cost and diminish the prospects of a successful output.

Economically, the capital investment required for research, development and production is enormous. Small scale operations are no longer viable; relatively massive volumes are required to support centres of manufacturing. Then, when all is said and done, there are no absolute guarantees of 100% efficacy or safety, nor can one expect in such circumstances that the products will be cheap.

However, the pharmaceutical industry remains one of the most exciting and creative of today's 'high tech' industries and no one should doubt that it will continue to succeed in its objectives of extending the boundaries of successful medicine.

4.2
Organization and Funding of Medical Research in the UK

NICHOLAS E. J. WELLS
Office of Health Economics, London, UK

4.2.1 INTRODUCTION

The wide range of medicines that is available today to provide effective treatment for many different diseases tends to overshadow the fact that therapeutic progress is a remarkably recent phenomenon. At the turn of the century, only four synthetic drugs were known—aspirin, phenacetin, Salvarsan and barbitone.[1] And in the 1920s, aspirin, phenacetin and caffeine were combined in various ways with codeine, quinine and belladonna to account for an estimated 60% of all prescriptions written by doctors.

The following decade, Domagk discovered the first sulfonamide and this development marked the beginning of the take-off into what has been termed 'the first pharmacological revolution'. Subsequent research efforts have yielded an ever-broadening spectrum of preparations for the treatment of infectious diseases but attention has not of course been confined to this particular therapeutic area alone. Medicines are now available to treat disorders affecting all bodily systems. As a result, it has been estimated from the *Monthly Index of Medical Specialties* (MIMS), a reference and prescribing guide for doctors in general practice, that there were 1507 discrete pharmaceutical products on the UK market in 1984.[2] The total number of branded products listed in MIMS at that time was 2125 and there were over 3900 formulations.

Medicines have come to play an extremely important role in health care provision. In the UK, for example, general practitioners issued, and chemist and appliance contractors dispensed, 393 million prescriptions—or 6.96 per head of population—in 1985 at a cost to the National Health Service (NHS), measured at manufacturers' prices, of £1344 million. Taking into account the cost of medicines prescribed by hospital and dispensing doctors raises the total to £1797 million in that year. At this level of expenditure, medicines' costs (excluding the dispensing and other fees paid to pharmacists) accounted for 9.8% of total spending on the NHS in 1985.

Table 1 shows that the NHS medicines bill has increased considerably in recent years. Between 1980 and 1985, for example, NHS purchases of medicines at manufacturers' prices increased by 74%.

Table 1 Pharmaceutical Sales to the NHS at Manufacturers' Prices, UK[16,a]

Year	*Pharmacists* (£ million)	*Doctors* (£ million)	*Hospitals* (£ million)	*Total NHS* (£ million)	Total NHS sales	
					As % GNP[b]	*As % gross NHS cost*
1970	131	7	33	171	0.38	8.3
1975	280	14	73	367	0.38	6.9
1980	777	43	213	1033	0.52	8.7
1981	903	52	247	1202	0.55	8.8
1982	1054	60	284	1398	0.59	9.6
1983	1191	69	317	1578	0.61	9.6
1984	1274	79	336	1689	0.61	9.7
1985	1344	88	365	1797	0.60	9.8

[a] Figures may not add up to total because of rounding. [b] At factor cost.

Even after adjustment has been made for the effects of price inflation, real growth amounted to 23% over the period. A number of factors underlie this trend and among these the ageing of the population is clearly of significance. The number of people aged 75 years and over increased by almost 500 000 between 1980 and 1985 and average pharmaceutical consumption in this age group, at 15 or more prescriptions *per capita*, is five times that in the population of working age.

Another important explanation is that the costs of researching and developing significant new medicines have continued to increase rapidly. The Centre for Medicines Research currently estimates that an investment of up to £93 million may be required to finance the discovery of a new chemical entity and its transition from the laboratory bench to the pharmaceutical market place.[3] The research and development content of such products coupled with the therapeutic advantages they offer over existing alternatives means that they command a price premium which is reflected in the medicines bill as consumption patterns switch in their favour.

The rising costs of research and development faced by pharmaceutical manufacturers have important implications for future therapeutic progress. Concurrently, there is widespread dissatisfaction among scientists working in the academic sector at the level of support being made available from the public purse for medical research. This chapter discusses the trends which have culminated in the present concerns of the research community and identifies the issues that need to be addressed if the record of success so far in new medicines innovation is to be sustained in the future.

4.2.2 THE RESPONSIBILITY FOR RESEARCH

Table 2 indicates that four separate bodies in the UK spent an annual total of £943 million on medical research in the mid-1980s.[5] Some degree of caution needs to be exercised, however, in interpreting this figure. In addition to the element of inconsistency arising out of the use of different accounting year ends within and between the organizational groupings, precise research expenditures in the university system are not known and have to be estimated. Second, it is axiomatic that the types of research supported by the different bodies differ considerably. Whereas much of the academic research effort will be directed towards gaining a better understanding of, for example, the fundamental mechanisms underlying disease processes, research expenditure by the pharmaceutical industry is predominantly aimed at the development of new medicines. Finally, the overall sum of £943 million understates the true level of medical research spending since it excludes the financial support made available by other organizations such as the Department of Health, the Health Authorities and the Science and Engineering Research Council. The latter, for example, allocated £26.6 million to biotechnology and biology in 1985–1986 but how much of this total might legitimately be categorized as medical research is unknown.

4.2.2.1 Universities

Under present arrangements in the UK, universities receive resources from the University Grants Committee (UGC) in the form of block grants to cover both research and teaching. The objective of the UGC's input into research is to provide the basic 'floor' of research capability in university

Table 2 Funding of Medical Research and Development[a]

	£million	*Year*
Universities	154	1985/86
Medical Research Council	129	1985/86
Medical Research Charities	110	1985
Pharmaceutical Industry	550	1985
Total	943	

[a] Sources of data include the Annual Review of Government Funded R and D, Annual Reports of the Medical Research Council and the Association of the British Pharmaceutical Industry and the Annual Handbook of the Association of Medical Research Charities.

departments which is necessary if speculative ideas are to be generated and developed to the stage where they may attract support from external sponsors, such as the Medical Research Council. However, there is now widespread concern that both sides of this dual-support system are inadequately funded by government and that this is exerting a detrimental impact on potential research achievement.

Substantial reductions in real terms in UGC funding—between 1981 and 1986, for example, there was a real total decline of about 20% in the resources provided by the UGC for medical schools[4]—have diminished the capacity to cover the overhead costs of research. In addition, these reductions coupled with a smaller volume of resources reaching the teaching hospitals from their local NHS authorities have resulted in cuts in the number of academic staff. Since 1980, 434 medical academic posts have been lost.[4] As a consequence, research has suffered since academics remaining in post have had to give greater priority to clinical service work and teaching to compensate for the shortfall in these two areas caused by the reduced staff numbers. In this regard it has been reported that in order to maintain patient care with fewer staff many academics are spending three-quarters of their time treating patients when it is intended they should devote only just over half their time to NHS work.[5]

It has also been argued that financial stringency is having the effect of altering to some extent the direction of the research carried out in the universities. This is happening because academic researchers are increasingly having to seek financial support from external agencies such as the medical charities and the pharmaceutical industry. In evidence to a Sub-Committee of the House of Lords Select Committee on Science and Technology, which is enquiring into priorities in medical research, the Committee of Vice-Chancellors and Principals has argued that 'there is a tendency for research to be distorted towards glamorous high tech projects which attract outside funding from medical charities and industry at the expense of work likely to lead to greater benefit for the greater number of people as the population ages.'[6] Similarly, the Association of Clinical Researchers in Medicine in evidence to the same committee has stated that 'while contacts with industry are clearly desirable, the consequences for research on illness and health are potentially very serious since industry is not in general prepared to invest in non-drug related research . . . '.

4.2.2.2 Medical Research Council

The Medical Research Council (MRC) seeks to improve the health of individuals in the community by supporting research into the cause, diagnosis and treatment of disease as well as studies of the social, environmental and preventive aspects of medicine. Slightly in excess of half of the MRC's annual expenditure is allocated to its own establishments including 53 research units of varying size, the National Institute for Medical Research and the Clinical Research Centre. The remainder of the MRC's income is spent on grants to universities and their medical schools and on training schemes.

In 1984–1985, the income of the MRC amounted to £123.7 million, most of which, £117.2 million or 95%, was provided by the government *via* the Parliamentary grant-in-aid. The latter sum represented an increase of 15.2% over the figure for 1981–1982 (the year when the full rigours of the government cash limits to control public expenditure began to be applied). When inflation is taken into account, however, it emerges that the period witnessed a real decline in the MRC's funds from the government of 1.7%. Coupled with pay awards in excess of the allowance within the cash limit, increased pension contributions and raised international subscription costs, this reduction meant in fact that the MRC experienced a fall in its purchasing power of 4% in 1984–1985 compared with the previous year.[7] Against this background, it has been observed that the MRC has passed from growth to level funding to decline in the space of about seven years.[8]

These deteriorating financial circumstances have inevitably inhibited the capacity to undertake medical research. During 1984–1985 most MRC establishments had their laboratory supplies allocation cut by 16% and received no funds for capital equipment. In addition the number of research grants awarded had to be reduced. In 1984–1985, the MRC was able to provide funding for only about 55% of alpha-quality grant applications (the remainder were categorized as 'approved but not funded'). And the number of programme grants (support for research designed to achieve broad objectives with a normal tenure of five years) has fallen since the beginning of the decade: at the start of 1985 the number of such grants in existence was approximately three-quarters the total recorded four years earlier.

The MRC's report for the 1985–1986 financial year notes that the increase in the parliamentary grant in aid failed fully to reflect price inflation. There was therefore a small decline in real terms in the MRC's funds from this source. As a result, the MRC's decision partially to restore some of its expenditure on laboratory supplies could only be accommodated by a further cutting back on grants and training awards. Project grant awards (usually up to three years' duration) were reduced by 7.5%, research studentships were cut from 230 in October 1984 to 160 in October 1985, advanced course studentships fell from 100 to 70 and intercalated awards were reduced from 380 to 340. Against this backdrop of developments, the MRC stated in the 1985–1986 annual report that 'at present funds are insufficient to maintain a healthy base of science for medical research in the United Kingdom and to maintain the confidence and morale of gifted workers, especially those now entering research.'

The funding of research in academic centres is predominantly the responsibility of government and the central objective is to advance scientific knowledge which may in turn pave the way, in medical research for example, for new and more effective means of therapeutic intervention. In 1985–1986, government funding of university research in medical science and of the MRC amounted to an estimated £276 million. This sum was equivalent to about 23% of total government funding of research undertaken in the universities and by the research councils, 13% of all civil research monies provided by the government and just 6% of the latter's total financial commitment to research. At this level, there is widespread concern that medical research in the UK is inadequately funded just at a time 'when the basic sciences are offering us the possibility of making the next 50 years the most exciting and productive for medical research.'[9]

4.2.2.3 Medical Research Charities

In sharp contrast to the developments described above, financial support for academic medical research provided by the medical charities has increased considerably in recent years. The Association of Medical Research Charities (AMRC) has reported that in 1985 its 35 members spent £110 million of their income on research.[10] Furthermore, this sum—which compares with £33 million in 1979, a doubling in real terms over the period—understates the overall charitable commitment to research funding because it excludes the contributions of many other charities which do not belong to the Association.

The medical charities are today performing an increasingly vital role in sustaining medical research in this country. At the end of the 1970s, the research funds available from the AMRC members were equivalent to approximately 47% of the MRC's expenditure on research. By the mid-1980s, this proportion had risen to 86%. The growth of the charities' role in this way has enabled many research initiatives to proceed which the academic sources of finance, despite approving the proposals, have not been able to fund. Yet these charitable funds should be seen as complementary to, and not substitutes for, public sector support. The resources that the charities channel into medical research are dependent on donations from the public and industry and are therefore potentially susceptible to sharp year-on-year fluctuations. In addition, a substantial proportion of the charities' total expenditure is directed at only a small number of diseases. In 1985, heart disease and cancer (*via* the British Heart Foundation, the Imperial Cancer Research Fund and the Cancer Research Campaign) accounted for more than half of the AMRC members' expenditure on research. This pattern of distribution reflects proper public concern at the volume of morbidity and mortality generated by these particular diseases but it also means that other major illnesses are relatively neglected by the charities sector.

Furthermore, the approaches to research funding adopted by the MRC and the medical charities are significantly different. The former has a major commitment to long term research in the sciences basic to medicine, whereas the charities apply most of their funds to clearly defined projects of limited duration.[11] Finally, the responsibility for research training rests substantially with the MRC and the universities and is not a direct objective of the charities' expenditure on research projects.

4.2.2.4 Pharmaceutical Industry

In contrast to the academic centres where research is directed principally towards increasing fundamental scientific understanding about disease processes, research activity in the pharmaceutical industry is concentrated 'on the discovery of new scientific knowledge necessary to facilitate the conception and development of a new product (that is, a new chemical entity or process) for the treatment of disease.'[12] It is not surprising, therefore, that as the industry and its commitment to research have grown, pharmaceutical manufacturers have increasingly become the major source of new medicines. It has been calculated that 54% of all new products coming onto the US market between 1935 and 1949 were products of industry research, rising to 68% between 1950 and 1962 and 82% between 1963 and 70.[13] In more recent years, this proportion has continued to increase and the Centre for Medicines Research in the UK estimates that 95% of the medicines marketed have been discovered in the industry's own laboratories.[14]

The UK is a major centre for pharmaceutical industry funded research and development, attracting about 8% of worldwide industry funds allocated to these endeavours. This position has been achieved because of the excellence of the academic research infrastructure in this country and the recognition by government of the benefits of a thriving pharmaceutical industry for the health of both the community and the nation's economic accounts. In this context, market research data indicate that four out of the five top-selling medicines worldwide in 1985 had been discovered in UK laboratories.[15] The record of innovative success achieved by this country's pharmaceutical manufacturers has resulted in the industry becoming one of the UK's most significant net earners of foreign currency. In 1980 pharmaceutical exports exceeded imports by £523 million. By 1985, this surplus had risen by 60% to £835 million, a sum exceeded only by the manufacturers of power generating machinery (£1080 million), other transport equipment (£1015 million) and organic chemicals (£850 million). From another perspective, only the USA, Switzerland and West Germany enjoyed larger trade surpluses in pharmaceuticals than the UK in 1984.[16]

4.2.2.4.1 Trends in pharmaceutical innovation

In 1985, the UK pharmaceutical industry spent £550 million, equivalent to 14.2% of its output value, on research and development. This sum was more than 18 times the figure recorded in 1970 (Figure 1). Even after the effects of inflation are taken into account, the period still witnessed an almost fourfold increase in spending in real terms. However, rising levels of industrial investment in research and development in the UK and elsewhere among the other major pharmaceutical nations have not been associated with increasing numbers of new medicines becoming available for use each year.

Trends in pharmaceutical innovation can be gauged in several different ways. The number of compounds synthesized, screening tests undertaken, patents filed and scientific articles published are all measures that have been employed in studies of innovation patterns. In many instances, however, such indicators reflect the level of research activity and shed little light on output. Consequently, attention needs to be focused on product introductions and more specifically on new chemical entities (NCE) rather than reformulations or modified presentations of existing pharmaceutical preparations. Yet even the use of NCEs is not straightforward since surveys have differed with regard to the inclusion or otherwise of, *inter alia*, drugs restricted to use in hospitals only, vaccines and semi-novel combination products. In addition, analyses based on simple counts of new chemical entities yield no information about therapeutic novelty and worth nor commercial success. Furthermore, focusing too narrowly on trends in the numbers of new chemical entities marketed in any given country may generate a misleading impression of the true state of pharmaceutical innovation. In this context, Figure 2 shows the number of new chemical entities marketed each year in the UK between 1970 and 1984. In broad terms the data suggest a reasonably steady introduction rate of about 20 products per annum. However, this trend, whilst obviously reflecting innovative progress

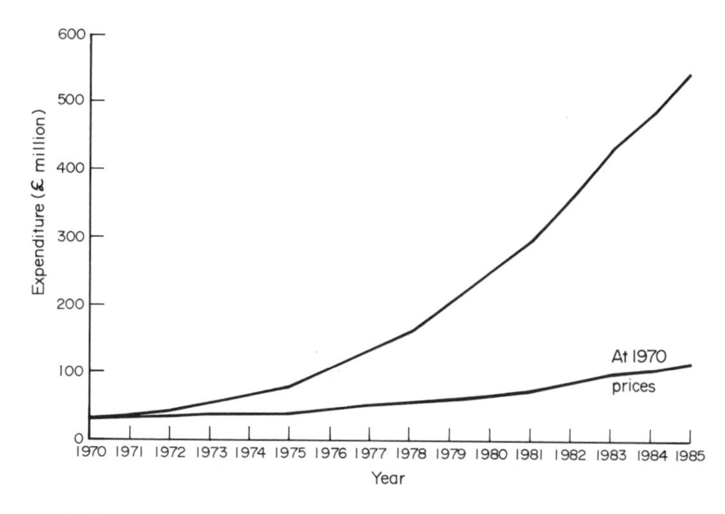

Figure 1 UK pharmaceutical industry expenditure on research and development, 1970–1985 (data from the Annual Reports of the Association of the British Pharmaceutical Industry)

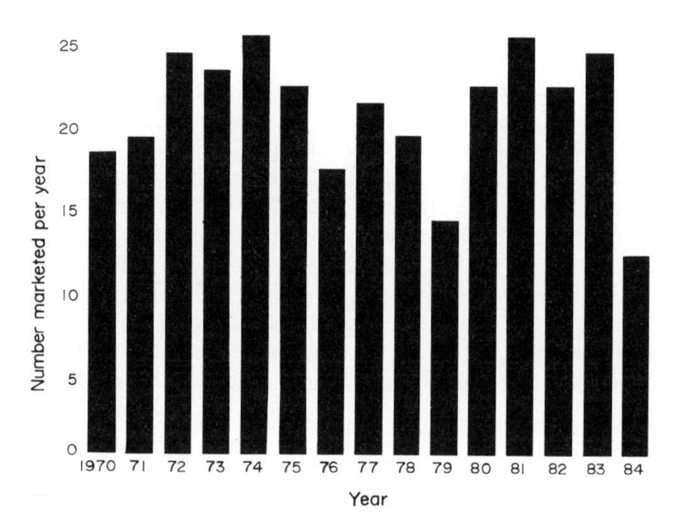

Figure 2 Number of new chemical entities marketed in the UK, 1970–1984 (reproduced from ref. 29 by permission of the Office of Health Economics)

to some degree, is also determined by the international marketing policies of the pharmaceutical manufacturers.

Some of the difficulties in using new chemical entities as a guide to innovation—notably the requirements of definition consistency and the need for a global overview—have been overcome in the time series study by Reis-Arndt.[17] Table 3 shows the number of new chemical entities introduced for the first time onto one or more of the world's markets each year between 1961 and 1985. The analysis is by country of origin of the discovering company. This means, for example, that British-owned companies introduced 86 new chemical entities over the period although the discovery need not necessarily have occurred in a laboratory based in Britain. (It follows that it is not possible therefore to associate these data with the research and development spending figures shown in Figure 1 which are the sums of the expenditures of British- and foreign-owned companies operating in this country.)

Several important points can be drawn from the data contained in Table 3. First, USA manufacturers have dominated NCE development throughout the period—almost one quarter of the output resulted from the research of companies with American origins. Second, the World's demand for new medicines is met by only a very small number of nations. Thus companies from the USA, France, West Germany, Japan, Italy, Switzerland and the UK were responsible for 86% of the NCE launches between 1961 and 1985. Third, manufacturers from the Eastern Bloc have played a negligible role in pharmaceutical innovation. The data in Table 3 indicate that, for the period as a

Table 3 Annual Number of New Chemical Entities Introduced on to the Pharmaceutical Market by Country of Origin of Introducing Company[17]

Year	*USA*[a]	*France*[a]	*West Germany*[a]	*Japan*[a]	*Italy*[a]	*Switzerland*[a]	*Eastern Bloc*	*UK*[a]	*Scandinavia*[a]	*Benelux*	*Spain*	*Austria*	*Others*[b]	*Total*
1961	31(1)	12	11	8	4	12(1)	4	6(1)	4(1)	2	—	3	—	95
1962	20	21	15	4	7	8	—	4	5	7	—	—	2	93
1963	22(1)	21	17(1)	12	2	7	2	9	4	2	1	1	—	99
1964	15	8	14	8	4	6	6	4	2	1	—	2	1	71
1965	13	14	10	13	6	7	2	4	1	1	—	2	—	73
1966	22	19	7	8	3	3	8	4	3	4	1	1	1	84
1967	19(1)	19	8	8	6	8	9	5(1)	3	—	1	3	—	88
1968	20(2)	17(1)	12(1)	7	7(1)	5(1)	5	4	3	3	2	1	1	84
1969	18(1)	23	11	5	9	3	9	3(1)	2	3	—	—	—	85
1970	21(1)	18	7(1)	7	2	6	4	2	2	1	1	2	—	72
1971	26	16	5	11	7	6	9	2	3	4	1	—	2	92
1972	14	13	4	9	9	3	8	3	1	2	—	—	1	67
1973	10	18	14(1)	1	8	9(1)	9	3	1	2	3	—	—	77
1974	17(1)	13	9(1)	14	4	5	5	4	3	2	2	1	—	78
1975	15	7	12	8	5	8(1)	12	4	1(1)	1	2	—	—	74
1976	16(1)	8	9(1)	2	7	2	5	3	4	1	1	1	—	58
1977	14	5	12	6	12	3	1	6	3	—	2	—	1	65
1978	12	9	7(1)	6	4	5(1)	5	4	2	2	1	—	—	56
1979	14	6	10	8	7	—	1	—	1	—	2	—	1	50
1980	14(1)	2	8(1)	10(1)	9	4(2)	1	—	2(1)	—	1	—	—	48
1981	14	5	15	16	3	6	5	4	1	1	2	—	2	74
1982	11	2	4	9	2	8	3	1	1	1	2	—	2	46
1983	14(1)	3	13	10	4	4	—	4(1)	2	1	—	1	—	55
1984	8(1)	3	4	14(1)	5	4	—	1	2	—	2	—	—	42
1985	22	6	9	12	6	1	—	2	1	1	1	—	—	61
Total	422(12)	288(1)	247(8)	216(2)	142(1)	133(7)	113	86(4)	57(3)	42	28	18	14	1787

[a] Figures in parentheses signify substances developed simultaneously in two countries and are included in preceding number shown for country and year. [b] Argentina, Australia, India, Israel, Canada, Mexico, New Zealand, Portugal.

whole, only 6% of NCEs originated from this source. Furthermore this average figure disguises the fact that the contribution from these countries has been declining since the start of the 1970s. Between 1971 and 1975, Eastern Bloc manufacturers accounted for 11% of worldwide introductions. This proportion fell to 5% for the 1976–80 period and to just 3% between 1981 and 1985. Finally, it is clear that the international rate of pharmaceutical innovation has fallen steadily over time. In 1961, 95 NCEs reached the world market compared to 61 in 1985. If the annual averages for periods of five years are considered in order to avoid the potential distortions inherent in single year data (this would appear to be especially necessary in the case of the figure for 1985), the study reveals a 35% reduction from 86.2 during 1961–65 to 55.6 for 1981–85.

4.2.2.4.2 Explaining the trends

There is no single reason for the declining trend in NCE introductions which has taken place over a period of time when national expenditures on research and development have risen. However, one of the most significant explanations lies in the changing nature and duration of the drug development process. In a review of UK trends in pharmaceutical innovation since 1960, it was reported that the development phase facing a major new chemical entity in its transition from initial discovery to marketing now extends over 10 years; in the early 1960s, it was not unusual for a corresponding project to be completed within a period of approximately three years.[18] Similarly, a study from West Germany has reported that the overall average duration of research and development work on new substances has increased from between 2.3 and 5 years in 1964 to between 9 and 13 years in 1981.[19]

Several factors have underpinned this trend and among these the proliferation of regulatory demands following the thalidomide tragedy of the early 1960s has perhaps attracted most attention. Over time, the duration and extent of safety evaluation has increased and as knowledge has advanced new safety tests have been devised and these have been added to the requirements already in existence. The direct impact of this trend on testing horizons has been compounded by delays within the regulatory authorities caused by the imbalance between the manpower available and the significantly increased volume of information that has to be analyzed.

It would be misleading to imply that in the absence of official intervention there would have been little change in the testing procedures applied to new candidate medicines. The survival of the industry depends on the production of safe and effective medicines and many of the tests demanded by the regulatory agencies would have been adopted by manufacturers had the option of self-regulation been available.[20] In addition, attention should also be drawn to the measures introduced in the UK at the start of the present decade by the Committee on Safety of Medicines and the licensing authority to promote greater flexibility in pre-clinical testing and to establish a clinical trial certificate exemption scheme respectively. Nevertheless, the fact remains that additional toxicity and clinical testing, coupled with the process of therapeutic transition away from antiinfective medicines towards those for the treatment of chronic diseases, has significantly lengthened development times and in so doing has played an important part in increasing the portion of research and development resources devoted to development from about 50% in 1970 to 70% in 1983.[12]

A number of other factors have been identified as potentially important contributors to the reduced innovative output of the pharmaceutical industry over the last two decades or so. It has been speculated, for example, that the trend represents a relative deceleration that had inevitably to follow the unprecedented levels of innovative activity of the 1950s and 1960s. In this context it has been argued that the drop in NCE productivity reflected a depletion of the stock of usable biological knowledge. Whilst there is undoubtedly some validity in the contention that the opportunities for discovering new antiinfectives by screening microorganisms in the soil were diminishing, new fields of potential advance were opening up. Consequently, any decline in innovation should, at worst, have manifested as a temporary interruption preceding the resumption of a high level of annual NCE introductions. However, as the data contained in Table 3 show, this did not occur.

Other possible explanations for the reduced level of research productivity since the early 1960s have included the direct and indirect effects of consumerism, sceptical attitudes towards science and technology, negative publicity, zero-risk philosophy, shortages of highly trained and motivated research manpower and the negative effects of national drug reimbursement policies.[12] In the final analysis, however, no one cause in isolation can account for the considerably reduced rate of NCE introduction of recent times. Instead, the trend is attributable to a combination of factors, the relative significance of each of which is, and seems likely to remain, a source of contention.

4.2.2.4.3 Research and development costs

One factor relevant to, but not included in, the discussion above is of course the cost of researching and developing a new medicine. Substantially extended development times, technological advance leading to improved and considerably more expensive research equipment, inflation in the prices of materials, buildings and energy as well as in wages have combined to increase significantly these costs over time. The nature of pharmaceutical research is such that it is impossible to isolate the costs attributable to a specific new medicine but it is clear that they are now substantially greater than the £2 to £3 million estimated for the first half of the 1960s. In a consultative document published in 1981, the Pharmaceutical Sector Committee of the Chemicals Economic Development Council reported that the cost of developing a successful major new drug can now be £50 million or more.[21] More recently, the Centre for Medicines Research in the UK has argued that the figure may now be as high as £93 million.

The escalation of costs towards this level is a further factor in the declining annual rate of NCE introduction described above. It also clearly has important implications for the development of new medicines in the future. As the minimum level of investment in research and development necessary to sustain a serious presence in the innovative pharmaceutical market has steadily increased, the number of manufacturers with the capacity to spend sums of this magnitude has fallen. As a result, there has been a trend towards the concentration of 'innovative responsibility' into fewer hands. But even the large heavily research-based companies are experiencing growing pressures on their innovative capabilities.

Investment in pharmaceutical research is fraught with risk. In addition to scientific risks, whereby potential medicines may suddenly fail at advanced stages in the development process having already incurred substantial costs, manufacturers face the possibility that their new product, once launched onto the market, may not recoup their investment in research. This may occur, for example, as a result of a competitor entering the market with a new product that renders those already available technologically obsolescent. Alternatively, unexpected side-effects may emerge during widespread clinical usage, causing the product to be withdrawn from the market. But perhaps the two most significant risks in recent times stem from government measures to contain the growth of public sector expenditure on health care, and the steady reduction in effective patent life.

Focusing on the first of these issues, pharmaceutical manufacturers require adequate prices for their products if they are to sustain an innovative research base. In the UK, the NHS is the industry's principal customer—purchasing 45% of its gross output in 1985—but the first half of the 1980s saw a series of initiatives designed to reduce public outlays on medicines.

The revenues that pharmaceutical manufacturers received from sales of medicines to the NHS totalled £1797 million in 1985. However, this sum accounted in fact for only 9.8% of overall NHS expenditure. In addition, the sum compares favourably with other familiar items of public and personal expenditure: the NHS medicines bill in 1985 was equivalent to nine pence per person per day, whereas corresponding expenditure on tobacco amounted to 34 pence, clothing 72 pence, alcohol 77 pence, education 85 pence, defence 89 pence and food 145 pence.[16] Furthermore, international comparisons reveal that less is spent on pharmaceuticals in the UK than in many other nations. The data contained in Table 4 indicate that pharmaceutical consumption *per capita* in the UK is about half that of Switzerland, West Germany and the USA—the other three members of the group of four nations with a pharmaceutical industry that has, to date, been highly successful.[22]

In spite of these observations, the UK Department of Health in 1983 introduced a 2.5% overall price reduction on NHS medicines and a price freeze which took effect from the beginning of August with the goal of cutting the cost of the pharmaceutical bill by £25 million in 1983–1984 and by £65 million in 1984–1985. In December 1983, the target rate of return on capital for pharmaceutical companies supplying the NHS—set under the Pharmaceutical Price Regulation Scheme (PPRS) which the government employs to regulate the industry's income from its NHS business—was lowered by four percentage points. Following this, a limited list prohibiting the prescribing of certain medicines at NHS expense was introduced and the level of permitted return on capital was cut again by another four percentage points.

In addition to the direct loss of sales revenue caused by these measures, the series of events served to undermine investment confidence. The serious implications of the latter for the future development of new medicines has, however, been recognized and the recently renewed PPRS seeks as one of its objectives to establish a more stable financial environment for the industry. In contrast, solutions to the industry's concerns surrounding the erosion of effective patent life are still awaited.

The lengthening of the various stages of the drug development process has significantly reduced the patent protection available to new medicines entering the market (Table 5). Figure 3 shows that

Table 4 Pharmaceutical Consumption per Person in 1983[a, b]

	£[c]	£[c]
Japan	94	(75)
USA	81	(55)
West Germany	66	(61)
Switzerland[d]	64	(58)
France	52	(51)
Belgium	49	(45)
Canada	46	(35)
Italy	37	
Sweden[d]	35	(34)
Finland	34	(32)
UK	32	(32)
Australia	28	(26)
Denmark[d]	28	(26)
Norway	27	(26)
New Zealand	25	(30)
Ireland	25	
Netherlands	25	(22)
Spain	24	(24)
Greece	17	
Portugal	12	

[a] D. G. Taylor and J. P. Griffin, *Pharm. J.*, 1985, **234** (23 February), 228. [b] Pharmaceutical consumption is equated with manufacturers' returns and therefore dispensing and other related expenditures are not included. [c] The figures in the first column were calculated using January 1985 exchange rates. Those in parentheses were based on 1983 rates. [d] 1982 data.

Table 5 The Stages of Medicine Development[25]

1	*2*	*3*	*4*	*5*
Academic observations and development of understanding of disease mechanisms Generation of therapeutic hypotheses	Discovery stage Candidate medicinal substances identified and selected	Development stage Assessment of activity and preclinical and clinical safety and efficacy testing Production establishment, demonstration of product quality	Product registration	Post marketing development Adverse reaction reporting, event monitoring, new use identification/ approval
Stages 1 and 2 may take from 3 years to over 20 Stage 1 often takes place outside the commercially funded research environment Patents normally filed towards the end of Stage 2		Stage 3 typically takes from 5 to 11 years Clinical work requires a strong national clinical medicine base	Product registration may take from 6 months to 2 years	Average effective patent term left now under 8 years of the full 20 year term

medicines first introduced in 1960 had an average of over 13 years of their patent life still to run but by the early 1980s true effective patent life had fallen to five or six years.[23] Even if account is taken of the licence of right period granted for new existing patents, the protection afforded to recently introduced products is extended only to around nine years. (In June 1978, the UK followed the ruling of the European Patent Convention and raised patent terms to 20 years—previously 16 years—from the date of filing. Existing patents at this time were not granted an extension of their terms with the exception of those filed after 1967—the 'new' existing patents. However, the latter had imposed upon them a licence of right endorsement which means that other companies can apply as of right to manufacture and/or sell the product concerned after the first 16 years of patent life.) Such sharp reductions in the period of time during which marketing can take place without the presence of price-lowering generic competition have increased the difficulties facing manufacturers of obtaining sufficient sales to cover research and development costs.

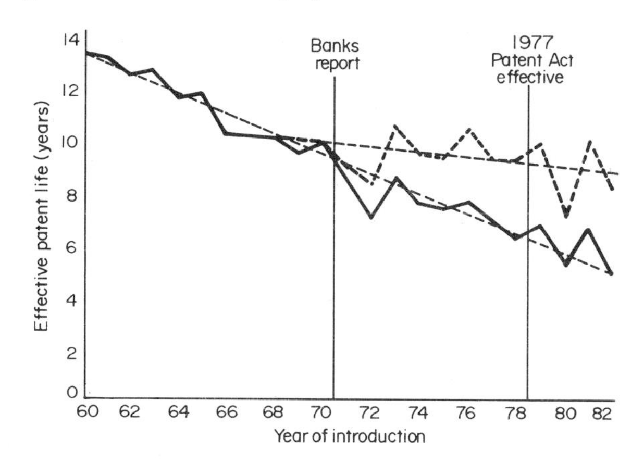

Figure 3 The erosion of effective patent life in the UK, 1960–1982. Total database = 617 NCEs. Patent data = 436 NCEs. ——, true effective patent life (EPL); ----, EPL plus four year licence of right; straight dashed lines indicate lines of best fit (reproduced from ref. 23 by permission of The Pharmaceutical Journal)

The impact of this process of patent erosion is illustrated by the finding that around 70% of the sales revenues generated by the top 100 medicines on the UK market in 1973 were protected by patents. A decade later, this proportion had halved to around 35%.[24] Furthermore, in the future, the effective pharmaceutical patent term may be expected to fall yet further if drug development times continue to lengthen in the absence of legislation to extend the current duration of patent life of 20 years.[25]

The concerns of the pharmaceutical industry at the developments described above have in part been acknowledged by government and in 1988 existing legislation was amended so as to exclude pharmaceutical patents from the effects of the licence of right provisions. More fundamentally, however, there is today a strong case for action aimed at the restoration of the patent life of pharmaceuticals, which in practice enjoy a very much shorter term of effective protection than most other patentable innovations. At the present time, setting aside the licence of right provision, a new chemical entity at launch has the prospect of about eight years of patent-protected commercial life before generic competition can enter the market. Yet recent research suggests that even the most successful of innovations—those large-selling medicines that effectively support the major companies' research and development expenditures—only succeed in achieving 'top ten' status after an average of about five and a half years marketing.[26] Consequently, after less than three years in this position, such products become subject to competition from potentially much cheaper copy products from generic and other manufacturers. In the present era in which governments are seeking ways to contain public expenditure, additional pressures on, and perhaps incentives for, prescribers to employ those cheaper products when they become available could serve seriously to diminish the original innovating company's revenues and thereby jeopardize the future income streams that are necessary to finance subsequent innovations.

4.2.3 THE FUTURE

Therapeutic progress in recent decades has made an important contribution to reductions in mortality and has extended effective control to the symptoms of many chronic diseases. Developments in chemotherapy and immunization have combined with environmental, social and economic improvement to restructure mortality profiles. The principal change has of course been the dramatic reduction in the mortality attributable to infectious diseases. In England and Wales in 1948, deaths from the latter cause accounted for 31.4% of mortality between the ages of one and 44 years and for this age group the death rate from infectious diseases was 542 per million population. The corresponding figures in 1984 were 1.5% and 10 per million respectively.

Within the infectious diseases grouping there have been particularly large reductions in mortality from respiratory tuberculosis. In 1948, deaths from this cause in England and Wales totalled 18 798. By 1984, the number had fallen to 376. Application of the 1948 age-specific mortality rates to population data for 1984 suggest that in the absence of improvement almost 22 000 respiratory tuberculosis fatalities might have been expected in the latter year. Furthermore, 35% of this total would have involved people aged between 25 and 44 years and 90% of cases would have been people of working age. These reductions coincided with the introduction and increasing use of anti-tubercular medicines, although it should also be pointed out that mortality was already on a well-established declining trend prior to the later 1940s.

Innovations in chemotherapy, especially new antiinfective medicines, have, in addition to contributing to reductions in mortality, also generated financial savings for the NHS by reducing admissions to, and the duration of stay in, the expensive hospital sector.[27] However, the major influence of medicines today is on the quality of life. Thus chemotherapy provides effective control of symptoms for individuals suffering from complaints as diverse as asthma, arthritis, epilepsy, glaucoma, anxiety, depression, angina, hypertension and Parkinsonism. Pharmacological advance has of course also played an important part in extending the role of surgical intervention in improving the quality of life for many patients.

It is nevertheless clear that there is a need for new and improved means of treatment in a large proportion of the diseases to which man is susceptible (Table 6). Major disorders in this context include senile dementia which affects about 20% of people aged 80 years and over. The significance of this observation lies in current demographic projections indicating that the number of people in this age group will increase by 30% between now and the early years of the next century by which time they will account for more than one in four of the elderly population (that is, people over 65 years of age). Other disorders awaiting significant therapeutic advance include schizophrenia, rheumatoid arthritis, multiple sclerosis and viral diseases including the acquired immune deficiency syndrome (AIDS).

Table 6 Progress in Drug Research During the Past 50 Years[29]

Disease or condition	*Progress*[a]	*Examples of drugs discovered*
Infection		
Bacteria	Excellent	Antibacterials such as ampicillin, ceftazidime, cephalexin and trimethoprim
Fungi	Good	Griseofulvin, ketoconazole
Animal parasites	Excellent	Ivermectin
Viruses	Good–excellent (vaccines)	Poliomyelitis, smallpox, rubella and others
Cardiovascular disease		
Hypertension	Moderate–good	Methyldopa, propranolol, captopril
Thrombosis	Poor	
Atheromatous vascular disease	Poor	
Ischaemic heart disease	Moderate (drugs–surgery)	Propranolol, atenolol
	Poor–moderate (infarction)	Timolol
Heart failure	Poor (selective inotropy)	
	Good	Thiazides, captopril
Alimentary tract		
Peptic ulcer	Good	Cimetidine, ranitidine
Skin diseases	Moderate	Betamethasone valerate, fluocinolone acetonide, retinoids
Immune diseases		
Allergy including bronchial asthma	Good	Glucocortoid steroids, oral (prednisone) and inhaled (beclomethasone dipropionate) Selective bronchodilators (salbutamol) Disodium cromoglycate Histamine H_1-antagonists
Others including rheumatoid arthritis	Moderate	Glucocortoid steroids, non-steroidal antiinflammatory drugs
Conditions involving the CNS		
Mental illness	Moderate	Neuroleptic and antidepressant agents
Pain	Moderate	Pentazocine, buprenorphine
Anaesthesia (and associated surgical practice)	Good	Halothane
Parkinson's disease	Moderate	Levodopa
Epilepsy	Good	Phenytoin and others
Cancers	Poor–moderate	Methotrexate, tamoxifen
Hormone and vitamin deficiencies	Good–excellent	Vitamins, steroid hormones, synacthen

[a] Scale = none, poor, moderate, good, excellent.

The likelihood and speed of therapeutic progress in these and other areas of ill-health are difficult to predict with any accuracy. It has been suggested that the many remaining targets for research are in increasingly difficult and complex areas.[12] More optimistically, however, it has been argued that continuing progress in the understanding of intracellular biochemistry, founded originally on the elucidation of the structure of DNA in the 1950s, should pave the way for a 'second therapeutic revolution' in which treatments will be developed for viral and autoimmune diseases as well as for cancers.[28] At the same time, ever more knowledge is being gained about the thousands of proteins—especially hormones, components of the immune system, enzymes which catalyze the body's chemical reactions and chemical transmitter substances—which regulate the functioning of different bodily systems. Research in this area raises the possibility of new treatments for a wide range of diseases which work either by enhancing or inhibiting the effects of the naturally occurring proteins. In addition, computer and related technologies are progressing rapidly and making an increasing contribution to the process of the design of new medicines.

The foregoing suggests that continuing pharmaceutical innovation is certainly feasible from a 'technical' point of view. This observation is borne out in a recent review of the pharmaceutical industry which, in the specific context of work related to receptors in the brain, for example, drew attention to several promising medicines currently in the research process, including non-addictive tranquillizers and novel analgesic compounds.[14] A steady stream of effective new treatments in the future is, however, a less certain prospect. Underlying this possibility are the difficulties currently facing all sections of the research community.

In the academic sector, the funding of scientific research in general is a source of major contemporary concern. As a result of insufficient finance large numbers of alpha-rated grant applications are unsupported and many researchers are seeking employment opportunities abroad.[29] Similar developments are occurring in medical research and these have potentially substantial long term costs. The innovative success of the pharmaceutical industry is becoming increasingly dependent on the academic scientific community for basic information about physiological and disease processes. Consequently, as fewer therapeutic advances are likely to result from chance observations and the screening of large numbers of candidate chemicals,[12] reductions in the volume of basic research work carried out in academic centres may be expected adversely to affect the flow of knowledge into and thus new medicines from the commercial sector. At the same time, inadequate funding of medical research in the universities and by the MRC will also have a detrimental impact on the future supply of research personnel. It has been noted already in this chapter that the MRC has reduced the number of training awards granted in recent years and the long term negative impact of this trend, embracing both quantitative and qualitative aspects of personnel supply, will impinge on industry as well as academia.

Focusing on the prospects for the pharmaceutical industry, recent trends in research costs and effective patent life erosion coupled with the growing concerns of governments to contain pharmaceutical and other public expenditures suggest that innovation will become increasingly confined to a small number of large highly creative research-based multinational companies. Most of the remaining medium- and small-sized firms will concentrate on generic manufacture, although some of the former may retain a research-based status by restricting attention to specific therapeutic groups.[20]

The productivity of the innovative sector will reflect, *inter alia*, confidence in the ability to achieve a reasonable return on research investments. A continuation of the pressures described in this chapter may be expected to lead to a concentration of research activities on large therapeutic markets where it has already been shown that an adequate reward can be obtained. Although this type of strategy may be genuinely beneficial in yielding improved alternatives to existing treatments, it is axiomatic that it severely inhibits the possible development of novel therapies for diseases that cannot be treated, or at best only very inadequately, at the present time. This risk-minimization approach may also negatively affect the academic research sector. Maxwell has argued that pharmaceutical developments can be of key importance to basic research: '. . . in more instances than not, the drugs come first, then the biochemical knowledge in animal models, then the theoretical constructions about drug actions in animals, followed by post-hoc rationalisation concerning the biochemistry of the human disease processes.'[30]

There are, of course, no easy solutions to the problems overshadowing the future of pharmaceutical innovation. However, key requirements must include, for the academic sector, sustained funding at levels consistent with the financial exigencies of contemporary research endeavours and, for the industrial sector, the creation and maintenance of an operating environment which is conducive to investment in the highly risk-laden activities of pharmaceutical research and development. The volume of financial resources that should appropriately be allocated to medical research raises many extremely complex issues to which there are no straightforward answers. It is nevertheless clear that if the preconditions for a viable national research effort noted above are to be met by government, then, in the current economic climate in which the competition for scarce resources continues to intensify, both the academic and commercial sectors will have to do more to demonstrate the efficiency and effectiveness of their separate and (perhaps increasingly) combined research activities.

4.2.4 REFERENCES

1. M. H. Cooper, 'Prices and Profits in the Pharmaceutical Industry', Pergamon Press, Oxford, 1966, p. 4.
2. S. R. Walker, L. Girling and R. A. Prentis, *Pharm. J.*, 1985, **234** (2 March), 264.
3. C. E. Lumley, R. A. Prentis and S. R. Walker, *Pharm. Med.*, 1987, **2**, 137.
4. Lord Prys-Davies, 'House of Lords Official Report', HMSO, London, 26 November 1986, col. 592.
5. *The Times*, 21 October 1985.
6. Committee of Vice-Chancellors and Principals, press statement issued on 18 May 1987.
7. Medical Research Council, 'Annual Report for 1984/85', MRC, London, 1985.
8. D. Noble, *Medical Research Council News*, 1985, **26** (March), 10.
9. D. Weatherall, quoted in an article entitled 'Medical research in crisis' in *The Times*, 1 May 1987.
10. Association of Medical Research Charities, 'Handbook 1986/87', AMRC, London, 1986.
11. D. Evered, 'Medical Research Today, What the Charities are doing', Association of Medical Research Charities, London, 1986.
12. National Economic Development Office, 'Pharmaceuticals, Focus on R and D', report by the Pharmaceuticals Economic Development Office, NEDO, London, 1987.
13. D. Swartzman, 'Innovation in the Pharmaceutical Industry', John Hopkins University Press, Baltimore, MD, 1976.
14. A. Wyke, *The Economist*, 1987 (7 February).
15. T. Jackson, 'Most top-selling drugs of 1985 discovered in UK', article in *The Financial Times*, 23 January 1987.
16. Association of the British Pharmaceutical Industry, 'The UK Pharmaceutical Industry, Facts '86', Association of the British Pharmaceutical Industry, London, 1986.
17. E. Reis-Arndt, *Pharm. Ind.*, 1987, **49**, 2, p. 136.
18. N. E. J. Wells, 'Pharmaceutical Innovation, Recent Trends Future Prospects', Office of Health Economics, London, 1983.
19. J. Thesing, 'Drug Industry Research Today', Medizinisch Pharmazeutische Studiengesellschaft, Mainz, 1984.
20. R. Rigoni, A. Griffiths and W. Laing, 'Pharmaceutical Multinationals, Polemics, Perceptions and Paradoxes. IRM Multinational Reports', Wiley, Chichester, 1985.
21. Pharmaceutical Sector Committee, 'Research and Development Costs, Patents and Regulatory Controls, A Consultative Document', National Economic Development Office, London, 1981.
22. R. Chew, G. Teeling Smith and N. E. J. Wells, 'Pharmaceuticals in Seven Nations', Office of Health Economics, London, 1985.
23. S. Walker and R. Prentis, *Pharm. J.*, 1985, **235** (5 January), 11.
24. D. G. Taylor, 'Patent Protection for New Medicines: The Case for its Restoration', Association of the British Pharmaceutical Industry, London, 1985.
25. 'Association of the British Pharmaceutical Industry Evidence to the House of Lords Science and Technology Committee (Sub Committee 2—Medical Research). House of Lords Session 1987/88. Third Report of the Select Committee on Science and Technology: Priorities in Medical Research', HMSO, London, 1988, vol. 2: oral evidence.
26. G. Teeling Smith, *Pharm. J.*, 1986, **237** (30 August), 242.
27. N. E. J. Wells, in 'The Costs and Benefits of Pharmaceutical Research', ed. G. Teeling Smith, Office of Health Economics, London, 1987.
28. G. Teeling Smith, 'The Future for Pharmaceuticals', Office of Health Economics, London, 1983.
29. N. E. J. Wells, 'Crisis in Research', Office of Health Economics, London, 1986.
30. R. A. Maxwell, *Drug Dev. Res.*, 1984, **4**, p. 375.

4.3 Organization and Funding of Medical Research in the USA

PATRICK S. CALLERY
University of Maryland, Baltimore, MD, USA

4.3.1 MEDICAL RESEARCH IN THE USA

The purpose of this chapter is to provide an overview of the mechanisms of funding medical research in the USA by way of a survey of funding sources, types of funding, how applications are made for funding and the prospects for future funding. Leading references for this chapter were derived from available government documents and surveys.[1–4]

The undisputed leader in medical research in the USA is embodied in the intramural and extramural programs of the National Institutes of Health (NIH). Starting with a $300 budget in a one-room attic laboratory over 100 years ago, the NIH budget has grown to over $6 billion per year (Table 1). Although this figure seems huge, it represents an investment in health-related research of less than 2% of the total spent by the nation on health. The investment in medical research appears smaller yet when compared to the estimated cost to society of disease (Table 2).

Table 1 Research and Development Budget Authority for Health[3]

	1987 actual ($ million)	*1988 estimate ($ million)*	*1989 estimate ($ million)*
Health programs (HHS)	6466	6991	7756
National Institutes of Health	5852	6320	6229
Acquired Immune Deficiency Syndrome (AIDS)	—[a]	—[a]	900
Alcohol, Drug Abuse and Mental Health Administration	509	558	522
Centers for Disease Control	65	72	63
Assistant Secretary for Health	20	21	20
Health Care Financing Administration	10	10	11
Health Resources and Services Administration	10	11	11
Food and Drug Administration (HHS)	85	91	88
Occupational Safety and Health Administration (Labor)	5	4	5
Total	6556	7087	7849

[a] In 1987 and 1988, AIDS activities were supported through funds appropriated directly to the Public Health Service agencies.

Table 2 Estimated Cost to Society of Disease and the Appropriations for Medical Research[a]

Disease	*Number of people afflicted (1986)*	*Cost to society (1986) ($ billion)*	*Research Appropriation (1987) ($ million)*
AIDS (virus)	1 500 000	1	271
Alcoholism/alcohol abuse	13 000 000	43	70
Alzheimer's Disease	3 000 000	50	64
Arthritis	37 000 000	25	118
Asthma	9 000 000	4	17
Blindness/visual impairment	13 000 000	15	217
Cancer (all forms)	1 000 000	72	1000
Cerebral Palsy	700 000	4	146
Cardiovascular disease	66 000 000	79	583
Diabetes	11 000 000	14	245
Drug abuse	12 000 000	16	101
Emphysema/bronchitis	14 000 000	10	24
Hearing loss	17 000 000	24	50
Influenza/pneumonia	97 000 000	14	32
Kidney disease	13 000 000	3	111
Mental illness	30 000 000	33	247
Multiple sclerosis	250 000	2	76
Stroke	500 000	13	38

[a] Adapted from the report of the 1988 Ad Hoc Group for Medical Research Funding, Washington, DC (sources = NIH, ADAMHA and various voluntary health agencies).

Under the direction of the Assistant Secretary of Health, the Public Health Service (PHS) administers a wide array of programs concerned with health. Of the six major PHS agencies, the most involved in medical research are the National Institutes of Health (NIH) and the Alcohol, Drug Abuse and Mental Health Administration (ADAMHA). The NIH mission is to carry out programs aimed at improving the nation's health by increasing knowledge related to health and disease through the conduct and support of research, research training and biomedical communication. ADAMHA's target is the reduction of health problems resulting from the abuse of alcohol and drugs and the fostering of improvements in the mental health of Americans.

The NIH intramural research program, centered on a campus setting in Bethesda, Maryland, represents an extremely creative environment for productive research in which 2000 scientists participate. The extramural research and training programs of NIH involve over 1300 non-Federal laboratories representing a working partnership of academia, industry and government. Industry and government depend on the universities to educate and train scientists and research administrators, while academia benefits from financial support and knowledge provided by government and industry in areas of interest held in common.

4.3.2 FUNDING SOURCES

Sources of funds for medical research are derived from government, non-government (non-profit institutions such as universities and foundations) and the private sector.

In general, current information on program availability and submission deadlines are announced by the individual programs. For government sources of funding, announcements are published in government documents such as the *Federal Register* or the *Commerce Business Daily.* Readily available current information on a broad base of biomedical funding is available in the form of abstracted summaries of potential funding source offerings in the *ARIS Biomedical Sciences Report* published on a subscription basis by Academic Research Information System, The Redstone Building, 2940 16th Street, Suite 314, San Francisco, CA 94103, USA. This report provides specified deadline dates along with program descriptions and requirements of organizations sponsoring biomedical research.

4.3.2.1 Government

Funds are awarded by NIH Institutes through their extramural programs for a variety of research projects, research career development, the training of new and established scientists and provision of medical library and research resources.

Another source of support for medicinal chemical research is the National Science Foundation (NSF). While NSF does not formally include medical research as part of its mission, there is some overlap with the mission of the NIH in the types of research funded, especially in the basic sciences.

The Veterans Administration (VA) funds meritorious projects consistent with defined needs.

Other funding sources which less directly or indirectly support medical or health-related research and development include the Food and Drug Administration, Centers for Disease Control, Occupational Safety and Health Administration, Environmental Protection Agency, Drug Enforcement Administration, Department of Defense, Department of Energy, National Bureau of Standards and various state and local agencies.

4.3.2.2 Foundations, Universities and Other Non-government Sources (Non-profit Institutions)

The USA has over 350 foundations which have giving programs of at least $1 million per year. According to summaries of the information found in foundation directories, such as 'The Foundation Directory',[5] 'The Foundation Grants Index',[6] and the 'Annual Register of Grant Support',[7] medical research is a popular target for funding by non-profit foundations and other non-government sources of funding. Foundation directories provide information on major foundation funding interests by subject area, geographical locality, types of support and an indication of the types of organizations that receive grants.

Academic medical centers, as a resource of scholarly research and training with goals of the discovery of new knowledge and the transmission of that knowledge to others, are major contributors to medical research.

4.3.2.3 Private Sector (Drug Industry)

Funding from the private sector, of which the pharmaceutical industry is the prime example for medical research, conforms for the most part closely with corporate directions, suggesting that investigator interests must be held closely in common with industry interests before a successful collaboration can be established. This funding is mostly in the form of research contracts with specific performance requirements made with other companies or academia, although, in some pharmaceutical companies, research institutes or foundations have been established which fund grants for basic biomedical research and training. Worldwide, USA pharmaceutical research laboratories invest more than $4.5 billion a year in biomedical research in its own laboratories and in support of academic and clinical research.

4.3.3 TYPES OF FUNDING

As the main sources of funding of medical research in the US, the NIH and ADAMHA also offer the largest variety of types of funding mechanisms. The major types of NIH funding are financial assistance awards including grants and cooperative agreements and acquisition awards, which are mostly in the form of contracts. Parallel types of funding, with some exceptions, are also offered by the Alcohol, Drug Abuse and Mental Health Administration (ADAMHA). Grants represent the traditional mode of funding of investigator-initiated ideas for the research or training project. Cooperative agreements are similar to grants except that there is more involvement of the program in assistance and support of the research. Acquisition awards or contracts are more restrictive awards for research and development which are targeted towards particular areas of research.

The largest category of NIH funding is in the form of research grants awarded to non-profit organizations and institutions, governments and their agencies and to individuals and for-profit organizations. Funding covers salaries, equipment, supplies, travel and other allowable direct costs and indirect costs. The traditional grant is for a research project awarded to an institution on behalf of a principal investigator. Program Project Grants are made to institutions on behalf of a principal investigator for the support of broadly based usually multidisciplinary long-term research programs. Center Grants are awarded to institutions on behalf of a program director heading a group of collaborating investigators and are more likely to have a clinical orientation than Program Projects.

Grants to small businesses with technological expertise pertaining to the research and development mission of the NIH are supported by the Small Business Innovation Research (SBIR) program.

4.3.3.1 Funding the Various Stages of a Medical Researcher's Career

Funding mechanisms are available to support each stage in the development of a medical research scientist, from student status to senior scientist.[1] Early support for college student training is often in the form of laboratory assistant work on an hourly basis including, perhaps, partial to full defrayment of educational expenses through scholarship programs. Predoctoral and postdoctoral training becomes more formalized and access to more funding programs results. As a researcher develops, career support is consumed as part of research grants and specialized fellowships. Support can also be found for scientists wishing to change fields or direction of research efforts.

4.3.3.1.1 Predoctoral training

Several types of National Research Services Awards (NRSA) have been established by NIH to increase the number of individuals trained for research and teaching in specifically designated biomedical areas and to improve the environment in which the biomedical training is conducted. Professional Student Short-Term Research Training Grants are made to predoctoral students in the health professional schools. For students in graduate programs, predoctoral training support is offered by NIH through Individual Predoctoral Fellowship Awards or Institutional Training Grants.

Funding for training in medical research is available for minority students through the NIH-sponsored Minority High School Student Research Apprentice Program and the Minority Access to Research Careers Program (MARC).

The NSF funds predoctoral and postdoctoral fellowships, although few if any are awarded for directly health-related research projects and training.

Other sources of predoctoral training support come from institutional graduate teaching assistantships which cover student living and school-related expenses and require some service to the university providing the funds. Similarly, research assistantships are often available from universities which usually require research activities closely related to the student's thesis project.

4.3.3.1.2 Postdoctoral fellowships

In addition to the NRSA predoctoral awards, Individual Postdoctoral Fellowship Awards are awarded to qualified individuals holding a doctorate or equivalent to support full-time research training in designated biomedical areas. Funding for postdoctoral fellows is often garnered from other sources not necessarily specifically awarded for training purposes, such as research grants. In general, a postdoctoral fellowship is offered on a one year basis, renewable for one or two additional years. The rate of pay for a postdoctoral fellow is about double that of a predoctoral fellow and roughly one-half the compensation earned by holders of an entry level industry or academic position.

4.3.3.1.3 Research career support

Most research project grants, such as the RO1 type of award made by NIH, provide partial to full salary support for awardees.

A mechanism to support the new independent investigators, awarded for five years to help effect the transition toward the traditional types of NIH research project grants, is available in the form of First Independent Research Support and Transition (FIRST) Awards (R29).

Several types of career awards are available to research and academic institutions on behalf of scientists with high research potential who require additional training and experience in a productive scientific environment. These include the Research Career Development Award (K04), Mid-career Award and Special Emphasis Research Award.

The National Cancer Institute supports an Outstanding Investigator Grant program to provide long-term funding for experienced investigators with outstanding records of research productivity.

The MERIT (Method to Extend Research in Time) award provides long-term support to investigators with research competence and productivity that are distinctly superior and who are likely to continue to perform in an outstanding manner. Investigators do not apply for this award, since candidates are nominated by NIH personnel.

4.3.3.1.4 Mid-career changes and senior fellowships

Investigators with at least seven years of relevant research or professional experience may apply to an NRSA program for senior fellowships. These awards are designed to fund major changes in the direction of their research careers, for the acquisition of new research capabilities, for the broadening of their scientific background, for the enlargement of their command of an allied research field or for time from regular responsibilities to increase their capabilities for engaging in medical research.

Scientists wishing to study abroad can apply to programs such as the Fogarty International Fellowships which are awarded to further international cooperation and health research.[8]

Sabbatical leaves are funded in part by many higher education institutions. Under these arrangements, qualified faculty can spend one year out of seven in off-campus research settings.

4.3.3.1.5 Distribution of new and renewal RO1 applications by age of principal investigator

In a study carried out by the NIH Division of Research Grants Statistics and Analysis Branch[9] on applications submitted in 1979, 1982 and 1985 (Table 3), it was found that the age group of 35 or younger consistently received the best priority scores and the highest success rates compared to

Table 3 Distribution of NIH Research Grant Applications by Age of Principal Investigator[9]

	Mean priority scores				*Success rates*[a]			
	New		*Renewal*		*New*		*Renewal*	
Age	*1979*	*1985*	*1979*	*1985*	*1979*	*1985*	*1979*	*1985*
26–30	250	222	—[b]	—[b]	40.2	30.4	—[b]	—[b]
31–35	261	230	229	193	37.7	28.0	58.9	50.3
36–40	275	238	234	194	32.0	24.3	55.7	50.1
41–45	287	249	233	192	26.3	21.8	56.3	51.5
46–50	290	254	234	198	24.3	17.9	58.4	48.4
51–55	294	263	237	196	21.6	15.4	54.9	51.5
56–60	295	271	235	199	23.3	13.4	53.8	48.6
61–65	297	263	233	206	20.7	14.2	53.9	44.7
66–70	308	254	235	217	20.8	21.3	51.6	37.9

[a] Awards as percent of number reviewed. [b] Less than 50 approved applications.

applicants in other age groups. While the proportion of traditional grant (RO1) applicants 35 or younger declined from 26.1% in 1979 to 13.4% in 1986, applications for New Investigator Research Awards and Research Career Development Awards (K04) increased. The New Investigator Award was replaced by the First Independent Research Support and Transition (FIRST) Award in 1986. Even though there was a decline in the proportion of applications from the 26 to 35 age group, these applicants were rated with the best scores and were the most successful in obtaining NIH funding, especially for new applications. For applications submitted in 1985, for each 10 years of age, the average score on a new application increased by about 16 points or 10%. There is less relationship between scores and applicant age with competing renewals, although applicants 31 to 45 received slightly better scores than those 46 and over. The success rate of obtaining NIH funding for new applications in 1985 ranged from a high of 30.4% for the 26 to 30 age group to a low of 14.8% for applicants over 50.

4.3.3.2 Contract Research

Academic institutions and other non-profit and commercial organizations can apply for NIH research and development contracts for specific scientific inquiry directed towards specific areas of research needed by the NIH or USA government.

Contracts, whether solicited or unsolicited, differ from traditional grants in that the NIH closely monitors for accomplishment of contract goals. Also, most NIH contracts are of the cost-reimbursement type for allowable, allocable and reasonable costs in performing the project. A fee may be paid where appropriate. The contract review process is based on offerer response to an NIH-defined statement of work outlined in a published Request for Proposal (RFP) and reviewed by a scientific peer review group. Peer reviewers' recommendations and separate NIH staff evaluations of the technical and costs proposal sections provide the information needed for negotiations with offerors in a competitive range. Offerors are then asked to submit a best and final offer.

Solicitation announcements and deadlines for contract proposals appear in the *Commerce Business Daily*, subscriptions of which are available from the Superintendent of Documents, Government Printing Office, Washington, DC 20402, USA. Availability of RFPs is published in the 'NIH Guide for Grants and Contracts' (NIH Guide Distribution Center, National Institutes of Health, Room B3BE07, Building 31, Bethesda, MD 20892, USA).

Included in the types of NIH contracts awarded are research contracts for specific research problems identified by NIH components, developmental contracts to develop substances, devices, systems, sophisticated instruments or other approaches to diagnose, prevent, treat or control disease, demonstration contracts and research and development support contracts.

Research contracts are also performed for the pharmaceutical industry, the Food and Drug Administration and the Military Establishment.

4.3.3.3 Equipment and Research Support

The NIH Division of Research Resources (DRR) programs offer funding mechanisms for biomedical research technology, animal resources, biomedical research support, research centers in minority institutions and general clinical research centers.[10] The Shared Instrumentation Program of the DRR has been especially useful for the updating of the aging instrumentation armamentarium found in most universities.

4.3.4 APPLYING FOR RESEARCH FUNDING

Funding agencies usually supply their own application forms or application kits on which proposals are submitted. These forms outline specific requirements such as order of presentation of information, page limitations, budget restrictions and overhead costs allowances. While some funding sources accept applications throughout the year, most sources announce specific deadlines. For example, Table 4 lists application receipt deadlines for most NIH and ADAMHA programs.

Table 4 Deadlines for NIH and ADAMHA Applications

Receipt dates	*Initial Review group meetings*	*Council Meetings*	*Earliest Possible start date*
February 1,[a] January 10,[b] March 1[c]	June/July	September/October	December 1
June 1,[a] May 10,[b] July 1[c]	October/November	January/February	April 1
October 1,[a] September 10,[b] November 1[c]	February/March	May/June	July 1

[a] Receipt dates for new regular research grant applications, both new and renewal research scientist development applications and new and competing renewal program project applications. [b] Receipt dates for all National Research Service Award applications. [c] Receipt dates for all competing renewal and supplemental research grant applications; revised research grants and career development applications (both new and renewal).

4.3.4.1 Proposal Review and the Funding Decision

NIH research grant and other assistance applications are submitted to the Division of Research Grants (DRG). A review of each application is carried out and assignment is made to the appropriate Initial Review Group (IRG) for review and to the appropriate NIH Institute for additional review and consideration of an award. The IRG, which is better known as the study section, provides peer review for scientific merit using the following criteria for assessment of applications: 'scientific, technical, or medical significance and originality of the proposed research; appropriateness and adequacy of the experimental approach and methods to be used; qualifications and experience of the principal investigator and staff in the area of the proposed research; reasonable availability of resources necessary to the proposed research; reasonableness of the proposed budget and duration in relation to the proposed research; and where an application involves activities that could have an adverse effect upon humans, animals, or the environment, the adequacy of the proposed means for protecting against such effects.'

The review of NIH proposals by the peer review system is based on two sequential levels of review, the first being the chartered initial review groups made up of experts in specific scientific disciplines as described above. The second level of review is conducted by the National Advisory Council or Boards of the Institutes that make awards. This review is based on the scientific merit, as judged by the scientific review groups. IRG-assigned priority scores and percentile ranking values calculated against a pool of favorably recommended applications are utilized as factors in the funding decision. Other factors include the relevance of the proposed study to the Institute's programs and priorities.

In contrast to the NIH grant review process, most proposals reviewed by the NSF are unsolicited and peer review is carried out by an alternate system. Proposals to the NSF are assigned to a program officer who oversees external review and evaluates comments made by the reviewers. The program officer makes recommendations to award or not to award taking into account other considerations such as the relationship of the work to the research area and other pending proposals and the program's mission and budget. These recommendations are reviewed at higher supervisory levels and checked by the Division of Grants and Contracts on non-scientific aspects of the award.

The NSF review procedure in the physical sciences is by mail to reviewers who respond individually. In the biological sciences, a panel reviews the proposal in addition to the mail reviewers. In the mail-out procedure, the proposal is sent to from three to ten reviewers identified by the project officer as knowledgeable on the topic. Instructions to reviewers are standardized and responses are submitted directly to the project officer on standard forms. Panels generally meet about three times a year to evaluate a group of proposals taking into account the prior mail reviews. NSF proposals on average are reviewed by 5.5 external reviewers.

The criteria for proposal selection are published by NSF in 'Grants for Scientific and Engineering Research and Education'. Among the NSF criteria are 'research performance competence pertaining to the technical capability of the investigator and adequacy of the institutional resources; intrinsic merit, the extent to which the proposed work is expected to lead to new discoveries or fundamental advances in its discipline or across disciplines; utility or relevance in terms of contributions to an extrinsic goal such as a new technology; and effects on the infrastructure of science and engineering regarding what the work will contribute to the Nation's research, education and human resource base.' The fourth criterion allows for the taking into account such matters as the participation of women and minorities, institutional distribution and the stimulation of important but under-developed research areas.

4.3.4.2 Costs of Peer Review of NIH Grant Applications

The costs of peer review of grant applications in the Division of Research Grants of the NIH have been tabulated for fiscal years 1972, 1983 and 1986 (Table 5).[11] The cost in actual dollars decreased 28% to a cost of $1010 per application. In constant 1986 dollars, the cost per application actually decreased 53%.

Table 5 Cost of Peer Review of NIH Grant Applications[11,a]

Fiscal year	Total estimated costs ($ actual)	Number of applications	Cost/application ($) Actual	Constant FY 1986[b]
1972	8 358 694	10 614	788	1912
1983	19 584 504	20 694	946	1064
1986	22 701 731	22 484	1010	1010

[a] Cost estimates were based on identifying the percentage of effort/budget directly related to peer review of each of the branches of the Division of Research Grants and of the Office of the Director. The total included costs of the Scientific Review and Evaluation Awards Office and the space rental and utilities costs incurred by house Division of Research Grants staff. [b] Base year of 1986 indexed at 100 using the GNP price index.

4.3.4.3 Perceptions of NSF Grant Applicants of the Peer Review Process

A recent study on the perceptions of applicants for NSF grants provided some interesting insights on the peer review process.[4] A 16-page survey was mailed to 14 282 applicants whose competitively reviewed proposals for research grants had been awarded or declined by NSF during fiscal year 1985. About two-thirds responded. As part of the survey, views were sought about the proposals review system.

Nearly half of the respondents were satisfied with the review process, although 38% were dissatisfied. Two out of three applicants that had been declined were dissatisfied. Among the reasons for dissatisfaction were that the reviewers selected by NSF were not sufficiently expert in the subject matter of the proposal or that the reviews were cursory, conflicting or did not seem to support NSF's decision. Other reasons cited were politics, cronyism or biases of various types.

Only one-third of all of the respondents felt that the reviewers' comments helped them to understand the NSF's decision a great deal. About 40% indicated that the reviewers' comments influenced the course of their subsequent research.

Applicants who contacted a program officer in writing, by phone or by a visit before submitting a proposal were somewhat more likely to be successful in being awarded a grant. Applicants declined on their first proposal were less likely to contact a program officer for an explanation of the decision, while those who had previously been awarded several grants were much more likely to make follow-up contact and to resubmit their proposals. Almost all of the consistently successful applicants, and three-fourths of all applicants, had served as a reviewer or review panelist for NSF at least once in the five years before the survey.

Of the applicant pool surveyed, 34% had proposals funded that year, 38% for chemistry applications. Of the applicants declined, about 25% subsequently resubmitted to NSF, 20% submitted to another funding source and nearly half had not taken further action at the time of the survey. Chemists declined funding by NSF more frequently than other scientists preferred the NIH as a source of funding.

4.3.4.4 Grantsmanship Overview

The mystique of obtaining a grant has led many scientists to make conscious efforts to develop skills in the art and science of grantsmanship. In short, grantsmanship can be looked at as learning about the funding process and following a set of simple common-sense instructions useful in the preparation of grant applications that would be more favorably received by reviewers and others participating in the funding decision. While the established investigator is likely to be a skilled grant writer, the new investigator, or the investigator with an application disapproved or approved but not funded, often seeks assistance in the form of published articles,[12–26] seminar series and workshops. The NIH has compiled selected articles written by NIH staff to assist researchers in the preparation of research grant applications.[26] Copies of these articles are available from the NIH Office of Grants Inquiries.

4.3.4.5 Obligations of the Recipient

Support for medical research on human subjects can be awarded by the NIH, PHS or Department of Health and Human Services (DHHS) only for research activities conducted in accordance with DHHS Regulations for the Protection of Human Subjects, Title 45, Part 46 of the Code of Federal Regulations (45 CFR 46). While certain research activities are specifically exempt, the regulations provide for protection for human subjects of research, including requirements for review and approval by an institutional review board established in accordance with the regulations, informed consent of subjects or the subject's legally authorized representative and adequate safeguards of the rights and welfare of individual subjects.

Research activities involving live vertebrate animals are not supported by the PHS unless conducted in accordance with the latest revision of the 'PHS Policy on Humane Care and Use of Laboratory Animals by Awardee Institutions.' Each recipient organization is required to establish an appropriate program for the care and use of animals by the standards described in the 'Guide for the Care and Use of Laboratory Animals' (NIH Pub. No. 85-23, 1985). Acceptable Animal Welfare Assurance must be provided as part of the application submission. Cited documents are available from the NIH Office for Protection from Human Risks.

Integrity of scientific research has recently come under careful scrutiny to the point where some groups are suggesting that the reporting of fraudulent scientific findings should be treated as a criminal offense. As stated by the PHS, implicit in the awarding of an application is the expectation that the investigator, institution and others associated with the project will adhere to commonly accepted norms of sound research design, accurate recording of data, unbiased interpretation of results, respect for the intellectual property of others and proper management of funds. Also stated is a responsibility of the awardee institution to deal forthrightly with instances of misconduct in science and to report such occurrences to the NIH. Furthermore, the accuracy and validity of all administrative, fiscal and scientific information in applications, proposals and progress reports is the joint responsibility of organizations and investigators. The number of cases is increasing where NIH has been forced to take administrative actions such as withdrawal of an application or the suspension and/or termination of an award resulting from deliberate witholding, falsification or misrepresentation of information.

Before a grant can be awarded, USA applicants must also certify compliance with laws regarding civil rights, handicapped individuals and sex discrimination.

Complete details on the obligations of awardees of PHS grants are outlined in the 'PHS Grants Policy Statement' (DHHS Publication No. OASH 82-50000, revised January 1, 1987).

4.3.5 FUNDING FUTURE MEDICAL RESEARCH

The USA spends more money annually on research and development than any other nation. A recent study by the NSF indicates a continuation in a 12-year pattern of expanding real research and development growth[27] with basic and applied health research most likely continuing to parallel the overall growth pattern.[28] For example, according to this study, the funds for the performance of applied research and development of drugs and medicines has increased dramatically from $605 million in 1973 to $2601 million in 1983.[27]

As of April 1988, the proposed budget authority for USA health research and development for 1989 represented an 11% increase over 1988. The breakdown by agency is listed in Table 1. Of particular note is special emphasis on funding for Acquired Immune Deficiency Syndrome (AIDS) research. In 1989, the PHS budget for AIDS will increase 42% and will no longer be appropriated directly to the PHS agencies.

In Table 6 is shown a slow steady increase in commitment to funding of NIH activities over the years in terms of budget requested and actually funded by the Administration and Congress.

Table 6 NIH Appropriations History ($ million)[a]

Year	Administration request	Approved	Percent above administration request	Percent change previous approval
1975	1835	2093	14.0	—
1976	1981	2302	16.2	10.0
1977	2165	2544	17.5	10.5
1978	2596	2843	9.5	11.7
1979	2885	3190	10.6	12.2
1980	3186	3429	7.6	7.5
1981	3512	3569	1.6	4.1
1982	3311	3642	10.0	2.0
1983	3749	4024	7.3	10.5
1984	4077	4494	10.2	11.7
1985	4567	5146	12.2	14.5
1986	4853	5501	13.4	6.9
1987	5080	6184	21.7	12.4
1988	5534	6667	20.5	7.8
1989[b]	7123	7199	1.1	8.0

[a] Figures provided by the Ad Hoc Committee for Medical Research Funding, Washington, DC.
[b] Estimated.

4.3.6 REFERENCES

1. 'National Institutes of Health Grants and Awards, NIH Funding Mechanisms', Grant Inquiries Office, Division of Research Grants, National Institutes of Health, Bethesda, MD, 1987, p. 1.
2. 'NIH Extramural Programs, Funding for Research and Research Training', NIH Publication No. 88-33, National Institutes of Health, Bethesda, MD, 1988, p. 1.
3. 'Federal R&D Funding by Budget Function, Fiscal Years 1987–89', National Science Foundation Publication NSF 88-315, Washington, DC, 1988, p. 17.
4. 'Proposal Review at NSF: Perceptions of Principal Investigators', National Science Foundation Report 88-4, 1988, p. 1.
5. L. Renz (ed.), 'The Foundation Directory', 10th edn., The Foundation Center, New York, 1985, p. 1.
6. E. Garonzik, (ed.), 'The Foundation Grants Index', 15th edn., The Foundation Center, New York, 1986, p. 1.
7. 'Annual Register of Grant Support. A Directory of Funding Sources 1987–1988' 21st edn., National Register Publishing Co., Wilmette, IL, 1987, p. 1.
8. 'Directory of International Opportunities in Biomedical and Behavioral Sciences', John E. Fogarty International Center for Advanced Study in the Health Sciences, National Institutes of Health, Bethesda, MD, 1988, p. 1.
9. 'NIH Peer Review Notes', Division of Research Grants, National Institutes of Health, Bethesda, MD, 1987, p. 4.
10. 'Division of Research Resources Program Highlights', NIH Publication No. 87-2309, National Institutes of Health, Bethesda, MD, 1987, p. 1.
11. 'NIH Peer Review Notes', Division of Research Grants, National Institutes of Health, Bethesda, MD, 1987, p. 8.
12. D. G. Murphy and D. J. Dean, in 'Institute Insight', National Institutes of Health, Bethesda, MD, 1984, p. 1.
13. D. G. Murphy and D. J. Dean, *Nutrition International*, 1986, **2**, 38.
14. A. C. Novello, *Miner. Electrolyte Metab.*, 1985, **11**, 281.
15. G. N. Eaves, *Grants Magazine*, 1984, **7**, 151.
16. G. N. Eaves, D. B. Rifkin, T. E. Malone, R. Ross and R. T. Schimke, *Fed. Proc., Fed. Am. Soc. Exp. Biol.*, 1973, **32**, 1541.
17. D. H. Merritt and G. N. Eaves, *Fed. Proc., Fed. Am. Soc. Exp. Biol.*, 1975, **34**, 131.
18. G. N. Eaves, J. M. Pike, and S. C. Bernard, *Grants Magazine*, 1978, **1**, 263.
19. C. Henley, *Fed. Proc., Fed. Am. Soc. Exp. Biol.*, 1977, **36**, 2066.
20. H. H. Fudenberg, (ed.), 'Biomedical Institutions, Biomedical Funding, and Public Policy', Plenum Press, New York, 1983, p. 1.
21. J. S. Greene (ed.), 'Grantsmanship: Money and How to Get It', Academic Media, Orange, NJ, 1973, p. 1.
22. J. Jagger, *Grants Magazine*, 1980, **3**, 216.
23. S. R. Loveland, *Respiratory Care*, 1980, **25**, 139.
24. J. Killingsworth, *IEEE Trans. Prof. Commun.*, 1983, **PC-26**, 79.
25. J. T. Dingle (ed.), 'How to Obtain Biomedical Research Funding', Elsevier, New York, 1986, p. 1.
26. 'Preparing a Research Grant Application to the National Institutes of Health: Selected Articles', Office of Grants Inquiries, National Institutes of Health, Bethesda, MD, 1987.
27. 'National Patterns of Science and Technology Resources: 1987', National Science Foundation Publication NSF 88-305, Washington, DC, 1988, p. 1.
28. 'Extramural Trends, FY 1978–1987', Information Systems Branch, Division of Research Grants, National Institutes of Health, Bethesda, MD, 1988, p. 1.

APPENDIX 1 ABBREVIATIONS

ADAMHA	Alcohol, Drug Abuse and Mental Health Administration
BID	Bureau, Institute or Division (of NIH)
DHHS	Department of Health and Human Services
DRG	Division of Research Grants
DRR	Division of Research Resources
F31	Predoctoral NRSA
F32	Postdoctoral NRSA
FIC	Fogarty International Center
FIRST	First Independent Research Support and Transition Award
IRG	Initial Review Group (grant and cooperative agreement applications)
KO1	Research Scientist Development Award, Level I
KO2	Research Scientist Development Award, Level II
KO4	RCDA
KO5	Research Scientist Award
MARC	Minority Access to Research Careers
MBRS	Minority Biomedical Research Support Program
MERIT	Method to Extend Research in Time Award
NCI	National Cancer Institute
NEI	National Eye Institute
NHLBI	National Heart, Lung and Blood Institute
NIA	National Institute of Aging
NIAAA	National Institute of Alcoholism, Alcohol Abuse
NIAID	National Institute of Allergy and Infectious Diseases
NIAMSD	National Institute of Arthritis and Musculoskeletal and Skin Diseases
NICHD	National Institute of Child Health and Human Development
NIDA	National Institute of Drug Abuse
NIDDK	National Institute of Diabetes and Digestive and Kidney Diseases
NIDR	National Institute of Dental Research
NIEHS	National Institute of Environmental Health Sciences
NIGMS	National Institute of General Medical Sciences
NIH	National Institutes of Health
NIMH	National Institute of Mental Health
NINCDS	National Institute of Neurological and Communicative Disorders and Stroke
NLM	National Library of Medicine
NCNR	National Center for Nursing Research
NRSA	National Research Service Award
NSF	National Science Foundation
P01	Program Project
P50	Research Centers
PHS	Public Health Service
RO1	Regular Research Grant (Traditional)
RO3	Small Grant Program
R13	Support of Scientific Meetings
R29	FIRST
R37	MERIT
RCDA	Research Career Development Award
RFA	Request for Applications (grants)
RFP	Request for Proposals (contracts)
SBIR	Small Business Innovation Research
SERCA	Special Emphasis Research Career Award
SRG	Scientific Review Group
T32	Institutional NRSA, Training Grant
VA	Veterans Administration

APPENDIX 2. SELECTED FUNDING INSTITUTION CONTACTS FOR NIH, ADAMHA AND NSF

Office of Grants Inquiries
Division of Research Grants
National Institutes of Health
Bethesda, MD 20892, USA

Division of Extramural Affairs
National Heart, Lung and Blood Institute, NIH
Room 7A17B, Westwood Building
Bethesda, MD 20892, USA

Division of Extramural Programs
National Library of Medicine, NIH
Room 5N505, Building 38A
Bethesda, MD 20894, USA

Division of Extramural Activities
National Institute of Diabetes and Digestive and Kidney Diseases, NIH
Room 657, Westwood Building
Bethesda, MD 20892, USA

Office of Extramural Affairs
National Institute on Aging, NIH
Room 5C05, Building 31
Bethesda, MD 20892, USA

Extramural Activities Program
National Institute of Allergy and Infectious Diseases, NIH
Room 703, Westwood Building
Bethesda, MD 20892, USA

Division of Extramural Activities
National Institute of Arthritis and Musculoskeletal and Skin Diseases, NIH
Room 2A04, Building 31
Bethesda, MD 20892, USA

Extramural Program
National Institute of Dental Research, NIH
Room 503, Westwood Building
Bethesda, MD 20892, USA

Extramural Program
National Institute of Environmental Health Sciences, NIH
Post Office Box 12233
Research Triangle Park
NC 27709, USA

Extramural and Collaborative Program
National Eye Institute, NIH
Room 6A51, Building 31
Bethesda, MD 20892, USA

Office of Program Activities
National Institute of General Medical Sciences, NIH
Room 953, Westwood Building
Bethesda, MD 20892, USA

Extramural Activities Program
National Institute of Neurological and Communicable Disorders and Stroke, NIH
Room 1016A, Federal Building
Bethesda, MD 20892, USA

Deputy Director
Division of Research Resources, NIH
Room 5B03, Building 31
Bethesda, MD 20892, USA

International Research and Awards Branch
Fogarty International Center
Room 613A, Building 38A
Bethesda, MD 20894, USA

Deputy Director
National Institute of Child Health and Human Development, NIH
Room 2A04, Building 31
Bethesda, MD 20892, USA

Small Business Innovation Research
Special Programs and Initiatives
Office of Extramural Research
National Institutes of Health
Bethesda, MD 20892, USA

Division of Extramural Activities
National Cancer Institute
Room 10A03, Building 31
National Institutes of Health
Bethesda, MD 20892, USA

Research Resources Information Center
1601 Research Boulevard
Rockville, MD 20850, USA

Office for Protection from Research Risks
Room 4B09, Building 31
National Institutes of Health
Bethesda, MD 20892, USA

Grants Management Branch
National Institute on Drug Abuse
Room 10-25, Parklawn Building
5600 Fishers Lane
Rockville, MD 20854, USA

Grants Management Branch
National Institute of Mental Health
Room 7C-05, Parklawn Building
5600 Fishers Lane
Rockville, MD 20854, USA

Division of Extramural Research
National Institute on Alcohol Abuse and Alcoholism
Room 14C-17, Parklawn Building
5600 Fishers Lane
Rockville, MD 20854, USA

Grants and Contracts Office
National Science Foundation
1800 G. Street NW
Washington, DC 20550, USA

4.4
Health Care in the UK

DAVID TAYLOR
ABPI, London, UK

4.4.1 INTRODUCTION

Since 1948 the National Health Service (NHS) has been the principal provider of health care in the UK. It currently (1987) spends in excess of £21 000 million, and employs in the order of 1.2 million full and part time staff. These figures represent around 6% of the nation's wealth and labour force, a resource share more than double that enjoyed by the NHS at around the start of its four decades of existence.

The NHS remains one of Britain's most popular, as well as its largest, institutions.[1] Yet alongside growth and frequent acclaim the last three to four decades have brought the service many new challenges. For example, the gradual emergence of a more individualistic, less professionally controlled, social environment has meant that British health care consumers have become less tolerant of inconvenient or inadequate patterns of service than they were in the 1950s.

Such trends, together with the tensions like those associated with the NHS industrial disputes of the 1970s and early 1980s and the 1974 and 1982 structural reorganizations, have led a number of commentators to believe that the NHS has reached a crisis point. Some even suggest that it is near to collapse.

The analysis on which this chapter is based does not support such views. Nevertheless, the efficiency and/or quality of current provisions could of course in some instances be improved: the widespread desire for better 'customer care', particularly in contexts such as waiting for hospital appointments and the quality of general practitioners' (GP) premises; certain unacceptable failures to match service levels offered abroad, as in the case of renal dialysis for older patients; and talk of 'NHS overmanning' on the one hand and of 'life-threatening' financial cuts on the other. All are examples of the type of issue about which there is genuine concern.

Against this background the primary goal of this chapter is to provide a balanced picture of the origins and continuing evolution of the British health care system with special reference to the structural arrangements applying in England (Scotland, Northern Ireland and to a lesser extent Wales have rather different NHS arrangements). It does not question the fundamental principle that the NHS exists to provide comprehensive health care for everyone in the population, regardless of their wealth or social status. Rather, it attempts to identify the ways in which a developed country's health care system based on the ideal of universal welfare can most efficiently achieve its aims.

To this end particular attention is paid to the administration of the NHS Family Practitioner Services (FPS). These include the General Medical and the Community Pharmaceutical Services, which in England and Wales are run quite independently of the hospital services (HCHS). They arguably embody some of the most distinctive and desirable facets of the British system. Indeed, despite criticisms from some quarters, the concepts underlying the FPS may eventually come to serve as a general model for the future development of the entire health service. For potentially they permit a more pragmatic combination of national planning and, where appropriate, economic competition between providers than the current structure of the NHS hospital service. If successfully extended throughout, the 'independent contractor' system at the heart of the FPS could well provide an important example for other national health services to follow.

4.4.2 BRITISH HEALTH CARE BEFORE THE TWENTIETH CENTURY

The origins of institutionally based health care in this country lie in provisions for the sick and destitute offered by the medieval monasteries and the first of the charitable hospitals. The oldest of the latter is St. Bartholomews, which was founded in London in 1123. State involvement cannot be said to have begun until the 17th century, when the Elizabethan Poor Laws gave the then nonelected local authorities the power to raise rates to finance the support of lame, blind or otherwise disadvantaged individuals who were unable to work.[2]

Following further legislation in the 1720s and the 1780s, the workhouse system was established, and soon became the main form of residential support for the impoverished sick. Between 1750 and 1800 the charitable, or voluntary, hospital sector also went through a notable period of growth as Britain increased her wealth through foreign trade and enhanced domestic productivity. The social problems generated by urbanization encouraged the establishment of further state services during the 19th century, most notably in the form of 'lunatic asylums' and extended Poor Law care. In 1875 the Disraeli government sponsored Public Health Act supplied a firm basis for the local provision of services like clean water and adequate drainage and sanitation. It also created an organizational kernel around which other local authority health functions, like mother and child care, would later group.

Towards the end of the 19th century there were further major extensions in the scale of both state and private/voluntary sector hospital care. Overall, the total number of beds available in England and Wales rose from just over 100 000 in 1891 to almost 200 000 in 1911. Also around that time there became established in Britain a clear distinction between medical specialists and general practitioners.

The latters' incomes were then entirely dependent on either fees earned directly from patients, or on payments made *via* the several thousand Friendly Societies which provided insurance cover for GP services and the costs of the medicines prescribed. The general practitioners were anxious that access to hospital-based services should only be obtained by way of referral by them. Otherwise, they feared, they would be destroyed by competition from out-patient departments, most of which offered free care.

4.4.3 THE IMPACT OF THE WORLD WARS

During the first half of the 20th century British medical care development was substantially influenced by the occurrence, or threat, of international conflict. Where previous social reformers argued simply that better health care would enhance national wealth creation, those of the early 1900s also pointed to the growing strength of Germany and the disturbingly poor physical condition of many of the would-be recruits to the army.

The School Medical Service, created in 1907, and Lloyd George's 1911 National Insurance Act, which required lower paid workers (but not their dependents) to be insured for basic general practitioner and pharmaceutical care, both partly stemmed from such pressures. The compulsory system then established, which was administered *via* county or county–borough wide Insurance Committees, was the forerunner of today's FPS administrative structure.

The next major piece of legislation to affect the health sphere was the 1929 Local Government Act. It transferred to the by then elected local authorities the responsibilities of the Poor Law Guardians, and permitted them to provide a full range of hospital services. Thus state sector institutional support for the sick (which by the mid 1930s constituted about 80% of the nation's total bed capacity) was linked, through local government control, with the expanding community services already under the Medical Officers of Health.

Standards were highly variable across the country during the short period between 1929 and 1939 when national government had no direct command over the development of health services. But in some localities at least the positive potential of a publicly financed, electorally accountable, system was demonstrated. Meanwhile, the independent sector had come under increasing financial pressure. In the interwar period the inefficiency of the myriad private health care insurance systems then operating in Britain began to be very apparent. Also 'middle class' patients found it more difficult to meet medical expenses directly, whilst viewing the prospect of having to use the existing public hospitals with some horror.

It is therefore possible that had World War 2 not begun, some form of government intervention designed to support further the voluntary sector, including the teaching institutions, would have taken place. But with the advent of fresh international conflict attitudes shifted. One important development was the introduction of the Emergency Medical Scheme (EMS) in 1939. Formed in part in expectation of enormous casualties from enemy bombing, this made use of both the public and private hospital services. The advantages of coordination and rationalization were made obvious to the public and to many professionals; hence the EMS paved the way for the formation of the NHS. Government failure to have taken note of its success might have created serious social discontent in postwar Britain.

Following the Beveridge report in 1942 (a year which also saw the publication of a BMA report urging the creation of a centrally planned, comprehensive public health service) the wartime administration produced in 1944 a White Paper proposing the formation of a National Health Service. Its detailed proposals met opposition, not least from the general practitioners' representative body, the Insurance Acts Committee of the British Medical Association (now known as the General Medical Services Committee).

When, after the war, Aneurin Bevan became Minister of Health he redrew the plans for the NHS substantially. The objections of the most powerful medical and other interest groups were accommodated and the tripartite NHS structure shown in Figure 1 emerged. The most radical element embodied in the 1946 legislation was that it involved the nationalization of most of the hospitals, although the teaching hospitals were given a special position within the new format.

The administration of the family doctor and allied services was achieved *via* bodies known as Executive Councils, which performed functions similar to those of the old Insurance Committees. The entire population became entitled to family practitioner care. Local authorities retained school and public health, together with community provisions such as health visiting and domiciliary nursing and mother and child care. They also ran the ambulance service, and were empowered to develop comprehensive health centre facilities.

The NHS as formed in 1948 was, therefore, by no means a complete break with past arrangements. Bevan compromised with and adapted existing structures in a public and political atmosphere strongly influenced by the disciplines and expectations engendered by the war. In some respects the result could be said to be victory for the 'values of equity, rationality and efficiency'.[3] But the NHS inherited certain constitutional weaknesses from its forebears, as well as some peculiarly British strengths.

For example, the problems surrounding the role of local authorities in what was intended as a national health service can in a sense be linked back to the difficulties seen in the interactions between local government and the Poor Law system almost a century before. The unsatisfactory linkage between the FPS and other community services and the hospital sector in the 1948 arrangements was also obviously related to previous structures.

4.4.4 THE EVOLUTION OF THE NHS

It is commonly argued that the main task confronting the NHS at the time of its formation in the 1940s related to the reduction of premature mortality, particularly amongst infants, children and working-age sufferers of conditions such as tuberculosis. But the success of the health service in bringing to an end the 'infectious disease' era resulted, this line of thought goes on to imply, in the emergence of a new set of problems related to chronic illness and disablement, particularly in those surviving to later life. This is why the NHS had to be reorganized in the early 1970s and greater emphasis given to the integration of community and hospital-based services, together with the establishment of fuller cooperative arrangements between the NHS and other care providers.

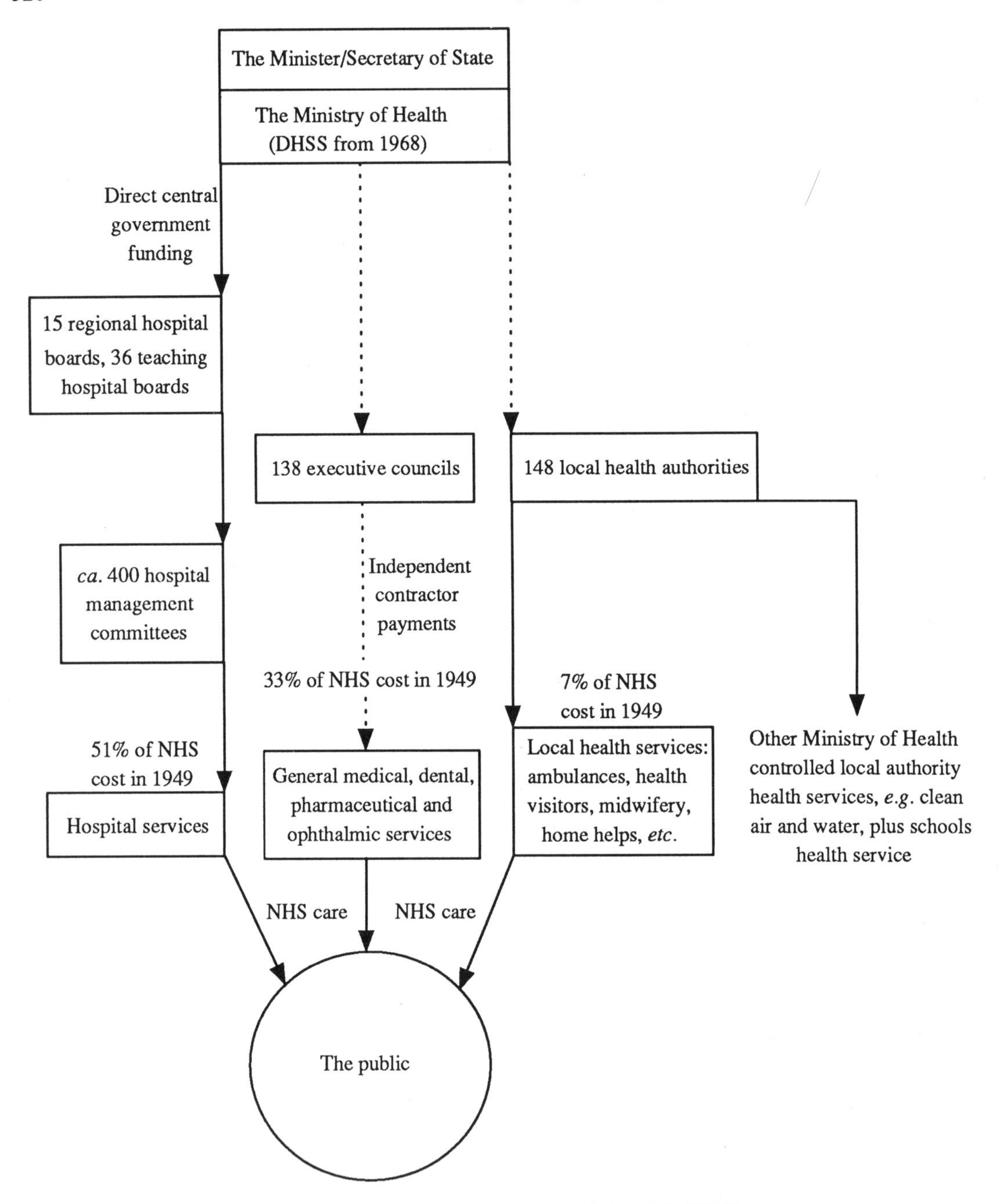

Figure 1 The NHS in England and Wales, 1948–1949[11]

This simplified picture of the developing challenge confronting the British health care system is of considerable validity and explanatory value. It is certainly true that antibiotics and vaccines such as that against polio were of importance in further reducing mortality from infections in the immediate postwar period, both in the UK and all other countries in the developed world. Indeed, life expectancy amongst elderly women has increased faster in the last three to four decades than at any other time in British history. For men over 65 the most marked advances have taken place since 1970.

However, the data in Figure 2 suggest that the mortality-reducing achievements of the NHS should in the main be seen as a consolidation and securing of past progress against the major infectious disease killers. This was initially gained by nutritional improvements and environmental changes. Life expectancy at birth for women, for instance, was already 70 years by 1948; in the same year female life expectancy at the age of 65 was 14–15 years. This compares with a little over 17 years

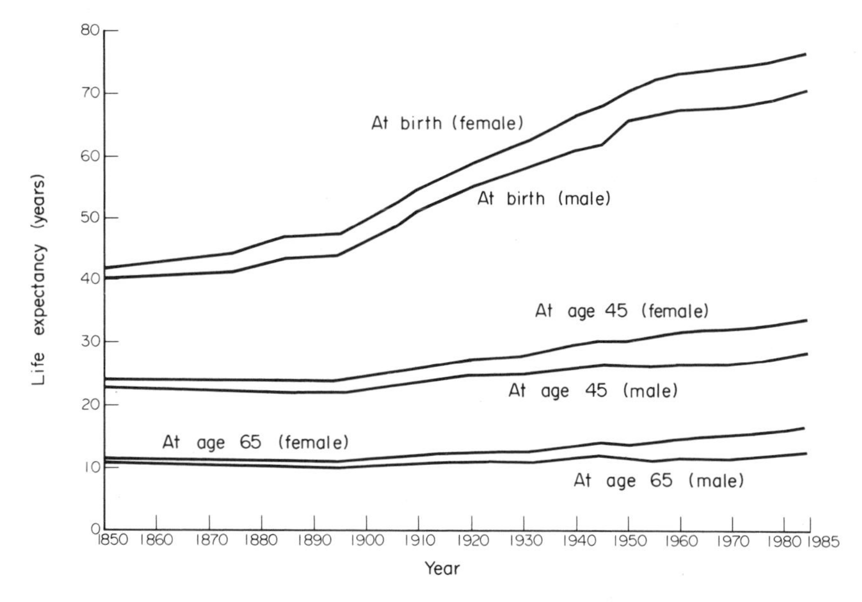

Figure 2 Life expectancy at birth, at age 45 and at age 65 in England and Wales, 1850–1985

today. Hence the 'problem' of old age was not created by modern medicine or the National Health Service. Neither was the pre-1948 prevalence of conditions such as severe intellectual impairment or long-term psychiatric disablement any less than that seen today—if anything, modern medicines, educational techniques and caring approaches have reduced the true level of such distress in society.

A key understanding to draw from these facts is that the NHS has always had as a central task the combined treatment and support of individuals with long-term sickness and disability. So too did the health and welfare systems which preceded it, right back to the days of the Victorian Poor Law.

Nevertheless, it is probably accurate to say that *greater awareness* of the needs of chronically ill and/or other disadvantaged people was the single major driving force behind the NHS reorganization of 1974, although throughout the 1950s and 1960s extensive NHS changes driven mainly by the advent of new medical technologies had already occurred. Enhanced understanding of the requirements of 'Cinderella' consumers was the result of an unmasking of a problem set which had to a degree been temporarily obscured in the 1950s. Such progress was in part due to the political leadership of men such as Richard Crossman and Sir Keith Joseph, and in part to the more vigorous public expression of consumer demands.

In general, the most important facets of the 1974 reorganization, which resulted in the reformed bipartite format shown in Figure 3, were:

(i) The creation of Regions and Areas as executive authorities, together with the establishment of a subarea tier of management, the Districts. Whilst the Areas coincided geographically with local authorities, in order to encourage service coordination, the smaller Districts (based essentially on hospital catchment areas) overlapped the local authority (LA)/area boundaries.

(ii) The effective preservation in England and Wales of the Executive Council system, with the boundaries of the new Family Practitioner Committees redrawn to match those of the areas.

(iii) The introduction of teams of managers at District, Area and Regional levels. These comprised individuals at the head of functional hierarchies like nursing and finance, together with representatives of local doctors in Districts and single-District areas. They were charged with reaching consensus management decisions. The former stayed unchanged in the 1982 reorganization of the NHS; the latter served as a model for the formation of the new District teams.

(iv) The integration of the teaching hospitals into the unified structure. Postgraduate teaching hospitals retained 'preserved' Boards of Governors, and have since come under the administrative control of a separate Special Health Authority.

(v) The formation of Community Health Councils (CHC) to represent consumer interests in the NHS at District level. Local authority members were also appointed to Areas and Regional Health Authorities (RHA).

(vi) The establishment of a comprehensive new planning system, better to equip the NHS to identify and pursue its priorities.

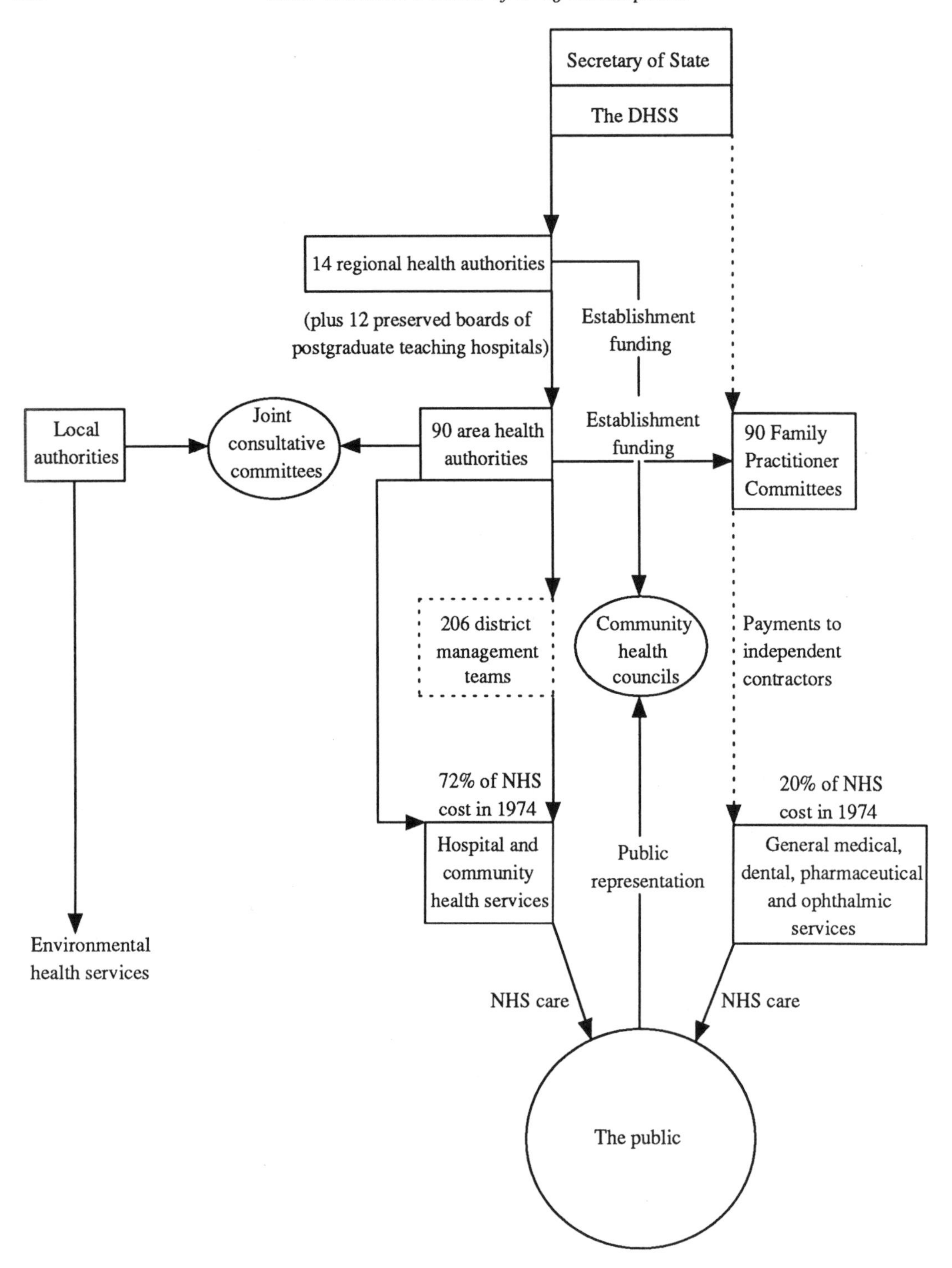

Figure 3 The NHS in England in 1974[11]

(vii) A heavy emphasis on the need to balance 'accountability upwards' by 'delegation downwards'. The upper tiers of the organization were to monitor the performance of lower ones, in order to ensure that agreed plans were being followed. But at the same time the members of the teams at District, Area and Regional levels were not in a linear relationship to one another. District and Area officers were jointly and (in some contexts) individually accountable to the Area Health Authorities (AHA), the Regional officers likewise to the RHAs.

4.4.5 FROM 1974 ONWARDS

The early 1970s were difficult years for Britain. The government began the decade with an aggressive search for growth, which in part involved heavy extra spending on public services. But in 1973 the oil crisis struck the Western economies, and Britain was particularly hard hit. The workforce, through the trade unions, resisted government attempts to limit state expenditure and cut back the rising rate of inflation, and a miners strike led to the three-day week and the eventual demise of the Heath government.

It thus fell on a Labour administration to supervise the final moments of the NHS's 1974 rebirth. The latter clearly had reservations regarding the new arrangements. This fact was reflected by the rapid introduction of reforms designed to include local government representatives on the RHAs and to increase their numbers on the AHAs. The CHCs were given powers regarding the approval of hospital closures.

During the five years of the Wilson and Callaghan governments the NHS was faced with new financial restraints (linked to the general economic situation) and an inevitable series of problems related to the disruption and changes in administrative and managerial structures and principles that the 1974 reorganization introduced. However, even in this 'teething' period a number of important steps were taken, including the publication of a seminal document on NHS priorities,[4] a new initiative on prevention[5] and the introduction of the Resource Allocation Working Party (RAWP) approach to regional resource reallocation in the NHS.

Following the return of a Conservative administration in 1979 the pressure for further structural and managerial changes within the NHS began once again to increase. The most important consequences of the new administration's determination to obtain better 'value for money' from the NHS (whilst also restraining after the start of the 1980s any further increases in the proportion of national wealth being utilized in public sector health care) were, in chronological order, as follows:

(i) In 1979 'Rayner scrutinies' of a number of NHS functions were initiated.

(ii) In 1982 the Area tier in England was eliminated, and local (unit) level management arrangements modified. This simplified the NHS structure. In the same year performance reviews were first introduced in the RHA/DHA structure, and innovations such as detailed option appraisals of capital investment projects introduced.

(iii) In 1983 the 'Griffiths' NHS Management Enquiry was set up, and new NHS audit and competitive tendering arrangements were first established.

(iv) From 1984 onwards the 'Griffiths' restructuring of NHS management began to have far-reaching effects on the way the health service is run. The introduction of a general management (as opposed to consensus team) function throughout the RHA/DHA/unit structure, the extension of accountability/performance reviews, the installation of management budgets for clinicians and, at the head, the establishment of an NHS Management Board, all stemmed from the 'Griffiths Report'.

(v) In 1985 the FPCs were made formally independent of the RHAs/DHAs, and significant managerial, planning and accountability developments followed in 1986. A discussion paper on primary health care was published in the latter year.

(vi) In the first half of 1987 a second (personal) Griffiths enquiry, this time into the provision of community care, was set up by ministerial invitation. A White Paper on primary health care is due for publication late in 1987, to be followed by legislation in 1988. At the time of writing (October 1987) it is expected to address four main areas, care standard maintenance, health promotion, consumer choice and value for money.

The revised English NHS structure resulting from this series of evolutionary changes is illustrated in Figure 4. The remainder of this chapter considers, against the background of the structural progress outlined above, two sets of issues of critical importance to the future of the NHS.

The first is the funding of the service, and the problems inherent in ensuring efficiency in the health care context. The second is the potential of the Family Practitioner Service branch of the NHS to contribute further to health promotion and community care in Britain; this last topic raises many vital questions about the pursuit of value for money and consumer satisfaction in the NHS.

4.4.6 THE FUNDING OF THE NHS

The difficulties inherent in, say, judging trends in the 'real' level of health service spending over time are considerable, as are those associated with measuring the true cost of many individual

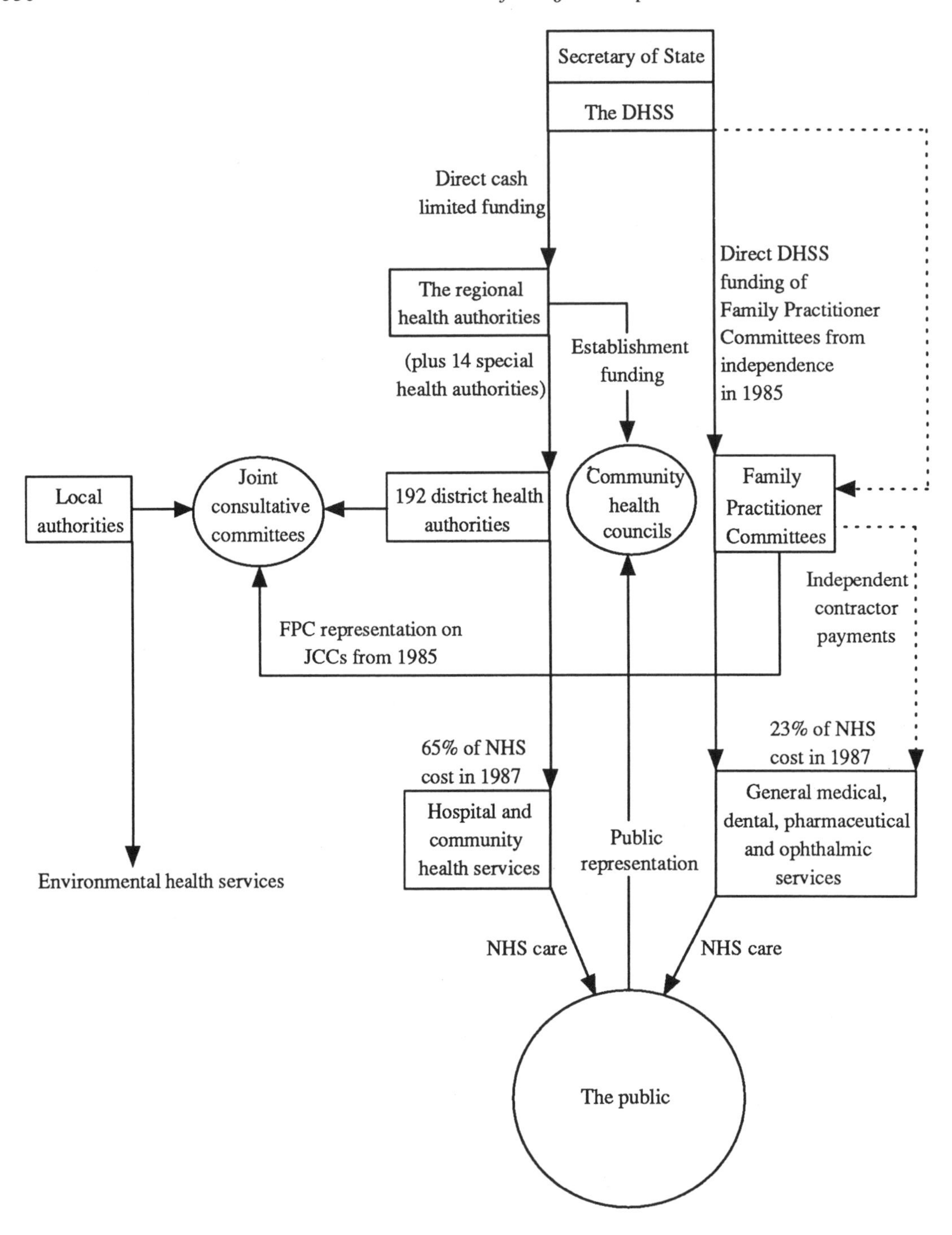

Figure 4 The NHS in England in 1987

service items and the value of the benefits they generate. This obviously complicates any attempt to assess the reasonableness or otherwise of current health service resourcing levels. However, recent work from a number of groups, including the newly established Kings Fund Institute for Health Policy Analysis[6] and the Office of Health Economics[7] has helped to provide an objective view of historical and planned NHS resourcing levels.

In cash terms NHS spending in 1987 will exceed, as noted in the introduction to this chapter, £21 000 million, approximately 6.2% of the gross national product (GNP). Figure 5 indicates that in general inflation-adjusted terms, the health service's resources multiplied by almost four times between 1949 and the present. Even after the onset of the 'oil crisis' in the early 1970s significant growth has been maintained, as measured by this scale. The available data show an 18% increase in

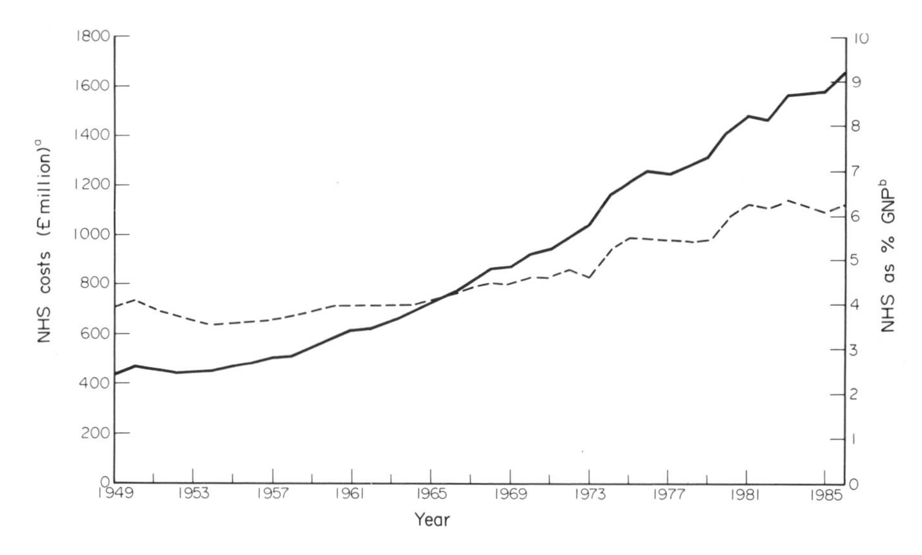

Figure 5 The cost of the NHS (at 1949 prices) (——)[a], also expressed as a percentage of GNP (- - - -)[b], in UK in 1949–1986. [a]As adjusted by the GDP deflator. [b]At factor cost

'real' NHS resources between 1982 and 1987, compared with 17% in the previous five year period and 26% between 1972 and 1977.

But adjustment by an index related specifically to NHS labour and other costs gives a much lower 'real' growth. Table 1 shows that during the 1980s the Family Practitioner Services enjoyed a rather greater expansion than the hospital and allied sector (HCHS)—in input cost-adjusted terms the latter's volume grew by about 6% between 1980/1981 and 1986/1987, compared with 14% for the FPS. Further modification of the figures to take account of increasing needs of elderly patients suggests that on the HCHS side of the NHS there has been in effect no 'real' growth in resource levels during the 1980s. Finally, adjustment to deflate the HCHS hospital spending for the extra cost of new treatments would indicate that there has in this sense been a marked short-fall in the hospital sector's 'real' funding.

Public discussion of such statistics has been strongly polarized. On the one hand those wishing to cast the recent record of NHS spending in a negative light have emphasized the latter set of calculations, taking them to be clear evidence of a significant 'cut' (of over 5% or approaching £1000 million) in NHS hospital funding since 1980. Data used to provide emotional support to this interpretation of the facts have included those relating to the falling number of NHS hospitals and hospital beds. Over the last 20 to 30 years the bed total has dropped by over a quarter, from almost 550 000 to under 400 000.

On the other hand those wishing to defend the NHS spending record point to the relatively strong growth in retail price index-adjusted outlays together with rises in indicators such as the number of in-patient and out-patient cases treated in the NHS hospitals. DHSS figures suggest a 15–20% increase in hospital and community health service activities since the start of the 1980s, together with a similar rise in the number of family doctors serving the British population. No British government has ever spent more on health and related services than the present one, either in absolute terms or as expressed as a percentage of the total national wealth.

NHS spending reached a record proportion of Britain's wealth in the early 1980s. It has been stable at just over 6% of the Gross National Product in 1981. Yet some parts of the NHS have been under very considerable stress in recent years, particularly those areas (like central London) adversely affected by resource reallocation programmes. Although talk of NHS 'cuts' is often misleading and on occasions irresponsible in that it serves only to undermine confidence in the standards of care available in the public sector, the reality is that, however measured, Britain's spending on health care is relatively low in international terms. The US, Sweden, France and Germany devote 50% more of their resources to the health sector.

At present it has to be admitted even by the most committed defenders of the NHS and/or of better health care generally that no adequate measures of the welfare benefits to be expected from higher national spending on health as opposed to, say, housing, education, holidays, or pensions for the elderly and disabled, exist. Hence any response to questions outlined above will probably always be controversial and questionable, although a report commissioned jointly by the Institute of

Table 1 Annual Percentage Increases in Public Spending[a] on Health 1980/81 to 1986/87[6]

	Hospital and community (current)			Family practitioners (current)			Total NHS		
	Cash	Cost	Volume	Cash	Cost	Volume	Cash	Cost	Volume
1980/1981	31.9	11.0	3.0	25.5	5.7	−1.2	30.5	9.9	2.8
1981/1982	10.4	−0.4	2.0	15.2	4.7	2.0	12.1	2.1	3.0
1982/1983	7.4	0.4	0.8	15.5	7.9	3.5	9.1	2.0	1.6
1983/1984	5.1	0.6	0.0	7.5	2.9	2.0	5.9	1.3	0.9
1984/1985	5.7	1.3	−0.1	9.9	4.8	2.8	7.4	2.4	1.4
1985/1986	5.4	−0.8	0.2	5.3	−0.6	−1.1	5.6	−0.4	0.1
1986/1987	7.2	4.1	0.3	8.5	5.4	2.4	7.7	4.6	1.1

[a] Cash expenditures are outlays at the prices prevailing in each year. Cost adjustments deflate the latter by a factor representing the changes in price of all goods and services over time. Volume expenditure expresses cash spending adjusted for price changes specifically within the health sector.

Health Services Management (IHSM), the BMA and the Royal College of Nursing (RCN)[8] makes one of the best possible overall economic cases for channelling more funds into the health sector.

It observes that internationally, demand for health care behaves as that for a 'luxury good', that is richer nations usually spend more of the greater wealth on health than poorer ones. The report then argues that 'the minimum expectation over the course of the next Parliament should be for health care spending to rise in line with national income, with separate provision for demographic pressures and any major new illness such as AIDS'. This would clearly have the effect of ensuring that over time a gradually increasing proportion of the national income will be devoted to health care. It would probably be in the order of 0.5% of GNP in a five year period, a reasonable expectation given the background of national and international experience over the past 40 years. Indeed, in as much as the logic of this case applies to the totality of health spending, NHS and private combined, some commentators within the public sector may regard it as being unduly modest given the rapid rate of private sector expansion (against a small base) in recent years.

As far as the outlook for the NHS is concerned, all that at present can be said with certainty is that current public expenditure plans (autumn 1987) envisage an increase in cash outlays of only about 4.5% per annum in the period 1988–1990. To what extent this will in practice prove sufficient to cover just cost inflation alone is unclear, a fact which must concern all those using, working in and supplying the health service. Recent government statements show no positive commitment to the IHSM/BMA/RCN approach, but have nevertheless recognized NHS funding problems.

However, it must be concluded that even if NHS spending in the years ahead is such that it enjoys a stable or marginally increasing share of the nation's wealth, the efficiency with which available resources are used will remain a matter of critical importance. This emphasizes the need for good management, cooperative attitudes and a commitment to productivity throughout the health care system.

4.4.7 THE INDEPENDENT CONTRACTOR SYSTEM—THE NEXT WAY FORWARD?

Throughout the 1970s and early 1980s public debate about improving the management and future performance of the British health care system concentrated on two main areas. First, the reform of the NHS hospital and related administrative system, and the development of the RHA/DHA system and, second, the options open for alternative, private sector provision.

However, since 1985 increasing attention has been paid to the provision of primary care, and the unique structural arrangements relating to the Family Practitioner Services. As described earlier, the latter are in England and Wales administered by 98 Family Practitioner Committees and in Scotland and Northern Ireland by the Health Boards. However, in all parts of the UK the community-based general practitioners, dentists, pharmacists and opticians are nearly all independent contractors. Their status differs fundamentally from that of those doctors, nurses, health visitors and other staff who are salaried employees of health authorities.

As Table 2 shows, the Family Practitioner Services currently account for about 23% of gross NHS (UK) spending. The pharmaceutical services are the principal component of this outlay, followed by the general medical services. Together these two elements represent some 80% of total FPS costs. Administrative expenditures in the form of FPC and allied budgets are not included in

Table 2 Family Practitioner Services Expenditure by Service 1949–1987[7]

	Family practitioner services									
	Pharmaceutical		*General medical*		*General dental*		*General ophthalmic*		*All FPS*	
Year	*£m*	*% total NHS costs*	*£m*	*% total NHS costs*	*£m*	*% total NHS costs*	*£m*	*% total NHS costs*	*£m*	*% total NHS costs*
1949	33	7.6	44	10.1	45	10.3	23	5.3	145	33.2
1955	58	9.5	62	10.2	38	6.3	15	2.5	173	28.5
1960	91	10.1	90	10.0	57	6.3	18	2.0	256	28.4
1965	145	11.1	102	7.8	67	5.1	21	1.6	335	25.7
1970	209	10.2	178	8.7	102	5.0	29	1.4	518	25.3
1975	453	8.5	345	6.5	217	4.1	72	1.4	1086	20.5
1980	1126	9.4	754	6.3	459	3.8	119	1.0	2457	20.6
1981	1278	9.3	890	6.5	534	3.9	141	1.0	2844	20.7
1982	1469	10.1	1006	6.9	598	4.1	165	1.1	3237	22.4
1983	1628	9.9	1100	6.7	650	4.0	264	1.6	3642	22.2
1984	1750	10.2	1231	7.1	710	4.1	190	1.1	3881	22.5
1985	1875	10.2	1345	7.3	779	4.2	159	0.9	4159	22.6
1986	2029	10.2	1460	7.4	845	4.3	149	0.8	4484	22.6
1987[a]	2192	10.2	1595	7.5	926	4.3	140	0.7	4853	22.7

[a] 1987 figures are estimates.

the figures given, but they were about £100 million (UK estimate) in 1986 compared to service outlays of approaching £4500 million.

The FPS budget is not, unlike most elements of government spending today, formally cash limited. Yet it would be incorrect to conclude from this that government is completely unable to control spending in this area. Complex negotiating machinery exists to determine both the target remuneration of the various groups of professional contractors and the return of capital to the manufacturers supplying pharmaceutical and allied products to the FPS/NHS. In many instances the fees and prices paid tend, over time, to vary inversely with the volume of items/services supplied, even though changes in fundamental determinants like professional manpower levels will have lasting cost consequences.

This understanding helps to draw out an important difference in the relationship between central government, the DHSS and the FPS, as compared to that which exists between the DHSS and the DHA/RHA-administered hospital and community provisions. It is relevant to the entire debate on how health care generally should or could be managed more efficiently.

In the case of the health authorities, there exists a recognizable chain of command between the centre and the periphery. Specific sums of money are allocated to the Regions and the Districts and the units within them, and the planning process allows local services to develop in the context of priorities and objectives laid down within the system overall. In introducing service developments involving investments in facilities like new hospitals (health 'factories') which demand not just large capital inputs but also adequate provision for the major recurrent costs associated with them, this is arguably vital.

But in the case of FPS primary care the system comprises many millions of individual consumers. There is, in reality, no directly interlinked chain of planning and priority/objective setting. Rather, the centrally negotiated payment system for the independent contractors has served to define a pattern of economic incentives against which they operate in the market-place as independent 'retailers'. Provided they comply with the basic terms of their NHS contracts, FPS contractors are traditionally free to a substantial degree to set their own objectives and standards, albeit that advisory interventions from FPCs together with their own professional bodies and expressed consumer preferences will guide their behaviour.

In looking to the future of the British health care system it may be argued that a careful balance has to be achieved between these two types of approach. Even accepting that a system devised to ensure efficiently universal access to appropriate care is the basic goal of the advocates of both, there are significant differences between them. While the former is driven primarily by political process and professional guidance, the latter model is—potentially at least—more sensitive to the direct influence of consumer decisions and preferences.

Although at present the nature of the FPS independent contractors agreements is such that they are not in competition with one another to any great degree, this situation could change. The White Paper on primary health care due for publication in November 1987 is expected to propose changes in the contracting arrangements together with initiatives in areas like prevention and, perhaps, the relationship between FPCs and regional health authorities. New measures to reward 'good practice' and improve practice premises are also likely to be introduced.

In the long term it could be that primary care in the NHS will evolve to become more like an 'HMO' system. In this system, independent 'firms' of family doctors and other professionals—including possibly health visitors and social workers—would compete to provide their clients with comprehensive care paid largely or entirely on a capitation fee basis by the government. Such practitioner groups could be in part responsible for buying in hospital or other social and health care services supplied by other independent, albeit NHS contracted, agencies.

In this way a market for both primary and secondary health care could be established within the NHS and linked to the private sector where appropriate. From the viewpoint of market economists this would have the advantage of incorporating stronger competitive motors in the NHS pursuit of value for money, while also offering the population the benefits of a planned, universally available health care service.[9]

4.4.8 CONCLUSION—THE NHS AS A BRITISH SUCCESS

In closing this chapter, it is important to emphasize that the record of the National Health Service is in the main one of outstanding success. Discussion of the possibilities for even better performance in the future should not be allowed to detract from, or mask, this fact. For example, despite occasional ill-informed criticism, Britain's mortality rates in fields ranging from perinatal survival to the longevity of the over 65s compare well with those of most other European and other comparable nations.

In terms of managerial reform, the provision of community care for disabled and other people and the cost-effective introduction of sophisticated medical technologies, the record of the NHS in the last 10 to 20 years is also relatively strong, despite this country's intentionally low level of expenditure on health. Fuller awareness of problems such as those existing in the community care areas[10] should not be taken as evidence of failure, but rather of the greater availability of the knowledge needed to continue the constructive evolution of Britain's health and welfare system.

The effective partnership between the NHS and the British pharmaceutical industry provides a further example of positive national achievement. The UK has in the past 40 years established a strongly innovative pharmaceutical industry, which contributes some £850 million net to the nation's balance of trade. But at the same time the country's per capita domestic volume use of and cash outlays on medicines and allied goods are significantly below those of its major trading rivals.

It is against the confidence that this background provides that British policy makers are currently examining ways of strengthening further the nation's primary health system and the linkages between it and the hospital and other welfare agencies, and creating greater direct consumer choice in the processes of care provision. Their ultimate task is to avoid ideological or professional polarities; maintain order and regulation in the health market-place while avoiding the false assumption that the appropriate future health care needs of the population can ever be predicted entirely accurately by bureaucratic planners. That is, the NHS today must strive pragmatically to combine 'market' and 'socialist' approaches to the equitable distribution of necessarily scarce national health care resources. It must also maintain complementary programmes of prevention, cure and care with the purpose of creating as robust and flexible a structure as possible to meet the uncertain challenges of tomorrow.

To the extent that Britain's health care system has already gone some considerable way to achieving such goals, and promises further success in the future, the continuing NHS experiment should be of interest to the entire worldwide health care community. This includes those sections of it concerned primarily with the development of new pharmaceutical treatments. For, in the final analysis, technical innovations in any area of medicine will achieve their full potential value to the community only if a health service structure exists to deliver them effectively and fairly to all those in need.

4.4.9 REFERENCES

1. M. Glendenning and W. A. Laing, 'The Politics of Health', Association of the British Pharmaceutical Industry, London, 1987.
2. J. E. Pater, 'The Making of the National Health Service', The Kings Fund, London, 1981.
3. R. Klein, 'The Politics of the National Health Service', Longman, Harlow, 1983.
4. Department of Health and Social Security, 'Priorities for Health and Social Services in England', HMSO, London, 1976.
5. Department of Health and Social Security, 'Prevention and Health: Everybodies Business', HMSO, London, 1976.
6. R. Robinson and K. Judge, 'Public Expenditure and the NHS: Trends and Prospects', Kings Fund Institute for Health Policy Analysis, London, 1987.
7. S. B. Chew, 'The Compendium of Health Service Statistics', Office of Health Economics, London, 1987.
8. M. O'Higgins, 'Health Spending—A Way to Sustainable Growth', The Institute of Health Services Management, London, 1987.
9. A. Maynard, personal communication.
10. Audit Commission, 'Making a Reality of Community Care', HMSO, London, 1986.
11. D. G. Taylor, 'Understanding the NHS', Office of Health Economics, London, 1984.

4.5
Health Care in the USA

Jack E. Fincham
Samford University, Birmingham, AL, USA

4.5.1 INTRODUCTION

The organization, format of delivery, financing and provision of health care services in the United States (US) are unique in comparison with the parallel components of other health care systems in the world. What separates the US system from systems elsewhere is the eclectic nature of impacts upon the delivery of US health care. The economics of US health care delivery are such that any service is available, provided money is available for payment. Independently wealthy patients are not terribly concerned about the inaccessibility of services to others. Individuals with health insurance have a range of treatments available to them. Individuals without health insurance have few options when ill and in need of health care. According to US governmental figures, the proportion of US citizens without health insurance is roughly 15% of the population.[1] These 37 000 000 Americans are not only without health care, but they also lack an advocacy group to represent their collective point of view in places where change could be instituted. Stark figures indicate the severity of the problem: one in five of American children is poor, the US infant mortality rate is higher than 17 other industrial nations, and on any given night there are 100 000 children among the homeless.

Since the federal and state governments, and employers are the major purchasers of health care for covered individuals, these groups have a major impact on the coverage limits, payment and delivery of US health care.

In 1984, fee-for-service payments for health care accounted for 89% of expended funds, while managed care payments accounted for the remaining 11%. In 1997, it is projected that managed care payment for services will approach 90% and fee-for-service payments will decrease to 10%.[2] After decades of the highest health care costs in the world, and obdurate increases in health care insurance premiums, this switch from fee-for-service to managed care provision of health care should not seem surprising. Managed care is a loosely defined construct representing an amalgamation of private health insurance, provision of health care and insurance through a Health Maintenance Organization (HMO) or a Preferred Provider Organization (PPO), or through an Exclusive Provider Organization (EPO). HMOs, PPOs, and EPOs have incorporated the delivery and financing of health care under umbrella structures as opposed to traditional insurance and delivery of health care through dichotomous units.[2] These private sector efforts in combination with governmental efforts such as Medicare and Medicaid have affected all involved in the delivery, financing and utilization of health care in the US.

Change has been an underlying force impacting upon the American health care system for many years. Although at present the health care system seems burdened with new impacts and forces of change, many current dilemmas, when examined closely, have been present for a long period of time.[4] The problems may not be new, but certainly specific facets have forced a revolution in US health care delivery. The questions of access, payment and allocation of scarce resources have been foci for change in the past, just as they are today, and most certainly will be tomorrow. The paradox of increasing technology coupled with decreasing resources allocated for payment has inserted political, social and philosophical aspects into health care delivery as never before.

4.5.2 US HEALTH CARE AND THE FORCES OF CHANGE

The functions, provision and utilization of US health care are rapidly changing. With the advent of governmental entitlement programs in the 1960s (Medicare, funded by the federal government, and Medicaid, funded jointly by the federal and state governments), political and economic overtones were thrust upon the health care system. Perceived shortages of physicians, pharmacists and other health care professionals in the 1960s and 1970s led to capitation grants from the US government to health professional schools. The results of this capitation have led to a projected surplus of physicians and others in the 1990s and beyond, and, additionally, to the emergence of new health professionals (physician assistants—PAs and nurse practitioners—NPs). Patterns of utilization will surely be affected by increasing technology and possible finite resources for payment for services delivered. Issues of indigent care, prospective payment for hospital services and competition have revolutionized the activities of hospitals in the past decade.[5] Patients have changed as well. Expectations from health care in some cases have exceeded possibilities. There is still only so much that can be done for patients even in the technologically intensive health care system of today. Even so, US patients have a long tradition of expecting the health care system to fix what is wrong with them.

The impacts affecting the health care system have included social, economic, professional, political and structural factors. The US system, perhaps as no other system, has been affected by, and subject to, political influence. Despite current spending of over 11% of the US Gross National Product (GNP) on health care, the resultant expectation of 'the best' health in the world has not come to pass. The percentage of the US health care expenditures in relation to the GNP, and the degree to which these percentages have precipitously increased since the 1960s, is depicted in Figure 1. The health care expenditure percentage of the GNP has increased by over 88% in slightly over 20 years. Despite the enormous expenditures for health care in the past century, rising income levels and advances in

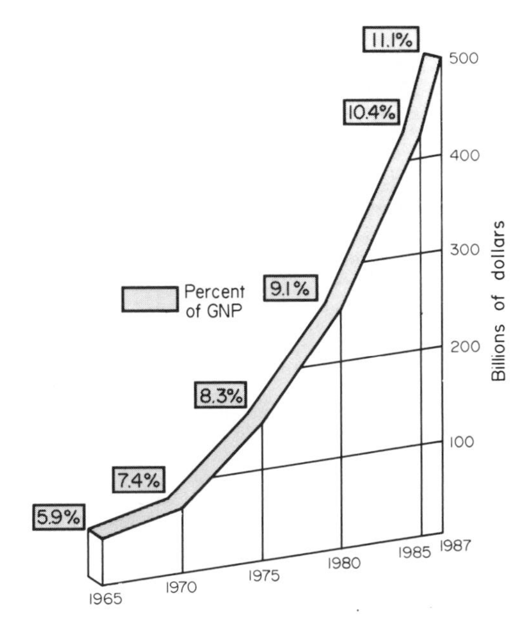

Figure 1 Total US health care expenditure (data from US Health Care Financing Administration, 1988)

public health practices have been suggested to be the most important reasons for the fall in infant mortality.[6]

The impact of autoimmune deficiency syndrome (AIDS) will also force added change on the US health care system. In a health care delivery system already burdened with limited resources, the treatment and care of the burgeoning AIDS population will undoubtedly impact upon the future of the health care system. An analysis of the cumulative lifetime medical care costs associated with treating the US AIDS population has been performed.[7] The analysis forecasts the costs to be $3.5 billion in 1989, $4.7 billion in 1990, $6.0 billion in 1991 and $7.5 billion in 1992. The figures do not include other costs borne by other patients that are preventive in nature. The bottom line of this and other projective studies is that all patients (those with AIDS and those without), providers, insurers and policy makers will need to factor the treatment of AIDS patients into any consideration of the future of the US health care system.

4.5.2.1 The Lack of Planning in the US Health Care System

Planning for health care needs of the future was an emphasis of the political climate of the 1970s.[8] There were many enabling legislative and regulatory efforts to institute a system of planning for future needs and regulation of the health care industry. These efforts were instituted to ensure any brick and mortar changes (buildings, hospitals, *etc.*) and/or equipment in the health care delivery system were truly needed in any one geographical area. Certificate of need (CON) approval was required for expansion of existing structural components or completion of new expansions. The intent of such efforts was to curtail expenditures in the entire system and thus save money, as well as to ensure that additions to the delivery system were truly necessary.

In the 1980s, this enabling legislation and regulatory thrust was dismantled in favor of competition. Competition was felt to be a more suitable way to deal with the costs of care in a much more appropriate fashion. This *laissez-faire* approach to health care did not achieve success in reducing the tide of expense outlays of the US health care system. In addition, without the necessary planning for the future, many efforts in the current health care climate are indeed reactive rather than proactive. The satisfaction of current needs takes precedence over any planning for the future. However, the situation for the future must be considered now, or the potential for future trouble will increase. As the system stands now, the health care system does not possess the structural or financial stability needed to deal with the future needs of providers or recipients of health care.

The influence of these competitive impacts upon physicians and pharmacists has been enormous. One need only consider preferred provider organizations (PPOs), health maintenance organizations

(HMOs), mail order pharmacy, physician dispensing and drug diversion to realize the scope and magnitude of the current health care delivery organizational status. We are where we are in the current system because of the lack of a continuous degree of proper planning for future health care needs and practice functions.

4.5.2.2 The Nature of Change, Present and Future

Although the US health care system has been characterized as a nonsystem, or as an amalgamation of entities without an organization structure, the components are definitely interrelated. Any impact affecting one segment or professional body has ramifications and reverberations affecting the entire system. This multidimensional interrelationship of the various health care system components has led to change that is also multidimensional in nature.

Examples abound to support this multidimensional nature of change. When the Diagnosis Related Group (DRG) system of prospective reimbursement for inpatient Medicare patient care was instituted, hospitals sought methods of increasing revenues in ways previously not pursued. Outpatient services took on added importance for hospitals since they were excluded from DRG reimbursement regulation. Thus outpatient prescription departments became potential areas for growth and expansion for institutions. Previously, these services were offered only on a small scale in most hospitals. These departments were in place in hospitals previously, but marketing efforts could best be characterized as passive. As well, and for similar reasons, home health care departments became prominent in institutions.

The involvement of hospitals in these two areas of nontraditional outpatient services placed hospitals in direct competition with community and chain pharmacies in an open fashion for the first time and at a competitive level hitherto unknown. Change is also present when considering physician dispensing and pharmacist prescribing. Pharmacists are averse to physician dispensing, but view pharmacist prescribing in a favorable light.[9] On the other hand, some physicians view dispensing as a function suitable for their practice repertoire and view pharmacist prescribing as a threat to their practice roles. Regardless of who possesses which point of view, both pharmacist prescribing and physician dispensing have roots in several facets of health care components—economics, practice responsibilities, numbers of professionals in the discipline and future desires for expanded role possibilities. These elements of change have affected the health care system and will affect all involved in the delivery and utilization of health care services.

4.5.2.3 The Business Climate in the Health Care System

One aspect of health care delivery and structure which recently has taken on added importance is the phenomenon of mergers. Whether it is the merging of brand name and generic drug companies, individual hospitals with hospital chains, combining of purchasing groups, the combination of academic health centers with hospital chains (Humana, Hospital Corporation of America), various practitioners (physicians, pharmacists, *etc.*) or any or all of the above, mergers have made an impact upon the health care system. Each of these previously mentioned mergers has the potential to influence many other points in the system. However, the merging and combination of structural components both have the potential to dramatically affect all in the health care system.[10]

Scenarios have been suggested whereby in the future a few very large conglomerates will be the major providers of US health care on a nationwide basis.[11] Under this nationwide tier of large scale providers will be a segment of smaller conglomerates with a lower level of a larger group of individuals and groups which feed into the components above them. Another possibility involves the formation of integrated health care clusters (IHC).[12] This integrated system would contain any and all services a defined population would need—from wellness or preventive services to acute and long term care. Also, this system would effectively 'lock out' nonparticipating providers. Which of these scenarios will succeed is debatable. What is significant is that they are currently being discussed and the subsequent attention the discussion garners. What of the patient and practitioners in such a system? Time will tell, but the health care system from utilization to payment will be irrevocably different if such scenarios play out.

4.5.3 QUALITY *VERSUS* COST IN HEALTH CARE

What should the paramount issue in health care be? Should quality be the goal or should cost determine the processes and outcomes of care? Perhaps the combination of quality and cost should be the driving force of assessment of health care. However, the patient, the provider or someone else may desire input into the decision making process. But, who should decide? An eclectic health care system leads to an eclectic series of decision making points. What has become apparent is the fact that economics drive the health care system, and, as such, many decisions come down to economics. Quality may be important, but only if the price is right.

4.5.3.1 Economics of Health Care

The purse strings of the health care system are controlled by many forces. These forces include private insurers, groups of private payers, state and local governments, and the federal government. A description of health care economics is beyond the scope of this chapter, however certain key points need to be addressed. Beyond the payment issue alone lies perhaps an even more pressing concern for the future. Namely, what health care should be paid for and for whom should the payments be made?

Expanding resources to secure health benefits is engaging in a game of chance. What works and what does not is not often easy to summarize. Zeckhauser and Shepard[13] analyzed age specific death rates for adult males from 1930–1970. Findings showed the older age group reduction in mortality was one-fourth as large as for younger age groups. The dramatic differences in mortality could perhaps be attributed to changes in sanitation, public health efforts or medical care. The implications are that whilst all segments are aided by health improvements, there are risk factors present in certain age groups not present in others. The interaction of risk factors and improvements in nutrition, medicine, public health, *etc.* has certainly benefited some. However, because of the larger number of risk factors in older individuals, the benefits may not have accrued in this age group as with younger individuals.

With increasing levels of technology present and projected in the future, it is possible that more and more health care will be available for fewer and fewer individuals. Funding may not be available to cover all interventions. Coverage for both expensive and inexpensive care possibly could be sparse.[14] Currently in the health care system, deductibles and coinsurance (varying amounts of money required to be paid by the patient before health insurance for coverage begins) force many needy individuals to delay or postpone care.

The financing of health care became a publicly debated topic in the 1930s with the emergence of the private insurance industry and discussion of health care as a 'right' similar to 'rights' of the population to police protection and public education. In the 1960s, the passage of Medicare and Medicaid provided millions of poor and elderly Americans with eligibility for unrestricted medical care for the first time. Public discussion of the future financial impact of these legislative efforts was negligible while tax coffers overflowed. However, when recession and inflation became simultaneously prominent in the 1970s, discussion and debate of the merits of public financing of medical care became commonplace.

Virtually every legislative impact upon the health care system since the early 1970s has sought to reduce expenditures for health, either through recipient benefit reduction, curtailing of covered services, or institution of beneficiary copayment for certain services (medical visits, prescriptions, *etc.*). What has characterized these past impacts upon the financing and payment for health care has been their reactive nature. An impending crisis must loom on the horizon before appropriate action is instituted. For example, DRG prospective reimbursement came into play in the mid-1980s after the financial solvency of the Social Security system was questioned. Dire forecasts of a bankrupt Social Security program occurred in the late 1970s and early 1980s. Whilst the nature of the next crisis is uncertain, the ramifications most certainly are clearer—more cutbacks and/or decreases in funding.

4.5.3.2 Quality of Care

The quality of medical care delivered has ramifications for all involved in health care, from user to provider to payer. Despite the ramifications, the determination of what constitutes 'good' quality has been mercurial.[15] However, care suspect in quality can profoundly affect all components of the health care system. These potential ramifications have led to governmental efforts to develop a method of ascertaining quality of care standardization. Although the face of quality assurance has changed with the demise of health planning regulations in the 1980s, other requirements associated with the Tax Equity and Fiscal Responsibility Act of 1984 (TEFRA) regulations[16] have put into place peer review organization (PRO) assessments of quality of care.[17] Despite the lubricious nature of quality of care assessment, the need and importance of assessing quality of care will no doubt continue in the future as governmental payment for health care services continues at present or perhaps enhanced levels in the years to come.

Corporate or managed health efforts at ensuring quality of care have also been recently enhanced. Managed systems have determined that the best method of attracting new subpopulations of patients is to provide health care that is cost effective, stable and of an ascertained certain level of quality. Health Maintenance Organizations have had quality assurance components built in their infrastructure from the onset of governmental regulation and involvement in the HMO movement. Thus, quality of care and economic constraints are not necessarily mutually exclusive constructs in the US health care system.

4.5.4 REIMBURSEMENT CONSIDERATIONS

4.5.4.1 Prospective Reimbursement

Prospective reimbursement has had a dramatic impact upon hospitals and related institutions. Formerly, hospitals were reimbursed on a retrospective basis for Medicare services provided. Because of the nature of 'after the fact billing', services such as pharmacy services were viewed as a revenue generating component of services provided. However, now that rates for hospital services are predetermined for Medicare patients, all expenses incurred are considered as costs as opposed to revenues. Thus, the lower the costs, the higher the amount hospitals can garner as the 'above cost–below payment level' profit buffer. On the other hand, costs which exceed maximum limits must be 'absorbed' by the institution. This has placed many hospitals in a most precarious economic posture. A four year phase in period somewhat softened the ultimate blow, but the full effects of the legislation currently are being felt. Because of the short term success of DRG reimbursement in reducing the rate of increase of health care expenditures in the Medicare program, it will no doubt be reatained for Medicare and perhaps instituted for other government programs such as Medicaid. In addition, private insurers see prospective reimbursement as a way to decrease costs associated with the care they collectively purchase.

4.5.4.2 Changes in Medicare

The Medicare Catastrophic Coverage Act of 1988 (P.L. 100-360) was signed into law in July of 1988. This law represents the largest expansion of the Medicare program since the passage of the Medicare enabling legislation in 1965. Changes mandated by the new law included the following:

Inpatient hospital coverage—no limits for enrollees.

Skilled nursing care—up to 150 days per nursing facility.

Hospice care—no limits.

Medical benefits—visits by skilled nurse practitioner of up to six days a week for 38 days, up to 80 hours a year of professional care of homebound patient to relieve unpaid family member or friend.

Prescription drug benefit—prescription drugs to minimize tissue rejection for the first year, intravenous drugs used at home (begins in 1990), all approved drugs eligible for reimbursement (begins in 1991).

The specific inclusion of a prescription drug benefit and added coverage for skilled nursing care have been seen as methods to reduce shortcomings of the current Medicare coverage limits and restrictions. What remains to be seen is how all of these added recipient benefits will be paid for in a system already stressed from economic and delivery standpoints.

4.5.4.3 Resource Limits

Who should decide what should be purchased in the way of health care services? Should it be the patient, physician, fiscal intermediaries for governmental programs or the various governments themselves (state or the federal government)? If it is to be a governmental entity, how much health care can and should governmental programs purchase? What limits should be placed upon recipients' use of health care services? These are questions previously not often asked in the US. The enormous increases in technology have been utilized by those who could afford it through their own payment or through insurance payments. This expensive and elaborate technology provides the uninsured or underinsured patient with little benefit for their health. Do the limits of the collective ability to pay for health care adversely affect one segment of society? Jonas[18] has suggested this situation has always existed with regard to US health care resource utilization. Previously, resource limits may not have been so segmented, but in the US there has always been an underserved segment in need of more care than could be obtained by them.

4.5.4.4 Who Shall be Insured?

If there are limits on how much technology can be purchased, who determines who obtains what? Disadvantaged segments of the population will no doubt increase numerically in the future. Bovbjerg[19] has noted: 'People on their own or in small groups, as well as increasing numbers of dependents of workers in large groups, are disadvantaged in comparison with workers in large groups.' Is it the insurer, the provider or the government who decides on coverage? The questions, although not answered easily, must nonetheless be addressed. Limits placed upon health care utilization have been in place in the past, but never to the extent they will be in the future. The experience of Great Britain sheds light on the problem and potential ways to manage the situation.[20] British patients and providers perhaps have dealt with limits on health care utilization and adoption of new technologies in better fashion than their American counterparts. Patients and provider alike understand the limits of the British system and as such do not place heroic demands related to what can and cannot be obtained.

4.5.5 PROFESSIONAL CONSIDERATIONS

The United States health care system has become a personnel intensive system. The array of specialists, subspecialists, technicians, technologists, consultants and attending individuals seems to expand yearly as technology and expectations for care have expanded. Despite different titles and hierarchical breakdown of assignment of duties within professional components, there have been overlaps of responsibilities and duties that no doubt will increase in the future. How the professions interact, overlap or challenge each other's efforts affects not only health care delivery, but also the utilization of care. In fact, the degree of cooperation or lack thereof will determine the future direction and control of the health care system.

4.5.5.1 Physicians

The practice of medicine has changed rapidly in the past decade. One aspect of change has been the dramatic shift from solo practice to group practice as the norm. Coupled with the change in practice location is the expansive increase in the number of practicing physicians. From the 1970s to the present, the scenario has changed from a projected lack of physicians to a projected surplus. If the glut has occurred, are there still underserved areas in desperate need for physicians? The answer is an explicit yes.[21] Despite the increase in numbers of physicians and specialities, there remain underserved populations in need of basic primary care practitioners, namely internists, general practitioners, obstetrician–gynecologists and pediatricians. To further complicate the picture, the medical malpractice liability crisis in medicine has forced many physicians from their chosen specialty to other specialties in the profession in medicine because of inexplicable increases in malpractice insurance premiums.

The physician glut disappears if patients live in a small community devoid of a physician. It conversely expands if patients live in a major metropolitan area with hundreds of possibilities for choosing a physician. The divergence of points of view pertaining to practice sites and under- and over-served areas will continue as long as physicians are free to choose when and where they will practice. This freedom has the potential to be impacted upon by many factors which may include: changing demographic shifts, payment for service alterations, future liability concerns, actual practice expansion due to technology or practice compression due to the expansion of other professionals' roles or responsibilities.

4.5.5.2 Nonphysician Providers

The so-called physician shortage of the 1960s and 1970s led to exploration of the concept of a 'physician extender'. The concept was proposed that an individual with requisite skills in patient triage, clinical assessment and patient education would ease the strain the lack of physicians had created. Generally, it was felt these physician extenders with proper training could serve patients and the system through easing the shortage of physicians.

Physician Assistant (PA) and Nurse Practitioner (NP) training programs were inaugurated and have flourished. The PA profession is male dominated and the NP profession is female dominated. This sexual segmentation of the two professions has led to interesting confrontations between the two physician extender components. Nurses in some cases have refused to act upon orders written by PAs. Apart from disagreements between themselves, NPs and PAs must potentially deal with the notion they both may compete for patients with physicians in the years to come. Physicians may view these extenders as potential prescribing replacements. Pharmacists may view these professions as potential unwelcome dispensers.[22]

As long as the shortage of physicians, coupled with a lower medical school enrollment, was perceived, the issue of overlap of services provided by physicians, NPs and PAs was not a major issue. As long as physicians were in short supply, any new health worker (in a subordinate position) was a welcome addition to the health care team. Where the key words in the late 1960s were 'shortage', 'crisis' and 'expansion of training capacity', in the 1980s the components of note have included 'oversupply', 'overtraining' and 'costs'. However, as medical school enrollment and graduates have increased, the overlap of services rendered will become a major issue for all health professions in the future.[23]

4.5.5.3 Pharmacists

The profession of pharmacy and practicing pharmacists have not been immune to the changes affecting physicians and others in the health care system. The changes in reimbursement and service delivery have forced the pharmacy profession to change and adapt. This is not a new phenomenon; the swirls of change have always fallen precipitously close to the pharmacy profession. The advent of DRG reimbursement in the institutional setting has impacted the range of services to be offered by hospital pharmacists. The emergence of mail order pharmacy on a large scale (obtaining prescriptions through the mail from in state or out-of-state pharmacies) has threatened the economic viability of community pharmacy practice. The provision of hospital based outpatient pharmacy services had further eroded the competitive stature of independent pharmacies. The competition within the chain pharmacy industry has impacted the chains themselves, as well as independent community pharmacies. The competition from chain pharmacies has also been a threat to the continued viability of many chain pharmacy operations themselves. Finally, the dispensing of prescriptions by *physicians* has served to indicate the traditional roles in the health care system are no longer so clearly defined.

Who shall perform what tasks will be a question to be answered in the years to come. Who prescribes, dispenses and administers medications may not be an easily definable series of questions.

4.5.6 PATIENTS

The patients treated in the US health care system collectively are aging. Physicians, pharmacists and others will be providing services in the future to a clientele on average older than any age group previously served by health care professionals. Current planning for future health needs of the elderly is lacking to say the least. American society, the health care system and health care professionals are not prepared to deal with the elderly population that awaits provision of care in the future. One in eight Americans is currently aged 65 or older. Projections are staggering for the future as well—one in three Americans will be aged 65 or over in the year 2030. According to the American Association of Retired Persons (AARP):[24] 'The old themselves are also getting older. In 1985, the 65–74 age group (17 million) was nearly eight times larger than in 1900, but the 75–84 age group (8.8 million) was eleven times larger, and the 85+ age group (2.7 million) was twenty-two times larger.' Because of age related complications of growing old, the amount of drugs consumed by the elderly at the turn of the 21st century will be staggering. Vestal has suggested the elderly will consume 40% of all medications, up from a projected 25% in 1986.[25]

4.5.6.1 Subpopulation Concerns

Certain subpopulations of the elderly will require the use of tremendous amounts of health care services and pharmaceuticals. This is attributable to the resultant morbidity associated with the aging process. Chronic treatment for chronic diseases and multiple pathologies will challenge the health care practitioner of the future. If and when provision of services becomes problematic because of resource scarcity, then the allocation of resources will become a crucial concern. Issues raised earlier in this chapter regarding decisions pertaining to who will receive what care will require attention. Will different subpopulations be in competition for health care? At present, 40% of the Medicaid budget for the poor of all ages is consumed by the elderly.[26] Competition between generations for health care services will challenge the moral and ethical fiber of the US health care system and professional components.

Despite the common sense assumption that the elderly of the future will be sicker with more and more infirmities, professionals may be providing more and more preventive or wellness based care for the well elderly subpopulation. These patients will be normal in every sense of the word; the only remarkable fact about them may be their advanced age. How professionals deal with this subpopulation will undoubtedly affect the success of all professionals and the health care system.

4.5.6.2 Changing Demographic Patterns

Often Americans are referred to as a mobile population. This overall assessment is particularly applicable to a number of facets of society, including the health care delivery system. It is particularly important from service provision, resource utilization and health professional training points of view. It is also important when analyzing the 'graying' of the US health care population.

Consider that in 1985 about 49% of persons 65 years and older lived in eight states. California, New York and Florida had over two million each, and Illinois, Michigan, Ohio, Pennsylvania and Texas had over one million each.[27] Even though previously the elderly have been less likely to change residence than any other group, in 1985 about 800 000 persons 65 and over had moved to a different state since 1980. Of these, over 35% had moved from the northeast or midwest regions to the south or west.[27] If indeed the 'well old' grow in numbers in the years to come, they may be more willing to move than their contemporaries of previous times. American society, the health care system, *etc.* are not prepared to deal with these future projected continuations of patient mobility and geographical change from delivery, financing and structural outlooks.

4.5.7 REGULATORY ASPECTS

Health care decisions have always had political ramifications. But, with the portion of health care consumed by governmental payers exceeding 40% of total payments, the interjection of politics into health care decisions has increased and will continue to do so. An example of increased political activity has been the formation and prominence of political action committees (PACs). These entities make campaign contributions to individuals running for state or national elected offices with the hope that the elected official will remember favorably the contribution when votes are taken after election which affect the sponsoring PAC. The lobbying and financial contributory efforts of PACs cannot be underestimated. If financial input equates with legislative decision outputs, does the greater contributor win? What if competing contributors have drastically different policy perspectives? Will pharmacy, medicine, consumer or manufacturing PACs have equivalent expectations for support from legislators, regardless of level of input (state, federal)? Chances are that the concerns and views may be quite different.

And what of the issues? From a regulatory or policy perspective, any number of 'hot' topics will be debated, acted upon or possibly changed from the *status quo* in the years to come. One such area of potential change is the drug milieu.

4.5.7.1 The Drug Milieu

Aspects pertaining to all components of medication consumption will be malleable in the future. These aspects include the switching of drugs from prescription to over-the-counter (OTC) status, the approval time and the approval process for new drugs, the drug lag and orphan drugs. What follows is a brief listing of each of these items and discussion pertaining to their importance in the health environment.

4.5.7.2 Switching of Drug from Prescription to Over-the-counter classification

Certainly the switching of medications from prescription to over-the-counter (OTC) status has had an effect on pharmacists and patients.[28] Pharmacists decry on one hand the lack of information consumers have with regard to the nonprescription drugs consumed, and yet are faced with more and more prescription drugs being switched from prescription to OTC status. The crucial consideration pertaining to appropriateness of the switching may revolve around consumer ability to self-medicate with these drugs. Self-monitoring of body functions is necessitated with the use of many of these drugs. Certainly different points of view may be espoused by OTC drug manufacturers, consumer groups and health professionals. OTC manufacturers wish to have their products available for purchase in as many locations as possible. Consumer groups have sought to have drugs available for purchase at the lowest possible price. Pharmacists have desired to have a monopoly on the purchase locale for nonprescription drugs, even though this is out of the realm of possibility with consumers able to purchase OTC drugs in supermarkets, discount stores, grocery stores, *etc.*

4.5.7.3 Drug Approval

The US Food and Drug Administration (FDA) is placed under tremendous pressure concerning the drug approval process. On the one hand, FDA efforts may be seen as too stringent by a pharmaceutical company considering the length of the drug approval process and the resultant number of years of patent protection remaining on a product. Also, consumers and their physicians desperate for the release of a product may see the extended drug approval process as being too lengthy. Other groups may view postmarketing emergent side effects (not discovered in clinical trials) as an indication that the drug approval process needs further scrutiny, since drugs may be allowed on the market too easily. Any effort to speed the approval process along for drugs for one group of patients (*e.g.* AIDS patients) may be resented by other groups of patients also requiring a miraculous drug but who perceive a longer wait for approval of their needed drug.

4.5.7.4 The Drug Lag

If a particular drug is available overseas before it is available in the US, reference is made to the problem of a drug lag for US prescribers and patients. Some opponents of US drug approval policies argue US patients are being deprived of available treatments in other countries. Depending upon varying points of view, the line between prudent patience before approval of a drug and the notion of a drug lag is a fine one indeed. The drug lag also can refer to prescription to OTC switched drugs which are made available in other countries before being made available as OTC drugs in the US. These deliberations are all the more interesting when professional, political and economic impacts are factored in the scenario.

4.5.7.5 Orphan Drugs

Drugs of importance in treating rarely occurring exotic diseases are referred to as orphan drugs. Often the costs associated with producing such drugs are prohibitive. The potential to recapture investment dollars often is limited and as such treatable diseases may be left untreated. Recently, the orphan drug issue has been partially diffused, and the manufacturer's burden has been reduced.[29] However, with costs skyrocketing throughout the health care system, orphan drugs and diseases may increase as resources to cover many health needs become stretched.

4.5.8 PRESENT AND FUTURE CONSIDERATIONS

Current issues of note in the health environment will also be issues of note in the future. What the professions provide to patients and what patients do for themselves will ultimately converge and allow for determination of how good the health of the system is or will be. Each practitioner must decide how to fit into the system and ultimately help patients achieve optimum health in the health care system.

4.5.8.1 Self-care

What role will self-care play in the years to come? With many more drugs available for self-medication and the promotion of self-care activities, more and more individuals will utilize self-managed treatments. The bombardment of consumers with advertisements promoting self-selection of products and therapies will undoubtedly stimulate further marketing of more products. The advertising, marketing and labeling of OTC products is complex and in some cases confusing. In addition, and to further compound the situation, more and more products will be targeted for switching from prescription to OTC status. There are debatable issues surrounding the switching process. This debate will not lessen in intensity in the years to come. What individuals should be allowed to do to, or for, themselves is not an issue likely to be resolved in the present or short or long term future.

4.5.8.2 Prevention *Versus* Treatment

Where should money be spent in the health care system? Is money better spent in treating the outcomes and sequelae of chronic disease or in the prevention of the disease? Certainly money will always need to be spent for acute and chronic care. There can be no question of the importance of the treatment of acute and chronic disease. But, is it better to spend enormous amounts of money in treating the outcomes of sometimes preventable conditions, for example cardiovascular disease? Or is money better spent in educational or other preventive programs aimed at reducing cigarette smoking, hypercholesterolemia or obesity? Individuals have suggested preventive efforts would be more cost effective in the long run.[30] Even so, there will always be a demand for technology to undo what it is we do to ourselves.

4.5.8.3 Elaborate Technology *Versus* Basic Needs

The US health care system provides some of the most elaborate and technologically advanced health care available anywhere in the world. Whether this technology is too elaborate is a debatable issue. The elaborate technology serves only those who can afford it; if it is unattainable for many patients, how usable is it? Questions have been raised surrounding the spending of money on technology which, by its design, has limited applicability to a broad spectrum of society, as opposed to expenditures for basic needs.[1] To a patient without access to a physician, access to elaborate medical devices and interventions becomes a secondary consideration.

4.5.9 THE FUTURE

No one can predict with certainty what the future will bring. However, some trends should increase in the years to come. The trends toward high drug consumption at play in the 1980s should continue to increase in the future. This projected increase is based upon projected morbidity that is chronic in nature, coupled with a population growing older and consuming even more medications. The drugs of the future may or may not be of the shape and form of today. But, there can be certainty that more medications will continue to be consumed by more and more US patients.

The switching of drugs from prescription to OTC status will increase in the future. The subsequent counseling as to proper use of the medications will challenge future pharmacists from both a content and delivery standpoint. The potential for misuse of OTCs will increase parallel to the potential for increased self-use. Pharmacists can impact upon both the use and misuse of OTC drugs in a positive fashion. Pharmacists must become better managers of the drugs patients consume. Aggressive management of patients' therapies needs to be undertaken in the ambulatory environment, just as it has been accomplished in the institutional setting (hospital and long term care facilities).

The US health care system is in a continual state of flux, more and better health care is provided through state of the art medical and institutional technology. Those most posed for success, either patient or provider, are the individuals prepared to deal with the current cascade of change.

Proposals have been suggested to deal with the current and future health care crises.[31] There is no guarantee that these or any other proposals will be put in place. These vary from reorganizing the American health care system in a similar fashion to the Canadian health care system to establishing a system of brokered care. The brokering system would establish the purchase of prudently priced health care in place of currently purchased overpriced services from hospitals, nursing homes, physicians or pharmacists.

What system will emerge or evolve in the future will be different from the current system of US health care delivery. In addition, it will no doubt be uniquely American. More than likely, any future changes or implementations will be scrutinized and continuously debated with many individuals and affected groups possessing differing points of view pertaining to a system invariably subject to the impacts of change.

4.5.10 REFERENCES

1. P. F. Short, J. C. Cantor and A. C. Monheit, *Inquiry*, 1989, **25**, 504.
2. F. R. Curtiss, *Am. J. Hosp. Pharm.*, 1987, **44**, 1797.
3. P. J. Kenkel, *Mod. Healthcare*, 1988, **18**, 31.
4. S. Jonas, Health Care Delivery in the United States,' 3rd edn., Springer, New York, 1986.
5. J. C. Merrill, and S.A. Somers, *Inquiry*, 1986, **23**, 316.
6. V. R. Fuchs, 'How We Live', Harvard University Press, Cambridge, MA, 1983.
7. F. J. Hellinger, *Inquiry*, 1988, **25**, 469.
8. J. T. Tierney and W. J. Waters, *N. Engl. J. Med.*, 1983, **308**, 95.
9. R. A. Angorn, *Legal Aspects of Pharmacy Practice*, 1986, **9** (3), 1.
10. E. Ginzberg, *N. Engl. J. Med.*, 1984, **310**, 1162.
11. J. R. Hagness, *AACP Newsletter*, June 1987, 6.
12. G. L. McManis, *Healthcare Executive*, 1987, **2** (1), 60.
13. R. J. Zeckhauser and D.S. Shepard, In 'Economic Aspects of Health,' ed. V.R. Fuchs, The University of Chicago Press, Chicago, 1982.
14. C. J. Schramm (ed.), 'Health Care and Its Costs', Norton, New York, 1987.
15. A. C. Einthoven, *N. Engl. J. Med.*, 1978, **298**, 1229.
16. *Prospective Payment System-Hospitals*, Social Security Amendments of 1983, Prospective Payments for Medicare Inpatient Hospital Services, Public Law 98-21, April 20, 1983.
17. P. E. Dans, J. P. Weiner and S. E. Otter, *N. Engl. J. Med.*, 1985, **313**, 1131.
18. S. Jonas, 'Health Care Delivery in the United States', 3rd edn., Springer, New York, 1986, p. 5.
19. R. R. Bovbjerg, *Inquiry*, 1986, **23**, 403.
20. H. J. Aaron and W. B. Schwartz, 'The Painful Prescription', The Brookings Institution, Washington, DC, 1984.
21. M. A. Fruen and J. R. Cantwell, *Inquiry*, 1982, **19**, 44.
22. B. Keith, *Drug Top.*, 1987, **131** (12), 82.
23. E. J. McTernan and A. M. Leiken, *J. Health Politics, Policy Law*, 1982, **6** (4), 739.
24. AARP, 'A Profile of Older Americans, 1986,' AARP, Washington, DC, 1987.
25. R. Vestal, 'Drugs in the Elderly', Adis Press, Boston, 1985.
26. J. Avorn, *Daedalus*, 1986, **115** (1), 211.
27. P. G. Clark, *Gerontologist*, 1985, **25** (2), 119.
28. The Proprietary Association, 'Rx-OTC: New Resources in Self-medication,' The Proprietary Association, Washington, DC, 1982.
29. T. H. Althuis, 'Contributions of the Pharmaceutical Industry', in 'Orphan Drugs', ed. F. E. Karch, Dekker, New York, 1982, p. 181.
30. A. V. Chobanian, *Am. J. Med.*, 1984, **77** (2B), 22.
31. R. Ruthen, *Sci. Am.*, 1989, **260** (3), 18.

4.6
Good Pharmaceutical Manufacturing Practice

JOHN SHARP
formerly of Waverley Pharmaceutical Ltd, Runcorn, UK

4.6.1 INTRODUCTION

'The quality of a product ultimately depends on the quality of those producing it.' (Sir Derrick Dunlop)[1]

Much time, effort and money is expended every year on the research and development of new, and in the investigation of existing, medicinal products. A lot of this effort is in the field of medicinal chemistry, although other disciplines (pharmacy, pharmacology, biophysics, toxicology, clinical medicine, engineering and so on) are, of course, involved.

The annual expenditure on research and development by the UK-based pharmaceutical industry alone (1984 figures) is of the order of £500 million.[2] (For comparison, the total national healthcare expenditure by both the public and private sectors, in the same year, was approximately £19 billion or an average of £335 per head of the population.)[3] All this energy and expense is wasted if the ultimate objective is not achieved. That objective is to deliver to the patient, prescription after prescription (throughout all of the nearly one million prescriptions which are dispensed in Britain daily), efficacious medicines, consistently manufactured, batch after routine batch, which conform to the appropriate standards of quality. All the costly research and development, all the medicinal chemistry, is unavailing if patients receive medicines which are incorrectly manufactured, contain wrong or poor quality materials, have deteriorated, become adulterated or have been incorrectly packaged or labelled. The purpose of good manufacturing practice (GMP) is to see these things do not happen, and since *bad* manufacturing practices can thwart all the, perhaps more sophisticated, effort that has gone before, it might be argued, not without justice, that GMP is the most important of all the many topics covered in this publication. It is also most appropriate that this chapter should be included under 'Socio-economic Factors . . .' as there are significant interactions with the economics of production (but *not* necessarily as an 'on cost', see later), since '. . . the object of GMP . . . . . is . . . . . ultimately the safety, well being and protection of the patient'[4] (*i.e.* of *people*), and since the people involved are the major influencing factor on the quality of manufactured medicines. Sharp[5] has stated many times that 'In the manufacture and quality control of medicines, the science and technology is the *easy* part. It is the "people" and communication problems that are difficult'.

In the paper from which the quotation at the head of this chapter was taken, Dunlop[1] went on to add 'The great majority of errors which occur in production are human errors on the shop floor of omission or commission caused often by carelessness or boredom'. With that one would almost entirely agree, regretting only the stress on error due to 'carelessness or boredom'. Many errors are, in fact, due to lack of complete understanding, or to well-meaning but misplaced enthusiasm leading to failure to follow established procedures, rather than to carelessness or boredom.

Information on the basic components of good manufacturing practice can be obtained from one or other of the various 'official', national and international, GMP codes, guides and regulations which are currently available. Amongst those most frequently consulted and referenced are the USA's 'Current Good Manufacturing Practice Regulations',[6] and the UK 'Guide to Good Pharmaceutical Manufacturing Practice',[7] the so-called 'Orange Guide'. At present there are some 20 national publications of this type, plus two with at least some degree of international status (WHO and the Pharmaceutical Inspection Convention). In addition, a European Economic Community GMP Guide appears to be drawing towards the end of a long gestation period. A most valuable compilation, which reprints all the available texts, with comparative cross-referencing, is 'International Drug GMP's'.[8] In the introduction to this latter compendium, the editor comments ' . . . . the United Kingdom . . . has played a pivotal role in the development of current trends in GMP philosophy . . . The style and scope of most national GMP Guidelines strongly reflect this British influence'.

Whilst the degree of detail in these various official publications does vary, there is a fair degree of agreement over the main topics covered. A brief summary of these topics, the major components of GMP, is as follows: (i) introductory matter, *e.g.* statement of purpose or intent, legal (or other) status and GMP/QA 'philosophy'; (ii) definition of terms used (or glossary); (iii) personnel and training; (iv) documentation (including records); (v) premises and equipment; (vi) control of starting materials and packaging materials; (vii) manufacturing operations; (viii) validation; (ix) recovered materials; (x) complaints procedure and product recall; (xi) good control laboratory practice; and (xii) operations carried out under contract, *e.g.* manufacture, analysis and servicing/maintenance. In addition, guidelines are usually required for various major categories of product. In view of the significance and hazards of the products, and of the specialized techniques involved, a *separate* section is often devoted to: (xiii) manufacture and control of sterile products. In addition, there may be sections on: (xiv) dry products and materials, *e.g.* tablets, capsules, powders and granules, *etc.*; (xv) liquids, creams and ointments, although these two latter categories, *i.e.* (xiv) and (xv), may be covered in the 'manufacturing operations' section. In addition, some include such additional sections as: medical gases, radiopharmaceuticals, veterinary medicines, computer systems and wholesaling.

4.6.2 THE NATURE AND PURPOSE OF GMP

4.6.2.1 Definition

The British 'Guide to Good Pharmaceutical Manufacturing Practice'[7] is the only publication of its kind which offers a formal definition of GMP, and it is worth quoting here: 'Good Manufacturing Practice is that part of Quality Assurance aimed at ensuring that products are consistently manufactured to a quality appropriate to their intended use. It is thus concerned with both Manufacturing and Quality Control procedures'.

For a full understanding, this requires a definition of 'Quality Assurance', which in the same publication is defined as follows: 'Quality Assurance is the sum total of the organised arrangements made with the object of ensuring that products will be of the quality required by their intended use. It is Good Manufacturing Practice *plus* factors outside the scope of this Guide (*e.g.* original product concept, design and development)'.

Quality control is further defined as: 'that part of Good Manufacturing Practice which is concerned with sampling, specification and testing, and with the organisation, documentation and release procedures which ensure that the necessary and relevant tests are, in fact, carried out and that materials are not released for use, nor products released for sale or supply, until their quality has been judged to be satisfactory'.

Thus, by these definitions quality assurance is the all-embracing concept, and includes factors outside day to day manufacturing considerations. It is, for example, necessary in the first instance to develop a product which it is *possible* to make consistently to appropriate quality standards. The primary objective of GMP is to render that possibility a *certainty*. (That such a 'certainty' cannot be absolutely achieved in practice is a philosophical idea which need not detain us. *Certainty* should be the *objective*.)

Although a review of the various 'official' statements on GMP, worldwide, will reveal a rather close agreement as to the *sort* of thing that good manufacturing practice is, there is considerable disagreement on the relationship between GMP, quality assurance and quality control. Furthermore, whilst the concept of 'quality' inevitably looms large in any discussion of GMP, there is a wide and confusing range of views on just what is meant by the word. There is also, perhaps, the additional complication of the specialized meaning of the term 'quality control' as applied to industrial production generally. This is exemplified by many of the published texts on quality control, which rarely, if at all, consider the special problems of medicines manufacture, and are concerned almost entirely with distribution theory, and the statistics of probability. Even texts ostensibly dealing with the *practical* application of mathematical quality control theory seem somewhat remote from the specialized world of medicines manufacture. For example, Caplan's 'A Practical Approach to Quality Control',[9] which in its introduction is offered as 'A thoroughly practical introduction to quality control which in particular . . . . . aims to show how quality control can be applied to *any* practical situation . . .', gives the object of quality control as 'The object of quality control is to produce a quality that:

1 Satisfies the customer
2 Is as cheap as possible
3 Can be achieved in time to meet delivery requirements'.

Doubtless, these are laudable objectives for almost any and every manufacturing industry. But medicines are different. As will be discussed later, even the question of satisfying the 'customer' is problematic, and whilst it would be hopelessly idealistic to deny that the pharmaceutical industry is in business to make a profit, one cannot quite so readily accept that the quality of manufactured medicines may be modified by considerations of cheapness and meeting of delivery dates. In his 'Quality Control Handbook', Juran[10] states that 'quality control' can have a variety of meanings. He gives examples of various limited interpretations of the term (*e.g.* mere inspection, or simply the collection and analysis of statistical data) and goes on to develop a definition as 'the totality of activities which must be carried out to achieve the quality objectives of the company'. For our purposes, such a definition is acceptable only in so far as it is recognized that the 'quality objectives' of a pharmaceutical manufacturer are (or ought to be) rather different from those of other industries, where economic and other considerations may reasonably be permitted to modify 'quality objectives'.

Similar points have been made elsewhere; for example, in reference to sampling tables, Jacobsen[11] stated '. . . . we know that they are made for other purposes, where a certain degree of error is easier to accept than for pharmaceutical products', and Setnikar[12] argued that' . . . statistical criteria are rarely applicable in pharmaceutical technology' and that '. . . the consumer of drugs wants a "zero defect" quality which is incompatible with the very theory of control based on sample inspection'. Now, it is possible to question the claim that statistical criteria are 'rarely' applicable, since clearly they do have a number of valuable applications in medicines manufacture. It is also entirely reasonable to argue that 'zero defect quality' is unattainable, although it should be noted that Setnikar does not suggest that it *is* attainable, only that it is what the consumer quite understandably *wants*.

More examples could be given, but sufficient has been said to indicate that: (i) in the manufacture of medicines something more is required than that which is indicated in 'standard' treatments of 'quality control'; and (ii) there should be, and indeed is, an awareness within the industry that this is so (for a more detailed discussion see Sharp[13]).

4.6.2.2 Quality

A point on which all publications on GMP are in complete accord is that it is concerned with the pursuit and attainment of 'quality'. But what is 'quality?' It is a word with a variety of meanings, some of which shade almost imperceptibly into one another. Amongst the various, and at times confused, meanings are: (i) the essential feature(s) which make something, concrete or abstract, what it is ('The quality of mercy . . . . '); (ii) peculiar excellence ('A. Butcher & Sons: Purveyors of Quality Meats for over a Century'); (iii) a measure of the conformity of a product or material with its specification (this, in whole or in part, is the definition adopted in texts on statistical quality control); and (iv) the totality of features and characteristics of a product or service that bear upon its ability to satisfy a given need.

The latter definition, which is offered by the Institute of Quality Assurance, is the one most closely applicable in the context of the manufacture of pharmaceuticals, if the 'given need' is considered to be the need to deliver to a patient the correct dosage of the right medicinal product, properly packaged and labelled, and without adulteration, contamination or deterioration. Clearly, some sense of *totality* of properties is required in relation to the quality of medicines. Equally clearly, quality as solely a measure of conformity with a specification may well be a valuable concept in some industries, but it is entirely inappropriate to the manufacture of medicines. It is perfectly possible for a medicinal product to comply absolutely with a well-founded specification and yet still fail to have the desired effect, and/or be lethal. The UK 'Guide'[7] in its introduction states, *à propos* 'quality' that 'In this guide the word is used in the sense both of the essential nature of a thing and also of the totality of its attributes and properties which render it fit for its intended purpose', and further adds that 'assurance of that required quality cannot be achieved by the testing of end-product samples alone'. It is the very special nature of medicines, as compared with most other commercial products, which make this broader view of quality and its assurance both appropriate and necessary.

4.6.2.3 Why GMP? The Special Nature of Medicines

Medicinal products generally have a great potential for good or ill. The right medicine correctly manufactured under controlled conditions can confer great benefit on a sick patient. A wrong, incorrectly manufactured or contaminated medicine can cause injury or death or fail to cure. Yet no other product is consumed so totally on trust. Indeed in some circumstances, the ultimate consumer may be totally unaware of what he is consuming, or even that he is consuming anything at all. The purchaser of a motor car is aware that he wants such a product. He knows what a car should be like, and has ways of deciding whether or not the particular model or specimen he has in mind has the attributes he requires or can afford. He can read the manufacturer's literature, and independent reports. He can hire an engineer to inspect and report. He can examine the car thoroughly himself, and he can take a test-drive. He is in a position to decide whether or not to be a consumer on the basis of a substantial quantity of data. The purchaser of food can still reject the actual or potential purchase on grounds of appearance, odour, texture or taste. The consumer of medicines is, however, 'taking his medicine' in the vast majority of cases entirely on trust. The consumer, for example, of an antiinflammatory tablet will usually be able to decide whether or not he has in fact got a tablet but cannot know whether or not it contains an antiinflammatory drug, whether it is the right antiinflammatory drug, whether it is the correct dosage quantity or whether or not there are any contaminants or degradation products present. Some 'consumers' may be unconscious when medicines are administered to them. The ultimate consumer, the patient, is very largely not in a position to recognize that a medicine is incorrect or defective. Nor, normally, is a penultimate 'consumer' (*e.g.* prescribing physician, administering nurse or dispensing pharmacist) in any better position. The patient is but one end of a chain of implicit trust which extends through the administration, dispensing, prescribing and distribution, back to those responsible for the manufacture of the medicine. The social and moral responsibility which this chain of trust imposes, quite apart from any economic and legal considerations (although it is obviously *not* good business to damage your customers or to be closed down by regulatory authorities) is one of the main reasons for GMP.

The other main reason for GMP is the problem of testing medicinal products, and of the potential hazards of a relatively small proportion of defectives. Most testing of medicinal products is performed only upon samples. This is inevitable, since the majority of the tests are destructive. One hundred per cent testing for all specified attributes might well be a worthy, instructive and even entertaining ideal to pursue. It is scarcely likely to prove economic.

Although the use of sampling plans and a consideration of the relevant statistics may well enable relatively reliable inferences to be made about the *probable* quality of the batch as a whole, the results of sample testing (especially on any scale that is economically viable) cannot provide totally reliable information on the *actual* quality of the unsampled portion of the batch. Whilst a small proportion of defectives may cause no more than irritation, anger or loss of sales in other industries, the consequences in medicines manufacture can be dire indeed. The problem is perhaps less acute when a production batch may reasonably be considered to be homogeneous (*e.g.* a bulk liquid). It is more acute when the product is in the form of discontinuous units (*e.g.* tablets, capsules) and it is most acute in the testing for that most critical of qualities, the sterility of a parenteral product, where the standard tests for sterility provide little in the way of reliable information on the batch as a whole.

A closely related problem is that testing is of necessity limited. No specification for a medicinal product, or the materials used in it, could possibly include a test for the absence of all possible and potentially hazardous contaminants. Further it is not usual, in testing, to 'account' for the constituents of the product 100%. Those portions of the product not accounted for by assay are normally only *assumed* to be excipient, diluent *etc.*

These then are the reasons for GMP; the social and moral responsibilities of the 'chain of trust' from manufacturer to patient (and the legal and economic implications of that 'chain of trust'), and the inevitable limitations of end-product sampling and testing.

4.6.2.4 The Growth of the GMP Concept

The French National Pharmacopoeia Commission's original publication on Good Manufacturing Practice (1978) contains in its preface a brief passage which summarizes, with elegant precision, the whole broad pattern of the historical development, in three main phases, of attitudes towards the maintenance of the quality of manufactured medicines. This paragraph describes the transition from early attitudes which perforce placed almost entire emphasis on following a formula and a preparative method, towards a later view (as test methodology and equipment became available) that material and end-product testing was all important, and thence to the modern view that true assurance of quality can only be achieved when such testing is integrated in the more general framework of a detailed understanding of the conditions of manufacture.

The earliest example, known to the author, of a medicinal product, produced on an industrial scale, and of a standardized potency confirmed by assay, dates from 1879 (it was a purified liquid extract of ergot, marketed by Parke-Davis & Co.). This may be said to have marked the beginning of the trend towards regarding end-product testing as the ultimate criterion of product quality; an attitude which pervaded generally for the next 50 years or more, and was still to be found in some quarters into the 1970s and which (for all one knows) may still exist in some farthest and least enlightened reaches of the industrial pharmaceutical world. However, by the 1940s, and certainly by the 1950s, attitudes were changing towards the present day view of quality assurance and GMP.

The earliest *official* publication which is clearly about GMP, although it does not in fact use the term, is the Canadian document on 'Manufacture Control and Distribution of Drugs'[14] published in 1957. It is also perhaps worthy of note that as far back as 1925, the UK Therapeutic Substances Act[15] displayed a recognition that, for at least some types of product, approaches other than end-product testing are necessary. However, perhaps the most interesting of early statements on GMP is a paper written in 1947 by Taylor.[16] The paper is entitled 'Quality Control', although Taylor does not employ the term in the text, referring only to 'control', 'control procedures', 'manufacturing control', 'packaging control' and so on. Nor does Taylor mention GMP or QA, although from a present day viewpoint, this is clearly what the paper is referring to, for in addition to discussing sampling and test methods, Taylor writes: 'Control procedures must encompass all things that may influence the quality of the completed medicinal preparation; they must permit inquiry into every phase of purchasing, manufacturing, packaging, storage and labelling . . .', and also discusses, amongst other things, master manufacturing formulas and methods, and batch manufacturing records.

The phrase 'Good Manufacturing Practice' first appeared in official print in the 1962 amendment to the US Food, Drugs and Cosmetic Act.[17] Twentyfive years on, the initial letters have come to be used and understood, wholly or partially, almost worldwide.

4.6.3 PERSONNEL

4.6.3.1 General

References to the 'four M's' (men, materials, machinery and methods) or to the 'four P's' (personnel, premises, plant and procedures) as the essential elements in any quality-orientated industrial enterprise are commonplace, and such generalizations do at least serve to focus discussion on basic requirements. There can be no doubt that of these elements, it is the people (the men or the personnel) that are the most important, in the assurance of quality. This is true of all levels within an organization from company president or managing director to the most junior employees. It may well be possible (if not altogether desirable) for high quality, dedicated personnel to compensate for some lack or deficiency in the other elements. Nothing, not even the finest premises, equipment, materials or procedures can compensate for the quality hazard represented by low standard, ill-trained or badly motivated staff.

4.6.3.2 Management and Organization

Many and various statements have been made about the object, or purpose of any commercial enterprise. The essential feature which may be extracted from such statements is, generally, that a business exists to deliver goods or services for which a demand exists (or for which a demand can be created) at a profit.[18] Although medicines may in some circumstances be manufactured without any aim or desire to yield a profit (*e.g.* in some hospitals), the great majority of medicines are produced, worldwide, in organizations where the profit motive is the driving force. To some this represents an ethical dilemma. Others see no special reason for considering making a profit from the manufacture of medicines any differently from making a profit from the manufacture of other basic human needs such as food, shelter or clothing. Discussion of this interesting point is something which is outside the scope of this chapter, as indeed is the even more thought-provoking question of the possibility of *creating* a demand for medicines (see above). Suffice it to say that the vast majority of medicines *are* manufactured in situations where profit is a prime mover.

It is possible to see this as inimical, as representing an opposing force, to the special dedication to quality which is required in the production of medicines. However, if ethical considerations do not provide a sufficiently powerful impetus, it would still be a foolhardy manufacturer who did not allocate sufficient resources to the prevention of the production of poor quality or defective medicines. The 'Guide to Good Pharmaceutical Manufacturing Practice'[7] states ' . . . . assurance that products will be of a quality appropriate to their intended use . . . requires the involvement and commitment of all concerned, at all stages'. This spirit of involvement, this commitment,* needs to begin at top management level and thence diffuse throughout the organization as a whole. In establishing an organizational structure for the manufacture and quality assurance of medicines, the most generally accepted view is that there should be two separate persons, each with overall responsibility for production *or* for quality control, neither of whom is responsible to the other. This organizational concept is explicitly stated, or is implicit, in a number of 'official' GMP publications. In some European countries the concept is overlaid by a statutory requirement that there shall be a single person designated as a 'responsible person' or some such similar title.

Aside from any statutory requirements, the separation of quality control from production has on occasions been challenged on grounds that it removes from production personnel the healthy sense of responsibility for product quality that they so very rightly should have. This is very largely to miss the point. The more generally accepted view is that it is sound sense that the person ultimately

* 'Think of ham and eggs. The chicken is involved. The pig is committed'. Attr. Ms Martina Navratilova, *c.* 1982.

responsible for quality control should be freed from the need to consider, or be influenced by, questions of *quantity* of production, meeting production schedules and sales estimates *etc.*, all of which are quite properly the province of a production manager. The quality control manager should thus be able to make decisions regarding quality standards and procedures and to approve or reject materials or products entirely uninfluenced or biased by such pressures. This is by no means to say that the production manager or the staff associated with production do not have a responsibility for implementing quality policies and procedures and for quality assurance. Objections to the managerial separation of production and quality control very largely disappear if it is understood that attainment of the required quality is everyone's responsibility, even though, within that general context, specific responsibilities may be assigned more especially to the quality control manager, or to the production manager. It is essential that these various responsibilities are defined and understood and this point is made in the UK Guide,[7] thus:

'The way in which the various key responsibilities which can influence product quality are distributed may vary with different manufacturers. These responsibilities should be clearly defined and allocated.

'The person responsible for Quality Control should have the authority to establish, verify and implement all quality control procedures. He should have the authority, independent of production, to approve materials and products, and to reject as he sees fit starting materials, packaging materials and intermediate, bulk and finished products which do not comply with the relevant specification, or which were not manufactured in accordance with the approved methods and under the prescribed conditions. (His authority in relation to packaging materials may be limited to those which may influence product quality and identity.)

'The Production Manager, in addition to his responsibilities for production areas, equipment, operations and records; for the management of production personnel, and for the manufacture of products in accordance with the appropriate Master Formula and Method, will have other responsibilities bearing on quality which he should share, or exercise jointly, with the person responsible for Quality Control.

'These shared or joint responsibilities may include monitoring and control of the manufacturing environment, plant hygiene, process capability studies, training of personnel, approval of suppliers of materials and of contract acceptors, protection of products and materials against spoilage and deterioration and retention of records. It is important that both direct and shared responsibilities are understood by those concerned.'

The way in which different medicine-manufacturing companies are structured can (and do) vary considerably on points of detail. It is essential that all concerned are made fully aware of both their functional and reporting responsibilities, by means of organization charts and written job descriptions.

4.6.3.3 Recruitment

It is clear that senior supervisory or managerial staff in production and quality control must have the education, qualifications and experience appropriate to the jobs they perform. In many countries this is mandatory, and not just a 'guideline' recommendation. A vital point which can be overlooked is that they should also have the *ability* to do the job. Specifically, they should have the ability to *manage*. In assessing a prospective senior employee's admirable 'paper' qualifications and their proven technical ability, it is perhaps a little too easy to forget that a major, perhaps most vital, part of the prospective manager's role will be to manage; to lead, direct and motivate the more junior staff. Management should bear in mind that whilst the possession of outstanding technical ability and exemplary professional qualifications *by no means* excludes the ability to manage, it does not guarantee that ability.

In all but the very smallest organizations, junior unqualified staff (production operators and maintenance men, cleaners, stores and service personnel) form the vast bulk of the workforce. It is these people and the way they are trained, directed and motivated that form perhaps *the* key element in the assurance of product quality. By far the largest proportion of reported defective medicinal products are the result of simple human error or human misunderstanding, and not of failure at a high level of technology. Advanced qualifications are not normally required at, for example, basic operator level. Nevertheless such personnel must, at minimum, have the education and intelligence to read and fully understand written instructions, and to carry them out. They must also be able to respond effectively to the very special challenges of medicines manufacture, and to understand the nature and purposes of good manufacturing practices. They must also have innately good standards of personal hygiene. Management needs therefore to be in a position to exercise some degree of selectivity in the recruitment of such staff. It follows that it should tailor its policy regarding rewards (monetary and otherwise), conditions of employment and prospects of advancement accordingly. It should also aim at creating a working environment in which staff turnover is reduced to a minimum, since it is difficult, if not impossible, to maintain a well-trained and motivated workforce where there is rapid staff turnover. The pharmaceutical industry is not really a job for those who are 'just passing through'!

4.6.3.4 Training

The necessity for training arises whenever there is any deficiency in the knowledge, understanding, attitudes and specific skills possessed by a person as compared with the knowledge, understanding, attitudes and specific skills that are required for the successful performance of any task assigned to that person. The need for training can, and does, arise at all levels in an organization from senior management to most junior employees. A new employee may have already acquired the basic skills required from previous employment, but still needs to learn how to exercise those skills in the new working environment, and to satisfy the new employers in this aspect. The new employee still needs to be made familiar with the new environment, company background, traditions, attitudes and policies, and, crucially, needs training in GMP, or to have any previous GMP training reinforced.

In no other industry can the need for sound training be more obviously apparent. The concept of 'sound training' implies a formal, systematic approach. Merely 'sitting with Nellie', on the grounds that this constitutes training 'on-the-job' is *not* sufficient.

Training needs to be directed at three groups of employees: (i) new employees; (ii) existing employees—when the nature or content of their job changes; and (iii) existing employees whose performance at a particular task declines below required standards.

Basic training should also be reinforced from time to time by ongoing training programmes designed to ensure that employee performance and attitudes remain up to standard. The fields in which training needs to be given may be considered as forming three basic training elements: (i) introductory, background (orientation or induction) training (for new employees); (ii) GMP training (including training in hygienic practices); and (iii) specific skills training. None of these elements should be neglected.

4.6.3.5 Motivation

'Self-responsible, motivated, activity is more efficient than commanded activity'.[19] Other things being equal, well-motivated staff will produce more goods which have a greater assurance of quality. Conversely, in the special context of medicines manufacture, ill-motivated staff can represent a hazard both to themselves, to the public and to the company profits.

Motivation and engendering of high morale in pharmaceutical workers should be a relatively easy task. Indeed it is difficult to understand why in any pharmaceutical manufacturing operation there should be ill-motivated workers. Work of any type, far from being a curse and a punishment to men for their sins is itself a motivator. By and large people *want* to work. Work defines a person's status. It places them in society and provides their major source of social interaction. A man *is* very much what he *does*. We all tend to ask of a new acquaintance 'What do you do?'

It has been argued that pay and conditions are less important, even insignificant, factors.[20] It is, however, difficult to see how a sense of being paid less than one's worth (or less than others who are doing comparable jobs), or how a feeling that the working environment and facilities are of a standard lower than the job requires, could not be demotivating factors. In any event, whilst remuneration is a matter of company policy, legislation and basic GMP requirements should guarantee that generally (apart from a few 'dirty jobs') the working environment will be more congenial than many. Even 'dirty jobs' are acceptable when it is realized that the 'dirt' is an inevitable part of the job (coal miners do not usually complain about coal!). It is when the working environment is seen to be more dirty or unpleasant than it need be, that it becomes demotivating. (A source of low morale which can arise in the modern, clean, air-conditioned, windowless, immaculately finished pharmaceutical factory, is the sense of isolation and lack of contact with others and with the outside world which can arise in such circumstances.)

In addition to the basic *need* to work, and the social satisfactions gained from working, the other main motivating factors may be summarized as sense of *purpose*, sense of *pride* and sense of *belonging*. The stimulation of such senses in the worker should be particularly easy in the pharmaceutical industry. Medicines serve a recognized, significant social purpose. Workers will easily understand this, and will readily take a pride in that purpose. From induction training onwards, they should be encouraged to see that they have a role to play, however marginal, in achieving the socially useful purpose of supplying medicines to cure or prevent illness. In the same way they can be made to feel that they *belong* to a team, the aim of which is to achieve that purpose. The most important word is 'communication', not just of facts but also of ideals, attitudes and objectives. Where morale and motivation are low, the blame must be laid squarely at the door of management.

4.6.3.6 Hygiene and Health

Most official guidelines or regulations on GMP stress the importance of personal hygiene. High standards of personal hygiene are clearly necessary for all involved in the manufacture of pharmaceuticals. This necessity is most crucial where the product is exposed, particularly when that product is intended to be sterile. Nevertheless, high standards should be demanded and achieved at all stages, and with all types of products. The primary reasons are to control contamination of product, materials or environment by that vigorous dispenser of microorganisms, the working human being, and also to prevent cross-contamination *via* workers' hands *etc.* There is a very good secondary reason. Persons with high regard for matters of hygiene will the more readily be able to adopt that special attitude of care and attention that the manufacture of medicines requires.

Microorganisms can abound on body surfaces, and in the nose, throat, mouth and intestines. They may be transferred by shedding from body surfaces, generally in association with inanimate particles (*e.g.* skin flakes), *via* sneezing and coughing or by direct contact with contaminated hands. The total number of microorganisms on the skin varies from person to person (and in accordance with their personal hygiene practices). It also varies in different parts of the skin surface. It can vary from less than one hundred organisms to several millions cm^{-2} of skin surface. The largest concentration of organisms are generally to be found on the head and neck, armpits, hands, feet (and beard, if worn). Saliva can contain up to 100 million organisms mL^{-1}, and nasal secretions up to 10 million. The number of coliform bacteria alone per gram of human faeces can be in the order of 100 million.

The human body continuously sheds inanimate particles, largely consisting of skin fragments. Dependent on skin type and level of activity (the more vigorous the activity, the greater the shedding) it has been stated[21] that the rate of shedding is of the order of 5–15 g of particles every 24 h. Skin microorganisms are most frequently shed in association with the skin particles. The extent and the hazard of microorganism dispersal increases where there is infection, especially of skin, the respiratory system or the alimentary canal. Steps need to be taken to control these hazards to product and environment, through the observation of hygienic practices by staff, and through the provision of suitable protective factory wear. Let it not be thought that the need to guard against bacterial contamination applies only to the manufacture of products intended to be sterile, or to certain creams, emulsions, suspensions, syrups *etc.*, which might 'grow'. There is evidence for example, of the adverse clinical significance of microbiologically contaminated tablet products.[22]

The first step towards ensuring hygienic practices amongst personnel is, of course, to recruit the right sort of people in the first place, people who will already observe high standards of personal hygiene. It then becomes a matter of providing training which emphasizes the especial risks and requirements of pharmaceutical manufacturing, and of providing the necessary facilities.

There should be a general medical examination prior employment, the extent of which may vary according to the nature of the work to be performed by the new employee. No person with a communicable disease, or with open lesions on exposed body surfaces should engage in the manufacture of medicinal products, most certainly not where the product may be exposed. Further, staff should be instructed to report any such conditions and supervisors to look out for them. Steps should be taken to encourage such reporting, and no person should suffer any loss (*e.g.* of remuneration) for so doing.

The 'Hygiene Recommendations' of the Association of Swedish Pharmaceutical Industry[21] contain a table giving suggestions for initial and follow-up medical checks for staff employed in different manufacturing environments. A full programme is given, based on three 'cleanliness' levels as follows: (i) environment where only closed units are handled; (ii) environment where open products are handled; and (iii) environment where aseptic work is carried out.

The implication that nonaseptic sterile production requires only the same standards of environmental cleanliness as, say, tablet production is perhaps questionable by the standards of more recent thinking. Nevertheless, this whole publication will repay careful reading. Points worth adding are:

(i) If high standards of cleanliness, hygiene and the wearing of protective clothing are to be observed and maintained by operators, it is necessary for supervisors and managers to set the appropriate good example.

(ii) If hand-washing facilities are to be used, not only must they be available, they must be *conveniently* available.

(iii) From the GMP angle, the protection of the operators and their 'normal' clothing is only a secondary consideration. The primary purpose of protective clothing is the protection of the product and thus the patient.

(iv) Protective garments are of no value if they are damaged, dirty or permitted to become themselves vehicles of contamination or cross-contamination. Suitable changing rooms should be provided, and the protective garments should not be worn outside the controlled factory environment or in any area where they could collect or distribute potential contaminants.

(v) Medical checks should include sight testing (including checks for colour blindness). This is something which is often neglected and it can assume great significance in jobs where visual acuity and/or distinction of colours is important.

4.6.4 BUILDINGS AND EQUIPMENT

4.6.4.1 Principles

The UK 'Guide to Good Pharmaceutical Manufacturing Practice'[7] states: 'Buildings should be located, designed, constructed, adapted and maintained to suit the operations carried out in them. Equipment should be designed, constructed, adapted, located and maintained to suit the processes and products for which it is used. Building construction and equipment layout, should ensure protection of the product from contamination, permit efficient cleaning and avoid the accumulation of dust and dirt'.

Similar statements appear in other GMP Guides, Codes and Regulations.

4.6.4.2 Basic Considerations

A pharmaceutical manufacturing building with equipment installed could be viewed as a 'black box' into which are fed: (i) raw and packaging materials (and/or part processed products); (ii) people ready to work; and (iii) services (air, heat, light, power, water *etc. plus* any additional support systems for (ii) above) and from which will emerge: (i) finished products (and/or part processed products); (ii) people leaving after their days work; and (iii) waste, scrap, rubbish, trash and effluent.

A primary consideration, therefore, for the siting and building of a pharmaceutical plant is that it must be possible (and, preferably, *conveniently* possible) to feed materials, people and services *to* the site, to distribute products issuing from it and to dispose of waste effluent. (People having been persuaded to come and work in this 'black box' will presumably find no greater difficulty in persuading themselves to leave it.)

Within the 'black box', there will be various flow patterns, principally of materials and products, and of personnel. Materials will be received and held pending tests, released for use, held in store, dispensed for manufacture and processed into products which are then packaged, tested and held in store pending distribution.

Working along with material/product flow patterns, and indeed allowing or causing them to happen are personnel flows, as people arrive for work, change into suitable protective clothing, carry out work, take breaks, change back to outdoor clothes and leave for home.

In addition to the material/product and personnel flows there will be flows of air of differing qualities (plain, conditioned and filtered), the flow of various services through pipework, ducting and conduit, and the disposal flow of waste, defective or contaminated material and of rubbish, sewerage and effluent.

Thus, the basic considerations for the design, structure and layout of a pharmaceutical manufacturing facility may be summarized as: (i) siting—including: (a) availability of staff, (b) availability of materials, (c) availability of services, (d) availability of waste disposal systems and (e) potential for distribution of products; (ii) size, scale and complexity of operation; (iii) material flows (and their interactions) including space for storage when material is *not* flowing; (iv) personnel flows; and (v) internal provision and distribution of services and waste disposal systems (including provision to personnel of such services as changing rooms, clean protective garments, refreshments and toilet facilities). Other strongly influencing factors will be the company's marketing strategies, and its inventory and physical distribution policies.

4.6.4.3 Siting

Sites for factories have been selected for many different, and not always the best, reasons. That the executive vice president likes the look of the local golf course ought *not* to be a major influence! Factors truly relevant are:

(i) Is the site suitable for the erection of a building of the size, shape and height proposed? Will the existing terrain allow the insertion of foundations which will support such a structure?
(ii) Is the site of sufficient area to accommodate not only the building, but also access roads, parking areas, hard standing for delivery and despatch vehicles and, one hopes, a certain amount of pleasing external landscaping and planting?
(iii) Do local regulations permit a building of the size, type and shape proposed?
(iv) What are the risks of water damage, flooding, pollution, pest/vermin infestation, or contamination and/or objectionable odours from other nearby activities? What control will the manufacturer have (*e.g. via* local regulatory authority) over any possible future development of such activities?
(v) Will it be possible to attract suitable staff?
(vi) Will local and personal transport allow convenient staff access to the site?
(vii) The convenience and economics of getting materials to the site, and distributing products from it.
(viii) The logistic and geographical relationships between the site and any other company-owned facilities, its subsidiaries, warehousing and distribution agents, wholesale and retail outlets.
(ix) Availability of services, water, power, electricity, waste and effluent disposal.
(x) Potential for future expansion.

4.6.4.4 The Structure

There are many ways to build factories. The most widely adopted current approach is based on the steel or reinforced concrete frame with fill-in external walls of bricks, building blocks, coated steel panels or combinations of these. Such structures provide a degree of flexibility in arranging internal nonload-bearing walls, which can be constructed from structural blocks rendered, made smooth and finished with a hard drying, smooth impervious surface finish; or from prefabricated partition panels of various types. Internal walls should be nonporous, nonshedding and be free of cracks, dirt-retaining holes and flaking paint. They should be washable and able to resist repeated applications of cleaning and disinfecting agents.

Floors should be even-surfaced, free from cracks and allow for ready cleaning and removal of any spillages. They should conform to the requirements similar to those indicated for walls above. They need also to be tough.

Where drains or drainage gullies are installed they should be able to be easily cleaned (and clean) and trapped to prevent reflux. Floors should fall to drains, not *vice versa*, although there still seem to be an extraordinary number of drain installers who seem convinced that, left to its own devices, a liquid will flow uphill.

A common approach is to employ suspended or false ceilings, with the void above the ceiling being used for pipework and services. Ceiling panels or tiles should be close-fitting and sealed or clamped together at joints. The entire ceiling should have a smooth impervious surface, and be easy to keep clean. Lighting should be fitted flush, or suspended from the ceiling in such a manner that the fittings may be kept clean. Acoustic tiling is generally inappropriate in areas where product is exposed.

4.6.4.5 Basic Design Features

Absolutely fundamental to good pharmaceutical building design are the concepts of segregation of different types of operation, grouping together of related types of activity or product, and smooth mainly undirectional flows of materials, in-process product and intermediates, with minimal crossing over of work flows or backtracking. In any but the simplest facility, manufacturing only one product (or a small range of closely similar products), it is perhaps not possible to achieve *ideal* segregation, grouping and flow, but the objective is clear. It is essentially to avoid mixup and contamination and additionally to create and maintain an orderly, efficient working environment in which supervision, rapid appraisal of just what is going on, and communication are all facilitated. In addition to the GMP aspects, it is hardly necessary to add that there are of course the economic benefits of more efficient production and higher productivity.

The immediate surrounds of the building should be such that they may be and are maintained in a clean, tidy and orderly condition. Around the entire perimeter there should be a width of concrete, tarmacadam or similar material, which should fall to drains and prevent water seepage into the building. All outside walls should be sealed to prevent entry of dust, damp and insects through cracks and gaps, as should all cut-outs for windows, piping and duct work. Windows direct to production areas should not normally be openable. All external loading and unloading points should be provided with protection from the weather.

The building should be secure, with access restricted to authorized personnel at authorized times only. The access of vermin, birds, insects and pests should be prevented.

The general external appearance should not be neglected. It might be argued that aesthetics have nothing to do with GMP, but a good, clean, attractive external appearance does help to encourage desirable operator attitudes.

A very basic block design showing a simple single-storey linear flow is given in Figure 1. Another popular layout is the horizontal U-flow, as shown in simplified form in Figure 2.

It is not difficult to picture a number of variations or combinations on just these two basic patterns alone. In practice *total* realization of the ideal layout is rarely found. The nature of the site, local conditions, availability and placement of services all tend to dictate modifications, and as business and product range change and develop and premises expand, supplementary flows (or subsidiary 'production loops') tend to be grafted on to the original pattern. Nevertheless the aim should always be to remain as close to the ideal as possible.

4.6.4.6 Equipment

The UK 'Guide to Good Pharmaceutical Manufacturing Practice'[7] contains eight paragraphs on 'Equipment'. They may be paraphrased as follows: 'Equipment should be designed and located to suit the processes and products for which it is to be used. It should have been shown to be capable of carrying out the processes for which it is used, and of being operated to the necessary hygienic standards. It should be maintained so as to be fit to perform its functions. It should be easily and conveniently cleanable, both inside and out. Parts which come into contact with materials being processed should be minimally reactive or absorptive with respect to those materials and there should be no hazard to a product through leaking glands, lubricant drips, and the like, or through inappropriate modifications or adaptations. It should be kept and stored in a clean condition and checked for cleanliness before each use'.

Within the last 20 years there has been a most significant trend away from machines 'borrowed' from other industries (such as bakery, confectionery or dairy) and from machines adapted or designed for pharmaceutical use with scant regard to GMP principles. The trend has been towards

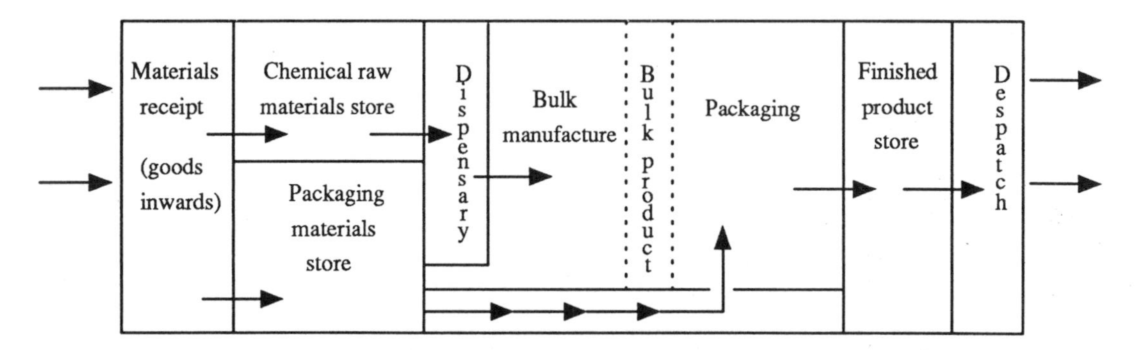

Figure 1 Simple single-storey linear flow pattern (schematic, not to scale, all sampling, quarantine and release stages *not* shown)

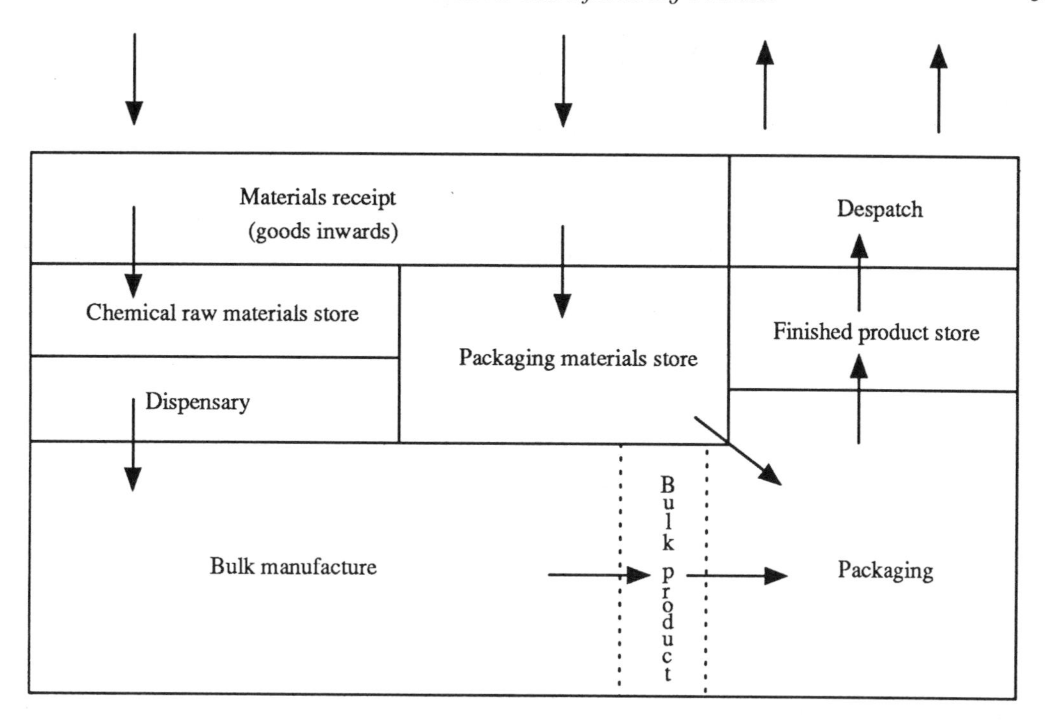

Figure 2 Simple horizontal U-flow pattern (schematic, not to scale, all sampling, quarantine and release stages *not* shown)

equipment designed, and built *specifically* for pharmaceutical use. To this end some particular trends have been towards: (i) more precision in control and adjustment; (ii) 'closed systems' of manufacture; (iii) more efficient, more automatic monitoring and recording; (iv) 'multisequential purpose' machinery, which performs a series of operations *in situ* (*e.g.* mixer/drier/granulators); (v) built-in local environmental and contamination control; and (vi) easy cleaning which can be readily confirmed.

Many of these trends are epitomized by the evolution of the 'hard' capsule-filling machine. For many years it seemed to be readily accepted that the *only* way to fill capsules was by using the early semiautomatic double ring plate machines with manually actuated powder hoppers. Those early machines were: (i) noisy; (ii) prodigiously dust generating; (iii) requiring of constant attention (one operator per machine); (iv) very demanding of operator skill and judgement; (v) much affected by the physical nature of powder; (vi) liable to produce wide-ranging fills; (vii) slow (say $<10\,000\ h^{-1}$); and (viii) unpleasant for operators.

By contrast, modern advanced capsule-filling machines are: (i) quiet in operation; (ii) enclosed and dust extracted; (iii) once set up, in need of little attention (one operator per several machines); and (iv) once set up, in need of little operator judgement. Furthermore: (i) they can be very precise, and fitted with automated weight check and adjustment; (ii) they are not so affected by powder nature and density; and (iii) they are fast ($\geqslant 150\,000\ h^{-1}$) and pleasant to work with.

4.6.5 DOCUMENTATION

In the context of GMP, 'documentation' is the term used to embrace the preparation, use, revision, application and retention of all forms, specifications, instructions, written procedures, manufacturing and test methods, formulas and records which have to do with manufacture of medicinal products, from the receipt of materials to the despatch of goods.

Much emphasis is placed on documentation in all GMP guides, and since most of these emanate from governmental regulatory agencies, it might well be concluded that this is but another example of the standard bureaucratic urge towards the proliferation of paper. This is not in fact so. Documentation is a vital part of any system of quality assurance. Its purposes are to define the system of control, to reduce the risk of error which purely oral communication introduces, to allow investigation and tracing of any defective products, and to permit appropriate, effective corrective action. The system of documentation should be such that the history of each batch of product, including the utilization and disposal of ingredients, packaging materials, intermediate, bulk and finished products may be determined.

The basic concept is to: (i) define what is to be done, (ii) do it, and (iii) record that it has been done—and that is good, sensible practice in any area of science and technology, rather than bureaucracy.

Documents usually considered necessary include: (i) raw materials specifications; (ii) packaging material specifications; (iii) intermediate and bulk product specifications; (iv) finished product specifications; (v) sampling and approval documentation; (vi) master formulae and manufacturing instructions; (vii) master packaging instructions; (viii) ingredients and packaging material receipt, test and issue records; (ix) batch manufacturing records; (x) batch packaging records; (xi) intermediate and bulk material release records; (xii) finished product release records; (xiii) distribution records; (xiv) complaints records; (xv) sampling procedures; (xvi) analytical methods; (xvii) analytical results records; and (xviii) standard procedures for operating machinery, equipment and instructions, complaint handling and product recall.

Perhaps the most comprehensive statement of the required content of such documents is to be found in the UK Guide.[7]

Attention should be given to careful and effective design of documents so as to facilitate their proper use. Of particular importance is the need for unambiguous instructions for the use and completion of a document and the need for revision and reissue when required. There should be a system for control of all documents to ensure that obsolete documents (instructions, specifications *etc.*) cannot be used.

All instructions to personnel (*e.g.* for sampling, testing, manufacturing, operating equipment, *etc.*) should be clear, precise, unambiguous and written as numbered steps, in the imperative. They should be written in a language and style that the user can readily understand.

Personnel should be trained in the use and completion of documents and should be in no doubt about where and how they should record information and sign for having completed an operation. All documents on which records are made should be double checked by a responsible person. In particular the release of product should be dependent upon a check to ensure the presence of all necessary information, and the satisfactory completion of all relevant records.

Many national regulatory authorities require that records specifically relating to batch manufacture and packaging, together with the analytical records of the testing of the materials, intermediates and final products, plus records of distribution, should be retained for specified minimum periods (British regulations, for example, require that batch manufacturing and packaging records should be retained for at least five years). Retained reference samples of finished products may be, and indeed often are, considered as 'records'. It is usually considered that they should be retained at least until the expiry date of the batch of product concerned or until such time as the batch may no longer be expected to be anywhere in stock or in use.

Electromagnetic and photographic recording techniques (microfilm and microfiche) are, of course, increasingly used. These are acceptable providing the system will reproduce the data clearly and accurately. Care needs to be taken where interpretation of an original document depends on colour distinction, *e.g.* multiple trace chart records. Original documents should be retained for at least six months, and not destroyed until any copies made from them have been checked for completeness and legibility. (The writer knows of one company, where, on converting to microfilm, several months records were lost because the originals were destroyed immediately on filming, and the film was destroyed through an accident in processing.)

4.6.6 OTHER ASPECTS OF GMP

In one short chapter it is not possible to cover, in any degree of detail, the entire field of GMP. The basic concepts and some key issues have been discussed. Space permits only a brief consideration of some other, but nevertheless important, issues.

4.6.6.1 Recovered Materials

In many manufacturing organizations there will be, on occasions, residues from production runs (*e.g.* powders or granules remaining in hoppers), or batches (or part batches) of product which fail to meet specification in some way that does not represent a specific or crucial patient hazard (*e.g.* tablets which are 'out of spec' with regard to weight check or hardness, slightly under- or over-filled capsules and so on). In instances such as these it may be both economic and feasible to recover or rework the material. It may also be acceptable from the GMP viewpoint, provided: (i) there is no adverse effect on product quality, safety and efficacy; (ii) the process is specifically authorized and documented; (iii) there are preset limits, established by quality control, for the amount of recovered material that may be added to subsequent batches; and (iv) batches containing product residues are not released until the batch(es) from which the residues were derived have been tested and approved.

With regard to the recovery and reuse of products returned after they have left the manufacturing site, the British Guide to GMP comments: 'A finished product returned from the Manufacturer's own stores or warehouse (because, for example, of soiled or damaged labels or outer packaging) may be relabelled or bulked for inclusion in subsequent batches, provided that there is no risk to product quality and the operation is specifically authorized and documented. If such products are relabelled, extra care is necessary to avoid mix up or mis-labelling.

Finished products returned from the market and which have left the control of the manufacturer should be considered for re-sale, relabelling or bulking with a subsequent batch only after they have been critically assessed by the person responsible for Quality Control . . . .'.

In general, it is doubtful if *any* material returned after it has left the control of the manufacturer should be considered for reprocessing or resale, other than chemical recovery and purification of valuable active ingredients. Any material returned as a subject of complaint, or reported adverse reaction should, of course, be effectively segregated from all other materials and products and *not* reused.

4.6.6.2 Complaints and Recall Procedure

Manufacturers should establish an organizational and recording system for dealing with complaints, or reports of defective products. Often such complaints *seem*, in isolation, to be trivial and unfounded and, on further investigation, may prove to be so. However, a full understanding of the nature and cause of a complaint, and a consideration of other possible reports of a similar nature might well lead to a conclusion that there really *is* something amiss, and that corrective action must be taken. That is why all complaints should be recorded and thoroughly investigated, with a decision on any necessary action being taken at an appropriately senior level. That decision could be to recall the product or batch.

All manufacturers should have a written recall procedure, with nominated persons responsible for implementing it as necessary, within, or outside of, normal working hours. Distribution records should be maintained which will facilitate effective recall, and the written procedure should include emergency and 'out-of-hours' contacts and telephone numbers.

The written procedure should state: (i) how the distribution or use of the product or batch should be halted; (ii) how contact should be made, as necessary, with the relevant regulatory authority; and (iii) how, and how widely, the recall should be notified and implemented. Any notification of a recall should include: (i) name of the product; (ii) batch number(s); (iii) nature of defect; (iv) action to be taken; (v) urgency of the action; and (vi) statement of *reasons* for the recall and of potential risks.

Account needs to be taken of any goods which may be in transit, and consideration must be given to the possibility that the fault may extend to other batches or products, for example due to a fault in equipment or machinery used in the manufacture of a number of different products or batches, or through the use of incorrect or faulty material in a number of different batches.

4.6.6.3 Good Control Laboratory Practice

The word 'control' is customarily inserted to provide a distinction from 'good laboratory practice' which has come specifically to imply reference to practices in toxicological or pharmacological testing laboratories.

Quite simply, indeed it *should* be 'quite obviously', good practices should be observed just as much in manufacturers' testing laboratories as in the production plant itself. Thus laboratory staff should be appropriately trained and experienced and properly managed and motivated. Design, layout and equipping of laboratories should be given the same level of attention.

All equipment and instruments should be appropriate to the test procedures carried out and should be properly serviced, maintained and calibrated at specified intervals, and records maintained of these operations. These records should clearly indicate when the next calibration or service is due.

Systems, procedures records and labelling (date, shelf-life, standardization factor, storage conditions) should be established to ensure the continuing validity of all reagents.

Samples for testing should be taken in accordance with a written procedure, so as to be representative of the bulk from which they are taken. The written procedure should include: (i) the method of sampling; (ii) the equipment to be used; (iii) the amount of sample to be taken; (iv) instructions for any required subdivision of the sample; (v) the type and condition of sample container to be used; (vi) any special precautions to be observed; and (vii) cleaning and storage of sampling equipment.

Records should be maintained of all the test procedures carried out. These records should include: (i) name of product or material and code reference; (ii) date of receipt and sampling; (iii) source of product or material; (iv) date of testing; (v) batch or lot number; (vi) indication of tests performed; (vii) reference to the method used; (viii) results; (ix) decision regarding release, rejection or other status; and (x) signature or initials of analyst, and signature of person taking the above decision.

4.6.6.4 Contracted Work

A number of manufacturers contract out part of their manufacturing or analytical work to other companies, and/or have servicing and maintenance work done for them by outside contractors. This can be perfectly acceptable provided both parties understand the specific technical requirements and observe the basic principles of GMP. Relative responsibilities should be fully understood and agreed in writing. It must always be remembered that ultimate moral and legal responsibility rests with the manufacturer for the quality of the products.

The various GMP guides and codes cover certain other topics in varying degree of detail. Two of the more important remain to be considered here: the special requirements of sterile product manufacture and the question of validation.

4.6.7 STERILE PRODUCTS

In the manufacture of sterile products, the application of the principles of GMP assumes a very special, indeed *vital* significance. These are products which, if they are faulty or defective in any way, probably represent the greatest of potential hazards to patients. Not only is the greatest care and attention to detail necessary to ensure that the product is 'right' in terms of its chemical composition and physical properties, that it is properly compounded from the correct ingredients of the appropriate quality, unadulterated and undeteriorated, it must also (i) *in fact* be sterile and (ii) be acceptably free of particulate matter.

The fallibility of the standard pharmacopoeial tests for sterility as a determinant of finished product quality has already been mentioned. Pharmacopoeias specify a maximum sample size of 20 containers per batch (ampoules, vials, bottles *etc.*) for sterility-testing purposes (smaller sample sizes are usually given for small batches). Assuming this sample size of 20 containers, it can be shown[23] that the probabilities of passing batches as 'sterile', with various levels of infected containers, are as shown in Table 1.

These probabilities are independent of batch size. Furthermore, most pharmacopoeias allow repeat tests, on further samples of 20 containers, if the first results in failure. Far from increasing the probability of detecting nonsterile units, the repeat testing serves only to reduce it!

Thus, whilst there can hardly be a more crucial quality characteristic than the sterility of an injectable (or other purportedly sterile) product, the only end-product test available is of strictly

Table 1 Probabilities of Passing Batches as 'Sterile', with Various Levels of Infected Containers

Contaminated containers in batch (%)	*Probability of passing batch as 'sterile'* (%)
0.1	98
1.0	82
5.0	36
10.0	12
20.0	1.2

limited (many would argue 'zero') value as a means of *assuring* that quality. Thus it is that GMP assumes a singular importance.

Particularly stringent standards are required of the premises in which the products are manufactured, their structure, layout and surface finishes, and of the quality of the air supplied to the different manufacturing areas.

Sterile products should be prepared in specially designed and constructed departments which are separate from other manufacturing areas, and where the different types of operation (*e.g.* container preparation, solution preparation, filling, sterilization, *etc.*) are strictly segregated from one another. All walls, floors and ceilings should be smooth finished, impervious and unbroken, so as to minimize the shedding or accumulation of particles, and to permit the application of cleaning and disinfecting agents. There should be no recesses which cannot be cleaned, and a minimum of projecting ledges, shelves, cupboards and fixtures. All equipment should be easily cleaned and disinfected (and *be* clean and disinfected!).

All sterile product-processing areas should be supplied and flushed with filtered air under positive pressure. There should be a positive pressure gradient, with respect to the surrounding environment, with the highest pressure at the zone of highest risk, *i.e.* where unsealed product is exposed.

Environmental standards for different levels (or classes) of sterile product processing rooms are given in Appendix I of the UK 'Guide to Good Manufacturing Practice' (see also British Standard BS 5295, 1976, and US Federal Standard 209B).

The greatest microbiological hazard to the products is presented by the personnel themselves. They should be kept to a minimum, and selected with great care to ensure that they may be relied upon to observe the necessary disciplines, and are not subject to any disease or condition which might hazard the microbiological quality of the product. They must maintain high standards of personal hygiene and cleanliness. They must be carefully trained in the principles and practice of sterile product processing. They should be clad in special protective clothing, appropriate to the type of processing area in which they are working.

A basic, fundamental concept is that it is far from good enough to manufacture sterile products to poor standards and in poor conditions and then rely on some terminal process or test to achieve sterility. The utmost care and attention is required throughout the entire production itself. It is here that the concept of process and equipment *validation* assumes its greatest significance.

4.6.8 VALIDATION

Validation is a subject which has become a major topic of discussion. Indeed, it could be said to be *the* industrial pharmaceutical 'buzz word' of the 1980s.

It is generally believed that it was in the USA, during the mid-1970s, that the term 'validation' first came into use in its specialized industrial pharmaceutical sense. The word, of course, is not new, nor indeed is the general concept of demonstrating that any thing, statement, argument or activity is sound and well-founded. Both word and concept are centuries old.

The first appearance of the *word* in written English dates from 1648. The general *concept* is hardly new. It is clearly not new in the general field of science and technology, nor indeed in the particular field of pharmaceutical production. Heat distribution and heat penetration studies on sterilizers, and 'media-fills' to check efficacy of aseptic procedures (the three topics most frequently discussed in recent papers on validation) have been practised in the British pharmaceutical industry for at least 30 years.

The first edition of the British 'Guide to Good Manufacturing Practice', published in 1971, has, in a section headed 'Verification of Procedures', the statement, 'Procedures should undergo a regular critical appraisal to ensure that they are, and remain, capable of achieving the results which they are intended to achieve'. That surely is 'validation', and it is thus reasonable to doubt if validation was, in fact, 'invented' in the USA in the 1970s.

A difficulty of validation has been that of deciding what precisely it is. Wordy definitions seem to abound, both of the term itself, and of the curious extensions of it ('total validation', 'narrow validation', 'retrospective validation', *etc.*).

Sharp[24,25] has discussed the general question of validation, and offered some examples of the more opaque definitions. Sharp concludes: 'Many such definitions serve only to confuse those who already have an idea of what the intention is, and would not enlighten those who do not'.

Fortunately, simple clarity has sometimes prevailed. In 1982, Fry of the US Food and Drug Administration, stated:[26] 'To prove that a process works is, in a nutshell, what we mean by the verb 'to validate'.

The 1983 UK GMP Guide,[7] a little more formally, but just as sensibly, defines 'validation' as 'The action of proving that any material, process, procedure, activity, equipment or mechanism used in manufacture or control, can, will and does achieve the desired and intended results'.

Fry[26] refers to some 'matters of concern', one of which is that 'the validation requirements have resulted in virtually a new industry being created, of consulting firms who perform validation studies for pharmaceutical manufacturing clients'. Fry points to the importance, and the difficulty, of ensuring that the consultant validators are competent to do the job. One might go further, and suggest that it is not just *outside* the manufacturing industry that the 'validation industry' has developed. It has also grown *within* the pharmaceutical industry, as a flourishing subculture. Is it all necessary? Is the expense justified? The answer is a cautious 'yes', provided that the validation is practised as, and to the extent that, it is really required. That is, it is *done* rather than endlessly theorized about, and that it is done in the interests of product quality and patient safety and not as an end in itself, or as a means of creating work. If validation is kept within bounds, and not pushed to some of the impractical, even ludicrous, extremes that have been suggested, then a useful purpose is served in focusing attention on the need to maintain a questioning attitude to the processes of manufacture, and a penetrating, systematic approach to their evaluation.

Validation is most crucially required in the manufacture of sterile products. Because of the limitations of the standard sterility test as a means of providing assurance of the sterility of a batch as a whole (see Section 4.6.7), there is an unquestioned need to validate the processes and equipment used to render and maintain the product sterile. In the validation of sterilization processes, an initial broad distinction may be made between 'physical validation', *i.e.* establishing by physical methods and controls, that a product or material *will* be subjected throughout a load to the desired and intended treatment during routine operation and 'biological validation', *i.e.* demonstrating that, with respect to an appropriate test organism (or organisms), the desired level of assurance of sterility will be achieved during routine operation.

The US Parenteral Drug Association has published three valuable 'Technical Monographs' on the validation of steam-sterilization cycles,[27] aseptic filling processes[28] and dry-heat processes.[29] Much useful information is also given in 'Hospital Technical Memorandum No 10—Sterilisers' (published by HMSO), although this document is specifically directed towards hospitals, and deals only with steam and dry-heat processes.

Discussions on validation have largely focused on sterilization processes. What of other manufacturing processes and other dosage forms? What about the validation of the manufacture of tablets, capsules, liquids, creams and ointments? On those topics, relatively little appears to have been said and written. Papers that do deal with the validation of such processes seem only to discuss matters that a few years ago would have been considered as embraced by product development, raw material quality control, in-process control, in-process testing or end-product testing. The only difference is a change in terminology. One is forced to conclude that the writers of those papers proceed from an uneasy assumption that since validation is accepted as a 'good thing' in sterile processing, it follows that it must be introduced into other manufacturing processes. But that is not necessarily so. Is the need for validation perhaps inversely proportional to the adequacy of product design, raw material control, in-process control and end-product testing to provide assurance of routine product quality? Such things are clearly *not* adequate to assure the quality of sterile products, but it may well be that they *are* adequate in, say, the production of tablets?

Computers are now increasingly being used to record, to monitor, to control and ultimately to direct and automate entire processes. Much careful thought needs to be given to the validation of computer hardware and software to ensure that the desired and intended results will indeed be achieved. However, in pharmaceutical manufacture the science and technology are the relatively easy part; it is the 'people problems' that are really difficult. Begg[30] has reminded us that 'people make products, people are required to follow production instructions . . . (and) to service and maintain equipment'. Begg's paper was on 'Validation' yet it was largely concerned with 'human aspects'—with personnel selection, training, motivation and management, with assessment of performance, and with avoidance of human errors. People are the most important factor in any manufacturing enterprise yet we cannot validate people. We can only direct our attention to those 'human aspects' emphasized by Begg.

It is difficult to say what precisely is the prime motive behind the vogue for what it seems, to some at least, is an excess of validation. It is however, not something which has been led or driven by regulatory bodies in Britain.

A reading of the four short paragraphs (5.9 to 5.12) on the subject in the UK Guide[7] will reveal them as distinctly 'cool' in comparison with other more fervent statements.

4.6.9 THE ECONOMICS OF GMP

On occasions one hears talk of 'the cost of GMP'. An appropriate reaction is to be concerned about the basic assumptions underlying such a question and to wonder if it is at all meaningful. To draw a simple analogy between a pharmaceutical factory and a motor car; then if the engine, providing as it does the basic drive, may be likened to the urge to run a sound and profitable business (and, one hopes, to provide a useful public service), then GMP may be likened to the steering wheel, the brakes and the other control mechanisms. These are *not* 'optional extras' which involve possibly unnecessary additional costs. They are basic essentials. The principles of GMP are basic and inherent components of medicines manufacture. Not only *should* you not — you *cannot* — make medicines without them.

The object is *quality*, not in any fancy philosophical sense of the word, nor as mere conformity with a predetermined specification, but quality as *fitness for purpose*, the purpose of being administered to a patient to cure them and not to harm them. As thus defined, 'quality is free'. Philip Crosby's book of that title concludes that 'The cost of quality is the expense of doing things wrong'.[31]

Even if quality failures are detected before the product is distributed, then there is still the cost of waste, of scrap, and of rework — and that's the good news. The bad news is the discovery of a quality failure after the product has been distributed and possibly consumed. There is then the very high cost of recall, possibly the very much higher cost of compensation for damage, and perhaps the incalculable cost of damage to the company's reputation — a blow from which, it is not fanciful to suggest, a company might never fully recover. Thought of in these terms, the market advantage of an unsullied reputation for quality as compared with the damaging effects of quality failure, then quality, and hence GMP, is indeed free and is certainly a bargain at that price. But that is far from the end of the story.

It has been said that GMP is not only good manufacturing practice — it is also good *management* practice.

GMP is about a planned, orderly approach to manufacture. It is about the provision of suitable premises and equipment, sufficient space for manufacture and storage; smooth efficient work flows; effective communication and supervision, efficient documentation systems, good stores records *etc.* It is about competent, well-trained, highly motivated staff at all levels. But all these things are not just good manufacturing practice, they are good sense, good management and good economics. For example, ensuring properly trained, well-directed, highly motivated staff with a full understanding of what they are doing, and why they are doing it, is good manufacturing practice. Such staff will also be happier, have a sense of purpose, work better and achieve higher productivity; *This is so* — it is not mere theoretical dreaming. To give just one example: a large pharmaceutical factory a few years ago engaged in a thorough overall GMP upgrade; buildings finishes, equipment, documentation, operator training, and so on. They spent a lot of money, but very rapidly they recorded over 25% increase in productivity, directly attributable to better operator attitudes and morale. There was a 50% reduction in work in progress, due to improved documentation resulting in more efficient, swifter testing and release mechanisms. That drop in work in progress, in turn resulted in a saving of many thousands of pounds per annum in inventory costs, and there was a dramatic drop in staff turnover.

The economics of GMP are, therefore, very *favourable* to the manufacturer, and with the ever-increasing complexity and potentially greater hazards of medicines manufacture, they are likely to remain, and indeed become even more, so.

4.6.10 REFERENCES

1. D. Dunlop, in 'Good Manufacturing Practices in the Pharmaceutical Industry, IFPMA Symposium, Geneva, 1971', IFPMA, Zurich, 1971, p. 19.
2. N. Wells, 'Crisis in Research', Office of Health Economics, London, 1986, p. 21.
3. R. Chew, 'Health Expenditure in the UK', Office of Health Economics, London, 1986, p. 3.
4. J. R. Sharp (ed.), in introduction to 'Guide to Good Pharmaceutical Manufacturing Practice', HMSO, London, 1983, p. 6.
5. J. R. Sharp, 'First Maxim', in many papers, published and unpublished, *c.* 1970 to present.
6. US Food and Drug Administration, 'Current Good Manufacturing Practices for Finished Pharmaceuticals' (Regulations Part 211), FDA, Washington, DC, 1988.
7. J. R. Sharp (ed.), 'Guide to Good Pharmaceutical Manufacturing Practice', HMSO, London, 1983.
8. M. Anisfled (ed.), 'International Drug GMP's', 2nd edn., Interpharm Press, Prairie View, IL, 1983.
9. R. H. Caplan, 'A Practical Approach to Quality Control', 3rd edn., Business Books, London, 1978.

10. J. M. Juran, 'Quality Control Handbook', McGraw-Hill, New York, 1962.
11. M. Jacobsen, 'Safety Aspects of Packaging', Proceedings EFTA Seminar on Safety Aspects in the Packaging and Labelling of Pharmaceuticals, Geneva, 1971, EFTA, Geneva, 1971, p. 10.
12. L. Setnikar, 'Reliability in Sampling and Control', Proceedings EFTA Seminar on Safety Aspects in the Packaging and Labelling of Pharmaceuticals, Geneva, 1971, EFTA, Geneva, 1971, p. 131.
13. J. R. Sharp, 'Quality Control, Present Problems, Future Possibilities in Collected Papers, PIC Seminar, Copenhagen', EFTA, Geneva, 1975.
14. 'Manufacture Control and Distribution of Drugs', Specifications Board, Supply and Services, Ontario, Canada, 1957.
15. 'Therapeutic Substances Act', HMSO, London, 1925.
16. F. O. Taylor, *J. Am. Pharm. Assoc.*, 1947, **8** (3), 149.
17. US Food, Drugs and Cosmetic Act, Kefauver–Harris amendment, 1962.
18. R. Falk, 'The Business of Management', 3rd edn., Penguin Books, London, 1963.
19. W. Zinn, in 'Collected papers of Seminar on Good Manufacturing Practice in Tablet Manufacture, PIC Seminar, Sunningdale', EFTA, Geneva, 1978.
20. J. A. C. Brown, 'The Social Psychology of Industry', Penguin Books, London, 1964.
21. 'Hygiene Recommendations, Association of the Swedish Pharmaceutical Industry', Stockholm, 1972.
22. M. Fors, 'Microbiological and Other Potential Contaminants of Tablet Products', PIC Seminar, Sunningdale', EFTA, Geneva, 1978.
23. A. D. Russell, in 'Pharmaceutical Microbiology', ed. W. D. Hugo and A. D. Russell, 3rd edn., Blackwell, Oxford, 1983.
24. J. R. Sharp, 'Pharmaceutical Process Validation', in 'Proceedings of the International Colloquium on Industrial Pharmacy, Rijksuniversiteit, Gent, Belgium, 1985', University of Gent, Belgium, 1985, p. 26.
25. J. R. Sharp, *Pharm. J.*, 11th January 1986, 43.
26. E. Fry, 'Validation, Theory and Concepts', in collected papers of a seminar on Validation, Dublin, EFTA, Geneva, 1982.
27. 'Validation of Steam Sterilisation Cycles', Technical Monograph No. 1, Parenteral Drug Association, Philadelphia, PA, 1978.
28. 'Validation of Aseptic Filling for Solution Drug Products', Technical Monograph No. 2, Parenteral Drug Association, Philadelphia, PA, 1980.
29. 'Validation of Dry Heat Processes used for Sterilisation and Depyrogenation', Technical Report No. 3, Parenteral Drug Association, Philadelphia, PA, 1981.
30. D. I. R. Begg, *Manuf. Chem. Aerosol News*, 1984, **55** (12), 31.
31. P. Crosby, 'Quality is Free', McGraw-Hill, New York, 1979.

4.7 Toxicological Evaluation of New Drugs

PETER MILLER
Roussel Laboratories Ltd, Swindon, UK

4.7.1 INTRODUCTION

'In the last analysis, it is only clinical trials which can reveal whether or not an effect observed in animal experiments is also relevant to man, and it is only from extensive clinical use of a new compound that valid conclusions can be drawn as to its safety when employed for a specific therapeutic purpose.'[1]

Although this statement is to a large extent a truism, it nevertheless is a reminder to all involved in drug discovery and development of the limitations of their biological testing schemes, both *in vitro* and *in vivo*. Despite these reservations, however, it is the role of the toxicologist, as it is of the pharmacologist and biochemist, to design test systems capable of predicting effects in man. For the toxicologist these test systems should be designed to identify and evaluate possible adverse effects which allow conclusions to be made as to their likely significance in man. In this context, the quotation above makes two further important points: in clinical usage a drug will be offered to many more patients than it has ever been evaluated in animals, either pharmacologically or toxicologically, and thus the opportunity of identifying rare adverse effects within the limitations of toxicological testing are restricted; and, secondly, the adverse effects of a drug are not an absolute but must be interpreted against the specific therapeutic purpose for which the agent is intended—side effects are perhaps tolerable in an agent intended for a life-threatening disease for which there is no present treatment, whereas they might not be for the banal treatment of a disease which is already well controlled by existing agents. In interpreting toxicological results we must, therefore, bear in mind both the predictivity and limitations of the methods available and the concept of a risk–benefit balance in extrapolating possible adverse effects to man.

In the public eye, the history of drug development is highlighted too often by its failures rather than its successes. Thus, the recent history of the pharmaceutical industry is marred, particularly in Europe, by the spectre of thalidomide and we must not forget that the fledgling pharmaceutical industry in the USA was shaken in the 1930s by the deaths caused by administration of Elixir of Sulfanilimide-Massengill, an effect later attributed to the diethylene glycol in the preparation.[2] In both these cases, as in others which could be cited, the response of both industry and regulatory authorities was to strengthen and increase the available and required toxicity testing procedures. Despite these greater demands, however, it is not always possible to identify all likely risks, either because of their rare occurrence or, even when they have been observed in man and with the benefits of hindsight, our inability to reproduce them in animals. A recent example of a failure of this type is provided by the β-blocking agent practolol, which induced in man a combination of symptoms particular to itself which came to be called oculomucocutaneous syndrome and for which suitable animal models have not been found.[3]

This then is the background against which the pharmaceutical toxicologist works. It is easy to see the role as one of an automaton, pushing a potential drug through a checklist of tests—this, however, will not ensure the safety of those receiving the drug. It is only by taking an interpretative proactive role that the safety of pharmaceuticals—as far as possible—can be assured: 'Testing procedures outlined by authority whether national or international will either be so vague as to be not worth disseminating or become by necessity more precise. The recommended tests would then be carried out by scores of unthinking technicians who will supply a mass of data eventually to be pushed under official noses. The scientific study of toxicology will atrophy and the hazards from new drugs remain as much of a problem in the future as it is today.'[4]

4.7.2 AIMS AND LIMITATIONS OF TOXICITY TESTING

Within the environment of the pharmaceutical industry, the aims of the toxicologist are basically twofold, both deriving from a knowledge of the toxicity of a particular compound in animals and *in vitro* systems. These are the exclusion from development of molecules which are too toxic and the definition of likely risk in man of molecules going forward to clinical trials and ultimately being marketed.

Many compounds are nominated within the industry for entry into the development process but only a small proportion of these reach the market. There are many reasons why they should fall along the way and toxicology is only one of them. The toxicologist is therefore just one of those with an input to the decision as to whether or not the development should continue. There are several clear-cut decision points within the toxicological evaluation of a compound and it is particularly, but not solely, at these points that the results of toxicology should be addressed. The phasing of toxicological testing is discussed in Section 4.7.4.

Determining the potential risk in a compound destined for the market is a process which requires judgement and a knowledge of the likely benefit to be gained from the compound, together with information on the toxicity of other agents used for treatment of the same disease(s). As the development of a compound unfolds and it is given an increasingly wider exposure in man, the toxicologist gains further information on its clinical use which allows the predictions from the animal studies to be tested. The process is thus one of feedback of information and checking of predictions. For example, on the basis of the initial one or three month studies in animals, the toxicologist may have warned that there could be a liver toxicity problem on sub-acute administration to man and recommended that particular attention be paid to liver function testing during the early clinical trials. The results of this testing could then confirm or put in context the predictive value of the animal studies.

Although the example just cited allows for a relatively quick feedback to the toxicologist on predictions made from animal studies, there are other elements of a toxicology programme upon which it is not so easy to obtain early information. The aim of sub-acute and chronic toxicity studies is to predict the likely effects of administration of the compound to man—the length of administration to man being determined by the availability of toxicological data according to Table 1, which cites as an example the national requirements of the UK and other members of the European Economic Community (EEC). The requirements of other countries can, and do, vary from these and, in addition, carcinogenicity studies will be required for longer term use in man (generally greater than six months).

Table 1 The Permitted Length of Exposure to a Drug in Man Based on Available Sub-acute and Chronic Toxicological Data (According to EEC Requirements)[5]

Intended duration of dosing in man	*Duration of toxicity tests required* (d)	*Intended duration of dosing in man*	*Duration of toxicity tests required* (d)
Single dose or several doses on 1 day	14	Repeated dose up to 30 days	90
Repeated dose up to 7 days	28	Repeated dose beyond 30 days[a]	180

[a] Long term administration may also require carcinogenicity data.

When one comes to consider the predictive value and feedback on reproductive and particularly carcinogenic studies, the toxicologist is faced with a different problem. The gradual build-up in a normal clinical trial programme, as increasing numbers of patients are treated for greater lengths of time, permits a constant monitoring of adverse effects and comparison of these with predictions made from toxicological testing. The extrapolation of reproductive studies, particularly teratology, to man requires a sudden rather than gentle approach as has been previously adopted and the toxicologists must be surer of their ground—there is less room for error in the predictivity of the studies. However, the predictions are limited by the tests used. Thalidomide was not predicted to be teratogenic on the basis of the studies done in rats; had rabbits been routinely used at this time for teratology studies then its teratogenic potential would have been identified. Toxicologists are thus limited by the test systems available and must be constantly on the search for meaningful tests to add to their armoury or to replace less valuable tests. The carcinogenic potential of a compound poses real problems to a toxicologist. The problems inherent in this type of study and their interpretation are discussed in greater detail below (Section 4.7.6.5) but as the purpose of such a study is to predict events that will occur during the total lifetime of the species being studied, its predictive value in man is not likely to be tested until many years of widespread drug usage in the clinic have been completed.

Further limitations on the predictive value of toxicity testing are the serendipitous reaction in man and adverse effects of rare occurrence. As has been mentioned, the oculomucocutaneous syndrome induced by practolol remains as an example of a complex reaction of unknown aetiology which defies modelling in animals. Adverse reactions in man with an occurrence of, for example, 1 in 10 000, are impossible to model reliably in animals; even studies designed to detect a treatment-related increase in the occurrence of an adverse effect are difficult to design and draw conclusions from. This is particularly a problem for carcinogenic studies, where dosing is for the lifetime of the animal and one of the parameters examined is the increase in background tumour incidence. This point will be discussed further in Section 4.7.6.5, but reference to Table 2 shows the size of the problem being examined. In this example, the number of animals required to demonstrate a statistically significant increase in tumour incidence above the 10% level found in background control groups is given for two levels of power—90% and 95%—and presuming a 5% level of significance. Thus, 105 animals are required in both control and treatment groups to detect an increase of 15% due to treatment from a background incidence of 10%, with 90% power. It is obvious that the demonstration of slight increases in incidence requires so many animals as to make the study untenable and, moreover, as most carcinogenic studies are carried out with group sizes of 50 animals of each sex per dose level and 100 of each sex in the control group, the limitations of these studies can clearly be seen. This is not to say that such studies are flawed and must be redesigned and repeated, but merely to underline their limitations.

Table 2 Group Sizes of Animals—Control and Test—Required to Demonstrate an Increase on a 10% Incidence of a Side Effect in Control Animals

Incidence (%)	*Number of animals required at*	
	90% power	*95% power*
15	751	950
20	214	271
25	105	133
30	65	82
40	33	42
50	20	26
60	14	17
70	10	13
80	7	9
90	5	7

4.7.3 THE ROLE OF KINETICS IN TOXICOLOGY

As the role of toxicology is to predict and investigate likely toxicity in man, it follows that the species and strain of animal used for testing should handle the drug in a manner similar, if not identical, to man. This should then be one of the criteria of choice. In making the choice of species and strain the toxicologist must have knowledge of the comparative kinetic parameters in man and the available species and the dosage regimen employed should reflect that used clinically. This is of particular importance for any long term studies being carried out as these will, to an extent, supersede earlier studies not carried out in the correct species. This ideal requirement poses great practical problems as the full knowledge of kinetic and metabolic parameters may not be available—in animals or in man—at the time the choice of species must be made. Because of the length of time involved in carrying out a long term toxicity or carcinogenicity study, it is often necessary to commit to those studies before all data are available. (The phasing of studies and their impact on development times are discussed later in Section 4.7.4.)

The first sub-acute studies of two or four weeks duration are carried out in a rodent (usually the rat) and non-rodent (usually the dog or primate) species. On the basis of this knowledge, the first administration to man can take place and, after suitable tolerance work, studies can be undertaken to investigate pharmacokinetics and metabolism. At the same time, similar studies should be undertaken in the available toxicity species to define the same parameters. Particular attention should be paid to peak plasma levels, time to peak plasma level, elimination half-life, area under the plasma level–time curve, routes of elimination and metabolic pattern in each of the species, including man. Whilst all these data are obtained after acute administration, it is also necessary to study pharmacokinetics after repeat dosing in animals under the protocol used in the toxicology studies (and preferably within these same studies) and also in man under the regimen employed in clinical usage. In this study, not only can the comparison be made of acute parameters and similarities in metabolic pathways, but evidence of accumulation or enhanced excretion can be sought.

Not only does knowledge of pharmacokinetics and metabolism permit validation of toxicology studies by demonstrating the similarities in the way in which a drug is handled in man and in the species chosen for toxicity testing, it also allows a further interpretation of the toxicology studies themselves. Thus, knowledge of the distribution pattern of drug gained from experiments with radiolabelled molecules will identify any organ(s) in which the compound is concentrated or particularly long-lived, urging the toxicologist to pay particular attention to them. However, if plasma levels of a compound are found to fall with repeated dosing, then liver enzyme induction might be suspected, alerting the toxicologist to perhaps include measurement of this parameter in the studies or at the least examine carefully the liver and measures of its function and toxicity in greater detail.

In examining the toxicity of a compound within animal studies, it is necessary to define the dose at which adverse effects are seen. This permits extrapolation to man based on the therapeutically active dose and allows a safety margin—based purely on weight of drug administered—to be established. An alternative and complementary approach is by comparison of the plasma levels achieved in normal clinical use with those achieved in the toxicology studies, allowing a safety margin to be established on maximum plasma levels or areas under the plasma level–time curve. This is an approach increasingly looked for by regulatory authorities in their evaluation of applications for licence.

4.7.4 PHASING OF TOXICITY TESTING

The choice of which toxicity studies should be performed and the design of those studies is a function not only of the intended therapeutic use of the compound but also of the particular questions being asked of a compound at each stage of its development. Whilst a certain core of data is required for an ultimate product licence application, the exact phasing of these studies will depend on the development pathway planned. As shown in Table 1, which cites the example of the EEC demands, the length of permitted administration to man is governed by the duration of the available toxicology. Thus, for example, if it is thought necessary that an initial clinical trial, given sufficient tolerance data, needs to be of three months duration in order to demonstrate an unequivocal effect, then the decision made at the outset of the toxicology should be to complete a six month study as rapidly as possible. In contrast, a toxicology study of only 28 days duration is required to evaluate an antibiotic, for example, requiring only seven days administration to demonstrate efficacy. Other examples of tailored toxicology studies can be cited—an antibiotic for short term administration does not require carcinogenicity studies, a drug solely for treatment for prostate cancer does not need teratology studies. In short, the planned toxicology should be designed to find the risk in the intended therapeutic usage. Also, these studies must take into account the known pharmacology of the compound—it is not possible, for example, to carry out a fertility and general reproductive study with a proposed contraceptive agent, except at a meaningless dose.

The phasing of testing must also take into account the other element of the role of toxicologists within the pharmaceutical industry, their input into the commercial decision as to whether or not the development of a particular compound should continue. It is thus necessary to ask the crucial toxicological questions of a compound as early as possible within the overall scheme. These can be general, as can be asked of any compound, or can be generated by a knowledge of the biological activities of the compound or the known toxicity of chemical analogues. A further limiting factor on the generation of toxicological data is financial. The longer term and carcinogenicity studies are extremely expensive and many companies require some evidence of clinical efficacy before they are willing to commit to them. To counterbalance this caution, however, is the knowledge that these studies take a long time to complete and their lack can delay the clinical development programme or the eventual product licence application.

There are thus many factors and pressures which the toxicologist must consider when planning the phasing of all the studies envisaged on a particular compound and the overall plan is one which has to be decided on a company-wide basis as its impact is widespread.

At the early stages of a compound's development, the first questions asked of it are normally with regard to its acute toxicity and mutagenic potential. Both of these are vitally important questions, particularly the latter, and can serve as an absolute barrier which has to be cleared. At this time it is usual to rely on only one mutagenicity test—often the Ames test (see Section 4.7.6.7.2)—with other tests of mutagenic potential to follow later. Neither of these tests, on their own, will allow administration of the compound to man and this can only follow studies of 14 or 28 days duration. The place of these first studies within a typical planning schedule is shown in Figure 1 (see later). It must be emphasized that this is an example of a typical phasing for a compound intended for long term use in man and does not represent any legislatively required order. It does, however, show the way in which studies can be phased to allow clinical development to continue and in which some time can be saved by running studies in parallel. The aim, design and objectives of these different studies are discussed later in Section 4.7.6. The exact order and degree of overlap which will be employed is a matter of planning, clinical development schedules and financial commitment. It can also be influenced by the choice of species for particular studies. For example, carcinogenicity studies in dogs or large mammals are, by definition, extremely long—frequently 12–15 years—and if the chemical nature or pharmacological action of the compound demands the choice of these species then it may well be decided to commit to these studies at an early stage.

4.7.5 THE PARAMETERS FOR MEASURING TOXIC EFFECTS

4.7.5.1 The Available Parameters

There are many observations during the course of and at the end of a toxicology study. Together they can give a lot of information on the compound under test and allow conclusions to be drawn as to its effects in the test species and, from these, predictions can be made as to the likely toxic effects in man. Within sub-acute and chronic toxicity studies, the areas of these observations can be grouped together under the categories listed in Table 3. Other specialized studies, *e.g.* teratology, require different protocols, but the list in Table 3 includes most parameters examined within a repeat-dose study in rodents or larger animals. The particular requirements of other studies are discussed later.

Table 3 Parameters for Observations in Sub-acute and Chronic Toxicity Studies

Mortality	Urinalysis
Clinical symptoms	Haematology
Body weight	Clinical chemistry
Food consumption	Organ weights
Water consumption	Gross pathology
Physical examinations (*e.g.* ECG)	Histopathology

All toxicology studies rely on a comparison of the observed effects in control and test groups. All studies include controls; in some studies, *e.g.* carcinogenicity, these can be twice the size of the test groups, whilst in others, *e.g.* genotoxicity, positive control groups can also be included to demonstrate the sensitivity of the test system. Further use of control data occurs within any one study where comparisons can be made for some but not all parameters of their value in each animal before treatment with those found during and at the end of treatment.

4.7.5.2 Mortality and Clinical Symptoms

It is vital to note the reactions of the animals to treatment, particularly to distinguish those expected as a consequence of the known pharmacological properties of the test compound and those arising as the result of an unexpected or toxic effect. In many ways, death can be viewed as the ultimate reaction to treatment, but many other more minor alterations in the behaviour or appearance of the animals provide information necessary to interpret the overall effect of the study. Symptoms often recorded include somnolence, loss of muscle tone, salivation, vomiting (in higher animals), altered posture. Once again, however, it is important to emphasize that interpretation of these observations should be made within the context of the whole study and also the total knowledge available about the compound.

4.7.5.3 Body Weight

This is an extremely sensitive parameter for the detection of a toxic effect. It can be used to define the maximum dose within a carcinogenicity study—in this situation, the maximum dose employed should be only mildly toxic, defined so as to produce at least a 10% loss in body weight gain as compared to controls. Within a study in rodents, which is normally carried out in animals during their growth phase—and can extend beyond this in longer term studies—the parameter examined is body weight gain, whereas in primates and dogs, where the study can be carried out on mature or more slowly growing animals, the parameter is often body weight itself.

4.7.5.4 Food and Water Consumption

These again are sensitive indicators of the health of the animals, though particular care must be taken that a toxic effect assumed to exist because of an effect on either of these parameters is not simply a consequence of the pharmacology of the compound—it may be necessary to take such an action into account when designing the study. As a simple example of this, a sleep-inducing agent will reduce the time available for an animal to eat and drink and it may thus be necessary to alter the normal feeding times to take account of this.

Measurement of food consumption and comparison with body weight, particularly in rodents in their growth phase, allows calculation of food conversion ratios. Comparison of food and water consumption is essential as it is possible for water consumption to change as a consequence of alteration in food intake as well as in its own right.

4.7.5.5 Urinalysis

It is normal during the course of a study to place animals from each group within metabolism cages to collect samples of urine, usually overnight. These are then examined for volume, pH, specific gravity, protein content, content of glucose, ketone, bile pigments, hemoglobin and also for the presence of deposits after centrifugation. Any deposits found must then be characterized.

4.7.5.6 Haematology

The effect of a compound on parameters of cellular distribution in the blood is an extremely important potential cause and symptom of toxicity and it is possible that the compound under test may have a selective effect in reducing the numbers of one particular type of white cell or alter parameters associated with erythrocyte function. This is normally investigated with each animal acting as its own control, blood samples being taken at the beginning and end of the study as well, in longer studies, as at predetermined intervals during the study. The parameters measured include cell counting—including all cell types, erythrocytes, platelets, neutrophils, lymphocytes, eosinophils, basophils and monocytes—as well as measures of erythrocyte function and tests of clotting efficiency. Measures of erythrocyte function include hemoglobin concentration, packed cell volume, mean cell volume and mean corpuscular hemoglobin concentration.

4.7.5.7 Clinical Chemistry

This again is an example of each animal being able to act as its own control, with blood samples being taken at the beginning of, end of and, if appropriate, during the study. Plasma is separated from the red cells and assayed for its content of various factors and serum can also be assayed for its electrolyte content. The factors assayed in plasma include glucose, proteins (total and sub-fractions), urea nitrogen, creatinine, cholesterol, fatty acids and various enzymes indicative of the general state of the animal, particularly the liver, *e.g.* alkaline phosphatase, glutamic–pyruvic transferase, 5^1-nucleotidase and γ-glutamyl transferase. Fortunately, the assay of a number of these factors has been automated as it has in the hospital laboratory, thus easing the work involved.

In addition to these tests, further evaluation of the clotting system can also be carried out with plasma prothrombin time determinations.

4.7.5.8 Physical Examinations

Ophthalmological examination of animals, particularly in the high dose and control groups, is carried out at the beginning and end of a study and also at times during the longer term studies. If problems are perceived, the examinations are extended to other groups and can be undertaken more frequently so as to determine more specifically the sequence of events as they unfold. The functional aspects of other organs and systems can also be investigated. Of these, the easiest to examine and the most common is the cardiovascular system where electrocardiograms (ECG) are routinely examined, particularly in the larger non-rodent species where other parameters, *e.g.* blood pressure, can also be monitored. The effect of a compound after acute or very short term dosing will have been investigated as part of the pharmacological investigation into the compound's effects. The toxicity studies, however, present an opportunity not only to examine effects on the cardiovascular system as a consequence of toxicity, but also the pharmacological consequences of such long term dosing.

Within carcinogenicity studies, the animals should be regularly examined externally and palpated for signs of tumour masses.

4.7.5.9 Organ Weights

At the termination of the study and the sacrifice of the animals, a number of organs—listed in Table 4—are excised and weighed before being preserved for histological examination. This is done for all animals in all groups. The analysis of differences in weights between control and test groups can be done either in absolute terms or as weights related to overall body weight. Where organs are in pairs, *e.g.* the adrenals, these should be weighed separately, although their joint weights can be used in calculations.

4.7.5.10 Gross Pathology

All animals should be examined at autopsy for abnormal signs, both externally and internally. Thus, records of colour changes of organs, growths, changes in size of organs, *etc.* are recorded for later correlation with other effects noted. Within these examinations, it is not sufficient merely to examine the outside of an organ, *e.g.* the liver, but it should also be examined after sectioning at intervals of a few millimetres—later histological studies will examine the appearance of the organs under the microscope as sections of only a few micrometres. Where possible, photographs should be taken of any abnormalities.

Within carcinogenicity studies, particular care should be taken at this stage to identify all tissue masses and separate and preserve them for future histopathology.

4.7.5.11 Histopathology

The tissues listed in Table 4, including any gross lesions noted during the gross pathological examination, should be preserved in an appropriate medium, normally 10% formalin, though some tissues require special consideration, *e.g.* eyes are normally preserved in Davidson's fixative, and it is sometimes required to prepare samples of some tissues for immediate electron microscopy. It is important to plan at this stage for all eventualities. It is obviously not possible to predetermine the results of the microscopic examination nor of the correlation which can be built from this with other

Table 4 List of Tissues to be Studied Histologically in a Repeat-dose Toxicity or Carcinogenicity Study[5]

Gross lesions	Colon
Tissue masses or tumours	Liver[a]
Lymph nodes	Gall bladder
Mammary glands	Pancreas
Salivary glands	Spleen[a]
Sternebrae, femur or vertebrae (including bone-marrow)	Kidneys[a]
Pituitary	Adrenals[a]
Thymus[a]	Bladder
Trachea	Prostate[a]
Lungs	Testes[a]
Heart[a]	Ovaries[a]
Thyroid[a]	Uterus
Oesophagus	Brain[a] (sectioned at 3 levels)
Stomach	Eyes
Small intestine	Spinal cord
Blood smears (in cases of anaemia, enlarged thymus, lymphadenopathy)	

[a] Organs which should also be weighed.

facets of the study and it is thus necessary to have tissues available for further study without the need to dose more animals or repeat the study.

It is not normally essential to examine all the tissues in all the animals in all the groups. In rodent studies, where group sizes are relatively large, histopathology should be performed on all organs and tissues of the high dose and control groups—from other dose levels, it is necessary to examine only those tissues in which gross abnormalities have been detected, or in which changes have been observed in the high dose groups. In other species, where smaller numbers of animals have been used at each dose level, all tissues should be examined. Where tissues have not been examined, the wax blocks containing them should be kept both for archival purposes and in case it later proves necessary to examine them as other facets of a drug's properties, for example particular distribution patterns or previously unrecognized toxic effects, become known.

4.7.5.12 Special Parameters

As mentioned above, a toxicology study also provides an opportunity to examine the effect of chronic dosing of a compound on particular body systems. Because of the chemical nature of the compound, or as a result of observations made in other experimental investigations, it may be necessary to include other parameters than the more generally applicable ones discussed above. For example, if the compound is suspected of causing liver enzyme induction, a study of this parameter could be incorporated by taking a sample of liver from the test animals or by including extra animals.

A further example is the study of the immune toxicology of a compound. This is an area receiving growing recognition in terms of importance, although no accepted way of investigating it has yet emerged. However, if a compound is suspected of exerting effects on the immune system these should be investigated. It is not sufficient to count the number of cells in the animals, but it is also necessary to measure their function. This can be done in a number of *in vitro* test systems using separated blood cells and is even possible with cells derived from a portion of spleen or thymus. Any *in vivo* investigations of immune function, however, would have to be carried out separately as their introduction into the toxicity study could thwart the prime intention of such a study—the definition of toxicity. This point is discussed further in Section 4.7.6.8.

Mention has already been made in Section 4.7.3, but it is important to emphasize once again, that pharmacokinetic determinations within a toxicity study allow direct correlation of toxic effects and aid greatly the interpretation of these studies and hence their power and predictive value.

4.7.6 INDIVIDUAL TOXICITY STUDIES

4.7.6.1 Acute Toxicity

The initial experiment which a toxicologist undertakes with a compound under development is the determination of its acute toxicity—a study which is normally carried out in rats and mice and often in animals of both sexes to meet regulatory requirements. Apart from chance observations made by the trained observer during the pharmacological evaluation of the compound, these acute toxicity determinations are the first formal attempt to gain information on the toxicity of the compound. The parameter most often associated with acute toxicity is the LD_{50} value, the dose killing 50% of the population, but an acute toxicity experiment can give far more information than this. A properly conducted acute toxicity experiment can satisfy a number of objectives: (i) to give information that allows assessment of the toxic potential of the compound and hence predict the likely hazard involved in its use; (ii) to provide information on one half of the risk–benefit equation by considering the observed toxicity within the framework of likely clinical advantages; (iii) to provide information on the mechanism of the toxic reaction; and (iv) to provide information of use in designing sub-acute toxicity studies. The LD_{50} value is not the only numerical parameter which can be derived from an acute experiment. Other data of value are the minimum lethal dose and the maximum no-effect dose. In some cases, the compounds may be so acutely non-toxic that it is not possible to derive an LD_{50} and the other datum points can then be used; they do, however, have an important value in their own right and not merely as substitutes for the LD_{50}.

Acute toxicity experiments provide more information than mere figures—LD_{50}, minimum lethal dose or maximum no-effect dose. Careful observation of the dosed animals provides vital information on the mechanism of any toxicity which can be coupled with post-mortem examination of animals dying. The route of administration chosen for these studies must always be that intended for use in man, but in addition a parenteral route, preferably intravenous, is also chosen to provide information on toxicity by a route which ensures complete absorption. Comparison of toxicity between the intravenous and, say, oral routes can also provide information on the degree of absorption.

In addition to studies in rodents, some regulatory authorities also demand acute toxicity investigations in a non-rodent species. Here, the group sizes are much smaller than the 10 animals per group routinely employed in rodent studies, but the objectives remain the same, although stringent derivation of the LD_{50} or other datum points is not required.

4.7.6.2 Dose-ranging Studies

The next stage in the development of the toxicology profile of a new compound is to ask the question—what happens on repeated administration? As shown in Table 1, administration of a new drug to man requires repeat-dose studies whose duration determines the permitted length of exposure. To progress from acute studies to formal sub-acute two week, one month or three month studies is a large step, even with the maximum amount of information derived from the acute studies, and it is normal to first carry out dose-ranging studies. The objective of these is to determine the type of toxicity which is encountered on repeat dosing, to identify target organs and to provide information on the choice of doses for the formal studies to follow. They also provide information which can relate to the decision as to whether or not to proceed with the development of a particular compound—the results of the first repeat-dose studies provide important information to input into this decision as further progress with development can involve considerable expense, not only in the cost of the toxicology studies themselves but also in attendant areas such as industrial synthesis, formulation development, *etc.*

Although two species are chosen for subsequent repeat-dose work, it is common to carry out dose-ranging in three species, one rodent—the rat—and two non-rodent—the dog and primate. In this way, the maximum information can be obtained as to variation between species in toxicity, the best choice of species for the formal studies which are to follow and, if repeat-dose kinetics are included within the protocols, information on plasma levels with regard to accumulation or not, metabolic induction or not and differences between the species.

4.7.6.3 Sub-acute Studies

The objective of the early toxicity studies remains twofold—to determine the toxicity of the compound and to allow as rapid as possible administration of the test compound to man. The length of the early sub-acute studies—14, 28 or 90 days—is thus determined by the duration of the first administration to man for the determination of clinical efficacy. Whatever the choice of duration, the amount of ancillary studies generated once dosing of the animals is complete is a function of the number of animals in the study and thus it may not be considered worthwhile to carry out the shorter studies, particularly those of two weeks, except in particular circumstances. For example, it may take two to four months to analyze and report a study of 10 animals per sex at each dose level and thus the difference in time gained by carrying out a two week rather than four week study is only marginal. Within the early stages of a drug's development, however, it is necessary to investigate bioavailability in volunteers and for this an intravenous formulation is ideally required. This is used to determine kinetic parameters by this route in comparison to those seen after, for example, oral dosing, if this is the intended route in man. For this work to be carried out in volunteers, two week intravenous toxicity studies in two species are required. This remains, however, a very specific and restricted example of the value and need for a two week study.

Four week and three month studies remain the normal choice facing the toxicologist for the initial sub-acute studies; which is chosen depends on the initial clinical trials. Whereas a three month study presents a large step if it is the first repeat-dose study undertaken and may only be started as a result of a four week dose-ranging study, it is not usual to carry out both four week and three month studies unless the clinical development programme demands this. Rather, it is the norm to follow a four week study with one of six months.

Whatever the duration chosen, the objectives of sub-acute studies—and of other repeat-dose studies—remain the same and are summarized in Table 5. Of these, the most important is the determination and identification of the target(s) for toxicity and this can be done both during the

Table 5 Objectives of Repeat-dose Studies

To determine the nature of toxic events	To identify differences in sensitivity between the sexes
To determine the target for toxicity, *e.g.* specific organs (liver, kidney, *etc.*)	To investigate accumulative effects
To establish if a dose–response relationship exists	To investigate any tolerance
To identify species differences in toxic response	To correlate findings with any other known effects, *e.g.* specific pharmacological action, pharmacokinetic properties

dosing period of the study by observation and measurement of haemotology and clinical chemistry parameters as discussed above in Section 4.7.5, at autopsy and particularly during histopathological examination of the preserved tissues (Table 4) by an expert pathologist.

The choice of dose is crucial to a successful toxicity test. Three dose levels are normally chosen based on knowledge and experience gained previously during the dose-ranging studies and also the pharmacology. The low dose should be asymptomatic in toxicity terms, but should be a multiple of that required for pharmacological action in that species (frequently a 5–10 times multiple is chosen). The top dose must be chosen to produce toxicity, or to be the maximum practically attainable. The value of a toxic top dose underlies the whole rationale of a sub-acute study. It must be emphasized that at this stage of the development of a new compound, the objective is to define its toxicity. Administration of a compound in sufficiently high doses allows a number of criteria to be met: (i) the load of the drug within the animal will increase until a steady state is reached; (ii) the animal may adapt to this load by altering its way of handling the drug, either in pharmacological response, metabolic response or excretory mechanisms; (iii) continued high doses may overload the body systems such that different metabolic pathways are exposed, possibly leading to the formation of new metabolites which may be toxic; and (iv) a failure of physiological functions may occur, leading ultimately to pathological change. It is the object of this type of study that all these criteria should be met—if they are not, explanations should be sought and, if toxicity is not seen, evidence presented that the maximum attainable dose was used and that it was systemically absorbed. The intermediate dose should be symptomatic but not lethal and is normally the geometric mean between the high and low doses.

The choice of animal species is dictated by the need to satisfy regulatory requirements and a knowledge of the actions of the test compound in the dose-ranging studies, pharmacological testing and its kinetic and metabolic profile. Regulatory demand is for two species, one of which is a rodent and the other a non-rodent. Once again, the familiar choice is the rat and either the dog or primate. The choice of species should be clearly reasoned and capable of justification. The information on which the choice is made comes mainly from the dose-ranging studies. These will have investigated the response of each species to the compound and together with appropriate kinetic investigation, either within the dose-ranging studies themselves or in parallel experiments, allow a rational choice of species and dose. In an ideal situation, the handling of the compound in the species chosen will be the same as that in man. As one objective of the initial sub-acute studies is to allow the first administration to man, the information is obviously not available to allow comparisons to be made between animals and man in deciding the choice of species; similarities of metabolism and kinetics between man and the species chosen remain a criterion of choice for longer term studies.

Unless there are compelling scientific reasons why this should not be so, animals of both sexes are included in the experiment—any variation from this will require a clear and cogent justification. Equal numbers of animals of each sex are used in each group—test and control. Group sizes need not be large for studies of this length; in one and three month studies animals will not die due to natural causes and allowance need not therefore be made for this by the inclusion of extra animals. It is common to include 10–15 animals of each sex at each dose level.

4.7.6.4 Chronic Studies

Long term dosing studies come in two forms—chronic studies involving dosing for long periods, generally taken to be more than 90 days, and carcinogenicity studies. These latter studies are dealt with below in Section 4.7.6.5. Although Table 1 lists restrictions on the duration of clinical trials dependent on the available repeat-dose toxicology, this does not take into account the requirements of all countries nor the need for carcinogenicity studies. Carcinogenicity studies are generally required for clinical usage exceeding six months and this is amplified below.

The figures in Table 1 refer to EEC guidelines,[5] but other countries require further data. Japan, for example, requires a 28 day sub-acute study and a six month chronic study for administrations of between one and four week duration and a 90 day sub-acute study and one year chronic study for administration for over four weeks.[6] At first sight, the Japanese demands for a 28 day sub-acute plus a six month chronic study would appear similar to the EEC demands of 180 day toxicity to cover up to 28 days administration in man. However, this similarity misses a fundamental point in the differing objectives of sub-acute and chronic studies. In a sub-acute study, the principal objective is to induce toxicity and establish its nature; in a chronic study, the objective is to determine if this same, or some other manifestation of toxicity, becomes apparent on longer term less stressful exposure to the compound. Thus, whereas a sub-acute study is deliberately designed to overload the animals' metabolic systems, doses in a chronic study are chosen to be less severe. The high dose should be such as to induce toxic effects of only moderate severity and of relatively short duration, at least during the first half of the study when the animals should remain in good condition; this level is often chosen as being between the intermediate and high dose levels of the sub-acute study. The low dose in a chronic study should remain in excess of human dose, often three to five times, and the intermediate levels the geometric mean between these two dose levels.

Reference has already been made to the differing requirements of Japan and the EEC with regard to the need for chronic studies and the picture becomes more confused as other countries are considered. Table 6 lists the requirements of Canada, the EEC, the USA and Japan and highlights the differences between these four authorities. It is highly unlikely that any new drug would be developed for only one market and thus chronic toxicology must be undertaken to satisfy the most demanding of the international regulatory authorities. Registration, however, is obviously possible at different times in different countries depending on their demands; it is also limited by the need for carcinogenicity studies (discussed in Section 4.7.6.5).

Table 6 Duration of Toxicity Studies Required for Differing Durations of Exposure in Man

	Duration of studies (months)			
Duration of administration to man	*Canada*[7]	*EEC*[5]	*USA*[8]	*Japan*[6]
1 week	4–6	1	3	3
1 month	18	3	6	6
3 months	18	6	6	12
6 months or longer	18	6	12–18[a]	12

[a] USA figures for six months or longer administration to man refer to 12 months in a non-rodent and 18 months in a rodent.

There has been much debate about the value of prolonging chronic studies beyond six months, differing publications claiming that there is or is not extra information to be gained by studies of longer duration. A World Health Organization publication[9] advocates longer term studies because; '(a) chemicals may produce different toxic responses when administered over a prolonged period: and (b) during the aging process, factors such as altered tissue sensitivity, changing metabolic and physiological capability and spontaneous disease may influence the degree and nature of toxic responses.' Other papers have reinforced the need for longer term studies,[9,10] citing examples of toxic effects which would have been missed had toxicity studies been curtailed at six months. On the other hand, Lumley and Walker[11] have published an analysis of data on 45 compounds from the files of pharmaceutical companies with a research and development presence in the UK; they concluded that no new information was gained in studies of longer than six months duration. The economic attractions of restricting chronic toxicity studies to six months are obvious, but the scientific justification for doing this remains an active area of debate.

Whatever the length of study undertaken, it is important to realize the amount of information gained previously to this stage in a compound's development and the impact this can have on the protocol for the chronic toxicity. Not only are the results of the sub-acute studies available, but also information on kinetics and metabolism in animals and man, early clinical pharmacology and clinical trials in man, and also probably in teratology. Figure 1 describes a possible scheme for the phasing of toxicology studies. All the available information can be brought together to define the protocol, particularly with regard to two aspects—choice of species and choice of dose. One rodent

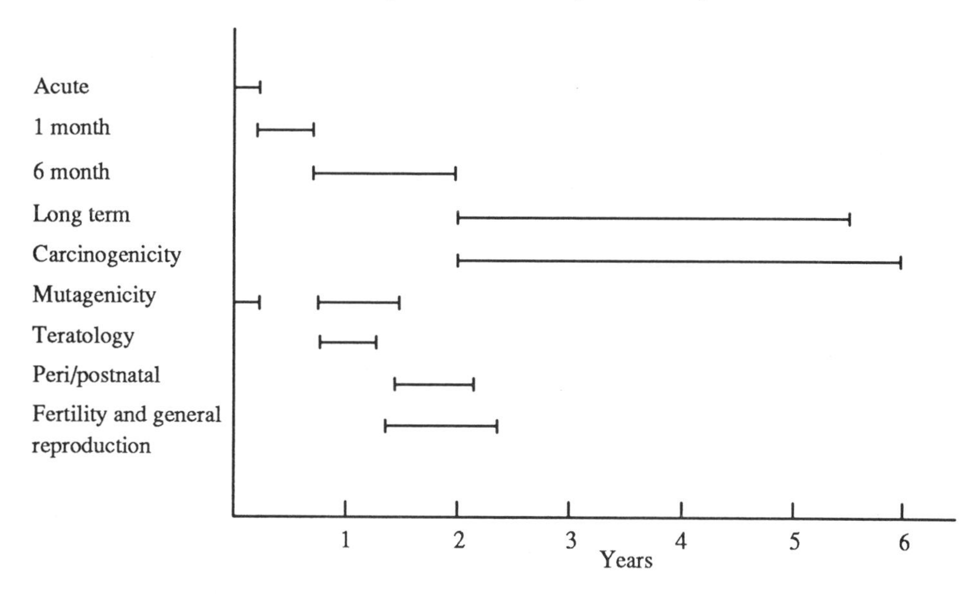

Figure 1 A possible phasing of toxicology studies on a typical compound requiring long term and carcinogenicity studies. The periods for each study include the time to establish the protocol, carry out the experimental phase, analyze the results and produce the report. The time axis gives a possible total of six years—this can be shortened if greater overlap is planned

(usually the rat) and one non-rodent species are required and these are often those used in the sub-acute studies. However, as has been discussed above, the sub-acute studies are conducted before any knowledge of the kinetic and metabolic handling of the compound in man is obtained. One of the aims of these sub-acute studies was to permit this knowledge, among others, to be gained and the species for the chronic studies should be chosen with full cognizance of these parameters.

The route chosen should be that intended for clinical use. If oral dosing is intended, it is normal in studies of longer than six months to administer the drug incorporated into the food of rodents, though individual dosing of the non-rodent species is maintained. If dosing *via* the diet is to be undertaken, ancillary studies should be established to investigate the mixing of food and drug, the stability of the mixture and the plasma kinetics of the compound.

In addition to knowledge of kinetic and metabolic parameters, other information may be available on the compound, either as a result of previous toxicity studies or from its use in man. It may be necessary to include within the protocol of the chronic studies specific investigations of parameters arising from observations in these earlier studies. This will of course influence the number of animals at each dose level. Interim investigations are normally included in longer term studies and these involve sacrifice of a number of animals in each group at intermediate times. Also, with six month studies, the recovery from any toxic effect is usually investigated in animals which are kept for a further six to eight weeks after the end of dosing before they are sacrificed—this allows the reversibility of any toxicity to be investigated. All these investigations require the inclusion of extra animals. In addition, it must be ensured that sufficient animals are started on the study so as to allow enough survivors at the end of a 12 or 18 month study, taking into account natural losses in what is, for a rodent, a study lasting a large part of its life.

Interim sacrifices, as have been mentioned, allow the development of any toxicity to be observed. Besides these sacrifices, toxic changes can also be monitored by blood, urine and faecal sampling at set intervals through the study.

The net result of these chronic studies—in rodent and non-rodent species—should be to allow predictions to be made on the likely toxic effects in man on chronic dosing. They often run parallel to clinical studies and as available toxicology limits the duration of clinical exposure, information gained at interim sacrifices (providing sufficient animals are included) can be fed back to allow more prolonged exposure of man to the compound.

4.7.6.5 Carcinogenicity Studies

Increasing awareness of the chemicals within our environment has led to heightened appreciation of the potential carcinogenicity of these agents and none more so than pharmaceuticals to which we have a controlled exposure of a given length and at a given level, *i.e.* dose. Appreciation of the control available over exposure to pharmaceuticals leads to a tighter definition of any risk–benefit equation liable to be employed in the evaluation of the results of a carcinogenicity study. Operationally, a carcinogen may be defined as any agent which significantly increases the incidence of malignant neoplasms, irrespective of its mechanism of action. In interpreting a carcinogenicity study, alterations in the incidence of tumours should be considered in terms of whether the test compound is an initiator or a promoter. Agents which are initiators are capable, either directly or after metabolic activation, of interacting with a cellular target, which is usually assumed to be DNA, to bring about changes in the cell leading to tumour growth. A promoter, on the other hand, is not itself capable of inducing tumour growth, but promotes tumours by bringing about an alteration in endocrine function or unmasking the effect of a virus or of a concurrently present genotoxic initiator.

The first successful carcinogenicity study was carried out in 1915 when tumours were produced experimentally by painting rabbits' ears with coal tar.[12] This led to the realization of the potential value of such studies because: (i) it showed the possibility of confirming experimentally observations made in humans; and (ii) the tumours obtained were comparable to those observed in people exposed to coal tar, lending extra weight to the parallel. Subsequent experimentation led to the identification of the constituents of coal tar responsible for the carcinogenic effect. Based on this and similar studies, long term tests were introduced: (i) to prove or disprove that chemicals or chemical mixtures suspected of producing cancers in humans could induce tumours in experimental animals; and (ii) to ascertain independently from any observation in humans that a chemical could produce cancers in an animal model. Validation of the predictivity of animal models came from the screening of known carcinogens—recognized from their effects in man—for their effects in animals.[13]

Rodents, the rat and mouse, are the species usually chosen for carcinogenicity studies and the general way in which such tests are conducted have been established for several decades. The principle of such a test is that it should involve lifetime exposure to the test compound and this, in itself, often dictates the choice of the rat and mouse, giving study lengths respectively of 24 and 18 months. Debate has taken place on longer duration of carcinogenicity studies, but the problem then arises from obfuscation of the interpretation because of loss of animals due to spontaneous death and subsequent autolysis and cannibalism. In a study of 179 chemicals, Grice and Burek[14] concluded that for over 95% there was evidence of carcinogenicity before 18 months in rats and for all there was evidence of carcinogenicity before 24 months. With improved animal husbandry, however, it is possible to maintain animals for longer periods and it is thus possible to consider prolonging studies if the results warrant it. On occasions where appreciation of the biological activities of the compound dictate the choice of alternative species the same principle should apply, *i.e.* exposure for a lifetime.

The choice of dose is crucial to a successful carcinogenicity study and the criteria employed differ fundamentally from those used in dose selection for chronic toxicology studies. The top dose should be one which exerts minimal overt toxicity, does not reduce lifespan except as a result of tumour induction and, as a broad rule, should not decrease body weight by more than 10%. In no case should the animals' metabolic/detoxification mechanisms be overwhelmed such that novel ways of handling the drug under test are exposed. It is common to choose a top dose which does not exceed 50 or 100 times the maximum daily dose in humans. This assumes a great deal of knowledge about pharmacokinetics and metabolism of the test compound both in animals and in man. The dosing regimen chosen should be based on such knowledge. As with chronic toxicity studies, drugs are often administered to rodents in the diet and, once again, particular care should be taken that suitable blood levels are achieved by this method. The lowest dose should be set at two or three times the maximum clinical dose or the dose that produces a pharmacologic effect: the intermediate dose will be the geometric mean between these two levels.

It is normal to start a carcinogenicity study with 50–75 animals of each sex in each test group with double this number in the control group. This is with the aim of having 15–20 animals per group alive at the termination of the study for full autopsy. All animals dying during the study should be autopsied and the cause of death established if possible. EEC guidelines[5] are typical with regard to which tissues should be subjected to histopathological examination. All tissues, basically in line with those given in Table 4, should be examined in the high dose and control groups and in animals dying or killed *in extremis* during the study. In addition, all grossly visible tumours and other lesions should be examined. Once any target organs have been identified, these should also be examined in the intermediate and lower dose groups. From these examinations, two possible positive effects of the test substance may be discerned, to increase the incidence of tumours or to decrease the latency to their appearance.

Reference has already been made in discussing Table 2 to the limitations in the power of a toxicology study placed on it by the choice of group size. Nowhere is this more important than in a carcinogenicity study. With a maximum of 150 animals, 75 male and 75 female, per dose level, reference to Table 2 shows the ability of the study to determine increases in tumour incidence over a 10% occurrence in the control group. This argument obviously lies in attributing causation to an apparent increase in incidence of a spontaneously occurring tumour in the species and strain under study. If the test compound induces a novel type of tumour which does not occur in control animals then causation is much easier to attribute, again as demonstrated in Table 2. Two principal factors assist in interpreting apparent increases in tumour incidence over background levels—the determination of whether or not any increase is dose related and the use of historical control data.

Any evaluation of apparent increases in tumour incidences is greatly aided if the effect is dose related or is related in incidence to both dose and duration of dose in that the higher doses cause changes earlier in the study than lower doses. Availability of historical data can similarly assist in evaluation of a carcinogenicity study, particularly when the differences in tumour incidence between the test and control groups within the particular study are small. However, if historical control data are to be included, several strict criteria should be met.[14] Animals should obviously be of the same strain, but also should have been housed under identical conditions, have been handled in the same way, have been fed with the same diet and generally should not differ from the study being evaluated.

The conduct of a long term carcinogenicity study should be particularly well controlled, as several factors, including diet, are known to be capable of affecting tumour incidence. Comprehensive guidelines for the conduct of such studies are available,[15] and for their interpretation based on statistical evaluation and an actuarial approach.[16,17]

4.7.6.6 Reproduction Studies

The aim of these studies is to investigate any effects a new drug may have on mating behaviour, foetal development seen as foetal loss and abnormalities, and any effect on the development of offspring in later life. These aims can be further particularized:[5] '(i) changes in fertility or in the production of normal young due to damage to the male and/or female gametes; (ii) interference with pre-implantation and implantation stages in the development of the conceptus; (iii) toxic effects on the embryo; (iv) toxic effects on the foetus; (v) changes in maternal physiology producing secondary effects on embryo or foetus; (vi) effects on uterine or placental growth or development; (vii) interference with parturition; (viii) effects with postnatal development and suckling of the progeny, and on maternal lactation; and (ix) late effects on the progeny.'

These aims are satisfied within three studies which are discussed in more detail below: (i) segment 1 or general reproduction and fertility involving commencement of dosing of both males and females before mating and observing effects on both offspring and parents; (ii) segment 2 is the teratogenicity/teratology study where any effects on the embryo are investigated by dosing the pregnant dams during specific periods of their gestation; and (iii) segment 3 or the perinatal or postnatal study aims to investigate effects on the suckling and lactating dam and upon development of the offspring.

Reproduction studies are generally carried out at various stages during the development pathway for a new drug. Often, however, the teratology study is carried out relatively early, as shown in Figure 1, as lack of this information can limit the scope of the envisaged clinical trials, as without it exposure of women of child-bearing age to the drug is not permitted. The other studies follow later, the exact timing depending upon the overall plan and the impact on this of knowledge gained from other studies. Reproduction studies should never be considered as a separate entity, but must always be viewed as part of the whole programme. Information gained in other studies—pharmacological, toxicological, kinetic and even clinical—should be considered in planning investigations of reproduction.

4.7.6.6.1 General reproduction and fertility

In this study, male and female rodents, usually rats, are dosed before mating and are then allowed to mate. The males have been dosed for 60 days and females for 14 days prior to this (Figure 2). When conception has occurred, the males are sacrificed and the rats examined histologically. The females continue to be dosed. Half the females are sacrificed during the pregnancy some days before the expected date of parturition and the uterus examined for number of corpora lutea, implantation

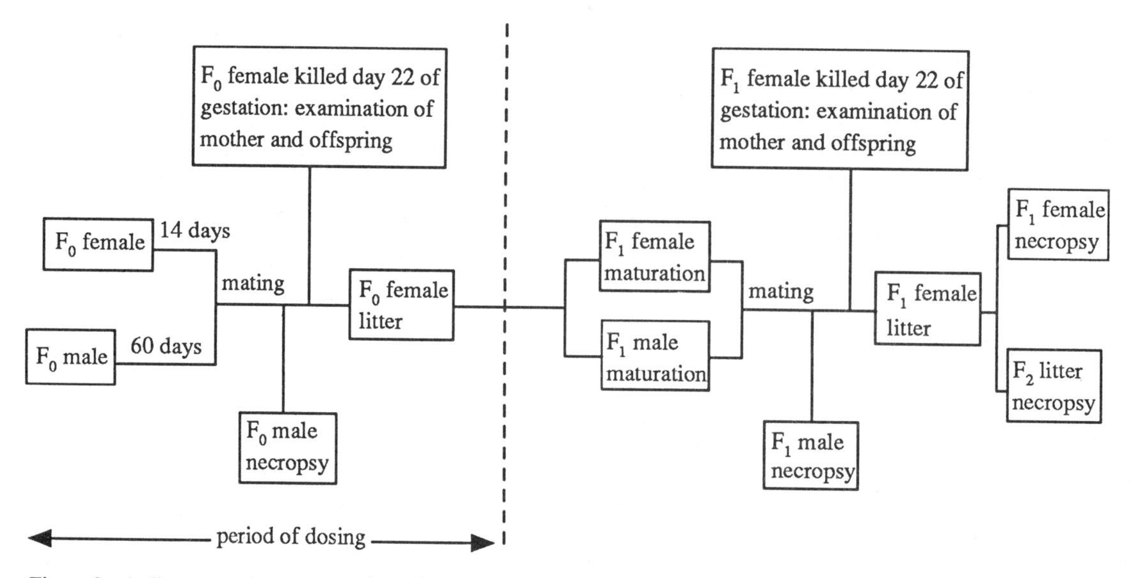

Figure 2 A diagrammatic representation of the protocol of a fertility and general reproduction study. Male and female animals of the F_0 generation are dosed prior to mating and dosing is continued in the females until weaning of the F_1 generation, part of which is then allowed to develop, mate and produce an F_2 generation

sites and resorptions as well as the foetuses for signs of abnormality. The remaining females continue to be dosed through pregnancy, delivery and lactation and are then sacrificed and examined. Part of the progeny should be raised to maturity so that any effect on auditory, visual and behavioural function can be examined. Finally, at least one litter should be produced (F_2 generation) by the progeny of the dosed parents.

This study thus potentially provides a wealth of information on all stages of the reproductive process, examining effects on both parents and offspring. The other types of reproductive studies focus on effects on the foetus and the newborn offspring.

4.7.6.6.2 Teratology

A teratogenic effect becoming apparent in an offspring may be the result of actions in the mother and/or the embryo. Whilst the general reproduction and fertility study examines to some extent the

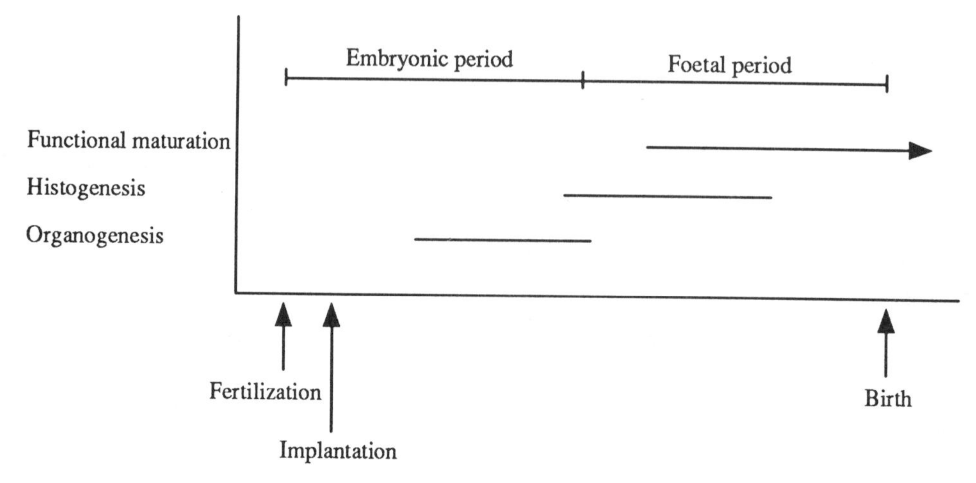

Figure 3 A representation of the different stages of development. The period covering organogenesis is that which is most sensitive to the effect of a teratogenic agent, although it is possible for minor structural defects to be generated in certain instances until after birth.

consequences of effects in the mother, it is the specific aim of the teratology studies to investigate effects of drugs on the embryo by dosing during the period of organogenesis. The sensitivity of an embryo to drug-induced malformations varies during gestation with the period of organogenesis providing the maximum sensitivity (Figure 3). Regulatory authorities require teratology studies in two species, one of which should be a non-rodent, and the rat and rabbit are the species usually chosen, though mice are also used as an alternative rodent. Whilst it is of fundamental importance to dose the mothers during the period of organogenesis, different organs develop at different rates and have differing susceptibilities with respect to exposure to a teratogen. For example, a single administration of a teratogen on day 10 of gestation of a rat would result in 35% brain defects, 33% eye defects, 24% heart defects, 18% skeletal defects, 6% urogenital defects and 0% palate defects.[18] Thus dosing of test agents is carried out not on a single occasion but over several days. Relative to conception, dosing is given on days 6–15 in the mouse and rat, though Japan requires dosing on days 7–17 in the rat, and days 6–18 in the rabbit. The normal gestation period in these species is 19, 22 and 30 days in the mouse, rat and rabbit respectively.

On the day prior to expected parturition, the foetuses are delivered by caesarean section and examined for the numbers living or dead, for gross visceral or skeletal abnormalities and the dams examined for evidence of resorptions. A certain percentage of foetuses will be spontaneously malformed and any drug-induced changes must be interpreted against this background. Moreover, the sensitivity of different species varies to different agents; thalidomide is not teratogenic in the rat, it is in the rabbit and this in itself has led to an increased need for teratological testing.[19]

4.7.6.6.3 Perinatal and postnatal study

The third segment of the battery of reproduction tests is the perinatal and postnatal study which examines effects of the test compound on late embryonic development, parturition and suckling up until weaning. The pregnant dam is given the compound during the final one-third of gestation and up until weaning, which in the rat—the species normally chosen for this study—is at three weeks. At the end of the weaning period, the offspring are sacrificed and examined at autopsy for signs of any abnormality—in addition to the various measures of their development made during suckling. In addition to the study of offspring at the end of weaning, it is also possible in certain circumstances to maintain some offspring to maturity so as to examine any effect on their later development and, by allowing them to mate, assess their reproductive capacity.

As with all the toxicity studies discussed in this chapter, the results of studies on reproductive function should be interpreted with care. The question, as always, is not what happens in an animal species, but what will happen in man in normal clinical use. As pointed out by Palmer,[20] almost any material may be teratogenic if given to the right species at the right time at the right dose. Choice of species for these studies is important and is based on experience and background knowledge. The rabbit is more susceptible to teratogenic agents than the rat and will identify thalidomide, for example, as a teratogen. However, the other side of this argument is that the rabbit is subject to a greater degree of 'noise', making extrapolations for risk assessment in human use more difficult.

The pharmacokinetics of a compound in man and in the test species are also important, as are the consequences of the pharmacological actions of the compound. Three examples will serve to illustrate this problem, all cited by Frohberg.[21] Comparative half-lives of compounds must be considered when interpreting the results of studies. Thus, in the rat, chloramphenicol shows an embryolethal effect at 300 to 1000 mg kg^{-1}, but from this it cannot be concluded that a therapeutic dose of 30 mg kg^{-1} would not involve a risk to the human foetus; comparative half-lives in the rat and humans are 30 min and 3 h and it follows that much higher doses are required in the rat to maintain exposure levels similar to those which would be experienced by the human foetus.

With regard to pharmacological activity, excess pharmacodynamic effects in the dams can lead to secondary impairments of development of the embryo. Thus, if the dose chosen is too high, administration of the sedative compounds chlorpromazine and phenobarbitone can lead to abortions, resorptions and underdeveloped foetuses simply because at the doses chosen the mothers were too sedated to eat properly. A final example concerns the evaluation of antibiotics, for example some cephalosporins, which are not teratogenic in the rat and mouse but are in the rabbit, often at doses approaching the human therapeutic level. This, however, is the consequence of the action of the cephalosporins in displacing the mainly Gram-positive intestinal flora in rabbits by Gram-negative coliform bacteria. This is not relevant for humans with their predominantly Gram-negative intestinal flora.

4.7.6.7 Genotoxicity Testing

Whilst many other types of toxicological investigations may seem rigid and well established, testing for genotoxicity is still a relatively new area and probably provides the greatest opportunity for confusion. It is only in the last 15 years that the evaluation of genotoxic potential has become a recognized and required part of the toxicological package for a new drug. However, this very rapid acceptance of the principle of genotoxicity testing has given rise to a plethora of potential testing methods offering a ready source of confusion. As with all toxicological testing, the choice of methods of testing and of the protocols employed should have a rational scientific basis with knowledge of the full toxicological activity of the compound and also its chemical nature. We have not yet reached a stage where the protocols to be used and the testing methods themselves are enshrined in legislature or guidelines, but if comparisons with other toxicity testing are made, this day cannot be far off—unless rational approaches to the validation and acceptance of both methods and protocols are adopted by regulatory authorities. With such a relatively young science, and with the development and evaluation of new and existing methods still proceeding, it is too early—if it is ever apposite—to establish rigid guidelines.

The origins of genotoxicity testing lie in many ways with the recognition of the damaging effects of ionizing radiation and the use of chromosome examination after the Hiroshima and Nagasaki atomic bombs. However, it was not until the work of Ames[22,23] that the area came to be incorporated routinely into the evaluation of new drugs—and also the evaluation of chemicals generally, based on growing public and regulatory awareness of the potential carcinogenic and mutagenic hazards posed by the chemicals introduced into our environment.

Just as the value of carcinogenicity testing is based empirically on the correlation of animal tests with clinical experience, so the validation of mutagenicity testing is based on correlations between animal carcinogenicity testing and short term genotoxic testing. Early estimates of the predictive value of the Ames test were high—85% of carcinogens being detected (135 out of 158 tested) and only less than 10% of 106 non-carcinogens[24]—but these correlations appear to have decreased as more data have accumulated. Thus, for example, in a study of 73 chemicals on which well-documented carcinogenicity data were available (44 positive and 29 negative carcinogens), Tennant *et al.*[25] found the Ames test to detect as positive only 45% of the carcinogens and 14% of the non-carcinogens. Over a battery of four genotoxicity tests, including the Ames test, 14% of the carcinogens were not detected by any test and 65% of the non-carcinogens were detected as positive in one or more tests. This once more underlines the necessity to approach the interpretation of toxicity and, above all, of genotoxicity with caution and total appreciation of all aspects of the biology of the compound. Closer examination of the apparent lack of correlation by examining individual compounds reveals scientific reasons for these discrepancies in a number of cases. Principal among the compounds which are positive carcinogens but negative mutagens are promoters. As already discussed in Section 4.7.6.5, both initiators and promoters can and will be detected in carcinogenicity testing—only initiators, however, will be detected in genotoxicity testing. Furthermore, genotoxic assays have been criticized for being oversensitive, in that they apparently throw up false negatives; these again can often be understood by greater knowledge of the compound and of the limitations of the particular assay systems used. An *in vitro* genotoxic assay can indicate whether or not a particular compound has genotoxic potential; whether or not that potential will be expressed can only be tested *in vivo* in a full carcinogenicity study and a further level of prediction involves the value of the carcinogenicity study to predict effects in man.

4.7.6.7.1 Metabolic activation

It has become apparent that a large proportion of genotoxic agents are detected in short term *in vitro* assays only after metabolic activation to the relevant active moiety. Most assays must, therefore, be conducted in the presence and absence of metabolic activation. The normal method of activation is by incorporation of the so-called 'S-9' mix. This is normally generated from a microsomal preparation of the livers of rats pretreated with an enzyme inducer; contained within the 'soup' are enzymes capable of generating electrophiles, while generally the detoxification enzymes are excluded as they are to be found in the cytosol supernatant fraction. The 'S-9' mix is not, therefore, entirely representative and results of its use should be interpreted with care. It is of particular value in tests such as the Ames test which employ bacteria; these generally lack elaborate metabolic capability. Although 'S-9' is also incorporated into assays involving mammalian cells, these cells have metabolic capability of their own and can give rise to a different set of metabolites.

Complete interpretation of an *in vitro* assay involving 'S-9' can include the determination of the metabolites produced by its usage.

4.7.6.7.2 Available methods

An exhaustive review of all the available methods—even an exhaustive listing—is beyond the scope of this chapter. Those interested in greater detail are referred to other references.[15,26] Rather, attention will be focussed on the tests more generally encountered in the evaluation of drugs. Regulatory requirements are not yet firmly fixed—and hopefully never will be—as to which tests should be performed, but rather give guidelines as to the properties to be examined. The EEC guidelines[27] are typical in requesting information on gene mutation and chromosome mutation and, if possible, genome mutation, referring respectively to alterations at the level of the individual gene, of chromosomal structure or of chromosomal number. Furthermore, the EEC asks for tests in both prokaryotes and eukaryotes as the organization of genetic material in these two types of organisms differ and further that one *in vivo* test should be included as well as inclusion of metabolic activation systems in *in vitro* systems.

To comply with these guidelines, the EEC suggests one test from each of the following categories be used: (i) gene mutation in bacteria, (ii) chromosome aberrations in mammalian cells *in vitro*, (iii) gene mutation in eukaryote systems, and (iv) an *in vivo* test.

The most commonly employed gene mutation assay in bacteria is the Ames test, based on reversion of strains of *Salmonella typhimurium* with a nutritional requirement for histidine. Bacteria of specific strains are plated with test compound, with and without metabolic activation, in medium with a limiting amount of histidine. The basis of the assay is the requirement of histidine for growth—only mutant strains will continue to grow. Several strains are used, each of which gives specific information. Amongst these, for example, TA 100 and TA 1535 give information on base changes (substitution of one base in DNA for another) while TA 98 and TA 1537 indicate frame shifts (deletion or addition of a base such that the sequence no longer reads correctly). Another similarly based test is established in *Escherichia coli* strain WP2 mutants, where base changes are measured by a mutation away from a requirement for tryptophan for growth.

Detection and analysis of chromosome aberrations relies on microscopic examination. This can be done after both *in vitro* and *in vivo* exposure to the test agent. A variety of different *in vitro* cell systems have been proposed, one advantage of these being that it is possible to choose a human cell system on which to experiment. As with the Ames test, however, full metabolic capability is not present and drugs should be tested both in the presence and absence of the 'S-9' mix. A common source of cell is the freshly obtained human peripheral blood lymphocyte. Cells are obtained from healthy donors, separated from blood and incubated with mitogen and treated with a spindle inhibitor (vinblastine) to arrest the cells in metaphase. After a further brief incubation the cells are separated and prepared for analysis. Other types of mammalian cells—cultured fibroblasts or Chinese hamster cell lines—can also be used with basically the same sort of protocol, though addition of a mitogen is not always necessary.

The most commonly used *in vivo* genotoxic test also relies on microscopic examination of possible chromosome damage. This is the micronucleus test which involves counting of micronuclei in polychromatic erythrocytes. The erythrocytes are obtained from rodents (usually mice, but rats are becoming more common) previously treated with the drug. For reasons that are not totally clear, maturing erythrocytes do not expel micronuclei along with the nuclei during their last mitosis in the bone-marrow. These remaining micronuclei, indicative of cytogenetic damage, can thus be easily counted using the microscope. Alternatives to the micronucleus test include the *in vivo* counterpart of the *in vitro* lymphocyte assay described above, where chromosome damage is sought in lymphocytes taken from rodents which have previously received the drug. A further alternative is the analysis of bone-marrow cells arrested in metaphase and taken from previously treated animals.

Several cell systems are available for the investigation of effects in inducing gene mutation in eukaryotic systems. Once again compounds are tested in the absence and presence of metabolic activation systems ('S-9' mix). The most commonly used cell lines are those from the Chinese hamster—the V79 lung fibroblast and the CHO cells—and the L5178Y mouse lymphoma cells. All these tests rely on the isolation of drug-resistant mutants. The V79 and CHO cells are resistant to 8-azaguanine and 6-thioguanine because of a mutation at the HGPRT (hypoxanthine-guanine phosphoribosyl transferase) locus. This enzyme is involved in the purine salvage pathway and transforms the 8-azaguanine and 6-thioguanine to lethal nucleotides. Mutants which lack the enzymes cannot do this and hence survive in culture—the basis of the assay is therefore the number of cells which survive. A similar principle exists in the L5178Y mouse lymphoma cell, where resistance to 5-bromodeoxyuridine is developed as a result of changes at the thymidine kinase locus.

These comprise the most commonly used test systems and a suitable choice from amongst them would satisfy regulatory authorities. Other test systems exist which employ tests in other cell lines, in yeasts and moulds, and in *Drosophila.* All of these test systems can be utilized and all address the problem of investigating potential genotoxicity. Which are chosen for a particular compound should be a matter of rational scientific decision, based on knowledge of the compound—both chemical and biological. The finding of positive effects in some tests may not necessarily be equated with direct genotoxic potential in man and further investigations may be required to elucidate this.

4.7.6.7.3 Interpretation

At its most simple, the result of a single genotoxic test indicates whether or not the test agent is genotoxic in that test system. Any further extrapolation of the likely consequences of this finding is a matter of correlation and experience. Before the ultimate extrapolation is made, *i.e.* to predict the likely genotoxic potential in man, several criteria should be met. Foremost amongst these is that the result of that particular experiment should be real. Most recommended protocols for the different genotoxic assays include a positive control, providing evidence of the sensitivity of the test system in that particular experiment. Similarly, negative controls are included in terms of untreated cells or animals; most assay systems have a degree of spontaneous mutation which is apparent in this control group. The degree of spontaneous mutation will vary from one experiment to another and will thus provide a range of background values against which any drug-induced changes should be judged. It is not, therefore, simply the data within one particular experiment which must be evaluated, but rather the results of that experiment should be considered against the historical control data for that assay in that laboratory. It is possible that a particular compound could show a statistically significant effect as compared to the negative control within that one experiment, but that the values obtained lie within the range of control values established over a number of experiments—the outcome of this is that a number of assays have a certain level, say two or five times background, which must be exceeded before a compound is considered to have a positive effect. Most tests involve several dose levels of the test compound and also replication of samples. It is important to recognize that true mutagens have positive effects which are both dose-related and reproducible. Furthermore, the need, or not, for metabolic activation for a positive effect should be evaluated and also the known reasons for likely false positives and false negatives excluded.

Taken together, a suitable battery of genotoxic tests, the results of which have been properly evaluated, can provide valuable information on the likely mutagenic potential of a compound both in animals and man. Inclusion of an Ames test is made early within the development plan of a compound and other genotoxicity tests are planned at different stages. Genotoxicity tests serve two functions within the overall development plan. They provide points at which go/no-go decisions can be made with regard to further development and they can also be considered as a part of the total data package, particularly alongside the carcinogenicity studies, in extrapolating risk to human usage.

4.7.6.8 Special Studies

Whilst the studies discussed in the previous sections provide the backbone of the toxicological evidence required for product licence application, special supplementary studies—relating to one or more particular aspect of the compound's toxicological profile—may also be required. These extra studies can arise from observations made during the other studies, for example the need to investigate further the observed toxic effects on a particular organ, or may be related to the chemical nature of the compound, its pharmacology or its intended route of administration. In each case, these extra studies should be undertaken only if required and should be specifically designed to address the problem being investigated.

These studies could include, for example, a special investigation of nephrotoxicity because the drug comes from a chemical class of compound known to have this problem. Although observations on nephrotoxic potential will have been made during other repeat-dose studies, it may be advisable to establish other tests to investigate this problem in isolation so that a greater range of parameters could be included and also comparisons run with other compounds already known to pose a problem in this area—often potential clinical comparators and competitors.

Examples of additional studies which could arise from observations made within the normal repeat-dose studies are the investigation of immune toxicological or behavioural teratological effects. Both these areas represent relatively new areas of interest to toxicologists and are under active investigation. Indications of potential problems in both these areas can be gained from the sub-acute and chronic toxicity and the teratology studies respectively. Indications of potential immune toxicity can be derived from effects on lymphoid organs and white cell counts, but the consequences of these events are not measured within a routine toxicity study. Measurement of immune function would have to be investigated in supplementary studies. Although no guidelines yet exist for the determination of immune toxicity, the envisaged study could include assessment of responsiveness of lymphocytes to mitogen, the ability to mount cell-mediated and antibody responses and the ability of the animals to withstand infection. Similarly, the routine teratology and, more particularly, the general reproduction and fertility study (described in Section 4.7.6.6.1) may indicate a potential problem in terms of behaviour of offspring and this may require further study. Again, no firm guidelines for doing this exist at the moment and the studies should be designed to test that element of behaviour which is thought to be abnormal. Both these examples cite relatively new areas of toxicology but of course the need for further studies to investigate in greater depth previously observed findings may fall into 'classic' lines, *e.g.* liver or kidney toxicity.

The intended route of administration of a compound can also provide certain demands in terms of additional requirements. Whilst all toxicological testing should be by the intended route of administration in man, be that oral, parenteral, inhalation or topical, it can often happen that a compound is considered for administration by two routes, only the major one of which is adequately covered by existing toxicological data—additional studies are thus required. As well as the appropriate minimum toxicity by the additional route, kinetic and metabolic studies should establish the validity of extrapolating results gained by another route to the present situation. Particular problems are posed by drugs to be administered topically where irritant and sensitization potential must also be investigated.

4.7.7 GOOD LABORATORY PRACTICE (GLP)

Today, the use of standard operating procedures is widespread throughout the pharmaceutical industry, both in drug development and in manufacturing. Compliance with these procedures is an integral part of GLP. Although the move towards regulations of this type was perhaps already underway, their introduction was expedited by the outcome of the Congressional hearings in the USA in the mid 1970s, in which faulty animal experimentation by one firm was investigated. Further investigations of other independent testing laboratories showed defects in design, conduct and reporting of studies. GLP became law in the USA in June 1979, bringing with it a system which introduced two major changes—the concept of the study director and the quality assurance unit. The latter operates independently to assure compliance with the regulations and the protocols for the study. The study director is the focal point within a study and it is through him or her that all information flows. Study directors have overall responsibility for carrying out the study. Their responsibilities start when they approve and sign the protocol and end when the final report is audited, approved and issued. Their function, however, is more than merely keeping track of paperwork; as well as being competent administrators they are first and foremost scientists. It is their role to check the events happening in a study, separating serendipitous events from drug-related events, making recommendations on necessary amendments to the protocol to deal with any eventuality.

Whilst GLP could be seen as a necessary evil which has increased the cost of toxicology studies and led to a checklist approach to safety evaluation, its impact has not been negative. It does ensure the maintenance of standards and that all events are documented and, in general, leads to a higher quality of study.

4.7.8 THE TOXICOLOGICAL EXPERT REPORT

A relatively recent addition to the toxicological evaluation of new drugs is the inclusion of expert opinion of the data within an application for a product licence. This has been introduced by the EEC—it remains to be seen how many other countries it will extend to in the future. This requirement holds both for submissions to individual countries within the EEC and also for applications using the multistate application system.

Within the EEC system, three expert reports are required in addition to the base data of the application. These three reports cover the pharmaceutical, pharmacotoxicological and clinical aspects of the data. The pharmacotoxicological report should address all the biological information—pharmacology, toxicology and pharmacokinetics. Unlike the other two expert reports—pharmaceutical and clinical—there are no formal qualifications required of the pharmacotoxicological expert, other than that he or she be competent. Moreover, given the range of topics covered within this report, it may be necessary to obtain expert opinion of specific areas within the data or even have more than one person prepare the report, providing that all sections and opinions are clearly attributed.

The report is accompanied by all the data, both in full report and summary table form, and the role of the expert is to tie together and interrelate all these diffuse pieces of information, providing an interpretation of their meaning and a critical evaluation of their worth—both from a scientific standpoint and as to the technical validity of the experiments performed. Finally, the expert should state what conclusions can be drawn as a result of all the experiments performed. The information should be discussed in the context of relevant literature, paying particular attention to the properties of already existing drugs for treatment of the same disease(s). Furthermore, to re-emphasize a point made in the introduction to this chapter, the data should be evaluated in relation to the proposed clinical usage. The report should be critical and comprehensive and should address the issues faced by the licensing authority in reviewing the application. Some of the points which could be considered in the preparation of the report are listed in Table 7. The report should, of course, be accompanied by an account of the experience and qualifications of the expert or experts.

Table 7 Points to be Considered in the Preparation of a Pharmacotoxicological Expert's Report

Choice of doses or effective concentrations	Shortcomings in the design and execution of experiments
Rationale for the selection of species, *e.g.* for kinetic and metabolic reasons	Shortcomings in the conclusions made on individual experiments
Assessment of possible accumulation and tachyphylaxis	Risk criteria (for compounds with a narrow therapeutic margin)
Development of tolerance and withdrawal	Relationship between results of animal experiments and clinical trial data
Assessment of pharmacological data for treatment of acute poisoning	Animal husbandry and good laboratory practice
Critical appraisal of the statistical methods used	

4.7.9 REGULATORY REQUIREMENTS IN DIFFERENT COUNTRIES

Regulatory requirements do differ between countries, but not greatly. They can differ both in terms of which studies are required and in terms of the protocols for a particular study. The pharmaceutical industry covers many nations and the potential market for a new drug covers the whole world. It is thus important that the toxicologist evaluating a new drug ensures that it is tested to the highest standard so as to comply with the requirements of all the regulatory bodies and licensing authorities. In general, this is possible by carrying out tests to the requirements of the country with the most stringent regulations. However, extra experiments may have to be performed for certain countries; for example rat teratology requires dosing on days 7–17 for Japan, but on days 6–15 for the rest of the world—it is difficult, if not impossible, to encompass both these schedules into one coherent experiment. Table 8 gives an outline of the differences in regulatory requirements for a selection of countries, based on the EEC guidelines as the comparator; the list is by no means exhaustive and other countries can have differing requirements. It can be seen that there is a large degree of homogeneity in the demands of different countries and that the differences are relatively minor, requiring in the case of acute toxicity extra species for some countries, but an extra teratology study in the rat for Japan. The situation is most confusing in the area of genotoxicity, where this reflects the relative novelty of this form of testing and, as yet, lack of guidelines multinationally for this type of work. However, as has already been discussed, it is hoped that guidelines and regulations will not be established, as it were, in stone, but that rational scientific approaches will be made both as to the recommended tests available and as to the choice to be made as to which tests should be applied to a particular compound.

Countries also differ in their requirements for the phasing of studies with regard to the evolution of clinical trials and the exposure of patients to a new drug as a function of the duration and type of

Table 8 Examples of Differences in Regulatory Requirement Between Different Countries

Test	*EEC*	*USA*	*Japan*	*Canada*	*Switzerland*
Acute toxicity	2 species	3–4 species (one a non-rodent)	As EEC	3 species (one a non-rodent)	As EEC
Sub-acute and chronic	2 species (one a non-rodent)	As EEC	As EEC	As EEC	As EEC
Carcinogenicity	2 species (mouse and rat)	As EEC	As EEC	As EEC	As EEC
General reproduction and fertility	1 species (rat)	As EEC	As EEC	As EEC	As EEC
Teratology	2 species (rat and mouse dosed days 6–15, rabbit days 6–18)	As EEC	2 species (rat dosed days 7–17, mouse days 6–15, rabbit days 6–18)	As EEC	As EEC
Peri/postnatal	1 species (rat)	As EEC	As EEC	As EEC	As EEC
Genotoxicity	Bacterial genome test	Not specified	Bacterial reversion	Tests for gene mutation and chromosome effects	As EEC
	In vitro chromosome damage *In vitro* gene mutation *In vivo* chromosome damage		*In vitro* chromosome damage		

toxicological studies available. Table 1 gave the situation as it exists in the EEC and, although this is typical of the phased approach, the detail of what information is required at each stage differs between countries.

4.7.10 THE FUTURE IN TOXICOLOGY

Despite the seeming regimentation of which tests are required and of the protocols to be adopted for them, the future of toxicology is not set in stone. New tests and approaches are constantly being evaluated and the established methods being questioned. In a recent review of the status of toxicological testing, Doull[28] discussed the difference between the science and the art of toxicology: the science of toxicology is the conduct of experiments and the gathering of data, whilst the art comes into existence where extrapolations have to be made from these data to predict likely toxic effects in man. It must remain as the major aim of toxicologists to maintain and strengthen the science or data gathering so as to be surer of the art of prediction.

Recent years have been characterized by a rethink of the value of the existing tests—how good are they at predicting risk? How many toxic drugs have been recognized during their evaluation and further development stopped, thus saving man from potential hazard as a result of exposure to them? Conversely, how many potentially beneficial medicines have been denied to man because of a misinterpretation of the predictive value of a toxicology study? The answers to both these questions will remain unknown. The challenge, as has been emphasized throughout this chapter, is to carry out the right tests to determine the potential toxicity of a compound within its proposed clinical usage. Restraints are present in the shape of guidelines and regulations as to which studies should be conducted and often as to the protocols that should be adopted. However, if the scientific basis is sound, dialogues should be established with the regulatory authorities to discuss why guidelines should not be adhered to and why alternative tests should be used in a particular situation. Also, thought must be given as to the follow-up studies which should be conducted to investigate toxicity seen in routine testing—it is not enough to observe that toxicity; its significance must be addressed. Presented, for example, with two species of carcinogenicity, one of which indicates a positive effect, a regulatory authority will righty opt for prudence and not accept the compound. It may be, however, that the negative finding is the more predictive for man and the reasons for this difference between the species should be sought.

New methodology is constantly being sought to investigate both existing and new aspects of toxicology. Reference has already been made to the increasing awareness of immunotoxic events and the ways in which these could be investigated. A further example is provided by the concept of behavioural teratology and also the possible prediction of central side effects, *e.g.* headache and depression in man. Both these latter examples depend on our ability to model these effects in animals and to design tests suitable for drug evaluation—either alone or as part of a chronic or sub-acute toxicity study.

Further examples of new tests which might be introduced are provided by *in vitro* techniques. The most obvious types of *in vitro* experiments in current use are the genotoxicity tests and these are constantly being modified and re-evaluated. Other types of *in vitro* tests, *e.g.* for teratology, are being investigated. Their precise value has yet to be determined. Early hopes for *in vitro* and short term *in vivo* genotoxicity tests was that they might replace the cumbersome carcinogenicity studies; this has not been the case and the two types of study tend to be used in conjunction to provide greater predictive capacity and often *in vitro* studies are used to clarify and investigate the results of the long term studies. The extent to which short term tests or *in vitro* studies will replace or be used alongside existing tests has yet to be determined. Whatever the new test considered, however, its introduction into testing should be made because it provides a scientific advance; it should not be introduced as merely another item to be ticked on the list of toxicity testing.

4.7.11 CONCLUSIONS

The trend towards increasing regulations covering the exposure of man to chemicals either as drugs, as agrochemicals or as industrial chemicals has increased the role of the toxicologist in industry. With the majority of chemicals being considered for their toxicity, the sheer volume of industrial chemicals does perhaps indicate the need for a more systematic approach to their evaluation and a checklist format is perhaps inevitable. With the time and money invested in new drugs, however, it is possible to judge each compound on its merits and to design a package of tests suitable for that compound, bearing in mind its chemical, pharmacological and kinetic properties. This package should, of course, fall within the guidelines issued by countries in which it is intended to be registered, but should also include other tests (or exclude tests) as required by the science of the molecule and findings in other experiments. In this way, the safety of patients can be ensured and new and effective drugs can come to the market.

The way forward in toxicology lies in the efficient utilization of the available tests and in the evaluation and introduction of new tests to investigate parameters not now examined, but whose importance is becoming more recognized. By adopting these approaches, the toxicologist can provide valuable information at each stage of a compound's development and share in this development in a proactive rather than reactive way.

4.7.12 REFERENCES

1. H. J. Bein, *Proc. Eur. Soc. Study Drug Toxic.*, 1963, **2**, 15.
2. E. M. K. Geiling and P. R. Cannon, *JAMA, J. Am. Med. Assoc.*, 1938, **111**, 919.
3. J. M. Cruikshank, J. D. Fitzgerald and M. Tucker, in 'Safety Testing of New Drugs: Laboratory Predictions and Clinical Performance', ed. D. R. Laurence, A. E. M. McLean and M. Weatherall, Academic Press, London, 1984, p. 93.
4. J. M. Barnes, *Proc. Eur. Soc. Study Drug Toxic.*, 1963, **2**, 57.
5. *Official Journal of the European Communities*, 1983, **26**, L332.
6. Kosheisho, 'Toxicity-Test Guidelines 1984', Yakugyo Jiho, Japan, 1984.
7. 'Health Protection Branch, Pre-clinical Toxicological Guidelines', Bureau of Human Prescription Drugs, Health Protection Branch, Health and Welfare, Canada, 1981.
8. J. W. Kesterson, *Drug. Inf. J.*, 1982, **16**, 22.
9. G. L. Frederick, in 'Long-Term Animal Studies. Their Predictive Value for Man', ed. S. R. Walker and A. D. Dayan, MTP Press, Lancaster, 1986, p. 65.
10. V. C. Glocklin, in 'Long-Term Animal Studies. Their Predictive Value for Man', ed. S. R. Walker and A. D. Dayan, MTP Press, Lancaster, 1986, p. 77.
11. C. E. Lumley and S. R. Walker, *Fundam. Appl. Toxicol.*, 1985, **5**, 1007.
12. K. Yamagiwa and K. Ickikawa, *Verh. Jpn. Pathol. Ges.*, 1915, **5**, 142.
13. L. Tomatis, *Annu. Rev. Pharmacol. Toxicol.*, 1979, **19**, 511.
14. V. J. Feron and R. Kroes, *JAT, J. Appl. Toxicol.*, 1986, **6**, 307.
15. 'IARC Monographs on the Evaluation of the Carcinogenic Risk of Chemicals to Humans', International Agency for Research on Cancer, Lyon, 1980, suppl. 2, p. 21.
16. R. Peto, *Br. J. Cancer*, 1974, **29**, 101.
17. 'IARC Monographs on the Evaluation of the Carcinogenic Risk of Chemicals to Humans', International Agency for Research on Cancer, Lyon, 1980, suppl. 2, p. 311.
18. F. Beck, in 'Advances in Pharmacology and Therapeutics II', ed. H. Yoshida, Y. Hagihara and S. Ebashi, Pergamon Press, Oxford, 1982, vol. 5, p. 17.
19. J. G. Wilson, *Teratology*, 1979, **20**, 205.
20. A. K. Palmer, in 'The Principles and Methods in Modern Toxicology', ed. C. L. Galli, S. D. Murphy and R. Paoletti, Elsevier/North-Holland Biomedical Press, 1980, p. 139.
21. H. Frohberg, in 'Advances in Pharmacology and Therapeutics II', ed. H. Yoshida, Y. Hagihara and S. Ebashi, Pergamon Press, Oxford, 1982, vol. 5, p. 41.
22. B. N. Ames, F. Lee and W. Durston, *Proc. Natl. Acad. Sci. USA*, 1973, **70**, 782.
23. J. McCann, E. Choi, E. Yamasaki and B. N. Ames, *Proc. Natl. Acad. Sci. USA*, 1975, **72**, 5135.
24. B. N. Ames, J. McCann and E. Yamasaki, *Mutat. Res.*, 1975, **31**, 347.
25. R. W. Tennant, B. H. Margolin, M. D. Shelby, E. Zeiger, J. K. Haseman, J. Spalding, W. Caspary, M. Resnick, S. Stasiewicz, B. Anderson and R. Minor, *Science (Washington, D.C.)*, 1987, **236**, 933.
26. B. J. Dean and P. Hodges, in 'Animals and Alternatives in Toxicity Testing', ed. M. Balls, R. J. Riddell and A. N. Worden, Academic Press, London, 1983, p. 381.
27. *Official Journal of the European Communities*, 1987, **30**, L73/4.
28. J. Doull, *Pharmacol. Rev.*, 1984, **36**, 15S.

4.8

Clinical Pharmacology and Clinical Trials

J. ANDREW SUTTON

Knoll Ltd, Maidenhead, UK

4.8.1 INTRODUCTION

In this chapter the nature of trials at different stages of drug development and the organization they require will be described. Of 10 000 compounds screened in pharmacology, 100 will arrive at the development phase to be tested in toxicology and clinical pharmacology, 10 will survive the initial screens and enter clinical trials and only one will emerge in the market place.[1] This process is extremely costly and time consuming, so much so that in most pharmaceutical companies twice the manpower is devoted to it as in the drug discovery phase. A new drug takes 10 to 12 years to arrive on the market and costs some £50 to 100 million.[2]

In 1977 the Food and Drug Administration (FDA) in the USA divided the trial program into four phases.[3] Phase I concerned the initial tolerance studies in volunteers, Phase II clinical pharmacology, Phase III the later trials up to product licence and, lastly, Phase IV comprised post-marketing studies. Each phase has its own objective, methods and indeed character. For example, Phase II is highly experimental and the compound does not have the highest priority within the sponsoring organization and trials are small scale. Phase III has a green light for an active compound which must be marketed without delay. The pharmaceutical company is eagerly anticipating the new compound and money spent on large scale trials is seen as well spent. In Phase IV the struggle to gain a market share is on, clinical practice has a habit of showing up the limitations of the compound, possibly uncommon but significant side effects emerge and any new indications seem yet a long way off. So, the objectives of the phases differ, as must their departmental organization. Accordingly, this chapter examines the clinical trial program phase by phase.

4.8.2 PHASE I

4.8.2.1 Definition

Originally the term was applied to studies in volunteers before patient studies began and consisted of tolerance testing and absorption studies. However, volunteer trials are needed throughout the life of a compound, mostly to test new formulations and so the term Phase I has expanded to include them also.

In the UK it is not subject to regulatory approval but as with all studies in man, it is subject to medical ethics. Accordingly, a clinician, usually a clinical pharmacologist, must be responsible for the trial. It is a time for wide cooperation within the company or academic department organizing it, since experience from pharmacology, toxicology, pharmacy and clinical trials must be combined. In a wider sense, medical experience itself must be included in the evaluation of the purpose and aims of volunteer trials.

4.8.2.2 Ethical Background

In a comprehensive yet readable review of the ethics of clinical trials Vere[4] illustrates the uncertainties with which ethics tries to grapple. On one hand, the trial which even aims at an unequivocal result is rare indeed, let alone the one which demonstrates it. On another, the definition of what is ethical is a function of the society concerned. The fundamental medical principle of protection of the health of patients applies to all clinical trials. The principle is codified in the Declaration of Helsinki[5] and by the Medical Research Council of the UK[6] and these directives apply to trials in both volunteers and patients.

The normal doctor–patient relationship is founded on trust and the same basis must exist in a volunteer–clinical trialist collaboration. Double or single blind studies are claimed by some to represent deception and hence the antithesis of this trust,[7–11] but Vere states the prevailing view: that when patients know in advance of the deceptions to be practised upon them, there can be no ethical problem.[4]

Acceptable risk is a crucial point, since abolition of all risk is unrealistic. One precondition is that the patient or volunteer knows and accepts the risk and must therefore be informed appropriately. However this is insufficient for the risk must also be acceptable to public peer review, and with the likelihood of benefit to society at large which genuinely offsets it. This means that there is a negligible increase compared with everyday life. This may be defined by comparison with risks attached to activities that the public engage in voluntarily, such as driving a car or commercial flying. In volunteer trials the risk must be balanced by the advantage to society at large and not to the individual, because there can be no therapeutic benefit to a healthy individual. This has led to the concept of ethical review of volunteer trials by ethics committees.

Ethics committee aims, constitution and organization are summarized in several guidelines. The report of the Royal College of Surgeons 1986 is an authoritative review.[12] It defines the objectives of ethics committees as follows: '.... to facilitate medical research in the interests of society, to protect subjects of research from possible harm, to preserve their rights and to provide reassurance to the public that this is being done. Committees also protect research workers from unjustified attack'. The report is comprehensive and deals with more difficult cases, such as trials on children. In 1988 the Association of the British Pharmaceutical Industry (ABPI) produced guidelines which are more focused on Phase I studies.[13] From these guidelines the constitution of the committee may be obtained. In the USA ethics committees are called Institutional Review Boards (IRBs) for which the Food and Drug Administration (FDA) has strict criteria and requires the *curriculum vitae* of all members of the boards.[14]

Choices must be made between a postal system or one in which decisions are taken only at meetings or some intermediate form. The third option may have the advantage of rapid responses to protocols with meetings frequent enough to discuss comments, results, ethical issues and possibly meet some of the investigators. An exclusively postal system will not generate enough detailed discussion of important issues such as these. Detailed ethical considerations of trials are considered in Sections 4.8.2 and 4.8.3.

In general, the history of Phase I studies indicates a small degree of risk to volunteers. Thus, while the recruitment of prisoners is not acceptable in Europe, a survey of 805 trials in volunteer prisoners in the state of Michigan by Zarafontis *et al.* illustrates this.[15] More than 30 000 volunteers in over 60 000 study days encountered 64 'significant medical events', of which 58 were considered drug or study related. On average this is one event per 26 volunteer years. Of the 64 events, two failed to regress completely: one was a stiff hip after sepsis at an injection site, the other was a myocardial infarction on a placebo treatment day. In the British Isles two fatal incidents have been reported. In Dublin a predictable drug interaction occurred when a volunteer failed to report that he had just been given a depot injection of a neuroleptic compound.[16] The other consisted of an aplastic anaemia, detected two months after one dose of a benzodiazepine, for which the cause was not proven.[17] The apparently low number of reactions is due to the care, vigilance and attention to detail practised by experienced staff. Dengler warns that only well-motivated, experienced staff will eliminate the carelessness that leads to injury.[18] Moreover, the same principle evidently applies to the volunteers themselves, so their vigilance also must be cultivated by clinical trial staff.

4.8.2.3 Organization of Studies

The major functions of a Phase I unit may be grouped as follows: organization of an ethics committee, recruitment and care of volunteers, conduct of trials and the analysis and reporting of data (Table 1).

Effective organization of an ethics committee requires excellent communications between the investigators and committee members. Whilst members must be fully independent of the institution running studies and its clinical pharmacologists, they will need to be sympathetic to clinical trials in general and to perceive their role in society. They cannot be paid for their time and effort in any currency other than the interest and satisfaction of helping with something worthwhile. Therefore, the Phase I unit staff or the university investigator must reciprocate by readily supplying all relevant information unasked. It is essential that the committee has a chairman to act as a focus for decisions on agendas, as an arbiter in disagreements within the committee itself and in running meetings. Either the chairman requires a secretariat or the Phase I unit itself must ensure that all documents are promptly circulated.

Organization of a volunteer panel requires meticulous record keeping. Individuals' participation in trials must be recorded, partly to prevent excessive recruitment into studies. Full medical histories must be recorded and kept absolutely confidential to the trial staff alone. For example, a volunteer's

Table 1 Major Functions of a Volunteer Trials Unit

1	*Maintenance of an ethics committee*
1.1	Secretarial organization, *e.g.* checklists of members, documents sent and received, taking and typing minutes
1.2	Recruitment of committee members; ensuring a balance of medical and lay members and that they are independently minded
1.3	Help with elections of chairmen or other officers
1.4	Contact with other departments, *e.g.* legal department on questions of indemnity
1.5	Contact with other organizations, *e.g.* ABPI for guidelines
2	*Recruitment and care of volunteers*
2.1	Pretrial tests for hepatitis antigens and in some cases HIV
2.2	Records of participation in trials, including blood taken
2.3	Records of honoraria paid
2.4	Contact with general practitioners
2.5	Contact with pathology laboratories *re* samples
2.6	Pretrial (group) discussions for each trial
2.7	Pre- and post-trial medical examinations
2.8	Instructions during trials, *e.g.* declaration of all concomitant treatments, avoidance of alcohol, strenuous exercise and driving a car
3	*Organization of trials*
3.1	Protocol development and circulation
3.2	Drug supply, labelling and dose allocation schedule
3.3	Instructions to nurses and technicians recording data
3.4	Dipstick pathology tests on blood and urine samples
3.5	Storage of samples for assay, *etc.*
3.6	Maintenance of equipment and quality control
3.7	Records of results, especially adverse reactions
3.8	Practice of emergency procedures
3.9	Organization, analysis and presentation of data
3.10	Report writing
3.11	Dissemination and interpretation of results

allergies must be established and subsequently checked before every study. Good communications with the volunteers will ensure that appointments and procedures are kept. These and the advantages of long term follow-up, closer contact and better understanding of procedures are easier to achieve with stable panels of volunteers than in systems that recruit them to single studies only.

Clinical expertise is required to judge the safety of each dose, to select appropriate tests and to devise practical study procedures. Physicians must be on the spot when a peak drug effect is possible and within 2–3 min call when the predicted, pharmacodynamic effects are declining. Emergency procedures must be defined clearly and practised regularly. Physicians must be well trained in them, for which regular contact with hospital specialists is advisable. The means and criteria for transferring a volunteer to hospital should be agreed with the topmost levels of the department or company.

Nurses and clinical monitoring staff have to be adaptable to the different tests, many of which are highly sophisticated. On the other hand they need to be meticulous keepers of schedules and records. They must provide well-organized, immediate, continuous supervision of volunteers throughout studies. They need the confidence of volunteers so that all unexpected effects are reported early.

There appears to be some debate on the necessary degree of volunteer supervision during Phase I studies when the effects of a dose have subsided.[19] Volunteers who are relatively unfamiliar to the unit staff may be retained overnight. However, units which recruit from an established panel frequently allow accompanied volunteers home several hours after a dose if effects were absent, slight or moderate and transient. This is because regular members of a well-organized panel can be relied on to telephone staff if needed, provided clear lines of communication are established. Otherwise, excessive arbitrary rules will negate good sense and discretion and eventually may even demotivate volunteers and staff who have to spend patently unnecessary nights in the department.

The special furniture needs of a department will include full resuscitation equipment: tipping beds, good suction, oxygen and intubation equipment. Overnight studies obviously need beds. Since these tend to be pharmacodynamic studies with groups of six or 12 volunteers, the number of beds needed should include these and enough for two staff. For most overnight studies showers or baths will be required.

Building plans will include a kitchen and dining room and adequate space for relaxation. Many volunteers want to work at books so a study area is useful and may even help recruitment, especially among senior staff. Toilets should accommodate a drug which causes considerable diarrhoea or vomiting in possibly six volunteers at a time. Volunteer traffic should be directed away from other functions such as staff offices, though it is efficient and improves surveillance of studies if these two functions are adjacent.

Special needs are best considered at the planning stage because subsequent alterations can prove expensive. Units performing radiolabelled compound studies will need washable working surfaces which have no cracks and crannies to accumulate spilt samples. Toilet outflows should be joined by other wastes to dilute labelled metabolites in urine or faeces. Several procedures require silence, including tests of vigilance, EEGs and pain thresholds, so an insulated room should be considered at the time of constructing the unit. Other specialized uses occur, such as metabolic balance studies which require a gas-tight room or airways compliance measurements by whole body plethysmography which need a room entirely free of pressure fluctuations caused by air-conditioning or even the bending of windows in the wind.

For data handling, computerized systems are becoming the only practical option as monitoring becomes ever more intense. For example, visual analogue scales are increasingly used for subjective assessments.[20] If 16 scales are used[21] four times per study day by 12 volunteers on four study days each, then the number of measurements is 3072. Automated data capture and delivery to computer analysis is faster and avoids transcription errors. A Phase I unit needs electronics and software expertise to harness computers to the monitoring equipment and to build into it the data packaging required.

Data analysis hinges on the correct use of statistical tests (see Section 4.8.4.3). Ready-made software systems such as SAS are widely accepted by regulatory authorities and they are capable of analyzing the distribution of data to determine which is the appropriate test and producing excellent graphs, *etc.* The RS 1 system has the advantage of user friendliness so that the clinical research personnel generating the data can use it. Having this degree of involvement with the analysis is good for motivation of staff running trials and improves the understanding of both input and results of analysis.

Correct and secure storage of data is essential. A question regarding effects seen in Phase I may well arise after the drug has reached the market or when the product licence is considered several years later. Computers again represent one practical way to store data and give ready access to it.

From the above it will be obvious that a team performing Phase I studies needs several kinds of expertise and personality. Flexibility is essential due to the constant appearance of compounds in different clinical areas. Furthermore, the 90% of drugs that fail, do so with very little warning. A flexible approach may be needed to interpret correctly an unpredicted effect. On the other hand, safety, good record keeping, report writing and data analysis all benefit from the meticulous, almost obsessional personality that tends to dislike change and uncertainty. Phase I studies are a progress-limiting bottleneck, so the team must expect criticism and this can be met constructively if they are prepared to explain their methods and share results with the other departments involved.

4.8.2.4 Scientific Preparation For Phase I Studies

The information needed to begin trials in man is summarized in guidelines prepared by the Association of the British Pharmaceutical Industry (ABPI),[22] the Department of Health and Social Security (DHSS)[23] and the Council for International Organizations of Medical Sciences (CIOMS), Geneva.[24]

Before a new drug is given to a volunteer there must be a positive reason for it in the form of unequivocal evidence of pharmacological activity which could be of clinical benefit. The clinician in charge has a duty to examine the animal data to satisfy himself that this is so. The experiments must have been well done, usually in two or more species, with positive controls and correct statistical analysis. For example, experiments using ratios of activity to compare active and placebo treatments frequently produce data requiring logarithmic transformation before the correct statistical analysis can be applied.

General pharmacology must indicate a lack of effect on vital systems other than those deliberately modified by the compound. For drugs likely to be given with other treatments, such as anaesthetics, and antiarrhythmics, it may be wise to determine their pharmacological safety in combination with other possible treatments at an early stage.

Animal pharmacokinetics must contain some evidence of suitable absorption–elimination characteristics.[25] There are no clear guidelines in this regard because interspecies differences are highly unpredictable. For example, Burns[26] reported the elimination half-life of phenylbutazone in plasma for seven species, as shown in Table 2.

The relationship between animal and human doses is not necessarily simple. Pitts recommends doses based on a unit surface area.[27] A commonly used rule of thumb states that the dose in mg kg^{-1} to a rat becomes a probable, active dose in man. Thus, 10 mg kg^{-1} translates into 10 mg total dose to man. However, this is most approximate and the dose–response curves are unlikely to run parallel in the two species, so much depends upon where the doses are on the curves. The choice of the first dose to man is discussed in Section 4.8.2.6.

Toxicological testing must have been adequate. Acute toxicity of single, high doses will attempt to indicate the likely effects of an overdose in man. The traditional requirement for full lethal dose in 50% of animals (LD_{50}) is now questioned, not least because it uses more animals than is scientifically necessary.[28]

Repeat dose studies (RDS) lasting two to four weeks are required in two species from different animal groups. Thus, two rodent species are insufficient and usually dogs and rats are used for logistic reasons and because they are relatively well characterized. Occasionally a primate species is included but this is expensive, provides data on few individuals and becomes ethically more difficult to justify as more becomes known of the sensitivities of primates. The main justification is when the canine species do not have appropriate metabolic or pharmacological responses to the drug.

Repeat dose toxicology requires at least one dose high enough to provoke clear adverse effects. If more doses will show that these are dose related, then so much the clearer is the relationship to the drug. Another aim is to establish a no-effect dose which is above the predicted dose to man. Hence, a well-designed study will include at least three dose levels: a lower dose to establish the no-effect level, an intermediate dose to obtain a wider range of tolerance or slight toxicologic effects, and a high dose to disclose them unequivocally.

Current requirements[24] for the duration of dosing in toxicology tests compared with clinical trials are discussed in Chapter 4.7. They are summarized in Table 3.

The interpretation of unexpected results requires considerable experience. For example the background incidence of benign tumours may vary according to the strain of rat used. If, by chance, the incidence is larger in treated groups than placebo, the tumours may appear to be drug induced. Some studies have even included two placebo groups to better identify the background incidence of such tumours and to exclude a drug-related effect. Occasionally the pharmacology of the compound may cause an unpredicted effect which would preclude human studies unless correctly interpreted. An example would be an opiate analgesic which in rat toxicology studies causes corneal opacities, not by an intrinsic, tissue-damaging mechanism but through a predictable loss of the protective, eyelash reflex. Hence intelligent interpretation of toxicology must include a possible gross exaggeration of the intended pharmacological effect.

Table 2 Elimination Half-lives of Phenylbutazone in Seven Species[26]

Species	*Elimination half-life* (h)
Rabbit	3
Guinea pig	5
Dog, rat, horse	6
Monkey	8
Man	72

Table 3 Guidelines for Duration of Chronic Toxicity Tests in Animals

Intended duration of dosing in man	*Duration of toxicity test in animals*
Single dose or 1 day	14 days
Repeated up to 10 days	28 days
Repeated up to 30 days	90 days
Repeated beyond 30 days	180 days

Source: Guidance Notes on Applications for Clinical Trials Certificates and Clinical Trial Exemptions, Department of Health and Social Security, 1984.

Mutagenicity tests are more contentious because their theoretical basis is 'inadequate and their practical validation limited.'[24] Increasingly, they are done before the first dose to man because a clearly positive result precludes further development unless the mechanism of the effect or its consequences are shown to be irrelevant, for example by showing that the cause was a metabolite not present in man. Considering that caffeine may produce positive tests, interpretation is difficult. Much depends on whether a dubious *in vitro* Ames test result is confirmed in other bacterial strains and tests *in vivo* such as the micronuclear test in mice. Overall, a lack of effect must be established before it is ethical to proceed to man.

Carcinogenicity testing normally is not required at this stage. However it will be indicated when mutagenicity or repeat dose toxicology produce doubtful results or when the compound and metabolites' structures resemble known carcinogens.

Pharmaceutical preparation requires that the stability of the dosage form be established before dosing. The methodology is examined fully in Volume 5 of this series and guidelines on the quality control aspects of new formulations acceptable to the regulatory authorities are available.[23] They include suitable stability in accelerated stability tests at various temperatures and, of course, a validated assay by which to measure it.

There is much in favour of liquid formulations because the drug is available from them at near optimum rates. They are more uniform than solids, which may have different particle sizes with different surface areas and thus different absorption rates. However, liquids may taste unpleasant or it may be impossible to match a placebo solution for taste and colour. Lastly, they may be less stable than solid dose forms, and in general the latter seem to be preferred.

4.8.2.5 Ethical Preparation For Phase I Studies

The selection of volunteers is ethically complex and much has been written on it.[4,29] The FDA in 1977 and the Royal College of Physicians' report on ethics committees are at pains to distinguish normal subjects from patients since the legal and ethical importance of the distinction is great.[3,13] In the UK patients fall within the jurisdiction of the health authorities, while volunteers do not. For example, a person who has entirely quiescent asthma until challenged by an allergen is always a patient for the purpose of trials involving allergen challenge, even when the predicted response is slight.[30]

Medical criteria will establish whether the volunteer should be considered in the first place. Females of child-bearing potential, the elderly or juvenile should not enter the earliest trials. Given healthy, normal recruits the essential points are: informed consent, confidentiality, indemnity, honoraria and follow-up.

Informed consent requires volunteers to be selected who can understand the aims, requirements and risks of the protocol. This is best achieved when volunteers belong to a stable panel organized by the unit because they become experienced in trial requirements and are less inhibited in asking questions. This can be turned to other volunteers' advantage in group pretrial discussions where everyone may consider each other's questions. Explanations of protocols must be both written and verbal, avoiding jargon and evasions. Whilst exhaustive detail is self defeating and it is accepted that not every imaginable risk can be itemized, it is essential that the investigator explains clearly all probable effects.

Failure by the recruit to declare disability or treatment being received is far less likely to occur in an atmosphere of understanding and complete confidentiality. Particularly following the Dublin fatality,[16] it has become prudent to obtain a signature confirming that all medical circumstances have been declared. Such formality also requires considerable mutual understanding and confidence.

Institutional indemnity for unpredicted or negligent harm to volunteers is essential. It should have formal approval from the insurers. It must be available without reference to legal liability if it is established that injury resulted from participation in a trial. The ethics committee should examine the details of cover and ensure that reference to it is included in information given to volunteers.

It is generally accepted that healthy volunteers may be paid when taking part in studies. Such honoraria have become established as compensation for scrupulous observation of protocol requirements and for inconvenience such as carrying bottles of urine or faeces and for the discomfort of venepuncture or fasting. They should relate to the financial circumstances of the volunteer. It is helpful to compare payments in similar institutions since peer practice largely determines ethical acceptability. A reasonable standard is the average wage of the community for the time spent away from work or family. Honoraria should not tempt volunteers into trials they would otherwise avoid. However, some will frankly admit that they would not even join the panel without them. So in practice a fine distinction has to be drawn between the principle of volunteering, for which payment is acceptable, and submission to a feature of a trial against personal better judgement, for which it is not. For example, if they discover that a trial makes unexpectedly onerous demands upon them, volunteers should withdraw yet not lose any part of their honorarium. Moreover, this should be stated in the information given to them before the trial.

4.8.2.6 The First Dose To Man

The first dose is usually oral because this is the preferred route in medical practice. However, the intravenous route is more precise, particularly in dose response and onset times of effects. There tends to be less variation between individuals because absorption processes and first pass metabolism are bypassed. Considering the low initial dose and the slow infusion rates which must be used (infusion times no less than 15 min) the route is safer than may first appear. Of course, the appropriate animal safety tests must also be done by the intravenous route.

The choice of the first dose to man is a matter of debate. Dollery and Davies[31] advocate 1–2% of the maximum well-tolerated dose in animals (MWTD). This means that for a therapeutic ratio ($MWTD/ED_{50}$) of 5 to 20, their dose is 1/20th to 1/5th of the ED_{50}. James[32] was more conservative in suggesting 2% of the ED_{50}, although he recommends progressive doubling thereafter, which can lead eventually to large increments. Pitts[27] describes a more holistic approach, influenced by effective pharmacodynamic dose levels (the ED_{25}) and well-tolerated levels in toxicology (10 to 20% of the MWTD). Rogers and Spector[19] also include reference to effective plasma concentrations, aiming for a starting dose that produces 10% of effective levels.

Certainly, these doses are suitable for first administration because they tend to fall well below the effective dose. For example, the most potent drug I have encountered was a β_2 agonist that produced significant increases in heart rate after 0.25 mg p.o. The effective dose range in rats was 10–100 $\mu g\,kg^{-1}$, so 1% of this range would give doses in a 70 kg man of 0.007 to 0.07 mg. The maximum well-tolerated dose proved to be 2.0 mg, 285 to 28 times greater than this.

In principle extra caution is required when a grossly exaggerated effect could produce a life-threatening event. Then it is wise to perform dose exploratory studies in a specialized hospital unit with physicians experienced in the use of the class of drug. Such a case might be a new cardiac antidysrhythmic agent, for example. An upper dose in the recommended range may be more appropriate when drugs of predictably wider tolerance, such as anxiolytics, are examined.

4.8.2.7 Integrated Tolerance And Pharmacodynamics

More than merely testing safety at arbitrarily chosen doses, the early studies will try to relate tolerance to pharmacodynamic effects and the projected dose in the clinic. For the majority of drugs entering Phase I it will be possible to monitor pharmacological effects and these will often approximate to the clinical dose. Thus asthma patients respond to spasmolytic treatments such as β_2 agonists at lower doses than volunteers, but not an order of magnitude less.[33] Similarly, to overcome more severe anxiety in patients a higher dose of an anxiolytic may be needed than one which produces mood changes in volunteers, but the threshold clinical doses for less severe anxiety will be similar.[34,35] Hence, before patients are treated it is often possible to foresee a dose range from the pharmacodynamic profile and to compare it with tolerance. For example, in the case of a B_2 agonist minimal increases in heart-rate or cardiac output can be identified and the dose producing them compared with the upper dose limit when marked tremor, jitteryness, nausea and changes in blood pressure supervene.

For some illnesses, such as depression, there is no volunteer counterpart, therefore no direct pharmacodynamic evidence can be obtained. However, indirect evidence may appear through secondary effects, such as in an electroencephalographic (EEG) profile, for example.[34] For other drugs, including antibiotics and thrombolytics, an *in vitro* test will show the plasma concentrations required. In principle, these can be matched in volunteer trials to establish the relevance of the tolerance data.

Placebo effects are surprisingly common.[36] Volunteers may appear outwardly calm despite inward turmoil, which can produce tachycardia, nausea, headache and loss of appetite. Overnight fasting and caffeine withdrawal headaches also contribute study-related effects, which must be distinguished from true drug-related effects.[37,38] Hence, placebo controls add to the statistical sensitivity of the design both for tolerance and dose response.

Designs should be kept simple and balanced, to produce an even distribution of doses to the different study days. When this is applied to the placebo doses, they may identify a dose order effect.

However, there is scope for special designs which fully exploit the opportunity which a trial presents. For example, the inclusion of two groups which receive alternate doses may widen the range of doses that can be explored. This is most useful for entirely new compounds, when the number of permissible doses is limited to three. This study design is illustrated in Table 4.

A placebo dose is distributed randomly within each group. Doses to each volunteer are given in strict ascending sequence. The next dose up can be given only when the tolerance has been established in at least two volunteers to the dose below. In this way group A prepares the way for B and *vice versa*. Note the uppermost increment is limited to 100 mg in this case as progressive doubling is likely to be excessive. The advantages of the design stem from the wide dose range overall and the large increments *within* individuals (larger than when two separate studies are performed, *i.e.* group A = 25, 50 and 100 mg, group B = 150, 200 and 300 mg). The wide range overall increases the likelihood of identifying the no-effect level, or the minimal effective dose because there is a greater contrast between upper and lower doses within subjects. Moreover, larger increments increase the chances of identifying effects in any one subject, which tends to reduce the numbers of volunteers needed. Group sizes of six subjects may be adequate for an initial pharmacodynamic study. This design will often establish activity within the first study, which, considering how small the first dose must be, can save many weeks' delay.

Interpretation of laboratory and clinical data may not be straightforward. Karch *et al.* reported a concordance rate of only 50% between physicians estimating the frequency of adverse reactions.[39] There can be wide differences between groups of volunteers for effects such as sedation.[36,40] Laboratory tests usually quote a normal range defined as two standard deviations from the mean. Therefore, by definition, almost 5% of normal results will lie outside this range. Such results must be compared with the dose to exclude a drug- or dose-related effect. Attempts are sometimes made to detect trends *within* the normal range; but these are usually defeated by normal variation and would be difficult to interpret. Moreover, several 'normal' conditions are associated with abnormal results. A classic example is the elevated bilirubin of Gilberts' syndrome.[41] Some extremely fit sportsmen have bradycardias and extrasystoles which may increase during the tension of dosing. Healthy flying personnel have been shown to exhibit an incidence of 1.2 per 1000 of complete bundle branch block in ECG tests.[42] In a study of 24 hour ambulatory ECGs, almost 25% of 120 normal subjects had episodes of changes in the ST segment resembling myocardial ischaemia.[42] Another quasi-normal effect is strenuous exercise, which will damage muscle and distort plasma creatine kinase enzymes. These may have to be differentiated from cardiac muscle damage of more serious origin. Exercise may also be the cause of microhaematuria in urine tests, which may be mistaken for damage to the kidney, although it seems to occur so frequently in healthy young adults that it must be included in the normal domain.[44]

Clearly, final interpretation of an apparently abnormal result should await a full history and follow-up which observes the return to baseline of the test. The numbers of volunteers recruited to this point are so small that if only one severe or serious reaction occurs, it has grave implications for tolerance in the wider population. Hence correct interpretation is crucial, indeed, the continued life of the new compound hangs by this thread.

Notwithstanding these difficulties, it is the physician's duty immediately to withdraw volunteers who have results that cannot be explained entirely by the history, or other obvious and innocuous cause. The physician must then search the records on the drug for any other event of a similar nature. A possible idiosyncratic reaction must be considered which may include an allergic mechanism.

To summarize, a logical programme begins at a dose below the predicted effective dose for safety reasons. The dose is then increased by modest increments until pharmacodynamic effects are encountered. Thereafter, doses are increased cautiously to the point where they provoke distinct but moderate effects which the clinician, nurses and volunteers themselves consider to be an acceptable limit. A dose-related incidence of effects increases the chances of positively identifying them as drug induced. If the maximum well-tolerated dose can be established then this becomes the first dose to

Table 4 A Useful Alternating Dose Group Design

	Doses					
Volunteer A	25		100		200	
Group B		50		150		300

use in patients. It is the most likely to produce the desired effect yet, by definition, it is also likely to be well tolerated.

4.8.2.8 Pharmacokinetics: Single Doses

During tolerance and pharmacodynamic studies it is usual to include drug assays. These may not be frequent enough to define the pharmacokinetic profile accurately, but they will indicate how rapidly the compound or metabolite plasma levels diminish. Care must be taken that the result is not assumed to show the elimination half-life when insufficient data are obtained to distinguish between the distribution and elimination phases because this will cause the elimination rate to be overestimated. Conversely, a lack of sensitivity of the assay could lead to an underestimate of the elimination half-life, as could a lack of specificity such that metabolites are mistaken for the parent compound.

A more defined plasma profile will be required as single dose studies progress. Peak concentration (C_{max}), time to peak (T_{max}), elimination and distribution curves and bioavailability will be needed. Peak characteristics require samples to be spaced at shorter intervals, of approximately 15 minutes, because the direction of the plasma concentration curve changes rapidly. By contrast, definition of the elimination phase requires samples spaced evenly, or such that its beginning and end are firmly established. Exact timings will therefore depend upon the elimination half-life.

If an intravenous dose has been given, the absolute bioavailability may be calculated, although an accurate result can only be obtained by comparison of oral and intravenous results from the same volunteer. It is perhaps the most important pharmacokinetic variable because it shows the efficiency of absorption, and whether it is advisable to look for a better-absorbed, successor compound. If low, there is more room for variation between subjects, particularly if due to a large first pass effect, because metabolic processes seem to differ between individuals more than elimination mechanisms.

For analysis of the results of assays, mathematical curve fitting procedures and compartment models are necessary. They are discussed in Volume 5 of this series.

4.8.2.9 Pharmacokinetics: Repeat Dose Studies

The exploration of ascending single doses will disclose the maximum well-tolerated dose (MWTD), *i.e.* the uppermost dose which is free from unwanted effects. Repeat dose studies will now examine the tolerance to a dose as near to the MWTD as medical prudence dictates. The purpose is to find the highest dose regime that is well tolerated so that the greatest room for therapeutic manoeuvre is obtained for clinical trials. Logistically, it avoids the question 'What if we tried a higher dose?' if, for any unforeseen reason, the MWTD proves inadequate.

The obvious risk is that drug or its metabolites will accumulate in plasma and produce higher concentrations which may exceed the well-tolerated limit. The most important determinant of accumulation is the frequency of dosing relative to elimination rates. Table 5 illustrates the effect of giving 10 mg of a drug at intervals equal to the $T_{1/2B}$.

A steady state is reached between five and eight half-lives. The same occurs when dosing intervals are half or twice the $T_{1/2B}$ as illustrated in Figure 1, but the plateau levels will differ in direct, inverse proportion. The levels may be calculated from formulae based upon the dose interval and the elimination half-life.[45] In practice a dose interval close to the $T_{1/2B}$ is frequently chosen in the knowledge that peak concentrations will reach approximately twice those of single doses.

Table 5 Amount in Body when 100 mg of the Drug is Given Every Half-life[45]

Dose	*1*	*2*	*3*	*4*	*5*	*6*	*7*	*8*
Maximum amount in body just after dosing	100	150	175	188	194	197	199	200
Minimum amount in body just before next dose	50	75	88	94	97	99	100	100

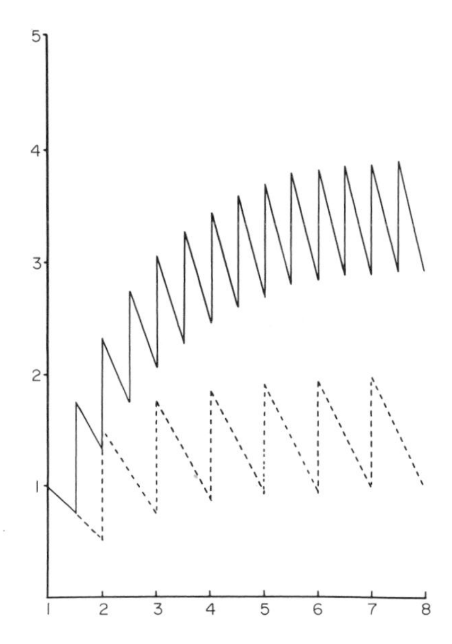

Figure 1 Units on the *y* axis represent amounts of drug in plasma expressed as the ratio of the maximum absorbed after the first dose. The *x* axis represents time expressed in elimination half-lives. A dose given every half-life (dashed line) produces a ratio of two at steady state, whilst a dose given twice per half-life (solid line) produces a ratio of four

Initially this is tested in a one day study for two to four times daily dosing, depending upon the rate of elimination. There follows a tolerance test of five, seven or 10 days dosing according to the elimination rates and pharmacodynamics of the compound. Such studies present logistic problems if volunteers are confined to the department for the entire duration of the study, for few normal people can afford to leave their families and occupation for so long. The reason why they are able to do so should be checked, along with their availability for follow-up when the trial has finished.

Repeat dose studies may discover increased plasma levels of the unbound compound if the first dose was largely bound to proteins which were saturated. Unbound drug is more available to sites of activity than bound material, so the drug effect may be unexpectedly enhanced. Alternatively, repeat dose studies may reveal low plasma levels because increasing amounts of drug are removed by the liver, *i.e.* the first pass effect increases progressively. This may be due to enzyme induction, the classic example of which is phenobarbitone. If pronounced, this effect may preclude definition of a clinical dose due to high within and between subject variation in bioavailability.

The linearity of absorption at different doses must be checked across the probable range of clinical doses. This either confirms or modifies the simple assumption that doubling the dose will also double the amount available to the tissues. Saturation of protein binding or metabolic pathways can lead to greater than expected bioavailability at upper doses, with the obvious risk of exceeding the limit of acceptable tolerance.

4.8.3 PHASE II

4.8.3.1 Definition

When adequate tolerance and pharmacokinetic data have been obtained from Phase I volunteer studies, the process moves on to trials in patients. These are known as Phase II, which is frequently divided into two subphases: IIA and IIB.

Phase IIA will often start with open pilot trials and then include two or three placebo-controlled studies that obtain statistically valid evidence that the new compound is effective. This usually requires about 50 patients. Phase IIB confirms efficacy and identifies the dose or doses for Phase III. This may need four or five double blind studies which enrol 50 to 100 more patients.

4.8.3.2 Regulatory Requirements

When patients are enrolled in trials the regulatory authorities become involved. The essential difference between volunteers and patients is that the latter can expect a beneficial effect from the experimental compound. However, this has been extended recently to include within the definition of a patient one who is 'in remission from a relapsing condition such as bronchial asthma (who) cannot be a volunteer for studies related to that condition, and should be regarded as a patient for studies involving the immune or respiratory systems'.[12] This is because a study which includes any risk of precipitating the condition is, by definition, a Phase II study.

The authorities will require a dossier which summarizes the chemistry, pharmacology, toxicology, Phase I trial results, the stability of the formulation and pharmacokinetics.[23] Whilst data from volunteers are not a *statutory* requirement, it greatly helps to justify and explain the proposed dose. Both investigators and regulatory authorities will need this summary, which should not exceed 30 pages. In the UK, a good dossier, with detailed protocols, will obtain permission for the trials to proceed in 35 days under the Clinical Trial Certificate Exemption (CTEx) scheme. If the CTEx dossier is inadequate the Department of Health and Social Security (DHSS) may require an extension of 28 days to consider it or for a resubmission of the dossier with suitable amendments.

The CTEx is a summary of data, a complete version of which used to be required for the previous, clinical trial certificate (CTC) procedure. The CTC was a considerable document and it took several months for the authorities to assess it. Accordingly, it is not surprising that the CTC method, though still available, is little used.

An investigator in the UK may apply for personal exemption from the full CTEx documentation. This will be granted only when it is clear that the sponsoring company did not make the first approach and if there is sufficient evidence of tolerance. An example might occur when a physician hears of a clinical trial of a new drug abroad.

4.8.3.3 Phase II Clinical Trial Programme

The first trial in patients is usually small scale, often a dozen or so patients. If efficacy is truly unknown, an open design may be used because it limits the number of doses, reduces the demands made on the patients and the duration of the trial. Open studies give the clinician a reasonable impression of efficacy and tolerance, good enough to indicate more precisely the correct dose for a double blind study.

The dose is related to the maximum well-tolerated dose (MWTD) determined in Phase I and, logically, should equal it. A single dose or a day's treatment is given first to patients with moderate states of the disease to be treated. In general it is difficult to justify giving an untried compound to severely affected patients at this stage unless the potential benefit is virtually certain. It is wise to open three or four centres because clinicians and patient populations vary from centre to centre. Moreover, an effect of recruiting moderately affected patients is to maximize the placebo response, a particular risk if the investigator is enthusiastic about the new treatment.

Pilot studies are followed by double blind trials. The comparator may be placebo, an established effective treatment, or both. Comparisons with placebo are aimed at measuring the drug effect. With established compounds an equivalent effect is the minimal aim, possibly with better tolerance. These trials commonly recruit 10 to 15 patients who crossover between treatments. They clear the path for similar but larger studies in Phase IIB. Larger studies will be needed to produce statistically valid results, but so great is the cost of these that a preliminary indication is necessary to assess numbers required and to ensure that they will be worthwhile. At the end of Phase IIB, if the desired result is obtained, Phase III studies are initiated.

4.8.3.4 Protocols and Study Design

Protocols once served the relatively simple function of stating the scientific methods of studies. They are now evolving towards medico–legal documents with contractual elements for the investigator and sponsor. Firstly, this is because ethics committees and regulatory bodies need to know and agree the basis of the study. Once they have done so, the protocol must be followed in detail. Secondly, the protocol is a ready reference for all the elements of a study, including contractual aspects such as indemnity, payment and publication. Thirdly, their inclusion minimizes mistakes when there are personnel changes within the organizations conducting or sponsoring the study. The elements of a protocol are listed in Table 6.

The function of each is as follows.

Front page. The front page contains the study reference number and title of the study. The title should include the disease and the treatments and whether the study is open or blind. The

Table 6 Elements of a Phase II Protocol

1.	Front page
2.	Introduction
3.	Objectives
4.	Ethical aspects including payments to patients
5.	Drug supply, labelling and randomization code
6.	Patient selection
7.	Methods
8.	Statistical analysis
9.	Dropouts
10.	Protocol amendments
11.	Adverse events
12.	Timetable
13.	Resources and personnel
14.	Financial agreement
15.	Indemnity
16.	Publication of results
17.	Record forms

addresses and names of the investigator and the contact in the sponsoring organization should be included also.

Introduction and rationale. This explains why the study is being done, describes briefly the new compound and states what information on it is available. It will include the projected therapeutic role of the compound. It may include references which are detailed at the end of the protocol.

Objectives. It is best to restrict each objective to one simple sentence.

Design. This states whether the study is open, single or double blind and the number of patients per group. The number of treatment periods per patient, run-in and washout periods between treatments will be defined.

In general it is best to avoid elaborate designs, no matter how elegant, because they tend to require larger numbers of patients. This includes stratifying patients into too many subgroups. If patients are stratified, the criteria must be stated clearly.

In crossover studies patients receive two or more treatments for within subject comparison. They tend to require smaller numbers of patients because more powerful statistical tests can be used. They assume that the two periods are comparable, so crossover designs become unsuitable for prolonged treatment periods in fluctuating illnesses. Parallel studies use two or more groups which receive different treatments. Sequential analysis designs compare matched pairs of patients on alternate therapies.[46,47] When a significant number of pairs has produced one result or the other, the trial can be concluded with the minimum number of recruits. The snag is that matched pairs are often as difficult to find as a much larger group of unmatched patients.

Latin square designs are useful for removing bias caused by treatment order. They ensure an equal distribution of treatments to first, middle and last treatment position, for example. Table 7 shows that an incomplete square supplies the three patients required to allocate three treatments to all three possible positions. If, in addition, all *sequences* are required to balance out a carry over effect from one treatment to the next, then a complete Latin square design will be needed and that involves a minimum of six patients.

Ethical requirements. The protocol now focuses on details of method by stating how the study conforms to accepted practice. It should say that the study conforms to the requirements of the Declaration of Helsinki (1964)[5] and its Tokyo amendment. It describes the ethics committee review procedure, the form of patient consent, honoraria and indemnity in case of misadventure or accident. Increasingly, for good clinical practice, the constitution of ethics committees and descriptions of persons serving on them are required. The written information given to patients to explain the purposes and foreseeable risks of the trial should comprise one of the protocol record forms, and an example is given at the end of this chapter. Simple, jargon free language will help to convince all who review the study that it is indeed well conceived and organized.

Honoraria. Occasionally honoraria are paid to patients in Phase II studies. They are usually restricted to reimbursement of expenses. Ethics committees will wish to see the honorarium stated in protocols.

Drug supply. This section states how and where drug supplies will be stored. It gives the dosage form and strength, the dose and the exact method of dosing. It describes the treatment order

Table 7 Complete and Incomplete Latin Square Designs for a Within Subject Comparison of Three Treatments A, B and C

	Dose			
Incomplete[a]				
Subject number	*Dose*	*1*	*2*	*3*
1		A	B	C
2		B	C	A
3		C	A	B
Complete[b]				
Subject number	*Dose*	*1*	*2*	*3*
1		A	B	C
2		A	C	B
3		B	A	C
4		B	C	A
5		C	A	B
6		C	B	A

[a] The incomplete design allocates each treatment to each dose equally, which balances any effect related purely to the order of treatment. [b] The complete design also ensures that all treatments are preceded by the others the same number of times and so balances out any carry-over effect.

randomization code, lists persons authorized to know it and details of the procedure to be followed if the code has to be revealed for any reason.

An example of the labelling of drug containers will be needed. In the USA a duplicate is torn off, to be included in the patient's record file to identify which container was allocated.

Patient selection. A comprehensive list of inclusion and exclusion criteria is necessary. The severity of the condition to be treated must be defined and how it is to be measured. Many conditions will be moderate in Phase II because it is not ethical to give an unknown treatment for a severe medical condition. This should be made clear because moderate conditions tend towards more placebo response than severe states, which increases the number of patients required. Women of child-bearing potential, children and the elderly are usually excluded from Phase II trials.

Methods. A methods section comprises detailed accounts of the items measured, how they are measured, when and by whom. Unless they are completely standard, the validity, reliability, sensitivity and specificity of the measurement should be defined. Checks on these should be included also. The aim is to enable any competent reader to reproduce the experiment if needed.

It is useful to list the tests in a table which has the timing of each in columns so that the number and time of every test is on one page. Table 8 is an example. If the text of the protocol states that 'tests will be done at the times shown in Table 8' rather than giving separate lists of timings in lines of text which are difficult to read, then the protocol is more likely to be understood fully. Moreover, amendments are frequently necessary and when they are made in the table it ensures that the impact of the amendment on other tests is seen. Restricting the list of tests to the table only has the advantage of avoiding confusion when, as often happens, the amendment is made in one place but not another. Table 8 can also be used as a chart of procedures on the ward. This has the advantages of ensuring that ward staff do exactly the tests intended and, as the standard layout becomes familiar, so omissions tend to be reduced.

For laboratory tests of tolerance the methods section identifies the laboratory, and its normal ranges. These tests must be done before treatment so that the individual patient baseline is established. In pharmacokinetic studies similar details of the assay laboratory, the assay technique and its lower limit of detection are required so that protocol reviewers may check that plasma sampling is justified and their timings appropriate.

Statistical analysis. A statistician should be consulted at the planning stage to ensure that appropriate statistical tests and procedures are adopted. These should be described briefly, giving reasons for the choices made. For example, non-parametric data are not normally distributed. Typically, it is obtained in 'yes–no' situations or from subjective classifications of the form: none, mild, moderate, severe, *etc.* Analysis requires non-parametric tests such as rank order tests of significance (*e.g.* the Wilcoxon Rank Sum Test.) When two sets of data are normally distributed and have similar variance around the mean, then comparison of their means is statistically valid. The 't' test is suitable for this, or the χ^2 test for comparing percentages.[48] Paired data are often preferred because they reduce variation about the mean or its source can be identified, which permits the use

Table 8 An Illustration of a Study Procedure Table[a]

Time	*Pre-dose*	*Dose*	*15 min*	*30 min*	*45 min*	*1 h*	*1.5 h*	*2 h*	*2.5 h*	*3 h*	*4 h*	*6 h*	*24 h*
Clinical pathology	*												*
Pulse and blood pressure	*	*	*	*		*		*		*	*	*	*
Electrocardiogram	*			*						*			
Assay blood sample	*		*	*	*	*	*	*	*	*	*	*	*
Reaction time	*			*						*		*	
Visual Analogue scales	*					*						*	*
Side effect questionnaire	*					*				*			*

[a] Items marked * must be completed by the ward observers. They may use copies of the table as a checklist. This table represents a simple study, designed to take assay samples of a new formulation of an established compound that is relatively rapidly eliminated and which may have sedative effects. The reaction time and mood-measuring visual analogue scales indicate fitness to leave the unit accompanied, but not necessarily fitness to operate machinery *etc*.

of more powerful statistical procedures, such as analysis of variance, which yield statistically significant differences within smaller treatment groups.

Thus having defined the nature of the data, including the variance within it, the most appropriate test can be identified. When the power of the design is decided, the number of patients can be calculated which fulfils the desired level of probability that a difference is, in fact, significant for that test (see Section 4.8.4).

Hence, the statistics section should at least state the expected form of data and nominate an appropriate test. This applies also to tests for treatment order effects, particularly if the study design is not a Latin square. However, Phase II studies may be so exploratory that the size of the effect may not be known, so precluding any calculation of the appropriate group size. Also, patient numbers are determined more by ethical and practical priorities and so simple tests of significance only are used.

A more detailed discussion of statistical tests will be found in Section 4.8.4.

Dropouts. The procedure must be listed for replacing patients who leave the trial. Consideration should be given to the possibility that a particular group of patients will drop out, so unbalancing the trial. However, it is usual to simply replace such patients as they occur by the next recruit.

Protocol amendments. Deviations and amendments to the protocol will be reported to the ethics committee. They require the agreement of the organization sponsoring the trial and a statement to this effect should be included. Amendments must be reported to the ethics committee and cannot proceed until the committee has sanctioned them in the approved manner. The date and implications of any changes must be included in the submission to the committee.

Adverse events. Procedures for recording and reporting adverse events will be listed in this section. If possible, methods of treating a predictable event should be included. All protocols must include a statement that 'adverse events will be recorded in the report forms of the protocol. All serious or life-threatening events must be reported immediately to the sponsor' (who will then report them to the regulatory authorities).

Timetable. The predicted recruitment rate and duration of the study should be included in the protocol. If this is clear from the beginning, difficulties in recruitment can be identified sooner. This may disclose that the selection criteria are too strict, for example, and therefore point to necessary amendments.

Resources and personnel. It is often helpful to list the various tasks the trial requires, and the personnel responsible. This includes the names and locations of medical, nursing pharmacy, secretarial and biometric assistants. A list of approved signatures is needed to identify those which appear on case record forms and validate them.

Financial agreement. It remains controversial to include in the protocol the exact sums of money paid to a department or investigator because the protocol has a wide circulation. However, it is required by some authorities for ethical reasons, in particular the FDA. The reason for including it is to check that investigator and sponsor have the same expectations. This may be served by stating the method and frequency of payment without giving the actual sum. Contingency plans should be made for an unexpected halt to the trial.

Indemnity. The form of insurance and indemnity for unforeseen and adverse events must be clarified in advance. An example is given in Appendix A.

Publication. An investigator with special expertise, as frequently chosen for Phase II studies, is likely to be interested in publication. Most companies also want their new compounds reported to the medical specialty in their journals but if not, this should be discussed early in the project. It is useful to include a statement of intent on publication in the protocol.

Record forms. No protocol is complete without its record forms. They must be designed with the procedures of the clinic in mind. The real need for each piece of information must be established because if the purpose is obscure, neither patient, nurse nor investigator will complete the form accurately. Design of record forms is discussed briefly in Section 4.8.8 and more comprehensively by Wright and Haybittle[55] or Grady.[56]

In summary, Phase II trials measure a specific drug effect on a carefully selected disease state. They attempt a precision which befits pharmacological studies in both animals and man, hence the title 'clinical pharmacology'. Once efficacy is so defined, the new medicine progresses to Phase III, where wider, less well-defined objectives predominate.

4.8.4 STATISTICS OF CLINICAL TRIALS

4.8.4.1 Introduction

The number of patients needed for a trial is calculated by statistical tests based upon the laws of probability and normal distribution. In this section the basic mathematical concepts of the 'likelihood' approach will be explained, since it is the more commonly used in clinical trial analysis. This assumes nothing can be concluded until strictly defined criteria are matched by the clinical trial data. The alternative, Bayesian approach increasingly has its proponents since it brings into account all the prior relevant knowledge as it accumulates and, by doing so, can achieve an earlier result. In a sense it is more flexible since it progressively adds information to the calculations until the probability of one outcome being true becomes acceptable. It may be used more in the future because it has potential for optimizing study designs.[48] For a detailed application of the two approaches reference may be made to the textbooks of Edwards[88] and Lindley[89] listed in Section 4.8.10.

Essentially, a trial is an attempt to measure the response of a representative sample of a larger population so that the overall response can be predicted with confidence. Statisticians can express the true random nature of a sample in the form of probability, which enables them to calculate how many patients it will take to find a representative group of patients. Since this is an essential part of planning a trial, this calculation should be made when the design of a trial is considered.

4.8.4.2 Basic Concepts: Power, Error and Non-parametric Tests

When the design of a study is discussed with the statistician the following questions will be asked. (1) What, exactly, is the question being asked (the hypothesis)? (2) What is the measurement which defines the result? (3) How small a difference is it required to detect? (4) With what degree of reliability? Now, the simplest form of the first question may be: 'Is treatment A different from B?' This may be resolved by straightforward 'yes–no' result. An example would be whether the number of successes or failures on treatment A is different from B. Such a question can also be resolved into the four items shown in Table 9.

As items three and four in Table 9 indicate, a trial involves the risk of two types of error: the false positive (Type 1) and the false negative (Type 2). Our trial populations are only a sample of the whole, which involves a risk that they will not be truly representative and so produce the wrong result without our knowledge. This is either the false conclusion that there is a difference between A and B (Type 1 error) or, contrarywise, that there is no difference when, in fact, the trial should have detected one (Type 2). Physicians tend to think of these two types of error as 'false positive'

Table 9 Statistical Decisions[49]

1. Percentage success on treatment 1 (usually the standard treatment)
2. Percentage success on treatment 2
3. α = level of significance in the χ^2 test used to determine a difference between treatments (usually set at $\alpha = 0.05$ or 1 in 20)
4. $1-\beta$ = the degree of certainty that a true difference would be detected (often set at $1-\beta = 0.9$ or 90%)

and 'false negative' because that is the kind of outcome they have to guard against when checking diagnostic test results.

α is used to quantify the risk of a Type 1 error. The value for α is usually set at $\alpha = 0.05$, *i.e.* a 1–20 probability that the result is due to chance alone. In other words, if the trial result is positive the odds are 19–1 in favour of a true effect, and 19–1 against a Type 1 error. If α is set at 0.01 then the odds rise to 99–1 against a Type 1 error.

$1-\beta$ is called the power of the trial. The value for β is often 0.1, which represents a 90% certainty that a difference, if it exists, will indeed be detected by the trial ($1 - \beta = 0.9$). In other words, the odds are 9–1 against a Type 2 error.

When the statistician's four questions have been answered, by estimating p_1 and p_2 and choosing values for α and β, the values can be substituted into the appropriate equations that determine the trial size. Thus, in the χ^2 tests the number of patients required in each group, n, is given by equation (1).

$$n = \frac{p_{\rm i} \times (100 - p_{\rm i}) + p_{\rm ii}(100 - p_{\rm ii})}{(p_{\rm ii} - p_{\rm i})^2} \times f(\alpha, \beta) \quad (1)$$

where $f(\alpha, \beta)$ is a function of α and β, which is looked up in a table such as Table 10. The table assumes a normal distribution of the data, and $f(\alpha, \beta)$ is actually given by equation (2)

$$f(\alpha,\beta) = [\Phi^{-1}(\alpha/2) + \Phi^{-1}(\beta)]^2 \quad (2)$$

in which Φ represents the cumulative distribútion of a standardized normal deviate, the origin of which will be discussed in the definition of normal distributions of data.

4.8.4.3 Statistical Aspects of Study Design

By working an example of this calculation we can study the interaction between clinical decisions and the numbers required for their different study designs. Pocock[49] quotes the calculation of patient group sizes for the anturane study of patients recovering from a myocardial infarction.[50] In this trial the percentage of patients surviving one year after an infarction on placebo was known to be 90%. The new treatment, anturane, was considered effective if it increased this rate to 95%. Values of 0.05 and 0.1 were chosen for α and β, as described above. Substituting in the formula for 'n' and using the table value for $f(\alpha, \beta)$ we have

$$n = \frac{90 \times 10 + 95 \times 5}{(95 - 90)^2} \times 10.5 \quad (3)$$

$$= 578 \text{ patients in both treatments or 1156 in the trial}$$

This is perhaps a surprisingly high number at first sight. The reason is that the difference between the treatments is small, only 5% in 90. Of course the trial planners wanted to be sure that if there were any effect, they would detect it, so the 5% difference is a minimum, the watershed between an ineffective treatment and a useful one. One reason why it is small is the choice of one year follow-up. Perhaps a longer period would produce a wider divergence between the treatments, although it might be negated by an increased patient dropout rate as treatment periods become prolonged.

Table 10 Values of $F(\alpha, \beta)$ for Calculating the Number of Patients Required in the χ^2 Test[49]

β[b] \ α[a]	0.1	0.05	0.02	0.01
0.05	10.8	13.0	15.8	17.8
1.10	8.6	10.5	13.0	14.9
0.20	6.2	7.9	10.0	11.7
0.50	2.7	3.8	5.4	6.6

[a] Type 1 error. [b] Type 2 error.

Supposing the actual improvement on anturane was 7% so that 97% of patients survived one year. How does this affect the necessary numbers? Substituting as before, we obtain

$$n = \frac{90 \times 10 + 97 \times 3}{(97 - 90)^2} \times 10.5 \quad (4)$$

$$= 256 \text{ patients per group or 512 in the trial}$$

The organizers could have chosen a less secure result because this needs less patients. A 1 in 10 probability of a Type 1 error, where $\alpha = 0.10$ instead of 0.05, reduces the group sizes to 473, or 946 patients in the trial. This figure is so close to the original 1156, the logistic and financial savings so small that it is easy to see why they took the more secure option. In fact a 1:20 probability is a more accepted level and 1:100 better still.

Increasing the risk of a Type 2 error, by replacing $1 - \beta = 0.9$ with $1 - \beta = 0.8$, which represents a 20% probability of this error, reduces the group sizes to 435 or 870 in the trial. Once again, setting up a trial of this size merits a more secure conclusion than a 1 in 5 probability that it is false.

Pocock[49] comments that an increase in power from 0.5 ($\beta = 0.5$) to 0.95 ($\beta = 0.05$) requires about a threefold increase in group size. Also, that much statistical work has elaborated other methods such as replacing the formula used above with a table of values (Fisher's exact test) which is suitable for trials with smaller numbers.[51] However, tables tend to be scanned for the values that fit preconceived ideas on the trial size. Berkson[52] has argued that Fisher's exact test is too conservative, but this is the very quality which causes others to recommend it for smaller scale studies.

4.8.4.4 Parametric Tests and Normal Distribution

When the data are in numerical form a good deal more information is available than the 'yes–no' option considered so far. This leads to more powerful statistical tests than χ^2, such as Student's 't' and analysis of variance. However, to compare groups of numbers something must be known of their distribution. If it conforms to a predictable pattern, that is a distribution curve with predictable parameters, then smaller random samples can be employed to predict behaviour of the whole population.

It so happens that the distribution of individual data points within a single class of random data often can be described by an equation or curve that closely resembles the curve for many other kinds of data. For example, it applies to mechanical data, such as lengths of wood cut by a machine and biological data such as heights of individuals within a population. This was discovered by Gauss (1777–1855) and it was first called a Gaussian distribution. Subsequently it was called a 'normal' distribution. It is illustrated in Figure 2.

If the curve is composed of actual measurements, the vertical axis represents the density of data points at each value within the range of data obtained along the horizontal axis. If the curve describes a hypothetical population, the vertical axis represents the probability density across the range of data. At the extremes of the range data points are few and the curve falls towards zero. Area under this curve represents the probability that data points will fall at the value shown.

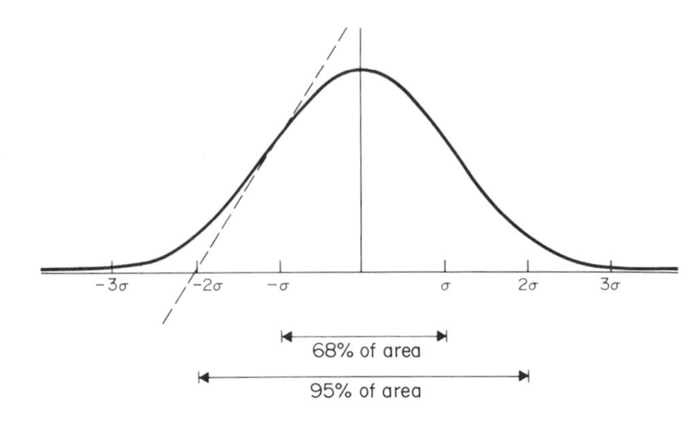

Figure 2 The curve of normal distribution. The y axis represents the number of observed values; the x axis the range of those values. In this case the peak, *i.e.* the most frequently observed value or the mode, coincides with both the mean of x and its median (medians have an equal number or values to either side). σ represents one standard deviation. The symmetry of the curve is such that a tangent at σ intersects the x axis at 2σ

The peak of the curve represents the most common value, known as the mode. In this rather special, symmetrical example it coincides with the mid-point of the range, where the vertical line is drawn. Due to the symmetrical distribution either side of the mid-point, showing that there is an equal amount of data to left and right, this line also shows the average or mean of the data. If the distribution were unequal to either side of the mode, the mean and mode would not coincide and the curve would be skewed.

A most useful parameter is the standard deviation. It is derived from the individual data points as follows: they are subtracted from the mean, the result is squared to remove negative values, these squared values are averaged and returned to the original scale of values *via* their square root. The result is the standard deviation, represented by σ in Figure 2, which has several useful properties.

Of most use to trial planners is that known proportions of the population are contained within one, two or three standard deviations from the mean. For example, 95% of the data points fall within two standard deviations of the mean. This provides a basis for comparing the location of two populations in a clinical trial, and calculating the probability that they do or do not differ.

It is quite possible to imagine many different samples taken from a normally distributed population. Each sample would have its own mean with its own distribution, similar to the individual data points described above. This produces a standard error of the mean which can be considered as a 'standard deviation of the component means'. The term 'error' is used to distinguish the two. The distinction is illustrated from the formula relating the two

$$\text{Standard error of mean} = \frac{\text{standard deviation}}{\sqrt{n}} \qquad (5)$$

where n is the number of data points under consideration.

The SEM can be used to test the hypothesis that no real difference exists between two mean values. This is the basis of the two-sample 't' test. The above formula is then adapted to give the standard error of the difference between two means as follows

$$\text{Standard error of difference} = \frac{\text{SD}_{\text{i}}^2 + \text{SD}_{\text{ii}}^2}{\sqrt{(n_{\text{i}} + n_{\text{ii}})}} \qquad (6)$$

where SD_{i}, SD_{ii} and n_{i}, n_{ii} denote the standard deviations and numbers of patients on each treatment. The probability of a true difference existing between the two populations is then looked up in a table.

Since these powerful predictions of the behaviour of populations can be made only if the data are normally distributed, what are the tests for a normal distribution of data? Firstly, the data should lie on one continuous, unimodal curve which is symmetrical about its mean. Secondly, the mean, standard deviation and the middlemost value, the median, should be close together. Note that the median is not the centre of the range, it is the value with an equal number of data points on either side. Thirdly, approximately 95% of the data points should lie within two standard deviations of the mean.

These conditions can be fulfilled without excluding a skewed distribution, wherein the mode, or peak, is displaced from the mean. Tests for skewness count the number of data points in equal, separate sections of the range, such as every inch in a survey of the heights of children. The cumulative distribution is calculated, or better still plotted on special probability graph paper; a normal distribution produces a straight line, a skewed distribution a curve. The slope of the straight line represents Φ, the cumulative distribution function, which is related to f in Table 10 by equation (2).

There is a great deal more to the statistics of normal distribution, probability and the appropriate use of statistical tests (see Section 4.8.10). It must be emphasized that so precise are the assumptions made in statistical tests based upon normal distribution that their misuse leads to most misleading conclusions. When data are not normally distributed or too few to demonstrate it, the non-parametric tests must be used.

4.8.5 PHASE III

4.8.5.1 Definition

Phase III trials provide a body of reliable data which justifies the launch of the new medicine onto the market. They attempt to answer the question 'What will be the role of the new compound in clinical practice?' This is a considerably wider purpose than the Phase II questions 'Is the drug effective and at what dose?'

The total number of patients required will depend upon the frequency of the medical condition, but a common ailment such as rheumatoid arthritis will need over 1000 patients actually on the treatment, *i.e.* over 2000 patients enrolled in the trial programme. If treatment is to be for life, as in rheumatoid arthritis, then at least 100 patient records must document good tolerance to one year of continuous treatment. For more rare indications or when the benefit to risk ratio is higher, this number may fall to several hundred.

The range of patients included must reflect the population that will be treated. For example many elderly patients suffer from rheumatoid arthritis and there is a version of the disease which occurs in childhood, so absorption, tolerance and efficacy will be needed in such patients. In chronic conditions it is inevitable that many other treatments will be given concurrently, so possible interactions must be studied. Any alterations in the formulation, or alternatives, must be tested for bioequivalence. If they should produce higher peak levels or slower elimination then particular attention must be paid to the tolerance.

Hence it is clear that 1000 patients represent several smaller subgroups, and as the section on statistics shows, it is from some points of view a relatively low number overall.

4.8.5.2 Strategies for a Phase III Programme

In preparation for clinical practice longer treatment periods in larger groups of patients are used in Phase III trials compared with Phase II. These simple changes markedly influence the design of studies and thence the organization they require.

Illness and the patient's perception of it usually fluctuate, which increases the variation within and between treatment groups in proportion to the duration of the trial. For a statistically valid result this increases the numbers required in each group, so removing the reason for crossover designs (*i.e.* valid comparisons in small patient groups). Moreover, prolonged treatment periods produce higher dropout rates, an effect which is all the greater if there has to be a washout period between treatments (as often occurs in crossover studies). Hence, there is a tendency to use parallel design studies in Phase III.

Comparisons with established treatments provide more information about the potential role of a novel drug than trials *versus* placebo. This reflects the change of emphasis in Phase III away from pure measurement of drug effects. However, comparison with another active treatment reduces the differences between the treatments, which again has the effect of increasing the numbers needed to show a difference. Thus, hundreds rather than tens of patients may be needed.

Thus the logistic aspects of Phase III studies greatly exceed those of Phase II. The protocol is an essential focus for planning these, a process described in some detail by Fidler and Koch.[53] Each study becomes a major effort of recruitment, follow-up and document keeping. When this has to be done to the exacting, detailed standards of the FDA, when six copies of all records have to be made, it means that a clinical trial monitor may be fully occupied by three or four trials. If gaps in data are numerous, the FDA may consider the entire study unsuitable for the registration dossier. In Europe the task is bureaucratically less demanding, but, even then, the cost is so high that most pharmaceutical companies try to ensure that the data are reliable enough for world wide registration dossiers. The reasons for so doing and a description of what it involves are given in Section 4.8.8.

4.8.5.3 Aspects of Study Design in Phase III

The choice of investigator is governed by their knowledge of the disease and its treatment, the number of patients available in their clinic, their experience of clinical trials and personal commitment to run them. Experts present their credentials in the form of published papers, which demonstrate their ability to marshall and analyze appropriate data. Publications suggest that the clinic has an effective team able to cope with all aspects of trials, although, naturally, this must never be taken for granted. 'Only an effective working relationship, leading to frequent access to the clinic, its facilities and staff will ensure that the trial is suitably located.'[54]

Patient inclusion–exclusion criteria should not admit too widely diversant patients nor should they be so strict that they select a group which is unrepresentative. If they are too strict, there will be dropouts, when the investigator is forced to recruit patients who are unsuitable for purely non-medical reasons.

Drug formulations must look and feel identical if the double blind nature of the trial is to be maintained. This requires a special form of the reference treatment. A matched placebo may be expensive. Another pharmaceutical company may be asked to supply the second treatment, a favour usually conferred against the need for a return favour in the future.

It is usual to requisition three or four times the calculated minimum number of tablets because supplies may be mislaid, badly stored or the trial may be extended. Not infrequently the trial may prompt a follow-up study, when a reserve of drug supplies will save considerable time and cost.

The drug allocation code for an individual patient must be held where the investigator has access to it in case of emergency. However, if the code is broken it must be recorded. The information may have to be obtained from a third party, such as a hospital pharmacist, or by breaking the seal of an envelope dedicated to the patient. Every dose of the trial medication material must be accounted for at the completion of the trial.

Randomization of treatment allocations aims to counteract bias introduced by external circumstances. Patients recruited early may be exposed to more uncertainty and even disorganization than later recruits. The season of the year may influence the illness and hence drug efficacy. Therefore equal numbers must be evenly distributed in the recruitment sequence.

Stratification of treatment allocations will be required if different classes of patients can be identified in advance. For example, if it is known that younger and more elderly patients will respond differently, they should have separate randomization sequences. Otherwise there will be a risk of uneven allocation of the two treatments to the different age groups which will distort the result.

Record forms have to be planned for a busy clinic and for use by people less versed in them so they must be especially clear and well organized. If the data demanded do not fit the long-evolved practices of the clinic, there will be omissions and errors. Hence a degree of flexibility is required within the Phase III programme overall. For guidelines on layouts, type formats and methods of transferring data from forms to systems of analysis the reader is referred to Wright and Haybittle[55] or Grady.[56]

Data will be collected by the monitor while the trial continues, although copies must be left with the investigator for reference and further visits by the patient. Analysis of a steady stream of data is obviously more efficient than waiting until the study finishes and, more important, it may detect significant omissions at a remediable stage. Care must be taken if interim data are analyzed because if they are replaced by new data the risk is increased of finding a spurious positive result (Type 1 error).[57,58] This occurs when the datum point can move anywhere within the total population range at random so that if it is measured enough times it will, by chance, lie outwith the proposed limits of probability. The correct statistical remedy is to use sequential analysis methods, which require fewer patients on the average.[58,59] Interim analyses should be planned in advance, not carried out when the data look 'interesting', which tends to select moments when chance has thrown up differences which the full number of datum points irons out.

Recruitment rates should be monitored and compared with predictions. An unexpectedly low rate may be one of the first signs that something is wrong with the protocol or the study centre. The inclusion–exclusion criteria may be too strict. Patients may not be available as predicted and the reason must be discovered. Dropouts must be expected and plans made for them. It is often possible to estimate these in advance by examining records of previous trials or asking the investigator. Dropouts require good statistical planning and extra supplies of drugs and record forms.

If patients fail to take the trial medicine, they may not admit it and the omission may not be evident. Non-compliance with the protocol is difficult, often impossible, to measure.[60,61] The topic is well aired in the literature and there are no clear rules on combatting it. The only practical countermeasure appears to be to count tablets remaining in trial canisters and possibly to measure metabolites in random collections of plasma or urine.[61,62] Compliance tends to weaken as treatment periods are prolonged.

There are several contractual aspects of protocols and it is essential that they are clearly established before the study begins (see Section 4.8.3.4). Their importance is increasing due to the enormous effort and time demanded of investigators and their staff. The method and frequency of payment by the sponsor must appear in or with the protocol since they are part of the good relationship that must be established. Budgeting must consider all the expenses at the centre, such as fees to junior medical staff, secretaries and pharmacists.[53,63] Patient reimbursements are usually small for they may receive out of pocket expenses and taxi fares but not honoraria. Finally, though not least in importance, the form of indemnity for unforeseen events must be established and agreed in advance.

To summarize, a good working relationship must be established between the sponsor and the investigator. This requires frequent visits which have the agreed purpose of monitoring progress, the quality of data and compliance with the protocol. In fact, good clinical practice requires investigators to know their commitments exactly and the status of the new drug. The origin of all data must be clear, the originator signing all observations, procedures and accounting for drug supplies. Each deviation from the protocol must be clearly stated and explained, the reasons for it and its consequences equally clearly established. Hence the role of the clinical research associate has increased in importance, as has the planning of the study from its conception.

4.8.6 PHASE IV

4.8.6.1 Definition

Phase IV refers to trials performed after a new therapy is marketed. They continue the process begun in Phase III, which is to determine the true clinical profile of the compound. This involves comparisons with established treatments, investigation of efficacy against other diseases and surveillance of undesired effects.

4.8.6.2 Strategies in Phase IV Trials

Comparisons with established treatments in Phase IV are identical to those of Phase III, and they entail the same large patient groups and their logistic problems. Multicentre studies may be required to obtain sufficient numbers of patients representative of subgroups such as children or the elderly.[63,64] Investigators may be general practitioners since the majority of patients may come from general practice, particularly when the condition can be treated adequately without referral to hospital clinics.[65] Current ABPI guidelines on general practice trials are aimed at establishing their correct scientific and statistical basis and will help to dispel the aura of 'promotional studies' that sometimes accompanies them.[66]

Trials extend further into subgroups such as the elderly. The aged have reduced metabolic clearance of drugs such that the average octogenarian requires only 70% of the full adult dose. Combined with their tendency to lower body weight, this reduces their dose requirements substantially. At the other extreme of age, children have increased dose requirements compared with adults on a simple body weight basis, and body surface area is the more accurate yardstick. Other important subpopulations include patients on therapies likely to be combined with the new treatment, such as surgery and anaesthesia or special treatments such as anticoagulation, anticonvulsant or antiasthma therapy. Efficacy in new therapeutic roles may be investigated in Phase IV. Initially this may explored by smaller scale studies resembling the original Phase II trials. Subsequently, comparison with established treatments probably will be required before suitability for the indication can be established.

The pharmacokinetic profile of new formulations will be established in volunteer studies, but acceptability in the clinical population must be confirmed if there is any tendency to more rapid absorption and higher peak levels. Absorption from slow release formulations is more prone to disruption by gastrointestinal disturbances such as diarrhoea than standard formulations.

Even the most extensive programme of clinical trials is unlikely to determine the incidence of severe adverse reactions, simply because such reactions are too infrequent. For example, after some 10 years on the market the intravenous anaesthetic Althesin was eventually withdrawn because spontaneous reports suggested an incidence of 1 : 10–20 000 of severe anaphylactic reactions which was not shown in clinical trials.[67] Wardell has calculated that a reaction occurring in 1 : 5000 patients against a background incidence of 1 : 10 000 would need 612 000 patients in a study of power 90%, to properly compare treated and non-treated populations.[68]

Dollery has suggested that the limitations of trials should be recognized so that post-marketing surveillance (PMS) for adverse reactions will be confined to surveys designed for the purpose.[69] Lasagna found evidence to support this view in a survey of laevodopa.[70] He examined reactions in 1500 patients followed closely for six years and found an incidence no different from that predicted in clinical trials. He concluded that the high cost of Phase IV studies large enough to define the incidence of reactions precluded any net benefit to society. Hence, Phase IV trials may establish many positive aspects of a new therapy but it does not seem appropriate to expect that they should exclude possible negative effects.

4.8.7 POST-MARKETING SURVEILLANCE (PMS)

4.8.7.1 Definition

Post-marketing surveillance (PMS) is more fully discussed in Chapter 4.9 of this volume, but consideration of its aims is relevant to the planning of Phase IV programmes since the two are complementary. PMS has been defined as the close observation of drug effects following marketing.[70] It includes: voluntary reporting schemes, case control studies, Phase IV trials, cohort or hospital monitoring surveys, and medical record linkage programmes.[72,73] Prescription event monitoring has been described as possible, although it requires considerable organization.[74]

4.8.7.2 Aims of Post-marketing Surveillance

There is wide agreement that both beneficial and adverse effects should be documented. The complete evaluation of a new medicine should include its full effect on the quality of life and the overall cost of treating the illness.[75] Thus if a new medicine avoids a surgical operation, this may be taken into account.

Griffin has remarked that the role of the UK Committee on Safety of Medicines (CSM) is to advise government and the medical profession on aspects of drug safety, and to do so it attempts to establish the risk–benefit ratio of a treatment.[77] However, this is no easy task since many adverse events occur for reasons unrelated to drugs and with an unknown background incidence.[72] Hence, validation is required of the link between a drug and a suspected adverse effect.

4.8.7.3 Outline of Methods of Post-marketing Surveillance

Such validation is based on epidemiological principles. Prominent among these are: any unusual features of the reaction either pathological or in its time course or geographical distribution. If they can be correlated with the number of patients exposed, then a kind of dose response will be obtained which increases the likelihood of a link. Here, the difficulty is that no individual practitioner will use a new compound often enough to obtain sufficient data. In fact, they are unlikely to suspect a link unless the reaction is highly unusual. This may be the main reason for under-reporting reactions in voluntary schemes such as the UK 'yellow card' system, although Inman has listed seven other more culpable reasons.[76] Hence, the incidence of adverse effects is known to be underestimated and the CSM has occasionally asked manufacturers to undertake further research or done so itself.

The CSM and the FDA have tried to prevent this difficulty by 'monitored release' of drugs, when Phase IV studies or their extension to larger groups of patients have been attempted. These include cohort studies and case control surveys of patients newly diagnosed. The more intensive the monitoring the more often they encounter the problems already described: that the numbers of patients required are enormous, and the value of the result questionable.[70,78]

A variant, termed 'recorded release' may prove more practical, less costly and actually monitor enough patients. It involves a monitoring centre which alerts local physicians to the need to record all unwanted effects in patients receiving new compounds. By clarifying with them exactly which drugs are involved and by follow-up discussions, the level of surveillance is likely to be raised. Prescription event monitoring is a similar scheme which attempts to link the reports to numbers of prescriptions.[72,73] It has been done for 50 000 prescriptions for ranitidine, for example.

To summarize, the majority of trials in Phase IV are comparative studies with existing treatments, in common with Phase III. Some multicentre trials may be needed in general practice or in important subgroups of patients. To investigate new indications smaller size trials will be appropriate initially and will resemble Phase II studies. Thereafter a few larger comparative trials will be needed but less than for the first clinical indication because the predictable tolerance to the new medicine will then be established.

However, the incidence of unpredicted, unwanted effects can be clarified only when numbers of patients reach thousands, as is typical of medical practice, numbers too great to monitor with the intensity of acceptable clinical trials. A 'halfway house' which monitors with less intensity is being developed under the banner of post-marketing surveillance. Its social, medicolegal and planning importance is likely to increase in the foreseeable future.

4.8.8 GOOD CLINICAL PRACTICE (GCP)

4.8.8.1 Definition

Good clinical practice (GCP) is the application of quality assurance concepts to clinical trials. It requires evidence that the trial complies with the protocol and that data have been correctly recorded and collected. It checks that legal requirements and acceptable scientific practices have been fulfilled.

A key issue is that the results do indeed mean what they say. For example, a measurement as simple as blood pressure is more often erroneous than expected because the sphygmomanometers are inaccurate. In GCP the data when the instrument was last checked will be recorded, as will the result and action taken. For laboratory tests normal ranges and comparisons with other results will be made so that the results can be interpreted more fully. Quality control data of assay procedures are particularly important, given the need to establish the assay sensitivity and specificity when metabolites similar to the parent compound are present.

4.8.8.2 Origins and Aims of Good Clinical Practice

The origin of GCP is well described by Farrell.[79] In principle the duty of the regulatory authorities is to interpret the data provided. This requires an assumption that the result is due to appropriate procedures, faithfully and fully reported. However, these assumptions have been questioned from time to time.[80,81,82] In 1974 the inspectorate of the FDA found the following incidence of failure to comply with the procedures stated in the protocol: drug accountability 50%, patient consent 35%, records accuracy 23%, records availability 22% and the role of the investigator in the study 12%. While this may be due in part to impractical procedures, there is always the option of amending them, which is one of the purposes of interim visits. Such amendments will be acceptable provided they have been documented and circulated appropriately.

The survey led to the GCP proposals which the American pharmaceutical industry applied, but without much success. Another survey (Arnold[83]) of 680 trials produced much the same result. Hence it is probable that the proposals will be made more practical and will be given legally enforceable status.

Their objective may be summarized as follows. Firstly, the rights and health of the patients should be fully protected, by informed consent and reviews of protocols by institutional ethics committees (Institutional Review Boards or IRBs). Secondly, monitoring the study should be organized by the sponsor (the pharmaceutical company) who must designate a monitor and provide written procedures for the monitor to follow. Thirdly, and incorporated from earlier guidelines, the principles already described in this chapter should apply to the trial design and analysis of the data obtained.

4.8.8.3 Obligations of Sponsors Complying With Good Clinical Practice

The obligations of sponsors and monitors include the following.

(a) Pre-investigation visits. A signed record will be required stating that the investigator knows the regulatory status of the new drug, the nature of the proposed trial and accepts their obligations to comply with FDA regulations. The monitor will ensure that the centre has adequate facilities and that the investigator will devote a proper amount of their personal attention to running the study.

(b) Guidelines for the constitution of ethics committees must be followed; the membership of the committee being a careful balance of medical and lay persons. Their *curriculum vitae* will be included in the evidence provided.

(c) Interim visits will ensure that facilities remain adequate. They will focus upon continued compliance with the protocol, regulations and completion of record forms. Detailed medical histories of each patient must be provided on entry and during the trial. Informed consent must be fully documented. Omissions in data must be repaired or explained, illegible entries repaired and patient dropouts listed, with the reasons for dropout and remedial action taken. Full records must be kept of these visits. Any attempt to promote or commercialize the new therapy must be reported with the assurance that it will not be repeated.

(d) Similar details of the obligations of investigators are included. They are to document the procedure described and to facilitate inspection by the FDA when asked.

4.8.8.4 Impact of Good Clinical Practice on Clinical Trials Programmes

Full implementation is time consuming and expensive. Completion of the documents often demands a full time assistant to the investigator on the spot and this requires a salary. Investigators have to be reimbursed for their time, which in the USA will be equivalent to their private practice fees. The procedures are so detailed that monitors have their hands full with three or four trials on one product only, which is less stimulating than a job which includes two or more new treatments. The cost of analyzing the mass of data is considerable, and when much of it is concerned with checking signatures, dates and requisition forms of one kind and another it is relatively non-productive.

These are not the only reasons why other regulatory authorities have not adopted the GCP proposals in full. For example, hospital records do not belong to the investigator but to the local health authority in the UK. Confidentiality of information is more jealously guarded in the UK and many other European countries and may have legal status. There are national differences in the definition of ethically acceptable trials. Thus in the Federal Republic of Germany placebo-controlled trials are frequently considered less than ethical and the German law of 1978 specifically does not make them essential for a grant of a product licence.[84] The close monitoring and duplication of records demanded is not acceptable to the medical professions of most European countries, particularly the rigorous inspection procedures that have the threat of penalties in the background. Perhaps the European view is most succinctly expressed by the WHO Group (1968), who considered that 'peer group review may actually be more effective than laws in protecting both the patient and the investigator'.[85] However, it is now widely accepted that a competent ethics committee must be included in the peer group, for which the insistence of the FDA has been a contributory factor.

For these reasons most European pharmaceutical companies attempt to comply with the spirit of GCP in European trials without incurring the full cost. This approach is suitable because data on a product known to be effective will be acceptable for a product licence submission in the USA and Japan provided that protocol violations remain few and minor.

4.8.8.5 USA Licence Submissions Using Foreign Data

The FDA issued regulations in 1975 on the acceptability of foreign data which state: 'It is in the interest of public health, whenever possible, to have access to and to consider detailed information resulting from those studies performed abroad which are well-conceived, well-controlled, performed by qualified experts and conducted in accordance with ethical principles acceptable to the world community'.[86]

Indeed, it has been known for a product to be licensed solely on the basis of foreign data (the Norwegian Multicenter Study Group trial on timolol for myocardial infarction). The FDA's conditions for this were published in 1982, and added the following to the existing requirements: (i) the foreign data must be applicable to populations in the USA and medical practice in the USA; (ii) the studies must be performed 'by clinical investigators of recognized competence'; and (iii) the data must remain 'valid' without an inspection of the centre, otherwise the investigator must agree to an inspection.[87] Farrell considers that the FDA is using 'valid' in the sense of reliable as much as in the scientific sense 'meaningful'.[79]

These proposals now influence greatly the practice of clinical trials throughout the pharmaceutical industry because data conforming to them will be acceptable to the largest market in the world. Moreover, the standards set in many different countries seem to be converging towards good clinical practice so it seems likely to consume more manpower and resources in the foreseeable future.

4.8.9 CONCLUSION

From the first dose in volunteers to a marketed product is a long campaign of clinical trials in three distinct phases. The first phase obtains proof of good tolerance to the expected clinical dose and evidence of adequate absorption and suitable pharmacokinetics in volunteers. Phase IIA begins with small-scale studies, sometimes of single doses only, which demonstrate that the compound has the desired pharmacological activity in patients. Proof of efficacy requires rigorous scientific and statistical techniques. Phase IIB confirms efficacy in approximately 50 to 100 patients, dosing usually from two to four weeks so that the sponsor of the new compound can reasonably expect that investment in Phase III and marketing will be worthwhile. Trials in Phase III recruit larger numbers of patients such that logistic problems predominate and the trials consume considerable resources.

After marketing the product will achieve its full potential only with the support of further trials, known as Phase IV, which amplify its primary clinical role, clarify questions of longer term tolerance and drug interactions, and may explore new indications. Finally, post-marketing surveillance is increasingly applied, to detect the side effect which is so rare that it eludes even the most extensive clinical trial programme.

4.8.10 SUGGESTIONS FOR FURTHER READING

(i) Management of Clinical Trials

1. H. Glenny and P. Nelmes (eds.), 'A Handbook of Clinical Drug Research', Blackwell Scientific Publications, Oxford, 1987. A practical book of chapters on main aspects of trials written by a specialist in each. The two on ethics and development of regulatory control of pharmaceutical research are essential reading.
2. S. J. Pocock, 'Clinical Trials', John Wiley and Sons, Chichester, England, 1983. A clear description of the statistical basis of trials that explains statistics in as much depth as most non-statisticians could want.
3. C. Maxwell, 'Clinical Research for All', Cambridge Medical Publications Ltd., England, 1973. A pocket-sized, very readable introduction to trials written in an engaging, chatty style.
4. C. S. Good (ed.), 'The Principles and Practice of Clinical Trials', Churchill Livingstone, Edinburgh, 1976. A symposium proceedings which has some uniquely useful chapters by experts on trials in specialized areas such as NSAIDs, antidepressants and antibiotics.
5. B. Spilker, 'A Guide to Clinical Studies and Developing Protocols', Raven Press, New York, 1984. A practical approach, clearly written by an experienced trialist in the pharmaceutical industry.

(ii) Pharmacokinetics

1. W. A. Ritschel, 'A Handbook of Pharmacokinetics', 2nd edn., Drug Intelligence Publications Inc., Hamilton, IL, 1980. A popular pocket-sized reference book which explains all the most used variables most concisely.
2. M. Rowland and T. N. Tozer, 'Clinical Pharmacokinetics. Concepts and Applications', Lea and Feibiger, Philadelphia, 1980. A book by experienced teachers of the subject which has remained popular for its clear explanations of difficult concepts.

(iii) Statistics of Trials

1. T. D. V. Swinscow, 'Statistics at Square One', the British Medical Association, London, 1977. Perhaps the most popular introduction to medical statistics with good illustrative examples.
2. S. M. Gore and D. G. Altman, 'Statistics in Practice', the British Medical Association, London, 1982. A collection of the most pertinent series of articles on statistics in trials from the British Medical Journal.
3. R. E. Parker, 'Introductory Statistics for Biology', the Institute of Biology's Studies in Biology No. 43, Edward Arnold, London 1976. A good introduction which by simple examples explains quite difficult statistical procedures.
4. G. J. Bourke, E. D. Daly and J. McGilvray, 'Interpretation and Uses of Medical Statistics', 3rd edn., Blackwell Scientific Publications, Oxford, 1985. A sound, readable account with good appendices on individual statistical tests.

(iv) Clinical Pharmacology of Drugs

1. B. G. Katzung, 'Basic and Clinical Pharmacology', Appleton and Lange, Norwalk, CT, 1987. A comprehensive, readable compendium of drug effects by several specialist authors. It includes valuable information that is otherwise difficult to find, such as the effects of renal failure on pharmacokinetics of antibiotics as a class of drugs.
2. D. R. Laurence and P. N. Bennett, 'Clinical Pharmacology', 6th edn. Churchill Livingstone, Edinburgh, 1987. A well-illustrated guide, much favoured by medical students and their teachers.

APPENDIX A: A FORM OF INDEMNITY FOR CLINICAL TRIALS

INDEMNITY

(Sponsoring organization Name and Address) will indemnify and hold harmless the (Hospital or Area Health Authority) and also authorized investigators employed in the service of (Area Health Authority) from and against any and all Third Party claims and suits for injuries and damages caused by the administration of (investigational new drug) during the course of the clinical investigation and in accordance with the protocol number (Protocol Number) or (Sponsor's) product (drug) and supplied to the authorized investigator for the purpose of the clinical investigation described in the protocol, provided that:

A Nothing herein shall be interpreted as an indemnity by (Sponsor) in respect of claims and suits for injury and damages caused by the failure of the Health Authority and/or the Investigator to observe all proper and reasonable standard in the use of (drug) according to the protocol.
B Immediately upon notice of any kind whatsoever of any claim or suit the Health Authority will notify (Sponsor) in writing and will permit (Sponsor's) lawyers at the discretion and expense of (Sponsor) to handle and control the defence of such claims.
C Such claims or suits will not be settled without the prior written consent of (Sponsor).

The authorized investigator(s) shall be: (Investigator's name(s)) together with such other qualified and competent persons as they may nominate to assist him/them in the said investigation and notify to (Sponsor).

In as much as the (Health Authority) and the said authorized investigators have their own insurance covering their own legal liability arising out of conducting clinical trials such insurance will prevail. The indemnity provided herein is deemed to operate over and above and in excess of any other insurance arrangement.

This indemnity shall be effective from the date of commencement of the clinical investigation described in the protocol and will remain effective until the date of the conclusion of that investigation.

Signed for and on behalf of (Sponsor)

Authorized Signatory.. Name..

Position in organization.. Date...

4.8.11 REFERENCES

1. C. E. Lumley, R. A. Prentis and S. R. Walker, *Pharm. Med.*, 1987, **2**, 137.
2. S. R. Walker, R. A. Prentis and M. K. Ravenscroft, *Pharm. Int.*, June 1986, 135.
3. *D. HEW Publ. (FDA) (US)*, 1977, 77-3040.
4. D. W. Vere, in 'Handbook of Clinical Drug Research', ed. H. Glenny and P. Nelmes, Blackwell, Oxford, 1986, p. 1.
5. World Health Organization, *Br. Med. J.*, 1964, **1**, 177.
6. The Medical Research Council, 'Responsibility in Investigations on Human Subjects', Command 2382, HMSO, London, p. 21.
7. *Lancet*, 1979, **1**, 534.
8. E. Davies, *Proc. R. Soc. Med.*, 1972, **66**, 533.
9. B. Simmons, *J. Med. Ethics*, 1978, **4**, 172.
10. M. D. Kirby, *J. Med. Ethics*, 1983, **9**, 69.
11. O. Gillie, 'Secret Drug Tests used on Patients', *The Sunday Times*, 27th July, 1975.
12. 'Guidelines on the Practice of Ethics Committees in Medical Research', Royal College of Physicians, London, 1975.
13. 'Guidelines on the Constitution and Function of Ethics Committees in the Pharmaceutical Industry', the Association of the British Pharmaceutical Industry (ABPI), London, 1988.
14. Food and Drug Administration (FDA), Code of Federal Regulations 50, *Fed. Regist.*, 1981, **46**, 8975.
15. C. J. D. Zarafonetis, P. A. Riley, P. W. Willis, III, L. H. Power, J. Werbelow, L. Farhat, W. Beckwith and B. H. Marks, *Clin. Pharmacol. Ther.*, 1978, **24**, 127.
16. A. Darragh, M. Kenny, R. Lambe and I. Brick, *Lancet*, 1985, 93.
17. Editorial: 'Death of a Volunteer', *Br. Med. J.*, 11th May, 1985, 1369.
18. H. J. Dengler, in 'Clinical Pharmacological Evaluation in Drug Control', World Health Organization, Geneva, 1973, p. 41.
19. H. J. Rogers and R. G. Spector, in 'Handbook of Clinical Drug Research', ed. H. Glenny and P. Nelmes, Blackwell, Oxford, 1986, p. 43.
20. C. Maxwell, *Br. J. Clin. Pharmacol.*, 1978, **6**, 15.
21. A. Bond and M. Lader, *Br. J. Med. Psychol.*, 1974, **47**, 211.

22. 'Investigations in Man', part 2 in 'Guidelines for Preclinical and Clinical testing of New Medicinal Products', Association of the British Pharmaceutical Industry (ABPI), London, 1977.
23. Department of Health and Social Security, 'Guidance Notes on Applications for, Clinical trials Certificates and Clinical Trial Exemptions', 1984, HMSO, London.
24. 'Safety Requirements for the First Use of New Drugs and Diagnostic Agents in Man', Council for International Organizations of Medical Sciences (COIMS), Geneva, 1983.
25. D. R. Laurence, in 'Clinical Pharmacological Evaluation in Drug Control', World Health Organization Heidelburg Symposium, 1972, p. 37.
26. J. J. Burns, *Ann. N. Y. Acad. Sci.*, 1968, **151**, 959.
27. N. E. Pitts, 'Principles and Techniques of Human Research and Therapeutics', ed. F. MacMahon, Futura, New York, 1974, p. 19.
28. A. D. Dayan, B. Clark, M. Jackson, H. Morgan and F. A. Charlesworth, *Lancet*, 1984, **1**, 555.
29. P. Joubert, L. Rivera-Calimlim and L. Lasagna, *Clin. Pharmacol. Ther.*, 1975, **17**, 253.
30. S. T. Holgate, 'How Good are Healthy Volunteers for Studying Drug Action', discussion in proceedings of a Society for Drug Research Symposium, *Pharm. Med. Suppl. 1*, 1988.
31. C. T. Dollery and D. S. Davies, *Br. Med. Bull.*, 1970, **26**, 233.
32. I. M. James, in 'The Principles and Practice of Clinical Trials', ed. C. S. Good, Churchill Livingstone, Edinburgh, 1976, p. 20.
33. S. T. Holgate, 'Volunteer Studies of Antiasthma Drugs', in 'How Good are Healthy Volunteers for Studying Drug Action?', proceedings of a Society for Drug Research symposium, *Pharm. Med. Suppl. 1*, 1988.
34. B. Saletu, in 'Human Psychopharmacology, Measures and Methods', ed. I. Hindmarch and P. D. Stonier, Wiley, Chichester, 1987, vol. 1, p. 180.
35. J. R. Wittenborn, in 'Human Psychopharmacology, Measures and Methods', ed. I. Hindmarch and P. D. Stonier, Wiley, Chichester, 1987, vol. 1, p. 73.
36. A. K. Shapiro, in 'Principles of Psychopharmacology', 2nd edn., Academic Press, New York, 1978, p. 441.
37. A. Goldstein, S. Kaizer and O. Whitby, *Clin. Pharmacol. Ther.*, 1969, **10**, 489.
38. R. Newton, L. J. Broughton, M. J. Lind, P. J. Morrison, H. J. Rogers and I. D. Bradbrook, *Eur. J. Clin. Pharmacol.*, 1981, **21**, 45.
39. F. E. Karch, C. L. Smith, B. Kerzner, J. M. Mazzulo, M. Weintraub and L. Lasagna, *Clin. Pharmacol. Ther.*, 1976, **19**, 489.
40. L. Lasagna and J. M. von Felsinger, *Science (Washington, D.C.)*, 1954, **120**, 359.
41. Gilbert's Syndrome; More Questions than Answers. *Lancet*, 1987, 1071.
42. R. G. Hiss and L. E. Lamb, *Circulation*, 1962, **25**, 947.
43. A. A. Quyyumi, C. Wright and K. Fox, *Br. Heart J.*, 1983, **50**, 460.
44. P. Froom, J. Ribak and J. Benbassat, *Br. Med. J.*, 1984, **288**, 20.
45. M. Rowland and T. N. Tozer, 'Clinical Pharmacokinetics: Concepts and Applications', Lea and Febiger, Philadelphia, 1980.
46. B. E. Rodda, in 'Importance of Experimental Design and Biostatistics', ed. F. G. McMahon, Futura, Mount Kisco, 1974, p. 19.
47. P. Armitage, 'Sequential Medical Trials', Blackwell, Oxford, 1975.
48. D. G. Clayton, *Br. J. Clin. Pharmacol.*, 1982, **13**, 469.
49. S. J. Pocock, 'Clinical Trials: A Practical Approach', Wiley, Chichester, 1983, p. 125.
50. Anturane Reinfarction Trial Research Group, *N. Engl. J. Med.*, 1980, **302**, 250.
51. J. L. Fleiss, 'Statistical Methods for Rates and Proportions', Wiley, New York, 1981.
52. J. Berkson, *J. Stat. Planning Inference*, 1978, **2**, 27.
53. K. Fidler and I. Koch, in 'Handbook of Clinical Drug Research', ed. H. Glenny and P. Nelmes, Blackwell, Oxford, 1986, p. 196.
54. B. Spilker, 'Guide to Clinical Studies and Developing Protocols', Raven Press, New York, 1984.
55. P. Wright and J. Haybittle, in 'Handbook of Clinical Research', ed. H. Glenny and P. Nelmes, Blackwell, Oxford, 1986, p. 247.
56. F. Grady, in 'The Principles and Practice of Clinical Trials', ed. C. S. Good, Churchill Livingstone, Edinburgh, 1976, p. 60.
57. P. Armitage, K. McPherson and B. C. Rowe, *J. R. Stat. Soc.*, 1969, **132**, 235.
58. K. McPherson, *Stat. Med.*, **1**, 25; *Experimentia (Suppl.)*, 1982, **41**, 454.
59. S. J. Pocock, 'Clinical Trials: A Practical Approach', Wiley, Chichester, 1983, p. 148.
60. A. Spriet and P. Simon, in 'Methodology of Clinical Drug Trials', translated by R. Edelstein and M. Weintraub, Kerger, Basle, 1985, p. 127.
61. H. P. Roth, H. S. Caron and B. P. Hsi, *Clin. Pharmacol. Ther.*, 1970, **11**, 228.
62. D. T. Lowenthal, W. A. Briggs, R. Mutterperl, B. Adelman and M. A. Creditor, *Curr. Ther. Res.*, 1976, **19**, 405.
63. P. A. Nicholson, in 'The Principles and Practice of Clinical Trials', ed. C. S. Good, Churchill Livingstone, Edinburgh, 1976, p. 81.
64. B. J. Culliton and W. K. Waterfall, *Br. Med. J.*, 1980, **280**, 1175.
65. J. E. Murphy, in 'Clinical Trials', ed. F. H. Johnson and S. Johnson, Blackwell, Oxford, 1977, p. 176.
66. 'Code of Practice for the Clinical Assessment of Licensed Medical Products in General Practice', the Association of the British Pharmaceutical Industry (ABPI), London, 1983.
67. R. S. J. Clarke, D. W. Dundee, R. T. Garrett, G. K. McCardle and J. A. Sutton, *Br. J. Anaesth.*, 1975, **47**, 575.
68. W. M. Wardell, M. C. Tsianco, N. A. Sadanand and H. T. Davis, *J. Clin. Pharmacol.*, 1979, **19**, 85.
69. C. T. Dollery, *Ciba Found. Symp.*, 1976, **44**, 73.
70. L. Lasagna, *JAMA, J. Am. Med. Assoc.*, 1983, **249**, 2224.
71. W. M. Castle, J. T. Nicholl and C. C. Downie, *Br. J. Clin. Pharmacol.*, 1983, **16**, 581.
72. G. R. Venning, *Br. Med. J.*, 1983, **286**, 544.
73. M. D. Rawlins, *Br. Med. J.*, **288**, 879.
74. P. Turner, *J. R. Soc. Med.*, 1984, **77**, 93.
75. 'Costs and Benefits of Pharmaceutical Research', seminar proceedings, ed. G. Teeling-Smith, Office of Health Economics, London.

76. W. H. W. Inman, in 'Monitoring for Drug Safety', M. T. P. Press, Lancaster, 1980, p. 9.
77. J. P. Griffin, in 'Pharmaceutical Medicine', ed. N. Macleod, Churchill Livingstone, Edinburgh, 1979, p. 108.
78. S. F. Sullman, in 'Post-marketing Surveillance of Adverse Reactions to New Medicines', Medico-Pharmaceutical Forum Publication No. 7, London, 1978.
79. F. G. Farrell, in 'Handbook of Clinical Drug Research', ed. H. Glenny and P. Nelmes, Blackwell, Oxford, 1986, p. 320.
80. L. Altman and L. Melcher, *Br. Med. J.*, **286**, 1983, 2003.
81. General Accounting Office, 'Supervision of Investigational Use of Selected Drugs', a report to the US Senate Committee on government operations, B-164031 (2), 23rd July 1973.
82. General Accounting Office, 'Federal Control of New Drug Testing is not Adequately Protecting Human Test Subjects and the Public', a report to Congress, HRD-76-96, 15th July 1976.
83. J. S. Arnold, *BIRA J.*, 1984, **3**,78.
84. R. Burkhardt and G. Kienle, *Lancet*, 1978, **2**, 1356.
85. 'Principles for the Clinical Evaluation of Drugs', a report of the WHO Scientific Group, Tech. Rep. Ser. No. 403, World Health Organization, Geneva, 1968.
86. FDA 21. CFR 312.20 at 40, Federal Regulation 16056, 19th April 1975.
87. FDA 21. CFR 314 proposed at 47, Federal Regulation 46622, 19th October, 1982.
88. A. W. F. Edward, 'Likelihood', Cambridge University Press, Cambridge, 1972.
89. D. V. Lindley, 'Making Decisions', Wiley, Chichester, 1973.

4.9
Post-marketing Surveillance

NIGEL S. B. RAWSON
Drug Safety Research Unit, Southampton, UK

4.9.1 HISTORICAL INTRODUCTION

4.9.1.1 Early Drug Therapy

The human race has long sought to ease its sufferings and to cure its diseases. At first sight, the methods that have been used seem to be both numerous and dissimilar, but they are all variations of three basic types: faith healing, hygienic measures and drug therapy. With the exception of a relatively short period during the classical Greek civilization when hygienic therapy prevailed, faith healing in the form of magic, religion or folklore was the dominant method of treatment until the Renaissance. Nevertheless, some form of drug therapy was undoubtedly practised throughout these millennia.

Although the magical and religious forms of faith healing slowly declined from the time of the Renaissance, hygienic therapy was not immediately revived. Excessive and irrational drug therapy became the dominant method of treatment in the 17th and 18th centuries. The extensive treatment administered to King Charles II and George Washington immediately prior to their deaths shows the type of therapy that was prevalent during this period.[1,2] Although the drugs were generally useless, some were harmful and even fatal.

The 19th century saw the decline of the excessive use of drugs as the prevailing form of treatment, the pendulum swinging so far in the opposite direction that 'therapeutic nihilism' became rife.[3] In England in particular hygienic measures were resurrected, largely due to the efforts of such men as Sir Edwin Chadwick, Sir John Simon, John Snow and William Farr; their work led to major improvements in sanitation.[4] Many important scientific discoveries were made during the 19th century, perhaps the greatest in medicine being those of Louis Pasteur. Pharmacology became a science seeking to identify the active substances in crude medicaments and to establish their actions in the human body, and important new pharmacopoeias, which for the first time laid down standards of drug purity, appeared in several countries.

There was some awareness that not all drug effects are therapeutically advantageous when the first formal enquiry into a suspected adverse drug reaction (ADR) took place in the closing years of the 19th century. Chloroform was first used as an anaesthetic in humans by Sir James Simpson in 1847 but, as its use increased, it was found that occasionally patients died suddenly and unexpectedly during the induction of anaesthesia. In 1877, the British Medical Association appointed a committee to investigate these deaths. Reports of fatal accidents in patients were analyzed and experiments were carried out in animals. In its final report,[5] the committee found that chloroform was hazardous not only because in large doses it caused respiratory depression but also because it had harmful effects on the heart; even in small doses, it could cause cardiac arrest at the beginning of induction. However, this effect did not occur in dogs,[6] giving one of the earliest demonstrations that pharmacological studies in animals are not necessarily applicable to humans. No other investigation of a suspected ADR of a comparable scientific calibre was to be carried out for another 40 years.

4.9.1.2 Drug Therapy in the 20th Century

Most major therapeutic developments have occurred in this the 20th century. When it began, physicians had little that was effective to offer their patients. Apart from their dedication, they had chloroform, cocaine and ether as anaesthetics, a few specifics such as digitalis, ipecacuanha, morphine and quinine, and little else. The situation soon changed, however, with a series of dramatic discoveries, *e.g.* the sulfonamides, insulin and penicillin, turning this century into the 'golden age of therapeutics'.[7]

These advances were not, however, free of unwanted effects. The second formal investigation into a suspected ADR was conducted at the end of World War I, when an epidemic of jaundice and fatal hepatic necrosis occurred among soldiers treated for syphilis with organic arsenicals at a military hospital near Cambridge. The epidemic was so serious that it was the subject of a special report by the Medical Research Committee, the predecessor to the present Medical Research Council (MRC).[8] The report stated that the most probable cause of the outbreak was the hepatic toxicity of the organoarsenical compounds. It is interesting to note that in both 1880 and 1922 medical and public opinion were sympathetic to the concept of setting up an independent body to assess the safety of drugs.[9]

Because they were associated with a wide variety of adverse effects, the sulfonamides, which were introduced in the 1930s, brought familiarity with ADRs to all doctors but the awareness and concern were only transient. Discoveries such as penicillin, streptomycin and the corticosteroids led to such dramatic advances in the efficacy of medical treatment that adverse effects, although recognized, were overshadowed and caused no great anxiety. By 1960, the discoveries made during the preceding 30–40 years and the associated improvements in the treatment of disease were well accepted, but ADRs were still of relatively little concern to either doctors or patients. The benefits of the new drugs were obvious; the adverse effects, although sometimes severe enough to demand that treatment should cease and even occasionally fatal, were deemed acceptable. Furthermore, when an ADR was discovered, it was often long after the drug was first marketed. A degree of concern similar to that existing about chloroform in the 1880s and arsenicals in the 1920s had not arisen again. The world was unprepared for the tragedy of thalidomide.

Thalidomide was introduced in the late 1950s as a 'safe' hypnotic and was soon used widely. It was not long before an 'epidemic' of phocomelia ('seal-like' limbs) and other congenital abnormalities was observed. Although tests for embryotoxicity were available, it was not standard practice at that time to perform such tests on new drugs before marketing and thalidomide was no exception. Nevertheless, the manufacturer claimed specifically that the drug was safe for use in pregnant women and nursing mothers, but this was pure speculation. In fact, it could have been predicted from elementary principles concerning the chemistry of the drug that thalidomide would gain access to the foetus.[10] Tragically, thousands of babies born to mothers who had taken thalidomide during pregnancy provided the missing data.

In the UK, about 400 thalidomide children survived, many terribly deformed, but it is thought that at least twice as many died around birth, mainly due to internal abnormalities.[10] It has been estimated that throughout the world 8000–10 000 babies were deformed as a result of their mothers taking thalidomide during the early stages of pregnancy. The vast majority (6000–7000) were born in West Germany, where thalidomide was not only widely prescribed by doctors but also available in non-prescription medicines.

Thalidomide was a watershed in the history of drug safety. Patients, doctors and the authorities had all been lulled into a false sense of security by the previous relatively safe years. Consequently, they were shocked by the tragedy which left its mark not only on the unfortunate children but also on the medical profession, the pharmaceutical industry, the public and governments. As a result of the catastrophe, many countries established agencies concerned with drug safety during the 1960s and, in 1968, the World Health Organization set up an international bureau to collect and collate information from national drug monitoring organizations.[11] The two most influential countries in the monitoring of drug safety, both then and now, are the UK and the USA.

4.9.2 NATIONAL DRUG MONITORING IN THE UK

4.9.2.1 The Committee on Safety of Medicines

In the UK, the Committee on Safety of Drugs (CSD) was established in 1964 by the health ministers in consultation with the medical and pharmaceutical professions and the pharmaceutical industry. The CSD had no legal powers but worked within a voluntary agreement with the trade associations of the pharmaceutical industry. These organizations agreed that their members would submit for approval data on the chemistry and pharmacy of new drugs and also the results of animal experiments and of subsequent clinical trials. Sub-Committees on Toxicity, Clinical Trials and Adverse Reactions were established in order to assist the CSD with decisions about pre-clinical laboratory studies, clinical trial results and post-marketing surveillance (PMS) for ADRs. Therefore, the CSD attempted to provide a voluntary registration system covering the examination of drugs both before and after marketing. The lack of legal powers to enforce its decisions was no great disadvantage but, as the scale and complexity of the operation increased, it was considered necessary to formulate regulations and to introduce a licensing system.[12]

Under the Medicines Act of 1968, the health ministers established a Licensing Authority to control the marketing and importation of medicinal products intended for human or veterinary use. The CSD was reconstituted as the Committee on Safety of Medicines (CSM) whose functions were: (a) to give advice with respect to safety, quality and efficacy in relation to human use of any substance or article (not being an instrument, apparatus or appliance) to which any provision of the Medicines Act is applicable; and (b) to promote the collection and investigation of information relating to ADRs for the purpose of enabling such advice to be given.[13] In September 1971, the CSM began to advise the Licensing Authority on the granting of clinical trial certificates and product licences for new drugs. Drugs already on the market were granted a temporary Product Licence of Right and a separate Committee on the Review of Medicines was subsequently established with the purpose of reviewing all Product Licences of Right by 1990. A further committee, the Committee on Dental and Surgical Materials, was set up to process applications for materials other than drugs, *e.g.* contact lenses. The work of these committees has been described by Cuthbert *et al.*[9]

The duty of the CSM is to monitor the safety of both established and new products, the assessment of the latter being perhaps the more difficult. Great emphasis is placed on pre-marketing tests of new drugs, but they alone are not considered to be sufficient because most serious ADRs are rare events, which are unlikely to be detected in clinical trials involving a few hundred patients. An effective PMS scheme based on thousands of patients taking drugs under general conditions of use is regarded as being more useful in the detection of ADRs than extensive clinical trials.[9] Consequently, the Sub-Committee on Adverse Reactions of the CSD and subsequently of the CSM has played an important role in drug safety in the UK. In 1982, the Sub-Committees on Toxicity, Clinical Trials and Adverse Reactions were amalgamated into the Sub-Committee on Safety, Efficacy and Adverse Reactions. A detailed account of the activities of this sub-committee has been given by Inman and Weber.[12]

4.9.2.2 The Yellow Card System

Until recently, the principal method for detecting unsuspected drug hazards after marketing was a voluntary reporting scheme known as the yellow card system. In May 1964, Dunlop (as chairman of the CSD) wrote to all doctors and dentists in the UK requesting reports of 'any untoward condition in a patient which might be the result of drug treatment'.[12] He announced the establishment of a Register of Adverse Reactions and enclosed a number of yellow reply-paid postcards for reporting suspected ADRs; all replies were to be treated with complete professional confidence. This system has been continued by the CSM. The Committee wished to obtain information about 'events'[14] or 'adversities'[15] experienced by the patients even when the doctor was uncertain about the exact role of the drug. In practice, however, doctors rarely report unless they strongly suspect that a drug has been responsible for an adverse event.[16] Doctors are asked to report serious or unusual reactions to all drugs; well-known minor ADRs, such as rashes with ampicillin, need not be reported but all events, however trivial, are requested for recently introduced drugs. There has deliberately been no attempt to lay down precise definitions of terms such as 'serious', 'unusual' or 'unexpected'.[12] The motto 'when in doubt—report' has been encouraged since the system began.[17]

During the four year period ending 30th June 1982, 78% of all reports of suspected ADRs received by the CSM came from the yellow card system.[12] A further 16% of the reports came from the pharmaceutical industry, a considerably larger proportion than was usual before the Medicines Act became effective in 1971,[18] since the Act imposed a statutory requirement on the pharmaceutical companies to report all adverse effects known to them. The rest of the reports came from correspondence, death certificates and articles published in medical journals. For the first few years, the number of yellow cards submitted to the Committee remained more or less constant at 3000–4000 a year but, as Table 1 shows, since 1975 the number has increased considerably to almost 15 000 a year.[19] These changes have not been completely explained, but possible reasons include increased concern about drug safety among the medical profession following the practolol incident (see Section 4.9.2.3), the introduction of new leaflets in the *Current Problems* information series published by the CSM, the introduction of prescription pads containing a slip of yellow paper reminding doctors to report ADRs, and the inclusion of 'tear-out' yellow cards in recent editions of the *British National Formulary*.

4.9.2.3 Underreporting of Adverse Drug Reactions

Although the figures in Table 1 are quite large, they reveal only a small proportion of ADRs; the rest are never reported. In a study of the association of blood dyscrasias with phenylbutazone and oxyphenbutazone,[20] it was found that only five of 44 deaths (11%) for which phenylbutazone or oxyphenbutazone had probably been responsible had been reported to the CSM. During the Committee's study of fatal thromboembolism in 1966–1967,[21] only two of the 53 general practitioners who had been aware that their patient had been taking oral contraceptives when she died had reported the death to the CSM prior to the special investigation. Another six were notified by consultants bringing the total reporting rate to about 15%, a low rate in the light of the public interest and concern at the time. When an 'epidemic' in which an excess of 3500 deaths from asthma were attributed to the excessive use of pressurized aerosols occurred in the mid-1960s, only six

Table 1 Numbers of Suspected Adverse Drug Reaction Reports Received by the Committee on Safety of Medicines, 1964–83

Year	*Number*	*Year*	*Number*
1964	1415	1974	4815
1965	3987	1975	5052
1966	2386	1976	6490
1967	3503	1977	11 255
1968	3486	1978	11 873
1969	4306	1979	10 880
1970	3563	1980	10 179
1971	2851	1981	12 357
1972	3638	1982	14 701
1973	3619	1983	13 974

doctors had considered the possibility that these deaths were related to treatment; they reported about a dozen deaths.[22]

Perhaps the most striking example of underreporting, however, was that associated with the β-adrenergic receptor blocking agent practolol, which was first marketed in 1970.[23] The syndrome produced by practolol is an excellent illustration of the type of tragedy which the yellow card system was designed to detect. Practolol was found to produce an unusual combination of signs and symptoms in some patients following periods of treatment ranging from one month to several years. The characteristic symptoms included psoriasiform rashes with hyperkeratosis, a reduction in tear secretion leading sometimes to conjunctival irritation and occasionally to corneal ulceration and blindness, a form of otitis media resulting in deafness, and unusual changes in the peritoneum (described as sclerosing or plastic peritonitis) causing intestinal obstruction with some loss of life. Not all the patients with reactions to practolol exhibited all these symptoms, but many had at least two of them and a few developed all four. More than three years elapsed and over 100 000 patients were treated with practolol before three independent groups of physicians published their observations. In May 1974, a report from dermatologists working in Newcastle-upon-Tyne suggested a probable association between practolol and psoriasiform rashes.[24,25] This was followed a month later by an ophthalmic surgeon's observation that patients treated with practolol could develop a dry-eye syndrome leading to serious loss of vision.[26,27] In December 1974, a group of Bristol physicians reported that a small number of patients had developed an unusual form of peritonitis with intestinal obstruction.[28]

All the symptoms were similar if not identical to naturally occurring ones, which was perhaps why only four reports of psoriasis and a single report of conjunctivitis associated with practolol had been submitted to the CSM during the first four years after the drug had been marketed.[29] The publication of the first two papers was followed by a warning letter distributed to all doctors by the manufacturer; later, three further letters from the manufacturer and a warning statement from the CSM were circulated. As a result, the Committee received reports of more than 200 patients with eye reactions (a small proportion of which had caused blindness), a larger number of reports of skin reactions, about 30 reports of deafness and almost 40 of sclerosing peritonitis.[30] Most of these reports related to observations made many months, even years, previously. The most disturbing feature of the peritonitis was that patients were developing the first abdominal symptoms up to three years after stopping the drug. In July 1975, the drug was restricted to hospital use and finally, in September 1976, oral practolol was completely withdrawn.[23] It is thought that several thousand reactions were caused by practolol.

4.9.2.4 Reasons for the Underreporting of Adverse Drug Reactions

In Sweden, where the reporting of ADRs is mandatory, it has been estimated that one-third of all major reactions are reported[31] but, in the UK, all the available evidence points to a much lower reporting rate, probably less than 10% and possibly as low as 1%.[18] Furthermore, of the doctors eligible to use the yellow card system between 1972 and 1980, only 16% are reported to have done so.[19] A recent study showed a similar low reporting rate, suggesting that the situation is unlikely to change in the near future.[32,33]

The reasons for underreporting are numerous and complex. Even if doctors do recognize the possibility of a connection between an adverse event and treatment, they will not necessarily report their suspicions. Among the many reasons for the apparent reluctance on the part of doctors to cooperate with the Committee, at least seven, described by Inman[30] as the 'seven deadly sins', have been identified: (1) *complacency*, the mistaken belief that only safe drugs are allowed on to the market; (2) *fear* of involvement in litigation, perhaps a lesser problem in the UK than in the USA; (3) *guilt* for having prescribed a treatment which may have harmed the patient; (4) *ambition* to be the first to collect and publish a personal series of cases; (5) *ignorance* of the requirements for reporting; (6) *diffidence* about reporting mere suspicions which might perhaps lead to ridicule; and (7) *lethargy*—an amalgam of indifference on the part of doctors to their role as clinical investigators, procrastination, lack of interest or 'time', inability to find a report card and other excuses.

To interest doctors in its work and to encourage them to report adverse events, the CSM provides an ADR information service, circulates letters and pamphlets in the *Adverse Reactions* and *Current Problems* series to all doctors, contributes a regular monthly column (*CSM Update*) to the *British Medical Journal*, and periodically issues a complete list of all reports made to the Register of Adverse Reactions. In addition, each doctor who reports a suspected ADR or enquires about one receives an up-to-date computer printout which lists all the reactions to the relevant drug that have been reported previously.[12]

4.9.2.5 The Value of the Yellow Card System

In spite of the low level of reporting, the yellow card system has proved to be of considerable value. Much of this is not obvious to the medical profession as a whole, let alone the public, because action to modify the manufacturers' claims or to insert suitable warnings in the promotional literature has often been all that was required. On several occasions, drugs have been removed voluntarily from the market on the advice of the CSM.[34] Action to remind doctors of the dangers of chloramphenicol, in the form of a pamphlet in the *Adverse Reactions* series, was reflected in an immediate reduction in the number of reported deaths from an average of one per month before the publication to only one death due to treatment in the UK in the following 14 years.[12]

Early reports of thromboembolism in women using oral contraceptives enabled the CSM to influence the MRC and the Royal College of General Practitioners towards investigating what appeared to be a potentially important problem.[35] The Committee's own mortality study provided the first positive evidence of a causal relationship between the taking of oral contraceptives and thromboembolic disease.[34,35] A study of the reports, together with comparable data obtained from the Danish and Swedish drug regulatory agencies, also revealed the importance of the dose of estrogen in establishing the risk of thromboembolism.[36] Later, it was possible to show that oral contraceptives also increase the risk of myocardial infarction, especially in older women.[37,38] Other hazardous relationships identified by the yellow card system include oxyphenbutazone and fatal blood dyscrasias,[20] overuse of pressurized aerosol bronchodilators and sudden death from asthma,[22] nalidixic acid and bizarre disturbances of the central nervous system,[34] halothane and jaundice,[39,40] and methyldopa and hepatitis.[41]

4.9.3 NATIONAL DRUG MONITORING IN THE USA

4.9.3.1 The Delayed Introduction of Drugs on to the American Market

In the USA, even greater emphasis is placed on stringent pre-marketing controls. The Kefauver–Harris amendments to the Food, Drug and Cosmetic Act passed in 1962 significantly increased the required pre-marketing tests. Pharmaceutical companies are required to carry out an enormous number of animal studies and clinical trials to show that a new drug is both safe and effective before it may be released onto the market. Consequently, many years can elapse before a useful product is available to the patients who will benefit from it; in 1976, the average time required from submitting a new drug application to gaining approval for marketing was nearly nine years.[42] In 1973, Wardell compared the pattern of introduction of new drugs in the USA over the decade 1962–1971 with the corresponding pattern in the UK.[43] A considerable difference was found between the two countries;

in particular, large differences of clinical importance were found between the two countries in antibacterial, cardiovascular, diuretic, gastrointestinal and respiratory drugs. Wardell concluded that, in comparison with the USA, the UK appeared to have gained from its less restrictive policy towards the marketing of new drugs.[44]

Many of the new drugs that have been delayed are therapeutically important. For example, the USA was the 15th country in which indomethacin was marketed and the 39th to approve the first oral cephalosporin, although both drugs were developed by companies based in the USA. The delay before the acceptance of drugs developed outside the USA is generally even longer; the USA was the 51st country to approve the antitubercular drug rifampicin, the 65th to approve the antiasthma drug sodium cromoglycate, and the 106th to approve the antibacterial cotrimoxazole.[42] Inevitably, such delays have led to important differences in the practice of medicine on the two sides of the Atlantic; in many respects, American 'therapeutics' in 1980 were equivalent to European 'therapeutics' in 1974.[45] However, it is important to remember that the thalidomide and practolol tragedies did not occur in the USA because their introduction on to the American market was delayed long enough for the adverse effects to be observed in other countries.

In 1978, Wardell compared the pattern of introduction of new drugs in the USA with the corresponding pattern in the UK for the five year period 1972–1976.[46] He found that there had been a perceptible change since his previous study. The change was not so much in the relative numbers of new drugs that became available, in which the UK still substantially exceeded the USA, but in the narrowing of the most obvious therapeutic differences between the two countries.

4.9.3.2 The Food and Drug Administration's PMS System

Efforts towards establishing an effective national PMS system in the USA have achieved little success.[47] In 1960, an ADR reporting system was started as a pilot study to obtain information about adverse events in five hospitals; by 1968, the number of hospitals had increased to around 200.[48] They reported on a contractual basis and produced several thousand reports each year. A voluntary reporting system was established in 1968 by the Division of Drug Experience, the ADR monitoring unit of the Food and Drug Administration (FDA).[48,49] Reports from pharmaceutical companies also began to be routinely submitted to the Division. The total number of reports received in 1971 was almost 20 000 (Table 2); for economic reasons, the hospital contract programme was phased out in 1971, which explains the marked drop in the number of reports from hospitals from over 7000 in 1971 to 2800 in 1972. Between 1972 and 1981 the number of reports varied from just under 10 000 to just over 13 000 in each year, with between 60% and 80% of the reports coming from pharmaceutical companies.[48] There has been a consistent rise in the number of reports during recent years. The reasons for this trend are not completely understood, but it is thought to be due to a number of factors, including an increased awareness of ADRs and the need to report them, and recent withdrawals of drugs from the market. In spite of the increasing numbers, the reports received by the FDA are considered to represent only a small proportion (perhaps as little as 2%) of the total number of ADRs in the USA.[48] Nevertheless, the scheme has, like the yellow card system, proved to be of some value.[50,51]

The FDA also uses other information sources such as special registries, *e.g.* the Armed Forces Institute of Pathology's Registry of Tissue Reactions to Drugs,[49,52] and intensive hospital monitoring schemes (see Section 4.9.4). Recently, the FDA has started to use the drug and diagnosis-linked data of the Medicaid Management Information System (MMIS).[53–55] Financed jointly by federal and state funds, Medicaid is a programme of medical assistance for certain low-income individuals and families and is the primary source of health care coverage for the 'economically disadvantaged'.[53] MMIS is divided into six sub-systems:[54] (i) recipient (of the medical service), (ii) provider (of the medical service), (iii) claims processing, (iv) reference file, (v) surveillance and utilization review, and (vi) management and administrative reporting. The first four sub-systems function together to process each claim and pay the provider of the medical service. The other two are used to consolidate, organize and present the data in a form that will assist managerial control over the programme. Once claims are paid, the surveillance and utilization review sub-system analyzes the 'payment characteristics' of the providers and recipients and identifies any patterns that may represent fraud or abuse; this is the major intended use of the MMIS database.[54]

In 1978, a pilot project was begun by Health Information Designs Inc. (HID) in Washington, DC using part of the claims processing sub-system data from the states of Michigan and Minnesota. After 18 months HID had amassed a database of almost 1.5 million patients; the scheme now

Table 2 Numbers of Suspected Adverse Drug Reaction Reports Received by the Food and Drug Administration, 1971–86

Year	*Number of reports from* *Manufacturers*	*Physicians*	*Others*	*Total*
1971	12217	226	7186	19629
1972	9750	231	2856	12837
1973	6700	2189	1872	10761
1974	6807	985	1917	9709
1975	8097	699	1960	10756
1976	7664	2138	2210	12012
1977	8945	1862	1652	12459
1978	9143	1335	781	11259
1979	9907	1389	2048	13344
1980	8642	1244	1416	11302
1981	9520	983	2314	12817
1982	19642	1806	5146	26594
1983	28221	1255	5030	34506
1984	27781	1401	4437	33619
1985	39985	1089	2500	43574
1986	51261	2010	2045	55316

includes 10 states and approximately six million patients.[54] This database, which is known as the Computerized On-line Medicaid Pharmaceutical Analysis and Surveillance System (COMPASS), links patient characteristics, details of outpatient drugs prescribed (including dates), inpatient or outpatient disorders treated [coded using the International Classification of Diseases (ICD)],[56] procedures provided (laboratory, radiological, *etc.*), and details of the provider of the service. In spite of the fact that the 'economically disadvantaged' are not necessarily representative of the general population, COMPASS has a number of features which the FDA have found useful for PMS.[53–55,57]

4.9.4 INTENSIVE HOSPITAL MONITORING SCHEMES

In the field of PMS, intensive hospital monitoring schemes, which have been developed mainly in the USA, currently represent the principal alternative to the voluntary reporting system. The disadvantages of voluntary reporting systems are that adverse effects are rarely reported until doctors are almost certain that they are caused by the drug and, even when an ADR has been identified, its incidence cannot be reliably estimated because reporting is incomplete and the number of people using the drug is unknown. Intensive hospital monitoring schemes have the advantages that adverse events can be recorded whether or not they are suspected to be due to drugs and accurate estimates of incidence are possible. However, they can usually only provide information about relatively common early reactions to drugs causing admission to the type of wards being monitored and they exclude certain important events (notably sudden death) that do not result in hospital admission. Furthermore, patients are not usually in hospital long enough for the detection of delayed effects.

Intensive hospital monitoring was pioneered at the Johns Hopkins Hospital in Baltimore in the early 1960s by Cluff.[58] Moving to the University of Florida in 1966, he further developed the scheme at Shands Teaching Hospital to include medical, surgical, paediatric and obstetric wards[59–61] and attempted outpatient monitoring in a community setting[62] before the system was discontinued in 1976. A similar system was developed by the Clinical Pharmacology Unit at Tufts University, Boston,[63] which subsequently took the title of the Boston Collaborative Drug Surveillance Program (BCDSP).[64]

4.9.4.1 The Boston Collaborative Drug Surveillance Program

The BCDSP, which began in 1966, has been the most successful intensive hospital monitoring scheme. The main aims of the BCDSP are: (a) to detect previously unknown ADRs, particularly serious ones; (b) to quantify and evaluate known ADRs; (c) to obtain an estimate of the efficacy of individual drugs; and (d) to evaluate the role of factors that influence efficacy and toxicity.

In the basic design of the BCDSP, specially trained 'nurse-monitors' were assigned to particular hospital wards and collected data on all patients admitted to them. At an initial interview, information was obtained from the patient about alcohol, coffee, tea and tobacco consumption, drugs used during the month prior to admission, any previous ADRs and characteristics such as age and sex. Subsequently, information about each drug prescription, including indication, dosage, frequency and route of administration, was collected. When a drug was stopped, the reasons for discontinuation and data on efficacy and any suspected ADRs (defined as any unintended or undesired effects of the drug) were obtained from the clinician during ward rounds. At the time of the patient's discharge, information about the occurrence of certain events, such as convulsions, deafness, gastrointestinal bleeding and jaundice, was recorded, irrespective of whether or not the events were attributed by the physician to a drug. Discharge diagnoses and certain laboratory data were also noted. At various times, the BCDSP has collected and analyzed information from over 40 hospitals in seven countries.

The results of the BCDSP can be divided into four types of studies:

(i) Descriptive studies of the patterns of use, toxicity and factors influencing toxicity for particular drugs, *e.g.* tricyclic antidepressants,[65] intravenous diazepam,[66] digoxin,[67] potassium chloride,[68] propranolol[69] and procainamide.[70] A major publication has provided some of the most reliable incidence rates for acute adverse events due to commonly used drugs in the hospital setting.[71]

(ii) Descriptive studies of particular ADRs, such as anaphylaxis, convulsions, deafness, extra-pyramidal symptoms,[72] gastrointestinal bleeding[73] and liver disease.[74] Studies on ADRs leading to hospital admission[75] and on fatal reactions among medical inpatients[76] have also been published.

(iii) Detailed studies of factors influencing the occurrence of particular ADRs or particular clinical events, including age, sex, weight, dosage, route of administration,[66,70,77] serum albumin,[78] renal function,[79] cigarette smoking[80] and genetic factors.[81]

(iv) Drug interactions, *e.g.* between chloral hydrate and warfarin,[82] tetracycline and diuretics,[83] spironolactone and potassium chloride,[84] and allopurinol and ampicillin.[85]

The BCDSP has published a book[71] and over 200 papers.

In the period 1966–1968, the BCDSP was exclusively a prospective study in which hospitalized patients were entered into the Program on admission and followed for drug usage, medical events and short-term drug effects on a daily basis until discharge. In order to overcome the limitations of the short-term follow-up and the lack of information on previous drug consumption, additional data in the form of a history of regular drug use during the month before admission began to be collected in 1968. From this time, the Program was not only prospective in nature but, since it obtained information concerning prior drug use, also contained the elements of a retrospective study permitting case-control comparisons to be accomplished. In 1971, when sufficient information concerning pre-admission drug use had been accumulated, an analysis of the data for validity was performed; the data were judged to be valid by assessing whether known ADRs would have been 'signalled'.[86] During a subsequent series of screening comparisons, in which each reasonably common diagnosis was evaluated in turn, a significant inverse association between the diagnosis of myocardial infarction and prior regular aspirin use emerged. This result was of particular interest because there were biological grounds for believing that aspirin might reduce the risk of myocardial infarction.[87]

A large-scale multipurpose case-control study (known as the Special Study) was initiated to verify the hypothesis.[88] Over the first 10 months of 1972, some 25 000 patients admitted to 24 hospitals in the Boston area were interviewed about the regular use of drugs during the three months prior to admission and categorized according to some 26 indications for treatment. The study provided confirmatory evidence of the inverse relationship between the regular use of aspirin and myocardial infarction.[89] Several studies of the regular use of aspirin by persons having suffered a myocardial infarction have been performed since then,[90–95] and, although their results have not always been consistent,[90] the general consensus of opinion now appears to agree with the findings of the BCDSP.[96]

The most controversial results produced by the BCDSP were those published in 1974 which showed an association between reserpine and breast cancer.[97] Over 20 studies have been carried out around the world trying to confirm or refute this finding;[98] the results of the majority have disagreed with the BCDSP and one of the authors of the original report has stated that the apparent association was due to chance.[99] In PMS studies, it is essential to examine the possibility that an association is in fact spurious in order to avoid alarming doctors and their patients; once aroused, such fears are almost impossible to calm.

In 1975, the BCDSP began to use the files of the Commission on Professional and Hospital Activities—Professional Activity Study (CPHA-PAS), a non-profit-making research and education centre.[100] The CPHA-PAS, which was established in 1955 in Ann Arbor, MI, records on computer files some 35% of all hospital discharges in the USA, including information from the hospitals of the Group Health Cooperative of Puget Sound (GHCPS). Founded in Seattle in 1947, the GHCPS is a consumer-owned health care plan providing virtually complete pre-paid medical coverage for outpatient care, drugs and hospital services. Almost all members of the plan, of which there are around 300 000,[101] are admitted (when necessary) to hospitals in the Seattle area maintained by the Cooperative. Information, such as discharge diagnoses, surgical procedures performed and routine demographic data, on all discharges from GHCPS hospitals has been recorded and computerized by the CPHA-PAS since 1972. In 1975, the GHCPS began to computerize records of prescriptions dispensed by its nine regional outpatient pharmacies. Most members normally use the Co-operative's pharmacies to obtain their prescriptions; in a study of the users and non-users of replacement estrogens,[102] it was found that 98% of the women interviewed stated that they routinely had their prescriptions dispensed at the GHCPS pharmacies. Therefore, the drug use records of virtually all members are accessible in machine-readable form and can be readily linked to data concerning hospital admissions. The BCDSP has used this source of information to conduct studies on endometrial cancer,[102] vascular diseases,[103] renal stones,[104] stillbirths[105] and perforated peptic ulcers.[106]

In 1981, the BCDSP ceased its original data collection. However, the data collected over the previous 15 years are retrospectively analyzed for specific purposes. A similar method is being utilized to study recently marketed drugs in inpatient populations.[107] The work of the BCDSP has been reviewed by Lawson.[108]

4.9.4.2 The Drug Epidemiology Unit

Before 1972, it was thought that a cohort approach was required to identify unsuspected associations between drug exposure and various conditions. However, the BCDSP's Special Study showed that, by interviewing patients with a wide range of conditions and by ascertaining a wide range of drug exposures, a relatively non-specific case-control approach could provide information on possible ADRs. In order to pursue these ideas further, Shapiro and Slone left the BCDSP in 1975 to establish the Drug Epidemiology Unit (DEU) within the Boston University Medical Center. The DEU has been renamed the Slone Epidemiology Unit following Slone's death.

Initially, the DEU undertook a critical review of the Special Study to determine how it could have been improved. The study was discontinued after an adequate number of patients with myocardial infarction had been entered and, although a number of associations between various conditions and various drugs were observed, some were also missed, *e.g.* the relationship between endometrial cancer and conjugated estrogens which was later reported by Ziel and Finkle.[109] This failure was partly due to a lack of numbers and partly due to comprehensive 'life-time histories' of drug use not being obtained; for instance, if the use of a drug had ceased more than three months before admission even though a patient may have taken the drug for many years in the more remote past, such use was not recorded. While this proved to be satisfactory for the evaluation of acute conditions in relation to a variety of drugs, it was inadequate for the evaluation of drugs that may have long-term adverse effects, such as carcinogenicity. Therefore, it seemed desirable in redesigning a case-control system to obtain the most comprehensive histories of prior drug use as possible.

In 1976, the DEU began the new multipurpose system, now known as case-control surveillance;[88,110] this has two broad objectives: (i) to discover serious drug-induced illnesses and (ii) to provide for the rapid testing of hypotheses, whether generated from within the system or from other sources. The scheme operates in some 15 hospitals in the USA, Canada and Israel, where patients having a wide variety of conditions are interviewed by specially trained 'nurse-monitors'. The information recorded includes:

(i) 'Life-time histories' of drug use including non-prescription drugs; these are recorded as fully as possible. The DEU acknowledges that short-term or sporadic use of a drug (particularly if it took place many years ago) is not amenable to adequate recall, but it does not appear to have performed any study to validate this method of collecting information about drug use. Recall concerning the timing of past drug exposure is known to be prone to error[111] and some DEU researchers have shown in a recent study that asking about past use of drugs by name yields significantly higher positive responses than either open-ended questions or questions about drugs taken for specific indications.[112]

(ii) Patient characteristics. Age, sex, weight and marital status are recorded routinely. Other information is recorded on an *ad hoc* basis depending on the hypotheses being studied.

(iii) The patient's consumption of cigarettes, alcohol, coffee and artificial sweeteners.

Subsequently, a copy of the patient's discharge summary is obtained and the diagnoses are recorded. The data are subjected to a series of standard analyses as well as *ad hoc* evaluations.

The DEU has carried out studies within the overall framework of case-control surveillance to investigate the relationship between maternal drug exposure and birth defects,[113,114] the role of oral contraceptives and other factors in the aetiology of premature myocardial infarction,[115-122] and the role of conjugated estrogens and other factors in relation to breast,[123-126] endometrial[127,128] and epithelial ovarian cancers,[129] as well as *ad hoc* studies.[130-132] The staff of the DEU have published two books[113,133] and over 60 papers.

The DEU is also the coordinating centre for the International Agranulocytosis and Aplastic Anaemia Study.[134] This study is an attempt to obtain a complete population-based ascertainment of all cases of agranulocytosis and aplastic anaemia occurring over a four year period in eight geographical regions (three of the regions were added after the beginning of the four years and thus did not participate for the whole period). The first report on this study showed widely differing results, in that there was a strong association between dipyrone and the occurrence of agranulocytosis in three of the regions and none in two others. It is not clear if this difference is real or the result of some bias in the methodology. The methodology of studies of this type has been criticized by Kramer *et al.*[135]

4.9.4.3 Other Hospital Monitoring Schemes

Attempts have been made to monitor ADRs in hospital outpatients,[62] but little success has been achieved. Probably the best known system was set up almost 20 years ago in San Francisco by the Kaiser-Permanente Group.[136] Between July and December 1969, information was obtained from over 220 000 clinic and pharmacy visits to the Kaiser-Permanente Medical Center made by about 75 000 patients. Although the extreme complexity of the approach, together with the difficulty of handling such large quantities of data, did not allow the system to develop beyond the experimental stage, the data have been used to investigate possible long-term carcinogenetic effects of commonly used drugs.[137,138]

Although large-scale intensive hospital monitoring has tended to be concentrated in the USA, other smaller schemes have been set up elsewhere. An example of this type of monitoring in the UK is the drug information system operated in hospitals in Aberdeen and Dundee by the Medicines Evaluation and Monitoring Group (MEMO), which was established almost 20 years ago.[139] This system incorporates both an accurate record of all drugs administered to inpatients and the information routinely collected for the Scottish morbidity returns, such as age, sex, diagnosis, and duration and outcome of hospital stay. Data on some 525 000 patient discharges and more than two million prescriptions have been collected.[139] Although it was not originally intended as a means of detecting previously unsuspected ADRs, the system has been used to investigate a number of problems, including the relationship between methyldopa and haemolytic anaemia, tricyclic antidepressants and sudden death, and frusemide and thrombocytopenia.[140]

4.9.5 MEDICAL RECORD LINKAGE

Other efforts in PMS have included the use of medical record linkage. In the original project of this type (the Oxford Record Linkage Study) which began in 1962, Acheson recognized the possibility of using record linkage for PMS.[141] He outlined a scheme for monitoring the effects of selected new drugs but considered that the standard of patient-identifying data on prescriptions would have to be improved for this to be possible. However, in 1974, a pilot study was begun in Oxford by Skegg and Doll to assess the feasibility of using record linkage to monitor major adverse effects of drugs.[142,143] The aim of the study was to develop methods that could be used in the future on a larger scale. The population included in the study consisted of some 43 000 people registered with 20 general practitioners in six practices. For a period of two years, three types of record were obtained for every person in this population: (i) basic information, such as name, address, sex and date of birth, obtained from the Oxford Community Health Project computer records; (ii) details of prescriptions dispensed were obtained as photocopies from the Prescription Pricing Authority (see

Section 4.9.7.2); and (iii) records of morbidity and mortality. The information about adverse events were obtained by two methods which were compared. In the first of these, records of illnesses seen by the general practitioners were obtained for 10 000 people in one practice. In the second, details of all hospital admissions, outpatient care, obstetric deliveries and deaths (in and out of hospital) were obtained for the other 33 000 people in the other five practices from the Oxford Record Linkage Study.

The results of this study suggested that prospective PMS using record linkage would be both feasible and effective. The conclusions were that the method of choice would be to link prescription data with records of deaths, hospital admissions and obstetric deliveries, that a full-scale project should cover a population of at least 500 000 people (preferably 10 times that number), and that it should continue for an indefinite period.[142,143] At the present time, no action has been taken on these recommendations.

More recently, the feasibility of linking hospital medical records to general practitioner prescribing for the whole of Tayside in Scotland has also been investigated.[144] All patients residing in the Tayside Health Board area are identified by a unique community health number and this number is used on all computerized records of hospital inpatient morbidity statistics and outpatient contacts. The scheme currently covers some 400 000 people, but similar systems are being established in two other health board areas so that potentially it could be extended to 1.2 million people. The conclusions drawn from the study were that record linkage in Tayside can provide a simple and inexpensive method of PMS and that it is particularly suited for the detection of serious ADRs which may take several years to develop. Skegg has recently reviewed record linkage schemes and their use in PMS.[145]

4.9.6 CONCERN OVER THE EFFECTIVENESS OF POST-MARKETING SURVEILLANCE SCHEMES AND PROPOSALS FOR NEW SCHEMES

With the establishment of agencies to monitor drug safety following the thalidomide disaster, there was a tendency for not only governments and the public but also the medical and pharmaceutical professions to slip back into the pre-thalidomide feeling of security. The drug safety regulations with their requirements for pre-marketing animal tests and clinical trials were there to prevent harmful drugs reaching the market. If one did reach the market, it was thought that it would be quickly picked up by the voluntary reporting systems or possibly by one of the intensive monitoring schemes. These feelings of security were severely damaged by the practolol incident. The practolol animal studies did not predict that the syndrome would occur in humans and, although the symptoms were undoubtedly seen in some patients during the clinical trials, their significance was unrecognized at the time.[30] Furthermore, the yellow card system failed to reveal the harmful effects of practolol. These failures led to a wide debate on the need to review current procedures and to consider new methods for more quickly detecting and, if possible, quantifying medium to low frequency adverse effects of new medicines. Results from the intensive hospital monitoring schemes and from other studies, *e.g.* those indicating relationships between the Japanese 'epidemic' of subacute myelo-optic neuropathy (SMON) and iodochlorhydroxyquin (clioquinol)[146,147] and between the occurrence of vaginal carcinoma in young women and exposure to diethylstilbestrol *in utero*,[148] added impetus to this debate.

Concern over the effectiveness of current PMS schemes was expressed in many countries. In the USA, where PMS covers a wider range of activities than in most other countries, encompassing drug use and efficacy as well as adverse and beneficial effects, there has been considerable pressure for the development and implementation of an organized PMS system. For instance, at an interdisciplinary conference on PMS held in 1976, Lasagna reported that in the discussions 'the conferees agreed that the question of whether or not PMS should be done is idle; that a relatively unstructured (open) system of PMS is now in existence; that PMS is an absolute necessity for assessing the benefit–risk ratio of a new drug; and that the proper object of discussion is the question of *whether* the present system should give way to some of the new (structured) proposals now under discussion, and to what extent'.[149]

4.9.6.1 Proposals for New PMS Schemes in the USA

The concern in the scientific community of the USA moved into the political arena when Senator Edward Kennedy became interested in PMS as a major theme in his review of drug law. In May

1976, at the annual meeting of the Pharmaceutical Manufacturers Association (PMA) he challenged the industry to create and fund an independent non-profit-making organization to collect adequate drug use data after the marketing of new drugs. The PMA decided to take up the senator's challenge and undertook the development of such an organization in conjunction with a number of potentially interested professional bodies. The formation of the Joint Commission on Prescription Drug Use (JCPDU) was announced by the PMA and Senator Kennedy on 30th November 1976. Given a three year life, the JCPDU was commissioned to design and recommend the details of a PMS system for the gathering of data on ADRs and new drug uses, and to develop a format, using available data, for reporting regularly on trends in drug prescribing and usage.

In February 1978, a contract to review PMS schemes in the USA and other countries and to develop new methods for possible use in the USA was awarded by the JCPDU to IMS America, a multinational organization which obtains information and prepares regular reports on the patterns of drug use and physician prescribing practice. Four tasks were set out in the contract: (a) an assessment of past and present efforts in PMS around the world; (b) the identification of other possible methods of PMS; (c) an evaluation of PMS methods and recommendations for methods to be experimentally tested; and (d) the preparation of a detailed system design and an evaluation design for the method(s) selected from those proposed in task (c).

In the final report of task (c), IMS America made three principal recommendations:[150]

(i) A national facility for the identification and formation of cohorts of users of new drugs should be established. For this role, emphasis was placed on the potential of a 'pharmacy panel', in which pharmacists working in consenting pharmacy chains and/or large hospitals would identify patients when they came to have their prescriptions dispensed and would obtain their permission to enrol them into the cohort; it was also considered that there might be instances in which a 'physician panel' would have to be formed and used instead. Some drugs are used too rarely to provide a sufficiently large cohort *via* a panel of a reasonable size and, in this situation, it was suggested that consideration be given to a scheme in which the drug could not be legally given without the enrolment of the patient into the cohort.

(ii) A national centre for the follow-up of the cohort and the subsequent data analysis should be set up.

(iii) The 'cohort scheme' should be supplemented by case-control surveillance of drug-induced conditions.

In the USA, 1980 was a presidential election year and, consequently, the final report of the JCPDU[151] did not receive the publicity it deserved.[47] Based on the work carried out by IMS America, the JCPDU made 31 major conclusions and/or recommendations and many lesser ones; three of the most important were that: (a) a systematic and comprehensive PMS system should be developed in the USA; (b) such a system should be able to detect important ADRs that occur more frequently than one per 1000 and to develop methods to detect both less frequent and delayed reactions; and (c) a non-profit-making centre for drug surveillance should be established to encourage cooperation between existing PMS programmes, to develop new methods for carrying out surveillance, to train scientists in the disciplines needed for PMS and to educate both providers and recipients of prescription drugs about the effects of these drugs. Unfortunately, the recommendations of the JCPDU and IMS America have not been implemented.

4.9.6.2 Proposals for New PMS Schemes in the UK

Serious concern over the effectiveness of current PMS schemes was also expressed in the UK and new schemes were proposed. Each of these was an attempt to improve the technique of 'monitored release', which is the release of a drug onto the market by the Licensing Authority under certain restrictions. The restrictions are usually an obligatory feedback of information gathered from its large-scale usage and often include the requirement that the company monitor a certain number of patients for a particular length of time after the drug is released. About 16–18 drugs have been subjected to 'monitored release' with the approval of the CSM, *e.g.* ketamine, buprenorphine and sodium valproate.[152,153] Varying degrees of success have been achieved with 'monitored release' but, in general, companies have found it expensive and difficult to run.[153,154]

The first new PMS scheme was proposed by Inman, then principal medical officer at the CSM. He gave a brief outline of his ideas at a symposium held in Honolulu in 1976[30] and a year later, in January 1977, at another international meeting in Honolulu he described his 'recorded release' scheme in more detail.[155] In the same month, Dollery and Rawlins published their proposal for 'registered release'.[156] Two months later, in March 1977, Lawson and Henry criticized 'registered release' and suggested their own scheme.[157] In October 1977, Wilson, then medical director of the Association of the British Pharmaceutical Industry, discussed the three proposed schemes in a paper published in the *British Medical Journal* and presented the industry's suggestions.[158] All four proposals, with some modifications, were reviewed at a meeting in London in December 1977.[159–162] In 1978, the CSM put forward its own version of 'recorded release' and later suggested a similar scheme known as 'Retrospective Assessment of Drug Safety'; both closely resembled Wilson's proposals.[158,162] Wilson has since reviewed all the proposed schemes and clarified his own suggestions.[163]

All the schemes involved the registration of patients by a central monitoring agency (Figure 1). In 'recorded release' and 'registered release', the prescribing doctor would have completed the registration form and sent it to the monitoring agency, while in Lawson and Henry's scheme the dispensing pharmacist would have recorded details from the patient's prescription and sent them to the agency. The Prescription Pricing Authority (PPA) would have been used in Wilson's scheme to identify treated patients in the normal course of its work and to send the relevant prescriptions to the monitoring agency.

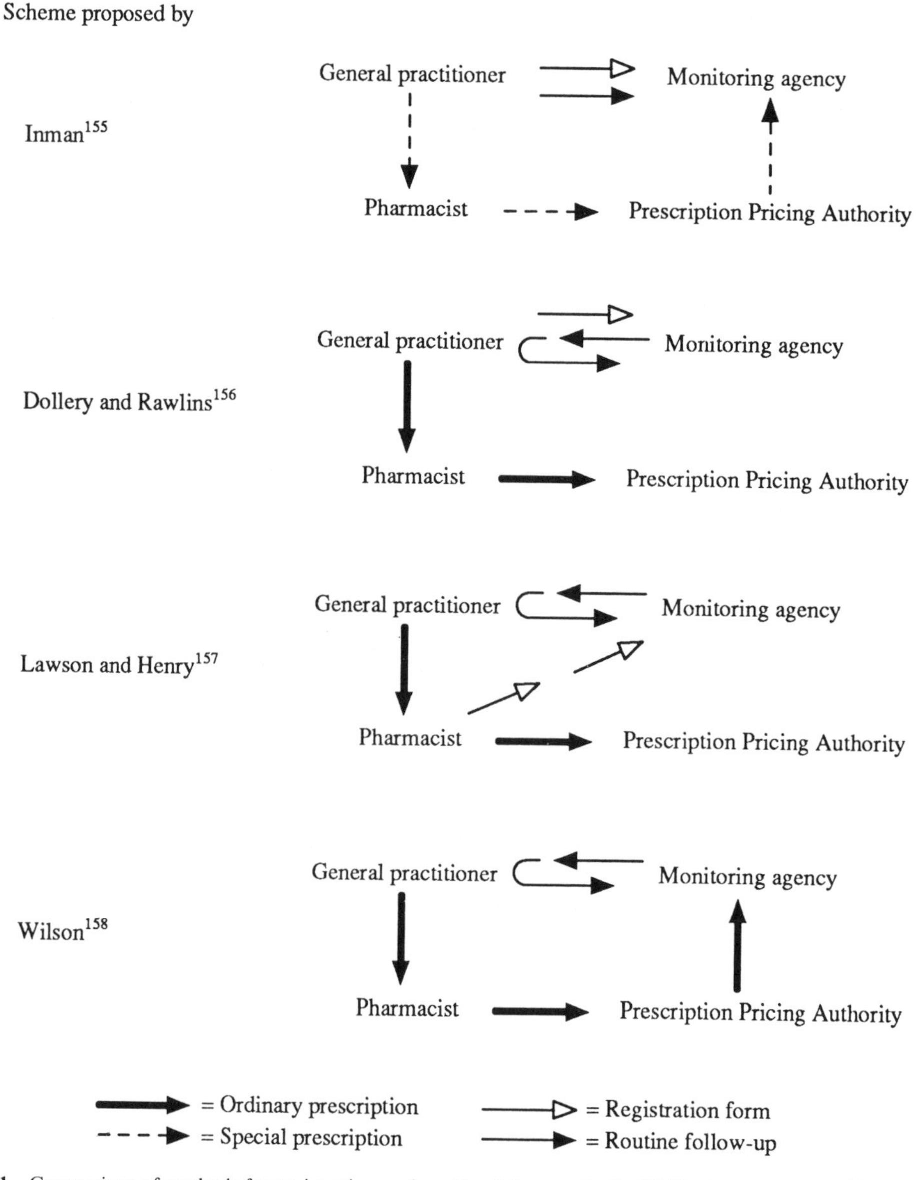

Figure 1 Comparison of methods for registration and routine follow-up in the PMS schemes proposed in 1977–1978

With the exception of Inman, all the proposers suggested that, when a specified period of time had elapsed since the registration of the patients, the monitoring agency should send a questionnaire to the prescribing doctors for them to complete and return (Figure 1). In Inman's scheme, the prescribing doctor would already have the questionnaire in the 'recorded drug package', which would have been supplied to all doctors.[155] This package would also contain the special prescription forms, without which the drug would not be dispensed to the patient, and the registration form, which would include a promise by the doctor to complete and post the follow-up questionnaire at the agreed time. All the proposers agreed both that the questionnaire should be kept as simple as possible and that some form of event reporting[14,15] (instead of ADR reporting) was essential. Inman and the CSM thought that all adverse events should be reported, while the others were satisfied with the reporting of 'serious' events. In order to monitor any long-term adverse effects, all but one of the schemes incorporated a request to the doctor for the patient's National Health Service (NHS) number in the follow-up questionnaire so that patients could be identified automatically by the Office of Population Censuses and Surveys if they were entered on a cancer register or when they died. In Dollery and Rawlins' scheme, an additional and important part of the routine follow-up was a questionnaire to the patient, sent either directly or *via* the patient's doctor, asking about 'many different bodily systems and symptoms'.[156]

All the schemes were intended for drugs which were new chemical entities (NCEs). Dollery and Rawlins suggested that all NCEs should be monitored, but the others were less ambitious suggesting that their schemes should be limited to 'selected' NCEs[155,157] or NCEs whose clinical indications required long-term administration.[158] The number of patients to be registered and followed-up by the monitoring agency varied from around 5000 to 100 000 but, unless there were exceptional circumstances, the number would generally have been around 10 000. Dollery and Rawlins proposed that for a new drug to receive a product licence the pharmaceutical company concerned should be required to complete a 'quota' of registrations. In all the schemes, there would have been a routine follow-up one year after the initial registration of the patient and then annually for three to five years.

Most of the proposers thought that a suitable 'control' drug was necessary but were vague as to how this should be implemented; suitable controls were part of Lawson and Henry's scheme in March 1977[157] but, by the following December, Lawson had changed his views and considered that they 'would be a positive hindrance'.[160] There would be no interference with the normal promotion of the monitored drug in any of the schemes, except 'registered release', in which it would only be possible after the required quota of registrations had been achieved. If their schemes were implemented, all the proposers, with the exception of Lawson and Henry, thought that the regulatory authority would be able to release drugs on to the market comparatively earlier than is currently possible.

None of the proposers had any clear opinion as to who should bear the responsibility for running the central monitoring agency, although three suggestions were made: the pharmaceutical industry,[156-158] the CSM,[156,157] and an independent group administered by one of the Royal Colleges[156-158] or a university.[157] It was suggested that the pharmaceutical industry has the resources for this work, but some companies have been suspected of 'dragging their feet' in reporting ADRs, especially when the reports are unconfirmed.[156] Some doctors have reservations about divulging information to a government agency[156] and, in addition, the CSM is sometimes subjected to political pressure from parliamentary questions to undertake premature analysis of its data. The independent agency does not suffer from these disadvantages.

The proposal in both 'recorded release' and 'registered release' that the prescribing doctors should register the patients was criticized because it is likely that their prescribing habits and patient selection criteria would differ from their normal practice. As a result, the study cohort would not be representative of the population that would subsequently be exposed to the new drug.[157,158] Dollery and Rawlins' suggestion that the pharmaceutical company should complete a quota of registrations for a new drug before it could be marketed normally was criticized for the same reasons.[157,158] Patient compliance could be affected if they became aware that they were receiving a new drug either because of special prescription forms or because of information from the doctor.[163] Patients could also become aware that they were receiving a new drug if the pharmacist was to perform the task of registration and again compliance could be affected. These drawbacks would be overcome by using the PPA to identify prescriptions, because the first involvement of the prescribing doctor would be at the follow-up stage, by which time a representative cohort of treated patients would have already been registered by the monitoring agency.

There was general agreement that the method of routine follow-up should be a questionnaire direct to the prescribing doctor. Dollery and Rawlins' proposal that a questionnaire should be sent to the patient as part of the routine follow-up was strongly criticized by Lawson and Henry, Wilson and Drury,[164] on the grounds that its value would 'be outweighed by the potential harm to the doctor–patient relationship'.[158]

On balance, the scheme proposed by Wilson was probably the best. It would have needed no new legislation which, if required to enforce 'registered release' or 'recorded release', would have been cumbersome,[157,159] and its cost was expected to have been lower than that of the other schemes. The extra workload on the doctors would have been minimal and could have been regarded as falling within their normal duties. In spite of the successful completion of a small-scale experiment of prospective monitoring in the West Midlands,[165] none of the proposed schemes has been implemented on a national scale in the UK.

4.9.7 THE DRUG SAFETY RESEARCH UNIT

4.9.7.1 Prescription–Event Monitoring (PEM)

With similar ideas to the recommendations made by the JCPDU (see Section 4.9.6.1), Inman left the CSM in June 1980 to set up the Drug Surveillance Research Unit (DSRU) as an independent non-regulatory centre within the Faculty of Medicine at the University of Southampton. The principal objectives of the DSRU are: (a) to develop and assess new techniques for PMS; (b) to provide a suitable environment for research and training in the epidemiology of drug-induced illnesses; (c) to review the 'state of the art' of drug monitoring; and (d) to produce impartial feedback in the form of publications and advice in order to improve communications concerning drug safety among patients, those responsible for health care, the pharmaceutical industry and the media.[166–168] In 1986, the DSRU ceased to be part of the University of Southampton and became the Drug Safety Research Unit managed by the Drug Safety Research Trust, which is an independent charity working in cooperation with the University.

The first PMS scheme to be developed and assessed by the DSRU is known as Prescription–Event Monitoring (PEM), which is similar to the scheme proposed by Wilson (see Section 4.9.6.2). It is considered to be most suitable for new medicines containing a single active ingredient, which are prescribed by one brand name for medium- to long-term treatment of chronic diseases and which are used on a scale sufficient to provide an adequate sample of patients within a reasonable period of time.[167]

PEM brings together two concepts, the monitoring of prescriptions and the reporting of events, which were both originally suggested as being of importance in PMS over 20 years ago. In the early 1960s, the MRC proposed a scheme to the CSD that was similar to Wilson's;[159] later, the value of prescription monitoring was recognized by Acheson in his record linkage system (see Section 4.9.5). Others have suggested that the facilities of the PPA might be used to advantage in PMS[169,170] and the feasibility of using the Authority for this purpose has been supported by several studies.[20,142,143,165,171,172] Event reporting was first proposed in 1965 by a statistician, Finney, who was a founding member of the CSD's Sub-Committee on Adverse Reactions.[14] He argued that if doctors could be persuaded to record significant events that had occurred after a drug had been prescribed rather than restrict their reports to those events for which they strongly suspect that drugs have been responsible, a clearer pattern of adverse experiences would emerge. Doctors would not be inhibited by the need to make up their minds about causation and there would also be less medicolegal risk, since they would be asked to do no more than copy details of events that had already been recorded in the patient's notes. Event reporting, which was an essential part of the proposals made in 1977–1978, has been shown to be practicable by studies carried out in the UK,[165] USA[173] and New Zealand.[174]

4.9.7.2 Methodology of PEM

Patients who have been prescribed drugs chosen for PEM are identified by the PPA during the normal course of its work. The PPA is a special health authority established to remunerate pharmacists who have dispensed NHS prescriptions issued by general practitioners working within the 90 English Family Practitioner Committee areas. Scotland, Wales and Northern Ireland, which have their own pricing authorities, have not been included in PEM. In those countries, many patients with the same surname may be found in a single practice. Since PEM depends on a general practitioner's ability to identify an individual patient rapidly, it was thought that the additional number of patients that might be obtained by including Scotland, Wales and Northern Ireland would not justify the extra workload that would be imposed upon doctors in those countries.

The PPA employs almost 2000 clerks to process over 350 million prescriptions each year. Until recently, this work has been done manually but now all the divisions of the PPA are computerized. The number of drugs that could be monitored at any one time by the non-computerized divisions was limited to four. It was found, by trial and error, that this was the maximum number of drug names that the clerks could memorize and thus reliably extract all prescriptions for these drugs that passed through their hands. More drugs can be monitored following computerization so that in the future the limiting factor will be the volume of prescriptions issued rather than the total number of drugs being studied. The prescriptions for the drugs selected for PEM are photocopied and the originals returned to the normal pricing procedure. The photocopies are sent to the DSRU in monthly batches.

When the photocopies of the prescriptions reach the unit, information from them is transferred to the DSRU's computer. The recorded data consist of the patient's surname, forename(s) or initials, sex (if deducible from the prescription) and abbreviated address, the date of the prescription, a practice code (shared by all the partners in the practice) and a code which identifies the individual doctor. Patients do not always see the same doctor within the practice at each consultation; consequently, in order to ensure that only one record is created for each patient, the program allows the data-entry clerk to check the list of patients already stored under the shared practice code before entering the patient's details.

The prescriptions are a means of identifying patients and doctors so that information about the patients may be obtained from their doctors. In order to obtain this information, follow-up questionnaires, which because of their colour are known as 'green forms', are generated by the computer and mailed to the relevant general practitioners. In the early studies, the green forms were mailed in batches at varying periods (usually 12 months) after the date of the earliest collected prescriptions. With the installation of an improved computer system, each questionnaire is now sent to the relevant doctor during the month after the first anniversary of the date of the first prescription for the patient. The green forms are designed so that each one is personalized for the individual patient and doctor by printing the patient's details and the name and address of the practitioner on the top section of the form. This part of the questionnaire is detachable so that any information about the patient that is subsequently written on the lower section is anonymous if the form falls into the wrong hands. A unique reference number is printed on both sections.

On each questionnaire, the doctor is asked to record the indication for which the drug was prescribed, the patient's date of birth and sex (if not already entered) and whether the patient is still taking the drug. If the patient has discontinued treatment with the drug, the doctor is asked to enter the date on which the treatment stopped and the reason for stopping. Finally, the doctor is asked to record the dates and details of any events that were experienced by the patient after the first prescription for the drug. An event is defined as any new diagnosis, any reason for referral to a consultant or admission to hospital, any unexpected deterioration (or improvement) in a concurrent illness, any suspected ADR or any other complaint which was considered of sufficient importance to enter in the patient's notes; this definition is printed on each green form. In order to emphasize that doctors should not restrict their reports to those events which are suspected ADRs, the simple example of a broken leg is given. The event-recording section of the green form is deliberately unstructured both to encourage the practitioners to reply and to allow them as much flexibility as possible in their replies. It is thought that if doctors were forced to categorize the events, they might be deterred from completing the questionnaires and important signals might be missed.

The following data from the returned green forms are added to the original computer records: (i) the patient's date of birth and sex; (ii) the indication for treatment with the drug; (iii) the dates on which the treatment began and (if applicable) stopped; (iv) whether the period of treatment was continuous; and (v) the dates of occurrence and the events experienced by the patient after the first prescription (although in the early studies, reported below, the dates of occurrence of the events were not recorded). Since the standard disease and diagnosis coding systems, such as ICD,[56] are unsuitable for some of the non-specific events reported by some doctors, a text-recording system has been developed so that events can be recorded as closely as possible to the description supplied by the practitioner. Further details of PEM have been given by Inman *et al.*[175] and Rawson,[176] and a detailed description of the methodology of PEM is in preparation.

4.9.7.3 Some Results from PEM

The DSRU began PEM with a pilot study to monitor two non-steroidal antiinflammatory drugs (NSAIDs), benoxaprofen (Opren) and fenbufen (Lederfen).[176] The study was designed to test the methodology of PEM and these drugs were selected by the DSRU because it was thought that they would provide sufficient prescriptions over a relatively short period. They were not chosen because of any particular concern about their safety. The pilot study showed that general practitioners were reasonably cooperative with PEM (a response rate of 51%) and, as in several previous studies,[165,173,174] that event reporting is highly practicable. The known adverse effects of benoxaprofen and fenbufen stood out against a background of unrelated events and one previously unknown adverse effect of benoxaprofen (irritation of the bladder) was also observed. It was concluded from the pilot study that PEM should prove to be an effective method for the detection of ADRs occurring with medium frequency.[176]

Subsequently, a second study of benoxaprofen and fenbufen was performed and studies of other NSAIDs, namely piroxicam (Feldene), zomepirac (Zomax) and two formulations of indomethacin (Indocid and Osmosin, the latter being an osmotic pump form of indomethacin), were also carried out.[175-179] The response rates in these subsequent studies ranged between 58% and 68% (Table 3). Since piroxicam and zomepirac had been marketed for about two years and Indocid for around 20 years before the relevant prescriptions were collected, patients who had commenced treatment more than a year before the collection of the relevant prescriptions were excluded in an attempt to avoid 'survivor bias'. Patients who have been on treatment for a considerable period prior to a study are 'survivors' of an earlier cohort of patients who had probably found the drug to be effective and free of adverse effects. Patients in the same cohort who had found the drug to be ineffective or who had experienced ADRs probably ceased to take the drug before the collection of the prescriptions. The number of patients included in the analyses are shown in Table 3.

The sex and age distributions of the patients taking the NSAIDs were similar, although both the zomepirac and the Indocid patients were generally younger (Table 4). The indications for treatment showed a greater variation (Table 5). Proportionally more benoxaprofen patients were suffering from rheumatoid arthritis, a not unexpected finding since claims were made when benoxaprofen was first marketed that it had disease-modifying properties in patients suffering from this disease. The Indocid and zomepirac patients were prescribed the drugs for a wider range of indications; in fact, zomepirac was promoted as a general purpose analgesic rather than as a NSAID and this is reflected in the 188 different indications for treatment with this drug. 142 patients were prescribed zomepirac to alleviate pain caused by cancer. These patients, especially those with bone secondaries, generally found the drug to be effective; when zomepirac was withdrawn following reports of fatal anaphylaxis associated with the drug, much suffering was caused among them.

In these PEM studies, the green forms were mailed at varying times after the dates of the earliest collected prescriptions. Consequently, the distributions of the periods of treatment to which the patients were exposed varied (Table 6); note that a treatment period of less than 15 days is recorded as zero months in this table. The average treatment periods of the zomepirac and Osmosin patients were particularly short because the studies of these drugs were only initiated when the DSRU heard that they were being withdrawn from the market. The collected prescriptions were those that were still being processed by the PPA at the time of the withdrawal.

The failure to record the dates of occurrence of the events in these early studies and the varying treatment period distributions present difficulties in the method of analysis of the recorded events. In an attempt to take account of both the varying numbers of patients and treatment periods, rates of

Table 3 Response to the PEM Questionnaires

	Benoxaprofen		*Fenbufen*		*Piroxicam*	*Zomepirac*	*Indomethacin*	
	Pilot study	*Second study*	*Pilot study*	*Second study*			*Indocid*	*Osmosin*
Questionnaires mailed	11 646	18 127	4113	8204	20 646	19 207	26 407	21 886
Questionnaires returned	5954 (51.1%)	11 171 (61.6%)	2126 (51.7%)	4893 (59.6%)	14 084 (68.2%)	11 091 (57.7%)	15 563 (58.9%)	13 325 (60.9%)
Questionnaires used in the analysis	5526 (47.4%)	9574 (52.8%)	1962 (47.7%)	4191 (51.1%)	7139 (34.6%)	6865 (35.7%)	4368 (16.5%)	12 265 (56.0%)

Table 4 Distributions of the Ages of the Patients

Age group	*Benoxaprofen* Pilot study	Second study	*Fenbufen* Pilot study	Second study	*Piroxicam*	*Zomepirac*	*Indomethacin* Indocid	Osmosin
<55	1269 (23.9%)	1653 (21.4%)	442 (23.5%)	828 (25.0%)	1607 (27.6%)	2655 (41.9%)	1565 (40.4%)	2723 (25.2%)
55–64	1327 (25.0%)	1990 (25.7%)	409 (21.8%)	787 (23.7%)	1297 (22.3%)	1189 (18.8%)	877 (22.6%)	2647 (24.5%)
65–74	1608 (30.2%)	2210 (28.6%)	568 (30.2%)	898 (27.1%)	1500 (25.8%)	1250 (19.7%)	825 (21.3%)	3000 (27.8%)
75+	1114 (20.9%)	1879 (24.3%)	459 (24.4%)	805 (24.3%)	1414 (24.3%)	1244 (19.6%)	606 (15.6%)	2439 (22.6%)
Total	5318 (100.0%)	7732 (100.0%)	1878 (100.0%)	3318 (100.0%)	5818 (100.0%)	6338 (100.0%)	3873 (100.0%)	10 809 (100.0%)
Age not known	208	1842	84	873	1321	527	495	1456
Average age (range)	63.2 (14–101)	64.4 (13–102)	64.1 (17–101)	63.6 (15–102)	62.6 (13–105)	57.0 (10–101)	57.1 (13–100)	63.0 (12–100)

events per 1000 patient-years of treatment exposure were compared. This is an unsatisfactory measure, since it is not the same as the incidence per 1000 patients per year.[176]

Nevertheless, the results of these studies showed the expected adverse effects. For example, the recorded rates of skin events per 1000 patient-years of treatment (Table 7) clearly showed the tendency of benoxaprofen to produce photosensitivity, onycholysis and other nail changes. To a lesser extent, it can be seen that patients taking benoxaprofen, fenbufen and zomepirac generally experienced more acute rashes. There was little difference between the patient groups in the other categories of skin events.

NSAIDs are often said to cause a large number of adverse gastrointestinal effects (especially serious ones such as peptic ulceration), although the evidence for this is scanty.[180] When the gastrointestinal events recorded in the PEM studies were examined (Table 8), it was seen that there was little difference between the drugs in the serious event categories. However, there were differences in the less serious categories, with the Osmosin rates being particularly high. It should be noted that Osmosin was designed to produce *less* adverse gastrointestinal effects. The zomepirac and Osmosin patients had generally been on treatment for shorter periods than the patients in the other studies and the high rates seen in Table 8 are partly due to this. Nevertheless, when rates *per 1000 patients* were also calculated, the Osmosin rates were again higher than one would expect. Therefore, it was concluded that patients with poor 'gastric tolerability' of NSAIDs were given Osmosin and they continued to experience gastrointestinal problems both during treatment with Osmosin and after they ceased to take the drug.[176,178,179] This is an example of a more general confounding factor affecting the interpretation of the results; it is usually impossible to control for such factors.

Analyses of the gastrointestinal events experienced by patients in four age groups (under 55, 55–64, 65–74 and over 74 years) have been presented by Rawson.[176] The main result was that the gastrointestinal haemorrhage rate increased with age in the fenbufen, piroxicam and indomethacin studies, with a suggestion of a similar trend in the benoxaprofen studies.

It should be noted here that the rates shown in Tables 7 and 8 are not the same as those presented in previous DSRU publications.[175,178,179] The results in these earlier publications included all patients in each study and, therefore, did not attempt to take account of 'survivor bias'.

4.9.7.4 Comments on PEM

Although the general methodology used to collect the information in these studies was essentially the same, there were important differences between them. The pilot study used a slightly different questionnaire and suffered from some logistic problems. The green forms in the second benoxaprofen and fenbufen studies were sent after benoxaprofen had been suspended. Similarly, zomepirac and Osmosin had been withdrawn from the market before the questionnaires in these studies were sent. Piroxicam and zomepirac had been available for around two years and Indocid for almost 20 years before the studies of these drugs were begun, although an attempt was made to reduce the possibility of 'survivor bias'. Finally, the time between the prescriptions being written and the green forms being mailed varied between three to four months in the zomepirac and Osmosin

Table 5 Grouped Indications for Treatment

Indication group	*Benoxaprofen*		*Fenbufen*		*Piroxicam*	*Zomepirac*	*Indomethacin*	
	Pilot study	*Second study*	*Pilot study*	*Second study*			*Indocid*	*Osmosin*
Osteoarthritis	46.8%	46.7%	50.2%	43.3%	41.7%	15.9%	26.2%	47.7%
Rheumatoid arthritis	20.1%	16.9%	9.3%	9.6%	6.6%	3.2%	3.4%	10.1%
'Arthritis'	10.7%	12.7%	10.0%	10.8%	8.8%	5.3%	8.0%	11.6%
Other arthropathies	0.9%	0.5%	0.5%	0.3%	0.4%	0.2%	0.6%	0.5%
Non-specific joint disorders	6.5%	6.9%	7.8%	9.2%	11.2%	7.5%	12.5%	7.6%
Spondylosis and allied disorders	4.4%	4.9%	5.4%	4.7%	5.4%	3.9%	4.4%	4.9%
Non-specific back and neck disorders	3.8%	4.1%	6.2%	7.5%	9.4%	17.7%	8.8%	6.6%
Other dorsopathies	1.5%	1.5%	2.1%	3.1%	2.9%	6.5%	3.7%	2.9%
Peripheral enthesopathies	1.3%	1.4%	2.4%	2.3%	3.3%	1.3%	5.1%	1.8%
Disorders of muscle, ligament and fascia	1.7%	2.0%	2.3%	4.5%	2.9%	6.0%	3.2%	1.9%
Other indications	2.4%	2.4%	3.7%	4.6%	7.4%	32.4%	24.1%	4.3%
All indications	5220	7176	1821	2695	4945	4749	3423	9927
	(100.0%)	(100.0%)	(100.0%)	(100.0%)	(100.0%)	(100.0%)	(100.0%)	(100.0%)
Indication not known	306	2398	141	1496	2194	2116	945	2338

Table 6 Distributions of the Duration of Treatment

Duration of treatment (months)	*Benoxaprofen*		*Fenbufen*		*Piroxicam*	*Zomepirac*	*Indomethacin*	
	Pilot study	*Second study*	*Pilot study*	*Second study*			*Indocid*	*Osmosin*
0–2	1423 (29.4%)	3042 (38.2%)	559 (33.5%)	1635 (46.8%)	2581 (38.4%)	4108 (73.2%)	2071 (52.3%)	5196 (47.3%)
3–5	775 (16.0%)	1453 (18.2%)	201 (12.0%)	459 (13.1%)	903 (13.4%)	896 (16.0%)	433 (10.9%)	3450 (31.4%)
6–8	470 (9.7%)	796 (10.0%)	128 (7.7%)	245 (7.0%)	602 (9.0%)	479 (8.5%)	285 (7.2%)	1964 (17.9%)
9–11	299 (6.2%)	499 (6.3%)	77 (4.6%)	174 (5.0%)	363 (5.4%)	122 (2.2%)	463 (11.7%)	342 (3.1%)
12–14	1395 (28.9%)	463 (5.8%)	353 (21.1%)	131 (3.7%)	309 (4.6%)	8 (0.1%)	350 (8.8%)	30 (0.3%)
15+	471 (9.7%)	1719 (21.6%)	352 (21.1%)	851 (24.3%)	1954 (29.1%)	0 (0.0%)	356 (9.0%)	6 (0.1%)
Total	4833 (100.0%)	7972 (100.0%)	1670 (100.0%)	3495 (100.0%)	6712 (100.0%)	5613 (100.0%)	3958 (100.0%)	10 988 (100.0%)
Duration not known	693	1602	292	696	427	1252	410	1277
Average period of treatment	7.6	7.1	8.5	9.0	8.3	2.0	5.1	3.3
(range)	(0–33)	(0–35)	(0–37)	(0–46)	(0–37)	(0–14)	(0–20)	(0–18)

Table 7 Comparison of the Recorded Rates of Disorders of the Skin, Nails and Hair per 1000 Patient-years of Treatment

Skin disorders	*Benoxaprofen*		*Fenbufen*		*Piroxicam*	*Zomepirac*	*Indomethacin*	
	Pilot study	*Second study*	*Pilot study*	*Second study*			*Indocid*	*Osmosin*
Photosensitivity	83.8	116.1	0.0	1.0	0.4	1.7	0.0	0.9
Onycholysis	18.1	19.1	0.0	0.0	0.6	0.0	0.0	0.3
Nail change	9.2	15.2	1.5	1.4	0.4	0.0	0.0	0.9
Rash	51.1	43.7	43.9	36.3	18.0	48.9	12.1	18.2
Pruritus without rash	15.4	16.3	12.1	7.1	7.2	6.1	4.9	4.5
Other inflammatory disorders	14.6	13.0	12.9	9.8	16.9	11.4	12.1	13.8
Other disorders	37.1	37.6	34.8	27.8	36.5	36.7	23.0	37.7
All skin disorders	229.3	261.0	105.2	83.5	80.1	104.9	52.1	76.3

studies to 22–27 months in the second benoxaprofen and fenbufen studies, which may have introduced bias.

In spite of these differences, the rates of events in the eight studies showed a remarkable consistency, with the exception of known ADRs and one or two surprising results.[176,178,179,181] In particular, the PEM studies indicated that there is little difference between these NSAIDs with regard to serious gastrointestinal effects. Nevertheless, it is clear that these drugs should be prescribed only after careful consideration has been given to other forms of treatment, especially in the elderly or patients with a previous history of digestive problems.

The questionnaires returned in the piroxicam study have been reprocessed on the DSRU's new computer using a flexible database management system. In the work presented above, an event was only recorded once in the relevant time period (*i.e.* during or after stopping treatment) irrespective of how many times the event occurred during the period. All occurrences of an event were recorded when the data were reprocessed. A rate per 1000 patient-years for events occurring within the first year of treatment and rates of events per 1000 patients occurring while on treatment during each month of the year after commencing treatment have been estimated; rates for various categories of digestive disorders are shown in Table 9. These rates are shown to illustrate the more detailed analyses that are possible now that the dates of the events are recorded; they are not directly comparable with the piroxicam results shown in Table 8.

In addition to the NSAID studies, PEM has been performed on the H_2 antagonist ranitidine[175,182] and the angiotensin-converting enzyme inhibitor enalapril.[183] PEM is now established as the second national drug monitoring system in England with 13 new drugs presently being studied.[184] Although in its usual form PEM is a hypothesis-*generating* method, it can be used to *test* hypotheses. The first two studies of this type were to test the hypothesis that erythromycin estolate is more commonly associated with jaundice than other preparations of this antibiotic[185] and to test whether oesophageal symptoms are associated with emepronium bromide.[186] In the former, no case of estolate-related jaundice was observed among more than 3300 patients, but three patients out of more than 4000 who were prescribed erythromycin stearate were thought to have suffered jaundice as an adverse effect. It was concluded that selective reporting of jaundice following treatment with the estolate had occurred. In the second hypothesis-testing study, there were relatively few cases of oesophageal symptoms that were thought to have been probably due to emepronium bromide.

4.9.8 PHARMACEUTICAL COMPANY POST-MARKETING SURVEILLANCE

Some pharmaceutical companies have undertaken post-marketing studies. These have generally been phase IV clinical trials, which assess the efficacy of a product after it has been marketed and usually include data on any adverse effects that occur during the study. However, some PMS studies have been performed. The distinction between the two types of studies appears to be unclear to some in the pharmaceutical industry.

Examples of PMS studies carried out or directly sponsored by pharmaceutical manufacturers are Reckitt and Colmán's study of buprenorphine,[187] Smith Kline and French's study of cimetidine,[188–190] Sandoz's study of ketotifen[191] and Ciba-Geigy's study of diclofenac.[192] In addition, Merck Sharp and Dohme have carried out a study of acute adverse reactions to enalapril in a highly selected group of patients.[193] Event reporting was included in the majority of these

Table 8 Comparison of the Recorded Rates of Disorders of the Digestive System per 1000 Patient-years of Treatment

Digestive disorders	*Benoxaprofen*		*Fenbufen*		*Piroxicam*	*Zomepirac*	*Indomethacin*	
	Pilot study	*Second study*	*Pilot study*	*Second study*			*Indocid*	*Osmosin*
Dyspepsia and gastritis	84.1	67.0	65.9	53.3	63.8	70.8	91.5	151.4
Peptic ulceration without complication	4.4	4.8	1.5	2.4	6.0	4.4	3.3	6.0
Peptic ulceration with haemorrhage or perforation	1.2	1.7	0.0	0.3	1.4	1.7	2.7	1.2
Gastrointestinal haemorrhage NOS[a]	7.4	3.0	8.3	2.0	9.9	8.7	8.8	8.1
Liver, gall bladder and pancreatic diseases	5.6	5.4	5.3	1.0	2.1	3.5	2.7	1.8
Other diseases	31.2	27.2	31.0	21.1	39.4	49.8	29.0	53.3
Nausea and vomiting	29.7	26.1	21.2	20.0	24.1	56.8	42.2	64.9
Abdominal pain	22.9	24.1	23.5	11.9	28.5	29.7	25.8	62.5
Other signs and symptoms	34.2	28.3	37.1	23.1	30.5	75.2	35.1	94.5
All digestive disorders	220.7	187.6	193.8	135.2	205.7	300.7	241.2	443.7

[a] NOS = not otherwise specified.

Table 9 Rates of Disorders of the Digestive System During the First 12 Months of Treatment with Piroxicam

Digestive disorders	*Rate per 1000 patient-years of exposure for year after first prescription*	*Rates per 1000 patients during each month of the first year*											
		1	*2*	*3*	*4*	*5*	*6*	*7*	*8*	*9*	*10*	*11*	*12*
Dyspepsia and gastritis	73.6	10.6	7.2	6.5	4.9	7.0	3.5	7.8	6.5	2.7	4.4	3.0	2.2
Peptic ulceration without complication	3.3	0.2	0.2	0.8	0.0	0.3	0.6	0.3	0.4	0.4	0.0	0.0	0.0
Peptic ulceration with haemorrhage or perforation	6.1	0.5	0.5	0.5	0.5	0.3	0.3	0.0	0.4	0.4	0.8	1.3	0.9
Gastrointestinal haemorrhage NOS[a]	5.1	0.5	0.2	0.8	0.8	0.3	0.0	0.0	0.4	0.0	1.2	0.8	0.0
Other upper gastrointestinal diseases	27.0	1.9	3.2	0.8	3.8	2.3	1.6	3.4	1.8	1.5	2.4	2.1	1.8
Lower gastrointestinal diseases	10.9	0.9	1.4	0.5	0.5	0.6	0.6	1.4	1.4	1.5	0.8	0.4	0.9
Liver, gall bladder and pancreatic diseases	3.0	0.3	0.2	0.2	0.3	0.3	0.0	0.7	0.4	0.0	0.0	0.0	0.4
Abdominal pain	37.9	3.1	4.5	4.0	1.9	3.8	2.9	2.7	2.2	4.2	3.2	2.5	1.3
Diarrhoea	16.7	1.9	1.1	1.0	1.6	1.2	1.3	1.7	1.4	0.4	1.6	1.3	1.8
Nausea and vomiting	27.0	4.5	2.9	2.8	1.6	1.8	2.2	1.0	1.4	0.8	1.6	1.3	1.8
Other signs and symptoms	22.7	1.9	2.3	2.5	1.6	2.3	2.2	2.4	1.1	0.8	1.6	1.7	1.3
All digestive disorders	233.3	26.4	23.7	20.3	17.7	20.2	15.2	21.4	17.4	12.6	17.7	14.4	12.4

[a] NOS = not otherwise specified.

studies, the method of analysis in the diclofenac study being similar to that employed in the early PEM studies. No previously unknown ADR was discovered in these studies.

Most pharmaceutical industry PMS studies have been expensive[187-191] and often difficult to run. The cimetidine study, for instance, was very expensive in both time and money. Similar results to those found for cimetidine were obtained from the DSRU's study of ranitidine more rapidly and at a considerably lower cost.[175,182] Company studies are also usually uncontrolled because competitors are unwilling to agree to the use of their products for what could turn out to be adverse comparisons. In addition, some companies appear to have performed studies for promotional purposes and this has tended to diminish the credibility of all company PMS research.

4.9.9 SOME GENERAL POST-MARKETING SURVEILLANCE PROBLEMS

Some of the problems that arise in PMS have been illustrated in the descriptions of the various methods in the previous sections. Nevertheless, in this section, some of the more general problems are considered. Several of these have been reviewed previously.[194,195]

4.9.9.1 Heterogeneity of the Data and Problems of Analysis

In many other types of drug studies, the patients are a specially selected group but this is not so in PMS. For example, many patients in PMS studies will often be chronically ill with multiple pathology and considerably older than patients selected for clinical trials. Since the patients often have multiple pathology, many will be receiving other drugs in addition to the one being studied; for instance, in the pilot study of PEM, more than 50% were prescribed at least one other drug in addition to the study drug and almost 10% were prescribed three or more additional products.[176] Many patients will also be taking medicines bought 'over-the-counter'. Therefore, the analysis and interpretation of the results of a PMS study can be difficult due to both the heterogeneity of the study population and the other drugs that are taken.

The special problems of observational studies and spontaneously submitted records have received little attention from statisticians and, consequently, such data are often analyzed as though no difficulties exist. The work that has been done has tended to be in what might be called 'more structured' observational studies, such as social surveys and case-control studies; the latter have played an important role in the detection of ADRs, but they were not specifically developed for this purpose. The statistical problems of PMS have been reviewed,[196,197] but few statisticians have attempted to derive methods of analysis for data collected in PMS studies. In the UK, only Finney has made a significant contribution to the philosophical basis and the methods of analysis of drug monitoring data.[14,15,198-202]

The biases in the collection of data in voluntary reporting systems and other observational PMS studies will usually prevent any probabilistic interpretation. If statistical tests of significance are used in hypothesis-generating PMS studies, there is always a risk of false-positive results when testing a large number of hypotheses. An example of this problem was the apparent association between reserpine and breast cancer reported by the BCDSP (see Section 4.9.4.1). Tests of significance are, therefore, generally inappropriate in such studies.[203] Lane and Rawson have reviewed some of the statistical problems in PMS.[204]

There is one further point of crucial importance concerning statistical significance that should be appreciated. Conclusions of reports which may affect the future of a drug should not be based on statistical significance alone because it does not necessarily equate with medical significance.[195,203] Medical significance depends not only on the number of adverse events and their severity, but also on the nature of the disease being treated and the risk–benefit ratio of both the drug being studied and others available. For instance, just one case of aplastic anaemia or Stevens–Johnson syndrome would be medically but not necessarily statistically significant. In PMS studies, medical significance usually has greater importance than statistical significance.

4.9.9.2 Numbers of Patients

It is generally recognized that large numbers of patients need to be studied in PMS since the adverse effects that PMS is designed to monitor are uncommon; common ADRs should be observed during clinical trials. Several authors have presented statistical estimates of the numbers of patients required based on the background incidence of the reaction in the general population, the magnitude of the drug's effects, and the degree of certainty with which it is desired that the drug's adverse effects (if present) will be detected.[194,195,205,206]

However, a number of assumptions are made in the calculation of these estimates that are often invalid in PMS. These include: (a) patients being assigned randomly to the different drug groups; (b) patients all being on treatment for the same period of time; (c) patients in all drug groups being equally susceptible to the adverse effect; and (d) unbiased reporting by doctors. Furthermore, they do not take into account events which were undetected by or unreported to the doctor, patients lost to follow-up or the detection of multiple ADRs. Therefore, such estimates have only a limited value.

4.9.9.3 Control Data

One of the main design problems in any PMS scheme is the provision of a control group or a measuring device with which to compare the drug being studied. Control data are important because many symptoms of adverse effects caused by drugs can occur for other reasons.[207-209] Furthermore, decisions about drugs are often taken in a heated sociopolitical atmosphere and, as Lord Rothschild commented in his 1978 Richard Dimbleby lecture on risk, 'comparisons, far from being odious, are the best antidote to panic'.[210] However, there is no ideal solution as to what the best controls are. In an affluent society, almost all patients with any significant disease or symptom receive some form of treatment and, therefore, it is almost impossible to establish the incidence of events occurring in untreated control populations because there are none. Even if patients with untreated disease could be found, they would be atypical for several reasons, *e.g.* they would be likely to have less severe disease.

One approach to the problem is to use national morbidity and mortality statistics as control data. These statistics describe the situation in the general population which is mainly healthy. In PMS, however, the patients are ill and the incidence of many medical events in such patients would be expected to be higher than the national statistics. A more common approach is to contrast events occurring in the same individuals before and during treatment with the study drug. However, the pre-treatment measurements may significantly affect the patients' subsequent responses to the drug or the doctors' decision to treat them, and the patients may have been taking drugs similar to that being studied during the pre-treatment period. In PEM and other studies, events occurring in the same individuals during and after treatment with the study drug have been compared. This also has a number of disadvantages including 'carry-over' effects into the post-treatment period and the problem that patients ceasing to take the study drug often continue treatment with a similar type of product which may produce similar adverse effects.[176] The only remaining approach is to compare patients who receive different drugs for the same medical indications. Almost all decisions made by drug-monitoring agencies are based on this type of comparison. Nevertheless, such comparisons are not entirely satisfactory because there may be reasons for doctors selecting one treatment as opposed to another for a particular group of patients and this could bias the results. Furthermore, when a drug is the first of a new series of 'breakthrough' drugs, comparable event profiles may not be available.

4.9.9.4 Relationship between the Drug and the Adverse Event

In PMS studies, it is almost impossible to *prove* that the drug caused the adverse effect. At best, all one can do is make a judgement about the probability that the drug caused the event. In an endeavour to improve upon 'expert judgement', which has been shown to be unreliable for this task, 16 ADR causality assessment methods (algorithms) have been devised during the last 15 years.[211] These have varied depending on the types of physician that designed them and the areas in which their originators intended them to be used. For instance, the algorithm may be required to provide the maximum amount of information in an attempt to decide the causal relationship (perhaps for litigation purposes) or the aim may be to use the details about the patient so that a prediction can be made about the likelihood of a similar patient experiencing the same type of event in the future.

In general, most of the algorithms include the following evaluation criteria: (a) previous general experience with the drug; (b) any alternative causes of the event; (c) an appropriate time sequence between the administration of the drug and the onset of the event; (d) any evidence of overdose; (e) the effects of stopping or reducing the dose (dechallenge); and (f) the effects of restarting the drug (rechallenge). These criteria are used to arrive at a point on a probability scale, *e.g.* 'definite', 'probable', 'possible', 'unrelated', which estimates the certainty of the link between the event and the drug.

Three well-known algorithms were used to assess the possible relationship between benoxaprofen and jaundice and renal disorders in patients who were reported in the pilot study of PEM to have suffered from these conditions.[176] The causality assessments obtained from the algorithms were compared with each other and with the 'expert judgements' of the director of the DSRU. In general, there was good agreement between the algorithms and with the expert's judgements. However, the cases tended to be either those in which there was clearly no relationship with the drug or those in which there was insufficient information for either the expert or the algorithms to make any decision other than 'possibly associated'. These algorithms are, as Louik and her colleagues have also found,[212] difficult to apply to the results of event-monitoring systems because the reports are usually brief and are often based on incomplete information.

Recently, Lane and Hutchinson[213] established six criteria by which to judge the various causality assessment methods. These were that they should have good repeatability, be explicit, have an explanatory capability, make complete use of the additional information available, indicate the alternative diagnoses and allow strong evidence to prevail over weak evidence . All the algorithms failed at least one of these criteria. In an attempt to overcome these problems, Lane and his colleagues have proposed another method based on Bayesian statistical techniques,[214–216] which provides the most detailed assessment of causality yet devised. However, this method requires further refinement, especially if it is to be used for the everyday assessment of a large number of adverse event reports.[211]

Other problems that affect PMS studies include the possibility that litigation may be based on the individual report or study and the 'sociopolitical climate' prevailing at the time. An example of the latter is the climate which surrounded the withdrawal of several NSAIDs (especially benoxaprofen) in the early 1980s. It is difficult to take account of these factors but they cannot be forgotten.

4.9.10 FINAL REMARKS

Although several PMS methods have been described in this chapter, voluntary reporting systems and intensive hospital monitoring schemes have so far been the most successful; PEM can now be added to these. Medical record linkage schemes offer one of the best possibilities for PMS but until the problems of patient confidentiality can be overcome, it is unlikely that a national scheme will be introduced in the UK within the foreseeable future.

Although modern drugs are remarkably safe, there are many misconceptions about the safety and efficacy of drugs, which have led to a public expectation that all drugs should be totally safe; this is not possible if drugs are to be effective. PMS can lead to a better understanding of the risks involved which, together with the efficacy data available from clinical trials, should provide the information required to make sensible decisions about the future of a drug based on a careful balancing of the adverse and beneficial effects. Without such information, drug-licensing authorities will be stampeded into erroneous decisions by alarmist anecdotal reports.

ACKNOWLEDGEMENTS

The majority of this work formed part of a thesis presented for the degree of Doctor of Philosophy in the Faculty of Medicine at the University of Southampton.[176] I therefore wish to thank my academic supervisor, Professor Charles George, and his cosupervisor, Professor Bill Inman, for their advice and guidance. I would also like to thank my colleagues at the DSRU for their work on the PEM results. In addition, I am indebted to Dr Allen Rossi of the FDA for supplying information for Table 2 and to Drs Richard Alderslade, David Lane, Jørgen Seldrup and Lynda Wilton for their helpful comments on an earlier draft of this chapter.

4.9.11 REFERENCES

1. H. W. Haggard, 'Devils, Drugs and Doctors: the Story of the Science of Healing from Medicine-Man to Doctor', River, Boston, 1980.
2. N. E. Davies, G. H. Davies and E. D. Sanders, *JAMA, J. Am. Med. Assoc.*, 1983, **249**, 912.
3. D. Dunlop, 'Medicines in Our Time', Nuffield Provincial Hospitals Trust, London, 1973.
4. F. F. Cartwright, 'A Social History of Medicine', Longman, London, 1977.
5. J. G. McKendrick, J. Coats and D. Newman, *Br. Med. J.*, 1880, **2**, 957.
6. E. Lawrie, T. L. Brunton, G. Bomford, R. D. Hakim, *Lancet*, 1890, **1**, 149.
7. C. F. George, 'Prescriptions—Placebo, Poison or Panacea?', Inaugural Lecture, University of Southampton, 1978.

8. *Med. Res. Counc. (G. B.), Spec. Rep. Ser.*, 1922, **66**.
9. M. F. Cuthbert, J. P. Griffin and W. H. W. Inman, in 'Controlling the Use of Therapeutic Drugs: an International Comparison', ed. W. M. Wardell, American Enterprise Institute for Public Policy Research, Washington, DC, 1978, p. 99.
10. Sunday Times Insight Team, 'Suffer the Children: the Story of Thalidomide', Deutsch, London, 1979.
11. W. H. W. Inman (ed.), 'Monitoring for Drug Safety', 2nd edn., MTP Press, Lancaster, 1986.
12. W. H. W. Inman and J. C. P. Weber, in 'Monitoring for Drug Safety', 2nd edn., ed. W. H. W. Inman, MTP Press, Lancaster, 1986, p. 13.
13. J. Leahy Taylor, in 'Textbook of Adverse Drug Reactions', 2nd edn., ed. D. M. Davies, Oxford University Press, Oxford, 1981, p. 602.
14. D. J. Finney, *J. Chronic Dis.*, 1965, **18**, 77.
15. D. J. Finney, in 'Epidemiological Issues in Reported Drug Induced Illnesses: SMON and other Examples', ed. M. Gent and I. Shigematsu, McMaster University Library Press, Hamilton, Ontario, 1978, p. 32.
16. W. H. W. Inman and D. A. Price Evans, *Br. Med. J.*, 1972, **3**, 746.
17. Committee on Safety of Drugs, 'Adverse Reaction Series No. 7', HMSO, London, 1968.
18. W. H. W. Inman, in 'Adverse Drug Reactions: their Prediction, Detection and Assessment', ed. D. J. Richards and R. K. Rondel, Churchill Livingstone, Edinburgh, 1972, p. 86.
19. C. J. Speirs, J. P. Griffin, J. C. P. Weber, M. Glen-Bott and C. Twomey, *Health Trends*, 1984, **16**, 49.
20. W. H. W. Inman, *Br. Med. J.*, 1977, **1**, 1500.
21. W. H. W. Inman and M. P. Vessey, *Br. Med. J.*, 1968, **2**, 193.
22. W. H. W. Inman and A. M. Adelstein, *Lancet*, 1969, **2**, 279.
23. J. T. Nicholls, in 'Post-marketing Surveillance of Adverse Reactions to New Medicines', Medico-Pharmaceutical Forum, London, 1978, publication no. 7, p. 4.
24. R. H. Felix and F. A. Ive, *Br. Med. J.*, 1974, **2**, 333.
25. R. H. Felix, F. A. Ive and M. G. C. Dahl, *Br. Med. J.*, 1974, **4**, 321.
26. P. Wright, *Br. Med. J.*, 1974, **2**, 560.
27. P. Wright, *Br. Med. J.*, 1975, **1**, 595.
28. P. Brown, H. Baddeley, A. E. Read, J. D. Davies and J. McGarry, *Lancet*, 1974, **2**, 1477.
29. W. H. W. Inman, in 'Epidemiological Evaluation of Drugs', ed. F. Colombo, S. Shapiro, D. Slone and G. Tognoni, Elsevier, Amsterdam, 1977, p. 231.
30. W. H. W. Inman, in 'Epidemiological Issues in Reported Drug Induced Illnesses: SMON and other Examples', ed. M. Gent and I. Shigematsu, McMaster University Library Press, Hamilton, Ontario, 1978, p. 17.
31. L. E. Bottiger and B. Westerholm, *Br. Med. J.*, 1973, **3**, 339.
32. C. E. Lumley, S. R. Walker, G. C. Hall, N. Staunton and P. Grob, *Pharm. Med.*, 1986, **1**, 205.
33. S. R. Walker and C. E. Lumley, *Pharm. Med.*, 1986, **1**, 195.
34. W. H. W. Inman and M. P. Vessey, in 'Recent Advances in Community Medicine', ed. A. E. Bennett, Churchill Livingstone, Edinburgh, 1978, no. 1, p. 215.
35. W. H. W. Inman, *Br. Med. Bull.*, 1970, **26**, 248.
36. W. H. W. Inman, M. P. Vessey, B. Westerholm and A. Engelund, *Br. Med. J.*, 1970, **2**, 203.
37. J. I. Mann and W. H. W. Inman, *Br. Med. J.*, 1975, **2**, 245.
38. J. I. Mann, W. H. W. Inman and M. Thorogood, *Br. Med. J.*, 1976, **2**, 445.
39. W. H. W. Inman and W. W. Mushin, *Br. Med. J.*, 1974, **1**, 5.
40. W. H. W. Inman and W. W. Mushin, *Br. Med. J.*, 1978, **2**, 1455.
41. P. J. Toghill, P. G. Smith, P. Benton, R. C. Brown and H. L. Matthews, *Br. Med. J.*, 1974, **3**, 545.
42. W. M. Wardell, *Regulation*, 1979, **3**, 25.
43. W. M. Wardell, *Clin. Pharmacol. Ther.*, 1973, **14**, 773.
44. W. M. Wardell, *Clin. Pharmacol. Ther.*, 1974, **15**, 73.
45. C. F. George, *Br. Med. J.*, 1980, **281**, 507.
46. W. M. Wardell, *Clin. Pharmacol. Ther.*, 1978, **24**, 499.
47. B. J. Culliton and W. K. Waterfall, *Br. Med. J.*, 1980, **280**, 1175.
48. A. Ruskin and C. Anello, in 'Monitoring for Drug Safety', ed. W. H. W. Inman, MTP Press, Lancaster, 1980, p. 115.
49. B. Lee and W. M. Turner, *Am. J. Hosp. Pharm.*, 1978, **35**, 929.
50. D. E. Knapp, B. B. Zax, A. C. Rossi and R. T. O'Neill, *Drug Intell. Clin. Pharm.*, 1980, **14**, 23.
51. A. C. Rossi, D. E. Knapp, C. Anello, R. T. O'Neill, C. F. Graham, P. S. Mendelis and G. R. Stanley, *JAMA, J. Am. Med. Assoc.*, 1983, **249**, 2226.
52. J. K. Jones, G. A. Faich and C. Anello, in 'Monitoring for Drug Safety', 2nd edn., ed. W. H. W. Inman, MTP Press, Lancaster, 1986, p. 153.
53. J. K. Jones, S. W. Van de Carr, F. Rosa, M. L. Morse and A. A. LeRoy, *Acta Med. Scand., Suppl.*, 1984, **683**, 127.
54. M. L. Morse, A. A. LeRoy and B. L. Strom, in 'Monitoring for Drug Safety', 2nd edn., ed. W. H. W. Inman, MTP Press, Lancaster, 1986, p. 237.
55. B. L. Strom, J. L. Carson, M. L. Morse and A. A. LeRoy, *Clin. Pharmacol. Ther.*, 1985, **38**, 359.
56. World Health Organization, 'Manual of the International Statistical Classification of Diseases, Injuries and Causes of Death', 9th revision, World Health Organization, Geneva, 1977.
57. S. W. Van de Carr, D. L. Kennedy, F. W. Rosa, C. Anello and J. K. Jones, *Am. J. Epidemiol.*, 1983, **117**, 153.
58. L. E. Cluff, G. F. Thornton and L. G. Seidl, *JAMA, J. Am. Med. Assoc.*, 1964, **188**, 976.
59. P. L. Doering and R. B. Stewart, *JAMA, J. Am. Med. Assoc.*, 1978, **239**, 843.
60. M. W. McKenzie, G. L. Marchall, M. L. Netzloff and L. E. Cluff, *J. Pediatr. (St. Louis)*, 1976, **89**, 487.
61. R. B. Stewart, L. E. Cluff and J. R. Philp (eds.), 'Drug Monitoring: a Requirement for Responsible Drug Use', Williams and Wilkins, Baltimore, 1977.
62. R. B. Stewart and L. E. Cluff, *Johns Hopkins Med. J.*, 1971, **129**, 319.
63. D. Slone, H. Jick, I. Borda, T. C. Chalmers, M. Feinleib, H. Muench, L. Lipworth, C. Bellotti and B. Gilman, *Lancet*, 1966, **2**, 901.
64. H. Jick, O. S. Miettinen, S. Shapiro, G. P. Lewis, V. Siskind and D. Slone, *JAMA, J. Am. Med. Assoc.*, 1970, **213**, 1455.

65. Boston Collaborative Drug Surveillance Program, *Lancet*, 1972, **1**, 529.
66. D. J. Greenblatt and J. Koch-Weser, *Am. J. Med. Sci.*, 1973, **266**, 261.
67. D. W. Duhme, D. J. Greenblatt and J. Koch-Weser, *Ann. Intern. Med.*, 1974, **80**, 516.
68. D. H. Lawson, *Q. J. Med.*, 1974, **43**, 433.
69. D. J. Greenblatt and J. Koch-Weser, *Drugs*, 1974, **7**, 118.
70. D. H. Lawson and H. Jick, *Br. J. Clin. Pharmacol.*, 1977, **4**, 507.
71. R. R. Miller and D. J. Greenblatt (eds.), 'Drug Effects in Hospitalized Patients: Experiences of the Boston Collaborative Drug Surveillance Program, 1966–75', Wiley, New York, 1976.
72. J. Porter and H. Jick, *Lancet*, 1977, **1**, 587.
73. H. Jick and J. Porter, *Lancet*, 1978, **2**, 87.
74. H. Jick, A. M. Walker and J. Porter, *J. Clin. Pharmacol.*, 1981, **21**, 359.
75. R. R. Miller, *Arch. Intern. Med.*, 1974, **134**, 219.
76. J. Porter and H. Jick, *JAMA, J. Am. Med. Assoc.*, 1977, **237**, 879.
77. D. J. Greenblatt and M. D. Allen, *Br. J. Clin. Pharmacol.*, 1978, **5**, 407.
78. D. J. Greenblatt, *J. Am. Geriatr. Soc.*, 1979, **27**, 20.
79. H. Jick, *Am. J. Med.*, 1977, **62**, 514.
80. H. J. Pfeifer and D. J. Greenblatt, *Chest*, 1978, **73**, 455.
81. H. Jick and J. Porter, *Arch. Intern. Med.*, 1978, **138**, 1566.
82. Boston Collaborative Drug Surveillance Program, *N. Engl. J. Med.*, 1972, **286**, 53.
83. Boston Collaborative Drug Surveillance Program, *JAMA, J. Am. Med. Assoc.*, 1972, **220**, 377.
84. D. J. Greenblatt and J. Koch-Weser, *JAMA, J. Am. Med. Assoc.*, 1973, **225**, 40.
85. H. Jick and J. B. Porter, *J. Clin. Pharmacol.*, 1981, **21**, 456.
86. Boston Collaborative Drug Surveillance Program, *Lancet*, 1973, **1**, 1399.
87. J. R. O'Brien, *Lancet*, 1968, **1**, 779.
88. S. Shapiro and D. Slone, in 'Drug Monitoring', ed. F. H. Gross and W. H. W. Inman, Academic Press, London, 1977, p. 33.
89. Boston Collaborative Drug Surveillance Program, *Br. Med. J.*, 1974, **1**, 440.
90. Aspirin Myocardial Infarction Study Research Group, *JAMA, J. Am. Med. Assoc.*, 1980, **243**, 661.
91. K. Breddin, D. Loew, K. Lechner, K. Uberla and E. Walter, *Thromb. Haemostasis*, 1979, **41**, 225.
92. Coronary Drug Project Research Group, *J. Chronic Dis.*, 1976, **29**, 625.
93. P. C. Elwood, A. L. Cochrane, M. L. Burr, P. M. Sweetnam, G. Williams, E. Welsby, S. J. Hughes and R. Renton, *Br. Med. J.*, 1974, **1**, 436.
94. P. C. Elwood and P. M. Sweetnam, *Lancet*, 1979, **2**, 1313.
95. Persantine–Aspirin Reinfarction Study Research Group, *Circulation*, 1980, **62**, 449.
96. P. C. Elwood, *Drugs*, 1984, **28**, 1.
97. Boston Collaborative Drug Surveillance Program, *Lancet*, 1974, **2**, 669.
98. C. R. B. Joyce, in 'Drugs between Research and Regulations', ed. C. Steichele, U. Abshagen and J. Koch-Weser, Steinkopff, Darmstadt, 1985, p. 89.
99. S. Shapiro, J. L. Parsells, L. Rosenberg, D. W. Kaufman, P. D. Stolley and D. Schottenfeld, *Eur. J. Clin. Pharmacol.*, 1984, **26**, 143.
100. H. Jick, *Am. J. Epidemiol.*, 1979, **109**, 625.
101. H. Jick, *Drug Inf. J.*, 1985, **19**, 237.
102. H. Jick, R. N. Watkins, J. R. Hunter, B. J. Dinan, S. Madsen, K. J. Rothman and A. M. Walker, *N. Engl. J. Med.*, 1979, **300**, 218.
103. H. Jick, B. Dinan, R. Herman and K. J. Rothman, *JAMA, J. Am. Med. Assoc.*, 1978, **240**, 2548.
104. H. Jick, B. J. Dinan and J. R. Hunter, *J. Urol.*, 1982, **127**, 224.
105. J. B. Porter, J. Hunter-Mitchell, H. Jick and A. M. Walker, *Am. J. Public Health*, 1986, **76**, 1428.
106. S. S. Jick, D. R. Perera, A. M. Walker and H. Jick, *Lancet*, 1987, **2**, 380.
107. J. B. Porter, K. Beard, A. M. Walker, D. H. Lawson, H. Jick and G. S. M. Kellaway, *Arch. Intern. Med.*, 1986, **146**, 2237.
108. D. H. Lawson, in 'Monitoring for Drug Safety', 2nd edn., ed. W. H. W. Inman, MTP Press, Lancaster, 1986, p. 255.
109. H. K. Ziel and W. D. Finkle, *N. Engl. J. Med.*, 1975, **293**, 1167.
110. D. Slone, S. Shapiro and O. S. Miettinen, in 'Epidemiological Evaluation of Drugs', ed. F. Colombo, S. Shapiro, D. Slone and G. Tognoni, Elsevier, Amsterdam, 1977, p. 59.
111. A. Klemetti and L. Saxen, *Am. J. Public Health*, 1967, **57**, 2071.
112. A. A. Mitchell, L. B. Cottler and S. Shapiro, *Am. J. Epidemiol.*, 1986, **123**, 670.
113. O. P. Heinonen, D. Slone and S. Shapiro, 'Birth Defects and Drugs in Pregnancy', Publishing Sciences Group, Littleton, MA, 1977.
114. S. Shapiro and D. Slone, *Epidemiol. Rev.*, 1979, **1**, 110.
115. L. Rosenberg, D. R. Miller, D. W. Kaufman, S. P. Helmrich, S. Van de Carr, P. D. Stolley and S. Shapiro, *JAMA, J. Am. Med. Assoc.*, 1983, **250**, 2801.
116. L. Rosenberg, S. Shapiro, D. W. Kaufman, D. Slone, O. S. Miettinen and P. D. Stolley, *Int. J. Epidemiol.*, 1980, **9**, 57.
117. L. Rosenberg, D. Slone, S. Shapiro, D. W. Kaufman, O. S. Miettinen and P. D. Stolley, *Am. J. Public Health*, 1981, **71**, 82.
118. L. Rosenberg, D. Slone, S. Shapiro, D. Kaufman, P. D. Stolley and O. S. Miettinen, *JAMA, J. Am. Med. Assoc.*, 1980, **244**, 339.
119. L. Rosenberg, D. Slone, S. Shapiro, D. W. Kaufman, P. D. Stolley and O. S. Miettinen, *Am. J. Epidemiol.*, 1980, **111**, 675.
120. S. Shapiro, D. Slone, L. Rosenberg, D. W. Kaufman, P. D. Stolley and O. S. Miettinen, *Lancet*, 1979, **1**, 743.
121. D. Slone, S. Shapiro, D. W. Kaufman, L. Rosenberg, O. S. Miettinen and P. D. Stolley, *N. Engl. J. Med.*, 1981, **305**, 420.
122. D. Slone, S. Shapiro, L. Rosenberg, D. W. Kaufman, S. C. Hartz, A. C. Rossi, P. D. Stolley and O. S. Miettinen, *N. Engl. J. Med.*, 1978, **298**, 1273.
123. S. P. Helmrich, S. Shapiro, L. Rosenberg, D. W. Kaufman, D. Slone, C. Bain, O. S. Miettinen, P. D. Stolley, N. B. Rosenshein, R. C. Knapp, T. Leavitt, D. Schottenfeld, R. L. Engle and M. Levy, *Am. J. Epidemiol.*, 1983, **117**, 35.
124. D. W. Kaufman, S. Shapiro, D. Slone, L. Rosenberg, S. P. Helmrich, O. S. Miettinen, P. D. Stolley, M. Levy and D. Schottenfeld, *Lancet*, 1982, **1**, 537.

125. L. Rosenberg, D. R. Miller, D. W. Kaufman, S. P. Helmrich, P. D. Stolley, D. Schottenfeld and S. Shapiro, *Am. J. Epidemiol.*, 1984, **119**, 167.
126. S. Shapiro, D. Slone, D. W. Kaufman, L. Rosenberg, O. S. Miettinen, P. D. Stolley, R. C. Knapp, T. Leavitt, W. G. Watring, N. B. Rosenshein and D. Schottenfeld, *JAMA, J. Am. Med. Assoc.*, 1980, **244**, 1685.
127. D. W. Kaufman, S. Shapiro, D. Slone, L. Rosenberg, O. S. Miettinen, P. D. Stolley, R. C. Knapp, T. Leavitt, W. G. Watring, N. B. Rosenshein, J. L. Lewis, D. Schottenfeld and R. L. Engle, *N. Engl. J. Med.*, 1980, **303**, 1045.
128. S. Shapiro, D. W. Kaufman, D. Slone, L. Rosenberg, O. S. Miettinen, P. D. Stolley, N. B. Rosenshein, W. G. Watring, T. Leavitt and R. C. Knapp, *N. Engl. J. Med.*, 1980, **303**, 485.
129. L. Rosenberg, S. Shapiro, D. Slone, D. W. Kaufman, S. P. Helmrich, O. S. Miettinen, P. D. Stolley, N. B. Rosenshein, D. Schottenfeld and R. L. Engle, *JAMA, J. Am. Med. Assoc.*, 1982, **247**, 3210.
130. A. A. Mitchell, P. Goldman, S. Shapiro and D. Slone, *Am. J. Epidemiol.*, 1979, **110**, 196.
131. A. A. Mitchell, C. Louik, P. Lacouture, D. Slone, P. Goldman and S. Shapiro, *JAMA, J. Am. Med. Assoc.*, 1982, **247**, 2385.
132. L. Rosenberg, S. Shapiro, D. Slone, D. W. Kaufman, O. S. Miettinen and P. D. Stolley, *N. Engl. J. Med.*, 1980, **303**, 546.
133. F. Colombo, S. Shapiro, D. Slone and G. Tognoni (eds.), 'Epidemiological Evaluation of Drugs', Elsevier, Amsterdam, 1977.
134. International Agranulocytosis and Aplastic Anaemia Study, *JAMA, J. Am. Med. Assoc.*, 1986, **256**, 1749.
135. M. S. Kramer, D. A. Lane and T. A. Hutchinson, *J. Chronic Dis.*, 1987, **40**, 1073.
136. G. D. Friedman, M. F. Collen, L. E. Harris, E. E. Van Brunt and L. S. Davis, *JAMA, J. Am. Med. Assoc.*, 1971, **217**, 567.
137. G. D. Friedman and H. K. Ury, *JNCI, J. Natl. Cancer Inst.*, 1980, **65**, 723.
138. G. D. Friedman and H. K. Ury, *JNCI, J. Natl. Cancer Inst.*, 1983, **71**, 1165.
139. D. C. Moir, in 'Monitoring for Drug Safety', 2nd edn., ed. W. H. W. Inman, MTP Press, Lancaster, 1986, p. 277.
140. D. Moir, in 'Adverse Drug Reactions: their Prediction, Detection and Assessment', ed. D. J. Richards and R. K. Rondel, Churchill Livingstone, Edinburgh, 1972, p. 42.
141. E. D. Acheson, 'Medical Record Linkage', Oxford University Press, London, 1967.
142. D. C. G. Skegg and R. Doll, *J. Epidemiol. Community Health*, 1981, **35**, 25.
143. D. C. G. Skegg, S. M. Richards and R. Doll, *J. Epidemiol. Community Health*, 1981, **35**, 32.
144. I. K. Crombie, S. V. Brown and J. G. Hamley, *J. Epidemiol. Community Health*, 1984, **38**, 226.
145. D. C. G. Skegg, in 'Monitoring for Drug Safety', 2nd edn., ed. W. H. W. Inman, MTP Press, Lancaster, 1986, p. 291.
146. G. P. Oakley, *JAMA, J. Am. Med. Assoc.*, 1973, **225**, 395.
147. M. Gent and I. Shigematsu (eds.), 'Epidemiological Issues in Reported Drug Induced Illnesses: SMON and other Examples', McMaster University Library Press, Hamilton, Ontario, 1978.
148. A. L. Herbst, H. Ulfelder and D. C. Poskanzer, *N. Engl. J. Med.*, 1971, **284**, 878.
149. L. Lasagna (ed.), 'Post-marketing Surveillance of Drugs', Medicine in the Public Interest, Washington, DC, 1977, p. 6.
150. IMS America, 'An Experiment in Early Post-marketing Surveillance of Drugs: Final Report — Task C', IMS America, Ambler, PA, 1980.
151. Joint Commission on Prescription Drug Use, 'Final Report', Joint Commission on Prescription Drug Use, Rockville, MD, 1980.
152. J. P. Griffin, *Health Trends*, 1981, **13**, 85.
153. A. W. Harcus, A. E. Ward and D. W. Smith, *Br. Med. J.*, 1979, **2**, 163.
154. S. F. Sullman, in 'Post-marketing Surveillance of Adverse Reactions to New Medicines', Medico-Pharmaceutical Forum, London, 1978, publication no. 7, p. 12.
155. W. H. W. Inman, in 'Drug Monitoring', ed. F. H. Gross and W. H. W. Inman, Academic Press, London, 1978, p. 65.
156. C. T. Dollery and M. D. Rawlins, *Br. Med. J.*, 1977, **1**, 96.
157. D. H. Lawson and D. A. Henry, *Br. Med. J.*, 1977, **1**, 691.
158. A. B. Wilson, *Br. Med. J.*, 1977, **2**, 1001.
159. W. H. W. Inman, in 'Post-marketing Surveillance of Adverse Reactions to New Medicines', Medico-Pharmaceutical Forum, London, 1978, publication no. 7, p. 45.
160. D. H. Lawson, in 'Post-marketing Surveillance of Adverse Reactions to New Medicines', Medico-Pharmaceutical Forum, London, 1978, publication no. 7, p. 50.
161. M. D. Rawlins and C. T. Dollery, in 'Post-marketing Surveillance of Adverse Reactions to New Medicines', Medico-Pharmaceutical Forum, London, 1978, publication no. 7, p. 40.
162. A. B. Wilson, in 'Post-marketing Surveillance of Adverse Reactions to New Medicines', Medico-Pharmaceutical Forum, London, 1978, publication no. 7, p. 56.
163. A. B. Wilson, in 'Monitoring for Drug Safety', ed. W. H. W. Inman, MTP Press, Lancaster, 1980, p. 189.
164. M. Drury, *Br. Med. J.*, 1977, **1**, 439.
165. M. Drury and F. M. Hull, *Br. Med. J.*, 1981, **283**, 1305.
166. W. H. W. Inman, *Br. Med. J.*, 1981, **282**, 1131.
167. W. H. W. Inman, *Br. Med. J.*, 1981, **282**, 1216.
168. W. H. W. Inman and N. S. B. Rawson, in 'The Impact of Computer Technology on Drug Information', ed. P. Manell and S. G. Johansson, North-Holland, Amsterdam, 1982, p. 153.
169. H. W. K. Acheson, *Br. Med. J.*, 1977, **1**, 439.
170. R. I. Tricker, 'Report of the Inquiry into the Prescription Pricing Authority', HMSO, London, 1977.
171. R. W. Smithells and E. R. Chinn, *Br. Med. J.*, 1964, **1**, 217.
172. R. W. Smithells and S. Sheppard, *Teratology*, 1978, **17**, 31.
173. E. K. Borden and J. G. Lee, *J. Chronic Dis.*, 1982, **35**, 803.
174. E. G. McQueen, in 'Advances in Pharmacology and Therapeutics II', ed. H. Yoshida, Y. Hagihara and S. Ebashi, Pergamon Press, Oxford, 1982, vol. 6, p. 79.
175. W. H. W. Inman, N. S. B. Rawson and L. V. Wilton, in 'Monitoring for Drug Safety', 2nd edn., ed. W. H. W. Inman, MTP Press, Lancaster, 1986, p. 213.
176. N. S. B. Rawson, Ph. D. Thesis, University of Southampton, 1987.
177. W. H. W. Inman, in 'Detection and Prevention of Adverse Drug Reactions', Almqvist and Wiksell International, Stockholm, 1984, p. 133.

178. N. S. B. Rawson and W. H. W. Inman, *Med. Toxicol.*, 1986, **1**, (suppl. 1), 79.
179. W. H. W. Inman and N. S. B. Rawson, in 'Side Effects of Anti-inflammatory Drugs, I: Clinical and Epidemiological Aspects', ed. K. D. Rainsford and G. P. Velo, MTP Press, Lancaster, 1987, p. 111.
180. J. H. Kurata, J. D. Elashoff and M. I. Grossman, *Gastroenterology*, 1982, **82**, 373.
181. W. H. W. Inman and N. S. B. Rawson, *Lancet*, 1983, **2**, 908.
182. N. S. B. Rawson and W. H. W. Inman, in 'Proceedings of the Second World Conference on Clinical Pharmacology and Therapeutics', ed. L. Lemberger and M. M. Reidenberg, American Society for Pharmacology and Experimental Therapeutics, Bethesda, MD, 1984, p. 939.
183. W. H. W. Inman, N. S. B. Rawson, L. V. Wilton, G. L. Pearce and C. J. Speirs, *Br. Med. J.*, 1988, **297**, 826.
184. Drug Safety Research Unit, 'PEM News', Drug Safety Research Unit, Southampton, 1988, no. 5, p. 5.
185. W. H. W. Inman and N. S. B. Rawson, *Br. Med. J.*, 1983, **286**, 1954.
186. Drug Surveillance Research Unit, 'PEM News', Drug Surveillance Research Unit, Southampton, 1984, no. 2, p. 11.
187. A. W. Harcus, A. E. Ward and D. W. Smith, *Br. Med. J.*, 1979, **2**, 163.
188. D. G. Colin-Jones, M. J. S. Langman, D. H. Lawson and M. P. Vessey, *Br. Med. J.*, 1982, **285**, 1311.
189. D. G. Colin-Jones, M. J. S. Langman, D. H. Lawson and M. P. Vessey, *Br. Med. J.*, 1983, **286**, 1713.
190. D. G. Colin-Jones, M. J. S. Langman, D. H. Lawson and M. P. Vessey, *Q. J. Med.*, 1985, **54**, 253.
191. W. P. Maclay, D. Crowder, S. Spiro and P. Turner, *Br. Med. J.*, 1984, **288**, 911.
192. J. Seldrup, *Biometrics Bull.*, 1987, **4**, 23 (paper presented at a conference on Statistical Methods in Medicine and Pharmacology held in Køge, Denmark, 6–8th October 1986).
193. W. D. Cooper, D. Sheldon, D. Brown, G. R. Kimber, V. L. Isitt and W. J. C. Currie, *J. R. Coll. Gen. Pract.*, 1987, **37**, 346.
194. W. M. Wardell, M. C. Tsianco, S. N. Anavekar and H. T. Davis, *J. Clin. Pharmacol.*, 1979, **19**, 85.
195. W. M. Castle, J. T. Nicholls and C. C. Downie, *Br. J. Clin. Pharmacol.*, 1983, **16**, 581.
196. A. R. Feinstein, *Clin. Pharmacol. Ther.*, 1974, **16**, 110.
197. H. Jesdinsky, in 'Drug Monitoring', ed. F. H. Gross and W. H. W. Inman, Academic Press, London, 1977, p. 91.
198. D. J. Finney, *J. Chronic Dis.*, 1964, **17**, 565.
199. D. J. Finney, *Proc. Eur. Soc. Study Drug Toxic.*, 1966, **7**, 198.
200. D. J. Finney, *Methods Inf. Med.*, 1971, **10**, 1.
201. D. J. Finney, *Methods Inf. Med.*, 1971, **10**, 237.
202. D. J. Finney, *Methods Inf. Med.*, 1974, **13**, 1.
203. D. R. Jones and L. Rushton, *Int. J. Epidemiol.*, 1982, **11**, 276.
204. D. A. Lane and N. S. B. Rawson, in 'Statistical Methodology in the Pharmaceutical Sciences', ed. D. A. Berry, Dekker, New York, 1989, p. 533.
205. J. A. Lewis, *Trends Pharmacol. Sci.*, 1981, **2**, 93.
206. D. L. Sackett, R. B. Haynes, M. Gent and D. W. Taylor, in 'Monitoring for Drug Safety', 2nd edn., ed. W. H. W. Inman, MTP Press, Lancaster, 1986, p. 471.
207. D. M. Green, *Ann. Intern. Med.*, 1964, **60**, 255.
208. M. M. Reidenberg and D. T. Lowenthal, *N. Engl. J. Med.*, 1968, **279**, 678.
209. P. H. Joubert, F. W. Jensen van Rijssen and J. P. Venter, *S. Afr. Med. J.*, 1977, **52**, 34.
210. Lord Rothschild, *Listener*, 30 November 1978, 715.
211. M. D. B. Stephens, *Adverse Drug React. Acute Poisoning Rev.*, 1987, **1**, 1.
212. C. Louik, P. G. Lacouture, A. A. Mitchell, R. Kaufman, F. H. Lovejoy, S. J. Yaffe and S. Shapiro, *Clin. Pharmacol. Ther.*, 1985, **38**, 183.
213. T. A. Hutchinson and D. A. Lane, *J. Clin. Epidemiol.*, 1989, **42**, 5 (University of Minnesota (School of Statistics) Technical Report No. 460, 1986).
214. D. A. Lane, *Drug Inf. J.*, 1986, **20**, 455.
215. D. A. Lane, M. S. Kramer, T. A. Hutchinson, J. K. Jones and C. Naranjo, *Pharm. Med.*, 1987, **2**, 265.
216. D. A. Lane, in 'Statistical Methodology in the Pharmaceutical Sciences', ed. D. A. Berry, Dekker, New York, 1989, p. 477.

4.10
Animal Experimentation

JOHN H. SEAMER and LYDIA A. BROWN
Salisbury, UK

4.10.1 INTRODUCTION

The use of animals by man for his own purposes has a very long history. The hunting of wild animals to provide man with food and clothing predates domestication. Following their domestication, animals have been used not only to provide food and clothing, but also for transport, carriage, companionship and sport. With man's own increasing civilization it was natural therefore to use animals for more sophisticated purposes such as the increase of knowledge. Thus it is recorded that Noah released a raven and a dove from the ark to see if the flood had abated; more certainly Pepys described several experiments on animals, including one on March 15th 1664–65 to observe the effect of 'the great poyson of Maccassa upon a dogg'.

The gradual evolution of scientific investigation, and particularly of the biological sciences, led to further usage of animals, until in the late 19th century animal use for scientific purposes had become quite commonplace in Western Europe. In Britain a spirit of protectionism had also developed and this had led to the passage, in 1822, of the Ill Treatment of Cattle Act (Martin's Act). Subsequently, despite the availability of early anaesthetics, surgical procedures were carried out by some scientists on fully conscious animals. Increasing concern among other scientists and humanitarians about these practices led to the development of antivivisection societies and the passage in Britain of the first law to protect animals in laboratories, the Cruelty to Animals Act, 1876.

Although the Judaeo–Christian tradition asserts man's domination over animals, albeit within humane limits, many people feel very strongly that animals should not be subordinated to human requirements. As in other situations current ethical thinking has to contend with current practice, but there is a formidable trend away from the use (some would call it exploitation) of animals by man in Western society. Concepts of animal rights have recently been advanced to support this view and the matter has become one of philosophical argument. Whether or not the case for animal rights can be sustained, few would deny the need for human obligations and responsibilities towards animals.

4.10.2 ANIMAL MODELS

Antivivisectionists and others often decry the use of animals in research, citing examples of results of animal experiments which have apparently proved to be misleading or inapplicable to man. The claim is, in effect, that the proper study of man is man himself. Leaving aside the question of advancing biological knowledge of animals, this argument would require observational and investigative studies of humans. Some advancement of knowledge can indeed be made by observational studies. However experimental scientific investigations sometimes require serious constraints and interferences which human subjects would not voluntarily accept. Moreover there are serious ethical difficulties in conducting trials in man, for example the selection of neonatal children for drug or placebo treatment. Human experimentation was formerly practised on criminals and lunatics, but this is now generally regarded as wrong, and in Britain only very limited human experiments are permitted. The risks of harm in these trials are very slight and there are rigorous safeguards.

While the biology of man clearly differs in some respects from that of the rat or pig, the three species do share many common processes. These similarities (and also occasionally differences) are utilized when animals are used as simulants for man. If the tenor of the scientific argument, that animals may be used to benefit man and other animals, is accepted, the question remaining is to define the limits of acceptability.

A general view of the welfare of animals based on the three Rs of Russell and Burch[1] is now prevalent among those involved in animal experiments. The three Rs are Replacement, Refinement and Reduction. Thus the use of animals is to be avoided if possible and they should be *replaced* by nonsentient systems. Techniques should be *refined* so that when animals are used they are properly selected, cared for and humanely treated. The numbers used should be *reduced* to the smallest necessary number to obtain a significant result in well-designed experiments. Great care should be taken to avoid, minimize or alleviate pain or distress up to, at least, the standards of current veterinary clinical practice by the use of analgesics or anaesthetics.[2] Extending these concepts still further is the fundamental principle underlying the new British Animals (Scientific Procedures) Act 1986 that the value of the results likely to be obtained from an animal experiment should be weighed against the pain, suffering, distress or lasting harm likely to be experienced by the animals. Stages of discomfort, distress, suffering and pain can be defined with increasing objectivity by veterinarians and others familiar with animal experiments who can recognize deviations from normality. Only particularly valuable results merit the infliction of more than trivial suffering.

4.10.3 REQUIREMENTS FOR THE USE OF ANIMALS IN RESEARCH

The requirements for the usage of animals in biomedical experiments are sufficiently various to defy simple categorization. For convenience they may be considered under the headings of fundamental research; applied research and testing; teaching and demonstration; and provision of biological materials. Fundamental medical research is concerned with finding out how the body functions. Early and simple examples involved the study of organ function by the removal of adrenals, pituitary, *etc.* from experimental animals. More recently extirpation of the Bursa of Fabricius in fowls and the thymus in mammals has led to a far wider understanding of immunology. The range of animals used in fundamental research is very wide. Species are frequently selected because they possess characteristics presumed to be particularly appropriate to a given line of work. The hamster cheek pouch has for example been used for transplanting tumours from other animal species because unusually it offers no immunological rejection. Similarly the susceptibility of the armadillo to human leprosy led to significant advances in the knowledge of this disease. Although a wide range of animal species may be used in fundamental biomedical research, the actual numbers of animals tend to be far less than those used in applied research.

Applied research follows on fundamental advances and sets out to solve a particular problem—how to cure a particular disease or to develop a vaccine to control it. In the case of human disease, the animals used frequently come from the 'common' laboratory species: mice, rats, guinea pigs, dogs, cats and monkeys. This may also be true, at least initially, of investigations into diseases of domestic animals, because of the inconvenience and cost of using larger animals such as pigs and cows. Applied research merges into the testing of substances in animals. Animal tests of one sort or another utilize by far the greatest number of animals used in experiments. Tests may be done to show that products for use in or by humans or their animals are effective and potent; that they are safe and not toxic; and that the batches and dosages are standard. In the last 20 years, national and international regulatory bodies have instituted standards for drugs, vaccines, food additives, domestic products, *etc.* to safeguard the health, safety and general interests of the public. These requirements, legal and quasi-legal, increased enormously the expense and complexity of animal tests and of course the numbers of animals used. Movement towards harmonization and mutual recognition of the various tests is now occurring with a consequent welcome reduction in expense and animal usage.

Although not strictly speaking a form of animal experimentation, the animals used in teaching and demonstration in universities and colleges should also be considered. In Britain relatively few animals are used in this way. However for many years some have been used in pharmacological and physiological demonstrations in medical and veterinary schools. Animals have not been used by students in British medical and veterinary schools for surgical practice, nor are experiments on animals allowed in primary or secondary schools. Furthermore the use of animals by qualified surgeons to acquire manual dexterity was until recently prohibited in Britain. This prohibition seriously inhibited surgeons wishing to become skilful in microsurgical techniques in which very small blood vessels are joined to benefit human patients. The prohibition was relaxed (but with stringent safeguards) when the Cruelty to Animals Act 1876 was replaced by the Animals (Scientific Procedures) Act 1986.

The last way in which animals may be used in biomedical research is in the provision of materials—cells, blood, antisera, *etc.* Many biomedical techniques require the use of blood or antisera almost as chemical reagents. These have to be obtained from live animals. In addition animal cells are used for various purposes, particularly for tissue cultures. While many tissue cell lines, which are virtually perpetual, exist, large quantities of tissue cultures are still prepared from the cells of freshly killed laboratory animals.

4.10.4 ALTERNATIVES TO ANIMALS

In earlier years there was a sharp debate between those who considered there were few, if any, alternatives to the use of animals in experiments and those who believed that alternative methods could and should be found quickly. Today in the face of quiet but significant progress, there is less polarization. Early proponents of alternatives argued strongly for the use of computers, perhaps without sufficient realization that the data to be used would probably have to come from animal sources. Nevertheless the greatly extended use of computers has probably reduced the numbers of animals used in research. Far greater reduction in animal usage has resulted from developments in tissue culture. Many virological studies, conducted originally in animals because there was no suitable alternative, are now carried out in cultures. Nevertheless it is still necessary for many tissue cultures to be obtained from animals and it should be recognized that for much research, for example into pregnancy or immunology, there is still no substitute for the living animal.

The reduction in animal use arising from computers and tissue cultures followed general developments in these areas. However there have been more specific developments which have led to a reduction in the use of animals. Pyrogenicity tests of substances used for injections have been traditionally carried out in rabbits, and indeed this is still a legal requirement. However, it has been found that pyrogenicity tests can also be made on amoebocytes of the crab *Limulus polyphemus* and although this test has not yet achieved official status, it is widely used for in-process testing. Similarly the development of the Ames test, in which bacteria are used instead of animals to screen compounds for mutagenicity, has led to an economic, effective screening test. There are other examples but for the time being the requirement for animal experimentation continues, particularly in pharmaceutical research. For a review of alternatives to animal experiments the reader is referred to Smyth.[3]

4.10.5 SUPPLY OF ANIMALS

In the early days of animal experimentation the relatively small numbers of animals required for research were acquired from a variety of sources—pet shops, animal fanciers and dealers. During World War 2, British scientists recognized that the quality of animals of uncertain origin could adversely affect research. As a result the Laboratory Animals Bureau was established in 1947 by the Medical Research Council to improve the source and supply of experimental animals. Today the vast majority of animals used in British laboratories are bred especially for the purpose. Clearly this is necessary on numerical grounds alone when more than a million rats and mice are required annually. Furthermore it is necessary for many mice and rats to be of a particular type or quality so that no 'ordinary' rat or mouse would suffice. Occasionally however it is necessary to use animals that have not been purpose bred—for example when studying the transmission of diseases in wildlife such as voles or squirrels.

Until recently the majority of monkeys used in research were caught and taken from their natural habitat and shipped to laboratories. This frequently caused suffering and distress to the animals concerned and in some cases seriously reduced the natural population. Many countries now prohibit the catching and export of monkeys on conservation grounds: breeding colonies have been established (Figure 1) and significant numbers of purpose-bred monkeys are now available for research projects.

Figure 1 Cynomolgus monkeys in a British breeding unit. Rhesus and cynomolgus monkeys were very extensively used in research in the 1950s and 1960s. They were imported from Asian countries where they were regarded as pests. Export of wild monkeys from many Asian countries has now ceased on conservationist grounds and purpose breeding for research has begun. This has resulted in better quality animals of known background (Crown copyright, reproduced with permission)

There is always public concern in Britain and other countries that stolen pets, particularly cats and dogs, are used in research. The Animals (Scientific Procedures) Act 1986 extends legislative control to the breeding and supply of animals for research, and enforcement of its provisions should prevent the supply of animals from any dubious sources to laboratories. In Britain the use of stray dogs for research has long been illegal; however, in some parts of the USA the use of 'pound' dogs is permitted although several states have prohibited it in recent years.

4.10.6 QUALITY OF ANIMALS

As indicated above, there was a gradual recognition that good quality animals were necessary to obtain good results in research. 'Good quality' means animals of known breeding that are free from pathological defects and have a defined microbiological status. Numerous strains of laboratory animals with defined characteristics are now recognized throughout the world (Festing).[4] These characteristics range from the physical, *e.g.* nude (Figure 2) or obesity, to the physiological, *e.g.* hypertension, to increased susceptibility to certain infections. These characteristics have been fixed in the strains by inbreeding—usually brother–sister mating—for 20 generations. The inbred strain is thus virtually homozygous—genetically uniform for the expression of certain characteristics. Selection of an appropriate strain can therefore be of great assistance in some lines of work. Microbiological quality is a more recent development. 'Germ-free' or 'axenic' animals are free of all microorganisms. These animals are born and raised in isolators (Figure 3). 'Gnotobiotic' means animals whose microbiological flora are known. 'SPF' (specific pathogen-free) refers to animals known to be free of certain, specified, pathogenic microorganisms. They are maintained under barrier conditions (Figure 4). 'Conventional' means animals which do not belong to any of the above categories and are maintained in less strictly controlled microbiological environments. Most large animal-breeding establishments in Britain run genetic and microbiological quality control schemes and conform to a self-policing microbiological-testing service which provides acceptable certification of the status of their laboratory animals.

Figure 2 Nude mouse. Found originally as a mutant in a laboratory colony, nude mice are now used extensively in research. They are particularly valuable in immunological research because they lack a thymus, which renders them immunologically incompetent. However, it also makes them very susceptible to intercurrent infections and they must therefore be maintained in strictly controlled microbiological environments (reproduced by permission of Harlan Olac Ltd)

4.10.7 ANIMAL FACILITIES

Animal experiments should not be undertaken lightly. Apart from the ethical considerations discussed earlier and the legal requirements to be described, animal experimentation is expensive, time consuming and difficult. The interrelated considerations for an experiment, namely selection of an appropriate species, appropriate numbers to provide a significant result and the source and quality of animals, may be readily decided. However, the husbandry and the management of the animals is much more demanding.

Figure 5 illustrates the Board of Agriculture's laboratory in Whitehall in 1904, complete with guinea pigs and rabbits. Modern animal accommodation is very different and is usually purpose built and of high quality. Walls, floors and ceilings must facilitate easy clearing and be impervious to water and disinfectants. The ventilation has to be controlled to provide sufficient air of the requisite temperature and humidity for the different animal species and is filtered to prevent noxious agents going to or coming from the animals. There must be an adequate water supply and drainage and the duration and intensity of light should be controlled. High quality (expensive) food to maintain

Figure 3 Germ-free isolators. Small animals can now be kept free of all microorganisms for long periods. The animals are obtained originally by Caesarian section and are maintained and bred in isolators. Filtered air is pumped in and the isolators are maintained under slight positive pressure. All food, bedding and other materials are sterilized before being passed into the isolator through airlocks. Waste is removed in a similar fashion. Work within the isolator is carried out by staff wearing long gloves sealed into the isolator walls. Isolators can also be used (under negative pressure) to contain safely animals infected with dangerous microorganisms (reproduced by permission of Harlan Olac Ltd)

health must be available for the animals as well as suitable bedding, and there must be means of collection and safe disposal of waste, soiled bedding and faeces. The cages used must contain the animals securely but must also permit ready observation, supply of food and water, be strong and durable but easy to handle and resistant to chemical or heat cleansing and sterilization. The supply must be sufficient to permit some to be cleansed while others are in use. Sometimes containment facilities such as isolators (Figure 3) are necessary, either to protect the animals or to prevent staff becoming infected with disease agents the animals may carry. Sufficient trained staff must be available to care for the animals every day of the year and to manage the ancillary facilities of provision of food, bedding, water, cleansing, sterilization, *etc.* It will be seen that the provision and maintenance of a modern animal facility is costly and labour intensive: a comprehensive manual on the care and management of laboratory animals has been prepared by UFAW.[5] Table 1 summarizes biological data of the common laboratory animals as well as environmental requirements for each species.

4.10.8 ANIMAL USAGE

The number of animals used for experiments throughout the world is not known. In the USA the Animal Plant Health and Inspection Service of the Department of Agriculture reported that some 1.8 million animals were used in experiments in 1986. However this figure does not include mice and rats, the animals most commonly used in research. Earlier, the Department of Agriculture had estimated that 11.1 million mice and rats were used in 1983. More accurate figures are available from Switzerland and The Netherlands. In Switzerland statistics from the Bundesamt fuer Veterinaerwesen show that 1.5 million animals were used in experiments in 1985. This total

Figure 4 Barrier mouse-breeding unit. The mice are kept in sterilizable plastic cages tiered in racks. Sterilized food and water are always available. The unit is sealed and maintained under slight positive pressure by the inflow of filtered air. Animal technicians enter and leave through an airlock. They take off their clothes, shower and put on sterilized clothes before they enter the animal rooms (reproduced by permission of Harlan Olac Ltd)

Figure 5 The laboratory of the Board of Agriculture at Whitehall Place in 1904. The caption to the original photograph, taken for the 'County Gentleman' read: 'Mr. J. McIntosh McCall M. B., M.R.C.V.S., Assistant Veterinary Officer and his Handy-man. Observe the pieces of pig from the provinces on the wire screens for investigation, and the labels above. The cages contain rabbits and guinea pigs for inoculation experiments when necessary' (reproduced from 'Animal Health—A Centenary 1865–1965' by permission of the Ministry of Agriculture, Fisheries and Food, Crown Copyright, 1987)

Table 1 Biological Data on Selected Laboratory Animals

	Rat	*Mouse*	*Rabbit*	*Guinea pig*	*Dog*	*Rhesus monkey*
Number of offspring	3–18	4–12	7–10	2–6	4–8	1
Weaning age (days)	21	18–21	42–56	14–21	42–56	
Lifespan (years)	2–3.5	1–3	5–7	2–5	10–15	20–30
Gestation (days)	21	19–21	31–32	62–72	63	162–170
Body temperature (°C)	38–39	36–37	38.5–40	36–40.5	38–39	37–39.1
Room temperature (°C)	20–22	20–22	15–20	18–22	15–21	20–24
Humidity (%)	50–60	50–60	45–65	45–70		

included 1.4 million mice and rats. Dutch Veterinary Public Health Inspectorate figures record that just under 1.2 million animals were used in 1985; almost 1 million of these were mice and rats.

Statistics of animals used in experiments have been kept in Britain for many years as part of the requirements of the Cruelty to Animals Act 1876. Apart from minor inconsistencies (for example the returns were technically 'experiments' and not of actual animals and some procedures were not classed as experiments) these data are very accurate and provide evidence of the ways in which animals are used. The returns for 1986 show that 442 research institutes, firms, universities and hospitals were registered for the conduct of animal experiments; some 20 000 people were permitted to carry out experiments although only about 9000 actually began an experiment in that year.[6] A total of 3 112 051 experiments were carried out. This is the lowest figure since 1956 and represents the tenth successive fall in the annual number of experiments. Table 2, summarized from the Home Office statistics, shows the numbers of animals used and the purposes for which the experiments were carried out. It will be seen that mice and rats were the most frequently used animals (1.6 and 0.8 million respectively) followed by fish (148 000), birds (145 000), guinea pigs (132 000) and rabbits (129 000). Dogs, cats and nonhuman primates receive special protection under the legislation and usage of these species was limited (12 901, 5251 and 5635 respectively). Horses and other *equidae* are also specially protected: 659 were used in 1986.

Table 2 Numbers and Species of Animals Used and Primary Purpose of Animal Experiments in Great Britain in 1986

Species	*1*[a]	*2*[b]	*3*[c]	*4*[d]	*Total*
Mouse	322 841	902 797	33 015	363 485	1 622 138
Rat	252 119	397 915	85 429	94 696	830 159
Guinea pig	12 831	67 513	38 333	13 759	132 436
Rabbit	11 050	95 326	15 692	7311	129 379
Primate	1163	4189	8	275	5635
Cat	2289	2584	52	326	5251
Dog	1599	9125	1446	731	12 901
Other mammals	23 934	31 191	1317	12 971	69 413
Bird	32 069	41 751	8126	63 642	145 588
Reptiles and amphibians	7880	257	1591	973	10 701
Fish	36 579	33 427	38 029	40 415	148 450
Total	704 354	1 586 075	223 038	598 584	3 112 051

[a] Study of normal or abnormal body structure or function. [b] To select, develop or study the use, hazards or safety of medical, dental and veterinary products. [c] To select, develop or study the use, hazards or safety of various substances. [d] Other purposes or more than one purpose.

It is perhaps surprising to note that fish are the third most frequently used experimental animal in the United Kingdom. There are over 26 000 species and a wide variety may be used by scientists in laboratories. Frequently salmonids are used as sentinel indicators for pollution studies. Some manufacturing companies place these fish in specially constructed tanks through which the effluent from a manufacturing process passes. Smaller tropical fish may be used in toxicological experiments for medicines destined in treat warm-blooded animals, for example the zebra fish (*Brachydanio rerio*) which has a relatively short life-cycle with a rapid breeding rate. Birds are also frequently used as laboratory animals. They are relatively easy to breed in a laboratory environment and cheap to maintain. They are used to study virally induced neoplasia, and in toxicology and immunology.

By far the greatest usage of animals (more than 1.5 million) was in the investigation of medical, dental and veterinary products, followed by 700 000 animals used in investigations of normal or abnormal bodily structure or function. Study of the use, hazards or safety of various substances accounted for the third largest total use of animals, more than 220 000. Taking the first and third of these figures together, it is evident that more than 1.8 million animals are used in the broad area of developing products for human and animal use or in studies of their safety. The discovery and development of a new drug may take as long as 8–12 years and cost £50–100 million. After fundamental research has led to its discovery, the drug has to be evaluated for safety and assessed for efficacy, while the formulation produced must be regularly checked for maintenance of quality. Before the drug can be marketed it has to be approved and registered by licensing authorities and even after marketing, postdevelopment surveillance is necessary. All of these stages require the usage of animals. For example a simple subacute toxicity test may use 8–12 male and 8–12 female rats and 3–4 male and 3–4 female dogs. Such a test could run for 30 days. The cost, without any pathological studies, is about £10–12 000.

A separate analysis of the Home Office returns shows that 58% of experiments were carried out by experimenters licensed in commercial concerns, although these licensees amounted to only 23% of the total . Conversely 56% of licensees were primarily licensed at universities but carried out only 23% of the experiments. (It should be noted that some experimenters are licensed to carry out experiments in more than one place.) Table 2 shows experiments listed for 'all other purposes': these included 161 941 to study cancer, 12 487 experiments on transplantation and 1028 'to demonstrate known facts' to students and others. Cosmetic and toiletry experiments (to which many people are antipathetic) accounted for 15 652 animals in 1986.

4.10.9 SAFETY

It must be appreciated that work with experimental animals can be very hazardous. The animal house will contain all the hazards of the workplace and laboratory as well as those hazards particularly associated with the animals themselves. Thus in addition to physical injuries from falls or sharp equipment there is the risk of bites, scratches and even kicks from animals. The larger the animal the more severe the injury is likely to be. The larger monkeys can be particularly dangerous and a wise general rule is that they should not be handled without prior sedation.

Animals can also carry infections transmissible to man—zoonoses. Although the risks from high quality, purpose-bred animals are generally less than those from wild trapped animals, accidental infection of laboratory colonies with organisms transmissible to man can occur. In microbiological laboratories animals are frequently infected deliberately with microorganisms and thus become a continuous source of infection which may be spread by their excretions as well as by other routes. Parenthetically it should be remembered that, to their detriment, animals may also pick up infections from humans. Recently the problem of allergy to laboratory animals has become prominent and certain forms are now recognized by the Department of Health and Social Security as a prescribed industrial disease. Repeated exposure to rodent urinary protein can result in severe allergic reactions in laboratory workers. The exposure occurs frequently in the form of aerosols, which are also a serious danger in the spread of infections. Where toxic, carcinogenic or radioactive work is carried out, particular attention must be given to the safety of animal technicians whose work may subject them to prolonged exposure, or to harmful residues in waste materials. Since exposure to various hazards in the first months of pregnancy can have disastrous effects on the foetus, special attention must also be given to the protection of young women. As elsewhere in the workplace, safety with animals now constitutes an important legal requirement under the provisions of the Health and Safety at Work Act 1974. Safety in the animal house is fully reviewed by Seamer and Wood.[7]

4.10.10 ANIMALS (SCIENTIFIC PROCEDURES) ACT 1986[8]

This is the principal legislation controlling animal experimentation in Britain. Under its provisions any interference carried out on a vertebrate animal for scientific purposes which is likely to cause pain, suffering, distress or lasting harm is subject to control. The breeding and supply of animals for scientific purposes is also controlled. With certain exceptions scientific procedures on animals may only be carried out in premises designated for the purpose by the Home Office. A certificate of designation is granted to someone holding a position in senior managment and also confers legal responsibility for seeing that its conditions are met. These conditions include the specification of rooms *etc.* for particular species and the requirement that there must be a named person legally responsible for the day to day care of the animals. Similarly there must be a named veterinarian responsible for the health and welfare of the animals. Scientific procedures are controlled by a dual licensing system. Project licences are granted to project leaders to cover a particular line of work. Individual licences are given to scientists who have satisfied the Home Office of their suitability and competence. These licences specify the particular procedures that the holder may carry out. Within the overall project individual licensees are legally responsible for the welfare of their animals and for seeing that the conditions of the Act are observed. Fundamental to the Act is a principal of cost benefit so that procedures causing more than trivial pain require greater justification in terms of the likely value of the results. Distress, suffering and pain have been banded as mild, moderate and substantial. Procedures likely to cause substantial pain require special justification. The Act is administered by Home Office inspectors with medical or veterinary qualifications. Additionally the Animal Procedures Committee considers problems of animal experimentation and may advise the Home Secretary on applications for licences about which there could be particular concern. The Act is accompanied by various guidelines which describe in greater detail the scope of the various provisions. Guidelines relating to the caging and care of the various animal species, some of which have yet to be issued, are particularly important. Although at the time of writing (summer 1987) all the provisions of the Act are not fully in force, no major problems appear to have arisen with its implementation. The provisions of the Act appear to comply fully with the recent European Economic Community (EEC) Directive 86/609 on animal experiments, while broadly similar provisions are under consideration to extend the Animal Welfare Act in the USA.

ACKNOWLEDGEMENTS

We are grateful to Dr M. Wood, Chemical Defence Establishment, Porton Down for assistance with the illustrations and for permission to publish Figure 1; to Mr. E. Bernard, Harlan Olac Ltd, Bicester for supplying Figures 2, 3 and 4 and permission to publish them; and to the MAFF Central Veterinary Laboratory, Weybridge for permission to publish Figure 5.

4.10.11 REFERENCES

1. W. M. S. Russell and R. L. Burch, 'The Principles of Humane Experimental Technique', Methuen, London, 1959.
2. C. J. Green, 'Animal Anaesthesia', Laboratory Animal Handbooks 8, Laboratory Animals Ltd, Newbury, 1979.
3. D. H. Smyth, 'Alternatives to Animal Experiments', Scolar Press, London, 1978.
4. M. F. W. Festing (ed.), 'International Index of Laboratory Animals', 5th edn., Laboratory Animals Handbooks 10, Laboratory Animals Ltd, Newbury, 1987.
5. T. B. Poole (ed.), 'The UFAW Handbook on the Care and Management of Laboratory Animals', 6th edn., Longman, Harlow, 1987.
6. Command 187, 'Statistics of Experiments on Living Animals Great Britain 1986', HMSO, London, 1987.
7. J. H. Seamer and M. Wood (eds.), 'Safety in the Animal House', 2nd edn., Laboratory Animal Handbooks 5, Laboratory Animals Ltd, Newbury, 1981.
8. Animals (Scientific Procedures) Act 1986, HMSO, London, 1986.

4.11
Orphan Drugs

BERT SPILKER
Burroughs Wellcome Co., Research Triangle Park, NC, USA

4.11.1 INTRODUCTION

Orphan drugs have been defined in the USA as those drugs intended to treat either a rare disease or a more common disease where the sponsor cannot expect to make any profit. A rare disease is defined as having a prevalence of under 200 000 patients in the USA. There are various estimates of the number of rare diseases, usually in the range of 2000 to 6000. Most patients with these diseases have either no therapy or inadequate therapy. The need for treatment of many rare diseases is great, yet the incentives for companies to concentrate research in these disease areas are small.

This chapter describes classifications and legislation in this area and then discusses how orphan drugs are discovered and developed. The chapter also describes factors that influence a company when deciding on whether to develop an orphan drug and the factors that influence an academic investigator in deciding whether to study an orphan drug.

4.11.2 CLASSIFICATION

When one peruses a list of the orphan drugs available to treat rare diseases, it is apparent that they are a heterogeneous group. Previous classifications of orphan drugs were based primarily on their stage of development and divided these drugs into four general categories[1] or 18 separate detailed categories.[2] There are various alternative classifications that could be developed, and Table 1 presents a classification consisting of five categories of orphan drugs based on their commercial potential and the availability of suitable treatment. This classification is closer to the way in which regulators have defined and described orphan drugs than the two former classifications. Pharmaceutical companies consider both the stage of drug development as well as commercial and medical potential when describing or classifying orphan drugs. Therefore, all three classifications are relevant for use by the drug industry. Practicing physicians think in terms of the drug's availability for patient use, and thus the former two categories are more meaningful to them than the last system.

The major difference between type I and type III orphan drugs in Table 1 is that the latter are used for diseases that already have at least some useful therapy available. A number of diseases have gone from the type I to type III categories over the previous few decades. These diseases include many rare bacterial diseases that can be treated with common antibiotics, plus Wilson's Disease, which can now be treated with penicillamine, zinc and triethylenetetramine.

Types I and III orphan drugs are generally the most difficult ones to find sponsors for if the drugs are known to have activity but are not yet marketed. There is often a hope that a type I, II or III drug will also be found useful for treating a common disease and become a profitable type V drug. This event is not common and should not usually be factored into the decision of whether to develop an orphan drug.

4.11.3 LEGISLATION

The Orphan Drug Act was officially signed by President Reagan in January 1983. This Act amended the 1938 Federal Food, Drug, and Cosmetic Act and had numerous provisions, including the following.

(i) Establishment of an Orphan Products Board in the Department of Health and Human Services (HHS).

(ii) Tax credits of 50% are granted for clinical studies conducted after the drug is given orphan drug status by the Food and Drug Administration (FDA) and prior to approval of the New Drug Application (NDA) or NDA supplement.

(iii) A seven year period of marketing exclusivity is given to manufacturers of nonpatentable orphan drugs. This period begins when the NDA is approved.

(iv) The Secretary of the HHS is authorized to request appropriations of four million dollars per year to support clinical testing of orphan drugs by investigators who cannot utilize the tax credits.

Additional details on the political, social, economic and other aspects of the history of this legislation are well described by Asbury[3] and Waxman.[4] Product liability aspects are described by Scharf.[5]

In 1985 the United States Congress passed some revisions to the Orphan Drug Act. These revisions extended the authorization for research grants (which expired in fiscal year 1985 according to the 1983 Act) for an additional three years, expanded the market protection for manufacturers of orphan drugs by allowing seven years of marketing exclusivity for patented, as well as unpatented drugs, and established a National Commission on Orphan Diseases to evaluate the government's research on rare diseases.

4.11.4 DISCOVERY OF ORPHAN DRUGS

Most drugs are used for multiple indications. These indications may be similar to each other and lie in the same therapeutic area, or they may vary widely across several therapeutic areas. One or more of these indications may be for a rare disease.

The processes used to discover new orphan drugs are essentially the same as those used to discover nonorphan drugs. The major discovery or hypothesis was primarily made either preclinically or in a clinical setting. Preclinical activities leading to discovery of an orphan drug include the following: (i) chemical synthesis targeted to finding a drug to treat a rare disease; (ii) biological testing of already synthesized compounds in an animal model of a rare disease targeted to finding an active compound; (iii) serendipitously found biological activity in animal studies that suggests potential clinical usefulness in treating a rare disease; and (iv) a new theory is advanced suggesting that a hypothetical chemical would have activity *versus* a rare disease.

Clinical activities leading to discovery of an orphan drug include the following: (i) serendipitously found clinical activity in human patients with the rare disease; (ii) a theory which suggests that a known drug would have clinical activity *versus* a rare disease. The theory is then tested clinically and found to be correct; and (iii) clinical testing conducted in patients with a rare disease, based on animal experiments suggesting activity.

Although the processes whereby orphan and nonorphan drugs are found are the same, the frequency of which processes are most common is not the same for these two groups. There are few

Table 1 Classification of Orphan Drugs

Categorization of a drug	*Description of the drug type*	*Number of patients in the USA (prevalence)*	*Anticipated profits for a new drug*	*Presence and adequacy of available treatment*
Type I	Therapeutic orphans with little or no commercial potential	$< 200\,000$	Poor to marginal	None or highly inadequate
Type II	Therapeutic orphans with commercial potential	$< 200\,000$	Good to excellent	None or highly inadequate
Type III	Orphan drug for a rare disease that can currently be treated	$< 200\,000$	Variable	Acceptable to excellent[a]
Type IV	Unprofitable drug for a common disease	$> 200\,000$	Poor to marginal	None or highly inadequate
Type V	Orphan drug used for both a rare and common disease	Both $>$ and $< 200\,000$	Variable[b]	Variable

[a] In most patients. [b] Each use of the drug, as well as total profits would be variable.

cases where a novel chemical was originally synthesized with the hope of its eventual use in treating a rare disease. This is an important difference in how orphan and nonorphan drug uses are found. The most common source of discovering drugs with activity *versus* a rare disease is probably serendipitously found clinical activity in human patients with that specific rare disease. These patients often have a second disease for which the drug was originally given. In other situations, their physicians tested the drug because of a personal belief or theory that suggested the drug might help the patient with the rare disease. Thus, most orphan drugs are 'discovered' as novel indications for known drugs.

While some chemists in academic or government laboratories may spend a significant portion of their time and efforts synthesizing a compound for a rare disease, this is extremely unusual in the drug industry. On the other hand, once a compound has been made which demonstrates an activity that suggests the compound will have clinical utility in a rare disease, many companies will pursue its development. Section 4.11.5 discusses some of the reasoning that would be used to support this decision.

Whether or not an academic chemist would synthesize compounds targeted towards an orphan disease would depend on (i) personal interest, (ii) the scientific and social environment of the workplace, (iii) whether there were collaborators who could test the compounds in biological test systems, (iv) perceived medical need, (v) support and encouragement from managers; and (vi) financial support. There are relatively few animal models that can mimic common human diseases and also provide reliable data for extrapolating results to clinical situations. Outside the area of infectious diseases there are even fewer reliable animal models of rare diseases. This means that there may not be any reasonable animal models of a rare disease in which a new chemical could be tested. This further complicates the potential value of synthesizing new compounds intended to treat a rare disease. If human patients with a rare disease are the only reliable model, then the number of hurdles to overcome before a newly synthesized compound may be adequately tested are enormous. Only with an extremely strong theoretical argument, financial support and corporate interest could this path of drug development be considered. At the opposite end of the rare disease spectrum are those diseases for which reasonable animal models exist and for which a good opportunity for profit also exists (*e.g.* several types of cancer).

After a drug is discovered to be useful in treating a rare disease, either during preclinical evaluation of a new compound or during clinical studies, the drug must be developed. Development of an already marketed drug may mean conducting clinical studies to understand the drug's profile of activity and preparing a regulatory application to obtain official labeling and permission to promote the drug for the new use. An application requesting official orphan status must also be filed with the FDA. Development of an investigational drug usually requires thousands of activities costing many millions of dollars.

The development program for many orphan drugs is often quite different than for nonorphan drugs. For instance, the number of available patients is less, which complicates drug testing. On the other hand, the need for phase III studies, *i.e.* those conducted in patient populations for which the drug is eventually intended, after efficacy has been established, is also less, which tends to speed up the development process. These studies generate additional data on both safety and efficacy in relatively large numbers of patients under normal use conditions in both controlled and uncontrolled studies. Several considerations for developing orphan drugs have been described elsewhere.[2] Examples of orphan drugs are listed in various publications[2,3,9,18–24] and officially approved orphan drugs under the Orphan Drug Act are listed in the Federal Register.

The case histories that describe the trials and tribulations of the discovery and development of orphan drugs make excellent reading. Table 2 lists some of the published case studies of orphan drug development. These case studies support the conclusion reached earlier, that orphan drugs are generally discovered in the same ways that new indications for nonorphans are often discovered, but the hurdles faced in their development are usually enormous.

4.11.5 PHARMACEUTICAL COMPANY'S PERSPECTIVE

Many years of preclinical and clinical testing are required before it is known whether a new compound possesses activity as a drug in humans. Companies naturally are reluctant to commit significant resources to develop a new compound for an orphan disease based only on results obtained in animal models. In some situations the models may be extrapolated to humans, but this is often not the case or it is only partially true.

Table 2 Selected References to Case Studies Discussing the Discovery and Development of Orphan Drugs

Drug	*Disease or medical problem*	*Ref.*
Alkylating local anesthetics	Local anesthetic	8
Antischistosomal	Schistosomiasis	9
Benzolamide	See ref. for potential uses	10
Carnitine	Carnitine deficiency, cardiac disease	11, 12
Cysteamine	Nephropathic cystinosis	13
Dopamine	Heart failure	12, 14
L-5-hydroxytryptophan	Myoclonus	12, 15
Penicillamine	Wilson's disease	16
Triethylenetetramine (trientine)	Wilson's disease	12, 17
Sodium valproate (valproic acid)	Epilepsy	9
Zinc	Wilson's disease	18, 19, 20

Many orphan drug indications are found by alert physicians who accidentally or through a hypothesis discover a new activity of a known drug. There are some well-known examples of this. But, what if these reports are incorrect or do not apply to all or even most patients with the rare disease? Most case reports of a drug's activity in a rare disease are not confirmed, and some drugs only work in a subpopulation of diseased patients that may not be easily identified. A company must exercise caution when it is reported that one of its marketed drugs is found to have activity in a patient who has a rare disease. Also what if marginally effective therapy already exists and it is uncertain whether the new drug offers important advantages?

When evidence of the drug's efficacy is overwhelming, uncertainty still remains about how extensively a company should develop the rare disease indication. Physicians are free to prescribe any marketed drug for their patients, even if the drug is not approved for the disease being treated. There would generally be strong interest in the medical community to conduct at least a few clinical studies to better understand how the drug could be optimally used to treat patients with the rare disease. This interest may or may not be present in the company. But, even if the company's interest is low, it will often supply the drug to an investigator who desires to study a new indication, whether for a rare or common disease. A high quality publication of a drug's activity in a rare disease might be the only development work needed to bring information on the new indication to the relevant group of physicians. Each set of circumstances would have to be evaluated separately by a company.

The usual flow of activities for a sponsor to follow in developing an orphan drug is shown in Figure 1. The actual series of events followed may be quite different, especially if more than one sponsor is developing the same orphan drug. This may appear at first to be a highly unlikely occurrence because of the way in which orphan drugs are often publicized (*e.g.* financial losers, hard to locate a sponsor to develop a promising drug). Nonetheless, there are several examples where two

Company A applies for orphan status of drug X for three indications

FDA grants orphan drug status to company A for two indications and asks for more information on third indication

Company A responds with more information

FDA approves or rejects orphan status for the third indication

Company A submits NDA on drug X for three indications

Company A requests marketing exclusivity (at least 120 days prior to approval of NDA) for all indications given orphan status

Company A's NDA is approved and the seven year period of marketing exclusivity begins (company A may or may not have a patent on the drug)

Time

Figure 1 Typical sequence of events as a sponsor develops an orphan drug

or more companies were each attempting to develop or had developed the same orphan drug. For example, the FAB antibody fragment of digoxin used to treat digitalis poisoning is an orphan drug developed by both Burroughs Wellcome Co. and Boehringer Mannheim. Ganciclovir or DHPG is an antiviral drug that is active against cytomegalovirus, and orphan status for this drug was granted to both Burroughs Wellcome Co. and Syntex. Some drugs differ slightly from each other (*e.g.* natural and artificial lung surfactant, natural and recombinant α-interferons) and it is not certain whether marketing exclusivity for one would preclude marketing exclusivity being granted for another.

One illustration of the various types of complexities that may arise in obtaining orphan drug status, NDA approval and marketing exclusivity for an orphan drug is shown in Figure 2. Marketing exclusivity for seven years is usually the prize desired and is granted to the first company whose NDA is approved. Of course, the company holding the patent (if any) is in a position of relative control. A situation may arise where the company with marketing exclusivity is not the company owning the patent. Presumably, if such a situation arose, both companies would meet and work out an agreement. Other considerations of pharmaceutical companies considering orphan drug development are described in ref. 1.

4.11.6 ACADEMIC INVESTIGATOR'S PERSPECTIVE

The term academician is used to include chemists who synthesize new compounds, physiologists, pharmacologists, biochemists and other scientists who test the biological activity of compounds, and clinicians who evaluate drugs in humans.

Many of the advantages for these individuals that result from developing orphan drugs are obvious and include (i) public recognition, (ii) publications, (iii) career enhancement, (iv) ability to obtain grants and new equipment, and (v) ability to treat their patients more effectively. Other advantages include the satisfaction of participating in preclinical research or clinical studies that lead to the development of a new drug or a new use for an established drug.

The major disadvantage an academic scientist or clinician faces in developing an orphan drug relates to the question of what other work and activities are not being pursued because of the time and money spent developing the orphan drug. It is likely that in many situations, the scientist or clinician would be more productive if alternate activities are pursued. This is easily appreciated when one realizes that most orphan drugs are not medical breakthroughs for patients with the specific disease being treated. These drugs represent modest advances, just as do most new drugs designed to

Company A applies for orphan status for drug X

Company B applies for orphan status for drug X

FDA grants orphan status to companies A and B

Company C applies for orphan status for drug X

Company C files NDA

FDA grants orphan status to company C

Company B files NDA

Company B requests marketing exclusivity

Company A files NDA and requests marketing exclusivity

Company C requests marketing exclusivity

FDA approves NDA for company C

FDA grants marketing exclusivity to company C

Patent is awarded to company A *etc.*

↓ Time

Figure 2 Potential sequence of events to illustrate some of the complexities involved when multiple sponsors are developing the same orphan drug

treat common diseases. In some situations, working on orphan drugs may take scientists away from the mainstream of their work to an area where the potential for progress is limited. One of the most serious problems facing researchers who want to work on orphan drugs is the small amount of money available through traditional funding mechanisms and through grants covered by the Orphan Drug Act.

4.11.7 NONTRADITIONAL ORPHAN DRUGS AND POPULATIONS

4.11.7.1 Drugs That Wouldn't Die

The title 'drugs that wouldn't die' is taken from an article by Weintraub and Northington.[6] They describe case histories of five unprofitable drugs that were taken off the market, but later reintroduced as a result of pressure from physicians, patients and other health professionals. Each of these drugs is an orphan, either based on the small size of the disease population or because the drug is unprofitable and only benefits a small segment of patients with a common disease. The five drugs discussed are tranylcypromine (Parnate), a monoamine oxidase inhibitor used for depressed patients; mecamylamine (Inversine), a ganglionic blocking agent used as an antihypertensive and reintroduced for patients with autonomic hyperreflexia due to spinal cord injuries; methoxamine (Vasoxyl), a nearly pure α receptor agonist that is used to avoid and reverse hypotension during anesthesia and for cardiopulmonary resuscitation; methotrimeprazine (Levoprome), parenterally used as a nonnarcotic analgesic, which provides an alternative for patients in whom physical dependence, respiratory depression, and/or other adverse reactions would be a significant problem; and alphaprodine (Nisentil), a short-acting narcotic analgesic that was particularly effective in the outpatient management of certain pediatric patients in dentistry.

Interestingly, tranylcypromine was far from an orphan drug when it was first introduced. Discovery of liver toxicity plus the 'famous' cheese and wine reaction in susceptible patients, however, caused a rapid change in its status. The cheese and wine reaction is a hypertensive crisis caused by tyramine in these and other foods, which is not destroyed at the normal rate because the drug inhibits the enzyme that normally destroys tyramine. Tyramine levels therefore increase and cause elevated blood pressure responses, which occasionally cause patient deaths.

4.11.7.2 Drugs That Were Killed By Publicity

The best example of this group of drugs is the combination of doxylamine and vitamin B_6 better known as Bendectin or Debendox in different countries. The media gave great attention to court trials concerning the possible teratogenicity of this drug. Nonscientist jurors often had to make decisions on scientific questions that went beyond their understanding, and they often reached verdicts against the manufacturer that were based largely on emotion. Enormous liability settlements made it unprofitable and unwise for this drug to be continued on the market and its manufacturer removed it in 1983. These court settlements were reached despite overwhelming scientific and medical evidence from numerous careful independent reviews that it is a safe combination drug and is not teratogenic.[7]

This drug was forced off the market for incorrect reasons. Its fate could be potentially shared by many other drugs under certain circumstances. The removal of this drug has left a void in the treatment of pregnant women who experience the nausea and vomiting of pregnancy.

Another useful drug for some patients that was killed by publicity is thalidomide. The therapeutic use in question is as an antileprosy drug, where its activity is great and the need is substantial. The threat of litigation, negative media attention to even this use, and the fact that most people believe thalidomide to be an unmitigated poison because of its teratogenic effects, however, are reasons why its manufacturer is reluctant to initiate clinical studies in patients with leprosy or to even supply the drug.

4.11.7.3 Orphan Populations

There are two major groups of patients that may be considered as orphan populations, children and pregnant women. Over 80% of all drugs prescribed for children contain warnings stating that the drug should not be used in children. This is usually stated in the drug's official labeling because insufficient data exist from clinical studies to determine appropriate recommendations. A similar situation exists for pregnant women, with one important difference. Whereas children are routinely treated with nonrecommended drugs, most pregnant women are advised not to take any drug during their pregnancy, unless their need is overwhelming.

4.11.7.4 Vaccines

The dramatic increase in the number and amount of liability settlements in vaccine-related cases in recent years has caused many vaccines to become virtual orphan drugs. Vaccines are orphans in the sense of companies being unable to make money from their sale. The number of companies producing vaccines in the United States decreased from about 15 in 1984 to only three in 1986. US Congress passed the National Childhood Vaccine Injury Act in November 1986, which was designed to decrease the liability problem by establishing a federal fund to help the families of children who are injured or die as a result of vaccination. This fund will receive money from an excise tax paid for every vaccination, but will only be available to parents who waive their rights to sue the manufacturer prior to the vaccine being given. Finally, the new law seeks to encourage future vaccine development by providing a five year period of marketing exclusivity for unpatentable vaccines.

4.11.8 REFERENCES

1. B. Spilker, in 'Orphan Diseases and Orphan Drugs', ed. I. H. Scheinberg and J. M. Walshe, Manchester University Press, Manchester, 1986, p. 119.
2. B. Spilker, *Trends Pharmacol. Sci.*, 1985, **6**, 185.
3. C. Asbury, 'Orphan Drugs', Lexington Books, Heath and Co., Lexington, MA, 1985.
4. H. A. Waxman, in 'Orphan Diseases and Orphan Drugs', ed. I. H. Scheinberg and J. M. Walshe, Manchester University Press, Manchester, 1986, p. 135.
5. S. F. Scharf, *Am. J. Law Med.*, 1985, **10**, 491.
6. M. Weintraub and F. K. Northington, *J. Am. Med. Assoc.*, 1986, **255**, 2327.
7. L. J. Sheffield and R. Batagol, *Med. J. Aust.*, 1985, **143**, 143.
8. J. F. Stubbins, in 'Orphan Drugs', ed. F. E. Karch, Dekker, New York, 1982, p. 73.
9. W. Sneader, 'Drug Discovery: The Evolution of Modern Medicines', Wiley, New York, 1985, p. 46, 275.
10. T. H. Maren, in 'Orphan Drugs', ed. F. E. Karch, Dekker, New York, 1982, p. 89.
11. S. L. De Felice, in 'Orphan Drugs', ed. F. E. Karch, Dekker, New York, 1982, p. 33.
12. L. Lasagna, *Regulation*, 1979, **3** (6), 27.
13. J. G. Thoene, in 'Cooperative Approaches to Research and Development of Orphan Drugs', ed. M. H. Van Woert and E. Chung, Liss, New York, 1985, p. 157.
14. L. I. Goldberg and J. F. Zaroslinski, in 'Orphan Drugs', ed. F. E. Karch, Dekker, New York, 1982, p. 117.
15. M. H. Van Woert, in 'Orphan Drugs', ed. F. E. Karch, Dekker, New York, 1982, p. 12.
16. M. E. Jaffe, in 'Orphan Diseases and Orphan Drugs', ed. I. H. Scheinberg and J. M. Walshe, Manchester University Press, Manchester, 1986, p. 43.
17. J. M. Walshe, in 'Orphan Drugs', ed. F. E. Karch, Dekker, New York, 1982, p. 57.
18. G. J. Brewer and G. M. Hill, in 'Cooperative Approaches to Research and Development of Orphan Drugs', ed. M. H. Van Woert and E. Chung, Liss, New York, 1985, p. 143.
19. G. Schouwink, in 'Orphan Diseases and Orphan Drugs', ed. I. H. Scheinberg and J. M. Walshe, Manchester University Press, Manchester, 1986, p. 56.
20. T. U. Hoogenraad, in 'Orphan Diseases and Orphan Drugs', ed. I. H. Scheinberg and J. M. Walshe, Manchester University Press, Manchester, 1986, p. 62.
21. T. H. Althuis, in 'Orphan Drugs', ed. F. E. Karch, Dekker, New York, 1982, p. 182.
22. F. E. Karch (ed.), 'Orphan Drugs', Dekker, New York, 1982.
23. M. H. Van Woert and E. Chung (eds.), 'Cooperative Approaches to Research and Development of Orphan Drugs', Liss, New York, 1985.
24. I. H. Scheinberg and J. M. Walshe (eds.), 'Orphan Diseases and Orphan Drugs', Manchester University Press, Manchester, 1986.

4.12
Patents

JEAN-CLAUDE VIEILLEFOSSE
Roussel-Uclaf, Romainville, France

4.12.1 INTRODUCTION

4.12.1.1 Background

With the development of commerce and industry, more and more importance has become attached to questions of industrial property, but this was not always so in the past. Thus, until the 15th century, the various states were scarcely concerned with furthering the development of industry and generally speaking manufacturing secrecy remained the rule.

Perceiving that the disclosure of inventions by their authors could be beneficial to the nation, governments in the Middle Ages granted certain manufacturing privileges in a discretionary way, by 'letters patent', to persons or to corporations when they had created things which were particularly significant for the city or for the country. These privileges were generally granted to span the whole life of the person in question and without taking any account of the new or inventive character of the manufactured object.

Thus under the Republic of Venice, in 1474, The Council of the Venetian Republic decreed a statute for inventors which is in fact the first law on patents. This statute for inventors stipulates that 'whoever in this city shall make any kind whatsoever of new and ingenious device will be able to ask the municipal authorities for protection against counterfeiting'. This first law concerning patents, before the letters patent, thus fixed two criteria for patentability which will be found later in the patent laws of certain countries: the criteria of novelty and of ingenuity, which in the patent laws of the 20th century were to become the criteria for an inventive step.

In England, it was not until 1623 under the reign of James I that the discretion of the sovereign was limited by the enactment of the 'Statute of Monopolies', which prescribed the cases for which there could be a monopoly (the invention had to be novel and nondetrimental to the general interest). This law on patents establishes a time limit for the duration of the monopoly.

At much the same time, in 1641, the Commonwealth of Massachusetts, which was later to become the State of Massachusetts in the USA, adopted a law which granted a short-term monopoly to inventors for novel inventions useful to the country. These early criteria for patentability were adopted in 1790 by Thomas Jefferson, who drafted the first American law on patents. Thomas Jefferson, besides being an inventor, and later President of the USA, was also one of the first three examiners in the United States Patent Office.

In France, the abolition of the privileges which had been the reward of inventors had to wait for the outbreak of the Revolution in 1789, but as from 7th January 1791 the first law on patents was passed. This law stipulated that 'every discovery or novel invention, in every branch of industry, is the property of its author'. In return for a disclosure of the invention by its author, the state granted the latter a 'monopoly' of manufacture for a period of time which was then 15 years.

In the 17th and 18th centuries, with the development of commerce and the beginnings of industry, a certain number of states thus adopted patent laws which gave their inventors protection against the infringement of their inventions. As each country then adopted its own law on patents, the criteria for patentability were liable to vary from one country to another.

In an attempt to secure uniformity, an international convention called the Paris Convention,[1,2] signed in 1883 and ratified since then by 99 states, established some major basic rules concerning industrial property. These basic rules have been adopted in almost all laws on patents. As a result of this convention, it is in particular possible to file patents in a foreign country while enjoying the same benefits as the nationals of the states where they are filed. The Paris Convention in particular grants a period of 12 months, called 'the priority period', during which applicants for a patent can file their patents in a foreign country without their own patent application or their own disclosures being held against them. This period is for 12 months and no extension, for any reason whatever, can be allowed.

Almost every country has now adopted a patent law. Until the last few years, the People's Republic of China had no patent law. Since March 1984, a patent law has existed and since 1st April 1985, it has become possible to file patent applications in the People's Republic of China, at least in certain fields. (Pharmaceutical and chemical products as such are not patentable, only the processes for the manufacture of these products are patentable.[2,3])

It was until quite recently necessary to file patent applications in as many countries as those in which it was desired to obtain protection. However, some years ago, a few systems of 'regional patents' made their appearance. An example of a regional patent is provided by the European patent which came into being on 1st June 1978, following the signing on 5th October 1973 of an international convention called the Munich Convention.[2,4] By means of a single filing at the European Patent Office (EPO) or at a 'receiving' office, it is possible to obtain national patents in the following countries: Austria, Belgium, Switzerland, Liechtenstein, FRG, France, Great Britain, Italy, Luxembourg, The Netherlands, Sweden and more recently Spain and Greece. It is anticipated that in 1992 Portugal will be added to these countries. The Munich Convention has set up a single examination system leading to the issue of a bundle of national patents. It should be noted that the European patent also covers countries which are not members of the European Economic Community (EEC).

The creation of a community patent is also envisaged at some, as yet uncertain, future point in time. The Luxembourg Convention[2,6] to this effect, which was signed on 15th December 1975, has not yet been ratified by the member states of the EEC. This patent will be a single instrument which will take effect in all the states of the EEC.

4.12.1.2 Why Is It Necessary To File Patents?—Definition of A Patent

A research worker who has just prepared a novel product, perfected a novel preparative process or found a novel use for a known product must decide what to do with the invention.

The first step would consist of jealously keeping the invention secret, without disclosing it or describing it. But from the moment when the product which is the subject of the invention leaves the research centre where it has been prepared in order to be tested, it can be analyzed and compared with other products by competitors. If the product has some therapeutic value, it is of course going to run the risk of being copied. It is the same for apparatus or devices; as soon as they have some economic importance, they are going to be taken apart, put together again, analyzed and finally copied. In the same way, once a product is marketed its preparation cannot indefinitely be kept secret. It is, in fact, well recognized that methods of analysis are becoming so precise that it is possible to detect traces of intermediate products remaining in the final product, indeed even to detect traces of solvents which can give important clues concerning the preparative process.

An industry, such as the chemical or pharmaceutical industry, which spends a lot of money on research and development, cannot allow itself to see its research efforts rendered worthless. In the pharmaceutical industry, the research and development of a product can easily occupy seven to ten years, and even longer. The costs involved are thus considerable. This research, which in the majority of cases is self-financed, cannot run the risk, a few months after the marketing of the product, of seeing a competitor put on the market an exact copy of this product which has just been perfected after such long and costly efforts. Not to protect oneself is therefore unacceptable and positive action is required. This positive action consists of filing an application for a patent. But what exactly is meant by a patent?

It is an official instrument issued by a state or a state organization which establishes, usually after examination for form and/or content, that at a given date, a person or a company has claimed an invention as being its own. In exchange for the disclosure of the invention by its author, the state grants to the latter an exclusive right, sometimes incorrectly called a 'monopoly'. This exclusive right given by the patent is always limited in duration.

In the majority of countries, the patent bestows the right of preventing any third party, without the consent of the owner of the patent, from manufacturing, offering, selling, using, importing or stocking the product which is the subject of the invention.

The exclusive rights given by this patent do not extend, in general, to acts carried out in private and for noncommercial purposes, nor usually to experimentation with a product when the final aim is not the marketing of the said product. On the other hand, in a reasonable number of countries, experimentation with a pharmaceutical product with the aim of marketing it just after the expiry of the patent may be regarded as an act of infringement and is liable to prosecution. It is therefore always necessary to bear in mind that the patent bestows a right to prevent others (and not the right to do so oneself, as inventors usually suppose).

As a consideration for this exclusive right to prevent others from copying an invention, given to the holder of the patent or their assignee, the inventor must describe the invention as honestly and as completely as possible. This description must be given in such a way that an expert could follow it without doing research or extensively perfecting it.

If the subject of the patent is a new chemical product, the author of the patent application must clearly indicate what are the starting materials and what means are used in order to arrive at the final product. If it is a pharmaceutical product, the author of the patent application must also indicate what are the pharmacological properties of the product, the dosage employed and the therapeutic indications.

As will be discussed subsequently in Section 4.12.2.3, the exclusive right given by the patent is limited in duration and is quite liable to vary from one country to another.

4.12.2 BASIC AND FORMAL REQUIREMENTS FOR FILING A PATENT APPLICATION—DURATION OF PATENTS

To be able to file a patent application, a certain number of basic requirements have to be met; in the main industrial countries these are quite similar to each other.

As regards the formal requirements, these vary a good deal from one country to another. The duration of patents is also very different, depending on the country. This last aspect will be dealt with separately at the end of this section.

4.12.2.1 Basic Requirements

Depending on the country, there are two or three basic requirements for patentability.

In most of the main industrial countries, there are three such requirements. This is the case with the European Patent Convention (EPC)[2,4] which in its Article 52 stipulates that: 'European patents shall be granted for any inventions which are susceptible of industrial application, which are new and which invlove an inventive step'.

We shall consider these three requirements for patentability in turn.

4.12.2.1.1 The invention must be novel

This requirement for novelty is the basic requirement as regards patentability. It is to be found in the oldest laws. The redisclosure of an already known technique is in effect of no value to the economy of a country and therefore does not justify granting an exclusive right of exploitation.

Article 54 of the European Patent Convention[2,4] stipulates:

'An invention shall be considered to be new if it does not form part of the state of the art'.

'The state of the art shall be held to comprise everything made available to the public by means of a written or oral description, by use, or in any other way, before the date of filing of the European patent application'.

'The content of European patent applications as filed, of which the dates of filing are prior to the date referred to in paragraph 2 and which were published on or after that date, shall be considered as comprised in the state of the art'.

From the requirements laid down it can thus be seen that the requirement for novelty is very severe since it has an absolute character and since any prior public knowledge is destructive of novelty.[7]

In order to ascertain if there is prior knowledge, three criteria will be taken into account: (i) public disclosure; (ii) sufficient disclosure and (iii) the date of disclosure. It will thus be investigated whether at the filing date of the patent application, the public could know about the invention claimed in the patent application. At this level, it is not necessary for the public actually to know about the invention. Thus, for example, the presence of a thesis in a university library can be destructive of novelty, simply because this thesis could be consulted by third parties.

It will then be taken into consideration whether the disclosure is sufficient for an expert to be able to perform the invention without an inventive act. Thus, for example, the disclosure of just the pharmaceutical properties of a product identified by a code name could only constitute prior knowledge if it is accompanied by an indication concerning the chemical nature or structure of the said product.

Finally the date of the disclosure will be considered, which must be certain and not in doubt. In order to destroy novelty, it must of course be prior to the filing date of the patent application.

From the criteria enunciated, it will thus be seen that the place of disclosure is irrelevant; the disclosure of the invention in the country where the patent application was filed has the same effect as a disclosure in any foreign country. It also scarcely matters whether the disclosure is written or oral as long as it fulfills the three criteria set out above. In the case of an oral disclosure, it may in some cases be difficult to prove the date. It also scarcely matters about the reception of the information as long as it could have been received. It is also irrelevant whether the applicant for the patent actually knew about this disclosure of the invention. Finally, it scarcely matters whether the prior disclosure springs from the actions of the applicant for the patent or of a third party.

The concept of absolute novelty at the filing date of the patent application is now to be found in practically all patent laws, with the exception, however, of American law. Article 102 of the American law on patents (US Code Title 35—Patents),[6] which establishes the criteria of novelty in the USA states, in effect, is as follows:

'A person shall be entitled to a patent unless—

(a) the invention was known or used by others in this country, or patented or described in a printed publication in this or a foreign country, before the invention thereof by the applicant for patent, or

(b) the invention was patented or described in a printed publication in this or a foreign country or in public use or on sale in this country, more than one year prior to the date of the application for patent in the United States, or

(c) he has abandoned the invention, or

(d) the invention was first patented or caused to be patented, or was the subject of an inventor's certificate, by the applicant or his legal representatives or assigns in a foreign country prior to the date of the application for patent in this country on an application for patent or inventor's certificate filed more than twelve months before the filing of the application in the United States, or

(e) the invention was described in a patent granted on an application for patent by another filed in the United States before the invention thereof by the applicant for patent, or on an international application by another who has fulfilled the requirements of paragraphs (1), (2), and (4) of section 371(c) of this title before the invention thereof by the applicant for patent, or

(f) he did not himself invent the subject matter sought to be patented, or

(g) before the applicant's invention thereof the invention was made in this country by another who had not abandoned, suppressed, or concealed it. In determining priority of invention there shall be considered not only the respective dates of conception and reduction to practice of the invention, but also the reasonable diligence of one who was first to conceive and last to reduce to practice, from a time prior to conception by the other (Amended July 28, 1972, Public Law 92-358, sec. 2, 86 Stat. 501; November 14, 1975, Public Law 94-131, sec. 5, 89 Stat. 691.).'

Without embarking on very lengthy commentaries which would necessitate editing this Article 102,[6] it is appropriate here to make a few comments.

Paragraphs (b) and (d) of this article provide a grace period of one year, which enables an American inventor to test an invention, without such testing being regarded as undermining novelty, which can therefore be freely developed without the possibility of personal disclosures being counted against the inventor. During this grace period of one year counting from the first disclosure, the inventor can then file the patent application. It is appropriate to emphasize here that Article 102, however, has another very important aspect to it. Since practically all the countries in the world have opted for the criterion of absolute novelty at the filing date of the patent application, an American inventor, taking advantage of this possibility will no longer have the chance to file in foreign countries since there is no longer novelty at the filing date of the patent application. It is therefore necessary to warn American research workers against the great risk that they run in disclosing their invention in the USA without having previously filed patents in foreign countries.

American law actually allows one to proceed in this manner, but it is then no longer possible to file in foreign countries. A cruel demonstration of this, for its authors, was the 'Boyer' case.[8,43,44] In this matter concerning the biotechnology field the patent matter was divulged in the USA prior to the USA filing date. As this divulgement was made before the USA filing date at the time of the foreign extensions, the inventor had no possibility to file a valid patent in any country.

Article 102 of the American law[6] refers moreover to the invention and not to the filing date of the patent application; this Article thus created the so-called 'first inventor' system as contrasted with the so-called 'first depositor' system used in almost all other countries (with the exception of Canada).

Without going into details, the previously quoted paragraph (g) of Article 102 of the US law lays down a ruling on the question of 'double patenting' quite different from that of other countries (in contrast, compare the wording of this paragraph with the third indented line of Article 54 of the European Patent Convention).[4] According to this paragraph (g) of Article 102,[6] in the case of 'double patenting', the patent will be granted to the inventor who not merely has the earliest dates for conception and for putting the invention into practice, but who also has displayed diligence in putting the invention to use. Thus an inventor who provides proof to be the first to have conceived the invention still risks losing the process of 'interference' which will involve the inventor if it is shown that diligence has not been displayed in putting the invention into practice. A few diagrams to illustrate the complexity of the problem are as follows:

```
Inventor

A        C———R

B                          C———R
```

Diagram 1

In this hypothesis, the inventor A who was the first to conceive the invention (C = 'conception date') and the first to achieve 'reduction to practice' (R) wins the interference. In this diagram, it does not matter whether the inventor has displayed diligence or not.

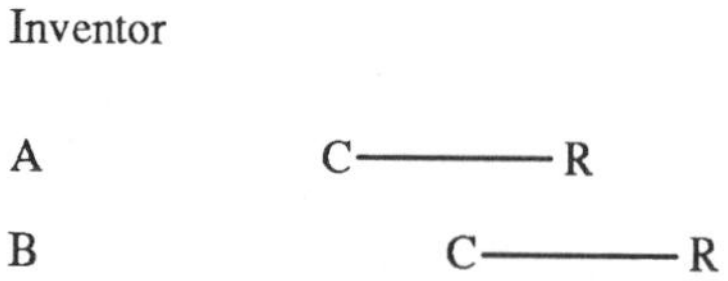

Inventor

A C——R

B C——R

Diagram 2

In this hypothesis, as in the previous diagram, inventor A also wins the interference.

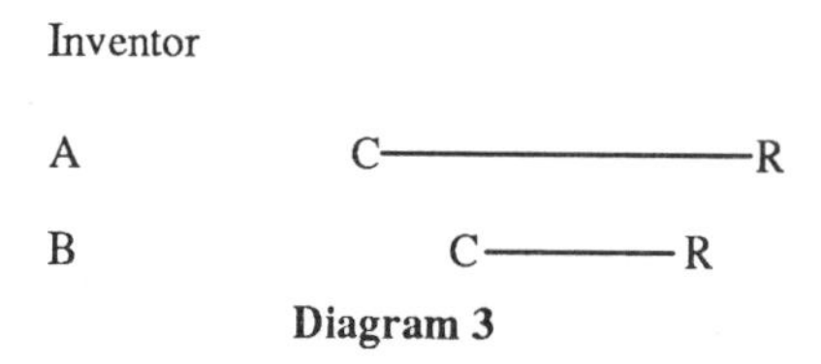

Diagram 3

In this hypothesis, inventor A is the first to conceive the invention but the last to achieve 'reduction to practice'. In this situation inventor A will have to show diligence during the period falling between B's conception date and A's own 'reduction to practice'. Only in this case will inventor A be able to win the interference.

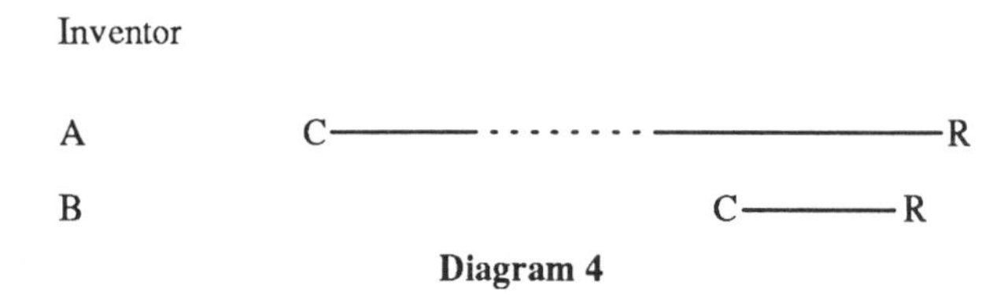

Diagram 4

In this hypothesis, inventor A was the first to conceive the invention, but the last to achieve 'reduction to practice'. Shortly after its conception, inventor A did not display diligence, but started to display diligence again just before the conception date of B. In this case, even so, inventor A will win the interference.

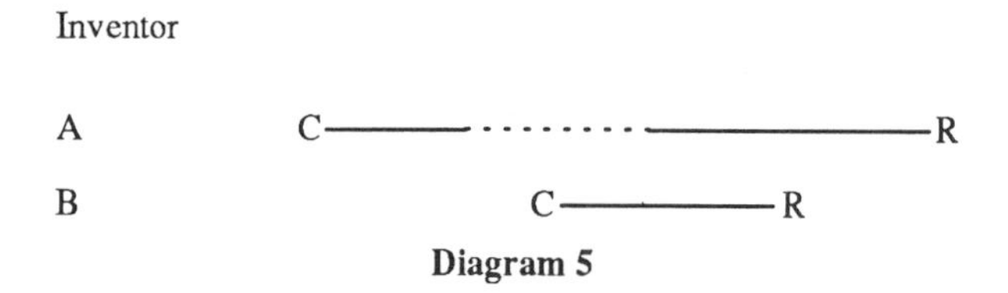

Diagram 5

In this hypothesis inventor A, who did not display diligence from the moment of conception of the invention by inventor B, will lose the interference.

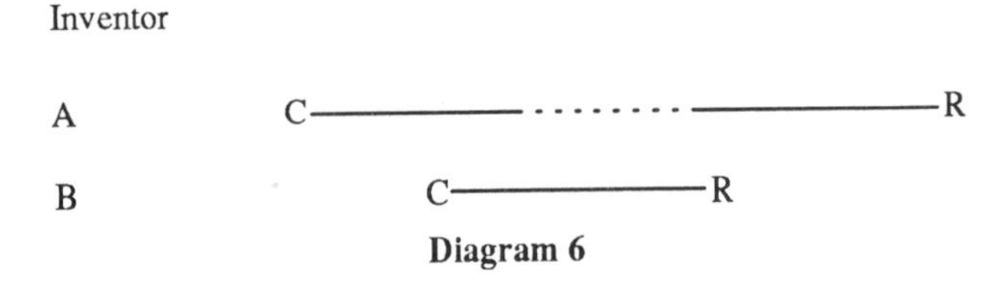

Diagram 6

In this hypothesis the interference will be lost by inventor A also, who for a certain period of time ceased to display diligence. This diagram therefore gives an outcome contrary to that of diagram number 3.

It is necessary to remark that it often happens that interferences arise between a greater number of inventors. In this situation, the outcome then becomes very complex.[9]

In 1931, an American patent attorney thought up an interference between three parties known as the 'three-party paradox', which was totally insoluble. This interference involving three inventors in interference two by two can be schematized as follows:

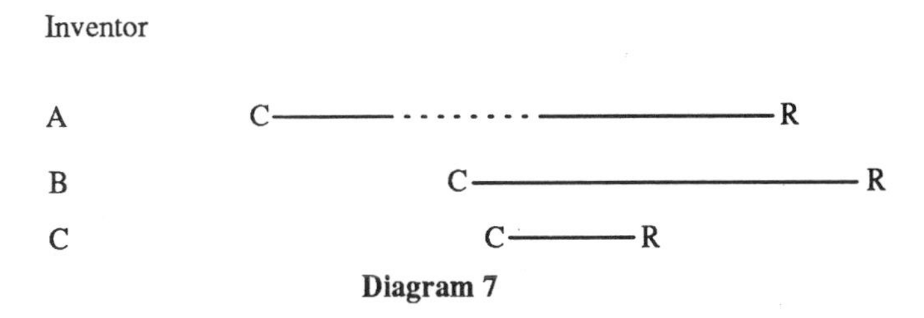

Diagram 7

In the interference between A and B, A, who was the first to conceive and to achieve 'reduction to practice', wins the interference (*cf.* diagram number 2). (In this case, the absence of diligence by inventor A does not matter.)

In the interference between B and C, B wins (*cf.* diagram number 3).

In the interference between C and A, C wins (*cf.* diagram number 5).

In such an interference, each party would be winning and losing at the same time. This case is offered as an anecdote to show the complexity of the problem.

As can be seen, the interference procedures before the 'Board of Patent Appeals and Interferences' are quite complex, and are the business of specialists. In a 'preliminary statement' each inventor must for each of the claims in issue (counts of inteference) indicate the dates of conception and the dates of putting into practice, and prove to have been diligent. As far as the filing of a patent application based on a patent application filed in a foreign country in the United States is concerned, it is not possible to go back beyond the earliest priority claimed at the time of filing; also, for foreign applicants, it is very strongly recommended to send all the research plans or documents to an American patent attorney, which would thus make it possible to establish a date of conception or a date of reduction to practice, earlier than that of the priority application filing.[10]

To conclude this discussion of novelty, it should be mentioned here that a country such as the FRG, which in its old law used to have a grace period of six months (and not of 12 months, as in the USA) has finally opted for the principle of 'absolute novelty'. Only Japan has retained this possibility in a very few cases. This question of a grace period is at present the subject of great controversy in the profession.

Finally, Canada, which used to have a so-called 'first inventor' system quite similar to that in the USA, is in the process of modifying its law on patents in order to adopt the so-called 'first depositor' system.[42]

4.12.2.1.2 The invention must involve an inventive step

From the inception of the first patent laws, the criterion of mere novelty had been regarded as insufficient; that is why it very quickly emerged that in order for there to be a patent, there should be something more than simple novelty in the strict sense.

The FRG therefore demanded that for a patent to be granted, it had to have a certain 'inventive height' or a certain 'technical advance'. For its part, the American law on patents speaks of nonobviousness.

For the European patent, the term of 'inventive step' was retained. This requirement is established by Article 56 of the European Patent Convention (EPC)[4] which stipulates:

'An invention is considered as involving an inventive step if, having regard to the state of the art, it is not obvious to a person skilled in the art. If the state of the art also includes documents within the meaning of Article 54, paragraph 3, these documents are not to be considered in deciding whether there has been an inventive step'.

From the requirements specified it will thus be seen that this Article refers to a 'person skilled in the art'. This question of a 'person skilled in the art' has been the subject of many commentaries regarding whom to choose. If in fact this person skilled in the art is taken from amongst lower-grade technicians, the invention which is assessed may be perceived as a great inventive step. If on the other hand the person skilled in the art is taken from amongst highly qualified experts in a particular field, the same invention may appear to be absolutely obvious. Where patents are concerned, and more particularly when the inventive step has to be judged, a person skilled in the art is defined as a good technician in the particular field. The entire subjective aspect of the assessment is thus sought to be eliminated.

There is also the question of knowing when exactly this assessment should be made; it should be made at the time of the filing of the patent application. In the case of conflict, the judge who has to determine whether or not there was an inventive step must, several years after the filing of the patent application, try the case according to the state of technical knowledge at the time when the patent was filed.[11–19]

It is finally necessary to note that, when taking inventive step into account, the second part of Article 56 of the European Patent Convention (EPC)[4] excludes from the state of the art European patent applications which were not published at the filing date of the patent application under consideration. It cannot really be expected that an invention should display an inventive step as compared with another invention which was totally unknown at the filing date of the application being considered. On the other hand the contents of European patent applications still unpublished at the filing date are retained as regards novelty, in order to avoid granting two patents for the same invention (double patenting).

In any given field, the state of the art is sometimes compared to a nucleus surrounded by a nebula. It is assumed that everything to be found in the nucleus is known, and that the knowledge contained in the nebula which surrounds the nucleus follows, in an obvious way, from the state of the art. Therefore it does not involve an inventive step. Only the knowledge which is to be discovered outside the nebula, and further and further away, can in this example involve an inventive step.

From a practical point of view, the substitution of a trifluoromethyl group on quinoline for a chlorine atom will be regarded as novel. (The trifluoromethyl radical is, in fact, different from a chlorine atom—the requirement for novelty is thus well fulfilled.) However, in a certain number of chemical series, the chlorine atom is quite often thought to be equivalent to a trifluoromethyl radical. There will in this case only be an inventive step if the pharmacological properties differ either as regards their nature, or as regards their level of activity.

In the same way, the preparative process for a known product can only be regarded as involving an inventive step if it leads to a yield which is clearly greater with a smaller number of stages, or if, for example, it enables a dangerous reaction to be avoided.

In the American law on patents, the criterion of inventive step is replaced by the criterion of 'nonobviousness'. Article 103 of the American law on patents (USA Code Title 35—Patents)[6] defines nonobviousness as follows:

'a patent may not be obtained though the invention is not identically disclosed or described as set forth in Section 102 of this title, if the differences between the subject matter sought to be patented and the prior art are such that the subject matter as a whole would have been obvious at the time the invention was made to a person having ordinary skill in the art to which said subject matter pertains.

Patentability shall not be negatived by the manner in which the invention was made'.

It is thus seen that in the American law on patents, the requirement of nonobviousness is a requirement of patentability which is added to the requirement of novelty. It is not sufficient for the subject of the invention to be novel, it is also necessary for it not to be obvious to a person having average competence in the art. As in the EPC, the American law refers to a person skilled in the art. This ideal person skilled in the art taken as a reference must have normal competence.

In order to assess 'the inventive step' in European law, or 'the nonobviousness' in American law, one can try to answer some questions which are equally indications of inventive step or of non-obviousness.

(i) For how long has the need been felt, to which the invention has responded?

(ii) Had there been many attempts in the prior art before reaching the solution given by the invention?

(iii) Was the invention immediately recognized as giving the solution to the problem posed? Has there been industrial and commercial success?

(iv) What are the unexpected results produced by the invention?

(v) Are there any prejudices in the prior art to overcome which could discourage the inventor from carrying out research in the field of the invention?

The answer to these various questions can enable the inventive step or nonobviousness to be assessed as objectively as possible.

4.12.2.1.3 The invention must be capable of industrial application

The laws of almost every country require that the invention should be capable of industrial application. Article 57 of the EPC[4] defines industrial application as follows:

'An invention is considered as susceptible of industrial application if its subject can be manufactured or used in any type of industry, including agriculture'.

The subject of the invention must therefore be able to be manufactured or used in some kind of industry. It is necessary to note here that the word 'industrial' should be taken in its broadest sense.

Article 52 of the EPC[4], which defines patentable inventions, uses the condition of industrial application to reject a category of invention which particularly interests us: methods of surgical or therapeutic treatment of the human or animal body.

Paragraph 4 of Article 52 of the EPC[4] reads as follows:

'Methods for treatment of the human or animal body by surgery or therapy and diagnostic methods practised on the human or animal body shall not be regarded as inventions which are susceptible of industrial application within the meaning of paragraph 1. This provision shall not apply to products, in particular substances or compositions, for use in any of these methods'.

We will return later to this question in Section 4.12.3.

In the American patent law, the criterion of 'industrial application' is replaced by the criterion of 'utility'. Article 101 of this law[6] defines patentable inventions:

'Whoever invents or discovers any new and *useful* process, machine, manufacture or composition of matter, or any new and useful improvement thereof, may obtain a patent thereof, subject to the conditions and requirements of this title'.

The criterion of utility which is found in this American law is quite close to the criterion of industrial application which is found in the laws of a good number of countries and in particular in the EPC.

As an example, it can be said that a chemical product for which no industrial application nor any pharmacological property had been found would not be patentable in terms of the EPC, because it would not have an industrial application.

Under American law, this product also would not be patentable, because it would not have any utility.

In chemistry or pharmacy, apart from the previously discussed point concerning the exclusion of methods of surgical or therapeutic treatment, this requirement for industrial application does not in general pose any particular problem.

4.12.2.2 Formal Requirements

4.12.2.2.1 The patent application must have a single subject

Mainly for reasons of documentary research, the patent application may relate to only one invention, or to a number of inventions linked together in such a way that they constitute only a single overall inventive concept.

In the countries under consideration, and especially in the USA, this requirement for a single subject per patent application is investigated with the greatest strictness. It can happen therefore that some patent offices may consider that a patent application contains 10, 20 or indeed 30 different inventions (say a patent application for a series of chemical products containing various heterocycles and various heteroatoms). In such a case, under penalty of rejection of the patent application, the applicant is given a certain amount of time to divide up the application.

4.12.2.2.2 The patent application must contain a description explaining the invention

In its preamble, the patent application must in general include the title of the invention and indicate the technical field to which the invention relates. So far as possible, the preamble of the patent application must also indicate the state of the prior art as far as this can be useful in understanding the invention.

The patent application must then explain the invention in a sufficiently clear and complete way for a person skilled in the art to be able to carry it out. Just reading the patent application must in fact enable a person skilled in the art to reproduce the invention without having to carry out research or lengthy and delicate operations.

A recent decision of the European Patent Office (Chamber of Technical Appeals 3.3.1.)[20] said on this subject that a document of the prior art 'would not constitute a sufficient disclosure of a chemical compound, even if it specifies its structure and the stages of the process by which it is produced, if the person skilled in the art is not in a position to find out, by basing himself on this document or by making use of his basic general knowledge, how to obtain the desired starting products or intermediate products. Information which could be obtained only by a very extensive research cannot be considered as being part of basic general knowledge'. The same applies to patents. The person skilled in the art must be able to reproduce the invention just by reading the patent application armed with the basic knowledge pertaining to the invention.

As regards patent applications filed in the USA, these must in addition give the best method of realizing the invention.

In the countries under consideration, it can also be useful to show comparative results between the products of the application and the products of the prior art, in order to demonstrate the advantages which the invention provides.

4.12.2.2.3 The patent application must contain claims

The claims define the subject of the protection demanded. They must be clear and concise and based on the description.

With the exception of 'provisional specifications' filed in Great Britain, which are in fact provisory patent applications and which are not allowed to contain claims, patent applications filed in all other countries must contain claims at the time of filing.

Generally speaking, the claims must have a preamble containing the designation of the subject of the invention and the technical characteristics which are necessary to the definition of the elements claimed but which, combined, are part of the state of the art. In a second part, which generally is preceded by the expressions 'characterized in' or 'characterized by', the claims must explain clearly the technical characteristics for which a protection is sought.

As will be seen in Section 4.12.3, the claims can be of several types. Depending on the country, the claims can cover products, processes, applications or compositions.

Every claim giving the essential characteristics of the invention can be followed by one or more subclaims concerning particular methods of performing the invention.

The claims must be of a reasonable number, having regard to the nature of the invention. From time to time patent applications are found containing more than a hundred claims.

Countries can to a certain extent manage to restrict the number of claims by instituting a tax on the number of claims.

Except in the last resort, the claims as regards their characteristic part must be based on the description, completed if appropriate with drawings. The claims cannot in any case extend more widely than the description, completed if appropriate with drawings.

4.12.2.2.4 The patent application must be filed in the language of the country in question

With exceptions, patent applications must always be filed in the language or one of the official languages of the country in which filing is contemplated.

Certain countries may, usually within the framework of reciprocity agreements between states, grant extensions for filing a translation into the language of the country concerned. Such agreements exist in particular between France and the USSR. These agreements usually give an extension of two months (or of 14 months from the priority date) for filing translations.

As regards European patent applications, these must be filed in one of the three official languages of the European Patent Office: German, English and French. Nevertheless, natural and moral persons having their domicile or headquarters in the territory of a contracting state with a language other than German, English or French as its official language, as well as nationals of such a state having their domicile abroad, can file European patent applications in an official language of such a state. In this case, a translation into one of the three languages of the European Office must be submitted within a period of three months, counting from the filing of the European patent application and in any case before the expiry of a period of 13 months from the priority date.

4.12.2.2.5 The patent application must designate the true inventors

This question will be examined in Section 4.12.4.

4.12.2.2.6 The patent application gives rise to payment of filing fees, examination fees and to payment of annuities

The cost of filing fees is very liable to variation from one country to another. The fees must generally be paid on the day of filing. Some states allow a period of one month for this payment to be effected. This is notably the case for patent applications filed at the European Patent Office. If payment is not made within the month following the filing of the patent application, the latter is regarded as nonexistent.

As an illustration, the filing of a European patent application designating all of the 12 states which it is now possible to designate, costs about US$ 3270 just in official fees ($ 320 filing fee, $ 1030 search fee, $ 160 designation fee per state). It is necessary where appropriate to add to this amount a fee of $ 40 per claim starting with the eleventh one, and in addition the fees of the European agent if one has been consulted to represent the inventor. At the beginning of the examination procedure before the European Office, it will be necessary to pay an examination fee ($ 1200), then at the time of issue a printing fee ($ 30 per page) and an issue fee ($ 220).[21] The agents fees are not included in these amounts. Finally, at the time of issue of the European patent, it will be necessary to pay for the translations into the languages of the various countries designated, and afterwards to pay for the annuities until the expiry of the patent.

On average, the filing of a patent application of about 20 pages costs of the order of $ 1400 per country. This amount in fact depends on the work that has to be carried out in order to proceed with the filing of the patent application (modification of text according to the requirements of the country and higher cost of translation into certain less common languages). For filings effected in Eastern bloc countries as well as in the USA, one should anticipate costs of the order of $ 1800 to $ 2000 per country.[2]

In the countries under consideration, in addition to payment of examination fees, one should anticipate fees for answering official letters. These amounts, which can be between $ 100 and $ 2000 for replying to one official letter, depend on the problems to be resolved.

Finally, in the majority of countries it is necessary over the whole life of the patent to pay annual or periodical fees. Until the last few years, the USA and Canada were exceptional in not requiring annuities. Since the 96-517 law of 12th December 1980, all American patents filed after this date are subject to payment of a renewal fee three and a half years, seven and a half years and eleven and a half years after the issue of the patent (of $245, $495 and $740 respectively). For American patents filed after 27th August 1982, these renewal fees have been set at $490, $990 and $1480 respectively. It should be noted that the amount of the renewal fee depends on the filing date of the patent application and not on the date of issue. A reduction of 50% in the amount of these fees is granted to small entities (independent inventors).[22,23]

As far as Canada is concerned, the government bill so far as known at present also envisages a system of annuities, the principles of which have not yet been decided.[10]

Table 1 gives an indication of the cost of maintaining a patent in force per country, up to the normal expiry of this patent.

4.12.2.3 Lifetime Of Patents—Patent Term Extension

At the start of the 1960s, the lifetime of patents varied enormously from one country to another. Thus, certain countries, like the USA and Canada, used to have (and still have) a lifetime of 17 years from issue. Countries like Great Britain, the Republic of Ireland and South Africa, as well as a

Table 1 Cost of Maintaining a Patent in Force up to the Normal Expiry Date

Country	*Cost ($ US)*	*Country*	*Cost ($ US)*
German Democratic Republic	23 000	USSR	5000
Rumania	17 000	UK	4800
Federal Republic of Germany	12 000	Japan	4300
Hungary	11 000	France	4100
The Netherlands	10 000	USA	3000
Austria	9100	South Africa	1000
Denmark	5500	Luxembourg	1000
Algeria	5000	Morocco	400

certain number of Commonwealth countries, used to have a lifetime of 16 years from the filing date. The FRG and GDR used to have a lifetime of 18 years from the filing date. Austria had a lifetime of 18 years from the publication date, but not exceeding a ceiling of 20 years from the filing date. Countries such as France had a lifetime of 20 years from the filing date. Japan had a lifetime of 15 years from the publication date for opposition by third parties, but this could not exceed 20 years from the filing date. Spain had a lifetime of 20 years from the date of issue. Some Eastern bloc countries such as Czechoslovakia, Rumania and the USSR had a lifetime of 15 years from the filing date.

Finally, countries such as The Netherlands had a system of calculation which was so complex that a foreigner could not with certainty determine the expiry of the patent without the risk of making a mistake.

With the introduction of the European patent, the countries ratifying this convention opted for a lifetime of 20 years from the filing date. Since then, some countries not participating in the EPC have also opted for this lifetime of 20 years. Table 2 shows however that most countries still have not adopted this lifetime of 20 years from the filing date.

It should, finally, be recognized that a few countries allow, in certain circumstances, the extension of the 'normal' lifetime of certain patents, particularly in the case of medicaments (Australia and South Africa). In these countries, the patentee is allowed to ask for an extension only when all the benefits that were expected from the patent have not been obtained. Usually the patentee blames administrative delays in the registration of the medicament. These procedures require preparation of voluminous files and can be subject to opposition by third parties.

The USA has recently adopted a law which enables the lifetime of certain American patents to be extended (law 98-417 of 24th September 1984). This law entitled 'Drug Price Competition and Patent Term Restoration Act',[24] linked with the names of senators Waxman and Hatch, bestows the possibility of obtaining a maximum extension of five years to holders of pharmaceutical patents when the marketing of a pharmaceutical product has been delayed in its registration by the FDA.

Patents affected by this law are product patents covering 'Human Drug Products' within the meaning of the Federal Food Drug and Cosmetic Act, as well as patents for processes and for methods of treatment embracing these products. The text of the law states that by 'Human Drug Products', is meant: 'The active ingredient of a new drug, antibiotic drug, or human biological product including any salt or ester of the active ingredient, as a single entity or in combination with another active ingredient'.

Also affected by this law are patents covering medical equipment, food additives or colorings subjected to examination with a view to registration by the FDA.

The extension which can be granted is a maximum of five years and takes into account the nature of the tests required by the FDA.

In order to work out the extension T that it is possible to obtain, the following formula should be applied:

$$T = (t_1/2) + t_2 - \text{(nondiligence)}$$

t_1 corresponds to the confirmatory stage, that is to say the period between the filing of the IND and the filing of the NDA; t_2 corresponds to the approval stage, that is to say to the period between the filing and the granting of the NDA.

As an example, let us take the case of an American patent which should have been issued on 31st August, 1977 (Table 3).

In this example, we assume that the applicant was diligent at all the stages up to the marketing of the product.

Before investigating whether an extension could be obtained, it should be noted that no extension is granted when the holder of the patent has a period of protection from the patent of 14 years or more. At the time when the product is marketed, the extension which can be granted added to the normal remaining life of the patent cannot confer on this 'extended patent' a remaining duration greater than 14 years.

Among the conditions which must be fulfilled in order to be able to ask for an extension it should be noted that: (i) the patent must not have expired at the time of filing the request for extension; (ii) the patent must never have been extended; (iii) the product must have been subjected to examination by the FDA; (iv) the request for extension must have been filed within 60 days of the granting of the NDA; (v) the request for extension must have been filed, at the latest, during the last three months of the life of the patent; and (vi) a fee of $750 must be paid.

To offset this possible extension of the lifetime of patents, the law, however, confers a certain number of advantages for the marketing of generic products. It is thus possible to file Abbreviated

Table 2 Lifetime of Patents in Europe, North America, Central and South America, Asia/Oceania and Africa

Country	*Years of validity from the date of:*			*Remarks*
	Filing	*Publication*	*Grant*	
Europe				
European patent designating:				
Austria	20			
Belgium	20			
FRG	20			
France	20			
Great Britain	20			
Italy	20			
Luxembourg	20			
The Netherlands	20			
Sweden	20			
Switzerland	20			
Spain	20			From 1.10.87 for Spain and Greece
Greece	20			
Austrian national patent		18		Cannot exceed 20 years from filing
Denmark	20			
Hungary	20			
Norway	20			
Portugal			15	20 years from filing after 1.10.92
Poland	15			
Rumania	15			
GDR	18			
Finland	20			
Turkey	15			
Yugoslavia		7		
North America				
USA			17	
Canada			17	From 1.1.89, 20 years from filing[42]
Central/South America				
Mexico			10	From 16.1.87, 14 years from grant
Honduras			20	
Costa Rica			20	
Panama			20	
Brazil	15			
Argentina	15			
Peru			10	
Paraguay	15			
Bolivia			15	
Colombia			5	Renewable once if exploited
Venezuela			10	
Uruguay			15	
Asia/Oceania				
Japan		15		15 years from publication but cannot exceed 20 years from filing
South Korea		15		From 1.7.87
People's Republic of China	15			
India	14			7 years for medicaments
Pakistan	15			
Taiwan		15		Not exceeding 18 years from filing
Australia	16			
New Zealand	16			
Africa				
Algeria	20			
Morocco	20			
Tunisia	20			
Egypt	15			10 years for medicaments
OAPI[a]	10			Renewable once 20 years
South Africa	20			From 1.1.79
ARIPO[b]	20			

[a] The African Intellectual Property Organization (OAPI) is an African regional patent-granting organization. The OAPI came into force on February 8, 1982.[36,37] The 13 contracting states are the following: Benin, Burkina Faso, Cameroon, Central African Republic, Congo, Gabon, Ivory Coast, Mali, Mauritania, Niger, Senegal, Togo and Tchad. [b] The African Regional Industrial Property Organization (ARIPO—formerly ESARIPO) is also an African regional patent-granting organization. The ARIPO came into force on April 25, 1984.[38,39] The nine contracting states are the following: Botswana, The Gambia, Ghana, Kenya, Malawi, Sudan, Uganda, Zambia and Zimbabwe.

Table 3

Filing of IND	1 January 1987	Confirmatory stage $t_1 = 15$ months
Filing of the NDA	1 April 1988	Approval stage $t_2 = 12$ months
Granting of the NDA	1 April 1989	
'Normal' end of US patent (before extension)	30 August 1994	
'Possible' end of US patent (after extension)		
$T = 15$ months/2 + 12 months – (nondiligence) = 19.5 months		
29 August 1994 + 19.5 months → 15 April 1996		

New Drug Applications (ANDA), which exempt manufacturers of generic products from making a certain number of clinical tests when the product has already been marketed. It is then not necessary to show the therapeutic activity and the harmlessness of the product when its therapeutic indications and its routes of administration are the same as those of the product already marketed.

Finally as regards this law, it should be pointed out that it makes provision for a minimum period between the granting of an NDA and of an ANDA.

Since the introduction of this law[24] in the USA, discussions have taken place in a certain number of countries in order to try to adopt similar measures. Leaving aside the beneficial effect that the extension of a patent can have for its holder, such measures have the disadvantage of giving a specialized character to patents for medicaments, an effect which practitioners have tried to suppress for many decades.

4.12.3 PATENTABLE INVENTIONS

Broadly speaking, inventions can be divided into four main categories, which will be examined in more detail subsequently, as follows: (i) inventions concerning products; (ii) inventions concerning processes; (iii) inventions concerning the use of products already known for other uses (primary therapeutic use and the case of inventions concerning secondary therapeutic uses); and (iv) inventions concerning compositions or combinations.

Depending on the country, it may be possible to obtain protection for each of these categories of invention. At the end of the present section, a table will provide a resume of the present situation country by country (Table 4).

4.12.3.1 Inventions Concerning Products

A product is something which has a defined constitution and which produces a determined effect.

In chemistry, it is a substance or a family of substances having a particular structure. The substance or family of substances must have a definite structure.

One cannot by means of a patent protect a substance having a hypothetical structure, but one can protect a substance which would have quite a short life.

The product or family of products claimed can be defined equally well by a general chemical formula as by specific chemical names.

A general claim covering a family of products, a subclaim covering a subfamily of preferred products and a claim directed to a single product are given as examples.

Claim 1: A 1,5-benzodiazepine-2,4-dione of the formula (**1**)

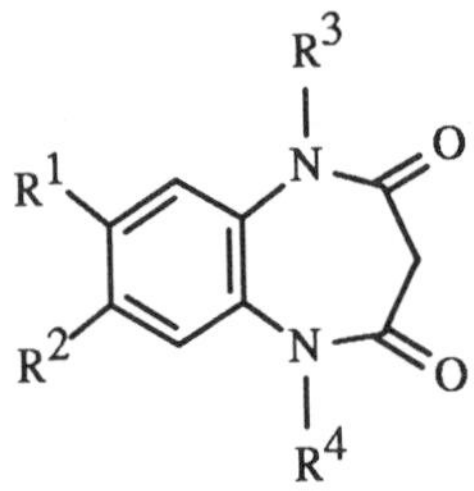

R^3 N O R^1 R^2 N O R^4

(**1**)

wherein: R^1 and R^2 are selected from the group consisting of hydrogen, hydroxy, methoxy, halogen and carbalkoxy wherein the alkyl group has from one to four carbon atoms; R^3 is selected from the group consisting of hydrogen, alkyl, alkenyl, aralkyl, cycloalkyl, (cycloalkyl) alkyl; and R^4 is selected from the group consisting of methyl and aryl.

Claim 2: A 1,5-benzodiazepine-2,4-dione as claimed in claim 1, wherein: R^1 is hydrogen; R^2 is selected from the group consisting of hydrogen, hydroxy, methoxy, halogen and carbalkoxy wherein the alkyl group has from one to four carbon atoms; R^3 is selected from the group consisting of hydrogen, aralkyl and cycloalkyl; and R^4 is phenyl.

Claim 3: 1-phenyl-5-methyl-8-chloro-1,2,4,5-tetrahydro-2,4-dioxo-3*H*-1,5-benzodiazepine.

The criteria for patentability discussed above apply to product claims. In comparison with the prior art, the product must therefore be novel, show an inventive step (or not be obvious) and finally display an industrial character.

As regards their effect, in the countries where they are permitted, product claims bestow a protection that can be regarded as 'absolute'. This protection prevents any third party from manufacturing the product thus protected or from using it, introducing it upon the market, or offering it for sale in the territory covered by the patent without the consent of the holder of the patent.

This protection bestowed by a product patent thus prevents a third party from using the claimed product for purposes which might not have been described or claimed by the holder of the patent.

A product patent also prevents a third party from using a novel preparative process for this product, perfected by them and which has not previously been described, without the agreement of the holder of the patent.

As regards inventions concerning products, it is appropriate to say a few words about so-called 'selection' inventions. This kind of invention has made its appearance in chemistry, where it sometimes happens that products may emerge within a huge general formula which show quite surprising properties.

In chemistry, by the permutation of various substituents, it often happens that chemical formulae embrace thousands, indeed tens of thousands of products. In such cases, it is quite obvious that not all the products have been prepared or tested.

A selection invention therefore lies in the act of isolating, from amongst a vast family of products, some subfamily of products or individual products which have not been described by name in the prior art, but which show very marked superior pharmacological properties. Products which are the subject of a selection invention may also be distinguished from the products of the prior art by the fact that the claimed subfamily does not show harmful side effects.[25,26]

A selection patent will be subservient to any general patent covering the products.

4.12.3.2 Inventions Concerning Processes

By a process invention is meant a system of chemical agents or mechanical parts which, when put into operation, leads to the obtention of a tangible object called a 'product' or an intangible effect called a 'result'.

In chemistry, a process patent thus describes a more or less prolonged series of reactions leading to the obtention of a given chemical product. When the process leads to a novel product 'X', one simply speaks of a preparative process for the product 'X'. When the process leads to a previously described product 'X', one speaks of a 'novel preparative process for the product 'X'.

As an illustration a general claim for the preparative process for a product, and a subclaim concerning the same process is given below.

Claim 1: A process for the preparation of a compound of the general formula (**2**), wherein: R^1 and R^2, which may be the same or different, are each a hydrogen or halogen atom, a hydroxy or methoxy group or a carboxyl group esterified with an alcohol having one to three carbon atoms; and R^3 is a methyl or aryl radical, in which a corresponding *N*-carbalkoxyacetyl-*o*-phenylenediamine, of the general formula (**3**), wherein R^1, R^2 and R^3 are as defined hereinbefore, and R' represents an alkyl radical having from one to four carbon atoms, is cyclized with acid to give a 1,5-benzodiazepine-2,4-dione having the formula given in (**4**).

Claim 2: A process as claimed in claim 1, in which the cyclization is effected by the action of a strong acid.

The criteria for patentability discussed above in Section 4.12.2.1 apply of course to inventions concerning processes. Even if it is easy enough to show that a process is novel in relation to a

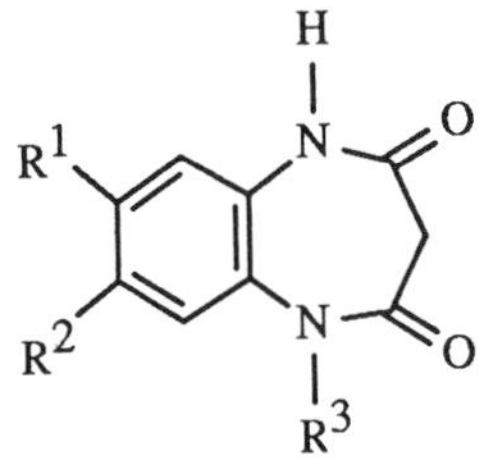

(2)

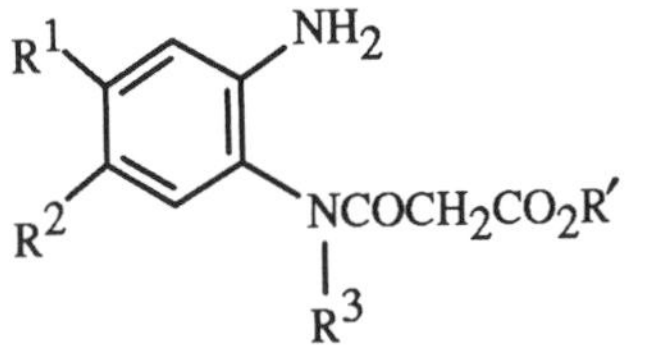

(3)

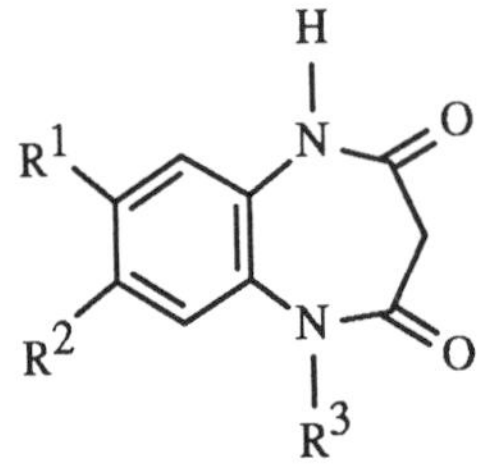

(4)

previously described process, it is on the other hand appreciably more difficult to demonstrate that this process displays inventive step as compared with a previously described process; a fair number of chemical reactions are in fact performed by analogy with reactions previously described with reference to other products.[27] To provide the necessary demonstration, one may look for anything which tends to show that it was not obvious to operate according to the claimed process. One may also demonstrate that the new process improves the yield to a significant extent.

As regards its effect, a process claim, in contrast to a product claim, gives only partial protection. In a country allowing only claims for processes, the process claim only has the effect of preventing a third party from using the same process. Such a process claim is ineffective against a third party who in that country might use a totally different process to make the same product.

It should therefore be borne in mind that in such countries, with only process patents, the protection given by the patent is only partial.

4.12.3.3 Inventions Concerning Novel Uses For Known Products

In the field of chemistry, it often happens that new uses are found for known products. Thus, for example, DDT was a known chemical product when its insecticidal properties were discovered. In the same way, sulfonamides were known chemical products and were used as dyestuffs when their antibiotic properties were discovered.

Generally speaking, it can be said that the use of a known product or a known process, so as to obtain effects which had not till then been thought about, can in many countries be the subject of an application for a use patent. When, however, the question is one of the pharmaceutical use of a known product, many countries have erected obstacles which prevent coverage of pharmaceutical uses.

In countries like the USA, the pharmaceutical use of a known chemical product has been accepted for a long time. In France, an edict of 4th February 1959 enacted by the decree of 30th May, 1960 instituting the 'Special Medicament Patent', recognized the patentability of pharmaceutical products by creating a special kind of industrial property, the famous BSM.[28] In Germany, the recognition of the patentability of pharmaceutical products (like that of chemical products) had to wait for the 1968 reform of the law. As regards the EPC, the patentability of medicaments has been recognized ever since the signing of the convention in 1973.

A certain number of countries, such as Austria and, more recently, Spain and Greece, have, however, devised restrictions so that the patentability of pharmaceutical products is ineffective in their territory for a certain time (Article 167, 2nd (a) of the EPC). In the case of Italy, it was necessary to wait for a decision from the Constitutional Court on 9th March 1978 for Article 14 of the Law on Patents excluding the patentability of medicaments to be declared unconstitutional.

As regards the European Patent Office, allowable claims[45,46] can have the following form: chemical compounds constituted by the derivatives of formula X for their use in a method of therapeutic treatment of the human or animal body.

If, as has just been briefly indicated, the patentability of any 'first therapeutic use'[29] is already strongly debatable, the patentability of any 'second therapeutic use' is still more so.

Until the last few years, the patentability of 'second therapeutic uses' was excluded everywhere, with the exception of the USA, Australia, Belgium, Japan, the Philippines and South Africa.

In the USA, the patentability of 'second therapeutic uses' has been recognized for a long time, provided that such uses answer the criteria of patentability set out previously. This moreover does not seem to raise any special problem.

On the contrary, as regards Europe, as was seen in Section 4.12.2.1, such 'second uses' are excluded from patentability by the effect of Article 52 (4) of the EPC[4] in default of industrial application.

Following the rejection of seven patent applications concerning second therapeutic uses, the Court of Appeals of the European Patent Office was moved, in accordance with Article 112 of the EPC, to put to the Supreme Court of Appeal the following question of law: 'Can a patent with claims directed to the use be granted for the use of a substance or composition for the treatment of the human or animal body by therapy?'

By a decision given on 5th December 1984, the Supreme Court of Appeal of the European Patent Office ruled as follows: 'It is decided that the question of law referred to the Enlarged Board of Appeal is to be answered as follows:

1) A European patent with claims directed to the use may not be granted for the use of a substance or composition for the treatment of the human or animal body by therapy.

2) A European patent may be granted with claims directed to the use of a substance or composition for the manufacture of a medicament for a specified new and inventive therapeutic application.'

The detail of this decision was published in the Official Journal of the European Patent Office dated 25th March 1985 (year 8/number 3).[30]

Following this very important decision, the President of the European Patent Office finally added the following paragraph to the Guidelines for Examination practiced in the EPO:[31]

'A claim in the form "Use of a substance or composition X for the manufacture of a medicament for therapeutic application Z" is allowable for either a first or "subsequent" (second or further) such application. In cases where an applicant simultaneously discloses more than one "subsequent" therapeutic use, claims of the above type directed to these different uses are allowable in the one application, but only if they form a single general inventive concept (Article 82 EPC).'

Thus, it is therefore now possible to protect 'second therapeutic uses'.

It is necessary to compare this very important Decision of the Supreme Court of Appeal of the European Patent Office, with the decisions made by the German, English and Swiss national courts in similar cases. The text of these decisions has been published in the Official Journal of the European Patent Office.[32]

The criteria for patentability discussed in Section 4.12.2.1 apply of course to inventions concerning the novel uses of known products. The use must be novel, but also inventive in relation to the uses described in the prior art. Finally the novel use must display an industrial character.

As regards its effect, a patent covering a novel use of known means (of known products) will only cover the claimed use. Other uses which might be discovered later will not be protected by a patent covering the first use. The use patent will be subservient to any product patent; thus an inventor who found a novel use for a product protected by a third party's patent would not be able freely to exploit the invention without the agreement of the holder of the product patent.

4.12.3.4 Inventions Concerning Compositions

Compositions, or combinations, in order to be patentable, must bring together means which have not previously been associated in the same way, so as to make them produce an overall result.

The mere juxtaposition of two means is not by itself patentable. Thus, in the mechanical field an example sometimes given is that of a furnace on wheels. This kind of invention is not patentable, because there is no operative interaction between the heating of the furnace and the turning of the wheels.

If inventions relating to compositions or combinations are to be patentable, they must yield a result of their own, or a technical effect of their own. Thus a combination or a composition containing both a diuretic and a hypertensive agent is only patentable if the properties of the combination are found to be modified as compared with those secured by the separate administration of each product.

The drafting of a composition claim does not raise any special problem. In the case of a pharmaceutical composition, the claim can be written as follows.

Pharmaceutical compositions, characterized in that they include, as an active principle, at least one chemical compound of general formula X, as well as an inert pharmaceutical excipient.

In the case of a combination of two known active principles, the claim would be written as follows.

A diuretic and antihypertensive composition comprising an antihypertensively and diuretically effective amount of A and B, in a weight ratio of 10 to 200 parts of A to one part of B, with the optional presence of an inert pharmaceutical carrier.

Just as with the other categories of invention, the criteria for patentability must be respected for any invention of a composition or combination.

As regards its effect, a patent covering a particular composition, or a given combination, will only cover the composition or combination effectively claimed. Thus, a single patent covering a composition containing active principles A and B will have no relevance to a composition containing A and C. As for use inventions, a patent covering a particular composition will be subservient to a patent for the corresponding product; thus an inventor who perfected a novel composition containing a product protected by the patent of a third party, would be unable to exploit the invention freely, without the agreement of the holder of the product patent.

Table 4 indicates the categories of patentable inventions according to the countries in question.

4.12.4 LIFE AND DEATH OF THE PATENT—JUDICIAL ACTIONS

Let us now look at the different phases in the life of a patent in succession: (i) the filing of the patent application in the country of origin of the invention (priority application); (ii) extensions of the priority application to foreign countries; (iii) procedures before the various patent offices; and (iv) judicial actions.

4.12.4.1 Filing The Patent Application

When an invention has just been made by an inventor, it is advisable to envisage the filing of a patent application. Since the majority of countries, with the exception of the USA and Canada, grant the patent rights to the one who first files the patent, the application should be filed as soon as possible.

In industry, the patent agent responsible for such a filing begins to assemble the chemical reports which have been written on the different products prepared. If applicable, the agent puts together the pharmacological results which it has been possible to obtain with the various products.

At the same time, the agent will have carried out a documentary search in order to locate the invention in the prior art; the agent then begins to draft the claims which are going to define the field of the invention which will be claimed. The patent engineer preferably drafts a certain number of subclaims like successive surrounding walls of a fortress. This form of drafting is very desirable because it enables a fall-back position to be retained in the event of partial anticipation. This is sometimes also spoken of as onion-peel structuring.

As has previously been shown, the patent application also includes a description. The patent engineer is going to have to describe the invention at this stage as completely as possible so that it can be reperformed by an expert in the field. In chemical cases the method of operation to secure a certain number of actually prepared products will be given by way of example, together with the constants enabling the products to be identified.

If the invention is concerned with pharmaceutical products, the description will indicate the pharmacological properties, the dosage of the products and the diseases treated. It is also desirable that the application should supply an indication of the pharmacological tests carried out, and the results obtained.[45,46]

If the invention is concerned with nonpharmaceutical chemical products, the patent application ought to give an indication of the utility of the products.

When the drafting of the patent application is finished, the true inventors should be identified, so that they can reread the patent application before it is filed. The true inventors will be the persons who have actually participated in achieving the invention.

People who have had only a supervisory function or who have only carried out routine tests (screening tests, for example) should not be regarded as true inventors. This task of identifying the true inventor is notably important, especially in the USA.

Table 4 Categories of Patentable Inventions for Europe, North America, Central and South America, Asia/Oceania and Africa

Country	*Products (pharmaceuticals)*	*Processes*	*'Pharmaceutical compositions'*	*'Second therapeutic use'*
Europe				
European patent designating:				
Austria	Yes (10/1987)	Yes	Yes (10/1987)	Yes (10/1987)
Belgium	Yes			
FRG	Yes			
France	Yes			
Great Britain	Yes			
Italy	Yes	Yes	Yes	Yes
Luxembourg	Yes			
The Netherlands	Yes			
Sweden	Yes			
Switzerland	Yes			
Spain	After 10/1992	Yes	After 10/1992	After 10/1992
Greece	After 10/1992			
Austrian national patent	Yes (10/1987)	Yes	Yes (10/1987)	Yes (10/1987)
Denmark	No	Yes	No	No
Hungary	No	Yes	No	No
Norway	No	Yes	No	No
Portugal	No	Yes	No	No
Poland	No	Yes	No	No
Rumania	No	Yes	No	No
GDR	No	Yes	No	No
Finland	No	Yes	No	No
Turkey	No	Yes	No	No
Yugoslavia	No	Yes	No	No
North America				
USA	Yes	Yes	Yes	Yes
Canada	Yes (1/1989)	Yes	Yes (1/1989)	No
Central/South America				
Mexico	No	Yes	No	No
Honduras	No	Yes	No	No
Costa Rica	No	Yes	No	No
Panama	No	Yes	No	No
Brazil	No	Yes	No	No
Argentina	No	Yes	No	No
Peru	No	Yes	No	No
Paraguay	No	Yes	No	No
Bolivia	No	Yes	No	No
Colombia	No	Yes	No	No
Venezuela	No	Yes	No	No
Uruguay	No	Yes	No	No
Asia Oceania				
Japan	Yes	Yes	Yes	Yes
South Korea	Yes (7/1987)	Yes	Yes	No
People's Republic of China	No	Yes	No	No
India	No	Yes	No	No
Pakistan	No	Yes	No	No
Taiwan	No	Yes	No	No
Australia	Yes	Yes	Yes	Yes
New Zealand	Yes	Yes	Yes	No
Africa				
Algeria	Yes	Yes	No	No
Morocco	No	Yes	No	No
Tunisia	No	Yes	No	No
Egypt	No	Yes	No	No
OAPI	Yes	Yes	Yes	No
South Africa	Yes	Yes	Yes	Yes
ARIPO	Yes	Yes	Yes	No

In the USA, in particular, any failure to designate the true inventor is a cause for invalidation of a patent. When this question has been settled, the filing of the patent application should be carried out as quickly as possible.

4.12.4.2 Extensions to Foreign Countries

During the eight or nine months which follow the filing of a patent application in its country of origin, the question arises as to the possible extension of the basic application to foreign countries. Generally, a detailed assessment is carried out with the research departments to ascertain if new products have been prepared since the filing of the patent application. An investigation is also performed as to whether any new results or new pharmacological tests have, in the meantime, redefined the profile of the invention which is the subject of the patent application. One then tries to forecast the prospects for the invention as objectively as possible. A list is drawn up of the countries in which it will be desirable to obtain protection, as a function of the profile of the products of its placing in foreign countries, and of the big potential sales in the area. Determination of the countries for foreign filing is most often done in collaboration with the commercial and marketing departments.

It is appropriate to emphasize the fact that the choice of countries for filing is a particularly difficult task which can have very serious consequences for the future marketing of the product. This choice must in fact be made at a very early stage in the life of an invention when it is not always easy to form any reliable judgement as to its importance.

As indicated above, by virtue of Article 4 of the Paris Convention of 1883,[1] the depositor of a 'priority' application has in fact a period of 12 months available in which to extend the patent application to foreign countries, without the possibility of any personal intervening disclosures being held against him. After this 12 months period the depositor can still with a certain amount of risk file in foreign countries, provided that there is no disclosure of the invention, but will then lose the advantage of the 'priority' date. After a period of 18 months from the priority date, the depositor has practically no further possibility to make foreign filings, because of the 'automatic' publication of patent applications at 18 months by the great majority of countries. As a more or less absolute rule, foreign filings therefore ought necessarily to be made within the 12 months which follow the filing of the 'priority' application.

When the list of countries is drawn up, for certain countries the depositor has a choice between several options. These are to either file national patent applications in each of the countries in question, or to file a regional patent such as the European patent, or finally to file an international patent application (PCT = patent cooperation treaty patent application).[2,33,34]

The filing of a national patent application is effected at the central industrial property department of that state. Such a filing is most often effected through a local patent agent who carries out this task in conformity with the rules of the country. The local patent agent makes a translation of the patent application into the language of the country and ensures that it conforms to the requirements of the country. In the case of large companies, the central industrial property department of such companies is able to prepare the text of the patent application so that it conforms to the requirements of the country concerned, and to file it direct. However, the national law of a country may insist on the appointment of a duly authorized local representative.

A European patent application may be filed either at the European Patent Office at Munich or at its branch at The Hague or, if the legislation of a contracting state allows this, at the central industrial property department or at other competent departments of this country (Article 75 of the EPC). The filing of a divisional application of a European patent, on the other hand, absolutely must be carried out at the European Patent Office at Munich or at its branch at The Hague.

When a European patent application is filed, one has to designate the contracting state or states in which it is hoped to obtain protection by means of a European patent. At least one state therefore should be designated amongst the following countries: Austria, Belgium, FRG, France, Great Britain, Italy, Luxembourg, The Netherlands, Sweden, Switzerland and Liechtenstein, Spain and Greece. One must pay as many designation fees as there are states designated.

When it is filed, a European patent application must contain at least a request for the issue of a European patent, a description of the invention, one or more claims, any drawings and a summary. The filing and examination costs have to be paid, at the latest, one month after the filing of the application. The designation fee is required to be paid within 12 months from the filing of the European patent application or, if priority has been claimed, from the date of priority. In the case of a filing after the priority period has expired, a supplementary period of one month is allowed for the payment to be made.

The European patent application must also include the designation of the inventor. If the applicant is not the inventor or the sole inventor, this designation must include a declaration indicating the basis for the acquisition of the right to the patent. Finally, if any priority is to be claimed, this must be done at the time of filing. The applicant must in addition supply a copy of the earlier 'priority' application, accompanied by translation thereof into one of the official languages of the European Patent Office (German, English or French), if the language of the earlier application is not an official language of the Office.

Any natural or juridical person and any company can apply for a European patent. A European patent application can also be filed either by coapplicants, or by several applicants designating different contracting states.

Natural and juridical persons who possess neither a domicile nor an office in the territory of any of the contracting states, must be represented by an approved representative and act *via* them as intermediary. On the other hand, natural and juridical persons who have their domicile or office in the territory of any of the contracting states can act *via* an employee as intermediary (Articles 133–134 of the EPC).[4] Since the procedures are relatively complex, it is strongly recommended always to operate *via* an approved agent as the intermediary.

As regards extensions to foreign countries, the latest possibility consists in filing a PCT patent application. An international convention called the PCT (patent cooperation treaty), signed on 19th June 1970[33] at Washington and recently (as at 1st January 1989) ratified by 41 states, has set up a common filing system including an international documentary search as to novelty and publication of the application at 18 months. After this 18 months period, the patent application is then transmitted to the industrial property departments of the designated states to be incorporated into the national issuing procedures of these states.

The main virtue of filing patent applications *via* the PCT procedure lies in the fact that the applicant has the opportunity at the expiry of the priority period to effect an extension of the basic application to a foreign country in the original language of filing.

The filing of a PCT patent application can be effected at the central industrial property department of any of the contracting states (receiving office).

At the 1st of January 1989, the contracting states were as follows: the FRG, Australia, Austria, Barbados, Belgium, Benin, Brazil, Bulgaria, Burkina Faso, Cameroon, Congo, Denmark, the USA, Finland, France, Gabon, Hungary, Italy, Japan, Liechtenstein, Luxembourg, Madagascar, Malawi, Mali, Mauritania, Monaco, Norway, The Netherlands, the Central African Republic, the Republic of Korea, the Democratic People's Republic of Korea, Rumania, the UK, Senegal, Sri Lanka, Sudan, Sweden, Switzerland, Chad, Togo and the USSR.

The documents needed for filing a PCT patent application are practically identical to those necessary for filing a European patent application.

By contrast, as regards the fees to be paid when a PCT patent application is filed, the matter is much more complex. It is necessary to pay a transmission fee, a research tax, an international fee including a basic fee, a designation fee per state designated for which national patents are requested, and a designation fee per regional patent application.

For further details on this point, reference should be made to the Applicant's Guide.

4.12.4.3 Procedures Before The Patent Offices

The procedures before the patent offices vary considerably from one country to another. There are countries which do not make any examination on the merits, as well as countries which, on the contrary, carry out an examination on the merits with varying degrees of severity.[2]

4.12.4.3.1 Countries not making an examination on the merits

These countries generally, though not necessarily, do undertake formal examination. The purpose of such an examination is, in particular, to make sure that the application is formally in order, and if appropriate that there is a certain unity of invention. These countries in fact register applications as filed, without checking if the criteria of patentability have been respected. Notable among these countries are Belgium, Luxembourg, Portugal, Greece, Tunisia, Morocco, South Africa, Peru, Argentina, Colombia *etc.*

4.12.4.3.2 Countries carrying out an examination on the merits

Such an examination may either commence within a few months following the filing of the patent application, or may, on the contrary, be deferred.

When examination commences within a few months following the filing of the patent application, the examiner charged with the examination sends to the depositor a first official notification which raises a certain number of objections as to the form and content of the application. The depositor generally is allowed two or three months, though capable of extension, in which to reply to the objections raised. The examiner may send a second official letter if it is thought that the depositor has not fully replied to the objections raised, and then decides whether to issue or reject the application. Rejection decisions are always subject to appeal. This is the typical procedure encountered in the major industrial countries which carry out an immediate examination, such as the USA, Denmark and the Nordic countries.

As regards the USA, the procedure is appreciably more complex because of the possibility of becoming involved in interference procedures, to determine who has the right to the patent (see Section 4.12.2.1.1).[35]

There is also a peculiarity in the USA due to the fact that it is possible, at almost any stage of the procedure, to refile the application by the subterfuge of the so-called 'continuation' or 'continuation in part' applications. The latter are basic applications which have been expanded by new examples or new results. The American procedure is very complex and would deserve a very lengthy explanation which cannot be offered here.

As indicated above, the examination can also be deferred. Countries such as the FRG, The Nertherlands and Japan have for almost 20 years operated a deferred examination system. In these countries, the depositor has a period of seven years counting from the date of filing in which to request examination. After this period of seven years, if the examination has not by then been requested, the patent application automatically falls into the public domain.

For several years, a certain number of countries such as South Korea and Brazil have also operated a deferred examination system, of respectively five years from the date of filing and of two years from the date of publication.

Finally, there are countries such as Great Britain and Australia where patent systems like the European patent or the PCT have established procedures in which the depositor has a period of six months from the publication of the search report within which to ask for examination. In these countries, the offices first of all carry out a documentary search (search report), the purpose of which is to inform the depositor, and then third parties, about the prior art considered to be the most pertinent. After this, and in some sense after a period of reflection, the applicant then has to ask for examination on the merits.

In all the countries which carry out such an examination on the merits, the examiners send out official letters (generally two or three per application) to the applicants. The applicant must then on each occasion refute the examiner's objections, if appropriate by comparative tests to show that there is indeed inventive activity and that the criteria of patentability have been respected. At any time, the applicant can abandon the application. Only when the examiner judges that the criteria of patentability have been respected is the application granted and then issued. A certain number of countries such as the FRG, The Netherlands and Japan, and more recently the European Patent Office, then publish the application as allowed by the examiner so that, if appropriate, third parties can oppose it. During the opposition period, varying from two to nine months according to the country, third parties can cite 'pertinent' references which have not been put forward by the examiner. The allowed patent then goes back before the examiners. It can then either be partially or wholly revoked or be finally issued definitively. By way of information, the examination and opposition procedures for a patent application in Japan can occupy about 10 years.

4.12.4.4 Judicial Actions

(i) An act of infringement can be defined as an attack on the rights of the patentee in the territory where the patent is filed. There is in fact no infringement in carrying out an invention in the FRG which has been protected by a patent filed only in Great Britain. For there to be an infringement of the rights of the patentee, there must therefore exist a patent. Thus, an action for infringement can only be launched as from the day the patent is granted. The reason for this rule is to protect third parties who might have infringed a patent of which they were unaware. The laws of a certain number of countries, on the other hand, do envisage the possibility of notifying a suspected infringer of the existence of a patent application. In such a case, when the patent has been issued and it has been decided that there was indeed infringement, the period between this notification and the issue of the patent could be taken into account in the calculation of the indemnity paid by way of reparation.[40,41]

When the patent has fallen into the public domain, either on normal expiry of the period of protection, or when the annuities have not been duly paid, or finally because it has been revoked by nullity *'erga omnes'*, there will no longer be infringement.

To determine if there has been infringement, one has to compare the subject matter protected by the patent with the act alleged to constitute an infringement. The general rules about this, which however are not necessarily applicable in every country, are as follows: (a) the slavish reproduction of the subject matter protected by the patent certainly constitutes an act of infringement; (b) according to the so-called theory of variants of execution, simple differences of detail do not take away the infringing character of the reproduction of a patented invention; (c) infringement is assessed by resemblances and not by differences; (d) according to the so-called theory of equivalents, there is infringement when the alleged infringement is a direct equivalent of the subject matter protected by the patent; and (e) infringement must be considered patent by patent.

Moreover, each country may indicate that certain acts constitute infringement while in other countries the same acts will not be regarded as acts of infringement.

(ii) An infringement action can be started either by the owner of a patent or by the licensee.

Each country[2] may establish a limited list of tribunals competent in matters of infringement. When there is a choice between several tribunals, the competent tribunal is generally the one where the infringement has taken place.

Proof of infringement must be provided by the owner of the patent. This proof can be provided by any means (witnesses, documents, publicity, purchase of products, *etc.*).

(iii) Relief for the infringing action may be provided by civil remedies or by criminal penalties. Among civil remedies there is the injunction of the infringer against continuing the infringement, under penalty of payment of a fine per article manufactured. There can also be confiscation of the infringing objects, or even of the means enabling the infringement to be carried out. Naturally, there is also the payment of compensation to make up for the loss of profit or for the loss suffered.

Finally, criminal penalties include the payment of a fine or a prison sentence.

In conclusion, patents have a first-order importance in protecting the fruits of research and development of every company which wishes to take its place at an international level.

Before making any disclosure of an invention, one absolutely must have the reflex thought of filing a patent application. At each stage of the research, one should investigate if there is patentable material before making any disclosure which might prove to be irreparable.

For companies who wish to market products on an international scale, the protection given by patents can only be contemplated on a worldwide scale, with every country at present having its own particular legislation.

During the last 20 years, very important progress has been made in the unification of the criteria for patentability and in the harmonization of the patent laws of many countries, but there still remains considerable room for progress. Between now and the end of the century, it can be hoped that the community patent will finally see the light of day.

4.12.5 REFERENCES

1. 'The Paris Convention for the Protection of Industrial Property', BIRPI, Geneva, 1969.
2. 'Manual for the Handling of Applications for Patents, Designs and Trade Marks Throughout the World', Octrooibureau Losen Stigter, Amsterdam, 1989.
3. 'Patent Law of the People's Republic of China' (also published in 'Patents and Licensing', 1984, p. 7).
4. 'European Patent Convention', European Patent Office, Munich, 1987.
5. 'Community Patent Convention', HMSO, London, 1975.
6. 'Patent and Trademark Laws', the Bureau of National Affairs, Washington, DC, 1989.
7. *Official Journal of the European Patent Office*, decision of the Technical Board of Appeal 3.3.1., 4 May 1985, T 31/84, 11/86, 369.
8. M. de Haas and Th. Mennessier, *Biofutur*, No. 38, September 1985, 63.
9. P. J. Federico, *Patent Interferences*, No. 1, 1971, 21.
10. M. B. Strefel, *J. Pat. Off. Soc.*, 1978, **60**, No. 9.
11. *Official Journal of the European Patent Office*, decision of the Technical Board of Appeal 3.5.1., 10 September 1982, T 21/81, 1/1983, 15.
12. *Official Journal of the European Patent Office*, decision of the Technical Board of Appeal 3.3.1., 13 October 1982, T 24/81, 4/1983, 133.

13. *Official Journal of the European Patent Office*, decision of the Technical Board of Appeal 3.3.1., 20 April 1983, T 65/82, 8/1983, 327.
14. *Official Journal of the European Patent Office*, decision of the Technical Board of Appeal 3.3.1., 17 March 1983, T 20/83, 10/1983, 419.
15. *Official Journal of the European Patent Office*, decision of the Technical Board of Appeal 3.3.1., 16 March 1983, T 04/83, 12/1983, 498.
16. *Official Journal of the European Patent Office*, decision of the Technical Board of Appeal 3.3.1., 25 February 1985, T 106/84, 5/1985, 132.
17. *Official Journal of the European Patent Office*, decision of the Technical Board of Appeal 3.3.1., 25 June 1985, T 205/83, 12/1985, 363.
18. *Official Journal of the European Patent Office*, decision of the Technical Board of Appeal 3.2.1., 10 October 1985, T 195/84, 5/1986, 121.
19. *Official Journal of the European Patent Office*, decision of the Technical Board of Appeal 3.3.1., 20 June 1986, T 94/84, 10/1986, 338.
20. *Official Journal of the European Patent Office*, decision of the Technical Board of Appeal 3.3.1., 26 March 1986, T 206/83, 1/1987, 5.
21. *Official Journal of the European Patent Office*, supplement to Official Journal 11/86, 1.
22. *Off. Gaz. US Pat. Trademark Office, Pat.*, 1982, **1023**, T MOG 31.
23. *Fed. Regist.*, 1985, **50**, No. 151.
24. 'Drug Price Competition and Patent Term Restoration Act', US Law 98-417, September 1984.
25. *Official Journal of the European Patent Office*, decision of the Technical Board of Appeal 3.3.1., 28 February 1985, T 198/84, 7/1985, 209.
26. *Official Journal of the European Patent Office*, decision of the Technical Board of Appeal, 6 June 1986, T 17/85, 12/1986, 406.
27. *Official Journal of the European Patent Office*, decision of the Technical Board of Appeal, 22 June 1982, T 22/82, 9/1982, 341.
28. *Journal Officiel de la République Française*, May 1960, No. 60-130 S.
29. *Official Journal of the European Patent Office*, decision of the Technical Board of Appeal 3.3.1., 14 May 1985, T 36/83, 9/1986, 295; 27 March 1986, T 144/83, 9/1986, 301.
30. *Official Journal of the European Patent Office*, decision of the Enlarged Board of Appeal, 3/1983, 59.
31. 'Guidelines for Examination in the European Patent Office', Part C IV 4.2, European Patent Office/Wila Verlag Wilhelm Lampl, Munich, 1987.
32. *Official Journal of the European Patent Office*, 1/1984, 26; 5/1984, 233; 11/84, 581.
33. 'Traité de coopération en matière de brevets (PCT) et Règlement d'execution du PCT','OMPI (Organisation Mondiale de la Properiété Intellectuelle), Genève, 1982.
34. 'Guidelines for the PCT', OMPI (Organisation Mondiale de la Propriété Intellectuelle), Genève, 1989.
35. 'Rules of Practice in Patent and Trademark Cases', Rules Service Company, Rockville, MD, 1989.
36. *Official Journal of the European Patent Office*, 5*1986, 149; 9*1985, 295.
37. '*La Propriété* Industrielle', January 1987, 20; 'Lois et traités', May 1987, texte 1-008, p. 001, Revue mensuelle de l'Organisation Mondiale de la Propriété Intellectuelle (OMPI), Genève.
38. *Official Journal of the European Patent Office*, 2/1983, 68.
39. 'La Propriété Industrielle', January 1987, 20, Revue mensuelle de l'Organisation Mondiale de la Propriété Intellectuelle (OMPI), Genève.
40. 'Le contentieux des brevets aux Etats-Unis d'Amerique', La Propriété Industrielle, Genève, May 1987, p. 192 (Revue mensuelle de l'Organisation Mondiale de la Propriété Intellectuelle, OMPI).
41. IIC (International Review of Industrial Property and Copyright Law), published by the Max Planck Institute for Foreign and International Patent Copyright and Competition Law Munich.
42. The House of Commons of Canada, 'Bill C', 'An Act to amend the Patent Act and to provide for certain matters in relation thereto', The Minister of Consumer and Corporate Affairs, 5.11.86. 22332 (assented to 19.11.87).
43. S. N. Cohen, *US Pat.* 4 237 224 (1980).
44. H. W. Boyer, *US Pat.* 4 468 464 (1984).
45. Roussel-Uclaf, *Eur. Pat.* 003 200 (1979).
46. Roussel-Uclaf, *Eur. Pat.* 223 701 (1986).

4.13
Trade Marks

ERICH A. HORAK
Ciba-Geigy Ltd, Basle, Switzerland

4.13.1 SCOPE OF CHAPTER

This chapter is primarily intended to provide a concise outline of the trade mark system for easy reference. Technical details are referred to only where necessary for the better understanding of the subject. The general view on trade marks presented in the chapter cannot, of course, be a substitute for the knowledge of the expert in the field. It is recommended that professional counsel be sought for advice whenever trademark-related issues arise in the course of business. As used in this chapter the words 'trade mark', 'mark', 'brand' and 'service mark' are synonymous. References to 'goods', 'products' and 'merchandise' relate also to 'services'.

4.13.2 THE FUNDAMENTALS OF TRADE MARKS

4.13.2.1 Tradition

The need of manufacturers and traders to make their goods distinguishable from each other by the addition of certain signs to the products is one that dates back far into the past. Personal symbols, signs identifying property and marks serving as an indication of product origin were the main groups of marks used throughout the ages. The closest to our present day marks were the symbols applied by craftsmen and artists to their products as identification of origin and the proprietary signs used by traders on their merchandise. It was only much later that trade mark regulations followed the use of marks. During the Middle Ages, trade mark use was dealt with in the statutes of the guilds, in the following centuries by way of privilege grants.

The first modern trade mark laws were enacted around the middle of the last century and today national trade mark statutes are effective in more than a hundred states. Multinational conventions serve to facilitate the obtaining of trade mark protection on a world-wide scale.

4.13.2.2 Definition of Marks

WIPO (the World Intellectual Property Organization) defines a trade mark as a sign which serves to distinguish the goods (as does the service mark with regard to services) of an industrial or commercial enterprise or a group of such enterprises. The sign may consist of one or more distinctive words, letters, numbers, drawings or pictures, emblems, monograms or signatures, colours or combinations of colours, and under some legislations it may also consist of the form or other special presentation of containers or packages for the product (provided they are not solely dictated by their function). The sign may, of course, consist also of combinations of any of these elements.

It is to be noted that in a number of states legislation does not yet provide for the protection of service marks.

4.13.2.3 Categories of Marks

In practice, the overwhelming number of marks used on goods and for services appear in three outward forms, *i.e.* as word marks, as pictorial marks and as combination marks. About 80% of the marks used today are word marks. They are signs where the word, independently of its graphic layout, forms the distinctive element of the mark. Pictorial marks consist of distinctive figurative representations of all kinds, including words, letters or numbers of any fancy design. Combination marks comprise general words and pictorial elements of distinctive character. Marks are predominantly used in commerce as individual marks, *i.e.* as marks distinguishing the goods or services of one enterprise. Unlike the individual mark, a collective mark serves to distinguish the origin or other common characteristics of goods or services of different enterprises. It indicates the membership of an enterprise in an association which is the owner of the collective mark, and permits its members to use the mark, subject to certain conditions. Certification marks, as distinguished from collective marks, do not indicate source but certify product characteristics, such as regional or other origin, material, quality, accuracy of manufacture or service, *etc.* A certification mark may be used by anyone as long as he complies with the conditions governing its use. Other forms of marks commonly used by enterprises are serial or derivative marks, umbrella marks and house marks. Serial marks are signs which, although distinguishable from each other, have a particular stem in common which indicates a common source. All other marks of the series are derived from that stem, too. Umbrella marks are used to distinguish a whole range of goods. In order to simplify the identification of individual products, umbrella marks are normally applied together with specific product marks or with generic names describing the individual article or service. House marks consist of the name, or of an abbreviated form of the name, of a manufacturing or trading enterprise. They are often employed as umbrella marks or as the characteristic stem of serial marks and play an important role in the creation of corporate identity. Some jurisdictions still distinguish between manufacturers' and traders' marks, although both enjoy equal legal protection. The definition of the so-called accompanying mark ties in with its name. It is a mark which accompanies the starting or finishing materials it designates through the successive steps of a manufacturing process. In the end, the accompanying mark appears on the final products together with the mark of their manufacturer. As a consequence, both the starting materials and the final products benefit mutually from the reputation their respective marks enjoy. Certain states endeavour to make the vinculum or linking of marks a mandatory requirement of their trade mark legislation. Linking means the associated use of a foreign trade mark, licensed to a domestic party, with a trade mark originating locally as property of the licensee. The goal of linking is to profit from the image of a high quality product, usually a pharmaceutical specialty bearing that foreign mark, in the case of its substitution by a product of local origin of a similar nature. Reserve marks are ordinary marks as to form and nature, and so are defensive marks. These two types of mark, however, are by definition not in use; hence, in this respect they are distinguished from the other marks which are used or for which a *bona fide* intention of use is assumed on registration. Reserve marks are usually contained in a portfolio of registered marks from which an enterprise may immediately draw in case a product is to be put on the market without delay. The inherent disadvantage of reserve marks derives from the use requirements embodied in the trade mark laws of many states. Generally, if a trade mark has not been used over an extended period of time in respect of the goods for which it is registered, it may become contestable, in the extreme case it may even automatically be removed from the Trade Mark Register. Consequently, ageing reserve marks which have never been put into use are of no particular value to their owners. While reserve marks are at least intended to be used sometime after their registration, defensive marks are not. They simply serve to build a defence around a very valuable mark of the same owner in order to prohibit marks of others coming too close to it. Some legislations provide expressly for defensive registrations which may be kept on the register without being used. The trade mark laws of several countries provide for the association of marks in the case where two or more marks of one owner resemble each other so closely that, if belonging to different owners, the same marks would not be registrable over each other. Such marks are entered into the register as so-called associated marks and can only be owned by one proprietor. In case of the assignment of one of the associated marks the association has either to be dissolved or, if that is not possible, the associated marks have to be assigned altogether.

Trade marks and trade names are different but closely related. A trade mark identifies goods, a service mark services and a trade name a business. Usually, but not necessarily, a trade name is also a trade mark. If so, it identifies the business as such, as well as the goods emanating from that business. A trade name can, for example, consist of a firm's name, any other lawfully adopted name, an individual's name, an abbreviation of a name, an invented fictitious name, *etc.* Since trade names can function as trade marks and *vice versa*, care has to be exercised with regard to existing trade mark rights when a trade name is adopted. Trade name protection differs from country to country. It is, therefore, advisable to register trade names as marks whenever possible.

4.13.2.4 Functions of Marks

Over the centuries the purpose of marking has continuously changed. For a long time, the indication of the origin of goods was the original and only function of marks. Today, in a world abounding with merchandise, a trade mark functions to guarantee quality as much as to indicate source. Furthermore, it helps to identify a product from similar goods of the same or of different origin. As to the origin function of a mark, it is sufficient that the public can expect the goods a particular mark designates to originate from the same, although anonymous, source from which similar goods under the same mark have always come. Nowadays, the most important functions of a mark are product identification and quality assurance. Both functions assist the purchaser to choose a particular product among the numerous articles competing on the market. The quality function helps to inform the consumer that he may reasonably expect the trade mark owner to maintain consistent quality standards with regard to the goods sold under the same trade mark. This is, however, not to be understood as a legally binding warranty against some default or change of the features of a given product. Being a means by which the purchaser recognizes a product, marks are important advertising tools, too. The advertising strength of a mark together with its reputation for quality linked to a particular source are the driving forces behind the selling power of marks.

4.13.2.5 Goodwill

Trade marks, owing to their particular role in sales and advertising, have increasingly become an important asset on their own, and consequently constitute a valuable part of the intangible assets of a business. Used for a long time on successful products, trade marks may symbolize the name, the reputation and the connections a business enjoys in the market. To preserve this goodwill, arising from the benefits acquired by a business beyond its mere value of property, particular care is exercised by manufacturers and traders by way of strict quality control of their products and by proper use of their respective marks.

4.13.2.6 Relations to other Intellectual Property Rights

In distinguishing goods in a competitive market, marks basically serve to safeguard fair competition. In the jurisdictions of continental Europe and in other areas, trade mark law is considered part of the broader concept of competition law which also provides protection against acts such as infringement of trade name, copying of style, discrimination of competitors and the like. In the UK and in countries having similar legal systems, actions in respect of unfair competition are not provided for. Unfair commercial acts are handled there within the concept of passing-off and the law of false trade description, respectively, depending on the underlying acts and the involved subjects. Trade marks, patents and copyrights, to name the more important fields of intellectual property, are independent special rights. Since patents are granted for new, unobvious and useful inventions, trade marks are by definition not patentable. As distinguished from patents, the publication of a trade mark is no obstacle to the later registration of the same mark for the same or similar goods anywhere. Whilst a patent owner is entitled to exclude others for a limited period of time from manufacturing, using and selling his invention, a trade mark owner can exclude others from using the same, or a confusingly similar mark, on the same or similar goods for an unlimited period of time, as long as his mark is properly used and periodically renewed. Copyrights differ from trade marks in a number of aspects. Like patents, they exist for a limited period of time only. They protect an author's exclusive right to publish, copy and sell his work, and to grant others the right to do any of these things. Trade marks, generally, cannot be copyrighted.

4.13.3 THE REGISTRATION OF TRADE MARKS

4.13.3.1 Selection of Marks

Designing a successful trade mark is no easy matter. It is an inherently risky business, given the commercial and legal implications it may entail. Marketing flops, damage to goodwill, costly litigation and other inconveniences can be results of a negligently exercised trade mark choice. The selection of the mark is substantially influenced by marketing and legal factors. Decisive marketing factors are the profile of the product and its positioning among competitive products in the market. The principal legal aspects are the distinctiveness of the prospective mark and its availability for use. Other determining factors are linguistic elements in connection with the use of the mark in foreign markets and the psychological aspects governing the impact of the mark on the consumer. Very important for the choice of the mark is the time factor. The selection process must be initiated in good time before the mark in question is put into actual use. Attention ought to be paid in this context to reserve marks or other existing marks which might be fitting for the product in question. Considerable amounts of time and money can be saved by proceeding along those lines.

The terms from which marks can be chosen are basically meaningless, suggestive or descriptive with regard to the product they designate. Meaningless marks bear no relationship to the product. They usually consist of an invented or fantasy word or of a dictionary word which is arbitrarily applied. The meaningless mark is inherently distinctive and, consequently, enjoys strong legal protection in this particular respect. The suggestive mark conveys a message with regard to the product and, on account of this, does provide certain marketing advantages. It can be reasonably protected, too. The suggestive mark is a mark which is frequently used in the pharmaceutical area. The descriptive mark, finally, gives straightforward information with regard to the product or its properties. Since it lacks distinctiveness it is not a mark in the sense of the law and is, at least initially, not protectable for the products it describes. It can, however, after extensive use in commerce acquire distinctiveness or 'secondary meaning', and hence become protectable when the public, eventually, associates the so-marked product with a particular source.

An important prerequisite for the choice of a 'good' mark is the consideration of a few guidelines. A good mark, in that sense, is short, that.is consists of two or a maximum of three syllables, is easily memorable, readable and writable, avoids unpleasant connotations and conveys a favourable impression of the products and its manufacturer. The coinage of marks may, for example, be accomplished by asking company employees for suggestions or may be achieved in discussions involving groups of specialists. Another possibility is the evaluation of computer runs based on special software. The ultimate selection of the mark may depend on preferences formed within the selecting group or may be left to surveys performed among the members of target groups. In pharmaceuticals, as in other areas, the coinage and the ultimate adoption of a mark are usually the responsibility of the product management and marketing departments, respectively, of the company. When a mark has been selected the next step is its clearance by the company's trade mark department as to availability for use. A mark which is likely to cause confusion or deceive consumers must not be adopted. Usually, the searches which are run to avoid confusing similarity of a proposed mark with prior marks are carried out in-house on a preliminary basis, whereas more accurate searches on a world-wide scale are done by specialized outside search institutes, patent offices and individual law firms.

4.13.3.2 Prior Use *versus* Prior Registration

Rights cannot be acquired in the trade mark *per se* but only in connection with the merchandise it is meant to distinguish. Trade mark protection is obtained by prior use and by registration, respectively. In roughly half of the world's jurisdictions, for example under the Common Law type of trade mark legislation in the UK, it is the first use of the mark in connection with a product which may confer proprietary rights to that mark for that product. In the legal systems of the other half of the world's countries, for example in most countries of continental Europe and in Japan, the exclusive right to the mark cannot be obtained by the mere use of it but only by way of the registration of the mark with the competent authorities. The registration of the mark is *prima facie* evidence of the ownership in the mark, unless proof to the contrary is produced. The registration of a mark provides a number of advantages. While the first user of an unregistered mark enjoys protection against the use or the registration of the identical or confusingly similar mark on the same or similar goods by a third party within the limits of unfair competition law or comparable legal remedies only, the first registrant may bring an action for infringement forthwith. Given the rather lengthy and cumbersome procedure in substantiating a claim to an unregistered mark, this possibility is one of the most beneficial aspects of registration. Equally advantageous is the early establishment of rights to the mark well before it is actually put into use. Together with the grace periods for non-use, which are up to five years after registration and are provided for in many legislations, this helps considerably in product development and marketing planning. The registration, furthermore, may simplify the assignment of the mark and is, in the country of origin, indispensable as a basis for obtaining registrations abroad under international agreements such as, for example, the Agreement of Madrid.

4.13.3.3 Requirements of Application

Before the application for the registration of the adopted mark is submitted to the relevant government office, the scope of the goods in respect of which the registration of the mark is requested must be completely clear. This is of importance insofar as subsequent changes in the list of goods initially submitted are not possible except for the restriction of goods or the insertion of disclaimers. Extensions of the list necessitate new applications and entail loss of time and additional cost. The classification of goods and services referred to in the Nice Agreement of 1957, to which many states adhere, serves as an excellent instrument for drawing up an appropriate list of goods, particularly when used together with the detailed guidelines provided for this purpose by various national industrial property offices. Since the international classification is a purely administrative tool, it is not a determining factor in the solution of trade mark conflicts involving an issue of similarity of goods. The decisive test in this respect is whether the disputed goods 'are of the same description' or, in other words, whether these goods would be considered in the market to belong to the same trade. Identical marks may thus be used or registered by different proprietors on different goods or in different classes as long as the above standard is met. As a rule of thumb in formulating a list of goods it is advisable to follow a middle of the road approach, that is to be not too selective but not too broad either.

4.13.3.4 Examination of Application

Throughout the world there are basically two systems of the examination of trade mark applications prior to acceptance. In a number of countries the application is examined only as to form, *i.e.* whether the filing documents conform to the legal requirements and the fees have been paid in full. The mark, then, is allotted a registration number and published in the country's trade mark journal as a registered trade mark. In this system it is left to the courts to judge on the validity of the registration if contested. The other system combines the examination of the mark with regard to formal requirements with an examination as to the material merits of the application and, consequently, does imply a presumption of validity of the ultimate registration, whereas the other system does not. There is, of course, no official 'guarantee' involved with regard to the enforceability of a trade mark right so granted. In some countries the *ex officio* examination of the mark is conducted as to absolute grounds of registrability as well as to possible conflicts with identical or confusingly similar marks registered earlier for the same or similar goods, *i.e.* as to relative grounds of registrability. In many of the jurisdictions following this system, the mark is published before registration to allow, within a prescribed period of time, interested third parties to oppose *ex publico* the grant of the registration. In other countries only absolute grounds for refusal are considered *ex officio*, with or without the possibility for interested parties to file opposition to the registration of the mark.

The most important absolute ground on which a sign cannot be registered as a mark is its lack of distinctiveness. Non-distinctive signs are not capable of distinguishing the goods of an enterprise due to their inherent characteristics indicating the kind, quality, quantity, value, place of origin, utility, *etc.*, of the goods they designate, in one way or the other. This also comprises descriptive signs including the commonly used and accepted names of chemical substances and their non-proprietary generic names. Signs which can only be registered by authorization of the competent authorities comprise representations of coats of arms, flags, emblems, insignia of armed forces, initials, names and abbreviations of states and international organizations, *etc.*, and include official signs, warranty seals, hall marks and the like. Equally not protectable are signs contrary to law and morality and signs which would be likely to deceive or to confuse the public with regard to the properties of the products or to their source, geographical terms included.

As far as the relative grounds for refusal are concerned, the registration of a mark for particular goods or services is not permitted if it is identical or so resembles a prior registered mark for the same or similar goods as to be likely to confuse the public with regard to the source of the goods or the connection between the concurrent users of the similar marks. The registration of a mark is also refused if it is similar to a so-called 'well-known mark'. This is a mark which is well known to the public in a given country, whether it is registered or in use in this country or not. Well-known marks enjoy particular protection under the Paris Convention and are protected in the national laws of most countries, too.

The similarity of two marks is judged in respect of their pronunciation, appearance and meaning, in this order, both marks being considered separately. Judging the similarity of marks is essentially influenced by the products the marks designate, by the relevant markets for which the marks and products are intended, by the respective trade channels, by the nature of the products and by how well the marks involved are known. The standards applied in determining the similarity of marks and the 'description' of goods vary from country to country. In the case where a mark is finally rejected after full examination by the industrial property office, most forms of legislation provide for administrative or judicial appeals against the decision. Conflicts coming to light during examination are generally settled between the involved parties on the basis of so-called prerights agreements. Such agreements simply state that the applicant for a registration undertakes to refrain from asserting rights against the trade mark of the proprietor of an earlier mark and also tolerates new registrations as well as registrations of modifications of the prior trade mark. The owner of the prior mark, then, in exchange for this undertaking, abstains from taking action against the applicant of the later mark. Any agreement of this kind must, of course, not reflect on the distinctiveness of the prior mark. In some jurisdictions so-called consents by registrants of earlier marks are accepted by the trade mark authorities to remove their objections to the registration of potentially confusingly similar marks.

It is to be noted that in some countries the health authorities may prohibit a trade mark from being used on a pharmaceutical specialty for reasons of public interest even though this mark has met all the relevant legal conditions for its registration as a trade mark.

4.13.3.5 Registration and Renewal

Once the application has been examined and the mark meets the criteria for its registration, the registration of the mark is recorded in the trade mark register and the mark is published together with all relevant particulars in the industrial property journal of the country or in any other official publication. The duration of trade mark registrations is nowadays in most countries fixed at 10 years from date of application, although in some areas longer or shorter periods are provided for. The trade mark registration may be renewed for similar periods; in many countries, however, only upon presentation of a declaration of use of the mark for goods for which it is registered within a certain period of time prior to the expiry of the registration. On average it takes between one and two years to get a mark on the register although shorter and longer periods of time are quite common depending on the country and, of course, the specific details of the case. It should also be noted that in a number of countries the latin script is not the norm, hence, transcriptions of latin type marks are recommended. It is quite useful, particularly in cases of infringement, to have both scripts of the mark registered separately.

4.13.4 THE EFFECTS OF REGISTERED TRADE MARKS

4.13.4.1 Rights to Use and to Forbid

The registration of the mark confers exclusive rights therein upon its owner, usually as of the publication date of the mark's registration. The proprietor of the mark is entitled to apply the mark to the goods in respect of which it is registered, to put the so-marked goods on the market and to make use of the mark in advertising, in correspondence, on labels, *etc.* Beyond his right to make use of the mark, the proprietor of the mark is entitled to prohibit a third party from using signs which are identical or confusingly similar to his mark in relation to goods which are identical or similar to those for which his mark is registered. The negative right of the owner of the mark to forbid thus extends beyond the limits of his positive right to use.

4.13.4.2 Limitations

The ownership of a mark does not bestow an absolute right with regard to the use or to the validity of the mark. Court actions throughout the entire lifetime of a mark are always possible for a variety of reasons, for example on grounds of infringement or invalidity, although in some jurisdictions the mark can become immune from attack on specific grounds after it has been indisputedly used over a certain period of time. The right to the mark must, moreover, not be exercised to prohibit others from using their surnames, *etc.*, geographical names, and all kinds of characteristics of their goods, such as indications as to quality, purpose, source, value and the like, provided, of course, that these indications are not used in some sort of trade mark sense.

4.13.5 ASSIGNMENT AND LICENSING

4.13.5.1 Change in Ownership

The right to a trade mark is an object of (intangible) property and, therefore, as any other property, is transferable from one proprietor to another. Although the conditions under which the trade marks can be assigned vary from country to country, depending on the interpretation of their functions, the mark, basically, can be transferred together with all or with part of the business to which the mark relates, or can even be freely assigned independently of the transfer of the business, in respect of all or of part of the goods for which it is registered. Since the goodwill of the business is symbolized by the mark, trade mark assignments with only part or without any part of the business are a delicate matter and special care must be taken, and is often legally required with regard to such transfers, in order to avoid the misleading of the public as to the source, quality, *etc.* of the products sold under the assigned mark. Assignments of marks are recorded and published in the usual way. Most countries provide for the transfer of the marks registered in respect of a certain business together with the said business in the case where it changes ownership.

4.13.5.2 Licensing

Licences under trade mark rights can generally be granted to third parties in respect of all or part of the goods for which the mark is registered. The licences may be exclusive or non-exclusive, may impose certain limitations on the licensee as to the use of the mark, for example with respect to the territorial scope, to the duration of the licence and to any acts endangering the validity of the mark, and the like. In order to be valid, licence agreements must also contain provisions ensuring quality consistency of the products for which the mark has been licensed and must further provide for the possibility of quality control by the licenser. Other provisions common to licence agreements regulate the rights and duties of licenser and licensee, for example in cases of infringement, and deal with the payments of royalties by the licensee. In a number of states, approval of licence agreements by the authorities is bound to certain conditions such as local investment by the licenser or the existence of technology transfer agreements, *etc.*, and recording of the trade mark licence is mandatory. In many jurisdictions, licence agreements will be held invalid if the stipulated conditions limit the licensee's possibilities of effectively competing with the licenser on the market. In other countries, such as in the UK and similar Common Law countries, 'user registrations' are provided for in the case that continuing control over the mark's further use is exercised. The thus permitted use of the mark by the 'registered user' is deemed to be use by the mark's proprietor and obviates the risk of losing the mark by negligently tolerating its use by somebody else.

4.13.6 TERMINATION OF REGISTRATION

Except for the regular expiration of its term, there are three possibilities by which a trade mark registration can terminate for all or part of the goods in respect of which the mark is registered. First, a mark can voluntarily be surrendered by its owner. A mark can, furthermore, be revoked from the register if it has not been used within a prescribed period of time preceeding the allegation of non-use. A trade mark may also be revoked from the register on the grounds that it has become a generic term for the product it identifies, or that it misleads the public with regard to the characteristics of the goods it marks. Finally, a trade mark registration may be declared invalid if the absolute and relative registration criteria provided by law have not been met. If revoked, a trade mark registration becomes legally ineffective as of the date when the grounds giving rise to the revocation existed. If declared invalid, a trade mark registration is deemed null and void within the tenor of the respective decision, as of the registration date. Some jurisdictions provide for a grace period, generally one year, after expiry of a trade mark registration, during which the former registrant only may apply for the reregistration of the mark.

4.13.7 INFRINGEMENT OF TRADE MARK RIGHTS

4.13.7.1 Acts

Anybody using a mark which is identical or confusingly similar to a mark registered in respect of identical or similar goods, in the course of trade and without the consent of the registered owner of the mark, is considered to commit trade mark infringement. Essential conditions of infringement are the use of the mark in writing and in a trade mark like fashion. Oral use of a mark does not constitute infringement, although unfair competition or passing-off actions might be brought in such a case, nor do acts where a mark is referred to in an explanatory fashion with regard to goods but not used in a trade mark like sense. The protection conveyed by the registration of the mark also extends to acts committed after the marked product has been lawfully put into circulation. It covers acts such as alterations of the mark by the purchaser, changes made with regard to the marked goods or their packaging, resale of marked goods in different containers and the like. The removal of the mark from the goods by the purchaser and resale under his own mark does not, however, constitute trade mark infringement, although quality issues, *etc.* may still be raised.

4.13.7.2 Counterfeiting

A specific form of trade mark infringement is the counterfeiting of the marked goods of a particular trade mark owner. It concerns mainly high quality products, such as pharmaceuticals.

Counterfeiting is the practice of using a trade mark which is identical or quasi-identical to the trade mark owner's original mark, in connection with falsified products which are at first sight indistinguishable from the genuine goods but are mostly of inferior quality. Counterfeiting deceives the consumer with regard to the object and to the quality of the products in question and deprives the trade mark owner of his legitimate business. Specific legislation is developing now in many jurisdictions providing for civil and criminal remedies, including the seizure and destruction of counterfeit products.

4.13.7.3 Remedies

Infringement actions are to be brought before the courts. The registered owner establishes a *prima facie* case of ownership of his mark, submitting as evidence his certificate of registration. It is then up to the defendant to oppose this proof by showing that the registration has been wrongfully obtained and that the rights of the registered owner have been lost, for example by degeneration of his mark to a generic term. If the court finds for the trade mark owner, civil sanctions can be imposed or also penal sanctions if the infringement was committed negligently or deliberately. The available remedies are usually the grant of an injunction and the award of damages. A precondition for bringing an action for infringement is the thorough policing of an enterprise's trade marks. A close watch of the market as to the appearance of conflicting trade marks and the installation of an appropriate surveillance operation with regard to newly published marks is indispensable in order to avoid the dilution of a company's trade mark rights by the marks of others coming too close to it. Specialized surveillance companies and, in pharmaceuticals, market-oriented publications facilitate the policing of a trade mark portfolio and contribute essentially to legal security.

4.13.8 THE USE OF TRADE MARKS

Trade marks are among the most valuable assets of a company. To avoid distinctive marks degenerating into a generic term, *i.e.* into an ordinary word of everday language, rules have to be set up and observed. Degeneration of a mark to a generic term can be brought about by its association on the part of the public with the product as a genus and by inactivity on the part of the trade mark owner to prevent such association and degeneration from taking place. Particularly endangered in this respect are marks of highly successful products.

A mark may be considered an adjective. One of the basic rules of proper trade mark use is to use the mark in such a way together with the common name of the product and to identify it clearly as a mark. This means that the 'adjective' mark is used together with the generic term of the product, but, of course, not written together, for example in one word or combined using a hyphen. It follows that a mark should always be written in capital letters and should be designated with the registration symbol ®, on the product itself, on the packing material and in all printed material relating to the product. The letters TM may be used to indicate that the respective mark is not yet registered. Footnotes referring to a mark, such as 'Registered Trade Mark of such and such a company', or similar remarks on the product will equally serve to clearly identify a word as a mark. The mark must never be declined, turned into a verb, conjugated, used together with other words to form composite terms or arbitrarily adapted to another language. There are no objections, however, to the simultaneous use of two marks, for example a product mark and a house mark or other umbrella mark of the same or of different ownership, if so indicated.

4.13.9 MISCELLANEOUS

4.13.9.1 Parallel Importation

Trade mark rights are in principle 'territorial rights', *i.e.* trade mark protection extends to the territory of the country in which the right was granted only. Once goods bearing a given mark have lawfully been put on the market of a country, the exclusive proprietary right to the mark is 'exhausted' in respect of his territory; in other words, the trade mark proprietor can no longer legally prevent consecutive resales of his marked goods on the national level, as long as the goods are resold under the original mark in 'unimpaired' condition. Relevant standards have been set in a number of court decisions. Precondition for so-called parallel importation is the disparity of prices for the same product in various states. This is particularly true for pharmaceuticals where prices, for one reason or another, are fixed at different levels in different countries. Parallel importation, then, arises when a product is sold by the trade mark owner, who is, usually, also the originator of the product, at a low price in one country, is purchased there by a third party, imported into a high price country and sold there without the authorization of, and often in competition with, the trade mark owner. Theoretically, the territorial rights bestowed by trade mark registrations are independent of each other, so that the exhaustion of one of the national rights should not affect parallel rights existing in other countries. Over the years, however, the doctrine of exhaustion was adopted by the courts of various states, including the European Court of Justice, in relation to the parallel importation of goods, too. It has constantly been held in this context that when the owner of a trade mark has parted from his proprietary right in one country by putting his marked product there into circulation, his rights to the mark in other countries, although they continue to exist, cease to be executable as a means of preventing parallel importation ('international exhaustion'). Under these rulings, therefore, parallel importation (or, as the case may be, reimportation of goods into the country of origin) does not constitute infringement of the trade mark rights of the owner of the mark in the respective territories. This philosophy is essentially followed in the European Community where the free flow of goods is a fundamental principle dominating the execution of industrial property rights (exceptions under certain conditions provided) and also in other areas, where, for example, the owner of the mark had expressly or implicitly consented to the use of his mark by subsidiaries or other parties.

4.13.9.2 Generic Names

As opposed to trade marks which serve to distinguish the goods of an enterprise, generic names implicitly do not. Basically, a generic name is the common name for a type of product and is freely available for use by anyone, hence cannot be monopolized. Although generic names lack distinctiveness and do not function as a mark in relation to the products they describe, a generic name if arbitrarily applied to an unrelated entity may very well serve as a mark. This is equally true for a generic name which has acquired distinctiveness in the course of trade.

In pharmaceuticals three types of nomenclature are in use. First, the trade mark which characterizes a formulated product in its entirety, including its inherent quality and properties. Then, the scientific name, which designates the pharmacologically active substance contained in the formulated drug, which embodies the systematic chemical nomenclature. Finally, there is the generic name which is used for convenient reference to the active principle of pharmaceuticals. Scientific names are usually long, complex and difficult to handle and to remember. Alternatively, the corresponding generic names are shorter, consist of one word generally composed of four to seven syllables and do not include any figures, brackets, commas, hyphens, *etc.* They are designed to convey some kind of use information to doctors and pharmacists and show often pharmacological group relationship. Recurring stems, prefixes and suffixes of generic names refer to recurring structural units in groups of structurally related compounds. In a number of countries generic names are devised and published by specially set up bodies according to established guidelines and procedures: USAN—United States Adopted Name in the US; BAN—British Approved Name in the UK; DCF—Dénomination Commune Française in France; and others.

International Non-proprietary Names (INN) are made available for world-wide use by WHO. They are selected in a two-step procedure. The possibility of commenting on the WHO's proposals or for filing formal objections by interested parties is provided for. Causes for objection are usually proposed non-proprietary names that are confusingly similar to existing trade marks. If there are no objections, or the parties agree on some compromise, the proposed or modified INN will be published by WHO as a so-called recommended INN. In some countries INNs are taken into account in the examination of applications for trade mark registration. Applications for trade marks which come too close to recommended INNs will officially be rebutted. Generic names are a quite useful tool for publishing, communicating and other activities in science and technology, for use in formularies and pharmacopoeias and everywhere where short designations for distinct chemical entities are needed. Their practicability as a means for identifying pharmaceutical products is controversial. In view of the enormous number of existing chemical compounds and their close structural relationship the number of non-proprietary names which closely resemble each other is equally very large. This, quite obviously, renders their handling in relation to pharmaceutical products rather difficult.

A pharmaceutical product which is marketed under one or more generic names is commonly referred to as 'generic'. A generic no longer enjoys patent protection with regard to its active ingredient and is usually, but not necessarily, available from a multitude of sources including the originator. To distinguish themselves from other generic manufacturers and to build up reputations for their generic lines some generic manufacturers market so-called 'branded generics'. As distinguished from 'commodity generics', these 'branded generics' are sold by reference to the generic names of their active principles but indicate their source by the display of an individualizing sign such as the trade name of the manufacturer or his specific product mark. In order to cut costs, while sustaining product quality, bioequivalence and bioavailability 'generic substitution', whereby a cheaper generic is dispensed instead of the prescribed trade marked product of the originator, is promoted in many countries.

4.13.9.3 International Agreements

There are essentially three agreements on the international level which form the basis for multinational trade mark work. The Paris Convention of 1883 to which at present 99 states are party provides that each national of a convention country is granted in all other convention countries the same right and protection as enjoyed by the nationals of these countries. It further provides for the right of priority: anybody who has filed a trade mark application in a convention country is entitled to claim the priority date of this first application when filing corresponding applications in any of the other convention countries. In the case of marks the priority period is six months.

The Madrid Agreement of 1891 to which at present 27 states are party provides that protection for a trade mark may be obtained, potentially in all member states, by virtue of one deposit of the mark at the International Bureau of WIPO provided that certain conditions for such 'international registration' are met. Under this system protection may be achieved by performing one single act of filing in one place with one set of application papers in one language and with one payment of fees. The effect of the international registration is that in each member State the same protection is obtained as if the mark had been applied for directly in the country in question. At present, the term of an international registration is 20 years, and renewable. The membership comprises most of the Western and Eastern industrialized countries, notable exceptions being the US, the UK and Japan.

The Nice Agreement of 1957 to which at present 33 states are party provides for a classification of goods and services in connection with the registration of marks. The classification system comprises 34 classes for goods and eight for services together with an alphabetical list tabling about 10 000 of such goods and services for easy reference.

4.14
Sources of Information

STEPHANIE A. NORTH and SANDRA E. WARD
Glaxo Group Research Ltd, Ware, UK
JANICE SKIDMORE
Smith Kline & French Research Ltd, Welwyn, UK
and
IAN J. TARR
PJB Publications, Richmond, UK

4.14.1 INTRODUCTION

Published information is the life blood of the medicinal chemist. State-of-the-art reviews set the scene for new projects, the literature yields clues to structure–activity relationships, provides the stimulus for developing new reaction sequences and is also the basis for the ultimate test for any new medicine—that of novelty.

The medicinal chemist requires information not only on chemistry but on biology, medicine, biochemistry and toxicology amongst other disciplines.[1–3] In all these sciences, the volumes of published information continue to grow dramatically. This growth in sources of information is matched by the pace of development in methods of accessing information, deriving principally from the advent of computerized versions of traditional sources. The application of the computer to scientific information handling has not only expanded the versatility and selectivity of techniques for searching published information but has increased currency of publication and has provided novel methods for delivering the latest information. The first public computer systems for searching 'Chemical Abstracts' (CA) appeared in 1973 and their use was largely the province of the information scientist. In 1989, the medicinal chemist can receive details of the latest journal articles on floppy disc for storage in a personal computer file, can search the full text of many journals on a remote computer and has access, at the desk, to a variety of abstracting services, encyclopaedias and numeric databanks *via* increasingly sophisticated searching tools which provide document request facilities if the original article is required. These new methods of access to information are encouraging their direct use by the individual chemist.

Chemical literature can be considered to fall into three groups.[2] The *primary* literature is composed of the first recordings of original research, *e.g.* conference proceedings, a journal article, a patent or a thesis. *Secondary* literature provides a summary of that first recording and includes indexes and abstracts to the primary literature and their publicly available computerized versions. Reaction indexes fall into this category. Reviews, reference works, handbooks and encyclopaedias which provide concise and evaluated summaries of this primary and secondary material are classified as *tertiary* sources.

In this chapter we intend to concentrate on the secondary and tertiary information sources which will support the medicinal chemist in two important areas: (i) using literature to develop ideas for target compounds, *i.e.* establishing what is known about the relationship of structure and biological activity including toxicology and metabolism, and (ii) acquiring information to facilitate the synthesis of target compounds and to confirm their identity. A variety of handbooks already exist which describe the printed sources of chemical and medicinal chemical information.[2,4–8] We will therefore concentrate on the electronic sources which offer much greater search flexibility, allow rapid re-searching where necessary and may be more current than printed sources. Computerized sources are increasingly the best way of identifying relevant information. In fact, researchers who cannot access published literature electronically are now severely disadvantaged.

Whether the medicinal chemist uses the tools we describe directly or *via* an intermediary, it is important that the potential of modern chemical and biomedical information services is clearly understood. This chapter will alert medicinal chemists to an exciting range of sources that lie within their grasp and will show how these can be exploited.

4.14.2 ONLINE AVAILABILITY OF MEDICINAL CHEMICAL INFORMATION

Many of the printed information tools familiar to the medicinal chemist are now publicly available as online databases that can be searched on remote computers *via* telecommunications networks. A few computerized sources are only available for use 'in-house', *i.e.* the source must be loaded on the user's own computer for searching. Some online databases are available, in whole or in part, for in-house use; criteria such as chemical structure, biological activity or date of publication can be used to produce relevant subsets.[9] Overall, the number of commercially available online databases has grown from 400 to more than 3500 in the last 10 years.[9] The number of databases of interest to the medicinal chemist continues to grow and the launch of BEILSTEIN ONLINE[11,12] will provide structure and data searching access to this valuable source. Although this service is not publicly available at the time of writing, BEILSTEIN ONLINE is treated in this chapter as an already available service because of its enormous potential value.

Online databases are made available by database vendors or hosts, organizations which provide access to a multiplicity of databases on their computers. The databases themselves originate from database producers who may be publishers, learned societies, research associations, governments, *etc.* Database vendors of importance for the medicinal chemist include: STN International (STN), a cooperative venture of the American Chemical Society and Fachinformationszentrum Energie, Physik, Mathematik GmbH (FIZ Karlsruhe) and the Japanese Information Centre of Science and Technology (JICST); Télésystèmes-Questel; Chemical Information Systems, Inc., a subsidiary of Fein-Marquart Associates, Inc.; DIALOG Information Services Inc. (DIALOG); BRS Information Technologies (BRS); Deutsches Institut für Medizinische Dokumentation und Information (DIMDI); National Library of Medicine (NLM); ORBIT Search Service; and Data-Star.

Many databases are available from more than one vendor. Searching software varies considerably between different vendors both in ease of use and in search facilities offered. For instance, only a few vendors (STN, Télésystèmes-Questel's DARC (Description, Acquisition, Retrieval and Correlation system) and Chemical Information Systems, Inc.) currently provide real facilities for structure searching. The size of the database offered also varies from vendor to vendor as well as, more rarely, the database content; for example abstracts from CA are only available on STN. To identify databases of interest and the vendors from which they are available database directories[10] can be used; these are available online as well as in printed form.

The first databases available online were simply computerized versions of abstracts journals containing bibliographic citations (title, author, journal reference) with the text of the abstracts and index terms produced as a by-product from the publishing process for the printed service. The number of these bibliographic databases, also known as reference databases, continues to grow.[10] A reference database provides only a method of identifying articles in which useful information may be found; the original article will then normally need to be consulted. Source databases are, in contrast, either databases containing the full text of the original source material, *e.g.* a journal article, or databanks containing textual–numeric or numeric data, *e.g.* the computerized versions of dictionaries and handbooks. Some source databanks have been specifically developed as machine-readable compilations.[2] Source databases thus provide immediate access to usable information.

The databases available for use by the medicinal chemist are of both types. Table 1 summarizes the secondary and tertiary published sources of principal interest for structure and activity searching. It indicates where these are available as online databases and the principal characteristics of the online version. Table 1 also includes some sources which are only available as computerized services. Details of the relative strengths of these services are explored in later sections. It should be noted that the timespan of many online databases is much less than their printed equivalents; this is less of a disadvantage for biomedical sources where older information is of less importance than for chemical information where, for instance, very old synthetic routes can still be of utility. Table 1 does not include reaction indexes which are discussed in Section 4.14.5.

Two services included in Table 1 are worth some expansion here. TOXLINE[4] is a computerized collection of bibliographic information taken from 13 secondary sources and covers the pharmacological, physiological, biochemical and toxicological effects of drugs and other chemicals. It includes 'International Pharmaceutical Abstracts' (IPA), which is also available *via* a number of other vendors, and the following files of interest to the medicinal chemist: (a) 'Abstracts on Health. Effects of Environmental Pollutants' (HEEP)—published effects of environmental chemicals and substances other than drugs that affect human health extracted from 'Biological Abstracts' (BA); (b) 'Chemical Biological Activities' (CBAC)—records from CA covering interactions of chemical substances with biological systems; (c) 'Environmental Mutagen Information Centre File' (EMIC)—worldwide literature on testing of mutagenicity and genetic toxicology of chemicals,

Table 1 Secondary and Tertiary Information Sources for the Medicinal Chemist

Printed source/publisher	*Hard copy start date*	*Subject coverage*	*Online availability and database content*	*Database classification*[a]	*Database size and growth rate*	*Utility*[b]
'Chemical Abstracts' (CA)/Chemical Abstracts Service (CAS)[4]	1907	Worldwide literature in chemistry including patents, reviews, monographs, theses, conference proceedings and technical reports; indexes all characterized compounds featured in publications	Four categories of online source exist: Registry Nomenclature and Structure service (RNSS); full CA nomenclature, synonyms, CAS registry number, molecular formula, ring system information, *e.g.* CANOM, CHEMDEX, CHEM NAME, CHEMICAL NOMENCLATURE	S	9 800 000 structures, 600 000 structures per year	C, B, D, E
			CAS ONLINE and EURECAS also contain connection tables	S		
			CA SEARCH; bibliographic and keyword index entries from printed CA with CAS registry number and additional subject entries	R	>8 200 000 records, 480 000 records per year	
			Abstracts from 1967 are held on STN only	R	6 300 000 abstracts	
'World Patents Index' (WPI), 'Chemical Patents Index' (CPI) and alerting bulletins including 'Farmdoc' and 'Agdoc'/ Derwent Publications Ltd[4]	1951	Chemical and chemically related patents issued by 33 major patenting authorities; indexing features all generic (Markush) structures featured in patent claims. Farmdoc file covers pharmaceuticals	WPI and WPIL (World Patents Index and World Patents Index Latest); bibliographic reference, abstract, title, patent assignee, inventor name, patent numbers, ICIREPAT priority and publication date. Classification and subject codes from 1963 for pharmaceuticals	R	2 000 000 citations from 1963, 82 000 new patents per year	C, E, B
No print equivalent/IFI/ Plenum Data company[13]	—	US patent literature	From 1950, CLAIMS/UNITERM and CLAIMS/COMPOUND REGISTRY contain citations to all US chemical and chemically related patents and a listing of all chemicals cited more than five times since 1950 in this database. Citation, index terms and fragment codes are included	R	1 700 000 citations, 14 000 compounds, updated weekly, 72 800 citations per year	C, R
'Index Chemicus'/Institute for Scientific Information (ISI)[4,14]	1960	New organic compounds and syntheses from 110 journals; compounds indexed include non-isolated intermediates	Not currently publicly available. Was available as Index Chemicus Online. Available for in-house use as Index Chemicus in-house database containing abstract, structure, molecular formula, number of atoms, abstract number, bibliographic reference, scientific data alerts, Wiswesser line notation (WLN) or CAS SDF-1 from 1962 to 1987	R, S	3 500 000 articles, 20 000 references and 192 000 compounds per year, Index Chemicus in-house database contains 4 200 000 structures	C, B, D

'Beilstein's Handbuch der Organischen Chemie' (Beilstein)/Beilstein Institute[5]	1830–1980	Worldwide chemical literature including journals, patents, reports relating to fully characterized compounds featured in source literature	BEILSTEIN ONLINE; structure and numerical data on preparation, physical properties, chemical behaviour, physiological behaviour and characterization data	S	1 600 000 compounds (all heterocyclic, some alicyclic). Eventually 3 500 000 compounds to be available	C, D, E
'Ringdoc Abstracts Journal'/Derwent Publications Ltd[15]	1964	Worldwide pharmaceutical literature	RINGDOC; bibliographic reference, abstract, keywords and fragment codes, from 1964	R	54 000 records per year	C, B, D
			Standard drug file (SDF); a companion dictionary to Ringdoc. Name, activities, fragment and ring codes of known drugs and other commonly occurring compounds	S	17 000 compounds	C, B
'Dictionary of Organic Compounds' (5th edn.), 'Dictionary of Organometallic Compounds', 'Amino Acids and Peptides', 'Carbohydrates'/Chapman and Hall Ltd[4,16]	5th edn. 1982, covering literature to 1980	Summarized information on fundamental chemicals extracted from worldwide chemical literature. Dictionaries of antibiotics and related compounds to be added	HEILBRON; molecular formula, molecular weight, chemical names, CAS registry number, physical and chemical properties, bibliographic source, compound type	S	215 000 compounds, 70 000 records, updated, biannually	C, D, E
'Excerpta Medica' (EM) abstract bulletins/Elsevier Science Publishers bv.[7,17]	1947	Worldwide biochemical li terature on human medicine and related areas of biological science. Comprehensive coverage of articles on drugs and potential drugs from 4500 journal sources	EMBASE is the full file; subsections exist, *e.g.* EMDRUGS which corresponds to the Pharmacology, Drug Literature Index and Adverse Reactions Section of EM. Files contain bibliographic citation, abstract, trade names, index terms, CAS registry number, from 1988	R	3 500 000 references, 300 000 references per year	B, D
'Biological Abstracts' (BA) and 'Biological Abstracts/RRM' (Reports, Reviews and Meetings)/Biosciences Information Service (BIOSIS)[17]	1926	Worldwide literature on life sciences from 9000 periodicals, books, patents, monographs, conference proceedings, symposia, research communications	BIOSIS PREVIEWS; from 1969, bibliographic citation and index terms, chemical name. Abstracts from 1976; the BIOCAS file contains references common to CA and BIOSIS PREVIEWS with CAS registry number	R	5 000 000 references, 480 000 records per year from BA and BA/RRM	B
'Index Medicus'/National Library of Medicine (NLM)[7,17]	1879	3200 worldwide journals from broad fields of biomedicine, research, clinical practice, policy and administration. Monographs included from 1981	MEDLINE; from 1964 bibliographic citation; author abstracts for 60% of papers from 1975, index terms, CAS registry number	R	5 200 000 references, 300 000 records per year	

Table 1 (*Contd.*)

Printed source/publisher	*Hard copy start date*	*Subject coverage*	*Online availability and database content*	*Database classification*[a]	*Database size and growth rate*	*Utility*[b]
No print equivalent/ National Library of Medicine (NLM)[4]	—	Nomenclature and structure information for compounds in NLM databases and TSCA inventory	CHEMLINE; CAS registry numbers, molecular formula, CA nomenclature, synonyms, ring descriptors	S	750 000 compounds, updated every two months	
No print equivalent/ National Library of Medicine (NLM)[4]	—	Worldwide literature on toxicology including chemicals, pharmaceuticals, pesticides, pollutants, mutagens and teratology compiled from 13 secondary sources	TOXLINE (a composite databank); bibliographic citations with abstracts	R	1 600 000 references, 144 000 records per year	B
No print equivalent/Fein-Marquart[17,18]	—	Commercially used chemicals including drugs included in the CIS (Chemical Information System), a collection of databases of physical property, safety, analytical and regulatory data on compounds represented in the Toxic Substances Control Act (TSCA) Inventory and in other US Federal agency files	SANSS (Structure and Nomenclature Search System); molecular formula, molecular weight name, synonyms, connection tables, CAS registry number	S	350 000 compounds	C
'Negwer'/Acadamie-Verlag Berlin[7]	1987	Drugs and organic chemicals; 10 000 compounds featured	—	—		
'Merck Index'/Merck and Company Ltd[4,19]	10th edn. 1983, covering literature from late 19th century to present	Summarized information on chemicals, drugs; and veterinary and agricultural products biologicals from worldwide literature including patents, journals, books and proceedings	THE MERCK INDEX ONLINE; molecular formula, molecular weight, systematic names including CA nomenclature, generic and trivial names, CAS registry number, physical and toxicity data, therapeutic and commercial use and bibliographic citations	S	10 000 monographs describing 30 000 substances, updated each six months	C, B, D
'Martindale: The Extra Pharmacopoeia'/The Pharmaceutical Society of	28th edn. 1983	Summarized/evaluated information on drugs used clinically from worldwide	MARTINDALE ONLINE; drug names, synonyms, codes chemical names, molecular formula, molecular weight, CAS registry	R, S	58 000 bibliographic citations on 5100 drugs and 20 000 preparations, 100 000 records	C, E, B, D

Great Britain[20]		scientific literature plus toxic substances and compounds used in manufacturing	number, physical and pharmaceutical properties, uses, preparations		in total, updated biannually	
'Pharmaprojects'/PJB Publications Ltd[21]	1979	New pharmaceutical formulations and compounds under development	PHARMAPROJECTS; generic and trade names, chemical name, CAS registry number, originating company, therapeutic activity, development status and references	R	5000 products including new formulations and compounds in development, 1200–1800 per year added	C, B, D
No print equivalent/IMS International[22]	—	Pharmaceutical products on the market worldwide or in research and development gleaned from a variety of sources plus full text of manuals, newsletters from IMS, *e.g.* 'Drug License Opportunities'	IMSBASE; chemical, brand generic names, planned launch date, trial status, company	S, R	> 50 000 compounds from 1977, updated weekly or periodically	B
'Registry of Toxic Effects of Chemical Substances' (RTECS)/US National Institute of Occupational Safety and Health (NIOSH)[23]	1970	Unevaluated literature covering toxicological measurements on chemicals	RTECS; CAS nomenclature and registry number, synonyms, molecular formula, name fragments, WLN, animal species, dosage methods, toxicity measures and values	S	135 000 measurements on 86 000 chemicals; quarterly update	B, C
National Library of Medicine (NLM)[4,24]	—	Substances of known toxicity to which substantial populations are exposed	TOXNET; composite Information Source System including RTECS and Hazardous Substances Databank (HSDB); substances identification (CAS nomenclature, registry number, synonyms, molecular formula), manufacturing/use, chemical and physicochemical properties; toxicity/biomedical effects, pharmacology, exposure standards, analytical techniques	S	4100 substances; updated periodically	B, D
US National Institute of Health (NIH)[4]	—	Worldwide literature (environmental surveys, IARC monographs, cancer journals) on substances which have undergone carcinogenicity and mutagenicity studies	Chemical Carcinogenesis Research Information System (CCRIS); bibliographic references and data on carcinogenicity and mutagenicity studies; includes positive and negative data	S/R	1100 substances; 200 chemicals per year	B
'International Pharmaceutical Abstracts'/American Society of Hospital Pharmacists[24]	1963	International Pharmaceutical Abstracts; literature pertaining to the development and use of drugs and to aspects of pharmaceutical practice; IPA covers over 800 journals	IPA; 1970 to date; abstracts, bibliographic citations, index terms	R	120 000 references; 14 400 per year	B

Table 1 (*Contd.*)

Printed source/publisher	*Hard copy start date*	*Subject coverage*	*Online availability and database content*	*Database classification*[a]	*Database size and growth rate*	*Utility*[b]
American Chemical Society (ACS) journals, Royal Society of Chemistry (RSC) journals/ACS, RSC[10]	Various	18 ACS and 10 RSC journals including 'Journal of Medicinal Chemistry', 'Journal of Organic Chemistry', 'Journal of Chemical Research', 'Faraday Transactions' 1 and 2 and 'Angewandte Chemie'	CJO (Chemical Journals Online); full text including captions, references and footnotes	S	67 000 articles, 9100 articles per year updated fortnightly	B, D
'Pascal' (formerly Bulletin Signaletique)/Centre Nationale de la Recherche Scientifique[25]	1940	Worldwide scientific literature including 4200 journals, books, theses, conference proceedings. Covers chemistry, biology and medical sciences	PASCAL M; from 1973 bibliographic citation and abstracts, English title and keywords	R	300 000 citations, 32 400 records per year	C, B, D
US Environmental Protection Agency (EPA)[10]	—	Worldwide scientific literature on mutagenicity	GENETIC TOXICITY; test data, bibliographic reference	S	2600 compounds, periodic update	B
No printed equivalent/ US Environmental Protection Agency (EPA)[26]	—	Published literature 1967–1984 examined to yield information on gastrointestinal absorption or excretion, metabolism distribution as part of the SPHERE project	GIABS (gastrointestinal absorption database)	S	>9200 references to >3400 studies of more than 2400 compounds	B
'Science Citation Index'/ISI[27]	1961	4500 scientific and technical journals	From 1974, SCISEARCH; ISI/BIOMED features citations to 1400 biomedical journals only and ISI/MULTISCI features 2500 scientific (including chemical) journals; content comprises author cited reference	R	8 500 000 record, 720 000 records per year	B, D

[a] R = reference database, S = source database. [b] C = facilities for structure and substructure retrieval, B = source provides access to biological data, D = source provides access to physicochemical data, E = source gives information on synthetic routes.

biological agents and selected physical agents; (d) 'Environmental Teratology Information Centre File' (ETIC)—teratogenic activity of chemical, biological and physical agents; (e) 'Pesticides Abstracts and Hayes File on Pesticides'—epidemiological effects of pesticides on humans; (f) 'Toxic Materials Information Centre File' (TMIC)—citations and abstracts on toxic materials prepared by the TMIC; (g) Toxicity Bibliography (TOXBIB)—subset of MEDLINE database covering adverse effects, toxicity, poisoning or environmental effects caused by drugs or chemicals; (h) 'Toxicology/ Epidemiology Research Projects' (RPROJ)—descriptions of US research projects; and (i) 'Toxicology Document and Data Depository' (TD3)—information on the toxicology reports literature from the National Technical Information Service (NTIS) database, a collection of more than a million government and government-sponsored technical research reports, predominantly from the USA.

The NIH-EPA Chemical Information System (CIS)[28] is an extremely comprehensive chemical data system developed under the auspices of the US government and including files of mass spectra, toxicity data and IR and NMR spectra. CIS also provides access to THE MERCK INDEX ONLINE (MERCK). A central structure file, the Structure and Nomenclature Search System (SANSS), provides a record of all structures included in the individual databanks and directs the user to which datafiles include information on a compound of interest. Not all the CIS datafiles are exclusive.

By bringing together information from a variety of sources these composite collections simplify access to online chemical information tremendously.

4.14.3 ACCESS TO CHEMICAL INFORMATION *VIA* CHEMICAL STRUCTURES

4.14.3.1 Introduction

Chemists enjoy an advantage over scientists in other disciplines in that much of their information is organized about a common language, namely the **structural** representation of a compound. The structural representation is used to index a compound or class of compounds described in the literature and so is of use in retrieving its associated information, *e.g.* preparation, reactions, physicochemical properties and biological activity.[29] A chemical structure is, therefore, an invaluable 'retrieval tool' for the medicinal chemist.

This section outlines the different structure-based indexes employed by the major information sources and shows how they can be used to locate either a single compound or a class of compounds containing a common structural component, *i.e.* a substructure.

4.14.3.2 Indexes Based on Chemical Structure

The methods developed for representing a chemical structure range from a simple atom count—the molecular formula—to sophisticated three-dimensional computer models. Some of these representations have been used to index the major secondary and tertiary chemical information sources; the most important indexing tools are: molecular formula, nomenclature ('trivial' and systematic), linear notations, fragment codes, connection tables/structure diagrams and registry numbers/codes.

A number of textbooks describe the development and utility of these systems;[2,4,5,30] the reader should refer to these and the additional references cited in the following sections where further detail is required.

4.14.3.2.1 Molecular formula

Although this has the advantage of being widely used, it is a fairly crude retrieval tool as there may often be several hundred compounds with the same molecular formula, even in a relatively modest compendium, in terms of compound numbers, such as MERCK.[19] In practice, almost all molecular formulae indexes include a cross-reference to a chemical name, in itself a source of ambiguity. Further problems arise from inconsistencies in the way that molecular formulae have been expressed over the years in a given source, and in the different formats used by different sources. For example, a salt may be dot-disconnected, as in $C_2H_4O_2 \cdot Na$, or included in the formula, as in $C_2H_4NaO_2$. Most molecular formulae indexes now employ the Hill system.[31]

4.14.3.2.2 Nomenclature

Fully systematic names can be used for precise retrieval of single compounds in some sources, but many sources only use non-systematic 'trivial' names in their indexes. Within medicine, the use of nomenclature is particularly complicated as there may be several different approved generic names for a medicinal substance, for example the INN (the International Non-proprietary Name approved by the World Health Organization), the BAN (British Approved Name) and the USAN (United States Adopted Name).[32] This proliferation of generic names, synonyms and trade names, rarely cross-referenced, often leads to incomplete or imprecise information retrieval. CA's nomenclature has emerged as the *de facto* standard for indexing. However, even systematic names are of little value for locating analogues containing particular substructures, as just a change in one functional group can result in a very different location for the systematic name in an alphabetically ordered index.[33] Some computerized dictionaries allow the combination of name fragments when searching and these, to a certain extent, enable a searcher to retrieve compounds containing, for example, the same functional groups or ring systems, *etc.*[30]

4.14.3.2.3 Linear notations

The most widely used notation is the Wiswesser line notation (WLN),[2,34] though its use is now declining. Linear notations provide unambiguous representations of structures as an alphanumeric code: essentially they are a means of translating a structure diagram into a computer-readable string of characters, numbers and symbols. Notations are superior to nomenclature because they allow retrieval of both single compounds and substructures. However, a specialized knowledge of the particular code used is required. Notations handle two-dimensional structures easily but cope with stereoisomerism with difficulty. The principal published online chemical source using WLN was ISI's Index Chemicus database which is no longer available.[35]

4.14.3.2.4 Fragment codes

Coding systems have been developed to represent common structural features or fragments such as functional groups, aromatic rings, *etc.* that appear in molecules. A partial description of a structure can, therefore, be built up simply by listing which of the fragments featured in the code are present, though this description does not specify the connections between the fragments.

Amongst the fragment coding systems used to index collections of compounds are the GREMAS[36,37] code developed by IDC (Internationale Dokumentationsgesellschaft für Chemie) and the New Chemical Code[38] used by Derwent Publications Ltd for their World Patent Index Online (WPI). Fragment codes are an ambiguous and partial description of a structure (*i.e.* very differently arranged molecules have the same fragments and not all atoms will be included in fragment codes). Like linear notations, they require a specialized knowledge of the coding systems used and, because of their ambiguity, are imprecise retrieval tools. Their use is now largely restricted to patent documentation systems and the Ringdoc database of pharmaceutical literature (RINGDOC). Fragment codes are cumbersome to use; a fragmentation code program 'TOPFRAG' has been developed to take a structure drawn using a personal computer and generate the fragments required for searching RINGDOC and other Derwent databases.[39] In connection table based structure searching systems, automatically generated fragment screens are used in the preliminary stage of structure searching.

4.14.3.2.5 Connection tables

A connection table describes a structure by defining the atoms it contains and the connections between them.[2] A number of connection-table-based systems have been developed including CROSSBOW (Computerized Retrieval of Organic Structures Based on Wiswesser),[30] Chemical Abstracts Service (CAS),[40] MACCS (Molecular Access System)[41] and SANSS,[18] and further work is in progress to develop connection tables that can accommodate Markush (generic) structures of the type found in patents.[39,42–45] Connection tables that include the stereochemical description for a compound are beginning to be seen in systems such as BEILSTEIN ONLINE.[46] There is also considerable interest in the use of connection tables for three-dimensional substructure retrieval as an aid to structure–property correlations in drug design. Connection tables provide a method for storing large computer-searchable files of structures and the systems developed to search them, such as CAS ONLINE (the structure searching software developed by CAS and usable *via* STN),[2,47,48] DARC,[2,48,49] or SANSS[18,50] allow input of a query in the familiar form of a structure diagram. Such systems can, therefore, be used to retrieve precisely a single compound or a class of compounds containing a common substructure. Their flexibility and ease of use are ensuring that they become the retrieval tools of first choice.

4.14.3.2.6 Registry numbers and codes

Each structure stored in a computer file or database is usually allocated a unique identification tag or number which can subsequently be used as a shorthand for the structure in indexing systems. The CAS registry number, for example, is used to index articles in several secondary sources as well as in CA itself. If a registry number is known it can obviously be used to retrieve information on a compound. However, as the number bears no relation to any particular chemical feature of the structure, closely related compounds may have widely different registry numbers. The CAS registry number cannot, therefore, be used to browse through related compounds in CA.[32] To overcome this problem, BEILSTEIN ONLINE includes an additional registry number, the Lawson number, which is structure based. Related compounds can, therefore, be retrieved using the Lawson number index, in a similar way to the use of the Beilstein system number in the printed 'Handbuch der Organischen Chemie' (Beilstein).[5]

4.14.3.2.7 The major index files

Table 2 identifies which of the major chemical information sources can be accessed using one or another of the common indexing techniques together with those biological information sources that use some form of chemical substance indexing. More general information on these entries is given in Table 1. The sources range from the major abstract collections to small specialist directories and encyclopaedias, such as MERCK and MARTINDALE ONLINE (MARTINDALE) or their printed versions, featuring only a few thousand compounds. It should be noted that indexes to sources such as CA or Beilstein are themselves so large that they are published as separate volumes to the main work or stored as separate files in their online versions. As CA and Beilstein are of particular relevance to medicinal chemists, further details of their online versions are provided below.

The online structure index to CA is known as the CAS Chemical Registry System. Each specific (*i.e.* characterized) compound identified by CAS in the literature is added to this registry. The registry system was started in 1965 and all compounds listed in the earlier hardcopy indexes are being added retrospectively. Eventually, all compounds indexed in CA since its inception in 1907 will be included in the registry.[51] The registry includes a molecular formula, connection table (which can be used to generate and display a structure diagram), the CA systematic name and any synonyms, and the CAS registry number for each compound. As mentioned in Section 4.14.2 and Table 1, CAS have not released all these components to every online vendor, with connection tables only being available from two vendors, STN and Télésystèmes-Questel.[52] With other vendors, sophisticated dictionary files have been developed to compensate for this lack of real substructure searching by providing facilities to search the systematic nomenclature using name strings and combinations of these. The CAS registry number is also included in a separate file containing details such as title, author, journal name and, on STN only, the abstracts for all papers describing the corresponding compounds held in the registry file. The CAS registry number therefore provides the link between the structural and the bibliographic information; the relationship between the two files is illustrated in Figure 1. Some of the uses of this linking feature are described in Section 4.14.4.

Using the registry number as a link to the relevant textual information in the online CA file is equivalent to physically turning from one of the hardcopy indexes to the appropriate entry in the printed abstracts collection. However, the process can be fully automated online—the registry numbers representing up to 5000 compounds retrieved by a substructure search (see Section 4.14.3.3.2) in the compound registry can be crossed over to the bibliographic files, using just one command, to retrieve any references describing them in CA.

Searching CA online has a number of other advantages, including the availability of a more useful chemical index in the form of the connection tables, allowing precise retrieval of information about a

Table 2 Chemical Indexing Techniques on Major Databases

File	*Type of compounds*	*Number of Compounds*	*Molecular formula*[a]	*Fragment code/ notation*[a]	*Connection table*[a]	*Nomenclature*	*Registry number*
CAS ONLINE (registry file)	All characterized compounds featured in selected publications	8 100 000	Y	CAS fragment screens	Y	CA nomenclature and synonyms	CAS registry number
DARC (CA registry file)	All characterized compounds featured in selected publications	8 100 000	Y	DARC fragment screens	Y		
BEILSTEIN ONLINE	All characterized compounds featured in source literature	700 000 (1988) to >3 500 00 (1991)	Y	—	Y	IUPAC-based systematic nomenclature	Lawson number, CAS registry number
WPI	Generic (Markush) structures featured in patent claims: Farmdoc subfile covers pharmaceuticals	Immeasurable	—	Fragment code	—	—	—
Index Chemicus in-house database	New organic compounds including non-isolated intermediates reported in the literature (1962–1987)	4 200 000	Y	WLN	Y	—	—
SDF (companion dictionary to Ringdoc)	Pharmaceuticals	17 000	—	Fragment code	—	Synonyms, trade names	—
HEILBRON	Fundamental organic compounds including pharmaceuticals and antibiotics	215 000	Y	—	—	Synonyms, trade names	CAS registry number, RTECS reference number
SANSS (CIS)	Commercially used chemicals including drugs	348 000	Y	—	Y	CA nomenclature	CAS registry number
Negwer	Drugs and organic compounds	10 000	Y	Y	—	Synonyms, trade names	CAS registry number
MERCK	Drugs, biologicals, veterinary and agricultural products	30 000	Y	—	—	CA nomenclature synonyms	CAS registry number
MARTINDALE	Drugs, and ancillary substances	5100	Y	—	—	Synonyms, trade names	CAS registry number

CHEMLINE	Substances featured in NLM databases and the TSCA inventory	750 000	Y	Ring descriptors	—	CA nomenclature synonyms	CAS registry number
EMBASE	Drugs, biologicals	—	—	—	—	Associated vocabulary file, synonyms, trade names	CAS registry number (from 1988)
BIOSIS	Drugs, biologicals	—	—	—	—	Synonyms, trade names	BIOCAS concordance
MEDLINE	Drugs, biologicals	—	—	—	—	Associated vocabulary file	CAS registry number (from 1980)
TOXLINE	Compounds tested for toxic effects	—	—	—	—	Synonyms, trade names	
PHARMAPROJECTS	Drugs in R & D	5000	—	—	—	CA nomenclature synonyms, trade names	CAS registry number
IMSBASE	Drugs, biologicals	> 50 000	—	—	—	Drugs and chemical names	CAS registry number
RTECS	Compounds tested for toxic effects	—	—	—	—	Drugs and chemical names	CAS registry number
HSDB	Compounds tested for toxic effects	—		—	—	Drugs and Chemical names	CAS registry number

[a] Y = featured, — = not featured.

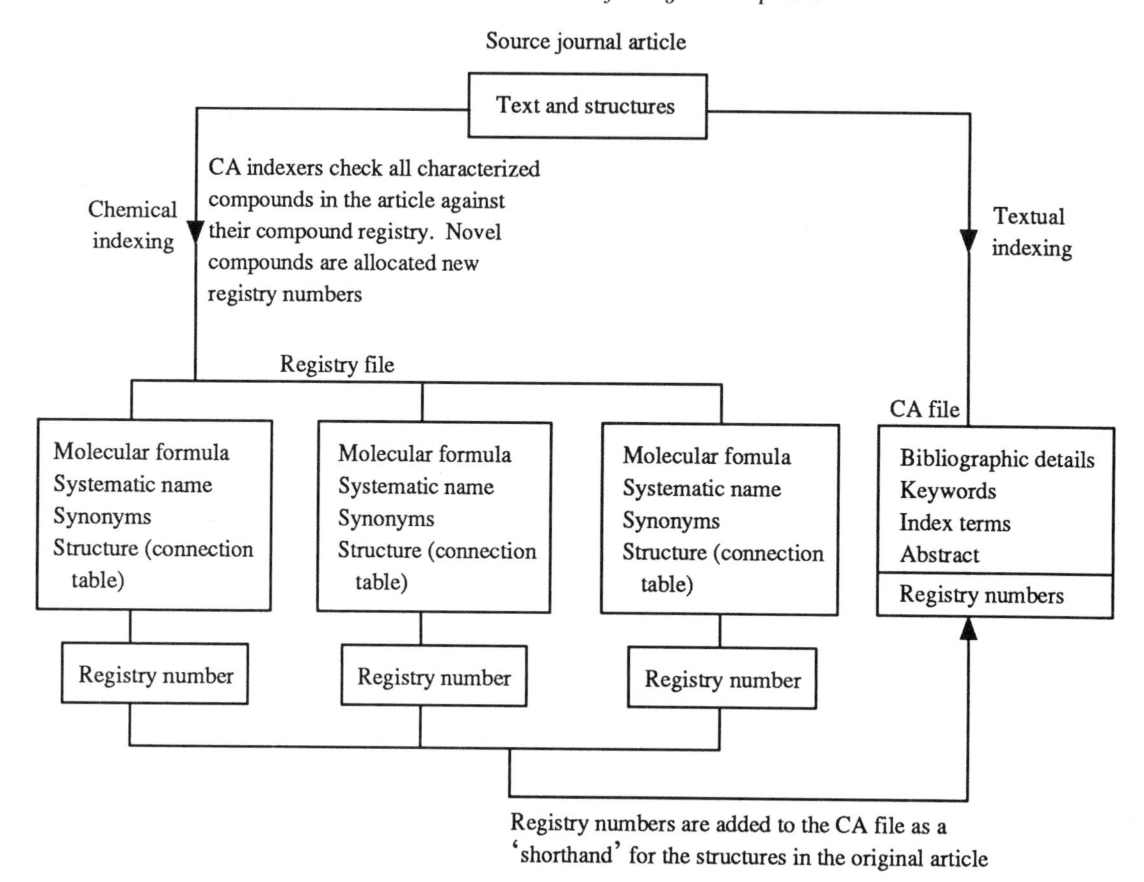

Figure 1 Relationship between structural and textual indexing in CAS ONLINE

particular substructure, as noted in Section 4.14.3.3. Parts of textual terms such as chemical names or biological concepts can also be used as search terms—a word can be truncated, as in 'metab' to retrieve references containing any word beginning with those letters, *e.g.* metabolic, metabolism, metabolised, metabolized, *etc.* Searching for each word individually would be extremely time-consuming using the printed indexes, and manually scanning each printed abstract would not be feasible. In contrast, the process only takes seconds online.

Searching CA online also allows search terms to be combined, for example to retrieve all references to compounds containing a particular substructure and a particular activity, or to be placed in a particular context, for example one term in the same sentence as another. Such retrieval techniques are not available when trying to find the same information in the equivalent printed sources. Many of these advantages of online searching also apply to information sources other than CA that are available in both electronic and printed forms.

The only other large, *i.e.* over 1 000 000 compounds, structural collection available online is BEILSTEIN ONLINE which can be searched *via* STN.[11] BEILSTEIN consists of a structure file and a factual data file and the structure file is also available for lease for in-house use. Eventually to contain more than 3 500 000 chemical structures, the Beilstein structure file is based on the Beilstein Registry Connection Table (BRCT) which, unlike CAS ONLINE, includes stereochemical information. The factual file includes:[46] structure-related data, *e.g.* purity and possible tautomers; preparative data, *e.g.* reactants, reagents, yield, conditions and by-products; physical properties, *e.g.* structure and energy parameters, melting point, optical properties and spectral information; chemical behaviour; physiological behaviour and applications; and characterization derivatives. On STN both the structure file and factual file have been combined as one database called the Beilstein File which contains data on 1 600 000 compounds from 1830 to 1980 (this includes all heterocyclic compounds and some alicyclic compounds). BEILSTEIN will eventually be available on other hosts including DIALOG and ORBIT.

BEILSTEIN ONLINE will make compound information available more quickly than has been possible with the printed version. The literature abstracted for the '5th Supplementary Handbook' series (1960–1979) and that from 1980 onwards will be made available initially in a non-evaluated form. For the critically evaluated data from the printed volumes, the data made available online will depend on whether the compound is classified as a large or small information compound (LIC or SIC). For SICs, all information in the handbook will be available; for LICs, *i.e.* those compounds which are regularly researched, not all the data can be made available online and the searcher will be referred to Beilstein for the residual information.

The flexibility of access promised—structure, data and keyword retrieval of critically evaluated data—will no doubt encourage the frequent use of this system by all medicinal chemists. For users of the hardcopy, a simple personal computer package, SANDRA,[11] allows the chemist to draw a structure and identify the system number and the area of the handbook where an entry will be found.

4.14.3.3 Using the Indexing Tools

4.14.3.3.1 Finding a single compound

Medicinal chemists may often need to find information about a single compound. Using the H_2 antagonist ranitidine as an example, the index entries using the various structural indexing tools would be as shown in Figure 2.

These various index entries provide the means for retrieving information about this single compound in different sources. In practice, a choice of indexing system and of index or search terms has to be made depending on the type of information required, and the amount of information already known about a given compound. For example, limited information about a known drug can easily be found in a pharmacopoeia using the molecular formula or trivial name as a search term. On the other hand, in order to determine whether or not a compound is novel a comprehensive search of the literature is required. This entails searching CA, Beilstein for the older literature and the WPI for compounds covered generically in patents. Unfortunately, the non-availability of Index Chemicus as a public online database removes easy access to a good source, which included the structures of intermediates not isolated during a reaction. The CHEMLINE dictionary is also a useful source. No one secondary source provides comprehensive coverage of the primary literature; each has different criteria for selecting which journals to cover, which papers to select from those journals and which compounds to index in each paper selected.

Structure (used to search files of connection tables)

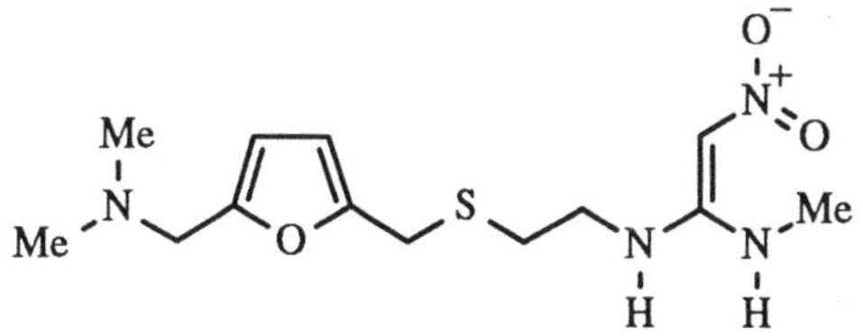

Approved name (BAN, USAN, INN):	Ranitidine
Synonyms/trade names:	Zantac, Zantic, Azantac, Ranacid, Ranidil, Raniplex, Sostril, AH 19065
Sytematic name:	*N*-[2-[[[5-[(dimethylamino)methyl]-2-furanyl]methyl]thio]-ethyl]-*N'*-methyl-2-nitro-1,1-ethenediamine
Molecular formula:	C13 H22 N4 O3 S
WLN:	T5OJ B1S2MYM1&U1NW E1N1&1
Derwent ringcode:	G7:02&;06&;067;075;10-;103;106;11&;11-;115;116;12&;122;123;124;125;13&;131;132;134;14-;151;158;17-;174;184;187;19&;19-;192;196
Chemical Abstracts registry number:	66357-35-5

Figure 2 Structural index entries for ranitidine

When searching for a specific type of information such as physical property data, it is advisable to turn initially to sources which may give the information directly, for example Beilstein or Heilbron's 'Dictionary of Organic Compounds',[16] as opposed to CA, which will only point you towards an appropriate reference. However, if a compound is not found in these, then a more comprehensive source such as CA should be consulted; getting at physical property data will be more tedious through this route but the information may be eventually located.

Effective retrieval, then, is a matter of choosing both the most appropriate source for the type of information required and the most effective indexing system for directing you to the information in that source. Tables 1 and 2 should assist in this choice.

Using an online retrieval system is generally thought to be more cost-effective than searching the same source manually and a number of 'user-friendly' aids are now being developed which make databases easier to search. For simple single-compound searches, it is expected that online searching will replace searching of the printed indexes in the near future, not only by information scientists but also by practising chemists.

4.14.3.3.2 Finding structural analogues or substructures

If a chemist is interested in compounds containing a particular common substructure for structure–activity studies or to find preparative methods for a series of close analogues, then the choice of viable methods for conducting a structure-based search becomes more limited. A single variable group in a structure can be enough to render a search by molecular formulae completely impractical. However, a search for the substructure depicted in Figure 3 could be achieved as described below.

If the variable (R groups) are limited then it remains possible to search by a combination of molecular formulae and systematic names. Thus, if the search was restricted to compounds where R = 1–3 carbon atoms the two structurally similar H_2 antagonists given in Figure 4 would be retrieved.

In effect, the chemist is merely looking repeatedly for several single compounds. However, such a technique rapidly becomes too tedious if the range of the enquiry broadens.

Some sources group compounds together by structural type, *e.g.* Beilstein, so allowing a certain degree of structural browsing. This is also true to a very limited extent in the CA Nomenclature Index for compounds indexed under the same 'Heading Parent'.

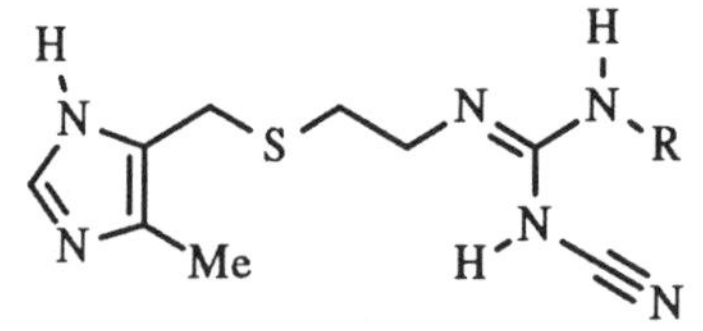

$R = C_1–C_3$

Figure 3 A typical substructure search enquiry

Cimetidine

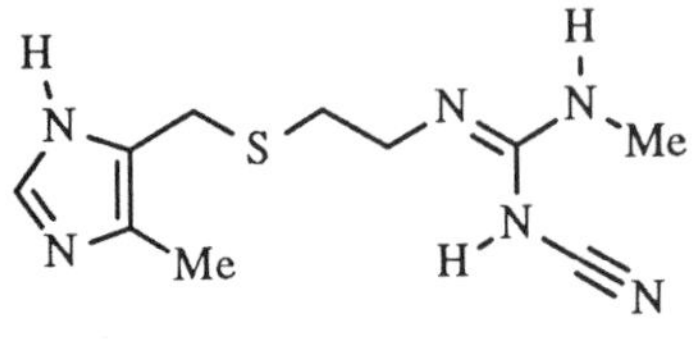

Molecular formula: C10 H16 N6 S

CAS name: Guanidine, *N*-cyano-*N'*-methyl-*N''*-[2-[[(5-methyl-1*H*-imidazol-4-yl)methyl]-thio]ethyl]-

Etintidine

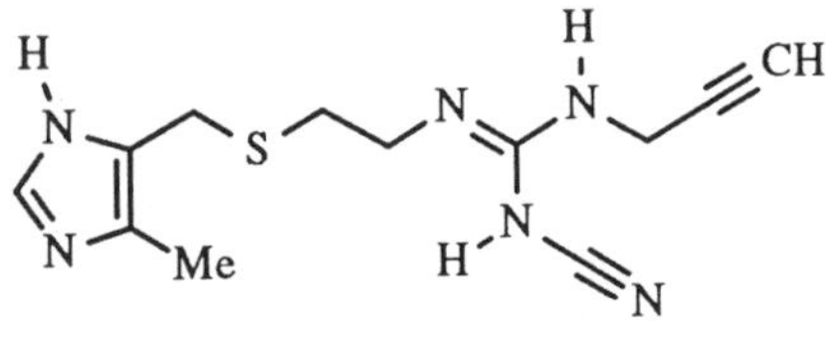

Molecular formula: C12 H16 N6 S

CAS name: Guanidine, *N*-cyano-*N'*-[2-[[(5-methyl-1*H*-imidazol-4-yl)methyl]thio]ethyl]-*N''*-2-propynyl-

Figure 4 Retrieved structures

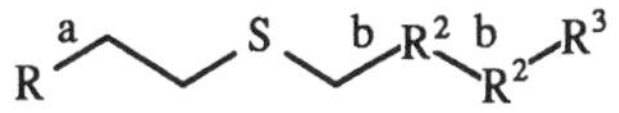

R^1 = O, N or S; R^2 = C or N; R^3 = C, N or S; a = any bond in a ring or chain, b = any bond

Figure 5 Complex substructure search

However, the most effective search technique is undoubtedly the use of a substructure search system such as CAS ONLINE; the less specific fragment-code systems providing another, poorer, option. Indeed, for broader substructural queries such as that shown in Figure 5 these are the only viable retrieval techniques.

This substructure search would retrieve several hundred compounds if used to search the CAS Chemical Registry System. The information associated with the compounds in CA's bibliographic files, *e.g.* How many are claimed as H_2 antagonists? How are they made?, could they be examined by linking appropriate keywords to the set of registry numbers (representing the compounds) retrieved by the substructure search. However, the retrieved abstracts and the original papers would then have to be examined for any available data.

Formulating a structural query such as the above calls for some expertise in using the vendor computer's command language together with an awareness of the general limitations of online substructure search systems. Choosing indexing terms to describe the context in which to place the retrieved structures calls for a detailed knowledge of the indexing policies employed by the various database producers (see Section 4.14.4).

4.14.3.3.3 Limitations of substructure search systems

Although online substructure search systems offer by far the most flexible and efficient means of retrieving compounds from large files of structures, they do have certain limitations.

Some systems allow stereochemical definitions to be included in a query as the connection tables that they search include this information (*e.g.* the Beilstein Registry connection table). However, others, most notably CAS ONLINE, do not. Consequently, in CAS ONLINE searches all stereoisomers are retrieved by a substructure search and identification of a particular stereoisomer requires scrutiny of their CA systematic names. Conventional connection tables do not include any information about the conformation of a molecule, though this area is being examined by several research groups.[29]

Substructure search systems such as CAS ONLINE search large files of connection tables in two stages: query structures are first analyzed for certain broad structural features (presence of rings of a given size, *etc.*) using a variety of screens to retrieve a more manageable number of structures which are then matched precisely (atom by atom) with the query structure drawn. If a query structure is particularly broad then many compounds will pass the initial screening stage; at present this number is restricted to a maximum of 50 000 on the CAS ONLINE system. For some types of query, this can be a severe limitation, *e.g.* a search for reactions involving the conversion of a common functional group to another.

Chemical patents often cover broad structural classes of compounds, frequently represented as Markush formulations. At present, patent databases which describe Markush structures are based on fragment codes as the connection-table-based systems can only represent specific compounds, but work is in progress to develop new generic connection table based systems.[42]

Proteins, peptides, gene sequences, *etc.* are not searchable by conventional substructure search systems as the molecules are too large to be stored as connection tables. Most systems have a limit to the number of atoms that can be stored per structure, *e.g.* 200 atoms on CAS ONLINE.

4.14.3.3.4 Summary: how and where to look

For a **single** compound, any structure-based indexing tool will lead you to information associated with that compound. As we have seen, however, many compounds will have the same molecular formula and a compound may be indexed under several different synonyms. In large compound registries such as CAS ONLINE and BEILSTEIN, formulae and nomenclature indexes should be treated with caution or avoided completely. Formulae and nomenclature do have their uses in the smaller tertiary sources such as the pharmacopoeias, and are the only tools available for searching some of these smaller sources. Even here the proliferation of trade names/synonyms may lead to some entries being missed by the searcher.

In general, if a compound is not found using one indexing tool, it cannot be assumed that it has not been described in the literature: another indexing tool or a more comprehensive source must be examined.

Where an information source is available online, it is generally simpler and quicker to find the same information using the online retrieval software than by searching the printed indexes. The examples given in this section have demonstrated how the choice of an appropriate chemical indexing tool becomes more limited as a query becomes broader. For retrieval of compound classes or substructures, molecular formulae and nomenclature cease to be viable search tools; most queries of this sort should be answered using an online structure search system such as CAS ONLINE or DARC. Such systems will eventually also encompass Markush formulations for greater retrieval from the patent literature.

We have described some of the limitations which apply to the use of chemical indexing tools and because of these they should be used with care. Nevertheless, any sort of chemical index can lead eventually to a wealth of information on a particular compound. In this respect structural based indexing tools provide an invaluable entry point to the published literature. In contrast, biological concepts are often extremely difficult to define and hence to index.

4.14.4 ACTIVITY SEARCHING OF THE MEDICINAL CHEMISTRY LITERATURE

4.14.4.1 Introduction

Rational development of drugs requires an understanding of the relationship between molecular structure and biological effect, both pharmacological and toxicological,[53] and hence 'the first foreign discipline a medicinal chemist is likely to encounter is biology.'[54] In this context the term biology embraces a number of subdisciplines, since the pharmacological activity of a drug has pharmaceutic, pharmacokinetic (including metabolic) as well as pharmacodynamic phases.[55]

Brown[56] has described the information needs of pharmaceutical research and development and Vinatzer-Chanal,[57] Franke[58] and Austel and Kutter[59] have discussed more specifically the first steps in the search for new drugs, namely the processes of lead generation and lead optimization, with which the medicinal chemist is intimately involved. With the obvious exception of random screening, many methods for lead generation depend upon published information such as compounds of known activity or observed experimental or clinical side effects. As compounds are synthesized and tested and structure–activity relationships (SAR) are established and refined, lead optimization follows with the dominant aim of optimizing activity and minimizing unwanted effects, while retaining patentability. SAR studies may be based on compounds synthesized in-house but it is also possible to make use of published data.[53] During both the periods of lead generation and lead optimization, indeed as soon as the research target has been defined, the published literature and patents provide essential information on competitor activity in the therapeutic or structural area. It is also important to monitor the literature for toxicity associated with compound types emerging as leads since, as Brown[56] has pointed out, chemists may be wary of early discards based on toxicological considerations but the liability for toxicity should always be borne in mind.

4.14.4.2 Biological Questions

During the initial period of lead generation, answers to questions such as the following may provide valuable starting points: What compounds are agonists/antagonists for receptor A (or a similar receptor)? What compounds inhibit/activate enzyme B (or a related enzyme)? What range of structures have shown therapeutic efficacy against disease C? Have any compounds been shown to affect both properties D and E? Then, as structural leads emerge, biological studies on similar compounds may indicate potential problem areas related to other activities or toxicity. Information on the metabolism of related structures may help in the design of a required pharmacokinetic profile or again warn of potential toxicological problems, and thus answers to the following questions may be relevant: What other activities/toxicities have been reported for this structure/substructure? How might this compound be metabolized?

4.14.4.3 Sources of Biological Information

A large number of secondary and tertiary services, including both reference databases and source databases, provide access to activity data, and Table 3 lists the major tools and describes how biological indexing is handled. Table 2 highlights the handling of compound information in these sources. In general the source databases listed in Table 3 provide activity and/or toxicity information on up to 80 000 chemicals, and marketed and/or investigational drugs. The records are based around a single compound and associated data.

MERCK contains over 10 000 monographs describing about 30 000 chemicals, drugs, biologicals and veterinary and agricultural products and includes patent, chemical, physical, use and toxicity data and literature references.

RTECS contains a wide range of both general and specific unevaluated toxicological data and references on approximately 80 000 standard chemicals.

Hazardous Substances Databank (HSDB) is a relatively new databank of evaluated data on potentially hazardous chemical substances; it includes chemical and physical properties, safety, handling, toxicity and pharmacology information.

PHARMAPROJECTS now contains information on over 5500 pharmaceutical products in development, and includes chemical, pharmacological, clinical and development status data and references; recently, products discontinued from development have been added to the file and products for which there is no evidence of continuing development may also be added shortly.

The Standard Drug File (SDF) contains chemical and pharmacological information on 17 000 drugs and other commonly occurring compounds.

Negwer is a compendium of 9000 drugs including synonyms and therapeutic uses. It is only available as a book but has indexes for synonyms, CAS registry number and chemical groups.

MARTINDALE contains physical, chemical, pharmaceutical and prescribing information on more than 5000 compounds used therapaeutically throughout the world; ancillary agents such as preservative, colouring and diagnostic agents are also included.

IMSBASE has been developed as a single online file from a number of different paper publications and contains not only prescribing information but also information on pharmaceutical compounds in research and development, licensing opportunities, company profiles and regulatory news relevant to the pharmaceutical industry. Unlike the majority of other sources, use of IMSBASE requires the payment of a costly subscription, currently $15 000 per year.*

These source databases also provide access to older information than the online reference databases and in general provide a simple means of locating standard activity and toxicity information on better known compounds. They are often exclusive to one database vendor. In contrast, the bibliographic databases are readily available from a number of vendors, and the major databases in the medicinal chemistry field are very large: CA, BIOSIS PREVIEWS (BIOSIS), MEDLINE and EMBASE each index about 250 000 papers, books, conference proceedings, *etc.* per year. CA has been described in detail in Section 4.14.3.2.7. BIOSIS provides a general coverage of the life sciences while MEDLINE and EMBASE are the major sources for biomedical literature. EMBASE is currently undergoing a radical restructuring of its indexing system. The coverage of TOXLINE has been described in Section 4.14.2. RINGDOC is a smaller database indexing about 50 000 documents per year, but it provides literature coverage of all aspects of drugs and detailed indexing of both chemical and biological concepts. As with IMSBASE, an annual subscription (currently £17 800) is a prerequisite for the use of RINGDOC.

Patents and patent databases are probably an underused resource since they can provide information on current technological developments and prior art as well as commercial intelligence, licensing opportunities and information on special technologies.

In addition to the handbooks mentioned earlier, further comprehensive lists of sources are found in Rimmer and Green,[60] Simmons (patent information),[61] Kruse and Snow (computerized databases covering the pharmaceutical literature),[62,63] Bawden and Brock,[64] Bawden and Kissman (toxicological information),[23,65] and Welch and King (general medical information).[66]

4.14.4.4 Retrieval of Biological Information

Retrieval of biological information differs from that of chemical information in two ways. Firstly, biological information tends to be dispersed amongst a number of both broad and specialist

* *Note added in proof*: IMS International has announced that the business intelligence database IMSBASE will be discontinued at some time after 31 December 1989.

Table 3 Biological Concept Indexing in the Principal Sources of Biological Data[a]

Name	*Broad sections or codes*	*Higher (generic) controlled index terms or codes*	*Controlled index terms or codes*	*Subheadings, roles or qualifiers*[b]	*Sentences*	*Abstracts*
CAS ONLINE	Y	S	S	Y	Y	Y
WPI	Y	Y	S	—	—	Y
Index Chemicus in-house database	Y	—	—	—	—	Y
RINGDOC (UDB)[c]	Y	Y	Y	Y	Y	Y
CIS (data and bibliographic files)	Y	S	S	—	S	S
EMBASE	Y	Y	Y	from 1988	—	S
BIOSIS PREVIEWS	Y	Y	S	—	—	S
MEDLINE	Y	Y	Y	Y	—	S
TOXLINE	Y	S	S	S	S	S
SDF	—	—	Y	—	—	—
Negwer	—	—	—	—	—	—
MERCK	—	S	S	—	—	Y
MARTINDALE	Y	Y	Y	Y	Y	Y
PHARMAPROJECTS	Y	Y	Y	—	Y	Y
IMSBASE	Y	Y	Y	—	—	Y
RTECS	—	Y	Y	—	Y	—
HSDB	—	Y	Y	—	Y	—
	General				Specific retrieval	

General ⟶ Specific retrieval

[a] Y = featured, — = not featured, S = featured to an extent. [b] Subheadings, roles or qualifiers indicate the context in which an index term is used; field qualifiers may be substituted, particularly in source databases. [c] Ringdoc—the older files have fewer generic index terms, do not have subheadings, or abstracts and searching within sentences is only possible using codes.

databases, and many questions will necessitate the use of multiple sources. Secondly, biological information is textual and cannot readily be translated into a coded form of general applicability such has been described for chemical compounds. For this reason a variety of approaches to retrieval of textual information have been developed, and frequently a combination of approaches and sources is still the most effective searching technique.

Information retrieval from textual records may depend upon the use of classification systems or assigned natural language keywords or phrases taken from a controlled list of terms.[17,67,68] Such systems have a number of advantages. These include simplifying the specification of single concepts by the provision of a standard list of index terms and allowing the specification of relationships between concepts by the provision of generic terms or grouping of index terms for related concepts. Such coordination may be achieved by the use of precoordinated terms (*e.g.* renal–vein), by the addition of standard qualifiers or roles to index terms (*e.g.* aspirin/pharmacology) or by combining index terms into sentences of logically related terms. Some of the disadvantages of such a controlled structured vocabulary, namely reliance on the accuracy of indexers and rigidity of terminology, may in future partly be overcome by the use of computer-assisted indexing using intelligent knowledge-based systems[69] designed to provide interactive assistance to indexers.

More recently, with the development of more powerful computers, it has become possible to store larger amounts of information (including the abstract or full text of a paper) and to search for single specific words or phrases within the abstract or text. Though searching in this way makes greater demands on the searcher in terms of formulating a strategy, such a system has considerable advantages in the retrieval of new biological concepts or minor aspects of a paper.[17,68]

A very different approach to indexing, using neither keyword nor free text searching, is that of citation indexing[27,70] as exemplified by the 'Science Citation Index'. This method uses the references cited by the author of a paper to define and represent the subject of the paper, thereby by-passing the semantic problems described above. An additional refinement has used the phenomenon of co-citation clustering to delineate the literature of individual subject areas (clusters), which are then assigned a subject descriptive name. An evaluation of this approach has been made by Bradshaw.[71]

Table 3 summarizes the availability of these various forms of textual indexing in the core databases containing biological information.

4.14.4.5 Retrieval of Associated Biological and Structural Information

Retrieval of activity or toxicity information on a single compound or group of similar compounds requires the capability both to specify and link the two types of information.

As has already been described in Section 4.14.3 there are a number of potential approaches to representing a chemical structure from molecular formula through nomenclature to fragment codes and connection tables. Chemical substance indexing in the databases containing activity information uses a variety of these approaches as defined in Table 2. The majority of the biological source databases and bibliographic databases have the non-proprietary or generic name of a drug and/or the brand name and/or the chemical name, and/or an investigational number and/or the CAS registry number. Some databases standardize on one or a number of these, others such as TOXLINE have no controlled indexing. These databases can thus be searched, with varying degrees of ease, for specific compounds. In contrast, RINGDOC, the chemical bibliographic databases and WPI have either fragment codes or connection tables which enable not only specific structures to be retrieved but also compounds containing defined substructures.

A further consideration in the choice of source for a structure/activity search is the ease with which the occurrence of generic activity or toxicity information can be specified. Source databases often have the advantage of restricted subject coverage—for example all data in RTECS is toxicological—and no further specification is necessary. In bibliographic databases a number of other techniques can be employed. Some databases are divided into subject sections, for example CA. Standard qualifiers may be useful, for example 'ST' for adverse effects or toxicology in RINGDOC and 'AE, TO, PO' for adverse effects, toxicity or poisoning in MEDLINE, but, in the absence of such aids, keyword or free text searching must be used. The availability of controlled keywords in the various databases is indicated in Table 3. However, free text and full text searching may be particularly useful for retrieval of toxicology information, since there is often poor standardization of nomenclature for toxic effects, or indeed a failure to index such effects at all.[64]

The final consideration is the ease and precision with which the two kinds of information can be linked, and this varies enormously from database to database. In the source databases a separate record is formed for each compound and all its associated properties and hence the link is both simple and precise. Searching for combined biological and structural information in bibliographic databases poses two main problems. Firstly, a direct link between a structure and associated activity or toxicity is not inherent in the format of the databases but depends on the use of grouped index terms. Secondly, many of the major bibliographic secondary sources are biased towards biology or chemistry, either in terms of coverage of the primary literature or ease of retrieval, the major exception being RINGDOC. This problem of subject bias can be overcome to some extent by combining, indirectly, a search of databases containing fragment codes or connection tables, with, for example, a search of bibliographic databases containing activity or toxicity information.[72] The search of the former can be used to yield a list of compounds, and information on these compounds can then be generated from the second set of databases by using whatever structural notation (names, CAS registry number) this second set of databases contains. As can be seen from Table 2, many biological databases now contain CAS registry numbers, which thereby provide a valuable 'link'. An agreement reached between CAS and BIOSIS provides, *via* STN, CAS registry numbers for compounds in BIOSIS records back to 1980. MEDLINE contains CAS registry numbers from 1980, and it is possible, using Télésystèmes-Questel, to generate a set of registry numbers from a structure search and use this set directly as a search parameter within MEDLINE. These techniques of tagged retrieval can, of course, work in reverse, *i.e.* following the identification of compounds of interest from a biological search, records containing the registry numbers can be used to retrieve further property information from other databases which include this link.

4.14.4.6 Sample Searches

To illustrate the differences between activity searching on these various databases a number of sample questions will be considered.

4.14.4.6.1 What compounds have been shown to have H_2 antagonist activity?

Appropriate source databases provide lists of compounds with the specified activity. In almost all the sources listed, use of controlled terminology allows retrieval of all H_2 antagonists with a single term, *e.g.* in MARTINDALE 'histamine-H_2 antagonist' and in PHARMAPROJECTS 'antiulcer-H2-antagonist'. Depending on the coverage of the database the retrieved list of compounds will be biased towards marketed products, *e.g.* MARTINDALE, or investigational drugs, *e.g.* PHARMAPROJECTS. The search could not be run on Negwer which is a hardcopy source with only name and substructure indexes.

A search of the biological bibliographic databases yields lists of papers in which H_2 antagonists have been studied. Again the use of a controlled vocabulary generally simplifies retrieval of all papers on this category of compounds, *e.g.* use of 'histamine H_2 receptor blockaders' in MEDLINE. Those bibliographic databases covering conference proceedings as a primary source such as RINGDOC and BIOSIS provide early mentions of new H_2 antagonists.

A search of the chemical bibliographic databases also retrieves papers in which compounds with this activity have been studied. The journal and subject coverage of CA (including research and preclinical studies of drugs, but not clinical) is usefully wide, but the semicontrolled semigeneric nature of the biological indexing necessitates use of both the term antihistaminics (covering studies of new antihistamines and studies of antihistamines as a class) as well as specific index terms for H_2 antagonists to retrieve all papers on H_2 antagonists. Use of the term antihistaminics, however, restricts the search to new (in contrast to known) compounds of this class, though, because of its lack of specificity, papers on new H_1 antagonists are also retrieved.

Index Chemicus covers papers reporting new organic compounds and includes any biological activities specified in the article. The activity descriptors are taken directly from the literature and a search for H_2 antagonists needs the input of all possible synonyms, from 'antihistaminic H2' to 'histamine H2 receptor antagonistic'. Finally, in WPI it is possible to search for new H_2 antagonists, from 1986, using a simple code (B12-D06A).

In all these databases it is possible that a new H_2 antagonist might have been categorized less specifically as an antiulcer compound, in which case it would not be retrieved.

The search illustrated above has demonstrated the value, for activity searching, of a controlled vocabulary to simplify retrieval of all papers reporting work on a standard activity. Nevertheless, if the receptor or enzyme of interest has recently been discovered, free text searching of titles, abstracts (and use of full text journals online) will be more valuable.

An alternative approach to the question would be a search of the 'Science Citation Index' for those papers citing a seminal paper in the field of histamine H_2 receptors or histamine H_2 receptor antagonists.

4.14.4.6.2 What toxicity has been reported for compound Y?

Source databases such as RTECS and HSDB are particularly useful, providing both a direct link between structure and biological effect and a controlled vocabulary for toxic effect, route of administration, species, *etc.*

The two most useful bibliographic databases from the point of view of toxicological coverage are CA and TOXLINE. CA has the advantages of ease of structure searching, *e.g.* by registry number, with grouped index terms increasing precision, but lack of controlled indexing makes searching for all toxicity or specific toxic effects quite difficult.[64] TOXLINE presents difficulties of both uncontrolled structure searching (a chemical dictionary file such as CHEMLINE can be used to generate synonyms for compound searching) and uncontrolled toxicity searching.

RINGDOC provides the advantages of controlled chemical and biological indexing, *e.g.* use of the role 'ST' for adverse effects or toxicology, and grouped index terms, but its coverage is limited to drugs. Similar limited coverage decreases the utility of the medical bibliographic databases, although the use of the MEDLINE subheadings 'AE, TO, PO' simplifies limiting the results to toxicity information.

4.14.4.6.3 What activities have compounds containing substructure X exhibited?

Of the source databases, Negwer can be searched by a printed index of chemical fragments, the SDF can be searched by both standard chemical structure text terms and by fragment codes, and MERCK can be searched for substructures *via* SANSS. The two former files yield activity data, MERCK yields both activity and toxicity information.

The chemical bibliographic databases are also substructurally searchable. Limiting the search to those papers in which activity data have been reported can be achieved by the use of section headings in CA (*e.g.* pharmacology) or the presence of the biological activity field in Index Chemicus. Both WPI and RINGDOC are substructure searchable by fragment codes. Location of a substructure in WPI (Farmdoc) infers that some associated activity information is probably present; in the RINGDOC UDB file, the search can subsequently be limited by the use of the role 'PH' to those papers reporting pharmacology information.

None of the other databases can be searched directly for substructures but, as already indicated, it is possible to link substructure searches in one file to activity searches in another using, for example, the CAS registry number as a link.

Two final points should be made concerning retrieval of biological information from published sources. Firstly, from this brief description it can readily be appreciated that, as yet, the retrieval of such information is not the province of medicinal chemists themselves, but requires the expertise of an information professional. Secondly, it must be emphasized that the problems of biological variability and the complexity of many tests for biological activity mean that it is difficult to make other than qualitative comparisons of activity information retrieved from different sources.

4.14.5 SYNTHESIS

4.14.5.1 Introduction

In their search for novel drug entities, medicinal chemists undertake many thousands of reactions, whilst in the chemical literature millions of reactions have been reported and new reactions are described every day. Thus efficient retrieval of reaction or synthesis information from this vast pool of knowledge is of fundamental importance. Consulting the literature will help stimulate ideas, encourage a diversity of methods and save the chemist time and money,[73,74] all of which are vital in the highly competitive field of medicinal chemistry.

This section discusses the types of synthesis questions which arise and the sources and systems used to retrieve the information. Sources of commercially available compounds are also mentioned, and a brief account of some expert systems is given.

4.14.5.2 Classification of Reaction Queries

A large proportion of compounds that medicinal chemists make are novel but a significant number of the intermediates and reagents will be known compounds. The information requirements therefore range between information about known reactions that can be applied to novel systems, and information on specific known compounds including their preparation, reactions and commercial availability.

The most typical questions that a chemist poses can be classified into several main categories.

(1) Synthesis of a compound from any starting material.

$$? \longrightarrow B$$

(2a) Conversion of one molecule into another.

$$A \longrightarrow B$$

(2b) Reactions between compounds.

$$A + B \longrightarrow C \cdots$$

Functional group transformations are generally of great interest as many reactions involve the manipulation of key functional groups to introduce the desired structural changes. In a number of instances these transformations need to be performed selectively, whilst leaving other functional group(s) in the molecule unaffected.

(3) The reactions of a given compound.

$$A \longrightarrow ?$$

(4) Questions on the utility of a given reagent or catalyst may be asked in a general sense, or in relation to specific compounds, or in conjunction with other reagents and solvents. The query may be described either by a structure, formula, keyword or keyphrase.

(5) The chemist may ask questions based on general reaction keywords (*e.g.* oxidation, reduction, photolysis) or reaction data (*e.g.* temperature, pressure, yield).

4.14.5.3 Sources

Having established the range of questions the medicinal chemist is likely to ask, the next consideration is the location of the answers. The information gathered by the secondary and tertiary sources such as CA, Index Chemicus or Beilstein is not primarily geared toward reaction information retrieval. Consequently there still remains the problem of sifting through the millions of reactions they feature.

Of potentially greater value are those services which selectively abstract synthesis information. Different services impose different selection criteria and the set of journals scrutinized varies between the services, as does the number of articles selected. For example, the selection criteria for ISI's 'Current Chemical Reactions' (CCR) are new and newly modified reactions and syntheses taken from approximately 112 source journals and 13 review journals or books. As well as novel chemistry, CCR includes known reactions, extensions of reactions giving new classes of products, modified known syntheses and new uses for old reagents. Approximately 5000 articles containing multistep reactions are abstracted into CCR each year. Derwent describe their selection criteria for 'Journal of Synthetic Methods' (JSM) as 'novel or unusual reactions and new synthetic methods; known processes effected by new reagents or improved synthetic methods; and interesting applications and extensions of known reactions.' Over 150 journals and patents are covered and from these around 3000 new reactions are selected annually.

Some of the sources of synthetic information are listed in Table 4. These works are a distillation of information on preparations, reactions, synthetic sequences and reagents reported in the primary literature. Manual searching and scanning of these sources can be a lengthy and tedious process, whereas the lifestyle of the modern medicinal chemist necessitates efficient and effective application of prior knowledge. Fortunately, a breakthrough in the methods available for searching these familiar sources conveniently and efficiently has been achieved over the last few years[87] and many of the secondary sources and reference works are now available as computer databases. These systems give the chemist the advantage of graphical input and output of chemical structures, *i.e.* the natural language of the chemist.

Table 4 Sources of Reactions or Synthesis Information[a]

Source (producer or publisher)	*Hardcopy*	*Online vendor*	*In-house*	*Ref.*
Compendium of Organic Synthetic Methods (Wiley)	Y	—	—	75
Current Chemical Reactions (CCR; ISI)	Y	—	Y	76
Current Literature File (CLF; Molecular Design Ltd)	—	—	Y	77
Journal of Synthetic Methods (JSM; Derwent)	Y	Y	Y	78
Methoden der Organischen Chemie (Georg Thieme Verlag)	Y	—	—	79
Organic Reactions (W. Dauben, Wiley)	Y	—	—	80
Organic Reactions Accessed by Computer (ORAC; ORAC Ltd)	—	Y	Y	81
Organic Synthesis (Wiley)	Y	—	Y	82
Reagents for Organic Synthesis (Fieser, Wiley)	Y	—	—	83
Survey of Organic Synthesis (Wiley)	Y	—	—	84
Synthesis Library (SYNLIB; Distributed Chemical Graphics Inc.)	—	—	Y	85
Synthetic Methods of Organic Chemistry (W. Theilheimer, Karger)	Y	Y	Y	86

[a] Y = available, — = not available.

4.14.5.4 Commercially Available Reaction Indexing Systems

This area of reaction information retrieval has witnessed great advances during recent years,[2,87] and the user requirements for such services have been reviewed.[88,89] Systems have been designed and developed which have largely overcome the earlier major problems of chemical reaction searching. These difficulties stemmed from the inherent nature of the chemical reaction, which is not a static entity, but rather it represents a series of changes and transformations with which many complex parameters and characteristics are associated. Hence the efficient storage and retrieval of graphical reaction information has posed a great challenge. Willet has described this as the 'multi-faceted nature of reaction information' in a review of reaction searching.[90]

The method by which the chemist may gain access to the information depends upon the database and the database producer. Some of the databases are publicly available on database hosts. Others can be purchased and run on an in-house computer, as noted in Table 4. Several commercial software packages are now available for reaction retrieval including REACCS, ORAC, SYNLIB and DARC-RMS detailed in Table 5. These systems, with the exception of DARC-RMS, have an accompanying library of literature reactions. An interesting comparison of REACCS, SYNLIB and ORAC has been published by Zass and Mueller.[91] Some reaction databases have been produced purely as computer sources of reaction information and do not have exact hardcopy equivalents, *e.g.* the ORAC and SYNLIB databases and the Current Literature File of REACCS. These databases are assembled from selected journals of past and current literature by international consortia of synthetic chemists who have their own set of selection criteria, depending on the database. Thus large groups of chemists are influencing and shaping the databases that they and their colleagues will use.

It is important to evaluate how the reaction retrieval systems can cope with everyday reaction queries. The ease of operation by the chemist and the flexibility of the systems are also essential factors to be considered.

4.14.5.4.1 Overview

The REactions ACCess System (REACCS) is a product of Molecular Design Ltd (MDL) and was first introduced at the ACS National Meeting in New York in August 1981. REACCS was installed at Eastman Kodak later that year to search their corporate reactions database.[92] A repertoire of approximately 100 000 literature reactions is provided with REACCS.[77] Included are the computerized versions of 'Theilheimer', 'Journal of Synthetic Methods' and 'Organic Synthesis'. MDL has created their own Current Literature File (CLF) which contains recently published novel reactions, new synthetic methods and improved known procedures. In addition MDL has an agreement with ISI to run a Current Chemical Reactions database on REACCS, which contains information from 1985 and which will have *ca.* 30 000 reactions added per year. Many major companies have collaborated to support this venture and have formed a Reactions Data Club.[93]

Organic Reactions Accessed by Computer (ORAC) was developed at the University of Leeds Wolfson CADOS Unit. A separate company, ORAC Ltd, has been formed to handle marketing and

Table 5 Computerized Reaction Indexing Systems

System	*Supplier*	*Database*	*Number of reactions (to July 1988)*
REACCS (Reaction Access System)	Molecular Design Ltd	Theilheimer (vols. 1–35)[a]	46 000
		Journal of Synthetic Methods[a] (JSM; 1980–1986)	25 000
		Current Literature File[b] (CLF)	21 000
		Organic Synthesis[a] (vols. 1–64)	5000
	ISI	Current Chemical Reactions[c] 1986	28 000
		1987	28 000
ORAC (Organic Reactions Accessed by Computer)	ORAC Ltd	ORAC[b]	40 000
		Theilheimer[a, d]	46 000
SYNLIB (Synthesis Library)	Distributed Chemical Graphics Inc.	SYNLIB	44 000[b]
DARC-RMS (DARC Reaction Management System)	Télésystèmes-DARC	None	

[a] From existing hardcopy sources. [b] Compiled by a consortium of chemists. [c] Available by agreement with ISI. [d] Complete database release scheduled for January 1989.

further development. (Maxwell Communications Corporation has recently acquired both ORAC Ltd. and MDL) The system was launched in early 1984[81] with a database of 5000 reactions. Since then, the database has grown rapidly to reach 40 000 reactions by July 1988. The Theilheimer database of around 40 000 reactions is also available on ORAC.

SYNthesis LIBrary (SYNLIB) has been available since 1981, and has evolved as a joint effort between Professor W. Clark Still (University of Columbia) and Smith Kline & French Laboratories.[85,94,95] SYNLIB is now marketed through Distributed Chemical Graphics Inc. A total of 44 000 reactions are available with SYNLIB (to February 1988).

The Questel-DARC system is an online structure search system.[28] Télésystèmes-Questel provide DARC software for in-house use including the Reaction Management System (DARC-RMS), which is a reaction storage and retrieval package.[96] At present there is not a commercially available database of literature reactions in DARC-RMS format and hence this system will not be considered further.

4.14.5.4.2 Using REACCS, ORAC and SYNLIB

All of these reaction retrieval packages function interactively with the chemist. A complex system command language does not need to be remembered as the systems are menu driven. The menu presentations of REACCS, ORAC and SYNLIB are different but all perform the functions of query creation searching and answer display. Generally, the menus are well formatted and the commands easily understood. A light pen or mouse is required to activate the menu commands and draw the chemical structures.

In all the systems the reaction query is constructed using a structure input menu (*e.g.* Figure 6). Templates of common chemical fragments are available to aid drawing and all the systems allow creation of user-defined templates. Variable bonds and atom lists can be assigned to structures, thus more general types of questions may be asked. REACCS, ORAC and SYNLIB also offer the facility to refine the query by assigning the bonds which are broken and formed during the reaction. Such refinements allow for greater accuracy in searching.

REACCS and ORAC offer substructure and exact structure search facilities for reaction queries containing the reactant(s) and/or the product(s). Queries are constructed in REACCS in the

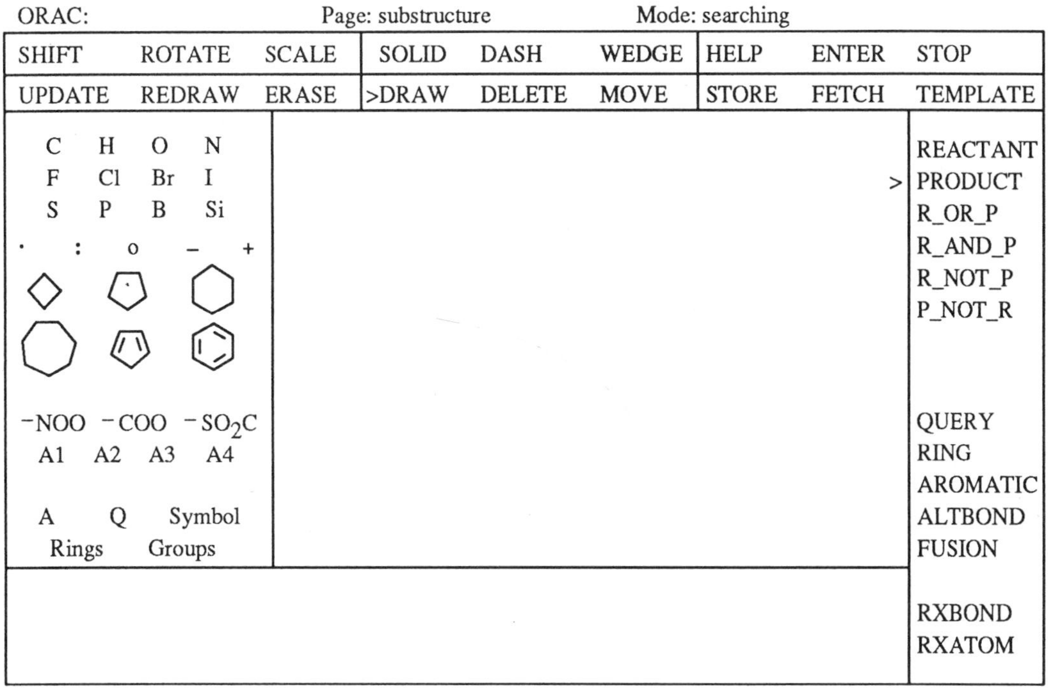

ORAC: Page: substructure Mode: searching
SHIFT ROTATE SCALE SOLID DASH WEDGE HELP ENTER STOP
UPDATE REDRAW ERASE >DRAW DELETE MOVE STORE FETCH TEMPLATE
C H O N
F Cl Br I
S P B Si
· : o − +
$-NOO$ $-COO$ $-SO_2C$
A1 A2 A3 A4
A Q Symbol
Rings Groups
REACTANT
> PRODUCT
R_OR_P
R_AND_P
R_NOT_P
P_NOT_R
QUERY
RING
AROMATIC
ALTBOND
FUSION
RXBOND
RXATOM

Figure 6 ORAC structure input menu (reproduced by permission of ORAC Ltd)

traditional way, with reactant and/or product either side of an arrow (Figure 7). Alternatively, the query molecule is searched using a target qualifier 'as reactant' or 'as product'. In ORAC the reactant and/or product are entered and searched separately and a reaction is retrieved from the combination of both searches.

Associated with each reaction in REACCS and ORAC are a wide range of text and data, including reaction conditions, yield, solvent, catalysts, reagents, keywords, journal, year, *etc.* The text and data terms in both systems may be searched by selecting the appropriate item from a menu.

Searches performed during a current session in REACCS and ORAC can be combined using logical operators. This allows for quite complex questions to be answered, for example the chemist can search for reactions occurring in a particular solvent and within a given temperature and yield range.

SYNLIB searching capabilities are somewhat limited when compared to REACCS and ORAC. Only substructure searches for products are allowed, *i.e.* searches for reactants are not possible. SYNLIB operates a two-tier search mode, a broad generic search and a narrow more precise search. The opportunities for text or data searches in SYNLIB *per se* are limited owing to the paucity of available textual/numerical information. Nevertheless it is possible to restrict substructure searches by applying some text or data parameters.

The results are displayed in the 'classical' way by all three systems, *i.e.* the reactant and product either side of a reaction arrow, with reagents and conditions shown above the arrow (Figure 8). All three systems also display the yield and reference citation. As explained earlier, both REACCS and ORAC have a number of text and data fields for each reaction. ORAC handles the answer display by treating the output as a reaction card (Figure 9), a concept with which all chemists are familiar. There are two 'sides' to each card, the 'upper' side is called the data card which contains all the reaction details as outlined above. The 'reverse' side is called the keys card which shows all the text and data indexing terms. A menu is available in REACCS which allows users to design their own output format.

In REACCS all the datatypes may be listed for a specific reaction; alternatively, specific datatypes (*e.g.* yield, temperature, solvent) may be listed for all the reactions in a given answer list. This facility has been elegantly extended to provide the user with a table display for which any of the datatypes may be selected thus providing the chemist with a highly effective way to summarize and compare reaction conditions.

This graphical storage and retrieval of chemical reactions represents a revolutionary move forward in reaction searching. These systems can convert the seemingly troublesome and tedious

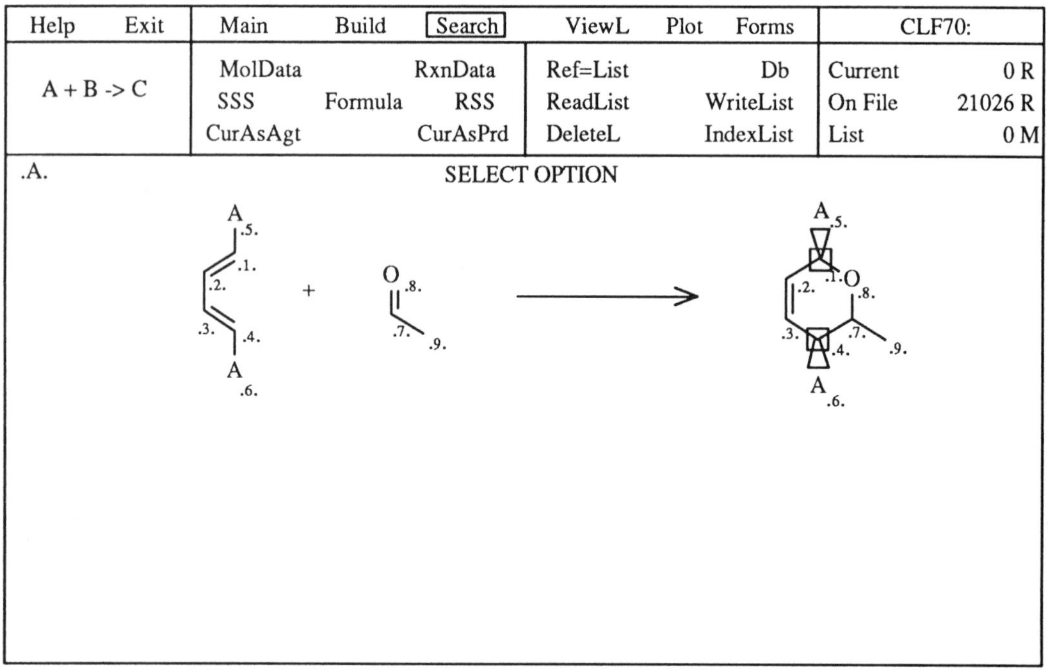

Help Exit	Main Build Search	ViewL Plot Forms	CLF70:
A + B -> C	MolData RxnData SSS Formula RSS CurAsAgt CurAsPrd	Ref=List Db ReadList WriteList DeleteL IndexList	Current 0 R On File 21026 R List 0 M

.A. SELECT OPTION

Figure 7 REACCS search mode (reproduced by permission of MDL)

Help Exit	Main Build Search	ViewL Plot Forms	CLF70:
A + B -> C	First Next Prev Item List Table Data Query	Ref=List Db ReadList WriteList DeleteL IndexList	Current 6676 R On File 21026 R List 2 R

M Bednarski, S Danishefsky, J Amer Chem Soc, 105(11), p. 3716-3717, 1983

Less endo selectivity observed for CH2C(OSiMe3)CHCHOMe and aliphatic aldehydes.

Keyphrases: Hetero Diels-Alder. Enol silyl ether. EU(FOD)3.

Keywords: Cycloaddition(CYA) Heterocyclic chemistry(HET) Regioselective(REG) Ring_closure(RCL)

regno: 6676	Extreg - folder - path - step 9100867 11 A 1 OF 2	VARIATION 1 OF 2

Figure 8 REACCS viewL mode (reproduced by permission of MDL)

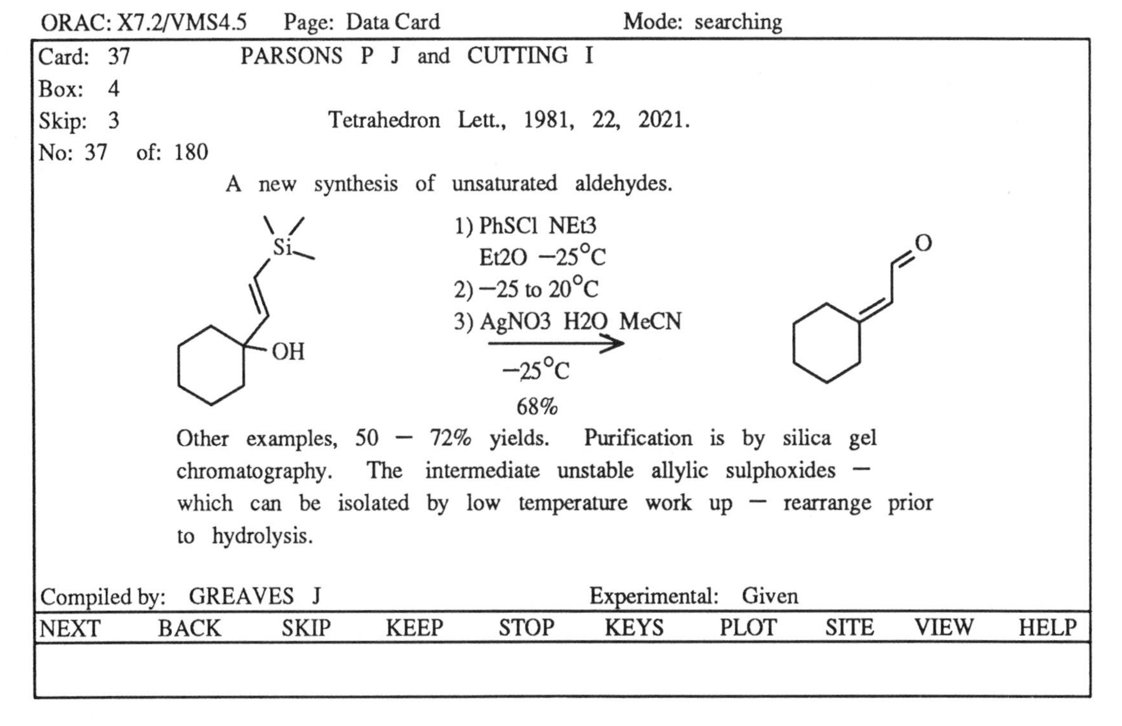

ORAC: X7.2/VMS4.5 Page: Data Card Mode: searching

Card: 37 PARSONS P J and CUTTING I

Box: 4

Skip: 3 Tetrahedron Lett., 1981, 22, 2021.

No: 37 of: 180

A new synthesis of unsaturated aldehydes.

1) PhSCl NEt3
Et2O −25°C
2) −25 to 20°C
3) AgNO3 H2O MeCN
−25°C
68%

Other examples, 50 — 72% yields. Purification is by silica gel chromatography. The intermediate unstable allylic sulphoxides — which can be isolated by low temperature work up — rearrange prior to hydrolysis.

Compiled by: GREAVES J Experimental: Given

NEXT BACK SKIP KEEP STOP KEYS PLOT SITE VIEW HELP

Figure 9 ORAC data card (reproduced by permission of ORAC Ltd)

task of reaction searching into an enjoyable, educational and fruitful experience. Chemists are at liberty to consult these systems in their own language, that of structures, and to interact freely with the knowledge base. Broad or general reaction queries may be performed in all the systems and the results scanned either sequentially or at random, thus providing the chemist with the opportunity to browse and take advantage of the serendipity factor.

On the whole, SYNLIB is less advanced and flexible than REACCS or ORAC. Those features lacking include the ability to search for reactants, efficient text and data retrieval, automatic storage of current session searches and the use of logical operators. SYNLIB is therefore unable to cope with the full range of reaction queries described earlier in this section. However, SYNLIB's reaction database is very good and has continued to grow steadily; furthermore its simplicity renders the system very attractive to some chemists.

In essence, both REACCS and ORAC offer the chemist a sophisticated and versatile approach to reaction retrieval; they are able to meet the criteria presented earlier and can deal with the majority of possible reaction queries. A comparison of the three systems' capabilities, with regard to the types of questions which can be answered, is given in Table 6.

The REACCS and ORAC programs are constantly undergoing improvement and development by their producers. Furthermore, the databases are expanding vigorously, for example reactions are being added to the ORAC database at the rate of *ca.* 2000 per month.

Other database developments include the conversion of FIZ Chemie's ChemInform (Chemischer Informationsdienst) to machine-readable format compatible with REACCS, ORAC and DARC-RMS.

4.14.5.5 Publicly Available Sources of Reaction Information

Access to reaction information may also be gained through the public networks. Indeed, ORAC is also available to online users. An interesting account is given by Bysouth and Hardwick on the user experience of retrieving chemical reaction information from publicly available online services.[97] Outlined below are some of the facilities available to the online user.

4.14.5.5.1 The Chemical Abstracts Reaction Service (CASREACT)

CA launched their reaction service in April 1988. Called CASREACT, the service is available *via* STN and is offered as an integrated part of the existing CAS ONLINE system with access to other CA information.[98]

The database of reactions drawn from over 100 core journals covers the literature since 1985, and it is anticipated that the database will increase by over 150 000 reactions each year. The subject coverage includes all single-step and multistep reactions from the journals indexed in the organic sections of CA (sections 21–34).

The system gives details of all reaction participants, with registry numbers assigned to the individual reaction components. Single-step and multistep reactions, yields and reaction sites are also indexed although, unfortunately, it is not possible to assign reaction centres to the query. A reaction role indicator allows the searcher to define the query substance as a reactant, product, reagent, catalyst or solvent. Yields, safety information and reaction conditions are also available.

The reaction retrieval system is structure driven with graphic input and output of reactants and products. Reaction substructure searching is performed using the existing CAS ONLINE Registry File, and a special CASREACT screen has been created to minimize any problems associated with existing search limits. Thus a substructure reaction search requires two full file searches, one for the reactant, the other for the product. The resulting answer sets are carried over into the CASREACT file and linked using qualifiers for reactant and product. However, only one full file search may be necessary when a low number of answers are obtained from the first search. In these cases it may be easier or cheaper to scan through the answers than to perform a second search.

The CA reaction service is the largest publicly available graphical online reaction retrieval system. As CASREACT has been introduced only recently, it is too early to comment on the general utility of the system, especially in comparison with the commercially available in-house reaction software packages. It is likely though, that the broad database of CASREACT will be used to complement the selective databases of systems such as REACCS and ORAC, for example when REACCS and ORAC have not provided suitable information.

4.14.5.5.2 CA

CA Online contains about 9 800 000 compounds with several million more reactions associated with them, and certain types of synthetic questions may be answered using the online systems. Two vendors currently offer graphical input and output of structures, namely STN and Télésystèmes-Questel.

Table 6 Comparison of Search Facilities of REACCS, ORAC and SYNLIB

	Class of search								
	? ⟶ B		A ⟶ B		A ⟶ ?		Structure search for reagents/catalysts/solvents		
System	*Exact*[a]	*Substructure*[a]	*Exact*[a]	*Substructure*[a]	*Exact*[a]	*Substructure*[a]	*Exact*[a]	*Substructure*[a]	*Text/data*
REACCS (version 7.0)	Y	Y	Y	Y	Y	Y	Y	Y	Molecular formula, authors, journal, year, reaction text, keywords, keyphrases, solvent, yield, temperature, pressure, work-up, comments
ORAC (version 7.0)	Y	Y	Y	Y	Y	Y	—	—	Authors, journal, year, reaction key, reagent key, solvent key, actual reagents, name of reaction, yield, temperature
SYNLIB (version 2.2)	Y	Y	—	Y	—	—	—	—	Minimum yield, date, text-string search

[a] Y = featured, — = not featured.

In addition to establishing chemical novelty (as discussed earlier), substructure searches may be used to find the preparations of a series of analogues. The query substructure is searched in the dictionary file and the references to the preparations are retrieved from the bibliographic file by a qualifier for preparation. This type of search may be extended to a substructure reaction search involving both reactant and product, but it is costly and not always precise. Another drawback is that some of the very simple structural queries such as those involving common functional groups cannot get through the limits set by the CAS substructure search system.

Two full substructure searches are required, one for the product, the other for the reactant. However, only the role of the product may be specified and not that of the reactant. Thus, when the two searches are combined, a list of references is obtained where the two substructures appear in the same paper, the preparation of the required product is given, but the compounds are not necessarily involved in the same reaction procedure. The chemist has to check the original reference to determine this.

The introduction of CASREACT has largely superseded this method of reaction searching. However, if information is required prior to 1985, the above method still needs to be employed.

Very general queries concerning functional group transformations can often be searched using keywords in the CA bibliographic file, since certain functional group transformations or named reactions are indexed if they are the main theme of a reference.[99] For example, general references to the preparation of amines from amides may be retrieved by employing in the search statement controlled terms for 'amines', 'amides', 'preparation' and 'reactions'. Named reactions such as the Diels–Alder reaction and the Wittig reaction may also be searched as keyword terms.

4.14.5.5.3 The Chemical Reactions Documentation Service (CRDS)

Both Theilheimer's yearbook 'Synthetic Methods of Organic Chemistry' (vols. 1–30, 1942–1974) and its continuing publication Derwent's 'Journal of Synthetic Methods' (JSM; 1975 to date) are available online as the Chemical Reactions Documentation Service (CRDS). This service, offered by Derwent Publications Ltd, provides a database of over 60 000 reactions. The history and features of CRDS are outlined in a review by Finch.[78]

Access to information within CRDS is not through graphical input and output of reactions. Instead, two different computer-based retrieval methods are operated, namely coding and keywording, and detailed offline preparation is required to formulate codes and determine keywords for query input. Although the online search itself is very quick, finding the answers can be troublesome, since the answers are displayed in a bibliographic format which may be too ambiguous to determine their relevance to the original reaction query. The searcher therefore has to scrutinize the hardcopy abstracts which contain the reaction diagrams before deciding whether the appropriate reactions have been retrieved.

The coding or 'multipunch' method covers the whole database and is derived from the old 80-column IBM punch card. The reactants, products, reagents, bond breakages, bond formations and reaction conditions are encoded into three-digit fragmentation codes. When input into CRDS, and linked together, the codes provide an overall description of the reaction taking place.

The newer keyword search method only applies to the latter half of the database (Theilheimer vols. 21–30 and all of JSM). A comprehensive list of controlled vocabulary terms (keywords) are employed to describe the reactants, products, reaction conditions, reaction types and generic reagent classes. It is advisable to search with both keywords and codes to be sure that all possible answers are retrieved.

The Theilheimer and JSM databases are now more easily searched using REACCS and ORAC.

4.14.5.6 Starting the Synthesis: Buying the Chemicals

Medicinal chemists need to acquire a range of chemicals to employ as starting materials, intermediates or reagents. The preparation of such compounds is rarely time- or cost-beneficial since many substances can be purchased readily. A host of suppliers exist, each with their own catalogue, version of chemical nomenclature, indexes and purchasing terms. Consequently, shopping around to find the right chemical from the best supplier can be monotonous and time consuming.

ChemQuest is a readily available source of information on chemicals and suppliers. It is a comprehensive database of international commercially available chemicals including organics, biochemicals, dyes and stains. The suppliers range from large major companies which produce thousands of compounds to the smaller specialist chemical manufacturers. ChemQuest contains over 180 000 catalogue entries from over 50 suppliers providing information on some 60 000 substances. Orbit Search Service own the ChemQuest database which is marketed through their subsidiary company Pergabase. The database is maintained and updated by Fraser Williams (Scientific Systems) Ltd.

Graphical substructure, exact structure, CAS registry number, formula, catalogue number or name searches can be performed online *via* ORBIT. Furthermore, the online version quotes the suppliers' prices. The ChemQuest database may be purchased and searched in-house using commercially available computer retrieval systems such as MACCS, REACCS, OSAC (Organic Structures Accessed by Computer) or DARC-SMS. A WLN version of ChemQuest is also available for searching using CROSSBOW.

ChemQuest was originally designed in the early 1970s by a consortium of pharmaceutical companies.[100] The group of companies decided to call their project CAOCI (Commercially Available Organic Chemicals Index). In 1979 responsibility for the file was passed to Fraser Williams (Scientific Systems) Ltd and the name changed to FCD (Fine Chemicals Directory). Then, in 1986, the database became available through Orbit Search Service and the system was renamed ChemQuest.

A number of hardcopy buyer's guides and chemical directories are available. A comprehensive list of buyer's guides and related tools is given by Maizell.[4] Two examples are the 'SRII Directory of Chemical Producers—Western Europe' and the 'Chemistry and Industry Buyer's Guide', published by the Society of Chemistry in Industry.

4.14.5.7 Computer-assisted Synthesis Design

4.14.5.7.1 Overview

Planning the synthesis of a potential drug candidate requires creativity and innovation from the chemist. Many ideas are based on existing knowledge of the literature. This knowledge can be efficiently acquired and greatly enriched by consulting computerized sources of information as outlined in the earlier sections. However, the chemist can further exploit the capabilities of the computer, not only to store and retrieve chemical information, but also to manipulate and apply information (both theoretical and empirical) in a logical manner.

To this end, computer programs have been developed which can assist the chemist in designing new synthetic routes. These systems are often referred to as 'knowledge-based' or 'expert' systems. There are many definitions of such systems; Johnson[81] aptly describes them as: 'Systems that are able to manipulate and refine knowledge in order to propose solutions that require logical or deductive inference .... They characteristically include the ability to organise and use knowledge that is often inexact and incomplete and operate on concepts, models and generalised rules as well as specific facts.'

The methods which can be applied to solve a complex synthetic problem and thus form the basis of an expert system need to be considered. A synthetic analysis may be approached from either the 'synthetic' or 'forward' direction, which proceeds in the same direction as a laboratory synthesis, or from the reverse direction, which is described as 'retrosynthetic or 'antithetic'.[101] Retrosynthetic analysis involves modification of the target structure in a stepwise fashion to give simpler precursors which in turn are further transformed until a suitable starting point is reached. These two distinct approaches are employed by expert systems, though the majority operate essentially in the retrosynthetic direction.

Expert systems may be classified into those which are reaction library based (essentially empirical) or theoretically based (non-empirical); Table 7 summarizes some of the expert systems available and their features. Some excellent reviews of computer-aided synthesis design are available,[102,103] and the influence of expert systems on the inventive process has been discussed elsewhere.[104]

4.14.5.7.2 Reaction library based systems

These depend upon a library of literature reaction transformations which are applied to the analysis of the target molecule using a set of heuristics, namely experimentally determined rules which guide a program, and are applied sequentially to the analysis of a problem, with any unfeasible solutions being discarded automatically. Corey and Wipke pioneered this approach in the late 1960s with the evolution of OCSS (Organic Chemical Simulation of Synthesis),[105] from which

Table 7 Chemical Expert Systems

Expert system	*Synthetic direction*[a]	*Information base*[b]	*Interactive*[c]
LHASA	R	L	Y
SECS/CASP	R	L	Y
PASCOP	R	L	Y
SYNCHEM	R or S	L	—
CAMEO	S	T	Y
EROS	R or S	T	—
SYNGEN	R	T	—

[a] R = retrosynthetic, S = synthetic (forward), [b] L = reaction library, T = theoretical. [c] Y = yes, — = no.

emerged the well-known LHASA (Logic and Heuristics Applied to Synthetic Analysis).[101,106–110] Wipke then went on to develop SECS (Simulation and Evaluation of Chemical Synthesis) from LHASA.[111–115] A consortium of Swiss and German companies are now developing CASP (Computer-Aided Synthesis Program) which is an offshoot of SECS.[101]

PASCOP (Programme d'Aide a la Synthèse de Composes Organo Phosphores) specializes in organophosphorus chemistry.[116–119] The SYNCHEM2 (Synthetic Chemistry) program also relies on a database of known reactions.[120–122] Programs such as LHASA and SECS are interactive, *i.e.* the chemist directs the search and is involved in each stage of the retrosynthetic analysis. The target structure is input by the chemist, who then selects a suitable retrosynthetic strategy from a menu (Figure 10). The program searches its database of reaction descriptions (transforms) for those which satisfy the given strategic requirements. Welford[102] describes a transform as 'a self-contained statement of the structural changes which occur when a particular reaction is applied to a certain substructure, and of the conditions which govern the success of that reaction.' A number of precursors are thus generated, each of which is examined by the chemist. Having established the first level of precursors, the chemist chooses the most suitable precursor(s) for further processing and selects another strategy option. The retrosynthetic analysis continues in this way until the chemist arrives at a stage where a suitable starting material is identified. Thus a 'synthesis tree' is constructed, where the target molecule is at the apex of the tree and the structures generated by the program are represented as numbered nodes on the tree (Figure 11).

Unlike LHASA and SECS, SYNCHEM2 is a 'non-interactive' program, *i.e.* the chemist is precluded from interacting with the system. The SYNCHEM2 program selects the intermediates for processing, and a synthesis tree is automatically generated.[123,124] The program selects the routes that lead to some available starting materials (a subset of the Aldrich catalogue). The chemist can examine the results only when the program has finished processing. SYNCHEM is also capable of operating in the synthetic direction.

SHORT-RANGE SEARCHES	BOND-MODE SEARCHES	LONG-RANGE SEARCHES
> UNCONSTRAINED	CYCLIC STRATEGIC	STEREOSPECIFIC C=C
DISCONNECTIVE	POLYFUSED STRATEGIC	DIELS-ALDER
RECONNECTIVE	APPENDAGES	ROBINSON ANNULATION
UNMASKING	RING APNDG ONLY	CYCLOPROPANES
STEREOSELECTIVE	BRANCH APNDG ONLY	HALOLACTONIZATION
	MANUAL DESIGNATION	QUINONE DIELS-ALDER
		BIRCH REDUCTION

OPTIONS	PERCEPTION ONLY	CONTROL
ALIPHATIC ONLY	CYCLIC STRATEGIC	SKETCH
> AROMATIC ONLY	POLYFUSED STRATEGIC	TREE
PRESERVED BONDS	6-RING CONFORMATIONS	DEBUG
PRESERVED STEREO	STEREOCHEMISTRY	TEST
STEREOCONSERVING	HUCKEL CALCULATION	EXIT
		HELP

Figure 10 LHASA process menu (reproduced by permission of LHASA UK)

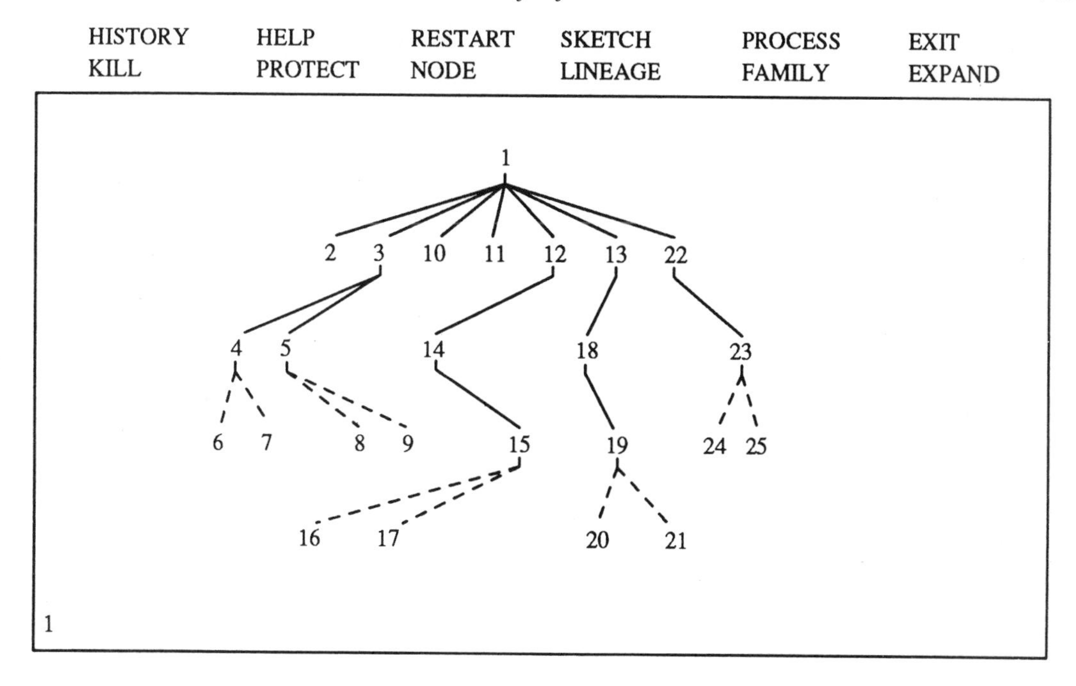

Figure 11 LHASA synthesis tree (reproduced by permission of LHASA UK)

4.14.5.7.3 Theory based systems

These systems are independent of a reaction library; instead, they are generally based on mathematical models and/or physicochemical parameters.

CAMEO (Computer Assisted Mechanistic Evaluation of Organic reactions)[125–130] is a mechanistically based interactive program developed by Jorgensen who describes the system as 'mimicking the traditional mechanistic reasoning of chemists.' The system operates in the forward synthetic direction only and predicts the products of a given reaction under given conditions, *e.g.* acid-catalyzed and electrophilic reactions. Hence CAMEO can be utilized to predict possible side products from a reaction, propose new reactions which are mechanistically sound and provide new ideas on reaction mechanisms. The program can also be used to calculate the enthalpy of each predicted reaction, and the heats of formation of products. CAMEO should therefore find most application in process research where the rigorous examination of reaction conditions and anticipation of the likely side products is of great importance.

The EROS program (Elaboration of Reactions for Organic Synthesis)[131–135] has the capability of operating both forward and retrosynthetic search strategies. This non-interactive system derives its reaction predictions from mapping electron movements. EROS is based on a mathematical model which is not restricted by known chemistry. Gasteiger[134] describes EROS as using a number of reaction generators which 'make and break bonds and shift free electrons according to simple well-defined schemes in order to generate new structures from molecules of interest.' Thus EROS has the ability to propose unknown chemical reactions which may or may not be chemically viable.

Hendrickson[136–139] has taken yet another view to synthesis design with the SYNGEN program. The strategic concept embodied in this program is based on the dissection of the target molecule's skeleton into the fewest number of fragments or 'bondset families' which correspond to available starting materials (part of the Aldrich catalogue). A synthesis is thus regarded as an assembly or convergence of small starting material pieces. The skeletal bonds which are constructed in a synthesis are called a bondset. The required construction reactions (for the bondset) are formulated from broad mechanistic guidelines, and therefore the necessity for a library of reactions is avoided. The SYNOUT program is used to display the output.

4.14.5.7.4 Conclusions and future trends

The programs and databases of expert systems need constant improvement and development to meet the ever-increasing demands of the synthetic chemist, and this is indeed happening for many of these systems. A future trend, which is already in evidence, is that of the 'bidirectional' approach.[81] This may be applied where certain starting material(s) can be perceived as candidates for the synthesis of particular target compounds. The starting materials may be based, for example, on ChemQuest. The synthesis can then be approached in parallel both from the target in the retrosynthetic direction and the starting materials in the synthetic direction, thereby dramatically reducing the number of potential intermediates (*i.e.* nodes on the tree) and, as Johnson[81] remarks, 'provide a short cut to the discovery of a good synthetic plan.' The LHASA program is presently being developed to incorporate the bidirectional approach. Wipke[140] has also described a program, SST, which allows the interactive selection of potential starting materials from a given target molecule.

Expert systems may assist the chemist in designing new synthetic pathways by helping to broaden, diversify and provide fresh insights into synthetic analysis.[141] However, while providing ideas for chemical synthesis, these systems are not intended to supply information on the various experimental procedures for the individual reactions within the synthesis. Here it can be seen that the reaction retrieval systems, *e.g.* REACCS, ORAC, SYNLIB, can be used in tandem with the expert systems to provide the chemist with a complete resource of synthesis information.

4.14.6 CHEMICAL DATA

During the course of their research, medicinal chemists may need access to a wide range of physicochemical data. This extends from the physical characteristics of a chemical such as melting point, optical rotation and various thermodynamic properties, through to spectral analysis for structure elucidation and confirmation, and ultimately to the physical behaviour of a drug in biological systems. This type of information is notoriously difficult to locate because of its many diverse sources, ranging from the original journals to hardcopy reference manuals and online databanks. Each source often contains different types of data for a specific chemical, *i.e.* many sources may need to be consulted if a complete data profile is required for a specific chemical. In some cases, the data are incomplete or not fully evaluated, and extrapolations, calculations or verifications may need to be employed. Hawkins[143] has commented that 'the scientist or engineer who wishes to obtain and use numeric data faces a monumental challenge.'

A comprehensive account of 'locating and using physical property and related data' is given by Maizell[4] and numeric databases in the sciences have been reviewed.[142,143] Another useful guide to sources has been written by J.M. Sweeney.[144]

Handbooks and manuals containing collections of data are frequently consulted. Some of the well-known hardcopy sources of chemical and physical data include the 'CRC Handbook of Chemistry and Physics', often referred to as 'the Rubber book',[145] 'The Chemists Companion' by Gordon and Ford,[146] 'Kirk-Othmer Encyclopedia of Chemical Technology'[147] and others already mentioned. In addition to these, the chemical catalogues (*e.g.* the Aldrich catalogue) often supply useful physical property data.

Some of the major hardcopy reference works described above are also available online, *e.g.* MERCK, HEILBRON, Kirk-Othmer and BEILSTEIN ONLINE.[148] A computer-readable source of numeric data is sometimes referred to as a 'databank' rather than a 'database'. A number of databanks may be accessed online from the major hosts or purchased for in-house use. A main advantage of the computerized sources over the hardcopy reference works is that the information is easily and regularly updated, whereas revised reference works may take years to appear and it can be costly to replace the older works. Generally, the access points to search the databanks may include chemical structure, CAS registry number, chemical name, molecular formula, property name or property constants. Several of the major sources of numeric data and the properties which they contain are listed in Table 8. Some of the sources include a wide selection of property types for a relatively small range of compounds, for example the DIPPR File (commercially important chemicals); other sources such as HEILBRON encompass a large number of organic compounds but may offer a more limited range of property data. Sources such as THERMO, TRC Thermophysical Property Datafile 1 and the Physical Property Data Service deal with specific classes of property information.

CA has not been listed as a source of information on chemical data because chemical properties are only indexed when they are the main theme of an article and, even then, actual data values are rarely indexed. Searching the CA abstract field uncovers little extra information, as nowadays the abstracts have become much less detailed. Often the best that the user can do is to search for the specific compound in CA, then examine the source document for any physical data. An experiment

Table 8 Numeric Databanks

Source	m.p./b.p.	$[\alpha]_D$	Refractive index	pK_a	Log P	Property[a] Solubility	Vapour pressure	Thermodynamic	Transport	Chemical technology	Other properties/comments
MERCK	Y	Y	Y	Y	—	Y	—	—	—	—	Full text, > 10 000 compounds
HEILBRON (Dictionary of organic compounds; Chapman and Hall	Y	Y	Y (common compounds)	Y	—	Y	—	—	—	—	Full text, > 175 000 compounds
Kirk-Othmer Online (John Wiley & Sons Inc.)	Y	Y	Y	Y	Y	Y	Y	Y	—	Y	Full text, > 1200 articles. 'common compounds'
DIPPR, Design Institute of Physical Property Data (American Institute of Chemical Engineers)	Y	—	Y	—	—	Y	Y	Y	Y	—	Evaluated data > 800 compounds
BEILSTEIN ONLINE	Y	Y	Y	Y	—	Y	Y	Y	Y	—	> 1 500 000 compounds. Indicates where spectral determination was carried out
DECHEMA, Dechema Chemical Engineering and Biotechnology Abstracts databank. (DECHEMA Deutche Gesellschaft für Chemisches Apparalewesen Chemische Technik und Biotechnologie e.v.)	—	—	—	—	—	—	—	—	—	Y	Includes information on laboratory methods, process and reaction engineering > 3000 compounds
DETHERM-SDR, DECHEMA Thermophysical Properties Databank— Data Retrieval System, DECHEMA	Y	—	—	—	—	—	Y	Y	—	—	Thermophysical properties, > 3000 industrially important chemicals, vapour pressure of > 2000 substances
THERMO, Thermodynamic Property Values Database, [NBS (National Bureau of Standards) and TRC (Thermodynamics Research Centre)]	—	—	—	—	—	—	—	Y	—	—	34 thermodynamic properties covered, no single entry contains information for all properties, > 15 000 compounds

Table 8 Numeric Databanks

Source	*m.p./b.p.*	$[\alpha]_D$	*Refractive index*	pK_a	*Log P*	*Property*[a] *Solubility*	*Vapour pressure*	*Thermodynamic*	*Transport*	*Chemical technology*	*Other properties/comments*
TRC Thermophysical Property Datafile 1: Vapour Pressure, (Thermodynamics Research Centre (TRC))	Y	—	—	—	—	—	Y	—	—	—	Also calculated vapour pressures, >20 000 compounds
Physical property Data Service (PPDS >800 components, (National Engineering Laboratory and the Institution of Chemical Engineers	Y	—	—	—	—	—	Y	Y	Y	—	Can calculate certain constant and variable properties >800 compounds, also available on magnetic tape
Log *P* and Related Parameters Database, (Pomona College) 'Medchem log *P* database'	—	—	—	Y	Y	—	—	—	—	—	Octanol/water distribution standard. >30 000 log *P* values in >300 solvent systems, also available on magnetic tape

[a] Y = featured, — = not featured.

on data tagging in CA has been conducted,[149] but the matter has not been taken any further. (Data tags are codes that indicate the presence in a primary document of specific types of numerical data.) ISI's Index Chemicus does include the data tags but it is no longer available online.

Chemical data are compiled by a substantial number of different producers, many of which are specialist data centres[150] or agencies who generate and handle specific types of data. Among these, the National Bureau of Standards (NBS) of the US Department of Commerce administers a group of data centres and projects called the National Standard Reference Data System (NSDRS),[151] the main purpose of which is to 'provide critically evaluated numerical data in convenient and accessible form, to the scientific and technical community.'[4] Two of the centres affiliated with NSDRS are the Thermodynamics Research Centre and DIPPR Data Projects, American Institute of Chemical Engineers.[152]

Certain online hosts are now developing their services to include clusters of numeric datafiles. The NIH-EPA CIS system has long provided a large number of numeric and spectral databases, among them the thermodynamic properties databank THERMO, MERCK and the Wiley Mass Spectral Search System.[153,154] STN plan to develop further their numeric data service and will include spectral, crystallographic, materials data and thermophysical/thermodynamic databases. Currently accessible *via* STN are DIPPR, DECHEMA, BEILSTEIN and the C13NMR database. The Technical Database Services Inc. (TDS) offer a service called Numerica which comprises several databanks including log *P* and related parameters, TRC Thermophysical Property Datafile and PPDS (Physical Property Data Service).

Spectroscopic data are widely available from many databanks, libraries and manuals. For example, Aldrich offer an NMR library, a library of IR spectra and a library of FT-IR Spectra. The major spectral databanks are given in Table 9. These are used either for the identification of an unknown substance or to obtain the spectrum of a known compound. Indeed, the area of structure elucidation/confirmation may be considered as part of analytical chemistry and many medicinal chemists rely on their specialist analytical colleagues for the appropriate information.

When it is not possible to locate reliable data, or perhaps not to find any data at all, the chemist may opt to calculate the required values. Several programs exist which can calculate properties from existing data. Depending on the system, the data used for the calculations are available either as a databank accompanying the program or are retrieved from another source. PPDS has a main package which contains the databank of thermophysical properties. In addition, the PPDS software can utilize the databank to calculate constant and variable properties for mixtures of components or to generate phase equilibrium data within a wide range of conditions. CHEMTRAN[6] calculates vapour–liquid equilibria for substances either using data from its own databank or from a two-dimensional chemical structure for compounds not in the file. Thermodynamic properties can also be estimated using EPIC (Estimate of Properties for Industrial Chemistry) produced by the University of Liege.[155,156] The CHEMEST system, developed by Arthur D. Little Inc., contains 36 methods for estimating 11 properties including melting point/boiling point, vapour pressure and acid dissociation constant.[157]

A substantial number of the numeric databanks and calculation programs may be purchased and run in-house, for example PPDS, CHEMEST, TRC thermophysical property datafile, MedChem

Table 9 Spectral Databanks

Databank	*Producer*	*Contents*
C-13 Nuclear Magnetic Resonance Database	BASF	>80 000 ^{13}C NMR spectra
CNMR (Carbon-13 Nuclear Resonance Search System	Netherlands Information Combine	>6500 ^{13}C NMR spectra
NMRLIT (Nuclear Magnetic Resonance Literature Search System)	NIH in conjunction with Preston Publications Inc.	>43 000 citations to literature on NMR
IRSS (Infrared Search System)	EPA and Boris Kidric Institute	>5000 complete IR spectra
SPIR (Search Program for Infrared Spectra)	Canadian Scientific Numeric Database Service	>140 000 IR spectra
WMSS (Wiley Mass Spectral Search System)	John Wiley & Sons	>130 000 mass spectra
MSSS (Mass Spectral Search System)	NBS	Mass spectral data for >42 000 compounds
Mass Spectrometry Bulletin	RSC	>160 000 citations
Cambridge Structural Database	Crystallographic Data Centre, Cambridge	Connectivity information on >41 000 compounds

log P database and the Cambridge crystallographic structural database. Molecular Design Ltd offer a pK file (from data compiled by Rhône-Poulenc Recherches) in MACCS-II and REACCS format. The pK file contains over 3400 molecules with over 10 500 dissociation constants.

The application of physicochemical parameters to the determination of quantitative structure–activity relationships (QSAR) and the prediction of the physical behaviour of drugs in biological systems with the resulting application to drug design is discussed in Volume 4 of this series. The Pomona College Medchem log P and parameter database, together with the Cambridge crystallographic structural database are important information resources in the area of QSAR.

4.14.7 KEEPING UP-TO-DATE

Monitoring the literature for new information of interest is an important stimulus. The volume of material published permits only a core group of primary journals to be read regularly and additional methods must be employed if the medicinal chemist wishes to ensure that all new information in a particular subject area is identified.

The printed secondary sources such as 'Chemical Abstracts Section Groupings'[4] and 'Index Chemicus'[4] can be skimmed through quickly. 'Index Chemicus', with its biological activity indicator which allows the quick identification of articles referencing new compounds with a particular biological activity, is particularly relevant. A more selective service for the medicinal chemist is the Ringdoc Profile booklets which are available on a variety of therapeutic topics, *e.g.* antiallergics, psychotropic agents, antiarrhythmics, drugs acting on enzymes and dermatological agents; these cover the pharmacology and therapeutic use of drugs. The service was developed by the Pharmadokumentationsring, a consortium of European pharmaceutical companies, prior to its transfer to Derwent Publications Ltd.[15] Less useful are the 'Current Contents' life sciences and physical, chemical and earth sciences booklets published by ISI and 'Chemical Titles' published by CAS which do not provide access to abstracts of articles.[4]

A number of services provide information on new synthetic routes. 'Methods in Organic Syntheses' is a monthly publication from the Royal Society of Chemistry which abstracts all the major European and American journals and covers functional group changes, carbon–carbon bond formation, new reagents, enzymatic and biological transformation, asymmetric synthesis and protecting groups.[158] ISI's 'Current Chemical Reactions' aims to provide a guide to novel and newly modified reactions and syntheses.[76] The 'Journal of Synthetic Methods'[78] is a monthly publication produced by Derwent's Chemical Reactions Documentation Service. Sources include both journal and patent literature and new reactions and synthetic methods, new reagents and improved methods, applications and extensions of known reactions are covered. The Ringdoc thematic booklet 'Chemistry' focusses on pharmaceutical chemistry and the reaction and synthesis of biologically active molecules. A less useful service because of its general coverage is 'ChemInform'[159] produced by FIZ Chemie; this does include heterocyclic compounds.

Searching the online databases can provide fairly selective alerts to new references. The currency of services of this type, known generally as 'selective dissemination of information' or SDI, obviously depends on the frequency with which a database is updated by the vendor in addition to the time lapse between the publication of an article and its processing by a database producer. The medicinal chemist has the option of undertaking such searches locally; if local facilities are not available, both publishers, online vendors and other organizations offer services of this type. Perhaps the best known of these are ASCA (automatic subject citation alert) profiles available from ISI. The user's profile, used to search weekly for new information added to the Scisearch database, can include author(s), journal(s), keywords from the article title or authors and journals to be cited by the selected article. The service is extremely current; its major disadvantage is that its output does not include abstracts but only comprehensive bibliographic details and brief details of the cited papers which have retrieved the item.[14] A further service from ISI has been the launch of all five 'Current Contents' publications as a three-month rolling online file which is updated weekly and whose tables of contents can be scanned easily.

SDI services which provide a regular list of articles on topics of interest (known as 'macroprofiles') to a large number of scientists are also produced from online databases. In 1987, more than 300 weekly ASCATOPICS were available from ISI including more than 40 on pharmacology and medicinal chemistry. Titles include 'Anti-ulcer agents', 'Prostaglandins' and 'Anti-viral agents'. The level of information on references included in ASCATOPICS parallels that in ASCAPROFILES.[4] Less current but containing full CA abstracts are the biweekly 'CA Selects' produced by CAS.[160] Only a few titles, however, are aimed specifically at the medicinal chemist;

these are β-lactam antibiotics, steroids (biochemical aspects), steroids (chemical aspects), antiinflammatory agents and arthritis, animal longevity and ageing, antitumour agents, atherosclerosis and heart disease, and biogenic amines and the nervous system. 'BIOSIS/CA Selects'[161] are biweekly abstracts booklets derived from both databases and include such topics as allergy and antiallergy, neuroreceptors and peptide and protein sequences.

Structure-based current awareness can best be carried out on CAS ONLINE's registry file which can be searched for the appearance of new compounds containing a particular substructure or for new references on structures or substructures of interest.[162] The use of Index Chemicus as an in-house current awareness tool has been described.[14] However, ISI introduced in 1986[163] Index Chemicus personal databases.* 16 databases on floppy disc (including amino acids/peptides; antibiotics; herbicidal compounds; DNA, RNA, protein synthesis inhibitors; antiinflammatory agents; anti-hypertensives/-hypotensives; heterocycles, macrocyclic compounds, natural products and derivatives; nucleosides/nucleotides; prostaglandins; steroids/antifertility compounds) provide direct access to articles on recently synthesized compounds. Each topic contains 200 selective compounds derived from the recent literature, with the compound name, molecular formula, molecular weight, biological activity, if indicated, and journal source. Quarterly updates are provided for use with MDL's ChemBase and annual updates for Scott Gould's ChemSmart software. Current Chemical Reactions personal databases are also available;* each database contains 100 reactions with quarterly updates derived from Current Chemical Reactions. Although these services do not provide comprehensive current awareness, they do provide the scope for analyzing the information in a novel way. Index Chemicus in-house databases derived, *inter alia*, using biological activity indicators or structure can also be made available for use on mainframe computers within organizations.

Monitoring current developments is not easy; medicinal chemists can best rely on a combination of core journal scanning, a few well-chosen SDI profiles or a relevant commercial SDI bulletin, and their 'invisible' college.

4.14.8 TRENDS AND DEVELOPMENTS

4.14.8.1 Sources and Services

The most exciting current developments for the medicinal chemist are the launch of BEILSTEIN ONLINE in 1988 and the recent introduction of CASREACT.

BEILSTEIN ONLINE and CASREACT are developments of enormous significance. Other developments worth noting here are the emergence of full text journals online, the increasing accessibility of Japanese information, the growth in the number of databanks and the introduction of sources on CD-ROM (compact disc read-only memory).

Full text journals can be made available as online databases as a by-product of the publication process. Both the American Chemical Society (ACS) and, more recently, the Royal Society of Chemistry (RSC) are making journals available in this form and, for instance, the 'Journal of Medicinal Chemistry' can now be read at a terminal. The present state of technology allows the text, footnotes, figure captions and references in each journal article to be included in the online file with the journal title, author(s), author's affiliation and abstract. Graphic information cannot yet be included. To date the journals that are most likely to be of use to the medicinal chemist are all available on STN. STN are actively increasing their collection of full text databases and 'Angewandte Chemie' will be available in 1989.

Every significant word in the text of the original document is searchable in a full text system. Full text therefore offers the potential advantage of being able to search in natural language and of not missing an item in the original which was insufficiently important to feature in the terms available for searching on reference databases, *i.e.* the abstract, title or keywords. Because they have no added indexing, searching full text databases involves specifying words likely to occur next or near to each other in the original article, *e.g.* 'melting' next to 'point'. Specifying words or phrases to retrieve information from articles that are often written in wildly disparate styles can be cumbersome, but a carefully constructed search will often retrieve the detailed information that is simply not available elsewhere. New search techniques and software are also under development which attempt to overcome the imprecise retrieval that can occur when searching full text databases. The practical utility of full text chemical journals online has been reported.[164–166] This preliminary experience does suggest that these sources are useful in searching for chemical, spectral and property data and

* *Note added in proof*: the Index Chemicus and Current Chemical Reactions personal databases have now been discontinued.

other information in the methodology sections of papers. However, their coverage is limited and searching requires considerable expertise. They can also provide a rapid copy of the full text of a paper albeit without the graphical information. Within Europe, for instance, ACS journals can often be read online before they can be read in local libraries.

The Japanese Association for International Chemical Information has been working with CAS for some time to provide input from Japanese chemical literature to CA. Recently other initiatives are extending the availability of Japanese information. In 1985 the Japan Information Centre for Science and Technology (JICST) began producing an English language version of its abstracts database of Japanese science and technology literature; the JICST file on science, technology and medicine in Japan (JOIS) is now available worldwide.[167,168] A databank system, the Networked Compound Information System, is also being developed by STA, the Science and Technology Agency; this will make available databanks of spectral, biochemical and drug information.[167] Japan Technology is the online version of Japan Technical Abstracts, an abstracting service based in the USA which prepares English language abstracts from almost 600 Japanese technical and business journals.[168] It is likely that initiatives to make available scientific information from China will be seen in the next few years.

Optical recording technology is making a significant impact in the publishing industry. Videodiscs and digital optical discs are available and the number of CD-ROM information products are growing.[169] A disc typically holds 270000 typed A4 pages and applications include subsets of bibliographic databases, encyclopaedias and combinations of databases. The potential user requires disc, reader and personal computer to conduct a search. Search software may be resident on the disc or within the personal computer. Only products of peripheral importance to the medicinal chemist have been produced so far; in the scientific field, initial services have been principally derived from the life sciences area.[170] The entire MEDLINE database is available as Compact Med-Base; OSH-ROM (Occupational Health and Safety Information) includes HSELINE, NIOSH and CIS DOC databases. The 'Kirk-Othmer Encyclopedia of Chemical Technology' and the 'John Wiley 1987 Registry of Mass Spectral Data' are also available.[171] EMBASE, Silver Platter Information Services, Yearbook Medical Press and the NLM are participating in a joint CD-ROM project to produce Ca-CD, a collection of cancer literature. ISI has recently launched the 'Science Citation Index' on CD-ROM; with this service, once a paper has been identified as being of interest, a single keystroke can retrieve other items which share common references.

The relevance of optical products to information retrieval for the medicinal chemist has still to be established. Once the product has been purchased, all searching is 'free' so costs may be cheaper. Searching techniques are simpler, albeit slower, than those of online searching and CD-ROMs can contain figures and tables. CD-ROMs also provide the facility to group information from more than one source. However, the information is not so current—most CD-ROM products are updated quarterly and often to search over several year's data requires separate searches of several discs. The software available for use with DIALOG's CD-ROM products (DIALOG on Disc) allows a CD-ROM search to be repeated on the online database to retrieve the most recent information.

A feature of the 1980s has been the growth in the number of source databanks,[172] many of these being released from government sources. GIABS (Gastrointestinal Absorption) is a new component of the NIH-EPA CIS which provides information from literature articles dealing with the absorption, distribution, metabolism or excretion of *ca.* 2400 specific chemical substances in laboratory animals or humans—experimental conditions, strain or species, rate of application, duration and purity of test substance are available for searching.[26] The database is an outgrowth of the SPHERE (Scientific Parameters in Health and Environment, Retrieval and Estimation) project sponsored by the US Environmental Protection Agency, Office of Toxic Substances.

The trend in the growth of databanks seems likely to continue as electronic publishing is applied to the publishing of handbooks and as further government data are made commercially available.

4.14.8.2 Methods of Access

Improved access methods for online systems are directed principally at encouraging scientists, known as 'end-users' to search for themselves. These 'front-end' tools include personal computer software for interfacing with selected external vendors and databases, simpler menu interfaces on the vendor's computer which may offer help in search composition and database selection, and gateways.[39,173,174] Gateways provide one point of access to many database vendors normally *via* menu-driven systems, *e.g.* Easynet.[175,176] Bridges between vendors are becoming more common; two vendors will agree to provide windows into each others' systems thereby extending the range of databases available to the users, *e.g.* ORBIT Search Service with ESA-IRS. The application of expert systems to assist the user to select relevant sources and develop search strategies is an active research and development area.[177,178,179]

Most of the developments listed above have focussed on simplifying textual searching. DIALOG'S Medical Connection[180] is an example of well-thought-out vendor software which can be accessed using a menu. The user selects one of four libraries of databases—the medical reference library, the bioscience reference library, the general reference library or the science and technology library which includes CA. Once in the library, a particular database must be selected but a search can easily be repeated on another database in the same library. Online help can be obtained in constructing the search strategy. Other vendor services are designed to encourage use outside normal working hours, thus DIALOG's Knowledge Index and BRS/After Dark offer cheap rates at night and at weekends. The latter provides access to CA and ACS journals (full text) and 'Analytical Abstracts', amongst others. BIOSIS' new online service, BIOSIS CONNECTION, another user-friendly system, is aimed at the bench scientist and offers access to a variety of databases.

Personal computer software packages for literature searching are almost too numerous to mention by name here.[173,181] Such software provides facilities for automatic log-on to the database of choice, may offer some help in search construction and the preparation of the search offline to reduce costs and may provide tools for the sorting and editing of results. Med-Base[182] is an advanced interface which enables a novice to match their natural language query to the structured index used in MEDLINE. Grateful-Med is another microprocessor providing simple access to a variety of medical databases.[183] Recently STN have entered the personal computer software field. Beginning with STN Mentor, a series of programs which offer instructions on searching, *e.g.* 'Introduction to CAS ONLINE'.[184] STN Communicator is an automatic log-in package for connecting to STN with one command. Launched in 1988 was STN Express which includes a chemical structure interface.[185] This enables the user to draw a structure offline using commands which are much simpler than those needed for CAS ONLINE itself, submit the query to CAS ONLINE and to view and print the results. STN Express is also being developed to provide search facilities for CASREACT. Molkick, being developed for Beilstein, and Chemlink, being developed by Télésystèmes-Questel, are other graphics-based packages under development. Superstructure is a structure input interface to CIS.[186]

Even with improved drawing and uploading packages, the chemist is still faced with a number of graphic structure packages to learn—CAS ONLINE, DARC, ORAC, *etc.* The ideal searching system would, of course, allow all structural databases to be searched from a search query prepared once. This ideal seems far from realizable currently. Molkick is, however, suitable for use with BEILSTEIN and CAS ONLINE and the Psidom range of PC packages has already been developed as a front-end to DARC in-house systems, to CAS ONLINE and to ORAC; access to other systems is under development.[186]

Gateways for the serious chemical searcher are not yet in place. The CSIN (Chemical Substances Information Network) project aimed to enable the user to retrieve and relate information about chemicals from many different information sources *via* a totally coordinated network of online chemical information systems.[187] The prototype was tested in 1983.

The information goal for medicinal chemists must be to access the world's chemical literature from their desk, to retrieve compounds of interest from a structural and/or property search, to display associated data, to be able to manipulate and store the information required and to have immediate help and assistance in all of these jobs. Such 'truly integrated systems' would include data retrieval and provide access to primary, secondary and tertiary[187] sources. The term 'vertical gateways' has been used to describe ideal systems which will provide integrated access to the information sources in a single discipline whether these are bibliographic, full text, numeric or pictorial.[189] The medicinal chemist has the advantage of working in a field which contains already some of the most sophisticated information tools. Commercial competition is encouraging a rapid pace of information development; hopefully, commercial forces will not prevent the required integration of different sources and techniques in the longer term.

4.14.9 USEFUL ADDRESSES

Beilstein Institute
Varrentrappestrasse 40–42
Carl-Bosch-Haus
D-6000 Frankfurt (Main) 90
Federal Republic of Germany
Tel: 69 79 171

BRS Information Technologies
1200 Route 7
Latham
NY 12110
USA
Tel: 800 345 4278/518 783 1161
800 833 4707/518 783 7251

ISI
3501 Market Street
Philadelphia
PA 19104
USA
Tel: 800 523 1850

Chemical Abstracts Service
Marketing Department 30887
PO Box 3012
Columbus
OH 43210
USA
Tel: 614 447 3600

Elsevier Science Publishers
PO Box 1527
1000 BM Amsterdam
The Netherlands
Tel: 20 580 3507

STN International
2540 Olentangy River Road
PO Box 02228
Columbus
OH 43202
USA
Tel: 800 848 6533/614 421 3600

ORBIT Search Service
Achilles House
Western Avenue
London W3 OUA
UK
Tel: 01 993 7334

Télésystèmes-Questel
5201 Leesburg Pike
Suite 603
Falls Church
VA 22041
USA
Tel: 800 424 9600
703 845 1133

DIMDI (Deutsches Institut für Medizinische Dokumentation und Information)
PO Box 42 05 80
Weisshausstrasse 27
D-5000 Cologne 41
Federal Republic of Germany
Tel: 221 4724

Télésystèmes-Questel
83–85 Boulevard Vincent-Auriol
75013 Paris
France
Tel: 582 6464

National Library of Medicine
8600 Rockville Pike
Bethesda
MD 20894
USA
Tel: 301 496 6193

Fein-Marquart Associates Inc.
7215 York Road
Baltimore
MD 21212
USA
Tel: 301 321 8440

Data-Star
Plaza Suite
114 Jermyn Street
London SW1Y 6HJ
UK
Tel: 1 930 5503

DIALOG Information Services Inc.
3460 Hill View Avenue
Palo Alto
CA 94304
USA
Tel: 415 858 3785

ESA-IRS (European Space Agency Information Retrieval Service)
Esrin
Via Galileo Galilei
0044 Frascati
Rome
Italy
Tel: 6 94011

4.14.10 REFERENCES

1. H. D. Brown, *J. Chem. Inf. Comput. Sci.*, 1984, **24**, 155.
2. J. E. Ash, P. A. Chubb, S. E. Ward, S. M. Welford and P. Willett, 'Communication, Storage and Retrieval of Chemical Information', Horwood, Chichester, 1985.
3. C. Jochum and P. Moricz, *Database*, 1987, **10** (4), 41.
4. R. E. Maizell, 'How to Find Chemical Information', 2nd edn., Wiley, New York, 1987.

5. R. T. Bottle, 'Use of Chemical Literature', 3rd edn., Butterworths London, 1979.
6. D. T. Hawkins, *Database*, 1985, **8** (2), 31.
7. L. T. Morton and S. Godbolt, 'Information Sources in the Medical Sciences', 3rd edn., Butterworths London, 1984.
8. T. Andrews, 'Guide to the Literature of Pharmacy and the Pharmaceutical Sciences', Libraries Unlimited, Colorado, 1986.
9. D. K. Bawden and T. K. Devon, *J. Chem. Inf. Comput. Sci.*, 1980, **20**, 1.
10. 'Directory of Online Databases', Cuadra/Elsevier, New York, 1988, vol. 9, part 1.
11. S. R. Heller, *Database*, 1987, **10**, 47.
12. C. Jochum, in 'Modern Approaches to Chemical Reaction Searching', ed. P. Willett, Gower, Aldershot, 1986, p. 165.
13. S. M. Kaback, in 'Computer Handling of Generic Chemical Structures, ed. J. M. Barnard, Gower, Aldershot, 1984, p. 49.
14. W. A. Warr, *Chem. Inf. Bull.*, 1984, **36** (2), 31.
15. R. E. Hoover, *Online Rev.*, 1981, **5/6**, 453.
16. J. MacIntyre, in 'Proceedings of the 10th International Online Meeting, London', Learned Information, Oxford, 1986, p. 53.
17. B. Snow, in 'Manual of Online Search Strategies', ed. C. J. Armstrong and J. A. Large, Gower, Aldershot, 1988, p. 273.
18. G. W. A. Milne, S. R. Heller, A. E. Fein, E. F. Frees, R. G. Marquart, J. A. McGill, J. A. Miller and D. S. Spiers, *J. Chem. Inf. Comput. Sci.*, 1978, **18**, 181.
19. 'The Merck Index', 10th edn., Merck, Rahway, New Jersey, 1983.
20. B. Snow, *Database*, 1988, **11** (3), 90.
21. H. Sack, 'Online Searching Made Simple', PJB Publications, Richmond, 1988.
22. IMS International, 364 Euston Road, London, NW1 3BL, UK.
23. D. Bawden, *Aslib Proc.*, 1988, **40**, 79.
24. *Online Sci-Tech. Inf.*, 1987, 84.
25. P. Hassanaly and H. Dou, in 'Manual of Online Search Strategies', ed. C. J. Armstrong and J. A. Large, Gower, Aldershot, 1988, p. 157.
26. GIABS, personal communication, Fraser Williams (Scientific Systems) Ltd.
27. D. Bawden, in 'Manual of Online Search Strategies', ed. C. J. Armstrong and J. A. Large, Gower, Aldershot, 1988, p. 44.
28. G. W. A. Milne and S. R. Heller, *J. Chem. Inf. Comput. Sci.*, 1980, **20**, 204.
29. W. A. Warr (ed.), 'Chemical Structures: The International Language of Chemistry', Springer, Berlin, 1988.
30. J. E. Ash and E. Hyde, 'Chemical Information Systems', Horwood, Chichester, 1975.
31. H. M. Woodburn, 'Using the Chemical Literature: A Practical Guide', Dekker, New York, 1974, p. 41.
32. E. Hyde, W. A. Warr and S. E. Ward, in 'Chemical Nomenclature Usage', ed. R. Lees and A. Smith, Horwood, Chichester, 1983, p. 29.
33. 'Chemical Substance Index Names', Chemical Abstracts Index Guide, 1984, appendix 4; 'Name and Indexing of Substances for Chemical Abstracts', CAS Reprint 298.
34. J. Vollmer, *J. Chem. Educ.*, 1983, **60**, 192.
35. D. Meyer, Institute for Scientific Information, personal communication, 1988.
36. C. Suhr, E. Von Harsdorf and W. Dethlefsen, in 'Computer Handling of Generic Chemical Structures', ed. J. M. Barnard, Gower, Aldershot, 1984, p. 96.
37. R. Fugmann, in 'Chemical Information Systems', Horwood, Chichester, 1985, p. 195.
38. S. M. Kaback, *J. Chem. Inf. Comput. Sci.*, 1980, **20**, 1.
39. W. A. Warr, *Database*, 1987, **10** (3), 122.
40. P. G. Dittmar, R. E. Stobaugh and C. E. Watson, *J. Chem. Inf. Comput. Sci.*, 1976, **16** (3), 111.
41. S. Anderson, *J. Mol. Graphics*, 1984, **2**, 83.
42. J. M. Barnard, *Database*, 1987, **10** (3), 27.
43. J. M. Barnard, in 'Proceedings of the 11th International Online Meeting, London', Learned Information, Oxford, 1987, p. 45.
44. K. Shenton, in 'Proceedings of the 9th International Online Meeting, London', Learned Information, Oxford, 1985, p. 43.
45. V. J. Gillet, S. M. Welford, M. F. Lynch, P. Willett, J. M. Barnard and G. M. Downs, *J. Chem. Inf. Comput. Sci.*, 1986, **26**, 126.
46. C. Jochum, G. Wittig and S. Welford, in 'Proceedings of the 10th International Online Information Meeting, London', Learned Information, Oxford, 1986, p. 43.
47. P. G. Dittmar, N. A. Farmer, W. Fisanick, R. C. Haines and J. Mockus, *J. Chem. Inf. Comput. Sci.*, 1983, **23**, 93.
48. W. G. Town, in 'Proceedings of the 8th International Online Meeting, London', Learned Information, Oxford, 1984, p. 29.
49. R. Attias, *J. Chem. Inf. Comput. Sci.*, 1983, **23**, 102.
50. R. J. Feldman, G. W. A. Milne, S. R. Heller, A. Fein, J. A. Miller and B. Koch, *J. Chem. Inf. Comput. Sci.*, 1977, **17** (3), 157.
51. R. E. Stobaugh, *J. Chem. Inf. Comput. Sci.*, 1985, **25**, 271.
52. T. Novak, in 'Proceedings of the 10th International Online Information Meeting, London', Learned Information, Location, 1986, p. 353.
53. P. C. Jurs, T. R. Stouch, M. Czerwinski and J. N. Narvaez, *J. Chem. Inf. Comput. Sci.*, 1985, **25**, 296.
54. H. D. Brown, *Drug Inf. J.*, 1982, **16**, 170.
55. E. J. Lien, *Prog. Drug Res.*, 1985, **29**, 67.
56. H. D. Brown, *J. Chem. Inf. Comput. Sci.*, 1985, **25**, 218.
57. E. Vinatzer-Chanal, *Drug Inf. J.*, 1982, **16**, 227.
58. R. Franke, 'Theoretical Drug Design Methods', Elsevier, Amsterdam, 1984, p. 11.
59. V. Austel and E. Kutter, in 'Drug Design', ed. E. J. Ariens, Academic Press, New York, 1980, vol. X, p. 1.
60. B. M. Rimmer and A. Green, *J. Doc.*, 1985, **41**, 247.
61. E. S. Simmons, in 'Manual of Online Search Strategies', ed. C. J. Armstrong and J. A. Large, Gower, Aldershot, 1988, p. 84.
62. K. W. Kruse, *Am. J. Hosp. Pharm.*, 1983, **40**, 240.
63. B. Snow, *Database*, 1984, **7** (1), 12.

64. D. Bawden and A. M. Brock, *J. Chem. Inf. Comput. Sci.*, 1985, **25**, 31.
65. H. M. Kissman, *Annu. Rev. Pharmacol. Toxicol.*, 1980, **20**, 285.
66. J. Welch and T. A. King, 'Searching the Medical Literature. A Guide to Printed and Online Sources', Chapman and Hall Medical, London, 1985.
67. R. G. Dunn, *Online Rev.*, 1983, **7**, 399.
68. C. P. R. Dubois, *Online Rev.*, 1987, **11**, 243.
69. S. M. Humphrey and N. E. Miller, *J. Am. Soc. Inf. Sci.*, 1987, **38**, 184.
70. E. Garfield, 'Citation Indexing: Its Theory and Applications in Science, Technology and Humanities', Wiley, New York, 1979.
71. J. Bradshaw, *Online Rev.*, 1983, **7**, 221.
72. B. Snow, *Database*, 1987, **10** (3), 100.
73. R. Rothwell, 'British Library Research & Development Report', London, 1983, no. 5782.
74. ACS Report Rates Information System Efficiency, *Chem. Eng. News*, 1969, **47**, 31, 45.
75. L. G. Wade, 'Compendium of Organic Synthetic Methods', Wiley, New York, 1984.
76. 'Current Chemical Reactions', ISI Inc.
77. W. T. Wipke, J. Dill, D. Hounshell, T. Moock and D. Grier, in 'Modern Approaches to Chemical Reaction Searching', ed. P. Willett, Gower, Aldershot, 1986, p. 92.
78. A. F. Finch, *J. Chem. Inf. Comput. Sci.*, 1986, **26**, 17.
79. E. Müller, O. Bayer, H. Meerwein and K. Ziegler (eds.), 'Methoden der Organischen (Houben-Weyl)', 4th edn., ed. H. Kropt, Thieme Stuttgart, 1952 to date.
80. W. Dauben (ed.), 'Organic Reactions', Wiley, New York, 1942 to date.
81. A. P. Johnson, *Chem. Br.*, 1985, **21**, 59.
82. 'Organic Syntheses', Wiley, New York, 1941 to date.
83. M. Fieser and J. G. Smith (eds.), 'Fieser and Fieser's Reagents for Organic Synthesis', Wiley, New York, 1967 to date.
84. C. A. Buehler and D. E. Pearson (eds.), 'Survey of Organic Syntheses', Wiley, New York, 1970, 1977, vols. 1 and 2.
85. D. F. Chodosh, in 'Modern Approaches to Chemical Reaction Searching', ed. P. Willett, Gower, Aldershot, 1986, p. 118.
86. A. Finch or W. Theilheimer (ed.), 'Synthetic Methods of Organic Chemistry', Karger, Basel, 1946 to date.
87. P. Willett (ed.), 'Modern Approaches to Chemical Reaction Searching', Gower, Aldershot, 1986.
88. S. A. North, in '2nd Colloq. Inf. Chim.', ed. A. Deroulede, Centre National Inf. Chim., Paris, 1986, p. 204.
89. A. Deroulede, *Inf. Chim.*, 1987, **289**, 143.
90. P. Willett, in 'Communication, Storage and Retrieval of Chemical Information', ed. J. Ash *et al.*, Horwood, Chichester, 1985, p. 203.
91. E. Zass and S. Mueller, *Chimia*, 1986, **40**, 38.
92. S. E. French, *CHEMTECH*, 1987, **17**, 106.
93. E. Garfield, *Curr. Contents*, 1987, **13**, 3.
94. D. F. Chodosh and W. L. Mendelson, *Pharm. Technol.*, 1983, **7**, 90.
95. J. Boother, *Chem. Br.*, 1985, **21**, 68.
96. J. P. Gay, in 'Modern Approaches to Chemical Reaction Searching', ed. P. Willett, Gower, Aldershot, 1986, p. 87.
97. P. T. Bysouth and J. R. Hardwick, in 'Modern Approaches to Chemical Reaction Searching', ed. P. Willett, Gower, Aldershot, 1986, p. 51.
98. P. E. Blower and R. C. Dana, in 'Modern Approaches to Chemical Reaction Searching', ed. P. Willett, Gower, Aldershot, 1986, p. 146.
99. A. J. Beach, H. F. Dabek, Jr. and N. L. Hosansky, *J. Chem. Inf. Comput. Sci.*, 1979, **19**, 149.
100. S. B. Walker, *J. Chem. Inf. Comput. Sci.*, 1983, **23**, 3.
101. E. J. Corey, A. K. Long and S. D. Rubenstein, *Science (Washington, D.C.)*, 1985, **228**, 408.
102. S. M. Welford, in 'Communication, Storage and Retrieval of Chemical Information', ed. J. Ash *et al.*, Horwood, Chichester, 1985, p. 231.
103. N. J. Hrib, *Annu. Rep. Med. Chem.*, 1986, **21**, 303.
104. P. Bamfield, in 'Computer Handling of Generic Chemical Structures', ed. J. Barnard, Gower, Aldershot, 1985, p. 218.
105. E. J. Corey and W. T. Wipke, *Science (Washington, D.C.)*, 1969, **166**, 178.
106. A. K. Long, S. D. Rubenstein and L. J. Joncas, *Chem. Eng. News*, 1983, **61**, 22.
107. D. A. Pensak and E. J. Corey, *ACS Symp. Ser.*, 1977, **61**, 1.
108. E. J. Corey, W. T. Wipke, R. D. Cramer and W. J. Howe, *J. Am. Chem. Soc.*, 1972, **94**, 431.
109. E. J. Corey, W. T. Wipke, R. D. Cramer and W. J. Howe, *J. Am. Chem. Soc.*, 1972, **94**, 421.
110. E. J. Corey, *Q. Rev., Chem. Soc.*, 1971, **25**, 453.
111. E. J. Corey, A. K. Long, T. W. Greene and J. W. Miller, *J. Org. Chem.*, 1985, **50**, 1920.
112. E. J. Corey, A. K. Long, J. Mulzer, H. W. Orf, A. P. Johnson and A. P. W. Hewett, *J. Chem. Inf. Comput. Sci.*, 1980, **20**, 221.
113. H. Bruns, in 'Textes des Conferences presentees dans le Cadre du Congres International 'Contribution des Calculateurs Electroniques au Developpement du Genie Chimique et de la Chimique Industrielle', Paris', Societe de Chimique Industrielle, Paris, 1978, sect. A, p. 83.
114. W. T. Wipke, H. Braun, G. Smith, F. Choplin and W. Sieber, *ACS Symp. Ser.*, 1977, **61**, 97.
115. W. T. Wipke and P. Gund, *J. Am. Chem. Soc.*, 1976, **98**, 8107.
116. W. T. Wipke, G. I. Ouchi and S. Krishnan, *Artif. Intelligence*, 1978, **11**, 173.
117. W. T. Wipke and T. M. Dyott, *J. Am. Chem. Soc.*, 1974, **96**, 4825.
118. C. Laurenco, L. Villien and G. Kaufmann, *Tetrahedron*, 1984, **40**, 2721.
119. C. Laurenco, L. Villien and G. Kaufmann, *Tetrahedron*, 1984, **40**, 2731.
120. M. H. Zimmer, *J. Chem. Inf. Comput. Sci.*, **19**, 235.
121. F. Choplin, P. Bonnet, M. H. Zimmer and G. Kaufmann, *Nouv. J. Chim.*, 1979, **3**, 223.
122. K. K. Agararal, D. L. Larsen and H. L. Gelernter, *Comput. Chem.*, 1978, **2**, 75.
123. H. L. Gelernter, A. F. Saunders, D. L. Larsen, K. K. Agarwal, R. H. Boivie, G. A. Spritzer, J. E. Searleman, *Science (Washington, D.C.)*, 1977, **197** (4308), 1041.
124. H. L. Gelernter, N. S. Sridharan, A. J. Hart, S. C. Yen, F. W. Fowler and H. J. Shue, *Top. Curr. Chem.*, 1973, **41**, 113.

125. T. D. Salatin and W. L. Jorgensen, *J. Org. Chem.*, 1980, **45**, 2043.
126. P. Metivier, A. J. Gushurst, W. J. Jorgensen, *J. Org. Chem.*, 1987, **52**, 3724.
127. M. G. Bures, B. L. Roos-Kozel and W. L. Jorgensen, *J. Org. Chem.*, 1985, **50**, 4490.
128. J. S. Burnier and W. L. Jorgensen, *J. Org. Chem.*, 1983, **49**, 3001.
129. J. A. Schmidt and W. L. Jorgensen, *J. Org. Chem.*, 1983, **48**, 3923.
130. C. E. Peishoff and W. L. Jorgensen, *J. Org. Chem.*, 1983, **48**, 1970.
131. J. Gasteiger and C. Jochum, *Top. Curr. Chem.*, 1978, **74**, 93.
132. M. G. Hutchings, *Anal. Proc.*, 1986, **23**, 300.
133. J. Gasteiger, M. G. Hutchings, P. Loew and H. Saller, *ACS Symp. Ser.*, 1986, **306**, 258.
134. M. Marsili, J. Gasteiger and R. E. Carter, *Chim. Oggi*, 1984, September (9), 11.
135. J. Gasteiger, in 'Textes des Conferences presentees dans le Cadre du Congres International 'Contribution des Calculateurs Electroniques au Developpement du Genie Chimique et de la Chimique Industrielle', Paris', Societe de Chimique Industrielle, Paris, 1978, sect. A, p. 90.
136. J. B. Hendrickson, D. L. Grier and A. G. Toczko, *J. Am. Chem. Soc.*, 1985, **107**, 5228.
137. J. B. Hendrickson, E. Braun-Keller and G. A. Toczko, *Tetrahedron*, 1981, **37**, 359.
138. J. B. Hendrickson and E. Braunkeller, *J. Comput. Chem.*, 1980, **1**, 323.
139. J. B. Hendrickson, *J. Chem. Inf. Comput. Sci.*, 1979, **19**, 129.
140. W. T. Wipke and D. Rogers, *J. Chem. Inf. Comput. Sci.*, 1984, **24**, 71.
141. T. V. Lee, *Chemom. Intell. Lab. Syst.*, 1987, **2** (4), 259.
142. D. T. Hawkins, *J. Chem. Inf. Comput. Sci.*, 1980, **20**, 143.
143. S. V. Meschel, *Online Rev.*, 1984, **8**, 77.
144. J. M. Sweeney, 'A Guide to Sources of Physical Property Held by the Science Reference and Information Service , British Library, London, 1988.
145. R. C. Weast (ed.), 'CRC Handbook of Chemistry and Physics', 67th edn., CRC Press, Boca Raton, FL, 1986.
146. A. J. Gordon and R. A. Ford, 'The Chemist's Companion', Wiley, New York, 1973.
147. M. Grayson (ed.), 'Kirk-Othmer Encyclopedia of Chemical Technology', 3rd edn., Wiley, New York, 1978–1984.
148. L. Domokos, C. Jochum and G. Wittig, *Mikrochim. Acta*, 1986, **2** (1–6), 423.
149. D. F. Zaye, *J. Chem. Inf. Comput. Sci.*, 1981, **21**, 73.
150. J. Martyn, *Aslib Proc.*, 1983, **35**, 258.
151. S. P. Fivozinsky, *Drexel Library Quarterly*, 1982, **18**, 27.
152. T. B. Selover, *Chem. Eng. Prog.*, 1987, **18**.
153. S. R. Heller, *Drexel Library Quarterly*, 1982, **18**, 39.
154. S. R. Heller, *J. Chem. Inf. Comput. Sci.*, 1985, **25** (3), 224.
155. G. Heyen *et al.*, in 'Conference Internationale de Thermodynamique Chimique, [Compte Rendu], 4th, Montpellier', ed. J. Rouquerel and R. Sabbah, Centre de Recherches de Microcalorimetrie et de Thermochimie du C. N. R. S., Marseille, 1975, vol. 9, p. 149.
156. A. Germain, B. Kalitventzeff, G. Heyen and R. Lecorsais, in 'Textes des Conferences presentees dans le Cadre du Congres International 'Contribution des Calculateurs Electroniques au Developpement du Genie Chimique et de la Chimique Industrielle', Paris', Societe de Chimique Industrielle, Paris, 1978, sect. E, p. 47.
157. R. S. Boethling, S. E. Campbell, D. G. Lynch and G. D. La Veck, *Ecotoxicol. Environ. Saf.*, 1988, **15**, 21.
158. 'Methods in Organic Synthesis', The Royal Society of Chemistry, Cambridge.
159. ChemInform, FachInformationsZentrum Chemie GmbH, ICSTI, 100 Berlin 12, Steinplatz 2.
160. J. E. Blake, V. J. Mathias and J. Patton, *J. Chem. Inf. Comput. Sci.*, 1978, **18**, 187.
161. BIOSIS/CAS Selects. BIOSIS, 2100 Arch Street, Philadelphia, PA 19103–1399, USA.
162. R. E. Buntrock, *Database*, 1988, **11** (6), 111.
163. E. Garfield, *Curr. Contents*, 1987, **9**, 3.
164. S. W. Terrant, L. R. Garson and B. E. Meyers, *J. Chem. Inf. Comput. Sci.*, 1984, **24**, 230.
165. R. A. Love and L. R. Garson, in 'Proceedings of National Online Meeting, 1985, NY', ed. M. E. Williams and T. H. Hogan, Learned Information, Oxford, 1985, p. 273.
166. J. P. Abbott and C. R. Smith, in 'Proceedings of National Online Meeting, 1985, NY', ed. M. E. Williams and T. H. Hogan, Learned Information, Oxford, 1985, p. 5.
167. H. Chihara, *J. Chem. Inf. Comput. Sci.*, 1987, **27**, 59.
168. P. H. Dorman, *Database*, 1987, **10** (4), 15.
169. M. Rivett, *J. Inf. Sci.*, 1987, **13**, 25.
170. B. Snow, *Online*, 1987, **11** (2), 113.
171. M. Rasdall (ed.) 'The Introductional Directory of Information Products on CDROM', 2nd edn., TFPL Publishing, London, 1988.
172. P. A. Chubb, in 'Communication, Storage and Retrieval of Chemical Information', ed. J. Ash *et al.*, Horwood, Chichester, 1985, p. 96.
173. C. Tenopir, *Libr. J.*, 1986 (July), 56.
174. T. Kleivane, in 'Online Information 1987: Proceedings of the 11th International Online Meeting, London', Learned Information, Oxford, 1987, p. 157.
175. G. Larson and S. Villumsen, in 'Proceedings of the 10th International Online Meeting, London', Learned Information, Oxford, 1986, p. 131.
176. M. O'Leary, *Online*, 1985, **9** (4), 106.
177. C. M. Bowman, J. A. Nosal and A. E. Rogers, *J. Chem. Inf. Comput. Sci.*, 1987, **27**, 147.
178. A Vickery and H. Brooks, *Online Rev.*, 1987, **11** (3), 149.
179. D. T. Hawkins, *Online*, 1988, **12** (1), 31.
180. J. Kwan, J. L. Watson and K. Deeney, *Online*, 1987, **11** (6), 32.
181. D. T. Hawkins and L. R. Levy, *Online*, 1986, **10**, 49.
182. B. Snow, *Online*, 1987, **11/3**, 125.
183. B. Snow, A. L. Corbett and F. A. Brahmi, *Database*, 1986, **9** (6), 94.
184. R. E. Buntrock, *Database*, 1988, **11** (1), 87.

185. *STN News*, 1988, **4** (1), 1.
186. W. G. Town, in 'Online Information 87: Proceedings of the 11th International Online Meeting, London', Learned Information, Oxford, 1987, p. 33.
187. J. A. Page-Castell and C. Hollister, *J. Chem. Inf. Comput. Sci.*, 1985, **25**, 359.
188. J. A. Hearty and R. A. Love, in 'Online Information 87: Proceedings of the 11th International Online Meeting, London', Learned Information, Oxford, 1987, p. 123
189. M. E. Williams, *J. Am. Soc. Inf. Sci.*, 1986, **37** (4), 204.

Subject Index